Use and Abuse
of
America's Natural Resources

Use and Abuse
of
America's Natural Resources

TIMBERS
OF THE
NEW WORLD

BY

SAMUEL J. RECORD

AND

ROBERT W. HESS

ARNO PRESS
A NEW YORK TIMES COMPANY
New York • 1972

Reprint Edition 1972 by Arno Press Inc.

Use and Abuse of America's Natural Resources
ISBN for complete set: 0-405-04500-X
See last pages of this volume for titles.

Manufactured in the United States of America

Library of Congress Cataloging in Publication Data

Record, Samuel James, 1881-1945.
Timbers of the New World.

(Use and abuse of America's natural resources)
Bibliography: p.
1. Timber--America. 2. Trees--America.
3. Timber--Latin America. 4. Trees--Latin America.
I. Hess, Robert William, joint author. II. Title.
III. Series.
SD434.R4 1972 582'.1609'7 70-140611
ISBN 0-405-02806-7

TIMBERS OF THE NEW WORLD

PUBLISHED ON THE
CHARLES LATHROP PACK FOUNDATION
AT YALE UNIVERSITY

Photo by U.S. Forest Service.

PLATE I. Dense stand of Tabonuco trees (*Dacryodes excelsa*) in the Caribbean National Forest, Puerto Rico.

TIMBERS
OF THE
NEW WORLD

BY

SAMUEL J. RECORD

DEAN OF THE SCHOOL OF FORESTRY,
YALE UNIVERSITY

AND

ROBERT W. HESS

ASSISTANT PROFESSOR OF FOREST PRODUCTS,
YALE UNIVERSITY

NEW HAVEN

YALE UNIVERSITY PRESS

LONDON · GEOFFREY CUMBERLEGE · OXFORD UNIVERSITY PRESS

Printed in the United States of America

First published, March, 1943
Second printing, May, 1944
Third printing, October, 1947
Fourth printing, July, 1949

PREFACE

TIMBERS OF THE NEW WORLD is the successor to *Timbers of Tropical America* which was published by the Yale University Press in 1924 and has been out of print since 1934. The new work contains more than twice the amount of material in the old. It not only covers the same ground more fully but its field is extended north and south of the tropics so as to include the trees and larger shrubs of the entire Western Hemisphere, exclusive of the islands of the Pacific. The bulk of the treatise directly concerns Latin America, however, since about 90 per cent of the species occur there, but the timbers of the United States and Canada are brought in to complete the picture. This is the first attempt by anyone to deal with all the larger woody plants of the Americas and it represents the cumulative results of more than twenty-five years of continuous investigation.

The book is intended not only for scientists but also for everyone interested in the utilization of American forests. It contains descriptions of the trees, tells where they grow and the sizes they attain, and attempts to evaluate their present and potential economic importance. It supplies the basis for the identification of the trees and their woods, correlates the vernacular, trade, and scientific names, and gives technological data on various subjects. It indicates the present and possible sources of rubber, resins, oils, tannins, dyestuffs, drugs, and fibers, and classifies the timbers with respect to their properties and uses. Persons looking for substitutes for well-known timbers or in search of a wood for a special purpose will find suggestions in the classified lists. The survey of the field is fairly complete since more than a thousand genera, or about 85 per cent of those containing trees native to the New World, have been described. Most of those omitted do not, as one might suppose, inhabit the vast Amazon basin but occur in the subtropical regions of Mexico and Brazil. From what is known about the botanical relationship of the missing genera it can be confidently stated that very few, if any, of them have economic possibilities.

The trees considered are in two principal divisions; namely, the Gymnosperms or conifers, the source of the softwood timbers, and the Angiosperms or broad-leaved trees, which produce the so-called hardwood lumber. Within each group the families and genera are arranged alphabetically to facilitate reference to them. The treatment is fairly consistent throughout. Each family is introduced with respect to its size, geographical distribution, and economic value, first throughout the world and then in the Americas. This is followed by a family description of the American woods. The genera are treated in a similar manner, though the emphasis on individual species and specific groups is limited to those in which the differences are considered important.

Each generic description concludes with a list of the trade and vernacular names. Their arrangement is by countries in the following order: United States and Canada, the West Indies, Mexico, Central America, and South America. These names have been compiled from the publications listed in the bibliography and also from collectors' labels, memoranda, letters, and other original sources. Common names are unsatisfactory at best and too much reliance should not be placed upon them, but used with proper reservation they are often very helpful in identification.

The descriptions of the woods are based upon specimens in the collections of the Yale School of Forestry. The total number of samples available for comparative study is about 40,600, of which 22,000 are American, as follows: U.S.A. and Canada, 2400; West Indies, 2600; Mexico, 1000; Central America, 3000; South America, 13,000. A great many of these are accompanied by leaves, flowers, or fruits from the same trees and it has been through the identification of this herbarium material by competent botanists that the scientific names for the woods have been ascertained. Systematic study of these authentic wood samples has made it possible to identify the timbers of commerce and the woods employed in local industry. Once the scientific name for a wood has been determined, the confusion in trade and local names disappears, the original and possible new sources of the timber become known, and the accumulated knowledge about the same or related species can be made available to the dealer and consumer. Trees, owing to the difficulty of collecting herbarium specimens from them, are often the least known botanically, and for this reason the wood anatomist is frequently in a better position than the taxonomic botanist to determine the range of economic timbers. The maps showing the distribution of some of the best known tropical trees are based upon data from many different sources. It is hoped that their publication will bring to light much new information.

Assembling a comprehensive collection of authentic wood samples is a slow and difficult task. The high degree of success attending the efforts at Yale is the result of long-sustained efforts involving the collaboration of many institutions and individual scientists, explorers, local residents, timber dealers, and manufacturers. The identification of the herbarium specimens has been possible through the co-operation of botanists at Field Museum of Natural History, U.S. National Museum, New York Botanical Garden, Arnold Arboretum, and elsewhere in the United States, Mexico, South America, and Europe. As a result of these investigations, several new genera and many new species have been discovered and numerous long-standing mistakes in classification have been rectified. The scientific names of very few American timbers of commerce remain in doubt. The preparation of permanent mounts of the cross, radial, and tangential sections of the woods for study under the compound microscope has been greatly facilitated through co-operation with individual investigators and scientific institutions, notably members of the faculty and their students at Harvard University. The Yale wood collections have provided material for research work in many parts of the world and the results of these investigations have been freely drawn upon in the preparation of this book.

Taxonomic botanists base their concepts of families, genera, and species on morphological characters of the reproductive and vegetative parts of a plant, and are not always in agreement as to the constitution of particular groups. The present authors have considered these taxonomic problems from the standpoint of wood anatomy and, when confronted with alternative proposals by different botanists, have made their choice on the basis of the anatomical characteristics. Some of these classifications are adopted only tentatively because the material available is not considered adequate for a final decision, but in such instances the alternative names and important synonyms are given. Various inconsistencies in current classifications are called to the attention of specialists in those groups.

The histological descriptions of the woods are mostly limited to family groups and are in finer print since they contain highly technical information of interest primarily to wood anatomists. (For explanation of the wood descriptions see p. 561). Accounts of several of the families, with descriptions of the different genera, were originally published as separate articles in *Tropical Woods* and in some instances were accompanied by keys. Such publication resulted in numerous constructive criticisms and the addition of important new material.

Although much progress has been made in the study of American woods since the publication of *Timbers of Tropical America,* it still must be said, as in the preface of that book, that "the field covered is so vast and the material and literature often so fragmentary that the work is necessarily of the nature of a general reconnaissance, with such attention to details as the subject seemed to warrant or as conditions in a given case would permit. It is hoped, however, not only that the original contributions and the digest of the available published material will prove helpful, but also that the calling of attention to the many broad gaps in our knowledge and to the chaotic condition of some of the literature and information will serve the useful purpose of stimulating and directing exploration and research on the part of other investigators."

S. J. RECORD

December 1, 1942.

CONTENTS

LIST OF ILLUSTRATIONS

PHOTOMICROGRAPHS

following page 588

MAPS

GYMNOSPERMS

THE commercially important Gymnosperms are members of a natural order of plants known as the Coniferae, or cone-bearers. These are all resinous trees or shrubs, typically with evergreen needle-like or scale-like leaves and bearing their seeds on the faces of cone scales. Their stems consist of a cylinder of non-porous, fairly homogeneous wood surrounding a small core of pith near the center and completely enveloped by bark. Coniferous trees growing in the north temperate zone supply most of the structural lumber in the markets of the world. The acreage and volume of the accessible timber (about one-third of the total conifer area) and the amount of standing timber cut or destroyed annually are shown in the following table prepared by Fraser Story in 1928 (*Australian Forestry Journal* 11: 3: 108):

ACCESSIBLE CONIFER FOREST (ESTIMATED)

COUNTRY	AREA *Million acres*	PRESENT STAND *Million cubic feet*	ANNUAL DRAIN *Million cubic feet*
Canada	240	100,000	4,000
U.S.A.	190	390,000	12,000
Europe	240	285,000	8,000
Siberia	100	100,000	1,000
Others	110	110,000	1,500
Total	880	985,000	26,500

The Coniferae are divisible into seven families, namely, Araucariaceae, Cephalotaxaceae, Cupressaceae, Pinaceae, Podocarpaceae, Taxaceae, and Taxiodiaceae. All of them, except the Cephalotaxaceae, are represented in the New World. The families of greatest importance in North America are Pinaceae (*Abies, Larix, Picea, Pinus, Pseudotsuga,* and *Tsuga*), Taxodiaceae (*Sequoia* and *Taxodium*), and Cupressaceae (*Chamaecyparis, Juniperus, Libocedrus,* and *Thuja*); in South America, Araucariaceae (*Araucaria*), Podocarpaceae (*Podocarpus* and *Saxogothaea*), and Cupressaceae (*Fitzroya* and *Pilgerodendron*). The two genera with species on both continents are *Libocedrus* and *Podocarpus*.

ARAUCARIACEAE

THIS is a small family of two closely related genera (*Agathis* and *Araucaria*) with over 30 species of large and interesting trees, survivors of a group widely distributed in Tertiary times but now almost entirely confined to the southern hemisphere. *Agathis,* with about 20 species, extends from the Philippines through Cochin-China and Malaysia to Australia and New Zealand; the best known tree is the New Zealand Kauri, the source of valuable timber and of Kauri resin.

Araucaria, with a dozen species, is represented in New Guinea and New Caledonia; eastern Australia, where the principal trees are the Hoop Pine (*A. Cunninghamii* Sweet) and Bunya Pine (*A. Bidwillii* Hook.); Norfolk Island, with one species (*A. excelsa* [Lamb.] R. Br.) attaining

MAP I. General location of South American timbers of commerce.

great size, though young specimens are often cultivated as house plants; and southern South America with two species, described below.

Chilean Pine, *Araucaria araucana* (Molina) K. Koch (= *A. imbricata* Pavon), occupies a small area in the Coast Cordillera of the Chilean province of Arauco (from which the genus derives its name), but is more abundant on both sides of the Andean divide between 37° and 40° south latitude (Plate II and Map 2). According to E. L. Bernath (*Tropical Woods* 52: 22) it "attains a height of 90 to 135, occasionally up to 150, feet, and a diameter (breast high) of 45 to 60 inches. Trees growing in the open have a large crown and rather pendulous branches covering more than half of the stem, but those in the forest have a small spherical or ellipsoid crown which extends less than a third of the way down the long cylindrical trunk and is composed of branches that are nearly vertical near the stem but curve toward the extremities. The bark, which is thick and broken into polygons of various sizes, has a very distinctive appearance; it secretes a resin that is collected and used medicinally by the Araucanian Indians. The scale-like, imbricated leaves are very stiff and sharp-pointed. The cones are as large as a man's head, and each contains from 120 to 180 large seeds, called 'piñones,' which ripen in March and form an important article of commerce in all parts of Chile, as they have a delectable aroma and flavor upon roasting.

"The pale yellowish, fine-textured timber is of good quality and is used for general construction, ceiling, flooring, doors, window sash, furniture, and boxes, and is considered suitable for paper pulp, although there is no pulp industry in the region. The lumber is obtainable in the markets of Temuco, Mulchen, Victoria, and Angol, but the annual consumption is small because of the difficulties attending the logging and extraction of the timber. It has been roughly estimated that there are about 600,000 acres of Araucarian forest in Chile having an average of 36 trees per acre, with a total possible yield of 1,215,000,000 cubic feet of timber, but most of the trees are scattered or in open stands in high and hardly accessible mountainous regions. As the tree is readily grown from seed and is common in southern parks and gardens, it is probably suitable for forest planting within its natural range."

Paraná Pine, *Araucaria angustifolia* (Bert.) O. Ktze. (= *A. brasiliana* A. Rich.) is the most important timber tree in Brazil. Its commercial range, originally amounting to about 100,000 square miles centering in the State of Paraná, is rapidly being reduced. The botanical distribution covers parts of Paraguay and Argentina (Misiones) and the Brazilian plateau region of Rio Grande do Sul, Santa Catharina, and Paraná, and small areas of western São Paulo and southeastern Minas Geraes (Map 2). It also occurs scatteringly in the mountains of Rio de Janeiro, Espirito Santo, and Bahia, but it is not unlikely that this northern extension represents escape from cultivation by early settlers or to dissemination by aborigines who used the large seeds for food. The tree appears to be more exacting as to climate and other environmental factors than the Chilean species and is rarely grown outside its regular habitat.

The mature trees are from 80 to 120 feet tall, with long, clear boles bearing a flattened crown of upturned limbs tufted at the ends, giving the effect of giant candelabra (Plate III). They grow gregariously, but the stands, instead of being open and park-like, as is the case with some of the Pine (*Pinus*) forests, have a second story and undergrowth of hardwoods and some Podocarps (*Podocarpus Selloi* Klotzsch and *P. Lambertii* Klotzsch), except where cleared to make room for the cultivation of Mate bushes (*Ilex paraguariensis* St. Hil.), the source of Paraguay tea. The principal hardwood timber in this association is the Imbuia (*Phoebe porosa* [Nees] Mez).

Paraná Pine is the most extensively exploited species in South America and supplies many million feet of lumber for the cities and agricultural regions of southern Brazil, Uruguay, Paraguay, and Argentina. Some of the logging and milling operations are conducted on a large scale with the type of equipment used in the Pine forests of southern United States. The sapwood is

MAP 2. Ranges of Paraná and Chilean Pines and Quebracho.

Plate II. Chilean Pine (*Araucaria araucana*) in the southern Andes Mountains.

PLATE III. Paraná Pine (*Araucaria angustifolia*) in southern Brazil. The understory is composed largely of Mate bushes (*Ilex paraguensis*) whose leaves are used for tea.

yellowish, the heartwood of various shades of brown, sometimes with bright red streaks. Growth rings are fairly distinct, but without prominent alternating bands of early and late wood. The sp. gr. (air-dry) is usually between 0.50 and 0.60, occasionally considerably greater; weight 30 to 40 lbs. per cu. ft. The wood is unscented, fairly straight-grained, of uniform texture, easy to work, requires care in seasoning, holds paint well, but is not highly resistant to decay. Limited amounts of the lumber are being imported into the United States for general millwork and such special uses as slats for Venetian blinds and backing for electrotypes. Considerable quantities were exported to Germany in 1938 and the timber, when thoroughly seasoned, was considered suitable for the same purposes as European conifers; the demand was for uniformly light-colored lumber (see *Tropical Woods* 58: 41).

The following description applies to the woods of both species. Growth rings distinct to apparently absent. Tracheids long; walls moderately thick; pits on radial walls mostly near ends of cells, close together without crassulae, usually alternately arranged in two or more rows; pits to ray cells bordered, numerous; spirals absent; resinous plates common, opposite the rays. Rays uniseriate, variable in height, the walls thin and unpitted; tracheids absent. Wood parenchyma sparingly developed.

Common names: *Araucaria araucana:* Araucaria, pehuen, pino (Chile, Arg.); pino chileno, p. piñonero (Chile). *A. angustifolia:* Paraná pine (English); pinheiro do Paraná, pinho, p. branco, p. do Paraná, vermelho (Braz.); curi-y, pinheiro do Brasil, pino blanco, p. colorado (Par.); curiy, kuviy, pino, p. del Brasil (Arg.).

CUPRESSACEAE

The Cypress family, with about 16 genera and 130 species of shrubs and small to very large trees, is widely dispersed over the world, though mostly in the temperate zones. Many of the plants are grown for ornamental purposes. In North America there are members of five genera, namely, *Chamaecyparis, Cupressus, Juniperus, Libocedrus,* and *Thuja,* but only *Cupressus* and *Juniperus* are represented in the tropics. One species each of *Fitzroya, Libocedrus,* and *Pilgerodendron* occur in the southern Andes Mountains of South America. The leaves are decussate or in three ranks, often of two forms, usually scale-like and producing flattened or angled branchlets; the fruit of *Juniperus* is a berry, that of the other genera a small cone with few scales (peltate in *Chamaecyparis* and *Cupressus*).

The timbers of this family are noted for their fragrant scent, fine and uniform texture, straightness of grain, ease of working, and high resistance to decay and insect attacks. The color of the heartwood varies from pale yellow through various shades of brown to reddish or purplish; the sapwood is white, often sharply demarcated. There is considerable range in density, from very light and soft (*e.g., Thuja*) to comparatively heavy and hard (*e.g., Juniperus*). Some species, notably of *Chamaecyparis, Juniperus,* and *Thuja,* are highly important commercially.

Growth rings present, often very distinct. Tracheids rectangular, rounded, or variable in cross section; transition from early to late wood gradual to abrupt; radial pitting mostly uniseriate, occasionally biseriate and rarely multiseriate in early wood; pits to ray cells 2 to 4 per cross-field, often simple and almost circular in *Thuja,* but with elliptic to circular border and slit-like to narrowly elliptic, vertical to diagonal aperture in the others; spiral thickenings absent in American genera (special type of thickenings present in some species of *Callitris*). Rays typically uniseriate, occasionally biseriate in part; maximum heights, 10 to 40 cells; walls without secondary thickening, hence without true pits, but with occasional to abundant depressed primary pit-fields; resinous deposits abundant in dark-colored heartwood; ray tracheids of sporadic occurrence, rather common to rare and difficult to find. Wood parenchyma diffuse, scattered to loosely zonate. Resin ducts absent. (For further details see "Systematic anatomy of the woods of the Cupressaceae," by Alan S. Peirce, *Tropical Woods* 49: 5–21.)

Chamaecyparis, with six species of trees, is confined to Japan and Formosa in the Old World and to the Atlantic and Pacific coast regions of North America. The Japanese trees in various forms are extensively planted for decorative purposes and are usually known to horticulturists as *Retinospora.* There are three American species. Some botanists consider *Chamaecyparis* a subgenus or section only of *Cupressus.* The principal distinctions are that the species of *Chamaecyparis* produce small, early deciduous cones with not over five membranous-winged seeds under each scale, whereas those of *Cupressus* have characteristically large, long-persistent cones, each scale covering numerous seeds having no membranous wings. The fructification in *Cupressus* is biennial, that of *Chamaecyparis* is usually, but not always, annual.

Southern White Cedar, *Chamaecyparis thyoides* (L.) B.S. & P., is a small to large tree often in pure stands of irregular occurrence in the Atlantic and Gulf coastal swamps and estuaries from southern Maine to southern Florida and southwestern Louisiana. The present centers of timber production are southern New Jersey, eastern North Carolina, and the Gulf coast; timber buried for centuries in peat bogs is still being utilized commercially along the New Jersey coast. Fresh heartwood is pale brown with a decidedly pinkish tinge and has a mild fragrant scent. It is light and soft (weight about 23 lbs. per cu. ft.); rather fine-textured, straight-grained, easy to work, and durable. The principal uses, besides posts and poles, are for boat planking, tank staves, shingles, woodenware, planing mill products and millwork (including house siding and trim). (See "Southern white cedar," U.S. Dept. Agr. Technical Bull. no. 251.)

Alaska or Yellow Cedar, *Chamaecyparis nootkatensis* (Lamb.) Sudw. (= *Cupressus nootkatensis* Lamb.), is a tree often 120 feet high and five or six feet in diameter, growing singly or in scattered clumps in the coastal forest from southwestern Alaska through British Columbia into northern California, reaching its best development in British Columbia. At high elevations and in dry situations it is small and often very much stunted. The estimated stand of saw timber is about 10 billion board feet, but most of it is not readily accessible and consumption is generally local for interior trim, furniture, and hulls of small boats. Alaska Indians prefer it for canoe paddles. The heartwood is of a pale yellowish color, mildly but unpleasantly scented, moderately light and soft (av. sp. gr. about 0.45), fine-textured, straight-grained, easily worked, and durable.

Port Orford Cedar, *Chamaecyparis Lawsoniana* (A. Murr.) Parl. (= *Cupressus Lawsoniana* A. Murr.), is a large and important timber tree of very limited distribution in a strip several to 40 miles wide along the Pacific coast from Coos Bay, Oregon, to Eureka, California. It grows in mixture with Douglas Fir, Red Cedar, and Hemlock, but at its best toward the northern part of its range it may attain a density of 20,000 board feet per acre and single acres have produced five times that amount of lumber. Heights of 135 to 175 feet and diameters up to seven feet are common, while individuals 200 feet tall and 12 feet through are occasionally found, the deeply furrowed bark nearly a foot thick at the base of the trunk. The bright, pale yellow pungently scented timber has exceptionally good technical properties. It weighs (air-dry) about 30 lbs. per cu. ft., is comparatively firm and strong, of uniform, rather fine texture and straight grain. It seasons readily, is easy to work, finishes very smoothly, takes paint and enamel well, holds its place when manufactured, and is highly durable. Its many uses include sash and doors, house siding, flooring, boat construction, storage battery separators, Venetian blinds, match sticks, light furniture, all kinds of millwork, and posts, poles, and railway crossties.

Common names: *Chamaecyparis thyoides:* Cedar (post, southern white, swamp, white), juniper (U.S.A.). *C. nootkatensis:* Cedar (Alaska or yellow), cypress—Alaska, Nootka, or Sitka (U.S.A.). *C. Lawsoniana:* Cedar (Oregon, Port Orford, or white), ginger pine, Lawson's cypress (U.S.A.).

Cupressus. The true Cypresses are native to Asia, the eastern Mediterranean region, and western North America. Most noted of the dozen or so species is *C. sempervirens* L. of southeastern Europe and southwestern Asia, from whose fine-textured, fragrant, and durable timber were constructed the gates of Constantinople and the doors of St. Peter's which remained sound until their removal after eleven centuries of service. The trees closely resemble the Junipers, but the cones are larger and open to shed the seeds upon ripening. They vary in form from nearly prostrate shrubs on wind-swept sites to stately trees in protected valleys. They are frequently planted for decorative purposes, but are nowhere of much importance for their timber.

The several American species and varieties are typically mountain trees and shrubs and have a combined range from southwestern Oregon to California and New Mexico and southward through Mexico into Central America. The most northern form, *Cupressus Macnabiana* A. Murr., is a rare and local tree occasionally 80 feet high and over three feet in diameter in Oregon, becoming smaller in California. The Monterey Cypress, *C. macrocarpa* Gord., is a picturesque tree of very limited distribution along the coast of central California. Closely related to it and sometimes considered only a variety is the Tecate Cypress, *C. guadalupensis* S. Wats., a small tree of southern California and Baja California. The species with the most eastern range is *C. arizonica* Greene, a stout-boled tree, rarely five feet in diameter, in Arizona, western New Mexico, and adjacent regions in Mexico. The principal Mexican species appears to be *C. lusitanica* Mill., which, including the variety *Benthami* (Endl.) Carr. (= *C. Benthami* Endl.), grows from about 21° north latitude southward through the mountains into Guatemala and Honduras. It is reported in Costa Rica, but according to Standley (*Flora of Costa Rica*, p. 6) it occurs there only in cultivation. The timber is highly prized locally throughout its range, but the native trees occur scatteringly in poorly accessible places. (See *Tropical Woods* 65: 1.)

Heartwood yellowish, pale brown, or pinkish, sometimes more or less streaked or variegated; sapwood white, usually sharply demarcated. Luster rather high. Fragrantly scented. Light and soft to hard and heavy; texture very fine and uniform; grain straight to irregular; working properties excellent; durability high.

Common names: Cypress (English): ciprés (Spanish); cedro, c. amarillo, c. blanco, c. de la sierra, ciprés de Mexico, gretado amarillo, g. galán, pinabete, teatlale, tlascal, tlatzcán, tlazzcán (Mex.).

Fitzroya, with a single species, *F. cupressoides* (Molina) Johnston (= *F. patagonica* J. D. Hooker), is a Chilean tree called Lahuán by the Indians, but more commonly known as Alerce (Larch), the name given it by the Spanish Conquistadores, though having nothing to do with true Larch (*Larix*). The following information is taken from E. L. Bernath's "Coniferous forest trees of Chile" (*Tropical Woods* 52: 19: 24–26).

"So far as definitely known, the area of its distribution begins near the city of Valdivia (39° 45′) and ends near the Río Futalelfú (43° 29′). It is typically a species of low swamps, but grows at higher elevations on Isla de Chiloé and in the Territory of Aysen in Patagonia. It is unique among Chilean conifers in forming dense, nearly pure forests over thousands of acres.

"The tree bears considerable resemblance to the California Redwood (*Sequoia sempervirens*). On Chiloé and in the Cordillera de Piuchué the average height of maturity is about 100 feet, but near Puerto Montt the old trees are mostly from 130 to 150 feet tall and one specimen measured 240 feet. The famous 'Silla del Presidente,' which formerly stood near Puerto Montt, was 15 feet in diameter; others in the same region reach a diameter (breast high) of nine feet. On Chiloé the average diameter is about four feet. The tree grows slowly and attains an age of more than 1000 years. Young trees have a conic-pyramidal form and very dense, dark green foliage; old ones have a small crown at the top of the long, straight bole, and so many of the branches are dry that the forest, seen from a distance, appears to be dead. The leaves, which are ericoid and pointed, are limited to the young twigs. The cones are small, terminal, with six scales and nine winged seeds. The bark is several inches thick, corky, and, upon incision, exudes a resin of agreeable odor. The inner bark called 'estopa de Alerce,' is harvested in the summer by men, women, and children who make bales of the thin strips and carry them on their backs to the nearest towns or harbors for use in calking boats and ships. The same people also collect resin and burn it as incense. The wood is of fine and uniform texture, straight and fine grain, low density (less than that of *Populus nigra* L.), and is strong and elastic for its weight. The sapwood, which is thin and white, is not utilized; the heartwood is red and long-lived under exposure.

"Alerce is exploited throughout its range (Plate IV). Usually the logs are not hauled to mills and sawed, but are worked in the forest by natives who cut the big trees and split them into boards and shingles with wedges made of the hard wood of *Myrtus luma*. In the 'alercales,' or great Alerce forests of the South, the woodsmen are specialists in this timber and produce rived boards of uniform thickness by means of a single wedge inserted at one end of a log. Men carry the lumber and shingles to market on their shoulders, often for many miles over poor roads through the swamps. On the Isla de Chiloé similar packing is done from the Cordillera Piuchué to Ancud, a very old town founded by the Spaniards. The cities of Puerto Montt and Osorno get their supply of Alerce in the same way from the distant forests in the Cordillera. The woods operations are limited to the dry season, from spring to autumn, during which time the woodsman lives with his family in a temporary dwelling, or 'rancho' built entirely of rived Alerce. The timber is the finest produced in Chile and ranks with the best and most useful in the world. It is ideal for carpentry and light and durable construction of all kinds, and for honey barrels and musical instruments; limited quantities are exported for making pencils and cigar boxes. The lumbermen also make and sell troughs that are noted for their durability. The price of the lumber in the Province Llanquihue is higher than for any other native timber. The total consumption is not large, however, owing to the present inaccessibility of most of the alercales."

Juniperus. The Junipers, of which there are about 60 species and numerous varieties, have an exceptionally wide distribution throughout the northern hemisphere, extending from the Arctic Circle to the Tropic of Cancer and crossing the equator in the mountains of eastern Africa. They exhibit various forms, from prostrate or sprawling shrubs to small, medium-sized or occasionally large trees with bushy, pyramidal, or spreading crowns. They are pungently aromatic; the leaves are sharp-pointed and in whorls of three or, more often, small and scale-like; the fruit is a fleshy berry formed by the coalition of the flower scales; and the bark is typically thin and shreddy. The heartwood is brown in some species, reddish to purplish, often variegated, in others; it has a pleasant and lasting aroma; is of very fine and uniform texture, firm and fairly hard, fissile, easy to cut smoothly in any direction, holds its place well when manufactured, and is highly resistant to decay. Its outstanding use, for which no entirely satisfactory substitute has been found, is for the manufacture of lead pencils, and the species most commonly employed are the Red Cedars of eastern and southeastern United States, *Juniperus virginiana* L. and *J. lucayana* Britt. (the latter extending into the West Indies), and more recently the East African Cedar, *J. procera* Hochst. (see *Empire For. Journ.* 5: 1: 39–53. 1926). The Himalayan Pencil Cedar is *J. macropoda* Boiss.

The most widely distributed Juniper is *Juniperus communis* L. with various forms and varieties extending over most of the northern hemisphere. In North America it is a prostrate shrub or decumbent tree, only rarely erect. The aromatic fruit is used in large quantities in Europe to impart a special flavor to gin. The species having the greatest diameter is *J. occidentalis* Hook., which attains its best development in the California sierras at altitudes between 6000 and 10,000 feet above sea level. One specimen in Tuolumne County was measured by W. S. Glock of the Carnegie Institution of Washington and found to be 80 feet high and 14 feet in average diameter at breast height; maximum diameter at the ground, 21.5 feet. The outer 12-inch layer of wood contained 700 annual rings; the estimated age of the tree is at least 3000 years. In the

Photo by Max Rothkugel.

PLATE IV. Hauling split logs of Alerce (*Fitzroya patagonica*) in the southern Andes Mountains.

Photo by G. P. Lewis

PLATE V. Sabina or Red Cedar tree (*Juniperus gracilior*), 41 inches in diameter and 40 feet to the first limb, in Dominican Republic.

vicinity of this exceptionally large tree are many others 8 to 10 feet in diameter.

The common Juniper tree or Red Cedar of most of the eastern half of the United States and southern Ontario is *Juniperus virginiana* L. It is described by Sargent (*Manual of the trees of North America*, p. 88) as "a tree, occasionally 100 feet high, with a trunk three to four feet in diameter, often lobed and eccentric, and frequently buttressed toward the base, generally not more than 40 to 50 feet tall, with short slender branches horizontal on the lower part of the tree, erect above, forming a narrow compact pyramidal head, in old age usually becoming broad and round-topped or irregular, and slender branchlets terete after the disappearance of the leaves and covered with close dark brown bark tinged with red or gray; on exposed cliffs on the coast of Maine, sometimes only a few inches high with long branches forming broad dense mats." It does not ascend the mountains of New England, New York, nor the high southern Alleghenies; in middle Kentucky and Tennessee and northern Alabama and Mississippi it covers great areas of rolling limestone hills with nearly pure forests of small bushy trees. The best of the timber is cut into pencil slats; knotty lumber is used in making clothes chests and for lining wardrobes and closets and the sawdust is distilled for ethereal and other oils used in perfumes and medicines and in various industries.

The Juniper or Red Cedar in swamps and along streams in Florida and southern Georgia is *Juniperus lucayana* Britt. (= J. *barbadensis* Sarg., non L.). It also grows in coastal and high land of the West Indies. It is a spreading flat-topped tree in the open but attains a height of 50 feet with an erect trunk of timber size in the forest. The wood is used for the same purposes as that of *J. virginiana*, and formerly was in demand for cigar boxes. Some of the Cuban wood has proved satisfactory for archery bows (see *Tropical Woods* 3: 9).

There are other species in the West Indies, but the only likely source of commercial timber remaining appears to be in the Cordillera Central of Dominican Republic. The trees, probably *Juniperus gracilior* Pilger, attain large size, as one specimen measured by Mr. G. P. Lewis, of Meredithville, Va., had a straight slowly tapering trunk 41 inches in diameter three feet above the ground and was entirely free of branches for 40 feet (Plate V). The heartwood is dark red and is appreciably harder than *J. virginiana*.

There are four or five species in Mexico, the most widely distributed being *Juniperus mexicana* Spreng., which occurs from western Texas to the highlands of Guatemala. The brown wood is used locally for fence posts, railway crossties, telegraph poles, and general construction, but is not exported.

COMMON NAMES: Cedar, juniper, pencil cedar, red cedar, savin (U.S.A.); enebro (P.R.); enebro criollo, sabina, s. de costa (Cuba); sabina (Dom. R.); cyprès (Haiti); cedro, c. colorado, enebro, sabino, táscate, taxate, yutnu-itne (Mex.); ciprés, huitun (Guat.).

Libocedrus, with eight species of trees, occurs in southwestern China, the Moluccas, New Guinea, New Caledonia, New Zealand, western United States, and southern Chile and Patagonia.

The Incense Cedar, *Libocedrus decurrens* Torr., the only species in North America, is a medium-sized to very large tree (maximum recorded height 186 feet, diameter eight feet), growing singly or in small groves from Mount Hood, Oregon, through the mountains of California and western Nevada into Baja California. It attains its best development at elevations of 5000 to 7000 feet above sea level in the Sierra Nevada mountains of central California. It has a rapidly tapering bole, which in youth is covered with flaky, rather than shreddy, bark. The scale-like leaves are borne in flat sprays, suggesting Arbor-vitae or Northern White Cedar, *Thuja occidentalis* L. Most of the large trees are affected by dry rot or "peckiness," such as occurs in Southern Cypress (*Taxodium*), which seriously reduces the amount of clear lumber obtainable.

The heartwood is spicily scented and is light brown, tinged with red, darkening with age and exposure; the sapwood is white or cream-colored. It is light and soft, weight (air-dry) about 25 lbs. per cu. ft.; fine-textured, fissile, very easy to work, and is highly durable. Most of the timber is employed locally for building purposes, posts, and poles. Most of the finest grade is used as a substitute for Juniper (*Juniperus*) in making lead pencils.

As usually classified, there are two species of *Libocedrus* in South America, but Rudolf Florin (*Svensk Bot. Tidskrift* 24: 1: 132–135. 1930) has proposed a separate genus, *Pilgerodendron,* for one of them and it is considered under that designation here. The other, *L. chilensis* (Don) Endl., called Ciprés in southern Chile and Cedro in Patagonia, occurs, according to Bernath (*Tropical Woods* 52: 22), "on dry, sterile, rocky hillsides and mountains, where no other trees can thrive, in the Andes Cordillera between 34° 25′ and 44° south latitude. In the locality of the Río de los Cypreses (Province Colchahua), its northern limit, the species is found at elevations of 4500 to 4800 feet; at 35° in the valley of the Río Teno (Province Curicó), 3900 feet; at 36° 40′ in the Cordillera de Chillán, 2700 feet; at 38° 3′ south latitude and 72° 35′ west longitude, at Hacienda Nupangue (Province Malleco), it occurs at about 1000 feet, while still farther south it descends to 300 feet or less. It often grows gregariously, forming small, nearly pure stands, sometimes several acres in extent. Young trees are pyramidal and resemble *Chamaecyparis Lawsoniana* Parl.; old ones are flat-crowned and have a thick trunk usually clear of limbs for two-thirds of its length and covered with a thick, furrowed, grayish brown bark. The juvenile leaves are about half an inch long and sharp-pointed, but those developed later are comparatively broad and scale-like. The cones, which are small, ripen in February or March and soon fall to the ground; they are borne only on old trees. The seeds have a broad wing and germinate readily. The wood is fine-textured, straight-grained, light in weight, and easy to work. Growth rings are distinct and usually rather broad. The sapwood is thin and nearly colorless, the heartwood reddish brown, scented, and highly resistant to decay. Owing to the inaccessibility of the trees, the timber is not exploited commercially, but is suitable for a great many purposes requiring ease of working and durability rather than great strength."

Common names: Cedar (bastard, incense, post, white), juniper (U.S.A.); ciprés (Chile); cedro (Arg.).

Pilgerodendron is a Chilean genus with a single species, *P. uviferum* (Don) Florin (= *Juniperus uvifera* Don = *Thuja tetragona* Hooker = *Libocedrus tetragona* [Hook.] Endl. = *L. cupressoides* Sarg. = *L. uvifera* [Don] Pilg.) The wood structure indicates a very close relationship to *Libocedrus*. The tree is known locally as Ciprés, Ciprés de las Guaytecas, and Lahuán. Its range extends from latitude 40° (Río Valdivia) to Tierra del Fuego, including the various islands and archipelagos. Its vertical distribution is from near sea level to 3300 feet.

According to Bernath (*Tropical Woods* 52: 25), the tree is of slow growth but attains a height of 90 feet or more. The crown is small and pyramidal in youth, becoming open and irregular and tending to be dry-topped at maturity. The dull green, scale-like leaves are 4-ranked, making the twigs characteristically squarish in section. The cones, which ripen in April, consist of two pairs of scales, bearing four small-winged seeds. "The sapwood is creamy yellow, the heartwood brownish and very durable. The timber is of the first class and is much exploited in the South, particularly on the Isla de Chiloé and the Patagonian islands, for house construction, flooring, doors, window sash, and furniture. The smaller sizes are in demand for fence and vineyard posts and telephone poles and are noted for their lasting quality in moist soil."

Thuja (or *Thuya*), with five species of trees, is confined to northeastern Asia and northeastern and northwestern America. The Chinese *T. orientalis* L., with numerous horticultural varieties, is often planted in the United States for decorative purposes, as are the native species. The woods are yellowish to reddish brown, fragrantly scented, very light and soft (sp. gr. 0.30 to 0.45), of medium to coarse-texture, straight-grained, easily worked, and moderately to highly resistant to decay. The two species in the New World are confined to temperate North America.

Northern White Cedar, *Thuja occidentalis* L., is a tree 50 to 60 feet high and one to three, rarely up to six, feet in diameter or divided into two or three secondary stems, of common occurrence and frequently forming dense forests on swampy ground in Canada from Nova Scotia and New Brunswick, becoming less abundant and smaller in the Lake States and in the Atlantic region to New Jersey and along the mountains to North Carolina and Tennessee. The principal use for the timber is in the round for fencing, poles, and street railway crossties. When used for decorative planting the tree is usually called Arbor-vitae.

Western Red Cedar, *Thuja plicata* D. Don, is a very important timber tree 100 to 150, occasionally up to 200, feet tall with a massive trunk three to eight, rarely 12, feet in diameter about the buttressed base, growing principally in a belt along the Pacific coast from southern Alaska into northern California and extending inland to western Montana, being at its best in the Puget Sound region. The maximum age attained is about 1000 years. About two-thirds of all the timber cut is converted into shingles, supplying about 90 per cent of all used in the United States. The lumber is used for bevel siding, closet lining, interior trim, sash and doors, patterns, tank and pail staves, boat construction, porch columns, greenhouse sash and trays, and many other purposes where durability rather than strength is a requisite. Small trees are the source of large quantities of poles and fence posts.

COMMON NAMES: *Thuja occidentalis:* Arbor-vitae, cedar—northern white, white (U.S.A., Canada). *T. plicata:* Arbor-vitae, cedar (canoe, red, western, w. red), giant arbor-vitae, shinglewood (U.S.A.); red cedar (Canada).

EPHEDRACEAE

Ephedra, the only genus in this family, includes about 35 species of shrubs inhabiting warm, dry regions in both hemispheres. The stems are slender and jointed; the leaves are reduced to opposite or verticillate scales; the staminate flowers are in short catkins, the fertile inflorescence is cone-like; the fruit is nut-like, angled and somewhat fleshy. The only use for the plants is medicinal, particularly as a source of the drug ephedrine. The wood is described here because its anatomy is intermediate of the Gymnosperms and Angiosperms.

Growth rings present. Pores numerous, small to very small, typically solitary; fairly evenly distributed to more or less zonate in early wood (Plate XXXVII). Vessels with foraminate perforation plates; without spirals. Rays one to several cells wide and few to many cells high, greatly variable in the same specimen; large rays usually with herring-bone appearance on cross section due to intrusions of vertical tracheids; heterogeneous, the cells of many different sizes and shapes throughout and abundantly pitted; pits to vessels and tracheids small, rather few, the pit-pairs half-bordered; ray tracheids absent. Wood parenchyma very sparingly diffuse. Tracheids of the coniferous type compose the ground mass of the wood; radially flattened in late wood; bordered pits large and numerous in both radial and tangential walls; crassulae present in radial walls. No resin ducts seen.

COMMON NAMES: Joint fir (English); cañatilla, hintomo real, itamo real, pitamo real, popotillo, retama real, sanguinaria, tepopote (Mex.); cola de caballo, pfirco, pfiuco, pinco-pinco (Peru); cipó da areia, morango do campo (Braz.); pingó-pingó (Boliv., Par.); pico de loco (Urug.); tramontana (Arg.).

PINACEAE

THE Pine family includes such well-known and important timber trees of the north temperate zone as the true Pine (*Pinus*), Spruce (*Picea*), true Fir (*Abies*), Larch (*Larix*), Douglas Fir (*Pseudotsuga*), Hemlock (*Tsuga*), and the famed Cedar of Lebanon (*Cedrus*). Although some of them grow well in warm regions, only a few species of *Pinus* grow at low elevations in the tropics. No member of the family is native to South America, only three genera (*Pinus, Abies,* and *Pseudotsuga*) occur in Mexico, and only one (*Pinus*) is found in the West

Indies and as far southward as Nicaragua.

Heartwood yellowish or pale brown to pinkish or reddish brown; not always clearly differentiated from the sapwood in *Abies, Picea,* and *Tsuga.* Luster medium to high. Odor mild, but often distinctive; not fragrant (except in exotic genus *Cedrus*); unpleasant in fresh specimens of some species of *Abies* and *Tsuga.* Light and soft to heavy and hard; texture fine to rather coarse; grain generally straight; working properties good, though variable; durability low to high.

Growth rings fairly to very distinct; late wood thin to thick, with gradual to abrupt transition from early wood. Tracheids usually thin-walled in early wood, thick-walled in late wood; striations common in dense wood; spiral thickenings present in *Pseudotsuga;* pits on radial walls large, usually in a single row, with distinct crassulae; pits on tangential walls small, confined to late wood (absent in Pitch Pine group). Rays all uniseriate or locally biseriate, except those containing resin ducts in *Larix, Picea, Pinus,* and *Pseudotsuga;* ray tracheids of constant occurrence except in *Abies* where they are absent or sporadic; parenchyma cells with secondary walls and blind pits; pits to vertical tracheids mostly small, sometimes large (White Pines). Wood parenchyma not present (except around resin ducts) or limited to scattered cells on outer face of late wood or rarely diffuse (*Tsuga*). Vertical resin ducts of normal occurrence in *Larix, Picea, Pinus,* and *Pseudotsuga;* of traumatic occurrence in *Abies* and *Tsuga;* constricted at intervals or like vertical series of cysts except in *Pinus;* epithelial cells thin-walled in *Pinus;* typically thick-walled in the others; strand tracheids frequently present.

Abies, with about 40 species of evergreen trees, is widely distributed in Asia, Asia Minor, Central Europe, and sparingly in the highlands of northern Africa, and in the New World from Alaska through the Pacific and Rocky Mountain region to Honduras and through central and southern Canada to Labrador and southward into the Lake States and along the Appalachian Mountains to North Carolina. There are about 13 species in North America.

The range of the common Balsam Fir, *Abies balsamea* Mill., is much the same as that of the White and the Black Spruce, with the heaviest commercial stands in Quebec and Ontario, Canada. It often forms nearly pure forests on swampy land and the trees are generally small. The principal use for the timber is, in mixture with Spruce, for the manufacture of pulp. A resin, known as Canada balsam, is obtained from "blisters" in the bark. The Balsam Fir at elevations of 4000 to 6000 feet above sea level in the southern Appalachians is *A. Fraseri* Poir.

A group of species commonly known as White Firs attain large size in western North America and supply white, often coarse-grained, rather weak and brittle timber. *Abies grandis* Lindl. occurs in the coastal region from Vancouver to central California and inland through Washington to Idaho and western Montana. At its best near the coast it attains a height of nearly 300 feet and a diameter of four feet. *A. concolor* Lindl. & Gord. is scattered through the Rocky Mountains, but is most abundant and of largest size, up to 250 feet tall and six feet in diameter, in southern Oregon and the California Sierras; it is of limited occurrence and rather small size in the mountains of northwestern Mexico. *A. amabilis* Forbes, another very large tree, attains its best development at moderate elevations in British Columbia and Washington. The tree is frequently known as Silver Fir and local lumbermen sometimes call it Larch.

Some of the other western Firs have considerably darker and stronger timber, frequently designated as Red Fir, which is better adapted for general construction. Noble Fir, *Abies nobilis* Lindl., occurs in Washington and Oregon; California Red Fir, *A. magnifica* A. Murr., and Shasta Red Fir, *A. magnifica* var. *shastensis* Lemm., grow in the interior of California and northward into Oregon; some of these trees attain a height of 200 feet or more and diameters of six to eight feet. The lumber of Noble Fir is sometimes sold as Larch. Another species with similar wood is the Alpine Fir, *A. lasiocarpa* Nutt., of high mountainous regions from the Yukon to Arizona and New Mexico; the southern form with thick spongy bark (Corkbark Fir) is considered by some botanists to be a separate species, *A. arizonica* Merr.

The principal species of *Abies* south of

Photo by U.S. Forest Service.

PLATE VI. Stand of old Larch trees (*Larix occidentalls*) on Lolo National Forest, Montana.

Photo by U.S. Forest Service.

PLATE VII. Virgin forest of White Pine (*Pinus monticola*) on the Kaniksu National Forest in Idaho.

the United States is *A. religiosa* (H.B.K.) Schl. & Cham., a small to large tree growing at elevations of 4000 to 11,000 feet in central Mexico. The specific epithet (*religiosa*) refers to the common use of the evergreen branches for decorations in churches. Resin, similar to Canada balsam and known as "aceito de abieto," is obtained from "blisters" in the bark and employed for medicinal and industrial purposes. The wood has distinct growth rings; is whitish or brownish, of rather low density but firm and fairly strong, fine-textured, and nonresinous. It is used locally to a limited extent for common lumber and general construction not requiring great strength or durability.

A form from the mountains of Oaxaca has been described as *Abies Hickeli* Flous & Gaussen, and one from Guatemala as *A. guatemalensis* Rehder. Rehder says (*Journ. Arnold Arboretum* 20: 3: 281) that both species "are easily distinguished even without cones, from *A. religiosa* by the pectinately spreading emarginate leaves while in the latter they are on the upper surface of the branch, directed forward and more or less appressed to the branch and always acute or obtusish at the apex, never emarginate. *Abies guatemalensis,* which so far is known only from a restricted area near Lake Atitlan in the high mountain range along the western coast of Guatemala, marks the southernmost extension of the range of the whole genus, occurring as it does, between 14° and 15° N. lat., while in Asia and Africa it does not even reach the Tropic of Cancer." This is probably the species occurring in the high mountains of Honduras.

Wood nearly colorless to various shades of brown with reddish or pinkish tinge, the late wood often purplish; not clearly differentiated from the sapwood. Luster medium. Without distinctive taste; odor sometimes unpleasant in fresh wood. Very light, soft, and weak to moderately heavy, hard, and strong; sp. gr. (air-dry) 0.30 to 0.60; weight 19 to 38 lbs. per cu. ft.; easy to work; poorly resistant to decay.

COMMON NAMES: Balm-of-Gilead fir, balsam, b. fir, black gum, blister pine, fir (alpine, corkbark, noble, red, silver, white, etc.), f. pine, larch (trade), she balsam (U.S.A.); fir, sapin, single spruce (Canada); abeto, acxoyatl, bansú, guayame, huallame, jolacote, oyamel, oyametl, pinabete, pino, p. oyamel, xalócotl (Mex.); cipreso? (Guat.); pino (Hond.).

Larix, with 10 species of deciduous trees, is widely distributed over the subarctic and mountainous regions of the northern hemisphere. The European Larch, *L. decidua* Mill., is planted for timber and ornament in its native habitat and also to some extent in New England and the Middle West. There are three American species, two of them supplying commercial timber.

The Tamarack or Hackmatack, *Larix laricina* (DuRoi) Koch (*L. americana* Michx.), has a very extensive range from Alaska to Labrador and southward into the Lake States, New England, and the Appalachian Mountains, but the trees are generally small because of unfavorable site. The species is most abundant in sphagnum bogs and muskegs, either in pure stands or in association with Black Spruce (*Picea mariana* B.S.P.), but the largest trees grow on better-drained soils where they attain heights of 50 to 75 feet and diameters of 14 to 20 inches, occasionally more. The timber is used chiefly for fence posts, poles, and railway crossties.

Western Larch, *Larix occidentalis* Nutt., has its center of distribution in northern Idaho, with extensions into Washington, Oregon, Montana, and British Columbia. It is at its best on moist, but well-drained soils and attains a maximum height of about 200 feet and a diameter of six feet, though usually it is considerably smaller. It sometimes forms open, pure stands (Plate VI), but more often is associated with Idaho White Pine, Western Red Cedar, Douglas Fir (*Pseudotsuga*), and in some places with Ponderosa Pine. The timber is suitable for all kinds of construction and is often sold in mixture with Douglas Fir. At high elevations, near timber-line, the species is replaced by *L. Lyallii* Parl., a short-boled tree with woolly branchlets.

Heartwood yellowish brown, never very resinous (*Larix laricina*), or red or reddish brown, sometimes highly resinous (*L. occi-*

dentalis); clearly defined from the white sapwood. Not very lustrous. Without distinctive odor or taste. Moderately to decidedly hard and heavy; sp. gr. (air-dry) 0.55 to 0.85; weight 34 to 53 lbs. per cu. ft.; texture medium, feel rather harsh; grain straight to irregular; working properties fair to good; durability high.

COMMON NAMES: Hackmatack, larch (American, alpine, black, western), tamarack (U.S.A.); epinette rouge, juniper, arch, tamarack (Canada).

Picea, with about 40 species of evergreen trees, mostly Asiatic, is widely distributed in subarctic and temperate regions of the northern hemisphere. The Norway Spruce, *P. excelsa* (Lam.) Link (= *P. Abies* [L.] Karst.) is one of the most valuable timber trees of Europe and is often planted for ornament and shade in the United States. Seven species and a few varieties are recognized in North America, their combined range covering most of Alaska and Canada and extending through the Rocky Mountains to the Mexican border and along the Appalachians to western North Carolina.

Eastern Spruce is a general term used in the timber trade of the United States for the wood of the following three species. White, Cat, or Skunk Spruce, *Picea glauca* (Moench) Voss (= *P. canadensis* [Mill.] B.S.P. = *P. alba* [Ait.] Link), is a tree rarely more than 60 to 70 feet tall and two feet in diameter, usually much smaller, characterized by ill-smelling foliage. It is of common occurrence along water courses and lakes from Alaska to Labrador and southward into the Lake States and New England. Black Spruce, *P. mariana* (Mill.) B.S.P. (= *P. nigra* [Ait.] Link), has the same general distribution as the White Spruce, though extending into the southern Appalachians. In the northern part of its range it grows on well-drained soils and attains fairly large size, but near the United States border it occupies sphagnum bogs and muskegs and often is very small. Red Spruce, *P. rubra* (Lam.) Link (= *P. rubens* Sarg.), which is perhaps only a form of Black Spruce growing on better sites, extends from the Maritime Provinces and Eastern Quebec through New England and New York to western North Carolina and eastern Tennessee. It reaches a height of 100 feet and a diameter of three feet, but usually is not over 75 feet tall and 20 inches in diameter. The timber is commonly used, without distinction as to species, for pulp and for lumber for building purposes, boxes, food containers, and sounding boards for musical instruments, such as pianos and violins. (The Swiss Pine of the violin trade is European Spruce, *P. excelsa.*)

There are four western species. Blue Spruce, *Picea pungens* Engelm. (= *P. Parryana* Sarg.), is a medium-sized to large tree growing singly or forming small groves at high altitudes in the Rocky Mountain region, with its center of distribution in Colorado. Because of its bluish foliage it is often cultivated for ornament, the most popular form being from a grafted plant known as Koster's Blue Spruce. Engelmann Spruce, *P. Engelmannii* (Parry) Engelm., is also known as White Spruce and has unpleasantly scented foliage like its northern relative, *P. glauca*. It occurs in the Rocky Mountains from Canada to Arizona and New Mexico, attaining its best development in the inland mountainous region of British Columbia and western Alberta, sometimes attaining a height of 120 feet and a diameter of three feet. The timber is not extensively used.

All of the foregoing are characterized by 4-angled leaves, but there are two western species with flattened leaves. Weeping Spruce, *Picea Breweriana* S. Wats., is of very limited occurrence at high elevations in northern California and southern Oregon. The Sitka or Tideland Spruce, *Picea sitchensis* (Bong.) Carr., is a tree usually not over 100 feet high with a tapering buttressed trunk three to four feet through, but sometimes 200 feet tall and 16 feet in diameter at the base. It grows at low altitudes near the ocean from Alaska to northern California; usually in association with Western Hemlock, Douglas Fir, and Western Red Cedar. It is an important timber tree and the estimated stand is upward of 40 billion board feet about equally divided into three regions—Alaska, British Columbia, and the states of Washington, Oregon, and California. Instead of being straw-colored or nearly white, like the other commercial Spruces, the wood has a decidedly pinkish or reddish tinge. Some of it is very coarse-grained and saws rather woolly, but that from the outer part of old trees is of slow and uniform rate of growth and very easy to work, but likely to be brittle. Sitka Spruce came into prominence during the World War because of its use in airplane manufacture. It is also employed for building purposes (except heavy construction), planing-mill products, boxes, and crates.

Heartwood not clearly differentiated from the sapwood. Luster satiny; tangential surface finely dappled. Without distinctive odor or taste. Mostly light in weight, but firm and strong; sp. gr. (air-dry) 0.35 to 0.65; weight 22 to 41 lbs. per cu. ft.; contains little resin; texture fine to medium; grain generally straight; working properties good; poorly resistant to decay.

Common names: He balsam, juniper, spruce (black, blue, bog, cat, double, eastern, Engelmann, silver, single, skunk, tideland, water, western, white, yellow, weeping), s. pine, yew pine (U.S.A.); epinette jaune, spruce (Canada).

Pinus, the most important genus in the world for timber and resin production, includes about 100 species of evergreen trees, widely distributed in the northern hemisphere, with only a Malaysian species (*P. Merkusii* Jungh. & De Vr.) crossing the equator. The linear leaves are borne in fascicles of 1 to 8 (usually 2, 3, or 5), with a persistent or deciduous sheath at the base. There are about 35 species in the United States, the range of nine extending into Canada, eleven into Mexico, and one (*Pinus caribaea* Mor.) into West Indies and Central America. There are about 15 other species native to Mexico, mostly growing at elevations between 4000 and 15,000 feet above sea level; four of them follow the mountains into Central America, with the southern limit of the genus a little north of Bluefield Bluff, Nicaragua, in latitude 12° 5′ north. There are three species in the West Indies (one in the Bahamas, one in the Island of Haiti, three in Cuba, two in Isle of Pines). The genus is not represented in the native flora of Jamaica, Puerto Rico, Lesser Antilles, Costa Rica, Panama, and South America.

Botanists divide *Pinus* into two subgenera: *Haploxylon* (single xylem) and *Diploxylon* (double xylem), the names referring to the number of vascular bundles in the leaves. The terms are also appropriate for the secondary wood, since the species of the *Haploxylon* group, the Soft Pines, have homogeneous growth rings with gradual transition from early to late wood, whereas the growth rings of the *Diploxylon* group, the Pitch and Hard Pines, are two-layered, the soft early wood usually giving way abruptly to the dense late wood. The timber of the Soft Pines, particularly the White Pines, is noted for its mellow and uniform consistency and ease of working, and is suitable for all kinds of carpentry and joinery where great strength is not required. The timber of the Pitch Pines, particularly the Hard Pines, is typically harder, stronger, and more resinous, though exhibiting a wide range of variation due in part to the species and in part to the conditions of growth. The densest grades are especially adapted for heavy construction and when rich in resin are highly resistant to decay. At the other extreme are woods approaching the White Pines in properties and uses; this type of material is common in the outer part of old trees that are growing very slowly. Some of the Pitch Pines yield an oleo-resin that is the source of turpentine and other commercially important products.

Although 28 kinds of Pine contribute to the timber industry of the United States, the bulk of the lumber is obtained from three southern, two northern, and three western species. The outstanding members of the White Pine group are *Pinus Strobus* L., *P. monticola* Dougl., and *P. Lambertiana* Dougl. There are several related species in the mountainous regions of the United States and Mexico, for example, *P. albicaulis* Engelm., *P. ayacahuite* Ehrenb., and *P. flexilis* James. Unlike most but not all of the Pitch Pines (*e.g.*, *P. Torreyana* Carr. and *P. leiophylla* Schlecht. & Cham.), their leaves are in clusters of five. Some of the Old World trees in this group are

P. cembra L. of the Carpathian Alps and northeastern Russia and Siberia, *P. koraiensis* S. & Z. of Corea or Chosen, *P. parviflora* S. & Z. of Japan, and *P. excelsa* Wallich of the Himalaya Mountains. White Pines are subject to a serious disease caused by a heteroecious rust which requires some species of *Ribes* as an intermediate host.

Northern White Pine, *Pinus Strobus* L., is widely distributed from the southern Appalachians to Newfoundland and from the Lake States deep into Ontario and Quebec. It is estimated that the original stand of this timber was about 900 billion board feet, the amount equally divided between the United States and Canada. The timber was formerly used in great quantity for innumerable purposes of general construction, joinery, and planing-mill and factory products, being noted for its mellowness, uniform texture, ease of working, and ability to hold its place when manufactured. Old trees are no longer abundant and the best grade of lumber from them is chiefly employed for patterns, sash and doors, and other exacting purposes. Much of the timber now available is second-growth and its principal uses are for boxes, pail stock, and match sticks. It is known to trade as Northern White Pine, Northern Pine, White Pine, Soft White Pine, and Cork Pine, often with the name of a state or locality added. The species is extensively planted for ornamental and forestry purposes.

Western White Pine, *Pinus monticola* D. Don, is very closely related to the preceding species and the woods of the two are nearly identical in structure, properties, and uses. The trees are of scattered occurrence in the mountain forests from central California to Puget Sound and more abundant and better developed in northern Idaho (Plate VII) and adjacent regions of Montana and British Columbia. The lumber is commonly known in the eastern markets of the United States as Idaho White Pine.

Sugar Pine, *Pinus Lambertiana* Dougl., one of the largest and most valuable timber trees on the Pacific coast, is of irregular occurrence in mixture with other conifers in the mountains from western Oregon to northern Baja California, attaining its best development at elevations between 3500 to 6000 feet on the western slopes of the Sierra Nevada and parts of the Coast Range in California, where it grows to a maximum height of nearly 250 feet, with a well-formed trunk six to eight, occasionally up to 12, feet in diameter. The estimated stand is about 20 billion board feet. The wood is somewhat coarser-textured, has larger and more prominent resin ducts, and more pronounced sweetish-resinous scent than the other White Pines. Sugary exudations are common on fresh wood. The lumber is available in large sizes free of defects and the best quality is used for pattern-making. Planing mills work it into moldings, panels, and railings as well as sash, blinds, doors, and stair work. The common grades are consumed locally in the manufacture of fruit boxes and other containers.

In southwestern United States and northern Mexico is a small group of closely related species known as Piñon or Nut Pines. They are dwarf, scrubby trees characterized by wingless, edible seeds a half inch or more in length. The principal species is the Piñon Pine, *Pinus edulis* Engelm., and it is an important component of woodland forests covering about 28 per cent of the total area of New Mexico and Arizona, or more than 42 million acres, as well as smaller areas of adjacent states. The trunks of the trees are rarely suitable for lumber, but some of them are cut locally for fence posts, mine props, and fuel. The wood is fine-textured and firm, but is cross-grained and brittle; resinous pieces have the scent of beeswax. The chief value of the trees is in their seeds or "nuts." According to Elbert L. Little, Jr. (*Chronica Botanica* 6: 15: 348. 1941), Piñon nut crops are not annual, but irregular, infrequent, and usually local in area. In a particular locality the interval between crops may vary from two to five, or even ten, years, but owing to the vastness of the area there is a good crop somewhere nearly every year. The annual harvest in New Mexico, Arizona, and Colorado is usually between one and two million pounds, but the crop in 1936, the largest ever gathered, totaled about eight million pounds. At 10 cents a pound paid to pickers and 25 cents retail price, the value of each crop is between $250,000 and $500,000, occasionally more.

In the subgenus *Diploxylon* is a group of species whose woods are readily recognized by the fact that the ray tracheids are dentate or reticulate as in the other Pitch Pines, but the pits from the ray parenchyma cells to the vertical tracheids are large as in the White Pines. Here belong the Japanese Red Pine, *Pinus densiflora* S. & Z., the Scots Pine, *P. sylvestris* L., and Austrian Pine, *P. laricio* Poir., which sometimes are planted for decorative and forestry purposes in the United States. The only important American species is the Norway or Red Pine, *P. resinosa* Ait., which is associated with Northern White Pine, especially in the northern part of its range. It is a fairly large tree, rarely up to 120 feet high, with a slender bole three feet in diameter. The principal commercial stands are in Michigan, Minnesota, Wisconsin, and adjoining regions in Canada. First growth timber has a darker heartwood, more pronounced late wood, and is more resinous than young trees of rapid growth. The lumber is often sold in mixture with White Pine under the name of Northern Pine. Its principal use is in general construction; the lower grades are consumed largely by the box trade. The species is much used in forest planting within its range.

Jack Pine, *Pinus Banksiana* Lam., is a small to medium-sized tree of extensive range in Canada east of the Rocky Mountains and extending into the Lake States and New England. The wood is rather light and soft and is principally used for pulp and (after antiseptic treatment) for railway crossties. Lodgepole Pine, *P. contorta* Loud. and forms or varieties, has an extensive range from Alaska to northern New Mexico and Baja California, being at its best in the Rocky Mountains. Mature trees vary in height from only a few feet, when growing in sphagnum bogs, to 150 feet where conditions are favorable. The chief uses of the timber are local for mining timbers and railway crossties, sometimes under the name of Tamarack.

Ponderosa Pine is the name now in common use for the timber of *Pinus ponderosa* Laws. and its forms or varieties, though the trees are usually known as Yellow Pine or Bull Pine, the young ones sometimes as Black Jack. It is very widely distributed in mountainous regions from British Columbia far into Mexico and from California to Nebraska, in many places forming great forests. At its best it is a very large tree, sometimes 225 feet tall with a smooth bole five to eight feet in diameter, but in arid regions it is short-bodied and round-topped. The quality of the timber varies greatly from coarse, hard, and resinous to fine, soft, and free from pitch. For many years the white, slowly grown, uniform-textured sapwood of large trees was sold in the markets of the Middle West and north Atlantic states as California, Arizona, or New Mexico White Pine and was used for the same purposes as Northern or Idaho White Pine. Lumber from northwestern parts of the United States was for a time known to the trade as Pondosa Pine, but this was later changed to Ponderosa Pine for the species regardless of the source, and applies also to the wood of the closely related *P. Jeffreyi* Oreg. Com. of California, sometimes considered only a variety of *P. ponderosa*.

Southern Yellow Pine is a collective term for several species growing in southeastern United States from Virginia to Texas. The three most important are Longleaf, *Pinus palustris* Mill., Shortleaf, *P. echinata* Mill., and Loblolly, Old Field, or North Carolina Pine, *P. taeda* L. Others of more or less value are Slash Pine, *P. caribaea* Mor., Spruce Pine, *P. glabra* Watt. and *P. virginiana* Mill., and Pond Pine, *P. rigida* Mill., var. *serotina* (Michx.) Loud. In the export trade the timbers of this group are commonly known as Pitch Pine, but this name should not be confused with that of a tree, *P. rigida* Mill., distributed from New England through New York to Georgia and eastern Tennessee and supplying lumber of rather low quality which is for the most part consumed locally.

Longleaf Pine comprises about half of the stand of Southern Pine and is the chief occupant of a belt of forest 125 to 200 miles wide along the Atlantic and Gulf coasts from southern Virginia to eastern Texas. The mature tree is of excellent tim-

MAP 3. Range of Slash Pine in U.S.A., Cuba, and Central America.

Photo by M. Martinez.

Plate VIII. Mexican Pine tree in Durango, Mexico.

Photo by M. Martinez.

Plate IX. Interior of a Pine Forest in Durango, Mexico.

ber form, the trunk, being tall, straight, and cylindrical. The wood is the heaviest, hardest, and most resinous of the group. In the southeastern part of its range it is in part replaced by the Slash Pine, which extends into the tropics. These species are alike in having very long leaves and being rich in resin. They are the principal American sources of naval stores, the annual production being about 30 million gallons of turpentine and over two million barrels of rosin.

Shortleaf Pine is the name commonly applied in trade to the timber of all of the other Southern Pines with short leaves. The tree that foresters and botanists call Shortleaf Pine, *Pinus echinata,* occurs as second growth, both in pure stands and mixed with hardwoods on extensive areas of flat lands and foothills from Maryland southward and westward to the Mississippi River, while in the Gulf states and the central Mississippi basin it is abundant as second-growth and also composes a large proportion of the remaining virgin Southern Yellow Pine. The wood is more variable than Longleaf and less so than Loblolly. Some of it, particularly from the southern part of the range, is hard and heavy while that produced further north or at higher altitudes is moderately light and soft and may have a special designation in the trade, for example, Arkansas Soft Pine.

Loblolly or Old Field Pine has much the same general range as Shortleaf. East of the Mississippi River it is a tree of the low pine barrens and their swampy borders, and further back from the coast is found on the tablelands of northern Mississippi, Alabama, Georgia, and South Carolina. West of the Mississippi River there are heavily timbered areas in Arkansas and Louisiana, and in Texas as far as the Colorado River. The diverse conditions of growth affect the form and size of the tree and the quality of the wood. At its best it is an excellent timber tree producing lumber of high grade, though usually of coarser grain than Longleaf and Shortleaf. A large part of the cut is from second-growth stands which produce wood of mediocre quality, being of coarse grain and texture, knotty, and perishable when exposed to the weather. The lumber known as North Carolina Pine is mostly of this species.

The range of variation in the wood of a single species of Southern Yellow Pine in response to environmental factors is often as great as that exhibited by the woods of the different species, hence specific distinctions based on anatomy alone are often impossible. Wood of high density is preferred for heavy construction, while the lighter timber is used mostly for planing-mill products. According to the U.S. Forest Products Laboratory (*Technical Note* No. 214): "There are no fundamental differences among the Southern Pines which make all the wood of one species preferable to all of the wood of another for any given purpose. The dense wood of any Southern Pine has practically the same strength and other characteristics as the dense wood of any other Southern Pine, and the lighter-weight pieces are more or less alike. In tests at the Forest Products Laboratory, Longleaf and Slash Pines have been found to have somewhat higher average strength properties than Shortleaf, Loblolly, and Pond Pines, but dense pieces of the latter species were found to be stronger than the average pieces of the former. Aside from defects, density can be taken as the factor that determines the suitability of any piece of Southern Pine for structural purposes. It is easily determined by visual methods, and structural timbers are now being classified and sold under density specifications."

For Mexico the total volume of commercial Pine timber is estimated to be 430 million cubic meters. The annual consumption is about 1.8 million cubic meters of wood, 13 million kilograms of gum resin, and 210,000 kilograms of nuts (see *Tropical Woods* 58: 35). The principal forest region is a belt 40 to 50 miles wide and 800 miles long extending along the Sierra Madre Occidental from near Arizona and New Mexico to Jalisco. Much of this region, particularly the northern half, resembles the southwestern United States, being mostly rather open and predominantly coniferous, though there are many hardwoods, especially Oaks. There are also coniferous forests on the eastern slopes of the Sierra Madre Oriental, while the largest Pines grow on the Sierra Madre del Sur. Pine occurs, too, in northern Baja California and on cross ranges and volcanoes of southern Mexico.

The most valuable White Pine in Mexico is *Pinus ayacahuite* Ehrenb. (including *P. strobiformis* Engelm.) which is very widely distributed in the mountains from the United States border to Guatemala. It grows at elevations of 7000 to 10,000 feet in Chihuahua, Sonora, and Durango and is often 75 feet tall and two to three feet in diameter. Further south it occupies higher and moister regions and is sometimes 150 feet tall, with a well-formed trunk occasionally five feet through. It does not form pure stands and is not very abundant, but the timber is considered the best of the Mexican conifers. *Pinus Strobus* var. *chiapensis* Martínez is a large forest tree of limited occurrence in Chiapas (see *Anales Inst. Biol.* 11: 1: 79; *Tropical Woods* 60: 10).

The Western Yellow Pine (*Pinus ponderosa* and varieties) of the Rocky Mountain region, is also one of the most important timber trees of northern Sonora and Chihuahua at elevations of 6000 to 8000 feet, usually growing in rather open forests, but forming fairly dense stands on the higher slopes. *P. leiophylla* Schlecht. & Cham. is one of the most common pines in southern Mexico, occurring from subtropical elevations to about 9500 feet up the mountain slopes. Though occasionally 100 feet tall, it is usually rather small and is poorly suited for timber. In northern Mexico it is replaced by *P. chihuahuana* Engelm., which some authorities consider only a variety. *P. teocote* Schlecht. & Cham. is a common Pine on dry sites in the mountains of eastern and southern Mexico. Though frequently short and scrubby it is occasionally up to 100 feet high and three feet in diameter. The wood is hard and heavy and the tree produces turpentine. *P. montezumae* Lamb. is a variable species of general distribution from Durango to Guatemala, and is particularly abundant, often forming pure stands, in southern Mexico. The timber is similar to that of *P. ponderosa* and is used for the same purposes. *P. oocarpa* Schiede is a medium-sized, soft-wooded tree growing at subtropical elevations from Sinaloa and Zacatecas to Chiapas in Mexico and in the mountains of British Honduras, Guatemala, Honduras, Nicaragua, and Salvador, in the last country being the only native species. The timber finds many local uses but is not exported.

In Central America the principal species is *Pinus caribaea* Mor. (= *P. bahamensis* Gris. = *P. Elliottii* Engelm. = *P. heterophylla* Sudw.), which is the same as the Slash Pine of southeastern United States. The tropical range includes the Bahamas, western Cuba and the Isle of Pines, and Central America from northern British Honduras through eastern Guatemala and Honduras into Nicaragua (Map 3). In British Honduras Pine forests occupy about a third of the mainland, mostly on the sandy soils of the dry savannas and at an average elevation of 2000 feet on the interior Mountain Pine Ridge; the timber is merchantable, but it has never been extensively exploited. Regarding the Pine forests of the Republic of Honduras, William D. Durland says (*Tropical Woods* 10: 8):

"The northern Pine belt is about 50 miles inland from the coast and is composed of open, park-like stands on the slopes of the hills and mountains, and on the ridges and plateaus. This forest extends southward and eastward, reaching the coast at Cape Gracias in the southeastern part of the Republic. Hardwoods often are in association with the Pine, but in all cases Pine is the dominant tree. An undergrowth of herbaceous and grassy vegetation is plentiful and even dense in some places, and scanty or lacking entirely in others. No regularity was observed in the volume of the stand; 2000 to 10,000 board feet per acre being representative, though some acres will tally more and some less. The size of the trees is also variable. On the lower elevation 14 to 18 in. diameters and one or two 16 ft. logs per tree are representative, while on the higher elevations the diameters increase to 30 and 34 in. or more, with three or more 16 ft. logs per tree. At the heads of the draws, in the gully bottoms and stream beds where there is an accumulation of fertile top soil washed down from the higher slopes, patches of hardwoods grow to the exclusion of Pine. The soil on which the pine grows is for the most part a red subsoil or laterite located on the exposed ridges and slopes. With increase in elevation above 1500 feet the hardwood patches become fewer, and above 2500 feet the Pine is permitted to occupy more fertile sites than at lower altitudes within the particular type location."

In Nicaragua the best stands of Pine occur on scattered ridges in a belt about 30 miles wide near the northeastern coast and tapering to about half that wide at its western end some 55 miles inland. The trees vary in height up to 100 feet and in diameter up to 36 inches, rarely more, are often clear of branches for 70 feet, and yield about 3000 board feet per acre. The amount of lumber exported annually is less than 10 million board feet, mostly to Great Britain and the United States, some to the West Indies and Costa Rica. The New York demand is chiefly for dimension timbers, sometimes up to 50 feet in length, for use by railroads and shipyards. A small percentage of the timber is comparatively light and soft and is better suited for general millwork than for structural purposes. (For detailed report on Nicaraguan Pine see *Tropical Woods* 55: 1–16; Sept. 1, 1938.)

In the West Indies Pine is of little commercial importance except in Dominican Republic, the species being *Pinus occidentalis* Sw. (= *P. cubensis* Gris.). W. R. Durland says (*Geographical Review* 12: 2: 216; April 1922): "The Pine forest of the Dominican Republic is of more interest commercially than all of the other forest types put together. . . . The timber is resinous and heavy, barely floating in water. In past years a few attempts have been made to exploit this Pine, but operations have been on a small scale and of a temporary nature. . . . The boundaries of the Pine forest are usually sharply defined. As a type it occupies the northern slopes of the main range from the vicinity of Cotui west to the Haitian border. Botanically the formation is not complex, 80 per cent and over of the surface area being covered by one species of Pine identified as *Pinus occidentalis*. In appearance the type somewhat resembles the open park-like stands of pine in our own southern states, although it is far less abundant per acre. It ranges in diameter on the average from 12 to 35 inches breast-high and in total tree height from 40 to 60 feet. Eggers reports that at the higher elevations (4000 feet) individuals approaching 200 feet in height and four feet or more in diameter are not uncommon. The trees are scattered more or less singly over the area in a very open manner, on the ridges in particular. They are found at times, although exceptionally, in dense pure stands. A luxuriant growth of wild grass over which cattle graze at will forms a ground cover beneath the Pine. This ground cover is coincident with the occurrence of the Pine and is characteristic of the type." The local development of the Pine timber industry in Dominican Republic is of recent origin. More than seven million board feet of lumber was sawed in 1937, mostly from Santiago Province. During the same year about a million feet of lumber, mostly Pine, was imported, a reduction of 90 per cent during the previous decade.

In Haiti, according to Schiller Nicholas (*Caribbean Forester* 1: 3: 21; April 1940), Pine constitutes the forest resource of greatest potentiality. "Although more than three-fourths of its original stands have been displaced by agriculture or destroyed by fire, especially at lower altitudes, this species, *Pinus occidentalis,* still occurs in fairly dense stands over some 300 square miles. This total area is located in four distinct forests, as follows: a little-known, inaccessible area in the southern peninsula at elevations from 1200 to 2500 m. above sea level; an area of 212 square miles in the eastern center of Haiti at altitudes ranging from 450 to 1200 m.; a small area, not surveyed, near a road in the north center of Haiti, at elevations of 300 to 900 m.; and an area of 40 square miles, southeast of Port-au-Prince and accessible by car at high altitudes, between 1300 and 2500 m. All except the first area, where the precipitation is abundant, may be classified as semi-arid. . . . The volume of these forests is estimated conservatively at 350 million board feet."

There are two American species in the Pitch Pine group which are of minor importance in their native habitats but have proved well adapted for commercial planting in parts of the southern hemisphere of the Old World. One of these is the Monterey Pine, *Pinus radiata* D. Don (= *P. insignis* Dougl.), a medium-sized to large tree of very restricted natural range along the coast of southern California. It is being successfully grown in parts of Australia and all of New Zealand. According to the New Zealand State Forest Service (Circ. No. 3), Insignis Pine "grows with a clean straight trunk to a height of 120 to 150 feet, with a maximum diameter of from 40 to 60 inches. . . . As a rule a three-year-old tree measures about three feet in height. After that period the growth increases at a rapid rate, sometimes as much as six to eight feet vertical growth being made annually over a number of years. . . . Examination of the timber in open-grown specimens shows a diameter increase of two inches sustained over 10 to 12 years of the early period of the life of the tree. . . . It is invaluable for shelter and firewood, the manufacture of cases and boxes of all descriptions, and general building construction, and concrete casing." Tests on Australian-grown timber indicate that it is not brittle and devoid of strength, as is commonly believed, but has much the same mechanical properties as Oregon or Douglas Fir (see *Tropical Woods* 23: 34). The other species is *Pinus patula* Schl. & Cham., which apparently is confined to elevations of 6000 to 8000 feet on the moist mountains on the eastern side of the Mexican plateau between 19° and 21° north latitude. It is well adapted to conditions up to moderate elevations in the mountains of eastern South Africa and has been planted there since 1907. It grows rapidly and produces timber suitable for box-making and the general building trade. (See *Tropical Woods* 15: 55.)

COMMON NAMES: Pine—specific names in text (U.S.A.); pino, p. de Cuba, p. hembra, p. macho (Cuba); bois pin, b. chandelle (Haiti); pino (Dom. R.); acalocahuite, acanita, ayacahuite, a. colorado, guiri-biche, huiyoco, jalicote, ocotl, ocote, o. blanco, o. chino, o. hembro, o. macho, pinabete, pino acahuite, p. barbón, p. cahuite, p. de azúcar, p. Montezuma, p. piñon, p. prieto, p. real, p. triste, piñon, sacalacahuite, xalócotl, yutmusatnu (Mex.); pine, pino, huhub (Br. H.); ocote, pino (Guat.); pino (Salv.); ocote, pino, p. blanco, p. ocote, p. vete (Hond.); auas, ocote, pino (Nic.).

Pseudotsuga. There are four species in eastern Asia and two in western North America. The Bigcone Spruce, *P. macrocarpa* Mayr, is a medium-sized tree often forming open groves of considerable extent on steep mountain slopes in southern California and northern Baja California. Its principal use is for fuel. Douglas Fir, *P. taxifolia* (Lamb.) Britt. (= *Pseudotsuga Douglasii* [Lindl.] Carr. = *P. mucronata* [Raf.] Sudw.), is extensively distributed along the Pacific coast from British Columbia to central California and inland through the Rocky Mountains into northern Mexico, a range of more than 2000 miles north and south and almost 1000 miles east and west. Two forms of the tree are recognized, the Pacific Coast and the Rocky Mountain, which at their extremes exhibit such differences in the form of the tree, characters of the foliage and cones, and quality of the wood that they would be considered distinct species except that in portions of their range they intergrade and lose their identity.

The interior form, which some botanists designate *Pseudotsuga glauca* Mayr, occurs from latitude 55° north in British Columbia southward at progressively higher elevations throughout the Rocky Mountain system into Mexico as far as Hidalgo. The trees are comparatively small though sometimes up to 100 feet in height and two to three feet in diameter, and the stand per acre only occasionally amounts to 8000 board feet and generally is much less. Much of the wood is red, hard, more or less cross-grained and knotty, but durable in contact with the ground. Clear lumber is obtainable from the larger logs, but the uses are mostly for dimension timbers; large quantities are employed for mine timbers and railway crossties.

The coast form is essentially a different

Photo by U.S. Forest Service.

PLATE X. Douglas Fir trees (*Pseudotsuga taxifolia*) six to nine feet in diameter on the Olympic National Forest in Washington.

Photo by Llewelyn Williams.

PLATE XI. Merey Montañero or Caracolí (*Anacardium excelsum*) near Turén, Portuguesa, Venezuela.

tree and supplies most of the timber known to the American trade as Fir and abroad as Oregon Pine (often shortened to Oregon). It is at its best not far above sea level on Vancouver Island and adjacent mainland and in Washington and Oregon between the coast and the western foothills of the Cascade Mountains, where the humid climate and moist soil are conducive to tree growth. Douglas Fir usually grows in mixture with Hemlock, and to a greater or less extent with Spruce, Cedar (*Thuja*), and the true Firs (*Abies*), but sometimes is in nearly pure forest, the immense trunks close together and rising to great heights with scarcely a limb to mar their smoothness (Plate X). Trees 200 feet tall and six feet through are common, while heights of over 300 feet and diameters up to 15 feet are occasionally reached. Single trees have scaled 60,000 board feet and in one recorded instance half a million board feet of lumber was obtained from an acre. The usual yield, however, is between 35,000 and 60,000 board feet per acre. The remaining stand of merchantable timber in Oregon, Washington, and British Columbia is about 400 billion board feet, of which considerably more than half is in Oregon. The wood exhibits considerable variation as a result of conditions of growth, ranging from pinkish yellow, fine-grained, fairly uniform-textured, and moderately light and soft to reddish brown, with wide growth rings having open and weak early wood and very dense late wood. The timber resembles Southern Pine in properties and uses and now has the advantage of being available in greater abundance and larger dimensions.

COMMON NAMES: Douglas fir, D. spruce, fir, Oregon, O. fir, O. pine, O. spruce, Puget Sound pine, red fir, r. pine, r. spruce, yellow fir (U.S.A., Canada); abeto, acahuite, cahuite, hallarín, pinabete, pino de corcho (Mex.).

Tsuga comprises about 14 species of evergreen trees, 10 of them in Asia, from Japan to the Himalayas, and four in temperate North America. Eastern Hemlock, *T. canadensis* (L.) Carr. (or *T. americana* [Mill.] Farwell), is a medium-sized to large tree, sometimes 100 feet tall and four feet in diameter, widely distributed from Nova Scotia westward through southern Canada and the Lake States to Minnesota, southward through New England, New York, Pennsylvania and southern Appalachians to northern Georgia and Alabama. It frequently occurs in groves but usually is in mixture with other conifers and with hardwoods, such as Beech, Birch, and Maple. It is often planted for ornamental purposes. Formerly the only commercial use for the tree was as a source of tanbark, but later the timber came into the market for pulp and lumber for rough construction and boxes. The wood is brittle, splintery, often cross-grained, and when fresh has an objectionable scent. Near the southern limits of its range is another species, *T. caroliniana* Engelm., a medium-sized tree of very local occurrence in the mountains.

Alpine, Black, or Mountain Hemlock, *Tsuga Mertensiana* (Bong.) Sarg. (or *T. Pattoniana* [Jeffr.] Engelm.), is found in the timber-line forests of the Pacific coast from Alaska to central California, and further east in southern British Columbia, Montana, and Idaho. It commonly occurs in mixture with other species, but forms extensive and nearly pure stands in parts of the Cascade Mountains. It varies in size from 150 in height and five feet in diameter to short and stunted. The bark is rich in tannin and the wood is of fair quality, but the forests are difficult of access and are chiefly valuable in protecting the watersheds.

Western Hemlock, *Tsuga heterophylla* (Raf.) Sarg. (= *T. Mertensiana* [Lindl. & Gord.] Carr.), is a tall slender tree growing in humid regions along the coast from Alaska to California and inland from Puget Sound to western Montana. It is one of the principal trees in northwestern Oregon and western Washington, in parts of its range in nearly pure stands, but more often associated with Douglas Fir, Western Red Cedar, Sitka Spruce, White Pine, and true Firs (*Abies*). It is common as an understory of the dense forests of Douglas Fir in Oregon and Washington. Its usual mature size in forests of the coast and Cascade Mountains is from 150 to 200 feet tall and three to four feet in diameter, but exceptional trees reach 250 feet in height and eight feet in diameter; in the Idaho region it is two or three feet in diameter and 100 feet tall. Even-aged, fully-stocked, practically pure stands on the best sites may yield as much as 100,000 board feet per acre in 100 years. The total estimated stand is 233 billion board feet, located as follows: Alaska, 63; British Columbia, 64; Washington, 80; Oregon, 24; Idaho and Montana, 2. The bark contains from 12 to 15 per cent of tannin, but the amount consumed is not very great. The timber has suffered in competition with its associates, partly on account of the poor reputation of Eastern Hemlock. It is coming into more general use as rough lumber for house construction and also for planing-mill products such as flooring, interior trim, ceiling, and siding; other uses are boxes and crates, sash and doors, woodenware, slack cooperage, and mechanical and sulphite pulp.

Heartwood pale brown or grayish, with a tinge of red particularly in the late wood; not clearly differentiated from the sapwood. Luster medium. Without distinctive taste, but with a sour odor when fresh, but usually absent from dry wood. Rather light to moderately hard and heavy; sp. gr. (air-dry) 0.40 to 0.68; weight 25 to 42 lbs. per cu. ft.; not difficult to season, though containing considerable moisture when fresh; fairly easy to work, finishing smoothly; poorly resistant to decay. Western Hemlock is superior to the eastern species in nearly all technical properties.

COMMON NAMES: *Tsuga canadensis:* Hemlock (eastern, red, white), h. spruce, Huron pine, spruce, s. pine (U.S.A.). *T. heterophylla:* Alaska pine, hemlock (Pacific, west coast, western), h. spruce (U.S.A.); grey fir, hemlock (British Columbia, western), h. spruce, tsuga de l'ouest (Canada); Prince Albert's fir, western hemlock fir (England). *T. mertensiana:* Alpine spruce, hemlock (alpine, mountain), h. spruce, weeping spruce (U.S.A.).

PODOCARPACEAE

THIS family, which some botanists include with the Taxaceae, comprises seven genera and about 100 species of trees and shrubs distributed as follows: *Acmopyle* (1 sp., tree; New Caledonia), *Dacrydium* (20 spp., shrubs or trees; Oceania, Chile), *Microcachrys* (1 sp., shrub; Tasmania), *Pherosphaera* (2 spp., shrubs; Tasmania, Australia), *Phyllocladus* (6 spp., trees or shrubs; Oceania), *Podocarpus* (70 spp., trees or shrubs; pantropical and south temperate zone), and *Saxegothaea* (1 sp., tree; Chile).

Dacrydium. Some of the Australasian species are large and, particularly in New Zealand, of commercial importance for their timber. The only species native to the New World is *D. Fonkii* (Phil.) Benth., one of the smallest coniferous plants in the world. It is less than a foot high and its foliage has the general aspect of *Juniperus virginiana* L. It is a constituent of the ground cover of swamps and bogs in Chile from the coastal Cordillera of Valdivia to Tierra del Fuego.

Podocarpus is the most important genus of the family and some of the trees, especially in New Zealand and South Africa, are noted for their large size and the excellence of their timber. About a fourth of the species are American, with a combined range, mostly in mountainous regions, from the West Indies and southern Mexico to northern Argentina and southern Chile. The wood structure is very similar to that of *Saxegothaea*.

The best known northern species is *Podocarpus coriaceus* Rich. of the uplands of the West Indies, Venezuela, and Colombia; it has also been credited to southern Mexico, British Honduras, and Guatemala, but the form there inhabits lowlands and is now known as *P. guatemalensis* Standl. Two South American species, namely, *P. montanus* (Willd.) Lodd., with *Abies*-like branches, and *P. oleifolius* Don, an ill-scented tree with comparatively broad leaves, grows from Bolivia to Costa Rica. The latter species is known as Sisín in Ecuador and is the only indigenous conifer there. M. Acosta Solís says (*Tropical Woods* 57: 3): "It occurs scatteringly in the foothills of the Andes, for example at Leito (province of Tungurahua) and Saloya (province of Pichincha), where it attains a height of 8 to 16 m. and a diameter of 0.60 to 0.80 m. The wood is of a lustrous light clear yellow color, of fine and uniform texture, and easy to work. It is used in considerable quantity for carpentry, cabinet-making, and carving and usually is finished in natural color. It lacks figure and is not cut into veneers."

The following information about three Chilean species is from E. L. Bernath's paper on "Coniferous forest trees of Chile" (*Tropical Woods* 52: 19–20): "*Podocarpus nubigéns* Lindl., known as Mañío or Mañíu, or in the Araucanian language Manilihuán, is irregularly distributed from Río Toltén, about 39° 20′ south latitude, to Río Backer (Smith Canal) in the Territory of Magallanes, though the southern limit is poorly known. It occurs singly or in little groups on swampy land in the rain forests, commonly in association with various Myrtaceae (*Myrceugenia, Tepualia,* etc.) and Alerce (*Fitzroya*), with a ground cover of ferns, mosses, and marsh grasses. Mature trees vary in height from 30 to 80 feet and in diameter from 24 to 36 inches. The timber is logged in certain localities in Valdivia, Llanquihue, and Chiloé Provinces, especially on the Isle of Chiloé, and is one of the best and most valuable in the South. The pale yellow or whitish wood is fine-textured, comparatively heavy, and highly durable. The lumber is utilized in carpentry, cabinet and furniture making, flooring, ceiling, cooperage (particularly honey barrels), ship and boat building (masts, rudders, decks, and storerooms), and for railway crossties. The annual production is small because of the scarcity of the trees. There is no information available concerning the silvicultural requirements, methods of reproduction, or forest management of the species.

"*Podocarpus salignus* D. Don, also called Mañío and Mañíu, or Mañíu de la Frontera, extends from about 35° 30′ (Río Maule) to the Province Llanquihue, the southern limits not being exactly known. In the northern part of its range it occurs along streams in the understory of evergreen rain forests which are composed principally of *Persea lingue* Nees, *Laurelia aromatica* Juss., *Guevina avellana* Mol., *Myrceugenia apiculata* Med., and *Peumus boldus* Juss. In the Provinces Cautín, Valdivia, and Llanquihue the dominant trees in its forest association are *Weinmannia trichosperma* Cav., *Eucryphia cordifolia* Cav., and *Laurelia aromatica*. It grows singly or in small clumps on hillsides and along streams where the atmosphere is always humid. Nowhere does it attain large dimensions, the usual heights being between 30 and 60 feet. The leaves are long and narrow, suggesting Willow (*Salix*). The trunk, which frequently is crooked or twisted, rarely attains a diameter of 18 inches. Timber from well-formed trees is of excellent quality, but most of it is used locally for making doors, window sash, ceilings, and honey barrels. Only a small quantity of lumber is exported from the southern virgin forests because the logs are slender and the average number of exploitable trees is less than one per acre.

"*Podocarpus andinus* Poepp., Araucanian name Lleúque, is sparingly distributed in the Andes Cordillera from Río Maule to about 40° 30′, and in the Coast Cordillera from about 37° 30′ to 41° south latitude. It is a short, generally crooked tree, 20 to 30 feet high and rarely attaining a trunk diameter of 20 inches. It occurs as scattered individuals in association with *Nothofagus Dombeyi* Blume and *N. pumilio* Reiche on stony hillsides and in gulches at moderate elevations, usually remote from highways and railroads. Although the wood is similar to that of the other Podocarps, the timber has very little economic importance, its utilization being limited to occasional farm purposes."

Common names: Bois Lubin (Haiti); sabina cimarrón (Cuba); yacca (Jamaica); cypress (Br. H.); ciprés, c. de montaña (Guat., Hond.); cipricillo, cobola (C.R.); pino blanco (Pan.); chaquera pino, chaquiro, c. clavel, chaquito, pino, p. blanco, p. colorado, p. criollo, p. de pacho (Col.); castañeto, granadillo, pinabete, pino, p. aparrado, p. castañeto, p. de Castilla, p. romero, p. veti (Venez.); sisín sumí (Ec.); pinheirinho, p. bravo (Braz.); mañíu (Arg.); lleúque, manilihuán, mañío, mañíu (Chile).

Saxegothaea *conspicua* Lindl., the only species, is a Chilean tree commonly called Pino, also Mañío or Mañíu. According to Bernath (*loc. cit.*, p. 20), it "grows in swampy areas in the rain forests from Río Mauleto 45° (Río de los Maniuales) and probably farther south. In Provinces Cautín and Valdivia the chief associates are *Weinmannia trichosperma, Eucryphia cordifolia, Laurelia aromatica, Aextoxicum punctatum* R. & P., *Lomatia ferruginea* R. Br., and *Myrtus luma* Barn.; in Valdivia, Llanquihue, and Chiloé they are *Pilgerodendron uviferum* (Don) Florin, *Fitzroya cupressoides* (Mol.) Johnston, *Nothofagus Dombeyi,* and *N. antarctica* Oerst. The usual heights are between 30 and 60 feet and the number of exploitable trees per acre is too low in most localities to permit extensive utilization." The wood is of excellent quality, suggesting *Podocarpus* and *Taxodium* (Bald Cypress). Color yellowish brown, somewhat streaked with resinous lines; odorless; density medium; texture uniform; easy to work and fairly durable.

Growth rings distinct, transition from early to late wood gradual. Wood tracheids with pits in a single row or occasionally in pairs; without spirals; resin plates common; pits to ray cells small, bordered. Rays uniseriate; 1 to 30 cells high; without tracheids. Wood parenchyma cells numerous, diffuse or irregularly zonate. Resin ducts absent.

TAXACEAE

THE Yew family, in the restricted sense used here, includes only three genera and 13 species of evergreen shrubs and small to large trees, all of the north temperate zone, with the exception of *Austrotaxus,* a monotypic genus of New Caledonia. The woods of the other genera, *Taxus* and *Torreya,* are characterized by the absence of wood parenchyma, resin ducts, and ray tracheids, and the presence of spiral thickenings in the vertical tracheids.

Taxus, with seven species, which are distinguishable only by their leaf characters, habit of growth, and range, is widely distributed in Asia, Europe, northern Africa, and North America. The trees have spreading branches and brown or dark purple, scaly bark; the leaves are linear and flat, gradually narrowed at the base; the fruit consists of a nut-like seed seated in a fleshy scarlet cup. *T. baccata* L. and *T. cuspidata* S. & Z. are widely cultivated for ornamental purposes, especially for hedges.

There are four American species. *Taxus canadensis* Willd., often called Ground Hemlock, is a low straggling bush with diffusely spreading stems, common on moist soil, especially under coniferous forests in eastern Canada and northeastern United States. *F. floridana* Chapm. is a bushy tree, rarely 25 feet high with a short trunk a foot in diameter, of very local distribution in Florida. *T. globosa* Schlecht. is a small tree of scattered occurrence in the mountains of south-central Mexico. *T. brevifolia* Nutt. occurs singly or in clumps, usually under high evergreen forests from southern Alaska along the coast ranges to British Columbia, Washington, and Oregon (where it attains its largest dimensions) into Central California and eastward into northern Idaho and adjacent areas. It sometimes attains a height of 75 feet with a tall straight trunk frequently two, rarely four, feet in diameter, but more often it is of poor timber form, the bole being eccentric and irregular.

Heartwood bright orange to brown, sharply demarcated from the thin white sapwood. Luster rather high. Odorless and tasteless. Hard, moderately heavy, stiff, and elastic; sp. gr. (air-dry) 0.65 to 0.70; weight 40 to 44 lbs. per cu. ft.; texture very fine and uniform; grain straight to variable; working properties excellent; durability very high. Its best known use is for making archery bows.

Torreya (or *Tumion*) includes five species, two of which grow in China, one in Japan, and two in the United States. The plants are shrubs or small to large trees with a pungent, usually disagreeable scent. The inner layer of the seed coat is folded into the thick white albumen in a manner suggesting a nutmeg.

The Stinking Cedar, *Torreya taxifolia* Arn. (= *Tumion taxifolium* Greene), is a tree, rarely 40 feet high with a short trunk a foot or more in diameter, of rare and local occurrence on bluffs along the eastern bank of the Apalachicola River, Florida, and adjacent parts of Georgia. The California Nutmeg, *Torreya californica* Torr. (= *Tumion californicum* Greene), borders mountain streams in the coast ranges and the western slopes of the Sierra Nevada Mountains in California. It is nowhere common, but in the northern part of the state attains a maximum height of 100 feet with a trunk sometimes four feet in diameter. The wood is of a bright clear yellow color and has a mild disagreeable scent. It is rather light and soft; sp. gr. (air-dry) about 0.50; weight 31 lbs. per cu. ft.; of fine and uniform texture; easy to work and fairly durable. The timber is of no commercial importance because of its scarcity.

COMMON NAMES: California nutmeg, fetid yew, nutmeg tree, savin, stinking cedar, Torrey tree, yew (U.S.A.).

TAXODIACEAE

OF the eight genera and about a dozen species of trees making up the Taxodiaceae, one genus (*Athrotaxis*) is Tasmanian, five (*Cryptomeria, Glyptostrobus, Cunninghamia, Sciadopitys,* and *Taiwania*) are Chinese or Japanese, and two (*Sequoia* and *Taxodium*) are North American. *Cryptomeria* and *Cunninghamia* are of economic value in Asia, and *Sequoia* (Redwood) and *Taxodium* (Bald Cypress) are highly important as sources of timber in the United States.

The American woods are light in weight (sp. gr. 0.30 to 0.55), very soft to moderately hard, easy to work, and noted for their resistance to decay. They are much alike in structure but distinct in appearance. The heartwood of *Taxodium* is medium-textured, yellowish, reddish, light to dark brown, or irregularly striped or variegated, the lighter-colored material lustrous, the darker often with a waxy or oily appearance and feel and sometimes with a mildly rancid odor, whereas that of *Sequoia* is coarse-textured, light cherry-red to reddish brown, of rather dull dry appearance, and without distinctive odor.

Growth rings present. Tracheids thin-walled in early wood, thick-walled in outer late wood; without spiral thickenings; pits usually in two or more rows in radial walls in early wood; pits to ray cells 2 to 5 per cross field, in single row, the borders round, the apertures small-elliptic, horizontal to diagonal. Rays rarely to frequently biseriate in part, and 1 to 30, occasionally up to 60, cells high; ray tracheids rarely present in *Sequoia* only. Wood parenchyma abundant, diffuse or loosely aggregated into concentric zones; end walls uniform in *Sequoia,* irregularly thickened in *Taxodium;* dark resinous deposits abundant. Vertical traumatic resin ducts occasionally present in *Sequoia* only.

Sequoia, with two species of evergreen trees noted for their great size, is of limited distribution in California. The Bigtree, *S. Washingtonia* (Winslow) Sudw. or *S. gigantea* (Lindl.) Decne. (= *Wellingtonia gigantea* Lindl. = *S. Wellingtonia* Seem.), occurs in about 26 scattered groves, of a few hundred to several thousand trees each, at elevations of 5000 to 8400 feet above sea level on the western slope of the Sierra Nevada Mountains between 39° and 36° north latitude. It is described by Sargent (*Manual of the trees of North America,* p. 63) as being "at maturity usually about 275 feet high, with a trunk 20 feet in diameter near the ground, occasionally becoming 320 feet tall, with a trunk 35 feet in diameter, much enlarged and buttressed at base, fluted with broad, low, rounded ridges; in old age naked often for 150 feet, with short, thick, horizontal branches becoming, after the disappearance of the leaves, reddish brown more or less tinged with purple and covered with thin, close, scaly bark and naked buds. Bark one to two feet thick, divided into rounded lobes four to five feet wide, corresponding to the lobes of the trunk, separating into loose light cinnamon-red fibrous scales, the outer scales slightly tinged with purple." The maximum age, based on actual count of the annual rings, is about 4000 years. Unlike the other species, cones mature during the second season instead of the first, and reproduction is entirely by seed and not by sprouts. The wood is very brittle and light, weighing less than 20 lbs. per cu. ft. when dry, but highly durable. It is no longer of commercial importance, as the large trees are protected.

The Redwood, *Sequoia sempervirens* Endl., an important timber tree, is confined to about a million acres in a narrow, broken belt, 10 to 35 miles in width, extending southward from southern Oregon for nearly 500 miles along the Pacific slope of the Coast Range. It never reaches inland beyond the influence of the coast fogs, and the best stands are in Humboldt County and adjacent regions in northern California. The estimate of the total remaining timber is approximately 60 billion board feet, enough at the normal rate of cutting to last more than a century. The forest at its best is extremely dense and composed almost exclusively of the one species. The trees are very large, the average range in diameter being between five and ten feet, though many exceed this and a few measure more than 20 feet through. According to an account in *Hardwood Record* (October 25, 1914, p. 40), a Redwood tree near Eureka, California, was 380 feet high, 26 feet in diameter seven feet from the ground, 261 feet to the first limb, where the diameter was 11 feet, and scaled more than 344,000 board feet of lumber. The average yield per acre is between 60,000 and 75,000 board feet, though the best stands average about 100,000 feet; single acres have scaled 2,500,000 feet, but the excessive waste reduced the amount of sawed lumber to 1,500,000 feet. The maximum age, judging from actual ring counts, is about 2000 years. Reproduction is largely from sprouts, and early growth is very rapid. The old trees usually show considerable defect due to the action of an unknown fungus which causes the heart to become doty in places.

The quality of Redwood lumber is variable. Some of it is very soft, fine-grained, and uniform-textured, and some is coarse-grained with a flinty hardness to the summer wood. The butt logs contain the hardest and heaviest timber in a tree, the upper logs the lightest and softest. The wood is very strong for its weight, is resistant to decay, free from insect attacks, and may be had in all sizes free from defects. It exhibits its greatest strength in endwise compression and is well suited for columns; it is too soft and brittle for heavy beams. Freshly cut lumber contains a large amount of water, but dries without excessive shrinkage (unless so-called compression wood is present), and the damage from checking is small. Once well seasoned it is comparatively inert to changes in the humidity of the air.

Redwood may be used for the construction of practically every part of a house, though it is best suited for finish, foundations, and exterior work. For the places most exposed such as roofs, porches (columns, railings, and flooring), sills, outside steps, weatherboarding, sash and doors, and for pergolas, greenhouses, and fences, redwood ranks with the best of the Cedars and Cypress. When first exposed to water a portion of the coloring matter leaches out which interferes with the use of the wood for some purposes. Very wide panels are obtainable with the middle portion showing a pleasing slash grain which gradually merges into fine edge grain.

Although the grain is ordinarily straight, exceptions occur in the case of burls caused by local injuries to the tree and in curly and wavy grain from other causes. Burls of all sizes are common, some of them attaining a diameter of six feet and containing 10,000 feet of lumber. Small freshly cut burls find ready sale for sprouting, for when placed in water they develop into fern-like bouquets. The bark of mature trees is sometimes a foot thick and an average stand will produce nearly 25 cords per acre. It is used for making novelties and as a fibrous constituent of roofing paper.

Taxodium includes two or three closely related species of large deciduous trees. In the United States the generic common names for the tree is Bald Cypress, or Southern Cypress, but the timber is known simply as Cypress, though often with some qualifying term indicating its source, color, or other characteristic. It should not be confused with the true Cypress -(*Cupressus*) which belongs with the true Cedars and the Junipers and like them has fragrantly scented wood.

Taxodium distichum (L.) Rich. is by far the most important species commercially. It grows naturally in swamps and along streams throughout a wide coastal strip from southern New Jersey to Texas and up the Mississippi valley to southern Indiana and Illinois, including all of Florida, Mississippi, and Louisiana, the southern half of Georgia and Alabama, and lesser parts of 11 other states. The heaviest stands of timber are in deep coastal river swamps from the Carolinas to central Florida and the alluvial flood plain of the lower Mississippi River, especially south of Baton Rouge, Louisiana. The total amount of standing timber is estimated to be about 30 billion board feet.

From the Dismal Swamp in southeastern Virginia to western Florida and southern Alabama is a smaller form with acicular leaves. Some botanists consider this a variety, *Taxodium distichum* var. *imbricarium* (Nutt.) Sarg., but others regard it as a distinct species, *T. ascendens* Brongn. (= *T. imbricarium* [Nutt.] Harper). It is frequently cultivated and is generally known to horticulturists as *Glyptostrobus pendulus* Endl.

Bald Cypress attains a maximum height of 150 feet, a trunk diameter of 12 feet (above the basal swell), and an age of 2000 years, but the average for mature timber in fair situations is about 100 feet in height, three to five feet in diameter, and less than 500 years old. In deep swamps and locations subject to long inundation, conical leafless projections, called knees, are developed from the roots and serve not only as organs of aeration but also to reinforce the root system and provide stable anchorage in soft muck. The only serious injury to which the trees are subject is the attack of a fungus which tunnels into the heartwood of individuals after they have reached the age of about 200 years. It is estimated that about one-third of the trees are infected, while nearly 10 per cent of the lumber manufactured is more or less damaged or "pecky." The progress of the disease is arrested when the tree is felled and pecky Cypress is equal in durability, though not in strength, to sound wood.

Regarding the color of the heartwood, W. R. Mattoon says (U.S. Dept. Agr. Bull. No. 272, p. 6): "Cypress averages darker in color in the Gulf and South Atlantic region, and correspondingly lighter as one passes northward up the Mississippi and Atlantic coast. In any specified locality, however, wide color variations occur, although usually one color predominates. For example, the Cypress of the deep St. Johns River swamps, Florida, averages an amber or light orange-brown, and is referred to under the trade name of Yellow Cypress. Nearby, in shallower isolated swamps of different soil and moisture conditions, the wood is decidedly darker, with a more pronounced grain when sawed. Along the lower Apalachicola River and in similar deep alluvial river plains, the prevailing orange-brown of the heartwood is occasionally varied to the deepest brown or chocolate color, often streaked on a lighter background. This wood is very handsome in its color markings and mottling. Much of the lumber from the lower Mississippi delta and other southern regions shows prominently the harder and darker reddish colored bands of the heavier summerwood in each annual ring, and has come to be known commercially as Red Cypress. The annual rings of Cypress and a marked irregularity of growth give to the wood a richly grained effect, for which it is widely sought for interior finish. The wood from both of the northern regions is spoken of in the market as White Cypress. In logging, this term means that the wood floats high. Along the Atlantic coast Black Cypress refers to heavy wood which sinks, or floats very low if at all. In the Gulf regions Black Cypress refers more especially to distinctions in color rather than buoyancy, since in respect to weight dark wood is often not noticeably heavier than the lighter shades. The cause for variations in color can not now be completely pointed out."

In density and strength Southern Cypress is about midway between the White Pines and the hard Yellow Pines. Sp. gr. (air-dry) 0.35 to 0.60, av. about 0.45; weight 22 to 37, av. about 28, lbs. per cu.

ft. As a rule, trees growing in ponds and other swamps with acid soils produce considerably heavier timber than those along fresh, active streams. The wood of large roots, knees, and swollen butts is very light and soft (sometimes with a sp. gr. as low as 0.19, or 10 lbs. per cu. ft.), and is used locally as a substitute for cork in making small articles of buoyancy.

The timber is easily worked, though it requires care in seasoning and in the application of finishes. It is suitable for all parts of a building which are exposed to the weather or likely to remain moist. It makes excellent siding and shingles. One feature of a cypress roof that is sometimes considered important is that it does not impart taste to the water collected from it. When made into porch and portico columns it retains its shape, holds paint, and is strong enough for the purpose. It is also adapted for porch floors, steps, railings, cornice, gutters, outside blinds, sashes and doors. In the construction of greenhouses, where the conditions of heat and moisture test the mettle of the best of woods, Cypress is extensively and satisfactorily used. It is also used successfully in kitchens and pantries for drainboards, sinks, tables, cupboards, cabinets, and floor. The grain of the wood is often highly attractive and at least half of the high-grade lumber manufactured is used for interior trim, panels, and doors. Figured crotch wood from double stems is known to the veneer trade as Faux Satiné.

The Mexican Bald Cypress, *Taxodium mucronatum* Tenore (= *T. mexicanum* Carr.), is widely distributed in the Mexican tableland extending southward into Guatemala and northward probably into south-central Texas, chiefly in wet soils along streams at elevations between 4500 and 7500 feet. It is similar to *T. distichum* in habit of growth, including the formation of knees from submerged roots, and herbarium specimens of the two species are scarcely distinguishable. The southern form, however, has very much smaller seeds and it also grows more rapidly. Paul C. Standley says (*Contrib. U.S. Nat. Herb.* 23: 60) that it "is one of the best known trees of Mexico, being noted especially for its size. The largest individual reported is the famous tree at Santa María del Tule, Oaxaca, near the city of Oaxaca, which has a height of 38.6 meters and a trunk circumference of 51.8 meters; the greatest diameter of its trunk is 12 meters, and the spread of its branches about 42 meters. The Cypress of Montezuma, in the gardens of Chapultepec, has a height of 51 meters and a trunk circumference of 15 meters. It was a noted tree four centuries ago, and has been estimated to be about 700 years old. Other trees have been estimated to have attained a much greater age. A third famous tree is the 'Arbol de la Noche Triste,' in the village of Popatela, near the City of Mexico, which is noted for its association with Cortés."

The timber is similar in structure to that of the northern species but is generally lighter and softer and considered in every way inferior to that from Louisiana. It is used locally for carpentry and general construction where resistance to decay is more important than strength, but the supply is too limited or inaccessible to permit the development of an export trade.

Common names: Cypress—bald, buck, cow, pond, river, southern, swamp (U.S.A.); cypress—black, Gulf red, Gulf coast red, pecky, peggy, white (trade); ahuehuete, ahoehoetl, ahuehuetl, ciprés, c. de Montezuma, cipreso, pentamón, pentamú, sabino, tnuyucú, yaga-chichicino, yaga-guichi xiña, yucu-nda-tura (Mex.).

ANGIOSPERMS

THE Angiosperms far outnumber the Gymnosperms and comprise almost all of the seed-bearing plants in the real tropical flora. With a few minor exceptions, the food, forage, and medicinal plants of the world are Angiosperms, so named because their fruits contain the seeds in a closed cavity. They are of innumerable forms, from low herbs and shrubs to great timber trees, but are readily separable into two groups, namely, the Monocotyledons (plants with a single cotyledon or seed leaf), such as the palms and grasses, and the Dicotyledons (plants with two rudimentary leaves in the embryo of the seed), the source of the so-called hardwood timbers.

Monocotyledons are of incalculable value to man as sources of food, fiber, and special products, but not for lumber. Some of the best known plants are grass, bamboo, corn or maize, wheat, rye, oats, barley, rice, sorghum, millet, sedges, lilies (including onions, garlic, and leeks), orchids, iris, ginger, bananas, plantains, manila hemp, yucca, agave (sisal), rattan, and palms. The tree forms are almost entirely confined to the warm regions of the earth and are extensively employed for a great variety of purposes, including general construction. The bamboos have innumerable uses, especially in the Far East, and can be grown commercially for paper pulp; one of the special uses in the United States is, in a natural state, for ordinary fishing poles and, in manufactured form, for high-grade split-bamboo fishing rods. Other commercial uses for monocot woods, outside the region of their growth, are rakes, walking sticks, umbrella handles, "cane" for chair seats, "reeds" for wicker baskets and furniture, and various novelties.

The stems of Monocotyledons are fundamentally different in their structure from those of Gymnosperms and Dicotyledons in that they are typically unbranched and without an outer covering of bark, and the wood is confined to hard strands irregularly disposed in a mass of softer tissue (Plate XXXVII). Some stems are jointed, with the internodes either solid (as in maize) or hollow (as in bamboo), while others (such as the palms) are continuous and solid, although the tissue at the center may be soft and open. The leaves are parallel-veined and the parts of the flower are in threes. There are about 45 families, but the only one that will be given further consideration here is that of the Palms.

Dicotyledons are much more numerous and widely distributed than Monocotyledons. There are approximately 270 families, large and small, and while the plants of some are entirely herbaceous, there are shrubs and trees in at least 200 families, of which considerably more than half are represented in America. The timbers are commonly known as hardwoods, but the term has no reference to the density of the wood, since it applies to the lightest and softest as well as the hardest and heaviest kinds known.

In temperate North America the thirteen most important hardwood families and their principal timbers are as follows: Aceraceae (Maple), Betulaceae (Birch), Fagaceae (Oak, Beech, Chestnut), Hamamelidaceae (Red Gum), Juglandaceae (Walnut, Hickory), Magnoliaceae (Yellow Poplar), Nyssaceae (Tupelo), Oleaceae (Ash), Platanaceae (Sycamore), Rosaceae (Cherry), Salicaceae (Cottonwood, Willow), Tiliaceae (Basswood), and Ulmaceae (Elm). The tropical American hardwoods well established in foreign commerce are also of thirteen families, namely, Anacardiaceae (Quebracho), Anonaceae (Lancewood), Bignoniaceae (Bethabara), Bombacaceae (Balsa), Flacourtiaceae (Venezuelan Boxwood), Lauraceae (Greenheart),

Leguminosae (Rosewood, Cocobolo, Brazilette, Brazilwood, Cocus, Kingwood, Brazilian Tulipwood, Logwood), Meliaceae (Mahogany, Spanish Cedar, Andiroba), Moraceae (Fustic, Letterwood or Snakewood, Satiné), Rubiaceae (Degame or Lemonwood), Rutaceae (West Indian Satinwood), Sapotaceae (Massaranduba), and Zygophyllaceae (Lignum-vitae). The two lists differ, not only as to the families enumerated, but also as to the class of timbers, for the northern kinds are of general utility, while those from the tropics are either cabinetwoods or serve some special purposes. The bulk of the tropical American hardwoods are in the general utility class and belong to the following 20 families: Anacardiaceae, Apocynaceae, Bignoniaceae, Bombacaceae, Boraginaceae, Burseraceae, Combretaceae, Euphorbiaceae, Guttiferae, Lauraceae, Lecythidaceae, Leguminosae, Meliaceae, Moraceae, Myristicaceae, Rosaceae, Sapotaceae, Simarubaceae, Verbenaceae, and Vochysiaceae.

Hardwoods are more complex in structure and exhibit a much greater range of variation than the conifers. With rare exceptions (*e.g.*, *Drimys*) they contain vessels, and those growing in the tropics are nearly always diffuse-porous and the growth rings are not always distinct. All of the colors are represented except bright green and blue. Most of the woods are without distinctive odor, but some are very characteristically scented. A few have a pronounced taste, usually bitter. The recorded range in density for thoroughly air-dry specimens is from 0.044 to 1.42, or from 2.75 to 89 pounds per cubic foot. While many tropical woods are very heavy the majority of them are not. The great variety of tropical timber available permits the selection of material suitable for any purpose for which wood can be used.

ACANTHACEAE

THE Acanthus family is widely distributed and comprises about 200 genera and 2000 species of herbaceous and climbing plants, some shrubs, and a very few trees. The leaves are simple, opposite, and pubescent; the small to large, irregular, distinctively colored flowers, are borne in spikes, cymes, fascicles, or thyrses; the fruit is an elastically dehiscent capsule. There are many representatives of the family in the New World, but only a few species of two genera, *Bravaisia* and *Trichanthera*, can be classified as trees and they do not supply any economic products. The following description is based upon 23 wood specimens of 15 American species of *Anisacanthus*, *Aphelandra*, *Beloperone*, *Bravaisia*, *Pachystachys*, *Trichanthera*, and *Sanchezia*.

Heartwood absent or not distinguishable from the whitish, grayish, or oatmeal colored sapwood. Luster fairly high. Light and soft to moderately heavy and hard; texture fine to medium; grain straight; easy to work; poorly resistant to stain and decay. The pith is coarsely septate.

Growth rings absent or indistinct. Pores medium-sized in *Bravaisia* (exc. *B. tubiflora* Hemsl.) and usually so in *Trichanthera*, small in the others; in tangential to ring-porous arrangement in *Anisacanthus Thurberii* (Torr.) Gray and with local tendencies in *Beloperone californica* Benth.; often radially disposed in *Aphelandra*, *Beloperone*, and *Pachystachys Riedeliana* Nees; few to rather few in *Bravaisia* (exc. *B. tubiflora*), *Sanchezia*, and *Trichanthera*, very numerous (70 to 200 per sq. mm.) in the others. Vessels with simple perforations, often with wide rims; fine spiral thickenings observed in *Anisacanthus;* thin-walled tyloses present in *Pachystachys;* pitting alternate, typically fine to very fine. Rays decidedly heterogeneous, the cells variable in size and shape, often square or upright, infrequently procumbent; 1 to 4, sometimes up to 6, cells wide and ranging in height up to 40 cells in *Bravaisia* and *Sanchezia*, up to 80 in *Anisacanthus*, and up to 200 or more in *Tricanthera;* 1 or 2, sometimes 3, cells wide and less than 30 cells high in *Aphelandra* and *Pachystachys;* all uniseriate and not over 15 cells high in *Beloperone;* palisade and sheath cells common; pits to vessels either all very small (*Anisacanthus*, *Aphelandra*, *Beloperone*, *Pachystachys*) or varying, often in the same cross field, from small and rounded to elongated and in scalariform arrangement (*Bravaisia*, *Sanchezia*, *Tricanthera*). Wood parenchyma typically very sparse, scarcely visible with lens; mostly narrowly vasicentric, occasionally diffuse, or scattered terminal. Wood fibers with thin to medium walls and numerous, often irregularly distributed, small, simple or indistinctly bordered pits; abundantly septate except in *Aphelandra.* Ripple marks absent. No gum ducts seen.

Bravaisia. Several species have been described, but the best known is *B. floribunda* DC., or *B. integerrima* (Spreng.) Standl., which is distributed from southern Mexico through Central America into Colombia, Trinidad, Venezuela, and Peru. In the northern part of its range it appears to be only a small tree, rarely up to 50 feet in height, but in Colombia and Trinidad it attains larger dimensions, sometimes 75 feet tall and 18 to 36 inches in diameter, and often develops a great mass of stilt-like roots which suggest Mangrove (*Rhizophora*) except that they do not extend down from the branches. The timber has no special uses, but is suitable for boxboards and cheap lumber. *Bravaisia tubiflora* Hemsl. is a shrub or a slender tree occasionally 25 feet high in Yucatán, Mexico, and British Honduras. The wood is finer-textured than that of the other species, the pores not being visible without a lens.

COMMON NAMES: Cien-pies, palo blanco (Mex.); hulaba, hulabal, hulup (Br. H.); mangle blanco (Nic.); palo de agua (C.R.); manglé de agua (Pan.); palo de agua, sancho-araña (Col.); curte, naranjillo, n. bobo (Venez.), jiggerwood, white mangue (Trin.).

Trichanthera *gigantea* H.B.K., the only species, occurs sparingly along streams and in wet forests from Central America to Peru and the Amazon basin, becoming fairly common on certain islands in the Amazon estuary. It usually is a small, spreading tree only 15 to 25 feet high, but a height of 50 feet with a trunk diameter of 10 inches has been reported from Colombia. The British Guiana form, *T. gigantea* var. *guianensis* Gleason, is sometimes 50 feet high. The wood has about the consistency of Red Maple (*Acer rubrum* L.). The pith is large and septate.

COMMON NAMES: Tuno nacedero (Guat.); palo de agua (Pan., Col.); naranjillo (Venez.); beque, canella de Garca, pau santo (Braz.).

COMMON NAMES (other genera): *Anisacanthus:* Chuparosa (U.S.A.); muicle (Mex.). *Aphelandra:* Añil cimarrón, añilillo (Mex.); chuflete, cordoncillo, hierba del cadejo, oreja de coyote, palo de golpe (Salv.); chamoltaco (Nic.); cabellito, hueso de anta (Col.); sanguinaria (Venez.). *Beloperone:* Chuparosa (Mex.); chuchita (Col.). *Pachystachys:* Panache (Haiti). *Sanchezia:* Topumac (Peru).

ACERACEAE

THE Maple family, with two or three genera and perhaps 50 species of small to large trees, is an important source of timber in the north temperate zone. *Dipteronia,* with a single species, occurs in central China. The valuable genus is *Acer,* whose North American range extends into Mexico. *Negundo,* usually considered as a subgenus or section of *Acer,* is widely distributed in the United States and as far south as Guatemala.

Heartwood yellow or yellowish brown (*Negundo*) to pale brown, greenish brown, or decidedly reddish; sapwood yellowish to nearly white. Odorless and tasteless. Luster medium to high. Rather light (*Negundo*) to hard and heavy; sp. gr. (air-dry) 0.45 to 0.70; weight 28 to 44 lbs. per cu. ft.; texture medium to fine, uniform; grain usually straight, sometimes highly figured; easy to work, finishing very smoothly and taking a high polish; not highly resistant to decay.

Growth rings present; often limited by narrow layer of flattened wood fibers. Pores small to minute, not distinct without lens; numerous but not crowded; occurring singly or in radial multiples of 2 to 5, well distributed. Vessels with simple perforations; spiral thickenings present; intervascular pitting rather fine, alternate. Rays usually of two sizes, the larger 3 to 5, sometimes 7, cells wide and commonly less than 30 cells high; homogeneous or nearly so; pits to vessels small, rounded. Wood parenchyma very sparingly paratracheal; not distinct with lens; pith flecks common in some species. Wood fibers with rather thin to rather thick walls; sometimes in part filled with starch in sapwood; pits small, fairly numerous, bordered. Ripple marks and gum ducts absent.

Acer, with possibly 70 species of small to large deciduous trees, is widely distributed over the northern hemisphere, one form extending south of the equator to the mountains of Java. The leaves are opposite, simple, palmately lobed and nerved; the flowers are small, with or without petals, appearing in lateral clusters before the leaves or in terminal or lateral racemes or panicles with or later than the leaves; the fruit is a double samara. There are many kinds of Maple in Japan, some ornamental, others yielding valuable timber. There are three European species and the tree and timber of *A. pseudoplatanus* are usually called Sycamore in England, a name applied in the United States to *Platanus;* the wood when artificially stained gray is known as Harewood or Silver-grey wood.

The American Maples are of about a dozen species, with several varieties. On the basis of their woods they are divisible into two unequal groups, Hard Maple and Soft Maple. The first consists of the Sugar Maple, *Acer saccharum* Marsh., and the Black Maple, *A. nigrum* Michx., which some botanists consider only a variety of the first; for all practical purposes they may be treated as one species. Hard Maple is by far the most important of the Maples commercially as well as the most abundant. Its range includes most of the eastern hardwood region of the United States and Canada, with the best stands near the Great Lakes and in northern New England and the St. Lawrence valley. The tree is called Sugar Maple because its sap is sweet in early spring and is the source of sugar and syrup of considerable economic value. The timber is denser, stronger, and more resistant to wear than Soft Maple and accordingly is preferred for flooring in residences, bowling alleys, dance halls, and shops, and for the frames of implements, machinery, and vehicles where strength is essential; it is also of more attractive appearance than Soft Maple owing to its more conspicuous rays and to special types of figure, such as curly grain, fiddle-back mottle, and bird's-eye. The cause of bird's-eye is not definitely known, but it is not, as so frequently stated, attributable to buds.

The two principal Soft Maples are Red or Swamp Maple, *Acer rubrum* L., and Silver or White Maple, *A. saccharinum* L. They have much the same range as *A. saccharum,* but extend further south and attain their best development in the lowlands of the Ohio valley. The Bigleaf, Broad-leaved, or Oregon Maple, *A. macrophyllum* Pursh., grows along the Pacific coast from Alaska to southern California, attaining its greatest development in Washington and Oregon where it supplies a few million feet of lumber annually for local consumption. The wood is of the Soft Maple type, but is more deeply colored and more frequently figured. The wood of these species is used for furniture, interior trim, flooring, cooperage, woodenware, novelties, and many other purposes.

Two small Maples, of distinctive appearance but no economic value, occur in the undergrowth of northeastern hardwood forests; they are Striped Maple or Moosewood, *Acer pennsylvanicum* L., and Mountain Maple, *A. spicatum* La Marck. The Vine Maple, *A. circinatum* Pursh., is, according to Sargent (*Manual of the trees of North America,* p. 685), "one of the most abundant of the deciduous-leaved trees of western Washington and Oregon up to altitudes of 4000 ft. above the sea, and of its largest size on the rich alluvial soil of bottomlands, its vine-like stems in such situations springing four or five together from the ground, spreading in wide curves, and sending out long slender branches rooting when they touch the ground and forming impenetrable thickets of contorted and interlaced trunks, often many acres in extent."

Two Mexican species of *Acer* (*sens. str.*) have been described. They are small trees, or rarely up to 60 feet high and 28 inches in diameter, growing in the mountains in the northern part of the country. The largest is *A. brachypterium* Woot. & Standl. which, according to Standley (*Trees and shrubs of Mexico,* p. 690), "is closely related to *A. grandidentatum* Nutt. of the western United States and may be only a form of that species." The Spanish name for Maple is Arce.

Negundo, the Boxelder, has the characteristic Maple fruits but the leaves are pinnately compound with one terminal and one to three pairs of lateral leaflets. Its natural range covers almost the whole of the United States and extends through the mountains of Mexico into Guatemala. The principal and perhaps the only distinct species is *Negundo aceroides* Moench (= *Acer Negundo* L.), but it is highly variable and some of the extreme forms are recognized as varieties or even as distinct species, though the existence of connecting forms makes such segregation of doubtful value. The tree attains a maximum height of 70 feet and a trunk diameter of four feet, but it usually divides near the ground into several wide-spreading or erect branches and is of little importance as a source of timber.

The heartwood is an unattractive yellowish brown, merging rather gradually into the greenish yellow sapwood; sometimes it has coral-red streaks due to the presence of a soluble pigment produced by the colored hyphae of a fungus, ***Fusarium negundi*** Sherb. The timber is lighter in weight than commercial Maple, the average sp. gr. (air-dry) being about 0.45; weight about 28 lbs. per cu. ft. The uses, which are few and local, include cheap furniture, woodenware, cooperage, and fuel. The wood structure is practically identical with that of *Acer* (*sens. str.*).

COMMON NAMES: Ash-leaved maple, boxelder (U.S.A.); acecincle, acezintle (Mex.); palo de venagre, raxoch (Guat.).

ACHATOCARPACEAE

THIS unimportant family, often included in the Phytolaccaceae, is composed of two closely related genera and several species of armed or unarmed shrubs and small trees with a combined range from southwestern United States through Mexico, Central America, northern and western South America to Paraguay and Argentina. The leaves are simple and alternate; the flowers are dioecious, racemose or paniculate; the fruit is drupe-like.

Achatocarpus, with perhaps a dozen species distributed from Mexico to Paraguay, appears to be of largest size in northern Colombia and in Argentina, but at best is only 35 feet tall and 12 inches in diameter. Some of the plants are armed with axillary thorns. The gray or ashy yellow sapwood is hard, tough, strong, fine-textured, and of about the consistency of Lancewood (*Oxandra*). It is little used except for fuel and charcoal.

Pores numerous; small, invisible without lens; mostly in pairs, well distributed. Vessels with simple perforations; without spiral thickenings; pits very small. Rays fine, near limit of vision on cross section, fairly distinct on radial; mostly 1 to 3 cells wide, and rather low; heterogeneous; all pits very small. Wood parenchyma sparingly paratracheal, occasionally slightly confluent. Wood fibers with minute bordered pits. Ripple marks and gum ducts absent.

COMMON NAMES: Limoncillo (Salv.); limoncillo, l. moján, moján (Col.); zamurito (Venez.); cabo de lança, tintoreiro (Braz.); ibira-hú, ñuaté-curugú, palo mataco, perlitas, ruma-caspi (Arg.).

Phaulothamnus *spinescens* A. Gray, the only species, is a shrub 3 to 10 feet high in Texas and northwestern Mexico. It has a grayish bark, horizontal spine-tipped branches, and white translucent fruits. It is known in Sinaloa as Putia. The wood has not been studied.

AEXTOXICACEAE

Aextoxicon *punctatum* Ruiz & Pavon, the only genus and species of this family, is a medium-sized Chilean tree of local importance in the Provinces of Valdivia and northern Llanquihue. In some places it grows in such dense stands that all other vegetation is killed by the shade, and for this reason it is frequently known as Palo Muerto (tree of death). The leaves are simple, leathery, and alternate; the spherical buds are covered by a cap that later splits irregularly and falls away; the fruit is fleshy and one-seeded. The timber has many uses, such as general carpentry and construction and particularly for cooperage. The supply is too limited to permit export.

Heartwood pale brown, with a reddish

hue; merging gradually into the somewhat lighter-colored sapwood. Not highly lustrous. Has about the consistency and appearance of Red Gum (*Liquidambar*). Of medium density and weight, tough and strong; of fine and uniform texture; grain fairly straight; easy to work, finishes very smoothly; is rather durable.

Growth rings present but indistinct. Pores very small, angular, scarcely distinct with the lens; very numerous, crowded; well distributed without definite pattern. Vessels with scalariform perforation plates having many narrow and closely spaced bars; without spiral thickenings; pitting finely scalariform. Rays very numerous, inconspicuous; 1 to 4 cells wide and mostly less than 30 cells high; decidedly heterogeneous, the uniseriates with all cells upright or square, the others with procumbent body cells and few to several marginal rows of large cells; scattered rhombohedral crystals present, usually in enlarged cells; pits to vessels narrow-elongated and in scalariform arrangement. Wood parenchyma finely reticulate, barely visible with lens. Wood fibers with rather thick walls and numerous large bordered pits. Ripple marks and gum ducts absent.

Common names: Aceitunillo, olivillo, palo muerto, roble, teque, tique (Chile).

ANACARDIACEAE

This family consists of about 65 genera and over 500 species of trees, shrubs, and woody climbers, most abundant in tropical and subtropical regions, a few inhabiting the temperate zones. The bark is often resinous and many of the plants have a volatile oil that is caustic or poisonous. The leaves are typically alternate, without stipules, and simple, trifoliolate, or odd pinnate with few to many pairs of leaflets; the flowers are generally small and borne in axillary or terminal panicles; the fruits are drupaceous, sometimes edible, or dry and winged. Some of the best known members are the Mango (*Mangifera*), Pistachio (*Pistacia*), Sumachs (*Rhus*), Poison Ivy (*Toxicodendron*), and Quebracho (*Schinopsis*). There are arborescent species of about 25 genera in America, but only two genera (*Astronium* and *Schinopsis*) are of much commercial importance for their timber.

The woods of the family exhibit wide variation in appearance and properties. Color yellow with greenish cast to olive-green, often variegated or streaked, in *Cotinus, Mauria* (in part), *Rhus* (in part), and *Toxicodendron;* pink to light reddish orange in *Mosquitoxylum* and *Rhus* (in part); red darkening to brick red or deep brownish red in *Astronium* (*Myracrodruon* section), *Comocladia, Schinopsis,* and *Schinus* (in part); in other genera mostly light brown to dark reddish brown, sometimes whitish or light grayish brown, often with pinkish tinge in lighter-colored specimens, frequently striped or variegated. Luster low, medium, or high. Odor sometimes present but not very distinctive; taste of reddish woods sometimes astringent. Density 0.44 to 1.30; weight (air-dry) 27 to 81 lbs. per cu. ft.; texture coarse to fine; grain variable; working properties poor to excellent, generally good; durability very low to very high.

Growth rings present, but not always distinct. Ring-porous structure in some or all species of *Cotinus, Rhus,* and *Toxicodendron,* the early-wood pores small to medium-sized and in a narrow to wide band, those in late wood small to minute and tending to form diagonal or tangential bands; ulmiform pattern in *Schinus;* elsewhere the pores are mostly small, solitary or more often in radial multiples of two to several pores each. Vessels typically with simple perforations; scalariform plates with many narrow bars also present in some vessels of *Campnosperma;* spiral thickenings present in some or all of the vessels of *Cotinus, Lithraea, Rhus, Schinus* (in part), and *Toxicodendron;* vascular pitting coarse. Tyloses usually abundant in heartwood. Rays in part with small to large resin ducts, the epithelial layer composed of one to several rows of very small cells in some or all species of *Astronium, Campnosperma, Loxopterygium, Malosma, Metopium, Rhus, Schinopsis, Schinus, Spondias, Tapirira,* and *Toxicodendron;* other rays generally 1 or 2, occasionally 3, sometimes up to 5, rarely 8, cells wide and mostly less than 30, occasionally up to 50, rarely 100, cells high; weakly to decidedly heterogeneous; crystals common; pits to vessels large, oval to much elongated. Wood parenchyma apparently absent in *Campnosperma;* sparingly vasicentric in the others; sometimes also finely terminal and occasionally diffuse; crystalliferous strands present in a few genera, *e.g.*, *Lithraea* and *Mauria;* pith flecks common in soft woods. Wood fibers with thin to very thick walls; septate, at least in part, in *Anacardium, Astronium, Campnosperma, Comocladia, Lithraea, Loxopterygium, Mauria, Metopium, Schinopsis, Schinus, Spondias, Tapirira,* and *Toxicodendron;* pits small to minute, simple or indistinctly bordered. Ripple marks absent. Vertical resin ducts not known to occur in any member of the family; radial canals few to numerous, visible with lens, sometimes (*e.g.*, *Loxopterygium* and *Tapirira*) staining the surface of the wood with their dark-colored, oily exudations. For anatomy of the different genera see *Tropical Woods* 60: 16–45.

Actinocheita, with a single species, *A. filicina* (DC.) Barkley (= *Rhus filicina* DC. = *R. potentillaefolia* Turcz.), is a shrub or little tree, resembling Staghorn Sumac (*Rhus typhina* Torn.), limited to the mountains of southwestern Mexico, where it is known as Tetlazian. It is not represented in the Yale wood collections.

Anacardium, with several species of small to very large trees, is represented throughout tropical America, though most of the forms are Brazilian. The leaves are alternate, simple, entire, and petioled; the flowers are small and borne in large terminal panicles; the leathery, nut-like, indehiscent fruit is borne on an enlarged stalk. The timber of certain species has commercial possibilities.

The best known species is the Cashew, *Anacardium occidentale* L., a small tree indigenous to the Caribbean region and naturalized elsewhere in the tropics. Standley says (*Trees and shrubs of Mexico*, p. 659): "From the trunk there exudes a gum somewhat like gum arabic. This can be used for varnish, and in South America it is used for bookbinding in order to prevent the attacks of insects. The most important products of the tree are the fruit and receptacles, both of which are edible. The receptacle is pear-shaped, very fleshy, and yellow or reddish; it is astringent when green, but when ripe has a pleasantly acid flavor. In Mexico sweetmeats are sometimes made from the receptacles, and in some parts of tropical America a kind of wine is made from them, and this after fermentation is distilled to obtain brandy. The pericarp of the fruit proper contains an oil, cardol, which is acrid and caustic. This is driven off by heat, but the fumes which rise when the nuts are heated should not be allowed to reach the face or eyes. The roasted kernels are edible and have a pleasant milky flavor. The oil obtained from the nuts is applied in India to the floors and rafters of houses to preserve them from insects, but its use is dangerous. The ground kernels are sometimes used to flavor wine, and they are mixed with chocolate." (For an account of the Cashew nut industry in western India, including the harvesting and preparation of the nuts and the utilization of the by-products, see *Bull. Imperial Institute* [London] 36: 1: 44–52. 1938.) The French for Cashew is Acajou, the name generally applied to Mahogany (family Meliaceae). According to A. Chevalier (*Rev. Bot. Appl. & d'Agr. Tropicale* 17: 194: 713), the designation for Mahogany probably originated from the use of the Cashew resin to varnish the ends of *Swietenia* logs to prevent checking in shipment from the West Indies to France. The wood of the Cashew tree is of no commercial value and its uses are few and local.

Some of the other species of *Anacardium* are large trees. Among these are *A. microsepalum* Loes., *A. parvifolium* Ducke, *A. giganteum* Engl., and *A. Spruceanum* Benth. of the Amazon region, the last two extending northward into the Guianas; and *A. excelsum* (Bert. & Balb.) Skeels (= *Rhinocarpus excelsa* Bert. & Balb. = *A. Rhinocarpus* DC.) extending from Costa Rica to Ecuador and Venezuela (Plate XI). The following account of *A. excelsum*, known in Panama as Espavé, is by G. Proctor Cooper (*Tropical Woods* 22: 4–9):

"Espavé occurs in large quantities on the Pacific Coast of Costa Rica and Panama and on the Caribbean coast of Colombia and Venezuela. I failed to find it on the northern side of Panama from Bocas del Toro to the Colombian border and it is not known at Puerto Limón, Costa Rica. It seems to be confined to regions having distinct dry seasons. Although typical of the dense evergreen forest it does not grow at the higher elevations; at least this was found to be the case in Chiriquí, Panama, where the species finds its best development on the lower, well-drained soils. In that locality I found an average of 4 or 5 trees per acre, with a maximum of 10 or 12 massive trees 75 to 125 feet tall. In Darién Province it is said to compose almost pure stands over large areas. Its gregarious occurrence is presumably due to the facility with which it reproduces itself and to its ability to crowd out competing species. . . .

"Under favorable conditions in the forest, Espavé attains a height of 125 to 150 feet, with an unbuttressed trunk four to six feet in diameter above the basal swelling and free of limbs for 40 to 50, occasionally over 60, feet; the crown is often spreading or urn-shaped, with some of the branches large enough for small sawlogs. Grown in the open, the tree has a short, thick trunk and a full crown of low-spreading branches. The pale, grayish bark, which is scaly or roughly plated, but not deeply furrowed, serves readily to distinguish the tree from its associates in the forest. . . .

"The sapwood, which may be as much as six inches thick in large logs, is dingy gray, more or less streaked or blotched with yellow or purple. Fresh heartwood is dull brown, with purplish red streaks; the freshly exposed interior of thoroughly dry specimens has a lusterless lemon-yellow color, with darker striping. When a dry plank is exposed to the air and light for some time the color gradually changes into a rich golden brown, the striped effect becomes more pronounced, and the purplish tints change to reddish brown. On quarter-sawed boards, the rays are very distinct, for although not large they are darker than the fiber background and show effectively. The vessel lines are also dark-colored and distinct, especially on tangential surfaces. A newly felled log has a slight resinous or pungent scent, but this is probably attributable to the bark, as the dry wood is odorless and tasteless.

"The wood is in some ways easy to work, but radial surfaces have a tendency to become fuzzy when planed or sand-papered, due to the fact that the walls of the fibers are rather soft and yielding and do not cut off sharply and smoothly where the grain is alternating. This type of grain, often called roe or feather grain or cross-banding, is typical of many of the finest cabinet woods, but there the tools make a much sharper and cleaner cut over the crossed grain than is possible with Espavé. However, when varnish is applied, followed by light sanding and wax, a high polish is obtainable. The wood takes stains readily. Occasionally pin knots are found deeply buried in the log and their pith is likely to fleck out when the lumber is being worked. Espavé nails without splitting and holds its shape well in box form if kept dry; exposed to the weather it will warp and check. It is not very durable in contact with the ground.

"The natives use the wood for making kitchen utensils and dishes because it is easy to work and the articles made from it are light, but resistant to wear. The large trunks are well adapted for the making of dugout canoes as they can be fashioned with crude tools, are not easily split, and have a fairly long life. The timber is to be had at the local lumber yards in the various parts of the countries where the trees are available. It is used for general carpentry and construction, inexpensive furniture, and other purposes for which it is suited on account of its lightness, fair durability, and low cost. . . . Attempts to introduce the lumber into the markets of the United States have been made in a desultory way, but with little success, as the wood is not of sufficiently high grade to sell itself. It is none the less potentially a very useful timber and occurs in sufficient abundance to justify careful investigation by industries in this country."

Tests on a limited amount of Espavé timber gave the following results (*loc. cit.*, pp. 7–9): Sp. gr. (air-dry) 0.56; weight 35 lbs. per cu. ft. *Static bending* (lbs. per sq. in.): modulus of rupture, 7175; m. of elasticity 1,683,500; fiber stress at elastic limit, 4690. *Endwise compression* (lbs. per sq. in.): maximum crushing strength, 8230; modulus of elasticity, 1,800,000; fiber stress at e. l., 6190. *Tension* (perpendicular to grain): 586 to 718, av. 636, lbs. per sq. in. *Shearing strength* (along grain): 954 to 1388, av. 1178, lbs. per sq. in. *Cleavage strength:* 222 to 264, av. 247, lbs. per in. of width. *Hardness* (load in lbs. required to imbed 0.444-in. ball half its diam.): radial surface, 615 to 730, av. 680; tangential, 650 to 765, av. 700; end, 620 to 795, av. 710.

COMMON NAMES: *Anacardium occidentale:* Cashew, c. nut tree (Eng.); acajou (Fr.); marañón (Sp.); cajuil (P.R.); cacajuil, cajuil (Dom. R.); pomme cajou, p. d'acajou (Haiti); jocote marañón (Salv.); merey (Col.); caují, merei, merey, pajuil, paují, paujil (Venez.); boschkasjoe, kadjoe, kasjoe, kasjoen, mereke, olvi, orvi (Sur.); acaja-iba, acajou-iba, aloi, a. ichie, auloui, caschou (Fr. G.); cajú, casú (Peru); acajaiba, acajuiba, cajú, c. manso, cajueiro, oacajú (Braz.). *Other species:* Nariz (Cuba, int.); espavé, espavel, e. amarillo, e. rosado, quina (C.R.); espavé, wild cashew (Pan.); aspavé, caracolí (Col.); caracolí, c. blanco, chorote, lacre rosado, mija, mijaguo, mijao, paují (Venez.); hoobodia, wild cashew (Br. G.); akajoe, akoejoe, boesi kasjoen, bosch kasjoe, hoeboedie, kadjoe mattoe, merekeballi, oeboedi djamaroe (Sur.); caracolí (Ec.); cajú assú, c. da matta, c. do campo, c. do c. coberto, c. rastreiro, cajueiro do campo, cajuhy, cajurana, cajú-y (Braz.).

Astronium is an important genus with about a dozen species of medium-sized to large timber trees occurring in southern Mexico, Central America, Colombia, Venezuela, Trinidad, British Guiana, Ecuador, Bolivia, Argentina, Paraguay, and (most abundantly) in Brazil. The leaves are odd-pinnate, sometimes with few, commonly with many, entire or serrate leaflets; the flowers are small and paniculate; the sepals become enlarged and cover the small, dry, rounded or elongated fruits. Two sections of the genus are recognized, namely, *Euastronium,* typified by *Astronium graveolens* Jacq., and *Myracrodruon,* of which the best known species is *A. urundeuva* (Fr. Allem.) Engl.

Section *Euastronium,* according to Mattick (*Notizbl. Bot. Gart. Berlin-Dahlem* 11: 110: 991–1012. 1934), includes nine species, one variety, and six forms. *Astronium obliquum* Gris. is an evergreen tree confined to the hills in northern Trinidad, where it is known as Yoke. *A. Conzattii* Blake (= *A. zongolica* Reko) of southern Mexico is probably not specifically distinct from *A. graveolens.* The Muiracoatiara of the lower Amazon is *A. LeCointei* Ducke; some timber of this species from the Ford rubber plantations on the Rio Tapajoz has been sold in the United States under the name of Mura. *A. Ulei* Mattick, a tree sometimes 50 feet high, distinguished chiefly by the small number of its leaflets, occurs along the Rio Branco, a northern tributary of the Amazon; its wood has not been studied. *A. gracile* Engl. is a tree of medium height, but with a stout bole as much as 36 inches in diameter, known as Ubatan or Ubatão in Rio de Janeiro, and as Urunday-itá, Urunday-pará, and Urunday-pytá in Paraguay; the bark contains about 12 per cent of tannin; the black-striped brown wood is said to be only sparingly utilized because of its brittleness.

The two principal species are *Astronium graveolens,* including the variety *Planchonianum* Engl. (= *A. Planchonianum* Engl.), and *A. fraxinifolium* Schott, with four forms. Both species occur together throughout most of the range of the genus and their timbers are practically identical. The color varies from light to dark brown or reddish, more or less conspicuously marked with vertical blackish bands of variable spacing and often producing a very striking and beautiful figure. There is a considerable range of density in different specimens and also within the same sample, the dark zones being the heavier. The timber best known commercially is the Gonçalo Alves of eastern Brazil. Karl Schmieg, a New York manufacturer of fine furniture, says (*Tropical Woods* 5: 2): "Gonçalo Alves is obtainable in long logs 12 to 24 inches in diameter and very straight and sound. The wood is moderately hard and heavy, fairly close in texture, and stands very well. It is suitable for cutting into veneers and takes a beautiful polish. It has a rather pleasing stripe and bears some resemblance in figure and texture to Golden Ebony, or Coromandel, only it is of a warmer tone; the stripes are dark but not real black. Sometimes the wood exhibits a mottled figure and then approaches the Brazilian Rosewood."

Although a form of *Astronium fraxinifolium* grows in Colombia, most of the herbarium specimens from that country have been determined as *A. graveolens*. H. M. Curran says (*Timbers of Tropical America*, p. 389): "This is another form of the tree and timber known as Gonçalo Alves in Brazil. I am acquainted with it only in the Magdalena valley in Colombia, where it occurs scatteringly in the lowland forests. It attains a height of 100 feet or more, with commercial lengths of 50 feet and diameters up to three feet. It is a symmetrical tree, only slightly buttressed, and has a rather thin but somewhat rough bark of a brown or grayish color. It is considered the best timber in the Magdalena valley for house posts. The timber is commonly met with in the markets of both countries and is highly esteemed for cabinet work and fine furniture."

Section *Myracrodruon*, sometimes treated as a distinct genus, comprises, according to Mattick (*loc. cit.*), four species and one variety limited in distribution to tropical South America below the Amazon region. *Astronium macrocalyx* Engl. of Bahia and Rio de Janeiro is known in the latter region as Aroeira do Mucury. *A. concinnum* Schott (= *Myracrodruon concinnum* [Schott] Engl.) is apparently confined to the region about Rio de Janeiro. *A. Balansae* Engl. occurs in Brazil, Paraguay, and Argentina. It attains a height of about 50 feet and a trunk diameter of 24 inches and is important for its strong and durable timber, particularly in the Chaco; it is called Urunday in Misiones, Urunday Pardo in Corrientes, and Urunday-pichaí in the Chaco. The most widely distributed species is *A. urundeuva* (Fr. Allem.) Engl. (= *M. urundeuva* Fr. Allem. = *A. juglandifolium* Gris.). It is generally known in Brazil as Aroeira, also Chibatan and Ubatan; in Bolivia as Cuchi and Sotocolo; in Paraguay as Urunday-mí; as Urundel in Jujuy and Salta, Argentina, where it is perhaps at its best, with a maximum height of 100 feet, though usually it is not over 60 feet tall and 30 inches in diameter.

The woods of all species of this section, so far as studied, are similar. The yellowish sapwood is comparatively thin and is sharply demarcated from the heartwood, which has a fairly uniform cherry-red color deepening to dark brownish red upon exposure, but lacking the bold blackish striping characteristic of the woods of the section *Euastronium*. There is considerable resemblance to Quebracho (*Schinopsis*), but in general the grain is not so irregular, the consistency is not so flinty, the appearance is more oily, the demarcation between heartwood and sapwood is sharper. The timber is highly appreciated locally for railway crossties, posts, piling, bridge timbers, and other forms of heavy and durable construction, and for fuel.

The following description applies to all species of *Astronium*. Wood hard and heavy to extremely so; sp. gr. (air-dry) 0.85 to 1.28; weight 53 to 80 lbs. per cu. ft.; texture rather fine, uniform; grain variable; fairly easy to rather difficult to work, turns readily, finishes very smoothly and takes a high natural polish; noted for its durability. The softer grades with attractive figure are suitable for veneers for fine furniture; the denser and more deeply colored material, particularly specimens containing a large proportion of black wood, can be used for knife handles in place of Cocobolo (*Dalbergia retusa* Hemsl.).

COMMON NAMES: *Euastronium:* Yoke (Trin.); cero, copaiva, culebra, kulimche, palo de cera, p. de culebra, sangolica, sangualica, yaga-biche, zongolica (Mex.); glassy wood (Br. H.); ciruelo, jocote de fraile, palo mulato, p. obero (Guat.); ronrón (Salv.); ciruelillo, ciruelo, c. de montaña, foncontín, masicarán, palo obero, ronrón, pimientillo, sirguelillo (Hond.); ronrón (Nic., C.R.); zorro (Pan.); diomate, gusanero, quebracho, quiebrahacha (Col.); algarrobo barcino, diomate, gateado, g. barcino, gusanero, potro, roble gateado, tibigaro, tirigaro, yomate (Venez.); bauwana (Br. G.); guasango (Ec.); bolaquivo, palo de cruz (Peru); arantha, gomavel, gonçaleiro, gonçalo, g. alves, g. do matto, jejuira, muira-catiara, m.-coatiara, m.-quatiara, pau gonçalo, sangue-sugueira (Braz.); urunday-para (Arg.); urunday-ibá, u.-pytá, u.-para (Par.); almendro macho, cuchi blanco (Boliv.). *Myracrodruon:* Aderno, a. preto, a. vermelho, aroeira, a. do campo, d. do mucury, a. do sertão, a. preta, chibatan, gibatão, guarubú preto, orendauva, ubatan, ubatão, urundeuva (Braz.); urunday-mí (Par.); cuchi, sotocolo (Boliv.); urunday, u. blanco, u. crespo, u. del nordeste, u. del noroeste, u.-mí, u. pardo, u.-pichái, urundel (Arg.).

Campnosperma, with several species of medium-sized to large trees growing gregariously on wet lowlands near Mangrove formations, is best represented in the East Indies. There are two species in tropical America, namely, *C. gummifera* (Benth.) L. March., a tree about 65 feet high of common occurrence in marshy forests in parts of the lower Amazon, and *C. panamensis* Standl. of the Atlantic lowlands of northern Panama and adjacent localities in Costa Rica, where it is known as Orey. The following information is supplied by G. Proctor Cooper (*Tropical Woods* 12: 7):

"I first saw the Orey tree while I was on a trip from Almirante to Bocas by launch. There is a long neck of low swampy mainland off the western part of Columbus Island, on which the town of Bocas del Toro is situated, and viewed from a distance one sees an even, unbroken line of timber that contrasts noticeably with the appearance of the usual type of shore forest. This is due to the predominance of the Orey trees which comprise upward of 50 per cent of the whole stand and form groves here and there that are almost pure in so far as the larger trees are concerned. In this gregarious habit the Orey resembles the Cativo (*Prioria Copaifera* Gris.) and the Silica Palm. I was told by Mr. William Ponton, the British vice consul at Bocas, that Orey occurs in places all along the Caribbean coast from San Blas, Panama, to Puerto Limón, Costa Rica, and, according to reports by natives, was to be found also along the east coast of Nicaragua.

"The specimen in our collection (Yale 10500) was obtained on Columbus Island, near Bocas. It is from one of a group of trees growing just behind the fringe of Mangroves and almost at sea level where the ground is wet and during heavy rains is under water, site conditions said to be typical for the species. The trees are 12 to 18, occasionally 24, inches in diameter breast high, with low and stout buttresses, and a rather short bole covered with a thick, greenish gray bark that is rough, though not deeply furrowed. The crowns, which are large, forked, and spreading, have heavy branches and coarse, brittle, and blunt twigs which bear clusters of leaves at the ends and are marked with prominent leaf scars. From the tips of some of the twigs, beyond the leaves, extended spikes of small, yellow, faintly scented flowers. The new fruits, some of which were nearly mature before the tree was through blooming, are about half an inch in diameter and have a stony pit covered with a green fleshy exocarp from which, when cut into, a dirty gray juice exuded that stained the knife blade a bluish purple. A similar juice was found in cutting the bark of the trunk, but it was not very abundant. The wood is light and soft and is not 'sappy.' It has a rather distinctive, though not pronounced, odor when fresh, but there is nothing unpleasant about it. The color varies from white to grayish buff, without marked contrast between heartwood and sapwood. The dingy pink color that eventually covered the surface of the specimen was not in evidence at the time of cutting and did not penetrate deeply except along cracks. Some blue stain due to a fungus appeared in the sapwood while the sample was being seasoned. I did not observe any local use for the timber."

Orey wood has a somewhat silvery luster in proper light, but on the whole its appearance is not attractive. It is firm, rather fine-textured, tough, and strong for its weight; sp. gr. (air-dry) 0.44 to 0.48; weight 27 to 30 lbs. per cu. ft. Preliminary tests on pulping properties indicate that it is unsatisfactory for sulphite, kraft, or ground-wood because of the difficulty of removing the pinkish gray color, but can be used successfully with the soda process. (See *Tropical Woods* 12: 9.) The timber is not suitable for construction where exposed to decay or insect attack or for purposes requiring great strength or attractiveness of figure, grain, and color. It seems best adapted for making boxes and food containers, the principal use of the East Indian species, as it is easy to work, finishes smoothly, holds nails well, and is free from objectionable odor and taste. It is also worthy of trial for plywood for general utility purposes as it is somewhat like Okoumé (*Aucoumea*) of tropical West Africa.

Comocladia, with about 20 species of poisonous little trees, occasionally 30 feet high, with slender, usually unbranched stems, is confined to the West Indies and south-central Mexico. The leaves are odd-pinnate, with 3 to 18 pairs of entire or irregularly toothed, sometimes spinose, leaflets; the minute red flowers are borne in axillary panicles; the fruit is a small drupe. The bark contains a caustic sap that turns blackish and makes an indelible stain. The wood is of good quality and harmless, but is little used because of the small sizes available and the poisonous nature of the trees.

Heartwood uniform light red, deepening to brick-red upon exposure; sharply demarcated from the brownish or yellowish sapwood. Luster rather low. Without distinctive odor or taste. Very heavy, hard, and strong; sp. gr. (air-dry) 1.10; weight about 69 lbs. per cu. ft.; texture fine and uniform; grain irregular; rather difficult to work, but taking a glossy polish; highly resistant to decay.

COMMON NAMES: Black plum, maiden plum (Jam.); guao, g. común, g. de sabana, g. hediondo, g. prieto, g. real (Cuba); carrasco, maíz pelado, m. tostado, poison ash, prapa (P.R.); chicharrón, guao (Dom. R.); bois espagnol, b. pangnol, brésillet, b. franc (Haiti); chinil-té, hincha huevos, pata de pava, tatatil, tatatián, teclatilla (Mex.).

Cotinus consists of two closely related species of deciduous shrubs and little trees with simple, long-petioled leaves, scaly bark, fleshy roots, and strong-smelling juice. *C. coggrygria* Scop. (= *Rhus Cotinus* L.) is widely distributed through southern Europe and the Himalayas to central China. It is the Smoke-tree of gardens, so called because of the large, loose, finely plumose, terminal, gray fruiting panicles which may cover the crown. The American species, sometimes considered only a variety of the other, is *C. americanus* Nutt. (= *Rhus cotinoides* Nutt. = *C. cotinoides* [Nutt.] Britt.). It is of interrupted distribution in the lower Mississippi valley, attaining its largest size in northern Alabama and southern Tennessee, and in the Ozark mountains of southwestern Missouri, northwestern Arkansas, and eastern Oklahoma; it also occurs in Texas, often reduced to a low shrub abundant in mountain canyons and high hillsides in Kenyon County. At its best it rarely exceeds 30 feet in height with a trunk 12 to 14 inches in diameter. The timber is unimportant because of its scarcity, but makes good fence posts.

Heartwood greenish yellow, more or less striped; sharply demarcated from the thin, nearly white sapwood. Luster fairly high. Without distinctive odor or taste. Moderately hard and heavy; texture rather coarse, uneven; grain straight; easily worked, taking a glossy polish; highly resistant to decay. Of no commercial possibilities.

COMMON NAMES: Chittamwood, smoke-tree, yellow wood (U.S.A.).

Cyrtocarpa is a Mexican genus with two species of trees 20 to 30 feet high and 6 to 12 inches in diameter. *C. edulis* (T. S. Brandeg.) Standl. (= *Tapirira edulis* T. S. Brandeg.) is limited to southern Baja California, where it is known as Ciruelo because of its edible drupaceous fruit. *C. procera* H.B.K. occurs from Jalisco to Puebla and Oaxaca; Standley gives the following information about it (*Trees and shrubs of Mexico,* p. 659): "The wood is said to be soft and purplish, with a strong odor, and to be used for making trays, small images, and other articles. Goats are fond of the leaves. The fruit, which is much eaten, is yellow and the flesh resinous, with acid flavor. The fruits are said to be known at Jojutla (Morelos) as 'berracos' and 'chupandías.' The large seeds are eaten by pigs, and they have been used locally (taken internally) as a remedy for leprosy. The bark is employed as a substitute for soap." The genus is not represented in the Yale wood collections.

COMMON NAMES: Chupandía, ciruelo, copal, c. cocote, c. jocote, copalhi, maxocote, popoaqua (Mex.).

Haplorhus *peruviana* Engl. is an evergreen tree, sometimes 50 feet tall, having the appearance of a Weeping Willow (*Salix*), of very limited known distribution in the Peruvian Andes, where it is called Ccási. (See *Notizbl. Bot. Gart. Berlin-Dahlem* 11: 108: 719.) There are no specimens available for this study.

Lithraea, with three species of poisonous shrubs and small trees, is rather widely distributed in southern South America. The leaves are simple in *L. caustica* (Mol.) Miers, but odd-pinnate with 3 to 5 sessile leaflets and a winged rachis in *L. brasiliensis* March. and *L. molleoides* (Vell.) Engl. (= *Schinus molleoides* Vell.). The last species is of considerable local utility; the bark is rich in tannin and dyestuff; the leaves are aromatic and medicinal; the fruits contain an essential oil similar to turpentine; the timber, though available only in small sizes, is of good quality and is used for cabinet work and articles of turnery and carving as well as for fence posts, stakes, fuel, and ashes for soapmaking.

Heartwood blackish brown; sharply demarcated from the brownish, pinkish, or greenish tinged sapwood. Fairly lustrous. Odor and taste not distinctive. Hard and heavy, especially the heartwood; texture fine; grain straight to irregular; easy to work, finishing very smoothly; highly resistant to decay. Of no export possibilities.

COMMON NAMES: Aroeira branca, a. brava, a. da caapuera, a. de bugre, a. do matto, aroeirinha, a. preta, coração de bugre, molle, pau de bugre (Braz.); árbol malo, aruera, a. blanca, a. colorado, a. dura, a. nigra, arueriña, corazón de bugre, molle de beber, m. dulce, palo de bugre, quina (Urug.); aroeira blanca, a. negra, chicha, chichita, molle de beber (Arg.); litre, llithi (Chile).

Loxopterygium, with four species of trees, is irregularly distributed from Venezuela and the Guianas to Argentina. Best known is *L. Sagotti* Hook. f., usually known in Surinam as Slangenhout (Snakewood) and in British Guiana as Hububalli. The tree is said to be fairly common in the forests of those two countries and attains a height of 100 feet and a trunk diameter great enough to produce hewed timbers 20 inches square. The smoothish or wrinkled bark contains a sticky greenish yellow latex. The unequally pinnate leaves have several pairs of rather large leaflets suggesting Ash (*Fraxinus*); the small flowers are borne in axillary panicles; the fruit is a small samara resembling those of Maple (*Acer*), though not borne in pairs. The timber is of good quality and is used to a small extent locally for making boats and furniture. Attempts to increase its utilization in British Guiana have not been very successful, owing largely to the preference for Crabwood (*Carapa*).

Heartwood brown or reddish brown, with dark laminations or streaks of varying width and regularity; usually flecked with oil specks, distinct on lighter surfaces, especially the tangential; not sharply defined from the fairly thick, brownish gray sapwood. Luster medium. Odor absent or mildly unpleasant; taste not distinctive. Density variable; sp. gr. (air-dry) 0.60 to 0.75; weight 37 to 47 lbs. per cu. ft.; texture medium; grain straight to very irregular; easy to work, finishing smoothly and taking a good polish, though likely to show oil specks; appears durable. Commercial possibilities doubtful.

COMMON NAMES: Onotillo (Venez.); hoobooballi, hububalli (Br. G.); boesi mahonie, hoeboeballie, koeipjarie, slangenhout, snekie hoedoe (Sur.); kooel pialli (Fr. G.).

Malosma, with a single species, *M. laurina* Nutt. (= *Rhus laurina* Nutt.), is an innocuous evergreen shrub or small tree limited to southern California and Baja California. The leaves are simple and have long petioles; the flowers are borne in large terminal panicles; the fruits are small whitish drupes; the seeds contain a pungent oil.

The plant is said to have the odor of bitter almonds. It is known in California as Laurel Sumach.

Heartwood not seen; sapwood brownish with a tinge of pink. Luster medium. Odorless and tasteless. Moderately hard and heavy, suggesting Maple (*Acer*); texture fine and uniform, grain straight; easy to work, finishing very smoothly. Of no commercial possibilities.

Mauria, with about ten species of small to medium-sized trees, occurs from the central highlands of Honduras to Venezuela and Peru. The leaves are large and odd-pinnate or in part or wholly simple; the small whitish flowers are borne in terminal and axillary panicles; the fruit is a small flattened drupe.

The Chachique of Venezuela is *Mauria puberula* Tul., a rather small Andean tree which produces timber of good quality; the heartwood is pinkish brown or greenish, sometimes beautifully figured with irregular markings of black and highly lustrous, suitable for cabinet work, brush backs, and fancy articles, but apparently rare. *M. birringo* Tul. occurs in humid upland forests from the central region of Costa Rica through Panama to Ecuador and northern Peru. It was collected by A. Rimbach at elevations of 4000 to 4500 feet in the Cordilleras of Ecuador. According to his notes it is a small to fairly large tree, with a light brown warty bark containing a watery juice sometimes used as a hair tonic. The leaves and fruit have a peppery scent. The wood is plain but of good working qualities and is employed locally in construction and joinery. The tree is poisonous to some persons, producing dermatitis and fever. The specimens collected by L. Williams in northeastern Peru at elevations of 1500 to 3500 feet have been determined as *M. suaveolens* Poepp. & Endl. He says (*Woods of northeastern Peru,* p. 286) that it occasionally attains a height of 60 feet with a well-formed trunk 15 inches or more in diameter and free of branches for about 18 feet. The lustrous pinkish brown wood is sometimes used for the construction of native huts.

Heartwood brownish, pinkish, or greenish, more or less variegated and sometimes conspicuously streaked with dark brown or black; transition to sapwood usually gradual. Luster silky. Odor and taste not distinctive, though sometimes present. Rather light to moderately heavy, firm, and strong; texture fine and uniform; grain mostly straight; very easy to work, taking a fine natural polish, holds its place well when manufactured; probably fairly durable. Figured material attractive, but probably of no commercial possibilities because of its scarcity or inaccessibility.

COMMON NAMES: Cirrí amarillo, koró (C.R.); chocho (Col.); chachique (Venez.); alovillo (Ec.); ingaina blanca, itil, i. blanco, yurac ingaina (Peru).

Metopium, with two distinct species of shrubs and small to medium-sized, poisonous trees, occurs in the Greater Antilles, southern Florida, northern Guatemala, British Honduras, and from Yucatán to Vera Cruz, Mexico. *M. venosum* (Gris.) Engl. is a poorly known species of restricted range in eastern Cuba, where it is called Guao or Guao de Peladero. It has a slender stem and branches and lanceolate leaflets. There is no wood sample available for this study.

Metopium Brownei (Jacq.) Urban. (= *Rhus Metopium* L. = *M. toxiferum* [L.] Krug & Urb.) varies in size from a shrub to a tree 50 feet high. The odd-pinnate leaves have 3 to 7 large, round or obovate leaflets with long petiolules; the small yellow-green flowers are borne in large, long-stalked, axillary panicles; the fruit is an orange-colored drupe with resinous pulp. The thin, reddish brown bark contains a caustic juice. The wood is not poisonous. In British Honduras, where the timber is utilized locally to a limited extent for making furniture, the tree is often associated with Sapodilla (*Achras Zapota* L.) in swamp and intermediate forests on calcareous soil in the northern part of the Colony.

Heartwood variegated brown and red, with a greenish tinge and golden luster; rather sharply demarcated from yellowish white sapwood. Without distinctive odor or taste. Hard, heavy, and strong; sp. gr. (air-dry) 0.85; weight about 53 lbs. per cu. ft.; texture rather fine and uniform; grain variable; not easy to work, but capable of a high polish; durability high. An attractive wood of limited utility because of the small size of the trees.

Common names: Coral sumach, hog gum, poison wood (Florida); burn-wood, Jamaica sumach (Jam.); guao, g. de costa, g. de peladero (Cuba); cedro prieto, papayo, poison tree (P.R.); cochinillo (Dom. R.); bois mulâtre, mancenillier (Haiti); cabalchechem, chachin, chechem, cochinillo (Mex.); black poison wood, chechen (Br. H.).

Mosquitoxylon *jamaicense* Krug & Urb., the only species of the genus, is a tree sometimes 80 feet high with a straight smooth trunk occasionally over 24 inches in diameter, but usually much smaller. The generic name, meaning Mosquito Wood, is derived from the common designation in Jamaica where the tree was first discovered. The species also occurs sparingly on the mainland from southern Mexico to Panama. The leaves are odd-pinnate with 11 to 17 leathery leaflets; the little white or greenish flowers are borne in large axillary panicles; the fruits are small drupes, bright red when ripe and making the crowns of the trees conspicuous in the forest. The bark contains a copious odorous resin-like latex. The timber is of good quality for general construction, but is little used because of its scarcity.

Heartwood pink, deepening to red-orange, with yellowish streaks; rather sharply defined from the yellowish gray sapwood. Without distinctive taste, but with faint odor. Fairly lustrous. Moderately hard and heavy; texture rather fine, uniform; grain straight to very irregular; not very easy to work, but takes a smooth finish; not highly resistant to decay. Of no commercial possibilities.

Common names: Mosquito wood (Jam.); nictaa (Mex.); bastard mahogany, ridge redwood, wild mahogany (Br. H.); chichimeca (Guat.); ciruelo (Hond.); cirrí, c. blanco, c. colorado (C.R.); carbonero, jobillo (Pan.).

Pachycormus *discolor* (Benth.) Coville, the sole species, is a short-boled contorted tree growing in the arid central region of Baja California, where it is known as Copalquín and Torote Blanco. It is described by J. A. Veatch as follows (see *Contrib. U.S. Nat. Herb.* 16: 14: 345):

"The trunk divides into several ponderous branches that shoot off horizontally and are bent and contracted into grotesque resemblances of the flexed limbs of a corpulent human being. These huge branches often terminate suddenly in a few short twigs covered with a profusion of red flowers, reminding one of the proboscis of an elephant holding a nosegay. The resemblance is heightened by the peculiar brown skinlike epidermis that forms the outer bark, which splits and peels off annually, accommodating the increase of growth. This epidermis, when removed, exposes the smooth greenish-colored surface of the spongy inner bark, which is from 1 to 2 inches in thickness. When this bark is cut through, a milky juice exudes that soon hardens into a compact mass of gum and resin. The quantity furnished from a single cut is considerable.

"The branches of the larger trees often shoot out to a horizontal distance of 20 feet from the trunk, thus covering an area of 40 feet in diameter. Smaller subordinate limbs spring upward from the upper side of the large boughs, and in this way give a neat oval appearance to the outline of the tree. When loaded with its bright red flowers, the effect is strikingly beautiful, particularly where hundreds of the trees stand near each other, intertwining their boughs, and forbidding ingress to the mysterious space they cover and protect. The leaves are minute and fall off before the blossoms are fairly developed. The young tree looks a good deal like a huge radish protruding from the ground. On the mountain sides, from a little above seashore to an elevation of 1500 feet, these trees grow scatteringly, singly, and in small clumps, but in the narrow vales of the ravines they sometimes form groves of several acres in extent, presenting the impenetrable and compact form above described. From June till August seems to be their blooming season."

There is no specimen of the wood available for this study, but according to Standley (*Trees and shrubs of Mexico*, p. 672), it is soft and porous and soon decays. He adds that considerable quantities of the bark have been exported to Europe for use in tanning.

Pistacia. There are several species in the Old World and one in America. *P. vera* L., of the Mediterranean region and western Asia, is the source of Pistachio nuts of commerce and is extensively cultivated. Mastic is the resinous exudate from the branches of *P. lentiscus* L., another Mediterranean tree; it is used medicinally and for varnish. The Cyprus Turpentine tree is *P. Terebinthus* L. The American species is *P. mexicana* H.B.K., a small evergreen tree, occasionally 30 feet high, with a short trunk 15 to 18 inches in diameter, or more often a large shrub, growing from western Texas through Mexico to Guatemala. The leaves are odd-pinnate, with 9 to 29 small, nearly sessile leaflets; the little flowers are borne in axillary panicles; the fruits are small, nearly dry, purplish drupes with edible seeds. A resin exudes from the branches. There are no specimens of the wood in the Yale collections.

COMMON NAMES: Pistache (Texas); almacigo, lantrisco, lentisco, ramón, yagagueiguei (Mex.).

Poupartia, a genus allied to *Spondias*, consists of eight species, two of which occur in southeastern Asia, five in Madagascar and adjacent islands, and one in South America. The last, *P. amazonica* Ducke, is a large tree of the central and lower Amazon region of Brazil. According to Ducke (*Archiv. Jard. Bot. Rio de Janeiro* 3: 204), its fruit resembles that of *Spondias* and the wood is white and soft. The shape of the trunk and appearance of the long-fissured bark suggest Cedro (*Cedrela*). This genus is not represented in the Yale collections.

COMMON NAMES: Cedro blanco, c. rana, taperebá assú, yacayaca (Braz.).

Pseudosmodingium is a Mexican genus of shrubs and little trees allied to *Rhus*. Best known of the four species is *P. perniciosum* (H.B.K.) Engl. (= *Rhus perniciosa* H.B.K.) of southwest Mexico. Its resin is said to have a carrion-like odor and to be highly poisonous. The wood has not been studied.

COMMON NAMES: Copal jiote, cuajiote, c. blanco, quauxiotl, xiote, yaga-lache (Mex.).

Rhus. The *Rhus* complex, with many species of small trees, shrubs, and vines, differs from most of the Anacardiaceae in being extra-tropical, with centers of distribution in both the northern and the southern hemispheres. The classification used here follows that proposed by Fred A. Barkley (*Annals of the Missouri Botanical Garden* 24: 3. 1937), which segregates the American species into six genera, namely, *Actinocheita, Cotinus, Malosma, Metopium, Rhus,* and *Toxicodendron.*

Rhus, as here delimited, is divisible into two subgenera, *Sumac* and *Schmaltzia.* The first consists of about a dozen species and several varieties of erect shrubs and small trees widely distributed over much of North America, Asia, and the Mediterranean region. The plants are erect shrubs or small trees with comparatively few, coarse, staghorn-like branches, odd-pinnate (sometimes bipinnate) deciduous leaves with thin, sessile, commonly serrate-margined leaflets. Of the seven American species, the most widely distributed are *R. Copallina* L., *R. glabra* L., and *R. typhina* Torner. The last, known as Staghorn Sumach because the branchlets are covered with a thick brown velvety pubescence, is usually a tall shrub, spreading by underground shoots into thickets, but under favorable circumstances becomes a tree, occasionally 35 to 40 feet high, with a rather crooked trunk as much as 14 inches in diameter. *R. glabra,* typically a shrub, occasionally a tree 20 feet high, is widely dispersed from the mountains of Chihuahua, Mexico, throughout the United States and much of Canada. *R. Copallina,* widespread in eastern North America, is typically a low shrub, rarely a tree 30 feet high, characterized by leaves having a winged rachis; one variety extends into Cuba, and another, often considered a distinct species, *R. lanceolata* Gray, is common in Texas and northeastern Mexico. The woods of the group, so far as studied, are light and soft to moderately so, with a thin white sapwood and a lustrous olive-green heartwood more or less distinctly striped or variegated, becoming russet-brown superficially on exposure.

The subgenus *Schmaltzia,* which Barkley (*Am. Midland Naturalist* 24: 3: 647. 1940) considers worthy of generic rank, comprises about 30 species and many varieties of shrubs (rarely scandent) and small trees, with numerous comparatively slender branches and simple, trifoliate, or odd-pinnate leaves. With two exceptions the species have their center of distribution in Mexico, with extensions across the border into the United States and southward into Costa Rica. The two northern species are *R. aromatica* Ait. of eastern, and *R. trilobata* Nutt. of western, North America, with overlapping ranges in the Mississippi valley; they belong to the section *Lobadium* and their woods have the greenish color of the *Sumac* group, but are somewhat harder and heavier. Representatives of the southern species, *e.g., R. integrifolia* (Nutt.) B. & H. f. and *R. ovata* Wats., section *Styphonia,* and *R. virens* Lindh., section *Pseudoschmaltzia,* have heartwood of a uniform yellowish red or salmon color and for this reason are often called Mahogany Sumac; the sapwood may have a greenish tinge.

Various species of *Rhus* supply useful products. The acrid berries make a refreshing drink similar to lemonade. The leaves are rich in tannin and are of some commercial importance for the making of special types of leather. The bark and roots also yield a dye and have a part in native medicine. The pliable young stems of *R. trilobata* have long been used by the Indians for making baskets and are considered better than Willow (*Salix*) for that purpose. The wood, though attractive, very easy to work, and fairly durable, is little used because of the small sizes obtainable.

Common names: Lemonade berry, skunk bush, sumac, sumach (black, dwarf, laurel, mahogany, mountain, shining, smooth, staghorn, white), vinegar tree (U.S.A.); añil del Pinar, sumaque (Cuba); agrillo, agritos, capulín, correosa, hierba de temazcal, lambrisco, lantrisco, lemita, sumaco cimarrón, temezcal, tnu-ndé, yoga-biche, yucu-caya, zumaque (Mex.).

Schinopsis is the source of the valuable South American tanwood known as Quebracho (ax-breaker) in reference to the flinty hardness of the wood. There are two kinds of Quebracho—the red or Quebracho Colorado, the product of this genus, and the white or Quebracho Blanco supplied by *Aspidosperma quebracho-blanco* Schl. (Apocynaceae).

Although at least eight different species of *Schinopsis* have been described by botanists there are only two that are generally recognized as of commercial importance, namely, *S. Lorentzii* Engl., a deciduous tree with odd-pinnate leaves having numerous narrow leaflets, and *S. Balansae* Engl., a tree never entirely free of leaves, which in this case are simple and suggest those of Willow Oak (*Quercus phellos* L.), though showing transitions to pinnate with a few coarse leaflets. The former, which is more abundant in the drier western plains, is sometimes referred to as the Santiago type (Quebracho Colorado Santiagueño), while the latter, which extends into the swampy lands fringing the Paraná and Paraguay rivers, is known as the Santa Fé or Chaco type (Quebracho Colorado Chaqueño). The two kinds are much alike as to flowers, which are small and inconspicuous; fruits, which are samaras; and woods, which are of a light cherry red color at first, but deepening upon exposure to a rich dark red, often with blackish streaks, and are a third heavier than water. A form of dermatitis is said to be induced in sensitive persons by contact with the branches, leaves, or sawdust of the tree or even by close proximity to the foliage (see *Tropical Woods* 17: 7).

The botanical range of *Schinopsis* extends over northern Argentina, western Paraguay, a small portion of Bolivia, and an undetermined area in Brazil as far north as the interior of the state of Bahia (Map 2). The commercial range includes the northern half of the province of Santa Fé, the eastern part of the Province of Santiago del Estero, all of the Territories of Chaco and Formosa, and a narrow fringe east of the Río Paraná in the Province of Corrientes in Argentina, and throughout the explored country west of the Río Paraguay in Paraguay. The extent of this commercial range is approximately 200,000 square miles and embraces practically all of the country known as El Gran Chaco, a vast flat, poorly explored region of mingled jungles and open pampas, inhabited only by nomadic Indian tribes. Extensive areas have been destroyed by fire and wasteful logging. The trees grow singly or in small groups in mixture with other hardwoods in open stands or islands (Plate XII) and the space between the islands is usually covered with an almost impenetrable tangle of vines, thorn bushes, and cactus. They are from 30 to 50 feet high, one to three feet in diameter, and clear of branches for 20 to 30 feet. They are of scrubby growth and their trunks are often bent and twisted, and swollen at the base. The number of trees per acre rarely averages more than five.

The heartwood of Quebracho contains from 20 to 30 per cent of tannin and 3 or 4 per cent of soluble non-tannin. The sapwood, which is from one to three inches thick, contains only about 3 per cent of tannin and 8 per cent soluble non-tannin. The tannin and soluble non-tannin content of the bark are each about 10 per cent. The usefulness of Quebracho is almost entirely dependent on the tannin content of the heartwood, deriving therefrom its great durability as well as its high economic importance as a source of extract, *S. Balansae* being considered the most valuable for this purpose. The extract industry which started in a small way about 1890 is now one of the most important industries of Argentina and Paraguay.

The trees are felled by means of heavy axes, and the bark and white sapwood are hewn off to reduce the weight of the logs and also to prevent the attack of beetles which deposit their eggs in the bark of down timber and fire-damaged trees. The resulting larvae will within six months completely riddle the wood with their galleries. The larger limbs are sometimes used for tanwood and the smaller ones are cut into firewood if not too far from market. The logs are dragged by oxen through narrow paths to wider lanes, called "picadas," which had previously been hewn out from the railroad. They are then loaded on heavy carts and hauled out, two or three tons at a time, to the narrow-gauge logging road where they are swung onto cars and drawn out to the extract factory by means of small wood-burning locomotives. At the extract plant the logs are reduced to chips by mechanical means. These chips are run into copper vats, leached with hot water, and the resulting liquor evaporated in vacuum pans to the consistency of syrup. This is further concentrated by heating and stirring in another type of evaporator and is then run into bags and allowed to cool and harden. It is exported in this form and contains about 65 per cent tannin, 10 per cent soluble non-tannin (mostly coloring matter), and 25 per cent moisture. The spent chips are allowed to dry and are then used for fuel. Logs may be stored for long periods in the open with only superficial loss of tannin, but users prefer fresh wood because it is easier to chip.

Formerly the logs were exported in great quantities to the United States and European countries where the extract was prepared, but the local extracting industry has gradually developed until there are now about 25 factories in operation in northern Argentina and western Paraguay. The monthly capacities of these plants range from 250 to 2500 tons of solid extract, the total annual output being normally between 150,000 and 200,000 tons.

Quebracho timber has many other uses. It is seldom cut into boards because of the difficulty of sawing and also because thin lumber has a great tendency to check and warp. Practically all of the fence posts, telegraph poles, bridge timbers, and railway ties in the region are of Quebracho, which is one of the most durable woods known. It is extensively used for fuel not only for household purposes but also by factories, power plants, locomotives, and the smaller river steamers. It is also used for wood-block paving, heavy construction, and cart axles.

The Quebracho of Bahia, Brazil, commonly called Baraúna, although this name seems more properly applicable to *Melanoxylon* (Leguminosae), is *Schinopsis brasiliensis* Engl. According to data and specimens supplied by H. M. Curran, this is a common tree of the dry forest, and while

Photo by Max Rothkugel.

PLATE XII. Stand of young Quebracho trees (*Schinopsis*) in the Argentine Chaco.

PLATE XIII. Palo de Vaca or Cow tree (*Couma guatemalensis*) in eastern Guatemala.

its range has not been determined it probably extends southward until it meets or merges into other species. It occurs in nearly pure stands or in association with Aroeira (*Astronium* or *Schinus*) and a number of other dry-forest trees. The forest is thin and open, with a dense undergrowth of thorny shrubs and cactus. The mature tree is 40 to 50 feet high with a much-branched bushy crown suggesting an Oak tree grown in the open. The trunks are short, rarely over 20 feet, with diameters up to 24 inches. The bark is thick, rough, and black, suggesting certain Oaks. The leaves, which are 1 to 5 inches long, are odd-pinnate, with 2 to 8 pairs of leathery leaflets, the largest about an inch long and half as wide. The winged fruits suggest those of Maple (*Acer*), except they are not in pairs. The wood is practically identical in appearance and structure with the Quebracho of Argentina. It is confused locally with Aroeira and is little used.

Heartwood of all species light red, deepening to brick-red; uniform or with black streaks; distinct but not sharply demarcated from the yellowish sapwood. Luster low to medium. Odor not distinctive; taste astringent. Extremely hard, heavy, and strong, but brittle; sp. gr. (air-dry) 1.15 to 1.30; weight 70 to 80 lbs. per cu. ft.; texture fine and uniform; grain irregular, often roey; difficult to cut, becoming flinty when dry, though splitting readily; takes a high natural polish; exceptionally durable, though standing trees are often defective as a result of heart rot.

COMMON NAMES: Quebracho (Trade); baraúva, braúna, quebracho hembra (Braz.); quebracho, q. colorado, q. c. chaqueño, q. c. santiagueño, q. macho, q. moro, q. negro (Arg.); soto negro (Boliv.); quebracho, q. rubio (Par.).

Schinus, with 15 or more species of shrubs and small to medium-sized trees, has its center of distribution in the south-central part of South America. In the subgenus *Duvaua* (sometimes considered as a distinct genus) the leaves are simple and the flowers and small drupes are borne in little axillary clusters. The best known and most widely distributed member of this group is *S. dependens* Ortega, a tree, rarely over 20 feet high, with spinose branches, occurring in southern Brazil, Uruguay, Argentina, Chile, and Peru. Its resin is used medicinally.

The species of *Schinus* proper (subgenus *Euschinus*) have odd-pinnate leaves with few to many leaflets, and the flowers and drupaceous fruits are borne in panicles. The largest tree is *S. molle* L., a native of Peru, but widely planted for shade and ornamental purposes in regions with mild climates. It is grown abundantly in southern California and is known as Pepper-tree. Standley says (*Trees and shrubs of Mexico*, p. 662): "The tree ascends in the Andes to an altitude of 3600 to 3900 meters, but often occurs at much lower altitudes. It thrives in dry, sandy soil and can endure extended drought. The Pepper-tree is said to have been introduced into Mexico by Don Antonio de Mendoza, the first viceroy, who sent the seeds from Peru. The specific name, *molle*, is the name by which the tree is known in western South America, and is derived from Mulli, the old Peruvian name. *Schinus molle* is an excellent shade tree and a handsome one, remaining green throughout the year. The only objection to it is the fact that it harbors the black scale, which is a serious pest of Citrus fruits. The wood is useful for various purposes and the bark for tanning skins. When fragments of the leaves are placed in water they execute quick jerking movements, due to the sudden discharge of the oil which they contain. The fruit contains a volatile oil and has a flavor resembling that of a mixture of fennel and pepper. The seeds are sometimes used to adulterate pepper. In Mexico the fruit is ground and mixed with *atole* or other substances to form beverages. An intoxicating liquor, known as 'copalocle' or 'copalote,' is obtained by fermenting the fruit with pulque for one or two days. The Pepper-tree is much used in local medicine."

Heartwood dull light red, deepening upon exposure and becoming more or less purplish and rather oily looking; distinct but not sharply demarcated from the brownish gray sapwood which suggests Elm (*Ulmus*). Odor and taste not distinctive. Moderately hard and heavy; sp. gr. (air-dry) 0.54 to 0.68; weight per cu. ft. 34 to 43 lbs.; texture medium to fine, uniform; grain variable, often irregular; very easy to work, especially the heartwood of *Schinus weinmannifolius* Engl. which cuts like Red Cedar (*Juniperus*); durability high. Good timber, but of no commercial importance because of its small size or its scarcity.

Common names: *Schinus molle:* Pepper tree (U.S.A.); árbol del Perú, copalquahuitl, molle, Perú, pimiento de América, pirul, ttzacthumi, ttzacthunni, xaza (Mex.); árbol del Perú, pimiento, pimientillo (Salv.); pimiento de California (C.R.); muelle, pimiento (Col.); árbol de pimienta (Venez.); molle, mulli (Peru); aroeira molle (Braz.); aguaribay, bálsamo, curanguay, gualeguay, molle, pimiento, p. del diablo, terebinto (Arg.); aruera, arueriña, aguará-bay-mí, curanguay, molle, paráparay, pimientero, palo del diablo, pimiento del diablo (Par.); pímentero (Chile). *S. dependens:* Molho (Braz.); incienso, molle, m. blanco, m. colorado, m. de curtir, m. de incienso, m. de monte, m. falso, m. rastrero (Arg.); molle, m. de monte (Urug.); huingán (Chile). *Other species:* China-mulli (Peru); aroeira do campo, a. mansa, a. vermelha, aroeirinha, cabuy, cambuy, coração de negro (Braz.); aguará-ibá, aroeira colorado, chichitá mononi, corazón de bugre, laurel muchi, litre, litrecillo, molle negro, trementino (Arg.); arueriña del campo, molle ceniciento (Urug.); aguará-ibá-guazú (Par.).

Spondias, with five or six species of small to large deciduous trees, is widely dispersed throughout the tropics both naturally and through cultivation for the edible fruit. The leaves are large and odd-pinnate; the flowers are small and paniculate; the fruit is a plum-like drupe. The two species best known in tropical America are *S. purpurea* L., with red or purplish flowers and fruits borne in small panicles lateral on the old wood, and *S. mombin* L. (= *S. lutea* L.), a larger often prickly tree, with greenish flowers and yellow fruits in large terminal panicles. Sections of young stems and limbs of old trees readily take root and grow when planted in moist soil and are commonly used for living fence posts. The timber is so susceptible to decay and insect attack that it is considered nearly worthless for lumber, but if the logs could be sawed without delay and the boards kiln-dried the material would be suitable for making boxes and crates, as it is tough and strong for its weight and holds nails firmly. In some localities it is used for making match sticks.

Color nearly white throughout when fresh, but subject to blue stain. Luster medium. Odorless and tasteless. Rather light in weight, but firm; sp. gr. (air-dry) 0.50 to 0.60; weight 31 to 37 lbs. per cu. ft.; texture medium to coarse; grain fairly straight; not difficult to work, finishing smoothly. Not likely to be of any importance for export.

Common names: *Spondias mombin:* Hog plum (Eng.); jobo (Span.); jobito (Cuba); jobobán (Dom. R.); mombin, m. franc (Haiti); abal, chupandilla, ciruelo obo, cozticxocotl, coztilxocotl, hobo, jobo espino, j. roñoso, jovo, kanabal, kinin-jobo, mompin, obo de zopilote, pompoaque, pompoqua, xkinin-jobo, zuliabal (Mex.); pok (Guat.); jocote montero (Nic.); balá, bará, bra, braá, brakra, frap, mube, páalan, paran, xlúra (C.R.); jobito (Pan.); jobo blanco, j. de Castilla, j. macho, Pedro Hernández (Col.); caimito, jobo, marapa (Venez.); hooboo, hubu (Br. G.); hobbo, hoeboe, monbé, moppé (Sur.); mombin, monbin, prunier mombin (Fr. G.); acaíba, acajá, acajáiba, acayá-mirim, cajá, c. mirim, c. pequeno, cajáseiro or cajázeiro, c. miudo, imbuseiro, pau da tapéra, taperibá (Braz.); shungu, ubo, uvo, ushun (Peru). *S. purpurea:* Spanish plum (Eng.); ciruelo (Sp.); ciruelo del país, Jamaica plum, jobillo, jobo francés (P.R.); cirouèlle (Haiti); abal, atoyaxocotl, atoyaxotl, biaxhi, capuatl-cacao, chiabal, cupu, hobo, jocote, macaxocotl, xobo, xocot, xocotl (Mex.); palo de mulato (Guat.); jocote de jobo, pitarrillo (Salv.); jocote (C.R.); jobo colorado (Col.); ciruelo de hueso (Venez.); ajuelo (Peru).

Tapirira, with several closely allied species of small to rather large evergreen trees, occurs sparingly from southern Mexico to Peru and eastern Brazil. The best known species is *T. guianensis* Aubl. with its center of distribution in the Guianas. Its leaves are large and odd-pinnate, with 5 to 9 leaflets; the small yellow flowers are borne in terminal or axillary panicles; the fruit is a rather small green drupe. The trunks are well formed and have low buttresses. The bark is dark gray, shallowly and narrowly fissured; the inner bark is reddish and contains an oily resin. The pinkish or pale brownish timber is easy to work and is used for carpentry and general interior construction, but its appearance is marred by small dark-colored resinous exudations.

Heartwood light pinkish, becoming brownish upon exposure; not sharply differentiated from the sapwood. Fairly lustrous. Odorless and tasteless. Variable in consistency from light and fairly soft to rather hard, the denser kinds suggesting Birch (*Betula*); sp. gr. (air-dry) 0.50 to 0.75; weight 31 to 47 lbs. per cu. ft.; texture medium to fine, uniform; grain generally straight; finishes smoothly, holds nails firmly; poorly resistant to decay. Of no value for export.

COMMON NAMES: Bagamani, vagamani, vanamani (Pan.); gommier viande biche (Trin.); caoba de montaña, cedrillo, cedro nogal, corazón colorado, jobillo, jobo liso, tapaculo (Venez.); dooka, duka, waramia (Br. G.); ana-akara, anoema latti, ata-apiriri, atapiriri, basa mopé, danlieba, djedoe, doka, dokka, duka, kressi pisie, man krappa, massé, matawarie nengé, m. nengi-djedoe, saprieran, tamoene-nooitjano-atapiriri, warimia, w. balli, witte hoedoe (Sur.); bois tapiré, tapiriri (Fr. G.); cedrohy, guarúba, pau pombo, tatapiririca (Braz.); isaparitsi (Peru).

Thyrsodium consists of one West African and several Amazonian species of small to rather large resinous trees, apparently of infrequent occurrence. The leaves are very large, unequally pinnate, with several pairs of leathery leaflets; the flowers are borne in axillary panicled racemes; the fruit is a rather large ovoid, orange-colored drupe; the leaves, inflorescence, and twigs contain a white latex.

Wood grayish brown with a slight pinkish hue throughout specimen. Luster silky. Odorless and tasteless. Moderately hard, heavy, and compact; texture fine and uniform; grain fairly straight; working properties fair; durability doubtful. Presumably of no commercial possibilities.

COMMON NAMES: Ooluballi, uluballi (Br. G.); castanha de porco (Braz.).

Toxicodendron, with several species of poisonous vines, upright shrubs, and small trees, is widely distributed in America and Asia. The leaves are deciduous; the flowers are borne in pendent, axillary panicles; the fruits are small whitish drupes. The genus is often merged with *Rhus,* but Barkley says (*Annals of the Missouri Botanical Garden* 24: 3: 419): "The generic recognition of *Toxicodendron* as distinct from *Rhus* has long been a matter of controversy, and one which in the nature of such things can never be answered in an absolute manner. The non-glandular pubescence when present on the fruit-coat, the ceriferous mesocarp, the consistently poisonous effluvium, and the paniculate inflorescence in *Toxicodendron,* as contrasted with the glandular pubescence always present on the fruit-coat, the non-waxy mesocarp, the constantly innocuous effluvium, and the thyrsoidal inflorescence of *Rhus,* as well as many minor characters consistently different between these elements, are characters that seem to the author sufficiently well marked to separate the two as distinct genera."

The American species are divisible into two sections, namely, *Eutoxicodendron* and *Vernix.* The first comprises woody vines and small shrubs with slender branches and trifoliolate leaves. The best known species is the Poison Ivy or Poison Oak, *Toxicodendron radicans* (L.) Kuntze (= *Rhus radicans* L. = *R. Toxicodendron* L., in part) which grows from southern Canada throughout the United States to southern Mexico and the Bahamas. The section *Vernix* is composed of large shrubs and small trees having comparatively few rather stout branches and odd-pinnate leaves with 5 to 17 leaflets. *T. Vernix* (L.) Kuntze (= *Rhus Vernix* L.) occurs throughout the eastern half of the United States. *T. striata* (R. & P.) Kuntze (= *R. striata* R. & P. = *R. juglandifolia* Willd.), usually a small tree, sometimes up to 35 feet tall, is distributed from southern Mexico to Peru and eastern Brazil. Owing to the small size and the poisonous nature of the trees, the wood of *Toxicodendron* is not utilized.

Heartwood variegated olive and orange; sharply demarcated from the whitish sapwood. Luster low to medium. Without distinctive odor or taste. Of rather light weight, but firm; texture coarse to very coarse; grain mostly straight; easy to work, finishing smoothly; is probably durable. Of no commercial possibilities.

COMMON NAMES: Poison ash, p. dogwood, p. elder, p. ivy, p. oak, p. sumach, p. tree, p. wood, swamp sumach, thunderwood, three-leaved ivy (U.S.A.); bemberecua, chechén, guardalagua, guau, hiedra, h. mala, h. maligna, hincho-huevos, malamujer, mexye, yagalache (Mex.); amché (Guat.); hinchador (C.R.); ajicito, alicito, birringo, caspi, c. carracho, chiraco, fumo, manzanillo, Pedro Hernández (Col.); incati, itil (Peru).

ANONACEAE

THE Custard Apple family, with about 60 genera and many species of shrubs and slender trees, is well represented in the Indo-Malayan region, parts of Africa, and tropical America. The leaves are alternate, entire, and without stipules; the flowers typically have three sepals and six petals in two series, and are solitary or fasciculate in the axils of the leaves or opposite them; the fruit is baccate or compound; the seeds are inclosed in an aril.

Some species, particularly of *Anona*, are widely cultivated in tropical countries for their fruit, and one, *Canangium odoratum* Baill., for its strongly scented flowers, the source of ilang-ilang perfume of the Philippine Islands. The Papaw, *Asimina triloba* Dunal, of eastern United States, is the most northern member of the family, and its wood differs from the others in several ways, most noticeably in being ring-porous. There are more than 20 genera in tropical America, their combined range including the West Indies, southern Mexico, Central America, and most of South America, especially the Amazon basin. The only timber known to the export trade is Lancewood, *Oxandra lanceolata* (Sw.) Baill. Botanists have difficulty in finding reliable characters to separate some of the genera, and the various classifications proposed have resulted in much confusion in the nomenclature of even the commonest trees, Lancewood, for example, having been referred to four different genera.

The bark is smooth, finely laminated, and fibrous and is used to some extent locally for cordage. Slender hard-wooded stems, though not resistant to decay, are well suited for poles, whether for propelling boats or making the framework of dwellings, and for masts of small sailing vessels, handles of implements, brooms, and spears. The sapwood the part of the timber best known, is yellowish, pale brown, or olive, but the older trees develop an irregular mass of dark heartwood, the colors ranging from reddish brown and dark olive to almost black, sometimes attractively variegated, though tending to become dull upon exposure to light and air. The heartwood occasionally has a disagreeable odor when freshly cut. There is a great range in density, from light and spongy to very hard and strong. The outstanding structural feature is the occurrence of evenly spaced lines or narrow bands of parenchyma which, together with the rays, form a spider-web pattern distinct under the lens. The following anatomical description is based on samples of American species of 21 genera, namely, *Anaxagorea, Anona, Asimina, Bocageopsis, Cymbopetalum, Desmopsis, Diclinanona, Duguetia, Fusaea, Guatteria, Heteropetalum, Hornschuchia, Malmea, Onychopetalum, Oxandra, Pseudoxandra, Rollinia, Sapranthus, Stenanona, Unonopsis*, and *Xylopia*.

Growth rings present or absent, sometimes distinct; ring-porous structure in *Asimina*. Pores medium-sized to large in *Guatteria, Rollinia*, and *Unonopsis;* small to very small in *Cymbopetalum, Hornschuchia, Sapranthus*, and *Stenanona;* variable in *Malmea* and *Xylopia;* mostly small or medium-sized or both in the others; generally few to fairly numerous, sometimes very few (*Heteropetalum*) or very numerous (*Asimina*); solitary and in short, occasionally long, multiples and sometimes in small clusters; distribution fairly even. Vessels with exclusively simple perforations; spiral thickenings present in *Asimina;* fine striations observed in *Guatteria Slateri* Standl.; gum deposits common; pitting typically alternate, sometimes opposite (*e.g.*, *Cymbopetalum*); pits variable from minute (*e.g.*, *Desmopsis* and *Duguetia*) to moderately large (*e.g.*, *Guatteria*), but mostly small; apertures often coalescent. Rays heterogeneous (with many of the cells squarish) to nearly homogeneous; not over 5 cells wide in *Diclinanona, Fusaea, Heteropetalum, Hornschuchia, Malmea,* and *Xylopia,* up to 6 to 10 cells wide in many of the others, and up to 13 or 15 cells in *Anaxagorea, Anona, Guatteria,* and *Stenanona;* rays less than 4 cells wide rather few in genera with widest rays; heights variable sometimes only up to 50 or 75, but more often up to 100, or considerably more; scattered crystals occasionally present; large oil cells observed in species of *Duguetia* and near an injury in a variety of *Xylopia mirabilis* R. E. Fries; ray-vessel pitting very fine to rather coarse, occasionally unilaterally compound. Wood parenchyma mostly in uniseriate or biseriate metatracheal bands, 1 to 2 pore widths apart; also sparingly paratracheal in some instances; crystals rare; oil cells observed in *Cymbopetalum* and near an injury in a variety of *Xylopia mirabilis.* Wood fibers non-septate; walls thin to very thick, often varying in same growth rings; walls gelatinous in part in *Duguetia* and *Xylopia;* pits very small, simple or bordered. Ripple marks absent. No gum ducts seen.

Anaxagorea. There are a few species in the Indo-Malayan region and 20 in Central America and in South America to northeastern Peru and Rio de Janeiro, Brazil, though most frequent along the Amazon. The trees grow mostly near watercourses and are ordinarily less than 25 feet high and six inches in diameter. The wood is moderately hard, but tough and elastic; the rays are about as prominent as in Sycamore (*Platanus*).

Common names: Palanco (Guat.); majagua (Venez.); amoura, arara, lancewood, yarri-yarri (Br. G.); invireira (Braz.); espintana (Peru).

Anona (or *Annona*) is sparingly represented in Asia and there are about 10 species in tropical Africa, but it is in tropical America that it is most generously distributed, with about 100 species having a combined range including southern Florida, the West Indies, part of Mexico, Central America, and tropical and subtropical South America. The trees are chiefly valuable for their edible fruits, some of which are large and succulent, known to English-speaking people as custard apple, sugar apple, sweetsop, soursop, and bullock's heart. One of the best known species is *Anona glabra* L., a tree sometimes over 40 feet high, with a short trunk 18 inches in diameter above the swollen or buttressed base. In southern Florida and the West Indies it frequently forms thickets in swampy lands near the coast. The wood, particularly of the roots, is so light and spongy that it is used locally as a substitute for cork in floats for fishing nets, stoppers of bottles, etc. Some species are said to yield a good quality of timber of the Lancewood class, but such reports are probably incorrect.

Common names: Alligator apple, bobwood, corkwood, monkey apple, pond apple (Eng.); anón, anona, chirimolia, chirimoya, corcho, corazón, guanábano, palo de corcho, sencuyo, sincuyo, or soncoyo (Span.); attier, hattier, cachiman, cachimentier, corossolier, mammies, pommier canelle (Fr.); kaneelappel, zuurzak (Dutch); bousi soursakki (Neg. Eng.); ahate, cabeza de ilama, c. de negra, chacoop, chincua ilama, kanoop, mac, mak, oop, op, polbox, poox, pox, saramulla, takob, tsalmuy, tsulipox, tukib, xmaak, xmacoop, xmak, yaxoop, zacoop, zulipox (Mex.); cahuey, matacuy, pac, tsumuy (Guat.); camarón, negrito, toreta (Pan.); gallina gorda, guanábano pun, matimba (Col.); anón liso, anoncillo, cabeza de negro, catigüire, catoche, catuche, cayube, chilimoya, chirimoriñón, chirimoya cimarrona, guanábana cimarrona, manirite, manirote, riñón, turagua (Venez.); anonilla, huanábana, masamba (Peru); agoutytreva, araticú, araticum, ata, corticeira, envireira, graviola, imbira, jaca, jacáma, maca do cobra, pindaúba, restinga (Braz.); araticú (Arg.).

Asimina. There are six species, five of them low shrubs in the south Atlantic and Gulf regions, the other, the Papaw or Pawpaw, *A. triloba* (L.) Dunal, a shrub or small tree rarely 30 feet high and a foot in diameter, distributed over the entire eastern half of the United States with the exception of New England and eastern New York. It is common in the undergrowth of hardwood forests growing on rich soil, in the Mississippi valley, often forming thickets. The bark is smooth and emits a disagreeable odor when bruised; the smooth glossy leaves are 6 to 12 inches long; the maroon drooping flowers, which are nearly two inches across when fully expanded, are borne in the axils of the leaves of the previous year; the fruit is 2 to 6 inches long, with a smooth skin, custard-like, rich-flavored flesh, and large brown seeds. This fruit is of some commercial importance in some localities, but should not be confused with the tropical Papaw or Papaya (*Carica*). The wood is greenish yellow, with thin white sapwood; light, soft, brittle, and coarse-textured; it has no special uses.

COMMON NAMES: Banana (false, wild), custard apple, jasmine, jasminier, papaw, pawpaw (U.S.A.).

Bocageopsis, with a few species of shrubs and small trees, occurs in the Amazon basin. The only specimen available (Yale 16190; Gleason 880) is of *B. multiflora* (Mart.) R. E. Fries (= *Bocagea multiflora* Mart.) which was collected by H. A. Gleason on Mount Duida. The wood is greenish, moderately hard and heavy.

Cymbopetalum, with nine species of shrubs and small trees, is distributed from southern Mexico to Peru and Rio de Janeiro, Brazil. The yellowish or greenish, moderately dense, fine-textured wood has no special utility.

COMMON NAMES: Eirojuelo, orejuelo, oroguelo (Mex.); guanabanillo (Venez.); espintana (Peru).

Desmopsis includes several species of small trees, mostly Central American. The only wood sample available (Yale 14889; N. S. Stevenson 105) is of *D. stenopetala* (Donn. Sm.) R. E. Fries from British Honduras. The wood (sapwood) is yellowish with a greenish tinge, moderately dense, medium-textured, straight-grained, and showing the rays prominently on radial surface.

Diclinanona, with two species of small trees, 20 to 30 feet high, is limited to the upper Amazon. The hard, heavy, brownish wood is said to have a honey-like scent when fresh, but is odorless when dry. The timber is too scarce and small to be important.

Duguetia, with about 65 species, has its center of distribution in the Amazon-Guiana region, extending northward to Panama and the Island of St. Vincent, southward to Paraguay and São Paulo, Brazil. Many of the plants are only shrubs and small trees, but a few are reported to reach a height of 60 to 75 feet. The woods are pale olive, more or less streaked with brown; quarter-sawed material is attractively figured. Their density varies from rather light and soft (*D. Spixiana* Mart.) to decidedly hard and heavy (*D. vallicola* Macbride). The timber of the latter species is used in Colombia for handles of axes and other tools.

COMMON NAMES: Yaya, y. pino (Col.); yeshidan (Br. G.); anona, espintana, e. blanca, tortuga-caspi (Peru); ameijú, biribá-rana, caniço branco, c. preto, embiratai, envira condurú, enviratai, envireira forte, jurueira sangue, meijú, pinhão, taiassuba, tiassuba (Braz.).

Fusaea, with three species of shrubs and little trees, is limited to the Amazon basin. One species, *F. longifolia* (Aubl.) Saff. (= *Anona longifolia* Aubl.), grows from eastern Peru to British Guiana. The wood (Yale 35700; A. C. Smith 2779) is brown (with black specks due to injury), hard, heavy, and rather fine-textured.

Guatteria is an American genus of nearly 200 species of medium-sized to small trees, shrubs, and rarely lianas, widely distributed from southern Mexico to Brazil and Peru. The fruit is a cluster of numerous small berries on long slender stalks. The olive-colored wood exhibits about the same range in density and texture as that of *Duguetia* and has the same general uses as the other members of the family.

COMMON NAMES: Bois noir (Haiti); eklemuy, elemuy, e. box (Mex.); flor de Guineo (Salv.); anonillo (C.R.); malagueto, m. prieto (Pan.); solera (Col.); yalla, yaya (Venez.); arara, black kuyama, kadaburichi, karemero (Br. G.); ouregou (Fr. G.); ag-guio, anonilla, carahuasca, charahuaca, espintana negra, espintanal, envira, icoja, icoje, janahuasco, tortuga-caspi, yana-huasca, yuno (Peru); cipó-ira, cipó-uira, embuyu branco, envira, e. amarella, e. amargosa, e. preta, e. preta do igapo, enviratai, envireira, e. fraca, juruá cacáuo, laranjinha, pindaiba (Braz.).

Heteropetalum, with two species of small trees, is limited to the Amazon basin. The only wood sample available (Yale 33830; Ducke 303) is of *H. brasiliensis* Benth., collected by Adolpho Ducke in deeply inundated forest along the Rio Negro in Brazil. The wood is exceptionally light, soft, and spongy.

Hornschuchia, with five species of shrubs and little trees, occurs in Brazil and the Guianas. The only specimen available (Yale 35467; A. C. Smith 2130) is from the type tree of *H. caudata* R. E. Fries, collected by A. C. Smith in British Guiana. The greenish yellow wood is of medium density and rather fine-textured; the rays are narrow, not very conspicuous on radial surface.

Malmea, with 11 species of shrubs and trees, some of which are said to be large, is distributed from southern Mexico to the Amazon basin. The brownish, rather dense wood is inclined to be fibrous; apparently not utilized.

Common names: Carahuasca, espintana, sabina (Peru); envira (Braz.).

Onychopetalum. The only specimen of this small South American genus available (Yale 36971; Krukoff 6909) is from the type of *O. lanceolatum* R. E. Fries, a tree 100 feet tall in the upland forest of the State of Amazonas. The wood is fairly lustrous, pale olive-brown, with prominent ray markings on the radial surface. It is of medium weight, hardness, and texture, easy to work, and finishes smoothly. The timber apparently is not utilized.

Oxandra is a tropical American genus with 20 species in South America, mostly in the Amazon basin, and two others, *O. lanceolata* (Sw.) Baill. and *O. laurifolia* (Sw.) A. Rich., both slender forest trees common in the West Indies. The woods of these two are considerably alike, but that of the first species, which is the true Lancewood of commerce, is denser and finer-textured.

Lancewood enters the market in the form of poles with the bark on, known as Lancewood spars. They are straight, with little taper, mostly about 13 feet long and rarely over five inches in diameter at the small end. The trees grow large and develop a dark heartwood, but only the pale yellow sapwood is in demand and consequently the young trees are preferred for that reason, as well as because they are easier for the natives to transport from the interior to the seacoast. The wood is hard, heavy, compact, fine-textured, straight-grained, tough, strong, elastic, but not resistant to decay; sp. gr. (air-dry) about 1.00; weight about 62 lbs. per cu. ft. Most of the timber comes from Cuba and Jamaica. It is in little demand in the United States, much less so than formerly, and most of the spars entering New York are trans-shipped to Europe. At one time it had an important place in the vehicle industry for spokes, shafts, and side bars, but its principal uses now are for the manufacture of fishing rods, billiard cues, archery bows, and articles of turnery. It is often associated in the trade with Degame spars (*Calycophyllum candidissimum*) from the same region.

Common names: *Oxandra lanceolata:* Lancewood (Eng.); Lanzenholz (Germ.); bois de lance, b. d'arc (Fr.); palo de lanza (Span.); black lancewood, white lancewood (Jam.); yaya, y. comun, y. de monte, y. hembra, y. prieta (Cuba); yaya boba (Dom. R.); haya, yaya (P.R.). *Other species:* Purio (Cuba); yaya (Dom. R.); bufumo, bujumo (Venez.); envira preta (Braz.); espintano (Peru).

Pseudoxandra, with six species of shrubs and trees sometimes 65 feet high, is limited to the Amazon basin. The wood of *P. quianensis* R. E. Fries (Yale 35651; A. C. Smith 2665) is yellowish green, rather light and soft, with conspicuous rays.

Rollinia, with 56 species, has a very extensive range from Dominican Republic and Mexico to Paraguay, northern Argentina, and southern Brazil, or from 20° north to 30° south latitude. The trees resemble *Anona* in fruit and foliage and the woods are of the same general type as the rest of the family.

COMMON NAMES: Anón, candongo (Dom. R.); anona (Mex.); anona de montaña (Guat.); anona, toreto (Pan.); anoncillo, cuchuchi, ilama (Venez.); black mahoe, koyetchi (Br. G.); anona (Pèru); araticú do matto, beribá, biribá, embira, imbira, pindahyba (Braz.); arachichú, araticú, a. guazú (Arg.).

Sapranthus, with a few species of shrubs and small to medium-sized trees, with malodorous flowers, has its center of distribution in southern Mexico and Central America. In the only specimen available, *S. nicaraguensis* (Seem.) Standl. (Yale 10083), the heartwood is dark brown, sharply demarcated from the greenish yellow sapwood. The density is medium. The timber is not utilized.

COMMON NAMES: Chacmax, madre de cacao, murciélago, nitxmaxché, zopilotillo (Mex.); cojón de venado (Guat.); guineo, palanco (C.R.).

Stenanona *panamensis* Standl., the only species, is a slender tree, 20 to 30 feet high, collected by G. Proctor Cooper in the region of Almirante, Panama. The rather dense wood (Yale 12046; Cooper 427) is slightly greenish, with nearly black areas near wounds.

Unonopsis, with 24 species of shrubs and small to medium-sized trees, occurs from the West Indies and Central America to southern Brazil. The specimens available are greenish, light in weight, rather coarse-textured, easy to work, and of attractive appearance on radial surface.

COMMON NAMES: Yaya blanca (Pan.); black yarri-yarri (Br. G.); cara-caspi or cura-caspi, icoje, yana-varas (Peru).

Xylopia, a large genus, is represented in the Indo-Malayan region, New Caledonia, Madagascar, tropical West Africa, and tropical America. Of the 45 American species, one of the best known and most widely distributed is *X. frutescens* Aubl., a slender tree with narrow leaves crowded along the twigs. The fruit is a cluster of red berries which split open when ripe and are sometimes used for spice. According to Englesing (*Tropical Woods* 17: 21), the Mangalargo of Nicaragua, a variety of *X. mirabilis* R. E. Fries, is sometimes 80 feet tall, with a straight bole 10 inches in diameter above the buttressed base. The lowest branches are fastigiate, the upper ones radiating like the spokes of a wheel. The twigs have a fern-like appearance. The heartwood is blackish brown, with a fetid odor when freshly cut.

COMMON NAMES: Malagueta, m. colorado, panetela, pimienta de Guinea, yararey (Cuba); polewood (Br. H.); palanco (Guat.); mangalargo (Nic.); corobá, malagueto hembra, m. macho, rayado (Pan.); achón, burilico, fruta del burro macho, malagueto, zembé (Col.); burriquito, fruta de burro, f. de capuchin, guaruchi, yaya (Venez.); jejerecou, white kuyuma (Br. G.); pegrekoe (Sur.); anona, anonilla, matarro, omechuai-caspi, pichi-varilla, tortuga de tierra (Peru); envira, e. branca, e. preta, envireira condurá, envireirinha, facheiro, imbira, jérerecú, pindahiba, p. de folho pequena, pindahuba, pindá-uba (Braz.).

APOCYNACEAE

THE Apocynaceae comprise about 175 genera and 1500 species of perennial herbs, shrubs, vines, and small to large trees with a very wide distribution, but mostly tropical. The leaves are opposite or whorled, simple, and without stipules; the flowers have a tubular corolla and are usually in axillary or terminal cymes; the fruit is typically of two dry or fleshy carpels, dehiscent or indehiscent. Most of the plants have a milky juice, which in some instances is sweet and palatable, in others bitter and reputedly poisonous; some species are the source of limited quantities of rubber and chicle; many yield products of medicinal use; some are ornamental; very few are valuable for their timber.

Some of the Asiatic trees have a limited amount of colorless, fine-textured, easily worked wood suitable for turnery and carving. The yellow South African timber known to the world trade as Knysna or Kamassi Boxwood is used to some extent for shuttles but is not suited for fine engraving. (For detailed account see *Boxwoods*, Yale School of Forestry Bull. 14, pp. 47–53.)

Of the 30 or more arborescent genera in tropical America, 25 are represented in the Yale wood collections, as follows: *Ambelania, Anacampta, Aspidosperma, Bonafousia, Cameraria, Couma, Geissospermum, Hancornia, Himatanthus, Lacmellea, Macoubea, Malouetia, Microplumeria* (?), *Neocouma, Parahancornia, Peschiera, Plumeria, Plumeriopsis, Rauwolfia, Stemmadenia, Stenosolen, Tabernaemontana, Thevetia, Vallesia,* and *Zschokkea.* With the exception of *Aspidosperma,* a highly important genus, their timber is of minor utility.

Heartwood variable in color—different shades of yellow, brown, and rose, sometimes variegated or striped; frequently not clearly differentiated from the sapwood. Luster low to rather high. Odor not distinctive in dry material; taste often bitter. Density very low (*Ambelania*) to very high (*Cameraria*); texture very fine to rather coarse, mostly medium to fine; working properties poor to excellent, mostly good; durability variable, generally poor.

Growth rings usually present. Pores ranging from minute to moderately large, usually not distinct without lens; numerous or fairly so; in short to long radial multiples or rows in most genera, but nearly all solitary in *Aspidosperma, Geissospermum, Microplumeria,* and *Vallesia.* Vessel lines occasionally distinct, but mostly inconspicuous; tyloses sometimes present; perforations exclusively simple, typically with wide rims, except in *Anacampta* and *Rauwolfia Duckei* Mgf., which have some short scalariform plates; spiral thickenings absent; pits small to very small; vestured. Rays (without latex tubes) frequently all uniseriate or locally biseriate; sometimes up to 4, 5, or 6 cells wide in *Anartia, Aspidosperma* (in part), *Bonafousia, Microplumeria* (?), *Peschiera, Rauwolfia, Stemmadenia, Tabernaemontana,* and *Vallesia;* generally less than 20, but sometimes 50 or more, cells high; decidedly heterogeneous in most genera, but homogeneous or nearly so in *Aspidosperma, Ambelania* sp., *Cameraria, Geissospermum,* and *Hancornia,* and with few distinctly upright cells in *Couma, Parahancornia, Thevetia,* and *Vallesia;* distended cells (suggesting oil cells) seen in *Anartia, Bonafousia, Geissospermum, Microplumeria* (?), *Peschiera, Stemmadenia,* and *Tabernaemontana;* latex tubes in some of the rays of *Ambelania, Couma, Hancornia, Parahancornia, Peschiera,* and *Stemmadenia;* crystals sometimes present; pits to vessels small to minute or elongated (in unilateral compound pitting). Wood parenchyma absent or sparse in *Anartia, Bonafousia, Peschiera, Stemmadenia, Stenosolen,* and *Tabernaemontana;* rather abundant in the other genera, being mostly reticulate, sometimes in concentric lines or narrow bands, occasionally terminal; unilaterally paratracheal in some species of *Aspidosperma;* crystalliferous strands common. Wood fibers septate in *Anartia, Bonafousia, Peschiera, Stemmadenia, Stenosolen,* and *Tabernaemontana;* pits in other genera often distinctly bordered. Ripple marks absent. Large radial channels, suggesting disintegrated leaf traces, observed in *Ambelania, Aspidosperma subincanum* Mart., *Bonafousia, Couma, Himatanthus, Lacmellia, Macoubea, Malouetia, Microplumeria* (?), *Necrocouma, Parahancornia, Peschiera, Plumeria, Rauwolfia, Stemmadenia, Stenosolen, Tabernaemontana, Thevetia,* and *Zschokkea.*

Ambelania, with 10 species of small trees and shrubs, is limited to northern South America and the Amazon region of Brazil. The plants have no commercial importance. Some of the fruits are comestible or medicinal, and the latex is sometimes used as an adulterant of balata.

Ambelania laxa (Benth.) Muell. Arg., a small tree with a swollen trunk, inhabits the understory of forests subject to deep inundation in the Rio Negro basin. The wood is remarkable for its lightness, and its local use as a substitute for cork was noted by explorers a hundred years ago. In this prop-
another apocynaceous species, *Alstonia*
erty it suggests the aerial root wood of
spathulata Blume of Malaysia, which however weighs only from 3 to 4½ lbs. per cu. ft. (see *Tropical Woods* 32: 2).

Wood creamy white throughout. Luster rather low. Odorless and tasteless. Light, soft, and spongy; sp. gr. (air-dry) 0.15; weight 9 lbs. per cu. ft.; feels velvety; not easy to work, as very sharp tools are needed to produce a smooth surface; has low resistance to decay. Appears suitable for some of the same purposes of Balsa (*Ochroma*), but lacks the strength and toughness of that timber.

Ambelania acida Aubl. is a medium-sized forest tree growing on fairly well-drained soil in the Guianas. The only sample available (Yale 35613; A. C. Smith 2581) was collected in the interior of British Guiana by A. C. Smith and determined by R. E. Woodson. Heartwood pale grayish brown; sapwood whitish. Luster medium. Odorless and tasteless. Moderately heavy and hard, having about the consistency of Soft Maple (*Acer rubrum* L.); texture medium; grain slightly roey; very easy to work; probably perishable when exposed to decay. Apparently without commercial possibilities.

COMMON NAMES: *Ambelania laxa:* Molongó (Braz.). *Other species:* Makuriru (Br. G.); ambararie, bat batti, bati bati, makoriro, mapa (Sur.); ambelani, paraveris, quienbiendent (Fr. G.); angelica do igapó, molongó, pau de leite, pepino do matto (Braz.).

Anacampta, with 14 species of shrubs and little trees, commonly included in the genus *Tabernaemontana*, is of scattered occurrence throughout the Amazon basin. The only sample available (Yale 18939) is of *A. Muelleriana* (Mart.) Mgf., collected by L. Williams in northeastern Peru. Wood pale brownish throughout specimen. Luster low. Odorless and tasteless. Rather heavy, hard, compact, and strong; texture fine and uniform; poorly resistant to decay. Of no commercial possibilities.

COMMON NAMES: Siuca-sanango (Peru); paqueretė (Braz.).

Anartia, with four species of shrubs or small trees, commonly referred to *Tabernaemontana*, is apparently limited to the Amazon basin. The only specimen available (Yale 36004) is from a tree 20 feet high collected in the interior of British Guiana by A. C. Smith and determined by R. E. Woodson as *Anartia attenuata* (Miers) Mgf. Wood pale brownish yellow throughout sample. Luster medium. Odorless and tasteless. Moderately heavy and hard; texture medium; grain straight; readily worked; probably perishable in contact with the ground. Of no commercial possibilities.

Aspidosperma, with about 70 species of small, medium-sized, or large trees, is widely but unevenly distributed in tropical America, with most of the forms native to the drier regions of South America, particularly Brazil. The timber of some species is highly esteemed locally.

The principal and perhaps the only distinct species on the mainland of North America is *Aspidosperma megalocarpon* Muell. Arg. of southern Mexico and Central America; although not definitely known to occur south of Guatemala, the same type of timber has been found in Panama and Colombia. The only species reported from the West Indies is *A. domingense* Urban, a small, shrubby tree of the island of Haiti. Four species are credited to Colombia, the most important being *A. Curranii* Standl. and *A. Dugandii* Standl., which are the source of the locally valuable timber known as Carreto (see *Tropical Woods* 36: 13–15). Pittier (*Las plantas usuales de Venezuela*) lists only two species from Venezuela, namely, *A. Vargasii* A. DC., with a dense yellowish wood of the Boxwood class, and *A. cuspa* (H.B.K.) Blake (= *A. lucentivenium* Blake), whose leaves and bark are a source of a bitter principle used in native medicine as a febrifuge. Little is known about the Guiana species. According to Markgraf (Pulle's *Flora of Suriname* 4: 18–21 and *Notizbl. Bot. Gart. Berlin-Dahlem* 12: 115: 559), only three species have been found in Surinam, namely, *A. Woodsonianum* Mgf., *M. oblongum* A. DC., and *A. nitidum* Benth. Of the several Amazon species, the only one of any economic importance is the Araracanga, *A. desmanthum* Benth. It is in eastern and southern Brazil that *Aspidosperma* is of chief importance for its timber. Some of these (*e.g.*, *A. eburneum* Fr. Allem. and *A. tomentosum* Mart.) are in the Boxwood class; others of more general utility (*e.g.*, *A. peroba* Fr. Allem.) have a roseate wood known as Peroba, particularly Peroba Rosa. (The Peroba or Ipé Peroba of the coastal forest of Rio de Janeiro and Minas Geraes is *Paratecoma*, family Bignoniaceae.) The principal Argentine species is *A. quebracho-blanco* Schl.

The woods of *Aspidosperma* exhibit considerable range in appearance and properties, but have many anatomical features in common. The predominant colors are yellow and pink. Luster medium to high. Without distinctive odor when dry; taste sometimes bitter. Mostly hard, heavy, tough, and strong; sp. gr. (air-dry) 0.70 to 1.00; weight 44 to 63 lbs. per cu. ft.; texture rather to very fine; grain variable, sometimes very irregular; working properties variable from excellent to rather poor; wood of some species fairly resistant to decay.

The range of variation in different species permits their separation into five groups, though the transition from one to the next is usually gradual. These have been designated as follows: (1) Araracanga (type *Aspidosperma desmanthum*); (2) Quebracho Blanco (type *A. quebracho-blanco*); (3) Carreto (type *A. Curranii*); (4) Peroba Rosa (type *A. peroba*); (5) Boxwood (type *A. Vargasii*).

Araracanga group.—There are four or five species in this group. The Araracanga, *Aspidosperma desmanthum* Muell. Arg., is a large tree of the high forest on moist terra firma in the lower Amazon region, being of common occurrence in the estuary. The timber, which has a rich orange-brown color and a waxy feel, is used locally for all kinds of heavy and durable construction. Another Amazonian species with orange-colored wood is *A. centrale* Mgf.; a specimen (Yale 20688; Ducke 7) is labeled Pequiá Marfim, a name more commonly applied in southeastern Brazil to a timber of the Boxwood class (*A. eburneum*). Other South American species which, from their botanical classification, probably belong here are *A. Spruceanum* Muell. Arg. and *A. Woodsonianum* Mgf. The North American representative is *A. megalocarpon* Muell. Arg., a common tree of British Honduras, where it is called My Lady. Mature timber has a pinkish color and is used locally for railway crossties, house frames, scaffolding, and heavy construction; stems of young trees are frequently employed for rafting poles.

Wood very hard and heavy; sp. gr. (air-dry) about 0.90; weight about 56 lbs. per cu. ft.; texture medium; grain straight to irregular; not very difficult to work, finishing very smoothly and taking a high polish; durability medium to high. Not likely to enter the export trade.

Common names: Volador (Mex.); chichi, malady, my lady (Br. H.); chichica, chicique (Guat.); alcarreto, ocre (Pan.); copachi, costillo (Col.); sibadanni (Br. G.); araracanga, ararauba da terra firme, jacamim, maparaná, piquiá marfim (Braz.); gavetillo (Boliv.).

Quebracho Blanco group.—There are probably at least two species in this group. *Aspidosperma horco-kebracho* Speg. is a small tree of the Chaco region where it is known as Horco-quebracho and Quebrachillo Blanco. *A. quebracho-blanco* Schl. is a medium-sized to large timber tree of Argentina, Paraguay, and southwestern Brazil. It is said to attain a maximum height of 100 feet and a trunk diameter of nearly four feet, but usually it is much smaller. The limbs are stiff; the leaves are rather small, hard, and bristle-tipped; the bark is white and much broken. It grows on dry sandy soil and originally formed the bulk of the stand over hundreds of square miles with 10 to 20 trees of commercial size per acre, but much of the forest has been destroyed by fires set to improve the grazing. The timber is well known in Buenos Aires, but it has never commanded the highest price, partly because of its abundance and also because of the care necessary in seasoning the lumber to prevent warping. It is not considered suitable for use in damp situations, but otherwise serves for heavy construction. Special uses are for hubs and felloes of logging cart wheels and for bent work. It makes excellent firewood and is a source of tannin extract.

Heartwood yellowish or brownish to rose-red, sometimes more or less variegated; fading upon exposure; not sharply differentiated from the sapwood. Very hard and heavy, tough, strong, and resilient; sp. gr. (air-dry) 0.80 to 1.00; weight 50 to 63 lbs. per cu. ft.; texture medium; grain irregular; rather difficult to work, but takes a smooth finish. Of no importance for export.

Common names: Horco quebracho, quebrachillo, quebracho blanco, q. b. llorón (Arg.).

Carreto group.—There are two or three species of *Aspidosperma* known as Carreto in northern Colombia. The type of *A. Curranii* Standl. (Yale 405) was collected by H. N. Whitford and J. Pinzón in the non-flooded valley of the Río Magdalena in 1917, but was not named until 1933 (see *Tropical Woods* 36: 14). The tree has Willow-like leaves and velvety fruit pods. The wood, which in dry condition is nearly colorless, is described by the collectors as being reddish or roseate when freshly cut. H. M. Curran, who obtained wood (Yale 1524) and sterile herbarium material in the Department of Bolívar in 1916, submitted the following memorandum: "Carreto is perhaps the best known and most widely used timber for heavy bridges, railroad ties, and house posts in the Magdalena Valley, at least in the lower portions. The tree is a constant factor in the forests on the lower slopes of the hills on sites intermediate between those occupied by Tolú (*Bombacopsis*) and Cedro (*Cedrela*) and the higher slopes where the Albarco (*Cariniana*) is found. Over certain limited areas it is very abundant, being apparently gregarious in small groups. One may pass along a trail and count 10 to 15 trees per acre and a little further on not encounter any. The tree grows to a height of over 100 feet with a slender cylindrical bole with clear lengths of 50 to 60 feet and diameters of two to three feet. The bark is light-colored, smooth, and Beech-like; the crown is narrow and the branches are rather small. The wood is heavy and hard and highly durable in contact with the soil. Carreto is not so well known or so extensively used in Venezuela, but a certain amount of the timber is obtained from the forests of the Lake Maracaibo region and is used for the same purposes as in Colombia."

Aspidosperma Dugandii Standl., also known as Carreto, was collected by A. Dugand near Barranquillo and by the senior author near Santa Marta (*Tropical Woods* 36: 15). Dugand says: "The trees vary in size from 5 to 12 meters, the trunk generally straight, the bark greenish gray, peeling off in small patches, but sometimes reddish gray, smooth, and somewhat glossy on the branches. There is scarcely any latex, the juice being rather watery and not milky. Trunk diameter 20 to 40 cm., but I understand some trees have trunks averaging over 80 cm., being, of course, taller, 15 to 20 meters. The heartwood is of a handsome pink color, the sapwood being dull grayish white." Another Colombian species, *A. ellipticum* Rusby, commonly called Amargo, Chivato, and Macuiro, is sometimes known as Carreto. Included in this group also are *A. domingensis* Urb., called Madame Jean in Haiti and Guiconejo and Palo de Caja in Dominican Republic, and *A. inundatum* Ducke, known as Jacamim and Maparaná in the Amazon estuary.

In the Guianas and northern Brazil is a small group of species noted for the peculiar form of the trunk, which instead of being cylindrical is divided almost to the middle into irregular plank-like parts often only an inch thick. The Yaruru or Paddle Wood of British Guiana, *Aspidosperma excelsum* Benth., is described by L. S. Hohenkerk (*Journ. Bd. Agr. Brit. Guiana* 12: 3: 185) as deeply fluted, as though consisting of a bundle of boards standing on end, their inner edges forming a common center. The ribs are used locally for making paddles and tool handles, the cores of large trees for mill rollers. *A. aquaticum* Ducke, a small tree in the deep swamps of the lower Amazon region, is called Carapanaúba (mosquito tree) because the humid recesses of the bole harbor mosquitoes. *A. nitidum* Benth. is also placed in this category, but there are no authentic specimens available to determine the matter. *A. Kuhlmannii* Mgf., of Matto Grosso, is said to have a deeply sulcate stem; it also bears the name Carapanaúba.

Heartwood yellowish brown or roseate, sometimes striped with rose or purplish red; not sharply demarcated from the yellowish sapwood. Luster medium to high. Odor not distinctive; taste often bitter. Hard, heavy, tough, and strong; sp. gr. (air-dry) 0.80 to 0.96; weight 50 to 60 lbs. per cu. ft.; texture fine and uniform; grain straight to irregular; not very difficult to work, taking a lustrous polish; fairly resistant to decay.

Common names: Madame Jean (Haiti); guaconejo (Dom. R.); amargo, carretillo, carreto, chivato, macuiro (Col.); paddle wood, yaruru (Br. G.); apoekoetja, apohita, apokoita, hariroro-jaroeroe, jaroeroe, j. hiraro, parelhout, parihoedoe, tamoené-apoekoetja, witie apokita, w. parihoedoe, witte parelhout (Sur.); bois chapelle, b. pagaie (Fr. G.); carapanaúba, jacamim, maparaná, pau de remo, sapupéma (Braz.).

Peroba group.—Peroba is the name for a highly important group of Brazilian timbers comparable in general utility to Oak (*Quercus*) in the United States. It is used for all sorts of general construction, including sills, framing, flooring, interior trim, and sash and doors, as well as for furniture and cabinet work. Because of the roseate or rose-yellow color of the wood it is commonly called Peroba Rosa. While several species are probably concerned, the principal one is *Aspidosperma peroba* Fr. Allem. (= *A. polyneuron* Muell. Arg.), a large tree of southeastern Brazil, attaining a maximum height of 125 feet, with a well-formed trunk four or five feet in diameter. Huascar Pereira says (*Timber trees of the State of São Paulo, Brazil,* pp. 68–69): "The timber of the Perobas represents an enormous wealth to the state of São Paulo. It is used in all kinds of building construction, both civil and naval, and is in great abundance. The slender trees, with their sparse foliage and wrinkled bark, stand out in the landscape clearly marked by their majestic appearance." According to the same author (p. 67) various colors of the woods are indicated by their names, thus: Peroba Preta with conspicuous black veins; Peroba Miuda, red with darker patches; Peroba Poca, a white wood; Peroba Rajada, color light red with large black patches; Peroba Tremida, yellow with lighter patches almost golden; Peroba Reversa, dark-colored with contorted grain. The last, called also Peroba Revessa, owes its color and peculiar grain to an abnormal structure somewhat of the nature of bird's eye in Maple (*Acer*).

Heartwood rose-red to yellowish, often variegated or streaked with purple or brown, the surface becoming brownish yellow to dark brown upon exposure; not sharply demarcated from the yellowish sapwood. Luster low to medium. Odor not distinctive; taste bitter. Hard and heavy or moderately so; sp. gr. (air-dry) 0.70 to 0.85; weight 44 to 53 lbs. per cu. ft.; texture fine and uniform; grain straight to very irregular; working properties good; durability medium.

Common names: Amarello, amargoso, bucheiro, cainga, muirá-jussára, peroba, p. rosa (Braz.); palo rosa (Arg.).

Boxwood group.—These woods are characterized by a clear yellow color and a very fine and uniform texture which gives them value in carving and turnery. *Aspidosperma Vargasii* A. DC., a small tree of the coastal region of Venezuela, where it is commonly known as Amarillo and Yema de Huevo, is believed to be one of the first woods known to the trade as West Indian Boxwood. It is said to have been exported to Europe in considerable quantities during the last decades of the nineteenth century, but was superseded by the Zapatero (*Gossypiospermum*) of the Maracaibo Lake region and is no longer commercially important. (For further information, including detailed description of the wood, see *Boxwoods,* Yale School of Forestry Bulletin 14, pp. 54–57.) According to Sandwith (*Kew Bulletin,* 1939, p. 16) the species has been found along the Demerara River, British Guiana, where it attains a height of 100 feet.

A similar wood in Brazil is supplied by *Aspidosperma eburneum* Fr. Allem.; it is called Pequiá Marfim and Pau Setim, though these names are also applied to species of Rutaceae. The tree grows in São Paulo and Victoria and attains large size, but is of infrequent occurrence. The wood is highly esteemed locally for fancy articles of turnery, cabinet-work, and inlays, and appears suitable for many of the same purposes as Boxwood. The Pequiá of the Bahia region and the Guatambú of São Paulo are of doubtful specific identity, but a herbarium specimen (Yale 3298) collected by H. N. Whitford in Paraná has been determined as *A. tomentosum* Mart. He states that the trees were growing at an altitude of about 2500 feet and were from 65 to 100 feet tall, the largest trunks being about 36 inches in diameter. Muirá-jussára-y, from near Manáos, Brazil, belongs in this group. These woods exhibit considerable range in density, some specimens being only moderately heavy and hard. The color is of various shades of yellow, with lemon predominating, without the roseate hue of the other groups. The timber is fairly plentiful and is used for many of the purposes of Maple (*Acer*). The harder grades make excellent flooring.

Heartwood mostly bright clear yellow, sometimes with a slight greenish tinge; not sharply differentiated from the sapwood. Luster high in dense specimens, medium in others. Odorless and tasteless. Sp. gr. (air-dry) 0.75 to 0.94; weight 47 to 59 lbs. per cu. ft.; texture very fine and uniform; grain mostly straight; working properties excellent; poorly resistant to decay.

Common names: Amarillo, a. yema de huevo, candado, guariche, limoncillo, manzanillo, membrillo, palo amarillo, yema de huevo (Venez.); guatambú, muirá-jussará, pau setim, pequiá, p. amarilla, p. marfim, pereiro do campo, peroba amarella, p. setim (Braz.); pinshi-caspi, quillo-bordon (Peru); guatambú amarillo (Arg.).

Bonafousia, with nine species of shrubs and small trees, usually included in the genus *Tabernaemontana,* is widely distributed throughout northern South America. The five specimens available were collected by L. Williams in northeastern Peru and are of a single species, *B. sananho* (R. & P.) Mgf. Wood cream-colored or brownish. Luster rather low. Odorless and tasteless. Light in weight, but rather hard; texture fine, grain irregular; easy to work; of low resistance to decay. Of no commercial value.

Common names: Roi de mapipire, wild jasmine (Trin.); dog stones, peru-ishi-lukudu, tanta-ran-tang (Br. G.); abuti abud, battie battie, kapoea kraroen, manbati bati, melki tiki, wagoe ston, wako (Sur.); sanangillo, sanango, sananho, uchu-sanango (Peru).

Cameraria, with two species of small trees or shrubs, occurs in open woodlands and savannas in the West Indies and British Honduras. *C. angustifolia* L. has very narrow leaves, whereas they are rounded in the other species, *C. latifolia* L. The latter was not known to occur on the mainland until discovered in 1926 by the senior author in Orange Walk District, British Honduras; it was considered a new species and named *C. belizensis* Standley, but this name has been reduced to synonymy by Woodson (*North American Flora* 29: 2: 121). Both species are reputed to be highly poisonous if the sap or latex comes in contact with the body, producing bad burns and inflammation. The timber is sparingly utilized, principally because of the small size of the trees.

Heartwood very dark brown, or somewhat variegated; sharply demarcated from the yellowish or grayish sapwood. Luster medium. Odor and taste not distinctive. Very to extremely hard, heavy, tough, and strong; sp. gr. (air-dry) 1.05 to 1.33; weight 66 to 83 lbs. per cu. ft.; heartwood considerably heavier than sapwood because of infiltrated material; difficult to work; but finishing smoothly; durability presumably high. Without commercial possibilities.

Common names: Maboa, m. blanca, m. de loma, m. de montaña, m. de sabana (Cuba); laitier (Haiti); chechem de caballo, white poisonwood (Br. H.).

Couma. There are six species of medium-sized to large trees, as follows: *C. utilis* (Mart.) Muell. Arg., *C. amara* Mgf., and *C. macrocarpa* Barb. Rodr. are Amazonian trees; *C. guianensis* Aubl. occurs in the Guianas and near the mouth of the Amazon; *C. sapida* Pittier is limited to Venezuela; *C. guatemalensis* Standl. (Plate XIII), discovered by the senior author near Puerto Barrios, Guatemala, and later found in southern British Honduras. The trees have thick smooth bark with a copious latex which flows freely from incisions and is sweet and palatable (except in *C. amarga*), thus bringing the species into the group of so-called Cow trees. The latex of the Guatemalan species contains about 70 per cent water and about two-thirds of 1 per cent of rubber; the remainder of the organic solid matter is a waxy substance resembling chicle from *Achras Zapota* L. Although the tree grows much more rapidly than the Sapodillo, tapping experiments suggest that it may not be well adapted to plantation culture, as the yield of gum is low and the rate of recovery from incisions is slow (see *American Journal of Botany* 22: 580–593).

The latex from Amazonian species is the source of Sorva gum, a chicle substitute. The timber of *Couma* is suitable for interior work in place of coniferous lumber.

Wood cream-colored or pale brown, often with a pinkish tinge, throughout. Luster rather low. Odorless and tasteless. Light in weight but firm and strong; sp. gr. (air-dry) 0.45; weight 28 lbs. per cu. ft.; texture medium; grain fairly straight; very easy to work, finishing smoothly; is of low resistance to decay.

COMMON NAMES: Cow tree, milk tree (Eng.); barca, palacio (Br. H.); palo de vaca (Guat.); avichurí (Col.); guaimaro macho, vacahosca (Venez.); dukaballi, dukalliballi, dukataballi, karimein (Br. G.); akoema, ama-apa, amaparian, apalan, bauka-mapa, djakali, dokalli, mappa, péra (Sur.); couma, poirier (Fr. G.); leche-caspi, osurba (Peru); cumá assú, c. de catinga, mucujé, sorva, s. de catinga, s. grande, s. pequena, sorveira (Braz.).

Geissospermum, with three or four species of medium-sized to large trees with deeply sulcate trunks, occurs in the Amazon basin from the Guianas to Bolivia. *G. excelsum* Kuhlm, is a tall tree discovered by Adolpho Ducke in high forest in Amazonas, Brazil, where it is called Carapanaúba. *G. sericeum* (Sagot) Miers, with the range of the genus, is characterized by its spirally arranged, silky-haired leaves; the bark contains a bitter principle. The dense and strong timber is little used because of its scarcity and poor form, but is suitable for tool handles and wagon spokes.

Heartwood yellowish brown or orange; not sharply differentiated from the yellowish white sapwood. Luster medium. Odor and taste not distinctive. Very hard, heavy, tough, and strong; sp. gr. (air-dry) about 0.95; weight 59 lbs. per cu. ft.; texture fine; grain straight; takes a lustrous polish; durability rather low. Of no commercial possibilities.

COMMON NAMES: Maniokinaballi, marisoba, uataki (Br. G.); bergi bita, bitterhout (Sur.); naranjo (Ec.); acary-rana, carapanaúba, quina-quina, quina-rana (Braz.).

Hancornia *speciosa* Com., the only species, is a small tree of eastern Brazil, where it is known as Mangabeira. The branches are drooping and pliable; the succulent berries have a sweet acidulous taste and are esteemed for eating and for making wine. The latex is used as a local source of a poor grade of rubber. The timber is used to a minor extent locally for interior construction and boxes.

Wood uniform pale brownish throughout the two specimens available. Luster medium. Odorless and tasteless. Rather light in weight but moderately hard; medium-textured; straight-grained; very easily worked; finishing smoothly; poorly resistant to decay. Has no possibilities for export.

Himatanthus, with seven species of low to tall and slender trees, has its center of distribution in the Amazon basin. It is usually not considered distinct from *Plumeria,* but according to Woodson (*Annals of the Missouri Botanical Garden* 25: 1: 189–224), there are sufficient grounds for making two genera despite the "apparent absence of well-defined taxonomic criteria."

The most widely distributed species is *Himatanthus articulata* (Vahl) Woodson (= *Plumeria articulata* Vahl = *P. drastica* Mart.), a slender tree 15 to 80 feet high ranging from Panama through Venezuela, the Guianas, and the lower Amazon to southern Brazil. The timber is used to some extent in Brazil for interior construction, oars, and cooperage; from the latex is obtained a medicinal oil, called tiborna, and a residue of gutta percha. *H. sucuuba* (Spruce) Woodson occurs throughout the Amazon region of Brazil, Bolivia, and Peru; it is said to attain a maximum height of 100 feet. Some of the species appear to hybridize readily.

Heartwood yellowish brown, variable in shade; not sharply demarcated from the lighter colored sapwood. Luster medium to low. Odorless and tasteless. Density variable but mostly medium; texture fine to medium; easy to work, finishing smoothly; poorly resistant to decay. Of no value for export.

COMMON NAMES: Wild jasmine (Pan.);

balata blanc (Fr. G.); bellaco-caspi (Peru); leche-leche (Boliv.); carabina, cipoal, janahuba, malongó, molongó, sucúuba, s. rana (Braz.).

Macoubea, with four species of medium-sized to rather large trees, has its center of distribution in the Amazon region, with extensions to the Guianas and Bahia, Brazil (see *Notizbl. Bot. Gart. Berlin-Dahlem* 14: 122: 178–179). The species best known and having the greatest range is *M. guianensis* Aubl. (= *Tabernaemontana paucifolia* Spruce), a tall tree growing in Surinam, French Guiana, northern Brazil, and eastern Peru. The many-seeded fruit is as large as an orange. The bark is light gray and contains a copious latex. The timber is used to a limited extent locally for interior construction and carpentry.

Wood yellowish or brownish throughout. Luster medium. Odorless and tasteless when dry, but said to give off an agreeable scent while seasoning. Of rather low density and medium hardness and texture; sp. gr. (air-dry) 0.45; weight about 28 lbs. per cu. ft.; easy to work, but perishable when exposed.

COMMON NAMES: Liekapatoe, mapa, mappa, rokko-rokko, sokko-sokko (Sur.); amapá doce, molongó (Braz.); huapa-caspi, yaco-sanango (Peru).

Malouetia, with about 20 species of shrubs and small, rarely medium-sized, trees, is largely confined to the Amazon basin, but there is one Cuban species, *M. cubana* A. DC., and one in Central America, *A. guatemalensis* (Muell. Arg.) Standl., extending from British Honduras to Panama. Perhaps the most widely distributed of the South American species is *M. tamaquarina* (Aubl.) A. DC., ranging from the Guianas through the Brazilian Amazon region to Peru; according to L. Williams (*Woods of northeastern Peru,* p. 424), it attains a maximum height of 50 feet with a well-formed trunk 20 inches in diameter; the copious latex is fairly sweet. Apparently the trees of this genus are too scarce to be of any commercial value.

Wood creamy white or brownish. Luster medium. Odorless and tasteless. Rather light in weight but hard; texture fine; feel soft; very easily worked, finishing smoothly; poorly resistant to decay. Suitable for pattern-making if available in large enough pieces.

COMMON NAMES: Guachamacá, lirio de monte (Venez.); kirikihi, white haia, jaramiloerang (Br. G.); oonsé-balli (Sur.); cuchara-caspi, raya-caspi (Peru); molongó de colher, tamanqueira de leite (Braz.).

Microplumeria is the name of a genus proposed by H. Baillon in 1889 for a small tree collected by Richard Spruce in the Amazon region. According to Markgraf (*Notizbl. Bot. Gart. Berlin-Dahlem* 13: 119: 458, June 30, 1937), the same specimen (Spruce 1664) had in 1860 been made the type of *Aspidosperma anomalum* Muell. Arg. Ducke, in 1930 (*Archiv. Jard. Bot. Rio de Janeiro* 5: 180), made this species the type of a new genus, *Cylindrosperma;* he further found that the type of *A. sessilis* Huber was merely a poorly developed specimen of *C. anomalum* (Muell. Arg.) Ducke. Apparently the correct name for this plant is *Microplumeria anomala* (Muell. Arg.) Mgf. It is a small tree or large shrub with all parts very bitter, the latex white, the leaves opposite or verticillate, the flowers small for the family. It is common in the periodically inundated forest of western Pará and adjacent areas in Amazonas as far as the Rio Negro, and is known to the inhabitants of Obidos as Cururú. The young shoots, which are long, straight, and elastic, are used for making fish weirs and boys' archery bows.

No wood sample of *Microplumeria anomalum* is available, but there is authentic material (Yale 6900; Pittier 11062) of *Aspidosperma quadriovulatum* Pittier, a tree about 35 feet high in the open woodland near Caracas, which Pittier says (*Bol. Cient. y Tecn. Mus. Com. Venezuela* 1: 66. 1926) is closely related to it, though in absence of fruit it is not now possible to tell if the two are congeneric. The wood differs in important details, particularly of the rays, from all of the other species of *Aspidosperma* examined, and in view of the close botanical relationship is tentatively considered as a species of *Microplumeria.*

Wood clear yellow throughout; injured

places brown. Luster medium. Odor not distinctive; taste mildly bitter. Hard, heavy, and compact; of fine and uniform texture; grain fairly straight; easy to work, finishing smoothly. Apparently not utilized, but suitable for small articles of turnery, though not quite fine enough to substitute for Boxwood.

Neocouma, with two species of small trees, is sparingly distributed in the northern Amazon region of Brazil. *N. ternstroemiacea* (Muell. Arg.) Pierre (= *Tabernaemontana ternstroemiacea* Muell. Arg.), known locally as Sorva da Catinga, had not been seen since the time of Richard Spruce until rediscovered in 1929 by Adolpho Ducke on the upper Rio Negro. Ducke also found a related form with golden yellow flowers and bitter latex which has been described as a new species, *N. Duckei* Mgf. The timber is suitable for the same purposes as soft kinds of Pine but is not utilized.

Wood cream-colored or brownish yellow. Luster rather high. Without distinctive odor or taste. Light in weight but firm and tough; texture rather fine, uniform, grain straight; very easily worked; subject to blue stain and poorly resistant to decay. Of no commercial possibilities.

Parahancornia *amapa* (Huber) Ducke, the only species, is a large tree, sometimes more than 100 feet tall, with a short crown and long clear well-formed trunk sometimes 36 inches in diameter, occurring infrequently throughout most of the Amazon basin. The copious somewhat bitter latex is used medicinally and as a minor source of a kind of gutta percha. The fruit is savory and comestible. The timber is used to a limited extent for carpentry and interior construction.

Wood cream-colored to pinkish brown. Fairly lustrous. Odorless and tasteless. Of medium weight but hard; sp. gr. (air-dry) 0.65; weight about 40 lbs. per cu. ft.; easily worked, finishing smoothly; poorly resistant to decay. Apparently of no commercial possibilities.

COMMON NAMES: Amaapa, dokalli, mampa, mappa (Sur.); amapá (Braz.); naranja podrida (Peru).

Peschiera, with 26 species of small to medium-sized trees, commonly included in the genus *Tabernaemontana,* occurs throughout almost all of tropical America, but most abundantly in Brazil. *P. arborea* (Rose) Mgf. of Panama is a tree 35 to 50 feet high with a straight trunk 10 to 14 inches in diameter. *P. australis* (Muell. Arg.) Miers occurs in southern Brazil, Paraguay, and northern Argentina; the latex is used medicinally and the white wood is of some local utility. Other Brazilian species supply small timber for interior construction and charcoal.

Wood white, yellowish, or brownish. Luster medium. Odorless and tasteless. Rather light and soft to moderately heavy and hard; texture fine and uniform; easy to work, finishing very smoothly; perishable in contact with the ground.

COMMON NAMES: Borache, wild jasmine (Trin.); cojotón (Br. H.); huevo de verraco, cojón del toro, c. de fraile (Col.); berraco, cojón de berraco, c. de verraco, palo verraco, varraco (Venez.); leiteira, pau de colher (Braz.); palo de víbora, p. de vívora, sapiranguí, s. guazú (Arg.).

Plumeria (*Plumiera, Plumieria*) is a tropical American genus of several species and numerous varieties, forms, and hybrids, some of them widely planted because of their conspicuous, highly fragrant, waxy-looking flowers which are especially suited for making garlands and leis. The variability of the plants has led to the naming of many species. Woodson (*North American Flora* 29: 2: 115–119) has reduced the number of North American species from more than 50 to six. Regarding *P. sucuuba* R. Spruce, Markgraf says (Pulle's *Flora of Suriname* 4: 17): "The variability of the leaf form of this species is astonishing. The method of the Dutch Forestry Bureau to collect material from numbered trees in the forest reserves has furnished striking evidence in favor of Schumann's view that out of the species of *Plumeria* with leathery leaves one may have been described repeatedly under different names, for the twigs collected at different times from the same tree bear at first large, obovate, remotely nerved leaves (flowering), later on small, oblong, densely nerved ones (fruiting)."

Wood yellowish brown throughout, sometimes with faint purplish streaks. Fairly lustrous. Odorless and tasteless. Typically hard, heavy, and compact; fine-textured; easy to work, taking a high polish. Frequently used locally for small articles of turnery, but is of no importance for export.

COMMON NAMES: Frangipanni, jessamine, Spanish jasmine (B.W.I.); alelí, a. blanco, lirio, l. amarillo, l. chucho, l. de costa, l. de dulce, l. de playa, l. de sabana, l. del valle, l. tricolor, súcheli (Cuba); alelí, a. cimarrón, frangipani, paucipán, tabaiba, tapaiba (P.R.); alelí (Dom. R.); frangipane, frangipanier, f. epineux, f. marron, f. rose (Haiti); ahaipuih, Alejandría, cacalosúchil, cacaloxochitl, campechana, campotoneta, chacnicté, ensalada, flor del cuervo, f. de mayo, jacalosuchil blanco, nicté, n. chom, quie-chachi, sabanicté, suchlcahue, súchil, tizaxochitl, zacnicte (Mex.); zopilote (Br. H.); flor de la cruz (Guat.); flor de ensarta, f. de la cruz, f. de mayo, f. de señora (Salv.); flor de pan, f. del toro, sucuanjoche (Nic.); cacalojoche, flor blanca, ingerto, juche (C.R.); caracucha, c. blanca, c. colorada (Pan.); amancayo, azuceno, florón, zapotillo (Col.); amapola, a. blanca, atapáimo, tamáiba (Venez.); frangipanier (Fr. G.); jasmin de Cayenna (Braz.).

Plumeriopsis *Ahouai* (L.) Rusby & Woodson, the sole species, is a shrub or little tree growing from southern Mexico to Brazil and Paraguay. Because of its large glossy leaves and bright red and showy fruits, it is often planted for decorative purposes.

Wood brownish gray throughout. Luster medium. Odorless and tasteless. Rather light in weight but firm and tough; texture fine and uniform; saws woolly when fresh, but finishes smoothly; is perishable in contact with the ground. Of no commercial value.

COMMON NAMES: Lecherillo, ojo de venado, venenillo (Mex.); cogotone (Br. H.); cojón de mico, c. de perro, cojotón (Guat.); cojón de gato, huevo de tigre, lava perro (Pan.); tomate cimarrón, tomatillo (Col.); ahohai (Braz.).

Rauwolfia is a pantropical genus of about 50 species of shrubs and small, rarely medium-sized, trees, well represented throughout tropical America. The plants have few uses and are of no economic importance.

Wood cream-colored or brownish. Fairly lustrous. Odorless and tasteless. Moderately light and soft to rather hard and heavy; texture fine; feel somewhat velvety; readily worked, finishing smoothly; perishable in contact with the ground.

COMMON NAMES: Cagada de aura, canina de perro, corazón de paloma, fruta de aura, frutilla, huevo de gallo, h. del toro, lechuga, lirio, l. de costa, l. de paredón, l. de monte, malambo, palo boniato, p. de leche, munique, sarna dé perro, súcheli blanco, víbona, vívara amarilla (Cuba); bitter-bark, cachinho, milk bush, palo amargo (P.R.); palo de leche chiquito (Dom. R.); bois lait femelle (Haiti); veneno (Tobago); cabamuc, cabatmuc, chacmuc, cocotombo, sarna de perro (Mex.); amatillo, hierba de San José, matacoyote, señorita (Salv.); comida de culebra, guataco colorado, viborilla (Nic.); cohataco, guataco (C.R.); fruta del diablo (Pan.); anguito, cruceto, venenito (Col.); boboró (Venez.); chiric-sanango (Peru); itapeua grande, marfim, muirajussara falsa, pau marfim falso (Braz.); lecherón negro (Arg.).

Stemmadenia, with about 15 species of shrubs and small or rarely medium-sized trees, is distributed from southern Mexico through Central America to Surinam. The species with the widest range is *S. grandiflora* (Jacq.) Miers. The plants are of no commercial value.

Wood cream-colored to brownish. Fairly lustrous. Odorless and tasteless. Of medium weight but hard; of fine and uniform texture; easy to work; of low resistance to decay.

COMMON NAMES: Berraco, cojón de puerco, laurel, lecherillo, lechoso, tapaco, sicte, sictillo, xlaul (Mex.); cojotón (Br. H.); flor del día (Salv.); bijarro, cojón de caballo, guijarro, güitil, huevo de caballo (C.R.); huevo de gato, lechuga, mountain jasmine, venenillo (Pan.); cojón de fraile, laidre (Col.); huevo de burro, h. de cochino (Venez.); skijtnotto (Sur.).

Stenosolen, with five species of shrubs and little trees, usually included in the genus *Tabernaemontana*, occurs from Colombia and the Guianas to northern Brazil, Bolivia, eastern Peru and western Ecuador. The only sample available for study (Yale 20738; Rimbach 45) is of *S. heterophyllus* (Vahl) Mgf. collected in moist forest at the foot of the western Cordillera in Ecuador by A. Rimbach who described it as a slender, lactescent tree 20 feet high, bearing sweet edible fruits. The local name is Petaquilla.

Wood cream-colored throughout the specimen. Luster medium. Odorless and tasteless. Of rather light weight but hard; of fine and uniform texture; very easy to work; perishable in contact with the soil. Not known to be utilized.

Tabernaemontana, in a broad sense, is a large pantropical genus of shrubs and small to medium-sized trees. Groups of American species have been segregated into separate genera, namely, *Anacampta, Anartia, Bonafousia, Macoubea, Peschiera; Stenosolen, Taberna,* and *Stemmadenia,* and the treatment in this book follows the classification proposed by Markgraf (*Notizbl. Bot. Gart. Berlin-Dahlem* 14: 122: 151–184). Apparently the genera are fairly distinct, but the material studied is too limited to justify definite conclusions.

According to Markgraf (*loc. cit.*), there are 12 American species referable to *Tabernaemontana,* as thus restricted, and their combined range includes the West Indies, southern Mexico, Central America, and northern South America. The two species having the widest distribution are *T. amygdalaefolia* Jacq. and *T. citrifolia* L., the latter with several synonyms. The latex finds some application in native medicine, but there are no special uses for the timber.

Wood cream-colored or brownish. Fairly lustrous. Odorless and tasteless. Of rather light weight, moderately hard; of fine and uniform texture; very easily worked, finishing smoothly; subject to sapstain and perishable in contact with the soil. Apparently of no commercial possibilities.

Common names: Huevo de gallo, lechero, lechoso, pegojo (Cuba); palo de lechoso, pegoge (P.R.); palo de leche (Dom. R.); bois lait mâle (Haiti); berraco, b. de la costa, chusumpek, cojón de gato, c. de puerco, c. de toro, huevo del toro, jazmín de perro, lecheria, lecherillo, olfato de perro, palo de San Diego, rejalgar, utsubpek, utzupek (Mex.); cogotone, cojotone (Br. H.); chanchito de flores blancas, chapupo, cojón, c. de mico (Guat.); chilindrón, cojón macho, jazmín de monte, leche de perra, amatillo (Salv.); cojón de mico (Hond.); cachito (Nic.); cachitor, cojón (C.R.); huevo del gato, jazmín de monte, nuno (Pan.); cojón de cabrito, fraile cillo, guacharaco, jazmín de montaña, platanito, tornasol, turma de perro (Col.); azaharito, berracho, buril, cojón de verraco, lechero, verraco (Venez.).

Thevetia, with several species of shrubs and small trees, occurs throughout tropical America, though typically a Middle American genus. The best known and most widely distributed species, much planted for ornament, is *T. peruviana* (Pers.) K. Schum. It sometimes grows to a height of 30 feet, but is usually much smaller, sometimes a clambering shrub. It has thick fleshy narrow leaves and large showy sweet-scented flowers. People often carry the large seeds in the pocket as talismans and the kernels yield an oil and a glucocide called thevetine. The genus supplies no timber of value.

Wood brownish gray. Fairly lustrous. Odorless and tasteless. Rather light in weight but firm and strong; fine-textured, with a silky feel; readily worked; has low resistance to decay.

Common names: Trumpet flower, yellow oleander (Florida); lucky nut, l. seed (B.W.I.); cabalonga, cobalonga (Cuba); caballón, cabalonga, milk tree (P.R.); bois saisissement, serpent (Haiti); acitch, acitz, cabalonga, cabrito, calaveritas, campanilla, chirca, codo de fraile, fraile, huevo de gato, joyote, joyoto, meriendita, naranjo amarillo, narciso amarillo, regalgar, tzinacanytlacuatl, yambigo, yoyote, yoyotli, yucucaca (Mex.); good-luck seed, willow (Br. H.); chilca, chilindrón, chirca, cojón de gato (Guat.); campanilla amarilla, chilindrón, c. blanco (Salv.); chirca, c. venenosa

(C.R.); amancay, campanilla, chirco (Pan.); acedio, aje de monte, amancal, amancay, azuceno, cabalonga, castañeto, caucho, olivo, pepo de cruz (Col.); caruache, casabel, cruceta real, fruta de cascabel, lechero, lengua de gato, retama (Venez.); camache, caruache (Br. G.); jurri-jurri, tawai (Sur.); bella-quillo (Peru); ahohai mirim, chapeu de Napoleão, jorro-jorro, lauro rosa amarello, mamma de cachorro (Braz.).

Vallesia, with several species of shrubs and little trees rarely 25 feet high, is largely confined to southern Mexico and Central America, though one species, *V. glabra* (Cav.) Link, extends from Florida to Argentina. The plants appear to have no economic value.

Wood brownish yellow. Fairly lustrous. Odorless and tasteless. Of medium weight but firm and strong; texture fine; working properties excellent; durability low.

Common names: Palo boniato, p. moniato (Cuba); cacorahue, frutilla, huelatave, otatave (Mex.); San Pedro Martín (Guat.); amargoso (Venez.); chuviringana (Braz.); ancoche (Arg.).

Zschokkea (= *Lacmellea?*), with numerous species of shrubs and small trees rarely over 35 feet high, has its center of distribution in the Brazilian Amazon region, with extensions northward to Guatemala and British Honduras. The trunks of the trees are often beset with spine-tipped cones suggesting *Zanthoxylum*. The latex is abundant and sweet, though tacky, and the residue can be used for chewing gum; the berries are edible. The timber is utilized locally to a minor extent for interior construction. The most northern species, *Z. Standleyi* Woodson, has until recently been confused with *Lacmellea edulis* Karst. of Colombia and Venezuela. Standley and Steyermark (*Field Mus. Bot. Ser.* 22: 5: 371) say that "they will not be surprised if ultimately *Lacmellea* and *Zschokkea* prove to be congeneric."

Color yellowish or brownish, often with a tinge of pink; sometimes streaked. Luster fairly high. Rather light but firm to rather heavy and hard; texture fine and uniform; feel silky; grain straight; readily worked, finishing smoothly; of low resistance to decay. Apparently without commercial possibilities.

Common names: Cow tree, palo de vaca, vaca (Br. H.); cabrahosca, lechemiel, mutuculicu (Venez.); haia-haia, hya-hya (Br. G.); awaratalla, boeboeraballi, pritijari, schopsteelhout (Sur.); chicle, condurú de espinho, cumahy, molongó, pau de chicle, p. de colher, sobrerinha, sorvinha, tamanqueire de leite, tucujá (Braz.); chicle, huiqui-caspi, pajar-umu, quina, quina-quina (Peru).

AQUIFOLIACEAE

The Holly family is composed of four or five genera of shrubs and small to medium-sized, rarely large, trees, the principal genus being *Ilex*.

Ilex, with over 250 named species, is very widely distributed in temperate and tropical regions of the world. The leaves are alternate, simple, and without stipules; the small white or greenish flowers are axillary; the fruit is a yellow, red, or black, berry-like drupe. *I. aquifolium* L. is the Holly so frequently mentioned in literature. The American Holly is *I. opaca* Ait., a small tree of the eastern United States whose glossy spiny leaves and bright red berries are used so extensively for Christmas greens. Both supply good woods noted for their chalky whiteness and long use for inlay, marquetry, and, when dyed black, for imitation Ebony.

There are many species of *Ilex* in tropical America, their combined range including the West Indies, Mexico, Central America, and South America to Paraguay; the genus is especially well represented in Brazil. Mate, the dried leaves of certain species, principally *I. paraguensis* St. Hil., is a highly important article of commerce, ranking fourth in value of Brazilian exports. An infusion, similar to that from ordinary tea leaves, is in daily use by millions of South Americans. The center of the industry is in Paraná, Brazil, where the small trees are propagated under the open canopy of the Paraná Pine, *Araucaria angustifolia* (Bert.) O. Kuntze. Argentina consumes about

100,000 tons of mate annually, most of which was imported from Paraguay and Brazil until recently, the local production (chiefly in Misiones) having developed from 9000 tons in 1924 to more than 80,000 tons in 1936.

The largest tree is *Ilex guianensis* (Aubl.) Kuntze which grows in northern South America and Central America. In the Guianas it is the source of timber used locally and logs are obtainable that are 30 feet long, squaring 12 inches. In the limestone plateau of British Honduras it is a characteristic component of the advanced forest, attaining a maximum height of 130 feet, with a straight slender trunk 20 inches in diameter. The timber is not utilized for any special purpose and is inferior in texture and color to the Hollywood of commerce.

Wood chalky white to grayish or bluish white with little contrast between heartwood and sapwood. Luster low. Odorless and tasteless. Rather hard, but variable in density in different species, having about the consistency of Maple (*Acer*); sp. gr. (air-dry) 0.60 to 0.80; weight 37 to 50 lbs. per cu. ft.; texture uniform, fine to medium; grain generally straight; easy to work, finishing very smoothly; perishable in contact with the ground.

Pores numerous; very small to small; arranged in short to long radial rows or clusters. Vessels with many-barred scalariform perforation plates; spiral thickenings frequently present, particularly in woods grown in temperate zone; intervascular pitting rather fine, opposite to scalariform. Rays of two sizes, uniseriate with all cells upright or square, and multiseriate, 3 to 6, sometimes up to 10 cells, wide and 40 to over 100 cells high, the body cells procumbent; pits to vessels small, opposite or much elongated and parallel. Wood parenchyma rather sparingly developed in short tangential lines and diffuse; not distinct with lens. Wood fibers thick-walled, frequently with spiral thickenings; pits with distinct borders and narrow apertures. Ripple marks and gum ducts absent.

Common names: Holly (U.S.A.); acebo, a. cubano, a. de sierra, a. hojas de mirte, pinsapo, vigueta naranjo, yanilla blanca (Cuba); briqueta, cuero de sapo, gongolin, guerrero, hueso prieto, macoucoua (P.R.); junco serrano, limoncillo, naranjillo (Mex.); casada (Br. H.); cerezo (Guat.); garlic-wood (Pan.); cardenillo, naranjuello (Col.); jaque negro, matías, punte (Venez.); kakatara, kakataraballi, kookerite-balli? (Br. G.); pavier blanc (Fr. G.); huito quiro (Peru); azevinho, caáchí, caá-mí, caá-mina, caá-úna, caraúba do matto, cauminha, caúna, c. amargosa, chá do rio, congonha, cravo do matto, herva de anta, h. mate, h. matte, macucú, m. verdadeiro, mate, migueira, orelha de burro, o. mansa, pau macucú, vondeira (Braz.); caá, caá-rá, roble de Tucumán, yerba mate (Arg.); bacucú, caá, caá-guazú, caá-mí, caá-na, yerba mate (Par.).

ARALIACEAE

The Ginseng family, with over 60 genera and about 700 species of trees, shrubs, vines, and herbaceous plants, is of general distribution in the tropics and there are several representatives in temperate regions. The alternate stipulate leaves are simple, digitate, or once or twice pinnate; the small whitish or greenish flowers are borne in racemose or panicled umbels; the fruit is a berry containing 2 to 12 1-seeded nutlets. One of the best known herbs is the Ginseng, *Panax quinquefolium* L., cultivated for its fleshy roots which are exported in considerable quantity to China for medicinal purposes. The English Ivy, *Hedera helix* L., is another member of this family. The pith of some of the plants is very large, and that of *Tetrapanax papyriferum* (B. & H.) K. Koch, of eastern Asia, is rotary-cut into thin sheets, called rice paper, for making artificial flowers.

The principal genera represented in America are *Aralia, Dendropanax, Didymopanax, Oreopanax, Pseudopanax, Schefflera,* and *Sciadodendron*. Although there are many trees, some of them of large size, they are in no instance of more than local importance for timber. With proper handling, however, lumber can be obtained that is suitable for general carpentry, interior construction, shelving, boxes and crates, paper pulp, and match sticks.

Wood pale brown or grayish, with little or no contrast between heartwood and sapwood. Luster medium to rather high. Has about the consistency of certain species of Willow (*Salix*) and Cottonwood (*Populus*), and like them is tough and strong in proportion to its weight; difficult to dry in the open air without becoming badly stained; perishable when exposed to decay or wood-destroying insects.

Growth rings present or absent, sometimes very distinct. Tropical species diffuse-porous, the pores medium-sized (100 to 200μ) to very small, rather few to very numerous, occurring singly and, more often, in multiples or clusters fairly uniformly distributed without definite pattern; temperate species frequently ring-porous, the medium-sized early-wood pores in one to several rows, those in late wood much smaller and often aggregated in distinctive patterns. Vessels with either simple or multiple perforations or both together; plates with multiple perforations scalariform, with few to many, fine to coarse, closely to widely spaced, straight or anastomosing bars; plates with simple perforations often steeply inclined and with wide rims; no spiral thickenings seen in American species; intervascular pitting very coarse (12 to 20μ, occasionally up to 36μ), alternate to opposite with tendency to scalariform. Rays greatly variable in width and height, but infrequently uniseriate; generally of two distinct sizes, the wider mostly 2 to 4, sometimes up to 10, cells wide and often less than 25, sometimes up to 50, rarely 100 or more, cells high; heterogeneous, at least in part, though sometimes with only single marginal rows of squarish cells; pits to vessels variable in size and shape, generally very large, at least in part, and tending to scalariform arrangement. Wood parenchyma mostly sparingly paratracheal, not distinct with lens. Wood fibers with thin to moderately thick walls; commonly septate and containing starch in sapwood; pits very small, simple. Ripple marks absent. No vertical gum ducts seen; very small to very large intercellular canals observed in the rays of American specimens of *Dendropanax, Didymopanax, Oreopanax,* and *Sciadodendron* (also in Old World specimens of *Acanthopanax, Arthrophyllum, Cheirodendron, Cussonia, Dendropanax, Dyzygotheca, Myodocarpus, Nothopanax,* and *Schefflera*), but apparently not always constant in a species.

Aralia, with about 40 species of aromatic, spiny trees and shrubs with stout, pithy branchlets, or bristly or unarmed perennial herbs, occurs in both temperate and tropical regions of Asia and North America. Best known of the American species is the Prickly Ash or Hercules Club, *A. spinosa* L., a small and usually prickly tree with large doubly pinnate leaves clustered at the ends of the coarse branches. It occurs in bottomlands in the south-central hardwood region and is occasionally planted for ornamental purposes in eastern United States. According to Standley (*Trees and shrubs of Mexico,* p. 1080), there are five species in Mexico, but only one, *A. pubescens* DC., is arborescent. The genus does not supply any commercial timber and the uses of the plants are mostly medicinal.

The following description applies particularly to *Aralia spinosa.* Heartwood pale brownish; sapwood yellowish. Fairly lustrous. Odorless and tasteless. Light in weight, but firm; texture rather fine to rather coarse; grain straight; easily worked, but brittle; poorly resistant to decay.

Common names: *Aralia spinosa:* Angelica tree, Hercules club, prickly ash (U.S.A.). *Other species:* American spikenard, bristly sarsaparilla, wild elder, w. sarsaparilla (U.S.A.); cuajilotillo (Mex.); cinco dedos (Col.).

Dendropanax (or *Gilibertia*) is represented by numerous species in eastern Asia and throughout most of tropical America, but the trees are typically small, rarely up to 50 feet tall and two feet in diameter. In Puerto Rico they are frequently left for shade in coffee plantations. The grayish, light but firm and tough, easily worked timber has about the consistency of Yellow Poplar (*Liriodendron*) and is utilized locally to a small extent for common lumber.

Common names: Angelica tree, galipee (Jam.); ahorca jíbaro, bíbono, palo cachimba, ramón de vaca, víbona (Cuba); corcho blanco, muñeca, palo cachumba, p. de gangulin, p. de pollo, p. de vaca, pana, víbona (P.R.); lengua de vaca, ramón de vaca (Dom. R.); bois négresse (Haiti); mano de danta, m. de león, m. de oso, palo blanco, p. de agua, p. de danta, p. santo, sac-chachah, tamalcoahuite (Mex.); white gumbolimbo (Br. H.); palo de agua (Hond.); mano de león (Salv.); cacho de venado (C.R.); paloma, vaquero (Pan.); banco (Col.); María molle (Braz.); achcuisman (Peru).

Didymopanax is a tropical American genus and, while numerous species have been described, the one that is best known and most widely distributed is *D. morototoni* (Aubl.) D. & P. It is a medium-sized to large tree, sometimes over 100 feet tall and 30 inches in diameter, and is characteristic of open forests, the edges of savannas, and former clearings. The seedlings are easy to transplant, make very rapid growth, and are not exacting in soil requirements, but must have plenty of sunlight. The trees are free from disease and may be reproduced by coppicing. The light wood is used locally for general carpentry work not exposed to the weather and for match splints and boxes. In the Amazon region the timber is sometimes sold as Marupá (*Simaruba amara* Aubl.). The Mandioqueira of southern Brazil (*D. longipetiolatum* March.) is denser and finer-textured and is considered an excellent timber for making boxes that are subjected to rough handling.

Wood grayish or pale brownish throughout. Luster medium. Variable in density from rather light and soft to moderately heavy and hard; mean sp. gr. (air-dry) about 0.50; average weight 31 lbs. per cu. ft.; texture medium to rather fine; grain mostly straight; easy to work, finishing smoothly; poorly resistant to decay.

Common names: Arriero, badana, botijón, cordoban, cordován, guanillo, padero, yagrumo macho (Cuba); grayume, g. macho, grayumo, llagrume, pana cimarrona, yagrume, y. macho (P.R.); palo de viento (Dom. R.); bois trembler (Haiti); chancarro blanco, manuel-dante, roble blanco (Mex.); pava, pavilla, pavo (C.R.); gargorán, mangabé, pavo (Pan.); yarumero (Col.); higuerotón, orumo macho, pan de trigo, volador, yagrumo, y. macho, yarumo (Venez.); carahora (Br. H.); bigi boesie papajahoedoe, cassavehout, karohoro, karoro, khobbé, mollototto-oe, moretoto-oe (Sur.); mandioqueira, morototó (Braz.); anonilla, sacha-uva (Peru); ambaí-guazú, ambay-guazú (Arg.).

Oreopanax, with numerous species of trees and shrubs, the latter sometimes epiphytic, has an extensive range in tropical America, but the plants are commonest in the West Indies, southern Mexico, Central America, and northwestern South America. *O. argentatus* (H.B.K.) Dcne. & Planch. is a sparsely branched forest tree about 50 feet high and 10 inches in diameter at elevations of 7500 feet above sea level in the Western Cordillera of Ecuador where, because of the shape of the leaves, it is known as Puma-maquí (paw of the puma). Its white easily worked wood is used for small joinery work and, on account of its even texture and flexibility, is preferred for making candy boxes, wooden spoons, kitchen utensils, and guitars. There are eight Mexican species, chiefly shrubs, the largest tree reported, *O. Salvinii* Hemsl., being less than 40 feet tall.

Wood grayish throughout. Luster medium. Mostly of moderate density, but occasionally very light, soft, and "stringy"; texture rather fine, uniform; grain straight; working properties good; durability low.

Common names: Growing stick, three-fingered Jack, woman wood (Jam.); palo de viento, víbona (Dom. R.); bois d'anjou (Haiti); mano de león (Mex.); yash-hulup (Br. H.); brasil (Salv.); higuero, matapalo (C.R.); cinquo dedos (Col.); apio de monte, candelero, mosquito (Venez.); puma-maquí (Ec.); sacha-uvilla, maquí-maquí (Peru).

Pentapanax, with several, mostly Asiatic, species of trees and shrubs, has at least two representatives in Argentina, namely, *P. angelicifolius* Gris. and *P. Warmingiana* (March.) Harms. Both of these are said to be large trees with yellowish white, very soft to moderately hard wood, of no commercial importance. There are no authentic specimens available for study.

Common names: Albiche, caroba blanca, c. brava, pino, quino-quino, sacha paraíso, saúco silvestre, sauguero bravo (Arg.).

Pseudopanax, with a few species of trees and shrubs, occurs in New Zealand, Chile, and western Argentina. The only American wood specimen available (Yale 1760) is of *P. laetevirens* (Gay) Seem. from Patagonia, where it is known as Saúco del Diablo. Color light grayish brown throughout. Luster medium. Without distinctive odor or taste.

Moderately hard and heavy; texture fine; grain fairly straight; working properties good. Of no commercial possibilities.

Schefflera (= *Sciodaphyllum* or *Sciadophyllum* and *Actinophyllum*) is a large pantropical genus of trees and shrubs. It is represented in the West Indies, southern Central America, and throughout most of South America. *S. paraensis* Ducke is a tree about 65 feet high in the Lower Amazon basin. There are no authentic specimens of any American species available for study.

COMMON NAMES: Adonképau, atapé, biegi boesie papaje, karohoro, karohoroe, mallototo-oe, moretoto-oe, papaja hoedoe, p. hout, perewa moeroetoeto-oe, p. moetototo-oe, tobi-toetoe (Sur.); parapará (Braz.).

Sciadodendron *excelsum* Gris., the only species, grows in Haiti, Central America, and Colombia. It is a fairly large tree, sometimes 75 feet tall, with a stout, straight trunk 24 to 40 inches in diameter and free of branches for two-thirds of its length. It is sparingly utilized locally for common lumber. Wood gray or yellowish gray throughout. Luster rather low. Moderately light but firm, though brittle; texture rather coarse; feel harsh; grain mostly straight; easy to work; perishable in contact with the ground. Not likely to enter the export trade.

COMMON NAMES: Juan primero (Dom. R.); ouane primaire (Haiti); corroncha de lagarto, lagarto (Salv.); palo de lagarto (Nic.); jobo lagarto, mangabé (Pan.); madura plátano (Col.).

ARISTOLOCHIACEAE

THE members of this family are mostly perennial herbaceous or woody vines, a few erect plants. There are about 180 species of *Aristolochia,* widely distributed throughout the tropics and warmer regions of the world. Some of them have very conspicuous irregular flowers and are grown for ornamental purposes. The principal use is medicinal, especially as a remedy for snake bite, though there is no clinical proof of its efficacy. The wood of the larger stems is white or yellow, and coarse-textured.

Frequently ring-porous. Some of the pores very large, others much smaller. Vessels with simple perforations; spirals sometimes present in smallest vessels; intervascular pitting coarse, tending to scalariform. Rays all wide, height often equal to length of internodes, usually not in contact with the vessels; fairly homogeneous, cells thin-walled; crystals common. Wood parenchyma sparingly paratracheal and sometimes in irregular, uniseriate, tangential lines. Wood fibers with numerous conspicuous bordered pits.

AVICENNIACEAE

Avicennia, the only genus, is commonly included with the Verbenaceae. It consists of about 14 species of small to medium-sized evergreen trees of pantropical occurrence along the coast and in tidal marshes in association with *Rhizophora* and other plants of the Mangrove formation. Although the trees sometimes attain a height 60 to 75 feet and a trunk diameter of two feet, they are generally not over 25 feet high, and towards the northern and southern limits of range become shrubby. They have long heavy roots which grow in and out of the mud in an arched and entangled manner and also develop asparagus-like leafless branches (pneumatophores) which catch the silt from the rivers and the flotsam of the sea, thus protecting and extending the shore line (Plate XIV). The seed usually germinates on the tree and is growing and ready to take root when it falls into the soft mud. The only distinct species in the western hemisphere, *A. marina* (Forsk.) Vierhapper (= *A. nitida* Jacq.), occurs in southern Florida, the West Indies, the north coast of South America and both coasts of Mexico and Central America.

The timber of all species is practically identical and is of negligible economic importance, though used locally for fuel and miscellaneous purposes in the round. Experiments have demonstrated that it can be pulped successfully by the soda process, but because of the shortness of fiber is unsuited for use alone and must be mixed with coniferous pulp. The billets are subject to discoloration by fungi and the stain adds to the difficulty of bleaching. The pulp is not a promising source of viscose, since, aside from other considerations, the content

Photo by B. E. Dahlgren.

PLATE XIV. Black Mangrove trees (*Avicennia marina*) in a coastal marsh in British Guiana.

Photo by Max Rothkugel.

PLATE XV. Large specimen of Lapacho Amarillo (*Tabebuia ochracea*) left standing when the forest was cleared for a plantation of Yerba Mate (*Ilex*) in Argentina.

of alpha cellulose is too low, being only about 69 per cent.

Heartwood brown to nearly black, with oily appearance; yellowish powder (lapachol compound) usually present, as in some members of the Bignoniaceae. Sapwood thick, white when freshly cut, but staining gray or purplish. Odor and taste absent or not distinctive. Heavy and hard; sp. gr. (air-dry) about 0.95; weight nearly 60 lbs. per cu. ft.; texture coarse and uneven; grain irregular; wood laminated, difficult to work and not suitable for lumber; old heartwood resistant to decay, but the fiber layers tend to separate.

Structure anomalous, concentric type (Plate LVII, 3); strands of included phloem associated with bands and rays of conjunctive tissue, each band containing a regular row of stone cells. Pores small to minute, not visible without lens; occurring singly or more often in radial multiples or rows, and sometimes irregularly clustered; fairly evenly distributed. Vessels with simple perforations; spirals absent; pitting very fine, alternate. Rays of conjunctive tissue rather few; xylem rays numerous, mostly 1 to 3, sometimes up to 6, cells wide and few to 25, occasionally up to 50, cells high; heterogeneous, most of the cells square or upright; small crystals common; pits to vessels very small, rounded. Wood parenchyma rather sparingly paratracheal, sometimes slightly aliform. Wood fibers usually thick-walled; sometimes septate; pits few, very small, simple. Ripple marks and gum ducts absent; phloem strands upon disintegrating leave intercellular spaces resembling vertical gum ducts.

Common names: Black mangrove, black tree, blackwood (Florida); black mangrove, green turtle bough (Jam.); cativo bastardo, mangle negro, m. prieto (Cuba); black mangrove, chifle de vaca, mangle bobo, olive mangrove, salt-pond, white mangrove (P.R.); mangle prieto (Dom. R.); manglier noir, palétuvier (Haiti); black mangrove, limewood, olive mangrove (Trin., Grenada); bois de mèche (Guad.); mangle blanco, m. negro, m. prieto, puyeque (Mex.); black mangrove, mangle negro (Br. H.); palo de sal (Hond.); árbol de sal, ishtalén, istalén (Salv.); culumate, mangle salsa, palo de sal (C.R.); mangle salado (Pan.); manglecito (Col.); mangle amarillo, m. prieto (Venez.); pariva, parwa (Sur.); guapirá, palétuvier blanc (Fr. G.); ceriuba, ciriuba, siriubo, mangue amarello, m. branco (Braz.); mangle salado (Ec.).

BERBERIDACEAE

The Barberry family includes several genera of herbs, shrubs, and small trees, widely distributed over the world, but mostly in the north temperate zone. The principal woody genus is *Berberis* (including *Mahonia*).

Berberis, with over 100 species, occurs in Asia, Europe, United States, Mexico, Cuba, and the Andean region of South America to Tierra del Fuego. The plants are typically low shrubs, often spiny; the inner bark is yellow; the leaves are alternate or fascicled, 1–9-foliolate; the yellow flowers are borne in drooping racemes; the fruits are tart, those of certain species grapelike and edible. A few Mexican species are trees, 20 to 30 feet high, with a trunk over a foot in diameter. The roots are bitter and contain alkaloids. The hard, fine-textured wood is used as a source of a yellow dye, and sometimes also for small articles of turnery and carving. The heartwood is chocolate-brown, sharply demarcated from the sapwood, which is bright yellow with a greenish hue.

Pores small to minute, the large ones zonate (ring-porous), the others arranged in an irregular pattern of wavy and zig-zag bands and patches. Vessels with simple perforations; members storied; spirals present; pitting fine, alternate. Rays nearly all large (*Platanus* type), giving rise to conspicuous silver grain on radial surface; not storied; nearly homogeneous; pits to vessels small, circular. Wood parenchyma sparse or absent. Wood fibers small, thick-walled; pits very small, simple. Ripple marks fairly distinct under lens.

Common names: Barberry (Eng.); cerillo de loma, cera amarilla (Cuba); agrillo, agrito, cachisdá, camisdá, leña amarilla, palo amarillo, p. de tenir, p. jarilla, quisquirindin, quisquiringuin, retamilla, xoxoco, yagabuxe (Mex.); espino de oro (Col.); bollo, bosuga, uña de gato (Venez.); chchecche, ccjeshua-chcheeche (Peru); calafate, c. grande, c. pilpil-voqui, leña amarilla, mi-chay, m. chico, palo amarillo, quebrachillo, sacha-uva, salall (Arg.); espina amarilla (Urug.).

BETULACEAE

THE Birch family, in the restricted sense, comprises two well-known north temperate genera, *Betula* (Birch) and *Alnus* (Alder); *Carpinus, Corylus,* and *Ostrya,* which often are included here, are considered in a separate family, the Corylaceae. The Betulaceae are trees or shrubs with alternate, simple, prominently penninerved, mostly serrate leaves and free stipules; the flowers are monoecious, the male inflorescence a pendulous catkin, the female a cylindric cone-like spike which is solitary in *Betula,* racemose in *Alnus;* the fruits are very small winged nutlets borne on the ament scales which are deciduous in *Betula* and persistent in *Alnus.* Another distinction between the two genera is in the winter buds, those of *Betula* being covered by imbricated scales while those of *Alnus* are without scales.

Alnus, with about 20 species of shrubs and trees, inhabits swamps, river bottoms, and high mountains throughout the northern hemisphere and extends southward at high altitudes to upper Assam in the Old World and through Mexico and Central America, and the Andean region of South America to Peru, Bolivia, and northern Argentina. There are about 10 American species, mostly shrubs and small trees, but in a few instances attaining large dimensions. The White Alder, *A. rhombifolia* Nutt., is at its best in the valleys of central California, where it is frequently 80 feet tall, with a straight trunk 2 to 3 feet in diameter, but its timber is not extensively used. The Red Alder, *A. rubra* Bong., grows from Alaska to southern California, but its commercial range is limited mainly to moist situations in western Oregon and Washington, the estimated stand being about a billion board feet. The good quality of the timber combined with the scarcity of native hardwood species make it highly important locally, particularly to the furniture industry. The lack of attractive natural color is overcome by staining, mostly in imitation of Walnut or Mahogany.

There are a few species in Mexico and the bark is used as a source of tannin, dyestuff, and medicines. One species, *Alnus glabrata* Fern., is said to attain rather large dimensions. The Central American species is *A. acuminata* H.B.K., a shrub or tree up to 35 feet high of common occurrence in the mountains, ascending the higher slopes and forming small pure stands. The timber is used to a limited extent for general carpentry and construction.

According to M. Acosta Solís (*Tropical Woods* 57: 3), Alder is common throughout the subandine region of Ecuador and is planted about farms. The trees are 50 to 65 feet tall with trunks 16 to 20 feet long and 16 to 28 inches in diameter at the base. "The wood is pale roseate, sometimes striped with red, becoming somewhat brighter upon drying; it is odorless, of fine and uniform texture, and is easy to saw and to plane. It is used in carpentry, furniture-making, and cabinet work, and before Eucalypt lumber became available, was employed in general construction in the same way as Calupí, Sauce, Algarrobo, and Arrayán. In the province of Tungurahua it is used for making the boxes and packing cases in which fruit is exported." Apparently the South American Alders are all of one species, *Alnus jorullensis* H.B.K., but of several varieties, in part sometimes confused with *A. acuminata.* They ascend to high altitudes in the Peruvian Andes and are especially valuable in such regions for fuel and small construction. The bark and leaves are used for tanning and dyeing and their astringent properties are also useful in local medicine.

The woods of *Alnus* are very similar, being pale brown or superficially bronze-colored, with low luster. Odorless and tasteless when dry. Light in weight but firm; sp. gr. (air-dry) 0.45 to 0.60; weight 28 to 38 lbs. per cu. ft.; texture medium to fine; grain variable; seasons readily without excessive warping or splitting; easy to work, finishing smoothly; holds its place well when manufactured; not durable in contact with the ground.

Growth rings present. Pores small to minute, not visible without lens; thin-walled; very numerous; often in radial multiples, well distributed, without pattern. Vessels with many-barred scalariform perforation plates; without spiral thickenings; intervascular pitting mostly opposite. Rays homogeneous; uniseriate or locally biseriate and up to 50, mostly less than 25, cells high; often aggregated, especially in outer parts of older stems, being rather conspicuous on all sections; pits to vessels very small. Wood parenchyma sparingly diffuse; not distinct with lens; pith flecks common. Wood fibers with medium-thin walls and small bordered pits. Ripple marks and gum ducts absent.

Common names: Alder (U.S.A.); aile, abedul, aliso, alizo, ayle, ilite verde, olmo del país, palo de águila, yaga-bizie (Mex.); palo de lama (Guat.); alum, jaúl (C.R.); aliso, a. andino, chaquero (Col.); aliso (Venez., Ec., Peru); lambrán, ramram (Peru); aliso del cerro (Arg.).

Betula, with upward of 30 species of shrubs and small to large trees, is distributed from the Arctic circle to Japan, China, the Himalayas, and southern Europe in the Old World and throughout Canada and most of the United States in the New, some species forming extensive forests in the north or at high altitudes. There are about a dozen American species, of which nine are trees; several varieties and hybrids are also recognized. Most important commercially is the Yellow Birch, *B. lutea* Michx., a large tree of southeastern Canada, the Lake States, New England, and the Appalachian region to Georgia, being at its best in the United States near the Canadian border. Ranking second is the Sweet, Black, or Cherry Birch, *B. lenta* L., ranging from Newfoundland and Ontario through New England to the southern Appalachians, where the trees though less abundant attain their largest size. The timber of Sweet Birch is somewhat denser and more deeply colored than Yellow Birch, but the two kinds are sold together in the trade simply as Birch without distinction as to species, though the yellowish sapwood is frequently called White Birch and the light reddish brown heartwood, Red Birch. The lumber is much used in the manufacture of furniture, interior trim, flooring, and doors; also for fuel, destructive distillation, and railway crossties (after preservative treatment).

The Paper, Canoe, or White Birch, *Betula papyrifera* Marsh., with its four varieties, extends over Canada through New England southward to the southern Appalachians and westward through the Lake States and Dakotas into Oregon and Washington; its commercial center is in Maine. The demand is for the thick, nearly colorless sapwood, which is moderately hard, of fine and uniform texture, and particularly well suited for articles of turnery such as spools, bobbins, dowels, and small handles; it is also used for making toothpicks and shoepegs, and the bark for novelties and formerly by the Indians and early settlers for canoes. The western varieties of Paper Birch, var. *subcordata* (Rydberg) Sarg. and var. *occidentalis* (Hooker) Sarg., are of comparatively little economic importance at present; the best of the timber is in British Columbia. The Gray Birch, often called Wire Birch in Canada, *B. populifolia* Marsh., of northeastern United States and southeastern Canada, is a slender, often crooked tree with a smooth gray bark that does not exfoliate as freely as that of the Paper Birch. It grows on poor soil or on the margins of swamps and quickly takes possession of abandoned farm lands or burnt-over areas. The wood is of the type of Paper Birch and is used to some extent for the same purposes, though chiefly for fuel; its appearance is often marred by dark-colored pith flecks.

The River or Red Birch, *Betula nigra* L., is a common tree with shaggy, orange-colored bark and crooked bole, growing along streams in the Mississippi and Ohio valleys and along the Atlantic coast. It supplies some coarse-textured, brownish, tough and strong timber used chiefly for slack cooperage; pith flecks are very common in the wood. The Red or Black Birch, *B. fontinalis* Sarg., of the mountainous regions of the western United States, is a small tree with dark bronze bark and rather light, brownish wood of no commercial value.

Heartwood pale brown to rich reddish brown; distinct and sometimes rather sharply demarcated from the thick yellowish or white sapwood. Luster medium. Odorless and without pronounced taste. Density variable in different species; sp. gr. (air-dry) 0.50 to 0.80; weight 32 to 50 lbs. per cu. ft.; texture fine to coarse, uniform; grain usually straight, sometimes wavy; working properties good to excellent; sapwood perishable, heartwood not very durable in contact with the ground.

Growth rings present. Pores very small to medium-sized; numerous; mostly in multiples of 2 to 5 or in little clusters, well distributed. Vessels without spiral thickenings; perforation plates scalariform, with numerous narrow bars; pitting fine to very fine. Rays 1 to 4, sometimes 5, cells wide, the largest usually not over 25 cells high; rarely aggregated; homogeneous; pits to vessels minute to very small (*B. lenta, B. nigra*). Wood parenchyma in a uniseriate terminal layer and in few to fairly numerous short tangential rows or diffuse. Wood fibers with medium to rather thick walls and small bordered pits. Ripple marks and gum ducts absent.

Common names: Birch, boileau, merisier, m. rouge (Canada); birch—black, canoe, cherry, old-field, paper, poverty, red, river, silver, water, white, yellow (U.S.A.).

BIGNONIACEAE

The Bignoniaceae comprise more than 100 genera and several hundred species of general distribution in the tropics, a few in temperate regions. A great many of the plants are scandent shrubs or woody climbers, one of the best known being the Trumpet Creeper cultivated in gardens. The stems of many of the lianas are of anomalous structure, a common form having regularly spaced deep furrows filled with phloem wedges, the line of separation appearing on cross section like a miniature staircase (see Solereder's *Systematic anatomy of the dicotyledons*, pp. 605–610). The plants are noted for the beauty of their showy flowers and many are planted for decorative purposes. There are numerous arborescent genera and a few of them yield timber of commercial importance.

In the New World the trees are of 16 genera, namely, *Amphitecna, Astianthus, Catalpa, Chilopsis, Cotema, Crescentia, Cybistax, Ekmanianthe, Enallagma, Godmania, Jacaranda, Macrocatalpa, Paratecoma, Parmientiera, Tabebuia,* and *Tecoma*. They are all tropical except *Catalpa* and *Chilopsis*. The timbers exhibit a wide range of variation in appearance and properties. Many are light-colored throughout, others have a distinct heartwood in various shades of brown to olive or blackish, uniform or more or less variegated or striped; dark specimens frequently have an oily appearance and abundant sulphur-colored deposits (*Cotema, Godmania, Paratecoma,* and *Tabebuia* in part). Luster low to satiny. Taste bitter in *Tecoma* and in inner bark of *Godmania*, absent or not distinctive in the others; odor suggesting kerosene in *Catalpa*, absent or mild and not describable in the others. Density range very great; sp. gr. 0.40 to 1.25; texture fine to coarse; grain straight to roey or otherwise irregular; working properties fair to excellent; durability poor to very high.

Growth rings usually present, sometimes conspicuous; ring-porous structure occasionally present (*Catalpa* and *Chilopsis*). In diffuse-porous woods, pores medium-sized to minute; few to numerous; commonly solitary and in small multiples or little clusters, evenly distributed or tending to diagonal or concentric arrangement. Vessels with exclusively simple perforations (reticulate perforation plates sometimes present in certain Old World species; see *Forestry* 7: 1: 16–25); spiral thickenings characterize smallest vessels of *Catalpa* and *Chilopsis;* tyloses common; intervascular pitting alternate, rather fine to fairly coarse. Rays 1 to 8, generally 1 to 3, cells wide, often uniformly low, sometimes up to 50 or more cells high; distinctly heterogeneous in *Astianthus, Tecoma, Jacaranda* in part, weakly so in *Catalpa, Chilopsis,* and *Paratecoma*, mostly homogeneous in the others; small cubical crystals seen in *Tabebuia stenocalyx*, otherwise apparently absent; pits to vessels small and rounded to rather large, oval to irregular (*e.g., Catalpa*); pitting sometimes unilaterally compound. Wood parenchyma sparse to very abundant; paratracheal, narrow to coarse aliform to confluent in narrow or wide, short to concentric bands; often finely terminal; no crystals seen. Wood fibers with thin to very thick and gelatinous walls; septate in *Macrocatalpa* and in *Cybistax* in part; pits simple or more or less distinctly bordered. Ripple marks present in *Cotema, Crescentia, Cybistax, Godmania, Paratecoma,* and *Tabebuia;* 80 to 140 per inch; regular to irregular; all elements involved; secondary seriation often distinct with lens in the bands of 2-celled parenchyma strands. No gum ducts seen. For anatomy of the different genera see *Tropical Woods* 63: 10–38.

Amphitecna *macrophylla* (Seem.) Miers, the only species, is a shrub or small tree, closely related to *Enallagma*, of restricted occurrence in the mountainous forests of southern Mexico and Guatemala. The large, simple, alternate, nearly sessile leaves are clustered at the ends of the branches; the long-pedicellate greenish flowers are borne on short bracteate shoots on the old wood; the fruit resembles Cacao (*Theobroma*) and the seeds, which are without wings, are imbedded in pulp. There are no known uses for the timber. The wood has not been studied.

COMMON NAME: Huiro de montaña.

Astianthus *viminalis* (H.B.K.) Baill., the only species, is a tree sometimes 50 feet high growing on sand bars along streams from southern Mexico to Gualán, Guatemala. The simple, linear leaves are ternately whorled; the yellow flowers are borne in loose terminal panicled cymes; the small linear capsule contains numerous small broad-winged seeds. So far as known, the timber is not utilized for any special purpose.

COMMON NAME: Chilca (Guat.).

Catalpa, excluding the section *Macrocatalpa* as a separate genus, is composed of four species of deciduous trees of the north temperate zone, two occurring in Japan and northern China, the others of limited natural distribution in the southeastern quarter of the United States, though rather widely cultivated. They have stout twigs without a terminal bud; the large to very large leaves are typically in whorls of three; the flowers are conspicuous in terminal panicles; the fruit is a long, 2-valved capsule containing numerous flat, hairy-winged seeds.

Catalpa bignonioides Walt., often called Indian Bean, is typically a short, stout-boled tree, with a spreading crown, rarely attaining a height of 60 feet. It is planted for decorative purposes and is hardy as far north as eastern New England. It is of no importance as a source of timber. *C. speciosa* Warder, generally known as the Hardy Catalpa, is a forest tree, said to attain a maximum height of 120 feet with a tall straight trunk sometimes 50 inches in diameter on fertile bottomlands in the lower Ohio River region. The easily worked, durable timber has never been available in sufficient quantity to be a factor in the market, but it served as a background for the commercial exploitation of the species for forest plantations. Thirty to forty years ago no other tree was so generally and indiscriminately recommended for planting in the Middle West. Some of the plantations have proved successful, but a larger number have, for various reasons, been partial or complete failures. The principal use for the product is for fence posts and fuel, but young timber is not highly durable.

Heartwood light brown, sometimes tinged with olive; sapwood nearly white, tending to discolor and merging gradually into the heartwood. Luster fairly high. Odor suggesting kerosene; taste not distinctive. Light and soft; sp. gr. (air-dry) 0.40 to 0.45; weight 25 to 28 lbs. per cu. ft.; texture medium to coarse; grain generally straight; very easy to work, finishing smoothly, holds place well when manufactured; durability good to fair.

COMMON NAMES: Bean tree, bois puant, catalpa (common, hardy, western), catawba, cigar tree, Indian bean, I. cigar tree, smoking bean, Shawnee wood (U.S.A.).

Chilopsis *linearis* (Cav.) Sweet, the only species, is a small deciduous tree, sometimes 30 feet high with a short trunk a foot or more in diameter, growing along arroyos and in depressions in the desert in southwestern United States and northern Mexico. The showy, sweet-scented flowers, the long seed pods, and the rather soft brownish wood resemble *Catalpa*, but the narrow leaves and slender branches suggest *Salix*, hence the usual American name of Desert Willow or Flowering Willow. The pliable withes are used locally for making baskets and the stems supply a limited amount of post timber and fuel.

Heartwood golden brown and more or less streaked when fresh, and sometimes with a metallic luster; fairly distinct but not sharply demarcated from the thin, lighter-colored sapwood. Luster rather high. Slightly odorous, but without distinctive taste. Wood moderately light, but hard; sp. gr. (air-dry) 0.60; weight about 38 lbs. per cu. ft.; texture coarse; grain generally straight; working properties excellent; durability good. Of no commercial possibilities.

COMMON NAMES: Catalpa willow, desert willow, flowering willow, Spanish willow (U.S.A.); mimbre (Mex.).

Cotema, with four species of little trees rarely 25 feet high, is limited in distribution to eastern Cuba. The leaves are typically 3–5-foliolate, with slender petioles; the rather large, long-pedicelled flowers are solitary or in small axillary clusters; the slender terete capsule is curved or coiled; the thin, oblong seeds are winged at each end. The wood belongs in the Lapacho group of *Tabebuia;* it apparently is not utilized because of the small size and scarcity of the trees.

Heartwood brown, strongly marked with yellowish green vessel deposits (lapachol compound); distinct but not very sharply demarcated from the brownish sapwood. Luster rather low. Very hard and heavy; texture fine and uniform; grain fairly straight; not easy to work, but finishing very smoothly; a yellow dust arises in sawing; durability high.

Crescentia, with a few species of small to medium-sized, irregular, evergreen trees, is widely distributed in tropical America and commonly planted there and in the Old World for the gourd-like fruits. The best known species is *C. Cujete* L., the Calabash tree, with a short trunk, crooked limbs, and often drooping branches. The leaves are simple, obovate or spatulate (in *C. alata* H.B.K. they are trifoliolate, with winged petiole), and generally borne in irregularly spaced fascicles; the purple-blotched yellowish flowers appear singly or in clusters in the leaf axils or along the larger branches; the fruits, which are large and borne close to the trunk, have a thin skin over a woody shell inclosing a mass of pulp and seeds. The shells are used by the natives for making cups, dishes, and other household utensils and some of them are ornately carved or painted. The firm, tough timber is used for the ribs and knees of small boats (for which the natural shapes of the limbs are peculiarly suited), hubs and felloes of wheels, and for saddle-trees, firewood, and charcoal.

Wood nearly white when fresh, becoming pale brownish with fine darker markings of parenchyma; sapwood not clearly defined. Luster medium. Without distinctive odor or taste. Of medium density; sp. gr. (air-dry) 0.60 to 0.71; weight 37 to 44 lbs. per cu. ft.; texture medium, though appearing coarse; grain generally irregular; fairly easy to work, finishing smoothly; does not split readily; is poorly resistant to decay. Of no export possibilities.

COMMON NAMES: Calabash, packy, wild calabash (B.W.I.); güira, g. cimarrona, totuma (Cuba); higuero (P.R.); calabasse, c. marron (Haiti, Fr. W. I.); ayal, ayele, cirián, cuastecomate, cuastecomatl, cuautecomate, cuiro, cujete, guaje, g. cirial, g. cirián, güiro, huajericián, huaz, hüiro, jayacaste, jícaro, luch, morro, pog, tecomate, t. ayele, tecomatl (Mex.); calabash, wild calabash (Br. H.); hom, morro, m. guacalero (Guat.); cuchara, cutuco, guacal, huacal, jícaro, j. de cuchara, j. de guacal, morrito, morro (Salv.); jícaro (Hond.); calabacero, ferú-kó-uo, jícaro, kuakrá, me, mékuru, tamkrá (C.R.); calabazo, jícaro, palo de calabaza, p. de turtumas, totumbo, totumo (Pan.); calabazo, mate, totumo (Col.); camasa, cotumo, cucharo, güiro, taparito, taparo, taparón, totumo (Venez.); mate (Ec.); huingo, pati (Peru); cabaceira, coité, cuia maracá, c. pequena do igapó, cuieira, cuünha do igapó, cuité, cuiteseira, cujeté (Braz.); ibirá-acá-hiá (Par.); japacary (Arg.).

Cybistax, with four or five species of small to large trees, is widely but irregularly dispersed in tropical America. The leaves are opposite or sub-opposite, digitate, with 5 to 9, usually 7, leaflets; the golden yellow, rarely greenish, flowers are borne in contracted or large open terminal inflorescences; the capsule is linear-oblong, with 10 to 16 ridges on the coriaceous valves; the

Map 4. Range of Primavera in Mexico and Central America.

broadly heart-shaped seeds are surrounded with a large thin wing.

Cybistax Sprucei K. Sch., native to the Andean region of Peru, is cultivated for its leaves which are a source of a blue dye used as a substitute for indigo. *C. antisyphilitica* Mart. occurs in eastern Brazil and Misiones, Argentina. According to Pio Corrêa (*Diccionario dos plantes uteis do Brasil,* Vol. II, p. 61), it attains moderate size and supplies white, coarse-textured timber of medium density (sp. gr. 0.57 to 0.63) which is utilized locally for interior construction and boxes. The leaves and bark are employed medicinally. There are no wood samples available for this study.

Common names: Caroba de flôr verde, carobeira, cinco chagas, ipê branco, i. de flôr verde, i. mèrim, i. pardo, jacarandá (Braz.); lapacho blanco (Arg.); yangua (Peru).

Cybistax Donnell-Smithii (Rose) Seibert (= *Tabebuia Donnell-Smithii* Rose) is a large tree with a long smooth trunk often four feet in diameter, occurring in southern Mexico, Guatemala, Salvador, and Honduras (Map 4). The large, digitately compound leaves are deciduous; the yellow flowers appear before the new leaves and are borne in great pyramidal panicles, and the trees, "standing out against the sky like golden clouds," are considered the most beautiful in the region. The timber has been in the export trade for many years and was formerly known in the furniture industry by the incorrect designation of White Mahogany, but now by its proper name, Primavera. The wood finishes to resemble Ceylon Satinwood (*Chloroxylon*) and its principal use is in the form of veneers for cabinet work. The timber can be had in large logs, usually free of defects, but the annual imports into the United States are not very great.

Color yellowish white to light yellowish brown throughout, often more or less striped. Luster fairly high. Odorless and tasteless. Light in weight, but firm; sp. gr. (air-dry) 0.45 to 0.55; weight 28 to 34 lbs. per cu. ft.; texture medium to rather coarse; grain straight to finely and attractively roey; easy to work, finishing smoothly; does not check badly in drying; holds its place well when manufactured; is poorly resistant to decay.

Common names: Primavera (Trade); duranga, palo blanco, primavera (Mex.); copal, palo blanco (Guat.); cortez, c. blanco (Salv.); San Juan (Hond.).

Cybistax chrysea (Blake) Seibert (= *Tabebuia chrysea* Blake) is a common deciduous tree, up to 60, rarely to 80, feet tall, with a large crown supported by a stout, erect trunk sometimes three feet in diameter, apparently restricted in distribution to northern Colombia and northwestern Venezuela. The bark is rough and gray; the leaves are large, digitate, with five grayish green, rough, crenate leaflets covered with stellate pubescence; the profuse golden yellow flowers are borne in dense terminal racemes just before the advent of the new leaves; the capsules are 12 to 20 inches long and contain hundreds of white, flat, winged seeds. The timber is of good quality and according to L. Williams (*Maderas económicas de Venezuela,* p. 89) is used for cabinet work, carpentry, railway crossties, and naval construction. It is worthy of consideration for the same purposes as Primavera and as a substitute for the brownish shades of Satinwood.

Heartwood brownish yellow, with a golden luster; rather sharply demarcated from the thick whitish sapwood. Odorless and tasteless. Moderately hard and heavy; sp. gr. (air-dry) 0.65 to 0.70; weight 40 to 44 lbs. per cu. ft.; texture uniform, medium; grain finely roey; does not check badly in drying; easy to work, finishing smoothly and taking a lustrous polish. Suitable for fine furniture.

Common names: Roble, r. amarillo (Col.); araguán, cañada, penda (Venez.).

Ekmanianthe is a genus proposed by Urban in 1924 for two species of West Indian trees segregated from *Tecoma. E. longiflora* (Gris.) Urb. grows in the rocky uplands of central Cuba and in Haiti; *E. actinophylla* (Gris.) Urb. occurs in western Cuba, where it is known as Roble Caimán. The timber is of good quality, but appears to be scarce.

Heartwood brown, with lighter and darker stripes appearing as seasonal growth layers; has a slightly waxy appearance; sapwood not seen. Superficially dull, but with golden luster below. Tasteless, but with mild odor. Very hard, heavy, tough, and strong; texture medium; grain somewhat roey; not very difficult to work, taking a high polish; is probably durable under exposure.

Common names: Roble caimán, r. real (Cuba); chêne a glandes (Haiti).

Enallagma, with several species of small trees or shrubs, occurs in southern Florida, the West Indies, southern Mexico, Central America, and northern South America. The genus is closely related to *Crescentia,* but the simple leaves are of a different shape and are alternate instead of fascicled; the flowers are large, with long peduncles, and borne singly or in small clusters, terminal or axillary; the fruit is large and gourd-like. The most widely distributed species is *E. latifolia* (Mill.) Small, a tree sometimes 35 feet tall, common on lowlands, especially in tidal swamps. The tough and strong wood is used to a very limited extent in making ox yokes, plows, and handles.

Wood not distinctively colored when fresh, becoming pale brownish with an orange hue upon exposure. Luster medium. Odorless and tasteless. Moderately hard and heavy; texture coarse, at least in appearance; grain variable; easy to work; poorly resistant to decay. Of no commercial possibilities.

Common names: Black calabash (Florida, B.W.I.); güira, g. boba, g. de olor, magüira, masgüira (Cuba); higuerillo (P.R.); calebasse zombe (Haiti); calabash (Trin.); higuerillo, huiro de montaña, jicarillo (Mex.); calabash, wild calabash (Br. H.); morito del río (Guat.); cacao silvestre, gsií-krá (C.R.); tutumillo, tutumito (Pan.); camuro, taparito (Venez.).

Godmania, with two species of small trees, is distributed from southern Mexico to the Amazon region of Brazil. The common species is *G. aesculifolia* (H.B.K.) Standl. (= *G. macrocarpa* Hemsl.), a tree rarely over 25 feet high occurring from British Guiana to Mexico. The leaves are opposite, long-stalked, digitately compound with 5 to 9 leaflets; the rather small, variously colored flowers are borne in dense corymbs; the pods are very long, slender, and spirally curved and contain numerous flat seeds with a long thin wing at each end. The tree is not utilized for any special purposes.

Heartwood dull yellow-brown, becoming darker upon exposure; sharply demarcated from the lustrous pale brownish sapwood. Without distinctive odor, but with mild spicy taste (inner bark very bitter). Heartwood moderately hard and heavy, sapwood considerably lighter; texture medium; grain irregular to roey; easy to work, finishing smoothly; appears highly resistant to decay.

Common names: Cacho de toro, hoco, joco, roble (Mex.); palo blanco, señorita (Guat.); cortéz blanco (Salv.); corteza de chivo (C.R.); araguaney de sabana, cacho de chivo, c. de venado, cornicabro, cuerno de cabra, sabanero (Venez.).

Jacaranda, with numerous species of shrubs and small to large trees, is widely distributed in tropical America, though most of the representatives are Brazilian. The leaves are opposite, typically bipinnate with numerous small leaflets; the large blue or violet flowers are commonly borne in terminal panicles; the fruit is a rounded and compressed woody capsule with winged seeds. The plants are highly ornamental because of their foliage and flowers and a few of them supply useful timber. *Jacaranda coerulea* (L.) Gris. attains a maximum height of 40 feet in the Bahamas, where it is known as Boxwood, Cancer Tree, and What O'clock. It is also reported from Martinique, and from Cuba where it is called Abey and Abey Macho. The timber is yellowish white, hard, moderately heavy (sp. gr. 0.72), fine-textured, and easy to work, and is used for carpentry in rural districts. Other species producing woods of this type are *J. arborea* Urb., *J. mimosifolia* D. Don, and *J. rhombifolia* G. F. W. Mey.

The name Caroba is applied to various species of *Jacaranda* in Brazil and Argentina, notably *J. caroba* DC. and *J. semiserrata* Cham. They are said to be trees 30 to 60 feet tall and 16 to 30 inches in diameter, supplying some lumber for carpentry, though the principal uses are for fuel and charcoal. The woods are yellowish white or grayish, with fine brown parenchyma and vessel lines producing a distinct pattern on the tangential surface. They are rather light and soft (sp. gr. about 0.55), not very strong, brittle, straight-grained, fine-textured, easy to work, and finishing smoothly. A distinguishing feature of this group is that the rays, which are 1 to 3 cells wide, are definitely heterogeneous.

The best known and most widely distributed species is *Jacaranda copaia* (Aubl.) D. Don; it is a tall tree with pinnate leaves sometimes nearly five feet long and is very showy when covered with abundant panicles of blue or purplish flowers. It is distributed from British Honduras to Brazil. Huber (*Bol. Mus. Goeldi* 6: 202) says that it occurs in the upland forests of the Amazon region, where it is known as Caroba do Matto and Parapará, and supplies a soft white wood often mistaken for Marupá (*Simaruba*). According to Heyder (*Tropical Woods* 3: 7), the tree is common in the mixed hardwood forests throughout British Guiana, making its best growth on the coastal reefs and along the banks of the Essequibo, Demerara, Berbice, and Corentyne rivers, particularly near their mouths. A total height of 80 feet is common and is often exceeded. The distance from the root to the first large branch varies from 18 to 48 feet, occasionally more. The diameters range from 6 to 30 inches. Some of the tallest trees are often very slender, while those with the largest diameter may be comparatively short. The supply of the timber is abundant and the growth is rapid. Owing to the lightness of the wood the logs are used by the Indians and others to add buoyancy to rafts of heavier timber. Other uses are for making corials, or buck shells, cheap coffins, match sticks or splints, boxes, and houses. The wood is perishable in contact with the ground or exposed to the weather, but is suitable for all sorts of interior work where cheap material is required. A suggested use is paper pulp.

The following description applies particularly to the woods of *Jacaranda copaia* and *J. rhombifolia.* Color oatmeal to dingy white throughout, with prominent brown vessel lines. Luster rather high. Odorless and tasteless. Light in weight but firm; sp. gr. (air-dry) 0.40 to 0.50; weight 25 to 31 lbs. per cu. ft.; texture medium to coarse, uniform; grain straight; very easily worked, but saws woolly when fresh; finishes smoothly; holds nails firmly; perishable in contact with the ground.

COMMON NAMES: Boxwood, cancer tree, what o'clock (Bah.); abey, a. macho (Cuba); l'abbe blanc, l'a. franc (Haiti); samarapa (Br. H.); gallinazo (C.R.); palo de buba (Pan.); caballitos, caco, chingalí, gualanday, pavito (Col.); abey, amuscu, árbol de roseto, chingalí, cupay, guarupa, nogal blanco, saca-candela (Venez.); fotui, futi, phootee, sand trysil (Br. G.); alieskie-ie, a. wewe, diamilikie, fettejie, foete-ie, footee, futi, gobaia, jaivie, jawie, jassie-hoehoe, jessie-noedol, kabana, kandra-hoedoe, koepaja, kopaia, koroballi, maja-ariran, mari-mari, tjoekoenda, totui (Sur.); bois á pian, copaia, c. des chantiers, faux simarouba, onguent-pian (Fr. G.); barbatimão, camboatá pequena, camboté, caraúba, caroba, c. do campo, c. do carrasco, c. do matto, c. miúda, c. preta, c. roxa, carobeira, carobinho, c. do campo, c. do matto, c. guassú, carobossú, casco de cavalho, jacarandá caroba, j. da serra, j. mimosa, j. preto, marupaúba, parapará, p. guassú, simaruba copaia, s. falsa (Braz.); caroba, jacarandá, nazare, tarco (Arg.); amchiponga, ishpingo, ishtápi, paravisco, solimán de monte (Peru).

Macrocatalpa includes a few closely related species of small to medium-sized evergreen trees occurring in the West Indies. The group is generally included as a section of *Catalpa,* but Britton (*Journ. N.Y. Bot. Gard.* 19: 8. 1918) considers it a distinct genus and his classification is convenient for use in describing the woods.

Macrocatalpa punctata (Gris.) Britt. occurs in the Bahamas, where it is sometimes 35 feet tall, and in Cuba, where it is common in lowlands and river swamps, but is usually not over 25 feet high. It has highly fragrant yellow flowers, long, slender pods, and linear seeds. The lead-colored wood has a distinctive odor when fresh and is used to a minor extent for small structural work. *M. longissima* (Jacq.) Britt. also occurs in

Cuba, but is at its best in Haiti where the heavy, easily worked, durable lumber is highly esteemed locally for carpentry, flooring, and furniture. The logs available are 12 to 18 inches in diameter and the supply is too limited to permit development of an export trade. The species attains fairly large size in southern Jamaica and supplies one of the most useful native timbers for boatbuilding and general construction.

Common names: French oak, Jamaica oak, mast wood, yokewood (Jam.); roble de olor (Cuba); roble (Dom. R.); bois chêne, chêne (Haiti); chêne de Antilles (Mart.); randegonde (Guad.).

Paratecoma *peroba* (Record) Kuhlm., the only species, is a large and important forest tree of eastern Brazil (see Bol. 4, Serviço Florestal do Brasil, Rio de Janeiro, 1931). It attains a maximum height of about 130 feet, with a slender symmetrical bole about 95 feet long and 60 inches in diameter and covered with a distinctly ridged, yellowish bark. The leaves are opposite, digitately compound with 3 to 7, mostly 5, serrate leaflets; the long-tubed flowers are borne in terminal panicles; the fruit is a flattened woody capsule containing numerous flat, entire-winged seeds. The timber is one of the most important in the market of Rio de Janeiro and is known as a kind of Peroba, a name usually applied to species of *Aspidosperma*. It is used for interior trim in the better class of houses, banks, and stores and for making high-grade furniture. Small shipments have been made to New York and a cabinet-maker there says of it: "Peroba is a yellowish, moderately hard wood obtained in logs sometimes 30 inches through. It takes stain fairly well, is of about the color of Greenheart, and has a texture suggesting Santo Domingo Satinwood. If not properly dried it is likely to check when exposed to warm temperatures. Experienced workmen are careful to avoid the splinters which they believe to be poisonous."

Heartwood light olive, with a yellowish, greenish, or reddish hue, sometimes indistinctly striped; sharply demarcated from the white or yellowish sapwood. Fairly lustrous. Odor and taste not distinctive. Moderately hard and heavy; sp. gr. (air-dry) 0.70 to 0.83; weight 43 to 52 lbs. per. cu. ft.; texture medium; grain straight to roey or curly; not difficult to work, finishing smoothly and attractively; is highly durable. An excellent wood with export possibilities.

Common names: Ipê peroba, peroba, p. amarella, p. branca, p. do campo, p. manchada, p. reseca, p. tigrinha, p. tremida, p. verdadeira (Braz.).

Parmentiera, with five described species of shrubs or trees sometimes 40 feet high, is apparently limited to southern Mexico and Central America. The branches are often armed with short, incurved spines; the alternate or subopposite leaves are typically trifoliolate; the large greenish flowers are borne singly or in clusters along the main stem and larger branches; the fruit is gourdlike. There are two Mexican species, the better-known being *P. edulis* DC., often cultivated there and in Central America for its short, thick, yellowish green fruit which is sweet and edible either raw or cooked. There are three species in southern Central America. *P. cereifera* Seem. is of rather frequent occurrence in the forests on the Atlantic side of Panama, and suggests a very large shrub as it is usually branched almost to the base. The pendant, smooth and yellowish fruits look like long wax candles and have an apple-like odor; they are said to provide excellent feed for cattle. So far as known, the woods of the several species of *Parmentiera* are not utilized for any special purposes.

Sapwood pale yellowish brown with faint greenish tinge; heartwood not seen. Fairly lustrous. Odorless and tasteless. Hard, heavy, tough, and strong; texture fine; grain irregular; not difficult to work; durability presumably low. Of no commercial possibilities.

Common names: Cat, catcuuk, chote, cuachilote, cuajilote, guajilote, gueto-xiga, juajilote, kaat, kat, pepina, xkat-cunc (Mex.); cow okra (Br. H.); cuajilote (Cent. Am.); candle tree, palo de velas, wild calabash (Pan.).

Tabebuia is a large and important genus of trees of general distribution throughout the whole of tropical America. Some botanists include in it only the comparatively few species having simple leaves, placing those with trifoliolate and quinquefoliolate leaves in a different genus, *Tecoma,* but this is not a natural division and separates otherwise closely related forms. Moreover there are intermediate forms. Sprague and Sandwith (*Kew Bulletin,* 1932, p. 18) say: "Rehder (*Mitteil. Deutsch. Dendr. Ges.* 1913, 262) accepted the genus *Tabebuia* in the same sense as Bentham and Hooker, and showed that the genus *Tecoma* Juss. (1789), which originally covered species of three genera, namely, *Stenolobium* D. Don, *Campsis* Lour., and *Tabebuia* Gomes, should properly be applied to *Stenolobium.*"

Tabebuia, in the sense used here, includes species with simple, unifoliolate, and digitately compound leaves, while *Tecoma* (= *Stenolobium*) comprises species with pinnate leaves. The material at hand, though abundant, is not sufficient for a comprehensive systematic treatment, and the classification followed has an economic rather than a botanical basis. The herbarium vouchers for the woods have been critically examined by R. J. Seibert.

The designations and types of the groups treated are as follows: (1) White Cedar, *Tabebuia insignis,* var. *monophylla,* a pale brownish, medium-weight timber of the Guianas; (2) Roble, *T. pentaphylla,* a brownish, moderately hard wood resembling plain-sawn Oak, occurring from the West Indies and Mexico to northern South America; (3) Lapacho, a large, widely distributed group of species with olive-brown, more or less oily, very dense timbers containing an abundance of sulphur-like deposits (lapachol compound); (4) miscellaneous.

White Cedar group.—Apparently there are two species and one variety of South American trees involved in this group. *Tabebuia aquatalis* (E. Mey.) Spr. & Sandw. is said by Sandwith (Pulle's *Flora of Suriname* 4: 2: 70) to be a small tree or a shrub distributed from the Guianas to Pará and Maranhão, Brazil. According to Bertin (*Les bois de la Guyane française,* p. 73), the species, known as Bois Blanchet, attains large size, upward of three feet in diameter and over 100 feet in height, on inundated savannas in French Guiana. The leaves are alternate, digitate, with five leaflets; the white flowers are borne in dichotomous cymes on the old branches; the capsule is cylindrical, 2-valved, and contains numerous seeds with oblong wings. Since Sandwith (*loc. cit.*) says the seeds of this species are "almost sub-orbicular, wholly coriaceous," it may be that the Bois Blanchet is *Tabebuia insignis* (Miq.) Sandw. The wood (Yale 12735; Bertin 3011) closely resembles the White Cedar of British Guiana.

There are no authentic specimens of *Tabebuia insignis* in the Yale collections. The tree is said to be of small to medium size in swamp forests and wet savannas of the Amazonian region of Venezuela, the Guianas, and Brazil. The leaves are 3–5-foliolate; the flowers are white; the shortly linear capsule is densely lepidote; the seeds are transversely oblong, with broad membranous wings.

The tree commonly known as White Cedar in British Guiana, probably because of the resemblance of the bark to that of *Cedrela,* differs from *Tabebuia insignis* only in having constantly a single large leaflet articulated to the petiole. It is sometimes considered a distinct species (*T. longipes,* Baker), but more commonly now as a variety, namely, *T. insignis,* var. *monophylla* Sandw. The timber appears to be well known locally and is used to a limited extent for general carpentry and interior construction. The published statements that it is highly durable in contact with the ground are incorrect and indicate confusion with some other timber.

Heartwood brownish, with reddish or olive hue, varying in different specimens; distinct but not sharply demarcated from the lighter-colored sapwood. Luster fairly high. Odorless and tasteless. Rather hard and heavy; sp. gr. (air-dry) 0.68 to 0.74; weight 42 to 46 lbs. per cu. ft.; texture medium, uniform; grain fairly straight; not difficult to work, finishing very smoothly and with a glossy polish. A good wood for the same general purposes as Birch (*Betula lutea* L.) which it somewhat resembles. Results of tests at University of Michigan

(Bull. 7, School of Forestry and Conservation, 1938, table 1) on small clear specimens in green condition (in lbs. per sq. in. unless otherwise stated): *Static bending:* Fiber stress at elastic limit, 9700; modulus of rupture, 13,700; modulus of elasticity, 2,300,000. *Endwise compression:* Fiber stress at E.L., 5870; max. crushing strength, 6200; M. of E., 2,610,000. *Hardness* (in lbs.): End., 1310; side, 1230. *Shear:* Max. strength along grain, 1160. *Tension:* Max. strength across grain, 650.

COMMON NAMES: Woracoori, warakuri, warikuri, white cedar (Br. G.); alasoabo johoto, alas-waboe, anago-switie, courali, koepaia, mattoe, panda, pandorana, panta, p. hoedoe, waroekoelie, warokorie (Sur.); bois blanchet, cèdre blanc (Fr. G.).

Roble group.—Roble, a Spanish name for Oak (*Quercus*), is applied to several different kinds of trees, but to none more commonly than to *Tabebuia pentaphylla* (L.) Hemsl., whose wood bears a superficial resemblance to Oak, though lacking conspicuous rays. The species attains large size, inhabits various sites from wet lowlands to dry mountain sides, and occurs from the West Indies and southern Mexico to Venezuela and Ecuador. Other species should probably be included in this group, but their identities are uncertain.

In Puerto Rico, Roble or Roble Blanco is a tree 20 to 60 feet high, of general occurrence throughout the island and said to be most common in the limestone hills. The wood is used for ox yokes, house construction, boat-building, and piling. In Cuba, the Roble Blanco grows best on moist but not stagnantly wet soils in association with the Jucero (*Terminalia*), where it attains a height of 70 to 80 feet and a diameter of 24 to 30 inches, more commonly 14 to 20 inches. The trees are of good timber form, with clear lengths of 30 to 40 feet. The timber is used for ox yokes, carpentry, interior finish, and sheathing. The Roble de Yugo, with similar wood, is *T. leptoneura* Urb.

In Venezuela, *Tabebuia pentaphylla*, generally called Apamate, is a tree of the coastal forests and reaches a height of 80 to 90 feet, with a smooth-barked, symmetrical bole often clear of limbs for 50 feet. The tree is deciduous and bears great masses of rose-colored flowers before the new leaves appear, giving it value for ornamental planting. The fruits are Catalpa-like pods which later carpet the ground with their winged seeds. The timber is common in all the local sawmills and carpenter shops, and provides the native lumber most commonly used for rough structural purposes. It is also employed for flooring and interior finish. It is not very durable in contact with the ground and is liable to attack by insects. Regarding the Roble in Colombia, H. M. Curran says: "Scattered individuals were met with in the forests of Magdalena and it is a common second-growth tree about towns and along the waterways. As a rule it is short-boled, 18 to 20 inches in diameter, with heights under 75 feet and clear lengths not over 20 feet. It seemed to be well known and appreciated by the local carpenters for furniture and interior finish."

Northward the range of *Tabebuia pentaphylla* extends to Oaxaca and Tamaulipas, Mexico. Standley (*Flora of Costa Rica*, p. 1130) says: "This is one of the best known and most useful trees of Central America, common throughout most of the Pacific slope and occurring also on the Atlantic watershed. In beauty it has few rivals. The trees often form pure stands of considerable extent and when covered with their blossoms, during spring months, afford an exquisite display of color, suggestive of that of Japanese Cherries in variation of tints."

In Nayarit, Mexico, the tree is commonly called Amapa, and logs and heavy timbers have been shipped under that name into Nogales, Arizona, for manufacture into flooring, interior trim, paneling, and miscellaneous millwork. Emanuel Fritz (*Tropical Woods* 8: 8) has described the use of the lumber in two homes of moderate cost in Berkeley, California. In one it was employed as flooring in living room, dining room, and vestibule; in the other for an ornamental mantel and a heavy paneled front door, and for the interior door, baseboards, and the window and door trim of the principal rooms.

"Amapa wood is of medium density, light grayish brown in color, and has an attractive appearance. The rays are fine and inconspicuous, but the parenchyma layers give rise to a distinct pattern on the tangential faces of lumber. Sharp, clean edges are readily obtained and the surface planes to a glossy smoothness, except for the fine depressions of the vessel lines. It nails well, although in thin tongue-and-groove flooring it is advisable to pre-bore the nail holes in the tongues. In each of the Berkeley installations the wood is finished in natural color, only filler, white shellac, and wax being used. The effect is bright, but neat and dignified, harmonizing especially well with furnishings of a light color. The writer has finished some specimens with mahogany and oak stains with good results, particularly with moderately dark oak stain. As flooring, Amapa seems to be a good substitute for Oak where a lighter color and finer figure are desired. While the two installations described may not be the forerunners of wider applications of Amapa in the San Francisco Bay region, they, nevertheless, call attention to a comparatively new wood which possesses certain characteristics of color and beauty that should offer architects the means for obtaining certain effects not possible in other woods."

Heartwood pale brown, deepening somewhat upon exposure; finely striped with brown parenchyma lines or markings, conspicuous on tangential surface; not clearly differentiated from the sapwood. Luster low to medium. Odorless when dry, faint and suggesting watermelon when fresh; taste absent or bitter. Moderately light to fairly heavy; range of sp. gr. (air-dry) 0.62 to 0.80; weight 39 to 50 lbs. per cu. ft.; texture medium to rather coarse; grain straight to roey; working properties excellent; seasons without difficulty, holds its place well when manufactured; durability fair.

Common names: Roble (Span. Am., gen.); roble blanco, r. de yugo (Cuba); roble blanco, white cedar (P.R.); poirier, p. du pays (Mart.); white cedar (Grenada); pink pouí (Trin.); amapa, a. rosa, amapola, hokab, macuil, macuilixuatl, maculiz, m. prieto, maquile, palo de rosa, p. yugo, roble blanco, rosa morada (Mex.); Mayflower (Br. H.); mano de león, maqueliz, matilisquate (Guat., Hond.); macuilís, maculigua, maculís, maquiliqua (Salv.); roble blanco, r. de sabana (C.R.); roble del río, r. morado (Col.); apamate, orumo, roble blanco (Venez.).

Lapacho group.—This group is represented in all parts of continental tropical America and some of the Lesser Antilles. Many closely related species have been described, but from the standpoint of their woods the number could be greatly reduced. All are trees, usually of medium to large size with well-formed trunks. The leaves are typically quinquefolioate and deciduous; the flowers are mostly yellow, sometimes pink, red, or violet; the fruit is a long woody capsule with winged seeds. The timber is noted for its great strength and durability. The vessels contain an abundance of yellow powder (lapachol compound) which looks like sulphur but in the presence of alkaline solutions turns deep red. (For references to the properties of lapachol see *Tropical Woods* 48: 47.)

Lapacho is the name commonly applied in northeastern Argentina and adjoining regions to two or three species, of which the most important appears to be *Tabebuia ipe* (Mart.) Standl. The principal commercial operations are along the Paraná River in Misiones and across the river in Paraguay. The tree is one of the tallest in the Misiones forest, reaching a height of 100 to 125 feet, with a cylindrical though slightly buttressed bole, free of limbs for 60 to 70 feet, and often 3 or 4, sometimes 6, feet in diameter (Plate XV). The digitate leaves are 6 to 8 inches in diameter and are deciduous; the flowers, usually of a rose color, are borne in profusion just before the new leaves appear, making the trees a resplendent feature of the landscape. The trees occur gregariously on steep rocky hillsides and are also found in smaller numbers through the lowland forests. The timber reaches the market in the form of square-hewn logs, and between four and five million board feet are used annually for general construction, carpentry, cabinet work, turnery, and vehicles. It is considered one of the best woods for wagon spokes. It is highly resistant to decay and consequently is used for many purposes requiring a strong and durable material. It also yields a purplish dye.

The group is well represented in Brazil

and numerous species have been described. In the southern part of the country the common name is Ipê, with many qualifying terms in reference to the color of the flowers or appearance of the trees and woods. The principal species there is probably *Tabebuia ipe*, but *T. serratifolia* (Vahl) Nicholson is also present as far south as São Paulo and Bolivia. The latter species is most important of all in this group and its range extends northward through the lower Amazon country and the Guianas to Venezuela, Trinidad, and Colombia. It is probably the chief source of the Brazilian timber known as Pau d'Arco. Huber (*Bol. Mus. Goeldi* 6: 201) says that it is one of the tallest trees in the uplands of the Amazon and when, at the close of the dry season, it loses its leaves and covers itself with an infinite number of large golden yellow flowers the crowns look like gigantic bouquets in the midst of the forest. It is sometimes called Pau d'Arco Amarello to distinguish it from the Pau d'Arco Rôxo, *T. ipe*. The timber is used for railway crossties, fence and house posts, bridges, and all sorts of purposes requiring strength, toughness, resilience, and resistance to wear, insects, and decay.

There are several species in northern South America (see *Tropical Woods* 30: 44; 36: 16), but they are typified by *T. serratifolia* and the timber finds the same applications as in Brazil. Limited quantities are exported from Surinam to the United States for the manufacture of fishing rods; the usual trade names are Surinam Greenheart (not to be confused with Demerara Greenheart from British Guiana) and Bethabara (see *Tropical Woods* 19: 7). The trees attain a height of 120 feet, with trunks large enough to square 30 inches free of sapwood, but the largest logs are usually too heavy for the primitive methods of extraction.

The principal Central American species is *Tabebuia guayacan* (Seem.) Hemsl., a small to large, yellow-flowered tree having a range extending from Colombia to southern Mexico. In Panama it is called Guayacán, a name also given to Lignum-vitae (*Guaiacum*), and the timber is considered one of the best available for durable construction. During World War I some attempts were made to use Panama Guayacán for propeller-shaft bearings in steamships, but the wood, though dense and strong enough, lacks the oily resin which makes genuine Lignum-vitae self-lubricating and is not a satisfactory substitute for that timber. The northern form with pink or rose-colored flowers is *T. Palmeri* Rose, a medium-sized tree growing from southwestern Mexico to Panama. According to J. G. Ortega (see *Tropical Woods* 7: 37), it is found in all parts of Sinaloa at altitudes of 30 to 1300 feet and requires more than 100 years to reach maturity. The timber, which is obtainable in pieces 16 to 20 feet long and 16 inches square, is highly valued for house posts, beams, sills, door and window frames, and also for railway crossties, fence posts, fuel, and charcoal. The yellow dust arising in milling operations produces reddish stains on sweat-moistened parts of the laborers' clothing and also gives rise to a mild form of dermatitis. The usual Mexican name for this species is Amapa Prieta.

There are two species, namely, *Tabebuia serratifolia* and *T. rufescens* J. R. Johnston, in Trinidad, where they are known as Poui; the former, called Yellow Poui, is generally distributed, thriving on pure sands, but preferring the tops and sides of ridges and avoiding swampy areas. The other species, called Black Poui because of its very dark heartwood, is a smaller tree preferring heavier, clay soil; it is readily distinguished by its rough leaves. This species also grows in Tobago, where it is called Cogwood, and in Grenada, where it is known as Poui and Greenheart.

Heartwood olive-brown to blackish, often with lighter or darker striping; sometimes rather oily-looking; sawed surface often covered with yellow powder; sapwood whitish, yellowish, or pinkish, rather sharply demarcated. Luster low to medium. Odor and taste not distinctive; very to extremely hard, heavy, tough, and strong; sp. gr. (air-dry) 0.95 to 1.25; weight 59 to 78 lbs. per cu. ft.; texture fine to medium, uniform; grain straight to very irregular; rather difficult to work, inclined to be splintery, but takes a very smooth finish; is highly durable.

Common names: Arcwood, bastard lignum-vitae, bethabara, noibwood, Surinam greenheart, washiba, yellow guayacan (U.S. trade); greenheart, poui (Grenada); cogwood (Tobago); poui (black, white, yellow), p. vert, pui (Trin.); ahan-ché, amapa, a. prieta, hahuache, roble, r. cinero, verdecillo, xha-hua-ché (Mex.); cortez colorado (Guat.); cortez (Hond.); cortez, c. amarillo, c. coyote, c. negro, c. prieto (Salv.); aoka, cortez (Nic.); cortez de venado, corteza, c. amarilla, guayacán (C.R.); guayacán (Pan.); alumbre, arco, canaguate, c. amarillo, c. morado, chicalá, coralibe, c. de arco, curarí, curaride, guayacán, g. polvillo, lumbre, polvillo, roble, r. amarillo (Col.); acapro, alcapro, arabone, araguaney, aravaney, canãda, curarí, echahumo, flor amarillo, penda, pui, p. araguaney, p. chiripe, p. negro, puy (Venez.); arawnig-yek, bowwood, hackia, hackooya, hakkea, washiba, whoua-whoua (Br. G.); akkeja, akkekeja, alahorré, ala-oné, ala-onni, arowoné, arraoné, courali, enbotta-koenatjepie, gienhatti, grienharti, groenhart, maka-grien, makka groenhart, mangienhatti, wasieba, wassiba, wehete, woite (Sur.); arahoni, arrhonée, bois d'evilasse, d. d'ébène verte, ébène jaune, e. verte, guiraparíba (Fr. G.); madera negra (Ec.); tahuari, t. amarillo, t. negro (Peru); capitary, caraúba do campo, caraubeira do campo, caroba do campo, carobeira, guiraparíba, ipê, i. cascudo, i. jabotia, i. preto, i. tobaco, pau d'arco, p. d. amarello, p. d. rôxo, quiarapaíba, tamurá tuira, tauary, t. do gapo, urupariba (Braz.); lapacho, l. negro, peúva amarella, p. roxa (Par.); lapachillo, lapacho, l. amarillo, l. blanco, l. crespo, l. negro, l. rosa, tally, tayí, t. pichaí, t. pirurú, t. s̃aiyú (Arg.).

Miscellaneous.—Some wood specimens labeled Taipoca, from Bahia, Brazil, bear a superficial resemblance to the Roble group of *Tabebuia,* but the pores are very small and arranged in rather definite wavy bands and diagonal rows, suggesting *Jacaranda* in part. Wood parenchyma is sparingly developed and ripple marks are present.

Tabebuia nodosa Gris. is a small Argentine tree or shrub. The wood is light-colored, only moderately dense, contains no lapachol deposits, but is otherwise similar to those of the Lapacho group. It has no important uses other than for fuel.

Common names: Caspí-cruz, ibirá ti, i. curuzú, palo cruz, p. sinvergüenza, tororataí, uinaj, yaguá-rataí (Arg.).

In the larger islands of the West Indies there are numerous species whose woods differ in many details from those of the other groups described. The specimens examined have been determined as follows: *Tabebuia Cowelli* Britt., *T. dubia* (C. Wr.) Britt., *T. heterophylla* (DC.) Britt., *T. lepidota* (H.B.K.) Britt., *T. Maxonii* Urb., *T. mogotensis* Urb., *T. petrophylla* Greenm., and *T. Schurmanniana* Urb. They are shrubs or small trees with simple or digitately compound leaves and white, pink, red, or purplish flowers. The wood is pale brown or yellowish, sometimes with a faint greenish tinge; moderately hard and heavy; pores with tendency to diagonal or tangential arrangement; wood parenchyma abundant, aliform to confluent into concentric bands. There are no important special uses.

Common names: Cucharillo, ébano blanco, roble caimán, r. de costa, r. de playa, r. de sierra, r. macho, r. negro, r. sabanero, r. yanillo, roblecillo, rompe ropa (Cuba); roble blanco, r. colorado, r. de mona, r. de sierra, r. prieto (P.R.); bois nago, b. savane (Haiti).

Tabebuia barbata (E. Mey.) Sandw. (= *Couralia toxophora* [Mart.] C. & H.) is a small to medium-sized tree of fairly common occurrence on lowlands in the Amazon region of Brazil and Venezuela. The large leaves have five smooth, leathery leaflets; the large violet or roseate flowers are borne in terminal clusters; the long, thick, cylindrical capsule contains many large, wholly coriaceous seeds. The olive-brown heartwood shows distinct vessel lines. It is hard and heavy, coarse-textured, straight-grained, not difficult to work, and is probably durable. The presence of lapachol in many of the vessels of the heartwood indicates a relationship to the Lapacho group, but the pores are considerably larger (160μ) and the wood parenchyma is coarse-celled and abundant, aliform, confluent, and terminal; sclerotic cells are common. Ripple marks are uniform, about 120 per inch, with the parenchyma cells in secondary seriation.

Common names: Capitary, pau d'arco

blanco, p. d'a. rôxo, tauary do igapó (Braz.).

Tabebuia stenocalyx Spr. & Stapf is a large Guiana tree sometimes 115 feet tall with a low-buttressed trunk nearly three feet in diameter. The large, smooth, leathery leaves are simple; the flowers have a long, green calyx tube and a large white corolla. The wood of a single specimen (Yale 35960; A. C. Smith 3497) is pale brownish within, but with a golden sheen on long exposed surface; rather light in weight, but firm and crisp, medium-textured, straight-grained, and very easy to work. The pores are medium-sized (130μ) and joined by rather narrow confluent parenchyma into wavy or uniform concentric series, 1 or 2 pores wide and 3 to 6 pore-widths apart. The vessels do not contain lapachol. The rays, which are 1 to 3 cells wide and up to 30 cells high, are homogeneous and not storied; very small cubical crystals are common. The wood fibers have rather thin walls and are septate in part. Ripple marks are absent.

Tecoma. This name is often used in the sense of *Tabebuia,* but is here employed in place of *Stenolobium* (see *Kew Bull. Misc. Inf.* 1932, p. 18). It consists of a few species of shrubs and small trees, of which the best known and most widely distributed is *Tecoma stans* (L.) H.B.K. This is an ornamental plant growing naturally or as an escape from cultivation from southern United States to Argentina. It rarely attains a height of 25 feet; its leaves are opposite and pinnate, with 5 to 13 leaflets; the bright yellow flowers are borne in terminal racemes or panicles; the fruit is a dehiscent linear capsule containing broadly winged seeds. The genus does not supply any timber of value.

Heartwood light brown, somewhat variegated; rather sharply demarcated from the whitish sapwood. Luster medium. Odor not distinctive; taste bitter. Moderately hard and heavy; texture fine; grain fairly straight; not difficult to work, finishing smoothly; durability probably rather poor. Of no commercial possibilities.

COMMON NAMES: Ginger Thomas, trumpet-flower, yellow elder, yellow cedar (B. W.I.); saúco amarillo (Cuba); roble amarillo, ruibarba (P.R.); chevalier (Haiti); borla de San Pedro, candox, corneta amarilla, flor amarilla, f. de San Pedro, gloria, guie-bichi, hierba de San Nicolás, h. de San Pedro, huachacata, ichculili, kanlol, mazorca, miñona, nixtamaxochitl, retamo, San Pedro, sauco amarillo, trompeta, trompetilla, tronador, tulasúchil, xkanlol (Mex.); chakté, chanté timbogue (Guat.); flor de San Andrés, f. de San Sebastián, marchucha, San Andrés, tache, tagualaishte, tasto (Sal.); sardinillo (Hond.); chilca, sardinillo (Nic.); candelillo, carboncillo, vainilla (C.R.); copete (Pan.); caballito, chirlobirlos, fresno, palo hueso, roble amarillo (Col.); flor amarilla, fresnillo (Venez.); garrocha (Urug.); guarán amarillo, g. colorado, guaranguarán, guaranguay amarillo, g. colorado (Arg.).

BIXACEAE

Bixa, with only two distinct species of tropical American trees and shrubs, comprises this family, in its restricted sense. *B. arborea* Huber is a medium-sized to large forest tree of Amazonia, and the Anatto, *B. Orellana* L., is a shrub or small tree, 6 to 20 feet high, widely planted in tropical regions the world over because of the orange dye obtained from the seeds. The leaves are alternate and simple; the flowers are in terminal panicles; the fruit is a 2-valved reniform capsule an inch or more long, usually covered with spine-like bristles, and containing numerous round or flattened seeds. Covering the seeds is a thin, orange-red aril, which, after removal by maceration and washing, is dried and pressed into cakes or rolls for the market. Anatto dye was formerly used by the Indians of the Caribbean region for painting the body, partly for adornment and partly for relief from insects. It is used locally for coloring rice and other articles of food and in the United States and Europe for coloring butter, butter substitutes, cheese, candy, soap, and varnish. The wood has no special uses.

Heartwood yellowish brown or pinkish brown; not sharply demarcated from the sapwood. Luster medium. Without distinctive odor or taste. Light in weight and rather soft; of about the consistency of

Black Willow (*Salix nigra* L.); texture medium; grain fairly straight; easy to work, finishing smoothly; perishable in contact with the ground.

Growth rings absent or poorly defined. Pores numerous; small to medium-sized; near limit of vision; mostly in radial multiples of 2 or 3 pores each, well distributed. Vessels with simple perforatións; without spiral thickenings; pitting fine, alternate. Rays 1 or 2, occasionally 3 or 4, cells wide and usually of one tier-height (not over 15 cells), but sometimes occupying from 2 to 8 tiers; heterogeneous, at least in part; pits to vessels small. Wood parenchyma scarcely distinct with lens; finely reticulate and diffuse; cells thin-walled, generally 4 per strand. Wood fibers with thin walls and few, small, bordered pits. Ripple marks present, though often irregular; all elements storied, though some of the rays may occupy more than one tier. No gum ducts seen.

COMMON NAMES: Anatto, arnatto (Eng.); roucou (Fr.); Orlean-strauch (Germ.); achiote, achote (Sp. Am., gen.); bichet (Carib, women); emátabi (Carib, men); bija, bixa, cachicuto (Cuba); bijo (Dom. R.); achiotl, achiotillo, achotillo, arnato, chancanguarica, changuarica, kuxub, pumacua, urucú (Mex.); atta (Br. H.); xayan (Guat.); anatto (Hond.); cuajachote (Salv.); katsá, krikrá, so, songuó (Pan.); achihuiti, bija, color, onote (Col.); bija, bijo, caituco, onoto (Venez.); bosch koesoewé, koesoewee, koesoeweran, roekoe, schirabaeli, toenataletano koesoewé, urucú (Sur.); urucú, u. bravo, u. da matta, urucurana, u. da matta, urucú-uba, urucuy (Braz.); achihuiti, achiote colorado, achite amarillo, huantura, sacha achiote, shambú, s. huayo, s. quiro, s. shambú, urcú (Peru); achiote, urucú (Par.).

BOMBACACEAE

THIS pantropical family comprises about 25 genera and 150 species, mostly trees, often of giant stature and sometimes with bulging stems. The leaves are alternate, simple or digitate, with deciduous stipules; the flowers are large and showy; the fruit is a dehiscent or indehiscent capsule. The American genera may be grouped as follows: BOMBACINEAE: *Bombax, Bombacopsis, Pachira, Ceiba, Ochroma, Cavanillesia, Hampea,* and *Chorisia;* GYRANTHEREAE: *Gyranthera, Bernoullia,* and *Huberodendron;* MATISIEAE: *Matisia* and *Quararibea;* CATOSTEMMATEAE: *Catostemma, Scleronema,* and *Aguiaria.* Several genera have their seeds imbedded in a woolly or silky fiber which is utilized for stuffing pillows, cushions, and mattresses, and for insulating refrigerators; the commercial product known as kapok is obtained from plantations of *Ceiba pentandra* Gaertn. The inner bark is fibrous and frequently serves for making cordage and for clarifying sugar. The only regular timber of commerce is Balsa (*Ochroma*) and it is obtained from plantations; some of the other genera, however, particularly *Bombacopsis* and *Ceiba,* supply timber of local utility and occasional export.

The woods exhibit a wide range of variation in appearance and properties, though this family is no more heterogeneous than the other Malvales. Color typically light throughout, being white, yellowish, pinkish, brownish, or oatmeal; heartwood reddish brown in *Bombacopsis, Catostemma,* and *Aguiaria.* Luster generally low, sometimes high. Without distinctive odor and taste, except in some specimens of *Quararibea.* Sp. gr. 0.10 (*Cavanillesia*) to 1.14 (*Aguiaria*); weight 6 to 71 lbs. per cu. ft.; Bombacineae all light and soft, Catostemmateae hard and heavy, the others intermediate. Texture medium to coarse; feel velvety (particularly in *Ochroma*) to rather harsh; grain mostly straight; working properties variable; durability low, except in deeply colored material.

Pores small to very small, mostly indistinct without lens, in *Catostemma, Hampea, Quararibea,* and *Matisia;* larger, visible to very distinct, in the others; few to rather numerous, occurring singly or in short rows, generally well distributed. Vessels with simple perforations; without spirals; pits alternate, large in *Chorisia,* small in *Catostemma, Scleronema,* and *Hampea,* minute in *Matisia* and *Quararibea,* medium-sized in the others; tyloses often present, sometimes abundant. Rays heterogeneous; of two sizes, the larger frequently up to 6, sometimes up to 8 or 10, rarely more than 15, cells wide and of various heights up to 150 or more; large rays often very conspicuous, especially on radial surface, producing attractive silver grain on quarter-sawed lumber; sheath cells present in most genera, but absent or rare in *Gyranthera, Huberodendron, Scleronema,* and *Aguiaria;* low tile cells absent; *Pterospermum*-type of tile cells (see *New Phytologist* 32: 4: 262) present in *Ochroma* and *Hampea;* crystals common; pits to vessels large to very large except in *Hampea, Matisia,* and *Quararibea,* where they are small to minute. Wood parenchyma abundant; finely reticulate, at least in part, except in the Catostemmateae, where it is in distinct bands; cells in horizontal seriation in *Bombax, Bombacopsis, Pachira, Ceiba, Cavanillesia, Hampea,* and *Chorisia.* Wood fibers with thin to very thick walls; septate in *Bombax, Bombacopsis, Pachira,* and *Hampea;* pits simple or indistinctly bordered. Ripple marks present in all genera except *Ochroma, Huberodendron, Matisia,* and *Quararibea;* distinct to indistinct. Vertical traumatic gum ducts seen or reported to occur in *Bombax, Bombacopsis, Pachira, Ochroma, Cavanillesia, Catostemma,* and *Scleronema.*

Aguiaria *excelsa* Ducke, the only species, is one of the tallest trees in the region of the upper Rio Negro, a northern tributary of the Amazon, attaining a maximum height of 160 feet. The timber is used locally for construction purposes.

Heartwood rich reddish brown, with light-colored parenchyma markings; has a waxy appearance and feel; sapwood brownish, sharply demarcated. Odorless and tasteless. Very hard and heavy; sp. gr. (air-dry) 1.14; weight 71 lbs. per cu. ft.; rather coarse-textured; not difficult to work, finishing smoothly and with a high natural polish; durability rather high.

Common names: Duracá, duraque (Braz.).

Bernoullia *flammea* Oliv., the only species, is a rare tree having the massive proportions and general appearance of *Ceiba pentandra.* It is known to occur at elevations between 500 and 2500 feet in southern Mexico, British Honduras, Guatemala, Honduras, and Panama. A wood specimen (Yale 2187) apparently identical with those of this species from Central America was collected by Georges H. Barrel in the Peruvian Amazon region, the common name being given as Lupuna. The leaves of *Bernoullia* are digitately compound, usually with 5 or 6 leaflets, the inflorescence is scarlet, the fruit is a 5-valved woody capsule, 10 to 12 inches long, containing winged seeds suggesting those of Mahogany (*Swietenia*).

Wood dull, oatmeal-colored or brown, the rays showing prominently as dark flakes on radial surface and giving a fine, lace-like appearance to the tangential. Very light in weight, but rather firm when dry; very coarse-textured; does not finish smoothly; is perishable in contact with the ground.

Common names: Palo calabaza, p. de perdiz (Mex.); mapola, red mapola (Br. H.); yuca (Pan.).

Bombax, Bombacopsis, and **Pachira** are closely related genera, treated by some botanists as a single genus (*Bombax*), with a total of about 50 species of small to large tropical trees, mostly American. *Bombax,* in the restricted sense, has small, pea-like seeds imbedded in silky wool or kapok, the flowers are large and the stamen fascicles divide into a thousand or more simple filaments, which, with the short, thick stamen tube, give the appearance of a powder puff. *Bombacopsis* differs in having smaller flowers and an elongated, slender stamen tube and from 75 to 200 stamens. *Pachira* has large, chestnut-like seeds not inclosed in wool, the flowers are large, and the stamen fascicles are repeatedly branched.

Woods of *Bombax* and *Pachira* much alike, being dingy brown or grayish brown throughout, like *Ceiba;* light and soft; sp. gr. (air-dry) 0.15 to 0.40; weight 10 to 25 lbs. per cu. ft.; coarse-textured, tough and stringy, perishable, and poorly esteemed for timber. The wood of *Pachira* is somewhat darker, the texture is harsher, the density is greater, and the pores are usually larger and more numerous than that of *Bombax.* The early wood of *Bombax* is generally much softer than the late wood, whereas in *Pachira* the consistency is fairly uniform.

Common names: *Bombax:* Drago (Cuba); fromagier (Haiti); amapola, a. blanca, a. colorada, cabellos de ángel, ceiba, chacknyché, chichochuchi, chilochuchi, clavellina, c. de la barranca, coquito, cuajilote, disciplina, guajilote, itztamatl, jiquique, jumallo, kuyché, lele, pochote, pochotl, pongolote, tiati, titilamatl, xanocol, xcunché, xiloxochitl, xihuicxan, zackuyché (Mex.); mapola (Br. H.); doncella, muñeco (Guat.); jilinsuche, pilinsuchil, shilo, s. blanco, s. colorado (Salv.); brisakrá, kurí, pochote, psikrá, purí, zgun (C.R.); barrigón (Pan.); barrigón, ceiba de agua, c. de majagua, majagua, m. colorada (Col.); cachimbo, majagua, majumba, sibucara, tambor (Venez.); agonzé, boesiekatoen, boschkatoen, ieljoe-loetanokri, jacomini, kamakotie, katoen, kirikiri-maroro, klieklie-maloeloe, koenanaballi, konnanaballi, krie-kriemaroeroe, krikri-wosijono, kri-miauloeloe, mankatoen, mattoe-maauw, momo, parakatoen, sabana-katoen, seesabana-katoen, sienzon wekelau-manloeloe, wladilikoro, wosijono (Sur.); embirassú, embiratanha, imbira de folho, i. guassú, imbirassú, imbirussú, mamorana grande, mangubeira, monguba or munguba, m. rana, paineira, sumaúma de terra firme (Braz.); palo blanco (Par.); beldaco (Ec.); bellaco-caspi, huimba, huina-caspi, punga, p. blanca de chamisal, pungu (Peru). *Pachira:* Carolina, c. blanca, castaño silvestre, ceibón, c. de agua, c. de arroyo, c. silvestre (Cuba); colorade (Haiti); wild chataigne (Trin.); apombo, kuyche, zapote bobo, z. de agua, z. reventador, z. reventón (Mex.); provision tree, Santo Domingo (Br. H.); sunzapote (Guat.); pumpunjuche, zapatón (Hond.); shila blanca (Salv.); jelinjoche (C. R.); mamé de mono (Pan.); ceiba de agua, mahagua, majagua (Col.); castaño (Venez.); bosch-cacao, kaneriballi, momow (Sur.); châtaignier (Fr. G.); mamorana, m. grande (Braz.).

Bombacopsis, with three or four species of large trees, is apparently limited in its distribution to Venezuela, Colombia, and Central America as far north as Honduras. The trees are of two general types, one with smooth bark, the other very spiny. Best known of the latter group is *Bombacopsis quinatum* (Jacq.) Dugand (= *B. Fendleri* [Seem.] Pittier = *Pachira Fendleri* Seem. = *Bombax Fendleri* Hemsl. = *Bombax quinatum* Jacq.). Hugh M. Curran supplies the following information concerning it in Colombia, where it is called Tolú: "This is one of the very common timbers on the lower slopes in the northern part of the country, occupying the drier gravelly areas in a region of plentiful rainfall. It is associated with other trees of the Bombacaceae and also with *Cedrela.* It reaches a height of 100 feet, has a rather wide-spreading crown of heavy branches, and a somewhat irregular bole inclined to be buttressed and completely clothed with heavy prickles toward the base. It is deciduous and remains bare of leaves for many weeks. The fruits are dry capsules which upon bursting liberate a quantity of soft brown vegetable wool inclosing the small brown seeds. In the regions where it is the principal species there are five or six of the trees to the acre with a yield sometimes of 4000 to 5000 board feet. It reaches the local market either in the form of squared logs or as whip-sawn lumber and is commonly met with in every carpentry shop of the coastal region. It has a soft texture, works readily with all machinery and tools, and is remarkable for the difficulty with which it gives up moisture, carpenters saying that it never becomes thoroughly dry in that humid climate. It is used chiefly for the manufacture of boxes and as a substitute for Cedar (*Cedrela*), and is frequently referred to as false Cedar. The natives make large dugout canoes from the trunks. The same tree occurs, or at least is reported, in the Lake Maracaibo region of Venezuela, and the lumber is well known in all the local markets under the name of Saqui-saqui."

Pittier states (*Esbozo de las formaciones vegetales de Venezuela,* Caracas, 1920, p. 28) that the wood of the Saqui-saqui (*Bombacopsis sepium* Pitt.), which bears a general resemblance to Spanish Cedar, contains a hygroscopic gum which keeps the material perpetually moist. If the lumber is kept submerged, preferably in running water, it is claimed that this gum will be dissolved out and the quality of the timber thereby greatly improved. The local uses of the wood include general construction and cooperage, especially tanning vats and rum storage vats. The timber has not been exported, but there is a possible market for it for the staves of water conduits. According to Dugand (*Contrib. Hist. Nat. Colombiana*

Photo by A. Dugand.

PLATE XVI. Macondo (*Cavanillesia platanifolia*) during the dry season in northern Colombia.

Photo by Tom Gill.

Plate XVII. Large Ceiba tree (*Ceiba pentandra*) in southern Mexico.

1: 2, March 25, 1938), *B. sepium* is doubtfully distinct from *B. quinatum*.

Regarding Cedro Espinoso (*Bombacopsis quinatum*) on the Bayano River watershed of Panama, H. C. Kluge says (*Tropical Woods* 5: 5): "This well-known tree is distributed very generally over the region, growing very well on the barren hilltops. The trunk is very irregular, with prominent buttresses. It sheds its leaves during January and the flowers appear a few weeks later; the fruit pods ripen and fall before the new leaves. The wood dries with extreme slowness; it is much used by the natives because it is resistant to insect attacks and decay."

Heartwood light to dark reddish brown; sharply demarcated from the yellowish sapwood. Luster rather low. Odorless, but with a slightly astringent taste. Light and soft but firm and strong for its density; sp. gr. (air-dry) 0.40 to 0.60; weight 25 to 37 lbs. per cu. ft.; easy to work, finishing smoothly; some specimens of about as good a quality as *Cedrela*. Vertical traumatic gum ducts occur in some species. Tyloses are abundant. The other anatomical details are virtually the same as in *Pachira*.

COMMON NAMES: Cedro espino (Hond.); aba, cedro espino (Nic.); cedro macho, pochote (C.R.); cedro espinoso, ceiba (Pan., Col.); ceiba, c. colorada, c. tolú, c. tolúa, tolú (Col.); caoba bastardo, cedrillo, cedro dulce, c. colorado, ceiba, c. colorada, jaris, lanillo, masguara, murea, saqui-saqui (Venez.)..

Catostemma, with three species of trees, occurs in the Guianas and northern part of the Brazilian Amazon region. The best known species is *C. fragrans* Benth., which is the same as *Guenetia macrosperma* Sagot. It is a tall tree with a long, slender trunk, sometimes free of branches for 90 feet, in British and French Guiana, but has not been found in Surinam. Another British Guiana species is *C. Alstoni* Sandw., a tree sometimes 100 feet in height and two feet in diameter. The Brazilian species, *C. sclerophyllum* Ducke, is a medium-sized to tall tree rather common in the low inundated forests near Manáos and northward. The timber is sparingly utilized, although of fairly good quality and suitable for general construction.

Heartwood dull yellowish brown, distinct but not sharply demarcated from the yellowish brown sapwood.

COMMON NAMES: Baramalli (Br. G.); flambeau rouge (Fr. G.).

Cavanillesia, with two or three species of large trees, is of limited distribution in Central and South America. *C. arborea* (Willd.) K. Schum., of south-central Brazil, is said to attain a height of 65 feet and a trunk diameter of 12 feet. In the dry regions the bole becomes considerably swollen several feet from the ground, giving rise to the name Barriguda (pot-bellied). The extremely light and weak wood is not utilized

Cavanillesia platanifolia H.B.K. occurs in northwestern Colombia (Plate XVI) and southern Panama. It is a large tree, usually 65 to 80, sometimes over 100, feet tall, with a ponderous trunk six to nine feet through at the base, which frequently is considerably smaller than the flask-shaped part above. The tree sheds its lower branches, even when growing in the open, so that the crown is always small, open, and seemingly out of proportion to the size of the trunk. The bark is thick and fibrous, and smooth except for the old branch scars.

Wood yellowish white to pale brown throughout. Fairly lustrous. Without distinctive odor or taste. Extremely light in weight; sp. gr. (air-dry) 0.10; weight 6¼ lbs. per cu. ft.; very soft and fragile; of uniform texture; with rather harsh feel; perishable.

COMMON NAMES: Bongo, cuipo, hamati, quipo (Pan.); macondo (Col.); barrigudo (Braz.).

Ceiba. There are about 10 species of this genus, but the best known is *Ceiba pentandra* (L.) Gaertn. (= *Bombax pentandrum* L. = *Eriodendron anfractuosum* DC.), of pantropical distribution. It is a massive tree, often 80 to 100 feet high and sometimes much taller, with a cylindrical stem having high, thick plank-buttresses (Plate XVII). According to A. E. Navez (*Proc. Nat. Acad. Sci.* 16: 339), the tabular roots are best developed on the side corresponding to the direction of the strongest prevailing winds and are to be considered as traction-resistant rather than compression-resistant structures, *i.e.*, they serve as cables and not primarily as buttresses. The Ceiba is characteristically a tree of the open, with a large coarse crown which in some instances may have a spread of 150 feet. The branches of young trees are spiny. The seeds are rather widely disseminated and find ideal conditions for germination on abandoned agricultural land.

The commercial importance of the tree is in the floss about the seeds, known on the market as kapok. The following information is taken from *Bulletin of the Imperial Institute* (London) 24: 18, June 1926: "The hairs spring from the inner wall of the capsule and are not attached to the seed itself, thus rendering the separation of the seed much easier than in the case of cotton. The hairs are cylindrical, from 0.6 to 1.2 in. in length, with very thin walls. The cells are full of air and are very light; they also possess the property of being impermeable to moisture, and on this account are extremely buoyant. For this reason kapok is now used throughout the world for the manufacture of buoys, life belts, and life-saving jackets. Its chief use is for stuffing cushions, pillows, mattresses, and similar articles. In a natural condition the fibers lack cohesive force and are unsuitable as a textile material, but by roughening the surface by chemical treatment this difficulty can be overcome, though the yarns have poor wearing properties. Practically the entire commercial supply of kapok is obtained from Java, where the trees are for the most part cultivated as boundary trees and fences, along roadsides, and with other kinds in gardens. The trees begin to bear in three or four years, and 7-year-old trees will yield 350 to 400 pods, 10-year-old trees 600 pods or more. The average yield is about one pound of cleaned floss per hundred pods. In 1924, Java and Madura exported over 15,000 tons of kapok, of a total value exceeding 1½ million pounds sterling. Of this quantity the United States took nearly 8000 tons, the Netherlands and Austria about 3000 tons each."

Heartwood pinkish white to ashy brown, not clearly distinguished from the sapwood. Luster low. Odorless and tasteless. The wood is heavy with moisture when freshly cut but is very light, about 27 pounds per cubic foot, when dry. If seasoned in the log it is almost certain to discolor and is likely to rot. It is little used locally except for dug-out canoes and rafts. During the first half of 1939 about three-quarters of a million feet of logs were exported from Guatemala to Germany for use as a substitute for Douglas Fir in plywood corestock. Other possible uses are slack cooperage, packing cases and boxes, toys, and miscellaneous purposes requiring a soft, easily worked wood. It cannot compete with Balsa for purposes of buoyancy and insulation. It is doubtful if it is suited for matches. Regarding a special use of the trunks of Sumaúma in the Amazon region, Spruce says (*Notes of a botanist in the Amazon and Andes*, London, 1908, p. 186): "The softness and lightness of the wood render it suitable above all other trees for hollowing out the trunk into what are called cuchas or floating casks, which, being filled with turtle oil or capivi on the upper Amazon and securely calked, are floated down to the Barra do Rio Negro or Pará." Some of them had a capacity of over 2000 gallons.

Common names: Ceiba, corkwood, cotton tree, silk-cotton tree (B.W.I.); ceiba, ceibo (Sp. Am., gen.); mapou (Haiti); arbe à coton, fromagier (Fr. W.I.); árbol de algodón, cabellos de ángel, cho, kinim, mosmote, peem, piim, pitón, pochote, pochotl, xiloxochitl, yaxche (Mex.); inup, nup (Guat.); ceibón, paniki, panya, sisín, poxot (Nic.); bonga, ceiba, c. bonga, c. de lana (Col.); ceiba yuca (Venez.); kankantrie, koddobakkoe (Sur.); sumaúma, sumaumeira paineira (Braz.); toborochi (Boliv.).

Chorisia is typically a genus of southern South America, although one species is credited to Colombia. *Chorisia soluta* D. Smith, of the uplands of Guatemala, is a synonym for *Ceiba aesculifolia* (H.B.K.) Britt. & Baker. The Palo Borracho of Argentina, *Chorisia insignis* H.B.K., is a bottle-shaped tree, for that reason sometimes called Palo Botella, rarely found in the forest, but occurring singly or in small groups in the open, where it attains a height of about 50 feet and a diameter of six feet. The trunks are used for making durable dug-out canoes, the bast is good for cordage, and the cottony fiber about the seeds is said to be of sufficient length and strength for textile purposes as well as for stuffing. The Paineira of southern Brazil, *C. speciosa* St. Hil., also occurs in Argentina where it is generally known as Samohú, though that name appears to be shared by other species. From published descriptions the tree is taller, the trunk more slender, and the timber denser than that of *C. insignis*.

Heartwood light yellowish brown, with a reddish tinge. Not highly lustrous; unattractive. Sp. gr. (air-dry) 0.44; weight about 27 lbs. per cu. ft.; firm, tough, and strong; of moderately coarse texture and rather harsh feel; easy to work, finishing smoothly; not resistant to decay. It is suitable for general interior construction, common lumber, and slack cooperage.

COMMON NAMES: Barriguda, paineira (Braz.); lupuna (Peru); algodoneiro, lluchán, mandiyú-rá, painá, paineiro, palo borracho, p. botella, samohú, s. morotí, s. pytá, yuchán (Arg.); zamuhú (Par.); paineira (Urug.).

Gyranthera, with two species of medium-sized to rather large trees, is of limited distribution in Panama (*G. darienensis* Pittier) and Venezuela (*G. caribensis* Pittier). The latter, according to information supplied by Llewelyn Williams, is confined to the central part of the country and is the largest and most characteristic tree of the cloud forest at an altitude of about 1100 feet along the coast range between Maracay and Ocumore de la Costa, State of Arugua, where it attains a maximum height of about 90 feet, with an irregular crown supported by a straight, cylindrical trunk sometimes 36 inches in diameter. So far as known, the timber is not now utilized for any special purpose, but should be useful for general carpentry, boxes and crates, and other purposes requiring a tough and strong material light in weight and easy to work but not resistant to decay. Color rather lustrous yellow-brown, with purplish ray flecks on radial surface. Odorless and tasteless.

COMMON NAMES: Candelo, cucharón, niño (Venez.).

Hampea. Paul C. Standley says (*Journ. Wash. Acad. Sci.* 17: 15: 394): "The genus *Hampea* consists of a small group of American trees and shrubs which has been referred by most authors to the family Bombacaceae, although others have believed its more natural position to be in the Malvaceae. The genus was based by Schlechtendal upon a single species, *H. integerrima*, described from Veracruz. In 1862 Triana and Planchon described a second species, *H. thespesioides*, from Colombia, and in 1886 Sereno Watson a third from Guatemala. A variety of *H. integerrima* was described from Costa Rica in 1899 by Captain John Donnell Smith, and in 1923 I transferred to the genus a Mexican plant described as a *Thespesia* by Presl, and published a new species from the Yucatán Peninsula. Practically all the scanty herbarium material of the genus has been referred without question to the original *H. integerrima*. Our representation of the genus has increased rapidly in recent years and, when an attempt was made recently to name two Central American specimens, it became evident that the group was badly in need of revision. In the present treatment nine species are recognized, one indigenous in Colombia, the others ranging from Panama to southern Mexico."

Wood lustrous yellowish brown or pinkish. Light and soft; rather silky; of medium fine texture and soft feel; not durable. No special use because the trees are too small for timber.

There are two types of *Hampea* woods, although the differences are not pronounced. In specimens of *H. panamensis* Standl. and *H. integerrima* Schl., the color is yellowish white throughout, the feel is velvety, the tier junctions are distinct under lens, and secondary seriation is common, whereas in *H. euryphylla* Standl., *H. stipitata* Wats., and *H. trilobata* Standl. the general color is pinkish, the rays are reddish, the feel is somewhat harsh, the tier junctions are not sharply defined, and secondary seriation is indistinct.

COMMON NAMES: Jonote blanco,. majagua, too-hoob, zacitza (Mex.); broadleaf moho, kajana (Br. H.); majáo colorado (Hond.); buriogre (C.R.); azote, burillo (Pan.).

Huberodendron. There are two Amazonian species, namely, *H. swietenioides* (Gleason) Ducke (= *Bernoullia swietenioides* Ducke) and *B. ingens* Ducke, both very large forest trees, 100 to 175 feet tall, with thick, highly buttressed trunks. The only specimen in the Yale collections (Yale 23672; Ducke 210) is of the first named species (see *Arch. Inst. Biol. Veget.* 2: 1: 59, 72). Wood of a dull, light brown color. Moderately hard and heavy; rather coarse-textured; easy to work, finishing smoothly; probably perishable in contact with the ground.

Matisia, with about a dozen species of small, medium-sized, rarely large, trees, is most abundant in the Amazon basin, but with extensions to Panama and Costa Rica. The tallest tree is the Sapote of Peru and Ecuador, *M. cordata* H. & B., which is sometimes 100 feet high; it occurs in the forest and is also cultivated for its comestible fruits. Its wood is light and soft, one sample having a velvety feel. Uniseriate and biseriate rays are very numerous and the multiseriate are very large, the maximum size being about 25 cells wide and 300 cells high, with tall uniseriate margins; sheath cells abundant, though mostly rounded rather than upright (tang. sect.). Wood parenchyma strands composed mostly of short cells, usually eight per strand; crystals abundant. The other details are about the same as in *Quararibea.* The woods of three other species, namely, *M. dolichosiphon* Schum., *M. obliquifolia* Standl., and *M. ochrocalyx* Schum., are so similar to *Quararibea* that no reliable basis could be found for separating them from that genus. So far as known, the timber of *Matisia* is not utilized for any special purpose.

Common names: Cupúassú-rana, cupú-rana, inajá-rana (Braz.); sapote, sapotillo, zapote (Peru).

Ochroma, the source of Balsa, the lightest of commercial timbers, is a genus of large, very fast-growing trees widely distributed in tropical America. Some botanists claim to recognize 11 species, but for all practical purposes at least there is only one, *O. lagopus* Sw., of which the others are varieties or forms. (See *Tropical Woods* 70: 1.) The genus, like *Bombax* and *Ceiba,* bears seeds enveloped in silky floss or down, and the young seedlings appear promptly in every nearby opening where the soil is exposed. Given ample room on rich, well-drained soil at low elevations, the trees may attain a height of 50 to 60 feet and a trunk diameter of 24 to 30 inches in five or six years. The wood produced under such conditions is very light and soft, weighing only 6 to 8 pounds per cu. ft. when dry. Timber of slow growth, whether from the outer part of old trees or from young ones that developed under adverse conditions, may weigh 25 pounds per cu. ft. These natural variations affect the structure, appearance, properties, and uses of the wood. To secure a uniform product, the trees must be grown in plantations.

Regarding the culture of Balsa in Ecuador, the principal source of supply at present, Samuel Greenhouse says (*Journal of Forestry* 33: 10: 870, Oct. 1935): "Seeds, having proper moisture and temperature conditions, must be acted on by direct sunlight in order to germinate. This makes removal of all previous cover necessary. The density of natural reproduction makes unnecessary any artificial seeding or planting, but necessitates some means to prevent overcrowding. Until they are 6 months old and about 1½ inches in diameter, the plants are very little more than pith and cortex. They are very easily broken or injured, and even a slight injury often causes death. Even after the trees are fully mature, a slight hurt will cause the wood to develop a hard and fibrous texture, and thereby lose its commercial value. The tree is then called 'macho' ('burillo' in Central America) as against 'hembra' ('Balsa real' in Central America), the soft-textured tree. This makes it compulsory that the plantation be handled as little as possible, for the careless workman is the chief cause of injury. Until it is seven years old, the water needs of the Balsa tree are supplied by a shallow root system. About that time, due to increasing competition, it will develop a tap root. This causes the wood in the center of the tree, near the base, to become saturated. Decomposition takes place, and the wood in that area assumes a red color, and gradually becomes doty. With time this supersaturated area spreads laterally and in height, decreasing the value of the log. The tree must therefore be harvested before this development occurs. The bole above the first branch will not produce merchantable logs. Lumber with knots has no value. More emphasis must therefore be placed on growth in height than in diameter." (For method of growing Balsa in Costa Rica, see *Tropical Woods* 15: 34.)

Balsa timber is imported into the United States either in logs with the bark on or in the form of lumber. The latter has the advantage of eliminating obvious defects, such as knots and rotten or stained material, and also the bark for which no satisfactory use has been found as it makes poor fuel and its tannin content is very low (see *Tropical Woods* 2: 1). Although the weight of green wood is from two to three times the dry weight, the logs are comparatively light and easy to transport. The wood decays very quickly in contact with the ground and is subject to sapstain if not promptly dried. Seasoned material absorbs water readily when immersed, but this property can be largely overcome by waterproofing processes which do not add greatly to the weight.

Balsa came into prominence during the first World War, and large quantities of timber were consumed in manufacture of life preservers and rafts and for mine buoys; 80,000 Balsa floats were used in the 250-mile submarine mine barrage in the North Sea. It was also used to line refrigerator trucks for military use in France, and it is to its insulating properties there demonstrated that the timber owes most of its present commercial applications. For protected lining, wood of the lowest density is best, and its efficiency is about the same as that of cork. Cases made of a heavier grade of lumber, preferably with a weight of about 10 pounds per cu. ft., are used successfully for the shipment of pre-cooled perishable commodities such as yeast, dairy products, fresh fruits, meat, and fish. Such a container, with a capacity of 40 to 80 pounds, can be used for from 50 to over 100 round trips, and as it is much lighter than an ordinary wooden box there is material saving in transportation charges. Balsa is also employed in ceilings and partitions of rooms as a sound-deadener, and under heavy machinery to prevent transmission of vibration to other parts of the building. Other uses include stream-lining in airplanes, pads to protect furniture in shipment, novelties, toys, sea sleds, floats, hat blocks, and diaphragms for loudspeakers.

The resistance of Balsa timber to endwise compression and static bending is about half of that of the best quality of Spruce (*Picea*) having a density four to five times greater (see *Tropical Woods* 41: 45). There are other woods lighter than Balsa (see *Tropical Woods* 37: 52), but they lack its strength, and moreover are not available in sufficient size or quantity to be commercially important.

Heartwood pale brown or reddish; sapwood (comprising most of the commercial timber) nearly white or oatmeal-colored, often with a yellowish or pinkish hue. Luster usually rather high. Odorless and tasteless. Feel velvety.

Common names: Balsa, balsa wood (Trade); balsa (Cent. & So. Am., gen.); bombast mahoe, corkwood, down tree, dum (Jam.); balsa, corkwood, guano (P.R.); ceibón botija, c. lanero, lanero (Cuba); bois flot, mahaudème (Haiti); bois flot, patte de lièvre (Mart.); gonote real, maho, mo-ma-ah (Mex.); polak (Br. H.); lanilla, puh, tambor (Guat.); guano (Hond.); algodón (Salv.); gatillo, polak (Nic.); enea, piú, pung, urú (C.R.); lana, puero (Pan.); balso, lano, palo de lana, tacarigua (Venez.); pau de balsa (Braz.); palo de balsa, topa (Peru); tami (Boliv.).

Quararibea, with about a dozen species, is of rather common occurrence throughout tropical America. The trees are mostly small or medium-sized, usually with a slender, smooth-barked, gradually tapering bole and small branches that are in whorls on the young stems. The leaves and bark, at least in certain species, have a peculiar odor variously described as resembling curry powder, fenugreek, or licorice.

The only trees reaching a height of more than 50 or 60 feet belong to an undetermined species growing in the Rio Grongogy basin in southeastern Bahia, Brazil, where it is known as Veroity. According to Hugh M. Curran, it attains a maximum height of 150 feet, with a trunk 3 to 4 feet in diameter and usually free of limbs for 50 to 70 feet. It grows on rich soil in the rain forest and often comprises the bulk of the stand over considerable areas. The timber is held in poor esteem locally because of its lack of durability and its tendency to stain blue or black soon after cutting. When properly dried, however, it makes a clear white or ivory-white lumber resembling American Holly (*Ilex opaca* Ait.). The one serious defect is the presence of numerous large wormholes, rimmed with black, but the mature timber will probably be found to cut as much clear lumber as the average American hardwood, since it is generally free of knots and other defects.

All specimens of *Quararibea* without distinct heartwood; natural color white or yellowish; usually discolored by fungi. Luster low to fairly high. Tasteless and generally odorless, sometimes with the characteristic scent of the bark and leaves. Sp. gr. (airdry) 0.55 to 0.72; weight 34 to 45 lbs. per cu. ft.; texture medium to somewhat coarse; feel rather harsh; working properties good.

Common names: Garrocho, swizzle-stick tree (P.R.); cacahuaxochitl, cacaoxochitl, flor de cacao, madre de cacao, maha, rosa de cacao (Mex.); batidos, cincho, mahass, majahás, molenillo (Br. H.); cuyapo, moro (Guat.); coco mamá, moro (Hond.); garrocho, molenillo, pía (C.R.); bobito, cincodedos, guayabillo, guayabo (Pan.); botón (Col.); aspaí, mampuesto negro (Venez.); apezoeloe, cacao-oelie, kibiwara wesjilikodo, kirikiri, maipoelie, taga-hoedoe (Sur.); guarariba, inajá-rana, veroity (Braz.); huayuash-sapote, sapotillo (Peru).

Scleronema, with four species of moderately large trees, is rather widely distributed in the Amazon region of Brazil. The only wood sample in the Yale collection is of *S. micranthum* Ducke (Yale 20694; Ducke 13). Heartwood lustrous golden brown, with conspicuous ray flecks on radial surface; sapwood oatmeal-colored, sharply demarcated. Odorless and tasteless. Moderately heavy, hard, tough, strong; coarse-textured; easy to work; probably fairly durable; a good timber for general construction and medium grades of furniture.

BORAGINACEAE

The Borage family comprises about 90 genera and over 1600 species. They are chiefly rough-hairy herbaceous plants, widely distributed and most abundant in temperate regions, especially in the Mediterranean basin, central Asia, and western United States. Many of the tropical species are upright or scandent shrubs or lianas, and some are trees. The leaves are simple and alternate, rarely opposite; stipules are absent; the flowers are borne in normal or in unilateral (scorpioid) cymes; the fruit is a drupe or of four nutlets. Some of the herbs, such as the Forget-me-not, Hound's-tongue, and Gromwell, are cultivated in gardens; a few trees (*e.g., Cordia*) are planted in parks and along streets because of their attractive foliage and handsome fragrant flowers; and various kinds are of local account for their edible fruits and reputed medicinal properties of their leaves, flowers, or roots. The glutinous mesocarp of *Cordia* is used for bird-lime and glue. The most important genus for timber is *Cordia,* and the early Egyptians made extensive use of the wood of the Sebestan Plum, *C. Myxa* L., a medium-sized tree native to Egypt, Persia, Arabia, India, and the Malay Peninsula. The tropical American tree species belong to nine genera, namely, *Auxemma, Bourreria, Cordia, Ehretia, Lepidocardia, Patagonula, Rochefortia, Saccellium,* and *Tournefortia.* The two of value for their timber are *Cordia* and *Patagonula,* but only the former has possibilities for the export trade.

Heartwood typically light to dark brown, frequently streaked or variegated with olive, purple, or black; often sharply demarcated from the whitish sapwood. Luster usually golden in proper light, but sometimes obscured by parenchyma. Taste not distinctive, but some specimens, *e.g., Cordia,* have a spicy fragrance. Density low to high; texture fine to coarse; grain generally straight; working properties and durability fair to excellent.

Growth rings present. Pores rather few to numerous; sometimes large in part, but mostly small to very small; solitary in *Rochefortia;* in ring-porous arrangement in *Auxemma* and *Ehretia* (in part); in ulmiform pattern in *Patagonula* and *Saccellium;* in small multiples and clusters, sometimes joined by parenchyma, in the others. Vessels often with meandering course as seen on tangential surface; typically with simple, wide-rimmed perforations; occasional foraminate plates also present in *Cordia;* spiral thickenings seen in small vessels of *Ehritia* (in part); tyloses common; intervascular pitting very fine to medium, alternate; pit apertures commonly extended and coalescent. Rays 1 to 4, sometimes to 8, cells wide and of variable heights up to 100 cells or more; not distinctly two-sized; homogeneous to decidedly heterogeneous; outermost marginal cells frequently peaked; sheath cells sometimes present, *e.g.*, *Auxemma* and *Cordia;* crystals present or absent, sometimes numerous; gum deposits abundant in dark-colored woods; pits to vessels usually between very small and medium-sized, but elongated (10 to 18μ) in *Tournefortia* (in part). Wood parenchyma finely reticulate in *Bourreria, Ehretia, Rochefortia,* and *Tournefortia;* paratracheal and aliform to confluent in the others; in some instances terminal; crystals present or absent, numerous in *Auxemma.* Wood fibers with thin to very thick walls; pits very small to medium-sized, simple to distinctly bordered. Ripple marks absent. Small vertical traumatic gum ducts occasionally present in *Cordia.* For anatomy of the different genera see *Tropical Woods* 67: 20–32.

Auxemma. Two species of this Brazilian genus have been described, of which the better known is *Auxemma oncocalyx* (Fr. Allem.) Baill., a small to medium-sized deciduous tree with its center of distribution in Ceará where it grows gregariously and rather abundantly in regions with distinct wet and dry seasons. The olive-like fruit is completely inclosed in a 5-winged bladder formed by the inflated calyx tube. The common name is Pau Branco (white wood), but the heartwood is dark-colored and resembles Black Walnut (*Juglans nigra* L.); its principal use is for fuel, but it is suitable for furniture and durable construction. The following description is based on a specimen collected by H. M. Curran in Ceará.

Heartwood chocolate or purplish brown, somewhat variegated, the surface fading to light brown upon exposure; distinct but not sharply demarcated from the yellowish white sapwood. Luster fairly high. Without distinctive odor or taste. Heavy, hard, and strong; sp. gr. (air-dry) 0.70; weight about 43 lbs. per cu. ft.; texture coarse; grain straight; easily worked, finishing very smoothly and attractively; is highly resistant to decay. A good timber, but presumably of no commercial possibilities.

Bourreria (*Beureria, Beurreria*), with about 40 species of shrubs and small to medium-sized evergreen trees, occurs in the West Indies, southern Florida, Mexico, Central America, and northern Colombia and Venezuela. The trunks of the trees are usually sulcate or fluted; the fragrant white flowers are borne in terminal corymbose cymes; the fruit is a thin-fleshed drupe containing four bony nutlets. The timber is not utilized for any special purpose.

Two species reach the shores of southern Florida. One of them, *Bourreria revoluta* Miers, is a large shrub; the other, *B. ovata* Miers, is a tree sometimes 50 feet tall with a scaly, buttressed trunk a foot in diameter. The most interesting species is the Huanita tree of Mexico, *B. huanita* (Llave & Lex.) Hemsl., which ranges from Michoacán to Oaxaca and southward through Guatemala to Salvador. The flowers are an ingredient of a fermented beverage prepared from crude sugar and are also used for perfuming tobacco and flavoring sweetmeats and conserves. The tree was once the cause of a war. Moteczuma II wanted it for his celebrated botanical garden and, being unable to get it by peaceful means, sent an army to Tlaxiaco about 1496 and took some plants by force. According to some authorities the stock thus secured died before it could be planted, but other writers state that it lived and became one of the treasures of Moteczuma's gardens. (See *Tropical Woods* 28: 14, 24.)

Heartwood brown, with lighter streaks, merging gradually into the scarcely distinguishable sapwood. Luster medium. Odorless and tasteless. Hard and heavy to moderately so; texture fine; grain fairly straight; easily worked, finishing smoothly; durability fair. Has no commercial possibilities.

COMMON NAMES: Strong back, s. bark (Florida, B.W.I.); agalla, árbol de la frutica, ateje de sabana, cagón, cateicito, curaboca, fruta de catey, frutica de catey, guazumillo, hierro de costa, h. de sabana, jaguagüita, raspalengua, roble agalla (Cuba); palo de vaca, roble guayo, spoon tree (P.R.); muñeco (Dom. R.); café marron, mapou gris (Haiti); bacalché, beheck, esquisuchil, flor de palomita, guiexoba, huanita, izquixochitl, jazmín de Tehuantepec, kakalché, opay, sacbayeck, yaga guiexoba, ytayucuine, yzquixochitl (Mex.); black fiddlewood, laurel, roble (Br. H.); roble, sombra de ternero (Guat.); esquinsuche, listón (Salv.); guisjoche (C.R.); uvito macho (Col.); flor blanca, f. de ángel, grimanso, guatacare, g. blanco, semeruco (Venez.).

Cordia, the most important genus of the family, includes more than 200 species of unarmed shrubs and small to large trees widely distributed in tropical and warm extra-tropical regions of the world. There are many American species with a combined range extending from southern United States to Argentina. All of the larger trees supply useful timbers, but only a few are of commercial value.

Botanists are not in agreement as to what constitutes this genus, and one has gone so far as to discard the name *Cordia* and distribute the species among separate genera (see *Tropical Woods* 36: 51). The available wood samples, though numerous, are not sufficient for a comprehensive systematic study. They range in color from grayish yellow or yellowish brown to dark brown variegated with black, in density from light, soft, and spongy to decidedly heavy and hard, and in texture from medium to very coarse and fibrous. The fibers often have a golden luster but this is masked by the dullness of the parenchyma cells so that the surface of the wood has a mealy appearance but with a rich sub-luster in proper lighting. The commercial timbers are of two general classes: (1) dark-colored and hard-wooded, typified by *Cordia Sebestena* L.; (2) light-colored and soft-wooded, typified by *C. alliodora* (R. & P.) Cham. The distinctions are not always pronounced and there is also some question as to the effect of age and site on the kind of timber produced by the same species.

Hard-wooded, dark-colored group.—*Cordia Sebestena* is a small tree, rarely 35 feet high, in southern Florida, where it is called Geiger-tree, and in the West Indies, the northern coast of South America and Yucatán, Mexico; it is often planted for ornament on account of its orange or scarlet flowers. The beautiful faintly scented heartwood is used locally for small cabinet work and articles of turnery, but the trees are too small and scarce to be of commercial importance.

A larger tree of the same general range is *Cordia Gerascanthus* L. (= *C. gerascanthoides* H.B.K.), which is often confused in the literature with "*C. Gerascanthus* Jacq." (= *C. alliodora*). It (or a closely related species) is at its best in Venezuela, where it is known as Canalete. Its variegated reddish brown, strong, readily-worked, durable timber is highly esteemed locally for joinery, furniture, and house construction. The most important species of the group in Guatemala, British Honduras, and Yucatán and Chiapas, Mexico, is *C. dodecandra* DC., a tree sometimes 100 feet tall, with very rough leaves, orange-red flowers, and edible acid fruits. The timber is considered excellent for fine furniture and turned objects, but the supply is very limited. Other Mexican woods of this group have been identified as *C. elaeagnoides* DC. and *C. sonorae* Rose, both small to medium-sized trees. Some undetermined specimens of Peterebí from Argentina and Matto Grosso, Brazil, belong in this group, but others of that name are comparatively light and soft.

Heartwood tobacco-colored to reddish brown, with irregular dark brown or blackish streaks and variegations; with more or less of an oily or waxy appearance; rather sharply demarcated from the grayish or yellowish sapwood. Luster variable. Taste not distinctive; scent mildly fragrant, at least when fresh. Hard, heavy, and strong; sp. gr. (air-dry) 0.80 to 0.97; weight 50 to 60 lbs. per cu. ft.; texture medium and not always uniform; grain variable; not difficult to work, finishing very smoothly and attractively; holds its place well when manufactured; durability high. Of limited commercial possibilities because of the scarcity of the timber, but suitable for brush backs, turned articles, and cabinet work.

COMMON NAMES: Anaconda, geiger tree, sebestena (Florida); anacahuite, baría, b. carbonera, b. prieta, bomitel encarnado, cutiperí, platanillo, sebestena, varía, v. negra, vomitel colorado, v. encarnado (Cuba); aloe-wood, San Bartolomé, vomitel colorado (P.R.); coquelicot (Haiti); mapou (Dom.); amapa asta, a. boba, a. bola, anacuite, asta, baría, barl, bocote, bohom, bojón, cha-copté, coopté, copté, grisiño, gueramo, habeem, ocotillo meco, palo de asta, siricote, s. blanco, zac-copté (Mex.); ziricote (Br. H.); palo de asta (Guat.); canalete, c. prieta (Col.); canalete, candelo, no-me-olvides, pardillo negro, p. prieto (Venez.); loro negro, peterebí (Arg.); lauro pardo (Braz.).

Soft-wooded, light-colored group.—There are many species of trees and shrubs in this group, but apparently there are only two or three species of much importance for their timber. The most widely distributed is *Cordia alliodora* (R. & P.) Cham. (= *C. Gerascanthus* Jacq.), a medium-sized to large tree occurring in the West Indies and from southern Mexico to the southern edge of the tropics in South America. The forks of the young twigs almost always develop swellings which harbor fierce ants. The leaves are characterized by stellate pubescence in varying abundance and have a garlic-like odor when crushed. The fragrant white or yellowish corollas are borne in large panicles and remain on the tree almost unshriveled, finally serving as parachutes to the falling fruits. In Central America the usual name for the tree is Laurel, and two kinds are recognized, namely, Laurel Blanco and Laurel Negro, the names referring to the color of the heartwood. The latter might well be considered in a third group of soft, dark-colored woods, but after repeated failures to distinguish the two sorts in the forest the senior author is of the opinion that the differences noted are probably attributable to the age of the tree or to individual peculiarities. Laurel Blanco is light-colored throughout, is not scented, and though considered fairly resistant to termites it will not last long in contact with the ground. Laurel Negro has a nearly white sapwood and a rather dark, somewhat variegated, spicily scented, durable heartwood which suggests Walnut (*Juglans nigra* L.); it may prove to be *C. megalantha* Blake, as the one specimen available has the characteristic color. Both kinds are employed locally for general carpentry and construction, but the darker timber is preferred on account of its more attractive appearance and greater durability.

The usual Argentine name for timber of this group is Peterebí. The species is usually given as *Cordia trichotoma* (Vell.) Arrab., but is very closely related to *C. alliodora* and the differences are varietal rather than specific. The flowers are generally larger, the leaves more pubescent, and the twigs are rarely occupied by ants and are then never conspicuously deformed (see *Journ. Arnold Arboretum* (16: 1: 9). Castro (*Las maderas Argentinas*, p. 18) says that Peterebí or Loro Amarillo is found in Misiones and northern Corrientes, the trees being 50 to 80 feet high and about 20 inches in diameter. The lumber, which is easy to work and has a high luster, is used for general construction, door and window frames, and as a substitute for imported Oak (*Quercus*) for the manufacture of furniture. The following information is supplied by H. M. Curran:

"Peterebí is a symmetrical tree 100 to 125 feet high with a slender trunk usually 18 to 24 inches in diameter and free of branches for 40 to 50 feet. It is without buttresses, and the bark is dark brown and rough, suggesting Elm (*Ulmus*). The deciduous leaves are rather thin and about two inches long and one inch wide; the flowers are dark chocolate-brown and borne in great profusion at the extremities of the branches. Though nowhere abundant, this tree is common in the forests of Misiones, Argentina, and in Paraguay and the adjacent region of Brazil. It occupies the sandy or clay loam in the vicinity of the rivers and the best stands rarely contain more than two or three trees per acre. The timber reaches the market in the form of squared logs and the total amount consumed annually is between 500,000 and 1,000,000 board feet. The wood, which is of a golden brown color, is considered one of the best furniture woods in Argentina."

There are several kinds of *Cordia* in Brazil, and in the southern part of the country they are generally known as Louro. In the Amazon region, the principal species is the Frei Jorge or Freijo, *Cordia Goeldiana* Huber, closely related to *C. alliodora*. According to Huber (*Bol. Mus. Goeldi* 6: 90), it is a large tree in the high forest along the right-of-way of the railway between Belem (Pará) and Bragança, and probably elsewhere. He says that the wood is highly appreciated, particularly for cooperage. In this connection an American consul reported in 1925 that its importation into Portugal for the manufacture of staves had declined owing to the poor quality of timber received and to the fact that it imparts a flavor to wines (see *Tropical Woods* 4: 11). Le Cointe (*L'Amazonia Brésilienne* 2: 25) says that the lumber is of good quality, easy to work, much used in Belem (Pará) for carpentry and joinery, and in place of Teak in naval construction.

Freijo has been on the United States market in small quantities for more than 20 years, and has been known as Brazilian or South American Walnut, Jenny Wood, and Cordia Wood. Karl Schmeig, New York manufacturer of fine furniture, has used this timber with success. He says (*Tropical Woods* 9: 1): "The Cordia Wood or Jenny Wood is proving highly satisfactory and, what is always a matter of concern to the manufacturer, the supply is adequate. This wood is of a neutral color, suggesting Chestnut, has about the same density as American Walnut, takes a stain very well, and, on account of its close texture and even grain, receives a soft patina finish with comparatively little effort. It is especially well adapted for interiors of club rooms, for bank fittings, and for furniture of Spanish design. For many purposes it satisfactorily replaces Oak. We recently built a complete room, including the furniture, of Cordia Wood and were gratified with the results."

Heartwood yellowish to brown, uniform or more or less streaked and variegated; light-colored material not clearly differentiated from sapwood, dark-colored distinct but usually not sharply demarcated; brown rays make lightest kinds oatmeal colored. Luster of best grades rich and golden when viewed in proper lighting. Without distinctive taste; dark specimens spicy-scented. Light and soft to moderately hard and heavy; sp. gr. (air-dry) 0.40 to 0.70; weight 25 to 44 lbs. per cu. ft.; texture uniform, medium to coarse; grain generally straight; easily worked, though sawing woolly when fresh; seasons readily, finishes smoothly when dry, takes glue well, and holds its place remarkably well when manufactured. An attractive timber worthy of greater consideration by consumers everywhere.

COMMON NAMES: *Cordia Goeldiana:* Cordia wood, Jenny wood (U.S.A. trade); freijo, frei jorge (Braz.). *C. alliodora* and *C. trichotoma:* Prince wood, Spanish elm (B.W.I.); baría amarilla, capá roja, palo de rosa del país, varía, v. colorada (Cuba); capá, capaw (P.R.); bois de cipre, b. de cype (Mart.); cyp, cypre (Trin.); amapa prieta, buhún, bojón, b. blanco, b. prieto, hormiguero, nopo, palo de rosa, p. de María, suchicahue, tambor (Mex.); bohun, laurel blanco, salaam, salmwood (Br. H.); suchah (Guat.); laurel, l. blanco, l. negro (C.A., gen.); laurel hembre, l. macho (Nic.); dzeguí (C.R.); canalete de humo (Col.); alatrique, canjaro, cautaro, pardillo, p. de monte (Venez.); brown silverballi, taparai (Br.G.); laurel (Ec.); árbol del ajo (Peru); louro, l. amarello, pau cacharro, urúa, uruasinho, uruazeiro (Braz.); louro (Urug.); peafata grande, lapachillo, loro amarillo, peterebí, p. hú, p. saiyú (Arg.). *Other species:* Cocobey (Bah.); manjak (Grenada); anacahuita, ateje, a. amarillo, a. blanco, a. cimarrón, a. colorado, a. de costa, a. hembra, a. macho, atejillo, hierro de costa, palo de rosa, saúco, tabaco, uva gomosa, varía blanca, vomitel, v. amarillo, v. blanco (Cuba); basora, b. prieta, black sage, capá cimarrón, cerezo, c. blanco, copillo, cupillo, manjak (red, white), moral, muñeco, palo de muñeco, p. de perico, saraguaso, s. prieto, saraguero (P.R.); capá prieto (Dom. R.);

belle-belle, bois caparo, b. chique, b. soumis, bonbon captain, chêne caparo, fleur dent, parésol, trois pieds (Haiti); cipre a griver, c. balanic, c. oranger, mahot noir, mapou blanc (Guad.); black sage, bois lay-lay, hairy lay-lay, manjak, mapoo lay-lay (Trin.); anacahuite, azota caballo, babosa, bohonché, bubo, chavarobo, gonguipo, gulabere, huaché, huazimilla, kopché, koxolxek, macahuite, oreja de ratón, San Juanito, tacotillo, valozo, vara prieta, vavos, xcopché, zazamil (Mex.); bastard salmwood, jackwood (Br. H.); coralillo, manuno upay (Guat.); cebito, cuaja-tinta, escoba negra, escobilla negra, manuno, tigüilote, t. negro, tihuilote, varilla negra, zompopo (Salv.); carne asada, cachalaco, sombra de ternero, tigüilote (Hond.); buriogre, b. amarillo, b. de montaña, cuaja-tinta, escoba negra, jigüilote, muñeco, nigüito, salvilla cimarrón, tigüilote, varilla negra (C.R.); goma, lengua de buey, muñeco amarillo, nigüito, paico, sabto, tigüilote, ubero, uvillo (Pan.); canalete de humo, caujaro, guácimo, g. nogal, guasco, muneco, m. canalete, pata de gallina, solera, uvito, u. mocoso (Col.); achechibe, aguacatico, baboso, basura prieta, candilero, cariaco, cariaquito, c. de sabana, c. negro, caujarito, caujaro, celedonia, guapalo, majañe negro, majao negro, pardillo blanco, tarare blanco (Venez.); ants' plant, yuwanarow (Br. G.); alatoeloeka, aloeko uonoré, anakara, anoemalatti, aratroeka, arowtroeka, arreuonoe, awali emoeloe, baka eoma, berg tafraboom, blaka oema, b. ivintje, boggi lobbi, danlieba, dokka dokoa, hoereuerereroko, horowé jore lokko, h. j. roko, kaboejakoro diamaroe, kakoro, k. konokodikoro, makoeja pipa, manblaka oema, marribonsoehoedoe, mattoe toenbalobbi, omosé, tafelboom, tafraboom, toenbalobbi, waijanaka erepaloe, wakoekwatokon (Sur.); arbre parasol, montjoly, tiki topichi (Fr. G.); tigua balsosa (Ec.); almendrillo, añallio-caspi, bacurí, motemullaca, orccolauraimana, tahuampa-caspi (Peru); araticú, a. guazú, colita, gomita, mbuy-rembiú-guazú (Arg.); carbón (Chile); azaherero del monte (Urug.); achira-mourou, arvore de umbella, babosa branca, café do matto, carú-caá, catinga de barrão, cauarú-caá, cha de bugre, cha de frade, chapeu de sol, claraiba, jaguará murú, juruté, laranjeira do matto, María preta, mata fome, paraparà, pau de formige, p. de jangada, pinchíricóto (Braz.).

Ehretia, with some 50 species of shrubs and small to large trees, is most abundantly represented in the tropics of the Old World, about 10 species occurring in the West Indies and from Costa Rica through Mexico to western Texas. *E. Austin-Smithii* Standl., which grows in the mountains of Costa Rica, attains a maximum height of 100 feet, with a trunk six feet in diameter at the base, but often dividing near the ground into several stems. The leaves are very rough and are used locally for scrubbing and scouring. *E. tinifolia* L., of the West Indies and Mexico, ranges in size from a shrub to a tall tree; it is often planted for shade, and has an edible fruit. The most northern species, *E. anacuna* (Berl.) Johnston, of Mexico and the Texas borderland, is a shrub on barren soil but grows to a height of 50 feet and a diameter of 36 inches in river bottoms. It is frequently planted along streets, as it has a handsome dense-foliaged crown, panicles of small white fragrant flowers, and sweet edible fruits. The wood is used locally for tool handles, yokes, and the spokes of wheels.

Wood grayish brown throughout. Luster medium. Odorless and tasteless. Heavy and hard; texture rather fine; grain fairly straight; not difficult to work, taking a high polish; not very durable. Of no commercial possibilities.

COMMON NAMES: Anama, anaqua, knackaway, nockaway, sugarberry, yara (Texas); quiebrahacha, roble prieto (Cuba); arrayán (Dom. R.); chêne noir, filière (Haiti); bois de rose noir (Guad.); anacahuite, anacua, anagua, bec, beec, capulín cimarrona, manzana, manzanillo, manzanita, nandimbo, roble, saúco (Mex.); guarlo, manzanita (Salv.); laurel (C.R.).

Patagonula. There are two species, namely, *Patagonula americana* L. and *P. bahiensis* Moric, but only the first is at all well known. It is a medium-sized to large tree, sometimes 85 feet tall and 30 inches in diameter, growing in southern Brazil, northern Argentina, Paraguay, and Uruguay. After blooming, the five calyx lobes greatly elongate and provide wings for the small, pointed fruit. The timber, which is of excellent quality and attractive appearance, is so much in demand in Argentina that the available stands have been seriously depleted. The thick sapwood is used for making tool handles, oars, agricultural implements, yokes, and vehicles, and the Indians of Misiones prefer it for their bows. The variegated heartwood is highly esteemed for fine furniture, bentwood chairs, interior trim, and articles of turnery. Importation of chairs of the Vienna type has been greatly reduced through the establishment of local factories using timber from the Chaco forests. The common name in Argentina is Guayabí, and it is sometimes confused with the Guayabil of Salta (*Saccellium lanceolatum* H. & B.).

Heartwood usually variegated in various shades of brown to blackish purple; sometimes fairly uniform dark olive; distinct but not always sharply demarcated from the white to brownish sapwood. Luster medium. Odor and taste not distinctive. Hard, heavy, tough, and resilient; sp. gr. (air-dry) 0.80 to 0.95; weight 50 to 59 lbs. per cu. ft.; texture rather fine, uniform; grain fairly straight; fairly easy to work, taking a high polish; heartwood very resistant to decay. Not likely to become important for export because of the local demand.

COMMON NAMES: Guayabí, g. blanco, g. crespo, g. morotí, g. negro, g.-rá, guayaibí, guayubí, guayabil (Arg., Urug.); ipê branco, guajuvira, g. branca (Braz.).

Rochefortia, with about eight species of shrubs and small trees, mostly armed with spines, is limited to the West Indies and Colombia. The timber is utilized to a limited extent locally for fence posts, articles of turnery, fuel, and charcoal.

Heartwood very dark brown, uniform or variegated; usually with a greenish yellow hue; has an oily or waxy appearance; rather sharply demarcated from the thin yellowish brown sapwood. Luster low. Without distinctive odor or taste. Extremely hard and heavy; sp. gr. (air-dry) 1.25; weight about 79 lbs. per cu. ft.; texture fine and uniform; grain fairly straight; not difficult to work, taking a glossy polish; very resistant to decay. Of no commercial importance because of small size.

COMMON NAMES: Carbonera, cerillo de costa (Cuba); juzo (P.R.); bois d'ébène (Haiti); bois vert (Guad.).

Saccellium. Three species have been described, namely, *S. brasiliense* I. M. Johnston (Matto Grosso, Brazil), *S. Oliverii* Britt. (Amazonian Bolivia), and *S. lanceolatum* Humb. & Bondpl. The last, which is the only well-known species, occurs in the Amazon drainage of northeastern Peru and from the mountains of southern Bolivia southward along the mountains of northern Argentina to Tucumán (see *Journ. Arnold Arboretum* 16: 2: 181–183). The branches are elongate and stiff; the evergreen leaves are distinctly lanceolate; the flowers are white or yellowish; the fruit is a small drupe inclosed by the accrescent calyx. Though usually a small tree, it sometimes attains a height of 65 feet and a diameter of 20 inches. The wood resembles that of *Patagonula* and is used to a small extent for general carpentry and furniture. Only one specimen is available (Yale 32601; Imp. For. Inst. 9629) and it is all sapwood.

Heartwood said to be variegated dark brown; sapwood pale brownish, with fine stripes or markings due to higher luster of wood-fiber zones. Odorless and tasteless. Hard and moderately heavy; easy to work, finishing smoothly; heartwood said to be highly durable. Apparently without commercial possibilities, at least for export.

COMMON NAMES: Guayabil, g. negro, guayibil (Arg.).

Tournefortia, with about 100 variable and poorly defined species of woody vines, erect or scandent shrubs, and a few small trees, is of pantropical distribution, but most abundantly represented in tropical America. The leaves are typically broad; the small white or yellowish flowers are borne in scorpioid cymose spikes or racemes; the small drupaceous fruits contain four nutlets. The only reported uses of the plants are in local medicines.

Wood pale brownish throughout, except about wounds, where it is chocolate-brown.

Luster medium. Rather hard and heavy; texture fine and uniform; grain straight; easy to work, taking a glossy polish; is presumably perishable in contact with soil. Of no commercial possibilities.

Common names: Soldier bush (Bah.); alhucema de costa, balsamillo, bejuco cayaya, cayaya, inciensio de costa, i. de playa, nigua, n. de paredón, romero de costa (Cuba); bejuco de nigua, b. masa, chiggernit, mata de nigua, nigua, n. peluda, pringamoza, té del mar, temporana (P.R.); nigua, n. peluda (Dom. R.); kallaba, liane chique (Haiti); chacnichmax, confite coyote, hierba del burro, h. del negro, h. del zapo, h. rasposa, ortigüilla, perlas sicimay, tlachichiona, topoya, tlepatli, xulkin (Mex.); frutilla, tirica (Nic.); maíz de gallo (C.R.); lágrimas de San Pedro (Col.); pesso (Ec.); herva de jaboty, h. de lagarto (Braz.).

BRUNELLIACEAE

Brunellia, the only genus, with about 17 species of shrubs and small or occasionally medium-sized trees up to 65 feet in height, occurs in the mountains from Peru to southern Mexico and the West Indies. The only wood samples available are of *B. comocladiifolia* H. & B. from Cuba and Colombia, a tree with opposite, imparipinnate leaves, rusty-pubescent twigs, and smooth grayish brown bark. There are no special uses for the timber, as it is of poor quality, scarce, and of small size.

Wood pale brown throughout. Luster rather low. Odorless and tasteless. Light and soft to moderately so; rather fine-textured; easy to work, but saws woolly; perishable in contact with the ground.

Growth rings absent or indistinct. Pores numerous; small, not visible without lens; mostly in short to long radial multiples, well distributed without pattern. Vessels with both simple and multiple (scalariform type) perforations; spirals absent; intervascular pitting opposite or scalariform. Rays uniseriate or occasionally biseriate, and few to 25, sometimes up to 50, cells high; heterogeneous, nearly all of the cells square or upright; pits to vessels large, mostly elongated and in scalariform arrangement. Wood parenchyma apparently absent. Wood fibers thin-walled; septate in part; pits minute, simple or indistinctly bordered. Ripple marks absent. No gum ducts seen.

Common names: West Indian sumac (Jam.); palo bobo (P.R.); bois Mabel (Haiti); jobo macho de tierra fría, majaguito de tierra fría, riñón (Col.); caobillo, caóbano, cedro (Venez.).

BURSERACEAE

The family includes 17 genera and about 600 species of unarmed, mostly resinous shrubs and trees of general distribution in the tropics and to a small extent in subtropical regions of the world. The leaves are typically alternate, unequally pinnate, and without stipules; the small flowers are borne in clustered racemes or panicles; the fruit is drupaceous. The trees are best known as the source of true frankincense, from Arabian species of *Boswellia;* myrrh, from *Commiphora* (*Balsamodendron*) of Arabia and the African coast of the Red Sea; and gum elemi, from Philippine species of *Canarium* and American species of *Bursera* and *Protium*. The most important commercial timber is the West African Okoumé (*Aucoumea Klaineana* Pierre) of Gaboon and the coastal region of the Middle Congo and Spanish Guinea, which is used extensively in Europe for surface veneers, plywood, and cigar boxes (see *Tropical Woods* 17: 1–5). In tropical America there are about 200 species of six genera, namely, *Bursera, Crepidospermum, Dacryodes, Protium, Tetragastris,* and *Trattinickia*. Their timber is used to a limited extent locally but is unknown to the export trade.

Heartwood, if present and distinct, brown or reddish brown, with sharp or gradual transition to the yellowish, pinkish, or nearly colorless sapwood. Luster fairly high. Without distinctive taste but sometimes fragrantly scented. Density widely variable; texture fine to moderately coarse; grain straight to roey; working properties good; durability fair to very poor. The best woods are of the type of Birch (*Betula*).

Growth rings usually visible. Pores distinct to invisible; rather numerous but not crowded; solitary and in multiples, occasionally with tendency to diagonal or zonate arrangement. Vessels with exclusively simple perforations; without spiral thickenings; tyloses common; intervascular pitting coarse to very coarse, alternate. Rays (without ducts) mostly uniseriate and biseriate in *Crepidospermum, Dacryodes, Protium, Tetragastris,* and *Trattinickia;* 1 to 5 cells wide in *Bursera;* often less than 20, sometimes up to 30, occasionally (*Bursera*) up to 40, cells high; heterogeneous; crystals common; gum abundant in heartwood; pits to vessels large to very large, oval to much elongated; gum ducts present in some of the rays in *Bursera, Tetragastris,* and possibly *Protium.* Wood parenchyma sparingly paratracheal; pith flecks common. Wood fibers septate; walls thin to medium; pits simple or indistinctly bordered. Ripple marks absent, no vertical gum ducts seen.

Bursera (*Elaphrium*), with about 100 species of shrubs and small to rather large trees, is widely distributed in tropical and subtropical America but most abundantly in Mexico. The leaves are odd-pinnate or more or less bipinnate, with entire or toothed leaflets and winged or non-winged rachis. The small panicled flowers have 3 to 5 distinct petals, valvate in the bud. The fruit is capsular or somewhat fleshy with an aril-like pulp. Specific determinations are often difficult and the collection of a series of herbarium specimens from the same trees at different seasons of the year has shown that certain characters which have been used diagnostically are worthless. The genus is of minor importance as a source of timber, but supplies some other products of value. Bullock says (*Kew Bulletin* 1936, pp. 348–9):

"Several species of *Bursera* have been exploited commercially, mainly on account of the high percentage of fragrant essential oil in the wood and fruit, and also to a less extent for the sake of their resins. Many are used locally as a source of incense, and all Mexican works on *materia medica* contain references to their value in medicine, particularly in the treatment of uterine diseases; they are also used in the preparation of surgical dressings. The resin of some species is valued for the manufacture of varnishes; on dissolving the resin in turpentine, varnish of high quality is obtained. Perhaps owing to the unsettled nature of large areas in Mexico, the commercial possibilities of the individual species have never been thoroughly tested or fully exploited, and the taxonomic difficulties encountered in dealing with the genus render it impossible to refer with certainty any product to a definite species. There is no doubt that trade names apply to products derived from several species. . . . Elemi is the name given to a number of oleo-resins derived from different botanical sources. Probably all belong to the family Burseraceae. Manila elemi is the most important commercially, but some comes from Mexico, probably collected from several species, including *B. jorullensis* (H.B.K.) Engl. and *B. copallifera* (Sessé & Moc.) Bullock. It is doubtful if *B. elemifera* (Royle) Baill. can be regarded as a source of supply. The resin is used in the preparation of ointments, lithographic inks, and varnishes. On the whole, little is known concerning the great majority of species, and it seems likely that a careful examination of many of them, and controlled exploitation, would prove profitable. At present, apparently, no control is exercised, and the oil-yielding trees are being rapidly exterminated."

According to the same authority (*loc. cit.*, p. 348), the two species most concerned in the production of Mexican Linaloe oil are *Bursera penicillata* (Sessé & Moc.) Engl. and *B. glabrifolia* (H.B.K.) Engl. The distribution of the former is given as the States of Colima, Mexico, Morelos, and Puebla; of the latter, Mexico, Michocán, Morelos, Oaxaca, and Puebla. According to Standley (*Trees and shrubs of Mexico,* p. 549, 551), *B. penicillata* is a shrub 3 to 10 feet high; the other is a shrub or small tree whose yellow and very fragrant wood, said to have been exported to England, yields upon distillation an oil used in the manufacture of perfumes. Bullock (*loc. cit.* p. 377) suggests that *B. tecomaca* (DC.) Standl. "may be one of the species used for the production of linaloe oil; it is very resinous and may also be the source of Mexican elemi." The maximum height recorded for this "sweetly aromatic shrub or tree" is about 30 feet. Standley says (*loc. cit.*): "Sessé and Mociño give the vernacular name as Tacamahaca, and state that the gum is used for healing wounds."

The best known species is *Bursera Simaruba* (L.) Sarg. (= *B. gummifera* L.), a tree sometimes 80 feet tall and 36 inches in diameter, though usually much smaller, of common occurrence in southern Florida, the West Indies, southern Mexico, Central America, and northern South America. Among its many common names are Birch, Gumbo-limbo (a corruption of Goma Elemi), Almácigo (Mastic-tree), Indio Desnudo (naked Indian), and Gommier (gum-tree). The tree is readily recognized by its smooth, lustrous, copper-colored bark, which peels in papery shreds like some of the Birches (*Betula*). Growing in the open it develops a rather short thick trunk dividing into heavy, often crooked, wide-spreading branches; in the forest it is tall and slender, with a fairly straight bole. The red resin obtained from wounds is employed locally as a cement in mending crockery and as a crude varnish. Branches and small stems are commonly used for fence posts which usually take root and develop into trees. The timber is full of moisture when first cut and soon discolors and begins to decay if not sawed into lumber and promptly dried. On this account the wood is little used, although there is an abundant supply of it, particularly in the West Indies. Experience has shown that properly prepared lumber is suitable for crates, boxes, and light interior carpentry and construction work. In Jamaica it is used for match splints. The woods of the several species examined are similar in structure and properties.

Color white, yellowish, or slightly brownish throughout; specimens usually discolored from fungus. Luster rather high. Without distinctive taste; with or without fragrant scent. Light in weight but firm and tough; sp. gr. (air-dry) 0.30 to 0.40; weight 19 to 25 lbs. per cu. ft.; texture mostly medium; feel somewhat harsh; grain more or less irregular; easy to cut, can be finished smoothly, and holds nails firmly without splitting; perishable when exposed to the weather.

COMMON NAMES: *Bursera Simaruba:* Gum elemi, gumbo-limbo, West Indian birch (Florida, Bah.); birch (red, West Indian), budge, incense tree, mastic tree, mulatto tree, turpentine tree (Jam.); aceitero, almácigo, a. colorado (Cuba); almácigo, a. encarnado, gum tree, turpentine tree (P.R.); almácigo, a. blanco (Dom. R.); chiboué, gommier blanc (Haiti); archipén, cachibou, chibou, gomard, gommart, gommier, g. gris (Fr. W.I.); Indien nue, naked Indian (Trin.); chaca, chacah, hukup, jiote, mulato, palo chino, p. colorado, p. jiote, p. mulato, piocha, quiote, sac-chacah, torote, xioquauitl, zongalica (Mex.); birch, gumbo-limbo (Br. H.); jiote, palo jiote (Salv.); chinacahuite, chino, indio desnudo, jicote, jiote, palo chino, p. jiote (Guat., Hond.); jiñicüite, jiñocuabo (Nic.); almácigo, caraña, dibit-krá, dorí-tská, jiñocuave, jiñote, karmari (C.R.); almácigo, carate (Pan.); almácigo, carate, caratosa, chibú, curatero, indio desnudo, i. en cuero, jinote, resbalamono (Col.); almácigo, caraña, caricarito, cucheme, indio desnudo, mara, m. colorada, mararo, palo de incienso, pellejo de indio (Venez.). *Other species:* Black birch (Jam.); almácigo de costa, almaciguillo, copal (Cuba); chutama, cirujano, copal, c. amargo, c. amargoso, c. blanco, c. chino, c. c. colorado, c. de penca, c. santo, copalquahuitl, copalquin, copalxochitl, cuajiote, c. amarillo, c. chino, c. colorado, c. verde, guande, g. colorado, inanué, incienso del país, jaboncillo, linaloé or linalué, nabanché, ngedi, ngidi, palo copal, papelillo, quincanchiri, sisiote, sochicopal, suchicopal, tacamaca, tacamahaca, tecomaca, teponaxtle, teponaxtli, teponaztli, tetlate, tetlatia, tetlatián, tetlatín, torote, t. blanco, t. copal, t. prieto, xochicopal (Mex.); caraña, carate, caratero, crispín, sasafrás, tatamaco (Col.); bálsamo, b. incienso, caricarito, chachique, isicagua de burro, mara blanca, sasafrás (Venez.); imborana, imburana (Braz.).

Crepidospermum, with two species of small resinous trees, is of infrequent occurrence in Colombia, the Guianas, and the Brazilian and Peruvian Amazon regions. The leaves have a few pairs of thin pointed serrate leaflets; the small flowers are borne in axillary panicles; the fruit is a small drupe with very resinous mesocarp. The only authentic wood samples available are of *C. rhoifolium* (Benth.) Tr. Pl., collected by B. A. Krukoff in Amazonas, Brazil. The timber is of good quality and suitable for the same purposes as Birch (*Betula lenta* L.), but the quantity is probably very limited.

Heartwood pale brown, distinct but not sharply demarcated from the grayish sapwood. Luster medium. Odor and taste absent or not distinctive. Moderately heavy, hard, and strong; texture rather fine, uniform; grain straight; not difficult to work, finishing very smoothly; probably not highly durable.

COMMON NAMES: Breu branco (Braz.); isula-micuna?, trompetero-caspi? (Peru).

Dacryodes. There is only one well-known species, namely, *D. excelsa* Vahl, a medium-sized to large tree occurring in the upland forests of Puerto Rico (Plate I) and some of the Lesser Antilles. The flowers have three distinct petals, valvate in the bud; the calyx is adherent to the ovary; the drupe has a single cell and seed. The trunks are long, smooth, and white-barked, often scarred near the base from cuts made to secure a fragrant resin which is used locally for making candles and torches and for medicinal purposes. The timber is of good quality and is appreciated in Puerto Rico for carpentry, furniture, and general construction. The quantity available is too small to be of much economic importance.

Heartwood pale brown with a tinge of pink; not clearly differentiated from the sapwood. Luster high. Odor and taste absent or not distinctive. Moderately hard and heavy, tough and strong, suggesting certain kinds of Birch (*Betula*); texture medium, uniform; grain more or less roey; finishes very smoothly, but is said to dull planer knives quickly; holds its place well when manufactured; is probably low in resistance to decay and unsuited for treatment with creosote.

COMMON NAMES: Candle-tree, tabanuco, tabonuco (P.R.); mountain gommier (Grenada).

Protium (including *Icica*), with about 90 species of small to large trees, is of pantropical distribution but most abundantly represented in the Amazon basin. The flowers have distinct petals, which are valvate in the bud; the fruit is a drupe. The bark is rich in a fragrant whitish or yellowish resin or balsam which is collected and used for medicinal purposes and as incense in churches. The timber is used to a limited extent locally for general carpentry and construction, but is of little or no commercial importance.

The genus is widely but sparsely distributed in the West Indies. The principal Middle American species is *Protium copal* (S. & C.) Engl., with its northern limit in Oaxaca and Vera Cruz, Mexico, and merging southward into the doubtfully distinct *P. sessiliflorum* (Rose) Standl. In northeastern Nicaragua, according to F. C. Englesing (*Tropical Woods* 17: 23), it is a small or medium-sized tree growing in dense shade on deep well-drained soil along river courses. The resinous bark is light gray, tinged with pink. The leaves are compound, with 2 or 3 pairs of dark green coarse-textured leaflets. The flowers are of a rich cream color and are borne in axillary spikes or panicles. The sapwood is creamy white, the heartwood pinkish brown; the timber is used for kindling, as it ignites readily.

Two other species growing in Panama are *Protium panamense* (Rose) Johnston and *P. asperum* Standl. The latter is described as a tree 30 to 65 feet tall, with large, scabrous leaves. Standley says (*Tropical Woods* 8: 5): "The tree at once attracts attention because of the fact that from wounds in its trunk it exudes large quantities of a balsam or resin which has a penetrating, agreeable, and distinctive odor. A large deposit of the semi-liquid balsam frequently may be found at the base of the trunk. I was informed by a native guide that in Panama this resin is gathered and sold. It is used probably for incense, like the resins furnished by other members of this family."

Specimens of Anime Blanca obtained by H. M. Curran in northern Colombia have been determined as *Protium guianense* (Aubl.) March. According to the collector it is a slender tree, 75 to 100 feet tall and 18 to 24 inches in diameter. The reddish wood is moderately hard, fine-textured, easy to work, and suitable for the same purposes as Birch (*Betula*).

One of the most widely distributed South American species is *Protium heptaphyllum* March. (=*Icica tacamahaca* H.B.K.), ranging from Colombia and Venezuela to Brazil and Paraguay. It is said that where this tree grows the air in the vicinity seems pleasant and wholesome from the incense-like resin that drops from any wound in the bark and collects in masses on the ground. In the northern part of the tree's range the Indians use the resin, called "tacamahaca," to scent the oil with which they anoint themselves. In the State of Pará, Brazil, where the usual name is Breu Branco, the resin is collected and enters the world market under the designation of "elemi," a term applied also to other burseraceous gums.

Williams (*Woods of northeastern Peru*, pp. 232–236) found eight species of *Protium* in the Peruvian Amazon region, where they are commonly called Copal-caspi. They vary in height from 10 to 95 feet, and in trunk diameter up to 21 inches. The only uses recorded for the timber are fuel and some local construction.

The authentic wood samples available are much like those of *Tetragastris*. Some specimens provisionally determined as *Protium* are of doubtful authenticity and it is not now possible to prepare a satisfactory description of the genus as a whole. Heartwood brown or reddish brown, usually not sharply demarcated from the pinkish sapwood. Luster rather high. Dry specimens without distinctive odor or taste. Density variable from medium to rather high; texture uniform, medium to rather fine; grain straight to very irregular; not difficult to work, finishing very smoothly; durability probably low.

Common names: Copal, incienso (Cuba); bois gommier, gommier blanc, g. juane des carrières (Fr. W.I.); copal (Mex., Central Am.); latilla, pom (Mex.); copal macho (Br. H.); tontol (Guat.); alcanfor, fontole, frontón (Hond.); fosforito (Nic.); caraño, chutra, comido del mono (Pan.); anime, a. blanco, ate, caraño, currucay, elemi, guacamayo, incienso, San Matias, tacamahaca, tacamocha, vara blanca (Col.); bálsamo, caraño, guacharaco, tacamahaco, t. macho (Venez.); baradaballi, haiawa?, incense tree, kuraka, kurana, kurokai, noyeau (Br. G.); bois encens, caragne blanc, icicariba (Fr. G.); mantos (Ec.); copal, c. caspi, pihuicho-micuna, shitari-runti-caspi (Peru); almecega, almecegueira, almesca, a. ussú, arourou, arurú, breu almecega, b. branco, b. b. da matta, b. b. da varzea, b. b. do campo, b. preto, b. xinca, cabumbo de azeite, cajú bravo, cicantaá-ihuá, coquilheiro, cuaruba, cupuba, pau de pombas, sucuriúba, sucurubeiro, uxy (Braz.); caayci guazú (Par.).

Tetragastris, with four species of medium-sized to large trees, occurs in the West Indies, Central America, and northern South America. The flowers have 4 or 5 united petals, valvate in the bud; the calyx is free; the drupe has 2 to 4 cells; the mesocarp is resinous. A clear resin exudes from the incised bark and finds various domestic uses. The timber is in the general utility class and is employed locally for the same purposes as certain kinds of Birch (*e.g.*, *Betula lenta* L. and *B. lutea* Michx.).

The West Indian species is *Tetragastris balsamifera* (Sw.) Kuntze, a pale-barked tree, sometimes over 60 feet tall, growing in well-watered upland forests in Puerto Rico, Cuba, the Island of Haiti, and some of the Lesser Antilles. The timber is used for the best grade of interior construction and carpentry, furniture, oars, and formerly for staves for sugar barrels.

There are two Central American species, the most northern being *Tetragastris Stevensonii* Standley, a medium-sized tree common in hill forests from British Honduras to Nicaragua. The principal use for the timber is for fuel, as it will burn readily, without seasoning. *T. panamensis* (Engl.) Kuntze is, according to some authors, endemic to Panama, but others consider it the same as the later-named *T. balsamifera*. According to Sandwith (*Kew Bulletin*, 1933, p. 323), the Haiowaballi of British Guiana is *T. panamensis*.

The Surinam species is *Tetragastris Hostmanni* (Engl.) Kuntze. There is no specimen in the Yale wood collections, but from

Pfeiffer's account (*Houtsoorten van Suriname* 1: 319–322), it is similar to that of other species in structure, but has a higher density (sp. gr. 0.90 to 1.02). A sample (Yale 12748, Bertin 3018) of the Encens Rouge described by Bertin (*Les bois de la Guyane française,* pp. 93–94) under the scientific name of *Papayrola guianensis* Aubl. (fam. Violaceae) is a species of *Tetragastris,* but the sp. gr. is given as 0.77. The timber is scarce.

Heartwood light brown when fresh, becoming somewhat orange-brown upon exposure; rather sharply demarcated from the lighter-colored sapwood which turns yellowish brown on the surface. Luster medium, golden. Without distinctive taste, but with a mildly fragrant balsamic scent. Density variable, but mostly medium; texture rather fine; grain irregular to distinctly roey; not difficult to work, taking a high polish; said to be resistant to insects but not very durable in contact with the ground. A good timber, but presumably without commercial possibilities because of the limited supply available.

Common names: Azucarero, incienso, palo cochino, p. de cochino (Cuba); masa, palo de aceite (P.R.); amacey (Dom. R.); bois cochon, sucrier de montagne (Haiti); capache, carbón (Br. H.); aguarrás, palo de cerdo, trementino azucarero (Col.); haiowaballi (Br. G.); aloewau-oe, Corantijnsch mahonie, joeliballi, j. fieberoebana, joeriballi, pakiria-sipioli, pelika patiara sipioro, salie, witte salie (Sur.); bois cochon, b. gommier rouge, encens rouge, gommier de montagne, sucrier de montagne (Fr. G.); almesca (Braz.).

Trattinickia, with four species of medium-sized trees, is of limited occurrence in the Amazon basin. The leaves are typically congested at the ends of the branches. The flowers have three united petals, valvate in the bud. The stems have gray bark, rich in resin. Information concerning the timber is meager and confused, but it appears some of the Guiana woods commonly referred to the genus *Protium* belong here. This conclusion is reached from a study of a British Guiana specimen (Yale 32882; B.G. For. Dept. 915) of Ulo, *T. demererae* Sandw., and two (Yale 15886 and 15887) of Tingie Monnie, *Trattinickia* sp., supplied by the Landbouwhoogeschool, Wageningen, Netherlands. The other common names listed for this genus are those given by Pfeiffer (*Houtsoorten van Suriname* 1: 323) for the Tingie Monnie group.

Wood yellowish brown or oatmeal-colored, sometimes with a greenish tinge; without clear distinction between heartwood and sapwood. Luster medium to high. Odorless and tasteless. Density variable, mostly medium; texture medium to rather coarse; grain more or less roey; not difficult to work, though one specimen (Yale 32882) has a wiry feel after being planed; durability presumably low. According to Pfeiffer (*loc. cit.*), there are two classes of Tingie Monnie timber, the one just described and another with reddish brown heartwood. Yale specimen No. 15886 is denser and finer-textured than the other two, and has fairly distinct yellowish brown heartwood and gray sapwood.

Common names: Bastard cedar, haiawa?, olu, ulo, white cedar (Br. G.); ajawa, aloewa-oe, apoto ajawa, gange iesè, g. pisie, hajawa, h.-balli, h.-b. hororodikoro, khalemeroe, koeroekai, ollo, olo, o. karau-bandikorro, orokaike, pakiria sipiolie, sipio, sipo, siwaarwahajawa, tapoekjen ajawa, tiengi monni, tingie monnie, wooitano ajawa, wosiono sipio, witte ceder (Sur.); ayawa, bois encens, cedre blanc, encens gris (Fr. G.).

BUXACEAE

The Boxwood family, as usually classified, includes six genera and about 35 species of shrubs, small trees, and a few herbs, occurring in the tropical and mild temperate regions of the world. The important genus is *Buxus,* especially *B. sempervirens* L., the source of true Boxwood, the only material suitable for the finest wood engravings and formerly used much more extensively than now for shuttles and articles of turnery. In the broader sense, *Buxus* has several species in southern Europe, Asia, and the West Indies, and one in South Africa, but according to a different classification the African species is placed in a separate genus, *Buxella,* and the West Indian in another, *Tricera* (see Yale School of Forestry Bull. No. 14, *Boxwoods*). The other woody genera in America are *Styloceras,* with a few species in the tropical Andean region of South America, and *Simmondsia,* with one species, in northern Mexico and southwestern United States. In these three the leaves are opposite, simple, and persistent; stipules are absent; the small inconspicuous flowers are in axillary inflorescences; the fruit is a 3-celled capsule.

Simmondsia *californica* Nutt. is a shrub or little tree sometimes 15 feet high, growing on sandy hills in southern California and Arizona and in northwestern Mexico. The seeds, which are about the size of a hazelnut, have an oily, slightly bitter kernel and are eaten, raw or cooked, or used medicinally. The only other use for the plant is for fuel.

Wood lemon-yellow throughout. Without distinctive odor or taste. Hard and heavy; fine-textured; of irregular grain; not resistant to decay. Of anomalous structure, phloem strands being included in coarse, concentric, anastomosing bands of conjunctive tissue suggesting *Avicennia.*

COMMON NAMES: Goatnut, jojora nut, pignut, quinine plant, sheepnut, wild hazel (U.S.A.); jojoba (Mex.).

Styloceras, with three species of trees, usually less than 25, rarely up to 40, feet high, occurs at elevations of 8000 to 11,500 feet in the Andes Mountains in Colombia, Ecuador, Peru, and Bolivia. The only specimens available are of *S. laurifolium* H.B.K. collected by A. Rimbach in the valley of the Río Chimbo, Western Cordillera, Ecuador. The leathery, yellowish green, obovate leaves are about three inches long; the 2-horned globose fruits are solitary in the leaf axils. The bark is deeply furrowed and corky. The wood is highly esteemed locally for joinery.

Color uniform yellowish white throughout. Luster medium. Odor and taste absent or not distinctive. Of about the same consistency as Holly (*Ilex*); texture rather fine, uniform; grain straight to somewhat irregular; working properties excellent; finishes very smoothly, holds its place well when manufactured; probably perishable in contact with the ground. A first class timber but apparently without commercial possibilities.

Growth rings absent or poorly defined. Pores small to minute, angular, not visible without lens; fairly numerous; mostly solitary, sometimes in small radial multiples, well distributed. Vessels with long scalariform perforation plates with many narrow and closely spaced bars; spiral thickenings absent; intervascular pitting tending to scalariform. Rays 1 to 3, occasionally 4, cells wide and up to 75, often less than 25, cells high; decidedly heterogeneous, the cells greatly variable in form; pits abundant, medium-sized, oval to elongated. Wood parenchyma fairly abundant, diffuse to finely reticulate; barely visible with lens. Wood fibers large; rather thick-walled; distinctly bordered pits numerous in both radial and tangential walls. Ripple marks and gum ducts absent.

COMMON NAMES: Platuquero, guishcás (Ec.).

Tricera, either as a distinct genus or as a section of *Buxus,* includes about 15 species having their center of distribution in the West Indies, with known extensions into southern Mexico, British Honduras, Panama, and Venezuela. They are rather rare plants, nearly all evergreen shrubs, though occasionally trees 15 to 25 feet tall. The wood is light yellow throughout; odorless and tasteless; hard, heavy, compact, of very fine and uniform texture; closely resembles *Buxus sempervirens.* It is not of commercial value because of the scarcity of the larger sizes, but is suitable for articles of turnery and for engraving.

Growth rings present. Pores minute, not visible without lens; numerous; well distributed without pattern; always solitary (except for tangential pairs where vessel members overlap). Vessels with scalariform perforation plates with few to 15 bars; spirals absent; intervascular pitting (in radial walls where members overlap) very fine, alternate. Rays 1 to 4 cells wide and few to 25 cells high; somewhat heterogeneous; pits to vessels very small. Wood parenchyma sparingly vasicentric and diffuse; not visible with lens. Wood fibers with thick walls and very numerous distinctly bordered pits in both radial and tangential walls. Ripple marks and gum ducts absent.

CACTACEAE

THE Cactus family, according to Vaupel's classification in *Pflanzenfamilien* (2nd ed.) 21: 594, comprises 26 genera and about 1200 species of succulent and spiny shrubs and trees adapted to growth in a desert climate. The family is most abundantly represented in the arid region of southwestern United States and northern Mexico, but there are many forms in the West Indies and in South America to Patagonia and Chile. Species of *Pereskia* and *Pereskiopsis* look most like ordinary trees, but columnar forms 20 to 60 feet high and up to two feet in diameter occur in the genera *Cereus, Cephalocereus, Nopalea,* and *Opuntia,* and within the fleshy stems are skeletons or ribs of wood for which there is considerable local use because of the scarcity of other timber. In parts of northern Venezuela, for example, the Cardón (*Cereus* or *Cephalocereus*) is very common, in places occurring in pure stands, and supplies attractive, easily worked lumber for making chairs, small tables, and other household equipment; also for rafters, picket fences, and fuel; it is not exported. The woods exhibit considerable variation, but in general are creamy white to light brown, apparently without differentiation into heartwood and sapwood. Some are light and soft, others hard and heavy, but the kinds commonly used are of medium density.

Pores sometimes large enough to be seen, but more often very small; numerous to rather few; solitary or in small groups, without definite pattern. Vessels with simple perforations; without spirals; pits large, irregular, often elongated. Rays coarse, usually highly conspicuous and giving rise to silver grain on radial surface, suggesting *Platanus;* heterogeneous, the cells generally large and irregular; pits to vessels large, mostly elongated, tending to scalariform arrangement. Wood parenchyma sparse to abundant; paratracheal and diffuse, but without special pattern. Wood fibers often septate; pits simple, narrowing outward and sometimes appearing bordered in surface view. Ripple marks absent. Radial channels often present, giving a pitted appearance to the tangential surface of the wood.

CALYCANTHACEAE

THIS family consists of two genera and several species of aromatic shrubs. *Chimonanthus* is limited to eastern Asia. *Calycanthus* has three species in southeastern United States and one in the Coast Range of California. The plants are usually less than 10 feet tall. The coarse, usually pubescent leaves are simple, entire, opposite, and without stipules; the solitary, axillary flowers have the fragrance of strawberries; the fruit is an achene, many of them being borne together within a capsule-like pseudocarp. *C. floridus* L., which grows naturally on hillsides and along streams from southern Virginia to Florida, Alabama, and Mississippi, is often planted because of the rich fruity fragrance of its dull purple flowers. The following description is based on one specimen (Yale 20680) of *C. floridus* and two (Yale 23691 and 27296) of *C. occidentalis* H. & A.

Wood whitish or brownish throughout. Luster medium. Without distinctive odor or taste, although the bark has a strong camphor-like scent. Moderately hard, heavy, and strong; texture fine; grain straight. Without utility because of the small size of the plants.

Growth rings present; structure more or less definitely ring-porous. Pores small (50 to 65μ); those in late wood in short to rather long multiples and in clusters, with diagonal or wavy radial arrangement. Vessels with simple perforations; pits large (10 to 11μ), round, alternate, not crowded; many very slender vessels with spiral thickenings present. Rays 1 to 3 cells wide and up to 20 cells high in *Calycanthus floridus,* uniseriate or biseriate and up to 35, sometimes to 70, cells high in *C. occidentalis;* decidedly heterogeneous; ray-vessel pitting coarse, frequently unilaterally compound. Wood parenchyma very sparse. Wood fibers with small indistinctly bordered pits. Ripple marks absent. No gum ducts seen. Structure does not suggest that of the order Rosales.

COMMON NAMES: Carolina allspice, spicebush, strawberry shrub, sweet shrub, sweet-scented shrub (U.S.A.).

CANELLACEAE

THIS family, with five or six genera and about a dozen species of aromatic shrubs and trees, is of very limited distribution: *Cinnamosma* (2 spp.) in Madagascar; *Warburgia* (3 spp.) in East Africa; *Canella* (1 sp.) in West Indies; *Cinnamodendron* (4 spp.) in West Indies and eastern Brazil; *Pleodendron* (1 sp.) in Puerto Rico; *Capsicodendron* (1 sp.) in southeastern Brazil. The leaves are simple, entire, leathery, gland-dotted, alternate, and without stipules; the flowers are solitary in the leaf axils or in terminal corymbs; the fruit is a berry. The principal value of the plants is in the bark, which contains pungent spicy-resinous oils used for incense, perfumes, condiments, and medicines. The wood of *Warburgia*, and perhaps of *Cinnamosma* also, is used as a substitute for Sandalwood (*Santalum*) (see *Kew Bulletin*, 1933, pp. 3–15; 1935, pp. 185–186). There are no specimens of *Cinnamodendron* available for study. The woods of *Canella* and *Pleodendron* are very similar. *Capsicodendron* will be considered separately.

Canella, with one species, *C. alba* Murr. or *C. Winterana* (L.) Gaertn., occurs in southern Florida and southward through the West Indies to Barbados. It varies in size from a shrub to a tree 50 feet high with a straight trunk 8 to 10 inches thick. The yellowish inner bark has the scent of cinnamon and is used as an aromatic stimulant and tonic. In Cuba the wood is used to a limited extent for plow beams, house poles, and miscellaneous purposes requiring great strength.

Heartwood dark olive-brown, with oily appearance; rather sharply demarcated from the greenish yellow or nearly white sapwood. Mildly scented when fresh; no distinctive taste. Density high, the sp. gr. of the heartwood near or slightly greater than unity; hard and strong; not easy to work, but takes a smooth finish and high natural polish; is resistant to decay.

Growth rings poorly defined. Pores fairly numerous; small to medium-sized, the larger scarcely visible without lens; solitary; irregularly distributed without pattern, sometimes in short rows but not in actual contact radially. Vessels with scalariform perforation plates with numerous bordered bars; spirals absent. Rays numerous; mostly uniseriate or partly biseriate and less than 30, sometimes up to 50, cells high; more or less heterogeneous, many of the cells being squarish; crystals abundant; oil cells absent; ray-vessel pit-pairs half-bordered, rather large, rounded or elongated. Wood parenchyma not visible without lens; unilaterally paratracheal (forming a cap 2 to 5 cells wide on outer side of pore) and apotracheal as single cells and small aggregates; pits to vessels often elongated; oil cells absent. Wood fibers with thick to very thick walls; pits numerous in both radial and tangential walls, the apertures lenticular, the borders large and distinct. Ripple marks and gum ducts absent.

COMMON NAMES: Canella, cinnamon bark, whitewood, wild cinnamon (Fla.); barbasco, canella, pepper cinnamon, wild cinnamon, winter bark (P.R.); amansa guapo, canela blanca, cúrbana, malambo, palo malambo (Cuba); canella powree (Haiti); canelillo (Dom. R.).

Capsicodendron *pimenteira* Hoehne, the only species, is a small tree of southeastern Brazil, locally known as Pimenteira because the bark has a peppery taste suggesting *Capsicum*. There are no known uses for the wood, which in many ways is different from that of the other members of the family so far as known. Sapwood white; heartwood not seen. Odorless and tasteless. Rather hard and heavy, suggesting Birch (*Betula*); texture fine and uniform.

Growth rings distinct, owing to slight differences in density and in size of pores. Pores small to medium; very numerous; well distributed, solitary. Vessels with long scalariform perforation plates having very numerous, fine, closely spaced, frequently anastomosing bars; pitting fine, opposite to scalariform. Rays numerous; uniseriate or locally biseriate; few to 30 cells high; homogeneous to heterogeneous; no crystals or oil cells observed; ray-vessel pitting very fine, mostly unilaterally compound. Wood parenchyma abundant, diffuse or in short tangential lines, scarcely distinct with lens; oil cells present. Wood fibers with moderately thick walls and very numerous distinctly bordered pits larger than those in vessels; non-septate. Resin ducts and ripple marks absent. *Material:* Yale 23444; Hoehne 27074.

Cinnamodendron, with four species of shrubs or little trees sometimes 30 feet high, occurs in the Antilles, Surinam, and eastern Brazil. The bark is used medicinally. The wood has not been studied.

COMMON NAMES: Cúrbana (Cuba); canella branca, herva moura do sertão, paratudo aromatico, pau pimenta (Braz.).

Pleodendron *macranthum* (Baill.) Van Tiegh., the only species, is a tree, sometimes 30 feet high with a trunk 14 inches in diameter, of rare occurrence in the forests of Puerto Rico, where it is endemic. It is known as Chupagallo. The only wood sample (Yale 35444) in the Yale collections was obtained with fruiting herbarium material by the Supervisor of the Caribbean National Forest from a tree growing in the northeastern section of the Luquillo Range. The structure of the wood is essentially the same as that of *Canella*.

CAPPARIDACEAE

THE Caper family comprises about 44 genera and 1000 species of herbs, erect or scandent shrubs, and small to medium-sized trees distributed over the warmer parts of the world. The leaves are typically alternate, simple or digitate, sometimes with minute or spiny stipules; the flowers are variously arranged; the fruit is a capsule or a berry. A few of the plants are cultivated for ornament, several are supposed to have medicinal value, and some supply woods of local utility in places where other timber is scarce, but the only articles of commerce are capers, the dried or pickled flower buds of *Capparis spinosa* L., a shrub of the Mediterranean region and extending to India. Eight genera of trees and shrubs are confined to the New World and two others (*Capparis* and *Crataeva*) are well represented there. Some of the plants have a fetid odor. The following description is based upon specimens of *Atamisquea, Capparis, Crataeva, Isomeris, Morisonia, Steriphoma,* and *Stuebelia*. The wood of *Forchhammeria* is of anomalous structure, having laminations of included phloem and conjunctive tissue.

Wood typically yellowish, sometimes gray or brownish; usually without much distinction between heartwood and sapwood. Faint disagreeable odor sometimes perceptible when dry wood is moistened; taste absent or not distinctive. Variable from hard and heavy to rather light but firm; texture fine to medium; grain variable, often irregular; working properties fair to good; durability usually low.

Growth rings generally present. Pores medium-sized to very small; generally thick-walled and rounded; occurring singly and, more often, in little groups or short to long rows. Vessel lines characteristically crooked on tangential section, often changing their course abruptly; "elbows" consisting of few to several tangentially arranged members with perforations near the middle appear on cross section as tangential multiples of 3 to 5 pores for a single vessel; perforations simple; no spiral thickenings seen; pits small to minute, alternate, vestured. Rays all uniseriate or uniseriate and biseriate in *Capparis* (in part), *Steriphoma,* and *Stuebelia;* uniseriates few, the others 3 or 4 cells wide, in *Atamisquea, Capparis* (in part), and *Morisonia,* and up to 5 or 6 cells wide in *Crataeva* and *Isomeris;* maximum height variable from 15 to 200 cells; homogeneous or nearly so in *Capparis* and *Crataeva,* heterogeneous in the others; ray-vessel pitting fine to very fine. Wood parenchyma generally sparingly paratracheal or vasicentric with little to considerable vasicentric-confluent; rarely somewhat aliform or aliform-confluent; sometimes terminal; in numerous concentric bands in *Steriphoma;* cells often fusiform or in 2-celled strands. Wood fibers commonly very short; occasionally septate; bands of rather loosely aggregated thinner-walled fibers present in some specimens of *Capparis* and less distinct in other genera; pits small to minute, simple or indistinctly bordered, more or less clustered as in disjunctive parenchyma. Ripple marks present in *Isomeris* and *Atamisquea;* larger rays not storied. No gum ducts seen.

Atamisquea *emarginata* Miers, the only species, is a densely branched ill-scented shrub of dry plains and hillsides of northwestern Mexico and southern South America. The only specimen available (Yale 32088; Imp. For. Inst. Oxford 9631) is from Argentina. Unlike the other representatives of the family studied, it has light olive-brown heartwood sharply demarcated from the thin white sapwood. It is hard and strong, of very fine and uniform texture, and appears to be durable.

COMMON NAMES: Atamisquea (Chile); atamisco (Arg.).

Capparis. There are about 350 species of xerophytic shrubs, climbers, and small to medium-sized trees in this genus. At least two of the West Indian species reach the shores of southern Florida where one of them, *C. cynophallophora* L., is sometimes 20 feet high with a trunk six inches in diameter; elsewhere a maximum height of 50 feet has been reported. There are many shrubby species in the thorn forests of arid regions in South America, especially near the coast, and they are useful for fuel. The following description is based on 22 samples of 15 species.

Wood whitish or yellowish throughout, frequently with dark brown streaks resulting from wounds. Luster medium. Faint disagreeable odor occasionally noticeable. Mostly moderately hard and heavy; texture rather fine; grain straight to irregular; working properties good; durability poor. Of no commercial possibilities.

COMMON NAMES: Caper tree, man-of-war bush, mustard tree (Eng.); black willow (Bah.); bejuco inglés, black wattle, b. willow, burro, b. blanco, linguam, palinguam, palo de burro, p. de b. prieto (P.R.); alcaparro, bejuco inglés, carbonero, ciguarayo, mostacilla, mostaza, olivo, palo de perro, p. diablo, p. verraco (Cuba); avocat maroon, balai velours, bois caca, b. d'argent, b. rave, b. Sénégal (Haiti); frijol, mostazo, olivo, palo de maco (Dom. R.); bois d'olive, mabouya, olivewood (Trin.); hoerihoeri, jeerba mostera, kade besji, olieba maatsjóe, paaloe die loora, p. preetoe, raaba, stokki (Dutch W.I.); chochitam, coloc, colorín, coquito, limoncillo, mangle de la sierra, mimbre de monte, naranjillo, palo zapo, vara blanca, v. prieta, xbayumac, xcoche, xpayumac, zapote prieto (Mex.); castillo (Br. H.); naranjillo (Guat.); azaharillo, olivo (Hond.); curumo, guacoco, quitacaezón (Salv.); escremento (Nic.); ají, ajicito, alcaparro, arará, auso, calabacillo, calabacito, contra arará, c. prieta, cotorrito, cucarachero, frijolillo, frijolito, fruta de burro, f. de zorro, huevo de burro, lengua de venado, limpiadente, maretiro, mataratú, medialuna, naranjuelo, nudo, olivo, o. macho, palo de agua, rala de gallina, taparera, tinto, tobaco de burro, totumito, tunito (Col.); ajacito, ajito, cansa caballo, gatillo, guariare, guayabo de loro, juan blanco, lengua de vaca, mamatete, mosto, olivo, o. macho, o. negro, o. santo, o. tabasca, paniagua, pan-y-agua, quebebé, vela de muerto, zorrocloro (Venez.); gatillo, guariare, ojicito, olivo (Br. G.); oenbatappo, warimiaballi (Sur.); alcaparreira, a. cheirosa, cipó-taía, fructeira de burro (Braz.); intuto-caspi, mango-micuna, ninacaspi, quina-quina (Peru); amarguillo, azarrá, azucena de monte, cayampa, ibirápororó, meloncillo, muicuré-caá, sacha membrillo, yerba de comandreja (Arg.).

Crataeva. About 20 species have been described, five of them tropical American with a combined range from the West Indies and Mexico to Argentina. The American species are closely related and the one best known and most widely distributed is *C. tapia* L., a tree rarely 50 feet high and 20 inches in diameter, with trifoliolate leaves and globose berries. The fruit and, in lesser degree, the fresh wood has a garlic-like odor. The timber, which is white or yellow, moderately hard, of medium texture, and rather brittle, is little used, but is suitable for minor carpentry purposes where resistance to decay is not a factor.

COMMON NAMES: Garlic pear, tocque (Jam., Trin.); cascarón, kolokmax, perihuete, tres Marías, trompo, xkolocmas, yuy, zapotillo amarillo (Mex.); waika bead (Br. H.); tortugo (Guat.); cachimbo (Hond.); anonillo, granadillo macho (Salv.); manzana de playa (Nic.); estralla, palo de guaco (Pan.); naranjillo, naranjito, naronjuelo, sorrocloco, zorrocloco (Col.); isiro, tambor, toco (Venez.); soeroen die moondi (Sur.); yagua de lagarto (Ec.); nina-caspi, tamara, ynsira-maskan (Peru); catauary, pau d'alho, tapiá, trapiá (Braz.); flor de seda, naronjillo, papagua-naranjo, papaguanyan, payaguá-naranjo (Arg.).

Forchhammeria, with nine species, typically shrubs and small trees, occurs in the West Indies and from California to Guatemala and Salvador. The largest tree reported is 50 feet high and 10 inches in diameter. There are no known uses for the wood.

COMMON NAMES: Palo San Juan, tres Marías (Mex.); bastard dogwood (Br. H.).

Isomeris *arborea* Nutt., the sole species, is an ill-smelling shrub or little tree limited in distribution to southern California and northwestern Mexico. The yellowish white, hard, fine-textured wood has no special uses.

COMMON NAMES: Bladder-pod, caper bush (U.S.A.).

Morisonia. Of the five species inhabiting parts of the West Indies, Mexico, and northern South America, the most widely distributed is *M. americana* L., a shrub or a small tree sometimes 30 feet high. The pale yellow, bitter, faintly odorous wood is fine-textured and suitable for turnery and carving though not quite dense enough to serve as a substitute for Boxwood.

COMMON NAMES: Wild mespili (P.R.); bois decree, b. doré (Dom.); árbol del diablo, chico, chicozapote (Mex.); árbol del diablo, cacao cimarrón, calabacillo, calabasuero, níspero de saino, rabo de mico, tabaco de burro, toro, totumito (Col.); zorrocloco (Venez., Br. G.); toco, wild sapodilla (Trin.).

Steriphoma. There are six species in tropical South America and Panama but apparently they are all shrubs. The whitish, mildly odorous, rather fine-textured wood is without value.

Stuebelia *nemorosa* (Jacq.) Dugand, the only species, is a small evergreen tree, 12 to 15 feet high, having an irregularly shaped trunk 8 to 12 inches in diameter, with rough, malodorous bark. It is fairly common in dry situations in northern Colombia. (See *Tropical Woods* 43: 15.) The yellowish wood has a fetid odor and a slightly astringent taste. It is of medium density and texture and suitable for turnery and carving, though probably without commercial possibilities.

COMMON NAMES: Calabacito, calabasuero, cabalazuelo (Col.).

CAPRIFOLIACEAE

THE Honeysuckle family is composed of about 10 genera and 340 species of small trees, shrubs, and woody vines of wide distribution in temperate regions and at higher elevations in the tropics. The leaves are opposite, simple or compound, and without stipules; the flowers are usually white and cymose; the fruit is a berry or a drupe. Some of the plants are cultivated for ornament, common ones being the Honeysuckle (*Lonicera*), the Snowball (*Viburnum opulus* L.), the Elder (*Sambucus*), and the Snowberry and Coral-berry (*Symphoricarpus*). The family furnishes no commercial woods. The only genera with tree species in America are *Sambucus* and *Viburnum*. Their woods are so different that they will be described separately.

Sambucus, with about 20 species, typically shrubs, is represented in Asia, Europe, northern Africa, North America, the West Indies, and the Andean region of South America. The natural range of some species has been extended by planting, as the masses of small flowers are showy and fragrant and the berries are edible and suitable for making wine. The plants have odorous pinnate leaves and are characterized by large white or yellow pith which has some commercial applications. Previous to the introduction of metal spouts, small stems of Elder were used in the United States to make "spiles" for conducting the sap of the Sugar Maple to wooden pails in which it was collected in the manufacture of sirup and sugar. The wood of the European Elder is used to a limited extent for small articles of turnery. Although some of the trees in Mexico and southern South America attain a height of 35 to 50 feet and a trunk diameter of 12 to 18 inches, the timber apparently is not employed for any special purposes.

Heartwood yellowish brown, not sharply demarcated from the white sapwood. Luster rather low. Odorless and tasteless. Light in weight but firm; usually brittle; texture medium, with a rather harsh feel to sawed material; easy to work, finishing smoothly; is not durable.

Growth rings generally distinct. Pores very numerous; small to minute, usually reduced in size during seasonal growth; clustered and tending to form tangential bands, particularly in the late wood. Vessels mostly with simple perforations; scalariform plates with few narrow bars sometimes present in very small vessels; no spiral thickenings seen; thin-walled tyloses common; pitting rather coarse, alternate. Rays 1 to 5 cells wide and generally less than 40 cells high, though occasionally higher; heterogeneous, but often without conspicuously upright cells; sheath cells sometimes present; pits to vessels mostly medium-sized and oval, but tending to become elongated. Wood parenchyma sparingly paratracheal, not distinct with lens. Wood fibers with rather thin walls and numerous very small simple or indistinctly bordered pits. Ripple marks and gum ducts absent.

Common names: Elder (Eng.); saúco (Sp.); sureau (Fr.); saúco blanco (Cuba); azumiatl, coyopa, cumdemba, cumdumba, nttzirza, skiiksh, xumetl, yutnucate (Mex.); bahman, sakatsum, tzoloh, tzolohquen, tzolotché (Guat.); rayan (Peru); acaporá (Par.).

Viburnum, with about 120 species of shrubs and small trees, has about the same range as *Sambucus,* though extending to Polynesia, Australia, and South Africa. Of the 15 American species, four are small trees, occasionally attaining a height of 30 to 40 feet and a diameter of 10 to 18 inches. The heartwood of the Sheepberry or Nannyberry (*V. lentago* L.) and the Black Haw (*V. prunifolium* L.) is brownish yellow with a somewhat oily appearance and feel and a disgusting odor which persists interminably. The wood of a Cuban specimen of *V. villosum* Sw. is similarly, though not so strongly, scented, whereas that of *V. pichinchense* Benth., a tree 25 feet tall and four inches in diameter in the Western Cordillera of Ecuador, is odorless. Specimens from different parts of the world exhibit considerable variation in density, texture, number and size of pores, and abundance and arrangement of parenchyma.

Heartwood pale brown to brownish yellow; sharply demarcated from the whitish sapwood. Luster low to medium. Hard and heavy to moderately so; of fine and uniform texture; grain usually straight; easy to work, finishing very smoothly; deeply colored specimens durable. Of no commercial possibilities.

Growth rings usually present. Pores very small to small; rather to very numerous; rarely in contact radially. Vessels with very long, scalariform perforation plates having many closely spaced bars; fine spiral thickenings sometimes present; pitting in overlapping members rather coarse, opposite to scalariform. Rays 1 or 2, sometimes 3 or 4, cells wide and up to 50, occasionally over 100, cells high; decidedly heterogeneous, the uniseriate rays and ray-margins composed of tall, upright cells; pits to vessels medium-sized and oval to much elongated and in scalariform arrangement. Wood parenchyma finely reticulate, scarcely distinct with lens. Wood fibers with rather thick walls and numerous to very numerous, large, distinctly bordered pits. Ripple marks absent. No gum ducts seen.

Common names: Arrow wood, black haw, cranberry tree, dockmackie, highbush cranberry, laurestinus, nannyberry, pimbina, sheepberry, squashberry, wayfaring tree, wild raisin, withe rod (U.S.A.); black wattle (Jam.); achichil, carindapaz, sunda, tlamahuacatl (Mex.); conchudo, curá, surá, tirrá (C.R.); pitá (Col.).

CARICACEAE

The Papaya family comprises four genera and about 45 species with the following distribution: *Carica,* with 40 species of small, rarely large, trees, often with unbranched stems, occurs throughout tropical and subtropical America; *Cylicomorpha,* with two species of medium-sized to large trees, is limited to tropical Africa; *Jacaratia,* with seven or eight species of small to large trees, some of them armed with short spines, has a range from southern Mexico to Paraguay; *Mocinna* (= *Jarilla*), with one or two species of herbs, is confined to Mexico.

The leaves are typically large, alternate, often variously digitately lobed or foliolate; stipules are wanting; the flowers are either perfect or, more commonly, unisexual and dioecious; the fruit is a large berry. All parts of the plants have a milky latex and that of *Carica* contains papain which resembles animal pepsin in its digestive action. *Carica papaya* L. is commonly cultivated in the warmer parts of the world for its highly esteemed melon-like fruit, the papaya.

The stems of *Carica* and *Jacaratia* are herbaceous in character and owe most of their strength to their bark which contains many thick-walled bast fibers and shows prominent phloem rays. The watery xylem mass consists of wedges of soft parenchymatous tissue and numerous vessels. The vessel wall, where in contact with parenchyma, appears reticulately thickened, the pits being large, irregularly gash-like or lenticular, and simple; where two vessels are in contact the pits have distinct, subcircular borders. There are no wood fibers. Ripple marks and gum ducts are absent.

The family supplies no wood of value for any purpose. Even the larger trees can be felled with a few blows of a machete. Upon drying, the soft interior of the stem shrinks to a small proportion of its original size. *Carica dolichaula* Donn. Sm. of Central America develops a large trunk and the bark is so thick and strong that the natives use cylinders of it for drums in which to store grain.

Common names: *Carica:* Melon tree, papaw, pawpaw (Eng.); papaya (fruit), papayo (Sp.); lechosa (P.R.); papaye (Haiti); chick-put, papayero, put (Mex.); lerdo, palo de barril, papayillo, papayo de monte, tapaculo (C.R.); babaco, chamburo, chiluacán (Ec.); mamoeirinho, mamoeiro (Braz.); higuerón, manón (Arg.). *Jacaratia:* Bonete, coahuayote, coalsuayote, cuaguayote, kumché, kunché, orejona, papayo montés (Mex.); cuayote (Salv.); papayillo de venado (C.R.); mamoeiro bravo, m. do matto (Braz.); jacaratiá (Arg., Urug.).

CARYOCARACEAE

This family, sometimes included with the Theaceae, is composed of only two genera, *Anthodiscus* and *Caryocar,* with about 25 named species of medium-sized to tall trees, rarely shrubs, making their best development in the Guianas and the Amazon basin, though the entire range extends as far north as Costa Rica and as far south as Paraguay. *Caryocar* is important as a source of structural timber for local uses, but is better known commercially for its edible, fat-yielding nuts. The woods differ in so many details that they will be described separately. (See *Tropical Woods* 42: 1.)

Anthodiscus, with ten species, occurs in the Guianas, Colombia, and the upper Amazon region of Brazil and eastern Peru. The largest species, and the only one whose wood has been studied, is *A. montanus* Gleason, a tree 60 to 100 feet tall, growing at elevations of 3400 to 4000 feet in central Colombia, where it is known as Cheepo.

Heartwood yellowish brown, more or less streaked, and nearly black around knots; distinct but not sharply demarcated from the sapwood. Appears and feels rather waxy. Faintly but pleasantly scented, but without distinctive taste. Moderately hard and heavy, tough and strong; rather fine-textured; straight-grained; easy to work, finishing smoothly; apparently durable.

Growth rings absent or indistinct. Pores medium-sized, barely visible without lens; numerous but not crowded; well distributed, occurring singly or more often in radial pairs or short rows. Vessels with simple perforations; without spirals; pitting very fine and alternate; tyloses absent. Rays numerous; uniseriate and biseriate and generally 1 to 25 cells high, but sometimes vertically fused; distinctly heterogeneous; ray-vessel pit-pairs very small and numerous; crystals absent or rare. Wood parenchyma inconspicuous; paratracheal and sometimes aliform and confluent; no crystalliferous strands seen; pith flecks common. Wood fibers thick-walled, non-septate; pits very small, simple or indistinctly bordered. Ripple marks and gum ducts absent.

Caryocar, with 14 to 17 species of shrubs and medium-sized to very large trees, has its center of distribution in the Amazon region. The Piquiá of Brazil, *C. brasiliense* Camb., and the Sawarie or Souari of British Guiana, *C. nuciferum* L., are the principal sources of nuts, often called butternuts, which should not be confused with the North American butternuts from *Juglans cinerea* L. The fruits are spherical, sometimes six inches in diameter, with a yellow pulp containing one to four bean-shaped seeds covered with tubercles. The oily kernels have a pleasant taste and oil expressed from them is used for culinary purposes. Some of the other species also have edible nuts and the pulp of some kinds is used as a fish poison, though it is edible when cooked.

In the Guianas and eastern Brazil the trees are large and of good form, logs 70 feet long and squaring 24 inches being obtainable. The timber is well known locally, though not exported, and is used for the frames, knees, and decking of ships, joists and flooring of warehouses, hubs and felloes of wagon and cart wheels, and for cooperage. Because of the similarity of common names in parts of Brazil, the wood of *Caryocar* is sometimes confused in the literature with that of *Aspidosperma* (fam. Apocynaceae) which has entirely different structure, properties, and uses.

Heartwood yellowish or grayish, sometimes rather oily, scarcely separable from the sapwood. Fresh material with a mild vinegary scent; taste not distinctive. Sp. gr. (air-dry) 0.80 to 0.90; weight 50 to 56 lbs. per cu. ft.; texture rather coarse; feel harsh; grain roey; tough and strong; fairly easy to saw but requires sharp tools to finish smoothly; is difficult to section with a microtome; does not appear resistant to decay, though oily heartwood is said to be durable.

Growth rings often distinct because of narrow bands deficient in parenchyma. Pores variable in size, the largest distinct without lens; not very numerous; fairly well distributed without pattern, solitary and in small multiples. Vessels with simple perforations, though tendency to the formation of scalariform plates has been reported; without spirals; pitting rather coarse, typically alternate; tyloses present, often abundant, sometimes rather thick-walled; vessel lines generally very distinct. Rays numerous; decidedly heterogeneous; of two sizes, uniseriate and mostly less than 20 cells high, and biseriate or triseriate (rarely wider) and commonly less than 50 cells high, but occasionally much higher through vertical fusion; ray-vessel pit-pairs very large, simple, irregular in form and arrangement; gum deposits abundant; crystals common. Wood parenchyma incompletely paratracheal and in very numerous irregular metatracheal lines and diffuse, producing a fine meshwork visible with lens; crystalliferous strands common. Wood fibers with thick, commonly gelatinous, walls and small to minute cavities; often septate; pits small, simple or indistinctly bordered. Ripple marks and gum ducts absent.

Common names: Ají, ajillo, ajo, caballo-kup (C.R.); achiotillo, almendrón, a. de mariquita, caquí, maní (Col.); batsouari, butternut-tree, cola, pekia, sawarri, souari, suwarrow (Br. G.); bokkenoot, ningre notto, sawarie, saouarou, saouary, soeri, zeephout (Sur.); bois de tatajuba, b. de tatayouba, b. Marie, b. Mary, chavarie, chawari, pekea, peki, pekia, pequi, pequy, schwari (Br. G.); pequi, pequiá, p. brava, p. eté, p. rana, piquiá, p. rana, p. r. da terra firme, p. r. da varzea, p. rocha (Braz.); almendro, a. de bajo (Peru).

CELASTRACEAE

The Staff-tree family comprises about 50 genera and 400 species of trees, shrubs, and woody climbers of general distribution but of very little economic importance. The leaves are alternate or opposite, simple, evergreen or deciduous, and with or without stipules; the flowers are mostly cymose or fasciculate; the fruit is a capsule or a drupe; the seeds often are arillate. There are a few species of large trees in the Indo-Malayan region but the kinds used are small, supplying white wood of fine and uniform texture for carving, turnery, utensils, and combs.

There are about 25 genera in the New World, their combined range extending from central United States to Chile and Patagonia. One of the best known plants is the Bitter-sweet, *Celastrus scandens* L., a twining shrub with attractive orange-colored pods which open and disclose the scarlet covering of the seeds. Many species of *Evonymus* are planted for decorative purposes; they are mostly shrubs, although the American Wahoo or Burning Bush (*E. atropurpureus* Jacq.) and the European Spindle-tree (*E. europaeus* L.) are trees 20 to 25 feet high. Other arborescent genera in the United States are *Maytenus, Rhacoma,* and *Schaefferia* in southern Florida, and

Canotia in Arizona and southern California. There are 16 genera of shrubs and small trees in Mexico, of which 10 extend across the northern border. The family is well represented in the West Indies, but the plants are too small to furnish timber except for fuel or small articles. Of the 13 genera in South America, only two, *Goupia* and *Maytenus,* attain large dimensions and yield timbers of value, and the uses for these are entirely local. The following description of the wood is based on American species of *Austroplenckia, Canotia, Celastrus, Elaeodendron, Evonymus, Goupia, Maytenus, Microtropis, Mortonia, Pachystima, Rhacoma, Schaefferia, Torralbasia, Wimmeria,* and *Zinowiewia.* There is much variation in the anatomical details.

Color yellowish or nearly white in *Evonymus, Microtropis, Mortonia, Pachystima, Schaefferia, Torralbasia,* and *Zinowiewia;* light to dark brown, with a reddish or purplish hue, in the others. Odor and taste not distinctive. Mostly hard and heavy, sometimes only moderately so in *Canotia, Evonymus, Pachystima,* and *Torralbasia;* usually with fine and uniform texture, but occasionally coarse-textured and either rather soft (*e.g., Celastrus, Microtropis*) or dense (*e.g., Elaeodendron*).

Growth rings often distinct; ring-porous or with tendency in *Canotia, Celastrus, Mortonia,* and *Pachystima.* Pores mostly very small to minute, but sometimes medium-sized to large (*Celastrus* and *Goupia*); numerous or fairly so; either all solitary or in association with small multiples, clusters, or short radial rows, without definite pattern. Vessels generally with simple perforations, but scalariform plates characterize *Elaeodendron* and *Goupia;* spiral thickenings sometimes present in *Austroplenckia, Canotia Evonymus, Maytenus Boaria, Mortonia,* and *Pachystima;* fine striations common; vascular pits, when present, small to minute, typically alternate; fibriform vessel members, with spirals, rarely present in *Evonymus,* abundant in *Pachystima;* vessels may compose ground mass of wood (*e.g., Celastrus* and *Pachystima*). Rays varying in width from all uniseriate or partly biseriate (*Austroplenckia, Canotia, Evonymus, Microtropis, Mortonia, Pachystima, Rhacoma, Torralbasia,* and *Wimmeria*) to distinctly 2-sized, the larger 3 to 6 cells wide (*Celastrus, Elaeodendron, Goupia, Maytenus, Schaefferia,* and *Zinowiewia*), and in height from 1 to 15, less often up to 30, occasionally to 60, rarely (*Celastrus*) to 150, cells; often decidedly heterogeneous, occasionally nearly homogeneous (*Canotia, Evonymus,* and *Torralbasia*); crystals common; pits to vessels very small to minute. Wood parenchyma ranging from apparently absent or very sparse (*Austroplenckia, Celastrus, Elaeodendron, Evonymus, Maytenus, Mortonia,* and *Pachystima*), to moderately abundant, diffuse and in short tangential rows not distinct with lens (*Canotia, Goupia, Rhacoma,* and *Schaefferia*), to abundant and forming distinct, irregular rows or broken bands 3 to 7 cells wide and usually one band per growth ring (*Microtropis, Torralbasia, Wimmeria,* and *Zinowiewia*). Wood fibers with rather thin to very thick walls and, at least in part, with numerous, small to rather large, distinctly bordered pits; accompanied in *Austroplenckia, Elaeodendron,* and *Maytenus* by irregular but distinct bands of thin-walled, septate, libriform fibers resembling parenchyma; spiral thickenings rarely present in fiber-tracheids in *Evonymus.* Ripple marks absent. No gum ducts seen.

Austroplenckia (= *Plenckia* Reiss.) includes six species of shrubs or small trees rarely over 20 feet tall in southern Brazil and Paraguay, according to C. L. Lundell (*Lilloa* 4: 378). The hard, fine-textured, purplish brown wood of *A. populnea* (Reiss.) Lundell is utilized locally for small cabinet work, and the slender pliable branches for making wicker furniture.

COMMON NAMES: Marmeleiro do campo, marmelinho do campo, marmelo do campo (Braz.).

Canotia *holacantha* Torr., the only species, is a leafless sprawling shrub or a shrublike tree sometimes 20 to 30 feet high with a stout trunk rarely 12 inches in diameter occurring in dry mesas in the southwestern United States and perhaps in northern Mexico. The stout branches terminate in rigid spines. The bark of the trunk is deeply furrowed and contains white laminations. The timber is not utilized except for fuel. Botanists disagree as to the classification of this plant, some referring it to the Rutaceae, others to a monotypic family, Canotiaceae. Nothing was found in the structure of the wood to exclude it from the Celastraceae.

Heartwood grayish brown, with purplish

hue; sharply demarcated from the narrow whitish sapwood. Luster medium. Odorless and tasteless. Hard and moderately heavy; texture fine; grain straight; easy to work, finishing very smoothly; suitable for turnery; durability probably low; white pocket-rot common in standing tree. Of no commercial possibilities.

Celastrus, with about 50 species of twining shrubs, occurs in the Far East and in North America and northern South America. The common species in the United States is the Bitter-sweet, *Celastrus scandens* L., occurring from Maine to Manitoba and southward. The lasting fruits are used for decorative purposes, the bark is reputedly medicinal, and the stems are sometimes used for walking sticks. The wood is dark greenish brown, rather soft, and coarse-textured.

Common names: Bitter-sweet, staff tree, waxwork (U.S.A.).

Elaeodendron is best represented in South Africa. The principal American species, if not the only one, is *E. xylocarpum* (Vent.) DC., a shrub or less often a tree usually less than 30 feet high in the West Indies, but in the Tres Marías Islands off the west coast of Mexico, where it occurs scatteringly near arroyos, it develops a very crooked bole with a maximum diameter of 30 inches and a length of 20 to 60 feet. The pinkish brown, hard, heavy, tough, strong, fine-textured, fairly durable wood is seldom used, but is considered suitable for railway crossties.

Common names: Laurel de costa, mate prieto, palo blanco, piñipiñi, roñoso, sange de doncella (Cuba); corcorrón, coscorrán, guayarrote, marble tree, spoon tree, nut muscat (P.R.); mano de león, zacchechem (Mex.).

Evonymus (or *Euonymus*) with numerous species of upright, decumbent, or climbing shrubs and little trees, is most abundantly represented in the Far East, but there are a few species in Europe and in North America as far south as Costa Rica. The European Spindle-tree, *E. europaeus* L., supplies a clear yellowish white, fine-textured, easily worked wood used for spindles and manicure sticks and other small articles of turnery. The only arborescent American species is *E. atropurpureus* Jacq., the Wahoo, also known as Burning Bush because of the scarlet arils of the seeds exposed upon ripening. It occurs scatteringly throughout the eastern half of the United States, but it is rarely 20 to 25 feet high and 4 to 6 inches in diameter; its wood is much like that of the European species and suitable for the same purposes.

Common names: Arrow wood, bleeding heart, burning bush, Indian arrow, spindle tree, strawberry tree, wahoo, woohaw (U.S.A.).

Goniodiscus *elaeospermus* Kuhlmann, the only species, is a medium-sized tree of the State of Amazonas, Brazil, where it is known as Andirobinha, Cabeça de Cutia, and Mapiá. Its seeds contain a high percentage of oil which is the basis of a small local industry (see *Tropical Woods* 36: 58). The timber apparently is not utilized. No wood sample available for this study.

Goupia, with two or three species of small to very large trees, is of common occurrence in the lower Amazon region and the Guianas and infrequent in the hinterlands of Colombia. The best known species is *G. glabra* Aubl. which in British Guiana sometimes attains a height of 120 feet with a long trunk that will square 30 inches, while logs 60 feet long squaring 12 to 16 inches are common. When the trees are felled the stumps exude a gelatinous substance having a fetid odor; the fresh wood also has the same scent, but loses most of it upon drying. The timber is said to be suitable for heavy and durable construction and for furniture.

Heartwood light reddish brown, darkening superficially upon exposure; distinct but not sharply demarcated from thick brownish or pinkish sapwood. Luster medium to rather high. Without distinctive odor or taste when dry. Rather hard and heavy; sp. gr. (air-dry) 0.82 to 0.88; weight 51 to 55 lbs. per cu. ft.; texture rather coarse; feel harsh; grain straight to roey; easily worked, but coarser-textured material requires filler for smooth finish.

COMMON NAMES: Sapino (Col.); cabacalli, copi, copie, couepi, coupi, goupi, goupil, kabukalli, kaboekalli, koepie, kopie (Guianas); cupiúba, tento (Braz.).

Maytenus. There are over 100 named species, well distributed throughout Latin America, and while most of them are shrubs a few are medium-sized to large trees. The wood of the Cuban Nazareno Morado, *M. lineatus* Wr., is sometimes used by cabinetmakers. The Sombra de Toro of Argentina, *M. ilicifolia* Mart., is generally less than 20 feet high and 10 inches in diameter, but the wood serves for repairing furniture, vehicles, tools, and agricultural implements. The Maitén, *M. Boaria* Mol., is a well-known tree in Patagonia; in the forests of the Río Negro it reaches a height of 75 feet, though in other parts of its range it is usually less than 30 feet high and 15 inches in diameter, and on the cattle ranges is browsed down to a shrub. The wood, which is firm and elastic and does not split badly, is rather highly esteemed locally in all kinds of carpentry work and for making implements.

According to H. M. Curran, the Carne d'Anta, *Maytenus obtusifolia* Mart., is a common tree of the coast forests of the Bahia region of Brazil, where it occurs scatteringly, averaging rarely more than one or two trees per acre over large areas. It attains a height of from 75 to 100 feet and has a cylindrical bole, two to five feet in diameter, with little taper and without buttresses. The sapwood is white, but the heart is bright rose when first cut, though gradually fading upon seasoning. The logs are free of defects and are readily sawn into lumber, but thin boards have a tendency to warp badly if improperly dried. The wood is hard and heavy; sp. gr. (air-dry) 0.82; weight about 51 lbs. per cu. ft.; though not durable in contact with the ground, it finds many uses in rural carpentry and construction where not exposed. Material supplied to manufacturers in the United States has shown satisfactory working qualities, being well suited for turnery, though with a slight tendency to champ on sharp cuts across the grain. Mechanical tests gave the following results (in pounds per square inch): modulus of elasticity (in bending), 2,551,300; fiber stress at elastic limit, 10,240; endwise crushing strength, 10,240.

COMMON NAMES: Rockwood (Jam.); boje, carne de vaca, espinillo, nazareno morado, sangre de toro, tea (Cuba); cuero do sapo, bois flament (P.R.); acajou sauvage, bois formi (Haiti); aguabola, limoncillo, mangle, m. aguabola, m. dulce (Mex.); arizá, camarón, caney (Col.); cucharo, say (Venez.); kaiarineo (Br. G.); apiranga, cafésinho, carne d'anta, chuchuasca, pau de colher, coração de negro, sombra de tauro (Braz.); ivirá-yuguí, molle morotí (Par.); congorosa, colquiyuyu, ibirá-yuquí, leña dura, lengue maitén, maitén, m. chico, m. grande, naranjillo, naranjito, quebrachillo, sal de Indus, sombra de toro, tapia, yucurira, yuqui-rá (Arg.); cangorosa, carne gorda, congoña, palta, sombra de toro (Urug.).

Microtropis, with about 65 species of trees and erect, scandent, or epiphytic shrubs, is largely confined to the mountainous regions of India, Ceylon, and Java. Only four American species have been described, their combined range extending from southern Mexico to Costa Rica. The plants of the same species may vary from epiphytes less than six feet high to small trees. The only American specimen available (Yale 38418; H. E. Stork 4204) was collected in the fog zone at 7000 feet elevation near Palmira, Costa Rica, with flowering herbarium material determined by C. L. Lundell as *M. Standleyi* Lundell. According to the collector, this particular tree was 23 feet high and seven inches in diameter at the base, though others attain a height of 50 feet or more. The gray or grayish green bark is nearly smooth. "The wood is soft, but of fine texture, pure white when fresh, somewhat resembling White Pine." The trees are not rare in the mountain forest, but are always widely spaced.

Wood white throughout. Luster medium. Odorless and tasteless. Rather light in weight but firm, being somewhat heavier and harder than Basswood (*Tilia*); texture fine and uniform; grain straight; very easily worked, finishing very smoothly; low in resistance to decay. Suitable for spools and similar articles of turnery and as a general-purpose timber, but presumably of no commercial possibilities because of the small size and inaccessibility of the trees.

Mortonia, with five species of evergreen shrubs, is limited in distribution to southwestern United States and northern Mexico, where the plants are sometimes called Afinador. The only species studied is *M. scabrella* A. Gray. Its very hard, fine-textured, easily worked, yellow wood is of the Boxwood class and suitable for small engravings and articles of turnery.

Pachystima. There are two North American species of shrubs. *P. Canbyi* A. Gray is native to the mountains of Virginia and West Virginia; its wood has not been studied. *P. myrsinites* Raf. ranges through the Rocky Mountains to British Columbia and California and southward into Mexico. The plant is small, sometimes prostrate. It has white wood of fine and uniform texture, and is sometimes known as Oregon Boxwood.

Rhacoma (incl. *Gyminda* and *Myginda*), with about 15 species of shrubs and shrubby trees, occurs in the West Indies, southern Florida, Mexico, and Central America. The best known species are *R. latifolia* Sw. of the West Indies and *R. eucymosa* (Loes. & Pitt.) Standl. of Central America. The brown to dark brown, hard, heavy, fine-textured wood is apparently not utilized except for fuel.

COMMON NAMES: Poison cherry (Bah.); hierba maravedí, jinca pata (Cuba); coral, cocorroncito, mala mujer, manto, maravedí, poison cherry, wild cherry (P.R.); managuatillo (Mex.); carbón, limoncillo (Br. H.).

Schaefferia. Of the eight species, all but one are low shrubs growing in Mexico and the West Indies. *S. frutescens* Jacq., though often a shrub, attains tree size on favorable sites and is sometimes 35 to 45 feet high and 12 inches in diameter. Its range, which is the most extensive of the genus, is mostly in the West Indies and southern Florida, but it occurs sparingly in southern Mexico, Colombia, and Venezuela. Its bright yellow, hard, moderately heavy, fine-textured wood is said to have been used as a substitute for Boxwood (*Buxus*), but the quantity is too small to be a factor in the trade. (For detailed description see Yale School Forestry Bull. No. 14, pp. 75–77.)

COMMON NAMES: Florida boxwood (U.S.A.); amansa guapo, boj de Persia cimarrón, guairaje (Cuba); jiba (P.R.); cabra (Dom. R.); balai de la montagne, bois capable, b. petit garçon, b. pin marron, petit bois blanc, p. garçon (Haiti); limoncillo (Col.); fruta de paloma, limoncillo (Venez.).

Torralbasia *cuneifolia* (Wr.) Krug. & Urb., the only species, is a small Cuban tree, called Guairaje. Its yellowish, fine-textured, moderately dense wood is of no special utility.

Wimmeria is a Middle American genus with seven or eight species of shrubs and trees. The largest reported is *W. Bartlettii* Lundell of northern Guatemala and British Honduras where it is said to attain a height of about 85 feet and a diameter of 24 inches. The wood is used locally for making marimba keys. *W. concolor* S. & C. of Mexico is occasionally 40 feet tall and a foot in diameter and supplies some timber for railway crossties. The only wood sample available (Yale 9624) is of a smaller Mexican tree, *W. confusa* Hemsl.; it is hard, heavy, fine-textured, reddish or purplish brown, with thin layers of nearly black parenchyma; it is not very attractive, but finds various local applications.

COMMON NAMES: Acedilla, algodoncillo, cedilla, chapulizle, palo cadillo, pimientilla (Mex.); chintoc, ixolte ixnuc, quiebrahacha blanca (Guat.).

Zinowiewia was long considered a monotypic genus, with *Z. integerrima* Turcz., a small tree, extending from Mexico to Panama. In 1938, however, Lundell (*Bull. Torrey Bot. Club* 65: 469–476) described six new species, five of them Middle American, the other, *Z. australis* Lundell, Venezuelan. The last is said to be a tree 65 feet high, known locally as Canalete, a name also applied to species of *Cordia*. The largest tree of the group is *Z. rubra* Lundell, which is credited with a height of about 100 feet and a diameter of 36 inches in the uplands of Guatemala. The only wood sample of this genus available is of *Z. pallida* (Yale 37253; Nelson Smith 40) collected in British Honduras with sterile botanical material determined by C. L. Lundell.

Wood white throughout when fresh, becoming brownish gray upon drying; shows thin laminations of parenchyma, slightly darker than background. Luster rather low. Odorless and tasteless. Hard, moderately heavy, tough and strong; texture fine; grain irregular; not difficult to work; probably perishable in contact with the soil. Presumably without commercial possibilities.

COMMON NAMES: Gloria, palo blanco (Mex.); canalate (Venez.).

CHLORANTHACEAE

THIS small unimportant family consists of three genera and about 45 species of aromatic little trees, shrubs, and a few perennial herbs. *Chloranthus* occurs in eastern Asia and the Indo-Malayan region; *Ascarina* in the islands of the Pacific to New Zealand; *Hedyosmum* in tropical America. The plants have articulate branches which are enlarged at the nodes.

Hedyosmum, with about 25 species of shrubs and small trees, the largest rarely more than 25 feet high, is sparingly distributed in the uplands of the West Indies, southern Mexico, Central America, Colombia, Venezuela, Ecuador, Peru, Bolivia, and southern Brazil. The leaves are opposite and simple, the petioles united to form a sheath on which there are small marginal stipules; the staminate flowers are spicate, the pistillate in spikes or heads; the fruit is a small drupe. All parts of the plant, except the wood, are fragrantly scented, and exudations of resin are sometimes collected and used, as also are the leaves and bark, in native medicine. The twigs are brittle; the pith is large and mucilaginous, becoming chambered upon drying. The fruit of *H. Artocarpus* Solms is comestible and of agreeable flavor. The genus, like the other two, furnishes no timber of value.

Wood pale brown throughout, with Oak-like figure on radial surface. Luster silky in proper lighting. Odorless and tasteless, at least when dry. Of light weight, but firm and tough; medium-textured; fairly straight-grained; very easy to work; not highly resistant to decay.

Growth rings indistinct. Pores small to medium-sized; fairly numerous; solitary and in tangential pairs, sometimes in small clusters, well distributed without definite pattern; often in contact with the rays. Vessels with exceptionally long, fine-barred scalariform perforation plates; spiral thickenings absent; intervascular pitting coarse, frequently scalariform. Rays almost all wide, composing a third to a half of the cross section, and variable in height from 25 to over 200 cells, showing very conspicuously on radial surface; cells thin-walled, appearing square and larger than the wood fibers on cross section, mostly upright on the radial, and as irregular hexagons on the tangential; pits to vessels large, often elongated and in scalariform arrangement. Wood parenchyma rather sparingly paratracheal, not distinct with lens. Wood fibers with moderately thick walls and large lumen; sometimes septate; pits numerous on radial walls, simple or with vestigial borders. Ripple marks absent. No gum ducts seen.

COMMON NAMES: Headache bush (Jam.); azafrán (P.R.); vara blanca (C.R.); anisillo, malibú morado, motilón, sibato, silba-silba, silva-silva (Col.); corteza de Salomón, hojita de Díos (Venez.); guishcas, tarquí (Ec.); aitacupí, almáciga, anís, asar-guiro, asar-quiro, aytacupí, corpales, supinum (Peru); herva de soldado (Braz.).

CLETHRACEAE

THIS family, in a restricted sense, consists of a single genus, *Clethra*, which some botanists include with the Ericaceae. A few years ago a second genus, *Schizocardia*, was added, but its position is somewhat doubtful.

Clethra, with several species of shrubs and small trees, is represented in tropical and subtropical Asia, Madeira, southeastern and eastern United States, Mexico, Central America, and parts of northwestern and central South America. The leaves are alternate, simple, and without stipules; the small, fragrant flowers are borne in terminal racemes or panicles; the fruit is a small, 3-lobed, dehiscent capsule with many small seeds. The largest and most widely distributed species is *Clethra lanata* Mart. & Gal., usually a slender tree less than 40 feet tall and a foot in diameter, occasionally considerably larger, in the lowlands and along streams in the uplands from southern Mexico to northern South America. The timber is not utilized for any special purposes.

Heartwood brownish, with an olive tinge; merging gradually into the lighter-colored sapwood. Luster medium. Odorless and tasteless. Has about the consistency of Alder (*Alnus*); sp. gr. (air-dry) 0.65; weight 41 lbs. per cu. ft.; tough and strong; texture fine; feel rather harsh; grain straight to roey; easy to work, finishing smoothly; is poorly resistant to decay. Of no commercial possibilities.

Growth rings present or absent. Pores numerous; thin-walled; small (55 to 70μ), not visible without lens; close together, but rarely in contact radially; uniformly distributed without pattern. Vessels with scalariform perforation plates having very numerous, fine, closely spaced bars; spiral thickenings sometimes present in tips of members; intervascular pitting infrequent, fine, opposite. Rays numerous, of two sizes; uniseriates few, composed of 1 to 7 rows of upright cells; multiseriates 2 to 6 cells wide and up to 50, sometimes to 170, cells high, with stratum of slender procumbent cells and few marginal rows of large, squarish or upright cells; walls copiously pitted; gum deposits abundant; ray-vessel pitting half-bordered, fine, typically opposite. Wood parenchyma diffuse to reticulate, sparse to moderately abundant. Wood fibers with medium walls and large lumen; conspicuously bordered pits very numerous in both radial and tangential walls. Ripple marks absent. No gum ducts seen.

COMMON NAMES: *Clethra lanata:* Jicarillo, mameyito negro, mama malhuaztili (Mex.); tepezapote, terciopelo, zapotillo de montaña (Salv.); nance, n. macho (C.R.); nispero macho (Col.); sapo (Venez.). *Other species:* Sweet pepperbush, white alder (U.S.A.); jaboncillo (Mex.); azahar, cagalero (Col.); cangalheira falsa, mansillo, pau de cinzas (Braz.).

Schizocardia *belizensis* Sm. & Standl., the only species, is an apparently rare tree sometimes 60 feet tall and 36 inches in diameter discovered in 1932 by William A. Schipp in the mountain forests of southern British Honduras. According to Standley and Smith (*Tropical Woods* 32: 8), "at first glance, the plant has little resemblance to the rather uniform species of *Clethra*, the type of a family heretofore presumed to consist of a single genus. . . . *Schizocardia* differs from *Clethra* in having the racemes axillary rather than terminal, the sepals accrescent, persistent, and deeply laciniate rather than entire, and a 5-celled rather than a 3-celled ovary." The leaves are alternate and clustered at the ends of the branchlets, which have exfoliating bark; the fragrant flowers have pink petals and yellow anthers; the fruit is a small capsule.

Heartwood dull reddish brown, merging gradually into the lighter-colored sapwood. Without distinctive odor, but with mildly astringent taste. Hard, heavy, tough, and strong; texture fine; grain more or less interlocked; suggests Applewood (*Malus*) but is slightly coarser-textured; durability probably high. Of no commercial possibilities. (For minute anatomy see *Tropical Woods* 32: 11–14.)

CNEORACEAE

Cneorum, the sole genus in this family, includes three species of undershrubs, shrubs, or treelets native to the Mediterranean and Caribbean regions. The only American species is *C. trimerum* (Urb.) Chodat, a shrubby little tree growing in the mountains of eastern Cuba. The leaves are alternate, leathery, small, elongated, entire, and contain oil cells; stipules are absent; the flowers are borne in 3-flowered axillary racemes; the fruit consists of three globose drupes joined at the middle but eventually separating. The wood is not utilized because of the scarcity and small size of the trees. The only sample available (Yale 19295; Bucher 168) was collected by

G. C. Bucher in the Turquino region of Cuba.

Wood pale brownish gray throughout. Luster medium. Odorless and tasteless. Hard and rather heavy; texture fine; grain straight; working properties good. Without commercial possibilities.

Growth rings present. Pores small (80μ), not visible to the unaided eye; rather thick-walled; fairly numerous; mostly in small multiples and clusters, usually well distributed but sometimes with tendency to tangential arrangement in early wood. Vessels with simple perforations; without spirals; pits very small (3μ), alternate. Rays uniseriate or biseriate and up to 6, sometimes to 8, cells high; heterogeneous; procumbent cells short; tall upright cells few, sometimes pointed; ray-vessel pitting often unilaterally compound, the vascular pits very small. Wood parenchyma abundant; aliform and short-confluent; also finely terminal; cells in 2-celled strands or fusiform. Wood fibers not in radial rows; walls thick; pits small, simple or indistinctly bordered, most numerous near the "shoulders" of the fibers. Ripple marks present but very irregular and indistinct. No gum ducts seen.

COCHLOSPERMACEAE

THIS small pantropical family, sometimes included with the Bixaceae, includes three genera and about 25 species of trees, shrubs, and herbs. The sole arborescent genus represented in tropical America is *Cochlospermum*.

Cochlospermum. The best known and most widely distributed species and the only one on the North American continent is *C. vitifolium* (Willd.) Spreng. (= *Maximiliana vitifolia* K. & U.). This is a slender deciduous tree, sometimes 40 feet high, with long-stalked deeply lobed leaves and large bright yellow flowers resembling wild roses; the fruit is a pear-shaped capsule about 3 inches long, containing numerous reniform seeds covered with white cotton. The bark contains a tough fiber and yields a gum. Apparently no important use is made of this or the other tropical American species, but from the East Indian tree, *C. gossypifolium* DC., are obtained stuffing for pillows and mattresses, bast for cordage, and gum for use as a substitute for tragacanth. Some of the Amazonian trees attain large size, but their timber is practically worthless. The woods of four American species are similar, being very light, soft, spongy, and tending to disintegrate, upon drying, into a loose bundle of fiber layers. Color white at first, but turning brown; texture very coarse; durability very low.

Pores barely visible; well distributed, occurring singly and in radial multiples; not very numerous. Vessels with simple perforations; spirals absent; pitting coarse, alternate. Rays widely variable in size, mostly 1 to 4 cells wide, sometimes considerably wider, and few to over 100 cells high; heterogeneous, with many of the cells square; pits to vessels large, often elongated. Wood parenchyma abundantly developed; paratracheal and in metatracheal bands; cells very thin-walled. Wood fibers short, thin-walled, with small indistinctly bordered pits. Ripple marks present, though often irregular; all elements storied in some specimens, large rays occupying one to several tiers in others. Radial intercellular canals sometimes present.

COMMON NAMES: Botija (Cuba); apompo, chimu, chum, chuun, cocito, cojón de toro, flor izquierda, huarumbo, madera de pasta, palo amarillo, p. de rosa, panaco, pochote, pongolote, quie-quega, rosa amarilla, tecomarochitl, tecomasuche, tecomasuchl (Mex.); pochote, wild cotton (Br. H.); comasuche, tecomatillo, tecomasuche, tecomasúchil, tsuyuy (Guat.); jicarillo (Hond.); tecomasuche, tecomasúchil (Salv.); bombón, catamerisuche, poró-poró (Nic.); poró-poró (C.R.); binguá, minkrá, poró-poró (Pan.); batabana, bototo, carnestolendas, flechero, jurubai, papayote, poró-poró (Col.); bototo, botuto, carnestolendas, majagua (Venez.); kanakuchiballi (Br. G.); njoe fodoe (Sur.); bototillo, botulo (Ec.); algodão bravo, a. cravo, a. do campo, a. do matto, algodoeiro do campo, butuá de corvo, pacoté, periquiteira, p. do campo, p. grande de terra firme, rhuibarbo do campo, sumauma do igapó (Braz.); huimba, huiña, h. caspi, quillo-sisa (Peru); arbol de papel, palo papel (Arg.).

COLUMELLIACEAE

Columellia, the only genus, with two species of shrubs and small trees, inhabits the Andes Mountains of Ecuador, Peru, and Colombia. The leaves are opposite and simple; stipules are absent; the yellow flowers are borne in terminal cymes; the fruit is a small, dehiscent, many-seeded capsule. *C. obovata* R. & P. is a shrub apparently limited to Peru. The two wood samples in the Yale collections are of *C. oblonga* R. & P. and were obtained by Dr. A. Rimbach in the Eastern Cordillera of Ecuador at an elevation of about 10,000 feet, where it is known as Quinoa Blanco. The outer bark of young stems is brownish yellow and exfoliates in delicate thin papery flakes; the inner bark is olive-brown, very thin and compact. The wood, which suggests Tupelo (*Nyssa*), is not utilized, except for fuel.

Heartwood pale brown; sapwood nearly white. Not highly lustrous. Without distinctive odor or taste. Of medium density, firm but brittle; of fine and uniform texture; easy to work, finishing very smoothly; is probably not resistant to decay.

Growth rings absent. Pores very small, angular; very numerous; solitary, though often close together radially; uniformly distributed. Vessels with oval perforation plates having 10 to 20 narrow, frequently anastomosing, bars; spiral thickenings absent. Rays very numerous; uniseriate and 1 to 15 cells high; barely distinct with lens on cross section; heterogeneous, nearly all of the cells upright; pits to vessels very small, oval, mostly opposite. Wood parenchyma very scanty, occurring as single cells in contact with vessels. Wood fibers with rather thin walls and fairly large lumen; pits very numerous in both radial and tangential walls, conspicuous, bordered. Ripple marks absent. No gum ducts seen. Peculiar ingrowing of bark noted in one specimen.

COMBRETACEAE

This family comprises 17 genera and about 650 species of trees, shrubs, and woody vines of general distribution throughout the tropical and subtropical regions of the world. The simple, entire leaves are opposite, alternate, or verticillate; the flowers are borne in terminal or axillary, elongated or subcapitate spikes or in panicles; the one-seeded fruit is greatly variable in size and form, is fleshy or dry, usually indehiscent, and often is winged or ridged. Bark, leaves, and fruit of certain species are utilized in the tanning and dyeing industries and some of them are known for their medicinal properties. The largest trees are of the genus *Terminalia,* and timbers of a few species in India and west tropical Africa are fairly well known to the European trade. The American trees are of six genera, namely, *Buchenavia, Bucida, Conocarpus, Laguncularia, Ramatuela,* and *Terminalia* (incl. *Chuncoa*); there are three other woody genera, namely *Cacoucia, Combretum,* and *Quisqualis,* but they are typically woody climbers. The timber, particularly of *Terminalia,* is of good quality, but its consumption is almost entirely local.

Heartwood yellowish or grayish brown or occasionally deep reddish brown, frequently attractively striped or variegated with dark green, brown, or red; rather sharply demarcated from the sapwood, which is yellowish or grayish, with an olive hue. Luster usually rather high. Without distinctive taste, but often with a mild, indescribable odor. Density variable from medium to very high; sp. gr. (air-dry) 0.65 to 1.10, av. about 0.80; weight about 40 to 70, av. about 50, lbs. per cu. ft.; texture variable from rather fine to rather coarse, mostly medium; grain often roey, showing conspicuous striping on radial surface; working properties fair; durability variable, sometimes high.

Growth rings usually present. Pores mostly small and near limit of vision, sometimes distinct without lens; numerous but not crowded; well distributed, occurring singly and in pairs, less often in radial multiples of 3 to 6, without definite pattern though with slight tendency to tangential arrangement. Vessels with simple perforations; without spirals; gum deposits abundant; pits typically medium-sized, crowded, alternate, vestured. Rays uniseriate (except in *Bucida* where they are 1 to 4 cells wide); mostly less than 20, sometimes up to 40, cells high, the cells typically large; distinctly heterogeneous in *Ramatuella crispiolata* Ducke, homogeneous or weakly heterogeneous in the others, with none of the cells definitely upright, though often cubical where a ray passes through parenchyma bands, and sometimes also in interspersed rows containing large crystals of calcium oxalate visible with lens; gum deposits abundant; ray-vessel pitting similar in appearance to intervascular, the pits typically oval (occasionally elongated in *Ramatuella*). Parenchyma occasionally only sparingly developed, but usually abundant and distinct, suggesting some of the Leguminosae; in *Conocarpus,* vasicentric and vasicentric-confluent into short wavy bands; in *Laguncularia* and some species of *Terminalia,* vasicentric and aliform, sometimes confluent into wavy, usually broken, concentric bands; also in narrow, apparently terminal bands in *Buchenavia, Bucida,* and a few species of *Terminalia.* Wood fibers with thick to very thick walls and small to medium-sized, simple or indistinctly bordered pits. Ripple marks absent. Vertical traumatic gum ducts observed in specimens of *Buchenavia, Bucida,* and *Terminalia.*

Buchenavia, with about a dozen species of medium-sized to large trees and some erect shrubs, occurs in the West Indies and tropical South America. The leaves are alternate or crowded into false whorls; the spicate flowers are without petals; the calyx is toothed and soon falls, and the anthers are adnate to the filament and not versatile; the fruit is a 5-ridged or 5-angled drupe, the stone bony, suggesting an olive. The best known and most widely distributed species is *Buchenavia capitata* (Vahl.) Eich. (= *Terminalia obovata* Camb.), called Yellow Sanders in the British West Indies. It attains a height of 80 feet and a trunk diameter of three feet above the large buttresses. The yellowish wood is moderately hard, takes a beautiful satiny finish, and is used locally for making furniture. There are several species in the Amazon region, the largest being *B. grandis* Ducke which on good sites is from 100 to 145 feet tall, though only half so large on drier situations. The timber is of good quality and is frequently used.

Heartwood yellowish to golden brown, usually with an olive hue; sapwood somewhat lighter. Luster high. With faint odor and sometimes mildly bitter taste. Moderately to very hard and heavy; texture rather fine to rather coarse; grain more or less roey; not very easy to work, but taking a glossy finish; durability fair. Not likely to become of value for export.

Common names: Mountain wild olive, yellow sanders (Jam.); júcaro amarillo, j. mastelero (Cuba); granadillo (P.R.); bois gris-gris, b. Margot (Haiti); yellow olivier, y. sanders (Trin.); almendro (Col.); amarillo, a. boj, granadillo, olivo negro (Venez.); gemberhout, katoelima, matakki, toekadi, toekoeli (Sur.); birindiba, cuiarana, mirindiba, periquiteira, p. do igapó, piuna (Braz.).

Bucida, with two species of shrubs or small to medium-sized trees, is of common occurrence through the West Indies to the shores of the Caribbean Sea and the Bay of Panama. The leaves are clustered at the ends of the branches; the small flowers are without petals, the anthers are versatile, and the slightly toothed calyx is tardily deciduous; the fruits are small slightly fleshy drupes which are borne singly. *Bucida spinosa* (Northrop) Jennings, often considered only a form of the other species, is a shrub or flat-topped tree rarely 25 feet high, with spiny twigs and capitate flowers. *B. Buceras* L. is an unarmed tree, sometimes 80 feet tall and three feet in diameter. It frequently is abundant along the coast, and its olive-brown timber is highly valued locally for poles, posts, railway crossties, piling, and other durable construction, and for carpentry and furniture; it also makes a good grade of charcoal. The bark is used for tanning.

Heartwood dark yellowish brown, with an olive hue; sapwood yellowish. Luster rather high. Moderately to very heavy, hard, and strong; texture medium; grain often roey; rather difficult to work, but finishing smoothly; resistant to decay; not likely to be important for export.

Common names: Black olive tree (Florida); black olive, olive-bark tree (Jam.); black olive, brier tree, prickly tree, spiny black olive (Bah.); jucarillo, júcaro, j. bravo, j. común, j. de costa, j. espinoso, j. negro, j. prieto (Cuba); black olive, búcaro, Gregory wood, úcar, water gregre (P.R.); gregre (Virg. Is.); bois gli-gli, b. gris-gris (Guad.); caracoli de Puerto Rico (Dom. R.); bois gris-gris, gris-gris des montagnes

(Haiti); pucté, pukté (Mex.); bullet tree, bully tree (Br. H.); amarillo, marión (Pan.); leertouwarsboom (Sur.); grignon (Fr. G.).

Conocarpus *erecta* L., the only species, grows in Mangrove swamps along tropical American and West African shores. It varies in habit from a prostrate or trailing shrub to an erect tree that usually is rather small but sometimes 60 feet high and 30 inches in diameter. The alternate leaves are somewhat fleshy; the flowers are without petals and are densely aggregated into heads borne in axillary racemes or terminal panicles; the fruits are curved, scale-like, 2-winged, and in cone-like fruiting heads. The bark is rich in tannin. The timber is used to a minor extent for durable construction, but its chief value is for fuel and charcoal. The trees reproduce themselves readily from sprouts.

Heartwood olive-brown, sometimes with a reddish tinge; sapwood lighter. Luster rather high. Moderately heavy and hard; texture medium; grain straight to roey; not easy to work, but finishing smoothly; durability good. Of no promise for export.

COMMON NAMES: Button mangrove, b. tree, b. wood (Br. W.I.); mangle botón, saragosa, yana (Cuba); mangle (Dom. R.); palétuvier (Haiti); botoncahui, botoncillo, estachahuite, iztac-cuahuitl, kanché, mangle negro, m. prieto, taabché, xkanché, xtabché (Mex.); button mangrove (Br. H.); botoncillo (Salv.); maraquito (C.R.); mangle piñuelo (Pan.); mangle garbancillo, m. zaragoza (Col.); botoncillo, mangle blanco, m. botoncillo (Venez.); jelé (Ec.); mangue (Braz.).

Laguncularia *racemosa* (L.) Gaertn., the sole species, is another member of the Mangrove formation along the shores of tropical America and West Africa. It varies in form from a low shrub to a tree with a maximum height of 60 feet and a diameter of 20 inches. The thick coriaceous leaves are opposite; the small greenish flowers are borne in clustered spikes; the fruit is a leathery, ribbed or angled drupe which begins to germinate while on the tree. The bark, galls, and leaves contain from 10 to 17 per cent (of their dry weight) of tannin, and are used for tanning and medicinal purposes. The timber is used locally for fence posts, house-framing, and fuel; it is not considered very good for charcoal. In Cuba, withes from the tree are seasoned in salt water and twisted into cables (cujes) from which tobacco is suspended for curing.

Heartwood reddish olive-brown; sapwood yellowish. Fairly lustrous. Hard and heavy; texture medium; grain straight to roey; not very easy to work, but finishing smoothly; durability high. Of no commercial possibilities for export.

COMMON NAMES: White buttonwood, w. mangrove (Florida); gold wood, white mangrove (Jam.); bastard buttonwood, green turtle bough, white mangrove (Bah.); mangle amarillo, m. bobo, patabán (Cuba); mangle blanco (Dom. R.); manglier blanc (Haiti); mangle blanco, m. bobo, m. chino, zacolcom (Mex.); white mangrove (Br. H.); cincahuite (Salv.); palo de sal (C.R.); mangle amarillo, m. blanco (Col., Venez.); petit palétuvier (Fr. G.); ankira (Sur.); mangue branco, tinteira, t. dos mangues (Braz.).

Ramatuella, with three poorly known species of shrubs and trees, is apparently limited in distribution to the northwestern Amazon region and the hinterlands of Venezuela. The only wood sample available for this study is from the type of *R. crispialata* Ducke (Yale 31948; Ducke 221), a medium-sized tree with its leathery leaves clustered at the ends of the stout branchlets; the fruits have four crinkly wings and are congested at the ends of a long, solitary, axillary stalk.

Heartwood rich reddish brown; looks rather oily; distinct but not sharply demarcated from the brownish sapwood. Luster medium. Without distinctive odor or taste. Hard, heavy, tough, and strong; texture medium coarse; grain straight; not very difficult to work, finishing smoothly and taking a high polish; appears durable. Presumably without commercial possibilities.

Terminalia (including *Chuncoa*), with about 200 named species of shrubs and medium-sized to very large trees, is of pantropical distribution. The leaves are alternate or sub-opposite, frequently crowded near the ends of the branches; the flowers, which are commonly in elongated spikes, have versatile anthers and no petals; the fruit is variously winged, dry or fleshy, with a stony seed.

One of the best known species is the Indian Almond, *Terminalia catappa* L., extensively planted in tropical America for shade and decorative purposes and commonly called Almendro. According to Howard (*Timbers of the World,* pp. 264 and 537), the timber known in England as Indian Laurel is *T. tomentosa* W. & A.; White Chuglam and Silver-greywood are light and dark timber, respectively, of *T. bialata* Wall. *T. chebula* Ritz is the source of the myrobalan nuts of commerce, used in tanning and dyeing. There are several other important species in India and a few in tropical West Africa. Chief of the latter is the Limba or Limbo, *T. superba* Engl. & Diels, whose yellowish or pale olive-colored timber is used in France and Belgium as a substitute for Oak (*Quercus*), while the less common, darker, variegated or striped material, called Noyer du Mayombe (Congo Walnut) is used in place of real Walnut (*Juglans regia* L.) in cabinet work (see *Tropical Woods* 18: 26–28).

Numerous American forms of *Terminalia* have been described but most of them are imperfectly known and are probably only "herbarium" species. Their combined range extends from the West Indies and southern Mexico to northern Argentina and southeastern Brazil. The trees are often tall and well formed and the timber is of good quality, though virtually unknown to the export trade and not very extensively used locally. The woods exhibit much variation in their appearance as the color may be gray, yellow, brown, red, or nearly black, fairly uniform to variegated and striped. The figured material, which apparently may occur in any species, is suitable for the same class of uses as are Birch (*Betula*) and Maple (*Acer*). The anatomy is as described for the family.

The most widely distributed species is *Terminalia amazonia* (Gmel.) Exell (= *Terminalia obovata* [R. & P.] Steud.). It is a large, virtually evergreen tree, with a long, smooth, buttressed trunk, and grows from southern Mexico and Central America to the Guianas, Trinidad, Brazil, and Peru. Marshall studied the silvicultural characters of the tree in Trinidad and his results may be summarized as follows: Not exacting in its requirements, thriving on sands and clays, ridges and flats. Root system rather superficial. Flowers April–May. Fruit small, double samara, ripens almost immediately in May–June; often shed in clusters. Natural regeneration fairly abundant in some localities and could probably be obtained by a light clearing about seed tree. Growth moderately fast. Fairly tolerant of shade. Coppices exceptionally well. Apparently free from serious diseases and pests. Wood yellowish brown, often striped with red; strong and durable. (See *Tropical Woods* 27: 30.)

Along the lower slopes and in the valleys of the coastal region of Bahia, Brazil, is a tree known as Araça (the name commonly given to some of the Myrtaceae) and determined from herbarium material obtained by H. M. Curran as near, if not identical with, *Terminalia januarensis* DC. According to the collector it grows in almost pure stands or composes a large part of the forest over considerable areas. It attains a height of 100 feet, with a bole three feet or more in diameter above the wide-spreading buttresses and free of branches for 50 to 60 feet. It is conspicuous in the forest because of its smooth thin bark which suggests *Eucalyptus*. Although the timber is used locally there is no special market for it. The wood is similar to that of the preceding species.

The only well-defined species in Argentina, according to Exell (*Lilloa* 5: 1: 127–130), is *Terminalia triflora* (Gris.) Lillo. The tree is rarely over 50 feet tall and 20 inches in diameter, but the timber is of good quality and satisfactory for many purposes if not exposed to the weather; splints of the yellow wood are used for making hampers for handling corn, charcoal, and vegetables; the brown-veined material makes attractive furniture.

Apparently there are no native species of *Terminalia* in Puerto Rico and the Bahamas, but *T. catappa* has become naturalized

Photo by H. C. Kluge.

PLATE XVIII. Shaggy-barked Jucarillo tree (*Terminalia intermedia*) in Cuba.

Photo by Max Rothkugel.

PLATE XIX. Hauling a log of Rauli (*Nothofagus procera*) in Patagonia. Other Antarctic Beeches compose the forest in the background.

there as elsewhere in many parts of tropical America. In Jamaica, according to Fawcett and Rendle (*Flora of Jamaica* 5: 3: 304), *T. latifolia* Sw. is a common tree, 100 feet or more in height, with whorled and spreading branches. The fruit is a drupe, with a pulpy or fleshy greenish red pericarp and a white edible kernel tasting like a sweet almond. The tree "affords a valuable timber and splits readily into shingles." There is no specimen available for this study.

The two principal species in Cuba are *Terminalia Chicharronia* (Gris.) Wright and *T. intermedia* (A. Rich.) Urb., both called Chicharrón, a name applied also to several trees of other families (Plate XVIII). As in nearly all of the American species, the fruit is dry and winged. The principal distinction between the two species seems to be in the wood, that of the latter being yellowish throughout and less esteemed than that of the other, which is dark brown, durable, and employed for railway crossties, rollers in sugar mills, carriage-making, and various kinds of rural construction. *T. intermedia* grows also in Haiti.

According to Standley (*Forests and flora of British Honduras,* p. 279), "the Central American trees of this genus are in a chaotic state because the available material of them is quite inadequate for their proper understanding." In addition to *T. amazonia,* the names frequently applied are *T. chiriquensis* Pittier and *T. Hayesii* Pittier. It is not now possible to deal specifically with the woods.

Heartwood variable from yellowish olive or golden brown to dark reddish brown, often prominently striped. Luster rather high. Has a slight odor. Moderately to very hard, heavy, and strong; texture medium; grain often roey; working properties fair to good; durability usually high. May have commercial importance for export.

Common names: Broadleaf (Jam.); chicharrón, c. amarillo, c. de monte, c. prieto, guayo, jucarillo (Cuba); olivier mangue, poirier, white olivier (Trin.); canxun, cochun, pucte (Mex.); nargusta (Br. H.); bolador, guayavo, naranjo (Guat., Hond.); volador (Salv.); guayabo de montaña, guayabón, surrá (C.R.); amarillo, a. real, carboncillo, palo amarillo (Pan.); guayabo león, palo prieto (Col.); aceituno, amarillo caraqueno, a. carruajero, a. curragero, a. pijua, a. solido, arispin, arpino, chicharro, guayabito, mapurite blanco, nispero macho, querebere (Venez.); alasuhabu, coffee mortar, fujadi, naharu, swamp fukadi (Br. G.); ginja hoehoe, karalawai jakoenepele (Sur.); adamaram, langoussi, nangocy, nagosse, nagossi, tanibouca (Fr. G.); araça, a. d'agua, cinzeiro, cuia-rana, merendiba branca, mirindiba branca, pau mulato branco (Braz.); rifari, shapana (Peru); palo amarillo (Par.); amarillo del río, sarandí amarillo (Urug.); amarillo, a. del río, guayabí amarillo, g.-ré, g.-saiyu, lanza, l. amarilla, lapachillo, palo amarillo (Arg.).

COMPOSITAE

This, the largest of the plant families, comprises over 800 genera and 12,000 species, mostly herbs and erect or scandent shrubs, and is of world-wide distribution. The leaves are opposite or alternate, entire or dissected, never truly compound; the flowers are collected in a head on a receptacle surrounded by an involucre of bracts; the fruit is an achene, often with a pappus at the apex. Some of the best known plants of field and garden are the Aster, Daisy, Dahlia, Dandelion, Goldenrod, Lettuce, Sunflower, and Thistle. Some of the plants are laticiferous and one of them, the Guayule of Mexico, *Parthenium argentatum* A. Gray, is a commercial source of rubber.

The comparatively few trees in the family are limited to tropical regions and are commonly small and of little, if any, value. An exception is the genus *Brachylaena* with a few species of large trees in Madagascar which are a commercial source of a kind of Sandalwood and also supply excellent material for cabinet work as well as for durable construction. In tropical America there are about 20 genera containing some arborescent species, but their woods do not enter the larger markets, though some of them are used locally, particularly in localities where larger timber is not available.

In view of the wide range of forms of the plants and of the conditions attending their growth it is not surprising to find a very wide range of variation in the anatomy of their stems. The descriptions given below for several American genera are based on comparatively few specimens from arborescent forms only, and probably have very limited application. The woods vary from light and spongy to very heavy, hard, and strong; from fine to coarse in texture; and from perishable to fairly durable in contact with the ground.

Woods mostly diffuse-porous, sometimes more or less distinctly ring-porous; pores mostly small to minute, occasionally medium-sized; rather few and scattered, or more numerous and in radial rows and flame-like patches, or in concentric bands, or in ulmiform arrangement. Vessels with short members; only simple perforations seen by the authors, though scalariform plates with rounded contour and numerous thin bars have been reported in various members of the family; tyloses sometimes present; spiral thickenings common in temperate species; intervascular pitting alternate, very fine to coarse. Rays mostly 3 or more, sometimes 8 to 10, cells wide, and rather low to very high; occasionally nearly homogeneous in part in the denser woods, decidedly heterogeneous, with few procumbent cells, in the softer woods; no crystals seen; pits to vessels very small to very large, rounded to elongated and vertically or horizontally parallel. Wood parenchyma sparingly developed about the vessels; no crystal strands seen. Wood fibers often septate; pits very small, numerous, simple. Storied arrangement of some of the elements observed in a few instances. No gum ducts seen. Interxylary cork layers present in *Artemesia tridentata* Nutt. (see *Lloydia* 1: 39).

Baccharis, with some 300 species of shrubs and small trees, is widely distributed in tropical America, but is most abundantly represented in South America. The plants find various applications in local medicine and the strong stems are used for fuel and poles for hut construction.

Common names: Coyote bush, groundsel tree, salt bush, water wally (U.S.A.); bitter broom, mountain broom tree (Jam.); bajaquillo, chicoria, clavellina, espanta mosquitos, tapafrío, tres Marías (Cuba); palo de toro (Dom. R.); bois balai (Haiti); boshi, chamicillo, escoba de monte, escobillo, hierba del carbonero, h. del pasmo, holnuxib, huichín prieto, jara, j. dulce, jaral blanco, jarilla común, j. del río, limpia-tuna, popotillo, tepotote, tepopotl (Mex.); barba fina (Guat.); canutillo, chilca, guardabarranca, sauce, tapabarranca (Salv.); Santo Domingo (C.R.); Santa María (Pan.); algodoncillo, chilca de teñir, marucha (Col.); niquitauito (Venez.); chillca (Ec.); kuchu-kuchu, quimica-kuchu, tayanca (Peru); romerillo, vautro (Chile); chirca blanca, fumo bravo, junco, romerillo, suncho (Arg.); amargosa, carquejá, carquejiña, chilca blanca, c. negra, chirca de monte, yerba santa (Urug.); caá-nambuy-guazú, caapé-guazú, cañambi-guazú, carquejá, chilca, yerba santa (Par.); alecrim do campo, a. do matto, carquejá, carquejinha, carrasco do campo, charrúa pequena, charruinha, chilca, chirca do matto, vassoura (Braz.).

Clibadium, with numerous species of shrubs and occasional small trees sometimes up to 35 feet tall, is of fairly general distribution in tropical America. Some species are noted for their rough leaves. A few have toxic properties and are used by natives in Brazil to stupefy fish. There are no special uses for the soft, whitish woods.

Common names: Mastranzo de monte, zalagueña (Pan.); barbasco amarillo, juque (Venez.); conami, kunami (Br. G.); huaca, llama-caspi, ll. huasca, sacha-huaca, uchu-huaca (Peru); barbasco, conabi, conambin (Braz.).

Eupatorium, with about 500 species of herbs, shrubs, and a few small trees, is widely distributed throughout the tropics and warm regions of the world, but most abundantly in Mexico and southern South America. The only special uses are for dyes and medicines. The strong stems are used for fuel, props, hut frames, and minor construction.

Common names: Archangel, bitter bush, Christmas bush, cigar bush, jack-in-the-bush (Jam.); albahaca de sabana, celestrina, rompezaragüey, sebo-parra, trébol de olor, trebolillo (Cuba); curia, oreganillo, Santa María, yapana (P.R.); langa chata (Dom. R.); feuilles bounda, langue chatte (Haiti); árbol de Santa María, chamiso, chioplé, ciguapozle, cihuapatli, crucito, cuilotillo, hierba del ángel, h. del golpe, hillo, lengua de vaca, soscha, tokabal, tokaban, xiquite, xoltexnuc, xtokabal, yolochichitl, yoyochichil, zaçate minero, zactokaban (Mex.); old woman's walking stick (Br. H.); chilca, copalillo (Guat.); arnicacho, carrizo, chimaliote, chimuyo, coyontura, flor de plata, fregona, hierba de plata, suelda con suelda, taco, tamagua, tameagua, vara blanca, v. de cama, v. hueca (Salv.); carbatano, crucita, tiñe anzuelo, t. cordel (Hond.); crucita olorosa, garrapata (Nic.); chirrite, zbin-korgá (C.R.); hierba de chiva, paleca (Pan.); almoraduz, avellano, pulisa, puliza, varejón de caballo (Col.); churreto, niquitao, pernilla, pesebrito, yerba de pozo (Venez.); quela (Ec.); manca-ppaqui, manga-paqui (Peru); charrúa, chilca, c. de bañado, chirca, c. común, mió-mió, romerillo, romero, yerba del charrúa (Urug.); aliso (Arg.); anil-assú, ayapana, cambará, chilca, chirca, cruzeiro, herva de charrúa, h. de cobra, iapana, japana, perna de saracura, vassoura de ferro (Braz.).

Lychnophora, with several species of shrubs and small trees, occurs in southeastern Brazil, mostly in high mountains. No specimens are available for this study, but according to Pio Corrêa (*Diccionario das plantes uteis do Brasil,* p. 431) some of the trees attain a height of 30 feet and supply a limited amount of very dense, durable wood used for turnery, interior and exterior construction, cabinet work, fuel, and charcoal. The bark is rich in tannin and the aromatic leaves and flowers are medicinal.

Common names: Candeia, candieiro, paratudo, pau candeia (Braz.).

Montanoa, with about 35 species of shrubs and small trees, is chiefly a Mexican genus, but at least four species occur in the elevated regions of northwestern South America, from Venezuela and Colombia to northern Peru. Most of the Mexican and Central American species are shrubs, but *M. hexagona* Rob. & Greenm. of Chiapas and *M. Rekoi* Blake of Oaxaca are said to be large trees. The latter has a cork-like bark and a trunk up to 20 inches through; it contains a camphor-like substance which burns like pitch.

Montanoa quadrangularis Sch. Bip., the Tara Blanca of Venezuela, is described by Ernst as a tree up to 40 feet high and 8 to 10 inches in diameter, with spongy wood and very thick pith; the crowns are often broken off by the wind, whereupon numerous rapidly growing sprouts spring up from the roots. This species was described by Dawe (see *Tropical Woods* 7: 33) as being in western Colombia "a tree about 30 feet high, confined to volcanic lands north of Manizales in Caldas. The timber is very durable and is used for beams in the construction of buildings, for making billiard cues, etc. The bole is hollow, but the timber has a great reputation for its strength and durability, notwithstanding." Blake says (*loc. cit.*) that the species was incorrectly determined and that the correct name for it is *M. Lehmannii* (Hieron.) Blake. The tree in question is known as Arboloco and a wood specimen of that name, accompanied by fruits identified by Blake as *Montanoa* sp., was collected for the Yale School of Forestry by Jorge Pinzón de Castilla (Yale 5893); it is the only representative of the genus available for this study. The stem is not hollow, but has a very thick pith. The wood is lustrous, chestnut-brown, with very thin white sapwood; moderately hard and heavy, medium-textured, straight-grained; without figure except that the rays show distinctly on radial surface.

Common names: Cerbatana, ciguapacle, cihoapactli, homahak, ocotillo, paracua, singuapacle, sinhuapostle, tacote de flor, yagazeche, zihoapactli, zihuapatl, zoapatle, zuapatli (Mex.); flor de Santa Lucía, Margarita, palo de marimba, tatascame, tatascamite, t. blanco (Salv.); telecate blanco (Nic.); toona quirita, tora (C.R.); arboloco (Col.); tara blanca (Venez.).

Oliganthes, with several species of shrubs and small trees, sometimes up to 40 feet high, is distributed from southern Mexico to northern South America and Peru. The timber is used to a limited extent for fuel and small construction purposes.

Common names: Zamurito, zamuro (Venez.); ocuera negra, yana ocuera, y. varas, y. varra (Peru).

Perymenium, with about 40 species of herbs, shrubs, and an occasional tree, is most abundantly represented in Mexico, with a few species in northern Central America. The largest tree is the Tatascame of Salvador and the Con of western Honduras, *P. strigillosum* Greenm., which, at its best, is 30 to 40 feet tall, with a trunk 8 to 15 inches in diameter and free of branches for 12 to 15 feet. It is fairly abundant, sometimes growing gregariously, but it occurs in mountainous regions and is poorly accessible. The heartwood is light yellow or pale orange when fresh, but darkens upon exposure; sapwood thin, white. It is moderately hard, rather fine-textured, easy to work, takes a lustrous polish, and is moderately durable. It is used locally for house construction and fence posts and is suitable for furniture; it is said to be exceptionally good for charcoal for forges. The timber is not exported.

COMMON NAMES: Guisandira (Mex.); palo de tizate, tatascame, tatescamito, tisate (Salv.); con (Hond.).

Tessaria, with a few species of shrubs and trees rarely up to 50 feet high, is sparingly distributed from California to Argentina. The best known South American species is *T. integrifolia* Ruiz & Pav., a fast-growing tree common along streams and sometimes forming nearly pure stands like Alder (*Alnus*) or Willow (*Salix*). The plants are more useful in reclaiming soil or preventing erosion than for their timber. The wood is typically light and soft and not durable.

COMMON NAMES: Aliso, a. de tierra caliente, barredera, mimbre, sauce, s. blanco, s. playero (Col.); aliso, barredera (Venez.); huapariu, pájaro bobo (Peru); aliso, a. del río, bobo, pijaro bobo (Arg.); suncho rosado (Urug.); bobo, ocirana (Braz.).

Vernonia, with about 1000 species of annual or perennial herbs, erect or scandent shrubs, and a few tropical trees sometimes up to 25 feet high, is of almost world-wide distribution, though most abundantly represented in tropical America. A common herbaceous form in the United States is a coarse plant called Ironweed. Stem specimens of *V. patens* H.B.K. of Central America and *V. baccharoides* H.B.K. of Peru have large pith. The wood is yellowish or pale brownish, light and soft, medium-textured, and poorly resistant to decay. The plants are common in old clearings and are used to a minor extent for fuel and slender poles for making fences and huts.

COMMON NAMES: Lagaña de aura, rompezaragúey, r. macho, tapa-camino (Cuba); Santa María, tapa-camino (P.R.); ti baume marron (Haiti); carpanche, cihuapatl, flor de borla, f. de cuaresma, hoja-lisa, quiebramachete, tamanbub, tlamalacatilacotli, zitit (Mex.); palito de negro, semen, suquinay (Guat.); aroma, barreto, ciguapate, c. de parra, palo blanco, p. de asma, pie de zope, rajate bien, r. luego, sauquillo, suquinay, s. prieto, suquinayo (Salv.); suquinay (Hond.); caña de danto, quitirrí, tabaquillo, tubusí, tueta, tuete, t. blanco (C.R.); botón de pega-pega, hierba de San Juan, lengua de buey, l. de vaca, palo blanco, salvia, sanalego (Pan.); mata paja (Col.); palotal, pebetera (Venez.); ocuera, o. común, purma-caspi (Peru); quiebrarao (Urug.); assa-piexe, cambará, c. guazú, capichinguy de bicho, enxuga, pau candeia, p. de moquem (Braz.).

CONNARACEAE

THIS unimportant family comprises about 24 genera and 385 species of erect or scandent shrubs and small trees of pantropical distribution. The leaves are alternate, unequally pinnate or 1–3-foliolate, resembling some of the Leguminosae; stipules are absent; the flowers are small and paniculate; the fruit is a follicle, usually with a solitary arillate seed. The American genera supply no timber of economic value, and published statements to the contrary are apparently the result of erroneous identification.

Wood brown or pale reddish brown throughout. Luster rather low. Without distinctive odor or taste. Hard and heavy or moderately so; texture rather fine in upright, and the inner core of scandent, stems, becoming coarse in the outer part of the latter; grain typically irregular; not difficult to work; finishing smoothly, with a soapy feel; durability probably fair.

Growth rings generally present but not always distinct. Pores variable in size, number, and arrangement, even in the same specimen; mostly small to medium-sized, sometimes (especially in the lianas) distinct to the unaided eye; mostly without definite pattern but occasionally in radial or diagonal series. Vessels with simple perforations; fine spiral thickenings sometimes present in the very small vessels and fiber-tracheids associated with the larger vessels; tyloses common; intervascular pitting moderately coarse, alternate. Rays typically uniseriate and few to 25, sometimes up to 45, cells high; heterogeneous, usually with most of the cells square; gum deposits abundant; latex tubes observed in *Cnestidium* and *Connarus;* pits to vessels variable in size, form, and arrangement, typically large. Wood parenchyma absent or very sparse; sometimes associated with gum cysts. Wood fibers commonly septate, often finely chambered and crystalliferous; concentric bands or irregular patches of fibers with thinner walls and larger cavities or differing in contents may appear on cross section under lens like wood parenchyma; pits numerous, minute, simple; fiber-tracheids with large bordered pits and sometimes with delicate spiral thickenings frequently associated with the vessels. Ripple marks absent. Vertically elongated gum cysts with bright orange contents commonly present.

Bernardinia, with four species of scandent or erect shrubs and little trees, occurs in southeastern Brazil, the Peruvian Amazon region, and Costa Rica. The wood has not been studied.

Common names: Caja, café do mato, mata cachorro (Braz.).

Cnestidium. There are two species of scandent shrubs: *C. rufescens* Planch. is distributed from Cuba and southern Mexico to northern Colombia; *C. guianense* Schellenb. is only known in the Guianas. No uses are known for the plants.

Common names: Bejuco de sangre (Col.).

Connarus, with about 120 species of vines, upright shrubs, and small or rarely medium-sized trees, is of pantropical distribution. There are nearly 50 species in the New World, their combined range including Cuba, Central America, and nearly all of tropical South America. The published statements (see *Pflanzenreich* 3: 127: 220, 239, 242; Stone and Freeman's *Timbers of British Guiana,* p. 33) that certain species supply furniture woods are undoubtedly incorrect, and probably result from confusion with some leguminous timbers. The largest of the American trees appears to be *C. angustifolius* (Radlk.) Schellenb. of the lower Amazon region of Brazil. It attains medium size on non-inundated lands, but the light reddish brown wood is not attractive and has a soapy feel resulting from orange-colored exudations. The fruits of some species, perhaps *C. punctatus* Planch., are used by superstitious people in Brazil as a fetish or charm against witchcraft, for which reason the tree is called Arvore dos Feiticeiros.

Common names: Bejuco caoba, pico de judio, sangre de toro bejuco, tres puntas (Cuba); lian caco (Dominica); aceite macho (Venez.); wayamu menepulu (Sur.); aiaoua, karaouassira, pariki (Fr. G.); amarello, arariba do campo, arvore dos feiticeiros, azitona brava, cabello de negro, café grande do matto, mata cacharro, mão de gato, mara-çacaca, mauba do matto, meruana, merurana, muira-çacaca, pau ferro, vacca preta (Braz.); shitari-caspi (Peru). Some names wrongly applied to woods of this genus are: zebra wood (Br. G.); bois préfontaine, b. serpent (Fr. G.); pau de zebra (Braz.).

Pseudoconnarus, with four species of erect or scandent shrubs with trifoliolate leaves, is widely but sparsely distributed in the Amazon basin. The wood has not been studied.

Rourea (including *Eichleria*), with about 30 species of vines, shrubs, and little trees, is distributed throughout the whole of tropical America. The plants are noted chiefly for the toxic properties of the seeds which, though apparently harmless to birds, are said to be deadly to carnivorous animals. The brownish, moderately hard, coarse-textured wood is not available in large enough sizes to be utilized, except sometimes for fuel and charcoal.

Common names: Bejuco de Baracoa, b. prieto, bergajo, Juan caliente, mata negro

(Cuba); bejuco garrote, b. Juan caliente (P.R.); Luis Gomez (Dom. R.); chilillo, c. de la huasteca, c. venenoso, palo de chilillo (Mex.); canjuro (Salv.); granada de monte, megua (Col.); cajú bravo do campo, mata cacharro, pau do porco (Braz.).

CORIARIACEAE

Coriaria, the only genus, contains a few species of shrubs, widely but sparsely distributed in temperate regions and at high altitudes in the tropics. The only American species is *C. thymifolia* Humb. & Bonpl., a branched shrub 5 to 13 feet high occurring in the mountains of southern Mexico and Central America, and in the South American Andes to Peru. The pith is large. The bark has many corky lenticels as in Elder (*Sambucus*); it is somewhat odorous and has an astringent taste. The leaves are small, opposite, nearly sessile on slender lateral branches, suggesting a compound leaf, and terminating in racemes of very small flowers; the purplish fruit is composed of several laterally compressed carpels. The plants are poisonous and contain a toxic principle, coriarmirtine. The fruit is sometimes used to make an indelible ink. The only sample of this species available (Yale 33856) was collected by Dr. A. Rimbach at an elevation of nearly 10,000 feet in Ecuador.

Wood pale brownish throughout; said to have been white when freshly cut. Luster moderate. Odor and taste absent or not distinctive. Light in weight but firm; texture medium; grain straight; probably not durable. Stems too small to be useful.

Growth rings absent. Pores medium-sized (up to 140μ), barely visible; occurring singly and in numerous pairs and small clusters, well distributed. Vessels with simple perforations; intervascular pitting fine (5.5μ), alternate; apertures often coalescent. Rays all coarse, composing about one-third of the cross section; rarely in contact with the vessels; 5 to 15 cells wide and frequently over 4 inches high; heterogeneous, with virtually all cells upright or square; sheath cells numerous. Wood parenchyma narrowly vasicentric; parenchyma-vessel pitting fine, often unilaterally compound. Wood fibers with numerous, very small, indistinctly bordered pits. Ripple marks local; rays not storied. No gum ducts seen.

Common names: Tlalocopetate, tlalocopetatl (Mex.); tisés (Venez.).

CORNACEAE

The Dogwood family, in a restricted sense, includes 11 genera of trees and shrubs of wide distribution over the world, but more numerous in temperate than in tropical regions. The only genus represented in America is *Cornus*.

Cornus, with nearly 50 species, mostly shrubs, is widespread in the northern hemisphere. The leaves are opposite or rarely alternate, simple, deciduous, and without stipules; the small whitish or greenish flowers are borne in open naked cymes or compact heads surrounded by a corolla-like involucre; the fruit is a drupe.

There are about 16 species of *Cornus* in the United States, four of them arborescent. The best known, both as an ornamental tree and for its timber, the sapwood of which is used extensively for shuttles, is *C. florida* L., the Flowering Dogwood of eastern United States. The largest is *C. Nuttallii* Aud., growing from California to British Columbia and, under the most favorable conditions, attaining a height of 100 feet and a trunk diameter of 12 to 24 inches, though ordinarily much smaller. There are four species in Mexico, all shrubs or little trees, and one of them, *C. excelsa* H.B.K., extends as far south as Guatemala. The only species known to occur naturally south of the equator is *C. peruviana* Macbride of the mountainous regions of Peru, Bolivia, and Ecuador; it reaches a height of 35 feet and a diameter of a foot. Owing to the scarcity and small size of the trees, the tropical American species are not utilized for timber; moreover the quality of the wood is not so high as that of the Flowering Dogwood. The woods of the various species of *Cornus* are much alike in structure, though differing somewhat in their technical properties.

Heartwood brown or reddish brown, sometimes with a greenish hue; distinct and usually sharply demarcated from the white or roseate sapwood. Luster medium to high. Odorless and tasteless. Hard and heavy to moderately so, having about the consistency of Maple (*Acer*); of fine and uniform texture; mostly straight-grained; tough and strong; easy to work, finishing and wearing very smoothly; not highly resistant to decay.

Growth rings usually present. Pores numerous; small to minute, distinct with lens; rarely in contact radially; well distributed without pattern. Vessels with many-barred scalariform perforation plates; no spiral thickenings observed; intervascular pitting uncommon, opposite to scalariform. Rays decidedly heterogeneous; of two sizes, the uniseriate few to 35 cells high, all cells square or upright, the others up to 7 cells wide and up to 50 (occasionally to nearly 100) cells high, the interior cells procumbent; crystals sometimes present; pits to vessels small, round to elongated. Wood parenchyma sparingly diffuse and in rather numerous, short, irregular tangential lines scarcely visible with the lens. Wood fibers with thick walls and many distinctly bordered pits irregularly distributed in both radial and tangential walls. Ripple marks absent. No gum ducts seen.

COMMON NAMES: Cornel, dogwood (flowering, silky, red-osier), kinnikinnick (U.S.A.); aceitunillo, corona de Montezuma, c. de San Pedro, palo de membrillo, tepeacuilotl, tepecuilo, tepecuilote, topoza (Mex.).

CORYLACEAE

THE Hazel family, often included with the Betulaceae, comprises three north temperate genera of shrubs and small, rarely medium-sized, trees, namely, *Carpinus, Corylus,* and *Ostrya*. The branchlets lack terminal buds. The leaves are alternate, stipulate, mostly serrate; the male flowers are borne in catkins, the female paired in short spikes; the fruit is a nut or nutlet, partly or completely inclosed in the foliaceous accrescent involucre. The timber is of fair to good quality, but is not extensively used because of the small sizes and limited quantity.

Woods nearly colorless to brown with a roseate hue throughout. Luster medium to rather low. Odorless and tasteless. Mostly hard, heavy, tough, and strong; sp. gr. (air-dry) 0.75 to 0.85; weight 47 to 53 lbs. per cu. ft.; density considerably less in *Corylus;* texture fine and uniform; grain straight to irregular; working properties fair; not durable in contact with the ground.

Growth rings present. Pores mostly small to minute, decreasing in size during the growing season; numerous; usually in short radial rows, without definite pattern. Vessels with few-barred scalariform perforation plates in *Corylus;* mostly with simple perforations but with tendency to the formation of few-barred scalariform plates in the smallest vessels of the others; fine spiral thickenings present; intervascular pitting medium to coarse (8 to 13μ), alternate. Rays 1 to 3, occasionally also 4 to 6, cells wide and up to 30, infrequently to 50, cells high; conspicuous aggregates common in *Carpinus* and *Corylus;* weakly heterogeneous in *Corylus,* homogeneous in the others; large crystals in distended cells frequent in *Carpinus* and *Ostrya;* pits to vessels small to medium-sized, oval. Wood parenchyma in numerous irregular metatracheal lines, usually not distinct with lens. Wood fibers with rather thick walls and numerous small bordered pits. Ripple marks and gum ducts absent.

Carpinus. There are several species, all European and Asiatic but one, *C. caroliniana* Walt., of eastern United States, Mexico, Guatemala, and Honduras. This is a bushy tree, rarely 40 feet high, with a fluted smooth-barked trunk and white or pale brown wood of no particular value except for fuel and charcoal.

COMMON NAMES: Blue beech, hornbeam, ironwood, water beech (U.S.A.); lechillo, palo silo, p. barranco (Mex.).

Corylus, with a few species of shrubs, rarely trees, occurs in middle and southern Europe and northern Africa, central and eastern Asia, and eastern North America. The English name for the plants is Hazel and they are the source of hazelnuts and filberts. The American species are all shrubs, but there are two species in Europe and temperate Asia, namely, *C. avellana* L. and *C. colurna* L., which attain a height of 30 feet with a trunk diameter of a foot or more and supply some commercial timber which resembles Alder (*Alnus*) and is noted for its fine and uniform texture and flexibility. In the United States the name Hazel is often applied, particularly by architects, to the figured heartwood of Red Gum (*Liquidambar*).

Ostrya. There are two distinct species, namely, *O. carpinifolia* Scop., the Hop Hornbeam of southern Europe and Asia Minor, and *O. virginiana* (Mill.) Koch, the Ironwood or Hop Hornbeam of eastern United States, with southwestern varieties, sometimes considered species, extending from Arizona through Mexico into Central America. The trees are usually not over 25 feet tall and a foot in diameter, but sometimes attain a height of 50 to 60 feet with a stout trunk 18 to 24 inches through. The southernmost variety, sometimes known as *O. guatemalensis* (Winkl.) Rose, is typically a low and spreading tree of infrequent occurrence in uplands of Guatemala and Mexico. The fruits of the genus are borne in hop-like clusters and, unlike *Carpinus,* the bark is scaly and the wood lacks aggregate rays. The timber is brownish, often somewhat roseate, and noted for its strength and toughness. It has many local uses but is of little or no commercial importance.

COMMON NAMES: Hardhack, hop hornbeam, hornbeam, ironwood, leverwood (U.S.A.); guapaque (Mex.).

CROSSOSOMATACEAE

Crossosoma, the only genus, is composed of three or four species of unarmed or spinescent shrubs or small trees (*C. californicum* Nutt.) occurring in southern California, Arizona, and northwestern Mexico. The bark is rough and bitter; the leaves are small, simple, leathery, and often clustered on short branches or spurs; stipules are wanting; the showy white or purplish flowers are solitary and terminal; the fruit is a cluster of follicles with few to many globular or reniform seeds, each inclosed in a thin fimbriate aril. The plants apparently are not utilized and they are too small to be a source of wood. The only specimen available (Yale 14834) is of *C. Bigelovii* S. Wats., collected by S. B. Detwiler in Arizona.

Heartwood not seen; sapwood white. Luster fairly high. Without distinctive scent; taste mildly bitter. Moderately hard, heavy, tough and strong; texture very fine; grain mostly straight; easy to work; durability of heartwood unknown. Suitable for small articles of turnery if available in large enough pieces.

Growth rings distinct; structure more or less definitely ring-porous. Pores angular; small (55μ) to minute (15μ), gradually diminishing in size during a season's growth, the largest frequently in a single initial row; those in late wood nearly all solitary, well distributed without special pattern, though radially arranged because of the close spacing of the rays. Vessels with simple perforations; many very small vessels with fibriform members; pits small (6μ). Rays 1 to 6 cells wide and up to 55 cells high; heterogeneous, with most of the cells square or upright, the procumbent ones short; pits to vessels small (5 to 6μ), subcircular. Wood parenchyma sparingly paratracheal. Ground mass composed of thick-walled tracheids having abundant large bordered pits and scarcely distinguishable from the fibriform vessel members. Ripple marks absent. No gum ducts seen.

CUNONIACEAE

THIS family, treated by some taxonomists as a section of the Saxifragaceae, comprises 26 or 27 genera and about 250 species of trees and shrubs, mostly of temperate and subtropical regions of the southern hemisphere. The trifoliolate or pinnate, rarely simple, leaves are often glandular serrate and typically opposite; the stipules are sometimes large; the flowers are solitary or in panicles or heads; the fruit is dehiscent or indehiscent. Eight genera are of more or less commercial importance for their timber in eastern Australia and there are likewise a few in the East Indies, South Africa, and New Zealand. In Latin America there are only three genera, namely, *Belangera, Caldcluvia,* and *Weinmannia,* and their timbers, though of good quality so far as known and serviceable for the same purposes as Red Gum (*Liquidambar*), are of only local utility.

Heartwood uniform to slightly streaked, light brown (often with a pinkish hue) to light reddish brown; usually merging gradually into the sapwood. Luster rather high to medium. Odor and taste not distinctive. Light in weight to moderately so; firm and tough; texture fine and uniform; grain straight to roey; easy to work, finishing very smoothly; holds its place well when manufactured; is fairly durable.

Growth rings usually present, sometimes conspicuous because of the darker color of the late wood. Pores very numerous; small to minute, not visible without lens; well distributed without pattern, though diminishing in size and apparent abundance during seasonal growths in temperate regions; radial multiples common in *Belangera,* few or absent in the others. Vessels with many-barred scalariform perforation plates in *Caldcluvia* and *Weinmannia*; with both simple and multiple perforations in *Belangera;* spiral thickenings absent; pitting opposite to scalariform. Rays heterogeneous, often decidedly so; typically 2-sized, the larger ones commonly only 2 or 3 cells wide and less than 30 or 40 cells high; dark red gum deposits abundant in heartwood; pits to vessels large, elongated, often in scalariform arrangement. Wood parenchyma rather weakly developed and scarcely visible with lens; diffuse in early wood to finely reticulate in late wood. Wood fibers septate in *Belangera;* pits bordered and conspicuous in the others. Ripple marks and gum ducts absent.

Belangera, with several species of trees or large shrubs, occurs in southern Brazil, Paraguay, and in Argentina. The maximum height reported is about 50 feet. The bark is used in tanning. The light pinkish brown, easily worked wood has about the consistency of Alder (*Alnus*) and is suitable for the same purposes.

COMMON NAMES: Salgueiro do matto (Braz.); árbol Andrade (Arg.).

Caldcluvia *paniculata* (Cav.) Don, the only species, is a little tree growing along the coast of southern Chile, where it is known as Tiaca. The wood, which is similar to that of *Weinmannia,* is available only in small sizes and its principal use is for fuel.

Weinmannia, with about 140 species of trees and shrubs, has a very extensive range in the southern hemisphere and extends northward to the Philippines and Mexico. Some of the Old World species are large timber trees, but the Latin American species, though numerous and widely distributed, are of small size and confined to mountainous regions. Occasional individuals are 50 feet high, with a trunk diameter of a foot or more, but many are only shrubs. The bark is rich in tannin. The timber, when large enough, is suitable for common furniture and interior construction.

Heartwood brownish to light reddish brown; merging gradually into the lighter-colored sapwood. Luster medium. Odorless and tasteless. Rather light but firm to moderately heavy and hard; sp. gr. (air-dry) 0.59 to 0.65; weight 37 to 44 lbs. per cu. ft.; texture rather fine, uniform; grain variable, often roey; very readily worked, and useful for the same purposes as Red Gum (*Liquidambar*) or Birch (*Betula lutea* Michx.), depending on the density; durability fair.

COMMON NAMES: Oreganillo (P.R.); bastard brasiletto, wild brasiletto (Jam.); arrayán, lorito, loro (C.R.); arenillo, encinillo (Col.); curtidor, saí, saisai, say, saysay (Venez.); huichullu, machi, tiaca (Peru); tarco, tinco, tinel (Arg.); madén, palo santo, teneo, testui, tinal, tineo (Chile).

CYRILLACEAE

AN American family of four genera and 10 species of shrubs and small, rarely large, trees with a combined range including southeastern United States, the West Indies, southern Mexico, British Honduras, and South America to the Amazon region. The leaves are simple, entire, alternate, and without stipules; the small, regular flowers are borne in terminal or axillary racemes; the fruit is an indehiscent capsule. The timber is of no commercial importance.

Cliftonia *nitida* Gaertn., the sole species, commonly known as Ironwood or Titi, is a tree occasionally 40 to 50 feet high with a stout but often crooked or leaning trunk 15 to 18 inches in diameter, growing in swamps in the coastal region of southeastern United States. The wood, which is similar to that of *Cyrilla,* has no special uses except for fuel.

Cyrilla *racemiflora* L., the only well-defined species, has its northern limit in the coastal Pine belt from southeastern Virginia to Florida and westward to Texas. It is fairly common along streams and at the edge of swamps, only occasionally attaining a height of 35 feet and a trunk diameter of 10 to 12 inches, being often shrubby with many slender stems 15 to 20 feet high. It is sometimes called Leatherwood because the bark at the base of the trunk is spongy and pliable. The species attains its largest size in the upper mountain forests of Puerto Rico, being sometimes five feet or more in diameter, though the trees over 24 inches through are invariably hollow and provide the principal nesting places for the native parrots. The wood is chiefly used for charcoal; the lumber is not in demand because of its pronounced tendency to warp. In Cuba it is usually a swamp tree or shrub, but in Jamaica it is sometimes 80 feet tall. In southern Mexico and British Honduras it is a riparian species rarely 40 feet high.

Heartwood light to dark reddish brown, the most deeply colored specimens being rather oily; sapwood lighter, not very clearly demarcated. Luster rather low. Moderately hard and heavy; of fine and uniform texture; easy to work, finishing very smoothly; inclined to warp; apparently durable.

Growth rings usually present. Pores small to minute, not distinct without lens; numerous; well distributed; infrequently in contact radially. Vessels with scalariform perforation plates having many fine and very closely spaced bars; without spiral thickenings; intervascular pitting fine, opposite to scalariform. Rays 1 to 4, in some species up to 8, cells wide and few to 25 or 30 cells high; heterogeneous, the interior cells of the larger rays slender and procumbent; pits to vessels small, opposite or elongated and parallel. Wood parenchyma diffuse and in rather numerous short irregular tangential lines not distinct without lens. Wood fibers with very numerous distinctly bordered pits in both radial and tangential walls. Ripple marks and gum ducts absent.

Common names: Bloodwood, burning wood, ironwood, leatherwood, red titi (U.S.A.); bloodwood, beetwood (Jam.); colorado, palo colorado, southern leatherwood (P.R.); barril, clavellina, lirio de costa, llorona, palo de jutía, tranca de puerto, yanilla (Cuba).

Cyrillopsis *paraensis* Kuhlmann, the only species, is a small thin-barked tree of the lower Amazon region of Brazil. The wood (Yale 33826; Ducke 299) is decidedly unlike that of *Cyrilla* and *Cliftonia.* Heartwood brownish red, merging gradually into the sapwood; very hard, heavy, and compact; difficult to cut.

Pores small, near limit of vision; rather few; solitary; irregularly distributed without definite design. Vessels with simple perforations; without spirals. Rays uniseriate or less often biseriate and up to 25 cells high; heterogeneous; pits to vessels large, elongated, and in scalariform arrangement, radially, obliquely, or vertically. Wood parenchyma in numerous, distinct, continuous or interrupted bands, 2 to 4 cells wide and 1 to 4 pore-widths apart, touching most of the pores but not including them, the cells all short; also sparingly vasicentric. Wood fibers with very thick walls and numerous distinctly bordered pits; those next to vessels (vasicentric fiber-tracheids) not so thick-walled and with pits in several rows. Ripple marks and gum ducts absent.

Purdiaea (or *Costaea*), with seven species of rather large shrubs, occurs in the mountains of Cuba, Colombia, and eastern Peru. *P. nutans* Planch. is said to attain a maximum height of 25 feet at elevations of 7000 to 8000 feet in Colombia. Wood not seen.

DICHAPETALACEAE

The Dichapetalaceae, also known as the Chailletiaceae, are small trees and upright or climbing shrubs of wide distribution. There are four genera: *Dichapetalum* (about 200 spp., pantropical), *Stephanopodium* (4 spp., tropical South America), *Tapura* (12 spp., tropical America and Africa), and *Gonypetalum* (4 spp., Amazon region). The leaves are alternate, simple, and stipular; the small flowers are borne in axillary clusters; the fruit is a drupe, typically dry. The family is not the source of commercial timber. The following description applies to woods of American species of *Dichapetalum*, *Gonypetalum*, and *Ta-*

pura, which differ markedly in their general properties and in certain structural details.

Pores very small to minute or medium-sized, the larger (*Dichapetalum* and *Gonypetalum*) near limit of vision; numerous; mostly solitary, sometimes clustered; uniformly or rather unevenly distributed, or more or less diagonally arranged (*Gonypetalum*) without definite pattern. Vessels typically with simple perforations but sometimes with tendency to development of scalariform plates with thin bars; spiral thickenings absent; pitting very fine, alternate. Rays of two sizes, the larger mostly 3 to 5, occasionally up to 8, cells wide and 30 to 100 or more cells high, readily visible on cross section and rather conspicuous on the radial; decidedly heterogeneous; uniseriate rays and ray margins composed of square and upright cells; crystals common; ray-vessel pitting appearing very fine, but sometimes unilaterally compound. Wood parenchyma abundantly developed, but scarcely distinct without lens; paratracheal, confluent, and metatracheal. Wood fibers with thick to very thick walls. Ripple marks and gum ducts absent.

Dichapetalum (or *Chailletia*) is most abundantly represented in tropical Africa, less so in the Indo-Malayan region. There are seven species in tropical America, their combined range extending from the Amazon region into Central America and the West Indies. They are small trees or shrubs of infrequent occurrence and of no particular utility. Sapwood yellowish; heartwood not seen. Moderately hard and heavy; texture rather coarse.

Gonypetalum. The only wood sample available (Yale 34089; Ducke 325) is from a small tree of *G. lanceolatum* Ducke growing near Manáos in Brazil. Heartwood dull yellowish olive, with a slightly oily appearance, merging into the pale brown sapwood. Odorless and tasteless. Hard, heavy, and strong; texture rather coarse; feel of sawed material harsh; grain straight; easy to work, finishing smoothly; is probably durable. Presumably without commercial possibilities.

Tapura. Thirteen species have been described, two in tropical Africa, five in the West Indies, and six in South America, mostly in the Amazon region. The American representatives are little trees or shrubs of scattered occurrence in humid localities and are of no economic importance, although the stems of *T. cubensis* (Poepp. & Endl.) Gris. are used locally for house poles, cart bodies, and plow beams.

Heartwood yellow-brown, with an oily appearance and feel; sapwood lighter, not sharply demarcated. Hard and heavy; fine-textured; fairly straight-grained; easy to cut, finishing very smoothly; probably durable.

Common names: Aura, cagada de aura, vigueta naranjo, v. prieta (Cuba); waiaballi (Br. G.).

DICLIDANTHERACEAE

Diclidanthera, the only genus, is sometimes included with the Styracaceae. The two or three species are shrubs or small trees in eastern Brazil. The branches are more or less pendulous; the leaves are alternate, entire, and without stipules; the tubular white or yellow flowers are borne in axillary or terminal racemes or panicles; the 5-seeded globose fruits are indehiscent. The only specimen examined is a piece of a small branch of *Diclidanthera laurifolia* Mart. (Riedel 1158) in the herbarium of the New York Botanical Garden. The wood is very dense, fine-textured, and of anomalous structure, concentric type, with a few stone cells in the conjunctive tissue.

Pores very small; fairly numerous; virtually all solitary; rather irregularly distributed without definite pattern. Vessels with simple perforations; without spiral thickenings; pitting fine, alternate. Rays mostly uniseriate and fairly high, some (of conjunctive tissue) coarse; heterogeneous, many of the cells square; pits to vessels small, rounded. Wood parenchyma very sparingly paratracheal; not visible without lens. Wood fibers with very thick walls and small bordered pits. Ripple marks absent. No gum ducts seen.

DILLENIACEAE

A family of 11 genera and about 300 species of trees, erect or scandent shrubs, and a few perennial herbs. The simple, alternate leaves are usually scabrous; the flowers are borne in terminal or axillary panicles; the fruit is sometimes bacchate, more often capsular and dehiscent, with arillate seeds. The principal value of the plants is for decorative purposes. There are five arborescent genera in the Indo-Malayan region, but *Dillenia* is the only one with large trees. About 15 species are native to the Philippines, and their woods, known on the Manila market without distinction as Catmón, are used to some extent for interior finish, flooring, general construction, furniture, and cabinet work.

There are four genera in tropical America, namely, *Curatella, Davilla, Doliocarpus,* and *Tetracera;* they make up the section Tetracereae. The first three have rough leaves which are used locally as a substitute for sandpaper. The genera other than *Curatella* are typically lianas or climbing shrubs; the slender stems are used as "bush ropes," the thick ones as an occasional source of drinking water. These peculiarities account for many of the vernacular names, which are applied without discrimination as to genus.

Heartwood pinkish brown to dark reddish brown, merging gradually into the sapwood; flecked with large rays. Odor and taste not distinctive. Hard, heavy, and strong to moderately so; sp. gr. 0.70 to 0.85; texture medium to very coarse; grain often irregular; not very easy to work, but attractive when properly finished; durability rather high.

Pores small to large (in the lianas), in most cases visible without lens; typically solitary; not very numerous and scattered without pattern. Vessel perforation plates either predominantly simple or predominantly multiple, the scalariform plates with few to several bars; spirals absent; pitting coarse, with some tendency to scalariform. Rays of two sizes, the large ones of the Oak (*Quercus*) type; heterogeneous, the procumbent cells short; ray-vessel pit-pairs small to large, half-bordered, sometimes elongated; bundles of raphides common. Wood parenchyma in numerous, wavy, uniseriate, tangential lines or diffuse; not visible without lens and often indistinct with it. Wood fibers thick-walled; with numerous conspicuous bordered pits in both radial and tangential walls. Ripple marks and gum ducts absent. *Doliocarpus* differs from the others, so far as studied, in being of anomalous structure, the included phloem being associated with coarse bands and rays of conjunctive tissue.

Curatella, with one or two species of small trees, is widely distributed in tropical America. *C. americana* L. is a common tree of savannas (Plate XXXII) and open forests, sometimes forming thickets. It is occasionally 30 feet high with a crooked trunk 12 inches in diameter. The rough leaves, which contain silicon, are widely used for polishing wood and metals and scouring kitchen utensils. The red flaky bark is rich in tannin. The wood is attractive but is of no commercial importance because of the poor form of the timber; it is employed locally for fuel, charcoal, fence posts, and to a minor extent for cabinet work and articles of turnery.

COMMON NAMES: Rough-leaf tree (Eng.); chaparro (Span.); careicillo, peralejo macho, vaca-buey (Cuba); peralejo (Haiti); hoja-man, raspa-viejo, tlachicón, tla-chiquoni (Mex.); yaha (Br. H.); lenga de vaca, malcajaco (Salv.); chumico de palo, hoja chigüe, raspa-guacal (C.R.); chumico, c. de palo, curatela (Pan.); azufre, carne fiambre, paralejo (Col.); chaparro de sabana, curata, paricá (Venez.); manabadin, sandpaper tree (Br. G.); feuille rude, paricá (Fr. G.); bosch cachon, curatahie, wilde cachon (Sur.); racta-panga (Peru); lixeira (Par.); aderno, caimbé, cajueiro bravo, c. b. do campo, c. b. do matto, folha de lixa, lixeira, marajoára, pentieira, sambaiba, sobro (Braz.).

Davilla is a tropical American genus of climbers, with about 35 species, mostly Brazilian. The bark of one of the Brazilian species is the source of a black dye. The wood is of normal structure and resembles that of *Curatella* except that the rays are coarser (up to 40 cells wide) and the parenchyma is diffuse.

COMMON NAMES: Bejuco castaño, b. colorado, b. de cerca, b. guará, b. guarana, guaranillo (Cuba); bejuco de tachicón (Mex.); chaparro (Br. H.); bejuco chaparrón (Guat., Hond.); hoja chiqüe (Nic.); chumico de bejuco (C.R.); chumico, c. de bejuco, chumiquillo (Pan.); bejuco tomé

(Col.); capa-homem, cipo caboclo, c. capa-homem, c. de carijó, c. vermelho, fohla de lixa, muiraqueteca, muirateleca, sambaiba, sembaibinha (Braz.).

Doliocarpus is a tropical American genus with over 20 species of slender to coarse lianas, most of them Brazilian. It is a common occurrence for a traveler to get a drink by cutting a section of the stem, holding it aloft and letting a stream of clear water flow into his open mouth. No other uses are known.

Wood dull brownish to reddish brown throughout, with conjunctive tissue and coarse rays showing conspicuously. Odorless; taste slightly acrid. Rather light in weight to moderately heavy and hard, the bands of conjunctive tissue harder than the rest of the wood; texture very coarse; grain irregular. Wood suitable for small fancy articles.

COMMON NAMES: Bejuco de agua (P.R.); water tie-tie (Br. H.); cipó d'agua, c. caboclo venenoso, c. vermelho, muiraqueteca, murucutua, sambahiba (Braz.).

Tetracera is a pantropical genus of about 30 species, mostly woody vines of the water-liana class. The wood has not been studied.

COMMON NAMES: Bejuco carey, b. guará (Cuba); bejuco de agua (Mex.); lengua de vaca (Salv.); hoja-chigüe (Nic.); raspa, r. guacal (C.R.); chumico, pasmo de sol (Pan.); azraz, bejuco de agua, b. chaparro, b. tomé (Col.); liane rouge (Fr. G.); cipó de caboclo, c. mulatinho, c. vermelho, icipó, tigarea (Braz.).

EBENACEAE

THE Ebony family comprises four genera (*Diospyros, Euclea, Royena,* and *Tetraclis*) and about 300 species of trees and shrubs widely distributed in the tropical and mild temperate regions of the world. The leaves are alternate, entire, and without stipules; the dioecious or polygamous flowers are axillary; the fruit is a berry. The most important genus and the only one represented in the New World is *Diospyros.*

Diospyros (including *Maba*) contains over 250 species, some of which are important for their succulent fruits (*e.g.,* the date plum, kaki plum, and persimmon), several for their timber, particularly the heartwood, which is the true Ebony of commerce. The name Ebony suggests a black wood, but the heartwood of most species of *Diospyros* is only streaked or mottled with black, and in this category are Macassar Ebony (Celebes), Coromandel or Calamander (Ceylon), Camagon (Phil. Is.), Marblewood (Andaman Is.), and Persimmon (U.S.A.). The sapwood is white when fresh, becoming either bluish or reddish. Transition of sapwood into heartwood is believed to be accompanied by a transformation of lignin into ulmic substances (at first ulmic acid, which afterward may be decarbolixized), a process analogous to, if not identical with, fossilization (see *Tropical Woods* 41: 50–51). The heartwood is accordingly very brittle, breaks with a concoidal fracture, and is delicate to work, but it has long been esteemed and its use can be traced back to the Early Egyptians who probably obtained it in Abyssinia. Ebony from India was known to the Greeks prior to 350 B.C. (see *Tropical Woods* 42: 61–64). The principal present-day uses of black Ebony are for the handles of cutlery and instruments, finger-boards of violins, piano keys, carved objects, brush backs, inlays, and marquetry; figured wood is consumed chiefly in the manufacture of hair brushes. Some of the woods called Ebony in the trade belong to different families, especially the Leguminosae.

There are numerous American species of *Diospyros.* The largest tree and the only one of commercial importance for its timber is the Persimmon, *D. virginiana* L. of the southeastern United States, usually a small or medium-sized tree, but known to attain a height of 100 to 125 feet, with a long trunk 20 to 30 inches in diameter; the sapwood is in demand for the manufacture of large-sized weaving shuttles and also for the heads of golf clubs. The species in tropical America are too small or rare to be of economic value, although several of them have black heartwood used locally for making walking sticks, inlays, and miscellaneous articles of turnery and carving.

Heartwood black, brown, or variegated; sapwood white when fresh, sharply demarcated, subject to stain. Odorless and tasteless. Moderately to decidedly hard, heavy, and compact; heartwood brittle, sapwood tough and strong; texture medium to fine; grain fairly straight; finishes very smoothly; black heartwood highly resistant to decay.

Growth rings present or absent; tendency to ring-porous structure in *Diospyros virginiana.* Pores minute to large enough to be seen without lens; rather few to numerous; solitary and in small multiples or fairly long radial rows. Vessels with simple perforations; without spiral thickenings; pits very small, alternate. Rays mostly uniseriate, sometimes biseriate, and few to 30 cells high; heterogeneous, many of the cells square or upright; crystals often present; pits to vessels very small. Wood parenchyma sparingly developed about vessels and in numerous, fine, mostly uniseriate lines, sometimes regular and continuous, more often short and producing a fine meshwork barely visible with the lens; chambered crystalliferous strands sometimes abundant. Wood fibers thick-walled; pits numerous, minute, indistinctly bordered. Ripple marks present in *D. virginiana;* 70 to 80 per inch; all elements storied, but often irregularly. No gum ducts seen.

COMMON NAMES: Date plum, persimmon, plaqueminier, possum wood, simmon (U.S.A.); pigeon wood (Jam.); ébano, é. blanco, é. real, é. carbonero, manaté, tagua-tagua (Cuba); boa-wood, featherbed (Bah.); guayabota, g. nispero, tabeiba (P.R.); bois raide (Haiti); bois charbon, butterwood (Trin.); chapote, c. prieto, coacollatillo or coacolutillo, ébano, estrellito, guayaparín, pipinance, tauch, xkanché, xnobché, zapote enano, z. negro, z. n. montés, z. prieto, zapotillo (Mex.); cylil (Br. H.); pipinance (Salv.); bara-bara, barrabarra (Br. G.); piriquiteira (Braz.); baiuca-caspe (Peru).

ELAEOCARPACEAE

THIS family, often included in the Tiliaceae, consists of 10 genera and about 150 species of trees and shrubs of wide distribution in tropical and subtropical regions. The leaves are simple, alternate or subopposite; the flowers, which are with or without petals, are borne singly, in fascicles, or in racemes in the axils of the leaves; the fruit is baccate or capsular. Although some of the trees attain large size, the utilization of their timber is almost entirely local. The family is represented in Latin America by six genera, namely, *Aristotelia, Crinodendron* (*Tricuspidaria*), *Dicraspidia, Muntingia, Sloanea,* and *Vallea.*

Heartwood light to dark brown; often pinkish or reddish in *Sloanea;* sapwood lighter, sometimes sharply demarcated. Luster low to rather high. Odor and taste absent or not distinctive. Growth rings usually present. Density variable; *Sloanea* very hard and heavy, others typically of light weight, but firm and tenacious. Texture fine to medium; grain variable; excepting *Sloanea,* woods easy to work, finishing smoothly, nailing without splitting, not resistant to decay, suitable for general carpentry, interior construction, boxes, crates, and paper pulp; *Sloanea* not easy to work, darkest specimens appear durable, suitable for implement stock and heavy construction.

Growth rings present or absent; sometimes distinct. Pores medium-sized to small, the largest visible in *Dicraspidia, Muntingia,* and *Sloanea,* all very small to minute in the others; occurring singly and, more often, in radial multiples of few to many pores each; sometimes irregularly distributed, tending to form bands. Vessels typically with simple perforations; no spiral thickenings seen; intervascular pits medium-sized, often opposite. Rays 1 to 3 cells wide in *Crinodendron* and *Vallea,* occasionally also in *Aristotelia;* distinctly 2-sized in the others, the larger being up to 5 or 6, in *Muntingia* and *Sloanea* up to 8 or 10, cells wide; nearly homogeneous in *Muntingia;* heterogeneous, usually decidedly so in the others, the uniseriate rays and ray margins composed largely of square or upright cells; ray height variable, less than 50 cells in *Crinodendron* and *Vallea,* up to 100 or more in the others; sheath cells present in *Aristotelia, Dicraspidia,* and *Sloanea;* crystals sometimes present; pits to vessels small to very small in *Muntingia* and *Dicraspidia;* medium-sized to large, oval, and opposite to very large, elongated, and more or less in scalariform arrangement in the others. Wood parenchyma finely reticulate, barely visible with lens in *Dicraspidia* and *Muntingia;* very sparingly paratracheal and not distinct with lens in *Aristotelia, Crinodendron,* and *Vallea,* and some samples of *Sloanea;* in concentric bands of variable width, spacing, and distinctness, apparently demarcating seasonal growth rings, in other samples of *Sloanea;* pits to vessels large and often scalariformly arranged except in *Dicraspidia* and *Muntingia,* where they are small. Wood fibers with very thick walls in *Sloanea;* with thin to medium walls in the others, and sometimes septate; pits very small, simple or indistinctly bordered except in *Dicraspidia* and *Muntingia,* where they have small but distinct borders. Ripple marks present in *Dicraspidia* and *Muntingia,* all elements storied except the larger rays; metatracheal wood parenchyma cells in secondary seriation. Small vertical traumatic gum ducts occasionally present in *Sloanea.* Inner bark with distinct V-shaped patches except in *Crinodendron* and *Vallea.*

Aristotelia. There are seven species of shrubs and little trees, all Australasian but one, the Maqui, *A. maqui* L'Her., of Chile and Patagonia. The last is usually not over 15 feet high; its berries are edible and the wood is of good quality, suggesting Basswood or Linden (*Tilia*), though too small for commercial purposes. Color pale brownish throughout. Fairly lustrous. Odorless and tasteless. Light in weight, but firm and tough; texture fine; grain straight; very easy to work; holds its place well when manufactured; poorly resistant to decay.

Crinodendron (or *Tricuspidaria*), with three species of trees, is limited in distribution to southern South America. The serrate leaves are alternate or opposite; the white flowers are typically solitary in the leaf axils; the fruit is a leathery, dehiscent, few-sided capsule. *C. patagua* Mol. is a tree up to 50 feet tall with a stout bole; it is native to central Chile. *C. Hookerianum* Gay is usually not over 25 feet high and six inches in diameter; it occurs in Valdivia and Llanquihue, Chile. *C. tucumanum* Lillo is a large tree, sometimes attaining a height of 100 feet and a diameter of 40 inches, in Tucumán, Catamarca, and Salta, Argentina, and in Bolivia (see *Lilloa* 2: 341–352. 1938).

Heartwood not seen; sapwood nearly white, becoming brownish upon exposure. Luster rather high. Odorless and tasteless. Light in weight but firm and strong, suggesting *Tilia;* texture fine; grain irregular; very easily worked, finishing smoothly; not durable in contact with the ground. Suitable for the same purposes as Basswood.

Dicraspidia *Donnell-Smithii* Standl., the only species, is a shrub or little tree 10 to 15 feet high, native to Honduras, Costa Rica, and Panama. It is easy to recognize because of its large peltate stipules. Standley says (*Field Mus. Bot.* 4: 227, 1929): "The tree is a very curious and interesting one, and the excellent specimens have remained so long (over 30 years) in the herbarium only because it is difficult to determine the proper family to which it should be referred. Captain John Donnell Smith made a detailed study of the specimens and prepared a partial description many years ago. He recognized the fact that a new generic type was represented. . . . There is no doubt in the writer's mind that the plant is related to the genus *Muntingia* and that it should be referred, consequently, to the Elaeocarpaceae, or rather to the Tiliaceae, if these two families are to be combined. For general aspect, and in pubescence, leaf form, and gross appearance of the flowers, *Dicraspidia* and *Muntingia* are very similar, so much so that their relationship can scarcely be questioned." The woods of the two genera are very similar in all respects except that in *Dicraspidia* (Yale 15644) the rays are decidedly heterogeneous, whereas in *Muntingia* they are nearly homogeneous.

Wood pale brownish throughout. Luster rather low. Odorless and tasteless. Light in weight but firm and tough; texture medium; grain fairly straight; easy to work; poorly resistant to decay. Of no commercial possibilities.

Muntingia *calabura* L., the only species, is a small tree, rarely 40 to 50 feet high, sometimes only a shrub, very widely distributed in tropical America and naturalized in the Philippine Islands and Siam. The 3-nerved leaves are alternate; the long-pedicelled white flowers are solitary or fasciculate in the leaf axils; the sweet berries are edible. The tough fiber of the bark is used locally for making rope. The wood has few uses because of the small sizes obtainable.

Heartwood pale brown, merging gradually into the slightly lighter-colored sapwood; with oatmeal appearance on radial surface. Fairly lustrous. Odorless and tasteless. Light in weight but firm and tenacious; texture medium; grain irregular; very easily worked, finishing smoothly; poorly resistant to decay. Of no commercial possibilities.

COMMON NAMES: Capulé, capulinas, guácima boba, g. cereza, memiso (Cuba); memiso (Dom. R.); bois de soie marron (Haiti); bersilana, capolén, capulén, jonote, palman, puan (Mex.); capulén (Guat.); duiskap-krogró, tebekrá (C.R.); majaguillo, pacito (Pan.); acurucó, cedrillo, chirriador, chitotó, majaugua, majauguito, manguito, pasito, tapabotija, vijaguillo (Col.); cedrillo, mabaujo, majogua majaguillo (Venez.); bolina, ccoilor-ppanchu, iumanasa, mullaca-huayo, yumanazo (Peru); datilea, ratiles (Phil. Is.).

Sloanea, with about 60 species, is represented in both hemispheres. Its range in America includes the West Indies, southern Mexico, Central America, and South America to Brazil and Peru. Species in the mountains of the West Indies attain a maximum height of 100 feet and a trunk diameter of four feet, but the timber is too dense to be of much utility. Small lots have been exported under the name of Sloane's Greenheart. The Mameicillo Colorado, *S. megaphylla* Pittier, of the Atlantic coastal forests of Panama, is about 80 feet tall and 16 inches in diameter and is remarkable for its enormous leaves and fluted trunk; the wood is said to have been used in Colon for the understructure of the old wharves. *S. echinocarpa* Uitt. of Surinam and British Guiana is characterized by narrow plank-like buttresses which extend far up the stem. There are about 15 species in the Amazon region of Brazil, some of them very large trees.

Heartwood variable in color; grayish brown (sometimes with a reddish hue), grayish yellow-brown, or a uniform brownish pink; usually distinct from the lighter-colored sapwood, except in the pinkish type. Luster medium. Odorless and tasteless. Heavy to very heavy, hard and strong; texture fine in pinkish sorts, medium to coarse in the others; grain irregular; working properties fair to poor; durability doubtful. Not likely to prove of much value for export.

COMMON NAMES: Break-axe, ironwood, lignum-durum (Jam.); achotello, berijua, chicharrón, cocote de toro, cogote de toro, cresto de gallo, juba blanca, pite (Cuba); cacaillo, cacao motillo, c. otillo, c. roseto, motillo (P.R.); chicharrón (Dom. R.); bois coq, castor, chapeau carré (Haiti); chataigner, c. coco, c. franc, c. petit coco (Fr. W.I.); chantonier (Grenada); huesillo (Mex.); wild atta, w. grape (Br. H.); palo de peine, peine de mico (Guat.); picapica (Hond.); terciopelo (Salv.); abrojo, mano de león (C.R.); casaco, mameicillo colorado (Pan.); achiote de monte, guebo de gato (Col.); aruadan (Br. G.); akosai, boesi-koesoewé, bosch-koesoewé, firoberobana, kassa-bahoedoe, koeseweran, koesoeweerjan, koesoewé-pète, serewan, sierabolieballi (Sur.); chataigner, quapalier, goulougou-albani, oulougua-palou (Fr. G.); biacuí, quabiraba-branca, piqué-y, sapopemba, urucurana (Braz.); maquisapañaccha (Peru).

Vallea, with three scarcely distinguishable species of small trees, sometimes 30 feet tall and 10 inches in diameter, grows in the mountains of South America from Venezuela and Colombia to Peru. The leaves are alternate, mostly small and heart-shaped; the roseate flowers are borne in axillary cymes; the fruit is a small dehiscent capsule. The timber has about the consistency of Birch (*Betula*), but there are no special uses for it because of the small size and limited occurrence of the trees.

COMMON NAMES: Raque (Col.); achote, achotillo, companito, onotillo, roso (Venez.); guishmo, platuquero, sacha-peral (Ec.).

ERICACEAE

THE Heath family, with about 35 genera and over 600 species of shrubs, undershrubs, and small to occasionally large trees, is very generally distributed. The leaves are typically alternate, simple, evergreen, and without stipules; the flowers are often in terminal panicles or racemes, sometimes very showy; the fruit is a berry, drupe, or capsule.

The best known representatives in eastern United States are *Azalea, Gaultheria* (Wintergreen), *Kalmia* (Mountain Laurel), *Oxydendrum* (Sourwood or Sorrel tree), and *Rhododendron;* in the western states, *Arbutus* (Madroño) and *Arctostaphylos* (Manzanita). The only important commercial wood in the family is furnished by the root burls of Bruyère, *Erica arborea* L., of the Mediterranean region and much used for making French Briar or Brier pipes. Attempts to use other ericaceous burls, especially of *Kalmia* and *Rhododendron,* have been fairly successful in war times. Most of the woods are unattractive, but that of certain species of *Befaria* and *Arctostaphylos* is a rich deep red and is employed locally for articles of turnery and small cabinet work. The tropical American members of the family are all small and grow mostly in the mountains.

Heartwood pale to dark brown or deep red; sapwood white or brownish, sometimes sharply demarcated. Without distinctive odor or taste. Moderately to decidedly hard and heavy; texture fine and uniform; grain straight to very irregular; working qualities generally good.

Growth rings usually present, sometimes distinct. Pores medium-sized to minute; usually very numerous and well distributed, though sometimes zonate (more or less ring-porous) in temperate climate; solitary and in small multiples, sometimes clustered. Vessels typically with scalariform perforation plates having numerous fine bars; sometimes with simple perforations, at least in part; fine to rather coarse spiral thickenings often present, occasionally limited to tips of members; pitting fine, with tendency to scalariform. Rays heterogeneous, all very fine in some species, two-sized in many, the larger 3 to 8 cells wide and up to 50, occasionally over 100, cells high; large ray complexes observed in one specimen of *Kalmia;* pits to vessels variable, typically very small, though frequently elongated and tending to scalariform arrangement. Wood parenchyma usually very sparingly developed, sometimes (*e.g., Kalmia*) reticulate; mostly indistinct or invisible with lens. Wood fibers with rather thin to very thick walls; sometimes with spiral thickenings; septate in part in some species; pits very numerous, distinctly to indistinctly bordered. Ripple marks and gum ducts absent.

Arbutus, with about a dozen species, some of them doubtfully distinct, occurs in the Mediterranean region of the Old World and in North America from British Columbia to Guatemala. The American species, usually known as Madroño, are evergreen shrubs and small to large trees with smooth thin bark peeling off in sheets. The leaves are alternate; the pink or white flowers are borne in terminal panicles; the fruit is a globose 5-celled granular berry.

There are seven species in Mexico and southwestern United States, some of them trees up to 50 feet tall and 24 inches in diameter. Standley says (*Trees and shrubs of Mexico,* p. 1099): "The Mexican plants of this genus are extremely variable and seem not to possess a single constant character. It appears probable that ultimately all of them will have to be considered mere forms of *A. xalapensis* H.B.K."

The largest member of the Ericaceae is *Arbutus Menziesii* Pursh. of the coast region of northwestern United States and adjacent Canada, at its best in the Redwood (*Sequoia*) forests of California where it attains a maximum height of 125 feet with a tall straight trunk four to five feet in diameter. The pale reddish brown wood is fine-textured, hard, and heavy; sp. gr. (air-dry) about 0.80; weight about 50 lbs. per cu. ft. It is employed locally to a limited extent for furniture, articles of turnery, and fine charcoal. The bark is sometimes used as a source of tannin.

COMMON NAMES: Laurel, madroña, madrone, madroño, manzanita (U.S.A.); aile, madroño, manganita, nuzu-ndu (Mex.).

Arctostaphylos, with numerous species of evergreen low-spreading or erect shrubs and a few small trees, is widely distributed in North America, mostly in the western parts of the United States and Mexico. The leaves are alternate or rarely opposite; the small white or pink flowers are in terminal panicles or racemes; the fruit is a globose drupe, often with a granular pulp. The plants are common on dry slopes of the Pacific region, where they are commonly called Manzanita. The wood is hard, strong, fine-textured, and that of some species at least is a rich reddish brown suitable for turnery and cabinet work, though usually available in very limited amounts.

COMMON NAMES: Bearberry, manzanita (U.S.A.); frutilla, garambullo, guie-yana, gayuba del país, leño colorado, madroño, m. borracho, m. chino, manzana, manzanilla, niño en cuero, palo de pingüica, pingüeca or pingüica, pinquiqua, tepeizquintl, tepesquisuchil, tepetomate, tepezquite, tnu-ndido, tnu-tque, verdis, yaga nita, y. yana (Mex.); abril (Guat.).

Elliottia *racemosa* Muhl., the only species, is typically a shrub, but occasionally a tree 25 feet high and four inches in diameter, rarely up to 40 feet tall, of very limited distribution in eastern Georgia and adjacent South Carolina. The bark of the young shoots is brown, that of older stems gray. The simple, entire, alternate, ascending leaves are deciduous; the white or pinkish flowers are borne in long, erect, terminal panicles that are very showy; the fruit is a small capsule; the seeds are apparently sterile. The plant is chiefly of interest because of its rarity. Efforts are being made to propagate it vegetatively for ornamental purposes in southern gardens.

Kalmia, with several species of irregularly branched, mostly evergreen shrubs, rarely small trees, is widely distributed in the United States and Canada. The leathery leaves are alternate, opposite, or whorled; the pink, purple, or white flowers are borne in terminal or lateral corymbs or umbel-like clusters; the fruit is a small, woody, many-seeded, dehiscent capsule.

The largest member of the genus is the Mountain Laurel, *Kalmia latifolia* L., distributed from New Brunswick and Ontario to Florida and Louisiana. Throughout most of its range it is a dense spreading shrub with many crooked stems, sometimes forming almost impenetrable thickets. In a few places in the Carolinas it becomes a tree 30 to 40 feet high with a contorted trunk 18 to 20 inches in diameter. The hard, brownish, fine-textured wood has a rather cheesy consistency and is used for small objects of carving and turnery and the root burls for smokers' pipes.

COMMON NAMES: *Kalmia latifolia:* Calico bush, c. flower, c. tree, ivy, kalmia, laurel (mountain, sheep, small, wood), spoonwood (U.S.A.). *Other species:* Lambkill, laurel (pale, sheep), wicky (U.S.A.).

Lyonia (or *Xolisma*), with about 30 species of evergreen shrubs and little trees, occurs in the southeastern United States, Mexico, and more commonly in the West Indies. The leaves are alternate, leathery, and lepidote; the small whitish globular flowers are borne in clusters in the leaf axils; the fruit is a small dehiscent capsule. The only arborescent species in the United States is *L. ferruginea* (Walt.) Nutt. (= *Andromeda ferruginea* Walt. = *Xolisma ferruginea* [Walt.] Heller), growing along the coast from South Carolina to Florida; occasionally, in rich soil, a tree 25 to 30 feet high with a crooked trunk not over 10 inches in diameter; often a low shrub in barren soil. This, or a closely related form, extends into the West Indies and Mexico. The light reddish brown, hard, fine-textured wood is of good quality but of no special utility because of its scarcity.

COMMON NAMES: Clavellina, sangre de toro (Cuba).

Oxydendrum *arboreum* (L.) DC., the only species, is a deciduous tree, commonly called Sourwood or Sorrel tree, which is sometimes 50 to 60 feet high and up to 20 inches in diameter. It grows in the elevated regions of the area from southeastern Pennsylvania to Florida and west to southern Indiana and south to Louisiana, being at its best on the western slopes of the Big Smoky Mountains in Tennessee. The bark is thick and deeply furrowed; the leaves are alternate, oblong or lanceolate, and finely serrate, and have a pleasant acid taste; the small white flowers appear when the leaves are full grown and are borne in large panicles at the end of the season's growth; the fruit is a small capsule. The tan or flesh-colored timber is of about the same consistency as Maple (*Acer*) and has good technical properties, but it is little used because of the scarcity of the larger sizes. Slender sprouts were formerly employed for arrow shafts.

Common names: Arrow wood, elk tree, sorrel tree, sourwood, titi (U.S.A.).

Pernettya (or *Pernettia*), according to Sleumer's classification (*Notizbl. Bot. Gart. Berlin-Dahlem* 12: 115: 626–655), consists of 13 species and several varieties of mostly low shrubs occurring in Tasmania and New Zealand and the cooler parts of tropical America from Mexico to southernmost South America. Apparently the only useful products are the edible berries.

The one wood sample available (Yale 1753) is of *Pernettya mucronata* (L.f.) Gaud., a shrub or little tree, collected by Max Rothkugel in Patagonia, where it is known as Chaura Grande. Heartwood light grayish brown; not clearly differentiated from the slightly lighter-colored sapwood. Luster medium. Odorless and tasteless. Moderately hard and heavy; texture rather fine; grain fairly straight; working properties good. Presumably without commercial possibilities.

Common names: Arrayán, capulincillo (Mex.); borrachera, chivacú (Venez.); macha-macha (Peru); chaura, c. chicha, c. grande (Arg.).

Rhododendron, with about 500 species of shrubs and a few trees, is widely distributed in the mountainous regions of Asia and central Europe, especially in southwestern China and the Himalayas, and there are about 25 species in eastern and western North America. The leaves are alternate, leathery, persistent or deciduous, and either scattered or clustered at the ends of the branches; the white or colored flowers are usually borne in dense, often very showy, terminal clusters; the fruit is a small woody many-seeded capsule.

The only arborescent species in America is the Laurel or Rose Bay, *Rhododendron maximum* L., growing from Nova Scotia and Ontario to Ohio, Georgia, and Alabama. It is usually a spreading shrub with many crooked stems 8 to 12 feet high, but occasionally attains a height of 30 to 40 feet with an irregular trunk a foot in diameter. The brownish, fine-textured, hard wood is much like that of *Kalmia* and is used to small extent for tool handles and carved or turned objects. The plant is widely cultivated for ornament.

Common names: *Rhododendron maximum:* Banana bush, laurel (big, bigleaf, deer, deer-tongue, mountain, wild), rhododendron, rose bay, spoon hutch (U.S.A.).

Satyria, with about a dozen species of shrubs, is distributed from the mountains of southern Mexico through Central America to British Guiana and the western Amazon basin. The juicy acidulous berries are edible. *S. Warszewiczii* Kl. is a rather large epiphytic shrub of frequent occurrence in the Central American forests from sea level to 8000 feet elevation. The pale brown, moderately hard and heavy, fine-textured, easily worked wood is not utilized because of the small size of the plants.

Common names: Arrayán, colmillos (flowers), coralillo, matapalo de uva, muelas (fruits), palo de miel (C.R.).

ERYTHROXYLACEAE

The members of the Coca family are shrubs and small trees of two genera, namely, *Aneulophus,* with one unimportant species in tropical West Africa, and *Erythroxylon,* with many species of pantropical distribution.

Erythroxylon (or *Erythroxylum*). There are about 200 named species, most of them tropical American. The leaves are alternate, simple, and entire; the stipules are single and intrapetiolar; the flowers are in axillary fascicles; the fruit is a red drupe with acidulous flesh. The best known and most important species is *Erythroxylon coca* Lamk., a South American shrub growing naturally in moist wooded regions on the eastern slopes of the Andes from Colombia to Bolivia, and extensively cultivated for its leaves in plantations there and in the West Indies, Africa, and the Far East. Athletic prowess during the reign of the Incas was attributed in part to the strength and sustaining power derived from masticating dried Coca leaves. The idea still prevails and it is estimated that about eight million people, mostly in the Andean region, follow the ancient practice of chewing the leaves, usually in mixture with lime or ashes, and millions more partake of the properties of the alkaloids and salts as ingredients of beverages and medicine, particularly cocaine.

Some species of *Erythroxylon* are of tree size, and their heavy, strong, durable timber is used locally for posts, poles, fuel, and occasionally for wheelwright work and articles of turnery. In this category are the Redwood of British Honduras and the Aroba of Cuba. Some attempts have been made to find a market for it in the United States as a substitute for Cocobolo (*Dalbergia retusa* Hemsl.) in the knife-handle trade. Logs 6 to 12 feet long and 10 to 20 inches in diameter are available, but the wood must be seasoned carefully to avoid checking, and the finished material is rather dull and unattractive.

Heartwood reddish or purplish brown, usually sharply demarcated from the light brown sapwood. Without distinctive scent or taste. Texture fine and uniform; feel rather harsh; very dark specimens somewhat waxy and surfaces tend to become covered with fine white crystalline exudations; not difficult to work with tools, finishing very smoothly.

Growth rings present or absent. Pores small to minute; very numerous; well distributed in pairs or radial rows, without definite pattern. Vessels with simple perforations; tyloses common; no spiral thickenings observed, but striations sometimes present; pitting fine, alternate. Rays 1 or 2, sometimes 3, cells wide and up to 40, generally less than 25, cells high; decidedly heterogeneous; pits to vessels variable in the same cross-field from very small and round to large and irregular. Wood parenchyma developed in various amounts about the pores, sometimes confluent or diffused into a very fine, irregular network fairly distinct under lens; crystalliferous strands numerous, the cells squarish, the crystals heavily integumented. Wood fibers with thick walls, narrow cavities, and small indistinctly bordered pits. Gum deposits abundant in all cells of dark heartwood. Gum ducts and ripple marks absent.

COMMON NAMES: Redwood (B.W.I.); arobillo, aroba, a. carbonero, a. colorado, a. de piedro, a. jibá, a. prieto, a. real, careicillo, cubanicú, hija menuda, jibá, j. de costa, jivá, pluma gallina (Cuba); índio (P.R.); nagot, papellilo, poirier (Haiti); aroba, topillo (Dom. Rep.); brisselet (Virg. Is.); kakapol (Grenada); palillo, palo chino, pata de pájaro, zapotillo (Mex.); redwood, swamp redwood (Br. Hond.); coca, ébano, guayacancillo, pie de paloma (Salv.); ayuelo, coca, caguimo, hayo de montaña, hayuelo, hueso de negro, huevo de guidere, jayo de montaña, manzanita de rosa (Col.); cabo de asta, cerezo, clavito, hayito, hayo, h. de clavo, h. melado, h. negro, h. olivo, olivo, semeruco macho (Venez.); coca, coca-coca, cuca, motelacaspi, murcu-varilla-colorado, puca-llaja, shantona colorada, yutobanco (Peru); coca coquilla (Boliv.); arco de pipa, a. de p. miuda, bacapary, bacupary, cabello de negro, cataúba, catuaba, coca, cocão, fruta de pomba, f. de tucano, ipadú, mercurio do campo (Braz.); ajicillo, coca de monte, gato cama (Arg.).

ESCALLONIACEAE

A FAMILY of 22 genera and about 165 species of shrubs and trees, mostly small, of very wide distribution, especially in the southern hemisphere. The only American genus containing trees is *Escallonia*.

Escallonia. There are about 50 species, generally of small size and occurring chiefly in the Andes mountains from Venezuela to Peru at elevations of 6000 to 12,000 feet. The leaves are simple, alternate, with gland-tipped teeth; the flowers are mostly in terminal racemes; the fruit is capsular. The timber apparently has no commercial possibilities.

According to Pittier (*Tropical Woods:* 1: 13), *Escallonia floribunda* H.B.K. is a rather common tree in the mountains near Caracas, Venezuela, where it is known as Jarillo. "At Colonia Tovar the trees are 25 to 45 feet in height and 14 to 16 inches in diameter and supply timber used locally for building purposes and for fuel. Because of the peculiar stench of the fresh wood the vernacular name there is Cochinito, while in the Venezuelan Andes it is Puerquito, both names meaning 'little pig.' " *E. tortuosa* H.B.K. of Ecuador is a much-branched tree, sometimes 40 feet high, with a crooked, shreddy-barked trunk a foot in diameter. Regarding the species in Venezuela, Pittier says (*Tropical Woods* 26: 11): "The common name of this plant in the high valley of Mucuchíes is Quitasol (*i.e.*, sun-robber), and in the bare country around San Rafael, at altitudes between 3000 and 3500 m., it is, with the exception of *Alnus jorullensis* H.B.K., about the only real tree to be seen."

Heartwood uniform dull reddish brown, fading gradually into the lighter-colored sapwood. Odor and taste of dry material not distinctive. Moderately hard and heavy; of fine and uniform texture, being of about the consistency of Red Gum (*Liquidambar*); easy to cut, tough to split, finishes smoothly; appears fairly resistant to decay.

Growth rings absent or indistinct. Pores small (40 to 70μ), not distinct without lens; numerous, but not crowded; mostly solitary, sometimes in radial pairs, well distributed without definite pattern. Vessels with scalariform, occasionally reticulate, perforation plates, the fine bars few to 15; fine spiral thickenings present, though sometimes poorly developed; intervascular pitting rare, fine and opposite to scalariform. Rays 1 to 3, occasionally 4, cells wide and up to 25, sometimes to 40, cells high; heterogeneous; the tall uniseriates composed of square and upright cells, the others with low to fairly high margins of square, less often upright, cells; pits to vessels small. Wood parenchyma abundant, diffuse or in short to long, irregular, uniseriate rows. Wood fibers with rather thick walls; fine spiral thickenings often present; pits large, bordered, numerous in both radial and tangential walls. Ripple marks absent. No gum ducts seen.

Common names: Madroño (C.R.); corraleros (Col.); cochinito, cochino, jarillo, puerquito, quitasol (Venez.); chachacoma (Ec.); chacha, chachacoma, chachacuma, fassta, pauco, pumachilca, siuba, tassta, tiri encarnado (Peru); canudo de pito, esponja do matta (Braz.); chapel, seis camisas, seite çamisas (Arg.); lun, madroño (Chile).

EUCRYPHIACEAE

Eucryphia, the only genus, with five species of evergreen trees, sometimes reduced to shrubs, is native to Australia, Tasmania, southern Chile, and Patagonia, Argentina. The leaves are opposite, simple or pinnate, and evergreen; the stipules are small and intrapetiolar; the large flowers are borne singly in the leaf axils; the fruit is a leathery or woody capsule with several boat-shaped valves and numerous imbricated winged seeds. Some species are planted for ornamental purposes.

Eucryphia lucida (Labill.) Baill., the Tasmanian Leatherwood, varies in size from a low bush to a medium-sized, rarely large, tree with showy white flowers and supplies a fine-textured, tough, durable, brownish wood not in general use but suitable for tool handles, furniture, and heavy construction. *E. Moorei* F. v. M. of New South Wales is known as Coachwood because of its suitability for the frames and bodies of vehicles.

Three species are usually credited to southern South America, but according to J. Bausch (*Kew Bulletin,* 1938) there are only two. The Huinque of Neuquén, Argentina, is generally referred to as *Eucryphia patagonica* Speg., and is described by Carlos Spegazzini as being a tree about 50 feet tall and 16 inches in diameter, with a rosy white, fine-textured, light and soft wood suitable for ordinary carpentry. Bausch considers the species doubtful and thinks the wood described may be confused with that of some other tree. There are no specimens of it available for this study. *E. glutinosa* (P. & E.) Baill. is a shrubby tree characterized by imparipinnate leaves having one or two pairs of leaflets. It grows in the lowlands of Chile between 27° and 38° S. latitude. The species is not represented in the Yale collections.

The Ulmo, *Eucryphia cordifolia* Cav., occurs in Chile between 37° and 44° S. latitude, extending up the mountains to the edge of the glaciers. It has simple leaves and at its best is a large tree, sometimes 130 feet tall and 24 inches in diameter. Simmons says of it (Spl. Agents Ser. No. 117, Bu. For. and Dom. Commerce, Washington, D.C., p. 20): "In blossom the Ulmo is the

Dogwood [*Cornus florida* L.] of Chile. The spring brings forth prolific bloom of large white flowers which present an aspect of rare beauty. The wood of the tree is strong, very close-grained, hard, and not difficult to work and makes excellent flooring, which is its most exacting use in southern Chile. It is abundant south of the present principal lumbering region and is therefore not common on the markets. Ulmo is not durable exposed, but in the nitrate region, where there is no rain, it is in demand and, because it is not given to checking, is preferred to Roble Pellín (*Nothofagus obliqua* Mirb.) for crossties and telegraph poles. The Indians seek large Ulmo trees for making the best dugout canoes." Other uses given (*loc. cit.*, p. 18) are furniture, vehicle parts, yokes, oars, and piling. The bark contains about 15 per cent of tannin and is used for the manufacture of extract for commercial tanning. The flowers, which are rich in nectar, are an excellent source of honey.

The woods of the various species of *Eucryphia*, so far as studied, are similar in appearance, properties, and structure. Heartwood brown in several shades, sometimes variegated; not sharply demarcated for the lighter-colored sapwood. Luster rather high. Odorless and without distinctive taste. Sp. gr. (air-dry) 0.65 to 0.70; weight 40 to 45 lbs. per cu. ft.; texture fine and uniform; grain generally straight; working properties good; not very resistant to decay.

Growth rings usually distinct, often showing prominently on tangential surface. Pores small to minute, diminishing in size during seasonal growth; very numerous, sometimes crowded; well distributed without pattern, though with some tendency to concentric arrangement. Vessels typically with many-barred scalariform perforation plates, but simple perforations may also occur; delicate spiral thickenings present in the two Australian species, but none observed in *Eucryphia cordifolia*; pitting opposite or scalariform. Rays very numerous; of two sizes, the larger 2 or 3, sometimes 4 to 6, cells wide in the middle and up to 25, ordinarily less than 15, cells high; heterogeneous, with a few marginal rows of squarish, occasionally upright, cells; gum deposits abundant; pits to vessels large, simple, usually elongated and in scalariform arrangement; no crystals seen. Wood parenchyma very sparingly developed in early wood, becoming zonately diffuse or in irregular lines and in contact with vessels in late wood, and sometimes terminal; no crystalliferous strands observed. Wood fibers with thick walls and numerous, distinctly bordered pits. Ripple marks and gum ducts absent. Structure suggests Cunoniaceae.

COMMON NAMES: Gnulgu, muermo, nçulnço, ngulngo, nirrhe, roble de Chile, r. de Chiloë, ulmo (Chile); huinque? (Arg.).

EUPHORBIACEAE

THE Euphorbiaceae comprise about 285 genera and 7000 species of trees, shrubs, and herbaceous plants which are widely distributed over the world, though most abundant in the tropics. The outstanding product is Pará rubber, obtained from the latex of species of *Hevea*, especially *H. brasiliensis* (H.B.K.) Muell. Arg., now extensively cultivated in the Malayan Archipelago and to a less extent in other tropical countries. The latex of *Manihot Glaziovii* Muell. Arg. is the source of Ceará rubber, and the roots of *M. utilissima* Pohl supply a starchy substance for making cassava and tapioca. Castor oil is obtained from the seeds of *Ricinus communis* L., croton oil from species of *Croton* and *Jatropha*, and tung oil or Chinese wood oil from *Aleurites*. Various plants are sources of dyestuffs, poisons, resins, and ingredients of medicines, a few have edible fruits, and some are planted for decorative purposes. The woods are of comparatively little value.

Of the 57 woody genera in India, none produces first-class timber, the best being *Bischofia javanica* Blume, which is widely distributed throughout the Indo-Malayan region. The most important tropical African member of the family is *Oldfieldia africana* B. & H.f., a large tree with a very strong and durable wood generally known as African Oak, and formerly as African Teak because of its use as a substitute for Teak (*Tectona*) in ship construction.

There are about 65 arborescent genera in Latin America, but most of the trees are small, often scarcely more than shrubs. A few attain large dimensions, but their timber is of only local utility and almost wholly unknown to the export trade. *Hura crepitans* L. is one of the largest trees in the coastal region of Surinam, but persistent efforts for several years to market the timber in the United States, under the name of Rakuda, did not meet with success. From Cuba small logs of the attractively figured wood of *Gymnanthes lucida* Sw., known locally as Aite or Yaiti, are occasionally exported to New York and used for making fancy articles of turnery and backs of brushes and mirrors. The Walnut-like lumber of the Manchineel, *Hippomane Mancinella* L., is highly esteemed for furniture in Jamaica, but the supply is very limited. The reddish-brown timber of *Hieronyma alchorneoides* Fr. Allem. of South America is used locally for general and durable construction, but its appearance and technical properties are not of a high order. *Buxus*, the source of true Boxwood, formerly included with the Euphorbiaceae, has been made the type of a separate family, the Buxaceae. Woods somewhat resembling it in color, texture, and properties are to be found in the genera *Adelia, Bernardia, Lasiocroton, Leucocroton,* and *Securinega*, but the trees are too small to be important. The woods of *Cunuria, Glycydendron,* and *Bonania* appear to be of good quality for furniture, but the trees are probably too scarce for commercial purposes. Some of the euphorbiaceous woods have exceptionally long fibers and presumably would be good for paper pulp, if available in sufficient quantity. Many species are soft-wooded and their timber, which would be suitable for boxboards, crating, and interior work if kiln dried promptly after cutting, is almost impossible to air-season in a humid climate without becoming badly stained or decayed.

From the foregoing statements it is obvious that the woods exhibit great diversity in appearance, properties, and structure. The following description applies particularly to American species. Heartwood variable in color, being clear yellow, or pale to dark brown or red, or light to dark olive; the darker kinds are usually dull, with an oily appearance. Sapwood may be sharply defined, or gradually merge in the heartwood, or apparently compose the whole stem. Odor and taste are absent or not distinctive in dry material. There is a wide range in density from light, soft, and spongy to very heavy, hard, and strong; in texture, from very fine to coarse; in working properties, from good to bad; in durability, from quickly perishable to highly resistant to decay.

Growth rings present or absent, frequently distinct. Pores variable in maximum size in different genera and specimens from minute to small (less than 100μ), medium-sized (100–200μ), large (over 200μ) to very large (over 300μ); very numerous to few; occurring singly, in small clusters, and in radial multiples or series, without definite pattern but with a more or less pronounced tendency to radial or diagonal arrangement of the solitary pores and groups; in some instances the pores at one or both ends of a series are considerably larger than the others. Vessels usually with simple perforations; scalariform plates characterize *Securinega neopeltandra* and sometimes occur in association with simple perforations, *e.g.*, *Conceveiba, Drypetes, Hieronyma, Paradrypetes,* and *Securinega congesta;* spiral thickenings absent; intervascular pitting fine to coarse, typically alternate, but distinctly scalariform in *Cubanthus;* tyloses common, and in *Chaetocarpus, Cunuria, Pera,* and *Pogonophora* they are occasionally to predominantly sclerotic. Rays decidedly heterogeneous, with exception of *Celaenodendron, Hippomane, Hura,* and *Piranhea,* where they are nearly homogeneous; usually uniseriate or biseriate or occasionally triseriate, but maximum widths of 4, 6, or 8 cells are found in *Acalypha, Acidocroton, Hieronyma, Omphalea, Phyllanthus, Richeria,* and some specimens of *Aparisthmium* and *Securinega;* height usually less than 30 cells, but frequently up to 50; occasionally up to 100 or more in *Alchornea, Aparisthmium, Hieronyma, Phyllanthus, Sapium,* and *Securinega;* cells with very thin to very thick walls, sometimes disjunctive (*e.g.*, *Lasiocroton*); crystals abundant in many genera, especially in those without crystalliferous parenchyma strands; ray-vessel pitting occasionally very fine (*e.g.*, *Amanoa, Celaenodendron, Drypetes, Savia,* and *Securinega neopeltandra*), frequently medium, most often coarse to very coarse, tending to scalariform; some of the rays of *Anomalocalyx, Glycydendron, Hippomane,* and *Sebastiania pavoniana* contain small latex tubes. Wood parenchyma of three principal types; sparsely paratracheal in *Acalypha,*

Acidocroton, Phyllanthus, and some species of *Securinega;* mostly metatracheal, diffuse or in short or anastomosing lines, usually 1 to 3 fiber-diameters apart, making a fine reticulum often barely distinct with lens in many genera; in numerous, often poorly defined, sometimes distinct, concentric lines or bands 1 to 3 pore-widths apart in many other genera; cells often sclerotic in *Chaetocarpus* and *Cunuria;* crystals abundant in many genera; pits to vessels often large and gash-like or elongated. Wood fibers with very thin to very thick walls, the latter typically gelatinous; rarely septate (*e.g., Acalypha*); pits typically very small, with narrow, distinct to indistinct borders, or apparently simple, rarely rather large, with conspicuous borders (*e.g., Hieronyma*). Ripple marks absent. Gum ducts absent, but radial channels, frequently large and suggesting decomposed leaf traces, observed in certain species of *Alchornea, Conceveiba, Conceveibastrum, Croton, Euphorbia, Gavarretia, Mabea, Pera, Sapium,* and *Senefeldera.* Gum deposits common in all elements of the heartwood.

In summary, the outstanding features of most euphorbiaceous woods are as follows: Pores in multiples, with tendency to radial rows. Vessels with simple perforations and alternate pitting. Rays fine and inconspicuous, often uniseriate. Wood parenchyma reticulate or in concentric lines or bands, but rarely visible without lens and often difficult to see clearly with it. Wood fibers often very long, rarely septate, those with thick walls gelatinous, the pits minute, usually with small borders. Large radial channels present in several genera as in certain Apocynaceae. Crystals often abundant in either rays or wood parenchyma, occasionally in both. Ripple marks absent. (For key to American genera see *Tropical Woods* 54: 36–40.)

Acalypha, with about 430 species of trees, shrubs, or herbs, is widely distributed in the tropical and warmer regions of the world. The American trees are small or only medium-sized and are of no particular importance for their timber, which is yellowish or light brown and widely variable in density, from light and soft to moderately heavy and compact.

COMMON NAMES: Derrière gonflé, petit pompon (Haiti); chilibtux, hierba del cáncer (Mex.); chichicaste, sesik (Guat.); costilla de danto (Hond.); guasanillo, malva montañes, shuampa (Salv.); primavera (Nic.); salvia de monte (Col.); meona, tapacaminos (Venez.); chorão, rabo de macaco, tapiá-guassú (Braz.); canilla de venado (Ec.); pespita, yana ocuera, y. varilla (Peru).

Acidocroton, with 10 species of shrubs and trees, occurs in the West Indies. The wood is brownish, very heavy, hard, compact, and fine-textured.

COMMON NAMES: Diente de majá, rompe ropa (Cuba).

Actinostemon, with 34 species of trees and shrubs, occurs from the West Indies to southern South America. *A. lanceolatus* Sald. is said to be employed in Brazil for tool handles, interior work, and fuel. Recorded uses of other species are chiefly firewood and charcoal. The only specimen available is of *A. anisandrus* Pax from Argentina. The wood is yellowish, fine-textured, and moderately hard.

COMMON NAMES: Canella de veado, chifre de veado, laranjeira do matto (Braz.).

Adelia, with 11 species of shrubs and small trees, often spiny, is distributed from Texas to Paraguay. The woods are yellow, fine-textured, and moderately to decidedly dense; that of *A. Ricinella* L. is of the Boxwood type.

COMMON NAMES: Jiá, j. blanca, tarro de chivo (Cuba); escambrón, espinillo (P.R.); grenade marron (Haiti); chau, xtompac (Mex.); agajo, escambrón (Hond.); espino blanco, macagüite, tintorillo (Salv.); bagre (Col.); guayabo rosado, limoncillo, naranjillo, polegallo (Venez.); ñuatí-curuzú-morotí, tapiá-guazú-y (Arg.).

Alchornea, with about 60 species of trees and shrubs, is of pantropical distribution. The American trees are generally small, sometimes up to 65 feet high, with pale brown, light, soft, and perishable wood used locally to a limited extent for carpentry and crating.

COMMON NAMES: Aguacatillo, baconá, chote (Cuba); dove wood (Jam.); achiotillo, palo de catorro, yobillo (P.R.); bois crapaud, b. vache, grain d'or (Haiti); aguacatillo (Dom. R.); palo de puta, palo mujer (Mex.); canelito, cola de pava (Hond.);

pochote, pochetón, tambor, tepeachote (Salv.); sauso or sauzo (Venez.); cassava wood (Br. G.); basra bèbè, kannekedie-ballie, kasaba hoedoe, kjeraporan, koereroe, mattoe groègroè, moetoesirian, naporan (Sur.); amor secco, boleiro, oeirana, tapiá, t. guassú, urucurana (Braz.); cocapano, mojaro, mojaris-caspi, yoco-chihua (Peru); mora blanca, guampita, tapiá-guazú-y (Arg.).

Alchorneopsis. There are two species of small trees, *A. portoricensis* Urb. in the West Indies, and *A. floribunda* (Benth.) Muell. Arg. in the Amazon basin. The pale brown wood is light, soft, woolly, and perishable.

COMMON NAMES: Palo de gallino (P.R.); atapilio, atapiripio, danlieba, gire-gire oe-mattoe, hegon bèbè, hororadihoro, ieto-boro-balli, kanekediballi, kanoewaballi, kas-sabahoedoe, kassavehout, moe-toesirian, papantie-ie-apiesie, ware honne, wawa na-ton (Sur.).

Amanoa consists of nine known species of trees and shrubs, three of them African, five South American, and one in the Lesser Antilles. *A. grandiflora* Muell. Arg. of the Guianas also occurs in British Honduras in swamps and along streams, sometimes with a height of about 40 feet and a trunk diameter of 12 inches. The best known of the American species is *A. guianensis* Aubl., a medium-sized tree of the Guianas and Brazilian Amazon region, sometimes used locally for heavy and durable construction. The woods are reddish or purplish brown; moderately to very dense; difficult to work.

COMMON NAMES: Konoliebie, kwatto mo-pierie, tapoeripa (Sur.).

Anomalocalyx *Uleanus* Ducke, the sole species, is a small to medium-sized tree of the central Amazon region. The wood is pale brown, of rather light weight, woolly, and not durable.

Aparisthmium *cordatum* (Juss.) Baill., the only species, is a small to medium-sized, deciduous tree, of rather common occurrence in the understory of the mixed hardwood forests on lowlands in tropical South America. According to L. Williams (*Woods of northeastern Peru*, p. 270) it is abundant in the Peruvian Amazon region, but the timber is used only for fuel. The heartwood is dull brown, not sharply demarcated from the thick sapwood. It is tasteless and odorless when dry, but is said to have a spicy scent when fresh; of rather light weight, of fine and uniform texture, saws rather woolly but can be finished smoothly; probably not resistant to decay.

COMMON NAMES: Koesoewe-oe-mattoe, mababallie, sauoero nani, tossie kojo (Sur.); rucurana (Peru).

Aporosella, with one or two species of small trees, is of limited distribution in Argentina, Paraguay, and southern Brazil. *A. Hassleriana* Chod. is said to reach a height of about 20 feet, with a spreading crown and stout trunk; the soft, reddish wood is of no utility.

COMMON NAMES: Ibirá-rembé-y, malcoc, yacaré-pitó (Arg.).

Bernardia, with about 40 species of shrubs, half-shrubs, and a few small trees, is most abundantly represented in southern Brazil, but the northern range includes the West Indies and Mexico. The wood of *B. dichotoma* (Willd.) Muell. Arg. from eastern Cuba has intermingling shades of orange and red, rather lustrous and waxy, very fine-textured, hard and heavy, finishing very smoothly; an attractive wood suitable for small articles of turnery. The wood of *B. microphylla* (A. Rich.) Muell. Arg. is canary-yellow, very fine-textured, of the Boxwood type.

COMMON NAMES: Cacapul (Cuba); oreja de ratón, palo de taruga (Mex.); waika ribbon (Br. H.).

Bonania, with 10 species of shrubs or little trees, sometimes thorny, occurs on the larger islands of the West Indies. The only wood sample available for study is of *B. cubana* A. Rich., collected by G. C. Bucher in eastern Cuba (Yale 16284). The heartwood is a beautiful waxy olive-brown with fine veining of deep green or black, very fine-textured, hard, heavy, not difficult to work, and capable of a very high natural polish. It has no commercial possibilities, but is suitable for fancy articles of turnery.

Caryodendron, with three species of small trees, is distributed from Panama to Brazil. The wood is pale brown, light and fairly soft, rather fine-textured, perishable; said to be used sometimes to make charcoal for blasting powder.

COMMON NAMES: Tacay (Col.); nogal de Barquisimeto, palo de nuez, taque (Venez.).

Celaenodendron *mexicanum* Standl., the sole species, is a tree rarely over 50 feet tall with a trunk diameter of 18 to 24 inches, occasionally larger, growing along the western coast of Mexico in the region of Mazatlán where it is known as Palo Prieto. It makes up about 90 per cent of the merchantable timber on María Magdalena Island, off the coast from San Blas, Noyarit, the area covered being about 16,-500 acres. The timber is well known locally and highly appreciated for its great strength and long life; nearly all of the houses are roofed with it and it also serves for fence posts, railway crossties, and all sorts of heavy, durable construction. The heartwood is dark olive-brown, sometimes with alternating light and dark striping; has a waxy or oily appearance; rather sharply demarcated from the white sapwood; said to have a characteristic and agreeable scent when being worked; sp. gr. (air-dry) 1.07; weight 66 lbs. per cu. ft.; difficult to split, easy to cut, takes a very smooth finish but is not highly lustrous; not a cabinet or furniture wood.

Chaetocarpus consists of eight species of trees or shrubs, of which there are three in Indo-Malaya, one in tropical West Africa, and four in South America. The wood of *C. Schomburgkianus* (O. Ktze.) Pax & Hoffm. is hard and heavy to decidedly so, rather fine-textured, finishing very smoothly; said to be used for furniture in British Guiana.

COMMON NAMES: Boobooraballi, buburaballi, ruri? (Br. G.); basau botie-ie, boeloewé-balli, bokko bokkotakon, jappopalli, kantoballi, koesaljeppo, kwepiran, mammerieballi, mattoe swama, moraballi, toekoeleroe kanta, toepoeloe koesoliepo, tokovero kaumta, warakkajaroe harilaroe, witte apakwie-ie, w. djoebolletrie, w. djoebotrie (Sur.).

Cleidion is a pantropical genus with about 20 species of small to medium-sized trees or large shrubs, with four representatives distributed from Central America to Brazil and Peru. The yellowish brown wood is rather fine-textured, moderately light, woolly, perishable; local uses unknown.

Conceveiba, with a few species of slender trees, ,sometimes 70 feet tall, occurs in the Guianas and the Amazon region of Brazil. The best known species is *C. guianensis* Aubl., typically a small tree not over six inches in diameter; its seeds are said to be edible. The light grayish brown wood is of medium texture and density, odorless and tasteless, straight-grained, and easily worked, but apparently without commercial possibilities.

COMMON NAMES: Bakhie-bakhie, harimenango, jawareran, koesoewé vemattoe, mababalli, mabi, necoehoeda, péierjan, talemo méréhé, wadiehie koro, witte hoedoe (Sur.).

Conceveibastrum *Martianum* (Baill.) Pax & Hoffm., the only species, is a tree of the upper Amazon. The yellowish brown wood is light but firm, medium-textured, and rather woolly.

Croton, with more than 1000 species of trees, shrubs, and herbs, is of pantropical distribution. There are about 600 American species, but the woody plants are mostly shrubs and small trees, occasionally up to 65 feet high and 20 inches in diameter. Some of the plants are fragrantly scented, some supply limited amounts of dyestuffs, resins, and medicines, and a few furnish timber for miscellaneous local purposes. The woods exhibit great variation in appearance and properties, ranging in color from yellowish or brown to almost black, in texture from very fine to medium, and in consistency from light and soft to hard and heavy. There is a corresponding range in the type of uses, from match sticks and boxboards to general carpentry and construction. The timber apparently has no commercial future.

COMMON NAMES: Aceitillo, ají de costa, anis cimarrón, caobilla, ceranio, clavellina le laguna, cuabilla, cuabo de ingenio, frailecillo cimarrón, guasima de costa, g. roja, romero de costa, verraco (Cuba); corcho, ortega menuda, pringamoza (P.R.); ardormida, fire bush, guayacancillo, lechecillo, maran, pepper bush, sage, soldier whip, yellow balsam, yerba, y. bellaca (Virg. Is.); bois blanc, b. cabritte, b. guêpes, copahy, feuille père, paobelia, romarin (Haiti); palo bellaco, p. berraco (Dom. R.); jeerba kareeta, j. kraabo, j. de seeroe, j. tsjoebaatoe (Dutch W.I.); árbol de sangre, canelillo, caobilla, chul, copalchí, cuanaxonaxi, cuate, dominguillo, ecbalan, ekbalam, enchiladora, encinillo, epaxihuitl, ezquahuitl, hierba de la cruz, h. del gato, h. del zorrillo, h. loca, icaban, ocotillo, ortiguillo, palillo, palo muela, picosa, pinolillo, pozual or puzual, quina, q. blanca, robaldo, rosval, rubaldo, salvia, sangre de drago, sangregado, solimán, s. blanco, s. prieto, taanché, tlachinole, vara blanca, xa-balam, xic-gaban, xiximcoh, xonaxe, xunalixase, xunaxilase, yepaxihuitl (Mex.); sphere-skutch, wild cinnamon (Br. Hond.); chacolote, chirca, ciegavista, perescuch, sangre de drago (Guat.); barenillo, cascarilián, ciega ojo, c. vista, pela-nariz, quema-nariz (Hond.); copalchí (Nic.); artanto, cerro de la olla, copalchí, friega-plato, pan caliente, sasafrás, tostoncillo (Salv.); quizarrá copalchí, targuá, targuacillo, terré (C.R.); baquero, copalchí, coquillito, c. de cerro, sangrillo (Pan.); balsillo, barbasco montañero, berengeno, cascarilla, drago, jengibre arborescente, malabito, malambo, manteco, mosquiro, plateado, sangregao, sangro, tinte (Col.); amargosito, amargoso, barredero, canacanapire, carcanapire, c. macho, cáscara de lombrices, malambo, Matías, orejón, palo Matías, punta de lanza, quina blanca, sangre de drago, salvia muñeca, sarasaro, torco (Venez.); boko-boko wiwirie, koesapoelan, moeroewabbo, ojédiballi, smeri wiwirie, tassi, wakaladan hororodikoro, without (Sur.); adipate, adipati, alcamphoreira, caá-jussára, caixeta, cajuçara, capechingui, capixingui, casca gaivota, c. sacáca, castanheiro do brejo, catinga de porco, chá de periquito, cipo urtiguinha, curraleiro, gaivotinha, melmeleiro, mercurio do campo, pau caboclo, pé de perdiz, sacáca, sangue de drago, tapixingui, urucurana, velame do campo (Braz.); sangre de drago (Ec.); hingoquiro, loro callo, l. micuna, moena, m. amarilla, ruma sacha, rucurana, sangre de drago, sangre de dragón, yurac-siprana (Peru); carurumi (Par.); sangre de drago, urucurá (Arg.).

Cunuria, with three species of large trees in the Amazon basin, has a dark brown, waxy, fairly lustrous, hard and heavy wood, which is not difficult to work, finishes very smoothly, and is attractive though without figure. It is suitable for furniture.

COMMON NAMES: Cunurí (Venez.); ruri? (Br. G.); cunury (Braz.).

Didymocistus *chrysadenius* Kuhlm., the only species, is a little tree, 10 to 20 feet high, of the upper Amazon region (see *An. Prim. Reun. Sul-Am. Bot.* 3: 76–78). Only one wood sample is available (Yale 36739; Krukoff 6579). Color uniform brown with a pinkish hue throughout. Luster fairly high. Odorless and tasteless. Moderately heavy, hard, tough, and strong; texture rather fine; grain straight; readily worked, cutting very easily and smoothly; looks fairly resistant to decay. Presumably of no commercial importance because of the small size of the trees.

Ditta *myricoides* Gris., the single species, is a shrub or small tree of the West Indies. The wood is yellowish, of medium density, fine-textured, woolly, and of no special use.

Dodecastigma *amazonicum* Ducke, the only species, occurs in the Brazilian Amazon region. The wood is grayish olive-brown, of medium density and texture, and saws woolly.

Drypetes, with about 160 species, is widely distributed in the tropics of both hemispheres. There are about a dozen American species, with a combined range including southern Florida, the West Indies, Central America, and northern South America to the Amazon region. They are shrubs or small to medium-sized trees with light-colored, hard, heavy, tough, and strong woods used locally to some extent in rural construction, but chiefly for fuel and charcoal.

COMMON NAMES: Florida plum, Guiana plum, whitewood (Fla., Bah.); chicharrón, c. espinoso, cuero duro, huesillo, hueso, h. blanco, h. de costa, h. de monte, h. prieto, h. de tortuga, maco, ramón blanco (Cuba); cafeillo, hueso, h. amarillo, h. tortuga, palo blanco, p. de aceituna, p. de vaca blanca, varital (P.R.); bois côtelette, b. moussara, côtelette, laboue cochon (Haiti); bullhoof, b. macho (Br. H.); mula (Salv.).

Euphorbia, in a broad sense, comprises about 1600 species of herbs, shrubs, and small trees and is represented in nearly all parts of the world. There are many forms and some of them resemble Cacti. A tropical American group of about 15 species of shrubs and trees, confined chiefly to Mexico, Central America, and the West Indies, is sometimes segregated into a separate genus, *Euphorbiodendron.* The Mexican Palo Amarillo, *Euphorbia fulva* Stapf, is a yellow-barked tree 25 to 35 feet tall growing in poorly accessible places at altitudes of 5000–6000 feet in Michoacán, Guanajuato, Jalisco, and eastern Mexico. According to Juan Zinser (*Mexico Forestal* 19: 40–42. 1941), the latex contains 18 to 20 per cent of rubber and about 40 per cent of resin. The rubber is of good quality and the resin makes excellent varnish, but it is only recently that satisfactory methods have been devised for separating the two components of the latex. The tree appears suitable for culture in plantations, as it is readily propagated from cuttings and makes rapid growth. A secondary product is the seeds which contain about 30 per cent of a drying oil. The species is frequently confused with *E. calyculata* H.B.K., known locally as Chupiri or Tencuanete, which is much inferior as a source of rubber. The oil of its seeds has much the same properties as croton oil.

The woods of *Euphorbia,* judging from the several species studied, are similar in appearance and properties. No dependable criteria were found for distinguishing the species of *Euphorbiodendron* from the others. Color whitish or yellowish throughout. Odor and taste absent or not distinctive. Light, soft, and weak; texture fine; grain fairly straight; low in resistance to insects and decay. Of no commercial possibilities.

COMMON NAMES: Cardón, sanguinaria (Cuba); bon garçon, candélabre, désaison, dezhomme, malnommé, poinsetta, romain (Haiti); sanguinaria (Dom. R.); bandera, bebeta, boxchacah, candelilla, chupire, chupireni, chupiri, cuitla-xochitl, liga, malamujer, matagallina, mulatilla, noche-buena, palo amarillo, p. colorado, p. de cucaracha, piñoncillo, puno-puno, sac-chacah, tencuante, tenquanete, trompillo, zacchacah (Mex.); pastores, sapo (Nic.); barrabás, pastora (C.R.); lechero, sindarute, yuco (Venez.); gunapalu, koenaparoe, koenapoeloe, melki-wiwirie (Sur.); caá-cambuhy, coral, cumanan, maleiteira (Braz.); lechero (Arg.).

Garcia *nutans* Rohr, the sole species, is a small to medium-sized tree growing in the West Indies, southern Mexico, Central America, and Colombia. There are no special uses for the wood, which is pale brown, of medium density, rather fine-textured, and not difficult to work, though sawing rather woolly.

COMMON NAMES: Huevo de gato (Salv.); avellano (Col.); pascualito, pepita del indio (Venez.).

Gavarretia *terminalis* Baill., the only species, is a shrub or little tree of the Amazon region. Wood dark brown, oily looking, of medium density and texture, rather woolly.

Glycydendron *amazonicum* Ducke, the single species, is a large tree, sometimes 100 feet tall, in the high forests of the Amazon region, where it is called Mirindiba Doce. The moderately hard, rather attractive, somewhat waxy, orange-brown heartwood is sharply defined from the lustrous white sapwood.

Grimmeodendron, with two species of small trees, occurs in the larger islands of the West Indies. Wood not studied.

COMMON NAME: Manzanillo del morillo (Cuba).

Gymnanthes, with 12 species of shrubs and small trees, is limited to the Caribbean region. Best known and most widely distributed is *G. lucida* Sw., which grows in southern Florida, the West Indies, southern Mexico, and British Honduras. Its maximum height rarely exceeds 30 feet and its slender trunk, often irregularly ridged, is commonly less than 10 inches in diameter above the swollen base. The heartwood is attractively variegated olive and dark brown, sharply demarcated from the white sapwood. It is hard, heavy, and strong; sp. gr. (air-dry) 1.10 to 1.20; weight 68 to 75 lbs. per cu. ft.; texture very fine; not difficult to work, takes a high natural polish, and is durable. The timber is in local demand for poles, posts, stakes, tool handles, and small articles of turnery. Occasional shipments of small lots of the logs are sent to New York and used for backs of brushes and mirrors, walking sticks and umbrella handles, and veneers for marquetry.

COMMON NAMES: Crabwood, poisonwood (Fla.); aceitillo, aité, aití, nagrona, yaité, y. bobo, yaitecillo, yaya macho (Cuba); baboncillo, yaití (P.R.); bois marbré (Haiti); granadillo, palo de tabaco (Dom. R.); bois vert, casse haches, colas, ebène vert, e.v. brune (Fr. W.I.); false lignum-vitae (Br. H.); branquilho, capixava (Braz.).

Hevea is by far the most important genus of the Euphorbiaceae commercially, as its latex is the source of nearly all of the world's supply of rubber. There are 12 species in the Amazon basin. Some of the trees attain heights of 100 to 125 feet, the large cylindrical trunks with or without buttresses; others are small to medium-sized trees, and one is only a shrub. There are numerous varieties, forms, and hybrids. The tree producing the most and best latex and the only species cultivated in plantations is *Hevea brasiliensis* Muell. Arg. The timber of *Hevea* is pale brown, light in weight, brittle, medium coarse-textured; it stains readily and is perishable when exposed in a humid climate; accordingly it has few uses.

COMMON NAMES: Pará rubber tree (Eng.); árbol de caucho, jacía (Venez.); hatti, sibi-sibi (Br. G.); mapalapa, rappa-rappa, seue joeballi (Sur.); messigné (Fr. G.); seringa, seringuera, s. amarella, s. barriguda, s. branca, s. chicote, s. da catinga, s. da terra firme, s. folha de maniva, s. itaúba, s. itaúbarana, s. legitima, s. pescoço de veado, s. preta, s. roxa, s. tambaqui, s. torrada, s. vermelha (Braz.); capí, conorí, jéve, j. debil fino, j. d. muerto, j. fino, seringa, s. mapa, s. mashan, s. rana, shiringa amarilla, s. del cerro, urco-seringa (Peru).

Hieronyma, with about 25 species of trees and shrubs, is extensively distributed in tropical America from the West Indies and southern Mexico to southern Brazil. The best known species is *H. alchorneoides* Fr. Allem., a large tree sometimes 100 feet tall, with a trunk three feet in diameter above the buttressed base, of common occurrence in the Guianas and Brazil. The strong durable timber is used for posts, railway crossties, bridges, miscellaneous construction, and sometimes for furniture, but it is not exported.

Sapwood pinkish white; heartwood very dark brown or reddish brown, often exuding a blackish sap when freshly cut. Luster low. Rather hard and heavy; sp. gr. (air-dry) 0.70 to 0.80; weight 43 to 50 lbs. per cu. ft.; texture medium to coarse, sometimes uneven; grain more or less roey; not easy to cut or split, finishing smoothly; requires care in seasoning to prevent warping; has rather high durability.

COMMON NAMES: Cajuela (Cuba); cedro macho (P.R.); tapana, tapanare (Trin.); chac-te-cook (Br. H.); curtidor (Hond.); nancito (Nic.); pilón, Scotch ebo (C.R.); bully tree, palo chancho, pantano, pilón, zapatero (Pan.); aguacatillo, catatú, coral, florecillo morado, quindú canelo, torito, trompillo (Venez.); dalina, serdani, suradanni, surdina (Br. G.); ajono, ajowo, amapaia, anoniwana, katoelienja, makoeroerian, okotjo, piento-bolletrie, soeladan, soeradan, sorrodan, tapierin, tarroema, teloko-enoeroe, tokadie-ballie, troko-enoeroe (Sur.); aricurqua, mará-gonçalo, orocurana, urucurana, u. de leite, u. mirim (Braz.).

Hippomane *Mancinella* L., the only species, is widely distributed in the Caribbean region. Though usually small, the tree is sometimes 65 feet high with a trunk 36 inches in diameter. Its latex is caustic to the skin and poisonous if taken internally. The wood, which is a lustrous yellowish brown with markings of brown and black, suggesting Circassian Walnut, has long been appreciated in the West Indies for making good furniture, but is practically unknown to the export trade. Sp. gr. (air-dry) 0.60 to 0.68; weight 38 to 43 lbs. per cu. ft.; texture fine and uniform; easy to work, finishing very smoothly; holds its place well when manufactured; durability high.

COMMON NAMES: Manchineel (Eng.); manzanillo (Span.); mancenillier (Fr.); penipeniche, pinipiniche (Cuba); árbol de la muerte (P.R.); pomme zombi (Haiti); hincha-huevos (Mex.); limoncillo, manzanillo de playa (Venez.); arvore da morte, caximduba, mancenilheira (Braz.); ficha (Peru).

Hura, with two closely related species of large trees, occurs from the West Indies and southern Mexico to northern Brazil. The Mexican species, *H. polyandra* Baill., differs from the more widely distributed *H. crepitans* L. in the structure of the stamens, but otherwise the two are practically indistinguishable and have the same properties. The branches and trunk are often provided with sharp spines. The latex of the bark is caustic and is sometimes used for stupefying fish. The fruits, which resemble little pumpkins, explode with violence upon drying and scatter the sections and wafer-like seeds in all directions. The English name of Sandbox-tree is derived from the early practice of hollowing out the immature pods and using them as containers of blotting sand.

The trees are of common occurrence in many places and the timber is used locally for common lumber for interior construction, carpentry, boxes and crates. They attain their best development on the low narrow reefs of the coastal plain near Paramaribo, Surinam, and form nearly pure stands estimated to yield from 6000 to 100,000 board feet per acre, averaging at least 25,000 feet over an area of 20,000 acres. The trees reach a maximum height of 200 feet, with a straight, fairly regular trunk free of branches for from 50 to 100 feet and sometimes seven feet in diameter above the buttresses. The wood varies in color from a lustrous creamy white to yellowish brown or olive-gray, is light and soft, medium-textured, more or less woolly; sp. gr. (air-dry) 0.36 to 0.44; weighs 23 to 27 lbs. per cu. ft.; easy to cut, takes stains and glues well. Some of the material is very attractively roe-grained, though figure is not pronounced except in crotches. The principal objection to it as a cabinet wood is that it is too light and soft to withstand marring, though it would appear to be well suited for less exacting uses.

COMMON NAMES: Hura wood, possum wood, rakuda (U.S.A. trade); monkey's dinner bell, sandbox tree (B.W.I.); haba, habillo, javillo, salvadera (Cuba); javillo, javarillo, milinillo (P.R.); sablier (Haiti); javillo, seda blanca (Dom. R.); bois du diable, pet du diable, sablier (Fr. W.I.); árbol del diablo, cuatatachi, haba, h. de indio, habilla, jabillo, ovillo, quauhtlat-latzin, quauhayohuatli, solimanché (Mex.); jabillo (C. Am., gen.); teteretá (Guat.); nune, tronador (Pan.); acuapa, acuapar, arenillo, arenillero, castañeto, ceiba amarilla, c. blanca, c. de leche, c. lechosa, ceibo, habilla, mil-pesos, salvadera, trovador (Col.); ceiba blanca, c. habillo, habillo, jabillo, javilla (Venez.); possentrie, postentrie (Sur.); assacú, uassacú (Braz.); catahua, cataua, habilla, salvadera (Peru); ochohó (Boliv.).

Jatropha, with about 150 species of trees, shrubs, and herbs, often armed with stinging hairs, is extensively distributed in tropical Africa and America. The American species are chiefly useful as medicinal plants, and perhaps the best known is the Physic-nut, *J. curcas* L., whose seeds have purgative properties. The trees are small, with pale brown or nearly white, soft and spongy, coarse-textured, woolly, and perishable wood of no particular value.

COMMON NAMES: Frailecillo, frailejón, palo santo, peregrina, p. del Pinar, piñón, p. botija, primamoza, purga de fraile, sabrosa, seibilla, tártago, tíratíra (Cuba); physic nut, wild oil nut (Jam.); higuereta cimarrona, piñón, tartago, tíratíra (P.R.);

moussara, papaye sauvage, petit mapon (Haiti); juca cimarrona, piñón, pringa leche, tíratíra, yuca cimarrona (Dom. R.); cabalonga, chay, chaya, chipché, chulché, coatli, drago, jiotillo, mala mujer, m. m. lisa, mata-muchachos, piñoncillo, pomolché, sangre de drago, sangregado, sicilte, tecote prieto, telondilla, tlapalezpatli, tocote prieto, torote, t. amarillo, torotito, tsimtsimché, tzah, xalal, xcacalché, xkakalché (Mex.); piñón, wild physic nut (Br. H.); chichicaste, c. de burro, piñón, tempate (Guat., Hond.); chaidra, chairo, chayo, copapayo, frailecillo, papayilla, ruibarbo, tempate (Salv.); chicaquil, coquillo, frailecillo, tapate, tempate (C.R.); árbol santo, coquillo, ortiga, pringamoza, ruibarbo (Pan.); florón de montaña, juquillo, piñón, pringamoza de monte, p. de saino, purga de fraile, túa-túa (Col.); emético vegetal, frailecillo, guaritoto, piñón, sibidigua, tartora, túa-túa (Venez.); oejedi, roode schijtnooten, schnijtnooten, weroeto (Sur.); batata do inferno, cansanção, c. de leite, flor de coral, perna inchada, pião, p. roxo, pinhão, p. de purga, p. roxo, urtiga (Braz.); chotarpo, guaritoto, huanarpo, piñón, vanarpo (Peru); ortigón bravo macho, piñón, sacha-higuera (Arg.); mbaracayá-rai-nambí, pynó-guazú (Par.).

Joannesia, with two species of medium-sized to large trees, is limited in distribution to Brazil. *J. princeps* Vell. occurs in the eastern states and is cultivated in tropical regions of the Old World. It is a gnarly drought-resistant tree with a trunk 15 to 25 feet long and 16 to 24 inches in diameter supporting a spreading crown with very large digitately compound leaves thickly tufted at the ends of the coarse twigs. The fruit is used as a purgative and for stupefying fish. The white or yellowish wood is very coarse-textured, light and soft, and of poor quality but used locally for rough lumber. *J. heveoides* Ducke is a large tree, similar to the preceding, discovered along the Tapajoz River, a tributary of the Amazon, where it is called Castanha de Arára. It is remarkable chiefly for its huge fruits, as much as eight inches wide, each containing three large seeds rich in fatty oil. Wood brownish, light, soft, coarse-textured, woolly.

COMMON NAMES: *Joannesia princeps:* Andá assú, a. guassú, castanha de arára, coco de gentio, c. de purga, cotiero, fruta de arára, f. de cotia, indá assú, i. guassú, purga de cavallo, p. de gentio, p. dos Paulistas (Braz.). *J. heveoides:* Castanha de arára (Braz.).

Lasiocroton, with five species of shrubs and small trees, is limited to the West Indies. The clear, yellow, hard, heavy, very fine-textured wood might serve as a substitute for Boxwood, but the quantity is too small for commercial purposes.

Leucocroton, with 16 species of shrubs and little trees, occurs in Cuba. The wood resembles that of *Lasiocroton.*

Mabea, with about 40 species of trees and shrubs, is distributed from southern Mexico to southern Brazil, most of the species being Brazilian. Some of the woods are said to be light and soft, but those available for study are of medium density, brownish to olive-brown, somewhat oily-looking; useful for general construction perhaps, but not attractive.

COMMON NAMES: Nigüito (Venez.); bakaa poeirenga, baririe koli kakkekoro, koenbotassi, pajoelidan, pakjira emoeroe, wepelana-noe takalli, wepenjana atakarie (Sur.); canudo de pito, taquary, taquarizeiro (Braz.); aya-uma, manchuiga blanca (Peru).

Manihot, with 160 species of shrubs, small trees, and herbs, is widely distributed throughout tropical America. The most valuable species is *M. utilissima* Pohl, native of Brazil but extensively cultivated elsewhere for the meal, starch, and cassava or tapioca obtained from the tuberous roots. *M. Glaziovii* Muell. Arg. is the source of Ceará rubber, now of minor importance. The woods are light to dark brown, very soft, coarse-textured, perishable, and not utilized.

COMMON NAMES: Yuca, y. agria, y. blanca (Cuba); cassada, cassava, manioc, tapioca (Jam.); juca (Dom. R.); ayotectli, batul, cititsin, cuacamote, c. dulce, cuadrado, guacamote, guh-yaza, huacamote, huacamotli, matorral, pata de gallo, tsin, xcaché, xhac-ché, yuca, y. amarga, y. brava, y. cimarrona, y. de monte, y. dulce (Mex.); cassava, yuca (Br. H.); caxamote (Guat.); cassava, quiscamote, yuca (Hond.); caucho blanco, cerro de la olla (Salv.); yuca amarga, y. dulce, yuquilla (Venez.); alèpaipio, alesebie, alomie, alostiki, amoewanopo, arèsamoenè, jako-poekondrekasaba, kappasienjolo, koemakabo, koemè-repo, kolaroripio, komorepo, kwallabo, makkapo, oelanarè, oskiboe, pakoema, patakapio, piekieriepo, pittorolli, pulwapie, saniemè, seperalipo, sepiepabo, tapitjie, tapopirè, tapirin joepoe, tiana, tisiekiboe, tisiemoenè, tollokopo, topitjie, topitoe, towe sipio, walaloppo, walekopo, walemiepjo, wariri riepjo, wayaloepo, wolo-wolo, zoete cassave (Sur.); aipim, aipy, macaxera, mandioca, m. doce, m. mansa, maniçoba, maniva do campo, m. de veado, m. dos indios (Braz.); guaso mandió, higuerilla, mandió-guazú, mandioca brava, m. cimarrona (Arg.).

Maprounea, with four species of trees and shrubs, occurs in tropical Africa and South America, especially Brazil and the Guianas. The wood is pale brown, of medium density, and perishable.

COMMON NAMES: Awatie, dekie hatti, gingepan, ietjotono parapisi, kisi angoala, peihatti, pirapisi (Sur.); matadeira (Braz.); airána (Peru).

Micrandra, with four or five species of medium-sized to rather large trees, occurs from Venezuela through the Amazon region to southern Brazil. The seeds of *M. elata* (Didr.) Muell. Arg. are a source of industrial oil and the timber is said to be used to a limited extent for general construction. The bark of the different species is rich in a white latex which coagulates readily and makes an elastic rubber. The only wood sample studied (Williams 11481) is of *M. siphonioides* Benth., which was collected by L. Williams in the middle Caura region, Venezuela. He says that the tree attains a height of 65 feet or more, with a spreading crown and an erect cylindrical trunk 20 inches or over in diameter. It is fairly common in the high forest near streams and considerable quantities of the latex were formerly exported from Cuidad Bolívar. Specimen pale brownish throughout. Luster rather low. Without distinctive odor or taste. Of medium density, suggesting Birch (*Betula*); texture fine and uniform; grain fairly straight; working properties good; probably perishable in contact with the soil. Of no commercial possibilities.

COMMON NAMES: Caucho tomoro (Venez.); bartabalie-balli, koedi-biosé-balli, moereidam, topoeloe alomé (Sur.); arvore de mamona (Braz.).

Nealchornea *yapurensis* Huber, the sole species, is an upper Amazon tree with brownish wood of medium density, coarse texture, and low resistance to decay.

Omphalea, with 17 species, mostly shrubs, rarely trees, is of pantropical distribution. The several American species have a combined range including the West Indies, Central America, and South America to Brazil and Peru. The trees are small, with very light, soft, perishable, yellowish wood of no value. Apparently the only uses for the plants are medicinal.

COMMON NAMES: Cobnut, popnut (Jam.); noisetier (Haiti); liane à l'anse, l. papaya (Mart.); castañete, chirán, hoja de queso, palo de jabón, p. de queso, p. shilán, tambor (Salv.); ana, baboennoot, baboenotto, mekoekwaire (Sur.); omphalier, ouabé (Fr. G.); cayaté, castanha de cayaté, c. de cotia, c. de peixe, c. purgativa, comadre de azeite (Braz.).

Ophellantha *spinosa* Standl., the only species, is a little tree, 15 to 20 feet high, growing in Salvador, where it is called Limoncillo. Wood not seen.

Paradrypetes *ilicifolia* Kuhlm., the single species, is a little tree of frequent occurrence along the Rio Doce in Minas Geraes and Espirito Santo, Brazil. It has opposite, hard, lustrous, toothed leaves and a long-conical crown that makes it attractive for decorative planting.

COMMON NAMES: Ameixa, folha de serra (Braz.).

Pausandra, with nine species of small trees, occurs sparingly from Nicaragua, where it is called Jagua, to southern Brazil. The wood is pale brown, fine-textured, rather light to moderately heavy, not durable.

Pera, with about 35 species of trees or shrubs, is distributed from the West Indies and southern Mexico to southern Brazil. The wood varies in color from light to very dark brown, in texture from fine to coarse, in density from medium to very high. The principal local uses appear to be house poles and fuel.

COMMON NAMES: Jiquí, yayabacaná (Cuba); casser rage, cotelle (Haiti); palo damaso (Dom. R.); palo prieto (Mex.); black Maya (Br. H.); granadillo? (Guat.); felí, wild olive (Pan.); arguaco, cucharo (Col.); cenícero (Venez.); ruri (Br. G.); hatsiballi, koen boevienga, peprehoedoe, pirikraipio (Sur.); pereiro, talá-caá (Braz.); machu sacha mapiche (Peru).

Phyllanthus, with about 480 species of shrubs, small trees, and annual or biennial herbs, is widely distributed in the tropical and subtropical regions of the world. The woods of the American species examined are pinkish brown with yellowish sapwood, rather fine-textured, and variable from soft to rather hard. No special uses known.

COMMON NAMES: Azulejo, a. bejuco, a. de monte, grosella cimarrona, grosellero cimarrón, guaicaje, lloron, mirobalanos émblicos, panetela, raspalengua, sangre de toro, yerba de la niña (Cuba); amortiguado, avispillo, higuerillo, higuillo, millo, palo de millo, siete-cueros, yaquillo or yuquillo (P.R.); rock bush, seaside laurel, snap plant (Jam.); derrière-dos, Espagnol marron, neige (Haiti); perico (Dom. R.); karkidaaki, lokki-lokki, Surinaam bitter (Dutch W.I.); ciruello, kahyuo, panatela, xpbixtdon (Mex.); ciruello, clawberry, monkey rattle, pixton, wild plum, xmabalche (Br. H.); guinda, nistamal, pimienta, pimientilla (Salv.); carillo (Nic.); chilillo, galina (C.R.); jobitillo (Pan.); aceite, arito, barbancito, barbasco, barbasquillo, cedrito, chirrinchao, gabellón, garbanzo, llallo, madura-plátano, perla, pelolica, pinturero (Col.); barbascajo, cerezo agrio, chipito, lavandero (Venez.); kunaparu (Br. G.); ajakéballi, bita-wiwirie, boesi-kofi-tiki, djari-bita, finie-bita, hikoelitókong, manbita, pomitji, popóno, walaballi (Sur.); arranca pedras, canabi, catuaba, compadre de azeite, conabi, conambi, conami, conanu, conaui, conavi, ginja, herva pombinha, perola vegetal (Braz.); asnac-panga, gallinaso-panga, quinilla del tahuampa (Peru); ibirá-rembé-y, lenteja, malcoc, sarandí blanco (Arg.); pará-paray-mí (Par.); sarandí blanco (Ur.).

Piranhea *trifoliata* Baill., the only species, is a tree of the Brazilian Amazon region, where it is known as Piranheira. The wood is dull olive-brown, oily-looking, very dense, and of medium texture. No local uses are reported and the timber apparently has no commercial possibilities.

Pogonophora, with one or two species of trees or shrubs, occurs in northeastern South America. The dark reddish brown, oily-looking wood is very dense, durable, difficult to work, though taking a high natural polish.

COMMON NAMES: Hajokantoballi, pau-arangdja, poripio, sibidan, hojokantoballi (Sur.).

Richeria, with six species of small trees, is distributed from the Lesser Antilles to Peru and southern Brazil. The wood is rather dull yellowish or pinkish, fairly coarse-textured, moderately hard, and not durable.

Sapium, with about 100 species of trees or shrubs, is of pantropical distribution. There are many species in Latin America from Mexico and the West Indies to Uruguay. Most of the trees are small or medium-sized, but a few, such as *S. giganteum* Pittier and *S. pleiostachys* K. Schum. of Central America and *S. Jenmani* Hemsl. of British Guiana, attain maximum heights of 90–115 feet, with a well-formed trunk sometimes 36 inches in diameter. The woods are light and soft to moderately so, whitish or yellowish, but staining readily, easy to work but sawing woolly; suitable for boxboards, interior construction, sabots, and paper pulp.

COMMON NAMES: Tallow tree (Fla.); gum tree, milkwood (Jam.); pinichi (Cuba); hincha-huevos, lechesillo, manzanillo, tabeiba (P.R.); bois brûlant, b. lait (Haiti); bois lait (Grenada); lengua de vaca (Dom. R.); hierba mala, h. de la flecha, higuerillo bravo, hincha-huevos, hiza, mago, magot, ohol negro, palo de la flecha (Mex.); leche de María (Br. H.); chilicuate (Guat.); chilamate (Salv.); ñipa, olivo, o. macho (Pan.); floral, lechoso, palo de leche, piñico (Col.); caucho de Apure, lechero, marfil, pascualitas, pepo (Venez.); kuina-ek, ky-cher, mabwa, swamp mabua, touckpong (Br. G.); alekosine, amanopora, jarre nona, jawahedan, komaakaran, maboewa djamaro, mabowaballi (Sur.); burra-leiteira, caucho, caximduba, curupita, curupicahy, leiteira, murupita, pau de bicho, p. de leite, seringarana, tapurú, t. da vargem, tartaruguinha (Braz.); caucho blanco, palo de leche (Ec.); caucho-mashan, gutta-percha, pampa-caucho (Peru); palo de leche (Par.); árbol de leche, blanquillo?, curupícaí, c.-c. guazú, curupikí, ibirá-camby, i.-cambuí, lecherón, pega-pega, punuá (Arg.); árbol de leche, curupí (Ur.).

Savia, with about 25 species of trees and shrubs, is represented in Madagascar, South Africa, the West Indies, and Brazil. There are no special uses for the reddish brown, very hard, very fine-textured woods of the American species.

COMMON NAMES: Ahorca-jíbaro, aretillo, carbonero de costa, hicaquillo, maco (Cuba).

Sebastiania, with about 90 species of shrubs, trees, and herbs, a few in the tropics of the Old World, is abundantly represented in Latin America, especially Brazil. The plants are the source of various products used in native medicine. The seeds of a Mexican shrub, *S. pavoniana* Muell. Arg., are the famous "jumping beans," a common article in curio shops in southwestern United States; the larva of a small butterfly is responsible for the movement. The largest size reported for any American tree of the genus is for *S. Standleyana* Lundell, which is said (*Lloydia* 2: 2: 97) to attain a height of 130 feet, with a straight trunk 18 inches in diameter. The heartwood of the American species examined is variegated olive, sharply demarcated from the white sapwood; not very attractive; of medium density, fine-textured, easy to work, not highly durable. The timber finds various local applications and is said to make exceptionally good charcoal for metallurgical purposes and filters.

COMMON NAMES: Chechem blanco, hierba de la flecha, ichicheh, kanchunup, mincapatli, palo de la flecha, revantadillo, sacchechem (Mex.); ridge white poisonwood (Br. H.); ibirákambi (Guianas); branquilho, capixava, sarandy (Braz.); amarillo, blanquillo, espina de arroyo, ibirá-camby, lecheleche, lecheroncillo, palo de leche (Arg.); blanquillo, palo de leche (Urug.).

Securinega. Four wood samples of three species have been studied. *S. congesta* Muell Arg., a shrub or small tree of eastern Peru (Yale 17843; Williams 2234), has lustrous purplish brown wood of fine and uniform texture. *S. Acidoton* (L.) Fawcett or *S. Acidothamnus* Muell. Arg., a West Indian shrub, has a uniformly clear yellow, fine textured, hard and compact wood of the Boxwood class. *S. neopeltandra* (Gris. Urb. (= *Chascotheca neopeltandra* Gris. is a small Cuban tree or a shrub with a pal yellow, dense, very fine-textured wood, als of the Boxwood class.

Senefeldera, with seven species of tree and shrubs, occurs in South America fror Colombia to southern Brazil. The wood i dark orange-brown, streaked, waxy, hard heavy, tough, strong, and fibrous.

Tetrorchidium includes 10 species of trees and shrubs, four in tropical West Africa and six in tropical America from the West Indies and Central America to Peru and southern Brazil. One of the largest tre is *T. rotundatum* Standl. of Honduras ar Nicaragua. It attains a height of 100 fee with a slightly furrowed trunk 24 inches diameter above the high buttresses. Th timber is not utilized. The nearly whi woods of the American species are all ligh soft, woolly, and perishable.

FAGACEAE

THE Beech family, with six commonly recognized genera, namely, *Fagus* (Beech), *Nothofagus* (Antarctic Beech), *Castanea* (Chestnut), *Castanopsis, Lithocarpus,* and *Quercus* (Oak), is widely distributed in temperate and subtropical regions, except southern Africa. The leaves are simple, alternate, pinninerved, and stipulate; the flowers are monoecious, the male in heads or aments, the female solitary or in spikes or heads; the fruit is a nut partly or entirely inclosed by a woody involucre. The nuts are valuable as a source of food for man and animals. The timbers are among the most important in the world as their field of utility is exceptionally broad.

Heartwood light to dark brown or reddish; sapwood brownish or white, often sharply demarcated. Sp. gr. variable from less than 0.50 to greater than 1.00, mostly 0.65 to 0.90; texture fine, medium, or coarse; technical properties generally excellent, although densest material requires care in seasoning; durability good.

Growth rings usually present, often very distinct; ring-porous structure characteristic of some genera or species. Pores in *Fagus* and *Nothofagus* small, numerous, frequently in small multiples, and uniformly distributed without pattern; in *Castanea, Castanopsis, Lithocarpus,* and *Quercus* in part medium-sized to very large, mostly solitary, and radially arranged in single rows or groups in the late wood of ring-porous specimens and throughout in diffuse-porous material; zone of early-wood pores rather wide in *Castanea* and certain species of *Quercus,* uniseriate in *Castanopsis* and some specimens of *Lithocarpus;* decrease in size during seasonal growth gradual or abrupt. Vessels without spiral thickenings except occasionally in *Nothofagus;* tyloses often present, sometimes abundant; larger vessels with simple perforations and opposite pitting, smallest ones sometimes with scalariform plates and pitting. Rays homogeneous to weakly heterogeneous; pits to vessels medium-sized to large, often elongated and parallel; width of rays as follows: In *Castanea, Castanopsis,* and *Nothofagus* all uniseriate or biseriate; in *Lithocarpus* and *Quercus* sometimes all very fine (either regularly distributed or aggregated in part) or both fine and coarse, the latter either individually distinct or in complexes (aggregates of multiseriate rays); in *Fagus* 2-sized, the larger distinct but much smaller than their counterparts in *Quercus.* Wood parenchyma diffuse or in short tangential rows not distinct with lens in *Fagus* and *Nothofagus;* variable from diffuse or finely reticulate to rather open bands or fairly regular concentric lines distinct with lens in the others; crystals common except in *Fagus* and *Nothofagus.* Vasicentric tracheids common to abundant (except in *Fagus* and *Nothofagus*), appearing under lens like parenchyma and frequently combining with the pores to produce distinct flame-like or dendritic pattern; tracheids in contact with large vessels often irregular in shape, frequently vermiform. Wood fibers with rather thin to very thick walls; sometimes septate in *Nothofagus;* pits small to rather large; distinctly bordered except in *Nothofagus* where they are simple or have only vestigial borders. Ripple marks and gum ducts absent.

Castanea, with seven species of nut and timber trees of the north temperate zone, is widely distributed in eastern and southwestern Asia, northern Africa, southern Europe, and eastern North America. Old World species and varieties are cultivated for their nuts which are important articles of food.

The principal American species of Chestnut, *Castanea dentata* (Marsh.) Borkh., is a large tree, occasionally 100 feet tall with a straight trunk four feet in diameter in the forest, but with a much thicker and shorter bole and wide-spreading, round-topped crown in the open. It was formerly of great economic importance from southern New England to the southern Appalachians but in recent years it has been nearly eradicated in its northern range and threatened with destruction in the remainder as the result of a blight caused by a parasitic fungus. The chief uses for the brown, coarse-textured, easily worked, and durable timber are for railway crossties, poles, piling, fence posts, interior trim, common furniture, plywood cores, and shingles. The wood, particularly of southern-grown trees, is a commercial source of tannin and the spent chips are used for making paper pulp. The Chinquipin, *C. pumila* (L.) Mill., is a smaller tree of the southern states, at its best in southern Arkansas and eastern Texas, where it attains a maximum height of 50 feet with a stout trunk 36 inches in diameter. East of the Mississippi River it is often a sprawling shrub forming thickets of intricately branched stems. The wood is somewhat denser than that of the other species; its uses are local. *C. alnifolia* Nutt. is a coastal species forming low thickets by underground stems; an arborescent variety in Florida is occasionally 40 feet high and a foot in diameter.

Heartwood pale brown; usually sharply demarcated from the thin white sapwood. Luster rather low to medium. With mild indescribable odor and slightly astringent taste. Density variable, mostly medium; sp. gr. (air-dry) 0.40 to 0.60; weight 25 to 38 lbs. per cu. ft.; texture coarse, not uniform; grain straight; easy to work; highly durable.

Castanopsis, with about 25 species of trees and shrubs, occurs from southern China to Malaya and the eastern Himalayas, with two representatives in western United States. *C. sempervirens* Dudley is a low alpine shrub of the coast ranges and Sierra Nevada Mountains. *Castanopsis chrysophylla* (Hooker) A.DC., locally known as Chinquipin and Chestnut, inhabits the Pacific coast region from the Columbia River to southern California. In parts of its range it is reduced to a shrub, but in the humid valleys of northern California it is up to 100 feet tall with trunk three to six feet in diameter and free of branches for 50 feet. The timber is of excellent quality, but the supply is very limited and consumption is local.

Wood pale brown to roseate throughout. Luster medium. Without distinctive odor or taste, at least when dry. Moderately hard and heavy; texture medium to rather coarse; grain straight; very easy to work, finishing smoothly, and holding its place well when manufactured; fairly resistant to decay.

Fagus. There are five species of Beech in eastern Asia, one (*F. sylvatica* L.) in Europe, one (*F. grandifolia* Ehrh. = *F. ferruginea* Ait. = *F. americana* Sweet) distributed generally over the eastern half of the United States and southeastern Canada, and one (*F. mexicana* Martínez) in Mexico (see *Mexico Forestal* 17: 66). Regarding the Mexican Beech or Haya, the discoverer, Professor Maximino Martínez, says (*Tropical Woods* 60: 10) that the trees, which are 100 to 130 feet tall and 20 to 40 inches in diameter, occur in a forest near Zacualtipán, Hidalgo. The species differs from *F. grandifolia* in that the leaves are cuneate at the base and the fruits are longer. The wood is very hard, durable, and of beautiful appearance. The present stand of Beech timber in the United States is estimated to be over seven billion board feet, the bulk of it in New England, New York, and Pennsylvania. The principal uses are for boxes and crates, furniture (especially chairs), flooring, interior trim, turned handles, woodenware, laundry appliances, brushes, food containers, and antiseptically treated railway crossties. The terms White Beech and Red Beech are sometimes used in the trade to designate lighter and darker shades of the lumber.

Heartwood pale brown to rich reddish brown; transition from whitish sapwood gradual to abrupt. Luster medium. Odorless and tasteless. Moderately to decidedly heavy, hard, tough, and strong; sp. gr. (air-dry) 0.65 to 0.90; weight 40 to 56 lbs. per cu. ft.; texture medium to fine, uniform; grain straight; easily worked; durability low to rather high.

COMMON NAMES: Beech—American, red, ridge, white, winter (U.S.A.); haya (Mex.).

Lithocarpus, in the sense used here, is a genus intermediate of Oak and Chestnut, with about 100 species distributed from southern Japan and China through the Malay Peninsula to the Indian Archipelago, and a single representative in California, but according to O. Schwarz *Notizbl. Bot. Gart. Mus. Berlin-Dahlem* 13: 116: 6. 1936) there are three genera involved, namely, *Pasania* (with at least 60 species, including the one in California), *Cyclobalanus* (about 30 species), and *Lithocarpus* (about 10 species). According to Sargent and Sudworth, the American species should be designated *Lithocarpus densiflora* (H. & A.) Rehder instead of *Quercus densiflora* H. & A. or *Pasania densiflora* (H. & A.) Oerst. It varies in size from a shrub at high altitudes to a medium-sized or occasionally very large tree in the Redwood (*Sequoia*) forest of Napa and Mendocino Counties. In part of its range it is abundant and valuable as a source of tanbark, for which reason it is often called Tanbark Oak. The very hard and tough timber is difficult to season without warping and splitting, but is utilized to some extent.

COMMON NAMES: Oak—California chestnut, chestnut, live, peach, tan, tanbark (U.S.A.).

Nothofagus is one of the two most important genera of timber trees in the south temperate zone, the other being the Australian genus *Eucalyptus*. There are species in New Zealand, Tasmania, southeastern Australia, and southern South America. They are of especial interest to the forester because they grow in pure stands on lands unsuited for agriculture and can be regenerated readily and repeatedly without the invasion of different tree or shrub associations. In New Zealand, besides the six distinct species generally recognized, there are innumerable forms that are believed to result not only from first crosses between species but also from crosses between hybrids and species or between different hybrids or from self-fertilized hybrids.

There are eight or ten species of *Nothofagus* in the moister parts of the Andes Mountains in Chile and Patagonia, and six of them jointly and severally form hundreds of miles of dense forest, partly on arable soil but mostly in thinly populated hardly accessible regions. They are a great asset to the countries in which they grow, providing protection to the watersheds and supplying much timber for local consumption. The timber has never been, nor is it likely to be, exported in appreciable quantities to northern markets, for even if it were not needed at home it would have to compete with Beech (*Fagus*), Birch (*Betula*), and Red Gum (*Liquidambar*), of which there is ample supply.

The two species with the greatest range are the Ñire, *Nothofagus antarctica* Oerst., and the Lenga, *N. pumilio* Reiche, which occupy a narrow strip along the divide from Tierra del Fuego to upper Neuquén, or from 55° to 37° south latitude, being mostly on the Chilean side of the boundary south of 45° and mostly on the Argentine side thereafter. Ñire occurs as a tree at elevations of 2000 to 4000 feet above the sea, then gives way to the taller Lenga, only to reappear at about 6000 feet to form an almost impenetrable fringe to the forest for the next few hundred feet. An orange-yellow fungus (*Cyttaria*) often attacks the Ñire and produces clusters, of the size of a small apple, which are sweet and juicy and good to eat. The Myrtle of Tasmania, *Nothofagus Cunninghamia* Oerst., is likewise the host of a *Cyttaria* that was eaten by the aborigines. The fact that species of *Nothofagus*, widely separated geographically, have unusual similarities extending even to the parasites, lends support to the theory of the former continuity of the antarctic continents.

Guindo, *Nothofagus betuloides* Blume, is the characteristic species of the moist forests at elevations less than 1000 feet in southern Chile, extending northward to about 38°. To the east the forests fringe those of Lenga and in winter the line of demarcation is visible at great distances, as the Guindo is evergreen and the Lenga deciduous. The timber of both species is sawed in considerable quantities at local mills and the more readily accessible stands of Guindo are nearing exhaustion. It is used for the same purposes as Beech (*Fagus*) in the north temperate zone.

Coigüe or Coihue, *Nothofagus Dombeyi* Blume, begins at the northern limit of Guindo's range and extends along the Chilean coast and up the river valleys into the high cordilleras in northern Llanquihue. It is the dominant species in the south and on favorable sites attains a height of 130 feet or more and a trunk diameter of six to eight feet. For the most part, though, it occupies poor soils where settlements are few and the demand for lumber is not great, hence the forests of Coihue have not been heavily cut. The timber is not of the best quality, being inclined to warp and lacking durability, but the increasing scarcity of other kinds will stimulate its use.

Roble, *Nothofagus obliqua* Mirb., grows from the Province of Valparaiso to the Province of Valdivia, and occurs both in dense pure stands and in mixture with other species. It prefers good soil and much of the land it occupied has been cleared for wheat farming. Heights of 125 feet and diameters of four to five feet are common, while occasional trees are 160 feet tall. Immature trees produce light-colored, perishable timber called Hualle. The heartwood is known as Pellín; it is late in forming, but that from old trees is dark red, very hard and heavy, slow in drying, and highly resistant to decay. The principal use of Roble Pellín is for heavy construction, especially for railway crossties.

Raulí, *Nothofagus procera* (P. & E.) Oerst., has about the same range as Roble, but its timber is much more highly esteemed, being the principal hardwood used in Chile for furniture-making, cabinet work, flooring, stairways, doors, mouldings, cooperage, and innumerable other purposes. The wood is of a bright clear red or cherry color, of medium density, easy to dry and to work, finishing very smoothly and holding its place well when manufactured. Most of the original forests have been destroyed, partly through logging operations and burning, and partly through clearing for agriculture.

The woods of Roble and Raulí are the most deeply colored, varying from a rather dull reddish brown in Roble to bright cherry-red in Raulí. The others are pale brown, with a more or less distinct pinkish hue; they are less resistant to decay and more inclined to warp unless carefully seasoned. There is a considerable range in density in the same species, some specimens being comparatively light and soft, others decidedly hard and heavy. The texture is variable from fine to medium, sometimes uniform, though frequently there is a pronounced tendency to greater porosity in the early wood, some specimens (*e.g.*, Raulí) being almost ring-porous.

COMMON NAMES: Beech—antarctic, Chilean, South American (Eng.); haya, h. austral, h. antártica (Span.); anis, coigüe, coihue, coyam, coyan, guindo, hualle, hualo, lenga, ouchpaya, pellín, ñire, ñirre, raulí, roble, r. blanco, r. colorado, r. de Magallanes, r. de Santiago, r. pellín, roblí, ruil, ruilí (Chile); coihué, coihigué, coyan, lengue, ñiré, roble (Arg.).

Quercus, one of the most important genera in the world, is widely distributed in the north temperate zone and extends at high altitudes into the tropics as far south as the Indian Archipelago in the Old World and to the mountains of Colombia and Ecuador in the New. More than 500 forms of Oaks have been described as species and numerous varieties and hybrids have been named; they range in size from low shrubs, often forming thickets or chaparral, to stately forest trees. The acorns of many kinds are sweet and supply food for primitive people and mast for swine. Some species yield the best quality of tanbark for making harness and sole leather, while others are sources of dyes, ink, and medicinal products. The cork of commerce is from Cork Oak trees grown for that purpose in Spain, Portugal, and northern Africa. The timber is noted for its strength, durability, and beauty and is employed for innumerable purposes ranging from fuel, fence posts, and railway crossties to tight cooperage, heavy building construction, interior trim, flooring, and all grades of furniture to the finest cabinet work.

Oak woods exhibit considerable variation, but have many features in common. Anatomically they are readily separable into two groups, the White Oaks, with numerous thin-walled angular pores in the late wood, and the Black and Red Oaks, with thick-walled subcircular pores few and distinct enough to be counted under a hand lens. These divisions correspond to important botanical differences in the trees. The woods can also be grouped into ring-porous, from deciduous trees, and diffuse-porous, from the evergreen or so-called Live Oaks, but these distinctions are not fundamental. One of the most characteristic features of Oak wood is the presence of large rays ranging in height from half an inch to five inches and producing conspicuous figure on the radial section. Occasionally, however, large rays are absent or very sparingly developed.

There are a great many different kinds of Oak in Mexico. In the northern part of the country the trees occur at comparatively low altitudes in the arid mountains and also at high altitudes in the larger ranges such as the Sierra Madre, but in the southern

Photo by U.S. Forest Service.

PLATE XX. Stand of White Oak (*Quercus alba*) in the Monongahela National Forest, West Virginia.

Plate XXI. A reptile-like Matapalo (*Clusia* sp.) strangling a large tree in eastern Guatemal

part they are mostly confined to the high mountains, though a few grow near sea level. The timber is in demand locally for furniture and durable construction and small amounts are exported. There is one species of Oak in Cuba, several in British Honduras, Guatemala, and Salvador, and a few in Honduras, Costa Rica, Colombia, and Ecuador. Trees in the warmer regions produce very dense, often dark-colored, tough and strong wood. The pores are fairly large, rather few, and arranged in radial or diagonal rows. The rays often are very coarse and conspicuous, but frequently, especially near the center of the stem, they are only aggregates of very fine rays. Metatracheal wood parenchyma is all diffuse in some specimens, more or less distinctly zonate in others, and occasionally in fine concentric lines which may be very closely spaced.

There are about 50 species of Oak trees native to the United States. About 35 species and varieties are of more or less value for their timber, the bulk of it coming from the central hardwood region. The best of the first-growth stands now remaining are in the southern Appalachians.

The chief commercial species of the White Oak group are as follows: White Oak, *Quercus alba* L. (Plate XX), covering the entire eastern half of the United States; Bur Oak *Q. macrocarpa* Michx., with more northerly range and excluding most of the southeastern states; Post Oak, *Q. stellata* Wang., over most of the eastern half of the United States south of the Great Lakes and Nebraska; Swamp White Oak, *Q. bicolor* Willd., from New England westward to Kansas, avoiding the southern states; Swamp Chestnut Oak, *Q. prinus* L., in the Atlantic and Gulf states and up the Mississippi valley to the Ohio River; Chestnut Oak, *Q. montana* Willd., from central New England southwestward to the southern Appalachians; Chinquapin Oak, *Q. Muehlenbergii* Engelm., in the central hardwood region from western New England to central Texas, avoiding the Atlantic coast region. There are also the Valley White Oak, *Q. lobata* Née, and Blue Oak, *Q. Douglasii* H. & A., of western and central California; Oregon White Oak, *Q. Garryana* Douglas, from Puget Sound southward through western Washington, Oregon, and California; and Rocky Mountain White Oak, *Q. utahensis* (A. DC.) Rydberg, of scattered occurrence in the mountains of Utah, Arizona, and New Mexico. White Oak is considered the only satisfactory timber for tight cooperage to contain beer, wines, and spirits. Compared with Red Oak, the heartwood is much less porous (owing to abundance of tyloses), more resistant to decay, has a more attractive appearance, and usually is of finer texture and easier to work, though much depends upon the age of the tree and conditions of growth. Tyloses are poorly developed in *Q. montana*. Transition from the large pores of the early wood to the small ones of the late wood is usually abrupt, but in southern White Oaks of rapid growth the diminution in size is likely to be gradual.

The principal species of the Black and Red Oak group in the United States are distributed as follows: Black Oak, *Quercus velutina* La Marck, and Black Jack Oak, *Q. marilandica* Muench., throughout nearly all of the eastern half of the country; Red Oak, *Q. borealis* Michx. f. and var. *maxima* (Marsh.) Ashe, in the northern and central hardwood region; Southern Red Oak, *Q. rubra* L. and var. *pagodaefolia* (Ell.) Ashe, in the states of the Atlantic and Gulf coasts and lower Mississippi valley; Pin Oak, *Q. palustris* Muench., in the central hardwood region from southern New England to Kansas, avoiding the Appalachian Mountains; Scarlet Oak, *Q. coccinea* Muench., has the same general range, but includes the mountains; Water Oak, *Q. nigra* L., Willow Oak, *Q. phellos* L., Laurel Oak, *Q. laurifolia* Michx., and Live Oak, *Q. virginiana* Mill., occur in the southeastern states; California Black Oak, *Q. Kelloggii* Newb., and Canyon Live Oak, *Q. chrysolepis* Liebm., grow in southwestern Oregon and western and central California. The heartwood of this group usually has a reddish tinge, the vessels are generally open, and the natural resistance to decay is not great. The timber is generally not so highly esteemed as White Oak for fine furniture, interior trim, and flooring, and is too porous for tight cooperage. It is used extensively, after treatment with preservatives, for railway crossties, bridge timbers, and other heavy construction purposes.

COMMON NAMES: Oak (Eng.); encina, encino, roble (Span. Am., gen.); aguatle, ahoaquahuitl, ahoatl, ahuatl, aoatl, camaycuy, charrasquillo, cucharillo, encino amarillo, e. blanco, e. chino, e. colorado, e. de agua, e. prieto, e. roble, jiote, palo colorado, p. duro, manzanilla, maquilihuatl, mêtlza, papalote, pitan, sho, shokiup, roble blanco, roblecito, tarecuen, texmole, tnu-yáa, t.-yaha, tocuz, yaga-cino, y.-reche, y.-xoo, y.-yoo, zinuh (Mex.); roblecito (Guat.); encino negro (Hond.); bellota, roble encina, r. negro (Salv.); kos, koskrá, roble negro, sákira-kani (C.R.); roble, r. amarillo, r. antioqueño (Col.).

FLACOURTIACEAE

BOTANISTS are not in agreement as to the limits of this pantropical family. According to Gilg (*Pflanzenfamilien*, 2nd ed., 21: 377), it comprises about 84 genera and over 800 species, mostly shrubs and small to medium-sized trees. Hutchinson (*Families of flowering plants*, pp. 161–162, 166) distributes the genera among three families, namely, the Flacourtiaceae, Samydaceae, and Passifloraceae. From the standpoint of wood anatomy it appears advisable to follow Gilg's classification, though two genera (*Ancistrothyrsus* and *Peridiscus*) seem rather out of place. The plants have simple, alternate leaves; a few have edible fruits and some are sources of tanning materials and drugs, oils, and resins of more or less medicinal value. The seeds of *Taraktogenos Kurzii* King of the Far East furnish the Chaulmoogra oil used in the treatment of leprosy. The only commercial timber is the so-called West Indian Boxwood (*Gossypiospermum*) obtained almost entirely from the Maracaibo Lake region of Venezuela. There are about 275 species of 30 genera native to tropical America. The following description of the wood is based on a study of 142 specimens of 71 species of 23 genera. (For anatomy of the different genera, see *Tropical Woods* 68: 42–57.)

Heartwood pale to sulphur yellow, light brown, or reddish; in most instances unattractive; demarcation from sapwood usually gradual, sometimes sharp. Luster medium to high. Odor and taste absent or not distinctive. Hard, heavy, and strong to moderately so; texture typically fine; grain often straight; generally easy to work, taking a very smooth finish; durability usually low.

Growth rings present or absent, rarely distinct. Pores mostly small to minute, in a few woods medium-sized in part; variable in abundance from few to (commonly) numerous or very numerous; solitary and in small multiples or groups or in short to long radial rows, well distributed without pattern except for a tendency to radial arrangement because of the closely spaced rays. Vessels either with all perforations simple or with some to most of them multiple; scalariform plates with few to many bars predominate in *Carpotroche*, *Hasseltiopsis*, *Peridiscus*, and *Tetrathylacium;* both simple and scalariform types may occur together about equally in *Arechavaletaia*, *Azara*, and *Hasseltia*, but there is a preponderance of simple perforations in *Mayna*, *Ryania*, and *Zuelania* in part; spiral thickenings observed only in *Azara*, *Olmediella*, and *Xylosma*, striations in some species of *Casearia* and *Zuelania;* thin-walled tyloses fairly abundant in *Ancistrothyrsus* (sometimes also sclerotic), *Lindackeria*, and *Peridiscus;* pitting of two principal types: (1) coarse, the intervascular frequently opposite, the vessel-ray often scalariform, in *Ancistrothyrsus*, *Arechavaletaia*, *Azara*, *Carpotroche*, *Hasseltia*, *Hasseltiopsis*, *Lindackeria*, *Olmediella*, *Peridiscus*, and *Tetrathylacium;* (2) fine to very fine, the intervascular typically alternate, the vessel-ray frequently unilaterally compound, in *Banara*, *Casearia*, *Gossypiospermum*, *Hecatostemon*, *Homalium*, *Laetia*, *Lunania*, *Prockia*, *Ryania*, *Samyda*, *Xylosma*, and *Zuelania*. Rays very numerous, sometimes with only one or two rows of fibers between them and often composing half the volume of the wood; infrequently not over two cells wide (*e.g.*, *Peridiscus* and a few species of *Casearia*), but commonly of two sizes, the multiseriate up to 15, though usually not over 5, cells wide and greatly variable in height, sometimes up to several hundred cells; decidedly heterogeneous; multiseriate parts composed of either square or procumbent cells or both together; uniseriate rays and ray margins composed of square and upright cells (sometimes locally biseriate without increase in width) and variable in height from 1 to 50, occasionally to 150, cells; vertically fused rays common; sheath cells sporadic; perforated cells (by-pass vessel members) of frequent occurrence throughout the family; rhombohedral crystals often abundant, frequently 2 to 4 in apparently (but not actually) chambered upright cells; pits to vessels of two principal types: (1) medium-sized to large, either rounded or elongated and in scalariform arrangement suggesting Violaceae; (2) small to very small, often unilaterally compound and almost entirely so in *Ryania*. Wood parenchyma absent or very sparingly paratracheal, except in *Ancistrothyrsus* (irregularly paratracheal, short aliform, and diffuse) and *Peridiscus* (reticulate); crystalliferous strands observed in *Olmediella* only. Wood fibers with rather thin to exceedingly thick walls, usually with a gelatinous inner layer; gelatinous and non-gelatinous fibers may be intermixed or separately zonate; septate except in *Ancistrothyrsus* and *Peridiscus;* finely chambered and crystalliferous in *Banara;* pits typically small and indistinctly bordered, but large and distinctly bordered in *Ancistrothyrsus* and *Peridiscus* (apparently communicating only with ray and wood parenchyma cells). Ripple marks absent. No gum ducts seen.

Ancistrothyrsus *Tessmannii* Harms, the sole species, is a slender scandent shrub of the Amazon region of Brazil and eastern Peru (see *Notizbl. Bot. Gart. Berlin-Dahlem* 11: 146–9, 598–600). It belongs to the section Paropsieae which, according to Hutchinson, should be referred to the Passifloraceae. The inflorescence is provided with peculiar hooks like those in *Hugonia* (Linaceae). The wood is yellowish, hard, and coarse-textured.

Arechavaletaia *uruguayensis* Speg., the only species, is a shrub or little tree in the mountain forests of Uruguay (see *Revista Sudamericana de Botanica* 3: 105–109). The wood (sapwood) is brownish, rather dense, and fine-textured.

Azara, with about 20 species of shrubs and little trees, is confined to the southern Andes and Juan Fernandez Island. The flowers are highly scented and certain species are cultivated on that account. The pale brown, fine-textured wood has no special uses. It is said to have a bitter taste, but this is not noticeable in the specimens of three species available for this study.

Common names: Aromo, chinchin, corcolén, corrollea (Chile); chinchin, duraznillo (Patagonia).

Banara, with about 30 species of small trees, occurs throughout most of tropical America. The largest size reported is for *B. bernardinensis* Briq., which is said to attain a height of 50 feet and a diameter of a foot in Misiones and Formosa, Argentina.

Common names: Guayo blanco, machete (Cuba); cadenillo (Venez.); bimiti joelèkoko, moembo etase poeté, piekien fouroe dioifi (Sur.); lacre branco (Braz.); borracho sisa, galgaretama, linque, machinmangua, machu-mangua, oco cireyda, okuchi huasi, ratan-caspi, raya-caspi, tamararu, tamanara, teareo (Peru); Francisco Alvarez, ibirá-obi-rá, mbavi (Arg.).

Carpotroche, with several closely related species of small to moderately large trees, is distributed from Guatemala to southeastern Brazil and Peru. The aril-like covering of the seeds is edible and the kernels are the source of an oil sometimes used in the treatment of leprosy. The bark is rich in tannin. The yellowish or pale brown wood is fine-textured, hard, tough, and strong and is used locally to a limited extent for general construction in protected places and for fuel and charcoal.

Common names: Sucte (Guat.); achotillo crespo, morrocoy, tablón (Col.); taparaime (Sur.); cacao branco, canudeiro, canudo de pito, fruta de babado, f. de comona, f. de cotia, f. de lepra, f. de macaco, f. de sapucainha, mata-piolho, papo de anjo, pau canudo, p. d'anjo, p. de cachimbo, p. de cotia, p. de lepra, ruchuchú, sapucainha (Braz.); cacaoillo blanco, huira-huayo, sapote de mono (Peru).

Casearia. This large genus of shrubs and small trees is of pantropical distribution, there being about 25 species in Africa and Madagascar, 65 in the region from China to Australia, and about 70 in tropical America, especially in Brazil. The most widely distributed species is *C. sylvestris* Sw., a shrub or a little tree rarely over 30 feet high, with rough bark, slender branches, and small white flowers in axillary clusters; its fine-textured, hard, yellowish brown wood has no special uses. The American species as a whole are of minor importance and supply few useful products. The woods exhibit considerable range in color, density, texture, and properties. A few indicate a close relationship to the West Indian Boxwood, but most of them are brownish or reddish and without particular merit. They are of no commercial promise because of their poor timber form.

COMMON NAMES: Cafecillo cimarrón, cafeillo, c. cimarrón, cafetillo, cuero de sapo, falcon, palo blanco, raboratón, sarna de perro (P.R.); aguedita blanca, a. dulce, a. macho, grasimilla del Pinar, jía brava, j. colorada, j. peluda, j. prieta, j. morada, raspalengua, rompe-hueso, sarna de perro, sarnilla, s. cimarrona, yaná (Cuba); café cimarrón, c. de gallina, chicharrón, limoncillo, palo blanco, p. de yagua, yagua (Dom. R.); castor, piquant arada, p. carré (Haiti); geelhout, paaloe de Bonaire (Curaçao); cafetillo, capulincillo, chililillo, ciruela, crementinillo, garrapatilla, guayabillo, iximché, pochito-quillo, palo de piedra corteño (Mex.); paletilla, wild sage (Br. H.); guacuco, guayabillo, manocarpo, vara blanca, xiliché (Guat.); escambrón, manocarpo, sombra de armado, vara blanca (Hond.); barredera, canjurillo, canjuro, chililillo, comeculebra, cuculmico, limoncillo, palanco, raspa-lengua (Salv.); cerillo, cerito, huesillo, matacartago, puipute, raspa-lengua (C.R.); cerillo, comida de culebra (Nic.); caraña, comida de loro, corta-lengua, mauro, palo de la cruz, raspa-lengua (Pan.); cuchillo, limoncillo, mahajo, peloto, vara blanca, v. de piedra, anime, huesito, h. blanco (Col.); limoncillo, macapiritú, machacomo, naranjillo, palo Bonaire, punta de ral (Venez.); alawatta-moereroe, bassakandra, bastard kopie, bimiti jorelsko, boschkoffie, kibidan, kiebiehiedan, knorye, koembetadde, kojara tokon, kwassie-kwassietikie, manalliballi, marisiballi jotoh, tamoipio, zwart parelhout (Sur.); rompehato (Ec.); achu-caspi, capanca, chiric-sanango, cuipe-ey, espina cacha, espino del demonio, espuela casha, fortuga caspi, huactana, limón caspi, llajas, naranjilla, oje de tucunare, sasishy-ey, sishi-co-ey, supai-cashi, supiecacha, tambor huactana, titibeguisi-ey, tortuga-caspi, uchu-caspi, uchu-mullaca, ullu-mullaca, usico-ey (Peru).

Gossypiospermum, a recent segregate from *Casearia,* is the only American genus of the Flacourtiaceae that produces timber of commercial importance. Two species are recognized. *G. paraguariense Rehder* (= *Casearia gossypiosperma* Briquet), a slender tree with a straight, smooth trunk, sometimes a foot in diameter, and said to have a yellowish, hard, brittle wood, is native to Paraguay and northern Argentina. Its commercial possibilities are unknown.

Gossypiospermum praecox (Gris.) P. Wilson (= *Casearia praecox* Gris.) is a small to medium-sized tree of Dominican Republic, Cuba, the Maracaibo Lake region of Venezuela, where it is commonly called Zapatero, and in eastern Colombia. The timber which has been on the markets of the United States and Europe for half a century or more under the name of West Indian Boxwood, is almost exclusively of Venezuelan origin, with occasional small lots of logs from Cuba. The wood was first described (1880) under the name of *Aspidosperma Vargasii* DC. (fam. Apocynaceae), next (1904) as *Tabebuia* (*Tecoma*) *pentaphylla* B. & H. f. (fam. Bignoniaceae), and finally (1914) as *Caesaria praecox* Gris. The identification of the wood with *Gossypiospermum* was made by the senior author in 1932 (see *Tropical Woods* 32: 4).

Zapatero is by far the most important Boxwood of commerce and has very largely replaced Turkish Boxwood (*Buxus sempervirens* L.) for all purposes except the finest engraving blocks. The logs, which are well formed, round, and smooth, are from 6 to 12, occasionally up to 18, inches in diameter, and from 9 to 12 feet long. The normal consumption in the United States is between 2000 and 2500 tons annually, the logs averaging about 10 per ton. They are shipped without removal of the bark, which is reddish brown to grayish, often flaking off irregularly, and containing intercellular canals from which a dark gummy substance exudes when freshly cut. The principal use of the timber is for the manufacture of precision rules. Other purposes are veneers for cabinet work and marquetry, engravers' blocks, articles of carving and turnery such as combs, spoons, shuttles, and spindles for silk mills, jewelers' burnishing wheels, and (ebonized) for handles of cutlery, keyboards, piano keys, and inlay. (For further details see Record and Garratt's *Boxwoods,* Bull. 14, Yale School of Forestry, 1925.)

Wood lemon-yellow to nearly white, mostly uniform, with little or no difference between heart and sapwood; blue stain common in logs that have been stored in a warm humid atmosphere. Luster high. Odor and taste absent or not distinctive. Sp. gr. (air-dry) 0.80 to 0.90; weight 50 to 56 lbs. per cu. ft.; hard and compact; of very fine and uniform texture; grain generally straight; easy to carve and turn, finishing very smoothly and taking a high natural polish; splits more readily than real Boxwood (*Buxus*); is poorly resistant to decay.

COMMON NAMES: *Gossypiospermum paraguariense:* Catiguá-oby, mbavy (Arg.). *G. praecox:* Boxwood—Maracaibo, Venezuelan, or West Indian (U.S.A. trade); India boxwood (England); westindisches Buchs (Germ.); buis d'Amérique, b. des Antilles (France); agracejo, a. de monte, jía, j. de monte (Cuba); palo blanco (Dom. R.); agracejo, cuchillo (Col.); lima, limoncillo, manzanito de montaña, manzano, marfil, naranjillo, sapatero, zapatero, z. de Maracaibo (Venez.).

Hasseltia, with three species of shrubs and small trees, is distributed from Central America to Venezuela and southward in the Andes Mountains to Peru. The wood is cream-colored, of medium density, fine-textured, and easy to work, but has no special uses.

COMMON NAMES: Muñeca (Nic.); raspa-lengua (Pan.); pié de paloma (Venez.); okuchi-huasi, tamamara, ratón-caspi (Peru).

Hasseltiopsis, with three species of trees, occurs from southern Mexico to Colombia and along the Andes Mountains to Peru. According to Sleumer (*Notizbl. Bot. Gart. Berlin-Dahlem* 14: 121: 49–52), *H. leucothyrsa* Sleumer attains a height of 50 to 100 feet in eastern Peru and *H. albomicans* Sleumer is 40 to 85 feet tall near Bogotá, Colombia. The only species of which wood samples are available is *H. dioica* (Benth.) Sleumer. It is a small tree, rarely up to 45 feet in height, with a smooth thin-barked trunk 15 inches in diameter, and is apparently limited to southern Mexico and northern Central America. The wood is cream-colored, sometimes with brownish stripes, fairly hard and heavy, fine-textured, rather cross-grained, and splintery. There are no known uses other than for fuel.

COMMON NAMES: Pochitoquillo (Mex.); chichimi, quina (Guat.); guatuso (Hond.).

Hecatostemon *dasygynus* Blake, the only species, is a small Venezuelan tree known locally as Lagunero. The only authentic specimen of the genus available (Yale 10342; Pittier 12341) has bright yellow, hard, fine-textured wood. The tree apparently is too small and rare to be useful.

Homalium, with more than 200 species of trees and shrubs, is widely distributed in the tropics, especially in Africa. The several American species, all of the section *Racoubea,* occur in the West Indies, southern Mexico, Central America, and northern South America. Some of the Indo-Malayan species are large trees with durable and useful timbers, but those in America are small and of no value. The excellent Venezuelan timber called Angelino was formerly supposed to be of this genus, but it is a species of *Ocotea* (Lauraceae).

COMMON NAMES: White cogwood (Jam.); cerezo, coracolillo, tostado (P.R.); acoma blanc, a. sauvage, acomat, acouma, bois d'acouma (Fr. W.I.); corazón de paloma (Dom. R.); palo de piedra (Mex.); trebo (Col.); caramacate, marfil, naranjillo (Venez.); conageddiballi (Br. G.); bita hoedoe (Sur.); bois d'acouma, mavavé, mavévé, racoube de la Guiane (Fr. G.).

Laetia, with 10 to 15 species of shrubs and small trees, occurs in the West Indies, Mexico, Central America, and in South America to Brazil and Peru. The largest specimens rarely exceed 30 feet in height and 10 inches in diameter. The yellowish or brownish woods are rather fine-textured, tough and strong, but not durable. They have no special uses.

COMMON NAMES: Guagnací, mamoncillo, ranillo, raspa-lengua (Cuba); guácima, trompillo (Venez.); warakaiaro (Br. G.); agamoe kamma, aletepe, aloekoejoeroe, aroekoejoeroe, basra kopie, brakkahatti, jakarawa-sorroballijepo, kabisie, kasapa erepalli, majapo werie, mania-powerie, maré oelang, sabanapau, waikarra, wajakajaro, warakajaharoe, warakajaro (Sur.); bois Jacquot, b. lamende, b. Marie (Fr. G.); timaréhua (Peru); casinga cheirosa, muirapucú, resinouso, teareo (Braz.).

Lindackeria includes about a dozen species of trees and shrubs, half of them in tropical Africa, the others in America from southern Mexico to Brazil. The best known of the six American species is *L. laurina* Presl, a small to medium-sized tree, sometimes 50 feet high and 12 to 14 inches in diameter, with a range from southern Mexico through Central America into the mountains of western Colombia. The woods of the various American species are yellow-brown, hard, heavy, and fine-textured; they have no special uses.

COMMON NAMES: Achiote (Guat.); hugro, ugro (C.R.); carbonero, guavo cimarrón, uvre (Pan.); caracana, huacapú, lluicho-caspi, quinilla colorado (Peru).

Lunania, with about a dozen species, mostly small trees, is distributed sparingly in the West Indies, Mexico, Central America, and through northern South America into the Amazon basin. The largest tree reported is 45 feet tall with a trunk free of branches for 25 feet. There are no special uses for the hard, fine-textured, yellowish brown wood.

COMMON NAMES: Palo campeche, p. negro, pimiento (Cuba); charapa-huatana, palo negro, piña-quiro, rumo-caspi (Peru).

Mayna, with seven or eight species of shrubs and little trees, is of fairly common occurrence in the Amazon region and northward in the Guianas and Colombia. The wood is similar to the preceding in appearance and properties.

COMMON NAMES: Cacáo branco, canudo de pito (Braz.); congo-caspi, huira guayo, h. huara, sapote de mono, s. yacú, shamshu huayo (Peru).

Olmediella *Betschleriana* (Goepp.) Loes., the only species, is a rare plant of central Guatemala. Paul C. Standley gives the following account of its curious history (*Tropical Woods* 32: 17): "The tree or shrub has been in cultivation in Europe for 75 years or more, but there, apparently, it seldom flowers. Its origin has been unknown. Because of the Holly-like form of the handsome leaves it was first described as a species of *Ilex*. Rippa, who observed it in cultivation at Naples, described it as a new genus, *Licopolia,* disregarding the earlier name *Olmediella* of Baillon, who had referred the plant, fantastically enough, to the Moraceae. According to Loesener and Gilg, the genus is most closely related to *Dovyalis,* a group represented most extensively in Africa. It is altogether unexpected to discover such a genus in America, since there are no other close relatives in the western hemisphere. . . . The tree, which is reported to be a handsome one and well worthy of more extensive planting, is said to be cultivated frequently in the parks and plantations of central Guatemala." The pale brown, lustrous, fine-textured wood is not utilized because of its small size and scarcity.

COMMON NAMES: Cumbo de cerro, manzana, manzanote (Guat.).

Peridiscus *lucidus* Benth., the only species, is an apparently rare tree of the Brazilian Amazon region, where it is known as Pau Santo. The specimen studied (Yale No. 22573) was collected near Manáos by Adolpho Ducke, who says the tree is small, although in the original description the species is said to attain large size along the Brazil-Venezuela border. It has large, smooth, leathery, entire, alternate leaves and clusters of small white flowers in the axils. There is some doubt as to its taxonomic position, and its anatomy is unlike that of the other Flacourtiaceae, except the Paropsiae, which some botanists include with the Passifloraceae. The heartwood is of a dull sulphur-yellow color, and sharply demarcated from the sapwood, which in this specimen is dark brown. It is hard, heavy, and fine-textured. No special uses are known.

Prockia, with 10 species of shrubs and little trees, occurs throughout tropical America. The best known and most widely distributed species is *P. crucis* L., a shrub or a tree sometimes 35 feet high. The wood, which is of a light clear yellow color, hard, and very fine-textured, appears suitable for many of the same purposes as the West Indian Boxwood (*Gossypiospermum*), though it is not available in large enough sizes to compete with that timber.

COMMON NAMES: Guácima de costa, guacimilla (Cuba); huesito (Col.); guácima de montaña, huesito (Venez.); cuiteleiro (Braz.); charapilla, uchpú-aguajillo (Peru).

Ryania (or *Patrisia*), with 10 species of shrubs and little trees, occurs in northern South America and the Amazon basin. The wood specimens studied have coarse and conspicuous rays suggesting Myrsinaceae, though lacking the resinous cell complexes or cysts.

COMMON NAMES: Ciezo, guaricamo (Venez.); capança, matacachorro, malacalado (Braz.).

Samyda, with 10 species of shrubs and small trees up to 35 feet high, is limited in its distribution to the West Indies and southern Mexico. The wood is yellowish, very hard, fine-textured, and of the West Indian Boxwood (*Gossypiospermum*) type. The trees are too small to supply timber of commercial importance.

COMMON NAMES: Rosa cimarrón (Dom. R.); casser sèche (Haiti); wild guave (Virg. Is.); aguja de tórtola, habalkax, puus mucuy (Mex.).

Tetrathylacium, with four species of small trees sometimes 30 feet high and a foot in diameter, is distributed from Costa Rica to Peru. The woods are yellowish, fine-textured, moderately hard and heavy, tough and strong, but apparently are not utilized.

COMMON NAMES: Anonilla, llaja, mullahuayo (Peru).

Xylosma includes about 20 Asiatic and 40 American species of trees and shrubs, often with axillary thorns. The bark contains tannin and the fruit is sometimes the source of dyestuff. The woods are yellowish, roseate, or brownish, fine-textured, and moderately hard and heavy; they are little used except for fuel.

COMMON NAMES: Palo de candela, p. colorado, roseta (P.R.); huesillo, hueso de costa, h. espinosa, h. de sabana (Cuba); piquant rosie (Haiti); corona santa, coronilla, huichichiltemel, junco, malacate, manzanillo, num, numtzutzuy (Mex.); agua de árrea (Salv.); mata-cartago, puipute (C.R.); jobo de lagarto, needlewood, roseto (Pan.); aguja de árrea, corona, espino de cabra, puyón, quemacho (Col.); cunshicashán, diablo-casha, supai-caspi (Peru); auiba, auiuva, auui-uva, espinho de Judeu (Braz.); espino de corona (Urug.); caravánuatí, coronillo, c. blanco, espino colorado, e. de cabra, e. de corona, espinillo amarillo, irá-puitá, inkerí-rá, nuatí-puitá, quillay, yucaráo, yuguayú, yuquerí-rá (Arg.).

Zuelania, with three or four species of small to medium-sized trees, occurs in the West Indies, southern Mexico, and Central America. *Z. Roussoviae* Pittier of Panama is a deciduous tree 30 to 75 feet tall, with a gray-barked trunk occasionally 20 inches in diameter. A gum, called "caraña," is obtained from incisions in the bark and reputed to have medicinal virtue. The wood is yellow, fine-textured, hard and heavy, but is not used for any special purposes, presumably because of its scarcity. *Z. guidonia* (Sw.) Britt. & Millsp. occurs throughout the range of the genus. It is said to attain a height of 100 feet in Cuba and is similar to the preceding in its wood and resin. The timber is used locally to a minor extent for interior construction and carpentry.

COMMON NAMES: Guaguasí (Cuba); cachiman marron (Haiti); campanillo, manzanillo, manzano, tamay, tepecacao, volatín (Mex.); water wood (Br. H.); palacio, sangre de playa (Hond.); caraño (Pan.).

FOUQUIERIACEAE

This economically unimportant but scientifically interesting family consists of two genera of desert shrubs and trees of limited distribution in southwestern North America. R. R. Humphrey says (*Am. Journ. Bot.* 22: 184): "The bizarre appearance of the plants—*Fouquieria* with its branches radiating upward and outward from the base, each of these tipped in spring by a flame-colored inflorescence, and *Idria* with its thick, almost prehistoric-appearing trunk that looks not unlike a great inverted parsnip—has drawn comment from naturalists from the time of their discovery."

Fouquieria, with seven species, ranges from Oaxaca northward into southern California, Arizona, New Mexico, and western Texas. The best known and most widely distributed species is *F. splendens* Engelm. Its trunk is short, dividing near the base into few to many crooked stems sometimes 25 feet high which for most of the year appear as leafless, spiny, apparently dead sticks. The bark contains gum, resin, and wax of some local utility. The branches are used for local construction and fencing material and if planted in the ground will produce living hedges.

The following description of the wood is based on specimens from southwestern United States. Heartwood pinkish brown; sapwood white or yellowish brown, sharply demarcated. Luster rather low. Density medium; texture fine; grain irregular; durability probably low.

Growth rings present; some specimens distinctly ring-porous, at least in part. Pores variable from small but readily visible with lens to minute; the largest in short to long tangential or concentric rows, bands, or clusters, the others in irregular radial rows or groups, the arrangement materially affected by the rate of growth. Vessels with simple perforations; without spiral thickenings; pits small to medium-sized, often somewhat elongated, not very numerous, mostly more or less opposite. Rays 1 to 5, sometimes up to 8, cells wide and up to 100, occasionally more, cells high; heterogeneous, many of the cells square, some of them upright; pits to vessels small, rounded to long oval. Wood parenchyma abundantly developed; finely reticulate, scarcely visible with lens; mostly in irregular, uniseriate lines, 1 to 5 fibers apart, and diffuse. Wood fibers with thick walls, mostly gelatinous; pits small, inconspicuously bordered. Ripple marks and gum ducts absent.

Common names: Alabarda, albarda, barda, cirio, chunari, jaboncillo, ocotillo, palo de Adán, p. santo, p. verde, rosalillo, thapacón, torote verde, torotillo (Mex.); coach whip, Jacob's staff, ochotilla, ocotillo, vine cactus (U.S.A.).

Idria *columnaris* Kellogg, the only species, is virtually confined to Lower California between 27° 30′ and 30° north latitude. The trees consist of a single or divided stem, sometimes three feet in diameter at the base and attaining a height of 50 feet or more, clothed with a tangle of slender spiny branches. The main stem is succulent and consists very largely of soft parenchymatous tissue surrounded by a layer of wedge-shaped xylem strands. (For illustrations and detailed description see Humphrey, *loc. cit.*)

GARRYACEAE

Garrya, the only genus, often included with the Cornaceae, comprises about 10 species of shrubs and little trees in southwestern United States and Mexico, and one each in Guatemala and Jamaica. The twigs are 4-angled; the leaves are opposite, simple, evergreen, with the petioles united at their base by a ridge; stipules absent; the flowers are dioecious and borne in silky, catkin-like spikes or racemes; the fruit is a berry, crowned by the persistent styles. The tallest tree reported is only about 25 feet. The wood, though attractive, is not available in large enough sizes to be useful except for fuel.

Heartwood light chocolate, with irregular dark brown striping; sharply demarcated from the white sapwood. Luster medium. Odor and taste not distinctive. Hard, heavy, tough, and strong; fine-textured; easy to work, finishing very smoothly; appears durable.

Growth rings usually distinct. Pores very small, scarcely visible with lens; virtually all solitary; numerous but rarely in contact radially; well distributed without definite pattern. Vessels with few-barred scalariform perforation plates; spiral thickenings present, distinct. Rays decidedly heterogeneous, most of the cells upright or square; of two sizes, the uniseriate rather low, the others up to 6 cells wide and up to 50, sometimes nearly 100, cells high, showing conspicuously on radial surface; pits to vessels small, round to elongated. Wood parenchyma sparingly diffuse and in fairly numerous, short, irregular, tangential lines, not distinct with lens. Wood fibers thick-walled; delicate spiral thickenings present; bordered pits exceedingly numerous in both radial and tangential walls. Ripple marks and gum ducts absent.

Common names: Fever bush, quinine bush, skunk bush (U.S.A.); bois amér (Haiti); chichicuahuitl, cuahuchichi, guachichi, ovitano, zapotillo (Mex.).

GOMORTEGACEAE

Gomortega *nitida* Ruiz & Pav., the only genus and species in this family, is a forest tree in Chile where it is known as Queule. The leaves are simple, aromatic, leathery, evergreen, opposite, and without stipules; the flowers are borne in axillary or terminal racemes; the fruit is a drupe. The species is not represented in the Yale collections. The following brief description of the wood is from Solereder (*Systematic anatomy of the dicotyledons*, p. 709).

Pores very small. Vessels with exclusively scalariform perforation plates, mostly having numerous bars. Rays narrow, the cells "somewhat elongated in the vertical and radial directions"; pits to vessels "relatively large" and simple. Wood parenchyma sparingly developed. Wood fibers with bordered pits. Resin cells occur in the pith, primary cortex, and leaves, but are not reported for the wood.

GROSSULARIACEAE

The Gooseberry family, often included in the Saxifragaceae, consists of a single genus, *Ribes*, with about 140 species of erect or creeping shrubs of general distribution in the north temperate zone and in the Andes Mountains of South America. The leaves are simple, alternate, and without separate stipules; the flowers are small, subsolitary or racemose; the fruit is a pulpy berry, crowned by the persistent calyx. The plants of one group of species, the Currants, are unarmed; the others, the Gooseberries (sometimes considered in a separate genus, *Grossularia*), are typically armed with bristles and nodal spines. Some of the shrubs are decorative, others are cultivated for their edible fruit. The presence of either Currant or Gooseberry bushes in the vicinity of any of the White Pines is a source of danger to the trees, as both the wild and cultivated species of *Ribes* can serve as the intermediate host of the White Pine blister rust. The wood is not utilized, owing to the small size of the stems, the largest in the Yale collections being only two inches in diameter.

Heartwood brown; distinct but sharply demarcated from the brownish or nearly white sapwood. Luster rather low. Odorless and tasteless. Hard, heavy, tough, and strong; texture fine and uniform; grain straight to irregular; not difficult to work, finishing smoothly; durability probably low. Suitable perhaps for small articles of turnery.

Growth rings present; ring-porous structure common. Pores very small to small (40 to 80μ); larger ones often in a compact row in early wood; smaller, late-wood pores frequently in more or less definite ulmiform arrangement; small multiples and clusters few to numerous. Vessels with fine-barred scalariform perforation plates; intervascular pitting scalariform. Rays 2-sized, the larger ones often conspicuous; uniseriates low, composed mostly of square and upright cells; multiseriates up to 10, sometimes to 40, cells wide and 30 to 50 or more cells high; interior of large rays usually composed of slender procumbent cells completely surrounded by a sheath (1 to 4 cells thick) of square and upright cells; occasional rays with plural cores of fine cells; pits from larger cells to vessels medium-sized and oval to large, elongated, and in scalariform arrangement. Wood parenchyma apparently absent or very sparingly paratracheal. Wood fibers in part with very small simple pits and septate, in part with medium-sized distinctly bordered pits and nonseptate. Ripple marks absent. No gum ducts seen.

COMMON NAMES: Currant, gooseberry (U.S.A.); capalincillo, ciruelillo, saracuacho (Mex); grosellero, parilla (Arg.); parella (Chile).

GUTTIFERAE

THE Guttiferae, according to Engler's classification (*Pflanzenfamilien*, 2nd ed., 21: 154–237), are separable into five subfamilies (Kielmeyeroideae, Hypericoideae, Calophylloideae, Clusioideae, Moronoboideae) containing 46 genera and about 1000 species of trees, shrubs, and a few herbs, mostly tropical. In general they are characterized by opposite entire leaves and a resinous sap; some of them are useful for decorative purposes while others produce fruits, oils, resins, and timber of value locally and to a certain extent commercially. Among the best known members of the family are the Poon and Alexandrian Laurel of India and the Santa María of tropical America (*Calophyllum* spp.), the Butter Tree of West Africa (*Pentadesma butyracea* Sabine), the Gamboge of Siam (*Garcinia Hamburyi* Hook. f.), the Bitter Kola of West Africa (*G. kola* Heckel), and two that are cultivated throughout the tropics for their succulent fruits, namely, the Mammee Apple (*Mammea americana* L.) and the Mangosteen (*Garcinia mangostana* L.). The timbers are not an important factor in the trade, though some of them are well worthy of development.

The family is well represented in tropical America, especially in the Amazon basin. Following is a list of the arborescent genera (the asterisk indicating that the woods have been studied). **Caraipa*, **Haploclathra*, *Kielmeyera*, **Mahurea*, and **Marila* of the Kielmeyeroideae; **Hypericum* and **Vismia* of the Hypericoideae; **Calophyllum* and **Mammea* of the Calophylloideae; *Balboa*, **Chrysochlamys*, **Clusia*, *Havetia*, **Havetiopsis*, **Oedematopus*, *Pilosperma*, *Quapoia*, *Renggeria*, **Rheedia*, **Tovomita*, and **Tovomitopsis* of the Clusioideae; **Moronobea*, **Platonia*, and **Symphonia* of the Moronoboideae. To the foregoing Ducke (*Arch. Inst. Biol. Veg.* 1: 3: 210 and 2: 1: 61 RESD.) has added **Lorostemon* of a new subfamily Lorostemonoideae, and **Tovomitidium*, subfamily Clusioideae.

The woods exhibit a wide range of variation in appearance, structure, and properties. Heartwood in various shades of red, brown, or yellow (Moronoboideae), generally without sharp distinction from the sapwood. Luster usually rather low; sometimes silky. Odor and taste absent or not distinctive. Density greatly variable in different genera; sp. gr. range 0.50 to 1.10. Working properties fair to good. Durability mostly low, sometimes fair to good.

Growth rings present or absent. Pores generally medium-sized to minute, occasionally large in part; few to abundant; sometimes in ring-porous arrangement in *Hypericum*, diffuse in other genera; solitary and in multiples or clusters; arranged in irregular radial or oblique series in *Calophyllum* and *Vismia*, less distinctly so in *Haploclathera* and *Caraipa*, and sometimes with tendency in other genera. Vessels with exclusively simple perforations except in part in certain specimens of *Mammea* and *Marila;* tyloses often present; spirals characterize *Hypericum;* intervascular pitting coarse to very fine, typically alternate, but scalariform in all of the Clusioideae except *Rheedia*. Vasicentric tracheids and very small vessels (in association with larger ones) in *Calophyllum* and *Vismia;* a large part of the ground mass composed of minute vessels in *Hypericum laricifolium* Juss. Rays usually heterogeneous to decidedly heterogeneous, but sometimes homogeneous in *Hypericum, Moronobea, Platonia,* and *Symphonia;* uniseriate only or uniseriate and biseriate in *Calophyllum, Caraipa, Haploclathera, Hypericum, Lorostemon,* and occasionally in *Moronobea, Platonia,* and *Symphonia;* 1 to 3 or 4, sometimes 5 or 6, rarely 7, cells wide in the other genera, the larger rays varying in height from less than 30 to over 200 cells; gum deposits abundant; crystals absent or uncommon; cells often thick-walled and very abundantly pitted, but less so in Moronoboideae; tall palisade cells present in *Chrysochlamys, Clusia, Tovomita,* and *Tovomitopsis;* ray-vessel pitting typically very coarse, frequently scalariform, but fine in *Hypericum, Rheedia,* and often in *Vismia*. Parenchyma apparently absent in *Hypericum;* very sparse in *Marila, Mahurea,* and commonly so in *Chrysochlamys;* sparingly paratracheal in *Chrysochlamys, Clusia, Havetiopsis, Mahurea, Oedematopus, Tovomitidium,* and *Tovomitopsis;* unilaterally paratracheal, frequently with very short to fairly long lateral extensions, in *Caraipa, Haploclathra,* and in part in *Mammea* and *Tovomita;* aliform, also more or less vasicentric and confluent, in *Rheedia, Tovomita,* and *Tovomitidium;* in numerous narrow to fairly coarse concentric bands in *Calophyllum, Lorostemon, Moronobea, Platonia, Symphonia,* and *Vismia;* cells sometimes disjunctive, giving rise to clustered pits in radial walls; long crystalliferous strands with dilated cells common in *Clusia;* strands with septate crystalliferous cells observed in *Rheedia, Vismia,* and rarely in *Caraipa* and *Marila.* Wood fibers with rather thin to exceedingly thick walls, those with thick walls having a large to very small lumen; commonly septate in *Chrysochlamys, Clusia, Havetiopsis, Mahurea,* and *Tovomitopsis;* thin-walled septate fibers (intermingled with non-septate thick-walled fibers) resembling diffuse and aggregate-diffuse parenchyma on cross section present in *Marila;* pits typically indistinctly bordered, but ranging from apparently simple to conspicuously bordered (*Caraipa* and *Mammea*). Ripple marks absent. Small radial gum ducts sometimes present in *Moronobea;* small to large ones characterize *Mammea* and *Rheedia* and show as distinct oily specks on tangential surface.

Calophyllum, with about 100 described species of trees, mostly in tropical Asia and Oceania, a few in tropical America, is the most important timber-producing genus of the family. Best known of the Asiatic species are *C. Inophyllum* L., the Palo María of the Philippines and the Alexandrian Laurel of India, and *C. tomentosum* Wight of India which supplies the Poon spars of commerce. The principal American species is *C. brasiliense* Camb., with several varieties having a combined range covering nearly all of the tropical countries. When conditions are favorable the tree attains a height of 100 to 150 feet with a long, straight, clear bole three to six feet in diameter. The leaves are opposite, entire, with fine feather-veining; the flowers are white and scented; the fruit is a one-seeded drupe. A medicinal yellow gum-resin is obtained from the bark, and an oil, suitable for illuminating purposes, is expressed from the seeds.

The timber is well known in the regions where it grows and has long been employed for general construction, ship-building, shingles, furniture, and many other purposes where an attractive, strong, and durable material is needed. Small quantities have been exported from time to time to Europe and the United States, but no regular foreign market has yet been established, though there are good possibilities. Considerable attention has been given to the Central American species, generally known there as Santa María. Woodsmen in British Honduras recognize three classes of timber after the trees are felled, namely, white, red, and dark; the first two kinds are floatable, the other is denser and sinks in water, a matter of practical importance in logging operations. The lumber, which must be seasoned carefully to avoid warping and splitting, is recommended for flooring and furniture. Rotary-cut veneers are attractively figured, but their parenchyma layers, which show as delicate purplish traceries on a reddish background, are sometimes a source of weakness and may result in surface chipping or flaking.

Heartwood pink to brick-red; sapwood lighter, distinct, but often without sharp line of demarcation. Luster rather low. Hard and heavy to moderately so; sp. gr. (air-dry) 0.55 to 0.75; weight about 34 to 47 lbs. per cu. ft. Texture medium, fairly uniform; grain usually roey; rough lumber has a harsh feel; working qualities generally good, though logs from some localities are said to be hard to saw; lumber holds nails and screws firmly, and there are no special problems in gluing, staining, or painting, but thorough kiln-drying is desirable for material to be used in artificially heated rooms.

Tests made in the laboratories of the Yale School of Forestry on small clear specimens of Santa María from British Honduras gave the following average results in pounds per square inch (the first value is for unseasoned material, the second for air-dry, 12 per cent moisture): *Transverse bending* (*central loading*): Modulus of rupture, 9910 and 12,650; modulus of elasticity, 1,535,000 and 1,695,000; fiber stress at elastic limit, 6310 and 8740. *Compression along the grain:* Crushing strength, 5160 and 6670; modulus of elasticity, 1,507,000 and 1,619,000; fiber stress at elastic limit,

4850 and 5060. *Compression across the grain:* Fiber stress at elastic limit, 640 and 1210. The test specimens were comparatively low in density. (For further information see *Tropical Woods* 30: 9–16.)

COMMON NAMES: Ocuje, o. macho (Cuba); Marías, palo de María (P.R.); baría, María, Santa María, varilla (Dom. R.); dalemarie, Dame Marie, damage, mara (Haiti); aceito de María, crabwood, galba, galopa, galpa (Trin.); bará, baré, barí, b. oscuro, cedro cimarrón, leche de María, limoncillo de cordóba, ocú (Mex.); María, palo María, Santa María (Mex., C.A.); barillo, varillo (Salv.); krassa (Nic.); calaba (Pan.); aceite, a. de María, acuje, calambuca, chagualo, palo María (Col.); cachicamo, María, palo María, p. rey rosado (Venez.); edaballi, kurahara, kurahura, wild calabash (Br. G.); koerahara, koerali, lorahara, korakrie, mani kwaha, penoga (Sur.); aca, cupia, guamandy, jacareúba (Braz.); jacareúba, lagarto-caspi (Peru); balsamaría (Boliv.).

Caraipa, with about 30 species of trees and shrubs, occurs mostly in swampy areas in the Amazon basin. The leaves are alternate, pinnate-nerved, with the primary nerves connected by numerous parallel secondaries; the fragrant flowers are borne in terminal and axillary panicles, often large and conspicuous; the fruit is a capsule and in dehiscence the woody valves fall away from the 3-winged central column. The largest and most widely distributed species is *C. densiflora* Mart. which grows in British Guiana, Surinam, and the states of Amazonas, Pará, Maranhão, and Matto Grosso, Brazil, sometimes attaining a height of 100 feet. It is the source of Tamocoari balsam and the sap is said to be highly caustic. The timber is of good quality but apparently is too scarce to be of much importance.

Heartwood dull grayish brown with a slight to distinct reddish hue; not always clearly differentiated from the lighter-colored sapwood. Odorless and tasteless. Moderately hard and heavy; rather coarse-textured; grain variable; not difficult to work, finishing smoothly; durability uncertain.

COMMON NAMES: Inambú-quissáua?, tamacoaré, tamacoarí, tamaquare (Braz.); alakasieri, asasiballi, alakoe-seriballi, laksiri, matakki?, pakasa, sepeipjo, waloesji (Sur.).

Chrysochlamys, with about 10 species of shrubs and small to medium-sized trees, has its center of distribution in the Andes of Peru and Colombia, with extensions to British Guiana and Central America. It is closely related to *Tovomita* and should probably include *Tovomitopsis.* The glandular-punctate leaves are thin opposite and entire, the lateral nerves few and distant; the small flowers are borne in terminal panicles; the fruit is fleshy and contains four or five large seeds.

One of the best known species is *Chrysochlamys membranacea* Tr. & Pl. which attains a height of 65 feet and a diameter of 24 inches at elevations between 3000 and 9000 feet in northern Colombia. The commonest Peruvian species is *C. Weberbaueri* Engl., a small tree, rarely up to 40 feet, growing in the understory of dense forests at altitudes below 3000 feet. The same species was collected in the basin of Shodikae Creek, a tributary of the Essequebo River, British Guiana, by A. C. Smith (Yale 35719; Smith 2840). So far as known the timber of this genus is not used for any special purposes.

Heartwood pinkish brown to reddish brown, deepening upon exposure; not always clearly differentiated from the sapwood. Luster medium. Odorless and tasteless. Rather light and soft to moderately hard and heavy; texture medium; grain straight; easy to work, finishing smoothly; durability probably high.

COMMON NAMES: Rapadura (Pan.); huevo de tanga, sangre de toro (Col.); tapirero (Br. G.); maúba, sacha-indana (Peru).

Clusia, with 145 species of trees and shrubs, mostly epiphytic, is abundantly represented in tropical America, sparingly so in New Caledonia and Madagascar. The bark exudes a viscid resinous sap, yellow when fresh but becoming black upon drying. The opposite leaves have numerous parallel primary nerves that are scarcely visible when the blades are fleshy; the small to large, white, sometimes fragrant flowers are borne in terminal ternately branched inflorescences; the fruit is a dehiscent leathery or fleshy capsule with a winged central column and numerous arillate seeds.

The best known and most widely distributed tree is the Copey, *Clusia rosea* Jacq., which, like some species of *Ficus,* usually germinates on another tree, growing over, strangling, and eventually replacing it (Plate XXI). It attains a height of 60 feet and a trunk diameter of 24 inches. According to Oviedo (Lib. VIII, cap. 14): "In the early days of the conquest of Hispaniola and other islands, the Christians made playing cards of Copey leaves, and lost or gained much money with them, for lack of better ones; on the leaves they drew the kings, knights, knaves, and spots, and all the other figures and values that there usually are on cards, . . . As these leaves are very thick they held the drawings well and shuffling did not break them." The resin from the fruit is sometimes used medicinally. The wood is not attractive and has few uses other than fuel and fence posts.

Heartwood brownish, often with pinkish hue; not always clearly differentiated from the sapwood. Luster medium. Odorless and tasteless. Rather light but firm to moderately heavy and hard; texture medium to coarse; grain straight; easy to work; probably poorly resistant to decay. Of no commercial possibilities.

COMMON NAMES: Balsam fig, b. tree, pitch apple, West Indian gamboge, wild mammee (B.W.I.); copeicillo, c. de manglar, copey, a. amarillo, cupey (Cuba); bejuco de Castillo, b. de mona, copey grande, cupeillo, palo de cupey (P.R.); copey, cupey, cupeyejo, cupeyito (Dom. R.); figuier maudit, f. m. marron (Haiti); arali, figuier marron, liane meutrière, mille pieds, paletuvier montagne (Fr. W.I.); dam maatsjoe, koetsjoe, kopijk, tam maatsjoe (Dutch W.I.); aralee, arali, matapalo, copey grande, cupey, matapal, Scotchman tree (Trin.); chunup, higo-amate, matapalo, palo de águila, zapotillo (Mex.); matapalo (Br. H.); icaco montés, manzaro montés, matapalo (Salv.); areng-krá, azahar, a. de monte, copeicillo, copel, copey, seré, tispa (C.R.); cerillo, copé, copey, c. grande, cupey, poison dogwood, sambogum, tar-gum tree (Pan.); chagualito, chagualo, copé, copei, copey, c. de páramo, c. grande, c. negro, corocito de titi, cupay, gaque, mangle de montaña, rampacho, rapabalbo, tampacho (Col.); caota, caote, chaota, copey, cupey, isfuque, quiripití, tampaco, tampeque (Venez.); balsam tree, kufa, kupé, madaburi, muri kupé, pérépéré, votomite (Br. G.); abrasa, a. koemoedjoko, apotanare koenapolan, boschmammi, joeva-joeva, jowa-jowa, kaapove, kaatoetatai, katoetatai, kienboto, koapó, koeffa, koemoedjoko, koenapara epeliki, koenapolan, koenaporang, kofa, madaberie, madabrie, m. kofa djamaro, m. toliolo, mangro, pimpin, près-près, sabana mangro, savanne mangrove, swietie-watra mangro, watra mangro (Sur.); bois roi, b. serpent, figuier maudit, parcouri goupi, p. mani (Fr. G.); abano, abaneiro, apui, capororosa, cebole brava, criúba, criúva, cupay, mangue bravo, m. de praia, mata-pau, panapary, paugamelo, piré (Braz.); gáme, renaco, renaquillo, sacha-indana (Peru); curi-ibá (Par.); mandur, paccha (Ec.).

Haploclathra. According to Ducke (*Tropical Woods* 51: 17–18), this genus comprises three species and one variety of medium-sized to large trees in the Amazon region of Brazil. The leaves are opposite or verticillate, more or less leathery, medium-sized to large, with prominent lateral nerves; the white scented or inodorous flowers are borne in terminal panicles; the fruit is a 3-sided, dehiscent, few-seeded capsule.

Haploclathra paniculata (Mart.) Benth. is a tall tree of infrequent occurrence in forested swamps about Manáos; its leaves are always opposite. *H. verticillata* Ducke is a medium-sized to rather large tree, fairly common on rocky banks near the cataracts of the Rio Negro, and very attractive when in flower; the leaves are usually in whorls of three. A variety, *catingae* Ducke, is a form occurring in the low and often scrubby catinga which constitutes one of the most characteristic features of the vegetation in the upper Rio Negro basin. *H. leiantha* Benth. grows in the deeply inundated forests of the entire Rio Negro where it scarcely attains middle size; it has comparatively small leaves which are opposite or ternate. The common name for *Haploclathra* is Tamaquaré, the same as for the closely allied genus *Caraipa.* Another name, Muirapiranga, is sometimes heard, presumably because of the resemblance of the red wood to that of *Brosimum paraense* Huber. The timber, which is the most attractive of the family, is used locally for cabinet work and would probably find acceptance in the trade if available in sufficient quantity. The woods of all three species are much alike.

Heartwood deep orange-red; rather sharply demarcated from the light-colored sapwood. Luster medium. Odorless and tasteless. Hard, heavy, and strong; texture medium; grain fairly straight; rather easily worked, finishing smoothly and taking a high natural polish. Appears highly resistant to decay.

COMMON NAMES: Muirapiranga, tamaquaré (Braz.).

Havetiopsis, with five species of primarily epiphytic shrubs which eventually become independent shrubs and trees, has its center of distribution in the Amazon basin. The opposite, finely pinniveined, leathery leaves are mostly near the ends of the scandent branches; the small flowers are borne in terminal panicles; the fruit is capsular. There are no special uses for the timber. The only wood specimen available is of *Havetiopsis flavida* (Benth.) Pl. & Tr. (Yale 35635; A. C. Smith 2624) from the hinterlands of British Guiana. Heartwood absent; sapwood pinkish brown. Luster medium. Odorless and tasteless. Hard and heavy; texture fine; grain fairly straight; easy to work. Of no commercial possibilities.

Hypericum is a large genus of herbs and half-shrubs widely distributed in subtropical and to some extent in temperate regions, with occasional representatives becoming small trees. The leaves are small, entire, narrow, sessile, and usually black-punctate; the flowers are yellow; the fruit is a capsule with numerous seeds. The generic common name in the United States is St. John's-wort. Heartwood apparently absent; sapwood pinkish white. Luster rather low. Odorless and tasteless. Hard, moderately heavy; texture exceptionally fine; grain straight; very easily worked. Suitable for small articles of turnery.

Kielmeyera, with about 20 species of half-shrubs, shrubs, and small or rarely medium-sized trees, occurs for the most part in southeastern Brazil. The flowers of some species are large, fragrant, and attractive. The corky bark is ground and used for insulating refrigerators (see *Tropical Woods* 68: 11). The wood has not been studied.

COMMON NAMES: Folha santa, malva do campo, pau de S. José, p. santo, pinhão (Braz.).

Lorestemon *bombaciflorum* Ducke, the only species, is typically a small tree, rarely 50 feet tall, known only in the understory of the non-inundated forest in the vicinity of Manáos, Brazil (see *Arq. Inst. Biol. Veget.* 1: 3: 209–212). The bark contains a copious yellow latex. The large, smooth, leathery, pinniveined, opposite or subopposite leaves resemble those of *Platonia;* the large, single, terminal flowers suggest *Bombax;* the baccate fruit is oblong, stipitate, indehiscent, and contains a few seeds imbedded in a spongy pulp.

Heartwood rather dull reddish brown, with blackish streaks; distinct but not sharply demarcated from the pale brownish sapwood. Odorless and tasteless. Very hard, heavy, and strong; texture fine and uniform; grain straight; easy to work, considering its high density, finishing very smoothly; durability probably high. Presumably of no commercial possibilities.

Mahurea, with seven species of small to medium-sized resinous trees, occurs in southeastern Colombia, Venezuela, the Guianas, and the Amazon region of Brazil. The alternate, leathery, entire leaves have prominent lateral nerves; the roseate flowers are borne in conspicuous terminal racemes; the fruit is a dehiscent capsule. There are apparently no special uses for the timber.

Heartwood rather dull reddish brown with purplish hue, merging gradually into the sapwood. Odorless and tasteless. Moderately hard and heavy; texture fine; grain irregular; easily worked, finishing smoothly; durability probably high. Presumably without commercial possibilities.

Mammea has four species of trees, of which three are tropical African. The other, *M. americana* L., is native to the West Indies and northern South America but is now extensively cultivated for its fruits, the Mammee apple. It is an attractive medium-sized tree, rarely 60 feet tall, with a short main trunk, sometimes two feet in diameter, commonly dividing into several steeply ascending branches bearing an oval crown of dense dark green foliage. The firm but juicy yellow or reddish flesh of the large russet-colored leathery-skinned fruit is eaten raw or cooked and has an apricot flavor. The fragrant flowers are sometimes used in the French West Indies in distilling a liqueur, "eau de créole" or "crème de créole." The gum resin obtained from the bark is used to kill vermin on domestic animals. The timber is employed to some extent for fuel, fence posts, and miscellaneous domestic purposes, but has no possibilities for export.

Heartwood dull reddish or purplish brown, merging gradually into the slightly lighter-colored sapwood; surface frequently flecked with small oily exudations from radial ducts. Odorless and tasteless. Hard, rather heavy, strong; texture medium; grain fairly straight; not difficult to work; moderately resistant to decay.

Common names: Mamey (Span. gen.); mamey amarillo (Dom. R.); abricotier (Haiti); mammee, m. apple (B.W.I.); abricotier, mamey sapote (Trin.); chacalhaaz, zapote mamey, z. niño (Mex.); rurí (Nic.); shru, thsep (C.R.); mamaja, mamieboom (Sur.); abricó, abricotiero (Braz.).

Marila, with about a dozen species of shrubs and little trees, is sparsely distributed from the West Indies and Guatemala to the Guianas, Bolivia, and eastern Peru. The pinninerved leaves are opposite; the flowers are borne in axillary panicles; the fruit is an oblong dehiscent capsule with very numerous seeds. There are no special uses for the timber. Heartwood dull brown with a purplish hue, fading gradually into the sapwood. Odorless and tasteless. Moderately hard and heavy; texture fine; grain irregular; easily worked; appears durable. Has no commercial possibilities.

Common names: Cacao mangua, red mangue (Trin.).

Moronobea, with six species of small to large trees, occurs in the Guianas and northern Brazil. The bark exudes a yellowish sap. The leaves are opposite, finely pinninerved, and borne at the ends of the branches; the solitary terminal flowers are white or red; the fruit is a berry. The best known species is *M. coccinea* Aubl., which, owing to confusion in Aublet's illustrations, is sometimes considered the same as *Symphonia globulifera* L.f. It attains large dimensions in the forests of the Guianas and lower Amazon regions. The timber is suitable for heavy construction and machinery frames, but probably is too heavy for making furniture. The only other species represented in the Yale collections is *M. pulchra* Ducke, a small to medium-sized tree common near Manáos, Brazil. Its timber is of coarser texture and somewhat less dense than that of *M. coccinea.*

Heartwood streaked brownish yellow, with a fine pencil striping of white parenchyma lines on longitudinal surfaces; merging gradually into the lighter-colored sapwood. Luster medium. Odorless and tasteless. Mostly hard, heavy, and strong; sp. gr. (air-dry) 0.95; weight about 60 lbs. per cu. ft.; texture rather coarse; grain straight to irregular; not very difficult to work, though somewhat splintery; finishes very smoothly and takes a glossy polish; fairly resistant to decay. Probably of poor commercial possibilities.

Common names: Maniballi, manni, manniballi (Br. G.); anasey, a. de terra firme, bacury bravo (Braz.).

Oedematopus, with six species of shrubs and small trees having long pendulous branches, is widely but sparingly distributed from the subandean region of Colombia and Peru through the Amazon basin to Bahia, Brazil. There are no known local uses for the timber.

Wood brownish throughout; ray markings distinct on radial surface. Lustrous in proper light. Odorless and tasteless. Very hard, heavy, and strong; texture rather fine, uniform; grain straight; not difficult to work, finishing smoothly; durability doubtful. Presumably of no commercial possibilities because of the small size and scarcity of the trees.

Platonia. Two species have been described, but *P. grandiflora* Pl. & Tr. is probably only a form of *P. insignis* Mart., a large tree common in the Guianas and Brazil and occurring also in Ecuador and Paraguay. Under favorable conditions it attains a height of 100 feet, with a well-formed trunk two or three, rarely up to four, feet in diameter. The opposite pinniveined leaves are borne near the ends of the branches and in single pairs on short branchlets; the large roseate flowers are terminal and typically solitary; the fruit is a globose berry with an edible mesocarp usually containing a single seed. The resin from the bark is employed in veterinary medicine. The timber is well known in the region of its growth and is used for general construction, shipbuilding, carpentry, tight cooperage, and crating. Though not highly attractive, it is a useful general-purpose timber, and tests have shown that it makes satisfactory veneers for plywood. It is unknown to the markets of the United States, but some efforts have been made to introduce it into France along with Manil (*Symphonia*) which it rather closely resembles.

Heartwood dull yellow to orange-brown, with prominent grayish parenchyma markings, and sometimes with black streaks; distinct and rather sharply demarcated from the yellowish white sapwood. Odorless and tasteless. Moderately hard and heavy; sp. gr. (air-dry) 0.70 to 0.85; weight 44 to 53 lbs. per cu. ft.; texture coarse to medium; grain irregular; not difficult to work, finishing smoothly; durability fair to high.

Common names: Pahoorie, pakoorie, pakuri, wild mammee apple (Br. G.); apakwie-ie, bakoerie, geelhart, pakoeli, pakuri, papoelie, serapi-hout, mani-pau (Sur.); manilparcouri, parcouri, p. jaune, p. soufré (Fr. G.); bacoropary, bacury, b.-assu, b.-rana, b.-siero, b.-úba, ibacopary, ibacury, ibacurupary, pacuru, pacoury grande, p.-uva, ubacury (Braz.); matazama (Ec.); bacury-guazú (Par.).

Rheedia, with about 38 species of small to medium-sized or occasionally large trees, is widely distributed in tropical America and there are four species also in Madagascar. The leaves are pinninerved, opposite or verticillate; the small flowers are axillary, solitary or clustered; the fruits have a leathery skin and contain 1 to 5 seeds enveloped in a fleshy aril. The fruits of some species are sweet and edible and in certain localities are of some commercial importance. The bark is rich in tannin and exudes a yellowish sap which has some medicinal applications. The timber is used locally for handles of axes and other tools, general construction and carpentry, and for fence posts and railway crossties.

Heartwood dull grayish to pinkish brown, merging gradually into the sapwood; surface sometimes specked with resinous exudation from radial ducts. Odorless and tasteless. Hard, moderately heavy, tough and strong; texture coarse; feel harsh; grain irregular; not difficult to work but does not finish very smoothly; durability only fair. Of no commercial possibilities.

Common names: Espuela de caballero, e. de rey, manchi, managú, m. cimarron, m. de costa (Cuba); guayabacoa, palo de cruz, sebucán (P.R.); bois diou, b. de haut, b. du roux malle (Haiti); contrevent (Mart.); hat-stand tree, soiebo, wild rose apple (Trin.); limoncillo (Mex.); mamey ciruela, waika plum (Br. H.); caimito, c. de montaña (Hond.); chaparrón (Salv.); jorco (C.R.); cero, madroño, macharé, naranjuelo (Col.); cupí, madroño (Venez.); assachi (Br. G.); apakwie, aroome, asasie, a. hororodikoro, baaha manie pau, bigi boesi pakoerie, manieran, matakkie, nopietja, nopitja, pakorian, pakorie ibibero (Sur.); bacoparé, bacury miudo, b. pary, pacury (Braz.); achuni-caspi, brea-caspi, b.-huayo, charichuéla (Peru); pacuri (Arg.).

Symphonia, with about 16 species of trees and shrubs, is limited to Madagascar except for one species, *S. globulifera* L.f., which occurs in tropical West Africa and in the West Indies, Central America, and northern South America. This is a rather common tree in the mixed hardwood and palm forests of low humid regions, and occasionally is the dominant element of small stands. On favorable sites it is frequently 100 feet tall with a long, straight, slightly buttressed bole 20 to 30 inches through; exceptional specimens are 135 feet tall and over 40 inches in diameter. The bark contains a yellowish resin which, upon exposure, becomes black and pitchy and is used locally in calking boats. The leaves are opposite, narrow, finely pinninerved; the red flowers are borne in terminal or axillary cymes, rarely solitary; the fruit is a few-seeded edible berry. The timber is of the general-utility class and is employed locally for building purposes, carpentry, cooperage, crates and boxes, railway crossties, and fuel. Small shipments have been made to Europe and the United States, mostly for making rotary-cut veneers for plywood, for which it is well suited, although lacking figure.

Heartwood yellowish, grayish, or greenish brown, with a somewhat mealy appearance because of the abundance of coarse-celled wood parenchyma; sharply demarcated from the whitish sapwood. Odor and taste absent or not distinctive in dry material. Hard and moderately heavy; sp. gr. (air-dry) 0.65 to 0.78; weight 40 to 49 lbs. per cu. ft.; texture coarse; feel rather harsh; grain straight to irregular; working and finishing qualities good; durability fair. Mechanical tests made in the laboratories of the Yale School of Forestry on small, clear, air-dried specimens from British Honduras gave the following average strength values (adjusted to 12 per cent moisture content) in pounds per square inch: *Transverse bending* (*central loading*): Modulus of rupture, 18,450; modulus of elasticity, 2,642,000; fiber stress at elastic limit, 11,200. *Compression along the grain:* Crushing strength, 10,040; modulus of elasticity, 2,494,000; fiber stress at elastic limit, 7850. *Compression across the grain:* Fiber stress at elastic limit, 1530. These values exceed those of ordinary White Oak (*Quercus alba* L.). (For further information see *Tropical Woods* 45: 1–15.)

COMMON NAMES: Boarwood, doctor gum, hog gum (B.W.I.); yellow mangue (Trin.); chewstick, corbán, leche amarilla, mountain cow, waiki chewstick, whykee chewstick, wycot (Br. H.); barillo, pimientillo (Guat.); barillo, leche amarilla (Hond.); botoncillo, cerillo, sambogum (C.R.); barillo, bogum, cerillo, cero, sambogum (Pan.); machare (Col.); maní, paramán, peramán, peramancillo (Venez.); buck-wax tree, karimanni, manni, manniballi (Br. G.); manie, mannie, mannihororodikoro, mannipau, masagrie, matagrie, matakki, matatji, tapoekin mani (Sur.); bois cochon, maní, manil, manilparcouri (Fr. G.); anany, vanani (Braz.); brea-caspi (Peru).

Tovomita, with more than 30 species of shrubs and small or occasionally medium-sized trees, has its center of distribution in the Amazon basin but extends northward into the West Indies. The leaves are pinninerved, opposite; the flowers are borne in terminal panicles; the fruit is capsular. Some of the trees in upland forests develop stilt-roots suggesting Mangrove (*Rhizophora*). The timber is of good quality and when obtainable in large enough sizes is used locally for staves, shingles, and furniture.

Heartwood dull purplish brown; not sharply demarcated from the lighter sapwood. Without distinctive odor or taste. Density and texture medium; grain rather irregular; silver grain conspicuous on radial surface; working qualities fair; durability high.

COMMON NAMES: Mangue rouge (Dom.); paletuvier montagne (Mart.); red mangue (Trin.); chagualo, rapebaibo (Col.); wakome (Venez.); awaskuli, wild sapodilla (Br. G.); alakapoeli, arapori, awasakoele, etra kapoeri, koessapoli, kombotasje poté, konapoe, mangro, prasara, sabana mangro, watiepa oepadjare, w. oepajarèrè, w. opodjaré (Sur.); azedinho, pachiubarana, p. miuda, sangue de torro (Braz.).

Tovomitidium, with two species of small resinous stilt-rooted trees, is limited to the west-central part of the Amazon basin of Brazil (see *Archiv. Inst. Biol. Veget.* 2: 1:

61–62). The large leaves are pinninerved, opposite; the white flowers are borne in short terminal cymes; the fruit is capsular. *T. speciosum* Ducke of the Tapajoz region is noted for its very large thick leaves. The only wood sample of the genus available (Yale 34075; Ducke 311) is of the other species, *T. clusiiflorum* Ducke, growing in upland forest near São Paulo de Olivença (Amazonas). The timber is not utilized.

Heartwood reddish brown, with purplish hue; merging gradually into the sapwood. Dull superficially but with silky luster in proper light and displaying an attractive silver grain on radial surface owing to the contrast of the tall, deeply colored rays against the fiber background. Odorless and tasteless. Very hard and heavy; texture rather coarse; grain straight; moderately difficult to work, but finishing smoothly; durability probably high. Apparently of no commercial possibilities because of its scarcity and small size.

Tovomitopsis. In this genus there are about eight species of trees and shrubs ranging from Central America to eastern Brazil. It has proved impossible to separate the wood samples from those of *Chrysochlamys* and this confirms Standley's suggestion (*Flora of Costa Rica,* p. 710) "that it might be better to refer all species of this group to *Chrysochlamys,* as was done by Bentham and Hooker."

COMMON NAMES: Coloradito, mangle colorado, m. montanero, rapadura (Pan.); raizudo (Venez.); azedinho (Braz.).

Vismia. There are 25 species, mostly tropical American, a few in tropical Africa. They are shrubs and small trees, rarely 40 feet tall, common in second-growth on old clearings. The bark contains a resinous yellow sap which finds application in local medicine. The leaves are opposite, pinninerved, usually tomentose, and often punctate with numerous glands; the white, yellow, or rusty brown flowers are generally woolly and are borne in terminal or axillary panicles, in part on leafless branchlets; the fruit is a berry with elongated warty seeds. The timber is not utilized for any special purpose as the available sizes are small.

Heartwood light pinkish brown, merging gradually into the sapwood. Luster medium. Odorless and tasteless. Rather light and soft to fairly dense; texture medium; grain straight; easy to work, and suitable for general utility purposes; not highly resistant to decay. Of no commercial possibilities.

COMMON NAMES: Lacre (Cuba); bloodwood, kiskidee, lacre, lacraie (Trin.); broadleaf, old William, yellow sange (Br. H.); achiotillo, camparaguey, lengua de vaca (Guat.); achotillo, gurak, hunkri-krá, srin-gró (C.R.); mata rancha (Nic.); pinta-mozo, sangre de perro, sangregado de tierra fría, sangrillo (Pan.); anabaptista, caparrosa, carate, lance, punta de lanza (Col.); caparrosa, lacre blanco, lancetollo, motijo, onotillo, punta de lanza (Venez.); bloodwood, oralli (Br. G.); manpienya, moejè pienyapau, omemapienja, omipienjapau, pienja, p.-hoedoe, pienjapau, pinja, seebralala, seepjalala, seibralala, soeinjani, taloekwepe, tamoenè ajoewinani, tapirin soewinjani, tona, wajoelidan, wakare soewinjani, warahaje, w. iebi korobana, w. sarerokoena (Sur.); bois babtiste, b. cossais, b. d'acossais, b. d'arte, b. de la fièvre, b. de sang, b. sanglant, coaopia (Fr. G.); caopiá, caparosa, lacre, l. blanco, pau de lacre (Braz.); pichirina, yana pichirina (Peru).

HAMAMELIDACEAE

THE Witch Hazel family consists of about 21 genera and 100 species of trees and shrubs growing in the Far East, Madagascar, South Africa, and eastern North America. The two genera most important for their timber are *Altingia* in the Malay Peninsula and *Liquidambar* in the United States. The following description applies particularly to the three American genera, *Distylium, Hamamelis,* and *Liquidambar.*

Pores small to minute, numerous, diffuse; solitary in *Distylium* and *Hamamelis.* Tyloses common; perforation plates scalariform, with several to many bars; spirals present in tips of vessel members of *Liquidambar;* intervascular pitting opposite to scalariform. Rays fine, distinctly heterogeneous; pits to vessels round to much elongated and parallel. Wood parenchyma sparingly diffuse. Wood fibers with distinct to conspicuous bordered pits. Vertical traumatic gum ducts sometimes present in *Liquidambar*. Ripple marks absent.

Distylium is an Asiatic genus containing nine species of trees and shrubs distributed from Indo-Malaya to China and Japan. According to Harms (*Notizbl. Bot. Gart. Berlin-Dahlem* 11: 108: 714–718), there is another species, *D. guatemalense* Radlkofer, in Guatemala. The only material available for this study is a twig from a herbarium specimen collected by T. G. Yuncker (his No. 6377) near Siguatepeque, Honduras. From this it can be inferred that the timber is moderately dense, hard, and fine-textured.

Hamamelis, with five or six species of deciduous shrubs and small trees, occurs in eastern Asia and eastern North America. The flowers, with small, strap-shaped, yellow petals, appear in the fall or winter and the capsular fruits are elastically dehiscent. Best known of the three American species is the common Witch Hazel, *H. virginiana* L., usually a stout shrub with crooked branches, but under exceptional conditions developing into a tree 25 feet high with a short trunk a foot in diameter. It is widely distributed throughout the eastern half of the United States and neighboring regions of Canada. The species is of commercial importance in southern New England where for the past 75 years the twigs and brush have been distilled for the production of Witch Hazel extract. The southern form, *H. macrophylla* Pursh., is most abundant in Louisiana and Alabama and sometimes attains a height of 45 feet. The timber is not utilized.

Heartwood brownish, merging gradually into the lighter-colored sapwood, which may have a pinkish hue. Luster medium. Without distinctive odor or taste. Density medium; texture fine; grain fairly straight; rather easily worked; durability low.

Common names: Snapping hazel, spotted alder, winter bloom, witch hazel (U.S.A.).

Liquidambar. There are three species in eastern Asia and one, *L. styraciflua* L., in America, where it is commonly known as Red or Sweet Gum. This important timber tree is widely distributed throughout the southeastern part of the United States, its northern and western boundaries being a line from Connecticut westward through the Ohio valley to Kansas and southward through Oklahoma into Texas; it reappears on the mountains of southern Mexico and the highlands of southern British Honduras and eastern Guatemala and Honduras. It is most abundant and of largest size in river bottoms subject to inundations in the maritime region of the south Atlantic and Gulf states and in the lower Mississippi valley, where it is often 80 to 100, sometimes 150, feet high, with a long smooth bole two to five feet in diameter. The timber is extensively used for a wide range of purposes, such as furniture (often in combination with Walnut), interior trim, doors, and panels, and veneers for plywood, baskets, dishes, wire-bound boxes, and vegetable barrels. In Mexico and Central America the tree grows at elevations of 3000 to 5000 feet, and often is small or medium-sized. The timber is used to a limited extent for general carpentry, packing cases, match sticks, and toothpicks. The tree produces a resin or storax, called Xochicotzo by the ancient Mexicans, which was an article of tribute for the support of the central government and was used chiefly in medicine and perfumes.

Sapwood nearly white; heartwood brown or reddish brown, with a satiny luster, and sometimes beautifully figured with dark pigmented markings. Odorless and tasteless. Sp. gr. (air-dry) 0.50 to 0.65; average weight about 35 lbs. per cu. ft.; grain usually irregular; texture fine and uniform; easy to work, finishing very smoothly; likely to warp badly if not carefully dried; is not highly resistant to decay.

Common names: *Timber:* Gum, gumwood, red gum, hazel (U.S.A.); red gum, satin walnut (Gr. Brit.); Amberholz, Satinnussbaum (Germ.); noce satin, legno di noce satinato (Ital.); noyer satiné (Fr.); satijn noten (Hol.); saten (Sp.). *Tree:* Bilsted, copalm, gum (red, sweet, starleaved), gum tree, liquidambar (U.S.A.); liquidambar, liquidamber (Mex., C.A.); estoraque, maripenda, naba, ocotzotl, ocozol, ocozote, ocozotl, yaga-bito, xochiocotzoquáhuitl (Mex.).

HERNANDIACEAE

A pantropical family of four genera (*Gyrocarpus, Hernandia, Illigera,* and *Sparattanthelium*) and about 35 species of trees and erect or scandent shrubs. The leaves are alternate, simple or digitately compound, and without stipules; the inflorescences are cymose and axillary; the dry 1-seeded fruits are winged. *Illigera* is the only genus without representation in the New World. The following description applies particularly to American woods of *Gyrocarpus* and *Hernandia.*

Color whitish and lustrous when fresh, but subject to fungus staining to grayish, brownish, or oatmeal. Odorless and tasteless. Very light and soft; sp. gr. (air-dry) 0.21 to 0.43; weight 13 to 27 lbs. per cu. ft.; coarse-textured; grain straight to variable; easily worked with sharp tools, but saws woolly; perishable when exposed to decay.

Growth rings present or absent, sometimes distinct. Pores very few (about 3 per sq. mm.); large (225 to 300μ) or occasionally small; solitary and in small radial multiples or little clusters, unevenly distributed, sometimes locally zonate. Vessels often meandering tangentially; perforations simple; intervascular pitting coarse (12 to 15μ), alternate. Rays 1 to 3 cells wide and up to 20, usually less than 15, cells high in *Gyrocarpus;* sometimes up to 4 or 5 cells wide and 50 cells high in *Hernandia;* homogeneous; occasional oil cells present in some specimen of *Hernandia;* ray-vessel pitting very coarse and irregular, often distinctly half-bordered. Wood parenchyma variable in abundance and arrangement; coarsely vasicentric, short to long aliform, more or less confluent, occasionally terminal; increase in density generally accompanied by greater development of confluent parenchyma; pits to vessels very large, with tendency to scalariform arrangement. Wood fibers thin-walled, sometimes septate in part; pits numerous, very small, simple or with vestigial borders. Ripple marks absent. No gum ducts seen.

Gyrocarpus, with a single pantropical species, *G. americanus* Jacq., is a medium-sized tree with thick branches; large, simple, long-petioled leaves; and pendant clusters of shuttlecock-like fruits, each consisting of a nutlet with two long slender wings at one end. The timber is occasionally used locally for making toys and small boxes.

COMMON NAMES: Babá, ciis, palo hediondo, p. de zopilote, quitlacoctli, quitlacotli, volador, xkis (Mex.); titirillo (Guat.); corroncha de lagarto, lagarto, tambor (Salv.); caballitos, gallito, talalate (Nic.); banco, limoncillo, piñón? (Col.); volador (Col., Venez.).

Hernandia. There are about 14 species, mostly small evergreen trees, of infrequent occurrence in Indo-Malaya, Oceania, Madagascar, and the Caribbean region. The only species of any commercial importance is *H. Voyroni* J. Jum., the Hazomalana or Faux Camphrier of Madagascar; its soft, yellowish, waxy wood has a pronounced odor of camphor and is resistant to decay and insects; considerable quantities are exported to India as a source of ethereal oil. The American species appear not to be utilized.

COMMON NAMES: Mago (P.R.); palo de chicalpexte (Mex.); aguacatillo (Guat., C.R.); tambor (Guat., Hond.); hojal tamal, mano de león (Hond.); cebo burro, c. macho, lampa or lempa (Pan.); hua-hua, jack-in-the-box, takaruva (Br. G.); hernandier, myrobolan (Fr. G.); ajowo, foungou, kajoeballi, kassabahoedoe (Sur.); ventosa (Braz.).

Sparattanthelium comprises about 12 species of lianas, climbing shrubs, and less often small trees, sparingly distributed from southern Mexico to northern Brazil. One species is called Oneka in Surinam. The wood is white, lustrous, and soft, as in the other two genera.

HIPPOCASTANACEAE

THE Horsechestnut family is composed of two genera, namely, *Aesculus,* with about 20 species and several varieties of deciduous trees and shrubs widely distributed in the north temperate zone, and *Billia,* with two species of evergreen trees of limited occurrence from southern Mexico to Venezuela. The leaves are opposite, digitately compound, with 5 to 9 leaflets in *Aesculus* and 3 in *Billia;* the showy white, pale yellow, or red flowers are borne in terminal panicles; the fruit is a leathery 3-celled capsule, usually with a single lustrous seed in each cell. Numerous species, varieties, and hybrids of *Aesculus* are planted for shade and decorative purposes and a few species are the source of commercial timber.

Aesculus occurs in the Old World from Japan and China through the Himalaya region into Europe. The best known species is the Horsechestnut, *A. Hippocastanum* L., considered by many the handsomest tree in Europe and widely planted in lawns and parks and along roadways. The wood is similar to that of Willow (*Salix*) and Poplar (*Populus*) and used in limited quantity for the same purposes.

There is one species, *Aesculus Parryi* A. Gray, in Baja California, Mexico, but it is only a shrub. In the United States there are six arborescent species, all commonly known as Buckeye. Westernmost is *A. californica* Nutt., a low tree with a short but very thick trunk, growing along streams at elevations of 2000 feet or more in the coast ranges of California. The largest tree is the Yellow or Sweet Buckeye, *A. octandra* Marsh., of the south-central hardwood zone where in mixed hardwood stands on good soil it attains a height of 100 feet with a straight trunk up to 36 inches in diameter. Occurring in the same general region, but extending westward beyond the Mississippi River, is the Ohio Buckeye, *A. glabra* Willd., usually a small tree but sometimes up to 70 feet tall. Buckeye is of considerable commercial importance, especially in Tennessee, Kentucky, and Ohio, but the actual lumber production is unknown because it is often sold in mixture with other kinds, particularly as the sapwood of Yellow Poplar (*Liriodendron*). Its principal uses are for boxes and cases of all kinds, sugar and candy pails, fish kits, spice kegs, honey sections, piano and organ keys, drawer bottoms, interior trim that is to be painted or enameled, and various other purposes requiring a clean-appearing, easily worked, fairly tough material of fine and even texture.

The woods of the several species examined are much alike. Color whitish or yellowish throughout. Luster fairly high. Odorless and tasteless when dry; with mild, unpleasant scent when fresh. Of rather low density, but tough and strong for its weight; sp. gr. (air-dry) 0.45 to 0.60; weight 28 to 38 lbs. per cu. ft.; texture very fine; grain straight to wavy; easy to work, finishing smoothly and presenting a bright surface suitable for printing upon; of low resistance to decay.

Growth rings distinct. Pores thin-walled and angular; small to very small, not visible without lens; numerous; solitary and, more often, in short to long radial multiples; a single row of slightly larger pores common at beginning of the annual ring; pores in outer late wood often fewer and smaller. Vessels with fine spiral thickenings; perforations simple, with tendency to the formation of scalariform plates near pith; tyloses sometimes present; pitting alternate, rather fine. Rays uniseriate, rarely locally biseriate, and few to 20 cells high; weakly heterogeneous, with marginal cells squarish; pits to vessels usually limited to marginal cells, rounded to much elongated and in scalariform arrangement, the elongated pits subtending two or more vascular pits. Wood parenchyma finely terminal and as single cells in contact with vessels, often on tangential face only. Wood fibers with small, indistinctly bordered pits. Ripple marks sometimes present; 65 to 70 per inch; all elements more or less distinctly storied.

Billia. The Mexican species, *B. Hippocastanum* Peyr., is a tree apparently of limited occurrence in Vera Cruz and Oaxaca; no specimens are available for this study. *B. columbiana* Planch. & Lind. is a tree sometimes 65 feet tall growing on the mountains at elevations of 2000 to 7500 feet from Guatemala to Ecuador. It has bright red flowers and is very showy when growing in pastures. The only sample available (Yale 36035) was collected by Austin Smith near Zarcero, Costa Rica. Color silvery gray, with a slight pinkish tinge, throughout. Luster fairly high. Odorless and tasteless. Of medium density, fairly hard, tough, and strong; texture rather fine, uniform; grain variable; easy to work, finishing smoothly; poorly resistant to decay. (For anatomy of the wood see *Tropical Woods* 58: 1–2.)

COMMON NAMES: Cocora, cucaracho (C.R.).

HIPPOCRATEACEAE

An unimportant pantropical family of two conventional genera, *Hippocratea* and *Salacia,* and over 200 species of slender trees and scandent, climbing, or erect shrubs, about equally divided between the Old World and the New. The leaves are typically opposite, simple, and with small stipules or none; the flowers are small, normally have only three stamens, and are borne in fascicles or cymes; the fruit is drupaceous or capsular. A few species have edible pulp around the seeds and many contain a non-resinous latex which yields a kind of rubber and may have commercial possibilities (see *Bol. Min. Agr.*, Rio de Janeiro, 1934, pp. 29–31).

In his recent monograph on the American species of Hippocrateaceae (*Brittonia* 3: 341–555; November 1940) A. C. Smith recognizes 12 genera and 115 species. In the Yale collections are specimens of eight species of six of these genera, namely, *Cheiloclinium, Hemiangium, Hippocratea, Prionostemma, Pristimera,* and *Salacia.*

Wood yellow or brownish throughout; sometimes with prominent ray markings. Luster medium. Odor and taste not distinctive or undefinable. Moderately heavy and hard; texture medium to coarse; grain fairly straight; probably poorly resistant to decay. Of no economic value. Structure normal or anomalous; included phloem in narrow concentric bands in *Cheiloclinium Gleasonianum* A. C. Sm. and *Salacia megistophylla* Standl., and in numerous very small islands (cross section) in *C. cognatum* (Miers) A. C. Sm. and *Prionostemma aspera* (Lam.) Miers; steplike intrusions of bark present around the periphery of the stem of *Hippocratea* and with tendency to such formation in *Hemiangium.*

Growth rings poorly defined to distinct. Pores often thick-walled; variable in size, but mostly medium or large; rarely in radial contact except in *Pristimera celastroides* H.B.K.; distribution regular to irregular, sometimes with noticeably fewer pores in outer late wood. Vessels with simple perforations; intervascular pitting, when present, fine and alternate. Rays heterogeneous; all uniseriate or biseriate in *Cheiloclinium cognatum* and *Salacia megistophylla;* of two sizes in the others, the larger ones conspicuous (composed of conjunctive tissue in *C. Gleasonianum*); included patches of thin-walled cells common in large rays of some genera; crystals abundant; ray-vessel pit-pairs small (4 to 5μ) or unilaterally compound with a ray pit elongated and subtending two to several vascular complements (*C. cognatum, Pristimera,* and *Salacia*). Wood parenchyma sparingly paratracheal. Wood fibers in part septate and parenchyma-like in appearance and distribution, being aliform to confluent into narrow bands in *Cheiloclinium* and *Salacia,* diffuse to finely reticulate (not distinct with lens) in the others; in part non-septate, with thicker walls and smaller lumen, distinctly bordered pits, and making up the bulk of the ground mass and, near vessels, becoming more tracheid-like; vasicentric tracheids usually present, sometimes vermiform. Ripple marks absent. No gum ducts seen.

Cheiloclinium, with 20 species of lianas or more or less scandent shrubs and slender trees, occurs throughout Central America and tropical South America. Specimens of only two species are available for this study and they are very dissimilar. *C. Gleasonianum* A. C. Sm. is a liana known only from the upper Essequibo basin in British Guiana. The wood (Yale 35600; A. C. Smith 2547) is of anomalous structure, distinctly laminated like *Avicennia,* and having large (250μ), irregularly distributed pores. *C. cognatum* (Miers) A. C. Sm. is a shrub or a tree sometimes 40 feet high, often with scandent branches, of widespread and common occurrence from Tobago and Panama to Amazonian Peru and Bolivia and to southeastern Brazil. The inner part of the outer bark is dark orange. The brown, hard, rather heavy, medium-textured wood is not laminated, the included phloem appearing as numerous oval or tangentially elongated islands (mostly 250 to 400μ in diameter).

Common names: Goroeba branca, quina do matto (Braz.).

Cuervea. The only specimen available (Yale 12092) is of *C. Kappleriana* (Miq.) A. C. Sm., a liana collected in Panama by G. Proctor Cooper. It is about an inch in diameter and is of anomalous structure, the included phloem being in strands in a concentric band of conjunctive tissue. The wood is pale yellow, moderately hard, and coarse-textured.

Hemiangium *excelsum* (H.B.K.) A. C. Sm., the only species, is a liana, shrub, or a slender tree sometimes 30 feet high, of common occurrence in uplands from southern Mexico to Colombia and Venezuela, and reappearing in Paraguay and southern Brazil. Wood light brown, with a pinkish hue, throughout; ray markings conspicuous. Luster medium. Without distinctive odor or taste. Moderately heavy, hard, and strong; texture rather coarse; grain fairly straight; easy to work, finishing smoothly and attractively; durability probably low. Of no commercial possibilities because of the small size of the trees.

COMMON NAMES: Barajillo, chumloop (Mex.); cucaracho, mata-piojo (Salv.); fruto de rosa (Nic.).

Hippocratea, reduced by Smith (*loc. cit.*) to a single species, *H. volubilis* L., is a slender to stout liana of universal distribution in tropical and subtropical America. Young pliable stems are used by natives for binding; the seeds are edible and yield an oil sometimes employed medicinally. Stems of different thicknesses and from widely separated regions are alike in having numerous, shallow, broadly V-shaped intrusions of the bark, with the sides step-like, somewhat the same as in certain bignoniaceous lianas. There are no included bands or strands of phloem.

COMMON NAMES: Bejuco de vieja (Cuba); haquimey, jaiquimey (Dom. R.); liane blanc (Haiti); bejuco colorado (Mex.); barracuta tie-tie (Br. H.); rabo de mono (Col.).

Pristimera, with nine species of lianas, shrubs, and small trees with scandent branches, occurs throughout most of tropical America. The only reported uses are medicinal. The wood is yellowish, of medium density, and rather coarse-textured.

COMMON NAMES: Almendro (Cuba); liane mébi (Haiti); barajilla, bejuco del piojo, cuanabichi, hierba del piojo, mata-piojo, tatsi, tulubalam (Mex.); mata-piojo (Salv.); levanta-perro (Col.).

Salacia, with 29 American and several Old World species of lianas, shrubs, and slender trees, is of general distribution in the tropics of the New World. The only specimen at hand (Yale 12121) is from the type of *S. megistophylla* Standley, a liana collected by G. Proctor Cooper in Bocas del Toro, Panama. The structure of the stem is anomalous, included phloem being present in a narrow concentric band about 1 cm. from the pith.

COMMON NAMES: Bacupary, bochecha de velho, tuyué-tipi (Braz.); pacuri, sipota (Par.).

HUMIRIACEAE

THIS family comprises three genera, namely, *Humiria, Sacoglottis,* and *Vantanea,* and about 30 species of trees and shrubs, all confined to tropical Brazil and the Guianas, with the exception of one West African species. The leaves are alternate, simple, and sometimes gland-dotted; stipules are absent or small and deciduous; the flowers are in axillary panicles or racemes; the fruit is a drupe, sometimes with resin-filled cavities in the endocarp. In Winkler's classification (*Pflanzenfamilien,* 2nd ed., 19a, p. 126) the Humiriaceae are reduced to the status of a subfamily of the Linaceae, but this proposal has little support in the anatomy of the woods. Some of the trees are among the tallest in the forest, but their timbers are sparingly utilized even locally, presumably because of an abundance of better kinds.

Heartwood grayish brown to reddish or purplish brown; distinct, but not sharply demarcated, from the sapwood. Luster usually low. Odorless and tasteless when dry. Hard and heavy to decidedly so; sp. gr. (air-dry) 0.80 to 1.10; weight 50 to 69 lbs. per cu. ft.; texture medium to coarse; grain mostly irregular; not easy to work; is tough and strong; finishes smoothly; durability doubtful.

Growth rings absent or poorly defined. Pores often large enough, in part at least, to be seen readily without lens, but sometimes all small;

very numerous to moderately so, occasionally rather few; typically solitary; fairly well distributed without definite pattern, though with local tendencies to diagonal arrangement. Vessels with scalariform perforation plates having numerous widely spaced bars; gum deposits abundant; spiral thickenings absent; intervascular pitting rare. Rays very numerous; uniseriate and partly biseriate, rarely wider, and of various heights up to about 50 cells; decidedly heterogeneous; most of the cells square or upright, the marginal ones frequently forming a palisade; gum deposits abundant; cells thick-walled and abundantly pitted; pits to vessels small (*Humiria*), medium-sized (*Sacoglottis*), or very large (*Vantanea*), rounded to much elongated and in scalariform arrangement. Wood parenchyma sparsely to rather abundantly developed, barely visible with lens; diffuse and in short tangential lines that are in part metatracheal and in part in contact with the pores on the outer face only; crystalliferous strands common. Wood fibers in radial rows; cells often flattened tangentially; walls thick to extremely thick; pits numerous, often abundant, small to rather large, distinctly bordered. Ripple marks and gum ducts absent.

Humiria. There are three or four species, but the best known are *H. balsamifera* Aubl. and *H. floribunda* Mart. of Colombia, Venezuela, the Guianas, and Brazilian Amazon region. On unfavorable sites they are shrubby, but where conditions are favorable they are among the dominant trees of the forest, and there are reports from British Guiana of trunks 90 feet long that would square 20 inches free of sapwood. Some trees yield a small quantity of balsam, known in Brazil as "balsamo de umiry." The fruits are edible. The wood has the general appearance and properties of Bulletwood or Massaranduba (*Manilkara*), though coarser-textured and of poorer quality. The local uses include heavy construction timbers, spokes of wheels, and miscellaneous purposes where strength is the chief requisite.

Common names: Oloroso (Col.); bastard bulletwood, b. bully, hoorihea, hoorihee, hurilú, muri (Br. G.); bakabe-ie, basra-botrie, bastard-bolletrie, bhoso, kierie-ma, mérie, tauroniro, tawaranoe, tawanonero (Sur.); arbre à brai, bois à flambeau, b. d'encens, b. rouge, b. r. tisane, boume houmirí, houmirí, h. boumier, caramura, couranoura, gommier de montagne, homiry, omiry, racine, triane, turi, umiri, u. balsamo, umiry, toweroenierou (Fr. G.); couranira, nieri, tourameira, turamira, umiri, umiry, u. de casca cheiro, u. de cheiro (Braz.).

Sacoglottis has 16 to 18 species in northeastern South America and one in tropical West Africa. The fruits have a fleshy mesocarp which in some species is thick and good to eat. Some of the Brazilian trees attain a height of 150 feet, but their dense timber is rarely utilized.

Common names: Huriki (Br. G.); achuá, a. rana, axuá, cumaté, c. da catinga, parurú, uaxuá, uchy, u. corôa, u. curúa, u. pucú, u. rana, uixyseiro, uxy curúa, u. pucú (Braz.).

Vantanea, with seven or eight species, is distributed from the Guianas to the mountainous regions of Minas Geraes, Brazil. Some of the trees are small, but others, especially in the lower Amazon forests, are 100 feet tall. The dense, grayish brown timber has no important uses at present.

Common names: Louantan (Fr. G.); achuá-rana, aroeirana, uchy-rana (Braz.).

HYDRANGEACEAE

The Hydrangea family, often included in the Saxifragaceae, comprises 16 genera and about 200 species of erect or scandent shrubs and a few small trees, widely distributed in temperate and subtropical regions, mostly of the northern hemisphere. The leaves are simple, opposite, and without stipules; hairs, when present, are simple, stellate, or glandular; the flowers are small to large, often showy; the fruit is capsular. Some of the plants are well known in floral planting. Five small genera, namely, *Carpenteria, Fendlera, Fendlerella, Jamesia,* and *Whipplea,* are confined to North America, and four others, namely, *Decumaria, Deutzia, Hydrangea,* and *Philadelphus* are represented there. Their utility is limited to decorative purposes. The only American species represented in the Yale collections are of *Fendlera* and *Philadelphus*. Their hard, heavy, fine-textured woods are not used.

Growth rings present, with more or less pronounced tendency to ring-porous structure. Pores very small (45 to 60μ) in early wood to minute in late wood; solitary; numerous but not crowded. Vessels with scalariform perforation plates having few to several thick and widely spaced bars in *Fendlera*, or several to numerous rather fine bars in *Philadelphus;* vascular pitting infrequent, scalariform. Rays uniseriate or biseriate and up to 15 cells high in *Fendlera;* 2-sized in *Philadelphus*, the larger up to 4 or 5 cells wide and up to 70, sometimes over 100, cells high; all decidedly heterogeneous; pits to vessels small (5μ) to very small (3μ), oval. Wood parenchyma diffuse in *Fendlera*, sparingly paratracheal in *Philadelphus*. Wood fibers with thick walls and conspicuously bordered pits; fine spiral thickenings present. Ripple marks absent. No gum ducts seen.

COMMON NAMES: *Hydrangea:* Gray beard, hydrangea, old man's beard (U.S.A.). *Philadelphus:* False jasmin, mock orange, syringa (U.S.A.); acuilotl, cozticacuilotl, jazmín, j. de monte, jeringüilla, mosqueta (Mex.); mosqueta (Salv.); mosqueta, m. trepadora (C.R.).

HYDROPHYLLACEAE

THE Waterleaf family, with 17 genera and about 200 species of annual or perennial herbs, rarely shrubs or small trees, is widely distributed, but mostly in North America. The only genera represented in the Yale collections are *Eriodictyon, Nama*, and *Wigandia*. The leaves are alternate and simple; the flowers mostly in scorpioid cymes; the fruit is capsular. About the only uses for the plants are in local medicines.

Heartwood probably dark brown (judging from wound areas); sapwood white to brownish. Luster rather low. Odorless and tasteless. Rather light and soft to moderately heavy and hard; texture rather fine; grain straight; easy to work; durability doubtful. Of no commercial possibilities.

Growth rings distinct to indistinct; sometimes with definite tendency to ring-porous structure. Pores mostly small (60 to 75μ) to minute (20μ), but up to lower medium size (110 to 125μ) in *Wigandia;* virtually all solitary in *Eriodictyon* and *Nama;* a few small multiples present in *Wigandia;* tangential to broken concentric arrangement common. Vessels with simple perforations; pitting fine (6 to 7μ), alternate. Rays in *Eriodictyon* and *Nama* heterogeneous, 1 to 3 cells wide and up to 30 cells high; in *Wigandia* homogeneous or nearly so, 1 to 4, sometimes 5 or 6, cells wide and up to 35 cells high; ray-vessel pitting fine in all of these genera. Wood parenchyma abundantly diffuse or reticulate; not distinct with lens. Wood fibers with numerous distinctly bordered pits. Ripple marks absent. No gum ducts seen.

Wigandia, with a few species of shrubs, small trees, and large herbs, occurs in uplands of the West Indies and from Mexico to northern South America. The principal and most widely distributed species, perhaps the only one attaining a height of 15 feet, is *W. caracasana* H.B.K. It is covered with stiff, white, stinging hairs that penetrate the skin easily like those of a nettle. The flowers are large and purple, the seeds are minute and winged.

COMMON NAMES: Chichicaste, consuelda mayor, hoja de San Pablo, ortiga, o. grande, palo de San Pablo, quemadora, sosa, San Pablo, tabaco cimarrón (Mex.); chochón (Guat.); chichicaste (Hond.); mata-pulga, tabaquillo (Salv.); ortiga, o. de montaña (C.R.); pringamoza (Col.); borrajón (Venez.).

ICACINACEAE

THIS is a rather poorly defined family with about 60 genera and 200 species of trees, shrubs, and woody vines of wide distribution throughout the tropics but nowhere of importance commercially. The leaves are typically alternate, simple, and without stipules; the flowers are small, perfect or unisexual; the fruit is a 1-seeded drupe, rarely winged. There are 16 genera and about 50 species in tropical America, mostly small trees or shrubs, many of them rare and not well known. The following description is based upon specimens of the wood of *Calatola, Dendrobangia, Discophora* (*Kummeria*), *Emmotum, Mappia, Metteniusa* (*Aveledoa*), *Oecopetalum, Ottoschulzia, Poraqueiba*, and *Villaresia*.

Heartwood yellow or pale brown; sapwood yellowish. Luster medium to rather high. Odorless and tasteless. Moderately to very hard and heavy; texture fine to rather coarse; grain usually straight; generally not difficult to work, finishing very smoothly and taking a good polish; some specimens attractive when quartersawed, owing to the prominent rays; not highly resistant to decay.

Growth rings absent or not well defined. Pores very small to medium-sized, occasionally rather large; usually numerous but not crowded; in most cases all solitary, but with numerous small multiples and clusters in *Dendrobangia, Discophora,* and *Mappia;* well distributed, but with tendency toward tangential arrangement in *Ottoschulzia.* Vessels typically with scalariform (sometimes reticulate) perforation plates; exceptions, *Mappia* with exclusively, *Discophora* with predominantly, simple perforations; spiral thickenings observed only in *Villaresia mucronata* R. & P. (Yale 34058); tyloses common in *Emmotum* and *Poraqueiba;* intervascular pitting medium to coarse and alternate in *Discophora* and *Mappia,* absent or rare and opposite or scalariform in the others; pits to fiber-tracheids abundant, rather small. Rays heterogeneous and of two sizes; multiseriates greatly variable in width and height, becoming larger outwards, sometimes suggesting *Platanus,* sometimes *Quercus,* and frequently interrupted by fiber layers; crystals often present; pits to vessels variable: all or in part large to very large, elongated, and commonly in scalariform arrangement in *Discophora, Metteniusa, Oecopetalum,* and *Poraqueiba;* medium-sized and round in *Dendrobangia;* very small in *Ottoschulzia* and sometimes in *Villaresia;* medium-sized in the others. Wood parenchyma diffuse and unilaterally aliform in *Dendrobangia* and *Emmotum;* loosely aggregated into narrow concentric bands in *Mappia;* diffuse in *Oecopetalum;* reticulate in the others; strands often long and composed of many cells; disjunctive cells common. Wood fibers usually not in definite radial rows; walls thick to exceedingly thick and gelatinous; lumen very small in *Discophora, Mappia,* and *Ottoschulzia;* pits typically numerous, large, and conspicuously bordered, but fewer, smaller, and with small or no border in *Discophora* and *Mappia.* Ripple marks absent. No gum ducts seen.

Calatola, with three or four rather poorly differentiated species of small to medium-sized evergreen trees, occurs from southern Mexico through Central America to Venezuela and Peru. Standley says (*Field Mus. Bot. Series* 22: 1: 39; Jan. 26, 1940) that he proposed the genus "with considerable mental reservation, there being some doubt as to whether it had been referred to the proper family, but . . . study of the wood, particularly, indicates that *Calatola* is definitely referable to the Icacinaceae, where flower structure had seemed to necessitate its reference." *C. mollis* Standl. is known only from Mexico, and attains a height of about 65 feet at Pueblo. *C. laevigata* Standl., described originally from Mexico, is now known also from British Honduras, Guatemala, and the Atlantic coast of Honduras. *C. costaricensis* Standl. is a slender tree 25 to 40 feet high in Costa Rica and Panama. The Colombian and Venezuelan species is *C. venezuelana* Pitt., though Standley (*loc. cit.*, p. 40) doubts "whether there are any sure characters for separating *C. venezuelana* from *C. costaricensis*." Specimens from a small tree from Shapajilla, Department of Huánuco, Peru, "agree well with Venezuelan ones and, at least for the present, the Peruvian tree may be referred to *C. venezuelana*. The genus now may confidently be expected to grow in Ecuador, and perhaps also in eastern Bolivia."

Williams says (*Tropical Woods* 56: 7–8) that the Venezuelan tree "is easy to recognize by its curious fruits, which often are abundant on the ground and are noticed immediately because of the sharp crests suggesting walnuts in appearance. . . Pittier states that the nuts of *C. costaricensis* have a sweet and agreeable flavor but later leave a bitter sensation and produce violent pains, nausea, and vomiting, although the roasted seeds have no harmful effects and are much esteemed by the people living on the slopes of the volcanoes Barba and Poas. According to Standley they are also ground into coarse meal from which is made a kind of tortilla, the flavor resembling those prepared with grated cheese. The seeds of *C. mollis* are said to have vomitive-purgative properties."

Wood cream-colored throughout when freshly cut, but soon changing to gray and finally to purplish blue, suggesting *Genipa* though eventually fading upon exposure to sunlight. Luster medium. Odor and taste absent or not distinctive. Not very hard and heavy; sp. gr. (air-dry) 0.74; weight 46 lbs. per cu. ft.; texture medium, uniform; grain somewhat irregular; very easily worked; durability low. Of no commercial possibilities.

COMMON NAMES: Calatola, calatolazno, palo de tinto (Mex.); duraznilla, erepe, palo azul, p. de papa (C.R.); haguey (Pan.); venenito (Col.); orosul (Venez.).

Dendrobangia *boliviana* Rusby, the single species, is a medium-sized to large tree, sometimes 100 feet high, known to occur in Bolivia, the Guianas, and the estuary of the Amazon River (see *Lloydia* 2: 3: 193). According to Ducke (*Archiv. Jard. Bot. de Janeiro* 3: 207), the timber is of good quality and is exported from the region of Breves, Brazil, under the name of Pau de Cubiú. Heartwood grayish brown with prominent ray flecks; not very clearly demarcated from the sapwood. Luster medium; somewhat silvery in proper light. Moderately heavy, hard, tough, and strong; texture rather coarse; grain straight; easy to work, finishing smoothly and attractively; probably low in resistance to decay.

Discophora, with three species of small trees, is sparingly distributed in Panama, the Guianas, and Brazil. The wood apparently is not utilized. Color light grayish brown throughout, with distinct ray markings on radial surface. Luster medium. Odorless and tasteless. Moderately heavy and hard; texture medium; grain straight; working properties good; subject to stain. Of no commercial possibilities.

Emmotum, with seven species, occurs from the hinterlands of Venezuela to central Brazil. *E. argenteum* Gleason is a slender shrub found at an elevation of 4800 feet on Mount Duida, Venezuela. *E. fagifolium* Desv. is a common tree in the eastern part of the State of Pará, where it is known as Marachimbé or Muirachimbé, but the only recorded use of the timber is for fuel. The only wood specimen at hand (Yale 33816; Ducke 289) is from *E. holosericeum* Ducke, a rather large tree discovered near Manáos. Heartwood chestnut brown, with Oak-like ray markings; sapwood yellowish. Fairly lustrous. Odorless and tasteless. Very heavy, hard, and strong; texture medium; grain straight; not very difficult to work, but tending to split upon nailing; durability probably fair. Presumably of no commercial possibilities.

Mappia, with seven species of shrubs and little trees, mostly of tropical Asia, has two representatives in the New World. *M. mexicana* Rob. & Greenm. is a shrub in southern Mexico; its wood has not been studied. *M. racemosa* Jacq. is a tree, occasionally 25 to 35 feet high, in the West Indies and Central America. It is usually crooked and bushy and apparently does not supply any timber of value. Wood pale yellow throughout; rays somewhat lighter than background, producing distinct but not conspicuous markings on radial surface. Luster rather high. Odorless and tasteless. Moderately heavy, hard, and strong; texture medium; grain straight; working properties good; probably low in resistance to decay.

COMMON NAME: Palo de caña (Cuba).

Metteniusa, with three closely related and doubtfully distinct species of small to large trees, is of infrequent occurrence in northern Colombia and Venezuela, eastern Ecuador, and northeastern Peru. *M. edulis* Karsten, a tree about 25 feet high, was collected at an elevation of over 6000 feet near Santa Marta, Colombia. Karsten thought that it should be made the type of a new family, but Engler referred it to the Icacinaceae. *M. nucifera* (Pittier) Sleumer is a tree 25 to 40 feet tall growing at an elevation of about 3000 feet in the Distrito Federal, Venezuela. Pittier made it the type of a new genus, *Aveledoa,* and referred it to the tribe Opilieae of the Olacaceae (see *Bol. cient. y techn. Mus. Com. Venez.* 1: 45–47; 1925). *M. Tessmanniana* Sleumer was discovered at low elevation in eastern Peru (Marañón) by G. Tessmann who says that it varies in size from a little tree less than four inches in diameter to a large one 80 feet tall with a trunk 16 inches in diameter and free of branches for over 60 feet. (See *Notizbl. Bot. Gart. Berlin-Dahlem* 12: 112: 148–150; 13: 118: 359–361.) In a letter dated March 24, 1939, Dr. Sleumer said that he had identified as *M. Tessmanniana* a specimen (No. 2672) collected by Hertha Schultze-Rhonhof at an elevation of about 3000 feet in the region of the Río Pastaza, Ecuador. The tree was reputed to attain a height of 100 feet and to supply a hard timber of local utility. A wood sample, supposedly from the same source as the herbarium material, has proved to be of an entirely different family (Proteaceae).

The following description is based on one sample (Yale 456; Whitford 46) from the trunk of a small tree of *Metteniusa,* probably *M. edulis,* collected by H. N. Whitford near Puerto Cruz, Venezuela. Study was also made of pieces of twigs from herbarium sheets of *M. nucifera* (A. Jahn 1202) and *M. Tessmanniana* (G. Tessmann 4235). Insofar as the observed anatomical differences are concerned, all three specimens might have come from the same tree. In the senior author's opinion this genus rightly belongs with the Icacinaceae and not with the Olacaceae or Opiliaceae.

Wood yellowish, with a slight pinkish hue, throughout specimen; ray flecks prominent on radial surface in proper light. Luster rather golden. Odorless and tasteless. Moderately heavy, hard, and strong; texture medium, uniform; grain straight; not difficult to work, finishing smoothly and attractively; probably poorly resistant to decay. Suitable for furniture but not likely to be commercially important because of the scarcity or inaccessibility of merchantable sizes.

COMMON NAMES: Canchi (Col.); macagua, urupagua (Venez.); canelo amarillo, pilche (Ec.).

Oecopetalum, with three very similar species, occurs in the mountains in the State of Vera Cruz, Mexico, and southward into Central America. The following description is based on a specimen (Yale 35269; Skutch 2080) from the type of *O. guatemalense,* a tree sometimes 60 feet high, with a slender trunk 12 inches in diameter (see *Journ. Arnold Arboretum* 2: 4: 483). Heartwood yellowish brown; sapwood somewhat lighter; ray flecks suggesting *Platanus.* Luster medium. Odorless and tasteless. Moderately hard and heavy; texture rather fine; grain straight; working properties good; durability probably low. Presumably of no commercial possibilities.

Ottoschulzia, with three closely related species of small trees, is limited in its distribution to the West Indies. *O. rhodoxylon* Urban is a little-known endemic tree of Puerto Rico where the timber is used occasionally for making fancy articles of turnery. *O. cubensis* (C. Wr.) Urban grows in Cuba and is sometimes utilized for railway crossties; the name Rayo del Sol is given to it in Oriente Province in allusion to the figure produced by the rays on a cross section of the stem. *O. domingensis* Urban, of the Dominican Republic, has supplied a few logs for export to New York under the name of Palomino. The woods are practically identical and are suitable for brush backs, marquetry, small cabinet work, and turnery.

Color yellow or pale yellowish brown, with little contrast between heartwood and sapwood; ray markings conspicuous, though without decided color contrast with the fiber background. Luster medium. Odorless and tasteless. Moderately hard and heavy; sp. gr. (air-dry) 0.80 to 0.85; weight 50 to 53 lbs. per cu. ft.; texture medium; grain fairly straight; has high moisture content when fresh, but dries readily without serious checking and warping; brittle when dry and inclined to chip out when being worked.

COMMON NAMES: Cocote del toro, cogote del toro, frutón, rayo del sol (Cuba); palo de rosa (P.R.); palomino (Dom. R.).

Poraqueiba. There are three or four species in the Amazon basin, all small to medium-sized trees. The only one represented in the Yale collections is *P. sericea* Tul.; its timber is used to a limited extent in Brazil for carpentry and general construction. Wood light yellowish brown, with an orange hue, throughout; ray markings distinct. Luster rather low. Odorless and tasteless. Hard and moderately heavy; texture medium; grain straight; working properties good. Presumably of no commercial possibilities.

COMMON NAMES: Poraquebé (Fr. G.); mary, m. gordo, umary, u. amarello (Braz.); umarí, u. amarillo, u. negro (Peru).

Villaresia (or *Citronella*). About 20 species have been described, half of them native to the Indo-Malayan region and Australia, the others inhabiting southern Brazil, Paraguay, Argentina, and Chile. The principal timber tree is the so-called Maple, *V. Moorei* F. v. M. of New South Wales, but the use is local. In South America the plants are small trees or shrubs and their value is in their leaves, which are employed as a substitute for the true Mate or Paraguay tea, the produce of *Ilex,* a genus to which *Villaresia* seems to be fairly closely related. (See *Contrib. Gray Herb.* 142: 60–92. 1942.)

Wood whitish throughout; ray markings prominent. Odor and taste absent or not distinctive. Hard and moderately heavy; texture medium to rather fine; grain more or less irregular; easily worked; durability presumably low. Trees too small to yield commercial timber.

COMMON NAMES: Congonha, c. verdadeira, gongonha (Braz.); caá-guazú, caá-rá, congoña, mate, tarumá del pantano (Urug.); mate, yapón, yerva de palos (Par.); caá-rá, caona, congonha, mborevícaá, palo de anta, taruma del pantano, yerba (Arg.); guilli-patagua, naranjillo (Chile).

JUGLANDACEAE

THE Walnut family is composed of six genera, namely, *Alfaroa, Carya* or *Hicoria, Engelhardtia* (incl. *Oreomunnea*), *Juglans, Platycarya,* and *Pterocarya,* with about 100 described species of medium-sized to very large trees and a few shrubs widely distributed in the north temperate zone and in some of the mountainous regions in the tropics of both hemispheres. The leaves are mostly alternate, unequally pinnate, aromatic, deciduous, and without stipules; the flowers are monoecious, the male in axillary aments, the female in erect terminal spikes; the fruit is drupaceous, the husk dehiscent or indehiscent.

The best known genera are *Carya* (Hickory) and *Juglans* (Walnut), both of high commercial importance for timber and edible nuts. *Platycarya* and *Pterocarya* are small genera limited to eastern Asia, and most of the species of *Engelhardtia* are Indo-Malayan. In America there are many kinds of Walnut (southern Canada to Argentina) and Hickory (southeastern Canada to northeastern Mexico), one rare species of *Alfaroa* (Costa Rica), and two rare species comprising the *Oreomunnea* section of *Engelhardtia* (southern Mexico and Costa Rica). The only member of the family of any commercial importance in tropical America is Walnut, and the timber, owing to scarcity or inaccessibility, is rarely exported.

Heartwood light to dark brown, sometimes attractively streaked or variegated; often sharply demarcated from the white sapwood. Luster mostly high. Odor and taste distinctive but not pronounced in dry material of *Juglans;* absent or not distinctive in the others. Light and soft to hard, heavy, tough, and strong; sp. gr. (air-dry) 0.35 to 0.85, certain species of *Carya* being the densest; texture medium; grain generally straight; working properties excellent.

Growth rings generally distinct, frequently conspicuous. Pores greatly variable in size, especially in ring-porous woods (*e.g., Carya*), the larger ones in most species readily visible without lens; not very numerous; in diffuse-porous woods and in late wood of others, occurring singly and in small multiples, well distributed, though often with tendency to echelon arrangement. Vessels with simple perforations; scalariform perforation plates with few to several bars also present in *Alfaroa* and *Engelhardtia;* spiral thickenings absent in American genera (present in small vessels in *Platycarya*); tyloses often present; intervascular pits small, alternate. Rays nearly homogeneous to decidedly heterogeneous, varying within the genus; 1 to 7 cells wide, frequently mostly uniseriate or biseriate; few to 30, occasionally 50, infrequently up to 70, cells high; crystals common or rare; pits to vessels all small and rounded or in part small and in part large and gash-like (*Alfaroa, Engelhardtia*). Wood parenchyma diffuse and in numerous, fine, continuous or broken concentric lines or narrow bands 1 to 3 pore-widths apart, usually distinct with lens; large crystals sometimes present. Wood fibers with thin to thick walls; pits very small, not very numerous, bordered. Ripple marks and gum ducts absent. (See *Tropical Woods* 12: 16–21.)

Alfaroa *costaricensis* Standl., the only species, is a small or medium-sized tree endemic to Costa Rica, where it is known as Gaulín. It is said to be abundant in the wet mountains south of Cartego, especially at El Muñeco. The leaves are mostly opposite and have numerous narrow leaflets, handsomely colored when young with pink and dull red. The plants sometimes blossom when they are mere shrubs; the flowers are small, green, and arranged in terminal spikes. The fruits are not at all like those of *Engelhardtia,* but resemble Hickory nuts, while the aspect of the mature tree suggests *Juglans*. The only specimen of the wood available (Yale 9837) was collected by Paul C. Standley who discovered and named the tree; it is from a young stem without heartwood. Its structure indicates a closer relationship to *Engelhardtia* than to *Juglans.*

Carya (or *Hicoria*) is a very important genus with one species in southern China and 16 species and 20 varieties in eastern North America, the range extending from the valley of the St. Lawrence River throughout the eastern half of the United States into the highlands of Mexico. Selected varieties of *Carya pecan* Engl. & Graebn. are extensively cultivated in the southern United States for their nuts, the pecans of commerce. Hickory nuts are mostly from *C. ovata* K. Koch, called Shagbark or Shellbark Hickory because the bark separates into thick strips attached at the middle but free and more or less reflexed at the ends, giving the trunk a shaggy appearance. The largest nut, sometimes called Kingnut, is produced by the Big Shellbark, *C. laciniosa* Schn., which is sometimes 120 feet tall with a long trunk 24 to 36 inches through. The Mockernut Hickory, *C. tomentosa* (Lam.) Nutt. (= *C. alba* K. Koch), derives its common name from the fact that the amount of kernel in the nut is very small in proportion to the thick shell. Bitternut or Pignut Hickory, *C. cordiformis* K. Koch, has a small round thin-shelled nut with very bitter meat. This tree has a very extensive range and is the common species in Iowa, Nebraska, and Kansas, though at its best in the lower Ohio valley. The only species endemic in Mexico is *Carya mexicana* Engelm., a tree 50 to 60 feet high in the mountains of San Luis Potosí and Querétaro. Two others, both large trees of the Pecan group, are *C. myristicaeformis* Nutt. and *C. olivaeformis* Nutt. The timbers of all three are used locally to some extent, but are not exported.

Forest-grown Hickory, particularly from lowlands in the southern states, usually lacks the high degree of toughness and resilience required by the trade. Hickory timber most in demand is sapwood from second-growth trees, especially from the central and northern states, but density and soundness, rather than species or position in the tree, are the principal factors determining strength. The principal uses are handles of axes and hammers where resistance to impact or shock is very important; implements and machinery; spokes for carriage and automobile wheels; poles and shafts of horse-drawn vehicles; pump rods; shafts of golf clubs; also many other purposes where great strength, high elasticity, straight grain, and ease of working are requisites. Hickory is the standard for fuel wood and for smoking meat. Common defects are so-called birdpeck and rust streaks; the sapwood is liable to injury from powder-post beetles.

Heartwood brown to reddish brown, more or less variegated or streaked; sharply demarcated from the thick white sapwood. Luster medium. Odor and taste not distinctive. Sp. gr. (air-dry) 0.70 to 0.90; weight 44 to 56 lbs. per cu. ft.; texture rather coarse; grain usually straight, sometimes wavy or otherwise irregular; requires careful seasoning to prevent splitting; working properties excellent; resistance to decay is poor.

COMMON NAMES: Hickory (bitter, bitternut, blasted, bullnut, hognut, mockernut, nutmeg, pecan, pignut, scalybark, shagbark, shellbark, swamp), pecanier, p. sauvage, walnut (U.S.A.); nogal morado, n. de nuez chica, napacoma, nogalillo, pecanero, quauhcacoatl (Mex.).

PLATE XXII. Black Walnut tree (*Juglans nigra*) on the campus of Wabash College, Crawfordsville, Indiana. It was sold for $650 and utilized for gunstocks.

Photo by L. S. Hohenkirk.

PLATE XXIII. Interior of Greenheart (*Ocotea Rodiaei*) forest in British Guiana.

Engelhardtia. There are two distinct sections: *Pterilema,* with 15 species in southeastern Asia and Malaya, and *Oreomunnea* (sometimes considered a distinct genus), with two American species: *E. mexicana* Standl., a medium-sized tree in the mountains of Chiapas, Mexico, and *E. pterocarpa* (Oerst.) Standl., a very large tree apparently limited to the wet forests of Reventazón valley, between Naranjo and Tucurrique on the Atlantic slope of Costa Rica, where it is called Gavilán. (See *Tropical Woods* 12: 12–15.) The heartwood of the latter (Yale 10796) is pinkish brown streaked with olive.

Juglans, with about 50 named species, is indigenous to China, Japan, India, Persia, United States, Mexico, Central America, West Indies, and the Andean region of South America from Venezuela and Colombia to Argentina. The natural range of the Persian Walnut (*J. regia* L.) has been greatly extended through planting for its nuts, commonly known in the American markets as English walnuts; the lumber is usually designated with reference to its origin, as Circassian, Turkish, Italian, and French Walnut, and is rated as one of the world's finest cabinet woods. Some handsomely figured timber, called Claro Walnut, was obtained from Chico, California, in 1934, and is said to have grown from nuts from Spain planted by General John Bidwell about 1850.

There are five or six species of *Juglans* native to the United States. The Butternut or White Walnut, *J. cinerea* L., is widely distributed from southern New Brunswick and the St. Lawrence valley throughout most of the eastern half of the United States, occurring scattered in the hardwood forests where it sometimes attains a height of 100 feet with a tall straight trunk 24 to 36 inches in diameter; open-grown trees are low branched with a wide-spreading crown. The heartwood is light chestnut brown with darker zones, the sapwood thin and white. It is light and soft, weighing 25 to 28 lbs. per cu. ft., and is used for furniture, interior trim, and general carpentry. The supply is limited and diminishing.

Black Walnut, *Juglans nigra* L., has much the same range as Butternut and is one of the largest and most valuable trees in the United States, being at its best in fertile soil in the central hardwood region (Plate XXII). It attains a maximum height of 150 feet with a straight trunk six feet in diameter and free of branches for 60 feet. Nearly all of the forest-grown trees have been felled for lumber or in clearing for agriculture, but a continuous supply of younger timber is maintained through cultivation on farms, as the species grows rapidly, makes good shade, has few natural enemies, and yields edible nuts. From the time of the early colonists Black Walnut has been one of the foremost American woods for furniture, cabinet work, and gunstocks. It possesses excellent technical properties, but for certain uses it is not considered as desirable as European or Circassian Walnut because of its uniformly dark color which tends to deepen with age. Beautifully figured wood is obtained from burls and stumps.

There are three or four western species, notably *Juglans rupestris* Engelm., a small or medium-sized tree from Texas to Arizona, and *J. californica* S. Wats., usually low but frequently thick-boled trees of southern and central California. There are two varieties that some botanists consider worthy of specific rank. About 50 thousand board feet of Walnut was exported from Oregon to Europe in 1936, but it was obtained from trees of *Juglans nigra* planted for ornamental purposes by early settlers in the Willamette valley.

Woods of *Juglans* exhibit considerable variation in color and density, the two extremes being illustrated by American Black Walnut and Butternut. The woods of the West Indian species are of the latter type, while the others resemble Black Walnut in appearance, although some of them are rather light and soft. *J. jamaicensis* C. DC. occurs at high elevations in Puerto Rico and certain other of the West Indian islands and is sometimes 80 feet high and 24 inches in diameter; *J. insularis* Gris. has a limited distribution in Cuba; both are of only local importance. There are several species in Mexico, some of which are said to attain rather large dimensions on favorable sites, particularly on the lower slopes of the western Sierras, but the Walnut timber entering the trade from those localities is in small logs, presumably because of the difficulty of extracting the larger sizes. Walnut grows naturally in the Cordilleras of Colombia, and one undetermined species yields a wood of beautiful satiny luster and rich brown color, more or less variegated, having the general properties of Spanish Cedar (*Cedrela*), for which reason it is commonly known as Cedro Negro and the tree sometimes as Cedro Nogal. A similar wood, also known as Cedro Nogal, was obtained in Copáre, Honduras (Yale 300).

In Ecuador, according to M. Acosta Solís (*Tropical Woods* 57: 2), Walnut (Nogal) occurs scattered or in stands on the subandean spurs of both cordilleras where the climate is mild. It is exploited in Santo Domingo de los Colorados, the Corazón, Imbabura, Montañas de Maldonado, Baños, and elsewhere. The trees are 35 to 60 feet high with a trunk diameter of 36 to 60 inches. There are two classes of wood, dark-colored (café obscura) and light brown (café clara); the former is preferred for cabinet work and fine furniture.

According to some botanists there are three species of *Juglans* in Peru, but J. Francis Macbride (*Flora of Peru* 2: 264) feels "that differences noted represent only individual variations or races. . . . Practically, there is one acceptable name for the Walnut of Peru (*sens. lat.*) *Juglans neotropica* Diels." Llewelyn Williams (*Tropical Woods* 27: 16) says that it is the principal as well as the most useful tree growing in the vicinity of Chachapoyas, and adds: "It grows in sandy or dry medium loam at altitudes varying between 6000 and 8000 feet, mostly in ravines and valleys which are below the general elevation of the country. The lumber is highly prized by the natives for furniture, cabinet-making, and also for musical instruments. A decoction of the leaves, fruit, and bark is employed for dyeing purposes. Many of the trees encountered were 50 to 60 feet tall, with a spreading crown and a straight cylindrical trunk up to 36 inches in diameter, sometimes free from branches for half the height of the tree but more frequently divided 10 or 15 feet above the ground. The bark on the smaller limbs is smooth and gray, while that on the trunk and older branches is moderately thick, brown, and somewhat fissured. Although irregularly scattered, there is a fair abundance of large trees in this territory, principally along the banks of the river Urcu-bamba, an affluent of the Marañón; also at Bagua, two and a half days' journey from Chachapoyas. Due to the remoteness of the territory and physiographic obstacles, exploitation of this valuable wood is greatly hampered. One method of extraction is to raft the logs to Iquitos, but the journey is hazardous as a series of turbulent rapids, such as the Pongo de Manseriche, must be negotiated in the River Marañón. Another means is to transport small lots on pack mules over the western Andean range to Celendin, a six days' journey, and afterwards by road and railroad to Pacasmayo, the nearest port on the Pacific Coast. The entire distance from Chachapoyas to Pacasmayo is approximately 210 miles."

According to Georges H. Barrel (*Tropical Woods* 10: 51), Walnut is found on the western Andean slopes bordering the upper reaches of the Ucayali River and in the Chanchamayo valley. Along the Pichis trail, the principal overland route between Iquitos and Lima, he found many tall trees with straight boles more than three feet in diameter. "No less than 30 trees were counted along the trail within a few hundred yards. Their distribution is very erratic. On some slopes Walnut is decidedly abundant, on others rare, on still others totally wanting. . . . Although, as a whole, Walnut trees are abundant, logging them seems impossible in view of the topography of that remote land. While the vast net of streams furrowing through these immense forests is part of the big Amazon drainage, yet hundreds of miles of rapids would have to be negotiated before reaching rivers of sufficient depth and subdued turbulence to permit rafting. The Pacific Ocean, it is true, is only 300 miles away, but access to it is impeded by that most formidable of barriers, the Cordillera of the Andes, soaring to 20,000 feet and more."

The Argentine species, *Juglans australis* Gris., is found in considerable quantities in

Tucumán, Salta, Jujúy, and Santiago del Estero. The trees are mostly short, generally under 60 feet, but with thick trunks sometimes exceeding three feet in diameter. The wood is lighter in weight than Black Walnut, but otherwise closely resembles it and has the same uses. The bark and the sawdust are used locally as a source of dyestuff for giving a peculiar brown or coffee color to cloth.

Heartwood chestnut- to chocolate-brown, sometimes with a purplish tinge; light-colored wood often with blackish streaks; sharply demarcated from the white sapwood. Luster high, often golden. Odor and taste mild, but usually distinctive. Density greatly variable; sp. gr. (air-dry) 0.35 to 0.75; weight 22 to 47 lbs. per cu. ft.; texture rather coarse, fairly uniform; grain usually straight, sometimes wavy or curly; working properties excellent; durability of dark-colored heartwood high.

COMMON NAMES: *Juglans cinerea:* Butternut, buttnut, oil nut, white walnut (U.S.A.); *J. nigra:* Black walnut, gunwood, walnut (U.S.A.). *Other species:* Walnut (Eng., gen.); nogal (Span., gen.); nogal del país (Cuba); palo de nuez (P.R., Dom. R.); nogal silvestre, nuez meca, ttzatchu (Mex.); cedro nogal (Hond., Col.); nogal de Caracas (Venez.); nogal blanco, n. negro, tocte (Peru); nogal cayuré or cayurí, n. cimarrón, n. criollo, n. silvestre (Arg.).

JULIANIACEAE

THIS small tropical American family, which might well be included in the Anacardiaceae, contains only two genera, namely, *Orthopterygium,* with a single Peruvian species, and *Juliania* (or *Amphipterygium*), with four Mexican species. They are all resinous shrubs or small trees with alternate, deciduous, odd-pinnate (rarely simple) leaves having opposite, crenulate or dentate leaflets; the small flowers are dioecious, the male in axillary panicles, the female, consisting only of a pistil, crowded on a receptacle; the fruit is hard, indehiscent, and one-seeded and has a flat wing-like fruiting pedicel.

Juliania *adstringens* Schlecht., the only one of the four species that is represented in the Yale collections, is a tree, sometimes 25 feet high, growing in south-central Mexico. It is of some commercial value for its bark, which is a source of tannin, dyestuff, and medicinal decoctions. The following description applies to a specimen (Yale 35270) collected by Mrs. Gordon C. Abbott for Harvard University.

Heartwood pale olive-brown, more or less striped; sapwood lighter. Luster medium. Without distinctive odor or taste when dry. Rather light in weight, but firm; texture somewhat coarse, uniform; grain straight; easy to work, finishing smoothly; wood of old stems probably durable.

Growth rings poorly defined. Pores small to medium-sized (120μ); few to numerous, usually 15 to 25 per sq. mm.; mostly in multiples of 2 to 8, occasionally in little clusters, well distributed. Vessels with simple perforations; without spiral thickenings; thin-walled tyloses abundant; pits medium-sized, alternate to subopposite, hexagonal to elongated, the apertures narrow, sometimes coalescent. Rays 1 to 6, mostly 4 or 5, cells wide and up to 30, often less than 20, cells high; more or less distinctly heterogeneous; uniseriate rays composed of square and upright cells; multiseriates with definite stratum of procumbent cells and narrow margin of squarish or short-upright cells; outermost marginal cells often peaked, sometimes triangular; crystals frequent in squarish cells; pits to vessels medium-sized and oval to large and gash-like or much elongated. Wood parenchyma very sparingly vasicentric. Wood fibers septate; walls rather thin; pits simple or indistinctly bordered. Ripple marks absent. Radial resin ducts present in some of the rays, the epithelial layer composed of very small cells; vertical ducts absent from wood but occurring in the pith and cortex.

COMMON NAMES: Cuauchalalá, cuauchalote, matixeran, volador (Mex.).

Orthopterygium, with one species, *O. huaucui* (Gray) Hemsl. (= *Juliania huaucui* Gray), is a shrub or small tree growing on hillsides in southwestern Peru. According to one collector (MacLean), it is "seldom seen with leaves, and always black as if burned or blasted," and is known to the natives of Lima as Huaucui. Macbride

(*Flora of Peru* 1: 2: 266) says that the vernacular name has been improperly recorded and should be Huancui. The only sample available for this study is from a herbarium specimen. The structure is not essentially different from that of similar material of *Juliania*.

COMMON NAMES: Huanarpu, huancui, huaucui (Peru).

KOEBERLINIACEAE

Koeberlinia *spinosa* Zucc., the only species in the family, is an intricately branched, apparently leafless, spiny shrub or a tree sometimes 25 feet high with a trunk rarely a foot in diameter, occurring in Mexico from northwestern Sonora to Tamaulipas, and in the United States from western Texas to southern Arizona, often forming impenetrable thickets over wide areas; it is also present in a limited area of the dry Chaco of Bolivia. The leaves are alternate, minute, and early deciduous; the little flowers are in short umbel-like racemes below the spine tip of the branches; the fruit is a small black 2-celled berry with thin and succulent flesh. The outer bark is thin and flaky, the inner laminated and containing a viscous reddish brown resin. The wood is little used, even for fuel as it burns with a sputtering flame and a sooty, disagreeable smoke.

Heartwood variegated orange-brown to nearly black, resinous; sharply demarcated from the thin yellowish white sapwood. Luster rather low because of the oily nature of the wood. Without distinctive odor and taste when dry. Heartwood very hard and heavy; sp. gr. (air-dry) about 1.20; weight 75 lbs. per cu. ft.; texture rather fine; grain fairly straight to irregular; readily worked, finishing very smoothly and taking a high natural polish; appears very durable. Is suitable for small articles of turnery and carving and for inlay work.

Growth rings present. Often ring-porous, the larger pores in early wood in a single row, sometimes visible without lens in old stems; pores in late wood minute, often scarcely distinct with lens, usually solitary, well distributed. Vessels with simple perforations; spirals present in part; pitting very fine, alternate; resinous deposits abundant in heartwood. Rays 1 to 7 cells wide and few to 100 cells high; homogeneous; pits to vessels very small. Wood parenchyma diffuse or in broken, uniseriate, tangential rows, not distinct with lens. Wood fibers with very numerous distinctly bordered pits; spiral thickenings present but not always distinct. Tendency to ripple marks noted in one sample. True gum ducts absent, but radial channels were observed in one specimen. (For further details see *Tropical Woods* 8: 15.)

COMMON NAMES: All-thorn, crucifixion bush (English); abrojo, corona de Cristo, juanco, junco (Spanish).

LACISTEMACEAE

THIS unimportant family consists of two closely related genera, *Lacistema* and *Lozania*, and about 20 species of shrubs and small trees sparsely distributed throughout most of tropical America. The leaves are alternate, simple, and without stipules; the very small flowers are in short to long racemes or short-branched panicles which are clustered or solitary in the leaf axils; the fruit is a 3-valved dehiscent capsule, usually with a single seed. The woods of the two genera are very similar and apparently are not utilized.

Heartwood pale brown or dull orange; not sharply demarcated from the lighter-colored sapwood. Luster rather silky in proper light. Odorless and tasteless. Of rather low to medium density, firm and tough; some specimens have about the consistency of Willow (*Salix*); texture fine and uniform; grain straight; saws woolly when fresh, but is easy to cut and can be finished smoothly; durability presumably poor. An ordinary wood of no commercial possibilities.

Growth rings absent or poorly defined. Pores small, not visible without lens; numerous; arranged in short to long radial multiples in which the pores are all of about equal size and squarish in section. Vessels with scalariform perforation plates having numerous, narrow, closely spaced bars; no spiral thickenings observed; pitting fine, typically opposite. Rays numerous; 1 or 2, sometimes 3 or 4, cells wide and frequently very high as a result of vertical fusions; scarcely distinct with lens on cross section but fairly prominent on radial surface; decidedly heterogeneous, with definite strata of procumbent cells and uniseriate margins or interspersed strata of square to tall upright cells; perforated ray cells (by-pass members of a vessel) common (see *Tropical Woods* 36: 75); crystals frequent; ray-vessel pitting fine, sometimes unilaterally compound. Wood parenchyma diffuse and in irregular metatracheal lines, barely visible with lens. Wood fibers long; arranged in definite radial rows; walls thick; pits rather few, simple or indistinctly bordered. Ripple marks absent. No gum ducts seen.

Lacistema. The best known and most widely distributed species is *L. aggregatum* (Berg) Rusby, a slender tree sometimes 35 feet tall, with a smooth thin-barked stem 8 to 10 inches in diameter. It occurs in Jamaica, southern Mexico, Central America, and in South America to Brazil, Paraguay, Argentina, and northeastern Peru.

Common names: Guayparín (Mex.); palo mulato (Guat.); cera vegetal, palo mulato (Hond.); maíz cocido (Col.); huacupú-rana, nena, palo metahuayo, trompohuiayo (Peru).

Lozania. Five species are now recognized, ranging from Panama to Venezuela and northeastern Peru. They are typically little trees less than 25 feet high, but according to Mansfeld (*Notizbl. Bot. Gart. Berlin-Dahlem* 10: 98: 860), *Lozania Mutisiana* R. & S. attains a height of 60 feet in the mountainous forests near Santa Marta, Colombia. *L. pedicellata* (Standl.) L. B. Smith occurs in Panama. No common names are known for any of the species of this genus.

LACTORIDACEAE

Lactoris *fernandeziana* Phil., the only genus and species in this family, is a much-branched shrub indigenous to Juan Fernandez Island. The stems are jointed; the leaves are small, simple, alternate, entire, and punctate with numerous minute pellucid dots; the large membranous stipules are situated between the petioles; the flowers are small and axillary; the fruit is a beaked follicle containing 4 to 6 seeds having copious oily endosperm. The plant has no known utility but it is of scientific interest, particularly because of its supposed relationship to the Winteraceae. The following description is based largely on Robert P. McLaughlin's study of small stems (*Tropical Woods* 34: 27–28).

Pores small; not very numerous (up to 18 per sq. mm.); solitary and in radial multiples. Vessels with simple perforations; without spirals; vascular pits medium-sized and rounded to irregularly elongated; alternate to opposite arrangement. Medullary rays 2 or 3 cells wide; heterogeneous. (Rays absent from very small twig examined by the authors.) Wood parenchyma diffuse. Wood fibers with very numerous irregularly distributed pits having small borders and slit-like apertures. Ripple marks absent. No gum ducts seen.

LAURACEAE

This family consists of about 40 genera and 1000 species of trees and shrubs, widely distributed over the tropical and subtropical regions of the world, with a few representatives in the temperate zones. The leaves are simple, typically alternate, sometimes opposite, rarely lobed (*e.g., Sassafras*); the flowers are generally small and fragrant and borne in axillary panicles; the fruit is a one-seeded berry or drupe, with an enlarged and persistent flower tube surrounding the base, frequently suggesting an acorn (*Quercus*).

The best known members of this family in the United States are the Sassafras, *Sassafras albidum* (Nutt.) Nees, and the Spice Bush, *Lindera Benzoin* (L.) Meissn. Many species are rich in aromatic substances and are the source of familiar products such as camphor (*Camphora*) and cinnamon (*Cinnamonum*). The common European Laurel, *Laurus nobilis* L., is grown in many countries for shade and ornament; the Alligator Pear-tree or Avocado, *Persea americana* Mill., is extensively cultivated in tropical and sub-tropical countries for its fruit.

The woods of all of the trees are suitable for industrial purposes, but comparatively few are known to commerce. The principal uses are for carpentry, general construction, and ship-building, but some of them are valued because of their fragrant essential oils, attractive color and luster, great strength, or other special characteristics. The Chinese Nanmu tree, whose scented wood is highly prized for making caskets, is *Phoebe nanmu* (Oliv.) Gamble. The shavings of Pau Hoi, *Machilus* sp., are rich in mucilage and the watery extract is used by Chinese women to bandoline their hair. The South African Stinkwood or Cape Laurel, *Ocotea bullata* E. Mey., is considered the finest of native timbers for furniture. The Bois de Rose of French Guiana, *Aniba rosaeodora* Ducke, is the source of an essence used in the perfume industry. Demerara Greenheart, *Ocotea Rodiaei* (Rob. Schomb.) Mez, noted for its great strength and resistance to decay and marine borers, has been long and favorably known in the European markets. Some of the timbers, *e.g., Mezilaurus,* have possibilities as substitutes for Teak (*Tectona*). There are numerous other kinds with commercial possibilities, but many of the trees, even of different genera, look so much alike in the forest that selection of a particular sort for market is at present very difficult.

Taxonomists are not in agreement as to the bases for classification, and all of the older species have several synonyms, usually involving generic transfers. Flowers are generally necessary for determination, and as the two sexes are often borne on different individuals, the collection of complete herbarium material to accompany wood samples is frequently impossible. The wood anatomist is thereby seriously handicapped in dealing with closely related genera. The woods of the family as a whole exhibit much variation in appearance and properties, but they have a fairly consistent structural complex, which, in combination with other features, makes the recognition of a lauraceous specimen comparatively simple. Within the family it is often easy to identify a species, but most of the generic concepts are very hazy.

The following description applies particularly to American species of 17 genera, namely, *Aiouea, Anaueria, Aniba, Beilschmiedia, Cryptocarya, Dicypellium, Endlicheria, Licaria, Lindera, Mezilaurus, Nectandra, Ocotea, Persea, Phoebe, Pleurothyrium, Sassafras,* and *Umbellularia.*

Heartwood in various shades of yellow, olive, brown, and reddish brown to nearly black; greenish tinge common; light colors tend to become brown on exposure; sapwood whitish, yellowish, or greenish, often thick and not sharply demarcated. Luster typically high and satiny, silvery or golden; some specimens oily and superficially dull. Scent and taste often highly distinctive, being or suggesting camphor, cloves, cinnamon, anis, sassafras, or cedar. Light and soft to very hard and heavy, but mostly medium; sp. gr. (air-dry) 0.50 to 1.23; texture rather fine to coarse, commonly medium; grain mostly straight, sometimes roey or otherwise irregular; technical properties usually excellent; durability low to very high.

Growth rings generally present; some woods ring-porous (*e.g., Sassafras*). Pores small to large, typically medium-sized and fairly numerous; solitary and in small multiples, less often in clusters; mostly well distributed without pattern, sometimes in diagonal arrangement and ring-porous. Vessels with simple perforations exclusively or in part; scalariform plates with few to several bars fairly common; spiral thickenings apparently absent; tyloses common, often abundant, sometimes sclerotic in certain species of *Aniba, Licaria,* and *Ocotea;* pits rounded in outline, not crowded, alternate, mostly large, occasionally medium-sized (*e.g., Umbellularia*), rarely small (*Lindera*). Rays 1 to 5, mostly 2 or 3, cells wide and up to 60, commonly less than 30, cells high; distinctly heterogeneous to nearly homogeneous; oil cells common; small crystals rather to very frequent; pits to vessels typically large to very large, rounded to much elongated and in scalariform arrangement, but sometimes medium-sized to small. Wood parenchyma generally coarse-celled, sparse to abundant; paratracheal (often not completely surrounding pore), sometimes short aliform and connecting adjacent pores diagonally, occasionally in distinct but irregular tangential or concentric bands; oil cells often present, frequently very abundant. Wood fibers generally septate; walls thin to very thick and gelatinous; pits simple or with vestigial borders, typically very small. Ripple marks absent in American species (see *Cryptocarya*). No gum ducts seen. For anatomy of the different genera see *Tropical Woods* 69: 10–33.

Aiouea (or *Ajouea*), with about 25 species of trees, shrubs, and woody climbers, is rather widely distributed in tropical America from southeastern Brazil and northern Peru to Trinidad and Costa Rica, but apparently is nowhere abundant. The trees are rarely over 50 feet tall, but may have stout trunks occasionally up to 36 inches in diameter. The timber is used locally for construction and furniture.

Heartwood absent from samples, probably reddish brown, judging from knots; sapwood brownish. Luster rather high. Odorless and tasteless. Rather light but firm; texture medium; grain fairly straight. Very easy to work.

COMMON NAMES: Ira colorado, quizarrá (C.R.); laurier canelle, l. mango, longleaf pulcherro (Trin.); boradiea, wild calabash, w. currant (Br. G.); ajouvé (Fr. G.); ajuba, amajouva, anhauiná, canella, c. anhuiba, itaúba, louro (Braz.); ciruelillo, moena, m. colorado, yacu-moena (Peru).

Anaueria *brasiliensis* Kosterm., the only species, is a large forest tree discovered in 1931 by Adolpho Ducke on non-inundated land near São Paulo de Olivença, Amazonas, Brazil, where it is known as Anauerá.

Heartwood brownish, darkening upon exposure; rather sharply demarcated from the lighter-colored sapwood. Luster fairly high. With mild spicy resinous odor and taste. Moderately heavy, hard, tough, and strong; texture coarse; grain straight; saws woolly when fresh, but finishes smoothly when dry; not difficult to season, holds its place well when manufactured; durability probably fairly high. A useful wood for general construction, but presumably scarce.

Aniba, with about 55 species of trees and shrubs, has its center of distribution in the Guianas and the Brazilian Amazon region. There are few if any representatives on the North American continent and only two occur in the West Indies north of Trinidad. The useful products include medicinal seeds, fragrant bark, essential oil, and some lumber for furniture and building purposes.

The woods are typically yellowish with a greenish hue when fresh, becoming brown or olive on exposure, and range from rather light to moderately heavy. The one outstanding exception is *Aniba canelilla* (H.B.K.) Mez, an Amazonian tree sometimes over 100 feet tall, with a very dense, dark brown or blackish olive wood similar in structure and properties to *Licaria canella* (Meissn.) Kosterm. The timber is noted for its strength and durability, and the cinnamon-scented bark is used in powder form for perfuming linen and sometimes for making tea. The species is occasionally cultivated.

The best known timber accredited to this genus is the Comino or Laurel Comino of Colombia, *Aniba perutilis* Hemsl. This species, which ranges through the Andes to Bolivia, attains a height of 100 feet, though often it is considerably smaller. The wood has a satiny luster, is moderately dense, has excellent technical properties, and is esteemed locally for high-grade furniture, interior trim, and durable construction. Some of the trees produce attractively mottled wood, called Comino Crespo, which is in demand for veneers for cabinet work and is exported in limited quantities. (According to Jesús M. Duque, *Rev. Acad. Col.* 4: 14: 231, this tree is not *Aniba* but an undetermined species of *Ocotea,* near *O. pretiosa* Nees.)

One of the most valuable woods of French Guiana is the Bois de Rose Femelle, *Aniba rosaeodora* Ducke, which was used by furniture makers in France for more than a century before the species was classified and named. The lustrous yellow wood becomes brownish and is not particularly attractive in appearance, but it contains a fragrant volatile oil that made it desirable for chests and drawers in which linen and clothing were stored. In 1875 a Frenchman by the name of Samain succeeded in distilling the oil, known at first as "huile de linaloès" or "huile d'aloès" and later as "essence de bois de rose," with the result that the perfume industry eventually monopolized the timber supply. The distilling is done near the source of the timber to avoid loss of oil in transport. Billets are reduced to chips and placed in retorts of about 1000-liter capacity and the steam distillation process requires about 1½ hours. The distillate, which has a specific gravity of 0.86 to 0.88, is decanted from the surface of the water and shipped in tightly sealed copper-galvanized iron cans. On exposure to the air the essence soon loses its clarity and sweet fragrance, becoming yellow and syrupy and acquiring a turpentine odor. (See "Rosenhout, bois de rose femelle, uit Suriname," by J. W. Gonggrijp in *De Indische Mercuur,* April 23 and 30, 1920.)

Aniba rosaeodora is a tree as much as 100 feet tall with a straight cylindrical bole sometimes 36 inches in diameter. Though best known in French Guiana it also occurs in eastern Surinam and the lower Amazon region of Brazil, where it is now exploited to some extent for the oil, and possibly in Colombia and northeastern Peru. Ducke, who described the species, recognized a geographical variety, *amazonica,* which Kostermans considers a distinct species, *A. Duckei* Kosterm. The woods of the two are very similar in structure and properties and have the same taste and scent. The Pau Rosa or Louro Rosa of Santarem and Faro is *A. parviflora* (Meissn.) Mez, a little tree; the scent of the wood is mild and pleasant, but distinct from the others. The Pau Rosa of the Amazon estuary is *A. terminalis* Ducke, a medium-sized tree; its heartwood, which is said to be dark brown, aromatic, and highly durable, is used for construction but not for distillation.

COMMON NAMES: *Aniba canelilla:* Canelilla, guarimán (Venez.); arabaima, wabaima, waibaima (Br. G.); amapáima, canella, casca de Maranhão, c. preciosa, louro preciosa, pau preciosa, pereiorá, preciosa (Braz.). *A. rosaeodora:* Eclit rozenhout (Sur.); bois de rose, b. de r. femelle (Fr. G.); pau rosa (Braz.); Cayenne or Guiana linaloe wood (Eng., trade). *Other species:* Canelilla (P.R.); bois jaune, laurier falaise (Mart.); laurier canelle (Trin.); aceite de palo, caparrapí, comino, c. crespo, c. liso, laurel comino, picodocaspe (Col.); laurel capuchino (Venez.); keriti, silverballi—bastard, gale, ginger gale, longleaved yellow, and yellow (Br. G.); apisie-ie, banba apisie-ie, kanoaballi, kurallaballi, mannetjes rozenhout, pisie, tamuna wawae, topoprenwewe, valsch rozenhout, waikarra pisie, w. sipiropipo (Sur.); ayou-cut, bois de cèdre, b. de rose mäle, graine koumarou, muscadier de Pará (Fr. G.); ishpingo chico, moena amarilla, m. negra, sinchi-caspi (Peru); coto, c. fino, c. piquiante (Boliv.); aiúba, aniúba, ayuba, canella sassafras, cedro amarello, louro, l. abacate, l. amarello, l. puxury, l. rosa, macaca-poronga, pau rosa (Braz.).

Beilschmiedia (including *Hufelandia* and *Nesodaphne*), with numerous species of trees and shrubs, occurs in the tropics of both hemispheres and extends into the south temperate zone. The Australian and New Zealand species belong to the *Nesodaphne* group; some of them supply commercial timber.

There are 15 American species with a combined range extending from the West Indies and southern Mexico to southern Brazil and Chile. The best known is *Beilschmiedia pendula* (Sw.) Benth., a small to large tree sometimes up to 100 feet high and 36 inches in diameter, supplying some timber for local construction. Closely related to it is *B. mexicana* (Mez) Kosterm., a large tree ranging from Mexico to Colombia. *B. anay* (Blake) Kosterm. of Guatemala and Colombia has a fruit resembling the avocado (*Persea*). *B. sulcata* (R. & P.) Kosterm. extends from the highlands of Costa Rica to northeastern Peru, and is said to attain a maximum height of 130 feet. The only species in British Guiana is *B. curviramea* (Meissn.) Kosterm. The genus is not represented in the Amazon region of Brazil, but there are five species between Rio de Janeiro and São Paulo. There are two Chilean species: *Beilschmiedia Berteroana* (Gay) Kosterm., is a tree 40 to 65 feet high, known as Ulmo, a name also applied to *Eucryphia cordifolia* Cav.; and the Bellota, *B. Miersii* (Gay) Kosterm., a tree occasionally 80 feet tall, supplying some timber for general construction and shipbuilding. The only authentic wood samples available are of *Beilschmiedia pendula,* which is of general distribution in the West Indies.

Heartwood not seen; said to be yellowish brown to blackish; sapwood pale brownish. Luster medium. Moderately heavy, hard, and strong; texture rather fine, uniform; grain straight; has good working properties; said to be highly durable. Of local utility, but of no possibilities for export.

Common names: *Beilschmiedia pendula:* Slogwood, slugwood (Jam.); aceitunillo, aguacatillo, cabeza de toro, carne de doncella, curabara, curavara, mulato, m. de la sierra (Cuba); aguacate cimarrón, cedro macho, laurel, palo colorado (P.R.); carrasqueno (Dom. R.); bois noir, olive (Haiti); bois doux muscadier, b. de vert, b. negrésse, Isabelle blanc, laurier fourmi, l. madame (Fr. W.I.). *Other species:* Anay (Guat.); quizarrá (C.R.); aguacatillo cordillero, laurel canime, i. c. catentano (Col.); lana, l. balli (Br. G.); canella tapinha (Braz.); tashango colorado, ushun-moena (Peru); bellota, ulmo (Chile).

Cryptocarya, with many species of trees and shrubs, is a widely distributed genus having its center in the Indo-Malayan region. There are about seven American species with a combined range from French Guiana to southern Brazil and Chile. One of the best known is *C. moschata* Nees & Mart. of southeastern Brazil. It varies from a shrub to a tree 50 feet high, with a stout trunk 20 to 50 inches in diameter. The timber is useful for general construction, but chief interest is in the fruits which resemble nutmegs in appearance and pungent flavor.

The only wood sample at hand (Yale 5552) is of the Chilean Peumo, *Cryptocarya rubra* (Mol.) Skeels, a small to medium-sized tree with stiff branchlets and small, rigid, alternate to opposite leaves. The fruit is completely included in the reddish, thin and brittle enlarged flower tube. The tree occurs in the Andes Mountains from Santiago to Valdivia. The wood is used locally to a limited extent for interior construction and charcoal, and the bark is a minor source of tannin.

Color pale brownish throughout sample. Luster medium. Odorless and tasteless when dry. Of medium density hard, and strong; texture rather fine, uniform; grain straight; easy to work, finishing very smoothly. Of no commercial possibilities.

Common names: Cèdre canelle, c. jaune de marécage (Fr. G.); anhuvinha branca, caá-xió, caha-xió, caia-xió, cajaty, canella abacate, c. batalha, c. branca, c. cega, c. de brejo, c. de papagaio, c. lageana, c. sebosa, louro abacate (Braz.); peumo (Chile).

Dicypellium *caryophyllatum* Nees, the only species, is a small tree of the central Amazon region of Brazil and ranging northward into French Guiana. The bark has a strong flavor intermediate between cinnamon and cloves. It is esteemed in native medicine and cookery and the distillate is used by the perfume industry. The species does not supply any timber of importance. The name Cayenne Rosewood is sometimes applied to it, but the real Rosewood of French Guiana is *Aniba rosaeodora* Ducke.

Heartwood yellowish brown to dark olive; not sharply demarcated from the pale greenish yellow sapwood. Luster silky. Scent and taste spicy resinous. Moderately dense, hard, and strong; sp. gr. (air-dry) 0.78; weight 49 lbs. per. cu. ft.; texture medium fine; grain straight; has excellent technical properties; durability high. Without possibilities for export because of the small size of the trees.

Common names: Bois crabe, canelle giroflée (Fr. G.); cravo de Maranhão, c. do matto, canella falsa, ibirapitai, imira quinha, louro cravo, muiraquyia, pau cravo (Braz.).

Endlicheria, with about 40 species of shrubs and small, medium-sized, or rarely large trees, is widely distributed throughout tropical South America; there is one species, *B. Browniana* Mez, in Panama, and one, *E. sericea* Nees, extending from Bolivia into the Lesser Antilles. The last is said to attain a maximum height of 115 feet and to supply good timber for local construction and furniture. Another species with a wide range is *E. anomala* Nees, a low-branched tree rarely 50 feet tall, growing from British Guiana to Rio de Janeiro and northeastern Peru. *E. Williamsii* O. C. Schmidt is a medium-sized tree of common occurrence in lowlands of the Peruvian Amazon region; according to Williams (*Woods of northeastern Peru,* p. 151), the timber is sometimes employed in making canoes, furniture, and crating. *E. multiflora* (Miq.) Mez is a tree of moderate size in the Guianas and Amazonas, Brazil; the single wood specimen of this name (Yale 9460; Persaud 49), which was described by Kribs (*Tropical Woods* 13: 22), differs so much from the others of the genus that its identification is considered very doubtful.

Heartwood not sharply demarcated from the grayish sapwood. Luster rather silky. Without distinctive scent or taste when dry. Of rather low to moderate density; texture medium; grain straight to variable; easy to work; durability doubtful. Apparently of no commercial importance.

COMMON NAMES: Bois doux blanc, b. d. grand-feuille, b. montagne, cayali, cayari, laurier caillé, l. canelle, l. vache (Fr. W.I.); bastard silverballi, burhuda, serowa, water silverballi (Br. G.); apisie-ie, pisie, siroeaballi, sopo hudu, woko mapiri (Sur.); canela, c. moena, hioma cocuir-ey, ishpingo foetida, i. rufo, isma-moena, moena blanca, m. del agua, oberillo, pampa-moena (Peru); laurel blanco (Boliv.); cafeira do matto, canella branca, c. cernuta, c. ceroba, c. cheirosa, c. de folha miuda, c. de papagaio, c. guajaba, c. jacua, c. paluda, c. preta, louro (Braz.).

Licaria, or *Acrodiclidium* (including *Chenekia* and *Misanteca*), with about forty species of trees and shrubs, is distributed throughout tropical America. Judging from the wood samples available, the genus is in need of further revision.

The northernmost species is *Licaria triandra* (Sw.) Kosterm., a small to medium-sized tree of southern Florida and the West Indies. The greenish yellow wood is of medium density, very easy to work, and employed locally to a minor extent for general carpentry and interior construction.

There are about 10 species in Mexico and Central America, but the only one represented in the Yale collections is *Licaria campechiana* (Standl.) Kosterm. (previously included by Standley and others in *Ocotea, Chanekia, Misanteca,* or *Phoebe*). It is a tree sometimes 80 feet high and 18 inches in diameter. The wood is lustrous reddish brown, rather fine-textured, hard, heavy, and strong. The timber appears suited for general construction, but apparently it is not available in sufficient quantity to be commercially valuable.

There are many species in South America, but there is scant information concerning them. *Licaria armeniaca* (Nees) Kosterm. is a shrub or a tree up to 50 feet in height growing from Rio de Janeiro through the Amazon basin to northeastern Peru. The heartwood is brown to chocolate, sometimes streaked with black, the sapwood greenish yellow; of medium weight to rather heavy. No special uses are recorded. *L. limbosa* (R. & P.) Kosterm. is a small to medium-sized tree with a range extending from Bolivia and Peru to Venezuela and Costa Rica. The heartwood is probably brown (judging from knots), the sapwood grayish with a slight greenish hue; density rather low. The Aritú or Louro Aritú of the lower Amazon region, *L. Appelii* (Mez) Kosterm., occurs also in Matto Grosso and Minas Geraes, Brazil. The tree attains rather large size and yields a lustrous brown, more or less streaked, hard, strong, and durable timber suitable for furniture and heavy construction. The wood is similar to that of *L. polyphylla* (Nees) Kosterm., a large Amazonian tree known as Louro Chumbo. *Acrodiclidium caryophyllatum* Ducke, a medium-sized tree of the upper Rio Negro, where it is called Puchury, has a lustrous brown, more or less striped, spicy-scented wood of medium density and excellent working qualities, which would make attractive furniture.

The Kaneelhart of Surinam is said to be *Licaria canella* (Meissn.) Kosterm., and Yale specimens of the wood agree with Pfeiffer's description (*De houtsoorten van Suriname*, p. 176). It appears to be identical with the Waibaima of British Guiana. The heartwood is a dark brown with a tinge of red or violet, and very dense, sp. gr. (air-dry) up to 1.15; it is further characterized by sclerotic tyloses and parenchyma cells. The structure is very similar to that of the Amazonian Casca Preciosa, which is said to be *Aniba canelilla* (H.B.K.) Mez, but the heartwood lacks the distinctive violet hue, being more of an olive-brown. The wood takes a high polish, is very strong, and is noted for its resistance to decay.

Common names: *Licaria triandra:* Sweetwood (Jam.); guajané, laurel blanco, lebisa, levita, leviza (Cuba); cigua gorrita, c. prieta (Dom. R.); laurier jaune (Haiti). *L. canella* and *L. cayennensis:* Brown silverballi, wabaima, waibaima, waibama (Br. G.); ajoewa, apisie-ie, atjarie kjanari, a. tjanarie, badagos, banda, kaneelhart, kaneel-pisie, kaneerjoe, kanerie hoedoe, nagre hoedoe, siroeaballi tataroe, s. tjanaria, wajaaka (Sur.); bois canelle, cèdre flibustier (Fr. G.). *Other species:* Canel, canela, canelillo (P.R.); bois chique, bois colique, b. fourmi, b. modong, b. piant, laurier gris (Fr. W.I.); laurel, l. de la sierra, palo misanteco (Mex.); aguacatillo (Hond.); copalchi, dzol, ektit, zootní (Guat.); quizarrá (C.R.); akawai nutmeg, brown silverballi, camacou, cambara de cheiro, waccawai, yellow silverballi (Br. G.); apisie-ie, atjarie kanarie, bosoho apisie-ie, joroha pomoire, jorokan pomoire, kaneel-pisie, magre, nagrehoedoe, pisie, siroeaballi, s. karowatsiamaro, a. ojotak, topoprin wewe (Sur.); bois de licari, licari kanali, sassafras (Fr. G.); cucherín, efuina quirirafuina, moena, m. colorado, muena, puchery (Peru); louro preto (Boliv.); aritú, cambará de cheiro, canella, folha de ouro, f. dourada, itaúba camará, louro amarillo, l. aritú, l. chumbo, l. de casca preta, l. preto, mahuba, picherí, puchery, puchury, puxiri, puxirim (Braz.).

Lindera, with numerous species of shrubs and trees, is well represented in eastern Asia and the East Indies, but in America by only one or two species of aromatic shrubs. The Spice Bush, *Lindera Benzoin* (L.) Meissn., is common throughout most of the hardwood region of eastern United States. Its only use now is for ornamental planting, but pioneers in medicine prescribed tea made from the twigs, and during the Revolutionary War the fruit served as a substitute for allspice.

Common names: Allspice bush, Benjamin bush, spice bush (U.S.A.).

Mezilaurus (or *Silvia*), with eight species of small to very large trees, has its center of distribution in the eastern and northern part of the Amazon basin, but is lacking in the western regions. The timber is valuable for durable construction and shipbuilding.

The only important extra-Amazonian species is the Tapinhoan, *Mezilaurus navalium* (Fr. Allem.) Taub. (= *Silvia navalium* Fr. Allem.), a tree usually medium-sized but sometimes up to 80 feet high and 36 inches in diameter, growing in the subtropical mountain forests of the State of Rio de Janeiro. The timber is of excellent quality and is used locally for furniture making, boat-building, and all kinds of durable construction. The bark is a source of tannin.

The Amazon species are known generally as Itaúba, but the principal source of the timber of that name is *Mezilaurus itauba* (Meissn.) Taub. (= *S. itauba* [Meissn.] Pax), a large tree, sometimes over 100 feet tall and 30 inches in diameter, ranging from the Guianas and eastern Amazonas to northwestern Matto Grosso. According to Ducke (*Tropical Woods* 42: 19), it furnishes the most useful timber of the lower Amazon, especially for naval construction.

The other species are as follows: *Mezilaurus crossiramea* (Meissn.) Taub., a small tree, sometimes shrubby, in Goyaz, Brazil; *M. decurrens* (Ducke) Kosterm., a large tree on non-inundated lands of the upper Rio Negro; *M. Lindaviana* Schw. & Mez, a small to medium-sized tree on uplands in the lower Amazon, the upper Rio Branco, and on steep, rocky slopes along the Rupununi River, British Guiana; *M. Sprucei* (Meissn.) Taub., a small tree along the Rio Negro; *M. subcordata* (Ducke) Kosterm., a small tree in upland forest near Pará; and *M. synandra* (Mez) Kosterm., a medium-sized tree in the drier upland forests around Manáos. The following description is based upon samples of five species; the principal differences are in density and texture, *Mezilaurus itauba, M. navalium,* and *M. synandra* being the heaviest and finest.

Heartwood brownish yellow, becoming russet upon exposure; has oily appearance and feel; distinct but not sharply demarcated from the grayish sapwood. Luster generally low. Scent and taste mild and not distinctive in dry specimens. Moderately to decidedly hard and heavy; sp. gr. (air-dry) 0.75 to 1.00; weight 45 to 62 lbs. per cu. ft.; texture rather fine to coarse (*M. decurrens*); grain straight to somewhat roey; easy to work, seasons readily, holds its place well when manufactured; is highly durable. Has many of the properties of Teak (*Tectona*).

COMMON NAMES: Rukut?, tanwa-ye? (Br. G.); ajomtonto horoadihoro, a. diamoro, apisie-ie, kaneelhout, kaneerjoe, kaneeri-pisie, kjarie kjanarie, siroeaballi tataro (Sur.); itaúba, i. abacate, i. amarella, i. preta, i. verdadeira, tapinhoa, tapinhoam, tapinhoan, t. amarillo, t. olho de sapo (Braz.).

Nectandra, with about 100 species of trees and shrubs, is widely distributed throughout tropical America, most abundantly in South America. All of the trees produce timber of good quality for carpentry and general construction, but they are only of local importance and so far as known to the authors do not enter the export trade. The woods are so much like some of the species of *Ocotea, Aniba,* and *Phoebe* that it is not now possible to separate them. Taxonomists appear to have equal difficulty with herbarium specimens, particularly if flowers are lacking. Under these circumstances no attempt is made to describe individual species.

Heartwood greenish yellow to dark olive-brown, the color deepening upon exposure; in some species (if correctly identified) becoming blackish brown; transition to sapwood gradual except in darkest material. Luster usually silky or silvery. Scent spicy resinous, taste mild to pronounced. Rather light and soft to moderately hard and heavy; sp. gr. (air-dry) mostly between 0.60 and 0.75; weight between 37 and 47 lbs. per cu. ft.; texture medium to somewhat coarse; grain straight to roey; seasons readily without splitting, easy to work, holds its place well when manufactured; darker specimens durable. Suitable for many of the same purposes as Yellow Poplar (*Liriodendron*) and Birch (*Betula lutea* L.).

COMMON NAMES: Capberry, shingle wood, sweetwood, timber sweetwood, whitewood (Jam.); aguacatillo, baulluá, boniato aguacatillo, b. amarillo, b. cigua, b. prieto, cigua (Cuba); laurel, sigua blanco (Dom. R.); laurier blanc, l. grandes feuilles (Haiti); laurier, l. cypre, l. mattack, sweet myers (Trin.); aguacatillo, koyokiche, piesito de paloma (Mex.); laurel, sweetwood, wild pear (Br. H.); aguacatillo (Hond., Nic.); aguacate, amarillo, a. de mico, a. de monte, canelito, chipinahuaca, laurel, laurelillo, palo de chepina huaca, pimiento, tepeaguacate, trompillo, trompito (Salv.); aguacatillo, ira rosa, quina, quizarrá, q. hediondo, q. quina, sigua amarillo (C.R.); sigua, s. blanco, s. canelo, s. negro (Pan.); laurel, l. negro (Col.); angelino aceituno, capuchino, cobalonga, laurel, l. angelino, l. canelo, l. capuchino, l. chirimoyo, l. maestro, l. mangón, l. negro, l. rosado, pucherí, quina de la Colonia (Venez.); bouragie, burajie, mainap, shirua, silverballi (broad-leaved, sawarri-skin, white, yellow), sirua, tallifer (Br. G.); kanoaballi, pisie, p. oema, siroeaballi, waitjara, wajaaka, watjarang (Sur.); cèdre, c. gris, sassafras (Fr. G.); canelo (Ec.); hihuha, huarme tashango, isula micuna, moena, m. amarilla, m. blanca, muena, mundshuy, m. gateado, nomebe fuina, pishcu nahu muena, roble, tihua, tihuha (Peru); batalha, canella, c. amarella, c. branca, c. preta, inamuí, inhamuy, louro, l. amarello, l. de casca preta, l. de varzea, l. inhamuy, l. mamorim, l. mamory, l. nhamuy, l. preto, l. sassafras, mamorí, pau sassafraz (Braz.); ayuy-morotí, canela guaica, guaica, laurel, l. blanco, l. canelo (Arg.).

Ocotea. There are a few species of trees and shrubs in southern and eastern Africa and the Mascarenes and about 200 of wide distribution in tropical America. The most highly valued timbers supplied by the genus are the South African Stinkwood, *O. bullata* E. Mey. (see *Forest trees of the British Empire* 3: 43–50), and Demerara Greenheart, *O. Rodiaei* (Schomb.) Mez, the first prized for fine furniture, the second for heavy and durable construction work. Two other American species with distinctive timbers are *O. caracasana* (Nees) Mez, the Angelino of Venezuela, and *O. rubra* Mez, the Determa, Wane, and Grignon of British, Dutch, and French Guiana, respectively; they are practically unknown to the export trade. All of the woods have good technical qualities, but with a few exceptions their identities are merged with members of *Nectandra* and other genera so that at present it is impossible to deal with them specifically. They belong in a group commonly called Laurier in French, Laurel in Spanish, Silverballi in British Guiana, and Louro in northern, Canella in southeastern, Brazil.

Regarding the Brazilian Canellas, H. M. Curran says: "This group of trees includes a host of species, mostly of *Nectandra* and *Ocotea,* and in the present state of our botanical knowledge of the Brazilian forests it is impossible to specify which ones produce the timbers of commerce. The trees are very common, being most abundant perhaps in the transition zone between the tropical and higher sub-tropical regions. The trees as a rule occur in mixture with other hardwoods and sometimes comprise 50 per cent of the forest. They are generally of fairly good timber form, with diameters up to three feet or more and attaining heights frequently exceeding 100 feet, with clear lengths of 50 feet. The woods produced vary in color from very light to very dark, and in density from very soft to extremely hard. Many of them are highly durable and some are very resistant to insect attack, presumably on account of the presence of some resinous or oily deposits."

The Angelino of Venezuela is confused in the literature with a species of *Homalium,* but the commercial timber called Angelino or Laurel Angelino has been identified by Pittier as *Ocotea caracasana* (Nees) Mez (see *Bol. Cien. y Tech. Mus. Com. de Venez.* 1: 1: 14. 1927). The lustrous yellow to olive-colored wood is moderately heavy and hard; sp. gr. (air-dry) 0.75; weight about 47 lbs. per cu. ft.; without distinctive odor or taste; medium-textured, easy to work, finishes attractively, and is highly durable. It is considered one of the best furniture woods of the country and is also used in construction work.

Determa or Wane, *Ocotea rubra* Mez, which is widely distributed in the Guianas and lower Amazon region, is a tree 100 feet or more in height, with a long, well-formed trunk occasionally 50 inches in diameter. Timbers have been obtained that were 40 feet long and 30 inches square, without sapwood, and spars are available that are 70 to 80 feet long and 14 inches in diameter at the small end. The wood does not look like any of the other Lauraceae. It has a distinctive, rather light reddish brown color, with a subdued golden luster, suggesting some of the Meliaceae. It is rather coarse-textured and the vessel lines often show plainly; quartersawed lumber is sometimes attractively figured, owing to the fine but distinct rays and the presence of roe grain. It is fairly strong and hard, though its density is rather low; sp. gr. (air-dry) 0.55 to 0.65; weight 34 to 41 lbs. per cu. ft.; the working properties are excellent; it is said to be highly resistant to insects, moderately so to decay. It is used locally for furniture, interior construction, sugar boxes, and boat planking; the large trunks make good dug-out canoes.

Greenheart, *Ocotea Rodiaei* (Rob. Schomb.) Mez, is a large evergreen tree 75 to 125, rarely up to 150, feet high, with a straight, cylindrical, non-buttressed trunk, sometimes 40 inches in diameter and free of branches for 50 to 75 feet (Plate XXIII). It is essentially a British Guiana tree, although small stands of it have been discovered along the upper Maratakka River in Surinam. Reports of its occurrence in other countries are without corroboration. According to the Conservator of Forests of British Guiana (*Official Gazette,* Georgetown, July 17, 1926), Greenheart "occurs in commercial quantities in the north central portion of British Guiana, behind the coastlands, and principally in the area drained by the Cuyuni, Essequibo, Demerara, and Berbice Rivers. In these areas it avoids the drier and poorer soils, growing largely on the slopes leading down to the streams. It is also found in the damp ground near the streams, provided the conditions do not approach too closely to true swamp, but in such situations the tree is not so big nor the quality quite so good as on the slopes, nor are the stands per acre anything like so great as a rule. There is a very large quantity of this wood remaining in the Colony, the former workings having been almost all in the vicinity of the navigable streams, and this large quantity awaits development by more modern methods of logging than have been adopted in the past. . . . The area of forest over which Greenheart is known to occur is approximately 20,000 square miles, while strip valuation surveys . . . have disclosed, on an area of 400 square miles, commencing at six miles from a port at which steamers up to 16 feet draught can load, a total stand of 77,000,000 cubic feet of sound mill timber 16 inches and over in diameter at breast height. . . . The triangle of land having Bartica as its apex, the Essequibo and Mazaruni Rivers as its sides, and a line drawn from Tiboko to Potaro Mouth as its base, a total area of 2360 square miles, is estimated to contain above 300,000,000 cubic feet of sound merchantable Greenheart timber. . . . On the average, 19 per cent of the trees are unsound. Hewn logs of shipping specification are obtainable from 10 inches to 25 inches square, caliper measure, squares above 21 inches not being common and generally carrying a somewhat higher price. Log lengths are usually from 30 to 70 feet."

Through Bancroft's *History of the Guianas* the valuable properties of Greenheart first became known to timber merchants in England in 1769. Trade in the wood gradually developed until it became one of the foremost industries of British Guiana. The principal market has been in Europe, particularly the United Kingdom and, in smaller volumes, the Netherlands. There has also been a considerable demand in the West Indies, but almost none in the United States until recently. The exports to the U.S.A. for the years 1936 and 1937 were, respectively, as follows: hewn and round logs, 105,182 and 105,151 cu. ft.; sawn timber, 7501 and 21,541 cu. ft. The total exports during 1937 were 428,801 cu. ft. of logs and hewed timbers valued at $232,050, and 42,229 cu. ft. of sawed lumber worth $38,770; total, 471,030 cu. ft., value $270,820. There is no monopoly of the supply and exports of Greenheart are accompanied by a Government certificate of genuineness.

Greenheart is noted for its strength and durability. (For results of recent tests see *Mechanical properties of certain tropical woods, chiefly from South America*, by William Kynoch and N. A. Norton, Bul. 7, Univ. of Mich. School of Forestry and Conservation, 1938.) The heartwood varies in color from lustrous light to dark olive or blackish, often with intermingling of lighter and darker areas or irregular striping; not sharply defined from the thick, lighter-colored sapwood. It is very hard, heavy, and strong; sp. gr. (air-dry) 1.05 to 1.23; weight 66 to 77 lbs. per cu. ft. The texture is medium fine and uniform, the grain straight to roey. The fresh timber contains from 40 to 50 per cent moisture and requires careful seasoning, though it is less refractory than many other heavy structural timbers because the difference in radial and tangential shrinkage is exceptionally low, a fact which can perhaps be attributed to uniformity of texture and the gelatinous nature of the thick-walled fibers. The principal uses are in marine construction, especially for piling, piers, planking, lock and sluice gates, and heavy timbers in ship-building. Other uses are bench slats, picket fences, truck wheel spokes, board walks, paving blocks, and house posts. It is also employed in the manufacture of fishing rods and is sometimes confused with the Surinam Groenheart (*Tabebuia* sp.).

Large timbers are slow-burning because of their density. The high resistance of the heartwood to decay and insect injury is generally ascribed to the presence of certain alkaloids, but this has not been definitely determined. For a long time Greenheart had the reputation of being immune to marine borers, but experience has shown that it may be destroyed by certain species of teredo infesting brackish water, as in the Panama Canal. The Conservator of Forests of British Guiana (*loc. cit.*) says: "It can fairly be claimed that, whilst no timber in salt water is immune from attacks by teredo, the record of Greenheart is unsurpassed by any other timber, but it must be borne in mind that this applies in the tropics to *salt water only,* not to brackish water, and Greenheart has most undeservedly been given a reputation for failing in America in *salt* water because of a much advertised failure in *fresh and brackish water. . .*. It is seen that the teredos attacking Greenheart are both fresh and brackish water species, and are *not* the species which ordinarily occur in salt water, and that they are all tropical species. They cannot have any effect on Greenheart in salt water in the tropics, nor in water in temperate regions at all."

COMMON NAMES: *Ocotea rubra:* Determa (Br. G.); baaka wana, bewana, demma maata indold, taparin, teleloema, teteroma, tetroema, topoeroe, wana, wane, w. ilsie amain, wonae (Sur.); grignon, g. franc, g. rouge (Fr. G.); louro vermelho, puxuri-rana (Braz.). *O. Rodiaei:* Greenheart—black, brown, Demerara, yellow, etc. (Eng., trade); bebeere, bebeeru, bibira, bibiru, greenheart, rora-ek, sepira, sipiri, tugul, wainop (Br. G.); beeberoe, beberuboom, Demerara groenhart, sipiroe (Sur.). *Other species:* Lancewood (Florida); bastard torch, black candlewood, b. sweetwood, b. torch, lignum-dorum, loblolly sweetwood, spicewood, sweet torchwood, whitewood (B.W.I.); aguacatillo, boniato laurel, canelillo, canelón (Cuba); canela, canelón, granadillo, laurel sassafras, nemoca cimarrón, sassafras (P.R.); laurel, l. blanco (Dom. R.); laurier, l. puant (Haiti); duckwood, laurier, l. citron, l. cypre, l. mattack, l. zaboca, spicy mattack (Trin.); laurel de bejo, l. de chile, pububuk (Mex.); pubukuk (Guat.); canelo, pimiento (Salv.); aguacatillo (Hond.); palo colorado (Nic.); aguacatillo, canelillo, canelo, ira, i. mangle, quizarrá, q. amarillo, q. barceno, q. lantisco, singuo-gro, yaya (C.R.); bambita, sigua, s. canelo (Pan.); aguarras, caparrapí, palo de anís (Col.); canelito, laurel, l. blanco, l. negro, l. pardillo (Venez.); baradan, bouragie, silverballi (brown, white), waramia (Br. G.); ajoeroe, apisie-a-blaak-man, apisie-ie, baaka apiesie-ie, bamba pisie, bastard pisie, beradié, harde pisie, hohoradikoro, ingi siri, jekoeroe, joekoejapi, joro-joro pisie, keretiballí, keretie, koeratarie, koeroekai, krassi pisie, lolie hoedoe, pisie, rozenhout, sabana pisie, salie, sipoeroelan, siroeaballi, s. kheretie diamaro, s. k. wadilidiamaro, s. ojokto, spiroerian, tamoene wajaka, tetero sierwaballi, tokene mania potano wewe, tokkewe, topoporin wewe, waé, wajaaka, warilipipio, wawe eran, wawerjan, zachte pisie (Sur.); ishpingo, maraco-fuina, moena, m. aguarás, m. blanca, m. negra, muena, negrito, pampa muena, roble blanco, sipra-moena, tashango corazón negro (Peru); canella, cujumaryrana, folha de prata, f. prateada, louro, l. branco, l. camphora, l. da varzea, l. de cheiro, l. do igapó, l. pimenta, l. rosa, l. tamancão, l. tamanco, pau rosa (Braz.); ayuy-pará, a.-morotí, guaicá, laurel, l. amarillo, l. mestezo, l. negro, l. overo, l. saiyú (Arg.); laurel (Urug.).

Persea, with about 60 species of shrubs and small to large trees, is sparingly represented in the Far East, abundantly in tropical America, with a few extensions into the temperate zones. There are two species in southern United States, six in the West Indies, about a dozen in Mexico and Central America, two in Chile, the rest in tropical South America.

The only important tree is *Persea americana* Mill., probably native to Mexico and Central America but cultivated for centuries there and in other lands for its pear-shaped table fruit, which is known to Spanish-speaking people generally as Aguacate and anglicized to Avocado or Alligator Pear. It grows to a height of 30 to 60 feet, with a trunk 12 to 18 inches thick. The two principal horticultural types are the West Indian, with smooth, leathery-skinned fruit, and the Guatemalan, Mexican, or highland, with rough and warty, thin-skinned fruit, and leaves exuding an anise-like scent when bruised. The rich, pink or whitish pulp is eaten raw, usually as a salad with some dressing to compensate for the mildness of taste. The seeds are the source of a proprietary oil. The heartwood is pinkish to light reddish brown, not sharply demarcated from the thick, cream-colored or pale brownish sapwood; it is light in weight, 35 to 40 lbs. per cu. ft., of medium to coarse texture, very easy to work, not very durable. The timber has no special uses and is not commercially important.

Persea Schiedeana Nees is a medium-sized to large forest tree growing from Vera Cruz, Mexico, to Panama. The fruit is much like the common Avocado. The rather hard, coarse-textured, unattractive wood is of the general type of the preceding species. It is utilized to a limited extent locally for interior construction, but is not resistant to insects and is of little economic value. Another Central American tree is *P. amplifolia* Mez & D. Sm. A specimen collected by the senior author in northern Honduras has been identified with that species, but the lustrous greenish yellow, light and soft wood bears little resemblance in properties or structure to any other member of the genus so far as known. It would fit better in *Nectandra.*

Most attractive of the woods of this genus is the Ecuadorean Pacarcar, *P. sericea* H.B.K., judging from a specimen (Yale 24099; Rimbach 161) collected by A. Rimbach at an elevation of nearly 10,000 feet in the western Cordillera. He says that it is a forest tree 65 feet high, with a broad flat crown and a stout trunk sometimes 40 inches in diameter. The pale reddish brown wood has a silky golden luster, and is rather hard, medium-textured, and roe-grained. The timber is available in logs 20 feet long and is used locally in joinery and house construction.

The only species of *Persea* supplying commercial timber is *P. lingue* Nees, a Chilean tree called Lingue, Liñe, or Litchi. Its botanical range extends from Coquimbo to Valparaiso and Santiago, but it is most abundant in the Provinces of Malleco, Cautín, and Valdivia, sometimes forming nearly pure stands. Mature trees are said to have an average height of 55 to 60 feet with erect trunks attaining a diameter of 40 inches. The bark is an important local source of tannin. The pale brown heartwood has a golden sheen, is of about the same consistency and uses as Birch (*Betula lutea* L.), is easy to work, except that it is inclined to be knotty, and highly esteemed locally for joinery, furniture, and interior construction.

There are four species of trees and shrubs native to southern United States. The Red Bay, *Persea borbonia* (L.) Spreng., is usually a small to medium-sized tree, sometimes up to 75 feet tall and 36 inches in diameter, growing mostly near streams and in swamps along the Atlantic and Gulf coast regions from southern Delaware to Florida and eastern Texas, thence northward through Louisiana to southern Arkansas. The heartwood is bright red to reddish brown, sometimes with dark streaks, moderately hard, medium-textured, and easy to work. It is used to a minor extent for furniture and interior construction and formerly in ship-building, but the supply of large timber is very limited. The Swamp Bay, *P. palustris* (Raf.) Sarg., is a slender tree rarely 40 feet high and a foot in diameter, inhabiting Pine-barren swamps, often almost to the exclusion of other plants, from the Dismal Swamp in southeastern Virginia to the Everglades of Florida and westward along the Gulf to eastern Louisiana; also in the Bahamas. The wood, which is similar to that of the other species, though less attractively colored, is of no commercial importance.

The following general description of the woods of *Persea* includes several other species not mentioned above. Heartwood brown, reddish, or pinkish, the darkest-colored sharply demarcated from the gray or cream-colored sapwood. Luster medium to high. Odor and taste absent or not distinctive in dry specimens. Rather light to moderately heavy; sp. gr. 0.60 to 0.75; weight 37 to 48 lbs. per cu. ft.; texture medium to coarse; grain straight to irregular; working properties good; durability low to fairly high. Of no export possibilities.

COMMON NAMES: *Persea americana:* Alligator pear, avocado (Eng.); aguacate (Span.); avocatier (Fr.); zoboca (Haiti); aguacatillo, ahuacate, ahuacatl, ahuacaquahuitl, cupanda, koidium, koitum, kuitmkeip, on, pagua, pahuatl, tonalahuate, ttalzan, xinene (Mex.); butter pear, pear (Br. H.); aguacate de anís (Salv.); amó, bukrá, buvó, deborkor, dikóra, hamó, sútuh,

PLATE XXIV. Imbuia trees (*Phoebe porosa*) in a Paraná Pine forest of southern Brazil.

PLATE XXV. Spiny trunk of the Espina Corona (*Gleditsia amorphoides*) on the upper Paraná, Misiones, Argentina.

súutu (C.R.); cura, palto (Col.); advocaat (Sur.); ahuaca, aouacate, bois d'anis, laurier avocat (Fr. G.); palto (Ec.); huirapalta, palta, paltai, palto (Peru); abacate, louro abacate (Braz.). *P. Schiedeana:* Chinini (Mex.); wild pear (Br. H.); chaucte, chucte, coyó, coyocté, kiyau, kiyo, kotyó, shucte (Guat.); chucte (Salv.); yás (C.R.); aguacatillo (Pan.). *Other species:* Bay (red, swamp, sweet, white), Isabella wood, laurel tree (U.S.A.); bay (Bah.); aguacate cimarrón, boniato blanco (Cuba); canela (P.R.); pèche marron (Haiti); aguacate cimarrón, a. oloroso, laurel, l. de la sierra, tepehuacate (Mex.); aguacate de mico, cachimbo (Salv.); aguacatillo (Hond., C.R.); aguacate cimarrón, a. de anís, a. morado, aguacatillo, cuco (Col.); apisie-ie, basra pisie, boeradie, hegron pisie, kiesie-ma, koesapoi, wonojen-panda (Sur.); pacarcar (Ec.); aguacate cimarrón, paltojera, palton, piria (Peru); liñe, lingue, litchi (Chile); canella rosa (Braz.).

Phoebe. There are about 85 species of shrubs and small to very large trees, some of them in the East Indies and Malaya, the majority in tropical America with a combined range extending from the West Indies and southern Mexico southward through Central America and the Andes to Argentina and Brazil, avoiding the Guianas and the Amazon hylaea.

There are four species in the West Indies, the best known being *Phoebe montana* (Sw.) Gris., usually a small tree but sometimes up to 65 feet high. Its light-colored, moderately hard, medium-textured timber is used to a minor extent for interior construction. There are about 25 species in the region from Mexico to Colombia, some of them stately trees of scattered occurrence in the forest and supplying easily worked timber for rafters, siding, ceiling, and flooring of houses. The wood is typically light olive-colored, lustrous, of medium to coarse texture, and only moderately resistant to decay.

The only highly important species is *Phoebe porosa* (Nees & Mart.) Mez of southern Brazil, where it is known as Imbuia or Embuia. It grows in the moist *Araucaria* forest of Paraná and Santa Catharina, mostly at altitudes of 2500 to 4000 feet, and forms about 20 per cent of the stand (Plate XXIV). It attains a maximum height of 130 feet and a trunk diameter of about six feet. Although evergreen, it sheds most of its old leaves as the new ones appear in August to September. On the under side of the leaves in the axils of the median and certain secondary veins are little 2-lipped pockets (domatia) which are inhabited by minute insect parasites. The fruits mature in January and fall to the ground, where they provide mast for swine.

The Najer Lumber Company, Long Island City, N.Y., supplied the following information to the senior author in 1929 (*Tropical Woods* 18: 19): "This firm is distributing approximately 200,000 board feet of Imbuia or Brazilian Walnut annually. The first shipment of importance reached the United States about 1918 and amounted to 100,000 feet, although occasional small lots had come in mixed shipments before that time. In the States of Paraná and Santa Catharina and also in the cities of São Paulo and Santos, Imbuia is the most important wood for high-grade flooring, furniture, interior trim, and fixtures. Exports have been small in the past as the local demand for the lumber has been about equal to the supply. The logs are shipped to mills in the cities where they are sawed by frame saws of small capacity. Lumber is not carried in stock, being supplied direct from the saw to purchasers who select the logs and have them cut to order. An American band mill at Tres Barras, Santa Catharina, operating in Paraná Pine, accumulates about a million feet of Imbuia a year. This lumber is shipped by rail about 200 miles to San Francisco du Sul, a small open harbor, and lightered to vessels beyond the bank in the open sea. Coffee is the only other commodity exported and sailings are very irregular." According to Karl Schmieg (*Tropical Woods* 5: 3), Imbuia can be selected for color to match any kind of Walnut and "is today practically the only wood obtainable in large planks so much needed for heavy carved work, such as table trusses and chair legs."

The following description applies particularly to Imbuia. Heartwood yellowish or olive to chocolate-brown, either plain or beautifully variegated and figured. Luster medium. Has spicy resinous scent and taste, losing most of it in drying. Moderately hard and heavy; sp. gr. (air-dry) 0.70 to 0.76; weight 43 to 47 lbs. per cu. ft.; rather fine-textured; easy to work, taking a high polish, holds its place well when manufactured, and is durable. The fine dust arising in sawing is irritating to some workmen and may cause dermatitis, though tolerance is usually developed. (See *Public health reports* [Washington, D.C.] 46: 33: 1938–1942; Aug. 14, 1931.)

COMMON NAMES: *Phoebe porosa:* Canella imbuia, c. i. clara, c. i. escura, embuia, e. amarella, e. vermelha, embuya, imbuia (Brazil). *Other species:* Boniatillo, boniato blanco, b. bombo, b. del Pinar (Cuba); avispillo, laurel, l. bobo, l. geogeo (P.R.); laurel (Dom. R.); laurier rose (Haiti); bois doux, doux cypre (Guad.); laurier mama z'enfants (Trin.); aguacate cimarrón (Mex.); laurel, timber sweetwood (Br. H.); aguacatillo (Guat.); canelito, pimiento (Salv.); guambo (Hond.); aguacatillo, a. blanco, quizarrá, q. amarillo, sungro (C.R.); sigua blanca (Pan.); laurel higüito (Col.); laurel aguacatillo (Venez.); moena, muena, pishco-moena, pishcu-muena (Peru); ayuy-hú, a. pichaé, laurel crespo, l. de la falda, l. negro (Arg.).

Pleurothyrium, with about a dozen species of small to large trees, occurs in the upper Amazon region of Peru and Brazil. According to Macbride (*Flora of Peru* 2: 928), the genus is "doubtfully separable naturally from *Ocotea.*" The timber is said to be used locally for carpentry and interior construction. Heartwood olive; not sharply demarcated from the greenish yellow sapwood. Luster silky. Odor mildly fragrant; taste not distinctive. Light and soft to moderately so; texture medium; grain straight; very easy to work.

COMMON NAMES: Canella muena, moena (Peru); louro abacate (Brazil).

Sassafras, with a single species, *S. albidum* (Nutt.) Nees, is of general distribution in the eastern half of the United States and southern Ontario, Canada. Throughout much of its range it varies in size from little more than a crooked shrub to a slender tree less than 40 feet tall; at its best on rich well-drained soil it attains a height of 75 to 90 feet and a trunk diameter of two to five feet; the branches are short, stout, and more or less contorted. The leaves are entire, 3-lobed, or mitten-shaped. All parts of the tree are aromatic, and the bark of the roots is the source of the commercial oil of Sassafras used to scent or flavor various products; the small roots in asparagus-like bundles are a common article of local commerce and are used in making Sassafras tea, a beverage with reputed medicinal virtues, particularly as a spring tonic. The center of these industries is in the southern states where the tree often takes possession of abandoned fields and develops a wide-spreading root system. The wood has the general appearance of Ash (*Fraxinus*) or Chestnut (*Castanea*) and is used to a limited extent in inexpensive furniture, boxes, and slack cooperage, and on farms for posts, fence rails, and kindling.

Heartwood pale brown, deepening to dull orange-brown upon exposure; not very sharply demarcated from the thin yellowish white sapwood. Luster medium. With characteristic but mild scent and taste. Light, fairly soft, brittle; sp. gr. (air-dry) about 0.52; weight 33 lbs. per cu. ft.; texture coarse; grain straight; very easy to season and to work, holds its place well when manufactured; is rather highly resistant to insects and decay.

COMMON NAMES: Gumbo file, sassafrac, sassafras (red, white), saxifrax (U.S.A.).

Umbellularia *californica* (H. & A.) Nutt., the only species, is a pungently aromatic tree common along water courses in southwestern Oregon, where it is generally known as Myrtle, and southward through the mountains of western California, where its usual name is Laurel. On rich soils in the Coos River valley of Oregon it forms dense stands in association with Maple (*Acer macrophyllum* Pursh.), sometimes attaining a height of 150 to 175 feet with a straight bole three to six feet in diameter

Open-grown trees usually have a short stout trunk divided near the ground into several coarse, diverging stems forming a broad, round-topped crown. In southern California it is much smaller and at high altitudes becomes shrubby and sometimes forms broad mats of prostrate stems. The shoots lengthen and new leaves appear throughout a long growing season and some of the leaves persist for five or six years. When the leaves and green bark are crushed they give a volatile oil as pungent as camphor. The timber is of excellent quality and is locally important for making furniture. Burls and gnarly trunks are sliced into veneers for cabinet work, generally under the name of Oregon Myrtle. Lumbermen say that the light color of the wood can be darkened by placing the fresh logs in water and leaving them submerged for a long time, producing the so-called Black Myrtle so highly esteemed for furniture and interior trim.

Heartwood yellowish brown or olive, often variegated; not sharply demarcated from the very thick pale brownish sapwood. Luster medium. With mild scent and taste when fresh. Of medium density, hard, and strong; sp. gr. (air-dry) 0.55 to 0.66; weight 34 to 41 lbs. per cu. ft.; texture medium; grain straight to wavy or contorted; very easy to work, finishing very smoothly, holds its place well when manufactured; heartwood durable.

Common names: Oregon myrtle (trade); bay, b. laurel, b. tree, cajeput, California baytree, C. laurel, C. olive, laurel, mountain laurel, pepperwood (Calif.); myrtle (black, white, yellow), myrtle tree (Oregon).

LECYTHIDACEAE

This family, in its broadest sense, comprises 17 genera and about 200 species of small to gigantic tropical trees. Some botanists think that the Old World genera should be considered as a separate family, the Barringtoniaceae. Thus restricted, the Lecythidaceae consist of 10 tropical American genera, namely, *Allantoma, Asteranthos, Bertholletia, Cariniana, Couratari, Couroupita, Eschweilera, Grias, Gustavia,* and *Lecythis.* The family is best represented and the trees attain largest size in the Amazon basin; the northern outposts are the West Indies and British Honduras.

The leaves are simple, alternate, and without stipules; the flowers are usually large and showy; the fruits are very characteristic, being hard and woody, variable in size but often large, and nearly always with an apical opening which, until maturity, is closed by a lid-like structure. They either contain several winged seeds or, more often, few to many nut-like seeds, some of which, particularly the Brazil nuts (*Bertholletia*) and paradise or cream nuts (*Lecythis*), are important articles of commerce. The inner bark is composed of numerous very thin layers for which various important local uses are found, especially the calking of boats.

The timbers have considerable local utility in carpentry, general construction, boxes and crates, but are only slightly known to the export trade. They are not especially attractive in color or grain and some of the denser kinds contain enough silica to interfere with sawing into lumber though probably not with slicing into veneers. The Manbarklak of Surinam (*Eschweilera*) has proved highly resistant to marine borers, but the difficulty of extracting the heavy timber has so far prevented its exploitation on a commercial scale. The Colombian Albarco (*Cariniana*) has been tried in a small way for furniture in France and the United States, but without success. More recently there have been some shipments of Brazil-nut timber (*Bertholletia*) to the United States, but it has to compete with native hardwoods of the general utility class. Eventually, however, the Lecythidaceae should become highly important, for the trees are plentiful, many of them are noted for their very large and well-formed boles, and there is a wide range of choice in the technical properties of the timber from different species and genera.

Color of heartwood variable; yellow in *Grias* and *Gustavia;* yellowish or pinkish brown in *Couratari* and *Couroupita;* golden brown or light reddish brown in *Allantoma* and *Cariniana;* dark brown or reddish in *Asteranthos, Bertholletia, Eschweilera,* and

Lecythis; sapwood sharply demarcated only in woods with deeply colored heartwood. Luster low to medium. Odor and taste usually absent in dry specimens; a rancid or fetid scent characterizes some specimens of *Asteranthos, Couratari, Couroupita,* and *Gustavia.* Density low in *Couroupita;* high to very high in *Bertholletia, Lecythis, Asteranthos,* and *Eschweilera;* medium to rather high in the others; sp. gr. 0.50 to 1.21; weight 31 to 75 lbs. per cu. ft. Texture medium to rather fine; smooth surface of densest material has a slate-like feel; grain fairly straight; lumber mostly without figure except for the fine parenchyma markings on the tangential surface of dark-colored material.

Growth rings usually distinct, mostly because of differences in the spacing of parenchyma layers. Pores variable in size from large to very small; commonly solitary and in small multiples, well distributed without definite pattern; larger pores rather few. Vessels with exclusively simple perforations except in part in *Asteranthos* and *Grias;* spiral thickenings absent; tyloses usually abundant in heartwood; intervascular pitting alternate and rather coarse, the pit borders large and the apertures small and included, except in *Gustavia* where the pitting is exceedingly fine; pits to parenchyma cells bordered. Rays 1 to 6, mostly not over 3, cells wide and generally less than 50, sometimes up to 120, cells high, except in *Grias* and *Gustavia* where they are up to 15 cells wide and occasionally over 200 cells high; homogeneous to heterogeneous, though not conspicuously heterogeneous except in *Asteranthos* where nearly all of the cells are squarish; ray-vessel pit-pairs of two types (except in *Asteranthos* and *Grias*), one similar in appearance to the intervascular (*i.e.,* showing a wide border and small aperture), the other showing a narrow border and wide aperture and often much larger than the intervascular; pit-pairs in *Grias* and *Asteranthos* virtually all of the second type, some of them very large, oval to elongated, irregularly to scalariformly arranged. Wood parenchyma abundantly developed, distinct with lens and sometimes barely visible to conspicuous without it; mostly in regular to wavy concentric metatracheal lines or bands generally 1 or 2, occasionally up to 4, rarely up to 8, cells wide, the spacing uniform to variable, often in the same growth ring, the average being about one pore-width; sometimes finely reticulate; all genera characterized by crystalliferous parenchyma strands, of about the size of the wood fibers, typically at the margins of the concentric bands, the cells cubical and each containing a small solitary rhombohedral crystal of calcium oxalate, imbedded to half or more of its diameter in a thickened tangential wall; disjunctive parenchyma cells common in dense woods. Wood fibers with rather thin to very thick walls, the latter commonly gelatinous; pits small, simple or bordered. Vertical traumatic gum ducts, resembling the gum veins in *Eucalyptus,* found in several specimens of *Eschweilera,* a few of *Lecythis,* and one of *Cariniana.* Ripple marks absent.

In summary, the two principal characteristics of the American genera of the Lecythidaceae concern the crystalliferous parenchyma strands and the concentric arrangement of the metatracheal parenchyma. *Grias* and *Gustavia* are distinct from the others in many ways, such as yellow color, wide rays, and small and numerous pores. *Gustavia* differs from all the others in having minute pores and extremely fine intervascular pitting. *Asteranthos* is distinct in its possession of some many-barred scalariform perforation plates and scalariform ray-vessel pitting. All of the others are much alike in structural plan, though exhibiting considerable variation in details; study of much more material is necessary before reliable generic distinctions are possible. (See *Tropical Woods* 43: 1–15.)

Allantoma, with a few species of medium-sized trees, is apparently limited in distribution to the Amazon region. The best known species is *A. lineata* (Berg) Miers of the Amazon estuary. The leathery leaves are 5 to 11 inches long and 2 to 4 inches wide, the new ones being of a beautiful dark violet color; the white flowers have a truncate, not hooded, androecium and are borne in axillary racemes; the woody fruit, which is shaped like the bowl of a Turkish pipe, is 4- or 5-celled and contains a few oblong, verrucose, wingless seeds. The tree is said to bear flowers and fruit throughout the year, and during the month of June is rendered conspicuous by its brilliantly colored foliage (see *Arch. Jard. Bot. Rio de Janeiro* 4: 155). There is no information available regarding the uses of the timber, which is light reddish brown, of medium density, and not difficult to work, though considered of inferior quality.

COMMON NAMES: Castanheiro da serra, cerú, cherú, churú, tauary (Braz.).

Asteranthos *brasiliensis* Desf., the only species, is a medium-sized tree of the Rio Negro region in the north-central Amazon basin. The twigs are slender; the smooth leathery entire leaves are 2 to 7 inches long and 1 to 2 inches wide; the large, yellow, fragrant, apetalous, gamosepalous flowers are borne singly in rather long pedicels in the axils of the leaves; the fruit appears as a brown 6-ribbed little cone with a flaring leathery collar near the base and contains a bony nutlet. Botanists are not in agreement as to the taxonomic position of this genus and Knuth suggests (*Notizbl. Bot. Gart. Berlin-Dahlem* 11: 110: 1034) that it be segregated into a monotypic family, the Asteranthaceae. Although the wood differs in various anatomical details from the other members of the Lecythidaceae, a close alliance is indicated.

Heartwood dark reddish brown, rather waxy looking; sharply demarcated from the sapwood which, in dry material, is pale brown. Luster medium. Odor slightly rancid; taste not distinctive. Very hard, heavy, tough, and strong; texture medium; grain irregular; difficult to work, but finishing smoothly; appears highly durable. Presumably too rare or inaccessible to be commercially important.

Bertholletia *excelsa* H.B.K., the sole species, is one of the largest trees of northern South America and of prime importance economically in the Brazilian Amazon region, though the value is not in its timber but in its seeds, the Brazil nuts of commerce. The somewhat leathery leaves are 1 to 2 feet long and 2 to 6 inches broad; the large yellow flowers have a hooded androecium and are borne in sparsely branched robust terminal panicles; the fruit is a thick-walled woody spheroid capsule 4 to 6 inches in diameter, tightly packed with 12 to 24 large angular seeds with a hard black shell filled with a rich oily white kernel. The capsule does not open and liberate the seeds as in most of the other genera of this family but falls to the ground intact, and germination, often of several seeds simultaneously; is through the small opercular opening (see *Botanical Gazette* 52: 226). The kernels contain between 60 and 70 per cent of oil, composed of elaeine (74 per cent) and stearine (26 per cent), which can be used as a substitute for olive oil. Most of the seeds are used as confections and form a staple article of trade worth millions of dollars. The timber is sparingly utilized, partly on account of the abundance of equally good or better kinds available, but mostly because the trees are much more valuable for their fruit. Small quantities of lumber have been shipped to the United States as salvage from clearings for rubber plantations in Brazil.

Heartwood rather light pinkish brown, tending to fade on exposure; sharply demarcated from the yellowish brown sapwood. Luster medium to low. Odorless and tasteless. Moderately hard and heavy, tough and strong; texture medium; grain fairly straight; not difficult to work, finishing smoothly; holds its place well; is probably not highly resistant to decay. A general purpose wood for interior work, but lacking attractiveness of grain or color and of limited commercial possibilities.

COMMON NAMES: Brazil-nut tree (Eng.); almendro, castaña del Marañón, olla de mico (Col.); iubia, iuvia, juvia, tucá, turury, ya, yubia, yuvia (Venez.); yuvia (B.G.); Brazilnoot, Braziliaansche noot, ingie noto, kokeleko, Pará noot, tetoka, toeka, totoka (Sur.); castanha verdadeira, castanheiro, c. do Maranhão, c. do Pará, iniá, nha, niá, tocary, toucá, tucary (Braz.).

Cariniana, with 10 or more species of very large trees, occurs from Central Brazil to Venezuela and Colombia. The leaves of different species vary in length from an inch to a foot; the comparatively minute greenish or yellowish white flowers do not have a hooded androecium and are borne in abundance on slender widespread subterminal panicles; the pipe-like woody fruits contain numerous imbricated seeds narrowly winged at one end. The genus is closely related to *Couratari* and *Allantoma.* Some of the timbers are of local importance for general construction but are little known to the export trade.

The northernmost species, *Cariniana pyriformis* Miers, known as Bacú in Venezuela and Albarco in Colombia, has supplied limited amounts of timber for export from the latter country to Europe and the United States for the past 60 years. Selected material with roe grain has an attractive figure when finished and for a time was sold under the misleading name of Colombian Mahogany (see Circ. 185, U.S. Forest Service). Manufacturers objected to it on the ground that it quickly dulled the saws used in working it. This dulling action is attributed to the presence of silica deposits in the parenchyma cells, and experience with other timbers, notably Oriental wood from Australia (*Endiandra Palmerstoni*), has demonstrated that a high silica content does not interfere seriously with the manufacture of sliced or rotary-cut veneers.

According to H. M. Curran, "The 'albarco' is one of the most common trees in the forests of northern Colombia at elevations of from 200 to 2000 feet above the sea, the best stands lying between 500 and 1000 feet. It is confined to the lower slopes and well-watered valleys, and does not extend into the flood plains except occasionally in the better drained areas. Over much of the region lying between the Cauca and Magdalena Rivers it may be considered one of the principal timbers if not the one supplying the bulk of the cut. The bark is of a very fibrous nature and is used by the natives for rough cordage. The wood is known on the local markets, but on account of a plentiful supply of more accessible timbers, it is little used. The claim of certain mill men in the United States that this wood dulls the saws and other tools very quickly has not been borne out in my own experience and it seems likely that there are differences in the quality of the timber in this respect, depending upon the particular locality of growth or other factors not yet determined. Since this timber exists in large quantity where it can be readily logged, efforts to use it should not be abandoned until it is conclusively shown that the alleged difficulties in working are real and cannot be overcome."

Closely related to *Cariniana pyriformis* is *C. micrantha* Ducke of eastern Pará and western Amazonas, where it is known as Tauary, a name commonly applied in that region to species of *Couratari*. *C. decandra* Ducke and *C. integrifolia* Ducke, two other large trees of the same general locality, and *C. Kuhlmannii* Ducke, a medium-sized tree of northwestern Matto Grosso, form a group exhibiting considerable resemblance to *Allantoma*. (For description of the wood of *C. integrifolia* see *Tropical Woods* 31: 27.)

Regarding other Brazilian species, Miers (*Trans. Linn. Soc. London* 30: 285–289) gives the following information: *Cariniana rubra* (Gardn.) Miers is "a large tree, with an inflorescence greatly resembling that of *C. uahupensis*, but with very different leaves and different fruit; the latter is called Cachimbo de Macaca" (monkey's pipe). *C. uahupensis* (Spruce) Miers is "a tree of vast size, with a trunk 100 feet high, five feet in diameter, strengthened at its base by projecting buttresses, and called Charão or Choro by the natives, signifying magnificent." *C. domestica* (Mart.) Miers "is a lofty tree, bearing the name of Jequitibá, known to occur in the Province of Goyaz." *C. estrellensis* (Raddi) Kuntze "is a tree with a trunk of immense diameter and 120 feet high, not infrequent in the Province of Rio de Janeiro, affording a valuable timber and bearing the name of Jequitibá Branca, which is much employed in the campos for sugar cases." *C. legalis* (Mart.) Kuntze of Rio de Janeiro "is a tree of immense size, with very broad, lofty, spreading branches, its trunk producing a timber of much value, useful in works of construction and shipbuilding, which is known by the name of Jequitibá Rosa."

According to H. M. Curran, the Jequitibá trees are the giants of the Brazilian forests, especially in the coast region and São Paulo, though the largest he observed near Bahia were about 125 feet high with well-formed boles three or four feet in diameter and free of branches for from 60 to 80 feet. The trees occur in mixed hardwood forests and are a constant factor over vast areas. "The wood is well known in Brazil and used locally for interior construction, though it is not on the market in quantity. It is light brown, easily worked, and gives promise of becoming one of the most important timbers of South America for general construction and carpentry." The principal species of Jequitibá appears to be *Cariniana legalis*.

The following description is based upon authentic wood samples of five species of *Cariniana*. Heartwood yellowish, pinkish, or reddish brown, sometimes streaked; usually not sharply demarcated from the pale brown sapwood. Luster medium. Without distinctive odor or taste. Rather light to moderately heavy, always firm and tough; sp. gr. (air-dry) 0.50 to 0.70; weight 31 to 43 lbs. per cu. ft.; texture medium; grain straight to roey; easy to cut with a knife, but sometimes difficult to saw; finishes smoothly and holds its place well when manufactured; deeply colored specimens resistant to decay.

Common names: Abarco, albarco, caobono, cobano, coco albarco (Col.); bacú (Venez.); cerú, chorão, choro, chupa, churú, jequitibá, j. amarella, j. branca, j. rosa, j. vermelha, tauary (Braz.).

Couratari, with about a dozen species of medium-sized to large deciduous trees, has its center of distribution in the Guianas, with extensions southward to Rio de Janeiro and westward into Panama. The leaves are 2 to 8 inches long; the flowers are reddish and the hood of the androecium is spirally coiled; the fruit is pipe-like and the seeds are winged all around. Without complete herbarium specimens it is often difficult to separate the three genera *Allantoma, Cariniana,* and *Couratari,* consequently the terminology is still more or less confused.

The following notes are from Miers (*loc. cit.*, 280–284). *Couratari glabra* Camb. is a tree 20 to 30 feet high in the forests of Rio de Janeiro. *C. rufescens* Camb., of the same region, is "a noble tree, with a straight trunk 40 to 60 feet high, with a wide-spreading head." *C. Tauari* Berg of Pará is "a magnificent tree, with a straight trunk strengthened at its base by strong buttresses, and affords a valuable hard timber of a palish color, used in shipbuilding; yields logs 65 feet long and 18 inches square. The name Tauari is also given to its laminated inner bark, consisting of many thin whitish sheets of the substance of paper, well known in commerce, being extensively used for the covering of cigarettes." *C. guianensis* Aubl. of Guiana is "a tree of immense size, growing on the borders of rivers, with a straight trunk 60 feet long and four feet in diameter, affording an excellent timber; its inner bark, composed of several scores of laminae, is applied to many useful purposes."

Couratari panamensis Standl., the only Central American species, is a tree about 100 feet tall with a well-formed trunk three to four feet in diameter above the stout buttresses. It was discovered by G. Proctor Cooper in the Cricamola valley, region of Almirante, Panama, and as the herbarium material is poor and the fruits are likened by Standley to the illustration of those of *C. macrocarpa* Mart., which (according to Ducke) may be a synonym for *Allantoma lineata* (Berg) Miers, there is room for doubt as to the accuracy of the classification. The wood (Yale 12162) is very similar in appearance, properties, and structure to a specimen (Yale 21657; Ford Co. 354) of *C. fagifolia* (Miq.) Eyma from the Brazilian Amazon region.

The following description applies to the two wood samples mentioned, the only ones available that were collected with herbarium specimens. Heartwood absent but said to be reddish; sapwood yellowish with a tinge of pink. Luster rather high. Odor slightly fetid; taste not distinctive. Of medium density but firm, tough, and strong; easy to work, finishing smoothly; durability of heartwood unknown. A good timber for interior construction and carpentry.

Common names: Tabarí, tauarí (Venez.); mari-mori, wadara, w. kakawalli (Br. G.); brakka ingie pipa, djoemoe, ingie pipa, ipipjo, kabba matjauw djoemoe, kalienja tamere, kariodan, kiesiepolloe joelle malledie, oelemali, oelemarie, olemallie or olemellie, oremerie, oremerivadili, powassa djoemoe, sipoeloejoeroe maladi, towekin oremariri, wadala, wadara, wata dje, oe, wataala or watala, zwarte ingie pipa (Sur.); balalaboua, balata blanc, bourrac, caouroubara, couratari, maho cigarre, m. couratari, m. fer (Fr. G.); tauary (Braz.).

Couroupita, with about a dozen species of large deciduous trees, is sparsely distributed from the lower Amazon region northwestward into Nicaragua. The large leaves are more or less wedge-shaped and clustered at the ends of the branchlets; the flowers, which have a hooded androecium and are mostly large, fleshy, brightly colored, and highly fragrant, are borne on short to very long and vine-like peduncles issuing from the trunk or old branches; the fruits look like rusty cannon balls and contain several to many small flat seeds which at maturity are imbedded in a semi-liquid, nauseous pulp. The wood also has a repulsive odor when fresh, but loses most or all of it upon drying; it is of no commercial importance as the quality is not high and the trees are scarce.

Five Central American species have been described, namely, *Couroupita Cutteri* Mort. & Skutch, *C. darienensis* Pittier, *C. nicaraguensis* Pitt., *C. odoratissima* Pitt., and *C. parviflora* Standl., all very tall trees of rare occurrence. *C. darienensis* appears unique in that the flower racemes appear on the branches and branchlets and not on the trunk.

The following notes on South American species are from Miers (*loc. cit.*, pp. 189–194): *Couroupita peruviana* Miers of Peru is "a very large tree about 100 feet high, with a softish wood, known by the name of Aiaúma (*caput mortui*), probably from its large round fetid fruit. . . . The pulp on exposure to the air becomes black and fetid. It is used by the natives to cure skin diseases in animals." *C. odoratissima* Seem. of Panama is "a lofty tree, 60 to 80 feet high, growing in the forest, the trunk bare to the height of 20 feet and bearing many racemes with handsome flowers as large as those of a passion-flower, from a resemblance to which it bears the vernacular name of Granadillo; it is also called Palo de Paraíso. . . . The fragrance of the flowers can be perceived at a distance of half a mile." *C. membranacea* Miers of Colombia is "a tall tree occurring in plains (not in the forests) and, as in some other species of the genus, with the inflorescence issuing from the main trunk or its bare branches. . . . The petals . . . when fresh are succulent and yield a blue juice when squeezed."

The best known species is the Cannon-ball tree, *Couroupita guianensis* Aubl., which has its center of distribution in the Guianas. Dahlgren says (Field Museum botanical leaflet No. 6): "The Cannon-ball tree is one of the most curious of the many remarkable forest trees of the South American tropics. In general aspect and habit of growth it bears some resemblance to a large Elm [*Ulmus americana* L.], though with larger leaves massed at the tips of the slender twigs. It is, however, distinguished from all other trees by the tangle of crooked branches which surround the lower part of its trunk . . . [these] branches which have no direct connection with the wood of the tree, grow out from the bark and, increasing in length, bear the flowers from year to year for many years. Distinct as these are from the foliage branches, it happens occasionally that one of the fruiting limbs also puts forth leaves. . . . The ground around the base of the tree is generally littered with fallen fruits and their remains in various stages of decay. These give off an unmistakable corpse-like odor which on closer acquaintance with this tree is found to be characteristic also of the freshly cut wood. The flowers are unusual in shape and appear at first sight to be somewhat orchid-like. . . . Their fleshy petals are salmon pink to crimson madder in color on the inner surface, white with a dash of yellow on the outside. . . . The flowering of the Cannon-ball tree is said to be almost continuous, but in Guiana an abundance of flowers is found early in the year when fruits on the tree measure 6 to 8 inches in diameter. These are the fruits of the preceding year, which, requiring some 18 months to ripen, remain on the tree till the new crop of fruit is well advanced. . . . The Cannon-ball tree grows . . . in places over the entire northern part of South America from Brazil to Central

America and the lesser Antilles. A single tree of this kind exists in the United States, having been planted at Fort Myers, Florida, where it grows outside the tropics better than might be expected."

The following description is based on authentic wood samples of *Couroupita guianensis, C. odoratissima,* and *C. parviflora.* Heartwood absent or not distinguishable; sapwood yellowish brown, probably nearly white when fresh. Luster rather low. Scent slightly fetid; taste not distinctive. Of fairly light weight but firm, tough, and strong; texture moderately coarse; grain mostly straight; not difficult to work, finishing smoothly; sapwood perishable. Suitable for general interior construction and boxes.

COMMON NAMES: Bala de canón (Cuba, int.); coco de mono, c. zapote, granadillo, palo de paraíso, zapote de mono (Pan.); coco de mono, mamey hediondo, múco, mucurutú, nispero hediondo (Venez.); arbre à bombes, cannon ball, moke, múco (Trin.); cannon-ball tree (Br. G.); boesi kalabasi, bosch kalabas, iewadaballi, koppe, wilde abrikoos (Sur.); abricot de singe, boulet de canon, calabasse colin, couroupitoumou (Fr. G.); abricó de macaco, castanha de macaco, cuia de macaco, cuirana (Braz.); aiauma, avahuma (Peru).

Eschweilera (including *Chytroma* and *Jugastrum,* which some authorities consider valid genera) is closely allied to *Lecythis* and comprises about 80 species of medium-sized to very large trees distributed from eastern Brazil through the Amazon basin to Trinidad and Costa Rica. The smooth leathery leaves range in size from rather small to very large; the medium-sized to large white, yellow, red, or violet flowers are borne in axillary or terminal racemes or panicles; the hood of the androecium is coiled inward, with the apical part abruptly folded back; the fruit, which is about as long as broad, opens with a large operculum; the seeds are sessile, not winged, and in many species have very bitter kernels. *Eschweilera* is recognized as a difficult genus and many of the species are imperfectly known and not clearly demarcated.

There are 15 or more species of *Eschweilera* in British Guiana, all commonly known by the name of Kakeralli. According to T. A. W. Davis, in a memorandum supplied by the Conservator of Forests, the majority are small or medium-sized trees, but some of them are dominant or subdominant in climax rain forests. In valuation and reconnaissance surveys the trees are generally recorded as Kakeralli without distinction as to kind, but in most areas at least 75 per cent of them are Black Kakeralli, *E. Sagotiana* Miers. It attains a height of 100, rarely 120, feet and a diameter of 24 inches or more. Buttresses are usually very small or entirely absent but sometimes are strongly developed. Merchantable lengths average about 40 feet, maximum about 60 feet. Between the Venezuelan border and the Demerara River the average number of trees 16 inches and over at breast height varies from one to two, and the volume of merchantable timber in the round is from 50 to 120 cubic feet per acre. The species becomes less common and rather local in the eastern districts and is absent on white sands (Wallaba forest) and is replaced in the Rupununi district by *E. subglandulosa* (Steud.) Miers. The leaves are large and stiffly coriaceous; the flowers are small and creamy white. The bark is of medium thickness, with a pale flesh-colored slash.

Smooth-leaved Kakeralli, *Eschweilera decolorans* Sandw., is similar in appearance and habit to *E. Sagotiana* and has about the same range and abundance, but the number of trees exceeding 16 inches in diameter is only 0.1 to 0.2 per acre. The large white flowers are easily recognizable because they discolor blue-green instead of brown when bruised by falling and also by their short stout green pedicels and green calyx lobes. The bark is rather thin, with a whitish slash. When not in flower or fruit the species is scarcely distinguishable from *E. odora* (Poepp.) Miers, which has fairly large creamy white flowers and peach-colored calyx and pedicels, almost exactly like *E. subglandulosa,* the Black Kakeralli of the Rupununi district. (The name Black Kakeralli refers to the mottled blackish and brown bark.)

Wena Kakeralli, *Eschweilera corrugata* (Poit.) Miers, is a medium-sized unbuttressed tree with rather thick smooth pale yellowish bark usually conspicuously striated with shallow longitudinal fissures; the slash is flesh-colored. The leaves are large and stiff and the flowers are mauve-pink or purple. It is of general distribution throughout British Guiana, averaging 0.1 tree or less per acre in most acres, but more common in Wallaba (*Eperua*) forest on white sand. *E. confertiflora* A. C. Smith, though somewhat like the preceding in general appearance, can be distinguished readily in the field. It is a medium-sized unbuttressed tree with thick, light reddish brown, slightly granular bark and flesh-colored slash. It is abundant in certain localities, such as the Mazaruni-Kuribrong district west of the Kaburi River, in a small area on the left bank of the lower Mazaruni River, and near Great Fall, Demerara River. Swamp Wena Kakeralli, *E. longipes* (Poit.) Miers, is a comparatively small tree of the understory, rarely exceeding 60 feet in height and 12 inches in diameter. It is generally distributed but nowhere abundant, occurring mainly in Mora and other riparian types of forest. The Haudan or Howdan, *E. holocogyne* Sandw., is a fairly large tree of general but infrequent occurrence (0.02 to 0.04 per acre). The flowers are large and white, but do not discolor blue-green like those of *E. corrugata;* also the leaves are smaller and not so stiff as in that species.

Guava-skin Kakeralli, *Eschweilera alata* A. C. Smith, is a medium-sized tree fairly common in the Northwest District and in the Mazaruni-Kuribrong region west of the Kaburi River. It is easily recognized by its very smooth, thin, almost chocolate-colored bark which is shed in large thin flakes suggesting the common Guava (*Psidium guajava* L.). The leaves are large and stiff; the flowers are large and creamy, usually with a tinge of mauve-pink.

The only *Eschweilera* timber well known to the export trade is the Manbarklak of Surinam, sometimes called Toledo Wood (a name given to it more than 15 years ago by Mr. R. G. H. McArthur of Paramaribo). According to Pfeiffer (*De houtsoorten van Suriname*, pp. 404–408), there are two species of Manbarklak, namely *E. longipes* (Poit.) Miers and *E. subglandulosa* (Steud.) Miers. The same vernacular designation is applied sometimes to other species, such as *E. odora* (Poepp.) Miers, and *E. corrugata* (Poit.) Miers (see Pulle's *Flora of Surinam* 3: 138–142). The timber, owing to the grit (silica particles) in the cells, is highly resistant to marine borers and is coming into use for marine construction, especially in brackish water. Piles of Manbarklak supporting the railway bridge across the Saramacca Canal at Beekhuizen, Surinam, were placed in use in March 1904 and removed in March 1921, and after 17 years' service were perfectly sound and fit for further use. The *Neoteredo,* the most destructive of shipworm, infests the waters of this canal but did not attack this wood at all, though a piece of Demerara Greenheart (*Ocotea Rodiei* Mez) placed in the water of this canal in February 1920 was found upon removal in May 1921 to be completely riddled. The only damage to the Manbarklak was a slight superficial injury from marine stoneborers, *Martesia cuneiformis* Say. Specimens of these timbers have been deposited in the Yale collections (Nos. 4954–4958).

A comparison of authentic specimens of several species from British Guiana with Manbarklak from commercial sources indicates that the Common or Black Kakeralli, *Eschweilera Sagotiana,* is nearest to that timber, followed closely by *E. subglandulosa.* The following description applies particularly to the Manbarklak group from which the others differ more or less in density and color. Heartwood olive-brown to reddish brown, somewhat streaked; rather sharply demarcated from the yellowish sapwood. Luster rather low. Extremely hard, heavy, compact, tough, and strong; sp. gr. (air-dry) 1.10 to 1.25; weight 69 to 78 lbs. per cu. ft.; texture uniform and rather fine; grain generally straight; difficult to work, but easy to split; finishes very smoothly, with a slate-like feel; highly durable.

In Yale specimens of other species the colors observed are: pale olive or nearly white (perhaps all sapwood), *Eschweilera calyculata* Pitt., *E. panamensis* Pitt., and *E. Rimbachii* Standl.; light olive-brown, *E. grata* and *E. odora;* fairly uniformly salmon-pink, *E. corrugata* and *E. jarana* (Huber) Ducke; dull orange with conspicuous purplish striping, *E. confertiflora;* plain reddish brown, *E. holocogyne* Sandw.; reddish brown with inconspicuous lighter striping, *E. alata* and *E. decolorans;* dark reddish brown with inconspicuous blackish streaks, *E. coriacea* (DC.) Mart. In *E. jarana, E. corrugata, E. grata, E. holocogyne,* and *E. odora* the parenchyma bands are 1 to 8 cells wide and distinct without lens. Gum ducts were seen in *E. calyculata, E. coriacea, E. corrugata, E. jarana,* and *E. Rimbachii.*

Common names: Majagua de indio, mata cansada, olleto, ollito (Pan.); coco de mono, curtidor montañero, guatequero, hebrito, naranjillo (Venez.); haudan, howdan, kakaralli, kakeralli, wena, w. kakeralli (Br. G.); akakarrie, arepawana, atotoito, baakalaka, barklak, barkraki, booka, bosoho, bosso, engoso, haudan, hiroe kakalalli, hiarokoro, houdin, jawané-bolotin, kakaralie wadili, kakorelli, k. balli, k. hyaro, k. wadilikoro, koearda, komantie kwatie-ie, kottobon, kwateli, kwatie-ie, kwateri, kwaterie, kwattere, k. pipipjo, manbarklak, manbarkraki, mekoekoeware, oemanbarklak, oemanbarkraki, oriebina, pipina, senskiebieta, sienkwatta, snekie-bieta, tamoene kwatere, tapirin kwatere, tapoeroe kwatere, tekarajan kwatere, tetei hoedoe, teteimerie, tité barkraki, t. meri, toewasa korajan kwatere, toko, toledo wood, wadilie kakeralie, w. lanaballi diamaro, wadodorie, w. lanaballi diamaro, wadoedoeli (Sur.); idatimon (Fr. G.); aterebá, ateribá, atereua, biriba, biribi, castanheira das aguas, genipa-rana da terra firme, ibirá-rana, jarána, jatereua, macacarecuia, matá-matá, m.-m. da casca branca, morrão branco, m. vermelho, pau de macaco, ripeiro, sapucaia amargosa (Braz.); piñuela (Ec.); machinmango (Peru).

Grias. There are five species, namely, *G. cauliflora* L. of the West Indies. *G. Fendleri* Seem. of Central America. *G. foetidissima* Dugand of Colombia, *G. tetrapetala* (Aubl.) Ndz. of the Guianas, and *G. peruviana* Miers of Peru. Although sometimes described as big trees, the greatest dimensions recorded are 50 feet in height and about 14 inches in trunk diameter. They usually inhabit swampy places, sometimes composing nearly pure stands. The leaves are large, sometimes three feet long, and crowded at the ends of the branches; the rather small white regular flowers are borne in clusters along the main stem and branches; the indehiscent fruit has one or few cells, each containing a single seed. The yellow wood is of medium weight and density, rather coarse-textured, easy to work, but is not durable and has no special uses.

Common names: Anchovy pear (Jam.); genip (Br. H.); cayilla (Guat.); irayol, jagüillo (Hond.); jaguey, membrillo macho (Pan.); pepeguára, tokóro (Col.); sapote de perro (Ec.).

Gustavia, with several species of shrubs and small, or occasionally medium-sized, trees is distributed from Panama throughout northern South America to Peru and Brazil. The leaves, which vary in length from 4 to 40 inches, are frequently clustered at the ends of the branchlets; the fragrant white, yellowish, or pinkish flowers are in terminal or axillary panicles or clustered on the old wood below the leaves; the androecium is regular, with the terminal parts of the stamens involuted; the fruits are the shape of a quince or a pear, sometimes winged or ribbed, with a leathery pericarp and, when ripe, with a pulpy interior which escapes through the opercular opening and carries with it the few to numerous bean-like seeds. The timber is of minor importance locally and has no commercial possibilities.

Heartwood apparently absent from specimens studied; sapwood yellowish, sometimes with a tinge of pink. Luster low to high, depending on the lighting. Without distinctive taste, but usually with fetid odor. Rather hard, heavy, tough, and strong; texture medium; grain straight to irregular; not difficult to work, finishing very smoothly; with rather attractive appearance, especially on radial surface; durability doubtful.

Common names: Membrillo (Pan.);

baco, chupa membrillo (Col.); chupón, flor de muerto, mariche, motín, palo de muerto, ventoso (Venez.); apolo-uokomollo-kotelo, arepawana, aripawana, a. enekan, a. waton, dijoelano, haudan, hegron taproepa, hoogland taproepa, itjoetano aripawana, kapoeaeb, kokonibieta, laagland, lanaballi, l. diamaro, l. hororodikoro, mantapoepa, mantapoeripa, okajomoe manetare, omitapoepa or omitaproepa, pakassa, pirigaramepé, stinkhout, watramamí bobbi, wokomolo kotele (Sur.); bois puant (Fr. G.); chope, genipa-rana, janiparindiba, pau fedorente (Braz.); membrillo de monte (Ec.); chope, sacha chope, s. manga (Peru).

Lecythis, with numerous, mostly imperfectly known, species of medium-sized to exceptionally large trees, is widely distributed from southeastern Brazil through northern South America to Costa Rica. The leaves vary greatly in size and texture and frequently are serrate; the yellow to purplish flowers are borne in axillary or terminal racemes; the hood of the androecium is coiled inwards but its apical part is not abruptly folded back as in *Couratari*. Miers says (*loc. cit.*, p. 162): "The fruit is a pyxidium, usually of great size, very thick, densely ligneous, extremely variable in form, and always marked by two concentric lines. . . . [It] naturally hangs in an inverted position and, when ripe, a swelling is at first noted around the opacular zone; afterwards, by the rupture of the central column, the operculum falls off with a portion of the column attached to it. The numerous seeds, sustained by their funicles, now hang down in a bunch and soon fall to the ground, the main body of the shell remaining long suspended from the lofty tree; at this period the fleshy mass of the funicles ferments and exhales a very nauseous odor. . . . The seeds of the species abundant in the province of Pará [*L. paraensis* (Huber) Ducke] are exported to Europe in considerable quantities and sold here in the shops as Sapucaia-nuts."

Most of the Brazilian species are known as Sapucaia, the Tupa Indian word for hen. H. M. Curran says: "The Sapucaias belong to the group of monkey-pot trees common over a large part of tropical South America and constitute one of the most common features of the Brazilian forests, both in the Amazon lowlands and in the coastal mountain forests. It is common to find trees with somewhat buttressed boles five or six feet in diameter and free of limbs for 50 to 60 feet. The crowns are composed of stout wide-spreading branches and shed most of their leaves during the dry season, becoming almost bare. The light-colored flowers are borne in profusion and though scarcely noticeable while on the tree they carpet the ground for a week or two during their fall. The fruits are great balls, dark brown when ripe, with a circular opening closed by a lid which eventually falls away and lets the seeds escape. The seeds are of about the size of the Brazil-nuts of commerce, but differ from them in having a smooth and somewhat wrinkled brown shell that is not sharply angled. They have a sweet and creamy kernel and are known in the trade as paradise or cream nuts, but are not extensively exported, partly because of the local demand for them and also on account of the difficulty in collecting, the bulk being eaten by monkeys and other animals of the forest." Mr. Curran collected a wood specimen (Yale No. 4693) and herbarium material of the common Sapucaia of the Bahia region, which has been identified as *Lecythis ollaria* L. The trees are found along all of the waterways and though not very numerous are a very conspicuous feature of the forest because of their large size. The laminated inner bark supplies material for oakum, cigarette papers, and tinder and is also used in tanning. A reddish sap, obtained by tapping the trees, is said to make a refreshing drink. The large woody fruit shells are used for ornamental and various domestic purposes. The seeds are a valuable article of food and yield about 40 per cent of oil suitable for cooking and for industrial purposes. The timber is employed in heavy construction, bridges, railway ties, posts, carpentry, and cabinet work.

The Guatecare of Trinidad, according to R. C. Marshall (see *Tropical Woods* 27: 28), is *Lecythis laevifolia* Gris. It is a large evergreen tree widely distributed on poor soil and not exacting in its requirements except that it avoids swamp areas. It usually has a long straight trunk which often is irregularly swollen, especially near the base. Two varieties are recognized, the black and the white, depending on the color of the heartwood, but so far no reliable botanical distinction has been found. The timber is very hard and heavy, difficult to work, and fairly to highly durable. An undetermined species of similar appearance is of infrequent occurrence in Tobago, where it is known as Devilwood.

There is little information about the species in the Guianas. The common Wodaduri or Monkey-pot of British Guiana appears to be *Lecythis Davisii* Sandw., of which the Surinam form, called Kwattapot, is a variety (var. *gracilipes* Eyma). The tree reaches a height of about 100 feet, and the trunk is shallowly fluted and not buttressed. The dark salmon-red heartwood is of good quality, durable, and appropriate for the same uses as Bulletwood (*Manilkara*).

There are several species in Venezuela, Colombia, and Panama, and the most common names for them are Coco de Mono and Olla de Mono. *Lecythis Curranii* Pitt., discovered by H. M. Curran in Colombia, is a tree 70 to 100 feet tall, with a slender trunk 18 to 24 inches in diameter. The timber is cross-grained and coarsely laminated like that of *L. elliptica* H.B.K. of the same locality and *L. Melliana* Pitt. of Panama; it is used for exterior construction because of its strength and durability.

Concerning the Central American species, Pittier says (*Contr. U.S. Nat. Herb.* 26: 1: 11): "It is interesting to note that the three species found on the Atlantic watershed (*L. armilensis* Pitt., *L. ampla* Miers, and *L. costaricensis* Pitt.) are all very large trees with umbraculiform crown and small ovate serrate leaves, while the two [*L. Melliana* and *L. tuyrana* Pitt.] found on the Pacific side of the continental divide are of lesser dimensions, have radiating branches beginning low on the trunk, and have leaves relatively larger and always entire." According to Standley (*Flora of Costa Rica,* p. 765), *L. costaricensis* is a tall tree, the trunk 80 feet long and 40 inches in diameter, rather widely scattered in the Atlantic lowlands. "The hard wood is used in making carts. The seeds are said to be much like Brazil nuts in flavor and are much sought by squirrels, monkeys, and men, but the supply of course is insignificant."

Heartwood brown to dark salmon, the deeply colored material sharply demarcated from the yellowish sapwood. Luster mostly low. Without distinctive odor or taste. Typically very hard, heavy, compact, tough, and strong; sp. gr. (air-dry) 0.85 to 1.10; weight 53 to 69 lbs. per cu. ft.; texture medium fine and uniform; grain fairly straight; difficult to work, but finishing smoothly; durability fair to excellent. Suitable for heavy construction and purposes requiring a strong, resilient material.

COMMON NAMES: Cocobolo, olla de mono (C.R.); coco, negrito colorado, nomoncrí, olla de mono (Pan.); coco de mono, coquillo, guaycán coco, hoja de mono, olla de mono, ollito de mono (Col.); coco de mono, durote, olla de mono, olleto, ollito de mono (Venez.); guatacare (Trin.); devilwood (Tobago); monkey pot, wadaduri (Br. G.); groote potboom, kwattapatoe, kwattapot, quatopot (Sur.); maho jaune, marmite de singe, quatelé (Fr. G.); castanha sapucaia, jacapucaio, jequitabá rosa, sapucaia, s. branca, s. guassú, s. mirim, s. vermelha (Braz.); machin-mango (Peru).

LEGUMINOSAE

THIS great family is of world-wide distribution and comprises about 500 genera and 15,000 species of herbaceous plants, erect or climbing shrubs, and small to very large trees. The herbaceous forms include such well-known and useful plants as peas, beans, clover, and alfalfa. The trees are mostly tropical and subtropical, but there are several in the United States, *e.g.*, Black Locust (*Robinia*), Honey Locust (*Gleditsia*), Red Bud (*Cercis*), Kentucky Coffee Tree (*Gymnocladus*), and Mesquite (*Prosopis*). Many of the tropical timbers have been in the markets of the world for a very long time and are among the most valuable for cabinet work, carving, and other purposes where beautiful woods are appreciated. Logwood, Brazilwood, and other dyewoods were formerly in great demand for dyes. Many species are valuable as sources of food and forage, gums and resins, drugs and medicines, and some are cultivated for ornament and shade.

The Leguminosae are divisible into three fairly definite groups which some botanists recognize as distinct families, namely, Mimosaceae, Caesalpiniaceae, and Papilionaceae (or Fabaceae). There is no sharp line of demarcation between them (particularly the last two) and, from the standpoint of the woods at least, there are advantages in maintaining a single family. The woods of the different genera and species exhibit a great range of variation in appearance, properties, and structure and have few constant characteristics, and yet for every specimen there is usually some peculiar combination of features which relates it readily to the Leguminosae, though at present it is not always possible to determine the subdivision to which it belongs.

Much progress has been made during the past 25 years in clearing up the botanical identities of timbers which have been known to commerce for a long time. In this category are Cocobolo, the Rosewoods, Kingwood, Brazilwood and Brazilette, Angelim Rajado or Bois Serpent, Vinhatico, Sucupira, and several others. The senior author has had a share in this work, but despite long-sustained efforts it has not been possible to determine the specific identity of Brazilian Tulipwood, though there appears to be no doubt that it is a species of *Dalbergia* near *variabilis*. There are many other woods in eastern Brazil, some of them timbers of commerce, that are generically unknown because of lack of authentic specimens with which to compare them.

The following account of the American woods of the Leguminosae deals with 185 genera, but several of these are segregates from other genera and are not recognized by all botanists. Owing to inadequate material it is not always possible to form an opinion as to the validity of some of the proposed groups, and their treatment as separate genera does not necessarily imply indorsement of such concepts. Here, as in similar cases in other families, it is frequently considered worth while to keep the sorted data separate as a contribution to future investigations.

Heartwood usually distinctively colored with shades of red, brown, orange, yellow, violet, or black, uniform to variegated or streaked; sapwood white or yellowish, often sharply demarcated. Luster low, medium, or high. Scent more or less pronounced in *Coumarouna, Dalbergia, Eperua, Ferreirea, Haematoxylon, Machaerium, Martiodendron, Myrocarpus, Myroxylon, Schizolobium, Sclerolobium, Taralea, Torresia, Vouacapoua,* and *Zollernia;* taste usually distinctive in *Cynometra, Ferreirea, Haematoxylon, Mora, Pentaclethra, Robinia, Stahlia, Sweetia, Vatairea,* and *Vataireopsis,* being decidedly bitter in the last two genera. Density low in *Erythrina* and some specimens of *Parkia, Pterocarpus, Schizolobium,* and *Sclerolobium;* medium to very high in the others. Texture occasionally very fine, as in *Brya,* more often medium to coarse, sometimes very coarse as in *Andira, Bowdichia, Clathrotropis macrocarpa, Dimorphandra gigantea, Diplotropis, Erythrina, Gymnocladus, Ormosia, Ormosiopsis, Vatairea,* and *Vataireopsis.* Working properties and durability widely variable.

Growth rings present or absent. Ring-porous structure characteristic of *Asacara, Cercis, Gleditsia, Gymnocladus, Neomimosa, Robinia,* and *Sophora affinis;* tendency in some of the others. Diffuse pores variable in size, number, and arrangement, but mostly numerous but not crowded, solitary and in small multiples, well distributed. Vessels with exclusively simple perforations; spiral thickenings present (at least in small vessels) of *Asacara, Cercis, Edwardsia, Gleditsia triacanthos, Gymnocladus, Prosopis* (in part), *Robinia,* and *Zuccagnia;* striations present in *Sophora* and small vessels of *Cladrastis;* tyloses apparently limited to *Coursetia, Gliricidia, Hebestigma, Lennea, Notodon, Olneya, Robinia,* and *Sabinea;* gum deposits general; pits variable from very small to large, mostly medium to rather large; all pits vestured except in the Bauhinieae (*Bauhinia* and *Cercis*). Rays typically narrow and low, but in part 6 to 8, occasionally more, cells wide and 60 to 100 or more cells high in *Asacara, Cladrastis, Dalea, Dugandia, Edwardsia, Gleditsia, Haematoxylon, Hymenaea, Len-*

nea, Lucaya, Myrmecodendron, Piptadenia (occ.), *Piscidia, Robinia, Senegalia, Sophora,* and *Vachellia* (occ.); sometimes up to 20 cells wide in *Erythrina* (Plate XLIX, 3) and in part exceptionally large and Oak-like in *Monopteryx* (Plate XLII, 2); homogeneous in about three-fourths of all genera; weakly to distinctly heterogeneous with one to very few rows of squarish, rarely upright, cells in the others; homogeneous in about 65 per cent of genera with, and in about 80 per cent of those without, storied structure; crystals common; pits to vessels small to large, but mostly medium-sized; unilaterally compound pitting observed in *Batesia, Cynometra, Eperua, Jacqueshuberia,* and *Parkinsonia.* Wood parenchyma typically abundant, often conspicuous; principal types vasicentric, short to long aliform, confluent, and metatracheal bands, with or without terminal and diffuse; small patches of sclerotic cells observed in *Clathrotropis;* inflated cells present in *Batesia;* crystal strands common, often abundant; gum cysts characteristic of certain species of *Poincianella;* cells of ordinary strands frequently in horizontal seriation. Wood fibers mostly with thick walls and small simple pits; gelatinous walls fairly common; septate fibers observed in *Albizzia, Arthrosamanea, Batesia, Dugandia, Havardia, Inga, Leucaena, Peltophorum, Piptadenia, Plathymenia, Poeppigia, Poinciana, Schizolobium,* and *Senegalia tamarindifolia;* bands of thin-walled fibers resembling parenchyma on cross section present in *Poeppigia.* Ripple marks common in the Caesalpiniaceae and Papilionaceae. Normal vertical gum ducts present in *Copaifera, Eperua* (Plate LIV, 1), and *Prioria.* Vertical traumatic ducts observed in *Andira, Cercidiopsis, Senegalia glomerosa, Humboldtiella, Hymenaea,* and *Peltogyne.*

Abarema (or *Jupunba*) includes a group of tropical American shrubs and small to rather large trees of no commercial importance for their timber. Of the 16 species described by Britton and Rose (*North American Flora* 23: 24–28, 193), all but two are limited in their distribution to the West Indies. *Abarema pseudotamarindus* (Britton) (= *Jupunba pseudotamarindus* Britton = *Pith. pseudo-tamarindus* Standley) is a tree about 50 feet high discovered by G. Proctor Cooper in Bocas del Toro, Panama, where it is known as Wild Tamarind. The species with the greatest range is *A. Jupunba* (Willd.) Britt. & Killip (= *Pith. micradenium* Benth. = *P. trapezifolium* Benth.), a medium-sized tree occurring in Guadeloupe, Trinidad, Colombia, Venezuela, the Guianas, and Brazil. It is known as Soapwood in British Guiana because the natives use the bark and roots as a substitute for soap.

Heartwood of the various species, so far as examined, pale brown or flesh-colored; sapwood white. Odorless and tasteless. Texture rather fine to moderately coarse; grain variable from straight to interlocked; easily worked, finishing smoothly; probably not very durable. No special uses known.

Common names: Shadbark, shagbark, tamarind shadbark (Jam.); abey blanco, angelino, ciruelillo, encinillo (Cuba); caracoli (Dom. R.); wild tamarind (Pan.); hooroowassa, soapwood (Br. G.); angelim fraco, avaremotemo, lagrimas de nossa senhora, tento azul (Braz.).

Acaciella, with more than 50 species of unarmed shrubs, little trees, and a few herbaceous plants, has its center of distribution in Mexico, with extensions into southwestern United States, the West Indies, Central America, and northern South America. Leaves eglandular, bipinnate, with few to many pairs of leaflets; flowers with pedicels, in globose or oblong clusters; legume narrowly oblong, flat, splitting from the apex; seeds lenticular.

Heartwood lustrous pinkish brown, more or less striped, merging gradually into the yellowish sapwood. Without distinctive odor or taste. Moderately hard and heavy; rather fine-textured; of somewhat irregular grain; easy to work, finishing very smoothly and taking a high polish. Rarely utilized because of the small sizes available, but suitable for fancy articles and cabinet work.

Common names: Cantemó, guajillo, palo de pulque, timbe, vara colorada, xaas (Mex.); guaje, guajillo, quebracho liso (Salv.); carbonero (Col.); tiamo güire (Venez.); waabi, wata pana (Curaçao).

Acaciopsis is a Mexican genus with 14 species of shrubs and little trees, with a few extensions into Texas. They have bipinnate leaves, with small or large leaflets; stipular spines that usually are slender; small sessile flowers in axillary heads or spikes; and very narrow, thin legumes with oblong seeds.

Heartwood rather dull purplish brown, somewhat variegated; rather sharply demarcated from the thin white sapwood. Hard and heavy; medium-textured; of irregular grain; not difficult to work, finishing smoothly; probably durable. An attractive wood, but not utilized because of the small sizes available.

COMMON NAMES: Chaparro prieto, gavia, gigantillo, huisache, largoncillo, vara prieta (Mex.).

Adipera, with numerous species of shrubs and small trees, is well represented throughout tropical and subtropical America. Some of the shrubs have elongated, vine-like branches. The leaves are pinnate, with three or more pairs of rounded to acuminate leaflets; the large yellow flowers are corymbose or racemose; the legumes are slender, elongated, indehiscent or breaking open irregularly. The plants appear to have no value except possibly for decorative purposes.

Heartwood brown to blackish brown; rather sharply demarcated from the nearly white sapwood. Luster medium. Odorless and tasteless. Rather hard, heavy, tough, and strong; texture fairly coarse; grain irregular; easy to work, finishing smoothly; is probably durable. Has no commercial possibilities.

COMMON NAMES: Christmas bush, wild raisin (Jam.); hedionda macho, hoja de sen, sen del país, styver bush (P.R.); alcaparrillo, viche, vicho (Mex.); wild currant, wood creeper (Br. H.); brasilillo, caraguillo, carne asada, flor barbona amarillo, frijolillo, moco de chompipe, m. de gallo, m. de güegüecho (Salv.); bicho, stakuogró (C.R.); alcaparro, bombito, platanito (Col.); ormaco, rama negra (Venez.); caquera (Braz.); motuy, mutui (Peru).

Albizzia is a rather large genus of unarmed small to large trees widely distributed in tropical Asia, Africa, and America. The best known species is the Siris-tree, *A. Lebbeck* (L.) Benth., native of Asia, but planted for shade and ornament throughout tropical regions and often naturalized. The leaves, which are deciduous, are pinnately compound, with numerous leaflets. The masses of yellowish white flowers have a rather heavy fragrance. The fruits are flat straw-colored many-seeded pods 8 to 12 inches long which may remain on the tree for months and rustle in a breeze with a characteristic sound that has been likened to the sizzling of frying meat. Its hard, dark brown, somewhat streaked timber is known on the U.S. market as Koko (see *Tropical Woods* 18: 23–25). Two related species in the Far East are *A. procera* (Roxb.) Benth. and *A. acle* Merr. The sawdust has a peppery odor that causes sneezing. There are about 20 species native to tropical America and while some of them are rather large trees they are of scattered occurrence in the forest and their timber is not especially attractive. There are numerous local uses, but mostly for general construction.

Heartwood pale yellowish brown or brownish, often with pinkish streaks; sapwood yellowish, not sharply demarcated. Luster medium to fairly high. Odorless and tasteless. Mostly hard, heavy, tough, and strong; texture medium to coarse; not difficult to work; does not appear durable.

COMMON NAMES: Amaquí, bacona (Cuba); bois savanne (Haiti); arellano, bolillo, cico, palo escopeta, p. fierro, p. joso (Mex.); prickly yellow, salom, small-leaved prickly yellow, wild tamarind (Br. H.); cenícero macho, gavilana (C.R.); bayeto antioqueño, carbonero de sombrio, carita, dormilón, guacamayo, guamuche, muche blanco, piñón, pisquín (Col.); hueso de pescado (Venez.); canafistula, faviero do matto (Braz.); pacará blanco, palo flojo, sacha-monte, tarco, timbó blanco, t.-y-atá, t. morotí, t. verde (Arg.).

Aldina, with seven species of large unarmed trees, is limited to the northern Amazon region and the Guianas. The leaves are imparipinnate, with few large leathery leaflets, or sometimes with only one; the white flowers are borne in simple axillary or terminal panicled racemes; the legumes are woody and ovoid and contain a single, large seed. The trees apparently are rare, judging from the scarcity of information concerning them. The only wood sample available (Yale 21003) is of *A. heterophylla* Benth., collected with flowering herbarium material by Adolpho Ducke near Manáos, Brazil, where the tree is known as Macucú.

Heartwood pale olive, with conspicuous, light-colored vessel lines; rather sharply demarcated from the yellowish white sapwood. Luster low. Without distinctive odor or taste. Very hard and heavy; texture coarse; feel harsh; grain irregular; not very difficult to work, inclined to be splintery, finishes fairly smoothly; durability doubtful. An unattractive wood of no commercial possibilities.

COMMON NAMES: Macucú, m. da catinga (Braz.).

Alexa, with four species of unarmed, typically tall, poorly known trees, is limited in its distribution to wet lands in the Guianas and the Brazilian Amazon region. *A. Wachenheimi* R. Ben. grows in French Guiana. *A. imperatricis* (Schomb.) Baker is found in central British Guiana, where it is known as Haiariballi. *A. bauhiniaeflora* Ducke was discovered along the Rio Negro in Brazil and differs from the other species in being a large shrub or a little tree. *A. grandiflora* Ducke is fairly common in the lower, southern Amazon. The leaves are large, unequally pinnate, with several large leathery leaflets; the white or yellowish flowers are racemose; the pods are large, velvety, dark red, woody, 2-valved, with spongy substance about the seeds. The inner bark sometimes has a pronounced acrid odor. The timber is not utilized.

Heartwood nearly colorless or brownish yellow, sometimes with darker shades; not sharply differentiated from the thick dingy white sapwood. Luster medium to low. Odorless and tasteless when dry. Of medium density; rather coarse-textured; mostly straight-grained; easy to work, finishing very smoothly; durability doubtful. Appears suitable for interior construction and carpentry, but with little if any commercial possibilities.

Amburana, with two described but doubtfully distinct species of medium-sized to large unarmed trees, has its center of distribution in Brazil. The genus is commonly designated *Torresia*, but according to A. C. Smith (*Tropical Woods* 62: 28) this is an incorrect spelling of *Torresea*, the original name, which is excluded by the International Rules of Botanical Nomenclature because of confusion with the earlier use of *Torresia* for a different genus. The leaves are imparipinnate, with numerous alternate leaflets; the small yellowish white fragrant flowers are borne in axillary racemes; the legumes are dark brown, short, rounded at both ends, and very flat except over the single rugose seed, which is provided with a basal papery wing. (See *Phytologia* 3: 136.) All parts of the trees are redolent of Cumarin (tonka bean).

Amburana cearensis (Fr. Allem.) A. C. Sm. (= *Torresea cearensis* Fr. Allem. = *Amburana Claudii* Schw. & Taub.) is a medium-sized tree widely distributed in the dry regions of Brazil and northern Argentina. The resin in the bark contains a fragrant volatile oil and is used medicinally. The seeds are used in perfuming snuff and toilet soap. The wood, which is likewise scented, is considered excellent for furniture in Argentina, but the supply of the timber is small. In northeastern Brazil it is used like coniferous lumber for general construction, carpentry, window frames, crating, and cooperage. A second form, *Amburana acreana* (Ducke) A. C. Sm., was discovered in 1933 by Adolpho Ducke in high forest on non-inundated land along the Rio Acre in Acre Territory, but it does not extend to the mouth of the river. It is distinguished from *A. cearensis* by its somewhat more numerous leaflets, more lax inflorescences, and its larger size, being sometimes 100 feet tall. Ducke says (*Tropical Woods* 43: 20) that it is highly esteemed for its excellent timber and for its seeds, which are a source of a popular perfume. The vernacular names, Cumarú and Imburana de Cheiro, are the same for both species.

Heartwood yellowish or very light brown with a slight orange hue, deepening somewhat upon exposure, and marked with conspicuous vessel lines; not always uniform;

rather waxy or oily looking; not sharply demarcated from the grayish sapwood. Luster medium to high in proper light. With mild but distinct scent and taste of cumarin or vanilla. Of medium weight and hardness; sp. gr. (air-dry) 0.55 to 0.65; weight 34 to 41 lbs. per cu. ft.; texture coarse, but uniform; grain irregular to roey; cuts easily, saws woolly when fresh, finishes very smoothly, holds its place well when manufactured; is probably fairly durable. Tests at the U.S. Forest Products Laboratory on shrinkage from fresh to oven-dry condition of several samples from a log from Argentina gave the following results (in percentage of green dimension): Radial, 2.6; tangential, 5.7; volumetric, 6.4 (see *Tropical Woods* 14: 16).

COMMON NAMES: Amburana, cerejeira, cumaré, cumarú, c. de Ceará, c. de cheiro, cumbarú das catingas, imburana, i. de cheiro, trebol, umburana (Braz.); palo trébol, roble, r. del país, trébol (Arg.).

Andira, with about 30 species of small to large, generally medium-sized, unarmed trees, is sparingly represented in Africa but most of the species are tropical American, their combined range including the West Indies, Mexico, Central America, northern South America, and Brazil. The leaves are large, odd-pinnate, with one to several leaflets; the roseate or purplish fragrant flowers are borne in terminal panicles; the fruit is drupe-like, one-seeded, indehiscent. The best known and most widely distributed species is *A. inermis* (Sw.) H.B.K., an evergreen tree usually of moderate stature but occasionally exceeding 100 feet in height; the bark is ragged and has a disagreeable odor. The timber is used locally for heavy durable construction and spokes of logging carts and has been known to the export trade for a long time, the usual English name being Partridge-wood; the amounts consumed are small, mostly for canes, umbrella handles, billiard-cue butts, and other articles of turnery. The woods of the several species are similar in structure, though varying in details.

Heartwood yellowish, reddish, or brown, sometimes very dark; distinct but not always sharply demarcated from the yellowish sapwood. Luster rather low because of abundant parenchyma. Without distinctive odor or taste when dry. Moderately to extremely hard, heavy, tough, and strong; texture very coarse; grain fairly straight; not easy to work, but can be finished smoothly; is highly durable. Light-colored specimens sometimes bear a superficial resemblance to hard Pine, and others may suggest Palm wood, especially on tangential surface.

COMMON NAMES: Partridge wood, pheasant wood (trade); angelin, bastard cabbage, cabbage-bark, wormwood (Jam.); moca, yaba, y. amarilla, y. colorada, yava (Cuba); angelin, cabbage-bark, moca (P.R.); bois palmiste (Haiti); angelica, angelin (Mart.); angelin, bastard cabbage-bark, lombricero (Trin.); cuilimbuca, cuartololoti, iximche, moca, m. colorada, macallo, macayo, pacay, yaba, yabo (Mex.); black blossom berry, cabbage-bark, cornwood (Br. H.); almendro (Cent. Am., gen.); almendro del río, a. macho, a. montes, a. real (Salv.); chaperno (Guat.); carbón (Hond.); cujia (Nic.); carne asada (C.R.); arenillo, cocú, quira (Pan.); angelino, congo, majagua gallina, palo de seca, peloto, purga (Col.); chigo, chirai, pilón (Venez.); angelin, bat seed, kuraru, red cabbage tree, wild olive (Br. G.); akoelie kiererie, a. tjerere, kabbes, koeraroe, k. talaboe, redietjabesi, rere erepare, roode kabbes, vreemoesoehoedoe, zwarte kabbes (Sur.); Saint Martin, S. M. rouge (Fr. G.); quinillo colorado (Peru); acapúrana, acatrus, andirá jareua, a. uchy, angelim, a. do igapo, a. morcegueira, a. rana, aracuhy, avineira, cumarú-rana, jacarandá morcega, lombigueira, morcegueira, sapupira da varzea, uchy-rana (Braz.).

Apoplanesia *paniculata* Presl, the only species, is a small unarmed tree, sometimes 35 feet high, of limited distribution in the Pacific coastal region of Middle America from Colima, Mexico, to Guatemala. The leaves are imparipinnate, with 11 to 17 rather small black-dotted leaflets; the small white flowers are borne in axillary and terminal panicled racemes; the legumes, which are half included in the calyx, are flat, leathery, indehiscent, one-seeded, and conspicuously punctate. The bark is said to yield a dye. The attractive wood is used a little for articles of turnery and small cabinet work, formerly for archery bows, but is of minor importance because of its scarcity.

Heartwood rich deep brown with faint to distinct blackish brown striping; rather waxy looking; sharply demarcated from the thin, white sapwood. Surface rather dull, but with golden luster beneath. Very hard, heavy, compact, strong, and resilient; texture fine; grain straight to irregular; rather easily worked, taking a high natural polish; is very resistant to decay.

COMMON NAMES: Arco negro, cacanaguaste, cacanaquasle, chulul, ébano, guiebiche, matagallina, palo de arco, p. de a. negro, p. matagallina (Mex.); palo de arco (Guat.).

Apuleia, with a single well-defined species, *A. leiocarpa* (Vog.) Macbride (= *A. praecox* Mart.), is a large unarmed tree occurring from Corrientes and Misiones, Argentina, throughout most of Brazil to Venezuela and eastern Peru. It has imparipinnate leaves, with 5 to 11 large alternate leathery leaflets; the small white flowers are borne in axillary corymbs or on leafless branches; the legumes are oblique, oval or oblong, leathery, indehiscent. The trees in Argentine and southern and northeastern Brazil are usually less than 80 feet tall and three feet in diameter at maturity, but occasional individuals are 100 feet high, with a trunk four feet through and clear of branches for from 50 to 60 feet. The fine-textured, yellow timber is highly appreciated for heavy construction, flooring, door frames, wheelwright work, shafts of vehicles, and fence posts. In the Amazon basin the tree grows best on rich well-drained clay soil and is sometimes up to 160 feet tall. The form has been described as a different species, *A. molaris* Spruce, but the differences from the type are minor and inconstant. The timber is not used much, one special use being in making canoes for use in rapids.

Heartwood rather lustrous golden yellow to yellowish brown, tending to acquire a reddish or coppery hue upon exposure; sharply demarcated from the thin nearly white sapwood. Generally odorless and tasteless when dry; more deeply colored specimens may smell slightly rancid. Hard, heavy, tough, and strong, though variable; sp. gr. 0.80 to 0.95; weight 50 to 60 lbs. per cu. ft.; texture fine and uniform; grain usually roey; easy to work, finishing very smoothly; is said to be durable. A good timber for many local applications, but not likely to become important in the export trade.

COMMON NAMES: Gateado, mapurite (Venez.); barajuba, burajuba, cumarúrana, faveiro, garapa, g. amarella, grapiapunha, g. branca, jutahy amarello, mirajuba, muirajuba, muiraruira, muirataua, pau mulato, p. setim (Braz.); guarapiapunha (Urug.); grapiapuña, ibirá peré, i. piapuña, ivirá peré, madera manchada (Arg.).

Arthrosamanea *pistaciaefolia* (Willd.) Britt. & Rose, the sole species, which different authors have included in *Mimosa, Pithecolobium,* and *Samanea,* is an unarmed tree 20 to 50 feet tall with an erect trunk 12 to 22 inches in diameter, of common occurrence in the Magdalena valley, Colombia. According to Dugand (*Contrib. Hist. Nat. Colombiana* 1: 9, March 25, 1938), it is the same as *Samanea guajacifolia* Pittier (which grows in the dry forests about El Sombrero, Guárico, Venezuela) and *Pithecolobium daulense* Spruce of Guayaquil, Ecuador. The bipinnate leaves resemble those of *Guaiacum;* the slender flattened legumes have undulated margins and are septate between the seeds, ultimately breaking transversely.

Heartwood probably dark brown, judging from color of knots; sapwood nearly white. Very hard, heavy, tough, and strong; rather coarse-textured; grain irregular; not easy to work. No special uses for the timber have been reported.

COMMON NAMES: Guayacán, g. chaparro, g. ciénega (Col.).

Asacara *aquatica* (Marsh.) Raf. (= (*Gleditsia aquatica* Marsh. = *G. monosperma* Walt.), the only species, is a swamp tree rarely more than 50 feet tall with a short stout trunk which usually divides a few feet above the base into heavy crooked branches forming a spreading flat-topped crown. It is rather widely distributed in the southeastern quarter of the United States, infrequent east of the Mississippi River, but abundant westward, and often occupies extensive areas of long-inundated bottomlands in Louisiana and Arkansas. It is commonly known as Water Locust. Unlike *Gleditsia triacanthos* L., the pods are short, pulpless, dehiscent, and usually contain only one seed, rarely two or three. The woods of the two species are very much alike.

Ateleia, with about 10 closely related species of unarmed shrubs and little trees rarely 30 feet high and eight inches in diameter, is represented in the West Indies, Mexico, Central America, Colombia, Bolivia, and southern Brazil. The leaves are unevenly pinnate, with several rather small somewhat leathery leaflets; the very small yellowish faintly fragrant flowers have a single petal and are borne in axillary racemes; the usually abundant fruits are small, compressed, one-seeded, indehiscent, and the upper suture is narrowly winged. The wood is sparingly utilized because of the small sizes available.

Heartwood, which is poorly developed and apparently traumatic in the specimens studied, dull reddish brown, more or less streaked; sharply demarcated from the thick yellowish sapwood. Odorless and tasteless. Moderately to very hard, heavy, compact, tough, and strong; texture fine; grain fairly straight to irregular; not very difficult to work, finishing very smoothly; durability doubtful. Sapwood suitable for tool handles and other purposes requiring strong material in small sizes.

Common names: Stinking pea (Bah.); carne de vaca, guamá, guayancillo, g. bobo, mierda de gallina, m. de gato, palo cenizo, pico de gallo, rala de gallina (Cuba); bois sentir, caiman franc (Haiti); timbé (Braz.).

Bahamia *acuifera* (Benth.) Britt. & Rose (= *Acacia acuifera* Benth.), the only species, is a much-branched shrub or little tree less than 15 feet high, endemic to the Bahamas. The trunk is armed with clusters of long stiff sharp spines; the twigs are slender and warty; the stipules are spinescent. No wood sample available for study.

Common names: Bahama acacia, cassip, pork-and-doughboy, rosewood (Bah.).

Batesia *floribunda* Benth., the sole species, is a large unarmed forest tree growing on sandy non-inundated land near Pará, Brazil. Because of its resemblance to Acapú (*Vouacapoua*) it is known as Acapú-rana (false Acapú), usually with the qualifying term "da terra firme" to distinguish it from another Acapú-rana (*Campsiandra laurifolia* Benth.) which occurs on the periodically flooded regions. The leaves are imparipinnate, with a ridged or slightly winged rachis and 9 to 11 large leathery leaflets; the yellow flowers are borne in large terminal panicles; the legume is short and somewhat swollen, the woody valves ridged on the inside between the few, depressed, hard, nearly circular red seeds.

Heartwood of a rich chestnut-brown color with golden luster; rather sharply demarcated from the pinkish brown sapwood. Without distinctive odor or taste. Of medium density; sp. gr. (air-dry) 0.60; weight about 38 lbs. per cu. ft.; rather coarse-textured; commonly roe-grained; saws rather woolly, but is very easy to cut, finishing smoothly; holds its place well when manufactured; is probably fairly durable. A good wood used locally for carpentry, but apparently well suited for furniture.

Common names: Acapú, a.-rana da terra firme, tenteiro, tento (Braz.).

Bauhinia, with about 250 species of armed or unarmed shrubs, lianas, and small trees, is of pantropical distribution. Some botanists segregate several groups of species into other genera, *e.g.*, *Casparea* and *Schnella* (vines), but the woods studied are not readily separable along the proposed lines and will be treated as of one genus. One of the most characteristic features of the plants is the simple palmately-veined leaves which are commonly more or less deeply notched at the end and when partly folded along the middle suggest the hoof of a deer or a goat. The flowers are generally large and showy. The shrubs often form thickets and the largest trees are rarely over 25 feet high.

Heartwood probably dark brown or reddish brown, judging from wound areas; sapwood pale brown, sometimes with a tinge of pink. Luster medium to high. Odorless and tasteless. Hard, heavy, tough, and strong; rather fine-textured; grain straight to irregular. Not utilized for any special purpose.

COMMON NAMES: Bull hoof, mountain ebony (Jam.); bejuco de tortuga, casca de mulo, guacacoa, guacamaya americana, majagüilla, pata de vaca (Cuba); araña gato (P.R.); pato de chivo (Dom. R.); bois calecon, collègue matouri, matourin, ti coleçon (Haiti); calzoncillo, chactsulubtok, cibix quibix, guacimilla cimarrón, hierba de la vaca, huamúchil, mano de vaca, papalocuahuite, pata de cabra, p. de vaca, p. de venado, p. de res, pie de cabra, p. de venado, timbe, tsulubtok, utsomeltok, zactsulubtok (Mex.); cow foot, c. tongue, pata de vaca (Br. H.); casco de venado, vainilla (Hond.); casco de venado, garrabatillo, pata de venado, pie de cabra, p. de venado, tripas de vieja (Salv.); espino blanco (Nic.); bejuco de culebra, casco de venado, escalera de mono (C.R.); bejuco de cadena, b. de culebra, b. mono, cocia, espino blanco (Pan.); bejuco de cadena, escalera de mico, machete vaina, pata de vaca (Col.); bejuco de cadena, b. de corona, cadenillo, dibrito, guarapa, perichargua, urape (Venez.); cipó de jaboty, c. escada, c. florão, c. unha de boi, escada de jaboty, guarapa, matámatá, mororó, m. sinho, pé de boi, unha de boi, u. de vacca, urape (Braz.); machete, m. vaina, vaina, v. de machete (Peru); caí-escalera, corazón de negro, escalera de mono, pata de toro, p. de vaca, toro-pó (Arg.).

Behaimia *cubensis* Gris., the only species, is an unarmed tree or a shrub of general distribution in Cuba. The imparipinnate leaves have several narrow somewhat leathery leaflets; the flowers are borne in simple or branched, axillary or terminal racemes; the pods are very thin, almost membranous, usually one-seeded. In the eastern part of the island the timber is used for railway crossties and rural construction.

Heartwood brown or orange-brown; sharply demarcated from the thin yellowish white sapwood. Luster fairly low. Without distinctive odor or taste. Very hard, heavy, tough, and strong; texture rather fine; grain interwoven; difficult to work but finishes very smoothly; is highly durable. Has about the consistency of some of the dense species of *Tabebuia*. Apparently without commercial possibilities.

COMMON NAMES: Ciruelillo, guayacán blanco, guayacancillo, g. de costa, rana macho (Cuba).

Belairia, with two or three species of spiny shrubs and small trees, is common on poor savanna soils and rocky sites throughout Cuba and the Isle of Pines. The leaves are equally pinnate, with few to several small to very small leaflets; the little yellow flowers are solitary or clustered; the pods are small and compressed. The twigs bear numerous needle-like stipular thorns and the bark of the trunk separates into large red flakes. The wood is of good quality, but on account of the small sizes obtainable is used chiefly for fence posts.

Heartwood olive-brown, more or less streaked with black, deepening upon exposure; sharply demarcated from the thin whitish sapwood. Luster low. Odor and taste absent or not distinctive. Very hard, heavy, compact, tough, and strong; texture fine and uniform; grain irregular; rather difficult to work, but finishing very smoothly and taking a high polish; is very resistant to decay. Is suitable for articles of turnery.

COMMON NAMES: Alfiler, yamaquey, y. amarillo, y. blanco, y. común, y. de costa, y. de loma, y. de sabana, y. negro (Cuba).

Bergeronia *sericea* Micheli, the only species, is a shrub or small tree sometimes 35 feet tall with a fluted trunk a foot in diameter, of fairly common occurrence in the Chaco region of southern South America. The leaves are imparipinnate, with opposite leaflets; the yellow flowers are borne in simple axillary racemes; the pods when ripe break into indehiscent segments. The timber is considered of poor quality except for fuel and has no special value.

Heartwood (apparently absent in only available specimen) is said to be reddish yellow; sapwood yellowish. Luster low. Odor of heartwood said to be disagreeable. Hard, heavy, and strong but brittle; texture coarse; grain variable; not difficult to work, but requires care in seasoning to prevent splitting; durability poor. Has no commercial possibilities.

COMMON NAMES: Ibirá itá, i. saiyú (Arg.).

Bowdichia, in a strict sense, includes only two species of medium-sized to very large unarmed trees on non-inundated land in eastern and northern Brazil, the Guianas, and Venezuela. (For other species often referred to this genus see *Diplotropis*.) The leaves are odd-pinnate, with 5 to 21 leathery leaflets usually 1 to 2 inches long; the lilac-blue flowers are borne in large terminal panicles; the elongated flattened pods have a narrow wing along the upper suture and are indehiscent; the seeds are small, black, and hard.

Bowdichia nitida Benth. occurs in the forest of the Rio Negro and lower Amazon region; its leaves have 5 to 9 oval to somewhat acuminate leaflets. *B. virgilioides* H.B.K. has a greater range, Venezuela, the Guianas, the upper Rio Negro and lower Amazon, and southward into Minas Geraes and Espirito Santo; it has smaller, more numerous, and oval or rounded leaflets. In Venezuela it is of small size and characteristic of the savannas (Plate XXXII), being readily recognized by the abundance of its blue flowers, which appear when the tree is leafless, and its wine-red pods. In better situations elsewhere it attains a height of 150 feet and a trunk diameter of four feet.

The timber is well known in Brazil, where it is generally called Sucupira or Sapupira, though this name is also applied to similar woods. H. M. Curran states in a memorandum written in 1923: "Sucupira is the wood preferred above all others in Brazil for making hubs and felloes of cart wheels. The tree occurs in the coastal forests, but is not abundant, and the logs entering the market are mostly from 12 to 18 inches in diameter. This is not a criterion, however, of the maximum size of tree, since the native loggers usually bring out only the small and medium-sized logs of heavy timber. Another kind of Sucupira, known as Sucupira-assú, has been introduced into the market. It appears to be more abundant, but the wood is considered inferior." Two specimens of Sucupira-assú (Yale Nos. 4720 and 7181), the first supplied by Mr. Curran, have been identified by the authors as *Diplotropis* sp.

Heartwood dull, chocolate to reddish brown, with parenchyma striping; sharply demarcated from the whitish sapwood. Luster low. Without distinctive odor or taste. Very hard, heavy, tough, and strong; sp. gr. (air-dry) about 1.00; weight 62 lbs. per cu. ft.; texture very coarse; feel harsh; grain irregular and interwoven; difficult to work, but can be finished fairly smoothly; is highly resistant to decay. Useful for heavy, durable construction, but is not suitable for furniture, though it has the general figure of Partridge-wood (*Andira*).

COMMON NAMES: Alcornoque (Venez.); mach (Br. G.); cutiuba, cutiubeira, paricárana, sapupira, s. do campo, s. preta, sebipira, sicupira, sucupira, s. preta (Braz.).

Brasilettia, with several species of unarmed shrubs and small to medium-sized trees, often referred to *Caesalpinia* or *Peltophorum*, is fairly well represented in tropical America. The bipinnate leaves are large, with 2 to 4 pairs of pinnae and few or several large leaflets; the yellow flowers are borne in terminal or axillary racemes; the legumes are thin, flat, few-seeded, and indehiscent.

Brasilettia platyloba (S. Wats.) Britt. & Rose (= *Caesalpinia platyloba* S. Wats.) grows in the southwest coast region of Mexico where it is known as Arellano, Palo Colorado, and Quebracho; a similar or perhaps the same species, called Chacte, occurs in Yucatán. According to J. G. Ortega (see *Tropical Woods* 9: 20), the tree is usually less than 25 feet high, with a trunk

8 to 16 inches in diameter and covered with a smoothish silver-gray or pearl-gray bark finely pitted from exfoliating scales. The sapwood is of a canary-yellow color and about one inch thick. The heartwood is orange to orange-red; macerated in water, its extract is garnet; in alcohol, it is topaz. The timber is very hard and heavy (sp. gr. 1.05) and on account of its strength and resistance to decay is highly esteemed for fence posts, house posts, railway crossties, and mine props. It makes excellent fuel and charcoal.

The other woods available for this study are of three species: *Brasilettia velutina* Britt. & Rose, from Guatemala, where it is known as Aripín; *B. mollis* (H.B.K.) Britt. & Killip (= *Peltophorum Suringari* Urb.), a small tree in the understory of the forest in northern Colombia, where it is called Cañafistula de Monte; and *B. violacea* (Mill.) Britt. & Rose of Cuba. The last is a tree, generally small or medium-sized but occasionally large, occurring in Jamaica, Cuba, British Honduras, and Yucatán, Mexico. It is called Brasiletto in Jamaica, and Fawcett and Rendle (*Flora of Jamaica* 4: 2: 90) state that the timber is of excellent quality, being elastic, tough, durable, and capable of receiving a fine polish. "It is used for ornamental purposes in cabinet-making and is well adapted for making the spokes of wheels of carriages. It is of a beautiful orange color, full of resin, and yields a full tincture by infusion. It was largely exported in Sloane's time for the use of dyers." The species is common in Pinar del Río in Cuba and grows to large size. It is there known as Yarúa, and areas where the trees are numerous are called "yaruales"; from them are obtained many slender stems for telegraph and telephone poles and fence posts. The timber is also highly esteemed in wheelwright work and turnery.

The woods of all species examined are much alike, although the color varies with age. Heartwood bright orange to orange-red; sharply demarcated, at least in old trees, from the yellowish or nearly white sapwood. Luster medium to high. Without distinctive odor or taste. Very hard and heavy; texture fine and uniform; grain straight to irregular; not difficult to work, finishing very smoothly; is highly durable. Of little value, except locally, because of the limited supply.

COMMON NAMES (various species): Brasiletto (Jam.); yarúa (Cuba); arellano, brasil, brasilete, chacte, palo colorado, quebracho (Mex.); brasiletto (Br. H.); aripín (Guat.); cañafistula, c. de monte, guamito macho (Col.).

Brownea, with more than 20 species of unarmed shrubs and small to medium-sized or rarely large trees, has its center of distribution in Venezuela and Colombia. Three species have been segregated into a separate genus, *Browneopsis,* but the type, *Browneopsis ucayalina* Huber, has been transferred to *Brownea* by Ducke (*Arch. Jard. Bot. Rio de Janeiro* 4: 51). A study of the wood (Yale 2258; Pittier 5511) of *Browneopsis excelsa* Pittier, a forest tree 80 to 100 feet tall and 18 inches in diameter in southern Darién, Panama, failed to reveal any characters that would separate it from *Brownea*. The leaves of *Brownea* are simply pinnate, with one to several pairs of large leathery leaflets, often brilliantly colored when young; the rose or red, rarely white, flowers are large and in conspicuous heads or short racemes, terminal, axillary, or borne on the trunk or branches; the legumes are large, flat, leathery, and dehiscent, and contain few to several large flattened seeds. The timber is not utilized for any special purposes, as it is unattractive, scarce, and usually available only in small sizes.

Several of the species are grown for ornamental purposes. Regarding two of them in Panama, Pittier says (*Contrib. U.S. Nat. Herb.* 18: 4: 145–157; 1916): "*Brownea ariza* Benth. is indeed, as are several other species of the same genus, a beautiful ornament of tropical parks and gardens. . . . The dense foliage itself, with the new leaves brightly purple-colored and hanging in heavy bunches, never fails to attract the eye, and the crimson heads of the flowers are of a gorgeous beauty. . . . At blooming time *Brownea macrophylla* Linden is one of the most striking features of the foothill belt of the Sambú Valley. In the semi-darkness of the dense tropical forest, its erect stems, entirely covered by the red blossoms, and showing for an instant between the trunks of the larger trees, strike the eye of the traveller almost as would lightning. . . . In every investigated case the trunk of *Brownea macrophylla* was found to be hollow and inhabited by a medium-sized black ant."

Heartwood not normally developed in specimens, but probably dark brown to blackish, judging from knots and wounds; sapwood brownish gray. Luster medium. Odorless and tasteless. Sapwood hard, heavy, tough, and strong; sp. gr. (air-dry) 0.75 to 0.85; weight 48 to 53 lbs. per cu. ft.; traumatic heartwood considerably denser; texture medium; grain rather irregular; not difficult to work, but is rather splintery; finishes smoothly, but is not attractive; heartwood probably highly durable.

COMMON NAMES: Flor de Jesucristo (C.R., int.); arizá, cuchillito, unmancri (Pan.); arizá, arizal, clavellín, flor de la cruz, palo de cruz, p. de rosa (Col.); guarabo, palo de sangre, p. de cruz, rosa blanca, r. de belvería, r. de cruz, r. de montaña, r. de monte, r. guaraba, r. macho (Venez.); bimiti topong, rose of the mountain (Br. G.); monterillo (Ec.); rosa da montanha, sol da Bolivia (Braz.).

Brya, a West Indian genus of prickly shrubs and small trees, consists of two closely related species, namely, *B. buxifolia* (Murr.) Urb. of the Island of Haiti and *B. Ebenus* DC. of Jamaica and Cuba, and has long been known as the source of a commercial timber known as Cocus or Granadillo. Mature trees are generally not over 25 feet tall and eight inches in diameter, but occasional individuals are considerably larger. The branches are clustered and armed with short sharp foliar prickles. The imparipinnate leaves are deciduous in dry weather and often have the petiole so reduced that the little leaflets appear single and sessile; the bright yellow or orange flowers are in axillary or subterminal clusters; the pods are indehiscent, jointed, dividing into 2 or 3 segments, the upper one usually sterile.

The logs, which are exported from Jamaica, Cuba, and Dominican Republic, are usually 3 to 6 inches in diameter and 4 to 8 feet long. They are shipped entire to protect the heartwood, the only valuable part, from checking and splitting. The bark is dark gray furrowed and shaggy. The wood is used in limited quantities for making musical instruments, particularly flutes and clarinets, for the handles of table cutlery, small articles of turnery, brush backs, and inlay.

Heartwood rich brown, variegated or finely striped, sometimes with an olive hue when fresh, deepening upon exposure; appears waxy; sharply demarcated from the yellowish sapwood which is from half an inch to an inch thick. Very hard, heavy, compact, tough, and strong; sp. gr. (air-dry) 1.05 to 1.20; weight 69 to 78 lbs. per cu. ft.; texture very fine and uniform; grain straight to finely roey; not difficult to work, finishing very smoothly with a high natural luster; requires care in seasoning but holds its place well when manufactured; is highly durable.

COMMON NAMES: Cocos, cocus, c. wood, ebony (American, brown, green, Jamaican, West Indian), granadillo, torchwood (English); amerikanisches Ebenholz, Kuba-Grenadilleholz, Kokusholz (Germ.); legno granadillo (Ital.); espino de sabana, granadillo, ojo de perdiz (Cuba); galle-galle (Haiti).

Caesalpinia, in a broad sense, includes more than 125 species of small to large trees and upright or scandent shrubs of pantropical distribution though most abundant in comparative localities in eastern Asia, the West Indies, Middle America, and extra-Amazonian South America. Many of the species are spiny or prickly.

In place of a single large genus with numerous sections, some botanists recognize several distinct genera. From a study of the woods it appears that such segregations are justifiable and they are accordingly treated separately in this book. On a basis of the color two groups are readily recognizable: (1) with bright orange or orange-red heartwood, including *Brasilettia, Caesalpinia* (*sens. str.*), *Guilandina,*

Poinciana, and *Tara;* (2) with dark red to chocolate-brown heartwood, including *Libidibia, Nicarago, Poincianella* (in part olive-brown), and *Russelodendron.* Ripple marks characterize three of these genera, namely, *Libidibia, Brasilettia,* and *Guilandina* (in part), the only ones of any economic importance for timber. The orange-red class contains the coloring principle brasilin and might be designated the Brazilwood group, although the principal source of the Brazilwood used in the United States is *Haematoxylon brasiletto* Karst. of Central America. The wood from which the South American country, Brazil, derived its name is *Guilandina echinata* (Lam.) Spreng. (= *Caesalpinia echinata* Lam.).

There are 11 species of *Caesalpinia* (*sens. str.*) in the West Indies, another in Mexico, and a few in South America. They are prickly or unarmed shrubs or small trees rarely over 25 feet tall, though *C. barahonensis* Urb. is said to attain a height of 65 feet in the Dominican Republic. The leaves are evenly bipinnate, with 1 to 7 pairs of pinnae; the flowers are racemose and the petals are small, mostly red or yellow, sometimes white or greenish; the pods are flat or compressed, unarmed, and elastically dehiscent.

Caesalpinia brasiliensis L., which sometimes is mistakenly credited with supplying the Brazilian Brazilwood, is a shrub or little tree of very limited distribution in Haiti. *C. bahamensis* Lam. occurs in the Bahamas and Cuba. *C. Andreana* Mich. is a small low-branching tree native to Colombia and Peru. All of the foregoing woods were formerly used to some extent for dyes and were known as Brazil or Brasiletto (see Kew. *Bull. Misc. Inf.* for 1916, pp. 214–216).

Heartwood bright orange, deepening upon exposure; sharply demarcated from the thin whitish sapwood. Luster high. Without distinctive odor or taste when dry; sometimes slightly fragrant when fresh Hard and heavy to moderately so; texture fine and uniform; grain straight to roey; very easy to work, finishing smoothly, with high luster.

COMMON NAMES: Brasilete, b. colorado (Cuba); brasiletto, braziletto (Bah.); brazil (Dom. R.); brésillet (Haiti); palo brasil (Col.).

Calliandra (or *Anneslia*) is a widely distributed tropical American genus of more than 100 species of unarmed shrubs and small trees of no importance for their timber. The leaves are bipinnate; the flowers have numerous, long-exserted stamens; the legume is flat and straight and the valves recurve from the apex when the pod opens. Some of the shrubs are planted for decorative purposes. Heartwood not seen; sapwood thick, yellowish white, not highly lustrous. Hard, heavy, and strong; medium-textured; of straight to irregular grain; not difficult to work. Used to a minor extent locally for tool handles, implement frames, and fuel.

COMMON NAMES: Night-flowering acacia (Jam.); moruro de costa?, sopillo (Cuba); acacia, cojobillo, zarza boba (P.R.); granolina (Dom. R.); petit gaiac (Haiti); cabalpich, cabellito, c. de ángel, c. de una vara, cabellos de ángel, cabeza de ángel, canela, carboncillo, charamusco, chivato grande, coquito, day, gavia, hierba burro, h. de canella, h. de ángel, lele, pambotano, pich, plumita, tabordillo, tentzonoxochitl, tepachera, tepejiloxochitl, tepexiloxochitl, texoxochitl, timbrillo, tlacoxiloxochitl, tlamacazatzotl, tlamacazcatacotl, tzonoxochitl, uaylahaltsac, xiloxochitl, xiloxochicuahitl, yalahatsac, zapotillo (Mex.); cabello de ángel, capulín de corona, ichumpich, old man's beard (Br. H.); barbón moñtanés, guacamayo, montés, pelo de vieja, salitrero (Salv.); carboncillo, c. blanco, c. rojo, pelo de ángel (C.R.); aromo (Pan.); ballotica, chicharrón, clavellina pichindé, rayado, tamarindo de monte (Col.); andarillo, cansacaballo, cimbra-potro, clavellina, c. rosada, c. serrana, mangle, quebra foice (Venez.); angelim capucina, mandarave, pau de salsa, salsa (Braz.); bobansana, yaco-shapano (Peru); chicotedeniño, niño-azote (Arg.).

Callistylon, with a single species, *C. arboreum* (Gris.) Pitt. (= *Coursetia arborea* Gris. = *Humboldtiella ferruginea* Harms), is an unarmed shrub or slender tree rarely over 25 feet tall growing in hot dry regions in northern South America (see *Journ. Wash. Acad. Sci.* 18: 8: 210). The leaves are pinnate, with 10 to 18 thin oval light green leaflets; the purplish white flowers are borne mostly in terminal racemes; the legumes are slender and flattened, the valves constricted between the numerous small black seeds, and twisting in dehiscence. The only recorded use for the wood is for fuel.

Heartwood yellowish white, not distinguishable from sapwood except that the vessels are filled with tyloses; wound areas dark brown. Luster medium. Odorless and tasteless. Very hard, heavy, tough, and strong; texture medium coarse; grain fairly straight; not easy to work; checks badly in drying; is probably not durable. Has no commercial possibilities.

COMMON NAMES: Maquiro, ramocillo, ramoncillo (Col.); grifo blanco, jebe negro (Venez.).

Campsiandra, with four species of unarmed medium-sized to large trees, is limited in its distribution to the Amazon basin. The leaves are imparipinnate, with several large leathery leaflets; the yellow or red flowers are borne in panicled racemes; the legumes are large, flat, leathery, and two-valved. The seeds are large and are sometimes used by the natives as a source of starch. The timber is employed locally to a minor extent for heavy and durable construction.

Campsiandra surinamensis Kleinh. grows along the Corantyn River in Surinam. *C. comosa* Benth. occurs rather sparingly in British Guiana and the Orinoco basin. The two Amazonian species are *C. angustifolia* Spruce and *C. laurifolia* Benth. The latter is very common along the borders of clearwater streams and lakes and in northeastern Peru, according to Williams (*Woods of northeastern Peru,* p. 194), it is sometimes 90 feet tall with a fairly straight and cylindrical trunk 24 inches in diameter and free of branches for 40 feet. Usually, however, the trees are smaller and not of good timber form.

Heartwood rather dull reddish brown, deepening upon exposure; marked with narrow lighter-colored vessel lines which may be highly conspicuous on tangential surface; rather sharply demarcated from the brownish or pinkish sapwood. Without distinctive odor or taste. Very hard, heavy, tough, and strong; sp. gr. (air-dry) 0.90 to 1.10; weight 56 to 69 lbs. per cu. ft.; texture rather coarse; grain fairly straight; not easy to work, but finishing smoothly and attractively; is highly durable.

COMMON NAMES: Chigo (Venez.); huacapú-rana, pampa huacapú-rana (Peru); acapú-rana, a. vermelha, capoerana, comandá-assú, cumandá, gapo, manaiára (Braz.).

Cascaronia *astralagina* Gris., the sole species, is a large rue-scented tree of rather infrequent occurrence in Jujuy, Salta, and Tucumán, Argentina. The trunk, which sometimes is up to three feet in diameter, has a thick, gray, deeply grooved, corky bark containing a red resin. The leaves are odd-pinnate, with several alternate leaflets having glands on the under side; the yellow flowers are borne in axillary racemes; the flat, narrowly winged pod is indehiscent and usually contains a single seed. The timber is used locally to a minor extent for common lumber and fuel.

The following description is based on a single specimen (Yale 36033) supplied by the Instituto Miguel Lillo, Buenos Aires. Normal heartwood absent from sample; wound areas brown with black edges; sapwood yellowish. Luster fairly high. Odorless and tasteless. Moderately hard and heavy, the dark streaks much denser; texture rather fine; grain irregular to finely roey; working properties fair. Presumably of no possibilities for export.

COMMON NAMES: Cascarón, saúco hediondo, tipa amarilla (Arg.).

Cassia, in the broadest sense of the generic term, includes about 500 species of herbs, shrubs, woody vines, and small to medium-sized, occasionally large, trees of pantropical distribution. The principal uses are medicinal. Some of the plants are cultivated for ornamental purposes and *C. Fistula* L., a rather large Indian tree, has become locally naturalized in many parts of the world. The genus is abundantly represented in tropical America but supplies few timbers and they are of only local utility. The genus as a whole exhibits a wide range of variation in its morphological and anatomical characters, which some taxonomists consider sufficient to justify the setting up of several distinct genera. The authors have found it convenient to follow the classification proposed by Britton and Rose (*North American Flora* 23: 229–301), and descriptions of *Cassia* woods appear under the following generic names: *Adipera, Cassia, Chamaefistula, Chamaesenna, Cowellocassia, Isandrina, Peiranisia,* and *Pseudocassia.* The species of *Apoucouita,* a section of the subgenus *Psilorhegma,* are treated as a separate group of *Cassia.*

Cassia, in a restricted sense, is a pantropical genus with about 25 species, some of them shrubs but mostly medium-sized to large trees. They have pinnate leaves, usually with many pairs of leaflets; the flowers are showy and in various colors, yellow, white, rose, red, purplish, solid or mixed; the legumes are elongated, woody, indehiscent, and septate between the numerous transversely oriented seeds. The American timbers, so far as they have been studied, are generally unattractive and of no special value. An exception must be made in the case of *C. fastuosa* Willd., judging from a specimen (Yale 22073) collected on the Ford concession in the Amazon region and determined by P. C. Standley. The tree is said to be medium-sized, occurring mostly in second-growth and on abandoned fields. The heartwood is a rich light reddish brown with a golden luster, finely feather-grained, and with distinct vessel lines. It is moderately hard and heavy but easy to work and should make good furniture. The local name is given as Faveirinha; other names reported are Baratinha and Angico, though the latter is generally applied to certain species of *Piptadenia.*

The most widely distributed species is *Cassia grandis* L.f., usually a spreading tree of medium height in the open, but attaining a height of 100 feet in the forest. Standley says of it (*Flora of the Panama Canal Zone,* p. 200): "In some places along the Pacific slope of Central America it is abundant, the trees when loaded with their delicate pink blossoms having a striking resemblance to apple trees. The wood is used for construction purposes." It is common in the West Indies and on the mainland from southern Mexico to northeastern Brazil and is often planted for shade. It has long coarse leaves with 10 to 20 pairs of rather large leaflets, and heavy squarish pods more than an inch in diameter and up to two feet long.

Heartwood (of *Cassia grandis*) variegated brown with light and dark, sometimes purplish, streaks and patches and marked with prominent vessel lines; rather sharply demarcated from the thick brownish or nearly white sapwood. Luster medium. Without distinctive odor or taste. Of medium weight, but firm and tough; texture coarse; grain straight to very irregular; easy to cut, but saws rather woolly, and is inclined to be "stringy"; durability doubtful. Of no promise commercially.

COMMON NAMES: *Cassia grandis:* Horse cassia (Jam.); cañandonga (Cuba); cañafístula cimarrón (P.R.); cañafístula, c. grande, quauhuayo (Mex.); beef-feed, bookoot, bookut, stinking-toe (Br. H.); carao (Hond.); carago, caragua, caragüe, carao, cargo (Salv.); carámano (Nic.); carao, sándalo (C.R.); cañafístula, carao (Pan.); cañafístula gruesa, cañandonga (Col.); cañafistola, cañafistula, c. macho, cañaflote (Venez.); cañafistola, jeneuna, marimary grande da terra firme, m. preto, m. rana, m. saro (Braz.).

Apoucouita, a section of *Psilorhegma,* a subgenus of *Cassia,* includes four or five species of medium-sized to large unarmed trees, centering in the Brazilian Amazon region where, because of their blackish brown heartwood, they are generally known as Coração de Negro. The leaves are pinnate, with a few pairs of rather large leaflets or many pairs of small ones; the large yellow flowers are borne in clustered racemes on twigs of the previous year. The heavy timber is noted for its durability and is commercially important in certain localities.

The species with the greatest range is *Cassia apoucouita* Aubl., which occurs from

French Guiana to Rio de Janeiro, Brazil. H. M. Curran says that it is fairly common in the rain forest along the coast of Bahia and Espirito Santo and attains a height of 100 feet with a trunk two to three feet in diameter and free of branches for 40 feet or more. The timber is considered excellent for fine furniture but only small quantities appear on the market. It is called Pau Perola (pearl wood). According to Ducke (*Arch. Jard. Bot. Rio de Janeiro* 4: 281) the timber is in demand for posts in Gurupá but is not known in other cities of the lower Amazon.

Cassia scleroxylon Ducke is a medium-sized Amazonian tree with a deeply furrowed trunk. The leaflets, usually three pairs, are sessile and the rhachis is channeled or winged, suggesting certain species of *Inga*. The timber is much used locally for durable construction. *C. xinguensis* Ducke, called Fava de Bezouro, is a little tree and according to Ducke (*loc. cit.*) the wood is white and soft throughout, though elsewhere he states that traces of dark-colored heartwood were seen. *C. adiantifolia* Benth. is a medium-sized to rather large tree of handsome appearance because of abundant flowers and graceful fern-like foliage.

Heartwood dark brown, with blackish streaks; superficially dull, but often with a golden luster beneath; transition to sapwood gradual to abrupt. Without distinctive odor or taste. Very hard, heavy, tough, and strong; sp. gr. (air-dry) 0.95 to 1.20; weight 60 to 75 lbs. per cu. ft.; texture coarse; feel harsh; grain irregular; difficult to work, but can be finished smoothly; highly resistant to decay. Some specimens resemble *Melanoxylon*. The wood of *C. adiantifolia* (Yale 20721; Ducke 40) has a higher luster and is coarser-textured and more fibrous than the others, and also differs in structural details.

COMMON NAMES: Apoucouita (Fr. G.); coração de negro, fava de bezouro, irary, memby, muirá paxiúba, m. pixuna, pau perola, p. preto, pixuneira-rana (Braz.).

Cedrelinga *catenaeformis* Ducke, the single species, is a tree 100 to 160 feet tall with an enormous trunk five to nine feet in diameter, growing in the humid "terra firma" of the Brazilian Amazon region. It has bipinnate leaves with few large leaflets; small white flowers in little heads in large terminal inflorescence; flat, indehiscent, articulated legumes with one large seed in each segment. The aspects of the tree, particularly the appearance of the bark, suggests *Cedrela,* and for that reason it is often called Cedro-rana (false Cedar). (See *Archivos do Jardim Botanico do Rio de Janeiro,* 1922, p. 70.)

Heartwood pale brown, with a golden luster; prominently marked with dark red vessel lines; merging gradually into the lighter-colored sapwood. Odor and taste absent in dry specimens, but fresh timber is said to emit a disagreeable scent when worked. Of medium density; sp. gr. (air-dry) 0.65; weight about 40 lbs. per cu. ft.; firm and tough; coarse-textured; roe-grained; saws woolly but is easy to cut; can be finished smoothly; is probably fairly durable. Though not utilized, the timber is suitable for general construction and inexpensive furniture.

COMMON NAMES: Cedro-rana, iacaica, paricá, yacayacá (Braz.).

Centrolobium, with five or six species of medium-sized to large well-formed unarmed timber trees, is of rather infrequent occurrence from Panama to Ecuador and southern Brazil. The large to very large leaves are imparipinnate, with 7 to 17 opposite to alternate leaflets bearing numerous glands on the under surface; the yellow or purplish flowers are borne in terminal panicles; the large samara-like indehiscent pod contains 1 to 3 seeds and looks like a Chestnut bur with the wing of a gigantic Maple seed attached to it.

Two species are generally recognized in eastern Brazil, from Bahia to Paraná, namely *Centrolobium robustum* Mart. and *C. tomentosum* Guill., generally known as Araribá and Putumujú. The trees grow rapidly and are often planted for ornamental purposes. The bark of the stem and roots contains a reddish sap that is used as a dye, although the color is not permanent. The timber is of good quality and exhibits considerable range of color from yellow through orange to red, often distinctly variegated, but tending to become brown upon exposure. Though not abundant it has a recognized position in the market and is highly esteemed for furniture, interior trim, doors, flooring, tight cooperage, and durable construction.

Centrolobium paraense Tul. occurs in the Rio Branco region of the Amazon basin, but apparently not in the State of Pará as the specific name would indicate (see *Arch. Jard. Bot. Rio de Janeiro* 4: 312). The usual name is Pau Rainha, but it is sometimes confused with Muiráquatiara (*Astronium*), another variegated timber. The range of the species extends into the Guianas and Venezuela.

The principal species in Venezuela and Colombia is *Centrolobium orinocense* (Benth.) Pittier. According to Amando Dugand, the tree in northern Colombia "is fairly common in upland transition forests, but is generally less than 40 feet tall with a trunk not over 16 inches in diameter. Heartwood, which is of a beautiful reddish orange color, is formed early, and slender trunks have thin sapwood and are used for house posts. The timber is known as Balaústre in the State of Magdalena and in the market at Barranquilla and is growing in demand for cabinet work and fine furniture." In Venezuela, where it is called Cartán as well as Balaústre, the timber is used for the frames of buildings and to some extent for furniture.

The Ecuadorean species is *Centrolobium ochroxylon* Rose, known as Amarillo de Guayaquil. The wood is of a brilliant orange color with purplish streaks when fresh but soon turns to a fairly uniform dark red on the surface. *C. patinense* Pittier of Panama has a similar wood, but there is little information available concerning the species.

In *Timbers of Tropical America*, pp. 292–293, attempt was made to divide the woods of *Centrolobium* into two groups, but study of more specimens has failed to provide a specific basis for the differences noted. It appears likely that conditions of growth may be a greater factor than the species in determining color, porosity, density, and certain other characteristics. The anatomy of the woods of the genus as a whole is fairly consistent and distinctive.

Heartwood yellow or orange, typically variegated, sometimes "rainbow-hued," usually changing to red or brown; rather sharply demarcated from the yellowish sapwood. Luster medium to high. Without distinctive odor or taste. Variable in density, but mostly hard, heavy, and strong; sp. gr. (air-dry) 0.75 to 1.00; weight 47 to 63 lbs. per cu. ft.; texture fine to rather coarse; grain straight to irregular; easy to work, finishing very smoothly; holds its place well when manufactured; is highly durable. Probably not of much potential value for export, as the local demand appears equal to the supply.

COMMON NAMES: Canary wood, porcupine wood, zebra wood (trade, U.S.A.); amarillo, bateo? (Pan.); balaústre, colorado, guayacán hobo (Col.); balaústre, birote de montaña, caguaro, cartán, cartanyé (Venez.); kartang, redwood (Br. G.); amarillo, a. de Guayaquil, a. lagarto, palisandro claro (Ec.); araraúba, ararauva, araribá, a. amarello, a. branco, a. grande, a. piranga, a. preto, a. rosa, a. testa de bois, a. tinga, a. vermelho, ararúva, caáguassú, caá-mirim, carijó, cartanié, guararibá, iriribá, i. rosa, mutumujú, pau da rainha, p. rainha, potomujú, putumujú, p. amarello, p. vermelho, putumuyú, tipiri (Braz.); morosimo (Par.).

Cercidiopsis *microphylla* (Torr.) Britt. & Rose (= *Parkinsonia microphylla* Torr.), the only species, is a spiny shrub or a little tree less than 25 feet high and a foot in diameter, occurring on dry plains and hillsides in southern Arizona and adjacent regions of California and Mexico and northern Baja California. The twigs are spine-tipped, but without lateral spines. The bipinnate leaves are small, with few pairs of deciduous leaflets; the pale yellow flowers are in short racemes; the pods are short moniliform and are edible. The wood is used for small carved articles and for fuel.

Heartwood yellowish or orange brown, rather oily looking; sapwood nearly white. Luster medium. Odorless and tasteless when dry. Rather hard and heavy, but brittle; sp. gr. (air-dry) 0.75; weight 47 lbs. per cu. ft.; texture rather fine; grain irregular; easy to carve, finishes very smoothly; durability doubtful. Of no commercial possibilities.

COMMON NAMES: Horse bean, little-leaf horse bean, palo verde (U.S.A.); lebón, palo verde, retama (Mex.).

Cercidium, with about 10 species of shrubs and small trees, occurs in dry regions from the southwestern United States to Argentina. The younger branches are usually armed with short straight spines. The leaves are bipinnate, the leaflets few, small, often fugacious; the yellow flowers are borne in axillary racemes; the legumes are flat or swollen, tardily dehiscent, the seeds single or few. The most widely distributed species is *Cercidium praecox* (R. & P.) Harms, extending from Mexico to Argentina. It is sometimes 35 feet tall with a short trunk a foot in diameter, but more often is shrubby. During certain months of the year the stem and branches become covered with a semi-transparent greenish gum or lac which is soluble in alkali and is used locally for making soap. About the only use for the wood is for fuel.

Normal heartwood absent from specimens; injured areas and knots reddish brown; sapwood yellowish. Luster low to medium. Odorless and tasteless. Hard, moderately heavy, brittle; texture uniform; grain irregular; easy to work, finishing smoothly; is probably perishable. Of no commercial possibilities.

COMMON NAMES: Green-barked acacia, horse bean, palo verde (U.S.A.); baie à onde, printemps (Haiti); llave (Curaçao); mantecoso, palo de berria, p. de mantecoso, p. verde, retama (Mex.); quica (Col.); brea, cuica, palo de brea, p. de cuica, yabita, yabo, yavo (Venez.); brea (Arg.).

Cercis, with six or seven species of unarmed shrubs or small trees, is represented in southern Europe, Afghanistan, Japan, the United States, and northern Mexico. The best known species are *C. Siliquastrum* L., the European Judas-tree, and *C. canadensis* L., the American Redbud or Judas-tree, with a range from Ontario to Texas and probably into Mexico. According to Standley (*Trees and shrubs of Mexico*, p. 412), "*C. reniformis* Engelm. (*C. texensis* Sarg.) has been reported from Mexico. That species does not appear to be very clearly distinct from *C. canadensis*. The Mexican material at hand seems to represent a single species and agrees better with *C. canadensis* than with the material that has been referred to *C. renifórmis*." The leaves are simple, entire, prominently nerved, and deciduous; the flowers, which are usually pink or purplish, are borne in clusters on the old wood before the appearance of the leaves. Because of their attractive appearance the trees are frequently planted for decorative purposes. The wood is rarely used because of the small size and scarcity of the trees, but is suitable for small cabinet work and turned articles.

Heartwood light olive-brown, often with dark streaks, with a golden luster, deepening to russet-brown; sharply demarcated from the white sapwood. Odorless and tasteless. Hard, heavy, tough, and strong; sp. gr. (air-dry) 0.75; weight 47 lbs. per cu. ft.; texture medium; grain fairly regular; easy to work, finishing very smoothly; is highly durable.

Chamaefistula is a large genus (or a section of *Cassia*) of unarmed erect or scandent shrubs, woody vines, and small trees, widely distributed in tropical America. The leaves are pinnate, with one or two pairs of large leaflets; the large yellow flowers are in terminal and axillary racemes; the legumes are slender, cylindrical, rather leathery, dehiscent along the inner suture, the seeds arranged transversely in glutinous pulp. The plants are of no commercial importance.

Heartwood brown or yellow, sometimes highly lustrous; slow in forming in some species; not always sharply demarcated from the yellowish or brownish sapwood. Without distinctive odor or taste. Moderately hard, heavy, tough, and strong; texture rather fine to rather coarse; grain mostly straight; easy to work, finishing smoothly; durability doubtful. Sizes too small for the timber to be useful for many purposes.

COMMON NAMES: Sen del país, platanillo (Cuba); hediondilla (P.R.); bois d'anneau (Haiti); quelite, quitegato (Mex.); vainillo (Nic.); sen de palillo (C.R.); bombillo, yema de huevo, platanito (Col.); velero (Venez.); gadoboom (Sur.); alleluia, São João (Braz.); amargo-caspi, flor de caña, matáyo, tangarana, yana huira, y. urarú (Peru); pito-movéva, sen, yerba de burro (Arg.).

Chamaesenna is a large genus (or a section of *Cassia*) of mostly unarmed shrubs, some small trees, and a few herbaceous plants, with numerous representatives in tropical America. The most widely distributed American species is *C. reticulata* (Willd.) Pittier, usually a coarse spreading shrub but sometimes a tree 20 feet high, ranging from southern Mexico to Brazil and Bolivia. The leaves are pinnate, with 8 to 12 pairs of large leaflets; the showy yellow flowers are borne in long racemes in the upper axils; the dehiscent legumes are flat, 3 to 6 inches long and a half-inch wide, and contain numerous slender seeds. The species is of no economic importance.

Normal heartwood absent from specimens, but probably brown. Sapwood nearly white, lustrous. Odorless and tasteless. Rather light and soft, to medium; texture moderately coarse; grain straight; very easily worked, finishing smoothly; durability of heartwood unknown.

COMMON NAMES: Yaaxhabin (Mex.); barajo (Guat.); barajillo, barajo, b. negro, sambrán, s. de río, vainilla (Salv.); saracontil, sorocontil (Nic.); candelita, saragundín (C.R.); laureño, wild senna (Pan.); Martín Galvis, montes de oc, mucuteno (Col.); espino amarillo, majagüillo, tarantán (Venez.); slabriki (Sur.); matapasto, m. grande (Braz.); retama, sapechihua (Peru); cabello de indio, pichana, retama (Arg.).

Chloroleucon, with at least a dozen species of spiny shrubs and small or occasionally medium-sized to large trees, is well represented throughout tropical America. The best known and most widely distributed species is *C. tortum* (Mart.) Pittier (= *Pithecolobium tortum* Mart.), a low-branched crooked smooth-barked shrub or small tree with a range including parts of the West Indies, Baja California, Central America, and northern and eastern South America to northeastern Argentina. The sapwood is thin and white. The heartwood is distinctive because of its color, which varies from golden brown to greenish yellow-brown, with a rather waxy appearance, and suggesting *Isandrina* (Caesalpiniaceae). It is firm and hard, fine-textured, of interlocked grain, easy to cut, finishes very smoothly with a high natural polish, and is durable. Owing to the small size and poor timber form of the trees, the uses are very limited, although the straighter stems make excellent fence posts and some of the others are suitable for small articles of turnery. The fumes of the burning wood are said to be of unpleasant odor.

Chloroleucon vinhatico Record is a large well-formed timber tree of scattered occurrence, rarely more than two or three per acre, in association with the Vinhatico Castanho (*Plathymenia reticulata* Benth.) on the upper slopes of hills in the Bahia region of Brazil, according to H. M. Curran who collected wood samples and herbarium material of it. It grows to a height of 100 feet, with a nearly cylindrical slightly buttressed bole three to four feet in diameter and clear of branches for 40 to 50 feet. The bark is thin and smooth, becoming somewhat shaggy on old trees. The young branches are armed with stout, straight, sharp, stipular spines often more than an inch long, while young sprouts are a veritable mass of thorns three or four inches in length; the bipinnate leaves, which suggest those of the Honey Locust (*Gleditsia triacanthos* L.), have a gland on the petiole below the insertion of the lowest of the 4 to 6 pairs of pinnae; the leaflets are small and crowded and bear numerous short coarse hairs, such as occur also on the petioles, peduncles, and twigs. The flowers are yellow, nearly glabrous, small (the calyx tube 2 to 3 mm., the corolla about 1 cm. long), sessile and spicate, the peduncles occurring 1 to 3 together in the axils of the leaves. The pods are small and contorted. It is known as Vinhatico de Espinho, a name also applied to *C. tortum,* but the tree differs from that species in its size and habit of growth and in certain structural details of the wood.

Heartwood lustrous yellow, sometimes with lighter and darker shades, slowly turning brown superficially; injuries may produce roseate discoloration. Has a waxy feel. No distinctive odor or taste. Rather light in weight but firm and tough; sp. gr. (air-dry) 0.60; weight about 37 lbs. per cu. ft.; grain fairly straight to roey; texture medium. Seasons readily, is not difficult to work, turns fairly well, takes a high natural polish, and holds its place when manufactured; appears resistant to decay. It differs from *C. tortum* in being coarser-textured, having more distinct vessel lines, larger rays (1 to 3 cells wide and up to 35 cells high), more sharply defined parenchyma, and in absence of the greenish tinge to the heartwood. It resembles Vinhatico Castanho (*Plathymenia*), but has smaller pores, less distinct vessel lines, more parenchyma, and no pronounced tendency to ripple marks; it is much less common in the local markets, but is used for the same purposes, namely, carpentry and general construction.

No specimens of *Pithecolobium scalare* Gris. are available for study, but from descriptions of the tree and wood there apparently is no doubt that it is a species of *Chloroleucon.* It occurs sparingly in northwestern Argentina and in Paraguay. It attains a height of about 50 feet and a diameter of 30 inches. Its yellowish brown timber is considered excellent for carpentry, furniture, ship-building, and general construction.

Common names: Aroma, humo, h. de costa, h. de sabana, h. espinoso (Cuba); losange (Haiti); cucharo, palo fierro, poralana (Mex.); guayabo de montaña, guayacán (Hond.); espino amarillo (Pan.); angarillo, calentura, hoyo de zorro, olla de zorro (Col.); cañaflote, cuchibán, cuchivano, cují, c. amarillo, c. blanco, c. de las vegas, macagua, quiebrahacha, retuerto (Venez.); algarrobo (Peru); amarello, arapiraca, arvore de macaco, cacunda, esponjeira, jaruma, jurema branca, vinhatico de espinho, v. de macaco (Braz.); espinello, palo cascarudo, tataré, t. blanco (Arg.).

Cladrastis is a small genus of trees with a few species in China and Japan and one in the south-central hardwood region of the United States. The American species, *C. lutea* (Michx. f.) K. Koch, is a medium-sized deciduous tree, sometimes 50 or 60 feet in height, with a wide-spreading crown and a short smooth-barked trunk 18 to 24, rarely up to 40, inches in diameter. The branchlets are slender and, having no terminal bud, are rather zig-zag. The leaves are odd-pinnate, with 7 to 11 rather large, usually alternate leaflets, and the base of the petiole, which is enlarged and hollow, covers the naked axillary buds; the white slightly fragrant flowers are borne in showy terminal panicles; the pods are thin, flat, tardily dehiscent, and contain 4 to 6 compressed seeds. The tree is cultivated for ornamental purposes and is hardy as far north as Massachusetts. The timber is little used because of its scarcity and poor form; it has no commercial possibilities.

Heartwood bright yellow, becoming brownish upon exposure; interior of old trees sometimes orange-colored; sapwood thin, white. Luster rather high. Odor and taste absent or not distinctive. Hard, heavy, tough, and strong; texture moderately coarse; grain straight to irregular; not difficult to work, finishing smoothly, but not very attractively; durability medium.

Common names: Gopherwood, yellow ash, y. locust, y. wood (U.S.A.).

Clathrotropis, with two or three species of small to medium-sized unarmed trees, is apparently limited in distribution to the central and northern Amazon region. The group is sometimes included with *Diplotropis.* The leaves are very large, imparipinnate, with 5 to 7 leaflets; the white or pale violet flowers are borne in large panicles; the pods are large, woody, usually 1-seeded, and elastically dehiscent.

Wood specimens of two species are available, namely, *Clathrotropis nitida* (Benth.) Harms and *C. macrocarpa* Ducke. The two are so distinct that it appears doubtful if they are congeneric. (See *Tropical Woods* 31: 16.) The timber of *C. nitida* is used to a limited extent for heavy, durable construction under the name of *Acapú* (more commonly applied to *Vouacapoua*). No uses are recorded for the other species. In the first species the heartwood is dark brown, somewhat variegated, prominently veined with parenchyma; has a waxy appearance; sharply defined from the yellowish sapwood. Luster rather low. No distinctive odor or taste. Very hard, heavy, tough, and strong; sp. gr. (air-dry) 1.20; weight 75 lbs. per cu. ft.; texture medium coarse; grain mostly straight; not easy to work, but finishing very smoothly and taking a good natural polish; durability high. In the Partridge-wood group, but finer-textured and with higher rays than the others. In *Clathrotropis macrocarpa* the heartwood is light brown, with a pinkish hue, the unfinished surface dull and of a mealy appearance; rather sharply demarcated from the thick yellowish sapwood. Odorless and tasteless. Hard, heavy, tough, and strong; texture very coarse; feel harsh; grain fairly straight; not easy to work, but finishing smoothly; durability doubtful. An unattractive wood, apparently of no commercial possibilities.

COMMON NAMES: *Clathrotropis nitida:* Acapú, a. do igapó (Braz.); *C. macrocarpa:* Cabarí, cabory, timbó pau, t. rana.

Clitoria, with about 70 named species of erect or scandent herbs and shrubs and a few trees, is widely distributed in tropical and warm regions of the world. The leaves are odd-pinnate, usually with three, sometimes very large, leaflets; flowers are large and showy, often bicolored, solitary or in clusters or short racemes in the axils of the leaves; the pods are linear, compressed. Some of the species are cultivated for ornament. The blue flowers of the pantropical *C. ternatea* L. are the source of a dyestuff used like litmus as a sensitive indicator of acidity and alkalinity. *C. Fendleri* Rusby is said to be a fairly large tree in northern Colombia. *C. brachycalyx* Harms is a deciduous tree about 40 feet high in British Guiana. *C. Hoffmanseggii* Benth. of the Amazon region ranges in size from a large shrub to a medium-sized tree; it is known locally as Faviera. Wood samples of these last three species are similar.

Heartwood absent or not distinguishable from the grayish yellow sapwood; with distinct parenchyma markings. Luster rather low. Odorless and tasteless. Moderately hard and heavy; coarse-textured; straight-grained; splits readily, saws woolly or stringy and is not easy to finish smoothly as the grain tends to "rise"; durability probably low. Has no commercial possibilities.

Cojoba. There are at least 18 species of unarmed shrubs and trees in the West Indies, southern Mexico, Central America, and northern South America. The largest and best known, and the only one of any commercial value as a source of timber, is *C. arborea* (L.) Britt. & Rose (= *Pithecolobium arboreum* [L.] Urban). Growing in the open it is usually less than 60 feet tall, with a spreading top and a stout bole sometimes four feet in diameter. It is less common in dense forest, but attains a height of over 100 feet in Honduras. Its doubly pinnate leaves are bright green, often 16 inches long, with 8 to 16 pairs of pinnae and 20 to 40 pairs of small, narrow leaflets. The white flowers are sessile in a globular head; the pod is bright red, somewhat fleshy, becoming coiled, usually deeply constricted between the nearly round shiny black seeds. The range includes part of southern Mexico, the east coast of Central America, and all of the West Indies except the Bahamas, although the timber is sometimes called Bahama Sabicú in the New York trade. The fact that the name Sabicú is applied to other West Indian timbers, notably *Lysiloma latisiliqua* (L.) Benth., has resulted in some confusion in the literature, although the woods themselves are readily distinguishable. According to Fawcett and Rendle (*Flora of Jamaica* 4: 2: 148), the timber of *C. arborea* "saws freely, is not too hard for general work, is beautifully grained, takes a fine polish, and is in general use [in Jamaica] for flooring, ceilings, and ornamental work. Altogether it is an excellent timber and very useful in building." In Cuba it is used for wheel hubs, rollers in mills, and railway cross-ties. It is employed to a very limited extent in the United States for furniture and brush backs.

Heartwood dark red or reddish brown, sometimes with narrow stripes of deeper color; sapwood nearly white or grayish, without sharp line of demarcation. Luster medium. Without distinctive odor or taste. Hard and heavy; sp. gr. (air-dry) 0.80; weight about 50 lbs. per cu. ft.; grain somewhat roey; texture medium to coarse; working properties good; durability high.

COMMON NAMES: *Cojoba arborea:* Moruro rojo, sabicú, s. moruno, zapatero (Cuba); red tamarind (Jam.); cojoba, cojóbana (P.R.); collier, poison lasinette (Haiti); cañamazo, coralillo, frijolillo (Mex.); barba jolote, black tamarind, John crow, J. c. bead, red tamarind, turkey gill, wild tamarind (Br. H.); cola de mico, c. de marrano, frijol de mico (Guat., Hond.); conchudo, lorito (C.R.); coralillo (Pan.). *C. costaricensis* Britt. & Rose: Cocobola, lorito (C.R.).

Conzattia, with two or three species of small to medium-sized unarmed trees, is apparently confined to southwestern Mexico. The leaves are bipinnate, with small leaflets; the small yellow flowers are in axillary racemes; the legume is flat, acute at both ends, few-seeded, dehiscent. There is little information about the species, and they are of no commercial importance. The only wood sample available (Yale 1184) is of *C. sericea* Standl., collected and identified by J. G. Ortega in Sinaloa. The tree, locally known as Navío, attains a maximum height of about 50 feet and a trunk diameter of 18 to 24 inches.

Heartwood (if present) not distinct from sapwood; lustrous grayish brown. Odorless and tasteless. Of rather light weight, but firm and tough; texture rather fine; feel soft; grain roey; very easily worked, finishing smoothly; is not resistant to decay or insects.

Copaifera, with 35 to 40 species of unarmed trees in tropical Africa and South America, is better known as the source of gum or balsam than for timber. The leaves are pinnate, with several to many pairs of leaflets; the small flowers are borne on panicled spikes or racemes; the short two-valved legumes each contain a single seed nearly inclosed by an aril. The woods of all species of *Copaifera* contain gum or oil canals which are the source of a commercial product known as copaiba balsam, copaiva balsam, balsam capivi, balsamo de copaiba, aceite de palo, etc. This is an oily liquid which is gathered by tapping. The natives use it in a crude state for medicinal purposes and also for anointing their hair and bodies. It is used commercially as an ingredient in medicines and in the manufacture of varnishes. The principal sources of the oil are Venezuela and the Amazon region.

The most northern of the American species is *Copaifera chiriquensis* Pittier, a tree sometimes 100 feet tall and four feet in diameter in the forests on the low hills along the Pacific coast of Panama. Southernmost is *C. Langsdorffii* Desf. in the region of São Paulo, Brazil; its timber is used locally for furniture, carriages, turnery, and ship-building, and the bark is sometimes employed in tanning. The best known species is *C. officinalis* L. of northern South America and cultivated in the West Indies. It attains large size and the timber is common on the local markets, being used for interior trim, carpentry and general construction.

The species of widest distribution in the Amazon region and the source of almost all of the copaiba balsam exported from the State of Pará is *Copaifera reticulata* Ducke, commonly known as Copahiba Marimary or Copahiba Jutahy. The yellowish brown balsam is thick and has a strong and disagreeable odor. The greater part of the commercial balsam exported from Manáos is fluid, transparent, and less strongly scented; its source is *C. multijuga* Hayne of the State of Pará and northern Matto Grosso, where it is called Copahiba Angelim as well as Copahiba Marimary. The trees of both species have grayish timber with irregular patches of brown; the wood of the first species has the scent of cumarin when fresh, the other is resinous, according to Ducke. The gum appearing in the trade of Pará and Manáos under the name of jacaré copahiba is the product of *Eperua oleifera* Ducke; it is of a brown color, very thick and resinous, and is used for making varnish. *C. Martii* Hayne of Pará, Matto Grosso, and the Guianas, has a brownish red, hard, fine-textured, resinous heartwood. *E. purpurea* Benth. grows along the upper Rio Negro, where it is called Yébaro; its wood is highly resinous but does not furnish balsam. (See *Tropical Woods* 32: 27.)

In the literature the woods of *Copaifera* are frequently confused with those of certain other genera, particularly with *Peltogyne,* the source of Purpleheart or Amaranth, and *Hymenaea courbaril* L., the West Indian Locust. In the latter instance the confusion doubtless results from the fact that this tree has a rosin-like gum or copal.

Heartwood reddish brown, variable, often with a coppery hue, and sometimes streaked; luster rather silky and golden in proper light; oily exudations sometimes present; not very sharply demarcated from the pinkish gray or nearly white sapwood. Without distinctive odor or taste in dry material. Density variable; sp. gr. (air-dry) 0.70 to 0.90; weight 44 to 56 lbs. per cu. ft.; hard, tough and strong; medium-textured; usually straight-grained; not difficult to work, finishing very smoothly; is highly durable.

Common names: Bálsamo de copaiba (Cuba); amacei, copaiba (Dom. R.); balsam (Trin.); camiba, camibar, caniva (Pan.); canime, copaiba, pata de gallo (Col.); aceite, cabima, cabimbo, cabimo, copaiba, currucaí, maramo, palo de aceite, p. de aceitillo (Venez.); apaoewa, hoepelboom, hoepelhout, hoepfroe-hoedoe, hoeproe, koepajoewa, koepawa, pasoemoeti, passiemoetie (Sur.); copaiba (Peru); copahibera, copahiba, copahyba, c. angelim, c. cuiarana, c. jutahy, c. marimary, c. parda, c. vermelha, copahúva, copaiba, copaibeira, copaúba, copaúva, jaboti meutámeutá, jacaré copahiba, oleo, o. folha, o. pardo, olho, pau d'oleo, yébaro (Braz.); cupay (Par.); timbó-y-atá (Arg.).

Coumarouna (=*Dipteryx* in part; see *Taralea*), with about a dozen tropical American species of small to very large unarmed trees, of which nine occur in Brazil, ranges from Honduras to eastern Peru and southeastern Brazil. The leaves are alternate, generally very large, the rachis reflexed-winged, evenly or unevenly pinnate, with 3 to 14 leathery leaflets which are sometimes longitudinally grooved and frequently pellucidly punctate; the roseate or violet flowers are borne in conspicuous terminal panicles and in certain species are very fragrant; the fruit is drupaceous, with a fleshy and oily mesocarp and a single, black, elongated, oily seed. The species are divisible into two groups, one characterized by petaloid sepals and inodorous seeds, the other (four species) having a coriaceous calyx and cumarin-yielding seeds, the tonka beans of commerce. Cumarin is a crystalline substance which has a peculiar fragrance suggesting vanilla and is used to flavor snuff, cigarettes, cigars, cocoa, confectionery, and as an ingredient in perfumes, sachet powders, and cosmetics. The oil, known on the market as "cumarú," is used medicinally.

The species commonly credited with being the source of tonka beans in northern South America is *Coumarouna odorata* Aubl. (=*Dipteryx odorata* [Aubl.] Willd.), but according to Ducke (*Tropical Woods* 61: 5; March 1940) its occurrence in the Guianas and Venezuela has not been substantiated by authentic specimens. It is plentiful, however, in the upland rain forests of Pará and Amazonas and yields the bulk of the cumarú beans exported from the Amazon region. It is a large tree often over 100 feet tall in virgin forest, but scarcely of medium size in second growth and in cultivation. The pericarp of the fruit is bitter and inedible. Two closely related species are *C. rosea* (Spruce) Taub. and *C. punctata* Blake, small to medium-sized trees with scented seeds but contributing little to the commercial supply.

Largest of the true Tonka-bean trees is *Coumarouna trifoliolata* Ducke which attains a height of 130 feet. Unlike that of the other species, the pericarp of the fruit is sweet and edible. It grows in the upper

Rio Branco and is believed to be the same as the Serrapia of the lower Orinoco, the source of the best quality of cumarin.

Among the species that are of no commercial value for their seeds are *Coumarouna polyphylla* (Huber) Ducke, a slender tree of the Rio Negro and Rio Japurá (Río Caquetá in the Colombian part); *C. ferrea* Ducke of Acre and Amazonas, where it is known as Cumarú Ferro on account of the density of its timber and is conspicuous when in flower because the bright rose crown overtops the common level of the forest; *C. alata* (Vog.) Taub., of small size in the dry woodlands of central and northeastern Brazil, where it is called Barú; *C. magnifica* Ducke, which attains large size from the head of the estuary to the center of the Amazon plain; *C. micrantha* (Harms) Ducke, a medium-sized tree of eastern Peru; and *C. speciosa* Ducke of the Tapajoz region of Pará, an imperfectly known tree which the discoverer thinks possibly may prove to belong to a new genus. Some of the foregoing are known as Cumarú-rana (false Cumarú), a name applied also to other Leguminosae because of their resemblance in appearance, scent, or other properties to the true Cumarú.

The most northern species is *Coumarouna oleifera* (Benth.) Taub. (= *Dipteryx oleifera* Benth. = *C. panamensis* Pitt.), a tree attaining a height of 150 feet with a well-formed trunk sometimes three feet in diameter in the forests of the Atlantic coast of Central America from Honduras to Panama. The fresh fruits are covered with a grayish green pubescence and impregnated with a sweet-smelling oil; the seeds are edible when roasted. Ducke suggests that this species might well be made the type of a new genus intermediate between *Coumarouna* and *Pterodon*, but no reliable means of distinguishing the wood from *Coumarouna* has been found. The species does not supply any commercial products.

The timber of *Coumarouna* has never been extensively exploited, formerly on account of the difficulty of extracting it from the forest and later because some of the species have become valuable for their fruit. Moreover the wood is difficult to work, especially when seasoned, being very hard and roe-grained. It is attractive when finished, however, and has been used to some extent for veneers for interior trim and for various articles of turnery, such as walking sticks and umbrella handles. Because of its oily nature and great strength and toughness it has been used to a limited extent as a substitute for Lignum-vitae (*Guaiacum*) in bearings for propeller shafts of steamships. It is not likely to become an important factor in the export trade. The following description is based upon authentic specimens of five species. Heartwood in various shades of brown from yellowish to reddish, more or less distinctly streaked, and marked with narrow but prominent vessel lines; has a waxy or oily appearance and feel; rather sharply demarcated from the yellowish sapwood. Luster of unfinished specimens rather low. With somewhat rancid odor in very oily specimens, sometimes with faint scent of vanilla; without distinctive taste. Extremely hard, heavy, tough, and strong; sp. gr. (air-dry) 0.90 to 1.20; weight 56 to 75 lbs. per cu. ft.; texture medium fine; grain interwoven and roey; difficult to work, but capable of a high natural polish; is very durable.

Common names: Tonka bean; also tonca, tonga, tonkin, or tonguin bean (English); ebo (Hond.); almendro, ebo (C.R., Pan.); sarrapia (Col.); sarrapia, s. mona, yape (Venez.); cuamara, gomorrow, kumara, quamary, tonka bean (Br. G.); groot locus, kelappa-bosie katoelimia, koemaroe, komaro, krapabosie, tonka, t. boon, t. hout (Sur.); bois de coumarue, faux gaïac, fevrier tonka, gaïac, g. franc, g. male, gayac (Fr. G.); barú, barujo, coco feijão, cumarú, c. amarello, c. ferro, c.-rana, c.-zeiro, cumbarú, emburena brava, feijão coco, muirapayé, pau cumarú (Braz.); charapilla, cumarut (Peru).

Coursetia, with about a dozen species of spiny or unarmed shrubs and small trees, occurs from the southwestern United States to Brazil. The leaves are evenly or oddly pinnate, with small to large leaflets; the variously colored flowers are borne in axillary racemes; the fruit is a slender, flattened, dehiscent pod with nearly circular seeds. The largest species is *C. arborea*

Gris., distributed from the Guianas to Trinidad and Panama; it is sometimes 40 to 50 feet high, but supplies no timber of value. The best known plant is *C. glandulosa* A. Gray of southern Arizona and northern Mexico. It is an unarmed shrub or a tree rarely over 20 feet tall, the branches often covered with a transparent brownish lac produced by insects and known to the drug trade as "goma Sonora"; it is used in making cough syrup.

Wood bright yellow throughout, with fine whitish parenchyma markings. Luster rather high. Odorless and tasteless. Hard and heavy; fine-textured; fairly straight-grained; finishes smoothly. Of no commercial importance because of the small sizes obtainable.

COMMON NAMES: Chino, samo prieto (Mex.).

Cowellocassia, with two species of small unarmed trees, is apparently limited in distribution to Cuba and the Island of Haiti. The leaves are pinnate, with 6 to 8 pairs of narrowly lanceolate leaflets; the small flowers are in panicled racemes; the legumes are linear, flat, septate between the few to several seeds, and promptly dehiscent. The trees rarely exceed 25 feet in height and have no special utility. The only wood sample available is a small branch of *C. domingensis* (Spreng.) Britton (= *Cassia domingensis* Spreng.), collected in Dominican Republic by W. L. Abbott (Yale 7466).

Heartwood slightly developed; very dark-colored; rather sharply demarcated from the brownish yellow sapwood. Luster rather high. Odorless and tasteless. Hard, heavy, tough, and strong; texture rather fine; grain irregular; working properties good; heartwood probably durable.

Crudia (or *Apalatoa*), with 20 species of unarmed shrubs and trees, is sparingly represented by two species in the East Indies and two in tropical Africa; of the remaining 12, one occurs in Jamaica, one in Salvador, one in Nicaragua, the others in the Amazon basin with a single extension into northern Venezuela and Trinidad. The leaves are imparipinnate, with several large alternate leaflets; the small flowers are borne in simple racemes, terminal or lateral on the previous year's twigs; the flat, ovate or roundish, leathery legumes contain one or two large seeds. Most of the trees are rather small, but in the swampy areas in Charuma, Trinidad, according to Marshall (*Trees of Trinidad and Tabago,* p. 40), *C. obliqua* Gris. is a large evergreen tree with a trunk up to three feet in diameter. None of the species produces timber of commercial value.

Heartwood pale brown, often with a tinge of pink; scarcely distinguished from the sapwood in dry specimens. Luster low to medium. Odorless and tasteless. Very hard heavy, tough, and strong; sp. gr. (air-dry) 0.85 to 0.96; weight 53 to 60 lbs. per cu. ft.; texture medium; grain straight to roey irregular; rather easily worked, finishing smoothly, is rather splintery; probably not resistant to decay. A plain, unattractive wood without special merit.

COMMON NAMES: Cocoon (Jam.); rosomacho, water locust (Trin.); copataishte (Salv.); cascarillo, guamo (Col.); pisho (Peru); faveira do igapó, ipê, i.-rana, jutahy-rana (Braz.).

Cyclolobium, with six species of trees and shrubs, is distributed from southeastern Brazil northward through the coastal forests and the lower Amazon region into the Guianas. The medium-sized to rather large leaves are unifoliolate; the small, purplish, papilionaceous flowers are borne in fascicled axillary racemes; the flat rounded pods contain 1 to 3 seeds and are indehiscent. According to F. C. Hoehne (*Flora Brasilica* 25: 3: 37. 1941), *C. Vecchii* A. Samp. attains a height of 50 to 65 feet and a diameter of 10 to 20 inches in the State of São Paulo. The only wood sample in the Yale collections (Yale 1267) is without herbarium material, but it comes from Araras and bears the name Louveira which is applied to the species in that locality. Its structure agrees perfectly with that of a small stem (Yale 40314; Curran 85) of *C. Blanchetianum* Tul. collected by H. M. Curran in the basin of the Rio Grongogy, Bahia, Brazil.

Heartwood brown with a purplish tinge, and marked with numerous vertical lines or narrow zones of deeper color which resemble late wood of growth rings; sapwood yellowish. Luster rather low. Without distinctive taste, but with mild licorice-like scent. Hard, heavy, and compact; texture fine; grain straight; not difficult to work, finishing very smoothly; durability probably high. A good furniture wood, though apparently too scarce to be commercially important.

COMMON NAMES: Amarelinho, louveira (Braz.).

Cynometra, with about 40 species of shrubs and small to medium-sized, occasionally large, unarmed trees, is widely distributed in the tropics of both hemispheres. The American species are of no commercial importance, either because they are scarce or because of their poor timber form. The leaves are evenly pinnate, with one to a few pairs of oblique leathery leaflets; the flowers are small and clustered in the leaf axils or on leafless branches; the pods are usually oblique or curved, somewhat swollen and rounded, leathery, tardily dehiscent or indehiscent.

A species found by the senior author in eastern Guatemala and named *Cynometra retusa* Britt. & Rose, is one of the commonest trees in the understory of the lowland forests. The new leaves are very light-colored and on drooping twigs, in conspicuous contrast with the deep green of the old foliage. The bark is thin, smooth, brown, with prominent scattered lenticels. The local name, Pata de Cabro, refers to the fancied resemblance of the paired leaflets to the foot of a goat. The timber is hard, tough, and strong, but is not utilized, except occasionally for fuel and charcoal, as the trees are small, rarely up to 35 feet in height.

There are six other species in continental North America and the West Indies, some of them of very limited distribution. The genus is fairly well represented in South America, but the small trees occur in the undergrowth or in rather inaccessible places and are poorly known. The species credited with the widest range is *Cynometra bauhiniaefolia* Benth., a small or medium-sized tree common along the banks of streams in Argentina, Peru, Brazil, and the Guianas. The wood has no special uses, as the trees are mostly of a drooping habit with stems of poor form for timber.

Heartwood reddish brown, finely veined with parenchyma lines on radial surface and with a distinct pattern on the tangential; transition to the pinkish sapwood gradual. Luster medium. Without distinctive odor, but with a mildly astringent taste. Hard and heavy to decidedly so; texture medium to rather fine; grain irregular; difficult to work, but can be finished smoothly and is rather attractive.

COMMON NAMES: Pico de gallo (Cuba); courbaril (Haiti); pata de cabro (Guat.); fruta de danto (Hond.); cativo (C.R.); trementino (Col.); aracito, cobalonga (Venez.); herairo (Peru); guarabú amarello, iauaranami, jutahy-rana, pororoca (Braz.); ingá-puitá (Arg.).

Dalbergia (including *Amerimnon* and *Ecastophyllum*), with about 250 species of trees and scandent shrubs, is of pantropical distribution. There are about 15 species which yield timber noted for its good technical qualities combined with richness of color and grain and usually a pleasant fragrance. All of the true Rosewoods known to the cabinet-maker are Dalbergias from Asia, Madagascar, Brazil, or Central America. The Rosewood of Siam and Cochin-China is *D. cochinchinensis* Pierre; the Sissoo of India is *D. Sissoo* Roxb.; the Blackwood famed in oriental furniture is said to be the same as the Rosewood of southern India, *D. latifolia* Roxb.; the two principal Rosewoods of Madagascar are *D. Greveana* Baill., exported from Majunga and Morandava, and another, possibly *D. Baroni* Baker, from Tamatave; the African Blackwood or Senegal Ebony, used in making musical instruments and in carving, is *D. melanoxylon* G. & P.

There are at least seven, probably more, species of *Dalbergia* of economic importance for their timber in tropical America. (See *Tropical Woods* 72: 1–10.) The well-known Rosewood from the coastal forests of Bahia, Brazil, is *D. nigra* Fr. Allem.; that from the Amazon region is *D. Spruceana* Benth. Other kinds of Rosewood or

Jacarandá of southeastern Brazil, usually credited to species of *Machaerium,* are probably Dalbergias. Kingwood from Ceará is believed to be *D. cearensis* Ducke. Brazilian Tulipwood from the dry interior of northeastern Brazil has not been specifically determined, but appears to be a Dalbergia. Cocobolo of the west coast of Central America is produced by a closely related and doubtfully distinct group of species of which *D. retusa* Hemsl. is the type. The Rosewood of Guatemala and Honduras is *D. cubilquitzensis* (D. Sm.) Pittier, and that of southern British Honduras, used in the United States for the bars of xylophones and marimbas, is *D. Stevensonii* Standl. There are other species in Central America and southern Mexico whose woods are of good quality, but they seem to be rare and are not exported.

The arborescent American species have odd-pinnate leaves, usually with few to several medium-sized to rather large, leathery, alternate leaflets, or sometimes (*D. nigra*) with numerous small ones; the white or yellow flowers are borne in clustered or panicled axillary or terminal racemes; the indehiscent pods are flat, thin, somewhat leathery, and usually with a single centrally located seed.

The woods have many features in common, the most noticeable difference being in the variegated colors and the nature of the infiltrated gummy or resinous material of the heartwood (described separately for each species). Fresh heartwood has a mild, rose-like scent which may be scarcely noticeable in old specimens. The sapwood is white or yellowish, generally sharply demarcated. Mostly hard, heavy, and strong, sometimes brittle; sp. gr. (air-dry) 0.75 to 1.22, generally 0.85 to 1.10; weight 47 to 75, commonly 53 to 69, lbs. per cu. ft.; texture fine to rather coarse; grain straight to irregular; not difficult to work, finishing very smoothly with a high natural polish; holds its place well when manufactured; is very durable.

Brazilian Rosewood or Jacarandá (*Dalbergia nigra* Fr. Allem.) has been known to commerce for about 300 years. It has been used extensively in the United States for making pianos and other kinds of cabinet work and fine furniture, but at present the consumption is mostly limited to the making of carpenters' spirit levels and plane handles, and for brush backs, butcher-knife handles, and, to a minor extent, for billiard tables, radio cabinets, and marquetry. The old timber in the more accessible regions of the States of Bahia, Rio de Janeiro, Espirito Santo, and the adjoining rain forests of Minas Geraes has become very scarce, but there are still considerable quantities further inland. The tree is of scattered occurrence, sometimes attaining a height of 125 feet, but the old trunks are generally hollow. The trees are slow in forming heartwood and large logs lose much of their volume when the wide and commercially valueless sapwood is hewed off. The heartwood of young trees is brown and not attractive; in fact, it is only from old defective stems that fragrantly scented lumber with rich purplish black marking is obtainable. Many kinds are distinguished locally on account of the prevailing color, whether brown, purple, or nearly black, but these are probably individual variations depending on conditions of growth, age, relative soundness, or other factors. Second-growth trees that are nearly all sapwood are generally known as Jacarandá Branco (white Rosewood) although this name is also applied to trees of other genera. The leaves of *D. nigra* differ from those of the other species described here in having numerous small oval leaflets; the flowers on forest trees are borne in yellow masses above the foliage; the pods are small and narrow, frequently with more than one seed.

Heartwood in various shades of brown to chocolate or violet, irregularly and conspicuously streaked with black; dark specimens with oily or waxy appearance and feel; sharply but irregularly demarcated from the white sapwood. Luster medium. Odor rose-like, mild but lasting in deeply colored specimens; scarcely noticeable in young heartwood. Taste distinctive but not definable. Density materially affected by amount of infiltrated material; sp. gr. (air-dry) 0.75 to 0.90; weight 47 to 56 lbs. per cu. ft.; texture medium; grain generally straight; working properties excellent, although some specimens may be too oily to take a good polish.

COMMON NAMES: Brazilian rosewood, palisander (Eng.); palissandre (Fr.); Jacarandaholz, Palisanderholz (Germ.); madera de palisandro (Span.); legno di jacaranda, pallissandro (Ital.); caá-biuna, cabeuna, cabiuna, c. parda, c. preta, c. roxa, cabiuva, camboré, camboriuna, caviuna, jacarandá, j. branco, j. cabiuna, j. preto, pau preto, uraúna (Braz.).

The Rosewood of the lower Amazon or Jacarandá do Pará is *Dalbergia Spruceana* Benth., a small to medium-sized or occasionally rather large tree growing on fairly dry land and supplying a limited amount of timber of good quality. The leaves are of an entirely different appearance from those of *D. nigra,* the leaflets being fewer, larger, and leathery, and the pods are 1-seeded and much larger. The heartwood, in the only authentic specimen available (Yale 22610; Ducke 150), is a rich golden brown with numerous narrow stripes of red or violet, the color deepening upon exposure. Scent very mild; taste not distinctive. Hard, heavy, compact, and strong; sp. gr. (air-dry) about 1.00; weight 63 lbs. per cu. ft.; of rather fine and uniform texture and fairly straight grain; not difficult to work, finishing very attractively. An excellent cabinet wood which should interest the export trade, if available in commercial quantity.

COMMON NAMES: Jacarandá, j. do Pará, saborana (Braz.).

Brazilian Kingwood (*Dalbergia cearensis* Ducke, probably) has been known to the cabinet-makers of Europe and the United States for a long time, but the species producing it was not named until 1925. The tree grows in the dry country of Ceará and is slender and of low stature; the leaves resemble those of *D. Spruceana,* but the flowers and pods suggest *D. variabilis* Vog., a liana or scandent shrub. The timber is exported in the form of logs or bolts, 3 to 6 feet long and 3 to 8 inches in diameter, without sapwood. Because of the small sizes available, the uses abroad have always been restricted to such purposes as inlays and marquetry and occasional fancy articles of turnery. Inlaid borders of fine panels are frequently of Kingwood. The wood is finely striped, being composed of alternating concentric layers of violet-brown and black or blackish violet. The dark parts are generally narrow and so regular as to give the appearance of late wood in seasonal growth rings, thus differing from *D. nigra* which is very irregularly streaked. The scent is fragrant, but scarcely noticeable in dry material. The heartwood is very hard, heavy, and strong, but brittle; sp. gr. (air-dry) about 1.20; weight 75 lbs. per cu. ft. The texture is fine and uniform although the dark layers are slightly harder than the others, owing to the greater abundance of infiltrated gummy or resinous material. The grain varies from straight to finely roey. The wood is not difficult to work with sharp tools, takes a high, waxy natural polish, and holds its place well when manufactured.

COMMON NAMES: Kingwood, Brazilian kingwood (U.S. trade); violete (Braz.).

Brazilian Tulipwood has long been known to the cabinet-maker and was a favorite in the composition of French furniture, especially of the Empire period. It is still used in small amount for inlays, marquetry, brush backs, and turnery. The tree is small, with an irregular trunk, slender branchlets, and odd-pinnate leaves having about 7 leaflets, suggesting *Dalbergia variabilis* Vog. There is no doubt that it is a species of *Dalbergia* and not *Physocalymma* (Lythraceae) as so generally misstated. (See *Tropical Woods* 20: 23.) The timber is brought by rail to Bahia, the chief port of shipment for that region, in the form of round logs or billets, mostly under 6 feet in length and from 2 to 8, rarely up to 12, inches in diameter, free of sapwood. It is a beautiful wood when freshly manufactured, but gradually fades and loses much of its original attractiveness.

Heartwood irregularly striped, the prevailing colors being yellow, rose, and violet, usually with the yellow predominating; sapwood yellowish. Luster rather high. With a mildly fragrant scent when worked; taste not distinctive. Very hard, heavy, and strong; sp. gr. (air-dry) 90 to 1.10; weight 56 to 69 lbs. per cu. ft.; texture rather fine; grain fairly straight to roey; not very easy to work, inclined to be splintery, but takes a high natural polish.

Common names: Tulipwood, Brazilian tulipwood (Eng.); bois de rose (Fr.); Tulpenholz (Germ.); cego machado, grão do porco?, pau cravo, p. rosa, p. rosada, sebastião de arruda (Braz.).

Cocobolo is a valuable Central American timber (Map 5) which has been in use in the United States, particularly for handles of cutlery, for more than 65 years. (For a comprehensive report on this wood see Yale School of Forestry Bull. No. 8, *Cocobolo.*) It was first introduced to the trade from Panama, but the principal sources now are western Costa Rica and Nicaragua. The Panama species was described by Hemsley in 1878 and named *Dalbergia retusa,* but the fact that it was the source of Cocobolo timber was first discovered about 1911 by Henry Pittier and published in 1918 (*Journ. Forestry* 16: 1: 76–84). In 1922 (*Journ. Wash. Acad. Sci.* 12: 3: 62–64), Pittier described three other species in the same section (*Miscolobium*), namely, *D. hypoleuca* and *D. lineata* from Costa Rica, and *D. granadillo* (based on same type as *Amerimnon granadillo* Standl.) from southwestern Mexico. The first two of these are definitely known to produce Cocobolo which, for all practical purposes at least, is identical with that of *D. retusa;* according to Standley (*Flora of Costa Rica,* p. 532), *D. hypoleuca* "probably is not distinct from *D. lineata*" and "it may well be that *D. lineata* is merely a form of" *D. retusa. D. pacifica* Standl. & Steyerm. is the name recently given to the Cocobolo tree of western Guatemala. According to the authors (*Field Mus. Bot. Ser.* 22: 4: 237), its closest relative is *D. lineata,* "with which eventually it may have to be united." The matter of *D. granadillo* is not so clear. A wood sample accompanying herbarium material determined as that species is purple and black, suggesting Kingwood, but two sterile herbarium specimens coming from the same region in Mexico and identified by P. C. Standley as his *Amerimnon granadillo* are definitely of the Cocobolo type. The Funera of Salvador, determined as *D. cuscatlanica* Standl., is also of the Cocobolo type (see *Tropical Woods* 1: 6). The occurrence of Cocobolo in western Honduras has been ascertained from a study of wood samples in the Yale collection from that country. Insofar as the structure, properties, and utility of the timber is concerned, Cocobolo from Panama to Mexico may be considered as one species, *D. retusa.*

Cocobolo woods from various countries do not exhibit greater differences than are to be noted in several specimens from the same region. Dealers profess no difficulty in distinguishing the source of a given lot of logs, but they do so through familiarity with certain features such as the size, shape, defects, and manner of hewing the logs. Cocobolo from Panama is usually of poor form, often mere shells or split pieces with more or less of the white and worthless sapwood present, while that from farther north reaches the market in solid logs with the sapwood off. This difference is due to the fact that the methods of extraction in Panama are less efficient than in some other parts of Central America where the industry is better organized. Tractors are used in parts of Costa Rica and make a decided difference in the size and quality of the logs brought out.

The sapwood is a dingy white, very sharply defined, and varying in thickness, mainly as a result of age, from about three inches in a young tree a foot in diameter to only an inch in large old trees. It is difficult to describe the color of the heartwood because it is so variable. Some pieces of it might very well be referred to as rainbow-hued, but upon exposure to sunlight and air the lighter colors lose their brilliance and merge into a deep red with a black striping or mottling. If the fresh wood is shellacked the oxidizing processes are retarded but the brilliant hues will not last. It is often claimed that the coloring matter in Cocobolo will not leach out and stain, but this is not borne out by laboratory tests. Pieces of the wood from Panama, Costa Rica, Nicaragua, and Mexico were placed in contact with wet filter paper and in every case a pronounced reddish brown stain resulted. Water in which the wood has been soaked for several days or in which chips have been boiled assumes a

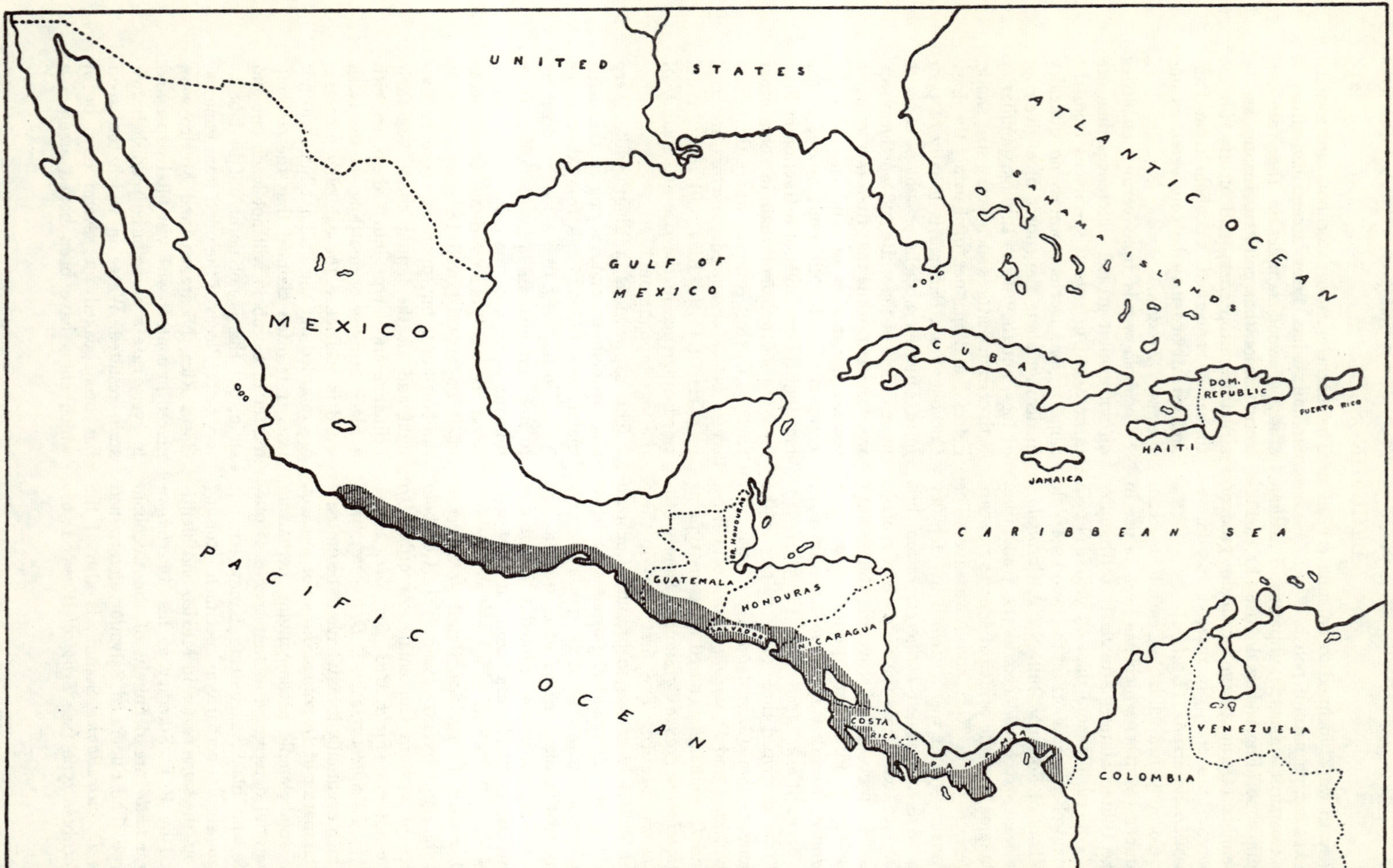

MAP 5. Range of Cocobolo in Mexico and Central America.

peculiar yellowish color. (For tests on the coloring matter see *Tropical Woods* 6: 23.) The odor is slightly pungent and mildly fragrant when worked; taste not distinctive. Sp. gr. (air-dry) 0.99 to 1.22; weight 62 to 76 lbs. per cu. ft.; grain fairly straight to interwoven; durability high.

Cocobolo is one of the most important woods in the cutlery trade, being extensively used for knife handles because of its beauty of color and grain, fine texture, dense structure, ease of working, and the presence of an oily substance which not only tends to waterproof the wood and keep it in shape after manufacture, but also makes it very easy to polish. If the smooth surface is rubbed with a cloth it acquires a waxy finish without the use of oil, wax, shellac, or filler. Prolonged or repeated immersing in soapy water has little effect on the wood except to darken its color, an important consideration in the case of kitchen and butcher knives. It is used also for small tool handles, brush backs, inlaying, Tunbridgeware, scales, musical and scientific instruments, launch-wheel drums, steering wheels, and for making small attachments to automobiles, launches, and expensive machinery. Some of the more beautifully figured or veined wood is manufactured into small jewelry boxes, canes, rosary beads, forks, spoons, buttons, chessmen, and numerous other articles. It is unsuited for gluing.

The fine dust arising in working Cocobolo may produce a rash or dermatitis resembling ivy poisoning. It may make its appearance on the fingers, forearms, face and neck, legs, or any part of the body, and is very annoying and irritating. Usually the infected employee has to discontinue work for a few days or a week and in most instances must permanently cease working in a room where he would be exposed to the dust. As in the case with poison ivy, many people are relatively immune while others are readily susceptible to the poison; and a person once infected, instead of acquiring immunity, seems to become all the more susceptible. Factories in which Cocobolo has been used for years have few cases of poisoning because they gradually acquire an immune working force in that department of the plant. Investigations by Dr. Ettore Ciampolini of the Graduate School of Yale University indicate that the irritating principle is contained in an oil which is liberated in fine particles when the wood is worked. This principle is soluble in alkalis but precipitated by acids. The theory, which seems to be substantiated by experiment, is that workmen whose perspiration gives an acid reaction are resistant, while an alkaline reaction indicates susceptibility, since the alkali dissolves the irritating agent and permits its entrance into the pores of the skin.

Common names: Cocobola, cocobolo, Nicaragua rosewood (Eng.); Cocoboloholz, Foseholz (Germ.); granadillo (Mex., Guat.); funera (Salv.); granadillo, palo negro (Hond.); ñambar, ñ. legitimo (Nic.); cocobolo, c. ñambar, c. negro, ñamba, ñambar, nnambar (C.R.); cocobolo, c. prieto (Pan.); cocobolo, palisandro (Col.).

Rosewood of British Honduras has been an article of export for about a century and there is a record of 118 pieces being shipped in 1841. It was classified as *Dalbergia* in *Timbers of Tropical America* (p. 285), but the specific identity of the tree was not established until December 1927 when Paul C. Standley described it under the name of *Dalbergia Stevensonii* in honor of Neil S. Stevenson who collected the herbarium material (see *Tropical Woods* 12: 4). According to Stevenson (*Tropical Woods* 12: 1): "It grows in damp 'broken ridge' of the riverain type from the Sarstoon River northward to Monkey River in the Toledo District of British Honduras. Between Deep River and the Temash it appears in fairly large patches, concentrated for the most part along the rivers though occurring also in the inter-riverain and drier areas. It attains a height of 50 to 100 feet, and the trunk, which often is fluted, commonly forks at about 20 or 25 feet from the ground. The papery disordered bark is about one-fourth inch thick, the scaly outer portion varying in color from a pale brownish gray to a dingy yellow-brownish gray. When freshly cut the bark (and sapwood as well) has a distinctive odor, suggesting stored apples, and the taste is slightly bitter. The dried bark separates readily into a thin and solid outer layer and a curiously matted inner portion.

"Honduras Rosewood is one of the best known timbers of the Colony. It is very hard and heavy, weighing from 58 to 68 lbs. per cu. ft. when thoroughly air-dry. The heartwood is of a pinkish-brown or purplish color, with alternating light and dark zones which are independent of the true growth rings; sapwood, which is 1 to 2 inches thick, is white with yellow vessel lines when first cut, but quickly turns yellow. The heartwood is highly durable, but the sapwood soon decays when in contact with the ground. The heart portion of a house post in use in Punta Gorda for 37 years was found to be as sound as when it was put in, but the sapwood, of course, had entirely disappeared.

"Honduras Rosewood is often well figured and used to a limited extent for cabinet work; it is chiefly employed for the bars of marimbas and xylophones manufactured in the United States. The requirements for the musical instrument trade are light-colored, straight-grained wood, in logs as nearly round as possible, hewn free of sap, mostly 4 to 6 feet in length and not less than 10 or 12 inches in diameter, although in times of shortage diameters as low as 5 inches may be taken. The exports of the timber, all to the United States, were 248 tons (valued at $5,362) in 1925, and 76 tons (valued at $2,315) in 1926."

The wood used in making marimbas in Guatemala is not Honduras Rosewood, but the product of one or more species of *Platymiscium*. In the United States, Brazilian Rosewood (*Dalbergia nigra*) was formerly employed for percussion bars, but it was found too weak to withstand the impact of professional players' mallets, and was superseded by Honduras Rosewood which is denser, tougher, and more highly resonant, and is not oily. In making the bars the logs are cut into quarters, then into edge-grain planks one inch thick, and from these the bar blanks of varying length are sawn. The blanks are ordinarily air-seasoned under cover for from two to seven years. There is considerable loss due to season checks and other defects as well as in manufacture, so that the weight of the finished bars represents only from 20 to 30 per cent of the weight of the logs.

COMMON NAMES: Honduras rosewood, nagaed (trade, U.S.A.); rosewood (Br. H.).

The Rosewood of eastern Guatemala and the Republic of Honduras is *Dalbergia cubilquitzensis* (D. Sm.) Pittier. It is a large tree, sometimes more than 100 feet tall, with a straight cylindrical trunk having at maturity an average diameter of 30 inches and being free of branches for three-fourths of its length. The bark is brown and fairly thick. The leaves are long, with rather numerous fairly large leaflets, brownish on the under side; the small flowers are borne in inconspicuous axillary panicles. The wood, which bears considerable resemblance to *D. Spruceana* of Brazil, is basically orange-colored, with more or less pronounced violet striping, deepening upon exposure to brown of various shades, sometimes purplish; looks somewhat waxy; sharply demarcated from the nearly white sapwood. It is nearly scentless. Moderately hard, heavy, tough, and strong; not difficult to work, taking a lustrous finish; deeply colored material is denser and more fragrant. The timber is considered one of the best in the country and is used for axles and tongues of wagons, spokes of truck wheels, and for durable construction as well as a limited amount of high-grade furniture. It is suitable for cabinet work, interior trim, and brush backs.

COMMON NAMES: Rosewood (Eng.); granadillo, junero (Guat.); granadillo, rosul (Hond.).

Dalea (or *Parosela*), a large genus of herbs, shrubs, and a few small trees, is distributed from southwestern United States through Mexico, Central America, and the Andean region of South America to Chile. Standley says (*Trees and shrubs of Mexico,* p. 445): "Many herbaceous species of the genus occur in Mexico. The shrubby forms [110 species are enumerated] are of scarcely any economic importance and very few vernacular names are reported for them." The best known arborescent species is *D. spinosa* A. Gray which is distributed from Arizona and southern California to

Sonora and Baja California. It is a densely branched shrub or a tree rarely more than 20 feet high, with a short contorted trunk 10 to 20 inches in diameter. The twigs are reduced to slender sharp thorns bearing at their base a few, irregularly scattered, early deciduous, simple, glandular, wedge-shaped leaves about an inch long; the dark blue flowers are borne in showy racemes, hence the name Indigo Bush; the ovoid, indehiscent, one-seeded pod is partially inclosed by the calyx. The timber is of no value because of the small sizes available but is suitable for brush backs and fancy articles.

Heartwood brownish gray, more or less conspicuously variegated or striped with dark brown; sharply demarcated from the moderately thick pure white sapwood. Luster medium. Odorless and tasteless when dry. Moderately hard and heavy; of medium to rather coarse texture; easy to work, taking a smooth finish; is probably fairly durable.

COMMON NAMES: Indigo bush, i. thorn, smoke tree (U.S.A.); mangle (Mex.).

Daubentonia, sometimes treated as a section of *Sesbania* (or *Sesban*), is represented in the New World by two species of shrubs and small trees. The leaves are evenly pinnate with many leaflets; the yellow, roseate, or purple flowers are borne in axillary racemes; the elongated leathery pods are 4-winged or 4-angled, with round seeds separated by cross partitions. *D. punicea* (Cav.) DC., sometimes called Purple Sesban, grows in the United States from Florida to Mississippi and also in Brazil and Argentina. *D. texana* Pierce (= *D. Drummondii* Rydb.) occurs from Florida to Texas and to San Luis Potosí, Mexico. There are apparently no special uses for the plants. The following description applies particularly to specimens of *D. punicea* (Yale 40114 and 40299) collected by J. L. Stearns in Florida, and Georgia, resp. (see *Tropical Woods* 72: 12–14).

Heartwood pinkish; rather sharply demarcated from the yellowish sapwood. Luster medium. Without distinctive odor or taste. Hard, heavy, and tough; has about the consistency of *Gleditsia;* texture coarse, and somewhat irregular; grain fairly straight; not difficult to work, finishing smoothly; durability doubtful.

The wood of *Daubentonia texana* bears little superficial resemblance to that of the other species. Heartwood (probably traumatic) grayish brown; sapwood grayish. Luster medium. Rather light in weight and of a "cheesy" consistency; texture coarse; grain irregular; saws woolly, but the feel of the roughened fibers is not harsh; durability probably low.

COMMON NAMES: *Daubentonia punicea:* Purple sesban (Georgia). *D. texana:* Coffee-bean weed, Drummond rattle-box (Texas).

Dialium includes about 30 species of tropical trees, but only one occurs in the Western Hemisphere. This is *D. guianense* (Aubl.) Sandw. (= *D. divaricatum* Vahl), a medium-sized to large unarmed tree ranging from southern Mexico through Central America to the Peruvian Amazon and Bahia and Matto Grosso, Brazil. The leaves are imparipinnate, with 3 to 7 long-tapering leaflets; the small flowers are borne in large panicled cymes; the indehiscent, edible, tamarind-like legumes are short and rounded, with a hard and brittle exocarp containing a single seed surrounded by a sweet pulp.

The tree is known to attain a height of 115 feet, with a long cylindrical smooth-barked bole, up to 24 inches in diameter above the narrow buttresses which sometimes are six feet high. It grows in dense evergreen forest on well-drained clay soils in Central America, but at the edge of virgin forests or in second growth on sandy soils in the Amazon region. In parts of its range it is very common and near the Ocongwas River in Nicaragua, for example, composes 15 per cent of the total stand. The dense dark-colored timber is noted for its strength and resistance to decay and termites. It is used locally for fence and house posts, bridge timbers, railway crossties, frames of heavy implements and vehicles, and for repairing logging-cart wheels.

Heartwood uniform brown or reddish brown, becoming darker upon exposure;

very distinct but not sharply demarcated from the thick whitish or yellowish sapwood. Luster medium. Without distinctive odor or taste. Very hard, heavy, tough, and strong; sp. gr. (air-dry) 0.90 to 1.10; weight 56 to 64 lbs. per cu. ft.; texture medium; grain interwoven; not very difficult to work, finishing smoothly; is highly durable. Not suitable for furniture because of its high density and lack of figure.

COMMON NAMES: Guapaque, paque, wapak (Mex.); ironwood, uhee-tee (Br. H.); tamarindo (Cent. Am., gen.); paleta (Guat., Hond.); tamarindo prieto (Hond.); comenegro, slim, tamarindo montero (Nic.); fría, monkey apple, tamarindo de montaña (Pan.); granadillo (Col.); cacho (Venez.); huitillo (Peru); azedinha, cururú, itú, jatahy, jetahy, jutahy, j. da varzea, j. mirim, j. peba, j. poca, j. pororoca, j. rana, parajuba, pororoca (Braz.).

Dicorynia is a small genus of large unarmed trees occurring in the Amazon region and the Guianas. The leaves are imparipinnate, with 7 to 13 large leaflets; the white flowers are small and borne in terminal panicles; the legumes are oval, leathery, winged along one edge, indehiscent, and contain one or two seeds. The best known and perhaps the only good species is *D. paraensis* Benth. It exhibits considerable variation in the size of the leaves, number and texture of the leaflets, and other inconstant morphological details, and these have been made the basis for proposing four or five other species. *D. ingens* Ducke, of the State of Pará, Brazil, is said to develop a very small amount of heartwood in comparison with the others so far known.

Dicorynia paraensis (or *D. guianensis* Amsh.) is most abundant and important in French Guiana and composes about 10 per cent of the stand in the western part of the Colony. Although it is common on lowlands it prefers well-drained soil and there attains a maximum height of 150 feet and a trunk diameter of five feet or more above the buttresses, which generally are not over six feet high. The local name is Angélique. Two kinds are recognized, namely, Angélique Rouge and Angélique Gris, the first being more abundant and supplying darker, denser, and stronger wood. The timber is considered a fairly satisfactory substitute for Oak and Teak, and is sometimes called Teck de la Guyane. It is used locally for joinery, furniture, carriage work, all sorts of construction, whether indoors, in the open, or in water, including posts, piling, and railway crossties. It is considered suitable for cooperage. The export trade has never been large, less than 2000 tons annually, but it may increase materially. The presence of silica particles in the wood, which interferes with sawing but comparatively little with other types of machining, adds to its usefulness in harbor works exposed to marine borers.

Heartwood said to be rose-colored when freshly cut, but quickly darkening to brown or purplish brown, solid or streaked; dull on surface but usually with a golden luster beneath, and showing parenchyma pattern on tangential surface; rather sharply defined from the sapwood which at first is nearly white, becoming light brown upon drying. Hard, heavy, tough, and strong, though variable in these respects; sp. gr. (air-dry) 0.70 to 0.90; weight 44 to 56 lbs. per cu. ft.; texture medium; grain usually straight; working properties variable with density and silica content; splits readily and finishes smoothly; is highly durable; lighter material holds its place well when manufactured and is preferable for joinery. A wood of general utility, rather than for show, as it lacks figure and the color is unattractive.

COMMON NAMES: Barakaroeballi, basra locus or lokus, b. loksi, basterd locus, kabakally, kieèrèoe, kièjère-oe, sienga pretoe, siengdia apeto, s. peto, tamoené-kereoe zand locus (Sur.); angélique, a. bâtard, a. franc, a. gris, a. rouge, teck de la Guyane (Fr. G.); angelica do Pará, tapaiuna (Braz.).

Dicymbe, with four species of medium-sized to rather large unarmed, apparently rare trees, occurs in the Amazon region of Brazil and the hinterlands of British Guiana. The only specimen studied (Yale 34076; Ducke 312) is of *D. corymbosa* Spruce, a moderately large tree in the forest on high sandy soil near São Paulo de

Olivença, State of Amazonas. The leaves are pinnate, with 2 to 4 pairs of rather large, leathery, long-pointed leaflets; the petals of the showy flowers are white with reddish bases; the inflorescence is large, terminal, and corymbose. There are no known uses for the timber.

Heartwood very pale brownish, merging gradually in the slightly lighter-colored sapwood. Luster medium. Odorless and tasteless. Moderately hard, heavy, tough, and strong; texture rather fine; grain fairly straight; not easy to work, being somewhat splintery or fibrous, but it can be finished very smoothly; durability probably low.

Dimorphandra, excluding the genus *Mora* which is sometimes merged with it, consists at present of about 25 species of small to large unarmed trees occurring in Brazil, the Guianas, and Amazonian Venezuela. According to Sandwith (*Kew Bulletin,* 1932, p. 395), the species in British Guiana, "unlike those of *Mora,* are always found in the 'white sand' patches of the coastal region or in the wide sandstone areas of the interior. . . . They may be small trees of the savannah type of vegetation (*D. conjugata* and *D. cuprea*), or very large trees with an immense girth (*D. Davisii*). One of them, *D. macrostachya,* is a conspicuous feature of the flora of the slopes of Roraima, but has not yet been met with on Duida, at the western extremity of the Pacaraima Range."

Ducke says (*Journ. Wash. Acad. Sci.* 25: 4: 193): "The genus *Dimorphandra* may be divided into two sections: *Eudimorphandra* and *Pocillum,* which are so natural that it would perhaps be better to consider them as subgenera. They are chiefly distinguished by the fruits, and each has a different geographical distribution. The species of *Eudimorphandra* occur throughout the Amazonian hylaea inclusive of Guiana and in tropical Brazil as far south as Rio de Janeiro and the State of S. Paulo. *Pocillum,* however, is strictly limited to the hylaea. This latter section now contains 15 species, 10 of which are found in the Brazilian States of Pará and Amazonas, and five in British Guiana. One of them, *D. macrostachya* of the slopes of Mount Roraima, must be included in the flora of both countries, as well as in that of Amazonian Venezuela. Here occurs also the Brazilian *D. pennigera.* The sole species found in French Guiana, *D. polyandra,* is, according to Sandwith, probably an anomalous form of *D. Hohenkerkii* of British Guiana. *D. pennigera,* collected in the Brazilian and Venezuelan upper Rio Negro, must certainly exist in the neighboring Colombian territories, and *D. gigantea* grows at the frontier of Peru."

Very little information is available concerning the woods of *Dimorphandra,* and no special uses are known for them. There are only four specimens in the Yale collections, and while all were obtained with herbarium material which has presumably been correctly determined, the differences in appearance and structure are decidedly pronounced. Four species are represented, two of *Eudimorphandra* and two of *Pocillum,* but there is no division conforming to the taxonomic classification. The anatomy is in many ways similar to that of *Mora.*

Dimorphandra Davisii Spr. & Sandw. is said to be a large tree about 120 feet high with a long, somewhat tapering, unbuttressed bole, growing gregariously on quartz sand only. The wood (Yale 32887; Davis 895) from the type tree is dark reddish brown, somewhat oily looking, and of the general type of *Mora excelsa. D. gigantea* Ducke attains a height of 160 feet or more in the non-inundated forest near Tabatinga in western Amazonas, Brazil, and the base of the trunk has high tabular buttresses. The wood (Yale 20703; Ducke 22) from the type tree is a fairly lustrous pale yellowish or grayish brown, marked with numerous very conspicuous dark-colored vessel lines. It is of medium density, very coarse-textured, roe-grained, and saws woolly. The pores are large and distinct and surrounded by abundant vasicentric, sometimes confluent, parenchyma which is not conspicuous on cross section because of a lack of color contrast with the background.

Dimorphandra conjugata (Splitg.) Sandw. is typically a small tree, but occasionally 40 feet high, growing abundantly in sandy places in the Guianas. The Arawak Indians call it Dakama, the Bush Negroes Anjama. It belongs to section *Eudimorphandra*. The wood (Yale 16011) is dark reddish brown, somewhat oily looking, very hard and heavy, coarse-textured, and cross-grained. Wood parenchyma is very abundant, distinct, and vasicentric-confluent into irregular diagonal bands. *D. parviflora* Benth. (of the same section) is a medium-sized forest tree growing on the non-inundated lands of the lower Amazon region. The heartwood (Yale 21342; Ducke 83) is yellowish with a tinge of pink, not sharply differentiated from the sapwood, and marked with fine but distinct vessel lines. The luster is satiny. According to Ducke (*Arch. Jard. Bot. Rio de Janeiro* 4: 258), *D. velutina* Ducke and *D. glabrifolia* Ducke, both of section *Pocillum,* have yellowish woods.

COMMON NAMES: Dakama, hawakaiyek (Br. G.); akajoeran, anjama, boesie kasjoe, bosch kasjol, dakama, kadjoe mattoe (Sur.); ataná (Braz.).

Dinizia *excelsa* Ducke, the only species, is one of the largest trees in the lower Amazon region, attaining a maximum height of nearly 200 feet and a trunk diameter of over six feet and having the aspect of *Hymenolobium*. It also occurs in the Guianas. The leaves are bipinnate, with numerous leaflets; the small green fragrant flowers are borne in large racemes; the legumes are leathery, flat, indehiscent, 10 to 14 inches long and about two inches wide; the reddish or sometimes grayish bark is broken into very numerous fine scales; the trunk is long and well formed. (See *Archivos do Jardim Botanico do Rio de Janeiro,* 1922, p. 76.)

Heartwood rather dull reddish brown, more or less streaked, and sometimes with an orange hue; distinct but not always sharply demarcated from the pinkish sapwood. Without distinctive odor or taste. Exceedingly hard, heavy, and strong; sp. gr. (air-dry) 1.15; weight 72 lbs. per cu. ft.; texture rather coarse; grain roey; difficult to work, but finishing smoothly and taking a high polish; is probably highly resistant to decay. The timber is sparingly used, but is suitable for heavy and durable construction.

COMMON NAMES: Kuraru, parakwa (Br. G.); angelim, a. falso, a. pedra, faveira grande (Braz.).

Diphysa, with several species varying in size from low shrubs to small or rarely medium-sized trees, is distributed, mostly in uplands, from southern Mexico through Central America to Venezuela. The leaves are odd-pinnate, with 5 to 25 small or rather small leaflets; the yellow flowers are borne in short axillary racemes; the pods are oblong, flattened, indehiscent, somewhat interrupted between the seeds, the pericarp separating into two layers, the papery exocarp becoming strongly inflated or bladder-like. The two best known and most widely distributed species are *D. carthagenensis* Jacq., characterized by numerous small leaflets, and *D. robinioides* Benth., with leaves suggesting Black Locust (*Robinia pseudacacia* L.). Both are poorly formed trees, usually less than 25, sometimes up to 50, feet tall. The timber is similar to Black Locust and is employed locally for fence and house posts and other purposes requiring a strong and durable material in small sizes. The heartwood yields a yellow dye, but it is not known to the trade. The trees are frequently planted to provide living fence posts. The following description applies particularly to the woods of the two species mentioned, which are practically identical.

Heartwood greenish yellow when fresh, changing to olive and eventually to russet-brown; of rather waxy or oily appearance; very distinct from the white or yellowish sapwood. Luster low to medium. Without distinctive odor or taste. Very hard, heavy, tough, and strong; texture medium; grain irregular, sometimes finely roey; not very difficult to work, finishing smoothly and taking a good polish; holds its place well when manufactured; is noted for its durability. Suitable for the same purposes as Black Locust, but apparently has no commercial possibilities.

COMMON NAMES: Cascabelillo, cuachepil, guiloche, huiloche, palo santo, retama, r.

de cerro, susuk, stutztzuk, tsutsuc, xbalché, xsusuc (Mex.); wild ruda (Br. G.); guachipelí, guachipelín (C.A., gen.); palo amarillo (Guat.); naguapate, sikró, singrá, tsikrá, urxk (C.R.); macano, m. amarillo (Pan.); arate, carate, negrito, vivaseca (Col.); achivare, bolsa de gato (Venez.).

Diplotropis comprises a group of five species of medium-sized to large unarmed South American trees which some botanists consider only a section of the genus *Bowdichia*. A study of a limited number of authentic wood specimens indicates that they are distinct. The species examined, in addition to the two of *Bowdichia* (*sens. str.*) are as follows: *Diplotropis racemosa* (Hoehne) Amsh. (= *B. racemosa* Hoehne), *D. brasiliensis* Benth. (and a variety), *D. guianensis* (Tul.) Benth., and *D. Martiusii* Benth. The range of the genus extends from the hinterlands of Colombia and Venezuela through the Guianas to Minas Geraes, Brazil. Only *D. Martiusii* inhabits overflowed lands and marshes.

The leaves are odd-pinnate, with 5 to 15 rather small to very large leathery leaflets; the roseate or white flowers are borne in terminal and axillary panicles and racemes which are usually inconspicuous because of the foliage; the legumes are typically thin, flat, indehiscent, disseminated by the wind, and contain 1 or 2 small, soft, black seeds; the fruit of *Diplotropis Martiusii* is large, thick, leathery, indehiscent, disseminated by water, and contains a single large seed with an oily pericarp. The trees are generally known as Sapupira in the Amazon region and the timber is used for heavy and durable construction of all kinds. In the Guianas the uses include furniture, house-framing, and shipbuilding.

Heartwood rich brown, with light striping of parenchyma; looks rather waxy; fairly uniform, except for variations in luster due to roe grain; sharply demarcated from the whitish or yellowish sapwood. Luster high and golden in proper light, though unpolished surface may appear dull as though oiled. Without distinctive odor or taste when dry. Very hard, heavy, tough, and strong; sp. gr. (air-dry) 0.95 to 1.15; weight 59 to 71 lbs. per cu. ft.; texture very coarse; feel harsh; grain irregular, usually roey; difficult to work, but finishing smoothly and taking a high natural polish; very resistant to decay. A timber of the Partridge-wood (*Andira*) type, fairly attractive when finished.

COMMON NAMES: Ají, zapán negro (Col.); botonallare, peonía (Venez.); aramatta, araumatta, tataboo, tatabu (Br. G.); levarte kabbes, zwarte kabbes (Sur.); coeur dehors (Fr. G.); sapupira, s. da varzea, sebipira, sucupira, s. assú (Braz.).

Drepanocarpus (frequently included in *Machaerium*), with ten species of armed or unarmed woody vines, rarely trees, has its center of distribution in the Amazon basin, though one species, *D. lunatus* (L.f.) G. F. W. Meyer occurs in the West Indies, Middle America, South America, and tropical Africa. The leaves are unequally pinnate with mostly alternate leaflets, the stipules sometimes spinescent; the purple or white, minutely bracted flowers are borne in small terminal panicles or axillary racemes; the 1-seeded leathery pod is broadly sickle-shaped or curved into a circle. The bark contains a red sap.

The only wood sample available (Yale 21250) is of *D. paludicola* Standl., a large shrub or small tree growing in inundated places in the State of Pará, Brazil, where it is known as Tucunaré Envira. (See *Tropical Woods* 33: 12.) Color very light grayish brown throughout; dark red stains common about wounds. Luster rather low. Odorless and tasteless. Fairly light and soft, brittle; texture medium, uniform; grain somewhat irregular; very easily worked, finishing smoothly. Of no commercial possibilities.

COMMON NAMES: Escambrón, palo de hoz (P.R.); chaperno, uña de gavilán (Venez.); andira uchy, aturiá, juquiry, tucunaré envira (Braz.).

Dugandia *rostrata* (H. & B.) Britt. & Killip, the sole species, which previously has been referred by various authors to *Acacia, Mimosa,* and *Lysiloma,* is a shrub or a small much-branched tree 15 to 20 feet high and 12 inches in diameter, growing in northern Colombia. The large bipinnate leaves have numerous very small leaflets; the young branches are armed with short recurved prickles; the legume is flat, septate between the seeds, the body separating from the margin, as in *Lysiloma*. The following description is based on a single specimen (Yale 29610; Dugand 529).

Heartwood rather dull reddish brown, more or less streaked; fading gradually into the greenish yellow sapwood. Without distinctive odor or taste. Hard, heavy, tough, and strong; rather coarse-textured; straight-grained; not very difficult to work. No special uses are recorded.

COMMON NAMES: Baranoa, iguanero de agua, rabo de iguano de agua (Col.).

Dussia (including *Cashalia* and *Vexillifera*), with ten doubtfully distinct species of medium-sized to large unarmed deciduous trees, is of infrequent occurrence from southern Mexico to the Guianas and the Amazon region of Brazil and Peru. The leaves are long, unequally pinnate, with 7 to 15 large, distinctly pinna-veined leaflets; the flowers are rose-lilac or variegated yellow and brown and are borne in long simple or panicled racemes; the leathery dehiscent pods contain 1 to 3 seeds.

The type of the genus is *Dussia martinicensis* Kr. & Urb., a tree about 50 feet high in the Lesser Antilles. The most northern form is *D. mexicana* (Standl.) Harms (= *Ormosia mexicana* Standl.) of Vera Cruz, Mexico. According to Standley (*Field Museum Bot. Ser.* 22: 5: 342), there are two Central American species, namely, *D. cuscatlanica* (Standl.) Standl. & Steyerm., a medium-sized to exceptionally large tree growing from Salvador to Colombia, and *D. macroprophyllata* (D. Sm.) Harms, a large tree of Costa Rica and Panama. The principal Amazonian species is *D. discolor* (Benth.) Amsh. which attains a height of 150 feet, with a large prominently buttressed trunk; the bark contains a red sap. The timber of *Dussia* is not utilized except in Salvador where it is said to be of some local importance for lumber.

Heartwood absent or not distinguishable from the creamy yellow sapwood; parenchyma markings distinct. Luster low. Odorless and tasteless. Of medium density, tough and strong; texture coarse; grain fairly straight; easy to work, finishing fairly smoothly; perishable in contact with the ground. So far as known, suitable only for common interior carpentry and construction; apparently too rare to be utilized extensively.

COMMON NAMES: Bois gamelle (Mart.); palo de tigre, soycol de monte (Guat.); cashal (Salv.); citrón (Pan.); peonía (Venez.).

Ebenopsis, with three described species of spiny shrubs and trees, is limited in distribution to Mexico and southwestern Texas. The largest is *E. flexicaulis* (Benth.) Britt. & Rose (= *Pithecolobium flexicaule* [Benth.] Coulter), ranging from Texas to Tamaulipas, Nuevo León, and Yucatán. It is often shrubby, but under favorable conditions becomes a tree, occasionally 50 feet high, with a stout trunk, two to four feet in diameter, and irregular, spreading branches. The leaves are bipinnate and lustrous; the flowers are yellow and fragrant; the woody pods are brown or black. Indians cook and eat the young seeds as a vegetable or roast the ripe ones for use as a substitute for coffee. The timber, which is noted for its strength and durability, is used for wheelwright work, fence posts, and miscellaneous farm purposes. It appears suitable for knife scales (handles) and brush backs.

Heartwood dark brown, somewhat variegated; has an oily or waxy appearance and feel; sharply demarcated from the thin yellow sapwood. Without distinctive odor or taste. Very hard and heavy; sp. gr. (air-dry) about 1.05; weight about 63 lbs. per cu. ft.; texture fine; grain fairly straight; not difficult to work, finishing very smoothly with a high natural polish and a golden luster.

COMMON NAMES: Ebony (Texas); ébano, guapinole, palo fierro, tempisque (Mex.).

Edwardsia, with about eight species of unarmed trees and shrubs, is often considered as a section of *Sophora,* but the woods appear distinct enough to justify segregation. The genus is represented in the Hawaiian Islands, New Zealand, south-central Chile, and Juan Fernández.

The most widely distributed species is *Edwardsia tetraptera* (Ait.) Salisb. (= *Sophora tetraptera* Ait.) which grows in New Zealand, where it is known as Kowhai, and in the Provinces of Maule, Llanquihue, and Chiloé, Chile, where the usual name is Pelú or Pilo. It varies in size from a low shrub to a small or rarely medium-sized tree. The average size of mature trees in Chile is about 20 feet high and 10 to 12 inches in trunk diameter. Young plants bear little resemblance to the old ones, for in the juvenile stage the branches are slender, flexuous, and interlaced into a rather compact mass, the bark is orange-colored, and the distantly spaced leaves have only a few pairs of rounded leaflets. Under favorable circumstances it develops an erect stem with a more open crown of drooping foliage; the odd-pinnate leaves are tardily deciduous and vary greatly in length and in the size, number, and shape of the paired leaflets; the bright yellow flowers are axillary, occurring singly or in few-flowered pendulous racemes, and appearing just before the advent of the new, silvery leaves; the legume is indehiscent, deeply constricted between the several seeds, and narrowly but distinctly 4-winged; the immature pods are covered with white silky hairs. The use of the timber is limited by the small sizes available, but it is considered of excellent quality for wheelwright work, implement handles, and other purposes requiring strength, resilience, and durability; also for cabinet work and ornamental turnery.

Heartwood light and dark brown in fine, irregular, alternating layers producing attractive figure on tangential surface; rather sharply demarcated from the nearly white sapwood. Luster rather low. Dry specimens without distinctive odor or taste. Very hard, heavy, tough, and strong; sp. gr. (air-dry) 0.90 to 1.05; weight 56 to 66 lbs. per cu. ft.; texture medium coarse; feel harsh; grain fairly straight; moderately difficult to work, but can be finished very smoothly; is highly resistant to decay.

COMMON NAMES: Mayu-monte-toromiro, pelú, pilo (Chile).

Elizabetha, with 10 species of small to medium-sized unarmed trees, occurs in the Amazon region of Brazil and the hinterlands of the Guianas, Venezuela, and Colombia. The leaves are pinnate, with few to many leaflets; the flowers are white or attractively colored, small but often with conspicuous stamens, and borne in short to very long inflorescences; the legumes are sickle-shaped, leathery or woody, and 2-valved. Ducke says (*Tropical Woods* 37: 18): "This genus may be considered as one of the least known of the Leguminosae of South America. . . . The plants are well worth the attention of botanists, not only because of their beauty, but also on account of certain characters of some of them, such as the sweet secretion in the vegetative buds of *E. Duckei* Huber, and the extraordinary hardness and density of the wood of *E. durissima* Ducke." He considers the latter species remarkable also because of its very numerous racemes of relatively small flowers which are pollinated by insects, whereas all the other species he knows are visited by humming birds. Apparently the only species well enough known to have a native name is *E. paraensis* Ducke, which is called Arapary Vermelho.

Heartwood reddish brown or chocolate, in solid color or intermingled with included sapwood; rather dull on the surface, but sometimes with a rich luster beneath; sharply demarcated from the white or pale brownish sapwood, which usually is very thick. Odorless and tasteless. Very hard and heavy to extremely so; texture medium to rather fine; grain straight to roey; difficult to work, but finishing very smoothly; probably highly durable. Of no commercial possibilities.

Enterolobium, with eight described species, occurs in tropical America from southern Mexico to northern Argentina. The best known species are *E. contortisiliquum* (Vell.) Morong (= *E. timbouva* Mart.)

in southern South America and *E. cyclocarpum* (Jacq.) Gris. in northern South America, Central America, West Indies, and Mexico. Both attain large size, with a wide-spreading crown and stout trunk in the open, much taller and longer-boled in the forest. The woods are of about the color of Walnut (*Juglans*) and resemble that of the Siris tree of India, *Albizzia Lebbeck* Benth.

Enterolobium contortisiliquum is common in northern Argentina, where it is called Timbó, and in the neighboring forests of Paraguay and Brazil, where it is usually known as Timboúba. It grows to a height of 75 to 100 feet, with a trunk sometimes six feet in diameter and free of branches for 30 to 40 feet, and with a crown spread of 50 to 100 feet. The large finely bipinnate leaves are deciduous and for a few months of each year the trees are entirely bare. The small white or yellowish flowers are borne in dense inconspicuous clusters. The ear-shaped pods are called Orejas de Negro by the Spanish and Orelhas de Negro by the Brazilians, names which are also applied to the tree. Timbó is well known to the timber trade in Argentina and because of its softness and ease of working it is used as a substitute for Cedar (*Cedrela*), being sometimes called Timbó Cedro. Since the tree grows rapidly and makes good shade, it is often planted in parks and along highways; some of the principal streets of Buenos Aires are lined with it. The same or a closely related species is common in São Paulo and eastern Brazil, where it is generally known as Tamboril, Tambory, Timboúba, or Timbuva. The principal use for the large trunks is for making dugout canoes. The timber rarely enters the market.

The Middle American species, *Enterolobium cyclocarpum,* is commonly called Guanacaste, a name of Nahuantl Indian origin meaning ear-tree. Standley says (*Trees and shrubs of Mexico,* p. 391): "The tree grows rapidly and makes an excellent shade tree because of its broad top. The large trunks are used for canoes, water troughs, etc., and the wood is very durable in water. It is employed in carpentry and cabinet work. The pods are said to be an excellent feed for cattle and the seeds as well as the young pods are sometimes cooked to be used for human food. The fruit and bark are rich in tannin. Rose reports that in Sinaloa the bark and fruit are used as a substitute for soap in washing woolen goods and that a syrup made from the bark is used for colds. The fruit is used as a soap substitute in Venezuela also. The gum which exudes from the trunk is employed in Sinaloa as a remedy for bronchitis."

The timber was introduced into the market of the United States about 1910 and for a while found considerable favor in the cities of the Pacific coast for paneling and interior trim. It was promoted for a time under the name of Juana Costa Mahogany. It is sold in small quantities in New York under the names Conacaste, Genizero or Jenisero, Pichwood, and (formerly) South American, Central American, or Mexican Walnut. Crotches are a source of figured wood used in matched panels. The heartwood is brown with various shadings, sometimes with a reddish tinge, and varies in consistency from light, soft, and spongy, to rather hard and heavy; sp. gr. (air-dry) 0.35 to 0.60; weight 22 to 37 lbs. per cu. ft. It is usually easy to work, finishes smoothly, holds its place well when manufactured, and is fairly durable. The solid dry wood is unscented, but the dust arising in working it has a disagreeable pungent odor and some workmen are allergic to it. The timber is useful and fairly attractive, but cannot be classed as a high-grade furniture wood.

Common names: Conacaste, genizero, guanacaste, jenisero, juana costa Mahogany, kelobra, kolobra (trade, U.S.A.); algarroba carretera, árbol de las orejas, cabellos de Venus, oreja de judío (Cuba); cascabel sonaja, conocaste, cuanacaztle, cuau-nacaztli, huanacaxtle, huinecaztle, nacaxtle, orejón, parota, perota, pich (Mex.); ear-tree, guanacaste, tubroos (Br. H.); conacaste, guanacaste, yaje (Guat.); árbol de orejas, conacaste, caro, c. hembra, cenícero, conacaste (Salv.); genécero or jenízero (Nic.); genícero, guanacaste, krukrá, kudsir, kurú, soró (C.R.); corotú, jarina (Pan.); anjero, campano, çarito, dormilón, orejero, oriera, piñón, samán (Col.); cara-caro, caro, c. hembra, hueso de pescada (Venez.); brown uya (Br. G.); acacia franc, poirier (Fr. G.); espina (Peru); chimbó, c. claro, c. escuro, fava de rosca, faviera grande, jacaré, monjolo, m. branco, m. cambuy, m. lizo, m. preto, m. vermelho, orelho de preto, tambor, tamboril, tamboriúva, tambory, tambureiro, tambuva, timbó, t. rana, t. úba, timbuva, vinhatico, ximbó (Braz.); timbó, t. colorado (Par.); orejá de negro, pacará, p. negra, p. ploma, timbó, t. atá, t. cedro, t. colorado, t. negro, t. puitá, t. pyitá, timboúba, timboúva (Arg.).

Eperua, with 11 species of small to large unarmed trees, has its center of distribution in the Guianas, with extensions into the northern Brazilian Amazon region and Venezuela. The leaves are mostly paripinnate, with few to several pairs of rather thin to thick and leathery leaflets; the large white, red, or purple flowers are borne in terminal panicled racemes; the legumes are large, rather woody, 2-valved, and few-seeded. The timber is of commercial importance and because of its abundance is worthy of efforts to increase its utilization. Its chief handicap is its oiliness.

Regarding certain Brazilian species, Ducke says (*Tropical Woods* 51: 16): "*Eperua bijuga* Mart. ex Benth. is a middle-sized tree with large purplish rose flowers and handsome red-brown but resinous timber found only in local commerce. It grows usually in moist or swampy places of the upland forest, chiefly on sandy soil with black humus, in the Amazonian estuary and from the western part of the lower Amazon (Lake of Faro) as far as Manáos, where it is very abundant. The common Muirapiranga of Manáos is this species, and not *Haploclathra paniculata* as stated by some authors. . . . This tree is called Muirapiranga in Soure, Island of Marajó, as in Manáos, but in the neighboring Breves Islands it is known as Ipé to Aipé, a name more generally applied in Amazonia to several species of *Macrolobium, Crudia,* and sometimes *Peltogyne,* and in extra-Amazonian Brazil to *Tecoma* [*Tabebuia*], fam. Bignoniaceae. In Faro, it is known as Espadeira, but the species so designated in the upper Trombetas is *Eperua falcata* Aubl., the common Wallaba of British Guiana. *Eperua Schomburgkiana* Benth. is rather like *E. bijuga,* but the flowers are white and the wood is darker and much more resinous. The tree grows in swampy upland rain forest on acid sandy soil, being frequent in some localities near Manáos, but otherwise only observed in the upper Trombetas region and the neighboring southern part of British Guiana." (For a detailed description of the wood of *E. Schomburgkiana,* see *Tropical Woods* 13: 25.)

The most southern species is *Eperua oleifera* Ducke, a large tree of the lower Rio Madeira, Amazonas, Brazil, where it is known as Jacaré Copahiba. It is the source of a commercial resin used in the manufacture of varnish. The wood is almost identical with that of *E. purpurea* Benth. of the upper Rio Negro, and both are of lighter color and considerably less dense than any of the other species examined. The wood of *E. leucantha* Benth., also of the upper Rio Negro, is a dull coppery brown, exceptionally hard and heavy, and, unlike all of the other species, is almost free from oily exudations. These three species are unique in the genus, so far as studied, in that the gum ducts are diffuse, as in *Prioria,* instead of in tangential or concentric series.

Concerning the species in British Guiana, Aitken says (*Tropical Woods* 23: 1): "There are several trees producing timber commonly known as Wallaba in British Guiana, but the best known are the Ituri Wallaba (*Eperua Jenmani* Oliver) and the Soft Wallaba (*E. falcata* Aubl.), the timbers derived from these two trees being indistinguishable commercially. Other trees known locally as Wallaba have been identified as *Eperua Schomburgkii* Benth. and *E. rubiginosa* Miq.; Yoboko Wallaba, *Eperua* sp., probably *E. Hohenkerkii* Sprague; Clump Wallaba, *Dicymbe corymbosa* Benth., and Water Wallaba, *Macrolobium bifolium* (Aubl.) Pers. The following statements refer principally to Soft and Ituri Wallabas, as the information available regarding the other species is scanty.

"The trees are small, diameters of over 24 inches not being common, and the heights average about 80 feet. They make fairly rapid growth and, under suitable conditions, when a Wallaba forest has been cut over the area is ready for cutting again in about 30 years. Wallaba occurs in almost pure stands and, while these are to be found throughout the greater part of the forest lands, the most extensive areas occur in the central and eastern part of the Colony and on the white residual quartz sands of the more elevated portions. This partiality of Wallaba for the white sands is very marked and, although scattered Wallaba trees may be found in almost every type of forest, there is a distinct line where one type of forest ends and Wallaba forest on white sand begins, the Wallaba type being obviously an edaphic climax. This white quartz sand is useless from an agricultural point of view, but it is capable of supporting Wallaba forest averaging 600 cubic feet of merchantable timber per acre. There are excellent Wallaba forests adjacent to good loading berths on some of the larger rivers of the Colony which are navigable for ocean-going steamers. The water has been analyzed and its suitability for use in the manufacture of paper pulp determined, while the head of water available offers a potential source of power. . . .

"Wallaba is in universal use throughout the Colony for house frames, vat staves, fence poles, telephone poles, and shingles. Its durability, due to its resinous character, makes it especially suitable for poles, and the engineering branch of the British Guiana Post Office Service reports that 'while the average life of an untreated pole is 20 years, poles have been found to be in excellent condition after 30 years.' There is an export trade to the West Indian Islands in shingles, staves, and poles, and the market, particularly of this last product, is worthy of further development. Most species of Wallaba burn well and the wood is used throughout British Guiana for fuel. An excellent type of charcoal is obtained."

The following information is taken by permission from a mimeograph report of tests made at the Forest Products Research Laboratory, Princes Risborough, England, on Soft Wallaba timber (*Eperua falcata*) from British Guiana: The three logs tested were obtained from virgin forest in the Demerara River basin, at an elevation of 150 feet. Fairly extensive splits developed in transit, although the ends were banded and covered with sacking. Ring shake was also present in one log. The weight at 50 per cent moisture was about 71 lbs. per cu. ft. During conversion of the logs it was necessary to apply a mixture of kerosene and water to the saw to keep the gummy exudation from clogging the teeth. An unpleasant sour odor was given off while the fresh wood was being worked, but this was less pronounced as the wood dried. Kiln-seasoning was difficult, and rather severe distortion, checking, and honey-combing resulted from efforts to obtain uniform drying to the center. The lumber machined fairly easily, but was somewhat hard to work by hand, and the gum adhered to the cutting edge of tools. From an examination of the wood structure and general physical properties, it is suggested that Soft Wallaba is likely to be most suitable for uses out-of-doors, where the sticky nature of the surface will not be an objection. For these purposes heartwood should be used, as the sapwood is not naturally durable. Following are average results of timber tests (at 58 per cent moisture content) in lbs. per sq. in. unless otherwise stated. *Static bending:* Fiber stress at elastic limit, 9860; same at max. load, 12,980; modulus of elasticity, 2,-172,000. *Impact bending* (50-lb. hammer): Fiber stress at E.L., 16,470; M. of E., 2,988,000. *Endwise compression:* Fiber stress at E.L., 6140; same at max. load, 8220; M. of E., 2,561,000. *Hardness:* Radial surface, 1551 lbs. (Western Red Cedar, 230 lbs.); tangential, 1304 lbs. (W.R.C., 260 lbs.); end, 1451 lbs. (W.R.C., 410 lbs.). *Cleavage,* in lbs. per inch of width: Radial, 277; tangential, 499.

Results of paper-making tests at the Imperial Institute on a log of Soft Wallaba 10.5 inches in diameter are briefly summarized as follows (*Bull. Imp. Inst.* 28: 4: 411–418, Jan. 1931): "This sample . . . gave a good yield of pulp which was of good quality but not easily bleached. It furnished paper of satisfactory strength and character, suitable for book or writing papers. The wood, however, cannot be recommended for the manufacture of kraft paper, as the product obtainable is lacking in strength for this class of paper and the cost of beating in the mill would be unusually high. Owing to the presence of a considerable quantity of resinous material in the wood it could not be satisfactorily employed for the production of pulp by the sulphite process."

Heartwood rather dull red to reddish or purplish brown; sharply demarcated from the gray sapwood; surface typically streaked and stained with oily or gummy exudations. Fresh wood with rather disagreeable odor; taste not distinctive. Moderately to decidedly hard and heavy; sp. gr. (air-dry) 0.75 to 1.04, mostly 0.85 to 0.95; weight 47 to 65 lbs. per cu. ft.; of medium to coarse texture; straight-grained; very readily split. An unattractive wood because of the viscous exudations.

Common names: Palo machete, uapa tabaco (Venez.); wallaba—ituri, soft, water, yoboko (Br. G.); baboen walaba, bie-ie-hoehoe, bijlhout, bili-hoedoe, birihoedoe, hariraro walaba, ietoeri walaba, pallewie, paparin bale-ie, pare-eh, parrewe, roode walaba, tamoene, toto amoté, wit bijlhout, witte birihoedoe, witte walaba, woapa, wouapa (Sur.); awapa, bois de sabre, eperu, wapa, w. blanc, w. gris, w. huileaux (Fr. G.); aipé, apá, apazeiro, copaiba-rana, cupaúba-rana, espadeira, ipé, jauácano, jacaré copahiba, jébaro, jebarú, muirapiranga, youácano, yébaro (Braz.).

Erythrina, with more than 130 described species of trees, shrubs, and some herbaceous plants with woody rootstocks, is widely distributed throughout the tropical and subtropical regions of the world. The stems and branches are usually armed with stout spines. The leaves are odd-pinnate, with three large leaflets; the scarlet or orange flowers, often scimitar-shaped, are borne in showy terminal or axillary panicles, commonly when the plants are leafless; the slender more or less moniliform pods are two-valved, generally dehiscent, and contain several red seeds.

The principal use is for live hedges and fence posts, as branches and sections of the stem take root readily, and for ornamental planting. The bark, seeds, and roots contain alkaloids which are employed medicinally and for making insecticides. Crushed stems and roots are sometimes used to stupefy fish. The succulent young flowers serve as seasoning in native cooking. Although some species attain large dimensions, especially in trunk diameter, the timber is sparingly utilized because it is light, coarse, unattractive, and lacks durability. It is tough and strong for its weight however, and is sometimes used for making woodenware, sabots, stable floors, floats, and toys. It apparently has no commercial possibilities.

Heartwood apparently absent; sapwood yellowish. Luster medium to rather high. Odorless and tasteless. Light to very light in weight, but comparatively firm and tough; av. sp. gr. (air-dry) about 0.25; weight 16 lbs. per cu. ft.; texture very coarse; feel usually harsh and wiry; grain straight to irregular; finishes fairly smoothly but cannot be polished; subject to blue-staining fungi and perishable in contact with the ground.

Common names: Coral bean (U.S.A.); coral tree, red bean tree (Jam.); búcare, b. espinoso, pinón, p. de costa, p. de Cuba, p. de pito, p. de sierra, p. de sombra, p. espinoso, p. florido, p. forastero, p. francés, p. milagroso, p. real (Cuba); baumortel, búcare, bucago, coral tree, machete, palo de boyo, pinón espinoso (P.R.); brucal (Dom. R.); bois immortel (Haiti); búcare, immortel, madre de cacao (Trin.); immortel du pays (Guad.); boontsji die paaloe, b. karta (D.W.I.); chacmaloche, chacmolche, chocolín, chottza, cochizquilitl, colorín, c. negro, coralina, cozquelite, demthy, equelite, iquimite, jiquimite, macayxtli, madre brava, m. cacao, patol, p. colorín, peonía, pichoco, piñón espinoso, pipal, pito, purenchequa, pureque, sumpancle, tzinacanquahuitl, tzompantle, tzompantli, xkolokmax, xoyo, zompancle, zompantle, zompantli, zumpantle (Mex.); coapma, sumpankle, tiger wood (Br. H.);

colorín, pito, tzintej (Guat.); guiliqueme (Hond.); ahuejote, ahuijote, pito, quilite (Salv.); elequene, palo machete (Nic.); boró, brukrá, burú-krá, elequeme, froro, fru, kueri-kanga, poro, p. colorado, p. de montaña (C.R.); bobo, gallito, immortal, machete, palo bobo, p. santo, pernilla de casa, pito (Pan.); anaco, balú, barbatuco, búcaro, cachimbo, cambulo, canto gallo, ceibo, chacafruto, choho, c. colorado, mata caimán, peronía, peronilla, písano (Col.); anauco, barisigua de cerro, b. de conuco, búcare, b. anauco, b. peronía, b. reinosa, ceibo, peonía, parichigûe, pericoa, pericoco (Venez.); palo prieto (Ec.); amasisa, pisonay (Peru); assacú, a. rana, assucará, bico de papagaio, corticeira, moxoxo, muchocho, mulungú, pau sabão, sanandú, sananduva, sapicuxava, suiña, suinan (Braz.); ceibo, c. del chaco, c. macho, chopo, seibo, zuinán, zuinandi (Arg.).

Etaballia *guianensis* Benth., the only described species, is a medium-sized unarmed tree of rather limited distribution along the Essequibo River in British Guiana and the Rio Branco, a large northern tributary of the Brazilian Amazon. The leaves are rather large, ovate, leathery, simple, and alternate; the dull-golden flowers are borne in short, dense, showy, clustered, axillary racemes; the fruit is flat, hard, one-seeded, suggesting *Pterocarpus*. The sap is red. A form with smaller leaves, which may be a new species, was collected by B. A. Krukoff on varzea land along the Rio Livramento in the State of Amazonas (No. 6215, a tree 120 feet tall; No. 6614, height 70 feet); the immature inflorescences suggest *Lacistema*. The timber is fairly attractive and is suitable for making good furniture, but apparently the supply is small or inaccessible.

Heartwood variegated red and brown, more or less streaked with black and darkening upon exposure; sharply demarcated from the rather thick, white sapwood. Luster medium to high; golden on polished specimens. Without distinctive odor or taste when dry. Very hard, heavy, compact, and strong, but somewhat brittle; texture fine; grain straight to irregular; not difficult to work, finishing very smoothly with a high natural polish; holds its place well when manufactured; is very durable.

COMMON NAMES: Etabally (Br. G.); mututy, m. vermelho (Braz.).

Eysenhardtia, with several species of shrubs and small trees, is distributed from western Texas and southern New Mexico and Arizona through Mexico to Guatemala. The leaves are equally pinnate, with numerous, small, glandular-punctate leaflets; the little creamy white flowers are in axillary and terminal long-spicate racemes; the legumes are small, compressed, indehiscent, and usually 1-seeded.

The best known species is *Eysenhardtia polystachia* (Ortega) Sargent, a shrub or a tree occasionally 25 feet tall, with a slender, gray, flaky-barked trunk six to eight inches in diameter, growing from southwestern United States to southern Mexico. The foliage is aromatic and the flowers are fragrant. The wood is of special interest because during the sixteenth, seventeenth, and early part of the eighteenth centuries it was celebrated throughout Europe for its reputed diuretic properties and for the remarkable fluorescence of its infusion in pure water. In the latter respect it is rivaled only by the Philippine Narra, *Pterocarpus indicus* Willd., and for a long time this and the Mexican wood were supposed to be the same and were called "lignum nephriticum." Precious cups were turned from lignum nephriticum, and water drunk from them or from bowls in which a few chips had been allowed to remain was supposed to work marvelous cures. This is not surprising in view of the beautiful opalescence imparted to the water (if alkaline), varying in sunlight and shadow from orange and yellow to peacock-blue. The sources of lignum nephriticum remained uncertain for centuries and, in fact, were not positively established until 1915. (See Smithsonian Institution Report for 1915, pp. 271–298.)

Heartwood brown or reddish brown, more or less streaked; looks somewhat waxy; sharply demarcated from the thin yellow sapwood. Luster fairly high. Without distinctive odor or taste. Hard, heavy, compact, tough, and strong; sp. gr. (air-dry) 0.90; weight 56 lbs. per cu. ft.; of rather fine texture; grain straight to irregular; not difficult to work, except that dark areas near wounds are flinty; finishes very smoothly and takes a high natural polish; very resistant to decay. Of no commercial possibilities.

COMMON NAMES: Coate, coatl, coatli, cuate, leña nefritica, palo cualo, p. dulce, p. d. blanco, quaté, rosilla, taray, tlapahoaxpatli, tlapalazpatli, vara dulce, varaduz (Mex.).

Fairchildia, with only one species, *F. panamensis* (Benth.) Britt. & Rose (= *Swartzia panamensis* Benth.), is a medium-sized deciduous tree, sometimes 65 feet high, with an erect or distorted trunk rarely over 16 inches in diameter, and an elongate crown of short ascending limbs. The imparipinnate leaves have five rather long and narrowly acuminate leaflets; the large yellow flowers have a single petal and are borne in long pendent racemes; the legumes, usually one to a raceme, are long, broad, flat, woody, and dehiscent, and contain 4 to 8 disk-like seeds which have no aril. (See *Journ. Wash. Acad. Sci.* 11: 7: 156, April 4, 1921.)

The only wood sample available (Yale 2964) was collected with flowers and fruit by H. Pittier on a rocky slope along the Chagres River, Panama. It is from an eccentric stem, 2 to 4 inches in radius, and is pale yellowish throughout; distinct heartwood absent. Luster medium. Very hard, heavy, and strong; texture rather coarse; feel harsh; grain irregular; moderately difficult to work but finishing smoothly; durability of heartwood unknown.

Feracacia. This is a Cuban genus with one or two species of slender trees 15 to 25 feet high, having stipular spines on the twigs and clusters of formidable stiff spines along the trunk. No wood sample available for study.

COMMON NAME: Abrojo, palo biajaca (Cuba).

Ferreirea *spectabilis* Fr. Allem., the only species, is a large unarmed tree in southeastern Brazil, Bolivia, Paraguay, and northeastern Argentina. At its best it is 125 feet tall with a long well-formed bole two to three feet in diameter and covered with a dark brown or nearly black, ridged bark. The leaves are odd-pinnate with 9 to 15 rather small, retuse, oval or elliptical leaflets; the small yellow flowers are borne in abundance in terminal panicles; the pod is 1-seeded and indehiscent. The bark is used for tanning and the timber for heavy, durable construction.

Heartwood streaked brownish yellow when fresh, becoming darker upon exposure; has a waxy appearance; sharply demarcated from the thin yellowish sapwood. Luster rather low. Odor mildly rancid; taste bitter owing to powder present known in Brazil as "pó de sulfato." Very hard, heavy, tough, and strong; sp. gr. (air-dry) 0.95 to 1.15; weight 59 to 72 lbs. per cu. ft.; texture medium coarse; feel of unfinished material harsh, of finished, waxy; grain interwoven; difficult to work and almost impossible to split radially, but can be finished smoothly and presents a fairly attractive appearance; is highly resistant to decay. Not likely to become of commercial importance for export.

COMMON NAMES: Angelim pedra, gracuhy, maracahyba, sucupira, s. amarella (Braz.); manisito (Boliv.); topó- riváǵuazú (Arg.).

Fishlockia *anegadensis* (Britt.) Britt. & Rose, the only species, formerly of *Acacia*, is described as an intricately branched tree, up to 25 feet high, the main branches widely spreading, the slender twigs tortuous and bearing paired, persistent, spinescent stipules and axillary heads of yellow flowers. The pods are short, swollen, curved, and tardily dehiscent. The plant is said to be abundant on the rocky plain and occasionally on the sandy plain near West End, Enegada, Puerto Rico, and is not known to occur elsewhere. No wood sample of it is available for study, but what is designated as an undescribed Cuban species of the genus has been supplied by Dr. Juan T. Roig (Yale 26538); the vernacular name is Erizo, which is also applied to a cactus.

Heartwood rather lustrous reddish brown, with fine and inconspicuous parenchyma markings; sapwood nearly white, becoming yellowish brown on the surface. Without distinctive odor or taste. Very hard and heavy; medium-textured; difficult to work; probably durable. No special uses are reported.

Fissicalyx *Fendleri* Benth., the only described species, is a small deciduous tree 25 to 40 feet high, of restricted distribution in Venezuela. The leaves are unequally pinnate with 4 or 5 pairs of leaflets; the orange-colored flowers are borne in conspicuous terminal panicles; the one-seeded fruit is winged. One of the two wood samples in the Yale collections (Yale 36021) was obtained at Yuma, Aragua, by L. Williams who states that the tree was about 35 feet tall, with a short, cylindrical, fairly straight bole 20 inches in diameter and covered with a light-brown, flaky bark. The local name is Tasajo, but this is applied also to *Platymiscium pinnatum* (Jacq.) Dugand and various species of *Vitex*. The timber is of no economic importance because of the small size and inaccessibility of the trees. The same or a closely related species of *Fissicalyx* was collected in March 1938 by A. C. Smith in dense forest in the western extremity of the Kanuku Mountains, drainage of the Takutu River, British Guiana. The wood (Yale 35787) is very similar to the Venezuelan material.

Heartwood absent from specimens or not distinguishable from the yellowish sapwood; areas about wounds dark brown. Luster rather high. Without distinctive odor or taste. Hard, heavy, tough, and strong; texture medium; grain finely roey; fairly easy to cut, but difficult to finish smoothly; durability probably poor.

Geoffraea (*Geoffroea* or *Geoffroya*), with four or five species of small trees, is limited in its distribution to northern South America. The leaves are rather large, odd-pinnate, with 15 to 25 alternate or subopposite oblong leaflets; the yellow flowers are rather small and are borne in single or clustered axillary racemes near the ends of the branchlets; the fruit is drupaceous, with a single, comestible seed. *G. superba* H. & B. occurs in Venezuela and in northern Brazil and Peru. *G. spinosa* Jacq. grows in northern Colombia, where it is known as Silbadero. It is sometimes 30 feet high, with a trunk 10 to 20 inches in diameter, covered with a gray finely furrowed bark. The wood, so far as known, is not utilized except casually for fuel.

Heartwood absent, but may be purplish brown, judging from color of small knots; sapwood thick, yellowish. Luster fairly high. Odorless and tasteless. Hard, heavy, tough, and strong; texture rather fine; grain irregular; saws woolly, but can be finished smoothly; durability of heartwood unknown. Has no commercial possibilities.

COMMON NAMES: Coa, silbadero (Col.); almendro, jigua, pasa del Río Negro, taque (Venez.).

Gleditsia (or *Gleditschia*), with about a dozen, mostly Asiatic, species of trees, usually with the stem and branches heavily armed with large simple or branched thorns, is represented in the western hemisphere by only two species, one in eastern North America, the other in southern South America. The leaves are simply or doubly pinnate (both forms common on the same tree), with numerous small leaflets; the small greenish flowers are borne in slender axillary drooping racemes; the legumes are broad, flat, indehiscent, often pulpy between the many seeds. Unarmed forms are frequently planted for ornament and shade. The timber is of good quality and finds many local uses where strong, tough, and fairly durable material is required.

Gleditsia triacanthos L., commonly called Honey Locust, is a large forest tree with a maximum height of 150 feet and a trunk diameter of six feet, though usually less than half those dimensions, is of scattered occurrence throughout most of the eastern half of the United States, being at its best along small streams in southern Indiana and Illinois. Growing in the open, it has a broad spreading crown of slender and somewhat pendulous branches. *G. texana* Sarg., a large tree known to occur in a single grove near Brazoria, Texas, is probably a hybrid between *G. triacanthos* and *Asacara aquatica* (Marsh.) Raf. (= *Gleditsia aquatica* Marsh.). Other trees believed to be hybrids of these two species have been found in Indiana, Louisiana, and Mississippi. Heartwood of *G. triacanthos* light reddish brown or bronze; rather sharply demarcated from the thick yellowish sapwood. Luster rather high. Without distinctive odor or taste. Hard, heavy, tough, and strong; sp. gr. (air-dry) 0.70 to 0.80; weight 44 to 50 lbs. per cu. ft.; texture moderately coarse; grain straight to irregular; not very easy to work, but finishing very smoothly; is fairly durable. Suitable for furniture, but the timber is too scarce to be of commercial value.

Gleditsia amorphoides (Gris.) Taub., the only species in the southern hemisphere, grows in northern Argentina, Bolivia, and southern Brazil. It is usually less than 50, occasionally up to 75, feet high, with a short thick bole 20 to 30 inches in diameter, and thickly beset with very long thorns (Plate XXV), though there is an unarmed form in Formosa, Argentina. The pods and bark contain saponin and are sometimes used by the aborigines as a substitute for soap. The principal use for the timber is for fuel, though the heartwood is employed to some extent for fence posts and durable construction. The thick sapwood is readily attacked by insects. Heartwood roseate to reddish brown, usually faintly streaked; sharply demarcated from the yellowish sapwood. Luster medium. Hard, heavy, tough, and strong; sp. gr. (air-dry) 0.80 to 0.95; weight 50 to 70 lbs. per cu. ft.; texture medium; grain straight to variable. Not a commercial timber.

Common names: *Gleditsia triacanthos:* Honey locust, sweet locust, thorntree (U.S.A.). *G. amorphoides:* Coronilla, espinho de Cristo, espinilho (Braz.); caá-nambí, cambá-nambí, coronillo, espina de corona, espina Cristi, espinillo, e. amarillo, ivó-pó, palo de corona, quillaí, sucará, yuguazú (Arg.).

Gliricidia. Although several species have been described, the only one that is well known is *G. sepium* (Jacq.) Steud. It is unarmed and short-boled, commonly less than 25 feet high and resembling an apple tree, of common occurrence in southern Mexico, Central America, and northern South America, and naturalized in the West Indies. The leaves are odd-pinnate, with 7 to 15 rather large leaflets which may be purple-blotched on the under side; the pink flowers are borne in great profusion after the leaves have fallen; the pods are linear, flat and dehiscent. The trees are commonly planted for living fence posts and to shade plantations of cacao and coffee. The leaves and roots are said to be poisonous to rats and mice. The timber is noted for its durability and is used locally for fence and house posts, stakes, and, when large enough, for railway crossties.

Heartwood light olive-brown, deepening upon exposure; with fine pencil-striping of parenchyma; rather sharply demarcated from the yellowish white sapwood. Luster rather low. Odorless and tasteless when dry. Very hard, heavy, tough, and strong; texture coarse; grain irregular; not easy to work, but finishing very smoothly; holds its place well when manufactured; highly resistant to decay. Is suitable for the same purposes as Black Locust (*Robinia*), but probably has no commercial possibilities in the export trade.

Common names: Acacia, amor y celos, bien vestida, desnudo florecido, júpiter, piñón amoroso, p. florido, p. francés, p. milagroso, p. violento (Cuba); lilas étranger (Haiti); madre cacao, m. de cacao (Mex., C.A., gen.); cacahuananche, cacahuanantl, cacahuanano, iaiti, lengua de perico, madera negra, sac-yeb, xak-yaab, yaguaguyt, zac-yab (Mex.); can-sim, kanté, yaite (Guat.); madre cacao, palo de hierro (Salv.); cacagua, foncontín, madre cacao, madrial (Hond.); madera negra, madreado (Nic.); aí-kra, bala, buri-kri, sangre de drago (C.R., Pan.); mata-ratón (Col., Venez.).

Goldmania *foetida* (Jacq.) Standl. is an unarmed Mexican shrub or small tree, sometimes up to 40 feet high. Its yellowish green flowers are ill-scented; the seed pods are short, flattened, curved, the valves hard and very slow in opening. There are apparently no special uses for the plant. Wood not seen.

Common names: Coyacate, cusa, huizache, pinzanguarimbo, sicoche, yondero (Mex.).

Goniorrhachis *marginata* Taub., the only described species, is said to be a little shrub of southeastern Brazil. The variety *elata* Kuhlmann is a common tree in the State of Bahia and Espirito Santo, where it forms almost pure stands over small areas in the transition zone between chaparral and rain forests. In such situations it is from 60 to 75 feet tall and up to 20 inches in diameter. It also extends along the streams into the rain forest belt and there attains a maximum height of 100 feet, with a well-formed bole sometimes three feet in diameter and clear of branches for 60 feet. The leaves are pinnate, with two pairs of rather large smooth leathery leaflets; the white flowers are borne in clustered terminal spikes; the grayish bark is thick and finely corrugated. The timber is used in Bahia for all kinds of heavy durable construction, but is rarely met with in the markets. In Espirito Santo it is considered one of the best woods of the region for railway crossties.

Heartwood dull sulphur-yellow when fresh, changing to rich purplish red, similar to *Peltogyne;* sometimes with blackish streaks; sharply demarcated from the thick white sapwood. Without distinctive odor or taste. Extremely hard, heavy, tough, and strong; sp. gr. (air-dry) 1.10; weight about 68 lbs. per cu. ft.; texture medium; grain straight to finely roey; not difficult to work, finishing very smoothly and taking a high polish; is very durable.

Common names: Guarabú amarello, itapicurú, i. amarello (Braz.).

Gourliea *spinosa* (Mol.) Skeels (= *G. decorticans* Gill.), the only species, is a shrub or a small tree usually less than 25 feet high, very widely distributed in dry open situations in Argentina where it frequently forms nearly pure thickets. The branches are thorny and the bark on the upper part of the tree exfoliates as in *Platanus*. The leaves are small, odd-pinnate, with 9 to 21 oval leaflets; the little golden yellow flowers are borne in short clustered racemes; the indehiscent pods provide forage for animals and medicine and a fermented beverage for the natives. The principal use of the wood is for fuel, but the larger trees supply some timber for rural construction.

The following description applies to the sapwood, as no normal heartwood is present in specimens. (Traumatic heartwood in Yale No. 6242 is purplish brown and flinty hard.) Color white or yellowish. Luster low. Variable in density from rather light to fairly heavy; sp. gr. (air-dry) 0.56 to 0.69; weight 35 to 43 lbs. per cu. ft.; texture medium fine to coarse; grain irregular; not difficult to work, finishing smoothly; is not durable. An unattractive wood of no commercial possibilities.

Common names: Chañar, c. breda, chañarcillo (Arg.).

Guilandina, commonly treated as a section of the genus *Caesalpinia,* is represented in tropical America by about 20 species of woody vines, scandent or straggling shrubs, and one important tree. Many of the plants are armed with stout recurved prickles. The leaves are bipinnate, generally large, the leaflets often punctate; the yellow flowers are borne in racemes or panicles; the prickly pods, which usually are about as wide as they are long, are dehiscent, and the smooth and shiny seeds are frequently nearly globose. Two or three species are scandent or straggling shrubs occurring along tropical and subtropical seacoasts the world over and sometimes forming impenetrable thickets. Their nomenclature is confused, but the specific names *crista, Bonduc, Bonducella, major,* and *nuga* generally refer to this group (see *Journ. Bot.* [London] 76: 173–180). Confusion with arborescent species is probably responsible for the statement sometimes made that these strand plants are the source of a dyewood of the Brazilwood group.

The only American species of commercial importance for its timber is Brazilwood or Pernambuco wood of the coastal forests of eastern Brazil, *Guilandina echinata* (Lam.) Spreng. (= *Caesalpinia echi-*

nata Lam.), a tree sometimes more than 100 feet tall, with a symmetrical bole as much as three feet in diameter and free of branches for 50 to 60 feet, though commonly much smaller. Its leaves are bipinnate, with numerous small, nearly sessile, punctate leaflets; the rather small yellow flowers are borne in axillary and terminal racemes which are rendered inconspicuous because of the foliage. The brown, bristly, elastically dehiscent pods are 2 to 3 inches long and an inch wide and contain 1 or 2 flat, brown seeds.

From very early times the name Brazil, Brasil, or Brésil has been applied to plants producing a red dye. Marco Polo, who traveled in the Far East about 1260, mentions two kinds, one believed to be *Morinda citrifolia* L., the other *Caesalpinia Sappan* L. The wood of the latter was an important article of commerce in Europe in the Middle Ages, but eventually became known as Sappan rather than as Brasil. In 1500 a similar wood was discovered in eastern South America and the region producing it was called Brazil. The fact that the great Republic of Brazil derived its name from a wood is often overlooked in the common terminology of timbers and the natural assumption is likely to be that all Brazilwood emanates from Brazil. In this instance, however, the name Brazil has no geographical significance but is a term applicable to a group of red or orange-colored dyewoods derived from several species of closely related caesalpiniaceous genera of pantropical distribution. At present the principal, if not the only, source of Brazilwood consumed by the dye trade of the United States is *Haematoxylon brasiletto* Karst. (see under *Haematoxylon*).

Brazilian Brazilwood was at one time in such great demand in Europe that the business was made a Portuguese royal monopoly in 1623, and private exploitation was strictly prohibited. Although the ports of Brazil were opened to foreign commerce in 1808 and exportation of most kinds of timber was permitted, the trade in Pau Brazil and Pau Rainha (*Brosimum paraense* Huber) remained a prerogative of the crown until about the middle of the nineteenth century when, with a great accumulation of wood and few buyers, the price had declined to a level at which it was no longer worth while to maintain a monopoly. (See *Tropical Woods* 40: 49.) The principal foreign demand for the timber at present is not for dyes but for the manufacture of violin bows and in this trade it is known as Pernambuco, taking its name from the Brazilian state where the best grades originate. It is to the bow what the Spruce or so-called Swiss Pine (*Picea*) is to the sounding board and Maple or European Sycamore (*Acer*) is to the back of a violin. It was formerly the custom for dealers to sort over dyewood stocks for pieces of the right quality to meet the exacting demands of the bow-maker, but most of the bow wood is now bought on contract. A typical French specification calls for genuine Pernambuco, freshly cut, of strictly straight grain, in faultless round pieces free of sapwood, 1 meter long and 10 to 20 centimeters (4 to 8 inches) in diameter, with a few diameters of 8 to 9 cm. admitted. Wood lacking in density, fineness of grain, brilliance and richness of coloration, or other characters detracting from its use for bows is often designated Bahia wood to distinguish it from Pernambuco wood, although both qualities are shipped from Bahia and presumably are derived from the same botanical species. The inference is that Pernambuco wood of the bow-maker is a special type or grade of Brazilian Brazilwood.

The following description applies only to the wood of *G. echinata*. Heartwood fairly uniform bright orange to orange-red when fresh, deepening to rich vinous red, with golden luster; sharply demarcated from the nearly white sapwood. Luster often very high. Dry specimens without distinctive odor or taste. Mostly very hard, heavy, compact, and strong; sp. gr. (air-dry) 0.90 to 1.25, rarely more than 1.10; weight 56 to 88 lbs. per cu. ft.; texture mostly fine and uniform; grain straight to irregular; not difficult to work, finishing very smoothly and taking a high natural polish; is very resistant to decay. The coloring principle, brasilin, is the same as that found in the heartwood of *Haematoxylon brasiletto* and responds to the same tests (see under *Haematoxylon*).

COMMON NAMES: *Guilandina echinata:* Bahia wood, brazilwood, Fernambuco wood, Pernambuco wood (trade); Brasillet de Fernambouc, bois de Brésil, Pernambuco (Fr.); Brasilienholz, Fernambukholz (Germ.); fernambucco, legno del Brasile, pernambucco, verzino del Brasile (Ital.); palo brasil (Span.); brasiletto, ibira-pitanga, pau brasil, p. b. claro, p. b. escuro, p. Pernambuco, p. rosado, ymira-piranga (Braz.). *Other species:* Bonduc, horse-eye, nicker tree, gray nickers, quasha, yellow nickers (B.W.I.); guacalote, g. amarillo, g. prieto, mate (Cuba); brier, gray nickers, mato azul, m. de playo, yellow nickers (P.R.); mate (Dom. R.); canique gris, oeil de chat, yeux de chat (Fr. W.I.); brasil, cojones de gato, guacolote, haba de San Antonio, jabilla, ojo de venado, taray, villa del mar (Mex.); alvellana (Salv.); mate, orilla del mar (Col.); garrapata de playa (Venez.).

Gymnocladus, with two species, is confined to central China and east-central North America. *G. dioicus* (L.) Koch is an unarmed deciduous tree of infrequent occurrence in hardwood forests on rich bottomlands of the Mississippi and Ohio valleys. It is often planted in parks and gardens in eastern United States and in Europe. It attains a maximum height of about 120 feet and a trunk diameter of four feet, but at a distance of 10 to 15 feet from the ground the main trunk commonly divides into three or four stems. The doubly pinnate leaves are large, sometimes three feet long, and the twigs are coarse and crooked and have a large orange-colored pith; the greenish white dioecious flowers are borne in terminal racemes; the legumes are large, thick, rather woody, dark brown, and indehiscent, and contain a few hard rounded seeds separated by a thick sweet pulp. The seeds are edible when roasted and in pioneer days were used as a substitute for coffee, hence the name Kentucky Coffee-tree. Planted trees in Maryland and New York are often called Mahogany. The timber has been used locally for fence posts, railway crossties, bridge timbers, sills, and occasionally for interior trim and furniture. It is of little commercial importance because of its scarcity.

Heartwood light cherry-red or reddish brown, varying in shade in same tree; sharply demarcated from the greenish white sapwood. Luster medium. Hard, heavy, tough, and strong; sp. gr. (air-dry) 0.65 to 0.70; weight 42 to 44 lbs. per cu. ft.; texture very coarse; grain usually straight; not difficult to work, but difficult to season without splitting; takes a high polish; is highly durable. Resembles *Gleditsia triacanthos*.

Haematoxylon, a tropical American genus of two species of bushy and usually thorny trees with channeled trunks, is well known as the source of two important dyewoods, Logwood and Brasilette (one of the Brazilwoods of commerce). The leaves are evenly pinnate, with few to several pairs of mostly obcordate leaflets; the small yellow flowers are borne in showy axillary racemes; the flat few-seeded legumes are dehiscent through the middle of the valves instead of along the sutures.

Haematoxylon campechianum L., the Logwood-tree, is typically small, rarely 50 feet high, with a short irregularly fluted or contorted trunk attaining a maximum diameter of about two feet, though usually much less. The natural range includes southern Mexico, Central America, Colombia, Venezuela, the Guianas, and probably parts of the West Indies. It was introduced into Jamaica in 1715 and has become completely naturalized there and on other islands where it did grow originally.

The value of the heartwood for dyeing was early recognized by the Europeans who began to import it, chiefly from Mexico, probably about the middle of the 16th century. The first attempts to use the dyes in England were not successful because no method was then known for making the colors permanent with mordants, and to protect the public from the bad work of the dyers an act of parliament in 1681 abolished the use of "a certain kind of ware or stuff called Logwood alias Blockwood" and ordered that any found should be forfeited and burned. The act was not repealed until 1662, when the colors produced from Logwood through improved methods were recognized as being as durable and serviceable as those of any other dyewoods. During the period of prohibition the wood was clandestinely imported and used. It acquired the names of Campeachy and Peachwood from Campeche, Mexico, one of the principal ports of shipment.

Logwood played a prominent part in the early history of British Honduras, which was first settled in 1638. Gibbs says (*British Honduras,* London, 1883): "It was at one time the practice of the class of privateers (almost identical with buccaneers) cruising against Spanish traders to set fire to all vessels they captured which might be laden with Logwood, having first stripped them of everything valuable. But it so happened that a Captain James, the master of a letter of Marque, having captured a Spanish vessel the cargo of which consisted of this wood, brought the ship and cargo into the Port of London. On endeavoring to dispose of the latter he was gratified as well as surprised to find for it a ready sale at an enormous price per ton. The crew, who had used up a portion of the precious freight to burn in the galley fire, had little idea that they were using fuel at a *hundred pounds* per ton during the voyage! The fame of this dyewood soon spread, and privateers were fitted out and dispatched to cruise off the Main, for the especial capture of Logwood-laden vessels, on their passage home to Spain from his Catholic Majesty's possessions in the 'Indies.' In course of time, as prizes became scarcer, protecting cruisers of the Spanish navy more abundant, the crews of the privateers found it more profitable to search for the wood on shore, cut it, and load their vessels with it."

The Logwood industry reached its peak during the latter half of the 19th century when the annual exports of the wood exceeded 100,000 tons. In 1872, Haiti exported about 90,000 tons, and in 1876 England imported over 64,000 tons, valued at nearly £416,000. Synthetic dyes have very largely replaced those obtained from wood, but for certain purposes the natural dyes still defy competition. Logwood cutting is now a minor industry. This is well illustrated in the case of British Honduras which now ships 125 tons annually, whereas in former times the amounts varied from 4000 to 13,000 tons a year.

Only the heart portion of the trunk is used, the thin, white sapwood being hewed off in the forest. In Mexico, British Honduras, and Honduras, the tree grows best in flat marshy lands often subject to overflow from the rivers, and the wood, in 3-foot lengths, is brought out in small boats or dugout canoes. In the West Indies the best timber is produced in interior valleys and moist coves in the lower slopes of the hills, and the trees attain merchantable size in about 12 years. Jamaica and Haiti are important sources of the wood, and some shipments have recently been made from Cuba. The length of logs or sticks from the islands varies from 3 to 12 feet. At the dyeworks the wood is reduced mechanically to small chips from which the dye is extracted by boiling in water. The coloring principle, called haematoxylin, forms an orange-red solution with boiling water, becoming yellowish as it cools, recovering its former hue if heated, and if left to itself it finally turns black. Various shades and colors can be obtained by the use of proper mordants, but the greatest consumption of Logwood is for blacks, which are obtained by alum and iron bases, and of any desired intensity. In addition to dyes and stains it is also used in the manufacture of inks and to a minor extent for medicinal purposes. The term "bastard Logwood" is applied rather indiscriminately, but if wood so called contains haematoxylin the presumption is that it comes from a physiological form of *Haematoxylon campechianum* deficient in the coloring principle, as no botanical characters have been found that distinguish the tree from typical or true Logwood.

Haematoxylon brasiletto Karst. is a shrub or a tree 15 to 35 feet high with an irregular trunk usually less than 12, rarely up to 24, inches in diameter, growing along the Pacific slope of North America from Baja California and Chihuahua, Mexico, to Nicaragua, and in northern Colombia and Venezuela. It appears to be closely related to the other species and the resemblance extends to the anatomy of the woods. From the standpoint of the dye obtained from the heartwood, however, the two are very distinct and are not used for the same purpose. *H. brasiletto* belongs in the category of Brazilwoods and the coloring principle is known as brasilin. According to the generally accepted chemical formulas, brasilin ($C_{16}H_{14}O_5$) differs from haematoxylin ($C_{16}H_{14}O_6$), the coloring principle of Logwood, only by an atom of oxygen. Both are adjective dyes (*i.e.*, they require a mordant), Logwood producing blues and blacks, Brazilwoods plum and purple shades and bright red. Haematoxylin apparently is found only in the heartwood of *H. campechianum*, but brasilin is more generally distributed, for example in various species of *Brasilettia* and *Caesalpinia*.

There is much confusion in the literature regarding the identities of the Brazilwoods of commerce. The original Brazilwood, an important article of commerce during the Middle Ages, was *Caesalpinia Sappan* L. of the East Indies. When, in 1500, an abundance of a similar wood (see under *Guilandina*) was discovered in eastern South America, the locality and eventually the whole country was called Brazil. The common name Brazil or Brasil has been applied to various red-wooded trees in tropical America and the scientific term *Caesalpinia brasiliensis* was given by Linnaeus in 1753 to a shrub or little tree in Haiti, but this is not the same as *C. brasiliensis* Sw. of Jamaica, Cuba, and Yucatán, which is now called *Brasilettia violacea* (Mill.) Britt. & Rose.

The foregoing are only a few of the problems in nomenclature that confront anyone who studies the literature. The authors have examined authentic specimens of the different tropical American woods used or said to be used for making dyes and in the present connection have had the co-operation of Roy H. Wisdom, chemist for the Taylor, White Extracting Company, manufacturers of dyewood and tannin extracts, Camden, New Jersey. As a result of these investigations it appears definitely settled that the tree supplying Brazilwood, Brazilette, Lima-wood, Nicaragua-wood, or Hypernic used in the American dye trade is *Haematoxylon brasiletto,* now mostly from Nicaragua, and that heartwood of this species growing in Colombia and Venezuela is also of commercial dyewood strength.

The wood is said to have been first cut and shipped to England from the region of the Gulf of Nicaragua, opposite Providence Island, believed to be the only place where the species grows naturally on the Atlantic side of North America. Exports from Mazatlan to Havre and Hamburg began about 1848, and as local supplies decreased other Mexican ports were used. Rail shipments from Mexico were made to the United States during the first World War when there was a temporary revival of the dyewood trade. At present the demand is small and the timber resources are ample to meet it.

No reliable anatomical distinctions between the woods of the two species of *Haematoxylon* have been found, as the range of variation within one species appears as great as between the two. They are, however, readily separated by chemical means. The coloring principle, which is soluble in water, gives with lime, baryta, and stannin chloride a blue precipitate for Logwood and a red precipitate for Brasilette. A simple method of identification is to apply a few drops of concentrated ammonia solution to a freshly exposed surface of the wood and note the stain imparted to white paper; that of Logwood is decidedly bluish, that of the other at first blood red, changing to reddish violet. Water stains, which are yellow or olive for both, instantly acquire their characteristic hue when exposed to ammonia fumes.

Heartwood bright orange changing, after exposure, to red and eventually to reddish black; sometimes with dark streaks; sapwood white or yellowish, sharply demarcated, thin to very thin in mature specimens. Luster high, golden. Odor mildly fragrant in fresh specimens, but may be lost in drying. Taste sweetish. Very hard heavy, and compact; sp. gr. (air-dry) 0.9[illegible] to 1.00 for Logwood, up to 1.10 for Brasilette; texture medium to fine; grain irregular; cuts rather flinty across the grain finishes very smoothly and takes a high polish; is strong but brittle; highly resistant to decay. Use for other purposes than dyewood largely prevented by the irregular shape of the stems.

COMMON NAMES: *Haematoxylon campechianum:* Blockwood, campeachy wood, campechy wood, campetch, logwood, peachwood, poachwood (Eng.); Allerheiligenholz, Blankholz, Blauholz, Blutholz, Campeschenholz (Germ.); bois bleu, b. de Campêche, b. de la Jamaïque, b. de sang, b. noir, b. sanglant, campêche, c. carmen, c. rouge, coeur rouge (Fr.); campeggio, legno azzurro, l. di campeggio, l. negro (Ital.); palo azul, p. de Campeche, p. de sange, p. de tinta, tinto (Span.); pau sanguinho (Port.); blauw-hout, campêche-hout (Dutch); ek (Mayan). *H. brasiletto:* Brazil, brazilette, brazilwood, hypernic, limawood, Nicaragua wood, redwood (Eng., trade); brasil (Span., Port.); azulillo, curaqua, palo de brasil, quamochitl, vitzquahuitl (Mex.); brasilito (Salv.); brasil de montaña (Hond.); brasileto, hala, palo de brasil (Col.); brasil zancudo, brasilete (Venez.); brazieja (Curaçao).

Harpalyce, with about a dozen species ranging in size from low shrubs to small trees, is represented in southern Mexico, Cuba, and Brazil. The leaves are odd-pinnate, with rather large leaflets bearing numerous yellowish or orange glands or gland-like scales on the lower surface; the flowers are large, in axillary racemes or terminal panicles; the pods are 2-valved and contain several seeds, which in some species are separated by spongy tissues. Only two wood samples are available for study, one of *H. cubensis* Gris., a little tree in the understory of the forest in western Cuba, the other of *Harpalyce* sp. from eastern Cuba.

Heartwood pale yellowish brown, sometimes with pinkish streaks; rather waxy; sharply demarcated from the thin yellowish white to canary sapwood. Odorless and tasteless. Very hard, heavy, compact, tough, and strong; texture fine; grain straight to irregular; not very difficult to work, finishing very smoothly and taking a high natural polish; inclined to split in drying; is probably highly durable. Suitable for small articles of turnery; uses limited by small sizes available; of no commercial possibilities.

COMMON NAMES: Cerillo de costa (Cuba); balché-ché (Mex.).

Havardia, with nine species of spiny shrubs and small or rarely medium-sized trees, is of limited distribution and of no commercial importance. Eight of the species occur in Mexico; one of them, *H. pallens* (Benth.) Britt. & Rose, a shrub or a tree up to 30 feet high, extends across the border into northeastern Texas, while another, *H. platyloba* (Spreng.) Britt. & Rose, grows in Yucatán, Aruba, Colombia, and Venezuela. There is one Cuban species, *H. prehensilis* (C. Wright) Britt. & Rose, which is a low shrub. The bipinnate leaves are petioled and the leaflets are numerous and usually small; the legume is flat, with thin valves and a few flattened seeds. According to Standley (*Trees and shrubs of Mexico*, p. 397), *H. pallens* is of some value for forage, as sheep and goats eat the leaves in winter, and for its wood, which has various local uses. *H. albicans*, which attains a maximum height of 60 feet, "is said to produce a gum similar to that of the Mezquite. The fruit is reported to contain 18 per cent of tannin and to yield a black dye."

The only wood sample available (Yale 16450) is of *H. platyloba*, collected by the senior author at Don Jaca, near Santa Marta, Colombia. Heartwood lustrous dark brown with a purplish tinge; sharply demarcated from the thin, yellowish white sapwood. Without distinctive odor or taste. Hard and heavy; rather fine-textured; durable.

COMMON NAMES: Guajillo (Texas); carbonero, chino, chucum, guajilla, huisache, manto, palo chino, p. gato, ramo de chivato, tenaza, uña de gato (Mex.); hoenja gato (Venez.).

Hebestigma *cubensis* (H.B.K.) Urb., the only species, is an unarmed small to medium-sized tree, sometimes 40 feet tall, growing in the forests of the mountains and rough lands throughout most of Cuba.

The leaves are large, odd-pinnate, with several large, mostly alternate, leaflets greatly variable in outline; the racemes of pink flowers make their appearance before and below the young leaves; the legumes are woody, linear, flat, 2-valved, with corky masses separating the several compressed seeds. The species has been referred to three other genera, namely, *Robinia, Lonchocarpus,* and *Gliricidia,* and from the standpoint of its wood a close relation to *Lonchocarpus* is indicated. The timber has a few rural uses, mostly for props and fence posts.

Heartwood not distinct in specimens studied; said to be orange or brown; sapwood yellow. Luster medium. Odor and taste apparently absent. Very hard, heavy, tough, and strong; texture medium; feel rather harsh; grain irregular; not easy to work; said to be durable. Has no commercial possibilities.

COMMON NAMES: Cucharilla, frijolillo, guamá candelón, g. piñón, jurabaina, yayabacaná amarilla (Cuba).

Heterostemon, with seven species of unarmed shrubs and small trees, is of limited occurrence in the northern part of the Brazilian Amazon region and the hinterlands of the Guianas, Venezuela, and Colombia. The leaves are pinnate, with one pair of large leathery leaves or an indefinite number of small ones; flowers large orchid-like and showy in short racemes which are terminal or borne on leafless branches; the legumes are large, straight or sickle-shaped, with leathery valves and several seeds. Apparently the only value of the plants is for decorative purposes. Ducke says (*Arch. Jard. Bot. Rio de Janeiro* 4: 271) that *H. mimosoides* Desf. is probably the most beautiful of all American Leguminosae because of its attractive foliage and gorgeous flowers; the Brazilian name is Hervão.

Heartwood of intermingling shades of light and rather dark brown; sharply demarcated from the whitish sapwood. Not highly lustrous. Odorless and tasteless. Of medium weight and hardness; rather fine-textured; grain straight to irregular; very easily worked, finishing smoothly; is probably durable. Without commercial possibilities because of the scarcity and small size of the trees.

Holocalyx, with two species of medium-sized to large unarmed trees, is common in the dry forest areas of southern South America. *H. balansae* Micheli grows in Argentina and Paraguay, *H. Glaziovii* Taub. in São Paulo, Brazil. The leaves are pinnate, with many pairs of dark green and shiny leaflets; the small flowers are borne in axillary racemes; the legumes are egg-shaped, fleshy, indehiscent, generally with a single large seed. The trunks are fluted and irregular, usually less than 18 inches in diameter, though occasionally considerably larger. The strong, resilient, durable timber is considered excellent for wheelwright work and turnery, heavy construction, fence and house posts, fuel and charcoal. The amount available is limited and there are no exports.

Heartwood purplish brown with numerous darker streaks and fine parenchyma markings; sharply demarcated from the thick yellowish white sapwood. Luster medium. Odorless and tasteless. Very hard, heavy, and tough; sp. gr. (air-dry) 0.95 to 1.05; weight 60 to 66 lbs. per cu. ft.; texture rather fine; grain nearly straight to roey; not difficult to work, finishes very smoothly; is highly resistant to decay. A good timber but of minor commercial importance.

COMMON NAMES: Alecrim (Braz.); ibirápepé, i.-p. colorado, i.-p. morotí (Par.); alecrín, ibirá-pepé, uirá-pepé (Arg.).

Humboldtiella *ferruginea* (H.B.K.) Harms, the only species, is a small unarmed deciduous tree 15 to 20 feet high, growing in northern South America and Trinidad and best known in Venezuela. According to Pittier (*Journ. Wash. Acad. Sci.* 18: 8: 208–210), the species should be retained in the genus *Robinia,* though he admits that, so far as general habit and macroscopic characters are concerned, there is little resemblance between it and *R. pseudacacia*. Since the woods of the two genera are distinct it is simpler to treat them separately.

Unlike *Robinia,* the tree has termina

buds, the branchlets are accordingly not zig-zag, and the axillary buds are scaly and not in covered depressions. The leaves are odd-pinnate, with several to 21 rather small, alternate to sub-opposite, oblong-lanceolate, sharp-pointed, pubescent leaflets, and without spiny stipules; the purplish mottled yellow flowers are borne in axillary racemes; the pod is slender, 4 to 5 inches long, and covered with a rusty tomentum. The wood apparently is not utilized because of the small size of the trees.

The following description is based upon two specimens, 2 to 3 inches in diameter, supplied by Pittier, one of them (Yale 8678) accompanied by flowering herbarium material. Normal heartwood absent; purplish black near small wounds; sapwood bright yellow. Fairly lustrous. Odorless and tasteless. Very hard, heavy, compact, tough, and strong; sp. gr. (air-dry) 0.90 to 1.05; weight 56 to 66 lbs. per cu. ft.; texture rather fine; grain fairly straight; not difficult to work, finishing very smoothly. Of no commercial possibilities.

Hymenaea, with 20 species and a few varieties of large unarmed trees, has its center of distribution in the Brazilian Amazon region, with one species, *H. courbaril* L., extending from Bolivia to Mexico and the West Indies. East African and Asiatic species of *Trachylobium* resemble *Hymenaea* and are sometimes included in the genus. The leaves are pinnate, with a single pair of rather large thick and glossy leaflets; the flowers are generally large, white or purplish, and borne in panicles; the legumes are large, rough, dark brown, woody, and indehiscent and contain few to several large seeds imbedded in a sweet and edible pulp.

The best known species, *Hymenaea courbaril,* is commonly called Locust in British America, Jutahy or Jatobá in Brazil, and Algarrobo or Guapinol in Spanish America. It prefers well-drained soil and attains a height of 100 feet or more, with a lofty spreading crown supported by an immense trunk which sometimes is more than six feet in diameter. The seeds are often slow to germinate, but the seedlings grow rapidly, though bushy and succulent at first and attractive to browsing animals. Plantations, unless protected from livestock, should be started at the beginning of the rainy season with one-year-old seedlings, pruned back to about a foot in height. The trees are taprooted when young but later develop a spreading root system.

One of the commercially valuable products of the tree is a pale yellow or reddish rosin-like gum usually known in the trade as South American copal to distinguish it from the various copals from other parts of the tropics. The exudations from the bark trickle to the ground and harden into lumps which eventually become buried in the soil. The gum gatherers dig around the roots and sometimes obtain a barrelful of copal lumps in a place, while the former site of a big tree, long since decayed, may yield several barrels of so-called fossil copal. Gum is also obtainable in small amounts by wounding the bark, but it is softer and less valuable and is known in the trade as balsam or soft anime. The copal is employed chiefly in the manufacture of special grades of varnish and to a minor extent for incense in churches, for medicinal purposes, and as cement for mending broken crockery.

The bark is used by the aborigines for making canoes. The process consists of removing the bark in one large piece, preferably from the standing tree, sewing it together at the ends with bush rope or bark-fiber cordage, waterproofing the seams with gum or resin, and then fitting in a few wooden cross-pieces to hold the shape. The bark of very old trees is often over an inch thick, and canoes with a carrying capacity of 25 to 30 men can be made from it. Locust bark is also a source of tannin.

The timber is of good quality and is used for carpentry, construction of all kinds, ship- and boat-building, treenails, furniture, cabinet work, felloes of wheels, and many other purposes. It is not considered highly durable in contact with the ground, though there is considerable variation in the wood in this respect. It was formerly well known in the European markets but is less common now. There is no likelihood of its becoming important in the United States, as the supply is inadequate. It is suitable for the same purposes as Black Locust (*Robinia*).

Heartwood russet to reddish brown, often more or less streaked; rather dull on the surface but with a subdued golden glow beneath; sharply demarcated from the white or gray sapwood. Without distinctive odor or taste. Mostly very hard, heavy, tough, and strong, though exhibiting much variation; sp. gr. (air-dry) 0.75 to 1.05; weight 47 to 66 lbs. per cu. ft.; texture medium to rather coarse; grain fairly straight to irregular; not easy to work but finishes smoothly; is moderately to highly durable. Tests made at the Forest Products Laboratories of Canada on British Guiana timber gave the following results (*Journ. Bd. Agr. Brit. Guiana* 13: 2: 91; April 1920): Weight at 15 per cent moisture, 47 lbs. per cu. ft. *Static bending* (in lbs. per sq. in.): Fiber stress at elastic limit, 8500; modulus of rupture, 15,690; modulus of elasticity, 2,153,000. *Endwise compression:* 8660 lbs. per sq. in. *Hardness:* Load in lbs. required to imbed a 0.444-inch steel ball to half its diameter in different surfaces: radial, 1630; tangential, 1670; end, 1896.

COMMON NAMES: Courbaril, locust (leather-leaved, South American, West Indian), stinking-toe (B.W.I.); curbaril (Cuba); algarrobo (Sp. Am., gen.); coapinol, copinole, cuapinol, cuapinole, guapinol, guapinole, nere, pak, quauhpinolli (Mex.); guapinol (C. Am., gen.); masaicarán, palito colorado (Hond.); copinol (Salv.); surixkrá, tema, tsi-tsi-ñaú (C.R.); corobore (Venez.); caouroubali, cimiri, kawanari, kwanari, locust, simiri, s. locust (Br. G.); jengi kanda, julchihout, kakanjan boesoe, karvanari, kawakanalli, locus, lokisi kaka, lokisie, loksi, semirie, simirie (Sur.); bois de courbaril, cacachien, caroubier, chimidida, copalier, courbaril, c. de savane, c. montagne, itaiba (Fr. G.); algarobo, comer de arara, jatahy, j. peba, j. roxo, jataiba, jatai, jatay, jatobá, j. amarello, j. roxa, jetahy, j. preta, jetay, jutahy, j. assú, j. café, j. catinga, j. da varzea, j. do campo, j. do igapó, j. miry, j. peba, j. pororoca, j. roxo, jutai, jutay branco, oleo jatahy, trapucá, yatayba, yutahy (Braz.); abatí timbary, avatí (Par.); asucar-huain, pampa estoraque (Peru).

Hymenolobium, with about a dozen species of medium-sized to exceedingly large unarmed trees, sometimes 150 feet tall and 10 feet in diameter, has its center of distribution in the upland rain forests of the central and eastern parts of the Brazilian Amazon region, with extensions northward into the Guianas and Venezuela and southward to Rio de Janeiro. The leaves are large, imparipinnate, with 5 to 49 rather small to large, oblong, more or less leathery leaflets; the flowers are pale roseate to red or violet, and are borne, generally at intervals of few to several years, in conspicuous terminal panicles nearly always after the fall of the leaves; the pods are large, typically elongate-oblong, membranous, in some species highly colored, and are adapted for wind dispersal, except in *H. heterocarpum* Ducke, which has a spongy fruit suggesting *Pterocarpus*. According to Ducke (*Tropical Woods* 47: 2), all species "yield hard and strong timber which is employed industrially at Pará, chiefly in naval construction." The usual Amazonian name is Angelim; the true Angelim of the Pará timber trade is *H. excelsum* Ducke.

Heartwood pale brown when fresh, deepening upon exposure and thus accentuating the differences between the darker fiber layers and the lighter parenchyma bands and stripes; with rather gradual transition to the white or grayish sapwood which is 1 to 2 inches thick. Luster low, chiefly because of parenchyma. Odor and taste not distinctive. Hard, heavy, tough, and strong; sp. gr. (air-dry) 0.70 to 1.00; weight 44 to 62 lbs. per cu. ft.; small patches and streaks having a density greater than unity sometimes encountered, the excessive weight resulting from deposits of hard gum or wax filling all cells. In general, the woods are all much alike, although that of *Hymenolobium nitidum* Benth. contains less parenchyma than the others and is considerably denser (sp. gr. 1.15) and more deeply colored. They have many features in common with *Andira* and *Vataireopsis*, some species of which are known as Angelim in Brazil. The timber is useful for strong and durable construction, but probably will not attain much commercial importance. It has a small figure of the Partridgewood (*Andira*) type, though hardly distinctive enough to create a demand for veneers and turnery.

Common names: Erejoeroe, kadjoesie awha, liadiadan koleroe, lialiadan koleroe, reejoeloe, r. erejoeroerian, riariadan hororodikoro, saandoe, s. awha, wormbast, wormboom (Sur.); angelim, a. do Pará, a. pedra, carámate, murarena, sapupira amarella (Braz.).

Inga, with about 250 species of unarmed trees and shrubs, is widely distributed throughout tropical America. The leaves are even-pinnate, the leaflets few and large, the rachis often winged; the flowers are rich in nectar and very attractive to bees and humming birds; the large fruit pods are round, flat, or square in section and the seeds are imbedded in an edible pulp. The trees are common in the forests, particularly along streams, and some of them attain rather large size, though they usually are low-branched. Many species are cultivated for decorative purposes or for their fruit, but their chief utility is for shade and protection to plantations of coffee and cacao. The timber is not of commercial importance for lumber but it is used locally for fuel, charcoal, and miscellaneous minor purposes where resistance to decay is not a factor.

Heartwood generally poorly defined, sometimes copper-colored or reddish brown, often streaked; sapwood nearly white when fresh, becoming gray or brownish, often with a greenish yellow, coppery, or even purplish tinge. Luster medium to rather high. Odorless and tasteless. Of medium density to hard and heavy; coarse-textured; mostly straight-grained; easy to work; durability poor.

Common names: Guaba, guama (Cuba); guaba, guama, guamo, guava (P.R.); guama, jina (Dom. R.); pois doux, p. sucrin, sucrier, sucrin (Haiti); cuajinicuil, cuajiniquil, guajiniquil (Mex., C.A., general); aguatope, avenillo, bitze, chalahuitl, chelele, guatope, guavo mecate, jinicuil, timbre, vainillo (Mex.); bri-bri, red fowl, tama-tama (Br. H.); cerel, cerelillo, cuje, cujín, guamá, guamó, g. macho, ixcapirol, paterno (Guat.); chapernillo, cujín, guama, nacaspilo, paternillo, paterno, pepete, pepetillo, pepeto, p. liso, p. negro, pepetón (Salv.); cafecito, guavo (Nic.); ataña, brudji-ua, gruók, guaróbua, guarómá, guava, guavillo, guavito, guavo, g. machete, g. mecate, g. peludo, g. p. verde, g. real, sauí, seuí, sotocaballo, suremmá, tsoeib-kra, tsuib-uá, ugurók, uxu-puru, ñaba, vainica, zuín (C.R.); bribri, buk-oro-mo, buk-udra, coralillo, guabito peludo, guavo, g. de monte, g. machete, g. torido, sabá (Pan.); amé, guamito, guamo, g. arroyero, g. bejuco, g. colorado, g. de arroyo, g. machete, g. macho, g. montañero, g. tapaculo (Col.); burzquillo, carbonero, guamo, g. bejuco, g. caraota, g. tapaculo, guatero, tasí (Venez.); awati, koroti waikey, maparakoni waikey, mapourokong, pairowa waikey, shirada waikey, waikey, whykee (Br. G.); aboonkini, adjawakie, alawata poesoeloekoeloe, anakara, apoeroekonnie, aprokonie, karoto, k. waikey, koelisiriwokoeloe, lokisie, l. wakie, maporokon, maprokon, maprokonie, moutouchie, pairawa itoelicki karoto, pairjawa, pairowa waikey, plakonie, proekoenie, prokonie, sepronie, shirada waikey, sierada, sipoeroeni, soepronie, sopronie, swietie boonkie, s. boontje, toerelie, tolelie, lorelie, waikey, w. ingooe, waljie, wekie, worisjeporo apotopo (Sur.); bois pagode, bougouni, pois doux, p. d. blanc, p. d. gris, p. sucré, p. s. rouge, sucre (Fr. G.); arapari, ayacshimbillo, cacahuillo, cotochupa, guava, kusillu-pacai, millua-shicshic, pacai, p. guava, palta-shimbillo, ramia, rosca-shimbillo, rufindi, rujiña-shimbillo, sapo-shimbillo, shimbillo, s. blanco, s. colorado, s. paca, s. rujinti, s. venenoso, yaco-shapona, yaco-shimbillo, yacoruma shimbillo (Peru); ca-ingá, imburana de cheiro, indá-uassú, ingá, i. assú, i. chichi, i. chichica, i. cipó, i. cururú, i. doce, i. ferradura, i. jero, i. mirim, i. peba, i. peua, i. rana, pacay, parápará, pau de vintem, tonga (Braz.); pacay, p. peludo (Boliv.); ingá, i. colorado, i. de comer, i. del agua, i. del cerro, i. de monte, i. guazú, i. puitá, i. pyitá (Arg.).

Isandrina. Four or five tropical American species of unarmed shrubs and trees have been referred to this genus, but the only one well known is *I. emarginata* (L.) Britt. & Rose (= *Cassia emarginata* L.), a small to medium-sized deciduous tree, sometimes up to 65 feet high, of rather common occurrence in the West Indies, northern South America, Central America, and southern Mexico. The leaves are pinnate, with 2 to 5 pairs of broad softly pubescent leaflets; the large yellow flowers are borne in axillary racemes; the legumes are narrow, thin, flat, often a foot long, indehiscent, eventually breaking transversely. The timber is utilized locally to a minor extent for fence posts, carpentry, and construction and formerly was exported from Jamaica as a yellow dyewood. It does not appear to have any commercial possibilities.

Heartwood of a rich golden olive color, sometimes with dark streaks, and with a waxy appearance and feel; rather sharply demarcated from the thin white sapwood. Odor and taste absent or not distinctive. Rather hard and heavy; texture medium to rather coarse; grain straight to roey; very easy to work, finishing smoothly and taking a glossy natural polish; holds its place well when manufactured; is highly durable. Heartwood tends to impart a yellow stain to objects in contact with it.

Common names: Senna tree, yellow candlewood (Jam.); frijolillo, guaraduro, guacamayo amarillo, Jupiter amarillo (Cuba); palo de chivo (Dom. R.); bois cabrite, casse marron (Haiti); alcaparro, chepile, chile perro, flor de San José, mora hedionda, palo de zorrillo, p. hediondo, retama, vara de San José, viche, xtuab (Mex.); barba jolote, Carib pine (Br. H.); vainillo (Guat., Nic.); arguchoco (Salv.); caranganito, chibato, chibuelo, chivato (Col.); brucha macho, brusco, carángano, mora, mote extranjero, mucuteno extranjero, platanillo (Venez.).

Jacqueshuberia, with two species of small to medium-sized unarmed trees, is apparently confined to the northern and eastern Amazon region of Brazil. The leaves are large, equally bipinnate, with many pinnae and very numerous, small, crowded leaflets, suggesting *Cojoba arborea;* the medium-sized yellow or purple flowers are racemose at the ends of slender nearly leafless branches; the legumes are linear, flat, ribbed along both margins, elastically dehiscent, and contain several flat seeds. The branches and young stems are distinctly 5-ribbed.

Jacqueshuberia quinquangulata Ducke is a little tree with long, often nearly scandent, branches, known only in the transition zone from forest to savanna near Gurupá at the beginning of the Amazon estuary (see *Arch. Jard. Bot. Rio de Janeiro,* 1922, pp. 118–120). The only wood samples of this genus in the Yale collections are of *J. purpurea* Ducke, a tree up to 50 feet high, of apparently rare occurrence in the forest on periodically overflowed banks of the Rio Curicuriary, a tributary of the upper Rio Negro (see *Tropical Woods* 31: 25).

Heartwood rather dull reddish or dark orange-brown, more or less streaked; rather sharply demarcated from the thin brownish gray sapwood. Without distinctive odor or taste when dry. Very hard and heavy; texture rather fine; grain irregular; not very difficult to work, finishing very smoothly and taking a high polish; appears very resistant to decay. Has no commercial possibilities because of the small size and scarcity of the trees.

Klugiodendron. There are two described species of large shrubs or small trees: *K. umbrianum* Britt. & Killip in Colombia and *K. laetum* (Poepp. & Endl.) Britt. & Killip in Peru. Williams (*Woods of northeastern Peru,* p. 190) describes the latter as follows: "Tall shrub or small tree up to 35 feet in height. Crown spreading, branches elongated. Trunk erect, fluted, 13 inches or more in diameter, and clear of limbs up to 25 feet. Bark light to dark brown with a grayish cast and fairly smooth; an infusion prepared by boiling the bark in water is reputed to be used as a remedy for tertiary fever. Flowers white, with greenish white petals and pale pink filaments; September–October. Fairly common in the lowland forest, especially in the lower Huallaga (alt. 400–500 ft.). Timber esteemed locally for tool handles and more particularly for canoe paddles, hence the local name [Remo-caspi], 'remo' = paddle, 'caspi' = wood."

Heartwood light brown, though not distinct in specimen (Yale 18470; Williams 4444); sapwood brownish gray (discolored). Of medium density and texture; straight-grained; easy to work, finishing smoothly, with a good polish. Probably has many of the properties of Ash (*Fraxinus*).

Lecointea *amazonica* Ducke, the only species, is a medium-sized to rather large unarmed tree of frequent occurrence on periodically overflowed sandy soils in the lower Amazon region of Brazil, where it is known as Pracuúba. The leaves are simple, rather large, alternate, leathery, with serrate margin; the small white flowers are borne in solitary or clustered axillary racemes; the legumes are thick and indehiscent, with fleshy pericarp and thin leathery mesocarp, and contain 1 or 2 large seeds. The sapwood is used for making ax handles and the heartwood for parts of fishing arrows. The trunks are so deeply scalloped or fluted that it is impossible to saw them into lumber. (See *Arch. Jard. Bot. Rio de Janeiro,* 1922, pp. 128–130.)

Heartwood reddish brown, with an orange hue; has an oily appearance; sharply demarcated from the slightly yellowish sapwood. Odorless and tasteless when dry, but said to be mildly rose-scented when fresh. Very heavy and hard; sp. gr. (air-dry) 1.20; weight 75 lbs. per cu. ft.; texture rather fine; grain irregular; rather easily worked, finishing smoothly with a high natural polish; appears very resistant to decay. Presumably of no commercial possibilities.

COMMON NAMES: Paracuhúba cheirosa da varzea, pracuúba, p. cheirosa, p. da varzea (Braz.).

Lennea, with four species of unarmed shrubs or small trees rarely 25 feet high, is sparingly distributed in southern Mexico, Salvador, and Panama. The only specimens available for study are of *L. salvadorensis* Standl. collected by S. Calderón in the type locality, where the tree is known as Polvo de Queso. The leaves are odd-pinnate, with 13 to 19 thin light green leaflets; the rather small greenish yellow flowers are borne in short dense racemes which are mostly below the leafy parts of the branchlets; the pods are flat, with few to several dark brown seeds, the thin but hard valves twisting in dehiscence. The outer bark on old stems is gray and corky-fibrous. There are apparently no special uses for the wood.

Normal heartwood absent from samples, but wound areas are purplish brown; sapwood yellow. Luster medium. Odorless and tasteless. Very hard, heavy, tough, and strong; texture medium; feel harsh; grain somewhat irregular; not easy to work, but finishing smoothly; durability doubtful. Has no commercial possibilities.

Leucaena, with more than 40 species of unarmed shrubs and small or occasionally medium-sized trees, has its center of distribution in Mexico. The leaves are bipinnate, with large or small leaflets; the flowers are white and in peduncled heads; the legumes are flat, with two membranous valves.

The best known species is *Leucaena glauca* (L.) Benth., native to America but widely distributed by planting in tropical and subtropical regions generally. It is common in the second-growth forests of the Philippines, where it is called Ipil-ipil. It is usually a shrub or a small tree 20 to 35 feet tall, but under very favorable conditions attains a height of 50 feet or more. It grows readily from cuttings, establishes itself firmly, and coppices freely, thus making it suitable for plantations for fuelwood. It has been used successfully as a nurse crop for more valuable species which lack its ability to control encroaching grasses. Attempts have been made to utilize the bark for tannin materials, but the extracts are dark-colored, producing a leather of unsatisfactory appearance for most purposes (see *Lignan Science Journal* 13: 2: 211–224). Standley says (*Trees and shrubs of Mexico,* p. 369): "There is a prevalent belief in tropical America that if horses, mules, or pigs eat any part of the plant their hair will fall out. Cattle are said not to be affected, and in Mauritius the plant is considered valuable as forage for goats. The seeds are sometimes used for making necklaces, bracelets, and other articles."

According to the same authority (*loc. cit.*), *Leucaena pulverulenta* (Schl.) Benth., of northern Mexico and southwestern Texas, is a tree sometimes 60 feet high, the tall straight trunk occasionally 20 inches in diameter, supplying a rich dark brown timber which is employed for general purposes. *L. esculenta* (Moc. & Sessé) Benth., a tree 20 to 50 feet high, is believed to be the plant of which Sahogún wrote: "There is a tree known as Uaxin. It is of medium size, its trunk is smooth, likewise the leaves, which are almost like those of the Arbol del Perú [*Schinus molle*]. It bears a fruit like that of the Carob, which is good to eat and is offered for sale in the markets." According to Robelo, the geographic name, Oaxaca, signifies "the place where the Huaxin begins to grow."

Among the Central American species are *Leucaena Shannoni* Donn. Smith, from Chiapas, Mexico, to Nicaragua; *L. salvadorensis* Standl. of Salvador; and *L. guatemalensis* Britt. & Rose of eastern Guatemala. All supply good wood of attractive appearance, strong and durable, but of limited amounts and only local utility. Apparently the only species native to the West Indies is *L. pseudotrichoides* (DC.) Britt. & Rose, a shrubby to rather tall tree in the Dominican Republic. *L. trichodes* (Jacq.) Benth. is a shrub or little tree of sporadic occurrence in the "tierra caliente" of Venezuela, Colombia, and Peru. There are also two other species in Colombia. They have no commercial possibilities.

Heartwood rich, fairly lustrous, medium brown, sometimes with a coppery hue; sapwood yellow, very distinct but without sharp line of demarcation. Odorless and tasteless. Hard and heavy to moderately so; medium-textured; mostly straight-grained; not difficult to work, finishing very smoothly and with a high polish.

COMMON NAMES: Cow-bush, jimbay, jumby bean, jumpy bean, lead-tree (B. W.I.); aroma blanca (Cuba); acacia palida, campeche, hediondilla, tantan, wild tamarind, zarza blanca (P.R.); granalino (Dom. R.); bois bourro, grains de lin pays (Haiti); macata blanc (Fr. W.I.); bolillo, guaje, guajillo, guaxi, hoatzin, hoaxin, huassi, huaxin, uachi blanco, uaxin (Mex.); wild tamarind (Br. H.); yaje (Guat.); cascahuite, guacamayo de montaña, guaje, jocoro (Salv.); cañafistula de monte, panelo, veranero (Col.); durote, ramón (Venez.); chamba (Peru).

Libidibia, with several species of medium-sized to large unarmed trees, usually included in the genus *Caesalpinia*, is distributed from Mexico to Argentina. The leaves are unevenly bipinnate, with numerous small leaflets; the small yellow flowers are borne in axillary or terminal racemes and panicles; the legumes are small, several-seeded, and indehiscent.

Libidibia sclerocarpa (Standl.) Britt. & Rose (= *Caesalpinia sclerocarpa* Standl.) is a tree 25 to 50 feet high, with a trunk 12 to 24 inches in diameter, the bark smooth and exfoliating. It grows at elevations between 30 and 650 feet above sea level in southwestern Mexico, where it is known as Ébano. The timber is obtainable in hewed pieces a foot square and up to 20 feet long, free of the thick white sapwood. The heartwood is exceedingly hard and heavy, of a deep coffee color, and highly durable; it is used locally for railway crossties, marine piling and harbor works, and heavy building construction. It also makes excellent fuel and charcoal. A jet black color is produced by treatment with lime and such timber is employed like Ebony for articles of turnery and cabinet work. (See *Tropical Woods* 6: 20–21.) Small quantities have been shipped from time to time to San Francisco and New York.

A timber known to the New York trade as Brown Ebony or Coffeewood and in European markets as a kind of Partridge wood is obtained in limited amount from Venezuela, chiefly for making umbrella handles. According to Pittier, the species is *Caesalpinia granadillo* Pittier, which is closely related to *L. ebano* (Karst.) Britt. & Killip of Colombia and to *L. corymbosa* (Benth.) Britt. & Killip of Colombia, Ecuador, and Peru. The local names are Ébano and Granadillo. Mature trees are from 50 to 75 feet tall, with a well-formed trunk sometimes 36 inches in diameter and clear of branches for 35 feet. The bark is thin and exfoliates in plates more or less like *Platanus*. The wood is similar to the Ébano of Mexico and has the same local uses.

In the rain forests of northeastern Brazil is a tree known, along with various other hard-wooded species, as Pau Ferro (iron wood), and believed to be *Caesalpinia ferrea* Mart. It occurs in small groups or as scattered individuals, reaches a height of 100 feet and has a cylindrical bole three feet or more in diameter and free of limbs for 50 feet. The thin hard bark is mottled and flaky. The timber is employed for heavy and durable building, but does not enter the larger markets.

The Argentine Guayacán or Ibirá-berá, *Caesalpinia melanocarpa* Gris., is a wide-spreading heavy-limbed tree, usually less than 65 feet high but with a trunk occasionally more than three feet thick and 20 feet long. It is common in the Quebracho (*Schinopsis*) forests of the Gran Chaco region, and the greenish or mottled gray and green trunks suggest Lignum-vitae (*Guaiacum sanctum* L.). The timber is of some commercial importance in the smaller local markets but is rarely found in the cities. It is used in hewed form for railway crossties, fence posts, telegraph poles, and as a substitute for Lapacho (*Tabebuia*) for spokes of heavy wheels. The chief objection to it is the difficulty of working it with ordinary tools. The leathery black pods are very rich in tannin.

The most important species in the group is *Libidibia coriaria* (Jacq.) Schl. (= *Caesalpinia coriaria* [Jacq.] Willd.), but its value is not in its reddish brown flinty-hard timber but in its curled or S-shaped pods which are a commercial source of tannin. It is a large shrub or a bushy tree 15 to 30 feet high, with a short trunk sometimes 16 inches in diameter, common in open semi-arid country, especially on the outskirts of the tide belt along the coasts of southern Mexico, Central America, northern South America, and the West Indies. It is also cultivated in Java, Ceylon, India, and parts of tropical Africa. The legumes, known to the trade as Divi-divi or Dibi-dibi pods, contain a yellowish powdery substance which yields as much as 50 per cent tannin of exceptional qualities. The principal source of supply for the United States is Colombia and Venezuela. A mature tree will yield about 100 pounds of pods annually. Considerable amounts are consumed in local leather industries and also for making a black dye. The wood is occasionally used as a source of a red dye, but is too hard and refractory for other purposes.

The woods of this group are all much alike. Heartwood dark red to chocolate-brown or nearly black, usually with fine pencil-striping of parenchyma; sharply demarcated from the yellowish or pinkish white sapwood. Luster medium to low. Without distinctive odor or taste. Exceedingly hard, heavy, and strong; sp. gr. (air-dry) 1.10 to 1.30; weight 68 to 81 lbs. per cu. ft.; texture medium; feel harsh; grain fairly straight to very irregular; difficult to work, but finishes very smoothly; is highly resistant to decay. Of limited commercial possibilities.

COMMON NAMES: *Libidibia coriaria:* Divi-divi, libi-dibi (Jam., P.R.); cacalote, dibi-dibi, divi-divi, guaracabuya, guatapaná (Cuba); guatapaná (Dom. R.); divi-divi (Haiti); cascalote, nacascolotl, nacascul (Mex.); nacascolote (Guat.); nacascolo, nacascolote, nacascolotl, tinaco (Salv.); nacascol, nacascolo (Nic.); nacascol (C.R.); agallo (Pan.); baranó, dibi-dibi, divi-divi, lumbre (Col.); divi-dive, guatapán, guatapanare (Venez.). *Other species:* Brown ebony, coffee wood, partridge wood (trade); ébano (Mex.); eleten (Nic.); ébano, granadillo (Col.); ébano, granadillo, macle, quiebrahacho (Venez.); catingueira, imirá-itá, jucá, muirá-pixuna, pau de rato, p. ferro (Braz.); guayacán (Par.); guayacán, g. blanco, g. hu, g. negro, ibirá-berá (Arg.).

Lonchocarpus, with about 150 species of small to medium-sized trees and scandent shrubs and lianas, occurs in tropical America, Africa, Madagascar, and Australia. The leaves are odd-pinnate, with 1 to 15 medium-sized to large leaflets; the white, pink, or purple flowers are borne in simple or branched racemes; the pods are flat, broadly ovate to narrowly elongate, stipitate, thin to leathery, indehiscent, and usually with 1 to 4 seeds. According to Pittier (*Contrib. U.S. Nat. Herb.* 20: 2: 38), specimens without fruit are difficult even for an experienced botanist to distinguish from *Robinia* and *Gliricidia*.

The trees are common on open hillsides and rather dry plains at low or moderate elevations in tropical America and have very little economic value for their timber, though the hardest kinds are used locally to a minor extent for implement frames, spokes of logging-cart wheels, and some heavy construction. The roots of some of the South American shrubs and climbers are used to stupefy fish and as a commercial source of rotenone for making insecticides (see *Tropical Woods* 23: 30; 33: 36). The bark of one or more Yucatán species has long been used to make a fermented beverage, called "balche" (see *Tropical Woods* 21: 7).

Heartwood yellowish brown to dark reddish brown, with parenchyma laminations of lighter color; apparently late in developing in most species and usually absent from small specimens; sharply demarcated from the thick yellowish sapwood. Luster medium. Odor and taste absent from dry material. Moderately to very hard and heavy, generally very tough and strong; sp. gr. (air-dry) 0.70 to 0.95; weight 44 to 59 lbs. per cu. ft.; texture coarse to very coarse; feel harsh; grain irregular; not easy to work, inclined to be splintery, though there is considerable range in the working properties; durability of dark-colored material good. Apparently without commercial possibilities.

Common names: Frijolillo amarillo, guama, g. bobo, g. candelón, g. cimarrón, g. común, g. de costa, g. de majagua, g. de soga, g. hediondo, g. hembra, g. macho, g. negro, g. piñón, guamao, guamaro (Cuba); fuerte ventura, geno, geno-geno, palo hediondo (P.R.); anón de majagua, a. del río, anoncillo, a. de majagua, biajama (Dom. R.); battre à caïman, bois caïman, caïman cimarron (Haiti); savonette (Guad., Mart.); aricuahue, balché, balchechi, beco, cabo de hacha, cajurica, canacin, cincho, gusano, gusavo, jumay, kantzin, lombrizero, machich, margarita, marinero, palo amarillo, palo de oro, pitorrilla, rosa morada, taliste, talistillo, xbalché, yaxbabin, zaayab, zopilacuage (Mex.); bitterwood, black cabbage-bark, dogwood, ridge dogwood, swamp dogwood (Br. H.); chaperno, cocorocho (Guat.); chapel, chilillo, cincho, guayacán, masicarán, sopilocuao, sisín (Hond.); chapulatapa, ikuapelo, sangre de chucho (Salv.); carao, chaperno, corteza de venado, siete cueros (C.R.); chaperno, comenegro, dogwood, gallote, malvecino (Pan.); barbasco de agua, chicharrón, macaratú, macarutú, majagua de gallina, majomo, meoparado, meoparao, miaparao, morocolo, murucutú, papo de zamba, papo-zamba, prieto (Col.); aco, acrorutú, acurutú, grifo, jebe, mahomo, majomo, m. amarillo, m. carretero, menudito, tocorito (Venez.); aya, haiari, white haiari (Br. G.); nekoe (Sur.); nicou (Fr. G.); aquíquy, facheiro, pau de boto, rabo macaco, timbo, t. assú, t. branco, t. b. pau, t. carajurú, t. commum, t. de massa, t. grande, t. legitimo, t. macaquinho, t. pau, t. rana, t. urucú, t. venenoso do Pará, t. vermelho (Braz.); aguíguy, barbasco, b. de monte, b. legítimo, coñapi, cube, huasca-barbasco, kumu, olla-vaja, pacai, sacha-barbasco (Peru); caá-buzú, guamé-piré, ibirá-itá, i.-i. amarillo, i.-i. blanco, i.-i. morotí, i.-i. saiyú, palo-maceta, quina blanca, rabo blanco, r. de macaco, r. molle, yerba de bugre (Arg.).

Lucaya *choriophylla* (Benth.) Britt. & Rose (= *Acacia choriophylla* Benth.), the only species, is an unarmed tree sometimes 30 feet tall, occurring in coppices and scrub lands throughout the Bahamas and the cays of northern Cuba. It has bipinnate leaves with broad leaflets; the yellow flowers are borne in dense globose heads; the compressed woody pods are tardily dehiscent.

Heartwood rather dull dark reddish brown, inconspicuously marked with fine parenchyma and vessel lines. Sapwood yellowish white, rather sharply demarcated. Odor and taste absent or not distinctive. Exceedingly hard and heavy; medium-textured; of irregular grain; difficult to work; durability high. No special uses known.

COMMON NAMES: Cinnecord (Bah.); alava-alava, frijolillo, tamarindillo (Cuba).

Luetzelburgia, with three species of medium-sized to large trees, occurs in Brazil south of the Amazon basin. The only wood sample available (Yale 37860) is from the type of *Luetzelburgia trialata* collected by Ducke in upland forest near Rio de Janeiro. The imparipinnate leaves of this species have 5 to 7 smooth leathery leaflets; the small reddish violet flowers are borne in panicled racemes; the fruit is a samara suggesting *Tipuana* and *Vatairea.*

Heartwood yellowish brown, with prominent parenchyma striping; rather waxy looking; sharply demarcated from the thick pale yellowish sapwood. Not highly lustrous. Without distinctive odor or taste. Very hard, heavy, tough, and strong; texture coarse; grain irregular; difficult to work; is probably durable. Of the general type of wood known in the New York trade as Partridge.

Lysiloma, with about 30 species of unarmed shrubs and small or occasionally rather large trees, is chiefly a Mexican genus with extensions into Central America, southernmost parts of the United States, and the Greater Antilles. The branches are slender; the leaves are bipinnate; the flowers are small and in racemes, spikes, or heads; the legumes are thin and flat, the valves at maturity separating from the undivided margins, their outer layer dark-colored and flaky.

The most widely distributed species is *Lysiloma bahamensis* Benth., a spreading tree sometimes 50 feet high with a short trunk three feet in diameter, occurring in southern Florida, the Bahamas, Cuba, Haiti, Dominican Republic, British Honduras, and Yucatán, Mexico. The lustrous brown wood has a sp. gr. (air-dry) of 0.60–0.65, is easily worked, durable, and useful for furniture and interior trim, but the trees are too scarce or too poorly formed to supply timber of commercial importance, though it is used to some extent in Cuba for railway crossties.

Lysiloma latisiliqua (L.) Benth. (= *L. sabicu* Benth.) occurs sparingly in the Bahamas and the Island of Haiti, but is at its best in Cuba, supplying one of the most valuable timbers of the island. It is known as Jigüe or more generally as Sabicú, which is also the trade name for the timber. Writers often confuse this species with *L. bahamensis,* but the latter has very numerous, small, narrow leaflets, suggesting the Tamarind, whereas the other has comparatively few and large, oval leaflets. The name Sabicú is sometimes applied to *Cojoba arborea* and *Peltophorum adnatum* Gris., but their woods are dark red, whereas those of *Lysiloma* are brown. The Cuban Sabicú is a spreading tree with a rather short shaggy-barked trunk, sometimes more than two feet in diameter and free of branches for 25 feet, though commonly much smaller. It is generally distributed and formerly rather abundant, but trees of commercial size are now confined mostly to moist upland coves and small valleys in localities where transportation facilities are poor. The principal local uses for the timber are for general construction, wheelwright work, mill rollers, and railway crossties. Considerable quantities have been exported to England and lesser amounts to the United States for furniture, interior trim, and shipbuilding, but the trade has practically ceased.

Heartwood lustrous brown, with a coppery tinge, sometimes faintly striped; sharply demarcated from the thin white sapwood. Without distinctive odor or taste. Rather hard and heavy; sp. gr. (air-dry) 0.77; weight about 48 lbs. per cu. ft.; texture medium; grain commonly roey; easy to work, finishing smoothly and taking a high natural polish; is highly durable.

COMMON NAMES: *Lysiloma bahamensis:* Wild tamarind (Fla., Bah.); abey, frijolillo, f. cuadrado, f. redondo, saplillo

(Cuba); tabernau, tavernon (Haiti); tzalam, tzucte, xiazek (Mex.); salom, tzalam (Br. H.). *L. latisiliqua:* Horseflesh, sabicú (Bah.); jigüe, sabicú (Cuba); sabicú, Cuban sabicú (trade, Eng.); Sabicuholz (trade, Germ.); bois de sabicú (trade, Fr.). *Other species:* Chicharrón, gumaga, laaguia, mauto, palo blanco, quiebracha, tepeguaje, tepehauge, tepemezquite, tepeoaxin (Mex.); cola de marrano, coyote, granadillo, quebracho de cerro, quiebrahacha (Hond.); carbón, cicahuite, quebracho, q. colorado (Salv.).

Machaerium. According to Hoehne (*Flora Brasilica* 25: 3; 1941), there are 121 valid species (excluding *Paramachaerium,* but including *Drepanocarpus*). They are scandent or upright shrubs and small, medium-sized, or rarely large trees, widely distributed throughout tropical America, but most abundant in Brazil. The leaves are unequally pinnate, typically with alternate leaflets which are either comparatively few (3 to 11) and medium-sized to rather large, or numerous (up to 70) and small; the stipules are either caducous, or persistent and transformed into spines; the flowers, which are purple, violet, or white, are generally small and borne in axillary or terminal racemes or panicles; the fruit is samara-like, stipitate, one-seeded, and indehiscent. The sap is red, as in *Pterocarpus.*

About a dozen species of *Machaerium* are said to attain large dimensions, and several of these, particularly in southeastern Brazil, are reputed to supply commercial timber of excellent quality and useful for the same purposes as Rosewood (*Dalbergia*). In the Yale collections there are 37 wood samples collected with herbarium specimens which have been identified with 22 different species of *Machaerium* and at least six others which have been determined generically, but only three of the entire lot contain normal heartwood. There are also numerous other samples obtained from commercial sources in southeastern Brazil under the names Caviúna and Jacarandá (various kinds). In no instance, however, has it been possible to identify the trade samples with authentic material from the same locality. According to Hoehne (*loc. cit.*), the species producing the best timber is *M. firmum* Benth., but the wood has not been studied.

Many of the woods studied are of anomalous structure, the anomaly consisting of concentric laminations of included phloem which usually are widely spaced (0.5 to 1 cm., occasionally much more) and thus are not a dependable feature when dealing with small samples. Such specimens have rather light and soft, fairly coarse-textured sapwood and only occasional traces of dense dark red heartwood.

The normal woods exhibit considerable variation. Most like the anomalous kinds, except that they lack included phloem, are the available specimens of *Machaerium aculeatum* Raddi, *M. brasiliense* Vog., *M. paraguariense* Hassler, *M. scleroxylon* Tul., and *M. stipitatum* (DC.) Vog. All of these species are trees, some of them attaining large size. Normal heartwood is absent from all of the samples. The specimens of another group are hard, heavy, fine-textured, and of a fairly uniform light yellow color throughout. These are from erect shrubs or little trees of the following species: *M. arboreum* (Jacq.) Vog., var. *latifolium* (Benth.) Hoehne, *M. intermedium* Pitt., *M. latialatum* Pitt., *M. melanophyllum* Standl., and *M. riparium* Brandeg. The specimen from the type tree of *M. Whitfordii* Macbr. belongs here except that normal purplish brown heartwood is abundant and the parenchyma bands are mostly widely spaced and apparently terminal (for detailed description see *Bull. Torrey Bot. Club* 47: 2: 73–79). Several commercial specimens of Jacarandá from southeastern Brazil are of the general type of this Colombian species. A single specimen (Yale 11070; Pittier 12484) of *M. acutifolium* Vog. was collected in Venezuela and is probably a piece of a branch. It has a core of normal, blackish violet, walnut-scented heartwood, surrounded by rather coarse-textured, hard, greenish yellow sapwood. Since the range of this species extends from Venezuela to Paraguay, Bolivia, and southern Brazil it may be the source of the Jacarandá timber that is most like *Dalbergia.* The following generic description is drawn from the authentic specimens only.

Heartwood dark violet-brown, more or less streaked; rather waxy looking; absent from most of the available specimens; sapwood whitish, grayish, or yellowish, often with an olive tinge. Luster medium to high. Heartwood sometimes walnut-scented; taste not distinctive. Sapwood rather light and soft to very hard, heavy, compact, tough and strong; sp. gr. (air-dry) 0.60 to 0.95; weight 37 to 59 lbs. per cu. ft.; heartwood always hard and heavy, but usually brittle, the sp. gr. ranging from about 0.90 to 1.12; texture coarse to fine; grain variable; working properties fair to excellent; heartwood highly resistant to decay.

COMMON NAMES: Guamecate prieto (Mex.); juruguay (Guat.); sangre bravo (Salv.); mata-piojo (Hond.); capote, coba, cumaricá, cuña, látigo, negrillo, palo de sangre, pico de loro, purgación, sangre, siete cascas, s. cueros, s. c. blanco, s. c. espinoso, tachuelo, uña de lobo, zarza (Col.); cascarón, chacaranday, cumaricá, ñaure (Venez.); chiche (Ec.); tuseque (Boliv.); agarra-saia, aranha dedo, a. gato, arranhadeiro, bico de pato, cabiuna, cascarón, caúva, cariuna, chofó, conconcuriuga, cumaricá, farinha sêca, jacarandá, j. antam, j. cipó, j. de espinho, j. da Bahia, j. do cerradão, j. do cerrado, j. do campo, j. preto, j. roseo, j. roxo, j. sangue, j. una, mamica de porco, ojo dezamuro, pau de malho, sangre de toro, sapúva, sapuvussú, siete cueros, timbó, unha de gato, vaina de espada, violete, ximbó (Braz.); canella de brejo, isapuí, isapuy, i. guazú, i.-hú, i. mini, i. morotí, paloma rembuí, rembe-ú, yaquerí-buzú-guazú (Arg.).

Macrolobium (or *Vouapa*) includes more than 50 species of unarmed shrubs and small to large trees, about equally divided between tropical Africa and tropical America. The center of distribution of the American species is the Amazon region of Brazil, with a few extensions southward into Matto Grosso, Bahia, and Espirito Santo, westward into Peru, and northward into the Guianas, Venezuela, Colombia, and Panama. Two sections are recognized, one having 2-foliolate leaves, the other, sometimes considered a separate genus (*Outea*), with few to many pairs of leaflets; the racemose inflorescence is axillary or terminal, sometimes borne on the old branches; the legumes are orbicular to oblong, 2-valved, and containing one or few large, flattened seeds. Some of the plants are decorative, and some of the trees are large enough and in sufficient supply for commercial purposes, but their timber is not in demand.

The best known species is *Macrolobium bifolium* (Aubl.) Pers. occurring from the Guianas to Bahia, Brazil. In British Guiana, where it is called Sarebebe or Water Wallaba, though it is not a true Wallaba (*Eperua*), its frequency of occurrence over large areas is from three to nine trees per acre, capable of supplying logs 16 to 20 feet long squaring from 8 to 12 inches. Foresters consider the timber suitable for furniture and for heavy construction not in contact with the ground. *M. acaiaefolium* Benth., a medium-sized to rather large tree of the upper Amazon region and the Guiana hinterlands, is common along the margins of lakes and streams. According to Ducke (*Arch. Jard. Bot. Rio de Janeiro* 4: 270), the light red, hard, highly durable timber is suitable for carpentry and furniture. Williams says (*Woods of northeastern Peru,* p. 203) that the species, which is not common in the Peruvian Amazon region, attains a height of 50 feet, with an erect trunk 14 inches in diameter; the principal use for the wood is for fuel.

Heartwood brown or rather light reddish brown, of solid color or streaked; sharply demarcated from the brownish sapwood. Luster medium. Odorless and tasteless. Moderately to very hard and heavy; sp. gr. (air-dry) 0.72 to 0.90; weight 45 to 55 lbs. per cu. ft.; of medium to coarse texture; grain rather irregular; not very difficult to work, finishing fairly smoothly and attractively; durability variable, mostly high. Some of the timber is suitable for furniture.

COMMON NAMES: Sarebebe, water wallaba (Br. G.); ata-apa, ataba, atapa, salebebe, sarebebe, watapa, waterbijlhout, watrabiriehoedoe, watrapa (Sur.); wapa-sec

(Fr. G.); aipé, arapary, a. da varzea, a. rana, campina-rana, fava de tambaqui, faveira, ingarana vermelho, i. xixy, ipé, i. da varzea, i. de folha miuda, i. verdadeiro, i. uba, jatobárana (Braz.); aripari, copal, pashaco, pashaquilla (Peru).

Macroule, with one species, *M. Coutinhoi* (Ducke) Pierce (= *Ormosia Coutinhoi* Ducke), is a rather large tree in the lower Amazon region of Brazil and in British Guiana. It has odd-pinnate leaves with 5 to 9 large, smooth, leathery leaflets; the flowers are violet, with white or yellow centers, and are borne in large open panicles; the fruit is an indehiscent woody pod, usually with a single large brownish red seed having a long hilar scar. According to Ducke (*As Leguminas da Amazonia brasileira,* p. 106), the bark has a strong acrid odor when freshly cut, and the grayish, coarse-textured, fibrous, moderately hard timber is not utilized.

The only authentic wood specimen available (Yale 40083; Ducke 352) was collected (with flowering and fruiting herbarium material) at Utinga near Belem do Para. The tree was small and the wood sample (3-inch radius) contains no colored heartwood. Two specimens of Korokororo or Kurukoruru (Yale 5126 and 32940) from the Forest Department of British Guiana, the latter labeled *Ormosia Coutinhoi* Ducke, are entirely heartwood. All three pieces have similar structure.

Heartwood brownish or yellowish brown, with darker striping, and waxy appearance, suggesting *Prioria;* sapwood gray. Luster golden in proper light. Odorless and tasteless. Moderately hard and heavy; texture coarse to moderately so; grain roey; saws woolly and is inclined to rough up in planing, but will take a high polish; probably resistant to decay. Appears suitable for good furniture.

COMMON NAMES: Korokororo, kurukoruru (Br. G.); boiussú, buiussú (Braz.).

Marmaroxylon, with a single species, *M. racemosum* (Ducke) Killip (= *Pithecolobium racemosum* Ducke), is a large unarmed tree of the lower Amazon region and the Guianas (see *Tropical Woods* 63: 3). The surface of the bark is very rough and broken into irregular plates. The leaves are bipinnate, with 3 to 5 pinnae, each bearing 8 to 14 pairs of small, sessile, asymmetrical, retuse, glabrous, leathery leaflets; the petioles and petiolules bear nectaries and are covered with brownish pubescence; the petiolules are ribbed; the stipules are caducous; the small sessile flowers are collected in numerous heads in pubescent racemes on the old wood; the corolla is salmon pink; the filaments white at the base, but yellow distally; the fruit is a narrow, curved, densely pubescent pod about three inches long.

The name *Marmaroxylon* (marble wood) refers to the characteristic appearance of the wood. Large irregular purplish brown streaks and patches against a background of brownish yellow give a highly distinctive appearance to specimens of sufficient size to display the figure. Although the timber is not extensively used it has long been known to the Brazilian trade as Angelim Rajado (streaked Angelim) and to the French as Bois Serpent. The presence of the species in British Guiana was first established in 1937 by A. C. Smith, of the New York Botanical Garden, who found a tree 100 feet tall growing in dense forest on high land in the basin of the Essequibo River. According to Bertin (*Les bois de la Guyana et du Brasil,* pp. 79–80), the tree is rare but reaches large size in the forests of French Guiana. The bole attains a length of 60 to 80 feet to the first large limb and a diameter of 18 to 24 inches just above the heavily buttressed base. The natives use the timber for wheelwright work, but it is considered suitable also for cabinetmaking, marquetry, and decorative panels.

Heartwood in varying proportions of brownish yellow and purplish brown; darker streaks due to deeply colored resin filling all of the cells; transition to yellowish sapwood gradual. Luster medium. Odorless and tasteless, at least when dry. Very hard, heavy, tough, and strong; sp. gr. (air-dry) 1.05 to 1.25; weight 66 to 78 lbs. per cu. ft.; texture medium; feel rather harsh; grain straight to irregular; difficult to work, but finishing smoothly; durability doubtful, but said to be fairly high.

COMMON NAMES: Bousi tamarin, puta locus, slang houdou, snecki housou (Sur.); bois serpent, b. zebra (Fr. G.); angelim rajado, ingá caetitú, urubuzeiro (Braz.).

Martiodendron (or *Martiusia*), with four species of medium-sized to large unarmed trees, is limited in its distribution to Brazil and the Guianas. (For nomenclatural notes see *Phytologia* 1: 3: 140.) The leaves are imparipinnate, with 7 to 10 alternate medium-sized to large leaflets; flowers yellow, in pyramidal panicles; legumes large, somewhat leathery, generally 1-seeded, indehiscent. The hard and durable timber is used locally to a very limited extent for heavy construction.

Martiodendron excelsa (Benth.) Gleason occurs in the Guianas and along the upper Rio Branco in Amazonas, Brazil. *M. parvifolia* (Benth.) Gleason is found in Maranhão, Piauhy, and Bahia. *M. elata* (Ducke) Gleason grows on fairly well-drained clay soil in the region below the cataracts of the Tapajoz, a tributary of the Amazon; it is from 80 to 145 feet high, with a heavily buttressed trunk, and is very ornamental because of its yellow flowers and purple fruits. *M. macrocarpon* Gleason ranges in height to nearly 150 feet in the upper Amazon.

Heartwood orange-brown, becoming reddish brown upon exposure; sharply demarcated from the nearly white sapwood. Luster medium. Tasteless, but sometimes with slight, unpleasant scent. Very hard, heavy, tough, and strong; sp. gr. (air-dry) 0.95 to 1.10; weight 59 to 69 lbs. per cu. ft.; texture coarse; feel harsh; grain fairly straight to very irregular, roey, and wavy; difficult to work, being inclined to chip out; appears highly durable. A refractory timber of little promise.

COMMON NAMES: Boschmahonie, dastan, witte purperhart (Sur.); bois d'amarante rouge (Fr. G.); jutahy-seca (Braz.).

Melanoxylon *brauna* Schott, the only definitely known species, is a large unarmed timber tree occurring in the coastal rain forests from Bahia to São Paulo, Brazil. Two other species have been described, but one of them, *M. amazonicum* Ducke, was later made the type of a new genus, *Recordoxylon;* whether the other, *M. speciosum* R. Benoist, a British Guiana tree, belongs to *Melanoxylon* or *Recordoxylon* cannot be decided without a study of the fruit and wood, both of which are unknown (see *Tropical Woods* 39: 18).

Melanoxylon brauna, generally known locally as Braúna or Graúna, attains a height of 100 feet or more and has a smooth, straight, well-formed, unbuttressed bole sometimes four feet in diameter. The leaves are imparipinnate, with numerous large leathery leaflets; the large yellow flowers are in showy clusters; the legumes are rather large, with depressions between the seeds, which have rectangular wings. The bark is a source of tanning material and a reddish brown or black dyestuff. The timber is well known in local trade and is noted for its durability. It is used chiefly for beams, sills, posts, bridge timbers, and railway crossties; also for spokes of wheels, flooring, and fine furniture.

Heartwood dark brown or blackish with brown streaks; not highly lustrous; sharply demarcated from the yellowish sapwood. Without distinctive odor; taste somewhat astringent. Very hard, heavy, tough, and strong; sp. gr. (air-dry) 0.95 to 1.05; weight 59 to 66 lbs. per cu. ft.; texture rather fine; grain usually irregular; rather difficult to work, but finishing very smoothly and attractively.

COMMON NAMES: Baraúna, braúna, b. preta, garaúna, graúna, g. parda, g. preta, guiraúna, ibirá-una, Maria preta, muiraúna, parovaúna, rabo de macaco (Braz.).

Mimosa is a genus of more than 400 species of herbs, shrubs, vines, or small trees widely distributed throughout the tropical and warm regions of the world. Most of the plants are prickly. The bipinnate leaves usually have numerous small leaflets, which often are sensitive; the small regular flowers have 4 to 10 stamens (in contrast with *Acacia,* which has many stamens) and are in peduncled heads or spikes which frequently are showy; the legumes are thin, flat, and wingless and break up at maturity into few to many joints. The plants have little value except for decorative purposes and some of them are pernicious weeds.

Apparently the only American species useful as a source of wood is *Mimosa bracaatinga* Hoehne, an unarmed shrub or a slender tree sometimes 50 feet high and 8 to 16 inches in diameter, native to southeastern Brazil, where it is known as Bracaátinga or Abarácaátinga. It forms natural, but short-lived, thickets and is also readily cultivated. Tests have demonstrated that it grows rapidly and in a few years produces fairly straight hard-wooded stems that are excellent as a source of fuel. (See F. C. Hoehne: *A bracaátinga ou abarácaátinga*, pub. by Sec. Agr. Ind. & Com., São Paulo, 1930.)

Mimosa Schomburgkii Benth. was discovered in British Guiana about 100 years ago, but there are only two records of its appearance there. In February 1926, the senior author collected it not far from the seashore along the right of way of the Truxillo Railroad Company midway between Corocito and Black River, Honduras. The trees were growing singly or in clumps at the edge of open woodland for a distance of a mile or more. They are graceful and ornamental and have orange-brown shreddy bark, fine spray, rusty brown pubescent twigs, and prominent spikes of very small tawny flowers terminating the branches. The largest specimen examined was about 25 feet tall and 8 inches in diameter; nearly all of them were in bloom. (See *Tropical Woods* 10: 26; 11: 24.)

Heartwood pale reddish brown with fine but distinct red striping of parenchyma; not particularly attractive; not sharply demarcated from the yellow or yellowish brown sapwood. Luster medium to fairly high. Odorless and tasteless. Hard and heavy to moderately so; tough and strong; texture variable, mostly medium; grain straight to irregular; not difficult to work; appears durable.

Mimosopsis, a genus with about 25 species of shrubs and little trees accredited to it, has its center of distribution in Mexico, with extensions in southwestern United States. There is at least one species in Colombia and Ecuador. The only wood samples available are of *M. biuncifera* (Benth.) Britt. & Rose from Arizona and New Mexico and *M. quitensis* (Benth.) Britt. & Rose from Ecuador. The structure of the former is like that of *Mimosa* (*sens str.*), but that of the other is very different. The specimen (Yale 19490) was collected by Dr. A. Rimbach in the dry highlands of the interior of Ecuador. The tree was only about 10 feet high and three inches in diameter. Wood pale yellow throughout, apparently all sapwood. Fairly lustrous. Odorless and tasteless. Hard, heavy, tough, and strong; rather fine-textured; straight-grained; easy to work; not durable. No special uses known.

Common names: Gatuño, uña de gato (Mex.); guarango (Col.).

Monopteryx, with two species of large unarmed trees, is confined to non-inundated places along streams in the north-central Brazilian Amazon region and the hinterlands of Venezuela. The trees attain a height of 100 feet, with a trunk four feet in diameter above the irregularly buttressed base. (For drawing of the base of *M. angustifolia* Spruce, see Spruce's *Notes of a botanist on the Amazon and Andes*, Vol. I, p. 335.) The leaves are imparipinnate, with 3 to 5 large leathery leaflets; the rose-colored flowers are borne in terminal panicled racemes and are not very conspicuous; the legumes are large, flat, two-valved, elastically dehiscent; the seeds contain a bitter oil, but are edible when roasted. The timber apparently is not utilized.

Heartwood yellowish or orange, with dark red vessel lines; sometimes more or less streaked; rather waxy looking; sharply defined from the yellowish sapwood. Luster high, golden. Without distinctive odor or taste when dry, but said to have a rather agreeable balsamic scent when fresh. Very hard, heavy, tough, and strong; texture coarse; feel harsh; grain irregular; saws rather woolly, but is not difficult to work, finishing very smoothly and taking a high natural polish; probably durable. A good timber for furniture, but probably

Photo by L. S. Hohenkirk.

PLATE XXVI. Strongly buttressed Mora trees (*Mora excelsa*) in British Guiana.

Photo by B. E. Dahlgren.

PLATE XXVII. Amazonian forest at edge of clearing for Ford Rubber Plantation, Rio Tapajoz, Brazil.

of no commercial possibilities because of its inaccessibility.

COMMON NAMES: Uacú, uaucú (Braz.).

Mora, with five or six species of large unarmed trees, is closely related to, and often included with, *Dimorphandra,* from which it is readily distinguished by having simple-pinnate rather than bipinnate leaves and very large and soft instead of small and hard seeds. Two species have been described from Hispaniola, namely, *Mora Ekmanii* (Urb.) Britt. & Rose of Haiti and *M. Abbottii* Rose & Leonard of Dominican Republic. There is little information concerning them and no wood samples are available for study.

Mora megistosperma (Pittier) Britt. & Rose (= *D. oleifera* Triana) is a characteristic tree of the inner tidal belt of the Pacific coast of Panama, where it is known as Alcornoque, and in western Colombia and northwestern Ecuador, where the local name is Nato. It usually grows in nearly pure stands extending for miles along some of the rivers. It attains a height of 150 feet with a straight trunk sometimes over three feet in diameter and free of branches for 35 to 50 feet. The buttresses at times form large wings, generally three in number, extending up the stem for five feet or more; in other instances they may be low and scarcely noticeable. The main roots are flattened laterally and run horizontally and can often be followed for a distance of 50 feet or more from the base of the tree, where there are likely to be clusters of small pneumatophores. The leaves have only two pairs of leaflets, which are large smooth and leathery. The flowers are white and in solitary terminal or axillary spikelets. Of the three ovules, only one reaches maturity, and the enormous pods, when ripe, open with a twist of the valves and remain for a time on the branchlets after liberating the kidney-shaped, lustrous, brownish red seed which is considered the largest of any dicotyledonous plant known. The seeds fall to the muddy ground and germinate immediately. The timber, though readily obtainable in large sizes, has never been extensively used. The reddish brown heartwood is strong, durable, and not difficult to work, and is suitable for various kinds of heavy construction not exposed to marine borers. Crossties of the timber from Colombia are used by the Peruvian government railways and the Lima street car company (see *Tropical Woods* 3: 3).

The best known species is the Mora, *Mora excelsa* Benth., of the Guianas, Trinidad, and the Orinoco delta of Venezuela, where it inhabits swamps and flat lands along streams and is generally considered an indicator of poor soil. Natural regeneration is abundant, trees of all age classes occur together, and the forest floor is covered with a dense mass of seedlings. The tree is tolerant of shade, coppices at an advanced age, and is comparatively free from pests and diseases, though the largest specimens are usually hollow.

Regarding the occurrence and importance of Mora in British Guiana, Miles Haman says (*Tropical Woods* 15: 6): "The Mora-Trysil type of forest occurs on low-lying, moist, but not very swampy sites that are inundated during the wet seasons, but above the level of high tide, and is not limited to any particular parts of the Colony. In one survey in this type near the Demerara River, Mora was found to constitute 70 per cent of the stand with an average of ten sound trees per acre. Trysil [*Pentaclethra filamentosa* Benth.] was next, making up 7 per cent of the stand. The average stand of Mora is from 10,000 to 15,000 board feet per acre. Where the mixture of other species is heavy, Mora will run only about 3000 feet, but where it occurs pure, as it does in limited areas that are low and moist, stands of 25,000 to 30,000 feet are not uncommon. Mora will average between 100 and 150 feet in height, and sometimes will attain 200 feet. It is unquestionably the most distinctive tree of the British Guiana forest. The huge buttresses and flanges at the base spread out, on the larger trees, a distance of 15 feet from all sides of the trunk, and diameters of eight feet, 20 feet above the ground, are not uncommon (Plate XXVI). The wood springs in sawing, does not splinter readily, planes well, and takes nails better than Greenheart. It is very resistant to decay, the heartwood being as durable as any wood in the Colony; it does not, however, resist the teredo and other marine borers. The worst defect is water cracks in the heart. It seasons well when properly stacked. It is excellent for shipbuilding, framing and knees for punts, railway ties, paving blocks, house frames and underpinning, fence posts, and all uses where resistance to decay is required. It is not a beautiful wood and hence is not in demand for furniture, but at present it is the most important wood in the Colony."

A closely related species, *Mora paraensis* Ducke, is a tree about 100 feet tall in the estuary of the Amazon, where it is called Pracuúba. It supplies a good construction timber of commercial importance in Pará.

Morabukea, *Mora Gonggrijpii* (Kleinh.) Sandwith (= *Dimorphandra Gonggrijpii* Sandw.), is another important timber tree in the Guianas. According to Davis and Richards (*Journal of Ecology* 221: 1: 113–116): "Morabukea forest has rather a local distribution in the Colony [British Guiana]. It covers very large areas in parts of the near interior, *e.g.*, the Ebini Hills, and does not reach much nearer the coast than the Moraballi district. In the North West district, Morabukea is rare. It is only known outside the Colony from Surinam, where it is said to be fairly abundant. The Morabukea tree is much smaller on the average than *Mora excelsa,* though occasional specimens are as tall as the largest Moras. It is also a strongly plank-buttressed tree, but the buttresses seldom reach more than two or three meters above the ground. Morabukea is much the darkest type of forest, and the deep gloom is one of its most striking features. Another very conspicuous characteristic is the dense thicket of seedlings of the dominant species, all about 1 to 1.5 m. long, which is usually present. Epiphytes seem to be quite absent, and the herbaceous ground cover, as might be expected, is very sparse. In this respect, as well as in the dense reddish brown carpet of dead leaves, the Morabukea forest is reminiscent of an English Beech wood. It is never found on flood plains, but on the lower slopes of the smaller, flat-topped hills (mostly under 50 m. high) and may extend over their summits if the soil is suitable. It only reaches the creek bank where the ground is high enough to be raised well clear of floods. In general the Morabukea land is damp but well drained. . . . The dominance of Morabukea seems more complete than it really is, owing to the extremely abundant regeneration. On the plot Morabukea formed 60.7 per cent of all trees over 16 in. (41 cm.) and 25.9 per cent of all trees over 4 in. (10 cm.) in diameter." According to the Conservator of Forests, there are large stands of Greenheart and Morabukea, averaging about 7000 board feet per acre, situated away from the rivers but offering no difficulties to railroad logging. He believes that the timber will prove the equal of Mora, if not its superior, and cites an instance where a log remained sound after lying in the forest for 26 years. At present it is practically unknown, except to the aboriginal Indians, who prefer it to Mora, claiming that it splits less and is fully as strong and durable. (See *Tropical Woods* 14: 32.)

The woods of the several species are much alike in general appearance and properties. Heartwood rather dull reddish brown, often with a slight purplish hue; plain or more or less prominently streaked; not especially attractive; very distinct but not sharply demarcated from the light-colored sapwood. Odorless, but with a slightly bitter taste. Hard, heavy, tough, and strong; sp. gr. (air-dry) 0.90 to 1.05; weight 56 to 66 lbs. per cu. ft.; texture rather coarse and harsh; grain variable from straight to very irregular; not difficult to work, though somewhat splintery; takes a smooth finish; requires care in seasoning; is highly resistant to decay.

Common names: Alcornoque (Pan.); nato, n. rojo (Col.); nato (Ec.); mora de Guayana (Venez.); mora, muro (Trin.); mora, morabukea (Br. G.); mora, morabucquia, palaloea, peto prakowa, roode mora (Sur.); pracuúba, p. branca, p. vermelha (Braz.).

Myrmecodendron, to which are now referred about a dozen species formerly of

Acacia, is best represented in Central America and southern Mexico. The leaves are bipinnate, with numerous small leaflets; the yellow or purplish flowers are spicate in some species, capitate in others; the promptly dehiscent legumes are compressed, have thickened valves, and the seeds are surrounded by a fleshy pulp. The stipular spines are often very large, hollow, and inhabited by ants. Standley says (*Flora of Lancetilla Valley,* p. 209): "The Bullhorn Acacias are plentiful in most parts of Central America, and they are among the most characteristic and curious plants of the region. It is difficult to find a plant whose large inflated spines are not inhabited by ants, although once in a great while one will find an isolated individual which has been overlooked by the insects. While the spines are still young and tender they are punctured near the apex by the ants, and occupied, each spine being inhabited by a separate colony. The ants feed, in part at least, on nectar bodies borne upon the young leaves. They protect the plants efficiently against animals which might molest them, and sally forth full of fight when the branches are jarred by any passing body. They bite severely and painfully and are perhaps the most ferocious of all the Central American ants, although there are others which are close competitors."

Heartwood in various shades of brown, often with a purplish or coppery hue, and irregularly streaked or variegated; with rather gradual transition to the whitish sapwood. Luster medium to high. No distinctive odor or taste. Very hard, heavy, tough, and strong; texture medium; grain irregular; not easy to work, but finishing smoothly; probably very durable. Timber not utilized much because of its small size and poor form, but suitable for small objects of turnery and inlay.

Common names: Carretadera, cornezuelo, guisache carteño, jarretadera, mauto blanco (Mex.); bullhorn acacia, cockspur (Br. H.); guascanal (Guat.); carnezuelo, cornezuelo, iscanal (Hond.); cachito, cutupito, guascanal, iscanal, i. negro (Salv.); cornezuelo (C.R.); cachito, cuernito (Pan.); cachito, casa de hormigas, cornizuelo (Col.).

Myrocarpus includes two closely related and doubtfully distinct species, namely, *M. frondosus* Fr. Allem. and *M. fastigiatus* Fr. Allem. They are medium-sized to large unarmed trees attaining a maximum height of 125 feet and a trunk diameter of four feet, though the usual commercial sizes of logs are 18 to 24 inches. They occur scatteringly in the forests of Corrientes and Misiones, Argentina, eastern Paraguay, and throughout southeastern Brazil to Bahia. The leaves are odd-pinnate, with 5 to 9 medium-sized ovate leathery leaflets characterized by pellucid lines and dots; the small white flowers are borne in rather small clustered axillary and terminal spicate racemes; the samara-like fruit has one or two seeds. The bark is rough and deeply fissured and contains a fragrant balsam which is used medicinally and formerly for incense in churches. The timber is highly esteemed locally for heavy and durable construction of all kinds, shipbuilding, carriage work, fine furniture and articles of turnery. Because of its spicy scent it is also used for lining clothes chests and wardrobes.

Heartwood rich brown, often with darker streaks and roe grain; has a rather waxy appearance; sharply demarcated from the whitish or yellowish sapwood. Luster golden in proper light. Taste not distinctive; scent pleasant, "cedary." Very hard, heavy, compact, tough, and strong; sp. gr. (air-dry) 0.90 to 1.10; weight 56 to 69 lbs. per cu. ft.; of uniform, medium-fine texture; feel rather harsh; grain more or less irregular; not easy to work, but finishing very smoothly with a lustrous polish; holds its place well when manufactured; is highly durable. A good timber, but not likely to become important in the export trade.

Common names: Cabreúva, c. amarella, c. do campo, caburé, caburehiba, oleo de macaco, o. pardo (Braz.); cabriuba, cabriuva, incienso (Arg.); incienso (Par.).

Myrospermum *frutescens* Jacq., a variable species but apparently the only valid one of the genus, is an unarmed shrub or a small or rarely medium-sized tree of limited distribution in Venezuela, Trinidad, Colombia, Costa Rica, Nicaragua, and Salvador. The deciduous leaves, which resemble those of *Robinia,* are imparipinnate, with numerous leaflets characterized by translucent lines; the racemose flowers are white tinged with pink, suggesting *Gliricidia;* the fruits are samara-like, usually with a single seed at the end of a broad wing. The wood is sparingly utilized, but is suitable for purposes requiring small sizes of very strong and tough material.

Heartwood brown, with greenish or purplish tinge, more or less striped; rather waxy looking; sharply demarcated from the whitish or yellowish sapwood. Luster medium to high. Without distinctive odor or taste when dry. Very to extremely hard, heavy, compact, tough, and strong; sp. gr. (air-dry) 0.95 to 1.20; weight 56 to 75 lbs. per cu. ft.; texture fine; grain straight to irregular; difficult to cut across, but fairly easy along the grain; fresh sapwood tends to saw woolly; heartwood finishes very smoothly with a lustrous natural polish; appears highly durable.

Common names: Guayacán (Salv.); chiriquirín (Nic.); balsamito, balsamo, ramoncillo (Col.); cereipo, estoraque, guatamare, macagua, macagüito, muji negro, pardillo, palo de olor, pui, roble negro (Venez.).

Myroxylon (or *Toluifera*), with probably only one distinct species, *M. balsamum* (L.) Harms (= *Toluifera balsamum* Cav. = *M. toluiferum* H.B.K.), although about a dozen have been proposed, is a medium-sized to large unarmed tree widely distributed in nearly all of continental tropical America except the Amazon basin. It has been planted for balsam production in West Africa, India, and Ceylon. The leaves are unequally pinnate, with 5 to 11 rather large leaflets bearing translucent oil glands as in *Myrocarpus* and *Myrospermum;* the whitish flowers are borne in simple axillary racemes; the large samara-like fruit is 1-seeded.

The form growing in North America is *Myroxylon balsamum,* var. *Pereirae* (Royle) Harms; its range extends from southern Mexico southward through Central America, mostly on the Pacific side. It is generally 50 to 65, occasionally up to 100, feet tall, with a trunk 18 to 30, rarely 36, inches in diameter. It is often planted, along with *Inga* and *Gliricidia,* to shade coffee, attaining a height of 30 feet in 10 or 12 years, 60 feet in 25 years. Its principal value is in its vanilla-scented resin, known as balsam of Peru, which was introduced into Europe by the Spaniards in the sixteenth century. Its source has always been the Balsam Coast, originally a part of Guatemala, now of Salvador, and the misleading name was derived from the fact that in colonial days the product was frequently sent to Callao, Peru, for transshipment to Spain.

Harvesting the balsam proceeds throughout the year, but chiefly during the dry season, from December to April. The method is crude and consists of removing a piece of bark, about 6 by 10 inches, beginning a foot above ground, and covering the wound with a piece of cotton cloth to absorb the liquid. When the natural flow ceases it is stimulated by scorching. The impregnated rags are boiled in water and pressed; the liquid thus obtained is called "balsamo de trapo," while that from pieces of bark is known as "balsamo de cascara." The best trees yield between four and five pounds of balsam a year. The annual production in Salvador is about 115,000 pounds, of which nearly half is exported to the United States. (See *Tropical Woods* 32: 26.) The balsam, though formerly credited with great medicinal value, is now chiefly used for its perfume in ointments and proprietary preparations. Many trees die as a result of tapping and their timber is used for furniture, interior trim, and railway crossties.

The South American trees also yield oleoresin, sometimes called "tolu balsam" or "balsamum tolutanum," but it is considered inferior to the Salvador product. The timber is of commercial importance in Brazil and Argentina and limited amounts are exported, usually under the name of Oleo Vermelho. Regarding the use of the timber for fine furniture, Karl Schmieg says (*Tropical Woods* 5: 2): "It is in the form of round logs of good size and quite sound, except for a small center defect. The wood is rather hard, somewhat more so than Cuban Mahogany, and is very firm and strong. It is fragrantly scented and its color varies from yellow-orange to purplish rose. It has a rich figure and the texture suggests Padauk, only the pores are smaller. When finished, Oleo Vermelho has a marked resemblance to Cuban Mahogany. The principal hindrance to its use is the fact that it does not respond well to staining and consequently is limited in its application."

Heartwood reddish brown, sometimes with a yellowish hue, when fresh; frequently becoming deep red or somewhat purplish upon exposure; fairly uniform to striped; sharply demarcated from the white sapwood. Luster medium to high and golden. Without distinctive taste, but with pronounced pleasant spicy or "cedary" scent. Very hard, heavy, tough, and strong; sp. gr. (air-dry) 0.90 to 1.10; weight 56 to 68 lbs. per cu. ft.; texture medium; feel rather harsh; grain often roey; moderately difficult to work, but finishing very smoothly with a high natural polish; is noted for its durability.

COMMON NAMES: Bálsamo, palo de bálsamo (Sp. Am., gen.); árbol de bálsamo, bálsamo, b. del Perú, b. de S. Salvador, b. negro, cedro chino, chucte, hoitziloxitl, nabá, nabal, yaga-guienite (Mex.); balsam (Br. H.); bálsamo de Matagalpa (Nic.); chirraca, sándalo (C.R.); bálsamo de tolú, tache, tolú (Col.); estoraque, olor (Venez.); sándalo (Ec.); balsamo de valle, estoraque, quina-quina (Peru); bálsamo, caboré, cabreúva, c. vermelha, oleo vermelho (Braz.); incienso colorado (Par.); incienso, quina, q. morada, quina-quina (Arg.).

Neomimosa, with six or seven species of clambering or erect shrubs and little trees, mostly Mexican, differs from *Mimosa* and *Pteromimosa* in having the valves of the legumes continuous instead of jointed. A few species are unarmed, the others prickly. The only wood specimen available is of *N. platycarpa* (Benth.) Britt. & Rose (= *Mimosa platycarpa* Benth.) which was collected by the senior author in the scrubby forest near Gualán, Guatemala. It was a small tree, with fibrous bark, prickly branches, and sparse, feathery foliage; the thin, flat, dark brown pods are smooth on the sides, but the margins are beset with sharp prickles. Heartwood pale brown, poorly developed; sapwood bright yellow. Hard, heavy, tough, and strong; rather fine-textured; easily worked; presumably not durable. Apparently not utilized.

COMMON NAMES: Rabo de iguana, sierilla, uña de gato (Mex.); zarza (Guat.).

Nicarago *vesicaria* (L.) Britt. & Rose (= *Caesalpinia vesicaria* L. = *C. bijuga* Sw.), the only species, is a much branched shrub or a tree rarely more than 20 feet tall, with or without prickles on the branches, occurring in the West Indies and in Yucatán, Mexico. The leaves are bipinnate, with 2 or 3 pairs of pinnae and 1 to 3 pairs of large and broad leaflets; the numerous yellow flowers are borne in terminal panicles; the legumes are somewhat swollen, succulent, few-seeded, and indehiscent. All parts of the tree have a balsamic scent when bruised.

Heartwood uniform dark chocolate, with a reddish hue; sharply demarcated from the nearly white sapwood. Fairly lustrous. Tasteless, but with a mildly unpleasant scent. Exceedingly hard, heavy, tough, and strong; texture fine and uniform; grain rather irregular; not easy to work, but taking a high polish; appears very resistant to decay. Has no commercial possibilities.

COMMON NAMES: Bastard nicarago, Indian savin tree, jack-fish wood (Jam.); brasil, cuaba rosa, guacamaya de costa, palo brasil (Cuba).

Niopa, with a few species of unarmed tropical American trees, is often considered as a section of the genus *Piptadenia*. The leaves are large, bipinnate, with numerous small and narrow leaflets; the flowers are small, in little heads; the pods are compressed, not transversely septate, readily dehiscent, the valves leathery. The most widely distributed species is *N. peregrina* (L.) Britt. & Rose, a small to medium-sized tree occurring on well-drained sites from central Brazil through the Guianas and Venezuela into the West Indies. The bark contains about 16 per cent of tannin and is employed locally both for making leather and for medicinal purposes; it also exudes a substance similar to gum arabic which is used in the treatment of pulmonary and bronchial diseases. From the seeds the natives prepare a narcotic snuff known in Brazil as "niopó" and in the West Indies as "cojoba." The reddish brown wood is of attractive appearance; hard, heavy, tough and strong; of medium texture; not difficult to work; apparently durable; of no commercial importance.

COMMON NAMES: Bastard tamarind (Jam.); cajoba, cojobana, cojobillo (P.R.); cojoba (Dom. R.); bois gallo, b. l'écorce, oeuf de poule (Haiti); savannah yoke (Trin.); niopó (Col.); cojoba, curuba, niopó, niupó (Venez.); angico niopó, paricá, p. de cortume, p. da terra firme (Braz.).

Notodon, with four species of prickly or unarmed shrubs and little trees, is apparently confined to Cuba. The leaves are pinnate, with 4 to 10 leaflets, the rachis margined to distinctly winged and extended into a fine free tip; the purplish or bluish flowers are borne in small fascicles in the axils of the old leaves; the small pods are pointed at each end. The only authentic specimen available (Yale 21451) is of *N. gracilis* (Gris.) Urb., a shrub or small tree with short recurved prickles on the branches.

Heartwood rich olive-brown, with occasional darker streaks; rather waxy looking; sharply demarcated from the thin yellowish sapwood. Luster fairly high. Odorless and tasteless. Very hard, heavy, and compact; texture fine and uniform; grain straight; not difficult to work, taking a high natural polish; very durable. Suitable for small articles of turnery.

COMMON NAMES: Granadillo, g. de Cuba, g. macho (Cuba).

Olneya *tesota* A. Gray, the only species, is confined to southwestern Arizona, southern California, and northwestern Mexico. It sometimes, particularly in Sonora, attains a height of 30 feet with a short scaly-barked trunk up to 18 inches in diameter. The twigs are armed with stiff sharp spines. The persistent odd-pinnate leaves have 11 to 15 rather small leaflets; the purplish white flowers are borne in short axillary racemes; the somewhat moniliform pods are glandular-hairy and contain one to several seeds. The wood is valued for fuel and is suitable for brush backs and articles of turnery.

Heartwood rich dark brown, somewhat variegated; sharply demarcated from the rather thick, yellowish white sapwood. Fairly lustrous. Without distinctive odor or taste when dry. Very hard, heavy, compact, and strong, but brittle; sp. gr. (air-dry) about 1.15; weight 72 lbs. per cu. ft.; texture medium coarse; grain irregular; not easy to work, but finishing very smoothly with a high natural polish; is very durable.

COMMON NAMES: Desert ironwood, ironwood (U.S.A.); árbol de hierro, palo de hierro, p. fierro, p. tinta, tésota, uña de gato (Mex.).

Ormosia, with more than 30 species of unarmed shrubs and trees, occurs in tropical Asia and America. There are about 20 American species, mostly Amazonian, but with extensions southward to São Paulo and northward to the West Indies, Central America, and southern Mexico.

Heartwood pinkish to reddish, mostly salmon-colored and more or less streaked; surface of unfinished material dull, with mealy appearance; transition to yellowish sapwood gradual. Without distinctive odor or taste. Hard and heavy to moderately so; sp. gr. (air-dry) 0.65 to 0.80; weight 40 to 50 lbs. per cu. ft.; texture very coarse; feel harsh; grain usually irregular; not easy to work, finishing poorly; durability doubtful. A coarse unattractive wood, apparently of no commercial possibilities.

COMMON NAMES: Nickel (Jam.); mato, mosongo, palo de matos, peronía (P.R.); colorín (Mex.); alcornoque, pernillo de monte (Pan.); peonía (Venez.); baracara (Br. G.); agipau, anakoko, awaakoko, barakaro, b. fieroberoe, b. ibikoro, b. korero ibibero iwi, basakanda, itjoe-ranano-anakoko, kokriki (Sur.); jatobáhy do igapó, olho de boi, o. de cabra, tenteiro, tento, t. amarello, t. da terra firme, t. de folhas grandes, t. dos compinas, t. grande da varzea (Braz.); huairúru (Peru).

Ormosiopsis, with two species of unarmed trees sometimes over 100 feet tall, occurs infrequently in the Amazon region, where it is called Tento. The leaves are odd-pinnate, with 3 to 7 rather large leaflets; the pale lilac or yellow flowers are borne in panicles; the legumes are red, leathery, elastically dehiscent, usually with one hard, globose, red or black seed marked with a small white hilum. The timber is of good quality, but is sparingly used because of its scarcity.

Heartwood lustrous brownish yellow, with narrow light-colored vessel lines; becomes russet upon prolonged exposure; transition to whitish sapwood rather gradual. Without distinctive odor or taste. Very hard, heavy, tough, and strong; texture very coarse; feel harsh; grain roey; not easy to work, but finishing smoothly with a golden luster; appears durable. Belongs in the class of Partridgewood (*Andira*).

Palovea (or *Paloue*), with four species of small to large unarmed trees, occurs along streams in the Guianas and the lower Amazon region of Brazil. The leaves are simple, alternate, large, leathery, and entire; the dark red long-stamened flowers are borne in axillary racemes or fascicled at the ends of the branchlets; the fruit is an elongated flat dehiscent pod with hard, compressed, non-arillate seeds. Apparently the trees are poorly known locally, as no common names and no uses are recorded. The following description is based upon a single specimen (Yale 35646; A. C. Smith 2652) of *P. induta* Sandw., a tall tree in dense forest in the basin of the Essequibo River, British Guiana.

Colored heartwood not seen, but according to Ducke (*As Leguminosas da Amazonia brasileira*, p. 63), there is a very small core of dark heartwood in *Palovea brasiliensis* Ducke; sapwood pale brownish, with fine but distinct vessel lines. Luster medium. Odorless and tasteless. Hard, moderately heavy, tough and strong, having about the consistency of Sugar Maple (*Acer saccharum* Marsh.); texture rather fine, uniform; grain straight; readily split, easy to work, finishing smoothly.

Paramachaerium, with two species of small to medium-sized unarmed trees, occurs on periodically inundated lands along streams in the northern Amazon region of Brazil and the hinterlands of the Guianas. The trunks are usually poorly formed and the bark contains a red sap. The odd-pinnate leaves have mostly 5, sometimes 3 or 4, large leaflets; the dark violet flowers are borne in dense spikes or lax racemes; the fruit is a thick, somewhat woody, indehiscent, one-seeded pod with a short to rather long terminal wing. The timber is not utilized. The only wood sample available for study (Yale 27153; Ducke 217) is of *P. Schomburgkii* (Benth.) Ducke (= *Machaerium Schomburgkii* Benth.). Ducke says (*Tropical Woods* 41: 6): "Rob. Schomburgk by mistake attributed to this tree the precious Tigerwood of British Guiana, but really the wood of this tree is valueless." The wood of the other species, *M. ormosioides* Ducke, is said to be white throughout (see *Arch. Jard. Bot. Rio de Janeiro* 3: 155).

Heartwood apparently absent; sapwood buff, with lemon-yellow streaks. Luster medium. Odorless and tasteless. Hard and heavy, fairly tough and strong; sp. gr. (air-dry) 0.86; weight 54 lbs. per cu. ft.; texture fine and uniform; grain irregular; not difficult to cut, finishing smoothly with a high polish; is presumably not resistant to decay. Apparently without commercial possibilities.

Parkia, with more than 35 species of medium-sized to very large unarmed trees, occurs in tropical Asia, Africa, and South America. There are about 20 species in the Amazon region of Brazil, and the range of a few of them extends into the Guianas. They have bipinnate leaves, with many small leaflets; the small flowers are in conspicuous heads; the legumes are leathery or woody, dehiscent or indehiscent, in some species very large, maximum length 26 inches. Many of the trees are beautiful and are planted for ornamental purposes. The bark is tanniniferous and is employed medicinally; in a few species the inner bark is strongly scented.

Heartwood pale brown, sometimes with pink streaks; not clearly differentiated from the thick white or grayish sapwood. Luster medium to silky. Odorless and tasteless. Variable in density from very light and soft to moderately heavy and hard; sp. gr. (air-dry) 0.25 to 0.55; weight 16 to 34 lbs. per cu. ft.; texture coarse to very coarse; grain straight to roey; easy to work; saws woolly, but finishes smoothly; is tough and strong for its weight; is perishable in contact with the ground. The timber is little used, but is suitable for boxboards and interior construction.

Common names: Uya (Br. G.); ajoewa, apa akaniran, a. kanilan, basra ajoewa, b. tontoawha, ipana, koejari itabatje, kwatta kamo, lialiadan tataroe, oeloeloe-oe, oeroeroe-oe, oja, saandoe (Sur.); acacia mâle, grignon fou (Fr. G.); arapary branco, arara petiú, a. tucupy, boloteiro, cipóuba, esponjeira, fava de belotas, faveira, f. branca, gipóuba, japacanim, jupuúba, manopé, m. da praia, paricá, p. branca, p. de esponjas, p. grande da terra firme, pau de arara, rabo de arara, sipóuba, tambury, visgueiro (Braz.).

Parkinsonia, with one or two species of thorny shrubs or small trees, is widely distributed naturally and through cultivation for ornament in the warm and tropical regions of the world. The American species, *P. aculeata* L., is found from Texas to Argentina; it occurs on low moist soil and will also thrive in regions of little rainfall. It is rarely more than 25 or 30 feet high, with a short smooth-barked trunk sometimes 15 inches in diameter, and a round-topped crown of slender, somewhat pendulous branches. The leaves are bipinnate, with one or two pairs of pinnae appearing like sessile pinnate leaves; the flat, persistent rachis is 8 to 16 inches long, with many rather widely spaced pairs of deciduous little leaflets; the showy yellow flowers are borne in axillary racemes; the few-seeded legumes are moniliform. About the only use for the wood is for fuel.

Heartwood reddish brown; slow in forming; sapwood yellowish white. Luster medium. Odorless and tasteless when dry. Moderately hard and heavy, but brittle; texture rather fine; grain irregular; easy to work, finishing smoothly; durability doubtful.

Common names: Horse bean (U.S.A., P.R.); Jerusalem thorn (Jam., P.R.); espinillo, junco marino, palo de rayo (Cuba); flor de rayo (P.R.); cambrón (Dom. R.); madan naiz (Haiti); bagoto, cacaporo, guichi-belle, guachóporo, mezquite extranjero, m. verde, palo verde, retama, r. de cerda (Mex.); sulfato (Salv.); acacia de agüijote, espino real de España (Nic.); goajiro, retama, sauce, yabo (Col.); cují extranjero, espinillo, e. de España, espinito (Venez.); cina-cina (Arg.).

Peiranisia, a large genus of unarmed shrubs and small trees, is widely distributed in tropical America. The leaves are pinnate, with few to many pairs of leaflets; the yellow flowers are large, mostly in few-flowered axillary peduncles; the legumes are elongated, thin, with thickened margins, and dehiscent along both sutures. Apparently there are no special uses for the plants. The two wood specimens available are of *Peiranisia biflora* (L.) Pittier (= *Cassia biflora* L.), one (Yale 9982) collected by the senior author in Honduras, the other (Yale 20541) by Ramón Espina near Santa Marta, Colombia. The little trees or shrubs frequently form thickets, which are very attractive when covered with golden yellow flowers.

Heartwood olive-brown, with a waxy or oily appearance; sharply demarcated from the yellowish white sapwood. Odorless and tasteless. Hard, heavy, tough, and strong; texture rather fine; grain straight to irregular; not difficult to work, finishing very smoothly; is probably durable.

Common names: Bushy senna, mosquito bush (Bah.); carbonero (Cuba); hediondilla, retama, r. prieta (P.R.); carga agua (Dom. R.); séné, briser menage (Haiti); abejón, biche silvertre, flor de San José (Mex.); comayagua (Hond.); ahumada, montenegrito (Nic.); bicho, bombito, escovito (Col.); brucha, flor amarilla (Venez.).

Peltogyne, with about 20 species of medium-sized to large unarmed trees, is represented from São Paulo, Brazil, to Trinidad and Panama, and possibly to Acapulco, Mexico, with the center of distribution in the north-middle part of the Brazilian Amazon region. The leaves are pinnate, with rather small to large, sometimes waxy, leaflets in single pairs as in *Hymenaea;* the small to medium-sized white or roseate flowers are borne in terminal panicled racemes; the legumes are short, flat, usually 1-seeded, dehiscent or indehiscent. One of the most distinctive characteristics of the genus is the rich purple color of the heartwood, which is known to the trade as Amaranth, Purpleheart, or Violet Wood.

The trees and timber are generally called Guarabú or Pau Roxo in Brazil. H. M. Curran says: "Pau Roxo is common in the Bahia coast forest and sometimes will yield 2000 to 3000 board feet per acre. At its best it attains a height of 125 feet and a diameter up to four feet, though the usual commercial diameters are between two and three feet. The trunk is well formed, no buttresses, is frequently clear of branches for 50 feet, and is covered with a smooth grayish black bark. The sapwood of mature trees is from two to four inches thick. The heartwood is of a dull horn color when freshly cut and turns deep purple on exposure to the air. Logs and planks exposed to sun and rain soon lose their purple color and become black. The wood is as well known locally as the Sucupira (*Bowdichia*) and is considered the best material available for making spokes for cart wheels; in fact one rarely sees spokes made of any other wood."

According to Ducke (*Tropical Woods* 54: 1), there are ten species in Amazonas and five in Pará. One of the most widely distributed is *Peltogyne paniculata* Benth., known locally as Pau Ferro and Coracy. Along the Rio Negro the trees are only moderately tall, whereas the form in the upland rain forest of the lower and middle Amazon attains heights of 100 to 120 feet. The heartwood is brown at first, changing to red-brown, eventually becoming purplish. Some specimens of *P. micrantha* Ducke, a small to medium-sized tree of frequent occurrence along the periodically inundated banks of black-water streams in the upper Rio Negro region, conserve their brown color, whereas others cut in the same locality quickly change to violet. *P. paradoxa* Ducke is a rather tall slender tree in the Campos regions from Monte Negro to Macapá in northeastern parts of the lower Amazon. The fertile branches are leafless and very flexible, like long whips, which accounts for the vernacular name Coatá-quiçáua (monkey hammock). The heartwood is dark grayish violet. *P. densiflora* Spruce is the commonest species in the eastern half of Amazonia and furnishes most of the Pau Roxo of the timber trade of Pará. It is typical of the "igapó" and flooded banks of fairly clear lakes and rivers and never occurs in true upland forest. The trees are generally short and crooked, but sometimes have rather long well-formed boles. The heartwood assumes a deep purple-violet on exposure. The most beautiful and richly colored woods are *P. catingae* Ducke, *P. excelsa* Ducke, *P. gracilipes* Ducke, *P. Lecointei* Ducke, and *P. maranhensis* Huber. The highly esteemed Pau Roxo da Terra Firme of the timber merchants of Pará is supplied by *P. Lecointei.*

There are a few, perhaps several, species of *Peltogyne* in the Guianas, but information concerning them is not exact. The two generally listed are *P. paniculata* and *P. pubescens,* but they may be merely two forms of the same species, as Ducke says (*loc. cit.*) that he cannot find any constant morphological characters to separate them. The common names are Purpleheart in British Guiana, Purperhart in Surinam, and Amarante or Violet in French Guiana. Pittier lists three species for Venezuela in his *Arboles y arbustos del orden de las Leguminosas* (pp. 138–140). They are *Peltogyne paniculata, P. parvifolia,* and *P. floribunda* (H.B.K.) Benth. The Trinidad species, *P. porphyrocardia* Gris., which probably is the same as *P. floribunda,* is a large evergreen tree with a trunk sometimes more than three feet in diameter above the low thin buttresses.

The only known Colombian species is *Peltogyne pubescens* Benth. which attains a height of from 75 to 100 feet and a trunk diameter of from 20 to 30 inches in the Magdalena valley, where it is known as Tananeo. The purple wood is noted for its durability in contact with the ground. *P. purpurea* Pittier is a large deciduous tree growing in the Province of Darién and probably elsewhere in Panama. According to Standley (*Flora of the Panama Canal Zone*, p. 201) it reaches a maximum height of 195 feet. The local names, Nazareno and Morado, refer to the heartwood, which becomes intense purple.

The only basis for stating that *Peltogyne* occurs in Mexico is a wood specimen (Yale 4401) collected by former American Consul John A. Gamon near Acapulco under the name of Palo Morado. This material is the densest of any species tested, although a sample (Yale 20692; Ducke 11) of *P. paniculata* is in the same class; sp. gr. (10 per cent moisture) 1.21; weight 70 lbs. per cu. ft.; sp. gr. (oven-dry wt. and vol.) 1.19. After soaking the oven-dry specimen for 15 days the sp. gr. (23 per cent moisture) was 1.26. The submergence in water changed the rich purple to almost jet black on the surface. The color and structure of the wood leave no room for doubt as the correctness of the identification with *Peltogyne.*

Purpleheart is well known to American timber dealers, but there has never been much demand for it. Its principal uses are for inlay and cabinet work and butts of billiard cues. Regarding a trial for interior trim and flooring, H. S. Clark, of the Caspar Ranger Lumber Company, Holyoke, Mass., writes as follows (*Tropical Woods* 25: 1–3): "Amaranth or Purpleheart from British Guiana was recently employed by our firm with good effect for all of the exposed woodwork of the library in the home of Mr. H. Halstead Lindsley in Lenox, Massachusetts. So far as we know it is the first time that this timber has been used for such purposes in this country. . . . We obtained our lumber from a Boston firm that imported the logs and sawed them to our order. The logs were about two years old when received and the fresh lumber weighed about eight pounds per board foot. We bought the lumber, log run, and kiln-dried it carefully for four weeks, thereby reducing the weight to about 4½ pounds per board foot. The four walls of the room were panelled with Amaranth, backed with Whitewood (*Liriodendron*). The floor boards are 3-ply, ¾ in. thick, and with a 10-in. finish face. The surface layer is Amaranth, the others Chestnut, with the grain of the middle ply at right angles to the other two. A strong *hot* glue was used, as it was found that it held the Amaranth better than a waterproof glue. Machining and hand work removes the dark purple color from the surface and exposes the much lighter wood beneath. Within 24 hours, however, the rich color commences to return and continues to deepen. On this account it was considered advantageous to allow the room to stand for three or four weeks before the application of any painter's finish. The finishing in this instance was simply the use of wax, as nothing further was needed to bring out the rich natural color. Amaranth is a hard, heavy, fine-textured wood requiring considerable labor and skill for producing fine cabinet work. It has to be run slowly through the machines and all cutter tools must be of high-speed steel. It is about as expensive a wood as any with which we have had experience. . . . The wood stays in place remarkably well when properly treated. There is enough feather-striping of the grain and sufficient variation in shade and luster to prevent monotony of color, and the general effect is as pleasant and attractive as it is unusual."

The following information is taken, with permission, from a mimeographed report by the Forest Products Research Laboratory at Princes Risborough, England, on tests of three logs of Purpleheart from British Guiana. Weight, at 10 per cent moisture, about 53 lbs., and at 50 per cent moisture, about 70 lbs., per cu. ft. Freshly sawed wood has a characteristic, somewhat unpleasant odor, which practically disappears upon drying. Thin lumber kiln-dries well and fairly readily; thicker stock requires special handling. Moderately hard to work with hand and machine tools; gum adhering to the back bevels of teeth and cutters tends to cause overheating of the cutting edges. Lumber with irregular grain requires care in handling to prevent "picking up," especially on radial surfaces. Can be turned readily with a smooth surface and clean edges requiring very little sanding to provide an excellent finish. Takes stains well, but natural color with oil or wax finish gives the best effect. Results of mechanical tests on small, clear specimens at 50 per cent moisture content were as follows (in lbs. per sq. in., except as otherwise stated): *Static bending:* Fiber stress at elastic limit, 9600; same at max. load, 14,980; modulus of elasticity, 2,237,000. *Impact bending* (50-lb. hammer): Fiber stress at E.L., 20,230; M. of E., 2,992,000. *Endwise compression:* Fiber stress at E.L., 5840; same at max. load, 7890; M. of E., 2,369,000. *Hardness* (in lbs.): Radial surface, 2184; tangential, 2022; end, 2080. *Shear:* Radial plane, 1764; tang., 1966. *Cleavage* (in lbs. per in. of width): Radial plane, 487; tangential, 563.

Heartwood dull brown or grayish brown when fresh, usually becoming uniformly purplish brown to intense purple or violet upon exposure to light, and black when soaked in water containing iron; sharply demarcated from the white or gray, eventually purple-streaked, sapwood. Odor and taste not distinctive in dry wood. Very hard, heavy, tough, and strong; sp. gr. (air-dry) mostly between 0.80 and 1.00; weight 50 to 63 lbs. per cu. ft., occasionally considerably more; texture medium to fine; grain straight to irregular or roey; feel of unplaned surface harsh and finely splintery; not especially difficult to work, finishing very smoothly; is highly resistant to decay.

COMMON NAMES: Amaranth, purpleheart, violet wood (Eng., trade); palo morado (Mex.); morado, nazareno (Pan.); tananeo (Col.); algarrobito, morado, nazareno, zapatero (Venez.); purpleheart, sapater, zapatero (Trin.); kooroobooelli, kooroobovelli, koroborelli, kouburelli, kuraburelli, purpleheart, sacka, saka, sakavalli (Br. G.); danstan, dastan, hoepelhout, kocolorelli, koeloeboeralli, kuruburelli, malako, mala-oko, marako, marawinazoo, poeprehati, porpratti, purperhart (Sur.); amarante, bois pourpre, b. violet, violet (Fr. G.); barabú, coatá-quiçáua, coracy, guarabú, g. branco, g. preto, g. rajado, g. roxinho, g. roxo, g. vermelho, guarabussú, ipê roxo, pau ferro, p. mulato da terra firme, p. roxo da catinga, p. r. da terra firme, p. r. da varzea, p. r. de campo, rosinha, roxinho, violeta (Braz.); morado (Boliv.).

Peltophorum. There are several species in tropical Asia and Africa, one or two in the West Indies, and one in southern South America. The American species are large unarmed trees and supply good timber for local uses. The leaves are large and doubly pinnate, with very numerous small leaflets; the bright yellow flowers are borne in showy terminal panicled racemes; the legumes are flat, narrowed at the ends, winged on the margins, few-seeded, and indehiscent. The trees are frequently grown for ornamental purposes.

Peltophorum adnatum Gris. is common on sandy soil in Pinar del Río, Cuba, where it is known as Moruro Abey and sometimes as Sabicú Colorado. Mature trees are frequently 100 feet high, sometimes considerably taller. The fern-like leaves are glossy above and rusty below, and the bark is smoothish and finely granular. The red hard durable wood is used for fence posts, railway crossties, and heavy construction. The timber is said to have been exported to the United States and England under the name of Sabicú, but the true Sabicú is *Lysiloma latisiliqua. P. Berteroanum* Urban, perhaps only a form of *P. adnatum,* occurs in Dominican Republic and Haiti.

Peltophorum Vogelianum Walp. is fairly plentiful in the forests of northeastern Argentina and of Paraguay, where it is generally called Cañafístula and Ibirá-puitá. It is at its best on the deep red clay loam on the terraces of the Paraná River and its tributaries where it attains a height of 125 feet, with a rather short irregular bole six feet through above the swollen or buttressed base. Under poorer conditions it is usually less than 75 feet tall and three feet in diameter. The bark is dark brown and rough. The timber is common on the Argentine markets and is employed in general construction, carpentry, and turnery, and for making furniture and vehicles, though it requires considerable care in seasoning to prevent warping and checking.

The woods are much alike in structure, but there is some difference in appearance. The heartwood of *Peltophorum adnatum* is rather dull reddish or purplish reddish brown, slightly streaked, but with little or no figure; that of *P. Vogelianum* is fairly lustrous, pinkish or reddish brown, often with dark purplish streaks and roe grain. Both marked with fine but distinct vessel lines. Transition to whitish sapwood gradual to abrupt. Odorless and tasteless. Hard, heavy, tough, and strong; sp. gr. (air-dry) 0.75 to 1.00; weight 47 to 63 lbs. per cu. ft.; texture medium to rather coarse; grain variable; easy to work, though roe-grained material is likely to "pick up" in planing; can be finished very smoothly and with a high polish; is strongly resistant to decay. Probably of no commercial possibilities except locally.

COMMON NAMES: Horse bush (Bah.); moruro abey, sabicú colorado, s. moruro, zapatero (Cuba); abey (Dom. R.); abé rouge (Haiti); cañafístula, ibirá-pitá, i.-puitá, i.-p. guazú, palo colorado, virapitá (Arg.).

Pentaclethra. There are two species of trees, namely, *P. macrophylla* Benth. in west tropical Africa and *P. macroloba* (Willd.) Kuntze (= *P. filamentosa* Benth.) in tropical America. The range of the latter extends from northwestern Brazil through the Guianas and Venezuela to Trinidad and St. Vincent in the West Indies and into Central America at least as far north as Bluefields, Nicaragua. It is a medium-sized to large tree, sometimes 125 feet high, with smooth thin bark and fine feathery foliage; the flowers are in conspicuous racemose spikes; the fruit is a flattish, brown, woody pod up to a foot long and two inches wide at one end, the valves at maturity springing open and curling back, liberating the few large rhomboid seeds.

It is common on poor soil along the east coast of Central America, but the timber is not well known, although used locally to some extent for heavy durable construction and for fuel. Regarding its occurrence in Trinidad, where it is called Bois Mulatre, Marshall says (*Trees of Trinidad and Tobago*, p. 46): "Bois Mulatre is a medium-sized evergreen tree with a maximum girth of eight to nine feet, but seldom exceeding four to five feet. . . . Large trees are commonly hollow. . . . It is one of the most abundant subdominant species in many rain-forest areas. The wood is fairly hard, brownish, and the heartwood of mature trees is reported to be durable." It is plentiful in British Guiana, with 4 to 16 trees of commercial size per acre, being more numerous but of smaller size on swampy coast lands where great quantities are cut for fuel. Larger trees supply timber for furniture, house frames, floor beams, and slender stems make good spars. The trees are not very common in Venezuela, but such timber as is available is considered of excellent quality for heavy construction; it is easy to saw but difficult to plane, though capable of receiving a high polish. In Brazil the wood is used for fuel on river boats, the bark is a source of tannin and a remedy for snake bite and dysentery, and the seeds supply an industrial oil for cooking, lubrication, and soap-making. (See *Annales du Musée Colonial de Marseille*, 1925, p. 23.)

Heartwood light to rather dark brown, uniform or somewhat striped; distinct but not sharply demarcated from the pinkish white sapwood. Luster medium to silky. Odorless; taste slightly astringent. Hard, heavy, tough, and strong; sp. gr. (air-dry) 0.75–0.85; weight 47 to 53 lbs. per cu. ft.; texture medium; grain straight to roey; easy to cut, inclined to chip in planing, but can be finished with a natural gloss. A clean-looking and fairly attractive wood, but usually without distinctive figure or grain.

COMMON NAMES: Bois mulatre, fine-leaf (Trin.); gabilán, palo de aceite, sangredo falso, wild tamarind (Nic.); gavilán (C.R., Pan.); carbonero, mulato, palo mulato (Venez.); kooroobaili, koroballi, trysil (Br. G.); koeroebahara, koeroeballi, koroballi, kroebara, trysil (Sur.); bois mulatre (Fr. G.); paraná-cachy, paracachy, paroa-caxy, pracachy, pracaxy (Braz.).

Phyllocarpus. There are two species of medium-sized to large unarmed apparently rare trees. *P. pterocarpus* Ried. (= ***P. Rie-***

delii Tul.) occurs in southeastern Brazil and Acre Territory. *P. septentrionalis* Donn. Sm. is apparently endemic to east-central Guatemala, where it is known as Flor de Mico. The leaves are simply pinnate, with several pairs of large leaflets; the flowers are in short racemes, often forming large panicles, unfolding before the leaves; the legumes are thin, few-seeded, winged on one suture, indehiscent. The only wood sample in the Yale collection is of the second species.

Specimen (apparently from a rather large tree) nearly white throughout except for brown vessel lines. Luster medium. Odorless and tasteless. Moderately hard and heavy; rather coarse-textured; straight-grained; easy to work, finishing very smoothly; is probably perishable in contact with the ground. There are no known uses, but it is suitable for general carpentry and interior construction.

Pictetia, with a few species of shrubs and little trees, is apparently limited in distribution to the West Indies and a part of Mexico. The leaflets are odd-pinnate, with 9 to 25 small, rounded to elongated, bristle-tipped leaflets and spiny or rigid stipules; the yellow flowers are borne in axillary racemes; the linear or oblong pods are few-jointed and indehiscent. *P. aculeata* (Vahl) Urb. grows in thickets in Puerto Rico, Haiti, and the Virgin Islands. It reaches a height of 20 to 30 feet, and the clustered stems have upright flexible branches. The wood is noted for its strength and durability and is used locally for stakes and posts. There are two other species and some varieties in Cuba and Haiti, but they are shrubs of the savannas.

Heartwood chocolate-brown, often with an olive hue; appears waxy; sharply demarcated from the thin yellowish sapwood. Luster medium. Without distinctive odor or taste. Very hard, heavy, compact, and strong; texture fine and uniform; grain roey; not difficult to work, finishing very smoothly and taking a glossy polish. It is suitable for small objects of turnery but has no commercial future.

Common names: Carrasquillo, yamaquey, zarcilla (Cuba); carrasbesu, tachuelo (P.R.); fustic (Virg. Is.); palo tobaco (Dom. R.); gratte galle (Haiti).

Piptadenia, with about 80 species of prickly or unarmed trees and erect or scandent shrubs, is abundantly represented in tropical South America, sparingly in tropical Africa and Asia and in New Guinea. The leaves are bipinnate; the small white or greenish flowers are spicate; the legumes are thin and flat, not transversely septate, and readily dehiscent. The larger trees supply good timber of local utility, but their principal value is in the bark which is an important source of tanning material. Some species are planted for shade and decorative purposes in parks and along highways.

The trees are at their best in northern Argentina and in Paraguay and Brazil, and several species attain heights of from 80 to 100 feet and diameters up to three feet or more. One of the best known species is *Piptadenia rigida* Benth., generally considered the true Angico of southern Brazil, although that name is rather indiscriminately applied to various members of the genus. The reddish brown wood bears a superficial resemblance to old Mahogany (*Swietenia*). It is very hard and heavy; sp. gr. (air-dry) 0.95; weight about 59 lbs. per cu. ft. It is used for heavy durable construction and for fuel. *P. macrocarpa* Benth., commonly known as Cebil Colorado and Curupay in Argentina, has a wider distribution in that country, but is generally smaller, rarely exceeding 65 feet in height and 24 inches in diameter. Timber from the Chaco region, presumably of this species, is considerably denser than that from the Province of Salta, and resembles Urunday (*Astronium*) in its red color with blackish stripes and somewhat oily feel. It enters the local markets and is preferred to Quebracho (*Schinopsis*) in places where the wood must be worked with tools. Although of attractive appearance it is too hard for furniture. There are several species in the Amazon region and three or four in Colombia, but apparently their timber is not utilized. One species, *P. flava* (Spreng.) Benth., a small prickly tree, occurs in Panama, Colombia, Ecuador, Venezuela, and Trinidad.

The tropical American woods of *Piptadenia* exhibit much variation in appearance, density, and certain anatomical features, but are of two general types. In one group they are very hard and heavy, of a definitely reddish color, and the pores are not distinct without lens. Included here are *P. cebil* Gris., *P. macrocarpa* Benth., and *P. speciosa* Britt. & Killip. In the other group the woods are considerably lighter in weight and in color and are of coarser texture. The heartwood is mostly light brown with a pinkish or greenish yellow hue, sometimes with a silky luster. Some of the species are *P. colubrina* Benth., *P. communis* Benth., *P. flava* (Spreng.) Benth., *P. Pittieri* Harms, and *P. tocantina* Ducke.

COMMON NAMES: Bocachico, chocolatillo, iguanero, rabo de iguana, zarza colorada (Col.); carbonero (Venez.); ckara-huilca, huilca, h. romana, tarahuilca (Peru); angico, a. amarello, a. branco, a. coco, a. escuro, a. preto, a. p. rajado, a. rosa, a. roxo, a. vermelho, arapiraca, cobí, curupay, cutiuba, espinho roxo, jacaré, paricá, p. branco, p. da terra firme, p. de cortume, p. grande da terra firme, p. rana, paricachí, pau secante, surucucú, timbaúba, timbó da matta, t. rana (Braz.); anchico, angico, a. blanco, a. colorado, cebil, c. blanco, c. colorado, c. moro, curupay, c. jatá, c.-ná, c.-rá, horco-cebil, ibirá-ré, sacha-cebil (Arg.); curupay, c.-atá, c.-ná (Par.).

Piscidia (or *Ichthyomethia*), with a few species of unarmed shrubs and small to medium-sized trees, occurs in southern Florida, the West Indies, Mexico, Central America, and northern South America. The leaves are mostly large, with 5 to 11 leaflets; the pink or white-and-red flowers are borne in panicles from the axils of the fallen leaves of the previous year; the indehiscent pods are slightly constricted between the numerous black seeds and have four thin longitudinal wings. The best known and most widely distributed species is *P. piscipula* (L.) Sarg. (=*Ichthyomethia piscipula* Hitch. = *I. communis* Blake), a tree sometimes 50 feet or more tall, with a stout trunk frequently 24 inches in diameter. The bark is used medicinally and for stupefying fish. The timber is employed locally for fence posts (if planted green they will take root and grow), railway crossties, heavy construction, vehicle frames, spokes of logging carts, and for fuel and charcoal. It apparently has no commercial possibilities.

Heartwood yellowish brown, with light-colored parenchyma layers; becomes dark brown upon exposure; sharply demarcated from the whitish sapwood. Luster medium to fairly high. Without distinctive odor or taste when dry. Hard, heavy, tough, and strong; sp. gr. (air-dry) 0.80 to 0.90; weight 50 to 56 lbs. per cu. ft.; texture coarse; feel harsh; grain often roey; not easy to work, but finishing smoothly; durability high.

COMMON NAMES: Jamaica dogwood (Fla.); dogwood, black dogwood (Jam.); guamá candelón, g. de costa, g. hediondo (Cuba); dogwood, fish poison, ventiera (P.R.); bois à énivrer (Guad.); alejo, anipak, chijol, cocuile, cocuite, colorín de peces, flor de papagallo, haabí, haabín, habí, habím, habín, jabín, javín, palo blanco, p. b. duro, talzungo, tuncuy, zatzumbo (Mex.); dogwood, may bush (Br. H.); palo de zope, tiaxab (Guat.); palo de zope, zopilocuavo (Salv.); arepo, baurá (Col.); barbasco, b. amarillo, jebe (Venez.).

Pithecolobium (or *Pithecellobium*) is in its broadest sense a large genus of about 200 species of shrubs and trees widely distributed throughout the tropical and subtropical regions of the world. Several groups are recognized and some botanists give them distinct generic rank. In the present instance the generic name is used in a restricted sense and is limited to what was designated in *Timbers of Tropical America* as the "Cat's-claw group." The plants are spiny and have bipinnate leaves with only 1 to 3 pairs of leaflets. They are not commercially valuable as a source of timber.

One of the best known species is *Pithecolobium dulce* (Roxb.) Benth., a small to medium-sized tree often confused with *P. unguis-cati* (L.) Mart. It grows naturally in Mexico, Central America, and northern South America, and is much planted for ornament, shade, and hedges in other tropical and subtropical regions and has become locally naturalized. Standley says (*Trees and shrubs of Mexico*, pp. 393–394): "The tree is very resistant to drought. It is nearly evergreen, but loses its old leaves as the new ones appear. The wood is widely employed for general construction purposes, for fence posts, and for fuel. The bark yields a yellow dye, and is much used for tanning skins; it is used in domestic medicine, also, because of its astringent properties. The gum exuding from the trunk is transparent and deep reddish brown; dissolved in water it makes good mucilage. The flowers are much frequented by bees and yield a good quality of honey. The fruit is highly esteemed in Mexico and is a common article in the markets. The acidulous aril surrounding the seeds is eaten and is used in the preparation of a beverage similar to lemonade. Stock of all kinds are fond of the pods, and in India monkeys are said to eat them greedily. The Nahuatl name for the tree, Coaca Machalli (and variants), signifies snake jaws, and refers to the fancied resemblance of the paired leaflets to the jaws of a snake. The leaves are often used medicinally. The tree is useful for hedges and will withstand clipping and chopping as well as the browsing of animals."

Heartwood reddish brown. Sapwood yellowish to nearly white, not always sharply demarcated. Luster medium. Odorless and tasteless. Hard, heavy, and strong but brittle; texture rather fine; grain straight to irregular; rather difficult to cut, but finishing smoothly; apparently durable.

COMMON NAMES: Bread-and-cheeses, blackbead, cat's claw, ramshorn (U.S.A.); manca mantero, uña de gato (Cuba); rolón, uña de gato (P.R.); sinaso (Dom. R.); campêche marron (Haiti); dabarouida (D.W.I.); auamúchil, caumóchil, caumúchil, conchi, guamáchi, guamúche, guamúchi, guamúchil, guamúchitl, guaymachile, huamuche, huamúchil, h. costeño, humo, h. cimarrón, muchite, pichejumo, pinzan, timúche, tsitsilché, tsiuché, tsuiché, tucuy, tzim-ché, tzin-ché, yaga-biguichi, yagubixihui (Mex.); red fowl (Br. H.); jaguay, j. cimarrón, j. de llano (Guat.); abracade, espino, guachimol, guayacán blanco, mangollano, mongollono (Salv.); espino de playa, huamúchil, uña de gato (Nic.); mochigüiste (C.R.); azabuche, buche, b. blanco, b. colorado, chancán, chiminango, dinde, espino, e. negro, gallinero, ojito de nena, payandé, payondé, pata de vaca, peronio, tiraco, uña de tigre (Col.); ahogagato, arranca-pellejo, dabarouida, espinuelo, espinito morado, guamo cabello de ángel, güichere, ñaure, orore, paují yacure, pinopinito, piquirgua, taguapire, vainita de iguana, yacure (Venez.).

Pityrocarpa, with a few unimportant species of tropical American trees, commonly treated as a section of *Piptadenia*, occurs in Mexico, northern Central America, the Guianas, and Brazil. The bipinnate leaves are large, with broad leaflets; the flowers are in elongated axillary spikes; the seed pods are flattened and moniliform. The only wood sample available (Yale 1192) is of *Pityrocarpa constricta* (Micheli & Rose) Britt. & Rose (= *Goldmania constricta* Mich. & Rose = *Piptadenia constricta* Macbride) and was collected near Mazatlán by Jesus Gonzáles Ortega. The tree is small, occasionally up to 40 feet tall, and unarmed or with a few short spines.

Specimen pale brownish (possibly all sapwood). Luster medium. Odor and taste not distinctive. Very hard, heavy, tough, and strong; rather fine-textured; of irregular grain; not easy to work, though finishing smoothly; apparently not durable. No special uses known.

COMMON NAMES: Iguano blanco (Mex.); lengua de vaca, quebracho (Salv.).

Plathymenia *reticulata* Benth., the only species, is widely distributed in eastern Brazil from the lower Amazon to São Paulo. It is an unarmed tree, of small to medium size at the northern limits of its range, but of large dimensions farther south, where it sometimes attains a height of 125 feet, with a cylindrical trunk three feet in diameter and free of branches for 60 to 70 feet and covered with a grayish or brownish bark which tends to become shaggy. The leaves are bipinnate, with numerous small leaflets; the flowers are small and spicate, the terminal inflorescence almost racemose; the legumes are rather large, thin, and flat, with several seeds having a rectangular membranous wing.

The trees are nowhere abundant, but are scattered through the forests, rarely as many as three of commercial size per acre, and more common on the upper slopes than along the rivers. The more accessible timber has been cut, along with Rosewood and Brazilwood, but there is still a considerable supply in the interior. The finest specimens are sought by the natives for making dug-out canoes. The lumber is highly esteemed in Brazil and is so much in demand that there is little export trade. It is common in the markets along the coast from Pernambuco to Rio de Janeiro. It is usually called Vinhatico, but this name is also applied to other yellowish woods of similar appearance and properties, especially Vinhatico Espinho, *Chloroleucon vinhatico* Record. The local uses of Vinhatico vary according to the density of the wood and include cabinet work and all grades of furniture, interior trim, doors, plain and parquet flooring, shipbuilding, and general construction and carpentry.

Heartwood lustrous yellow or pale orange, soon darkening superficially to a rich brown; often striped with lighter and darker shades; distinct but not very sharply demarcated from the yellowish white sapwood. Odorless and tasteless. Rather light to moderately heavy; sp. gr. (air-dry) 0.56 to 0.65; weight 35 to 40 lbs. per cu. ft.; texture medium; grain straight to decidedly roey; very easy to work, finishing very smoothly with a golden luster; seasons readily and holds its place well when manufactured; is fairly resistant to decay.

COMMON NAMES: Amarello, candeia, oiteira, paricá-zinho, pau amarello, p. de candeia, vinhatico, v. amarello, v. cabelleira, v. castanho, v. de campo, v. pé de boi, v. rôxo. (Braz.)

Platycyamus *Regnellii* Benth., the only well-defined species, is a medium-sized to large unarmed tree growing in the interior plateau region of Brazil from São Paulo to Minas Geraes and Espirito Santo. The leaves are odd-pinnate, with three leaflets which are frequently a foot long; the white or purplish flowers are borne in densely pubescent panicles; the pods are large and 2-valved, with prominent sutures and a narrow wing, and usually contain 2 or 3 kidney-shaped seeds. The bark and leaves are said to be poisonous to animals and are employed medicinally in treating fevers. The timber is used locally for tight cooperage, general construction, fence posts, and railway crossties. It is not considered suitable for fine furniture and cabinet work because of its coarse texture, lack of attractive figure, and the tendency of the color to fade.

Heartwood rose-red, more or less variegated, and with a yellowish hue; becoming yellowish brown upon exposure; rather sharply demarcated from the yellowish sapwood. Luster low to medium. Without distinctive odor or taste. Hard, heavy, tough, and strong; sp. gr. (air-dry) 0.75 to 0.85; weight 47 to 53 lbs. per cu. ft.; texture coarse; feel harsh; grain straight to irregular; rather difficult to work, but finishing smoothly; is highly durable. Probably of no commercial possibilities except locally.

COMMON NAMES: Angelim rosa, camará de bilro, cambará de bilro, cataguá, folha de bolo, f. larga, mangaló, m. rosa, pau pente, p. pereira, p. quente, pereira, p. amarella, p. vermelha (Braz.).

Platymiscium, with about 30 species of closely related, small to large, unarmed, nearly deciduous trees, is of common occurrence in continental tropical America from southern Mexico to the Brazilian Amazon region. The hollow branchlets are often inhabited by ants, which accounts for such names as Hormigo and Palo Santo. The leaves are opposite or whorled, odd-pinnate, with 3 to 7 large to very large, elongated or rounded, usually leathery leaflets, the lateral ones paired; simple racemes of yellow flowers are borne in abundance on the bare branchlets usually just before the advent of the new leaves; the pods are papery or slightly leathery, flat, rounded at both ends, 1-seeded, and indehiscent.

The two principal species in Brazil are *Platymiscium Ulei* Harms (= *P. paraense* Huber) and *P. trinitatis* Benth. Both attain fairly large size in the virgin forest and are generally known as Macacaúba. The first grows on lowlands of the lower Amazon, and its timber, which is moderately hard (sp. gr. 0.80) but easy to work, is used for joinery and good furniture. The other grows on higher ground toward the interior and the timber is considerably heavier and of a darker color. Two varieties are recognized, of which *P. trinitatis*, var. *nigrum* Ducke is the densest (sp. gr. 1.20). The color varies from plain red with occasional purplish stripes to brick-red with black markings. (See *Arch. Jard. Bot. Rio de Janeiro* 4: 314–316.)

There are several species in Central America. Six specimens from young trees of *Platymiscium dimorphandrum* Donn. Sm. have no heartwood except in a thin shell around the hollow center formerly occupied by ants. The samples are alike in being comparatively light, coarse-textured, and with abundant coarsely confluent parenchyma. At the other extreme is a specimen (Yale 8824) from British Honduras, determined only as to genus, that is very dense and fairly fine-textured; the heartwood is dark reddish brown with lighter streaks. It bears considerable resemblance to one of the varieties of *P. Duckei*. A Nicaraguan timber, known locally as Ñambar Bastardo, was introduced into the New York trade about 25 years ago in shipments of Cocobolo (*Dalbergia retusa*), and was given the name of Yama Cocobolo or Yama Rosewood. The wood is dark red and rather oily and has been used to a limited extent for brush backs, the handles of kitchen knives, and butts of billiard cues. Trifoliolate leaf specimens supplied by Messrs. E. Palazio & Company of Corinto, were originally identified as *P. trifoliatum* Benth., but Standley (*Flora of Costa Rica*, p. 553) states that Ñambar is *P. pleiostachyum* Benth., which also has three leaflets.

The best known and most widely distributed species is *Platymiscium pinnatum* (Jacq.) Dugand (= *P. polystachyum* Benth.; see *Cont. Hist. Nat. Col.* 1: 10–12), a small to fairly large tree of northern South America, Trinidad, and throughout the Atlantic coastal region of Central America. The leaves have 3 to 5 leaflets. The wood varies in color through various shades of red to reddish brown, irregularly and often attractively striped. It is used locally for furniture and marimba bars and for bridge planking, railway crossties, and other forms of heavy durable construction. Limited quantities of the logs are exported to the United States for veneers, brush backs, tool and knife handles, and various articles of turnery. It bears some resemblance to Padauk (*Pterocarpus*) and is known to the trade as Panama Redwood (if from Panama) and Roble, Roble Colorado, or Vencola (if from Venezuela).

The following description is based upon authentic specimens of six species of *Platymiscium*. Assuming that they have all been correctly determined, the range of variation within a species appears about as great as between different species. Heartwood bright red to reddish or purplish brown, more or less distinctly striped; darker specimens look waxy; sharply demarcated from the nearly white sapwood. Luster medium to high. No distinctive odor or taste. Moderately to extremely hard, heavy, tough, and strong; sp. gr. (air-dry) 0.75 to 1.20; weight 47 to 75 lbs. per cu. ft.; texture mostly rather fine, sometimes coarse; feel harsh; grain generally roey; not very difficult to work, finishing smoothly and taking a high natural polish; holds its place well when manufactured; is very resistant to decay.

Common names: Masa wood, monkey wood, Panama redwood, vencola, yama cocobolo, y. rosewood (trade, U.S.A.); chambico, granadillo, palo santo, subinché, tepezapote, zubinché (Mex.); granadillo (Br. H.); hormigo, mulato, ormigo (Guat.); cachimbo, cotín, foncontín, granadillo, hormigo, imera, junera (Hond.); aceituno montés, granadillo (Salv.); coyote, cristobal, ñambar, ñ. bastardo, quira, sinkrá, zrok (C.R.); quira (Pan.); trébo, trébol (Col.); roble, r. blanco, r. colorado,

r. montañero, r. negro, tasajo, uvedita (Venez.); roble (Trin.); dukalaballi (Br. G.); coenatepie, doekaliballi, koenapetie, konatepie (Sur.); macacahúba, macacaúba, m. amarella, m. da terra firme, m. da varzea, m. preta, m. vermelha (Braz.); cumaseba, moira-caoba (Peru).

Platypodium, with a single variable species, *P. elegans* Vogel (= *P. Maxonianum* Pittier), is a medium-sized to large unarmed tree distributed from southern Brazil to Panama. The bark contains a red sap. The rather large leaves are pinnate, with 10 to 20 ovate leathery leaflets; the yellow flowers are borne in pendent axillary racemes; the woody legume is samara-like, the basal part flat and membranous, the apex rounded, and bearing (usually) one seed. The tree is said to be common on the interior plateau of São Paulo and to supply timber for general construction and carpentry. In Panama it occurs frequently, sometimes gregariously, in the littoral forest, attaining a height of 80 feet and a trunk diameter of four feet. The bark is grayish and more or less shaggy. The large stems frequently have large cavities containing an oily liquid. The wood is little used.

Heartwood absent from authentic specimens; said to be light brown with darker streaks or patches; sapwood yellowish white to canary. Luster fairly high. Odorless and tasteless. Hard, heavy, tough, and strong; sp. gr. (air-dry) 0.75 to 0.90; weight 47 to 56 lbs. per cu. ft.; texture rather fine; grain somewhat roey; not difficult to work, finishing very smoothly and taking a good polish; durability of heartwood unknown. Appears to have no possibilities for the export trade.

COMMON NAMES: Carcuera (Pan.); cajón, cañoeto, jayo, lomo de caimán (Col.); drague (Venez.); jacarandá branco, j. do campo (Braz.).

Poecilanthe is a poorly known genus with six species of small to medium-sized unarmed trees, mostly Brazilian. *P. ovalifolia* Kleinh., a tree about 50 feet high, was discovered by J. W. Gonggrijp along the Tapanahoni River in Surinam. *P. amazonica* Ducke is a small tree in the understory of the forest on periodically inundated lands of the lower Rio Negro. *P. effusa* (Huber) Ducke is a little forest tree on non-inundated land in Amazonas and Pará; it is sometimes called Cumarú de Rato. Further south in Brazil are three other species, usually less than 35 feet tall, namely, *P. grandiflora* Benth. (common name Carrancuda), *P. parviflora* Benth., and *P. subcordata* Benth. The leaves are odd-pinnate, with 1 to 7 large leathery leaflets; the racemose flowers are white, yellow, red, blue, or variegated; the pods are flat, few-seeded, and dehiscent. There are apparently no special uses for the timber.

The only wood sample available for study is of *Poecilanthe effusa* (Yale 22595; Ducke 135). Heartwood orange-yellow, with fine, lighter-colored parenchyma markings; rather waxy looking; scarcely distinct from the thin, yellowish sapwood. Luster medium. Odorless and tasteless. Very hard, heavy, tough, and strong; fine-textured; has a rather harsh feel; grain straight to irregular; not easy to work; takes a high natural polish, but is inclined to sliver because of the parenchyma laminations; durability doubtful. No commercial possibilities indicated.

COMMON NAMES: Carrancuda, cumarú de rato (Braz.).

Poeppigia *procera* Presl., the only species, is an unarmed tree erratically distributed in tropical America from southern Mexico to Peru and Rio de Janeiro, Brazil. The leaves are pinnate, with very numerous small leaflets; the flowers are yellow and in racemose cymes; the fruits are flat, thin, narrowly winged along one margin, and indehiscent. It is usually of small stature, but in eastern Peru, according to Williams (*Woods of northeastern Peru,* p. 204), it is a flat-crowned tree sometimes 90 feet in height with a straight low-buttressed cylindrical trunk 16 inches in diameter and free of branches for 25 or 30 feet; it is fairly common in rather dense upland forest and supplies durable timber highly esteemed locally for beams of houses. In Salvador and southwestern Mexico it is rarely 40 feet tall, but the wood is used to a limited extent for spokes of wheels, small cabinet work, and railway crossties. The species is common in the calcareous hills throughout Cuba, but the trees are small and the principal use for the timber is for stakes, posts, and poles; the bark, though thin, is rich in tannin.

Heartwood light to dark brown, uniform or faintly striped with black; rather sharply demarcated from the white sapwood. Luster medium to silky. Without distinctive odor or taste. Density variable from medium in a Peruvian specimen to rather high in Cuban material; mostly tough and strong; fine-textured; fairly straight-grained; easy to work, finishing very smoothly. Although of good quality, the timber has no commercial possibilities because of its scarcity and small size.

COMMON NAMES: Abey hembra, tengue, t. amarillo, t. rojo, tengue-tengue (Cuba); bicho, hoja menudo, quiebrahacha (Mex.); frijolillo, memble, quebracho blanco, tepemiste (Salv.); cedro pashaco (Peru).

Poinciana *pulcherrima* L., the only species, is a shrub or small tree sometimes 20 feet high and eight inches in diameter, of unknown origin but now of pantropical distribution either naturally or through escape from cultivation for ornamental purposes. The branches are sometimes prickly. The evenly bipinnate leaves have 3 to 9 pairs of pinnae and 5 to 12 pairs of thin leaflets; the large and showy, sweet-scented, orange, yellow, or variegated flowers have long exserted stamens and are borne in panicles or racemes; the legumes are flat, the two valves twisting after dehiscence. Every part of the plant appears to have a place in native medicine. The wood, though attractive, has no commercial possibilities.

Heartwood lustrous orange, more or less variegated or striped; deepening upon exposure to coppery orange-brown; not very sharply demarcated from the thin whitish sapwood. Without distinctive odor or taste. Moderately hard and heavy, rather brittle; texture fine; grain fairly straight; easy to work, finishing very smoothly, and taking a high polish; is probably durable.

COMMON NAMES: Barbados flower, B. pride, bird-of-paradise flower, wild senna (B.W.I.); guacamaya (Cuba); clavellina, doodle-do, dul-dul (P.R.); carzazo (Dom. R.); francillade (Haiti); chacalxochitl, chacsicken, chalcasúchil, chamolxochitl, flor de camarón, f. de guacamaya, f. de San Francisco, kansickin, maravilla moreña, sirundaniqua, tabachil, tabachín, t. amarillo, tabachino, tabaquín, xiloxóchitl, zinkin (Mex.); flambeau flower, kansik (Br. H.); clavellina, espanto-lobos, flor de chapa, gallito, hierba del espanto, Santa Rosa, utsá (Guat.); barbón, clavellina, flor barbona, guacamay, guacamaya (Salv.); guacamaya pequeña (Nic.); clavellina, guaca ierón, g. zorón, hoja de sen, malinche (C.R.); gallito (Pan.); angelito, a. clavellina, clavellina, flor de ángel, f. de pavo, florito, guacamaya (Col.); clavellina colorado (Venez.); ángel sisa, flor del ángel (Peru); barba de barata (Braz.).

Poincianella includes 34 species (commonly referred to *Caesalpinia*), typically unarmed shrubs and small to medium-sized trees in Mexico, Central America, and the West Indies. With few exceptions the leaves are bipinnate and in one group bear sessile black glands beneath or on the margin; the red, yellow, or variegated flowers are borne in racemes; the legumes are flat and elastically dehiscent. The plants are of no commercial value.

The wood samples available are of six species and are readily separable into two very distinct groups, which will be designated A and B. The species in A are: *Poincianella affinis* (Hemsl.) Britt. & Rose of eastern Guatemala, where it is known as Carcaño; *P. Recordii* Britt. & Rose of British Honduras, local name Bastard Billy Webb; *P. yucatanensis* (Greenm.) Britt. & Rose, called Xpanpocolcum in Yucatán, Mexico, and Bastard Billy Webb in British Honduras. All are small trees or shrubby. The following description is based on specimens of the last two species, as the other (Yale 10055) lacks heartwood and the determination is questionable.

Heartwood rich chocolate-brown, with light brown parenchyma markings; rather sharply demarcated from the white sapwood. Luster medium. Odor and taste absent or not distinctive. Hard and heavy to moderately so; texture rather coarse, but uniform; grain straight to irregular; easy to work, finishing smoothly with a good luster; appears highly durable. An attractive wood, but of no commercial possibilities because of the small size of the trees.

COMMON NAMES: Guajillo, guete-regl, hojasén, h. del país, palo piojo, polilla, retamilla, sen, s. del país, tabachín de monte, trompetilla, viche, xkanpocolcum, yagati (Mex.); bastard Billy Webb, caromayo (Br. H.); carcaño (Guat.); camaroncillo, flor del mar, f. moreña (Salv.).

Included in group B are: *Poincianella eriostachys* (Benth.) Britt. & Rose, called Iguanero, Iguaco, and Palo Alejo in southern Mexico, Iguano and Pintadillo in Salvador, and Zahino in Nicaragua and Costa Rica; *P. Gaumeri* (Greenm.) Britt. & Rose, known as Xcitinché and Zinkin in Yucatán, and Bastard Logwood, Kitingché, Peccary-wood, and Warree-wood in British Honduras; *P. myabensis* (Britt.) Britt. & Rose, a small shrub of Cuba. The first two are trees up to 50 to 65 feet high and 18 to 24 inches in diameter.

Heartwood grayish olive-brown, with fine streaks of darker color; looks rather oily; with gradual to sharp transition to sapwood; gum specks distinct, especially in sapwood. Luster medium to low. Very hard, heavy, tough, and strong; texture rather fine; grain roey or irregular; not easy to work, inclined to be splintery, but finishes smoothly with a waxy polish; probably durable. An unattractive wood, apparently without commercial possibilities.

COMMON NAMES: Citinché, iguanero, iguano, palo alejo, xcitinché, zinkin (Mex.); bastard logwood, kiting-ché, peccary wood, warree wood, warri wood (Br. H.); iguano, pintadillo (Salv.); zahino, zajino (C.R., Nic.).

Poponax, with more than 15 species of shrubs and small bushy trees, usually classified with *Acacia* or *Mimosa,* are of common occurrence in dry regions throughout tropical America. The leaves are bipinnate, with numerous small leaflets; the twigs bear spinescent stipules, sometimes two inches long; the yellow flowers are in globose heads; the legumes are woody or leathery, more or less pulpy within and containing black round seeds in a single row.

Heartwood purplish brown with a reddish tinge, or rather light reddish brown, or reddish with a purplish tinge, and usually with streaks of yellow, dark brown, or purple; with gradual transition to the yellowish white sapwood. Luster medium. No distinctive odor or taste. Hard and heavy or moderately so; texture rather fine; grain mostly irregular; fairly difficult to work, but finishing smoothly; durability probably high. The wood is suitable for small articles of turnery and inlay, but is not available in sizes or quantity large enough to make it of any commercial importance.

COMMON NAMES: Long-spined acacia (Bah.); acacia bush, park nut, wild poponax, w. tamarind (Jam.); guatapaná (Cuba); acacia bush, cacia, stink casha, wild tamarind (P.R.); hoebada, hubada, waabi (Curaçao); algarroba, cucharitas, espino, huisache, h. chino, palo de cucharitas, quisache corteño, q. tepamo, q. tepano, tepame (Mex.); kuntich (Br. H.); tepame (Guat.); espino de playa (Hond.); espino jiote (Salv.); ambuca, ambuque, aroma real, bigarí, chucuncha, cují, guayacán trapiche trupí (Col.); cují, c. hediondo, c. negro, c. torcido, juba, úveda (Venez.); aroma (Ec.); algarrobillo, aromita, tusca (Arg.).

Prioria *Copaifera* Gris., the only species, is a large unarmed tree native to Jamaica, Nicaragua, Costa Rica, Panama, and Colombia. The leaves are pinnate, usually with two pairs of pellucid-dotted leathery leaflets; the small, cream-colored, scented flowers are borne on panicled spikes; the large, flat, woody, one-seeded pods hang in small clusters from the ends of the upper branches; the seeds germinate readily.

G. Proctor Cooper says (*Tropical Woods* 14: 6): "For miles along the Caribbean coast of western Panama and of Costa Rica, back of the Mangrove fringe and up to the banana and cacao cultivation, lies the 'cativale,' a belt of land from one to three miles wide and also running back into the lower valleys of some of the larger rivers. It is so called because the forests of these lowlands and valleys are composed mainly of Cativo, *Prioria Copaifera* Gris. It is noteworthy that the only other broadleaf tree in Central America occurring in large groves or making up the bulk of any large stand is the Orey [*Campnosperma*], which is also found in western Panama.

"Trees of all heights and diameters are found growing together, the bole always being clean and long, the crown confined to the upper third of the tree. The average sizes are from 24 to 36 inches in diameter and 75 to 100 feet in height, but larger specimens are not uncommon. Recently in a clearing of Cativo a stump was found which measured 64 inches in diameter five feet above the ground. There are no buttresses or basal swellings to interfere with low cutting. The bark is smooth or with fine warts and rather thick; it is mottled gray-green on the surface, but reddish brown beneath. The sapwood, which is very thick, is a light cream or buff color when freshly cut, with fine darker flecks from the vessels and gum ducts; there is a gradual transition to the brownish tan heartwood and an abrupt change to the almost black core. When the sapwood dries out it becomes tan on the surface, but the color below remains unchanged. It has a considerable amount of gum of a deep brown color in the large ducts which makes it difficult to saw. This gum exudes freely when the tree is freshly cut, soon covering the ax or machete. It has a sharp stinging taste, but attracts bees and insects of all kinds. The old stump soon becomes a veritable 'fly paper' and natives claim that bats and other small animals and birds will often become trapped in the gum. Some natives use the substances medicinally to soothe cuts and bites. The wood is soft and easy to cut and is not heavy when dry. The grain is somewhat uneven toward the center and when the lumber is planed it 'pulls' in long thread-like splinters. The large splinters are very flexible and will bend double without breaking. When sawed on the tangent surface at the mill it cuts rough and woolly, the fibers standing out almost half an inch above the surface. The stand of timber has been variously estimated at from 10,000 to over 100,000 board feet per acre, some trees having 2000 or 3000 feet, log scale. No use is made of the wood at present, not even for firewood, but it seems likely that a detailed study of the properties, with a possibility of removing and recovering the gum, might lead to its adoption as material for paper pulp. The woolly or stringy character of the grain also suggests its use for excelsior. Certainly there is an ample supply for many years' operation on a moderate scale. In the Estrella Valley in Costa Rica, the stand has been estimated at from 25 to 50 million feet, log scale, and from the Chiriquí Lagoon in Panama to the Costa Rican border there is probably an equal amount."

In a report on the trees on the watershed of the Bayano River, Panama, H. C. Kluge says (*Tropical Woods* 5: 4) that the species "occurs on the lower wet river bottoms, attaining a diameter of five feet and clear lengths of 40 feet. In the Atrato River region, as well as in the drainage areas of the rivers and creeks flowing into the Gulf of Darién or Uraba, it composes pure dense stands, the trunks averaging three feet in diameter and free of limbs for 50 feet. There are two kinds: one with a pink sapwood, the other white. A thick gum exudes from the freshly cut sapwood, hence the Colombian name Trementino. This gum is little in evidence on dry lumber and does not seem to interfere with the application of paint; in fact, the wood takes paint readily."

Although the timber is abundant, much of it is defective or contains too large a proportion of unattractive sapwood to be marketable. One American importer reports that a shipment of 50 logs, 16 to 30 inches in diameter, all had "rambling" and defective heartwood and many of them were damaged by pin worms. While the lumber was standing in the drying racks gum would exude and form ridges sometimes half an inch thick. An experienced operator in Panama says that he is convinced that Cativo logging must be highly selective, owing to the large percentage of plain wood, and he recommends using the primitive methods of native contractors in preference to expensive machinery.

The first commercial shipment of the timber to the United States appears to have been made from Panama early in 1925. It is now being used in limited amounts for making veneer, mostly 1/28 inch thick, and is known in the trade as Cautivo, Floresa, Spanish Walnut, Tabasara, and Taito. Mr. Karl Schmieg, president of Schmieg & Kotzian, Inc., New York City, makers of fine furniture, says (*Tropical Woods* 9: 2): "This wood is not of prepossessing appearance in unfinished condition and shows a number of narrow oily streaks and veins. With no other finish than wax, however, there are a subsurface luster and a rich striping of brown with many nicely blended tones and shades, such as one sees in English Brown Oak. Cautivo is somewhat lighter in weight than Walnut, is very uniform in texture, rather open-grained, easy to work, not inclined to warp or check, and the natural oil content seems to give no trouble in working or finishing the dry wood. It is worthy of attention from architects and interior decorators, as it would make beautiful woodwork for a hall, living room, or library."

Armando Dugand, of Bogotá, Colombia, states in a personal letter: "The wood of Cativo is generally reputed as very perishable, being subject to attack by termites and other insects and, therefore, not suitable for exterior construction. Its use in northern Colombia has thus been limited to scaffolding, coarse furniture, and for the making of boxes and crates. It is claimed, however, that soaking the timber in sea water for considerable time improves its resistance to insects and increases its durability in contact with the ground. Cativo so treated is now finding new use for the building of cottages near the seashore." A manufacturer in New Jersey says that part of a lot of Cativo logs from Panama were left in salt water where they were washed by the tides for six months. At the end of that time they were sawed and found to be in much better condition and to have considerably less gum than the others which had been manufactured immediately upon arrival. It is important, however, not to expose the timber to the attacks of teredo or other marine borers.

Heartwood medium to light brown, often attractively streaked; superficially dull, but with a golden luster beneath; sharply demarcated from the thick sapwood, which is pinkish to white when fresh, becoming dingy and unattractive on the surface because of oily exudations. Odorless and without distinctive taste when dry. Of medium density, but firm and strong; of rather fine texture; straight-grained; very easy to work, taking a smooth finish; holds its place well when manufactured; must be seasoned carefully to prevent collapse, especially along the darker streaks in the heartwood.

COMMON NAMES: Camíbar, cativo (C.R.); amansa-mujer, cativo, cautivo (Pan.); amansa-mujer, canime, cativo, c. blanco, c. negro, trementino (Col.); cautivo, floresa, tabasaro, taito (U.S.A., trade).

Prosopis, with perhaps 30 valid species though many more have been named, and numerous varieties, is a highly important genus in the regions of its growth but not elsewhere. One species occurs in India and Persia, one in the eastern Mediterranean region, two in tropical Africa, and the rest in the New World from Patagonia to southwestern United States, the largest number (about 15) being native to Argentina. They are small to large shrubs or spreading short-boled trees growing in the open or forming clumps and thickets in localities too dry for most other kinds to thrive. They are mostly characterized by spiny shoots, though some forms are unarmed; the leaves are bipinnate, with one or two pairs of pinnae and numerous small leaflets; the flowers are small and borne in spikes or heads; the pods are indehiscent, linear, sometimes constricted between the seeds, and either fairly straight or spirally coiled.

Prosopis is most abundantly represented in Argentina and Uruguay and supplies more commercial timber there than elsewhere. Although there are many species, the bulk of the wood on the market is the product of *Prosopis nigra* Hieron. and *P. alba* Gris., and is known as Algarrobo. It is not uncommon to find 10 to 25 Algarrobo trees per acre, but at best they are only 25 to 40 feet high with short twisted boles 12 to 18, rarely 24, inches in diameter. The trees have umbrella-like crowns and a group of them suggests an apple orchard. The pods are borne in profusion and make excellent fodder for mules, burros, and horses and are also an important source of food for the Indians. The wide distribution of the tree in Argentina is doubtless attributable to the Indians, since in the neighborhood of old settlements and about towns and villages it is the commonest and often the only tree. It occurs in commercial quantities throughout the Chaco and is exploited in the regions north and west of Buenos Aires, especially in the Province of Cordova. The bulk of the best timber accessible to the railways has been cut, but the tree renews itself readily from sprouts and is of fairly rapid growth so that it continues to hold a place in the trade. The timber enters the market in the form of round logs from four to eight feet long, and is used for a great variety of purposes because it is easy to work, highly inert to changes in moisture, and very resistant to decay and to wear. Many streets in Buenos Aires have been paved with blocks of it without preservative treatment. It is also used for hubs and felloes and is one of the best woods on the market for doors and window frames.

Extending further southward than the two species above mentioned is *Prosopis caldenia* Burkhart, commonly known as Caldén. The wood is of somewhat lighter weight than the others and has a tendency to become ring-porous (see Koutche & Carmelich's *Estudio forestal del caldén,* Buenos Aires, 1936). The densest wood of all is that of *P. Kuntzei* Harms, called Itín or Barba de Tigre, a low tree without leaves most of the year, beset with large sharp spines, and yielding a very dense dark purplish wood. The Nandubay or Espinillo, *P. nandubay* Lor., grows in pure stands over considerable areas of the drier forests of northern Argentina, though it rarely exceeds 15 to 20 feet in height. The wood is much like the Algarrobo and is common on the local market in the form of posts. One of the most widely distributed species in South America is *P. chilensis* (Mol.) Stuntz, which some botanists consider the same as the Mesquite of southwestern United States. In Uruguay it is the only timber available in many sections remote from forested belts along the rivers and supplies fuel, posts, stakes, and wood for many other purposes, and also food for man and beast.

In the corresponding zone north of the Equator are found various species and varieties of *Prosopis,* two of which are of considerable local importance. The Screw Bean or Screw-pod Mesquite of the southwestern United States and the Tornillo of Mexico, *P. odorata* Torr. & Frém. (= *P. pubescens* Benth.) is, as its names indicate, characterized by twisted pods which are in a narrow straight spiral with 12 to 20 turns. These pods are so sweet that they can be eaten as picked without any special preparation; they are also used by the Indians for making a syrup, a fermented beverage, and a kind of flour.

Best known of the northern species of *Prosopis* is the common Mesquite, *P. juliflora* (Sw.) DC. (with two or three varieties), growing in western Texas and eastern New Mexico in the United States, throughout most of Mexico, Central America, and the West Indies. It was introduced into Hawaii more than a century ago and has become a highly important timber in the drier regions of that country, supplying large quantities of fuel and stock feed and affording abundant nectar for bees. It has also been naturalized in the Philippine Islands, and has been planted in India, Australia, and South Africa. The tree is common in Mexico and the arid regions of the Southwestern United States. In the driest situations it is reduced to a shrub and in the northern part of its range the shrubby form develops enormous underground stems (known in Chihuahua as "cepas"), which make excellent firewood in localities where little else is obtainable, though involving a great deal of labor in removing them from the ground. The smaller roots penetrate to great depths in search of water, sometimes, it is said, to as much as 50, or even 75, feet. The largest trees are found in the river valleys where heights of 40 feet and diameters of four feet are not uncommon. On the plains and mesas the species frequently forms thickets or forests many miles in extent resembling orchards of peach or apple trees.

The wood is valuable for railway crossties, vehicle construction, fence posts, and especially for fuel, and in Texas is sometimes used for paving blocks. Its uses are limited only by the small sizes of the timber, but care is necessary to protect cordwood and fence posts from attacks by borers (see U.S. Dept. Agr. Farm. Bull. No. 1197). An amber-colored translucent gum, similar to gum arabic, exudes from the trunk and is used for making mucilage and as an ingredient in medicine and for various other purposes. The bark, as is also the wood, is used in tanning; the inner bark of the stem is employed medicinally, and that of the roots supplies fiber for cordage and coarse fabric. The flowers are much frequented by bees and yield a good grade of light-colored honey, and the Pima Indians sometimes eat the flower spikes after stripping off the flowers between their teeth. The pods are sweet and are an important source of forage for stock of all kinds (see U.S. Dept. Agr. Bull. No. 1194). They have long been an article of human food in America, and among some of the Indian tribes, as with the Pimas, they were the chief food staple, being ground into a meal after the seeds and coarser parts had been removed. The meal is made into cakes (known in Mexico as "mesquitamales") which are baked, or it is mixed with water to form a beverage known as "mesquitatole," or fermented to make a kind of beer. (See Standley's *Trees and shrubs of Mexico,* pp. 351–352.)

The following description applies particularly to the wood of *Prosopis chilensis* and *P. juliflora.* Heartwood rich dark brown, often with a purplish hue, marked with fine but distinct vessel lines; sapwood yellow, generally thin, rather sharply demarcated. Not highly lustrous. More or less fragrant, suggesting violets; taste not distinctive. Hard, heavy and strong; sp. gr. (air-dry) 0.80 to 0.92; weight 50 to 58 lbs. per cu. ft.; rather coarse-textured; mostly of irregular grain; easy to work, finishing smoothly and taking a high natural polish; is very resistant to decay.

Common names: Algaroba, honey locust, h. mesquite, h. pod, ironwood, mescrew, mesquite, screwbean, s. mesquite, screwpod mesquite (U.S.A.); cashaw (Jam.); algarroba del Brasil, cambrón, mezquite, plumo de oro (Cuba); bayahonda, bohahunda, vallahonda (Dom. R.); bayahon, bayarone (Haiti); cojí, indjoe, wawahi (Curaçao); algarroba, catzimek, chachaca, chúcata, guisache, mezquite, m. amarillo, m. blanco, m. chino, m. colorado, mizquitl, tornillo, ttahi, tziritzequa (Mex.); nacascol, nacascolote (Guat.); carbón (Salv.); acacia de Catarina (Nic.); aroma, carbón (C.R.); manca-caballa (Pan., Col.); trupillo (Col.); barbasco, caobano gateado, cejí, cojí, cují, c. negro, c. yaque, yaque (Venez.); algarobo, bate caixa (Braz.); algarroba, algarrobo (Ec., Peru, Arg., Urug., Hawaii); thacco (Peru); algarobillo, algarrobo blanco, a. amarillo, a. colorado, a. mezclado, a. morado, a. negro, a. pardo, alpataco, barba tigre, caldén, cama-tala, espinillo, e. ñandubay, ibopéhú, ibopé morotí, i. pyitá, i. saiyú, iscayante, ischilín, itín, jacarandá, j. palo lanza, lamar, ñandubay, quentitacú, quilín, retorton, tatané, tintatico, vinal, visnal, v. tartagal (Arg.); algarrobo, cupesi, churqui (Boliv.).

Pseudalbizzia *Berteriana* (Balbis) Britt. & Rose, the only species, is an unarmed tree 25 to 50 feet high, limited in its distribution to Jamaica, Cuba, Haiti, and Dominican Republic. It is closely related to *Albizzia,* but, unlike that genus, its legumes are indehiscent, at length breaking transversely between the seed-bearing parts. The timber is used locally to a limited extent for miscellaneous purposes, but is not highly esteemed.

Wood pale yellowish throughout; apparently not forming a distinctly colored heartwood. Fairly lustrous. Odorless and tasteless. Hard and heavy, but not very strong; rather coarse-textured; more or less roe-grained; not difficult to work, finishing smoothly and taking a good polish; is not resistant to decay.

Common names: Abey blanco (Cuba);

córbano, c. blanco (Dom. R.); bois savanne (Haiti).

Pseudocassia. Two or three tropical American species of unarmed shrubs and small trees have been referred to this genus, but the only one well known is *P. spectabilis* (DC.) Britt. & Rose (= *Cassia spectabilis* DC.) which occurs in northern South America, the West Indies, Central America, and southern Mexico. It has long pinnate leaves, with 6 to 15 pairs of rather large pointed leaflets; the showy yellow flowers are borne in large erect panicles; the legumes are long, black, leathery, and terete, with many narrow transverse partitions between the lenticular seeds. The tree rarely exceeds 25 feet in height and its only use is for decorative purposes and live fence posts.

Heartwood rather lustrous light olive to golden brown; fading gradually into the nearly white sapwood. Odorless and tasteless. Moderately hard and heavy; of medium to coarse texture; straight-grained; very easily worked; probably fairly durable. A good wood but of no commercial importance because of the limited quantity and small sizes available.

Common names: Casse marron (Haiti); canchén (Mex.); candelillo (Hond., C.R.); cañafístola macho (Col.); cañafístola cimarrón, c. macho, chiquichique, mucutena (Venez.).

Pseudocopaiva, with a single species, *P. hymenaefolia* (Moric.) Britt. & Wils. (= *Copaifera hymenaefolia* Moric.), is a small or medium-sized unarmed tree of Cuba and the Isle of Pines. The leaves are pinnate, with one pair of oblique, leathery leaflets; the flowers are small and in short axillary racemes; the legumes are short, flat, one-seeded, and dehiscent; the glossy black seeds are short arillate. (See *Tropical Woods* 20: 28.) The tree is well known locally because of the flinty hardness and great natural durability of its purplish red heartwood, which is used for posts, beams, and railway crossties. Its usual name in eastern Cuba is Caguairán, in the western part, Quiebrahacha (axbreaker).

Heartwood rather dull purplish red, with very fine parenchyma markings; sharply demarcated from the nearly white sapwood. Odor and taste absent or not distinctive. Exceedingly hard, heavy, and strong; sp. gr. (air-dry) 1.33; weight nearly 84 lbs. per cu. ft.; texture fine; grain irregular; difficult to work, but finishing very smoothly. Of no importance except locally.

Pseudosamanea *guachapele* (H.B.K.) Harms, the sole species, is a medium-sized to large unarmed tree known to occur from Guatemala and Salvador to Venezuela and Ecuador. The bark is rough and the pith is finely chambered. The leaves are large and doubly pinnate, with 5 or 6 pairs of pinnae and rather few obliquely oblong leaflets; the flowers are yellowish white, in long-stalked tufts; the pods are flat. The wood is of the general type of *Enterolobium* and *Samanea,* but is somewhat finer-textured; it is of good quality, but is little used, presumably because of its scarcity. (See *Notizbl. Bot. Gart. Berlin-Dahlem* 11: 101; 52–55.)

Heartwood rich brown, with golden luster; rather sharply demarcated from the thin whitish sapwood. Odorless and tasteless. Mostly moderately heavy and hard; texture medium to rather coarse; grain straight to roey; easy to work; saws rather woolly but finishes smoothly and attractively; fairly resistant to decay. Suitable for furniture.

Common names: Cadeno (Guat.); tabaca, guamarillo (Col.); samanigua (Venez.); guachapele (Ec.).

Pterocarpus, with about 70 species of small to large unarmed trees of pantropical distribution, is well represented in America from the West Indies and southern Mexico to northern Argentina. A few of the species of the Old World yield valuable cabinet woods, such as Narra of the Philippines and Amboyna-wood burls from the Malayan Archipelago (*P. indicus* Willd.); Andaman Padauk, Andaman Redwood, or Vermilion (*P. dalbergioides* Willd.); Burma Padauk (*P. macrocarpus* Kurz.); and African Padauk (*P. santalinoides* L'Her., and probably others). The woods of the

American trees have little if any normally developed heartwood and are of no commercial importance.

The leaves are odd-pinnate, with few to several medium-sized to large, alternate to subopposite leaflets; the flowers are large, mostly yellow; in *Pterocarpus Ulei* Harms the rachis of the inflorescence is inhabited by ants; the fruits are indehiscent and circular, the central part (containing 1 or 2 seeds) enlarged and leathery or woody, and bordered by a ring, narrow to wide depending upon whether adapted to transport by water or by air. Some of the trees inhabit swamp lands along water courses and lakes, others grow on uplands. The bark when cut exudes a blood-red juice which soon hardens into a dark odorless resin with somewhat astringent taste; it was formerly used in medicine under the name of "dragon's blood."

The best known and most widely distributed American species is *Pterocarpus officinalis* Jacq. which occurs in Jamaica, Puerto Rico, southern Mexico, throughout most of Central America, and northern South America to the estuary of the Amazon. Standley says (*Tropical Woods* 28: 12): "Along the Central American coast this species is confined to the lowlands. Although it often grows on hillsides, its favorite habitat is swamps, especially those inundated by the tide. There, in the spring months, the surface of the water sometimes is covered by a floating layer of the buoyant seed pods, which can be confused with those of no other plant." It often forms dense nearly pure stands back of the Mangrove formation. The rather slender trunk is strongly buttressed by many thin, irregular, frequently wide-spreading root spurs.

There are authentic wood specimens of eight American species of *Pterocarpus* in the Yale collections. They are all yellow or whitish throughout, except near beetle holes, fire scars, or other injuries. The traumatic heartwood is dark brown or purplish, more or less variegated, generally very hard and much denser than the sapwood; without distinctive odor or taste. The sapwood varies from light and soft to moderately hard and compact, differing in the same species, although trees of the lowland usually have softer wood than the others. Texture medium-fine to coarse; grain fairly straight to irregular; not difficult to work, finishing smoothly; is perishable when exposed to decay. Suitable for interior construction lumber and plain joinery.

Common names: Palo de pollo (P.R.); bois chatousieux, b. de corail tendre (Guad.); bois l'étang, lagunero (Trin.); drago, guayabillo, llora-sangre, sangre de drago, s. de torro, sangregado (Mex.); kaway, swamp kaway (Br. H.); sangre de drago, sangregado (Guat., Nic.); chajada amarilla (C.R.); bloodwood, huevos de gato (Pan.); bollo blanco, grau blanco, sabrosó, sangre, s. de drago, sangregado, sangregao, s. blanco (Col.); drago, lagunero, mejoro, mucunana, sangre drago, tasajo (Venez.); corkwood, hiburu, itchikiboura, itik-boura, mutushi, warau (Br. G.); bebe, b. hoedoe, b. hout, bigbe, djoekabebe, hegronbebe, hoogland bebe, itikiboeroe, itikiboro, i. hororadikoro, itjoeroe-tanomoetoesie, matosirian, mattoe gwegive, moetoesi, moetoesiran, waata gwegwe, waterbebe (Sur.); liège du pays, moutouchi, m. de savane, m. grand bois (Fr. G.); angú, cortiça, corticeira, mututi, mututy, m. da varzea, m. de terra firme, pau sangua, tachy, tachyzeiro, tinteira (Braz.); nogal falso (Bol.); ibá-rá, payaguá-manduvi-mí (Arg.).

Pterodon. Botanists have named four species, but the only one cited in descriptions of the timber is *P. pubescens* Benth. This is a medium-sized to rather tall, coarsely branched, sparsely foliaged, unarmed, gray-barked tree growing in the dry regions of south-central Brazil, where it is known as Faveiro. The leaves are pinnate, with numerous leaflets; the small roseate flowers are borne in large panicles; the fruit contains one seed within a thin oily sarcocarp separating at maturity from the rather woody, indehiscent, winged exocarp which contains a pungent balsam. The timber is well known locally and is used for railway crossties, bridges, and other durable construction; also for implement frames and spokes of cart wheels.

Two classes are recognized, namely, Faveiro Amarello and Faveiro Vermelho; the latter is preferred in the matter of durability.

Heartwood light brown, deepening somewhat upon exposure; usually merging gradually into the yellowish sapwood, sometimes (in darker specimens) distinct. Luster low to medium. Without distinctive odor or taste. Very hard, heavy, tough, and strong; sp. gr. (air-dry) 0.90 to 1.15; weight 56 to 69 lbs. per cu. ft.; texture coarse; feel harsh; grain usually finely roey; difficult to work, but can be finished smoothly; durability fair to excellent. Appears to have no commercial possibilities except locally.

COMMON NAMES: Faveiro, f. amarello, f. do campo, f. da matta, f. vermelho, sucupira liza (Braz.).

Pterogyne *nitens* Tul., the only species, is a large unarmed evergreen tree of scattered occurrence on the lower slopes of the Andes Mountains in the Provinces of Tucumán and Jujuy and also in the forests of Chaco and Misiones, Argentina, and in southern Paraguay and Brazil. It attains a maximum height of over 100 feet, but is usually not over 75 feet, with a well-formed trunk two to three, exceptionally four, feet in diameter. It is nowhere abundant and even where most common will not average more than one tree per acre. The leaves are pinnate, with small leaflets, suggesting *Robinia;* the small yellow flowers are borne in short axillary racemes; the legumes are leathery and winged. The timber is of commercial importance in Argentina, but the consumption is limited because of the poor supply. It is of excellent quality and is used for cabinet work and good furniture, interior trim, articles of turnery, general carpentry, wheelwright work, cooperage, railway crossties, and fuel. The aqueous extract of the sawdust is purple and is sometimes employed locally for dyeing. The bark and leaves have a bitter taste, hence the Indian name Ibiráro (bitterwood) and the Spanish Palo Amargo.

Heartwood light pinkish brown or flesh-colored when fresh, deepening upon exposure and suggesting Mahogany (*Swietenia*); often with darker striping; distinct but not sharply demarcated from the sapwood. Luster medium to high. Without distinctive odor or taste. Hard, heavy, tough and strong; sp. gr. (air-dry) 0.80 to 0.90; weight 50 to 56 lbs. per cu. ft.; texture medium; grain often roey; rather easily worked, finishing very smoothly; holds its place well when manufactured; is fairly durable.

COMMON NAMES: Amendoim, a. bravo, ebiráro, ibiráro, jacutinga, oleo branco, pau amendoim, p. fava (Braz.); guiráro, ibiráro, ibiráro-mí, ivaráro, palo amargo, p. mortero, p. rosa, tipa colorada, viráro (Arg.); ibiráro (Par.).

Pteromimosa, with two described species of prickly shrubs or little trees, is of limited distribution in tropical America. The principal distinction between it and *Mimosa* is in the pods, which in *Pteromimosa* have soft and narrowly-lacerate wings along the margins. *P. bahamensis* (Benth.) Britt. & Rose is a much-branched plant less than 15 feet tall, growing in scrublands and thickets in the Bahamas. *P. hemiendyta* (Rose & Rob.) Britt. occurs in southern Mexico and northern Central America. It is less than 25 feet tall, with gray bark and stiff spreading branches bearing slightly recurved claws at the nodes. The bushes bear some resemblance to Logwood (*Haematoxylon*), but the timber is not utilized. The following description applies to the wood of the latter species (Yale 8807).

Heartwood lustrous pale reddish brown with yellowish streaks; with gradual transition to the thin yellowish sapwood. Odorless and tasteless. Hard, heavy, tough, and strong; texture medium; grain irregular; easy to cut; apparently rather durable.

COMMON NAMES: Haul-back (Bah.); citsím, sascatsím, zaccatsím (Mex.); bastard logwood, catseem, logwood brush (Br. H.).

Recordoxylon, with two described species of large unarmed trees, is apparently limited in distribution to the State of Amazonas, Brazil. *R. amazonicum* Ducke (= *Melanoxylon amazonicum* Ducke) was discovered by Adolpho Ducke in 1929; it is known to occur in the northern and southern parts of Amazonas, but has not been found in the intermediate region. The leaves are unequally pinnate, with 5 to 11 medium-sized to large leaflets; the golden yellow flowers are borne in conspicuous terminal panicles; the legumes are narrow, flat, leathery, and dehiscent; the seeds are few, brown, flattened but not winged. Ducke says (*Tropical Woods* 39: 17): "This tree, in the flowering stage, looks like *Melanoxylon,* but the pods are entirely different; they recall those of some species of *Cassia, Acacia,* and *Piptadenia;* the wood, however, is similar to those of *Bowdichia.* . . . I have named it *Recordoxylon* in honor of Professor Record, since it was his observation regarding the wood structure that led to the revision of my original classification." A second species, *R. stenopetalum* Ducke, known locally as Manicó, was described in 1938 (*Archiv. Inst. Biol. Veget.* 4: 1: 16). It was discovered in non-flooded forest near São Paulo de Olivença in western Amazonas and "perhaps eventually will be considered as a geographical variety of *R. amazonicum.*" The wood (Yale 39385; Ducke 338) is practically identical with that of *R. amazonicum.*

Heartwood dark brown with a yellowish hue; rather waxy looking and marked with prominent light-colored vessel lines; rather sharply demarcated from the thin, yellowish white sapwood. Without distinctive odor or taste. Very hard and heavy, somewhat horn-like, tough and strong; texture coarse and harsh; grain fairly straight; not easy to work, inclined to be splintery; finishes smoothly but not lustrously; is highly resistant to decay and insects. It is suitable for heavy and durable construction.

Robinia. There are about 20 species of shrubs and a few small, medium-sized, or large trees, all limited in natural distribution to the United States, with two exceptions, namely, *R. luxuriana* (Dieck) Rydberg, which extends into Sonora from western Texas and southern Nevada, and *R. Pringlei* Rose, a medium-sized tree known only from the type locality (elevation 6600 feet) near Tula, Hidalgo. Various tropical species have been described in this genus, but were later transferred to other genera. *R. ferruginea* H.B.K. of northern South America is valid, according to Pittier (*Journ. Wash. Acad. Sci.* 18: 8: 208–210), but the wood differs so much from that of other species of *Robinia* studied that it is more convenient to describe it under *Humboldtiella* as proposed by Harms.

In *Robinia,* in the sense used here, the branchlets are zig-zag, without a terminal bud, and with the axillary buds naked, superimposed, and protected collectively in a depression by a scale-like hair-lined covering which opens in early spring. The deciduous leaves are odd-pinnate, with numerous rather small opposite leaflets and persistent spiny or bristly stipules; the fragrant white, pink, or purplish flowers are borne in drooping axillary racemes; the dehiscent pods are small, linear, flat, membranous, and contain several small hard seeds.

The only species of importance for its timber is the Black Locust, *Robinia pseudacacia* L., a tree usually 50 to 60 feet high at maturity, rarely up to 90 or 100 feet in forest, with a trunk diameter commonly less than 20 inches, exceptionally up to three feet or even more. The center of its natural range is in the southern Appalachian region, but it is widely planted and naturalized throughout the eastern half of the United States and parts of Europe. Though at its best in rich well-drained soil, it will grow and produce thickets on very poor sites, and on this account has an important place in forest planting to control erosion. Trees in the open and in pure plantations are especially subject to the attack of a wood-boring insect (*Cyllene robiniae* Forst.) which frequently kills them or renders them unsightly for shade and unfit for timber. The roots increase soil fertility through the formation of nodules containing nitrogen-fixing bacteria, but apparently they are harmful to conifers and the two kinds of trees do not grow well in mixture.

Two forms of the tree are recognized, though botanists have had difficulty in finding constant characters to separate them. A variety called Shipmast Locust, *Robinia pseudacacia*, var. *rectissima* Ruber, was described in 1936 (Circ. No. 379, U.S. Dept. Agr.) from specimens collected on Long Island. "The Shipmast Locust has a more erect stem with less spread at the base, a narrower crown, thicker bark with deeper furrows, more durable wood, and fewer flowers with greener calyxes than has the common Locust of the locality. . . . Shipmast Locust trees, unlike those of common Locust, very seldom if ever bear pods and seed; the trees must, therefore, be propagated vegetatively by sprouts or root cuttings." The better-formed trees are often called Yellow Locust; they have a comparatively thin sapwood and a canary yellow heartwood (when fresh) very similar in appearance to that of Osage Orange (*Maclura*), whereas the ordinary forms have thicker sapwood and a more deeply colored brownish heartwood, often with an olive hue.

The timber is noted for its great strength and durability. When manufactured it holds its place well with little shrinkage or swelling and in the days of wooden ships was in demand for treenails. It was also excellent for the hubs of wagon wheels. Its principal use in manufactured form now is for pins to hold the glass insulators for telegraph and telephone wires. Large quantities are consumed on farms for fence posts. Ordinarily it requires from 15 to 20 years to grow trees large enough to make posts 7 to 8 feet long with a top diameter of 4 to 6 inches, and 30 to 40 years to produce bolts for insulator pins.

Heartwood golden yellow to yellowish brown, becoming russet-brown upon exposure; fairly uniform in same specimen; sharply demarcated from the white sapwood. Luster fairly high. Odorless, but sometimes with slight "beany" taste. About one and one-half times as hard, tough, stiff, and strong as White Oak (*Quercus alba* L.); sp. gr. (air-dry) 0.75 to 0.80; weight 47 to 50 lbs. per cu. ft.; texture uneven, coarse in early wood, fine in late wood; grain fairly straight; not difficult to work, finishing very smoothly and taking a high polish; sapwood perishable, heartwood, particularly the yellow kind, very resistant to decay.

COMMON NAMES: *Robinia pseudacacia:* Acacia, false acacia, locust—black, green, red, shipmast, white, yellow (U.S.A.); acacia, loco (Mex., int.). *Other species:* Locust (clammy, honey, red, red-flowering, rose-flowering), rose acacia (U.S.A.); uña de gato (Mex.).

Russellodendron *cacalaco* (H. & B.) Britt. & Rose, the single species, is a spiny tree or a large shrub of southwestern Mexico. The leaves are large and bipinnate, with a prickly rachis, 2 to 5 pairs of pinnae, and a few rounded leaflets; the large yellow-and-red flowers are borne in long racemes; the legumes are slender, indehiscent, rather succulent, and somewhat constricted between the edible seeds. The bark is the source of a toothache remedy and the pods are rich in dyestuffs and tannin, but the wood is little used except for charcoal. The tree is frequently grown for shade and ornamental purposes.

Heartwood rich reddish brown, marked with very fine white vessel lines; sharply demarcated from the nearly white sapwood. Luster medium. Odor and taste absent or not distinctive. Very hard, heavy, tough, and strong; texture rather fine; grain rather irregular; not easy to work, but finishing very smoothly; appears highly durable. Of no commercial possibilities because of the scarcity of the trees.

COMMON NAMES: Cacalaco, cascalote, chalalá, guachalalá, huisache, h. bola, huizache, nacascul (Mex.).

Sabinea, with three species of unarmed shrubs and little trees, is of limited distribution in the West Indies, chiefly Puerto Rico. The leaves are equally pinnate, with numerous small leaflets, the rachis terminating in a tip; the reddish or purplish flowers are borne in small clusters in the axils of the old leaves; the stipitate legumes are narrowly linear and flat, the thin valves twisting upon dehiscence. The best known species is *S. florida* (Vahl) DC., a shrub or a tree 20 feet tall occurring in thickets on moist lands in Puerto Rico, Virgin

Islands, and Dominica. It is often planted for ornament, as the long slender branches bear an abundance of lavender or light purple flowers which are highly attractive though short-lived.

Heartwood chocolate-brown, somewhat streaked; sharply demarcated from the thin, yellowish sapwood. Luster medium. Odorless and tasteless when dry. Rather hard and heavy; texture medium-fine; grain straight to irregular; easy to work, taking a high polish; appears very durable. Of no commercial value, but suitable for small articles of turnery.

Common names: Caracolillo, retama, wattapana (P.R.).

Samanea. Several species have been described, but the only one that is well known is *S. saman* (Jacq.) Merrill. It is indigenous to Central America and parts of northern South America and has been widely planted and naturalized in other tropical regions of both hemispheres. It is one of the most beautiful of neotropical trees and its enormous wide-spreading crown makes it ideal for shade in parks, barnyards, and pastures. It serves not only to protect domestic animals from the hot sun but also supplies them with food, as the abundant pods are excellent fodder. It is occasionally found along forest streams, for example the Magdalena River in Colombia, where it attains a height of 100 feet, with a long trunk four feet thick. Grown in the open, the bole is short and frequently over six feet in diameter.

Young trees grow rapidly and produce light and soft, easily worked, brown wood, but later growth is slower and the wood is comparatively hard and heavy, cross-grained, fibrous, tough, rather refractory, and of considerably darker color, sometimes almost black. The timber is of no commercial importance because forest-grown trees are rather scarce and the others are too useful for shade or their trunks are too large to be transported to a sawmill. Local uses include dug-out canoes, split posts, and some construction timber. On the Pacific coast of Central America, cross sections of the massive trunks are frequently used as wheels of ox carts. Selected lumber has a rather attractive grain, suggesting *Enterolobium.*

Common names: Rain tree (Florida); guango (Jam.); algarroba, a. del país, almacigo blanco (Cuba); samán (P.R.); cow tamarind (Trin.); carreto, c. real (Hond.); carreto, cenícero, zorra (Salv.); árbol de la lluvia, cenísero, c. claro, c. oscuro, cenízaro, samán (C.R.); guango, samán (Pan.); carito, compano, samán, tabaca, t. de monte (Col.); carabali, cenícero, daugení, lara, samán, s. blanco, s. negro, urero, u. macho, u. negro (Venez.); huacamayo-chico (Peru); bordão de velho, gipio (Braz.).

Samanea pedicellaris (DC.) Killip is a medium-sized to large tree of the Guianas and, according to Ducke (*As Leguminosas da Amazonia brasileira,* p. 26), extends through the lower Amazon region of Brazil to Rio de Janeiro. Bertin says (*Les bois de la Guyane,* p. 77) that in Saint-Laurent, French Guiana, where it is known as Bois la Morue, it is a beautiful tree with ascending branches, a cylindrical or irregular trunk sometimes more than four feet in diameter and 90 to 100 feet to the first large limb. The yellowish brown bark exudes a gum. The leaves are alternate and bipinnate, with 6 or 8 glandular pinnae, each with 15 to 25 pairs of oblong leaflets; the flowers have dark red corollas and are borne in small clusters; the blackish brown pod is less than four inches long and an inch wide, the margins of the valves thick and woody; there are from 12 to 18 seeds. The timber is used locally to a limited extent for carpentry and interior construction, and is suitable for furniture.

Heartwood brownish, more or less striped; usually with a golden luster; sapwood whitish, distinct but not sharply demarcated. Odorless and tasteless. Density variable; sp. gr. (air-dry) 0.70 to 0.90; weight 44 to 56 lbs. per cu. ft.; texture coarse; grain roey; easy to cut, but difficult to finish smoothly on radial surface; takes a high polish; is probably fairly durable.

Common names: Red manariballi (Br. G.); aboonkini, alawatta-moelerie, arawata-moerere, asau, baboen-banjie, hoeroewassa kabana, manalliballi-hororodikoro, manariballi-koereroe, plokonie, proekoenie, prokonie, sera, tamalin, tamarinde plokonie (Sur.); bois ara, b. cerf, b. la morue, b. macaque rouge, b. pagode, bougouni, cèdre d'argent, préfontaine rouge (Fr. G.); cambuí, ingá-rana (Braz.).

Schizolobium *parahybum* (Vell.) Blake, the only distinct species although two or three others have been proposed, is an unarmed deciduous tree of sporadic occurrence in southern Mexico, Central America, Colombia, eastern Peru, and the Amazon and coastal region of Brazil to São Paulo. It prefers rich well-drained soil and, while often of only medium size and found principally in second-growth, it sometimes attains a height of 100 feet or more in climax forest and has a long, slender, smooth, well-formed bole 18 to 30 inches in diameter above the prominent buttresses. The bipinnate leaves are sometimes three feet long, with very numerous oblong leaflets and, especially on young trees, suggest the fronds of a tree fern. The trees are leafless at the time of flowering and are made exceptionally conspicuous by the great masses of lustrous golden blossoms. The one-seeded legumes are flat, broadly spatulate, leathery, and tardily dehiscent. Germination is good and early growth is rapid. Young trees have light and soft wood, but that formed later, when growth is slow, is considerably denser, sometimes decidedly hard, tough, and fibrous. The timber is rarely utilized, but it has been suggested that the trees might be grown commercially for the manufacture of paper pulp (see *Tropical Woods* 2: 2–5).

Heartwood pale brown or pinkish, sometimes with reddish brown streaks; rather slow in forming and usually not in conspicuous contrast with the nearly white sapwood. Luster medium to satiny. Odorless and tasteless when dry; fresh wood said to be unpleasantly scented (see *Tropical Woods* 17: 27). Density variable; ranging from light, soft, and rather spongy to rather heavy, hard, and splintery; texture medium to coarse; feel soft to harsh; grain straight to decidedly interwoven; light and intermediate grades easy to work, sawing rather woolly but finishing smoothly; holds nails firmly; is not difficult to season, though sapwood stains readily; heartwood not very resistant to decay or insect attacks, the sapwood perishable.

COMMON NAMES: Judío (Mex.); quam, quamwood (Br. H.); plumajillo, sora, zorra (Guat.); tambor (Hond.); chapulaltapa, melón (Salv.); quon (Nic.); tambor (Col.); pashaco (Peru); bacurubú, faveira, guapuruvú, paricá, pau de vintem (Braz.).

Sclerolobium, with 20 species of unarmed small to medium-sized or rarely large trees, occurs in tropical Brazil, eastern Peru, the Guianas, and Venezuela. The leaves are generally imparipinnate, with several rather large leathery leaflets; the small white or yellow flowers are borne in terminal panicles of dense racemes; the legumes contain one or two large seeds. The twigs of some species are hollow and inhabited by ants called "tachy" in the Amazon region. The timber apparently has few special uses, except for making a fine grade of charcoal. It is employed locally to a limited extent for general carpentry.

Heartwood pale to rather dark brown, with a pinkish or olive tinge; sometimes with a satiny luster; scarcely distinct from the sapwood. Tasteless, but some specimens have a mild scent. Variable in density from rather light and soft to moderately hard and heavy; sp. gr. (air-dry) 0.40 to 0.70; weight 25 to 44 lbs. per cu. ft.; texture medium to rather coarse; grain often distinctly roey; easy to cut, but the fibers are likely to pull out in surfacing; takes nails without splitting; holds its place well when manufactured; does not appear very resistant to decay.

COMMON NAMES: Yawarridana (Br. G.); alaoelama, ararama, araurama, bintoela, djakidja, jawaledan, jawaredan, juwaredan, koereroe-janaledan, tamoené araurana, tapirin-aurarama, redji-djedoe, roode djedoe, witte djedoe (Sur.); ucsha-quiro (Peru); carvão de ferreiro, passariúwa, passuaré, ritaugueira, tachy, t. da terra firme, t. brancho, t. b. da terra firme, tachyrana, tachyseiro branco, taxiseiro (Braz.).

Senegalia is a large pantropical genus with more than 75 tropical American species of shrubs, woody vines, and small to medium-sized trees of little or no commercial importance. The leaves are bipinnate; the twigs are prickly or unarmed; the white or yellow flowers are in heads or spikes, rarely racemes; the pods are flat, mostly thin, and dehiscent. The timber of the larger trees is used locally for construction and miscellaneous rural purposes. The available wood samples identified with this genus exhibit a wide range of variation. The following description is based on the specimens that seem most reliable. (*Senegalia angustifolia* is treated separately.)

Heartwood dark brown or olive-brown, uniform or streaked; sapwood yellowish white. Luster medium to rather high. Odor and taste absent or not distinctive. Hard and heavy to moderately so; of medium to rather coarse texture; often cross-grained; not difficult to work; appears durable.

The wood of *Senegalia Greggii* (A. Gray) Britt. & Rose, a tree sometimes 30 feet high in southwestern United States and northern Mexico, is ring-porous and the heartwood is a rich, but not highly lustrous, dark brown, with a faint purplish tinge, the sapwood thin and yellowish; it resembles *Acaciopsis*. An Argentine variety of *Senegalia riparia* (H.B.K.) Britt. & Rose (Yale 14980) is olive-brown, with very fine markings of yellow and red and occasional prominent stripes of dark brown.

The description below applies to a sample (Yale 2800; Curran and Haman 427) collected in Curaçao, under the name of Chawari, with herbarium material determined at Gray Herbarium as *Senegalia tamarindifolia* (L.) Britt. & Rose. The wood is very distinct from all other specimens of *Senegalia* examined. It resembles a Venezuelan specimen (Yale 8627) labeled Taguapire, except that the latter is diffuse-porous. The wood appears well suited for inlay, small cabinet work, and articles of turnery.

Heartwood rich lustrous purplish brown with narrow purple and yellow streaks; sharply demarcated from the very thin yellowish white sapwood. Very hard and heavy; fine-textured; fairly straight-grained; not difficult to work, taking a very high natural polish.

Common names: Cat's claw, uña de gato (U.S.A.); tocino (Cuba); acacia nudosa, amarat, catch-and-keep, spineless acacia, tamarindo cimarrón, zarza (P.R.); désota, espino, gatuño, g. blanco, guajillo, guayolote, huajillo, malicia, matorral, mimbre, palo blanco, p. de arco, p. liso, panelo, rabo de iguana, r. de lagarto, tepeguaje, tlahuitol, tocino, toldillo, uña de gato, u. de g. negra, yax-catzim, zarza (Mex.); bastard prickly yellow, chuckem, tamarind—black, white, wild (Br. H.); lagarto, orotaguaje, sanpadrano (Guat.); bisquite, espino blanco, lagarto, sanpadrano (Hond.); cagalero, guajillo, llora-sangre, malacaro, palhuishle, quebracho, zarza (Salv.); baranó, baranoa, guarango, mulato, panelo, rasga, rasga-rasga, sarsa, tabonuco, toldillo, zarza redonda (Col.); bejuco de tiamo, chaguare, hoje menuda, rabo de iguana, ringuerín, tiamo, t. bejuco, t. flecha, t. güire (Venez.); parika (Br. G.); acacia, gujuba, paricarana, unho de gato (Braz.); pashaco (Peru); garabato, g. manso, ñapindá, n. guazú, n. hú, n. negro, yuquerí, y. guazú (Arg.).

Senegalia angustifolia (Lam.) Britt. & Rose (= *Acacia scleroxyla* Tussac) is a medium-sized to large unarmed tree, sometimes with a trunk 40 feet long and 24 inches in diameter, of rather common occurrence in the open forests in the dry coastal belt of the island of Haiti. It has doubly pinnate leaves with numerous opposite pinnae and many small crowded linear leaflets; the flowers are borne in slender spikes; the flat leathery pods are 4 to 5 inches long and less than three-fourths of an inch wide. The tree is commonly known in Dominican Republic as Córbano or more often as Candelón, a name often given in literature to *Colubrina ferruginosa* Brongm. The timber is utilized locally for fence and house posts, picket fencing, railway crossties, piling, and billiard cues, and is considered suitable for furniture.

Heartwood of a clear plum color or light brownish red; sharply demarcated from the yellowish sapwood. Luster medium. Odorless and tasteless. Very hard, heavy, and strong; texture rather fine; grain irregular to finely roey; takes a high polish; splits easily; is very resistant to decay.

Sophora, after segregation of a group of its species into the genus *Edwardsia,* comprises 50 or more species of unarmed shrubs, small to medium-sized trees, and some perennial herbs, occurring mostly in the mild temperate regions of North America and Asia. One species, *S. tomentosa* L., occurs along tropical seacoasts of both hemispheres. The leaves are imparipinnate, with numerous small or few medium-sized leaflets; the white or purplish flowers are borne in terminal or axillary racemes; the woody or fleshy pods are moniliform and indehiscent. One of the best known species is *S. japonica* L., a large tree of eastern Asia and often planted elsewhere for ornamental purposes. Its brown, ring-porous wood resembles Chestnut (*Castanea*) and is used for the same purposes, but the principal product of the tree is a yellow dye from the buds and bark.

Of the six or seven North American species, at least two are arborescent. *Sophora affinis* T. & G. is a round-topped tree less than 25 feet high and 10 inches in diameter, of limited distribution in the United States west of the Mississippi River in Arkansas, Oklahoma, Louisiana, and Texas. Its white flowers are in axillary racemes; the legume is fleshy; the leaves are thin and deciduous; the wood is definitely ring-porous, suggesting *Gleditsia triacanthos* L. *S. secundiflora* (Ort.) Lag. is a shrub or a low-branched tree sometimes 35 feet tall and a foot in diameter, frequently forming thickets in lowlands in western Texas, southern New Mexico, and northern Mexico. Its flowers are violet-blue, in terminal racemes; the pod is woody; the leaves are leathery and persistent; the wood is diffuse-porous. The seeds contain a bitter poisonous alkaloid, sophorin, with strong narcotic properties. The woods of these two American species exhibit many differences.

Sophora affinis: Heartwood brown or reddish-brown; merging gradually into the yellowish sapwood. Luster rather low. Odorless and tasteless. Very hard, heavy, tough, and strong; of uneven texture; grain fairly straight. Of no special utility and without commercial possibilities.

Sophora secundiflora: Heartwood orange-brown, more or less finely streaked with reddish brown; has a waxy appearance; sharply demarcated from the nearly white sapwood. Luster low. Dry specimens odorless and tasteless. Very hard, heavy, compact and strong; sp. gr. (air-dry) about 1.05; weight about 66 lbs. per cu. ft.; texture fine and uniform; grain fairly straight; not difficult to work, finishing very smoothly, with a high natural polish; appears durable. The heartwood is said to be the source of a yellow dye; no other special uses known.

COMMON NAMES: *Sophora affinis:* Beaded locust, coral bean, pink locust (U.S.A.). *S. secundiflora:* Coral bean, frijolito (U.S.A.); colorín, frijolillo (Mex.).

Stahlia *monosperma* (Tul.) Urban, the only species, is an unarmed tree of limited occurrence along or near the borders of Mangrove swamps in Puerto Rico, where it is known as Cóbana Negra, and in Dominican Republic, where it is called Caobanilla. It attains a maximum height of 60 feet, with a bole sometimes three feet in diameter. The leaves are pinnate, with 3 to 6 pairs of large pointed leaflets which bear numerous small black glands on the under side; the flowers are cream-colored with a pinkish tinge and are borne in rather large racemose clusters; the legumes are ovoid leathery pods about an inch long which do not dehiscence until the single seed begins to germinate.

Prior to 1934 the range of the tree was supposed to be limited to a few colonies near the coasts of Puerto Rico and the neighboring island dependency Vieques, but its presence in Dominican Republic was determined from herbarium specimens collected by James C. Scarff near San Pedro de Macorís. It has been described as a variety, *Stahlia monosperma,* var. *domingensis* Standl. According to Mr. Scarff (*Tropical Woods* 40: 16), it is "confined to ridges 6 to 30 feet high in or near coastal swamps and marshy deltas. The tree has a short bole, rarely over 10 or 12 feet, and it is covered with sprouts which the inhabitants of the region chop down when about wrist size to use for rafters and other building purposes. The cuttings can be repeated at intervals without any seeming loss of vigor in the parent tree. The heartwood is very hard and heavy (weight about 80 pounds per cubic foot, unseasoned), but is not very difficult to work and takes a lustrous finish resembling Mahogany (*Swietenia mahagoni*). It is noted for its resistance to decay and, according to José Schiffino (*Riqueza forestal Dominicana,* Santa Domingo, 1927, p. 49), many of the larger trees in the Provinces of Seybo and San Pedro de Macorís have been cut for railway crossties and exported. An objection to the use of Caobanilla for lumber is that the wood cracks when sawed, even after the logs have been seasoning for several years."

Heartwood rather dull brown or brick-red with intermingling of shades when fresh, later becoming chocolate or purplish brown and, upon long exposure, nearly black; sharply demarcated from the pinkish white sapwood. Without distinctive odor; taste slightly astringent. Exceedingly hard, heavy, tough, and strong; texture medium; feel harsh; grain irregular, mostly finely roey; not easy to work when dry, but can be finished smoothly; is highly durable.

Stryphnodendron is a tropical American genus with about 14 species of unarmed trees in Brazil and the Guianas, one in Costa Rica. The last, *S. excelsum* Harms, is said to attain large size along the Río Hondo and in the region of El General; no wood specimens of it are available for this study.

The only species of present economic worth is the Barbatimão, *Stryphnodendron barbatimao* Mart., typically a shrubby thin-foliaged little tree, occasionally with a stem 15 feet long and 16 to 24 inches in diameter, of common occurrence in the dry regions of central and eastern Brazil, extending southward to São Paulo. Its leaves are bipinnate, with numerous pinnae and small rounded leaflets; the flowers are white and spicate; the legumes are elongate, woody, indehiscent or nearly so, chambered within, and contain several red seeds with thread-like attachments. The bark, called "casca de virgindade," contains up to 40 per cent of tannin and is an important commercial source of tanning material for local consumption; the extract is also used medicinally. The hard and heavy timber, which is characteristically of twisted and gnarly growth, is used in the round for posts and to a limited extent for making furniture and small articles of turnery.

Stryphnodendron paniculatum Poepp. & Endl. is a large tree growing on non-inundated land in the central Amazon region. The bipinnate leaves have large leaflets and the small purplish flowers are in axillary spikes. The wood (Yale 33815; Ducke 288) resembles that of *S. barbatimao* (Yale 35273, from Aranha Pereira), but is of finer texture, straighter grain, and less dense. Various publications state that the specific gravity of Barbatimão wood ranges from 1.04 to 1.30, but this seems excessive; the sp. gr. (air-dry) of the specimen in the Yale collections is 0.82. The following description applies particularly to *S. paniculatum.*

Heartwood salmon-colored; conspicuously marked with closely spaced dark-colored vessel lines; transition to the grayish sapwood gradual. Luster rather high. Without distinctive odor or taste. Moderately hard and heavy; coarse-textured; fairly straight-grained; rather easily worked, finishing smoothly; is probably durable. A good, general-purpose wood, but not especially attractive.

COMMON NAMES: Boschtamalen (Sur.); barba de timan, barbatimão, b. verdadeiro, charãosinho roxo, mocidade, paricá-rana, talapirinria, timbaúba, timbó da matta, t. rana, uábitemo, ybátimo (Braz.); pashaco (Peru).

Swartzia (or *Tounatia*), a heterogeneous and poorly defined genus with more than 90 species of unarmed shrubs and small, medium-sized, or rarely large trees, is sparingly represented in tropical Africa and abundantly in northern South America, especially the Guianas and Amazon region, with extensions to the West Indies and through Central America to southern

Mexico. The leaves are imparipinnate, sometimes with winged petioles, often with large leaflets which are few in number and in a few species reduced to one; generally the flowers are yellow, have only one petal, and are borne in small racemes on older leafless branches, but there are numerous exceptions; the legumes are variable, leathery, swollen, indehiscent to tardily dehiscent, with one to few kidney-shaped seeds which are partly covered by a large aril. Some of the trees have deeply fluted, sulcate, or otherwise poorly formed trunks. The dense timber is of very little or no commercial importance, but the thick sapwood is used in some localities for implement frames and spokes of wheels, the heartwood for posts, heavy and durable construction, and articles of turnery. The Yale collections include more than 20 specifically determined kinds of *Swartzia,* and while all have the same general type of structure, there are differences in color, texture, the regularity and size of the parenchyma bands, and other features.

Heartwood dark brown to nearly black, sometimes reddish brown, in solid color or somewhat variegated; sharply demarcated from the yellowish sapwood. Luster medium to low. Without distinctive odor or taste. Hard and heavy to exceedingly so; sp. gr. (air-dry) 0.95 to 1.30; weight 60 to 81 lbs. per cu. ft.; texture medium to very fine; grain straight to irregular; mostly difficult to work, but finishing very smoothly; heartwood, which may be slow in forming, highly resistant to decay.

COMMON NAMES: Pico de gallo (Cuba); bois pois, wild orange (Trin.); bolsa de Judas, catalox, copa de oro, floripondio de monte, gorro de Napoleón, naranjillo, naranjito, tecomaxochitl, tetona (Mex.); naranjillo, paterno (Hond.); cornudo, cutaro, cutarro, naranjillo (Pan.); alma negra, mulato (Col.); bania, baracarra, bastard yaruru, black paddlewood, clubwood, ebony, brown ebony, ironwood, kerunite, larakusana, parakusan, parakusani, serebedan, wanebala, womara (Br. G.); akwan siba, aliana-oeu, anakoko, apoetoe, blakka parihoedoe, boegoe-boegoe, boucara, gandoe, iesrihart, iesriharti, ietikiboralli, ietjoetawoe, ijzerhart, isrihati, itikie boeroeballi, itjoeroe-tanomoetoe-sie, jaroroballi, kakaboekoe, kapoekoeroe-japoekoetjare, karoekoro, kharemeroe-jaroro, matosirian, moetoesi, moetoesiran, oelawabeta, oronapeta, sepietoena, tjabisi, tjabisihoedoe, toepoera apoekoetja, wajewoe, wepetano tamoené, yzerhart, zwarte parelhout (Sur.); anacoco, boco, bois à flèches, b. boco, b. de coco, b. de fer, b. de pagaie blanc, b. de perdix, b. pagaies, ferréol, heistère rouge, montouchy, panacoco, p. noir, p. rouge, perdix, siki-siki-danni, tounou, yakelele (Fr. G.); cascol (Ec.); icoje, nahua, nina-caspi, shatona blanca, yahua (Peru); orori (Bol.); anacoco wanebala, arapary da terra firme, coquido, coração de negro, cumbeira, fereol, goncalare, hucuya, jacarandá, j. branco, j. do campo coberto, j. preto, jarana vermelha, moçatahyba, m. branca, mocitahyba, muirá cutáca, m. gibóia, m. pixuna, muracutáca, mututy, m. da terra firme, pacapeuá, pacora de macaco, panacoco, p. rouge, paracutáca, patapeuá, pau ferro, p. de remo, p. de Santa Maria, p. de sangue, p. preto, p. p. da terra firme, p. p. da varzea, p. santo, pitaíca, p. da terra firme, p. da varzea, p. do campo, potajuca, pracuúba, p. cheirosa, saboarana, saboeirana, saborana (Braz.).

Sweetia, credited with about a dozen species of small to large unarmed trees, is distributed from southern Mexico to northern Argentina and southern Brazil. The leaves are odd-pinnate, with 5 to 13 medium-sized glossy leaflets; the small whitish flowers are borne in axillary panicles, generally obscured by the foliage; the thin, flat, elongated pods contain 1 to 3 flat seeds and are indehiscent.

The southernmost species, *Sweetia elegans* Benth., is said to be a medium-sized tree in São Paulo, where it is known as Perobinha, and in Argentina, where the local names are Lapachillo and Lapacho do Campo. No wood samples are available for this study and published descriptions are meager of details. Recorded uses are carpentry and joinery, fence posts, fuel, and charcoal. The best known species in South America is *S. nitens* (Vog.) Benth., a small to medium-sized tree of the central Amazon region, extending northward into the Guianas and Venezuela, mostly along streams and the margins of lakes. In Brazil it is known as Itaúba-rana, on account of some resemblance of the wood to Itaúba (*Silvia itauba*), and in Surinam as Watergroenhart, because it suggests Groenhart (*Tabebuia*). The timber is of good quality, strong, durable, and finishing attractively, but its scarcity and small size render it of minor economic value.

The only species on the North American continent is *Sweetia panamensis* Benth., a medium-sized to large tree, sometimes 125 feet tall, with a rather poorly formed trunk 30, rarely up to 45, inches in diameter and free of limbs for 40 to 65 feet. It occurs from southern Mexico throughout Central America to northern Colombia and northwestern Venezuela. The bitter inner bark, known as "cascara amarga," contains an alkaloid, and a fluid extract has been used in the treatment of syphilis and scrofula since about 1880 (see *Tropical Woods* 14: 39). The timber is noted for its strength and durability and is employed locally for heavy construction, railway crossties, sills, spokes of log wagons, and implement frames. It is not likely to become important in the export trade. The woods of *S. panamensis* and *S. nitens* are similar.

Heartwood brown or olive, more or less streaked, the color deepening upon exposure; looks somewhat waxy; usually rather sharply demarcated from the yellowish sapwood. Luster medium to fairly high. Without distinctive odor; taste of sapwood rather bitter. Density typically high; sp. gr. (air-dry) 0.90 to 1.10; weight 56 to 69 lbs. per cu. ft.; texture medium-fine; feel harsh; grain usually roey; not very difficult to work, finishing smoothly with a lustrous polish; durability high.

COMMON NAMES: *Sweetia nitens:* Kamarakatta (Br. G.); hoerowassa, sierietjo, sonbogienhatti, wadili-hoeroewassa, wassiba-omera-kodikoro, watragrien, watergroenhart (Sur.); arapichuna, darura, itaúbarana, piranheira (Braz.). *S. panamensis:* Guayacán, huesillo, huesito (Mex.); Billy Webb (Br. H.); chichipate (Guat., Hond., Salv.); coyote, visapolollo (Hond.); carboncillo, guayacán, g. corriente (C.R.); malvecino (Pan.); rejo (Col.); vera de agua (Venez.). *Other species:* Peroba meuda, p. mirim, perobinha (Braz.); lapachillo, lapacho del campo (Arg.).

Tachigalia, with 15 species of small to large unarmed trees, has its center of distribution in the Amazon region of Brazil, with extensions into eastern Peru, the Guianas, and Venezuela. The leaves are pinnate, with leathery leaflets; the small flowers are in simple axillary or terminal panicled racemes; the legumes are membranous and indehiscent. In several of the species the twigs, petioles of the leaves, and rachises of the inflorescences are hollow and inhabited by ants, known in Brazil as "tachy." The bark of *T. myrmecophila* Ducke is used locally to some extent for tanning; the wood is said to have a very fetid odor when fresh. There are no special uses for the timber of any species.

Heartwood brown; not sharply defined (in available specimens) from the thick light olive sapwood. Luster medium to high. Dry material without distinctive odor or taste. Hard, heavy, tough, and strong; texture medium; grain fairly straight; rather easily worked, finishing smoothly with a high polish; dark heartwood said to be durable. Not likely to be of any commercial value.

COMMON NAMES: Caracha-caspi (Peru); tachí, tachizeiro, t. branco, tachy, t. branco, t. b. matta, t. do igapó, t. preto da matta (Braz.).

Tara *spinosa* (Mol.) Britt. & Rose, the only species, is a prickly shrub or a small tree 20 to 30 feet high indigenous to Colombia, Venezuela, Ecuador, Peru, and Bolivia, and cultivated there and also in the West Indies and Chile. The leaves are evenly bipinnate, with 2 to 4 pairs of pinnae and 4 to 7 pairs of large leaflets; the yellow flowers are borne in elongated racemes; the legumes are compressed, leathery, several-seeded, and indehiscent. The principal value of the species is in the pods, which are a regular article of trade in Lima, Peru, and are used for dyeing, tannage, and making ink. The plants are also grown as hedge fences. The durable wood is of good quality, but is little used.

Heartwood bright orange, deepening upon exposure; sapwood not seen. Luster fairly high. Odorless and tasteless. Hard, heavy, tough, and strong; texture fine; grain mostly irregular; not difficult to work, finishing very smoothly with a high polish; is highly resistant to decay. Of no commercial possibilities because of its scarcity and small size; looks like a dyewood.

COMMON NAMES: Brasil, divi-divi, guarango (Col.); divi-dive de los Andes (Venez.); guarango, huarango (Ec.); espino, guarango espino, tara, taya, tura, yara (Peru); tara (Boliv., Chile).

Taralea (= *Dipteryx* in part), with five species of unarmed shrubs and medium-sized to rather large trees, occurs in the Amazon basin, including eastern Peru and the hinterlands of Venezuela and the Guianas. The opposite or alternate leaves are evenly or unevenly pinnate, with 4 to 6 large leathery leaflets; the red flowers are borne in conspicuous terminal panicles; the pods are flat, rounded, 2-valved, elastically dehiscent, and contain a single seed which is rich in oil, but odorless. The seeds of *T. oppositifolia* Aubl., known as Cumarú-rana (false Cumarú), are used in Pará, Brazil, as the source of an unscented oil for industrial purposes; statements that they supply cumarin, obtained from the seeds of the Tonka-bean tree (*Coumarouna*), are erroneous (see *Notizbl. Bot. Gart. Berlin-Dahlem* 14: 121; 127). The timber is suitable for heavy durable construction, but apparently is not utilized for any purpose.

Heartwood light reddish brown, with fine parenchyma striping; looks somewhat oily or waxy; not sharply demarcated from the yellowish sapwood. Luster rather low. Without distinctive taste, but with mild slightly unpleasant scent. Extremely hard, heavy, compact, tough, and strong; sp. gr. (air-dry) 1.20; weight 75 lbs. per cu. ft.; texture rather fine; feel harsh; grain roey; difficult to work, but finishing very smoothly; durability probably high. An unattractive wood not likely to become of any commercial importance.

COMMON NAMES: Tarala (Fr. G.); cumarú-rana (Braz.).

Tauroceras, with two species of shrubs or slender trees sometimes 25 feet tall, is limited in its natural distribution to Mexico and Central America. Standley says (*Trees and shrubs of Mexico*, p. 374): "The seeds are imbedded in abundant pulp. In this, as in [the species of *Myrmecodendron*], the Bullhorn Acacias, the large spines are usually inhabited by ferocious ants, which enter the spines by a puncture near the apex. These ants subsist upon nectar of the large pale glands which are borne as appendages on the tips of the leaflets." No wood samples are available for study.

COMMON NAMES: Árbol del cuerno, cuernecillo, tarrito (Cuba, int.); árbol del cuerno, cornezuelo, cuernitos, cuernos del toro, espino blanco, huitzmaamaxalli, iscanal blanco, zubín, zubinché (Mex.); cornezuelo, unka-guá (C.R.).

Tipuana (in a restricted sense) is a monotypic genus, the single species being *T. tipu* (Benth.) Hubbard & Rehder, a medium-sized to large unarmed deciduous South American tree sometimes 100 feet in height, with a long clear bole occasionally five feet in diameter above the buttressed base. The leaves are long and odd-pinnate, with 11 to 25 alternate or subopposite oblong leaflets; the bright yellow flowers are borne in terminal panicles at the time of leafing; the abundant samara-like fruits remain on the tree for nearly a year. The natural range of the tree, commonly known as Tipa, is limited to southern Bolivia and the Argentine provinces of Tucumán, Salta, and Jujuhy, but it is planted for shade and ornament along streets and in parks in many cities in southern South America, United States, France, and in Algeria. Tipa is not very abundant, but the lumber is highly appreciated in Argentina for furniture and cabinet work and in the construction of vehicles. The leaves are employed in veterinary medicine and the bark serves locally for tanning and dyeing.

Heartwood pale yellow to brownish, sometimes with a roseate hue; finely striped with vessel lines which may be dark-colored; not clearly demarcated from the yellowish white sapwood. Luster medium to high. Odorless and tasteless. Moderately hard, heavy, and strong; sp. gr. (air-dry) 0.65 to 0.75; weight 40 to 47 lbs. per cu. ft.; texture medium; grain usually finely roey; fairly readily worked, finishing very smoothly and taking a high polish; probably poorly resistant to decay. A good timber suitable for many of the same purposes as Maple (*Acer*) and bearing a superficial appearance to Primavera (*Cybistax Donnell-Smithii*).

COMMON NAMES: Tipa, t. blanca (Arg.); tipa, tipú, tipuana (Braz.).

Uleanthus *erythrinoides* Harms, the only species, is a small tree of fairly common occurrence in the upland high forest above the cataracts on the Tapajoz and Madeira Rivers, tributaries of the Amazon. In 1941 it was discovered on the Rio Urubú, northeast of Manáos, by Adolpho Ducke who collected floriferous herbarium material and a wood sample (Yale 40315) upon which the description below is based. The beautiful flowers, with red calyx and blue or rose-purple corolla, are borne in few-flowered racemes; the pods, which are eight to ten inches long and less than an inch wide, contain three to five seeds and are elastically dehiscent, the valves coiling upon opening.

Heartwood dark brown, marked with very fine yellow vessel lines; sharply demarcated from the thin yellowish sapwood. Luster rather low. Scentless and tasteless. Extremely heavy, hard, and strong; sp. gr. (air-dry) 1.02; weight 75 lbs. per cu. ft.; texture fine; grain straight; easily split, otherwise difficult to work, but finishing very smoothly. Durability presumably high. Probably without commercial possibilities.

Vachellia, with six species of shrubs and small trees, is very widely distributed in tropical and subtropical America. Best known is *V. Farnesiana* (L.) Wight & Arn., more commonly considered a species of *Acacia*. It is usually a shrub, but sometimes a bushy tree up to 30 feet high, growing naturally or in cultivation throughout the warmer regions of the world. It has bipinnate leaves, with numerous very small leaflets; the twigs are armed with pairs of stiff light-colored spines; the small, sweet-scented, bright yellow flowers are in globose heads; the dark brown woody pods contain two rows of seeds surrounded by pulp. Standley says of it (*Trees and shrubs of Mexico*, p. 379): "In many parts of Mexico the plant is found chiefly about dwellings and seems to be naturalized, but in other regions it appears to be native. The wood is used for many purposes. The bark and fruit contain tannin and are used for tanning and dyeing, and the fruit is often used for making ink. The vicious juice of the pods is employed in some places for mending broken china. The gum which exudes from the trunk is employed locally in making mucilage; it is very similar to gum arabic. The leaves are of value as forage for stock, especially in winter. In southern Europe the plant is cultivated extensively for the flowers (known in commerce as cassie flowers), from which perfume is manufactured. As much as 100,000 pounds of them are harvested annually about Grasse, France. In tropical America the flowers are often laid among linen to impart their perfume to it. An ointment made from the flowers is used in Mexico as a remedy for headache, and their infusion for dyspepsia. The green fruit is very astringent, and a decoction is employed for dysentery, inflammation of the skin and mucous membrane, etc. Seler reports even that in San Luis Potosí a decoction of the roots is employed as a supposed remedy for tuberculosis. The pulverized dried leaves are sometimes applied as a dressing to wounds. This is probably the plant reported from Baja California by Clavigero as 'Huisache.' The pods, he states, were used there for making ink. Cattle, he says, are fond of the tender branches, but these impart a bad flavor to their flesh."

The following description of the wood of *Vachellia* is based on nine specimens of three species, namely *V. astringens* Speg., *V. Farnesiana*, and *V. lutea* Speg. Heartwood rather dull purplish brown, with light-colored parenchyma markings and fine dark-colored vessel lines; sapwood thin and white. Without distinctive odor or taste. Hard and heavy; rather coarse-textured; of irregular grain; rather easily worked, finishing smoothly; is highly durable. Though of no commercial importance because of the small size of the trees, the wood is suitable for small articles of turnery.

COMMON NAMES: Aroma, aromo (Trop. Am., general); cashia, opopanax (Bah.); cassie flower (Jam.); aroma amarilla, cují, tamarindillo (Cuba); casha (P.R.); bihi, binorama, espino, finisachi, gabia, gavia, guisache, g. yondiro, huisache, h. de la semilla, huixachin, huizache, matitas, quisache, subín, ũisatsín, vinorama, xkantiriz, zubin, zubin-ché (Mex.); espino blanco, subín (Guat.); cachito de aromo (Hond.); espino blanco, e. ruco (Salv.); cachito de aroma (Nic.); aroma-ierón (C.R.); cují cimarrón, pelá, uña de cabra (Col.); cují, c. aroma, paují (Venez.); coronha, esponjeira (Braz.); huaranga (Peru); aromita, cavén, churqui blanco, c. negro, espinillo, e. del bañado, subín, tusca (Arg.).

Vatairea, with eight species of large unarmed trees, sometimes 125 feet tall and six feet in diameter, occurs from Rio de Janeiro, Brazil, to Venezuela, and along the Atlantic region of Central America to southern Mexico. The genus is very closely related to *Vataireopsis* and belongs to the section *Aristobulia* of *Andira;* some of the species have been referred incorrectly by certain authors to *Pterocarpus* and *Tipuana.* The leaves, which are congested at the ends of the branches, are odd-pinnate, with 5 to 17 mostly large and leathery leaflets; the purplish flowers are borne in large terminal panicles; the fruits are one-seeded and provided with a long terminal wing (suggesting *Machaerium*), except in *V. guianensis* Aubl. (= *Andira amazonum* Mart.), which has a large woody pod and only a rudimentary wing.

The woods described in *Timbers of tropical America* (pp. 301–303) under the heading "The bitter Angelims" are of two genera, *Vatairea* and *Vataireopsis.* From such information as is now available, the specimens with the finer and more numerous rays belong to the first genus, those with fewer and coarser rays to the second. They are much alike in general appearance and properties and in having a lasting bitter taste. The timber is used for durable construction as it is considered highly resistant to decay and insect attacks. Being rather brittle it is better suited for posts and columns than for beams. The bitter principle of the wood and bark is employed medicinally.

Heartwood yellow, becoming orange-brown upon exposure; striped with parenchyma; sometimes with oily appearance; distinct but not sharply demarcated from the whitish or gray sapwood. Luster fairly high in some specimens to low in others. Without distinctive odor; taste very bitter. Moderately hard and heavy; sp. gr. 0.65 to 0.80; weight 41 to 50 lbs. per cu. ft.; texture very coarse; grain variable; easy to work, finishing smoothly; holds its place well when manufactured. It is an unattractive wood, probably with no commercial future.

COMMON NAMES: Picho (Mex.); bitterwood (Br. H.); frijolillo (Guat.); amargoso, guacamayo (Hond.); amargo (Pan.); arasaru, arisauru, arisoro, arisouroo, ourisoura, yaksaru, yaksauru (Br. G.); arisoeroe, arisower (Sur.); amargoso, angelim amargoso, araçuy, fava de bolacha, faveira, f. de empigem, sucupira amarella (Braz.).

Vataireopsis is a genus described by Adolpho Ducke in 1933; at present it includes two species of medium-sized to large unarmed Brazilian trees. The leaves, which are clustered at the ends of the branches, are very large and imparipinnate, with 30 to 50 oblong leathery leaflets; the beautiful violet or blue-violet flowers are borne in abundance in large panicles when the trees are leafless or nearly so; the one-seeded samaroid fruit is similar to *Vatairea,* to which the genus is closely related.

Vataireopsis speciosa Ducke is a medium-sized tree with a non-buttressed trunk covered with a scaly bark; it occurs in upland forests in the central Amazon region. *V. araroba* (Aguiar) Ducke grows in high forest from Bahia to Rio de Janeiro, where it attains a height of from 65 to 100 feet. It is the source of Araroba powder, known locally as "pó de Araroba," "pó da Bahia," or "pó de Goa," which is obtained from the bark and used medicinally. The wood also contains the same or similar material, sometimes in sufficient quantity to fill the vessels and to cover the surface of sawed boards with fine yellow dust. A specimen sent by the senior author to Samuel C. Hooker, who was examining yellow deposits in wood in search of lapachol, was reported upon in a personal letter in 1921 as follows: "Angelim Amargoso yields on extraction with benzine a beautiful compound, apparently in a condition of great purity, evidently closely related to chrysarobin (chrysophan-anthranol) but of higher melting point and different minor characteristics. A trace dissolved in a drop or two of 1 per cent solution of caustic soda in the cold and exposed on a watch glass to the air gradually darkens the solution, becoming crimson-colored, probably due to the formation of chrysophanic acid or a closely allied derivative." The timber is used locally for general construction and carpentry. The dust liberated in sawing inflames the eyes of the workmen.

The wood is similar in appearance and structure to *Vatairea*, but generally is coarser-textured, has fewer and higher rays (1 to 5 cells wide and up to 50 cells high), and ripple marks are absent or only local. The Yale collections contain several specimens believed to be of this genus and representing a range from the upper Amazon to Bahia, but the only one accompanied by herbarium material is of *Vataireopsis speciosa* Ducke collected by Dr. Ducke near Manáos (Yale No. 20684).

COMMON NAMES: Angelim amarello, a. amargoso, a. araroba, araroba, faveira, moina (Braz.).

Vouacapoua, with three species of medium-sized to tall and slender unarmed trees, is limited in distribution to non-inundated lands of the eastern and northern parts of the Amazon basin. The leaves are imparipinnate, with five to nine large leathery leaflets; the rather small scented yellow flowers are borne in erect terminal panicled racemes; the legumes are dehiscent, thus distinguishing this genus from *Andira* with which it has been confused.

Vouacapoua americana Aubl. is a tall tree, but the unbuttressed trunk is generally less than 24, rarely up to 36, inches in diameter and from 50 to 75 feet long; the crown is rather small and composed of ascending branches. It occurs scatteringly in the forests in parts of Surinam and French Guiana, but attains its best development in the State of Pará, Brazil, where it is known as Acapú. The handsome dark-colored timber is highly esteemed in Brazil for all kinds of heavy durable construction and for floors, especially in association with some light-colored wood such as Pau Amarello (*Euxylophora*, fam. Rutaceae). Much of the readily accessible timber has been cut, but there are still considerable supplies available, though the range is limited and apparently does not extend westward into the State of Amazonas. According to Bertin (*Les bois de la Guyane française*, pp. 140–142), the species produces the best timber in French Guiana, but is of infrequent occurrence. The local uses are furniture, carpentry, wheelwright work, and posts; suggested uses include cabinet-making and all kinds of civil and naval construction. Pfeiffer (*De houtsoorten van Suriname*, p. 272) says that the wood, known as Bruinhart in Surinam, is one of the finest growing in that Colony. Its attractive color and grain combine with hardness and good working qualities to produce material of general utility, but because of its scarcity and high price it is best adapted for cabinet work, interior trim, and parquet flooring.

A second species of *Vouacapoua* was described in 1932 and named *V. pallidior* Ducke (see *Tropical Woods* 31: 16). It is a medium-sized tree which so far has been found only in the forests near Manáos and the vicinity of San Gabriel in the upper Rio Negro region. Both species have the same vernacular name, Acapú, and the woods are similar in structure. They are distinct in color, however, that of *V. pallidior* being dull chestnut-brown with an olive hue, with fine parenchyma markings. A third species, *V. macropetala* Sandwith, was described in 1937 (see *Kew Bulletin*, 1937, pp. 104–106). It is a tree 85 to 90 feet high and 16 inches in diameter growing in mixed forest on sandy clay soil along the upper Demerara River, British Guiana. This is "apparently the first record of this genus for the Colony." The Arawak name is Sarabebeballi. There are no wood samples of it available.

The following description is based primarily on *Vouacapoua americana,* but applies also to *V. pallidior* except as to color. Heartwood dark brown or reddish brown, deepening upon exposure; marked with fine but very distinct vessel lines of lighter color; sharply demarcated from the nearly white sapwood. Luster medium to rather low. Odor slightly unpleasant or absent; taste not distinctive. Very hard, heavy, tough, and strong; sp. gr. (air-dry) 0.87 to 1.00; weight 54 to 62 lbs. per cu. ft.; texture rather coarse, but uniform; feel harsh; grain fairly straight to irregular; not very difficult to work, but requires care in seasoning; splits readily; finishes smoothly and takes a good polish; is highly resistant to insect attack and decay. Is of the type of wood known in the U.S. trade as Partridge.

COMMON NAMES: Brownheart, partridge wood (Eng.); sarabebeballi (Br. G.); bowhanti, broinharti, bruinhart, dakamaballi, kabbi, tjanaren wakapoe, tjatjaboetja, wakabo, wakapo (Sur.); wacapou (Fr. G.); acapú (Braz.).

Zollernia includes about eight species of unarmed shrubs and trees, one in Central America, one in the lower Amazon region, the others in eastern Brazil. The leaves are simple (unifoliolate), leathery, in some species remotely toothed; the yellow or roseate flowers are borne in terminal panicled racemes; the legumes are thick, egg-shaped or nearly round, 2-valved, usually with a single large seed.

The Central American species, *Zollernia tango* Standl., was described in 1929. Its range is apparently limited to the Lancetilla Valley, Honduras, Río Dulce, Guatemala, and the Toledo district of British Honduras; its only known name is Tango. It grows in wet forests and has a long deeply fluted and irregular trunk. The wood is used to a limited extent for ax handles and cabinet and construction work. (See *Tropical Woods* 19: 6–7.)

Zollernia ilicifolia Vog. occurs in the coastal forests from Rio de Janeiro to São Paulo, where it is called Mocitahyba, a name also applied to *Swartzia crocea* Benth. It attains a height of 60 feet, with a trunk 30 to 40 feet long and 24 to 30 inches in diameter. The bark is a source of tannin, the fruits are medicinal, and the dense dark-colored durable wood is used for posts, poles, beams, railway crossties, and cabinet work.

Zollernia paraensis Huber is a medium-sized to tall tree, with a long straight trunk, growing on non-inundated land of the lower Amazon. The heartwood is well developed and some of it is said to be more or less figured as in Letterwood (*Piratinera*) and for that reason called Muirapinima Preta, although the usual name is Pau Santo. The timber is scarce and is highly prized locally for articles of turnery and small cabinet work. Small quantities have been exported to England as a substitute for Lignum-vitae (*Guaiacum*) and to the United States for handles of cutlery, butts of billiard cues, and brush backs.

Heartwood dark reddish brown (*Zollernia tango*), dark olive, or almost black, sometimes streaked or mottled, and with or without distinct parenchyma markings; has an oily or waxy appearance; sharply demarcated from the yellowish sapwood. Odor mildly fragrant; taste not distinctive. Extremely hard, heavy, and compact, but brittle; sp. gr. 1.10 to 1.35; weight 72 to 83 lbs. per cu. ft.; texture medium; feel waxy; grain irregular to definitely roey; working properties similar to Lignum-vitae; finishes smoothly, with a high natural polish; is very resistant to decay. Of only minor importance commercially because of its scarcity.

COMMON NAMES: Enemy, tango (Br. H.); tango (Hond.); coração de negro, maçatahyba, missutahyba, mocitahyba, m. parda, m. preta, m. vermelha, mossotahyba, muirapinima preta, pau santo (Braz.).

Zuccagnia *punctata,* the only species, is a shrub or a little tree sometimes 15 feet high and eight inches in diameter, widely distributed in central Argentina and certain parts of Chile. The short pinnate leaves have numerous very small, conspicuously punctate, leathery leaflets; the yellow flowers are borne in terminal racemes;

the bristly pods are short, ovate, flattened, and dehiscent. The species is common in brushlands and thickets and has no use except to a minor extent for fuel.

Heartwood orange-brown, with a waxy appearance; sharply demarcated from the thin yellowish sapwood. Luster medium. Without distinctive odor or taste. Very hard, heavy, tough, and strong; texture medium; grain irregular; not easy to work but finishing smoothly, with a high natural polish; is very durable.

COMMON NAMES: Jarilla de la puna, j. de pispito, j. pispa, j. pispita, j. mscho, j. pus-pus, pus-pus (Arg.).

Zygia, with about 20 species of unarmed shrubs and small trees, is widely distributed in tropical America. The species with the greatest range is *Z. latifolia* (L.) Fawcett & Rendle, a bushy tree usually less than 25 feet high, common along streams from southern Mexico to Bolivia and Brazil, and in the West Indies. It has bipinnate leaves, the leaflets large, usually 5, the upper ones opposite, the others alternate; the purplish flowers are borne in small heads arising from the older branches. The following description is based on sapwood specimens of seven species.

Normal heartwood not seen but probably brown, judging from wounds and knots; sapwood yellowish white. Luster medium to rather low, owing to abundance of parenchyma. Hard and heavy, tough and strong; texture medium; grain usually irregular; not very easy to work, sawed surfaces being splintery; fibers tend to pull out under tools. No special uses are recorded.

COMMON NAMES: Hoopwood, horsewood (Jam.); jasmín del río (Cuba); bois ca (Haiti); amesquite, hoja-vinsha, quiebrahacha, reviento-machete (Mex.); chilillo, choc-ché, turtle bone, t. b. macho (Br. H.); pepe nance (Guat.); paleto blanco (Hond.); cafecito (Nic.); kitá, kueh-krá, kuer, sota-caballo (C.R.); azota-caballo, sota-caballo, riverwood (Pan.); guamo macho, g. prieto, guayacán chaparro (Col.); guamo macho (Venez.); aliki, aliku, benda (Br. G.); ingá-rana, jarandeua (Braz.); avarembó-timbó, granadillo del río, guarápere (Arg.).

LEITNERIACEAE

Leitneria *floridana* Chapm., the only species in this family, is a shrub or small tree, occasionally up to 20 feet, with a slender stem four or five inches in diameter above the swollen gradually tapering base, of very limited occurrence in muddy swamps in Florida, Georgia, Texas, and southeastern Missouri. Its only common name is Corkwood. The simple, entire, alternate, deciduous leaves are four to six inches long; the flowers are in aments, the two sexes borne on different plants; the fruit is an oblong, compressed, dry, wrinkled drupe.

Wood pale yellow throughout. Luster rather high. Odorless and tasteless when dry, but with a distinctive odor when fresh. Very light, soft, and brittle; sp. gr. (air-dry) 0.22; weight about 14 lbs. per cu. ft.; texture rather fine; grain straight; perishable when exposed to decay. Of no commercial uses because of its small size and scarcity.

Wood imperfectly ring-porous. Pores in early wood in part small or occasionally medium-sized in association with very small ones and arranged in uniseriate arcs 2 to 6 pores long or in interrupted concentric rows; other pores very small to minute, rather few, and arranged in scattered radial, diagonal, or zig-zag rows or patches often originating at one end of the larger pore arcs and extending across the growth ring. Vessels with simple perforations; spiral thickenings present, at least in smaller members; pits small to medium, circular to oval and mostly opposite or, in larger vessels, elongated and in scalariform arrangement. Rays uniseriate and 10 to 12, sometimes to 25, cells high; cells rather large, thin-walled, short-upright or oblong; pits to vessels small and oval to elongated and in scalariform arrangement. Wood parenchyma rather sparingly paratracheal, surrounding pore groups and conforming to their outlines; also finely terminal. Wood fibers with very thin walls and very numerous extremely small pits. Ripple marks absent. Resin ducts absent from wood but present in the margin of the pith.

LILIACEAE

The Lily family contains many species of herbs, shrubs, and some small trees with simple or branched trunks. The plants are extensively cultivated for ornament, food, and textile fiber. The woody forms are common in dry regions and sometimes are without stems; the leaves are either linear or dagger-shaped, usually stiff and rigid, sometimes with spiny margins; the flowers are sometimes small, but often large and showy; the fruits are capsular or baccate.

The only genera to be considered here are *Dracaena* and *Yucca*. The young stems have a typically monocotyledonous structure. The comparatively few fibrovascular bundles have a few larger pores and considerable strengthening tissue; the parenchymatous ground mass is soft and succulent and is easily broken down so that dried stems usually have a core of separated fibers which is an inch or more in diameter. Unlike most other monocots, these plants eventually develop a continuous cambium sheath which gives rise to true secondary wood and bark. The secondary wood is more compact than the core but maintains the same general structure, that is, there are separate fibro-vascular strands with parenchymatous tissue around them. The principal difference is that the strands, as seen on cross section, are more or less elongated in a radial direction, the vascular elements are all thick-walled fiber-tracheids, and the interfascicular parenchyma is firmer, has its cells in radial arrangement, and resembles coarse anastomosing rays. (See De Bary's *Comparative Anatomy*, pp. 618–622.)

Dracaena is primarily an Old World genus, but one species, *D. americana* D. Sm., occurs in Central America. It is a small tree, 10 to 30 feet high and 5 to 10 inches in diameter, with a few coarse branches bearing many smooth linear leaves 8 to 12 inches long; the very small creamy white flowers are borne in large terminal panicles; the fruit is fleshy, yellowish green, about ¾-inch in diameter, and contains one or two large seeds. The tree is fairly common in the moist forest of the Atlantic coast, but is not utilized. The rather smooth gray bark is thin and fibrous.

Secondary wood whitish or pale brownish throughout. Luster rather low. Odorless and tasteless. Moderately light in weight, but firm; texture coarse; feel harsh and hairy; grain finely interwoven to fairly straight; not difficult to cut, but not easy to finish smoothly; is poorly resistant to decay. The wood is not used locally and is apparently without commercial possibilities.

Growth rings absent. Xylem bundles very numerous; cross section elliptical to decidedly elongated radially; slightly darker than parenchymatous tissue surrounding them (Plate XXXVII, 3). Parenchyma cells with thin to moderately thick walls; longest axis vertical; pits numerous, greatly variable in size. Fibrovascular strands composed of long-pointed tracheids with rather large cavities, thick walls, and large circular (or locally elongated) bordered pits having lenticular apertures; the phloem strand is centrally located, but often has uniseriate extensions to the parenchymatous ground mass. Ripple marks absent or local; parenchyma cells in distinct horizontal seriation on radial section. No gum ducts or radial channels seen.

Common names: Cerbatana (Br. H.); izote de montaña (Guat.); isote, izote (Hond.).

Yucca, with about 30 species and many varieties of shrubs and small trees, occurs in the West Indies, southern and southwestern United States, and nearly throughout Mexico, being particularly abundant in the more arid regions east of the western Sierra Madre. The leaves are from 5 to 50 inches long, linear-lanceolate, with a thickened clasping base; the creamy white waxy flowers are borne in large terminal panicles; the fruit is baccate and indehiscent or capsular and 3-valved. The plants are of some economic importance as a source of fiber obtained from the leaves for cordage, mats, and coarse cloth. The roots have saponifying properties and are used for washing purposes or as an ingredient of toilet soap, and the extract has been employed to produce foam in beverages. The flowers are a common article of food for the Mexicans, being eaten raw as a salad or cooked. The fleshy fruits are also edible. Thin sheets can be cut from the stems for making light, tough, surgical splints or for sheathing fruit trees; small boards are sold for use in making fire by friction; there is also a local trade in novelties.

One of the best known species in the southwestern United States is the Joshua Tree, *Yucca brevifolia* Engelm. (= *Y. arborescens* Trel.). It is at its best in the rocky mesas of the mountains of San Bernardino and Riverside Counties, south of the Mojave Desert, California (Plate XXVIII). It attains a height of 30 to 40 feet and a diameter of two to three feet. The numerous stout branches are clothed with slender sharp-pointed leaves 5 to 10 inches long. The thick laminated corky bark is deeply divided by irregular fissures. The stems have the general structure of *Dracaena,* but are more fibrous and have a larger core. The secondary cylinder appears laminated owing to periodic changes in the direction of the vascular strands. The surface of the dry wood usually shows numerous small radial channels (leaf traces).

COMMON NAMES: *Yucca brevifolia:* Joshua tree, J. yucca, the Joshua, praying tree, tree yucca, yucca cactus (U.S.A.). *Other species:* Spanish bayonet, S. dagger, soapweed, yucca (U.S.A.); amole (roots), isote, izote, palma, p. china, p. corriente, p. criollo, p. de datiles, p. loca, p. pita, p. samandoca, p. San José, palmilla, palmita, pitilla, soyate (Mex.); izote (Guat., Hond.); itabo (C.R.).

LINACEAE

THE Flax family, with about 19 genera and 300 species of trees, shrubs, and herbs, is very widely distributed; most of the trees are tropical. The only commercially important genus is *Linum* (with about 200 species), the source of linen, flaxseed, and linseed oil. Winkler's classification (*Pflanzenfamilien,* 2nd ed., 19a: 82–130) includes three genera (*Humiria, Sacoglottis,* and *Vantanea*) which are here considered in a separate family, the Humiriaceae.

The three arborescent genera of Linaceae represented in the New World are *Hebepetalum, Ochthocosmus,* and *Roucheria.* They are small to medium-sized or rarely large trees and shrubs of limited occurrence in the Amazon region and northern South America. The leaves are alternate, simple, and entire or nearly so; stipules are small or absent; the small yellow or white flowers are borne mostly in axillary racemes; the fruit is a drupe or a capsule (*Ochthocosmos*). The hard, heavy, strong, fine-textured woods are not utilized and are not likely to become of any economic value because of their scarcity and inaccessibility. Apparently no vernacular names are recorded for these genera.

Pores solitary; mostly small to very small; few to rather numerous; irregularly scattered without definite pattern or tending to diagonal arrangement. Vessels with simple perforations; without spirals. Rays 1 to 3, mostly 2, cells wide and few to 40 cells high; decidedly heterogeneous; pits to vessels large, oval or elongated and in scalariform arrangement. Wood parenchyma abundantly metatracheal in *Ochthocosmus,* sparingly paratracheal in the others. Wood fibers with very thick walls and very numerous distinctly bordered pits. Ripple marks and gum ducts absent.

Hebepetalum *humiriifolium* (Planch.) Benth., the only species, is widely distributed in the Amazon basin. It occurs on non-inundated land and while usually only medium-sized occasionally attains a height of 100 feet or more. Heartwood grayish yellow; distinct but not always sharply demarcated from the lighter sapwood. Luster rather low. Odorless and tasteless. Extremely hard and heavy; texture moderately fine; grain more or less irregular; splits readily, but is otherwise difficult to work; probably fairly resistant to decay.

Ochthocosmus, with about a dozen species of shrubs and small trees rarely over 30 feet high, occurs in tropical Africa and (four species) in the Guianas and northern Brazil. Wood reddish brown throughout.

Photo by U.S. Forest Service.

PLATE XXVIII. Clump of Joshua trees (*Yucca arborescens*) on the Mohave Desert, California.

Photo by H. H. Bartlett.

PLATE XXIX. Large Mahogany tree (*Swietenia macrophylla*) in Petén, Guatemala.

Luster medium. Odorless and tasteless. Hard and heavy; texture rather fine; grain irregular; moderately difficult to work; probably fairly durable.

Roucheria, with at least six species of little trees, is distributed from Colombia and the Guianas through the Amazon region to Bolivia. Wood yellowish gray throughout; perhaps all sapwood. Luster medium. Without distinctive odor or taste. Hard and heavy, but not horn-like as in *Hebepetalum;* texture rather fine; grain irregular; rather difficult to work; durability probably low.

LISSOCARPACEAE

Lissocarpa, the only genus, is often included with the Styracaceae. There are two species, namely, *L. guianensis* Gleason of the upper Mazaruni River, British Guiana, and *L. Benthami* Gürke, a small Brazilian tree. The leaves are alternate, simple, entire, and without stipules; the 4-lobed campanulate flowers are in axillary cymes; the indehiscent fruit contains one or two 3-ribbed seeds.

Heartwood not seen; sapwood brownish. Luster medium. Odor and taste not distinctive. Hard, heavy, tough, and strong; texture rather coarse; grain fairly straight; not difficult to work, though inclined to be fibrous; durability of heartwood unknown. Presumably of no commercial possibilities.

Growth rings absent or poorly defined. Pores medium-sized to small, not distinct without lens; fairly numerous, but not crowded; some solitary, but mostly in multiples of 2 to 4, well distributed. Vessels with simple perforations in part; other plates scalariform, with numerous closely spaced, sometimes anastomosed, bars; no spiral thickenings seen; pitting very fine, alternate to opposite. Rays 1 or 2, occasionally 3, cells wide and varying in height up to about 100 cells; decidedly heterogeneous; pits to vessels very small. Wood parenchyma abundant in coarse-celled, concentric, metatracheal lines, or in bands 2, rarely 3, cells wide, spaced about one pore-width apart; distinct with lens and suggesting Anonaceae. Wood fibers with very thick walls and minute cavities; pits with slit-like apertures and very small borders. Ripple marks absent. No gum ducts seen.

LOASACEAE

A FAMILY of 15 genera and more than 200 species, nearly all bristly-haired or scabrous herbs and low shrubs, mainly of the New World from the United States to southern Chile. Apparently the only arborescent representatives are species of *Mentzelia.*

Mentzelia. The numerous species are typically scabrous herbs which cling tenaciously to clothing, for which reason some of them are known as Stickleaf in the United States and Pegarropa in Mexico. The few arborescent forms are Mexican and belong to the section *Dendromentzelia.* The only specimen available (Yale 16346; Conzatti 9) is of *Mentzelia Conzattii* Greenm. which was collected by Professor C. Conzatti at an elevation of 5000 feet in Oaxaca. The species is a tall shrub or a tree sometimes more than 20 feet high in southwestern Mexico. The leaves are mostly opposite, simple, finely dentate, and without stipules; the large, bright yellow flowers are borne in panicled cymes; the fruit is a one-celled, thin-walled, many-seeded capsule. The stems have a thin bark and a large pith. The only known uses for the plant are in native medicine.

Heartwood not seen; sapwood brownish yellow with a satiny luster; conspicuously marked by the coarse rays. Without distinctive odor or taste. Light and soft, but tenacious; texture medium; grain fairly straight; very easy to cut, finishing smoothly and attractively; probably perishable in contact with the ground. Apparently of no commercial possibilities because of the small size of the trees.

Growth rings present. Pores mostly small to medium-sized, a few rather large (180μ); gradually decreasing in size during seasonal growth, occurring mostly in widely spaced pairs or little groups which sometimes tend to tangential arrangement. Vessels with wide-rimmed simple perforations; without spiral thickenings; all pitting coarse, irregular, opposite or occasionally more or less scalariform. Rays nearly all multiseriate; 5 to 12 cells wide and up to 100 cells high; distinct on cross and tangential sections and highly conspicuous on radial surface; decidedly heterogeneous, most of the cells large, square or upright; sheath cells numerous; pits to vessels large, irregular, often elongated. Wood parenchyma narrowly vasicentric. Wood fibers thin-walled; pits numerous, small, simple or indistinctly bordered. Ripple marks absent. No gum ducts seen.

COMMON NAMES: Arnica, yagaduchi, yaganduchi (Mex.).

LOBELIACEAE

THIS family, sometimes included in the Campanulaceae, comprises more than 20 genera and several hundred species of herbs and small to large shrubs, mainly in tropical and subtropical regions. Most of the members contain a milky latex. The leaves are simple, alternate, and without stipules; the flowers are zygomorphic and frequently showy; the fruit is a dehiscent capsule or fleshy. The value of the plants is chiefly horticultural, but some of them may have possibilities as a commercial source of rubber. The woods are not utilized.

Siphocampylus is the only American genus represented in the Yale collection by a specimen other than a small shrub. There are more than 200 species of erect or climbing shrubs, half shrubs, and coarse herbs in Latin America, particularly in the Andes of Colombia, Ecuador, and Peru. Some of the plants are highly ornamental, and the latex of *S. caoutchouc* G. Don of Colombia is said to be used locally for making rubber. The only wood sample available (Yale 19494; Rimbach 28) was collected by Dr. A. Rimbach at an elevation of 11,000 feet on the inner slope of the Cordilleras with floriferous herbarium material determined as *S. umbellatus* (H.B.K.) G. Don. The plant is described as a shrub 20 feet high, with a milky latex in the bark. The slender leaves are subverticillate and crowded near the ends of the branches; the large white flowers are borne singly on long pedicels extending beyond the leaves; the fruit is a many-seeded capsule.

Wood yellowish white throughout; radial surface flecked with the rays which usually appear of lighter color than the background. Luster fairly high. Odor and taste absent or not distinctive. Moderately heavy and hard, but brittle; texture medium; grain straight; easy to work, finishing very smoothly; durability low. Of no commercial possibilities.

Growth rings absent. Pores small (80μ), indistinct without lens; occurring singly and in small multiples, well distributed; mostly not in contact with the rays. Vessels with wide-rimmed simple perforations; members constricted at the ends; pits rather large (10μ), alternate, not crowded. Rays virtually all coarse and conspicuous; 4 to 8, occasionally up to 15, cells wide and up to about 200 cells (1 cm.) high; decidedly heterogeneous, without definite stratification; most of the cells upright-hexagonal (tang. sec.); sheath cells and some procumbent cells present; pits to vessels medium-sized (9μ). Wood parenchyma very sparingly paratracheal. Wood fibers with rather thin walls and large lumen; pits numerous, very small. Ripple marks absent. No gum ducts seen. (The structure is fairly typical of the family, so far as studied.)

COMMON NAMES: Jamchic, lechero de monte (Ec.).

LOGANIACEAE

THIS family comprises more than 30 genera and about 460 species of woody vines, erect shrubs, small to medium-sized trees, and some herbaceous plants, generally distributed in tropical and subtropical parts of the world, sparingly so in temperate regions. The leaves are opposite, simple, and with or without stipules; the flowers have a funnelform, salverform, or tubular corolla; the fruit is a capsule, berry, or drupe. One of the shrubs commonly planted for decorative purposes is the Butterfly Bush (*Buddleia*). The alkaloids strychnine and brucine are obtained from the silvery coin-like seeds of *Strychnos nux-vomica* L. of southern India. The seeds of the Clearing-nut tree, *S. potatorum* L.f., another Indian species, are not poisonous and are used to clear muddy water, the albumen precipitating suspended matter. The family is negligible as a source of commercial timber, although a few species in the Indo-Malayan region and Africa furnish some lumber for local construction and carpentry. The American trees of the family are of six genera and are all small and valueless for timber.

The woods of *Antonia, Bonyunia,* and *Strychnos* are of anomalous structure, the included phloem being in strands, whereas those of *Buddleia, Pagamea,* and *Potalia* are of normal structure (see *Tropical Woods* 50: 16). Color light to dark brown, or gray. Luster generally low. Mostly scentless and tasteless, sometimes bitter. Variable in density and consistency, but generally hard, heavy, and fine-textured.

Pores sometimes medium-sized to large in the climbers, mostly small to minute in the erect trees and shrubs; diffuse-porous, except in some species of *Buddleia;* pores usually numerous; solitary and in pairs or in short to rather long radial rows, well distributed without distinctive pattern except in some species of *Buddleia* and *Strychnos.* Vessels typically with simple perforations; spiral thickenings occasionally present; pits very small to medium-sized, alternate, vestured. Rays typically heterogeneous, often decidedly so; most of them uniseriate or biseriate, sometimes 3 to 5 cells wide; often very low, sometimes up to 25, infrequently to 50, cells high; considerably wider and taller in some species of *Strychnos;* fusiform rays containing phloem bridges present in *Antonia* and *Bonyunia;* crystals sometimes present; pits to vessels very small to moderately large. Wood parenchyma sparingly to abundantly developed; sometimes in concentric bands that are narrow to wide, continuous or broken, closely to widely spaced. Wood fibers with thin to very thick walls; occasionally septate; pits small, simple or with borders of varying degrees of distinctness. Ripple marks absent. No gum ducts seen.

Antonia *ovata* Pohl, the sole species, is a shrub or a small tree occasionally 20 feet tall, occurring in scrub savannas in the Guianas and Brazil. The macerated leaves are sometimes used by the natives to stupefy fish. Wood yellowish white throughout. Fairly lustrous. Odorless and tasteless. Density rather low; texture medium; grain irregular. Structure anomalous, the included phloem appearing as small scattered islands on cross section, not distinct without lens.

Common names: Inacú (Br. G.); harraroe thoeraroe, ipoentrie, kasabahoedoe, kasave, licahout, melassiehoedoe, paloeloipio (Sur.).

Bonyunia. There are five named species of shrubs and little trees, of which three are known only on Mount Duida (Venezuela) and Mount Roraima (British Guiana), one on the open plains of eastern Brazil, and one, *B. aquatica* Ducke, inhabits swamps in the upper Rio Negro valley, Amazonas. The only wood sample available is of this last species (Yale 31951; Ducke 224). Sapwood creamy white; heartwood probably yellow-brown. Odorless and tasteless. Rather dense, hard, fine-textured, easy to work, finishing smoothly, does not appear durable.

Buddleia, with about 100 species of shrubs and little trees, is widely distributed in tropical and subtropical America and Asia and in South Africa. The wood is of normal structure, pale brown, hard, heavy, and fine-textured.

Common names: Tabaco cimarrón (Cuba); azafrán, azafrancillo, azafranillo, cargolinán, escobilla, golondrilla, hierba de las escobas, lengua de vaca, mispatle, quimixpatli, salvia, tepoja, teposana, tepoza, tepozán, tepozancillo, tepozu, tepuza, topozán, tzompantle (Mex.); salvia santa, s. sija (Guat.); salvia, salviona (Salv.); hoja de salbe (C.R.); tabaquillo (Pan.); salvia blanca, tabaco de monte (Col.); hoja de queso, mata de queso, queso, salvia de páramo, Santa María (Venez.); barbasco, calção de velho, calças de velho, camará, cambará, cardeal, tingui da praia, vassoura, vassourinha, verbasco do Brasil (Braz.); quishuar (Peru); mirto (Par.); pañil (Arg.).

Pagamea, with several species of shrubs and small trees in the Amazon basin, is referred by some authors to the Rubiaceae. The only wood specimen available (Yale 20704) is of *P. macrophylla* Spruce collected by Adolpho Ducke. It is of normal structure, the heartwood dark olive-brown with yellowish streaks, the sapwood pale brown, rather sharply differentiated. Odorless and tasteless when dry. Hard, heavy, tough, strong, and probably durable.

Potalia *amara* Aubl., the only species, is a shrub or a little tree rarely over 15 feet high occurring throughout the Amazon basin. The only wood sample in the Yale collections was obtained by L. Williams in eastern Peru (Yale 17730). It is of normal structure, very hard and heavy, and fine-textured.

COMMON NAMES: Anabi (Braz.); sachamangua (Peru).

Strychnos, with over 100 species of woody vines and erect shrubs and small to moderately large trees, is of pantropical distribution. Most notorious of American species is *S. toxifera* Benth., which ranges from Brazil to Costa Rica and is a source of curare, reputed to be one of the deadliest of poisons. According to Standley (*Contr. U.S. Nat. Herb.* 27: 302), "curare is obtained from the bark and roots. Introduced into the circulation in minute quantities it paralyzes the motor nerves almost instantly and soon causes death. It has been employed extensively by the South American Indians for poisoning their arrows, especially those shot from blow-guns, and it is reported that similar use of the plant is made by the Indians of Panama." The woods are gray or yellow, hard, heavy, strong, rather coarse-textured, and are not utilized. Included phloem is in cylindrical strands (Plate LV, 4), surrounded and sometimes linked together by parenchyma.

COMMON NAMES: Manca-montero, matamontero, zarza cimarrona (Cuba); cabalonga de Tabasco, mata-perros, veneno del diablo (Mex.); chicoloro, luch-maax, snakeseed tietie (Br. H.); guaco (Hond.); guacal de mico, huacal de mico, guacamico, huacamico (Salv.); canjura, fruta de murciélago (Pan.); cruceta real, curare, mavacure (Venez.); rouhahamon (Sur.); apui-rana, quassia, quina de cipó, q. do campo, urare, urareúva (Braz.); cunshuhuayo (Peru); casia amarga, coyón de gallo, espuelo de gallo, nuatí-curuzú, té amargo (Arg.).

LORANTHACEAE

THE Mistletoe family includes about 36 genera and a great many species, almost all of which are shrubs parasitic on trees. The simple, entire, thick and leathery leaves are mostly opposite or whorled, sometimes scale-like; stipules are absent; the flowers are often brightly colored; the fruit is a berry or drupe with very viscid pulp. Two genera of normal terrestrial plants are *Nuytsia,* with one species, a small tree in West Australia, and *Gaiadendron* in tropical America.

Gaiadendron, with six species of trees and shrubs, occurs in Costa Rica and in the mountains of British Guiana, Venezuela, Colombia, Ecuador, and Peru. The only wood sample available (Yale 33726) is of *G. tagua* (H.B.K.) G. Don, collected by W. Gehriger at an elevation of about 8000 feet near Mucurubá, Meridá, Venezuela. Heartwood not seen. Sapwood white when fresh, becoming pale brown. Rather lustrous. Of medium-low density; fine-textured; easy to work; consistency about that of Alder (*Alnus*).

Growth rings present. Pores small to minute, not individually distinct without lens; not very numerous; irregularly distributed, occurring in short tangential rows or in clusters. Vessels with very short members; perforations simple; spiral thickenings absent, but pit apertures are spirally coalescent; pitting fine, alternate. Rays 1 to 6, mostly 3 to 5, cells wide and commonly less than 25, sometimes 50 or more, cells high; slightly heterogeneous, the cells rather large; pits to vessels small, nearly circular. Wood parenchyma reticulate, the short tangential lines being fairly distinct with lens; cells often disjunctive. Wood fibers with rather thin walls and comparatively few fairly large and distinct bordered pits. Gum ducts absent. Ripple marks present, the rays not storied; very fine, about 130 per inch; irregular, but fairly distinct under lens.

COMMON NAMES: Tábano (Venez.); matopalo (Peru).

LYTHRACEAE

THE Loosestrife family includes about 20 genera and perhaps 400 species of herbs, shrubs, and trees, widely dispersed over the earth. The leaves are simple, entire, opposite or verticillate, rarely alternate; stipules minute or absent; the flowers usually have a tubular calyx and are borne singly or in cymes or panicles; the fruit is typically capsular, with many seeds. A few members of the family are sources of dyestuffs. The Egyptian Henna, *Lawsonia inermis* L., is a shrub cultivated as a hedge plant and for its leaves which yield the henna dye used from the time of the early Egyptians for imparting an orange or brownish yellow color to the nails, skin, and hair. As a source of timber the only important genus is *Lagerstroemia* of the Indo-Malayan region.

There are five tropical American genera containing small to medium-sized trees. Their woods are yellow or olive (*Lafoensia*), light brown or pinkish (*Adenaria* and *Physocaylmma*), or dark brown to blackish (*Ginoria*). Luster medium to satiny. With mild scent in *Ginoria*, otherwise without distinctive odor or taste. Density high in *Ginoria* and *Lafoensia*, medium in the others; texture fine to medium, the latter having about the consistency of Birch (*Betula lutea* Michw.) or Maple (*Acer*); grain straight to roey; mostly not difficult to work, finishing smoothly; not durable; suited for general carpentry.

Growth rings present or absent. Pores small to medium-sized, the largest near limit of vision; rather few to numerous; occurring singly or more often in pairs, generally well distributed without pattern. Vessels with simple perforations; without spirals; pits small, alternate, vestured. Rays fine and inconspicuous; 1 or 2, rarely 3, cells wide and up to 25, occasionally to 50, cells high; homogeneous or nearly so in *Physocalymma*, heterogeneous in the others; pits to vessels small, of same appearance as those in the vessels. Wood parenchyma very sparingly developed, though sometimes apparently abundant. Wood fibers often septate; pits small, simple; thin-walled, large-lumined fibers in numerous distinct parenchyma-like bands present in *Ginoria* and *Physocalymma*, the cells containing starch at first, later crystalliferous; similar starch-bearing but non-crystalliferous fibers indistinctly paratracheal in *Lafoensia*. Ripple marks absent. No gum ducts seen.

Adenaria *floribunda* H.B.K., the only species, is a shrub or a slender tree rarely 20 feet high with a smooth-barked trunk occasionally eight inches thick. The leaves are dotted with black glands. The wood is unattractive and of no special merit. Heartwood light pinkish brown, more or less streaked; sapwood somewhat lighter-colored. Luster high. Without distinctive scent or taste. Moderately hard and heavy; texture rather fine; grain straight; working properties good; durability probably low.

Common names: Fruta de pavo (Pan.); chaparral (Col.); guayabito (Venez.); gurima-ey, puca-varilla, rumo-caspi (Peru).

Ginoria. There are seven species of shrubs and little trees, one Mexican, the others West Indian. Some of the plants are cultivated because of their attractive flowers. The wood is suitable for small articles of turnery. Heartwood dark brown to blackish brown, more or less variegated and streaked; sharply demarcated from the thin yellowish sapwood; sometimes with fairly thick gray intermediate zone containing abundant gummy deposits but not deeply colored. Luster rather high. With faint pleasant scent, but without distinctive taste. Very hard and heavy; texture fine and uniform; grain irregular; not difficult to work, taking a high natural polish; appears very durable.

Common names: Clavellina, c. de paredón, c. del río, c. espinosa, cuaresmilla, c. de paredón, c. espinosa, guairaje spinosa, g. uña, Júpiter arbol, rosa del río, yema de huevo (Cuba).

Grislea *secunda* Loefl., the only species, is a shrub or a tree occasionally 20 or 25 feet tall with a trunk 6 to 12 inches in diameter, of limited occurrence in Venezuela and Colombia and discovered by the senior author in the Aguán valley of Honduras in 1927. The bark is thin and smooth, the leaves are brown punctate, and the axillary clusters of small purplish red flowers are attractive. There are no special uses for the wood. Heartwood not seen; sapwood nearly white. Luster rather high. Odorless and tasteless. Of medium density, having about the consistency of White Birch; texture fine; grain fairly straight; easy to work, finishing smoothly; durability of heartwood unknown. Of no commercial possibilities.

Common names: Coloradillo (Hond.); guayabito de cerro (Col.); bijo, indiecito (Venez.).

Lafoensia, with about a dozen species of shrubs and small to medium-sized trees, is widely distributed in tropical America from southern Mexico to southern Brazil, though nowhere abundant. The largest trees are about 60 feet tall with a trunk 24 inches in diameter and clear of branches for 30 feet. The bark is used in Brazil as a source of a yellow dye, and the yellowish rather hard timber is employed for carpentry, interior construction, staves, fuel, and charcoal. According to Octavio Silveira Mello (see *Tropical Woods* 58: 39), the Merindiba, *L. glyptocarpa* Hoehne, is a timber tree growing abundantly near Gavea, Brazil, and worthy of propagation for the production of firewood and construction timber, for the protection of watersheds, for ornament and shade in parks and along highways, and even for hedges, because of its vigorous growth, deep root system, and its high degree of indifference to environmental conditions.

Heartwood bright greenish yellow or olive; distinct but not sharply demarcated from the lighter-colored sapwood. Luster high. Without distinctive odor or taste. Hard and heavy or moderately so; texture rather fine, somewhat roe-grained; not difficult to work, taking a smooth finish and high polish; durability fair. Not likely to prove of importance for export.

Common names: Coquito, moreno (Mex.); palo de culebra (Guat.); cuyapo, trompillo (Salv.); amarillo, a. fruta (Pan.); guayacán (Col.); ariauá, cabeça de monge, candeia de cajú, copinho, dedal, dedaleira, dedaleiro, d. amarello, d. preto, mangaba brava, mangabeira brava, merindiba, m. rosa, pacari, pacuri, p. do matto, p. salvagem, pau terra (Braz.); amarillo, cabeza de monge, chusma, chuspo, pocol (Peru); mangá-ná, moré-cipo, moresino (Pan.).

Physocalymma, with only one species, *P. scaberrimum* Pohl (= *P. floridum* Pohl), is a small to medium-sized tree of fairly common occurrence in the upper Amazon region. It was long supposed to furnish the well-known Brazilian Tulipwood (*Dalbergia* aff. *variabilis* Vog.), but in reality its timber is of the general type of Birch (*Betula lutea* Michx.) and unlike the Tulipwood. The presumably well-authenticated specimen described in *Timbers of tropical America* (p. 455) under the name of *Physocalymma* was later found by the senior author to belong to the Lauraceae (probably *Ocotea*) and so reported in *Tropical Woods* 20: 23, December 1, 1929.

Wood light grayish brown with pinkish hue throughout. Luster rather high. Odorless and tasteless. Moderately hard and heavy; texture medium; grain straight to somewhat roey; not difficult to work, finishing smoothly; durability probably low. Of no commercial possibilities.

Common names: Cego machado?, gipío (Braz.); huainava, huainuma (Peru).

MAGNOLIACEAE

The Magnolia family, in the restricted sense used here, comprises four genera, namely, *Liriodendron, Magnolia* (incl. *Kmeria* and *Pachylarnax*), *Michelia* (incl. *Elmerillia* and *Alcimandra*), and *Talauma* (incl. *Aromadendron* and *Manglietia*), with about 150 species of trees and erect shrubs occurring in temperate and tropical regions of Asia and (excepting *Michelia*) America. (See *Tropical Woods* 34: 3–39.) The leaves are alternate simple and entire (lobed in *Liriodendron*), with large deciduous stipules inclosing the young buds; the flowers are large, perfect, solitary, terminal or axillary; the fruit is compound and cone-like, composed of carpels that are closely imbricated, dry, deciduous, winged, and indehiscent in *Liriodendron,* or fleshy, with brightly arillate seeds attached by a thread-like funicle, the carpels bivalvate at maturity in *Magnolia,* indehiscent in *Talauma.* Many of the species are planted for decorative purposes. The genus most important for timber is *Liriodendron.*

The woods of all four genera are much alike in appearance, structure, and properties. Heartwood yellowish to brown, sometimes olive to purplish brown; usually sharply demarcated from the white sapwood. Luster medium. Without distinctive odor or taste. Light and soft to moderately so; sp. gr. (air-dry) 0.35 to 0.55; weight 23 to 34 lbs. per cu. ft.; of fine and uniform texture; usually straight-grained; easy to work, nails without splitting, holds paint well, and is well suited for interior woodwork and general carpentry.

Growth rings distinct to indistinct, usually without much contrast between early and late wood. Pores small polygonal and numerous or medium-sized rounded and not very numerous (*Talauma*); solitary and in multiples, with no definite pattern. Vessels with scalariform perforation plates, except in certain species of *Magnolia* (*e.g., M. acuminata*); spiral thickenings present or absent; tyloses common; intervascular pitting opposite to scalariform. Rays rather fine and inconspicuous; heterogeneous; oil cells sometimes present in tropical species; ray-vessel pitting often unilaterally compound. Wood parenchyma mostly terminal, in solid or much broken bands; cells often very irregularly thickened, especially in *Liriodendron.* Wood fibers with small bordered pits.

Liriodendron, with two geographical species of tall trees, occurs in central China and throughout most of the hardwood region of eastern United States and southern Ontario. *L. tulipifera* L. is often called Tulip Tree, while the lumber is known as Poplar, Yellow Poplar, or, especially in New England and New York, as Whitewood. Optimum development occurs on fertile soils in the Ohio valley region and on mountain slopes in North Carolina and Tennessee. About three-fourths of the remaining first-growth timber is in the southern Appalachians. The tree is known to attain a maximum height of nearly 200 feet with a trunk eight to ten feet in diameter and free of branches for 100 feet. Though generally less than 100 feet tall, it is typically of excellent timber form, with a long smooth cylindrical bole and a short crown of comparatively small limbs. Second-growth timber has very thick white sapwood, hence the name Whitewood, and a small core of heartwood which may be deeply colored and more or less iridescent. Old forest trees have a large proportion of pale olive-brown wood, which turns brown superficially upon exposure; the sapwood is often sold under the name of Poplar saps. The lumber is used in large quantities for sash, doors, painted interior trim, house-siding, boxes and crates, drawers and frames of furniture, fixtures, shelving, moldings, plywood, and many other purposes. The present estimated stand of merchantable timber is about nine billion board feet.

COMMON NAMES: Canary wood, canoe wood, poplar (blue, hickory, tulip, white, yellow), tulip tree, whitewood (U.S.A.).

Magnolia. There are about 70 species of deciduous or evergreen shrubs or medium-sized to large trees, widely distributed in eastern continental Asia and Japan, southeastern United States, and to a limited extent in Mexico, Central America, and the West Indies. The plants are more useful for ornamental purposes than for lumber.

There are eight or nine species in the United States but only three are of any commercial importance. The Cucumber-tree, *Magnolia acuminata* L., is the most widely distributed, its range including most of the hardwood region from southern Ontario to Louisiana. It is most abundant and of largest size (up to 100 feet tall and 48 inches in diameter) in the mountains of the Carolinas and Tennessee, and its lumber closely resembles Yellow Poplar (*Liriodendron*) and is often sold in mixture with it. *M. grandiflora* L. (= *M. foetida* Sarg.) is a large evergreen tree of the southeastern states, attaining a maximum height of 130 feet and a diameter of 56 inches in the lower Mississippi valley. It is of less importance than the preceding species, but supplies considerable lumber generally known to the trade as Magnolia. It has a greenish gray color and has about the consistency of Maple (*Acer*); it is used as a substitute for *Liriodendron.* The tree is widely planted for ornament in countries of temperate subtropical climate. The Sweet Bay or Swamp Bay, *M. virginiana* L. (= *M. glauca* L.) is a smaller tree of coastal swamps from North Carolina to Florida and westward to Louisiana. In parts of its southeastern range it attains comparatively large size and supplies some timber of value, mostly for local consumption. The heartwood is more brownish and the texture is coarser than in the other two species.

Magnolia splendens Urb., endemic to Puerto Rico, is considered one of the best timber trees of the island, but it is now very scarce; the bark contains an essential oil. *M. cubensis* Urb. is a fairly large tree in the mountains of Cuba. There are two species in south-central Mexico and one of them, *M. Schiedeana* Schl., is said to attain large size. *M. poasana* (Pitt.) Dandy is endemic to Costa Rica and is abundant and often large on the upper slopes of Poás. *M. sororum* Seibert reaches a height of about 70 feet in a limited zone at altitudes above 5000 feet on the east and northwest sides of the Volcán de Chiriquí in Panama. These tropical trees are all too scarce or inaccessible to provide timber of more than local utility.

Common names: *Magnolia acuminata:* Cowcumber, cucumber, c. tree, elkwood, linn (black, yellow), magnolia (blue, cucumber, mountain), wahoo (U.S.A.). *M. grandiflora:* Bat tree, bay (bull, laurel), big laurel, magnolia (U.S.A.). *M. virginiana:* Bay (swamp, sweet, white), beaver tree, Indian-bark, laurel (swamp, white), magnolia, swamp magnolia, s. sassafras (U.S.A.). *Other species:* Elk-browse, elkwood, Indian bitters, I. physic, magnolia (various kinds), umbrella tree (U.S.A.); magnolia, mantequero, marañón de la maestra (Cuba); jaguilla, laurel, l. sabino, mauricio, sabino (P.R.); corpus, elosúchil, semíramis (Mex.); candelillo (C.R.).

Talauma, with about 50 species of trees and shrubs, is of general occurrence throughout the tropics, except in Africa. There is a close resemblance to Magnolia in leaves, flowers, and wood, but the carpels of the fruit are indehiscent and the leaf petiole bears a stipular scar. While most of the trees are small, some are large enough for timber, though too scarce to be of more than local value. The uses for the lumber include general carpentry, especially interior work, boxes of all kinds, shelves, drawers, and similar purposes. The bark and flowers are employed in native medicine.

The range of the several species in the western hemisphere is scatteringly from central Mexico through Central America to Colombia, Venezuela, the Guianas, and Brazil. According to Standley (*Trees and shrubs of Mexico*, p. 275), *Talauma mexicana* (DC.) Don is sometimes 100 feet high, with a trunk four to six feet in diameter and "is one of the best known of Mexican trees. It was highly esteemed by the early inhabitants because of the sweet odor of the blossoms, a single flower being sufficient to perfume a whole house. The tree was cultivated in gardens, and the flowers were reserved for the extensive use of the nobility." *T. gloriensis* Pitt. is a medium-sized tree endemic to Costa Rica. *T. sambuensis* Pittier of Panama is 90 to 120 feet high, with a firm straight-grained yellowish to purplish wood similar to *Liriodendron*, but of somewhat coarser texture. The principal species in Brazil is *T. ovata* St. Hil., but at best it is only a medium-sized tree, and, like the others of this genus, is too infrequent to add appreciably to the lumber supply.

Common names: Azulejo, magnolia, marañón, m. de Costa (Cuba); coeur de cachiman, c. de cachiment (Guad.); bois pin, cachiman de montagne (Mart.); flor de corazón, guielachi, hierba de las mataduras, hualhua, laurel tulipán, yoloxochitl, yoloxichitl (Mex.); yorocónte (Guat., Hond.); magnolia, kakuá-biuí (Pan.); araticum fruta de pau, canella do brejo, magnolia do brejo, m. do matto, pau pombo, pinha do brejo, pinheiro do brejo (Braz.).

MALPIGHIACEAE

This unimportant family comprises about 60 genera and 800 species of woody vines, upright or clambering shrubs, and small or rarely large trees distributed throughout tropical and subtropical regions but most abundant in tropical America. The leaves are simple, often glandular, mostly opposite; stipules common, sometimes large and connate; the flowers have clawed petals; fruit various, drupaceous, nutlets, capsular, or a samara (sometimes suggesting *Acer.*) There are about 45 American genera, and about one-third of them contain arborescent species, though with few exceptions the trees are less than 25 feet high. The bark is sometimes of local utility as a source of tannin and medicine. *Byrsonima* is the only genus producing timber and it is not extensively used. The following description is based on specimens of twelve genera, namely, *Banisteria, Banisteriopsis, Bunchosia, Burdachia, Byrsonima, Diacidia, Glandonia, Lophanthera, Malpighia, Spachea, Tetrapodenia,* and *Thryallis.* Some of the lianas, not included here, have anomalous structure.

Wood yellow throughout in *Malpighia;* heartwood grayish brown to dark red or reddish brown in the others; transition to sapwood gradual. Luster low to rather high. Without distinctive odor or taste. Moderately to decidedly hard and heavy; texture medium to fine; grain usually irregular; working properties fair to excellent; durability variable.

Growth rings present or absent. Pores typically thick-walled and numerous; minute to medium-sized, sometimes more or less distinctly 2-sized; mostly in small multiples, but with tendency to form radial rows occasionally of considerable length. Vessels with simple perforations; without spirals; pits vestured, small to minute, mostly alternate, sometimes opposite in part. Rays inconspicuous, numerous; sometimes weakly heterogeneous in *Bunchosia,* decidedly heterogeneous in the others; occasionally all narrow; more often of two sizes, the larger varying even in the same genus from 2 or 3 to 6 or 8 cells wide and from less than 25 to over 100 cells high, but generally less than 4 cells wide and 40 cells high; gum deposits often abundant; crystals common, sometimes large; pits to vessels in part large elongated and in scalariform arrangement (*Burdachia, Glandonia,* and *Tetrapodenia*), mostly small to very small in the others. Wood parenchyma greatly variable in abundance; limited to a few cells in contact with the vessels in *Byrsonima* and *Glandonia;* sparingly paratracheal, sometimes with considerable diffuse in *Lophanthera, Spachea, Thryallis,* and also terminal in *Diacidia;* sparingly vasicentric, short aliform, and with considerable diffuse, also finely terminal, in *Banisteria, Banisteriopsis, Burdachia,* and *Tetrapodenia;* finely reticulate, with tendency to concentric lines and terminal in *Malpighia;* abundantly paratracheal, short aliform sometimes confluent into distinct bands 3 to 10 cells wide, occasionally composing more than a third of the cross section in *Bunchosia;* strands with heavily integumented crystals numerous to very numerous, the crystals small to medium-sized in *Byrsonima* and *Malpighia,* large (at least in part) in *Banisteria, Banisteriopsis, Bunchosia, Burdachia, Lophanthera,* and *Spachea.* Wood fibers with rather large cavities and septate in *Banisteria, Banisteriopsis, Byrsonima, Glandonia, Lophanthera,* and *Spachea;* with very narrow lumen and non-septate in the others; pits very small, simple or with vestigial borders. Ripple marks absent. No gum ducts seen.

Banisteria, with about 85 species of erect or scandent shrubs and a few small trees, is of general distribution in tropical America. The grayish or pinkish brown, moderately dense, fine-textured wood is not utilized for any special purposes.

COMMON NAMES: Bergajo de toro, vergajo de toro (Cuba); bejuco de buey (P.R.); bejuco de caballo, b. huesillo, escobillo, pinsamillo (Mex.); coral, nance colorado (C.R.); cointura (Pan.); ayahuasca, lluasca (Peru); caá-pi, cipó de São João, c. prata, timbo branco (Braz.).

Banisteriopsis, with several species of erect and clambering shrubs, is widely distributed from southern Mexico to southern Brazil. The only specimen available (Yale 9735), collected by H. Pittier in Venezuela, is similar to *Banisteria* (Plate XLII, 3).

COMMON NAMES: Ayahuasca, caapi, changro-panga, pejí, yaco, y. borrachero, yagué, yajé (Col.).

Bunchosia, with about 50 species, occurs in the West Indies and from southern Mexico through Central America to Venezuela and Peru. Most of the plants are shrubs and low trees, but *B. argentea* (Jacq.) DC. is said to attain a height of 50 to 65 feet with a trunk diameter of 12 to 16 inches in Colombia, and *B. glanduli-*

fera (Jacq.) H.B.K. is reported to reach a maximum height of 100 feet and a basal diameter of nearly five feet in Venezuela. The rich dark brown heartwood is in sharp contrast with the nearly white sapwood, and would be good for furniture if available in sufficient quantity and large enough dimensions.

COMMON NAMES: Icaquillo, mierda de gallina (Cuba); café forastero (P.R.); cabra hedionda, cabrita (Dom. R.); bois senti, b. zammi, caiman franc, merde rouge de la montagne (Haiti); capulincillo, ciruelillo, garbancillo, sipché, zapotillo de San Juan, zipché (Mex.); cojón de fraile (Br. H.); cerezo (C.R.); cerezo, c. de monte (Pan.); bolita de perro, cerezito, ciruelo de perro, lengua de venado, muñeco (Col.); ciruelo de fraile (Venez.); ciruelo, sacha-indán (Peru).

Burdachia, with two species of low trees, is of limited distribution in the Brazilian Amazon region. The wood is of a rather dull brown or reddish brown color, hard, heavy, fine-textured, and probably durable; apparently not utilized.

Byrsonima is a widely distributed tropical American genus of about 100 species and numerous varieties of low to high shrubs and small, medium-sized, or occasionally large trees. The species with the widest range is *B. crassifolia* (L.) Rich., a shrubby or low-branched tree rarely 35 feet high occurring in savannas and semiarid districts of southern Mexico, eastern Central America, northern South America, and the West Indies. The wood is used locally for small construction timbers and for fuel and charcoal. Some of the species, such as *B. coriacea* (Sw.) H.B.K., *B. aerugo* Sagot, and *B. stipulacea* Juss., are 100 to 125 feet tall in the forests of the Guianas and lower Amazon basin. The moderately hard, rather fine-textured, grayish to reddish brown timber is in about the same class as Red Gum (*Liquidambar*) and used locally for general carpentry, house-framing, and furniture. Other species, though smaller, have wood of more attractive color employed for cabinet-making, turnery, and fancy flooring. The name Doncella is often associated with some of the West Indian species, but the timber sold as Doncella in the New York market is of a different family (Sapotaceae).

COMMON NAMES: Candle berry, golden spoon, guana berry, hogberry, locus berry, locust berry, l. tree, lotus berry, pigeon berry (B.W.I.); carne de doncella, peralejo, p. de costa, p. de monte, p. de pinares, p. de sabana, p. blanco, p. común, p. enano, p. macho, sangre de doncella, s. de vaca (Cuba); candle berry, guana berry, maricao, m. cimarrón, palo de doncella (P.R.); bois corne (Haiti); bois canne, maurice, moureiller (Guad.); massif (Grenada); ceresa de monte, ceretle, mureche, serrette, surette de grands bois (Trin.); changugo, chi, nananche, nance, n. agrio, nanche, n. de perro, n. dulce, nanchi, nantzinxocotl, nanzinquehuitl, zacpah (Mex.); craboo, crapoo, sac-pah, wild craboo (Br. H.); tapal (Guat.); nance verde, nancite (Salv., Nic.); bek, merdiera, nance, n. hembra, nancite, sikrá, skirko (C.R.); nance, n. blanco, n. colorado, wild cherry (Pan.); chaparro, nance, nancí, peralejo, yaca or yuco (Col.); candelo, chaparro, c. de chinche, c. de sabana, c. manteca, manteco, peralejo (Venez.); arakadako, hitchia, huria, idin, kamadanni, kanoaballi (Br. G.); hoelia, hoeliadiamaro, holia, hori, horia, madabrieballi, moelei, moeleki, moeleidan, moelera, moeli, moeréi, moeréiran, moeroei weneran, sabana kwarie, savanna kwarie, s. mango (Sur.); bois canne, b. de tani, b. dysentérique, b. tan, itchia, mauricif, moureiller, mourresif, surette (Fr. G.); mirichi, m. rasteiro, morocy, murecí, murecy, m. miudo, m. pitanga, m. vermelho, murei, mureila, murici, m. de folhas pequenas, pau de cortuma, percegueiro bravo (Braz.); chupi-cara, indano, i. colorato, murushi, quillo-sisa (Peru).

Diacidia, with two species of shrubs and small trees, is of limited distribution along the border between Brazil and the Guianas and Venezuela. The only wood sample available (Yale 16188; Tate 563) is of *D. vestita* (Benth.) Jacks. obtained at an elevation of 4800 feet on Mount Duida. The collector describes it (*Bull. Torrey Bot. Club* 58: 380) as "a good-sized tree with straight trunk six to eight inches in diameter" and rather common along streams. The species had previously been known only from the original collection of Schomburgk in the region of Mount Roraima. It has yellow flowers and the persistent calyx turns red when the fruit matures. The pith is whitish in the middle and surrounded by a brown layer of thick-walled cells which also form distinct septa about 1 cm. apart. The grayish-brown, hard, medium-textured wood is not utilized.

Glandonia *macrocarpa* Gris., the only species, is a small or medium-sized tree apparently of infrequent occurrence in the Amazon region of Brazil and British Guiana, being known in the latter country as Kamadanni. The only wood sample available (Yale 31966) was collected by Adolpho Ducke in swampy forest in the vicinity of Manáos. The heartwood is a variegated rich reddish brown, with a rather waxy feel, of fine and uniform texture, hard and heavy but not difficult to work and taking a high natural polish. It would doubtless have many important uses if available in large sizes, but apparently it has no possibilities for export.

Lophanthera, with three species of small to medium-sized trees, occurs in the low or second-growth forests of the central and northern Amazon region. *L. lactescens* Ducke and *L. pendula* Ducke are characterized by a bitter white latex. In *L. longifolia* (H.B.K.) Gris. this latex exists only in feeble traces. The grayish brown, moderately dense, medium-textured timber has no special uses or commercial possibilities.

Malpighia, with about 36 species of shrubs and small trees, is of common occurrence in the West Indies, southern Mexico, Central America, and northern South America. The cherry-like fruits of some species are edible. Certain plants are provided with stinging hairs, which accounts for such vernacular designations as Touch-me-not and Cow-itch. The yellow hard wood is of fine and uniform texture and some of it approaches the Boxwood class, but the trees are too small and bushy to furnish any timber of commercial value.

Common names: Touch-me-not, wild cherry (Bah.); cherry, cowage cherry, cow-itch cherry (Jam.); cerezo, c. del país, palo bronco, p. de gallina, p. hierro, pegojo de costa (Cuba); azota caballo, cerezo colorado, chereese, cowage cherry, jolago, olaga, palo bronco, stinging bush, West Indian cherry (P.R.); capitaine, cerisier (Haiti); cerezo, chi, escobillo, kanibinché, manzana, manzanita, mora del campo, nancén, nanche, uzte, xbec-ché (Mex.); hicatee plum, sim-ché (Br. H.); camaroncito (Salv.); locktotl, xocot (Nic.); acerola, guacuco, huacuco, Jupiter (C.R.); cerezo de Castilla, grosella (Pan.); arrayán, a. macho, arrayancito, cerezo, c. agrio, huesito (Col.); cerezo, cimaruco, semeruco (Venez.); kersenboom (Sur.); cerejeiro (Braz.).

Spachea, with a few species of scandent shrubs and small trees, occurs from the West Indies to the Amazon basin. The grayish brown, moderately dense, rather fine-textured timber is apparently not utilized.

Tetrapodenia *glandifera* Gleason, the only species, is a small British Guiana tree closely related to *Burdachia* and *Glandonia.* The only wood sample available (Yale 9464; Persaud 53) is reddish or grayish brown, hard and heavy, fine-textured, and of irregular grain. It apparently is not utilized and has no commercial possibilities.

Thryallis (or *Galphimia*), with about a dozen species of erect shrubs and little trees, occurs from the southwestern United States to Brazil, but the center of distribution is in Mexico. *T. brasiliensis* L. is said to have a bitter latex. The best known species is *T. glauca* (Cav.) Kuntze, a slender shrub sometimes 15 feet tall, native to Mexico and Central America and naturalized in the West Indies where it is cultivated on account of its showy racemes of large yellow flowers. The reddish, moderately hard, fine-textured wood is not utilized.

Common names: Grano de oro (Cuba);

consulita, lluvia de oro (P.R.); consulita (Dom. R.); calderona amarilla, flor de Diciembre, hierba del piojo, h. del venado, huachácata, nachácata, palo del muerto, palo de San Vicente, ramo de oro, vachácata (Mex.).

MALVACEAE

THE Mallow family, with about 80 genera and over 1000 species of annual and perennial herbs, small to large shrubs, and a few trees, is widely distributed over the world, though most of the species are tropical. The member of outstanding importance is the cotton plant, *Gossypium.* The bark is typically fibrous and in several genera is of value in making textiles and cordage. The leaves are alternate, entire or variously lobed, mostly palmately nerved, and provided with stipules; the pubescence is typically of stellate hairs; the flowers are solitary in the leaf axils or borne in racemes, clusters, or panicles; the fruit is dry, or rarely berry-like, usually breaking into cocci, sometimes capsular. Some of the herbs and shrubs (*e.g., Hibiscus*) are noted for the beauty of their flowers and foliage and are widely grown for decorative purposes. Very few species are large enough to furnish timber and they are too scarce to be of more than local utility. The woods of the family as a whole exhibit a wide range of variation (see *Tropical Woods* 38: 15–36).

In tropical America there are more than 25 genera with at least some of the species shrubby or arborescent. The largest plants are in the genera *Abutilon, Bastardiopsis,* and *Tetrasida* of the Malveae, and *Hibiscus, Montezuma, Thespesia,* and *Wercklea* of the Hibisceae. The number of species involved is probably not over 10, but in their structure and properties they illustrate most of the range of variation within the entire family.

Heartwood commonly brownish, frequently with a reddish or purplish tinge, or grayish and then often with a greenish hue, frequently streaked or variegated; considerable change may be brought about in seasoning or as a result of exposure to light. Luster low to silky. Heartwood of a few species slightly fragrant, otherwise odorless. Sp. gr. 0.24 to 0.80, mostly less than 0.50; weight 15 to 50 lbs. per cu. ft.; texture very fine to rather coarse, mostly fine; grain straight to roey; working properties generally good; durability low except for deeply colored heartwood.

Growth rings distinct to indistinct. Pores small to medium-sized, rarely distinct without lens; solitary and, more often, in small radial multiples or little clusters; fairly to very numerous; generally uniformly distributed, but sometimes restricted to bands of parenchyma alternating with layers of wood fibers. Vessels with simple perforations; spiral thickenings sometimes present; pits very small to medium-sized or occasionally large; typically alternate. Rays heterogeneous; all 1 to 3 cells wide in some genera, much wider in others, greatly variable in height from few to over 250 cells; some, all, or none of the rays definitely storied; sheath cells present or absent; true tile cells absent; druses and rhombohedral crystals sometimes present; ray-vessel pitting half-bordered, mostly rather fine, not infrequently unilaterally compound. Wood parenchyma scanty to abundant; paratracheal in all genera, varying from vasicentric to aliform and confluent; metatracheal often finely reticulate, sometimes in bands 3 to 6 cells wide, occasionally terminal; abundance and arrangement not always constant in same species; cells sometimes fusiform, more often in strands of 2 or 4, rarely 8, cells each; crystals common, wood fibers with very thin to thick walls; pits few to moderately numerous, usually simple or indistinctly bordered, occasionally with small but distinct borders. Ripple marks often present, regular to irregular, distinct to indistinct, 50 to 125 per inch. Vertical traumatic gum ducts observed in certain specimens of *Hibiscus tiliaceus* L.

Abutilon has more than 100 species in tropical and subtropical regions, but most of them are herbs and shrubs, some of which are useful for their jute-like fiber. The largest tree represented in the Yale collections is *A. Chittendenii* Standl., discovered by Record and Kuylen near Olanchito, Honduras (Yale 10009). It was about 30 feet high, with thin rounded or heart-shaped leaves three inches in diameter and bearing large bright yellow flowers. According to Standley (*Tropical Woods* 10:

5), it is related to *A. Peyritschii* Standl., a small tree of Vera Cruz, Mexico.

Heartwood apparently absent; sapwood cream-colored. Odorless and tasteless. Hard, heavy, tough, and strong; texture fine and uniform; grain rather irregular; not difficult to work, finishing very smoothly; durability presumably low. Of no commercial possibilities.

Bastardiopsis, with a single species, *B. densiflora* (H. & A.) Hassl. (= *Sida densiflora* Hook. & Arn.), is a small to medium-sized tree sometimes 50 feet tall with a stout trunk 24 inches in diameter, in southern Brazil, Paraguay, and Argentina. The timber has about the consistency of Maple (*Acer*) and is used to a limited extent for general carpentry and interior construction.

Color yellowish brown throughout. Luster rather low. Odorless and tasteless. Moderately hard, heavy, tough, and strong; of medium texture; grain fairly straight; working qualities good; not durable in contact with the ground.

COMMON NAMES: Loro blanco, peterebí-morotí (Arg.).

Hibiscus, with about 200 species of herbs and shrubs, is of general distribution in tropical and temperate regions. Some of the plants are in cultivation for their showy, handsome flowers, and others for use as a vegetable, *e.g.*, the Okra, *Hibiscus esculentus* L.

The only really arborescent species is *Hibiscus tiliaceus* L. (= *Paritium tiliaceum* A. Juss.), common in swampy thickets along beaches and estuaries throughout the tropics, often forming almost pure stands. In such locations it is typically small and shrubby, usually not over 15 to 20 feet tall, and its principal value is in the tough and pliable bast which is used locally and sometimes commercially in making cordage, mats, and coarse cloth. In the West Indies there is an upland form or variety, considered by some botanists as a distinct species, *H. elatus* Sw., which attains a height of 60, rarely up to 80, feet, with a long clear bole 12 to 18 inches in diameter, occasionally considerably larger. Trees in the mountains are credited with producing timber of greater density and durability and more attractively colored than that from lower and wetter situations. The heartwood is often richly variegated with shades of purple, metallic blue, and olive, and has always been in demand locally for furniture, interior trim, and gunstocks. Such timber is now scarce and that available is of the general type of Yellow Poplar (*Liriodendron*), being light olive when fresh, but tending to darken superficially upon exposure. Some specimens are rather soft and fine-textured, with a silky luster, while others are hard, of medium to coarse texture, and rather dull.

COMMON NAMES: *Hibiscus tiliaceus:* Mahoe—blue, mountain, seaside (Jam.); majagua (Sp. Am., gen.); majagua azul, m. blanca, m. común, m. de la maestra (Cuba); emajagua, e. excelsa (P.R.); cotón marrón, mahaut franc (Haiti); mahot franc, marofranc (Guad.); mahagua del mar, mahault du bord de mer, seaside mahoe (Trin.); holó, masahua, mazahua, pox, xholol, xtoló (Mex.); blue moho, majao, mahoe (Br. H.); kipkuó, krokua, stsá, trokró (C.R.); majagua de playa (Pan.); majagüito de playa (Col.); algodoncillo (Venez.); maho (Br. G., Sur.); mahot, m. blanc, m. forestier (Fr. G.); maho, uacima de praia (Braz.); huamoga (Ec.); damajagua, emajagua (Peru).

Montezuma. There are two described species, namely, *Montezuma speciosissima* Sessé & Moc. (= *Thespesia grandiflora* DC.), a tree sometimes 50 feet high and 16 inches in diameter in central and western Puerto Rico and often planted along roadsides, and *M. cubensis* (Brit. & Wils.) Urb., a fairly common small to medium-sized tree in Cuba. The timber is of attractive color and figure and is used locally for making furniture and musical instruments, and because of its durability it is also employed for fence posts and piling.

Heartwood uniform rich reddish brown; clearly but not sharply differentiated from the yellowish sapwood. Luster medium. Without distinctive odor or taste. Moderately heavy and hard; texture medium, uniform; grain fairly straight; very readily worked, taking a smooth finish; radial surface with silver grain; highly resistant to decay. Not likely to be important for export.

COMMON NAMES: Maga, magar (P.R.); majagua de Cuba, m. peluda, negra Cuba (Cuba).

Tetrasida *polyantha* Ullr., the only species, is a small to medium-sized tree of infrequent occurrence in the upper Amazon region. The only specimen available (Yale 19033) was collected by Llewelyn Williams in second-growth at an altitude of about 1400 feet in northeastern Peru. The timber is not utilized for any special purposes.

Color pale yellowish brown throughout. Luster medium. Odorless and tasteless. Moderately heavy, hard, tough, and strong; texture rather fine, uniform; grain fairly straight; not difficult to work, finishing smoothly; probably perishable in contact with the ground; has about the consistency of Maple (*Acer*).

Thespesia, with six or seven species of shrubs and trees, is native to tropical Asia and Africa, with one species, *T. populnea* (L.) Soland., extending to the Caribbean region, where, in favorable situations, it grows to a height of 60 feet. It is frequently planted along roadsides and in gardens. The inner bark of the branches and young stems contains a tough fiber useful for cordage. The timber, which is suitable for furniture and cabinet work, is sparingly employed in America because of its small size, but in the Far East it finds many applications.

Heartwood rich reddish brown or chocolate, more or less variegated and striped; distinct but not sharply demarcated from the yellowish or pinkish sapwood. Luster rather low. Odorless and tasteless. Moderately heavy, firm and strong; texture medium; grain straight to variable; easy to work, finishing very smoothly and attractively; durability high.

COMMON NAMES: Seaside mahoe (Florida, B.W.I.); Portia tree (Fla.); John Bull tree (Jam.); bendy tree, cork tree, emajaguilla, otahite, palo de jagueca, Santa María (P.R.); majagua de la Florida (Cuba); catalpa (Fr. W.I.); gros mahaut (Haiti); cork tree (Br. H.); clemón (Col., Venez.); boschkatoen (Sur.).

Wercklea, with two species of small trees, is apparently endemic to Costa Rica. Regarding *Wercklea insignis* Pittier & Standley, which was not described until 1918, Standley says (*Flora of Costa Rica*, p. 677): "It is one of the most interesting and showy trees of Costa Rica and, abounding as it does along the old cart road to the Atlantic coast, in some places with its branches extending over the road so as to brush a traveler on horseback, it is hard to understand why it was not collected 50 years ago or more, for it must have been seen by most of the botanists who have visited Costa Rica." The tree is said to be from 20 to 35 feet high and 12 to 16 inches in diameter. The other species, *W. lutea* Rolfe, is similar to the preceding species but has yellow flowers (instead of lilac-rose), suggesting those of the pumpkin. According to Standley (*loc. cit.*), it is abundant in the forest about Las Nubes, forming extensive and dense stands along small streams. No uses are given for the timber. The only specimen at hand (Yale 4370) is of *Wercklea insignis* collected near La Palma by W. W. Rowlee.

Wood yellowish throughout. Luster moderately high. Odorless and tasteless. Extremely light, but rather firm; sp. gr. 0.10; weight 6 lbs. per cu. ft.; texture fine and uniform; feel rather velvety; grain straight; very easily worked; perishable when exposed to decay. Appears suitable for the same purposes as Balsa (*Ochroma*).

COMMON NAMES: Burío extrangero, flor de día, panamá (C.R.).

MARCGRAVIACEAE

THIS is a tropical American family of five genera, namely, *Norantea, Souroubea, Ruyschia, Caracasia,* and *Marcgravia,* with over 100 species of shrubs, sometimes upright but mostly epiphytic climbers. The leathery leaves are alternate, simple, sometimes dimorphic; stipules are absent; the flowers are in terminal racemes or umbels, the sterile ones with spurred or pitcherlike bracts containing nectar; the globose fruits are leathery or fleshy and indehiscent. The plants do not supply products of any special value. The woods of *Marcgravia, Norantea,* and *Souroubea,* the only ones available, are brown to reddish brown, without distinctive odor or taste, rather hard, and very coarse-textured.

Pores large; fairly numerous; occurring singly or in small multiples, diffuse without pattern. Vessels with simple perforations, though scalariform plates in *Norantea* have been reported by other investigators; no spiral thickenings seen; pits medium-sized, alternate, the apertures conspicuously coalesced. Rays fine to coarse; 1 to 8 cells wide in *Marcgravia,* the largest very high; heterogeneous, the cells mostly upright or square; pits to vessels of same appearance in surface view as the vascular; bundles of raphides sometimes present (*Marcgravia* and *Souroubea*). Wood parenchyma sparingly paratracheal and diffuse. Wood fibers commonly septate; pits numerous, simple or indistinctly bordered. Ripple marks and gum ducts absent.

Common names: *Marcgravia:* Bejuco codicia, b. palmar (Cuba); gallitos (Mex.); cachimba (Hond.); guinda-guinda, tiquire (Venez.); katjoesi anjalali (Sur.); atepele, dragona, mão de onça (Braz.); murcu-huasca (Peru). *Norantea*: Karagalla (Sur.); rabo de arara (Braz.). *Souroubea:* Caimirito (Venez.).

MELASTOMACEAE

This family comprises about 145 genera and nearly 4000 species of herbs, shrubs, woody vines, and small to occasionally large trees of pantropical distribution, ascending to alpine heights but absent from deserts and rarely extending into the temperate zones. The leaves are simple, opposite, and usually characterized by 3 to 9 nerves curving from base to apex, or 3-plinerved with parallel transverse veins at right angles between them, or sometimes pinnate-nerved (*e.g., Alloneuron* and *Mouriria*); the small to large, pink, purple, violet, rarely yellow or white flowers are borne singly or in clusters or cymes, which are terminal, axillary, or on the defoliated branches; the fruit is either a berry with few to many seeds or capsular and dehiscent. Some of the plants are decorative, some have edible fruits, and a few yield dyes, tannins, and medicinal products. Only a few attain large size and their timber is of very little economic value, though employed locally to a minor extent for charcoal, fuel, small construction, packing cases, and carpentry.

Heartwood pale yellow in two species of *Mouriria,* pale to dark brown or reddish brown in the others; usually not clearly differentiated from the lighter-colored sapwood. Luster medium to fairly high. Odor and taste usually lacking or not distinctive. Density variable from medium to very high; texture typically rather fine; grain straight to very irregular; working properties generally good, occasionally poor; durability variable, usually rather low. Of no possibilities for export.

Growth rings usually present, often irregular and indistinct. Included phloem in strands present in *Mouriria* and probably *Meliandra* (also in Old World genera *Kibessia, Lijndenia, Memecylon,* and *Pternandra*). Pores very small to medium-sized; few to numerous, but rarely crowded laterally; mostly solitary in some species of *Mouriria,* but also in multiples of 2 to 6 or in small clusters in the other genera; distribution typically uniform. Vessels with exclusively simple perforations; without spiral thickenings; intervascular pitting very fine to coarse; pits vestured. Rays very numerous, mostly uniseriate or locally biseriate, but 1 to 3 cells wide in *Blakea* and *Topobea* and up to 4 cells in two species of *Mouriria;* commonly rather low but sometimes up to 50, rarely over 100, cells high; typically heterogeneous, generally with nearly all of the cells upright or square; homogeneous in *Axinaea;* blind pits large and in vertical rows opposite large interstitial spaces; pits to vessels very small and rounded to large oval or elongated and in scalariform arrangement. Wood parenchyma various; paratracheal, aliform and confluent, with a few narrow broken metatracheal bands in *Calyptrella, Clidemia,* and *Mouriria;* often with septate cells which are loosely aggregated into groups or irregular or broken metatracheal bands in many genera; sometimes sparingly paratracheal. Wood fibers often septate; frequently in parenchyma-like patches or irregular bands with large interstitial spaces; thin to thick gelatinous inner layer common; pits small to medium-sized, simple or bordered. Ripple marks absent. No gum ducts seen.

Alloneuron (= *Meiandra*) consists of three Peruvian species with pinnate-nerved leaves and numerous-seeded berries. *A. major* Mfg. is said to have a trunk near 100 feet tall without branches. No specimen of the mature wood is available for study, but the twigs have normal structure (see *Tropical Woods* 17: 13).

Axinaea, with about 20 species of shrubs and small to rather large trees, inhabits chiefly the Andean region from Peru to Colombia and Venezuela, with a single species in Central America. This last is *A. costaricensis* Cogn., a shrub or little tree growing in the uplands of Costa Rica. The grayish wood, which is of medium density and texture and readily worked, is not utilized.

Bellucia, with 15 closely related species of small to high trees, is widely distributed in tropical America. The fruits, which are somewhat like large gooseberries, are edible. The wood is not utilized. It is pale reddish brown, fading gradually outward. Luster medium to rather high. Odorless and tasteless when dry. Moderately hard and heavy; texture medium; grain straight; working properties good; durability probably low.

Common names: Barajo (Guat.); capirote (Nic.); coronillo, papaturro agrio, skuar-bón (C.R.); manzana de corona, pomarosa de montaña (Venez.); dohiva (Br. G.); boschmispel, mespel, mispel, papajahout (Sur.); bois mêle (Fr. G.); araçá de anta, a. rana, goyaba de anta, muluba (Braz.); níspero de monte (Peru).

Blakea, with about 40 species of little trees and erect or occasionally scandent shrubs, is best represented in Central America and the northern Andean region of South America. Several species are noted for the beauty of their large flowers. The only specimen available (Yale 38425; H. E. Stork 4211) is of *B. grandiflora* Hemsl. from the mountains of Costa Rica. The heartwood is light brown, of medium luster, density, and texture, straight-grained, and easily worked, but apparently not utilized because of the small size and inaccessibility of the trees.

Common names: Catarina, San Miguel, S. M. blanco (C.R.); tira-agua (Col.).

Brachyotum, with several species of shrubs and little trees, occurs in the Andean region of South America from Colombia to Bolivia. The only specimen available (Yale 16914; Rimbach 9) is of *B. radula* Triana, a very common plant at an elevation of about 10,000 feet in the eastern cordillera of Ecuador where it attains a height of 15 to 20 feet. The wood is light reddish brown throughout, moderately dense, and fine-textured.

Calycogonium. About 35 species, all native to the West Indies, have been described in this genus, but they are nearly all shrubs of no economic value. An exception is *C. squamulosum* Cogn., a forest tree sometimes 50 feet high and 20 inches in diameter endemic to Puerto Rico where it grows in the eastern mountains. It is known as Camasey Negro, and the timber is utilized locally for railway crossties, house posts, and general construction. The following description is based upon one sample of this species and one of *C. rhomboideum* Urb. & Ekm. from Cuba.

Wood pale brownish throughout. Luster medium. Odorless and tasteless. Moderately hard and heavy, having about the consistency of Birch (*Betula lutea* Michx.); texture fine to medium; grain straight to roey; readily worked and suitable for the same purposes as Birch and Maple; durability probably rather low.

Common names: Rodwood (Jam.); cordobancillo (Cuba); camasey negro, c. jusillo (P.R.).

Calyptrella, with six species of shrubs and slender trees, occurs sparingly from Oaxaca, Mexico, to Peru. The only specimens in the Yale collections are of *C. cucullata* (Don) Triana, which is said to be a tree 10 to 35 feet high, common in second-growth stands on the lower slopes of the Andes, where it is known as Díspero or Níspero. The wood is brownish, with a slight tinge of red; of medium density and texture; easy to work, but not utilized.

Centronia includes a few species of shrubs and little trees, one in Costa Rica, the others in the Andean region of South America. The samples available were collected by A. Rimbach in the eastern Cordillera of Ecuador; they are of *C. tomentosa* Cogn., locally known as Huala, and *C. excelsa* (Bonpl.) Triana, a slender forest tree sometimes 25 feet tall. The creamy white, moderately dense, straight-grained wood is not utilized.

Clidemia, with more than 100 species of hairy shrubs and a few little trees, is widely distributed throughout tropical America. The fruits are sweet and edible. The only authentic specimens available are of *C. naevula* (Naud.) Triana collected by L. Williams in the Peruvian Amazon region. The pinkish brown, hard, moderately dense, fine-textured wood has no special uses.

COMMON NAMES: Cordobán, cordobancillo, c. peludo (Cuba); camacey, c. peludo (P.R.); guerit vite (Haiti); colación (Mex.); peluda (Salv.); sirín (Hond.); grosella azulada (Nic.); purra (C.R.); mortiño (Col.); oema snekie wiwirie, tamakoesji (Sur.); mullaca, pajar-mullaca (Peru); pixirica (Braz.).

Conostegia, with about 50 species of shrubs and small or rarely up to medium-sized trees, is generally distributed in the West Indies, southern Mexico, and Central America, and sparingly in northern South America. *C. xalapensis* (Bonpl.) D. Don has the approximate range of the genus and is one of the commonest woody members of the family in Middle America. It varies in size from a mere shrub to a spreading tree 25 to 35 feet high. Its rather large, dark purple, sweet and juicy fruits suggest blueberries (*Vaccinium*). The brownish, moderately dense, rather fine-textured wood of it and other species has no special value.

COMMON NAMES: Cordobán (Cuba); capulín, c. de cotorro, capulincillo, chicab, mora, nigua, pupu, serita, tecapulín, teshuate (Mex.); guabón (Guat.); capiroto, sirín, uva (Hond.); sirín (Salv.); cantarillo, escobillo, leña gata, lengua de gato, l. de vaca, mariquita, pu, purré (C.R.); canillito, dos caras, fruta de pava, quiera-vangué, quita-manteca, raspa-lengua (Pan.).

Graffenrieda, with about 25 species of shrubs and little trees, is widely but sparingly distributed in tropical South America. The only specimen available (Yale 18853) is of *G. limbata* Triana, a tree about 20 feet tall, collected by L. Williams in the Peruvian Amazon region where it is known as Díspero-sacha or Níspero-sacha. The pale reddish brown, moderately dense, medium-textured, straight-grained wood is not utilized.

Henriettea, with about 15 species of shrubs and small trees, occurs, usually at low elevations, from the West Indies and British Honduras to Rio de Janeiro, Brazil. The most widely distributed species, having almost the range of the genus, is *H. succosa* (Aubl.) DC., a little tree rarely 25 feet high and five inches in diameter. The pale reddish brown, hard, heavy, rather fine-textured woods of this group have no economic value.

COMMON NAMES: Boschkers, kaboanama biltere, kaloewanama, koloeanama beletere, mespel, mispel, moesoepoe, nanapolang, nonapora, pakira joejoeroe, pakiria jojoro (Sur.).

Henriettella, with about 35 species of shrubs and small or rarely rather large trees, is widely distributed throughout tropical America. The wood is similar to that of *Henriettea* and has no special uses.

COMMON NAMES: Sirín (Hond.); camasey negro, c. peludo (Col.); canilla de venado (Venez.); sergeanteklooten (Sur.); uchpa-caspi (Peru).

Heterotrichum, with several species of hispid or glandular shrubs or shrubby trees, occurs in the West Indies, Mexico, Central America, and northern South America. The fine-textured, lustrous, brownish, easily worked wood is not utilized because of the small sizes available.

Common names: American gooseberry (Jam.); camasey, terciopolo (P.R.); hoja peluda, peluda (Salv.).

Huberia, with about 10 species of shrubs and little trees, has its center of distribution in southeastern Brazil, with one representative in Peru. The only specimen available (Yale 23835) is of the Jacatirão do Grande, *H. semiserrata* DC., supplied by F. C. Hoehne from the Botanical Garden of São Paulo. The heartwood is bright reddish brown, with darker streaks, not sharply demarcated from the lighter sapwood. Apparently the only use for the plant, which has lustrous leaves and rather large yellow flowers, is for decorative purposes.

Leandra, with over 200 species of herbs, shrubs, and a few little trees, is widely distributed throughout tropical America except the West Indies. The only specimen available (Yale 12133) is of *L. dichotoma* Cogn. collected by G. P. Cooper in Panama. The wood is light reddish brown, of medium density, and fine-textured; it is not utilized.

Common names: Teshuate (Mex.); lengua de vaca (C.R.); aperta-ruão, camara-domato, pichirica, pixirica (Braz.); yuto-banco (Peru).

Llewelynia *Williamsii* Pittier, the only species, is a tree 45 to 65 feet tall and 18 inches in diameter of infrequent occurrence in the cloud forest covering the steep mountain slopes at elevations of over 3000 feet in Parque Nacional, Aragua, Venezuela. The following description is based on two wood samples collected by Llewelyn Williams in whose honor the genus and species were named.

Specimens yellowish gray throughout. Luster rather high. Without distinctive odor or taste. Fairly light but firm and strong, having about the consistency of White Birch (*Betula alba* L.); texture fine; grain irregular; both samples show collapse in drying; easy to work, but sawing rather woolly when fresh; durability presumably low. Of no commercial possibilities.

Loreya, with nine species of small to medium-sized trees, is apparently limited to the Amazon basin. *L. mespiloides* Miq. is a Guiana tree, sometimes of large size, according to Gleason (Pulle's *Flora of Surinam* 3: 215), with large leaves having fine parallel nerves. *L. arborescens* (Aubl.) DC. is a tree up to 65 feet in height in French Guiana, where it is known as Mela. The only specimen available (Yale 32650) is of *L. nigricans* Triana, a small tree collected by Adolpho Ducke in a swamp near Manáos, Brazil. The leaves are pinninerved as in *Mouriria,* but the wood is of normal structure. The lustrous light brown wood has a pinkish tinge and some dark stripes. It is of medium density and texture, straight-grained, very easy to work, and takes a high polish. Apparently it is not utilized.

Common names: Akaotombo, mespel, mispel, pakirja, wesopotare, sakwosépére (Sur.); mêle (Fr. G.).

Macairea, with several species of shrubs and little trees, is limited to the Amazon basin. Some of the plants are decorative and the bark of certain trees is the source of a black dye. In Brazil the name Cumaté is applied to these trees as well as to others which have a soft outer bark. The hard, purplish brown, medium-textured wood is not utilized.

Mecranium, with about 10 species of glabrous shrubs and little trees, is limited to the larger islands of the West Indies. The best known species is *M. amygdalinum* (Desr.) Triana, which attains a maximum height of about 25 feet; it is known in Haiti as Bois Pigeon. The brownish, moderately hard, fine-textured wood has no special uses.

Meliandra *monadelpha* Ducke, the sole species, is a tree said to be about 100 feet tall in the lower Amazon region of Brazil. It has punctate-veined leaves and is closely related to *Mouriria* (see *Archiv. Jard. Bot. Rio de Janeiro* 4: 156–158). The wood has not been studied, but it is probably of anomalous structure.

Menendezia, with three species of trees 15 to 65 feet high, segregated from *Tetrazygia,* is endemic to Puerto Rico. The only specimen available (Yale 1364) is of *M. Urbanii* (Cogn.) Britt. collected by W. P. Kramer; the local name given for the plant is Camasey de Canal. The brownish yellow, hard, fine-textured wood is not used.

Meriania, with more than 40 species of shrubs and small trees rarely 30 feet high, is widely distributed in tropical South America, especially in mountainous regions, and there are three species in the West Indies. The woods, which are dull olive-gray, hard, strong, medium-textured, have no important uses.

Miconia, with nearly 800 species of herbs, shrubs, and small to occasionally large trees, occurs throughout tropical America. The dull brownish woods, sometimes with a tinge of red, are hard and heavy to moderately so, of fine to medium texture, mostly straight-grained, easy to work, a d not durable in contact with the ground. They are used locally to some extent for interior construction.

Common names: Cenizo, cordobán, c. peludo, cordobancillo de arroyo, quitasolillo (Cuba); camasey, c. de paloma (P.R.); macrioi, trois côtes (Haiti); miconia, sardine (Trin.); cenizo, ojo de gato, sábano, teshuate tezhuate, totopozole (Mex.); bastard water-wood, maya—red, white (Br. H.); cachito, sirín, s. morado, sirinón (Guat.); quina blanca, sirín, sirinón (Salv.); cenizo, negrito, sirín, s. blanco (Hond.); canilla de mula, canillito, hoja de pasmo, lengua de vaca, María, resino, Santa María, shorka-krá, terciopelo de Santa María, zorka-kró (C.R.); cainillo, canillo, c. de cerro, dos caras, friega-platos, gorgojillo, gorgojo, mancha-mancha, oreja de mula, palo negro (Pan.); arnica, azucarero, camasey, c. amarillo, c. esquinado, c. morado, canilla de venado, jayo macho, mortiño, nigüito, punta de sarvia, tintillo (Col.); canillita, canilla de venado, morita, mortiño oiito. oreja de tigre, pepita, taparón, terciopelo, tocino, trompillo (Venez.); tara, wakradanni (Br. G.); arikadakoeballi, bakaradanballi, basterd mispel, besseroe edan, boesi smeriwiwirie, jalipi, jeseredan, jorokan pomoire, kamara kojoeroe, kopie, kremoto, lotohoedoe, totohout, lottohoedoe, maiporan, mispel, moesoepoe, oemanbarklak, pirimia, piritjo, saipiara, sakwasapére bibéroe, salero khoena, sangafoetoe soela, santo, sarero khona, seepialála, selele beletere, wakaradan, watoewena katopa, watra-kanell, witti boeka, wonoe (Sur.); colca (Ec.); bucacuru-caspi, caracha-caspi, isula-micuna, mozo-mozo, mullaca, m. colorado, mullu-caspi, níspero sacha-blanca, ñucñu-mullaca, palo blanco, pichirina, purma-caspi, rifari, rupunia, sinchi-mullaca, ubiamba, uchu-mullaca, yana-panga, yutobanco (Peru); cabucú, canella de velho, capitihú, carvoeiro, guamerim-felpudo, jacatérão, j. branco, jaguatirão, lacre branco, mará-mará, mundururu, papaterra, quina brava, sapatarinha, sapateiro, tangaraca, tintureira, vassourinha, velame do cerrado (Braz.); caá-itá (Arg.).

Mouriria, with about 65 species of small to moderately large trees and tall shrubs, is widely distributed in tropical America. The plants are not typical of the family, since the leaves are 1-nerved, the fruits are few-seeded, and the wood is characterized by included phloem (see *Tropical Woods* 17: 14).

The woods of *Mouriria* are of two types, the yellow and the brown. Out of 13 species represented in the Yale collections, there are only two in the first group. Earliest known of these is *M. pseudo-geminata* Pittier, collected by H. Pittier in the forests of Guaramales, Carabobo, Venezuela in 1909, where it is known as Paují or Pata de Paují. It grows to a height of 65 feet and in May is covered with a profusion of yellow flowers; two months later there is an abundance of rounded fruits, about an inch in diameter, having a sweet and agreeable flavor. The wood moderately dense; sp. gr. (air-dry) 0.78; weight about 50 lbs. per cu. ft. The second species, *M. Marshallii* Burtt Davy & Sandwith, was described in 1931 from specimens collected by R. C. Marshall in Trinidad. The tree, called Bois Lissette, is up to 100 feet tall and 40 inches in diameter. Marshall says (*Trees of Trinidad and Tobago*, p. 62): "It has a limited distribution in the Colony, being confined mainly to the hard dry soils which occur on the ridges in parts of the Moruga districts; here it is often locally abundant. The wood is fairly hard, yellow with characteristic white markings [included phloem strands], and is used for interior constructional work." The wood (Yale 21165) is denser than that from Venezuela, but otherwise the two species are similar. They differ from brown woods not only in color but also in anatomical details; for example, the pores are often in contact and arranged in radial rows and patches instead of being mostly solitary and irregularly distributed; the rays are frequently biseriate and occasionally triseriate instead of all uniseriate; and wood parenchyma is more abundant.

One of the most widely distributed of the brown-wooded group is *Mouriria parvifolia* Benth., a small to medium-sized tree occurring from southern Mexico through Central America to Colombia, Ecuador, and Bolivia. It has slender twigs; the leaves are small, ovate-lanceolate, and sessile; the flowers are small and white; the fruit is a red berry; the brown or reddish brown wood is very hard, heavy, tough, often with very irregular grain, difficult to work, but durable in contact with the ground. Another Central American species, *M. cyphocarpa* Standl., was discovered by the senior author in eastern Guatemala in 1926; it has much larger leaves and resembles *M. Muelleri* Cogn. of Oaxaca, Mexico. The genus is well represented in the West Indies, the Guianas, and Brazil, but most of the species are small and the timber of the larger trees has little to recommend it.

Heartwood in various shades of brown from dark grayish to reddish; not sharply differentiated from the grayish or yellowish brown sapwood. Luster low to medium. Sometimes with mild but unpleasant scent; taste not distinctive. Density high to very high; sp. gr. (air-dry) 0.90 to 1.20; weight 56 to 75 lbs. per cu. ft.; texture medium; feel harsh; grain usually irregular; difficult to work; durability fair to good. Of no possibilities for export.

Common names: Chicharron de monte, lebrero, maceṭa, mano de pilón, mirto del país, palo torcido, torcido, vigueta, yaya cimarrona, y. macho (Cuba); caimitillo, guasavara, mameyuelo, murta (P.R.); cormier (Haiti); bois lisette, monkey bone (Trin.); frutillo, hoja-viushi, yaglancito amarillo (Mex.); cacho venado, granadillo, half-crown, jug, sul-sul (Br. H.); camarón, capulín verde (Salv.); arracheche, cierito, kenna, solacra (Pan.); viguaro (Col.); guayabo, María, pata de paují, paují, pipi (Venez.); mamuriballi (Br. G.); ameraé-amoelaoe, amerau, amoelau, kammotolie, kimoto, komotolie, komotorie, k. balli, kromokoe, mamoeriballi, marakanakoewa, mopie, orokaikoballi, paroekoeroepe, pauwies mofo, piritjalaipio, siemorowi, spiekriehoedoe, spijkerhout, topie, toppie, wakapopi, wilde kers, wokopopi (Sur.); bois de fer, mouriricheira (Fr. G.); apiranga, camutim, cafézinho, casca de assahy, creoula, criuri, criviri, cruili, curiri, goyabarana, miraúba, muraúba, muiraúba, muriry, murta, m. de parida, murteira, socoró, s. zeiro, tucunaré mereçá, uapiranga, urury, xiputa (Braz.); charachuela, lanca-caspi (Peru); yabe (Boliv.).

Myriospora, with two species of hairy shrubs and small trees not more than 25 feet high, is limited to the Amazon basin. The only specimen at hand (Yale 36901; B. A. Krukoff 6823) is of *M. egensis* DC. from Brazil. The grayish brown wood is moderately hard and heavy, rather fine-textured, and easily worked, but apparently has no important uses.

Pachyanthus, with a few species of shrubs and little trees, is mostly confined to Cuba and Haiti. The only specimen available (Yale 15866) is of *P. cubensis* A. Rich. collected by A. J. Fors, Pinar del Río, Cuba. Unlike most woods in this family the heartwood, which is dark reddish in color, is rather sharply demarcated from the yellowish sapwood. It is very hard and heavy, fine-textured, and wavy-grained.

Common names: Cordobán, cordobancillo, hierro, h. del Pinar (Cuba).

Tessmannianthus *heterostemon* Mfg., the only species, is said to attain a maximum height of about a hundred feet with a buttressed bole 24 inches in diameter in the primary forest along the upper Marañon River, Peru (see *Notizbl. Bot. Gart. Berlin-Dahlem* 9: 90: 1141). The wood has not been studied.

Tetrazygia, with about 16 species of scurfy shrubs and trees rarely over 30 feet high, is limited to the West Indies. The largest size reported is for *T. pallens* Cogn., of Jamaica, Cuba, Haiti, and Dominica, which is said to reach a maximum height of 60 feet, though usually it is very much smaller. The yellowish brown or pinkish brown wood is fine-textured, moderately hard and strong, but has no important uses.

COMMON NAMES: Cordobán, cordobancillo (Cuba); cenizo, kre-kre, verde-seco (P.R.); granadillo bobo (Dom. R.).

Tibouchina, with about 230 described species of shrubs and a few herbs and trees, is widely distributed in tropical America but is most abundantly represented in southern Brazil and the Andes Mountains. The trees are infrequent and generally small, sometimes up to 45 feet high. Some of the fruits are edible. The hard, strong, pale reddish brown wood is used in Brazil to a minor extent for fuel, charcoal, and small beams and posts.

COMMON NAMES: Bois dents marron (Haiti); entrodelia (Mex.); mosqueta silvestre (Guat.); hierba del tabardillo, largona, sirín, talchinol (Salv.); macu-sacha pichirina, Santa Rosa sisa (Peru); arvore do papel, cuipeúna, flor de Maio, f. de quaresma, jacatirão do capote, jaguatirão, malmequer do campo, orelha de onça, pau de flor, p. papel, quaresma, quaresmeira (Braz.).

Tococa, with about 60 described species of shrubs and little trees rarely 25 feet high, is widely distributed throughout continental tropical America. Some of the plants are very ornamental. The leaf petioles usually bear inflated vesicles which harbor small ants. The yellowish brown or pale reddish brown, moderately hard, fine-textured wood has no special uses.

COMMON NAMES: Ant's plant (Br. G.); maranio (Peru).

Topobea, with about 30 species of small trees and erect scandent or epiphytic shrubs, is best represented in the uplands of Central America and the northern Andean region of South America, with a few species in the Amazon basin. The only sample at hand (Yale 38378; H. E. Stork 4162) is of *T. Maurofernideziana* Cogn. from Costa Rica. The brownish fairly lustrous wood is of medium density and texture, straight-grained, and easily worked, but is of no economic value.

MELIACEAE

THE Mahogany family, comprising about 50 genera and more than a thousand species of evergreen or deciduous trees and shrubs, is best represented in tropical and subtropical regions of America, Africa, and Asia, but extends into New Zealand and along the eastern coast of Australia. The leaves are alternate, rarely opposite, pinnately or digitately compound or sometimes unifoliolate or simple, and without stipules; the flowers are typically small and borne in terminal or axillary panicles; the fruit varies from a berry or a drupe to a dehiscent capsule with each cavity containing one to many seeds which sometimes are winged and imbricate in double rows. The bark is bitter and astringent.

The woods exhibit a wide range of variation in appearance and properties, but those of most commercial value are reddish in color, usually with a golden luster, sometimes fragrantly scented, of low to medium density, readily seasoned, easy to work, hold their place well when manufactured, and present an attractive and often beautiful figure. Included here are Mahogany (*Swietenia*), Spanish Cedar (*Cedrela*), and Andiroba or Crabwood (*Carapa*), of tropical America; African Mahogany (*Khaya*) and its allies (*Entandrophragma*), Tigerwood or so-called African Walnut (*Lovoa*), and Bossé (*Trichilia*), of West Africa; Red Cedar (*Toona* or *Cedrela*) and Rose Mahogany (*Dysoxylum*), of Australia; the Neem (*Azadirachta*), Persian Lilac (*Melia*), Redwood (*Soymida*), Chittagong (*Chukrasia*), and Toon (*Toona* or *Cedrela*), of India; and Calantas (*Toona* or *Cedrela*) of the Philippines. Several other genera are of local importance.

The American Meliaceae are of seven genera, namely, *Cabralea, Carapa, Cedrela, Elutheria* (*Schmardaea*), *Guarea* (including *Ruagea*), *Swietenia*, and *Trichilia* (including *Odontandra*). *Guarea* and *Trichilia* are represented in the Old World, and some botanists include *Xylocarpus* with *Carapa* and *Toona* with *Cedrela*. The Asiatic *Melia azederach* L. is a shrubby tree widely planted for ornament in tropical and warm climates which has become naturalized in America from southeastern United States throughout the West Indies, Mexico, Central America, and parts of South America, being generally known as Bead Tree, Lilac, and Paraíso; it is distinguished from the other members of the family by its large doubly pinnate leaves; the wood is ring-porous as in *Cedrela*, but the vessels in the late wood are small and clustered and have spiral thickenings. The following general description applies to native American woods of six genera.

Diffuse-porous, except in *Cedrela*. Pores small to large, being smallest in *Guarea* and *Trichilia* and largest (in part) in *Cedrela;* few to fairly numerous; solitary and, more often, in small multiples, well distributed. Vessels with exclusively simple perforations; spiral thickenings absent; gum plugs common; intervascular pitting fine to very fine. Rays uniseriate or locally biseriate in *Cabralea, Elutheria, Guarea*, and *Trichilia;* 1 to 5 cells wide in the others; up to 50, usually less than 25, cells high; mostly homogeneous in *Guarea* and *Trichilia*, more or less distinctly heterogeneous in the others, usually with single marginal rows of square or upright cells; crystals sometimes present; pits to vessels minute (*Swietenia*) to medium-sized (*Cedrela*). Wood parenchyma various; in distinct terminal or initial bands in *Carapa, Cedrela, Elutheria*, and *Swietenia;* in numerous very coarse concentric bands in *Cabralea;* in numerous, unevenly spaced, tangential or broken concentric lines or narrow bands in *Trichilia;* abundantly paratracheal and confluent into short to long bands in *Guarea;* more or less paratracheal in all genera and sometimes diffuse; crystals common. Wood fibers septate, at least in part, except in *Trichilia;* pits small, simple or indistinctly bordered. Ripple marks usually present in *Swietenia;* all elements storied. Vertical traumatic gum ducts known to occur occasionally in *Cedrela* and *Swietenia*. For anatomy of the different genera see *Tropical Woods* 66: 9–33.

Cabralea, with about 40 species of shrubs and small to large trees, occurs in the southern part of tropical America, being most abundantly represented in central and southeastern Brazil. The common name is Cangerana in Brazil and Cancharana in Argentina and applies to a group of species, although the one generally referred to in the literature is *C. cangerana* Sald. This tree is usually of medium height, but with a large trunk up to four feet in diameter. The bark contains an aromatic resin and is bitter, hence is much used in local medicine. The leaves are large, with eight or more pairs of leathery leaflets; the small white fragrant flowers are borne in axillary panicles; the fruit is a dark red, 5-celled, ovoid, woody capsule about the size of a small marble, dehiscent at the apex, with each cell containing one or two seeds which are emerald-green before maturity. The wood exhibits considerable variation in color and density, but this is attributable more to the site and conditions of growth than to differences in species. The timber is highly esteemed locally, as it has about all of the advantages of Cedar (*Cedrela*) coupled with greater firmness and strength. It is accordingly used for all sorts of construction, both interior and exterior, and for joinery, furniture, and sculpture. It is sometimes called Pau de Santo in Brazil because images of the saints are usually carved from Cangerana wood. By soaking the sawdust in water it is possible to obtain a red dye of some local utility. The supply of good timber is rather limited and is readily consumed by the domestic demand.

Heartwood typically dull red or maroon, with fine parenchyma markings; sometimes lighter colored, with purplish streaks; not always sharply demarcated from the pinkish sapwood. Without distinctive taste;

with fragrant scent when fresh, but losing it eventually upon drying. Mostly of medium density, but variable; sp. gr. (air-dry) 0.65 to 0.85, av. about 0.70; weight 40 to 53, av. about 44, lbs. per cu. ft.; texture medium to coarse; grain generally straight, sometimes wavy; easy to work, finishing smoothly; holds its place well when manufactured; is rather brittle; dark-colored material highly resistant to decay and insects.

COMMON NAMES: Cajarana, cangerana, c. grande, c. mirim, canharana, canjarana, canxarana, cayarana, cedro cangerana, pau de santo (Braz.); cancharana, canxarana, cedro macho, chanchorena, chanchorona (Arg.); congerana (Urug.); cedro-rá (Par.).

Carapa, with several closely related species of evergreen trees, occurs in West Africa and in tropical America from the West Indies and Central America to Peru and Brazil. The principal African species, *C. procera* DC., is said to occur also in the West Indies and northern South America, but the best known and most widely distributed American tree is *C. guianensis* Aubl. The principal common names are Andiroba in Brazil and Crabwood in British Guiana. The tree is of limited occurrence in the overflow delta lands of the Orinoco in Venezuela and is very common in the Amazon flood plains and in the Guianas, sometimes growing in nearly pure stands. At its best it is said to attain a height of 170 feet and a diameter of six feet above the buttresses, and is often 100 feet tall and three feet through. It has very large pinnate leaves, with rather numerous leathery leaflets; the small white flowers are borne in axillary or terminal panicles; the fruit is a large globose dehiscent capsule containing several smooth pale brown angular seeds as large as a horse chestnut (*Aesculus*). These seeds are the source of an oil used industrially for making soap, and in order to protect the industry in the State of Pará, Brazil, felling of the trees for timber has been prohibited. The bark is employed to a minor extent in tanning and contains an alkaloid, carapina, of some medicinal application.

Crabwood is popular in British Guiana and is used there for furniture and all kinds of construction work. There is considerable difference in the quality of the wood, depending upon the locality of growth. Trees in Mangrove swamps are small and the timber is of such poor quality and splits so badly that it is not used commercially. Growth in riparian swamps is better, but the best timber comes from creek banks which are only periodically inundated. The last two kinds of Crabwood are sometimes referred to, respectively, as "lowland" or "white" and "upland" or "red," but such terms are inaccurate as the species is rare on hillsides and the color variations of the wood are to be found in trees from the same site and even in the same tree. It is said that the presence of short longitudinal grooves in the surface of a peeled log is an indication of good quality. Tests on Crabwood at the Forest Products Research Laboratory at Princes Risborough, England, indicate the following: "Seasoning must be done carefully to avoid warping and checking. In the matter of strength, Crabwood compares favorably with Black Walnut (*Juglans nigra* L.). Straight-grained lumber finishes well, but roe-grained material requires considerable sanding to produce a smooth finish, owing to local tearing out of the fibers. Crabwood should be useful for such cabinet work as the carcassing in furniture suites, chair and table lengths, instrument cases, drawer linings and possibly drawer fronts, and for general hardwood joinery, such as shop fittings, display cabinets, and cupboards. Quarter-sawn material would appear to be particularly suitable for mouldings and also for plain and automatic square turning."

Attempts to establish a regular market for Crabwood lumber in the United States and England have not been successful, but with more attention to the quality of the timber exported this situation may be remedied. Brazilian material of the same species has been more favorably received and on this account Crabwood has recently been rechristened Empire Andiroba. J. F. Müller & Sohn, Hamburg, Germany, imported Brazilian Andiroba for the first time in 1938 and report that it is a firm and very usable wood which evidently can only be supplied in small quantities and diameters. In Colombia, according to Armando Dugand (*Tropical Woods* 31: 48), the timber, known there as Masábalo, is generally considered somewhat inferior to Albarco (*Cariniana pyriformis* Miers), but is acceptable as a substitute; shoemakers prefer it for the making of heel pieces.

The following information regarding the silviculture of *Carapa guianensis* in Trinidad is condensed from a report by R. C. Marshall (see *Tropical Woods* 27: 26): Tree not exacting as to soil and site, provided they are not too dry. Young plants produce taproots, but the tree tends to become surface-rooted. Flowers about June. Fruit, which is about the size of a cricket ball and contains a dozen seeds, requires about a year to mature; falls throughout the year, though mostly at beginning of the rainy season. Seeds large and readily collected from under the trees; subject to insect damage and, therefore, should not be stored. High percentage of sound seeds germinate within six weeks. Early growth fairly rapid. For direct seeding, plant one or two seeds one-half inch deep in spots spaced five feet apart, without preliminary working of the soil. For transplanting, undercut the roots during rainy weather, leaving four or five inches of the taproot, and allow the plants to stand a few weeks until new roots are formed; in this way plants up to three feet high can be successfully transplanted. Young trees do best under partial shade; excessive cleaning should be avoided. Crappo coppices well when not too old. In plantations, a borer (*Hypsipyla grandella* Zell.) attacks the shoots and also infests the seeds toward the end of the dry season. Young trees are badly browsed by deer.

The continental range of *Carapa* extends as far north as British Honduras. The principal species is *C. guianensis*, but two others have been described, namely, *C. Slateri* Standl., of Panama and Costa Rica, and *C. nicaraguensis* C. DC., of Costa Rica and Nicaragua. Regarding the last, F. C. Englesing says (*Tropical Woods* 17: 29) that it is a large tree of the shady forest on low hills, its frequency of occurrence being about one tree per acre between the Rawawas and Kukalaya Rivers. Specimens collected for the Yale School of Forestry (Yale 1231; Englesing 47) were from a tree 100 feet tall, with a cylindrical trunk free of branches for 75 feet and 15 inches in diameter above the branched buttresses which extended to a height of six feet. The freshly cut sapwood is white near the bark, shading inwardly to pink; the heartwood is light sepia. The timber is used locally to some extent for constructing buildings. The heartwood of the type tree of *Carapa Slateri* (Yale 10157; Cooper & Slater 59; see *Tropical Woods* 10: 49) is lighter in color than most samples of *Carapa* and has a high golden luster, but this is probably not a specific character. In general the appearance of *Carapa* wood is dull and plain, but there are exceptions for all species and localities of growth. The Central American trees are not abundant, but the timber is of the same type and useful for the same purposes as that from the Guianas and Brazil. Though usually of less attractive appearance than Mahogany, it should otherwise be fully as serviceable in plywood construction, particularly for motor boats.

Heartwood rather light to dark reddish brown; not always sharply defined from the pale brown or oatmeal-colored sapwood. Luster frequently low, sometimes golden. Odor and taste absent or not distinctive. Mostly of medium-low density, but firm and strong; sp. gr. (air-dry) 0.60 to 0.75; weight 37 to 47 lbs. per cu. ft.; texture rather coarse; grain generally straight, sometimes roey; technical properties fair to very good; takes paint and glue well; is durable.

Common names: Najesí (Cuba); cabirma de Guiana (Dom. R.); bois rouge carapat (Guad.); crabwood, crappo (Trin.); bastard mahogany (Br. H.); caobilla, cedro macho (C.R.); bateo, cedro bateo, c. macho, saba (Pan.); masábalo (Col.); carapa (Venez.); caraba, crabwood (highland, lowland, upland, red, white) empire andiroba (Br. G.); ietjoenban ka raäpa, kaäpa, karaba, kelaba, keraba, ke rapa, krappa, Surinaamsch mahonie (Sur.) bois caille, cachipou, carapa, c. blanc, c rouge (Fr. G.); figueroa, tangaré (Ec.)

andiroba (Peru); andiroba, a. branco, a. do igapó, a. saruba, a. vermelha, andirova, angiroba, camaçari, nandiroba, yandiroba (Braz.).

Cedrela, in a restricted sense, comprises numerous closely related and doubtfully distinct species of medium-sized to very large and important timber trees occurring in every country south of the United States except Chile. It has its counterpart in the *Toona* of Asia and Australia, a genus so closely related that some botanists merge it with *Cedrela.* The principal distinction is in the way the seeds are winged; there are no fundamental differences in their woods, which are more or less ring-porous, of a pinkish or reddish color, fragrantly scented, soft and easily worked, and highly resistant to decay and insects. Because of their fragrance, *Cedrela* trees and woods are generally known to English-speaking people as Cedar or Spanish Cedar and as Cedro to most Latin Americans. (The name Cedar was originally applied to species of *Cedrus,* a small group of coniferous trees growing in northern Africa and western and southern Asia.)

The first species, *Cedrela odorata* L., was described by Linnaeus in 1759 and was formerly credited with a much wider range than now. It is usually confused with *C. mexicana* Roem., but Alfred Rehder, of Arnold Arboretum, in a letter of May 28, 1937, says: "The two species are very close and, judging from the material in our herbarium, I am doubtful if they are really specifically distinct." The principal species of southern South America is *C. fissilis* Vell. So far as the woods of *Cedrela* are concerned they might well be of a single species, for although they exhibit considerable range in their properties, the differences observed could all be attributed to the age and conditions of growth of individual trees. The wood of young trees, especially of those of very rapid growth in the open, is less fragrant, of lighter color, and softer, though tougher, than that of old forest-grown trees.

In an account of the Peruvian Cedar by F. L. Herrera (*Revista Sudamericana de Botánica* [Montevideo, Urug.] 1: 21–27, 1934) it is stated that two forms of a single species are recognized, namely, Atoc-cedro, 50 to 65 feet high, growing along streams and producing fibrous, light-colored, porous, slightly scented wood, and Cedro Virgen, 80 to 100 feet tall, in hillside forests, supplying reddish, compact, highly resinous timber having a pungent odor, and much more highly valued for making furniture. Similar evidence as to the effect of site on the quality of the timber grown in Ecuador is given by M. Acosta Solís (*Tropical Woods* 57: 2. 1939), who says that Cedro "makes its most rapid growth in sheltered places where the air is very humid, . . . but the wood is too porous and light-colored to suit the furniture industry, particularly for use as veneer. On the other hand, timber produced in drier regions is denser and more deeply colored and therefore is in greater demand and more costly."

Under favorable conditions in the forest *Cedrela* attains stately proportions, often with heights of 100 feet or more and a straight cylindrical bole three to six feet in diameter above the substantial buttresses and free of branches for 40 to 60 feet. The large pinnate leaves have numerous entire leaflets and are deciduous; the small flowers are borne in panicles at or near the ends of the branches; the fruit is a woody capsule, much smaller and thinner-shelled than that of *Swietenia,* opening by five valves and liberating numerous small seeds, each with a papery wing at the lower end. The aromatic astringent bark is used medicinally as a tonic and febrifuge.

Cedrela supplies the most important timber for domestic use in tropical America. It is very easy to work, dries readily without warping or splitting, is strong in proportion to its weight, holds its place well when manufactured, and the better grades are attractive in color, grain, and odor, and are highly durable. The characteristic figure consists of a series of dark lines on a red background, but sometimes the wood is roe-grained and that of buttresses and burls is attractively figured. The denser kinds usually have a golden luster like Mahogany and are suitable for fine furniture and interior trim. Cedro serves almost every purpose for which lumber is needed in the tropics, but the principal use in the United States has been for making cigar boxes, as it was claimed that the volatile oil imparted a desirable aroma to the tobacco. Comparatively few cigars are now packed in Cedar boxes and log imports into northern countries have dwindled to a small fraction of their former volume. The principal defect in the lumber is the presence of too much gum, which may exude and discolor the surface. The tendency of oily vapors to condense on cold surfaces renders the lumber unsuitable for the lining of cabinets or cases containing glassware or metallic instruments. Loss of a foreign market is not serious for the producing countries, for the available supply of good Cedro timber is scarcely sufficient for their local needs.

Cedrela trees are often planted along streets and sometimes for shade for coffee and cacao, but comparatively few forest plantations have been established and little has been done to assure natural regeneration. The following information regarding the silviculture is condensed from R. C. Marshall's report on his experience with *Cedrela mexicana* in Trinidad (see *Tropical Woods* 27: 25). The tree is at its best on rich, well-drained clay soils of the older territory formations in Trinidad and on igneous soils in Tobago; prefers calcareous soils in sheltered positions on slopes and hills; it is rather exacting in its requirements and highly intolerant of waterlogging. The root system is superficial. It flowers in July, and the fruit ripens the following April or May. Trees in the open bear fruit every year, those in the forest less frequently. Ripe pods should be collected from the tree and placed in the sun to open. Each pod contains about 40 seeds and there are about 16,000 seeds per pound. Germination is usually good, up to 90 per cent or so within a fortnight. From early sowing on good, well-drained soil, with side protection but full light overhead, seedlings attain a height of four feet, sometimes six to eight feet, the first year. Direct seeding is feasible, and sowing in strips two feet wide has given promising results. The seed should be covered lightly, if at all. Seedlings four feet high can be transplanted during the dry season when they are leafless; also at the beginning of the rains, if new growth is trimmed off. Cutting back to within a few inches of the ground has not given good results. Present indications are that Cedar should be grown in mixture with other trees which will give it the necessary side protection, and that an evergreen underwood is necessary to keep the soil in good condition. Natural regeneration is often possible by clearing around seed trees. Cedar does not coppice. Its principal enemy is the shoot borer.

Heartwood pink to red or reddish brown, sometimes with a purplish tinge, fairly uniform in a given specimen; sharply to rather poorly demarcated from the pinkish to white sapwood. Luster medium to high and golden. Scent and taste distinctive and pleasant; very mild to pronounced. Density greatly variable; sp. gr. (air-dry) 0.37 to 0.75 (*Cedrela salvadorensis* Standl. being the heaviest tested at Yale); weight 23 to 47, av. about 35, lbs. per cu. ft.; texture rather fine and uniform to coarse and uneven; grain usually straight; some specimens are crisp under tools, others tough and fibrous; other properties as indicated previously.

COMMON NAMES: Cedar (cigar-box, Cuban, Jamaican, Spanish, West Indian, etc.), cedrela wood (Eng.); cedro (Span., Port.); cedro caracolillo, c. de ramazón, c. hembra, c. macho (Cuba); c. hembra (P.R.); cèdre (Haiti); acajou amer, a. femelle, a. rouge, cajou senti, cèdre acajou (Fr. W.I.); leli (Curaçao); calicedra, cedro chino, c. colorado, c. fino, c. hembra, c. liso, c. macho, c. oloroso, cóbano, cuché, kuché, kuiché, kulché, nogal cimarrón, n. corriente (Mex.); cedro blanco, c. macho (Salv.); cedro real, yalam (Nic.); aluk, cedro amargo, c. blanco, c. colorado, c. cóbano, c. dulce, cóbano, rru-argá, rru-rrugá, rruk, runkrá, táali, tali, tirikrú, tirigú, uara-krá, uluk, uruk (C.R.); cedro dulce, c. colorado (Pan.); cedro caoba, c. colorado, c. oloroso, c. real (Col.); cedro amargo, c. dulce (Venez.); red cedar (Br. G.); akkojaarie, akoejallie, ceder, c. hout, cedoe, cedre, kurana, samariehout, semmarie-apo (Sur.); acajou femelle, bois de cèdre, cèdre acajou, c. odorant, cedrel (Fr. G.); cedro colorado

(Ec.); atoc-cedro, cedro colorado, c. virgen (Peru); acajù catinga, basákiva, cedro amarello, c. bordado, c. branco, c. cheiroso, c. da varzea, c. rosa, c. roxo, c. vermelho, uenkutanema, yaporaissib (Braz.); cedro colorado, c. menotti, c. pinta, igary (Par.); cedro de Misiones, c. de Salta, c. obscuro, c. rojo (Arg.).

Elutheria (or *Schmardaea*), with two doubtfully distinct species, *E. microphylla* (Hook.) Karst. and *E. nobilis* (Karst.) Tr. and Pl., is of infrequent occurrence in the Andes Mountains from Peru and Ecuador to northern Colombia and Venezuela. The plants are shrubs or small trees 15 to 30 feet high. The leaves are unequally pinnate with 3 to 6 pairs of rather small, hairy, dentate, nearly sessile leaflets; the comparatively large, greenish flowers are borne in few-flowered axillary racemes; the fruit is a fusiform, dehiscent, 4-valved, woody capsule, the separate, reddish, inner layers with fibrous connectives; the seeds, which are winged at one end, are imbricated in four double rows, one in each valve of the capsule. So far as known, no wood samples from a main stem have ever been collected. The following description is drawn from a twig of *Elutheria nobilis* collected by J. Saer (No. 475) at an elevation of 4000 feet in Lara, Venezuela. Heartwood probably brown (judging from a wound); sapwood gray. Hard and heavy; of fine and uniform texture. Other properties unknown. Presumably useless because of the scarcity, small size, and inaccessibility of the trees.

COMMON NAME: Curito (Col.).

Guarea lends itself readily to the making of new species and already about 200 have been described. They range in size from shrubs to large trees and are mostly tropical American, though there are several in West Africa. The timber known commercially as Bossé is produced by *G. cedrata* (A. Chev.) Pellegr. of the Ivory Coast (see *Tropical Woods* 20: 10–14). Other species with similar wood occur from Liberia to the Belgian Congo.

The numerous American species are of minor importance for timber because of the small size or scarcity of the trees. Best known and most widely distributed is *Guarea trichilioides* L., or *G. guara* (Jacq.) P. Wils., usually a small or medium-sized tree, but sometimes over 100 feet tall, distributed from the West Indies and Central America to Argentina and southern Brazil. The equally pinnate leaves have 4 to 10 pairs of large pellucid-lined leaflets and continue for some time to produce new leaflets at the apex; the small white fragrant flowers are borne in axillary panicles; the fruit is a small globular apically dehiscent capsule with 2 to 4 cells, each containing a single seed inclosed in a scarlet aril. All parts of the tree have a musk-like scent. The powdered bark is used as an emetic and a hemostatic. The timber was formerly employed locally in the West Indies for the same general purposes as Mahogany (*Swietenia*), but the supply was never abundant and is now practically exhausted.

Heartwood pinkish to deep reddish brown; distinct but not sharply demarcated from the thick whitish or brownish sapwood. Luster rather low. Odor and taste very mild or not distinctive in dry specimens. Rather light to moderately heavy; sp. gr. (air-dry) 0.58 to 0.70; weight 36 to 44 lbs. per cu. ft.; texture medium; grain straight; easy to work, finishing very smoothly; is strong for its weight; deeply colored heartwood is durable.

COMMON NAMES: *Guarea trichilioides:* Alligator wood, musk wood, wild akee (Jam.); guaragáo, yamagua, y. colorado, yamáo (Cuba); guaraguáo (P.R.); cabilma, cabirma (Dom. R.); bois rouge (Haiti); bois à balles, b. balle, b. pistolet, b. rouge de Dominique, pistolet (Fr. W.I.); bailador, bilibili, guanco, mestizo, trompete, trompillo, zambo-cedro (Col.); trompillo, trompito (Venez.); carababalli (Br. G.); bois balle, guaré (Fr. G.); latapi, l. caspi, requía (Peru); açafroa, bilreiro, camboatá, cangerana miúda, carrapeta, c. verdadeira, cedrão, cedro branco, c. rana, cedrohy, cedro-y, gitó, guaré, jatuaúba, j. branca, jitó, macaqueiro, marinheiro, pau bala, p. de sabão, taúva, yaguá ratai (Braz.); comboatá, c. blanco (Arg.). *Other species:* Guaraguadillo (P.R.); cedrillo, c. cimarrón, chichón de montana, chohalaté, guaraguao, nochocche, ocotillo blanco, trementino (Mex.); cramantee (Br. H.); cedrillo (Guat.); carbón (Hond.); quitacalzón (Salv.); prontolivia (Nic.); campano, caoba, cocora, sota-caballo, turubúk (C.R.); dorita, mamecillo blanco (Pan.); guacharaco de terra fria, guamo blanco, g. cimarrón, tigre (Col.); cabimbo, caóbano, cedrillo horcón, hojiancho, tortolito (Venez.); kufiballi (Br. G.); gomma, g. hout, jarreewé, joekoetoena, karaballi, kodjo oedoe, koejakè fehoeta, saffeka, siwaroewa (Sur.); paujil-ruru (Peru); trompillo de monte (Bol.); atauba, camboatá, café branco, calcanhar de cotia, carrapeta, cayrana, cedrilho, ciricó, jatuaúba, j. preta, macaqueiro, marinheiro, pieto do pombo, tuaiussú, utuapoca (Braz.); cedrillo, guaimiré (Par.); cedrillo, c. blanco, guaimipiré, guaré (Arg.).

Swietenia, the source of the original or true Mahogany, the premier cabinetwood of the world, occurs in southern Florida, the West Indies, Mexico, Central America, Colombia, Venezuela, and the upper Amazonian region. The leaves, which usually are evenly pinnate, have 2 to 6 pairs of leaflets; the small whitish or greenish flowers are borne in axillary panicles; the fruit is an ovoid 5-valved capsule, 2 to 6 inches long, with a thick woody exocarp and much thinner leathery endocarp, the valves and seeds eventually falling away and leaving a 5-winged receptacle; the numerous seeds are imbricate in two rows in each cell and have a more or less quadrangular body and a terminal oblong wing.

The genus was described by Jacquin in 1760, and his single species, *Swietenia mahagoni,* was based on plate 81 in the second volume of Catesby's *Natural History of Carolina* which shows the fruit, leaves, and some withered flowers of a tree in the Bahamas. This was not the first binomial for the tree, however, as Linnaeus, the year before, had made the same illustration the basis for the name *Cedrela mahagoni.* About 1836, a second species, *S. humilis,* was described by Zuccarini from specimens collected in southwestern Mexico. It is distinguished chiefly by the fact that the leaflets are nearly sessile instead of having distinct petiolules, and the seeds are light brown instead of dark brown. A third species, *S. macrophylla* King, described in 1886, was based on trees grown in the Botanic Garden at Calcutta, India, from seeds reputedly from Honduras. It is distinguished from *S. mahagoni* by its appreciably larger leaves, flowers, fruits, and seeds. The first two species are of slower growth and have denser wood than *S. macrophylla.* These three species are fairly distinct and can be assigned definite ranges, namely, *S. mahagoni,* in the West Indies and southern Florida; *S. humilis,* in dry places along the Pacific coast from southwestern Mexico to Costa Rica; *S. macrophylla,* in regions of abundant rainfall from the Yucatan Peninsula through Central America into Colombia and Venezuela, and also in Peru and extreme western Brazil (Maps 6 and 7).

Swietenia humilis is of very little commercial importance. A form of it, *S. cirrhata* Blake, was described by S. F. Blake (*Journ. Wash. Acad.* 10: 286. 1920) as a distinct species, but Standley says (*Tropical Woods* 21: 6; March 1930): "Blake, although admitting that their ranges overlapped, divided the Pacific coast Mahogany into two species, chiefly on differences in the size of the leaflets, obviously not a character of great weight in distinguishing units. It was suggested, also, that the leaflets of *S. cirrhata* had longer and more slender cusps than those of *S. humilis.* This, I think, is a matter of accident. In the young leaves the cusps are very long, but they are also fragile and soon are broken off by the wind. . . . One is forced to the conclusion that only a single species is represented and that we must regard the Mexican and Central American Mahoganies as referable to only two species, *S. macrophylla* of the Atlantic coast and *S. humilis* of the Pacific slope."

Three South American species of *Swietenia* have been proposed, but each has been based on specimens from a single tree and, for all practical purposes at least, may be considered mere forms of *S. macrophylla.* The first, a Venezuelan tree, was named *Swietenia Candollei* by Pittier in 1920, but in 1921 (*Bol. Com. & Ind.* 18:

MAP 6. Range of Mahogany in West Indies, Mexico, and Central America.

585) he admits the inadequacy of his material and states that the only thing that may be affirmed at present is that the genus contains three fundamental types represented respectively by *S. mahagoni*, *S. humilis*, and *S. macrophylla*.

The rediscovery of Mahogany in the upper Amazon region in 1923 was followed by the description of two supposedly new species. *Swietenia Tessmannii* Harms (*Notizbl. Bot. Gart. Berlin-Dahlem* 10: 180. 1927) had as its basis some specimens from Yarina Cocha on the middle Ucayali and is distinguished by its author from *S. macrophylla* by the longer leaflet-petioles and the looser inflorescence. Macbride says (*Tropical Woods* 16: 50. December 1928): "It seems to me that the characters relied upon to separate these species [*S. Tessmannii*, *S. Candollei*, and *S. macrophylla*] are not convincing and not unlikely may prove to be relative in nature and valueless for purposes of classification." Equally unconvincing are the features distinguishing *S. Krukovii* Gleason & Panshin (*Am. Jour. Bot.* 23: 21. 1936), for the diagnosis is concerned chiefly with the size and shape of the upper two pairs of leaflets from a single specimen.

The original scientific discovery of Peruvian Mahogany was made in 1784 by Hipólito Ruiz, the eminent Spanish naturalist who was in charge of a botanical expedition to Peru and Chile during the years 1777–1788. In his account of the many plants found near Pozuzo, in the region drained by the Río Pachita, a tributary of the Ucayali, appears a brief note on *Swietenia macrocarpa*, "a tall tree with a large trunk and valuable wood." The next reference was in 1878, when Alphonse de Candolle (*Monographiae Phanerogamarum* 1: 723) identified as *Swietenia Mahagoni* ("*Mahogani*") the specimens collected by Ruiz 94 years earlier. The determination of the Peruvian species as *S. macrophylla* was made by S. F. Blake, at first tentatively on the basis of leaf specimens obtained at the senior author's suggestion by Georges H. Barrel, President of the Aguna Mahogany and Timber Company of Boston, on the Río Itaya, some 50 miles from its confluence with the upper Amazon (see *Tropical Woods* 6: 1. 1926). Two years later Blake wrote (*Tropical Woods* 14: 33): "This identification is now confirmed by the receipt of a nearly complete pod containing seeds, with portions of another, collected at Nanay, Río Amazonas, Peru, for Professor Record under direction of Mr. Barrel. . . . De Candolle's record undoubtedly belongs to the same species."

It should not be inferred from the foregoing discussion that the wood of *Swietenia macrophylla* is uniform throughout its wide distribution. On the contrary there is much variation in appearance, density, texture, and technical properties. The differences, however, which occur throughout the entire range are not appreciably greater than can be found within the boundaries of one small country, such as British Honduras, and are therefore attributable to site and conditions of growth rather than to differences in species. For commercial purposes *Swietenia humilis* Zucc. can be practically ignored, and all of the Mahogany of continental North and South America can be considered as of one botanical species, *Swietenia macrophylla* King. Whether observed differences in the timber can be correlated with botanical varieties, forms, or races remains to be determined.

Mahogany is the most valuable timber tree in tropical America. Its use by European colonists and explorers dates back at least to the sixteenth century. According to George N. Lamb (*The Mahogany Book*, 2nd ed., pp. 8–10), "the earliest surviving use of Mahogany is that of a rough-hewn cross preserved in the Cathedral of St. Domingo and bearing the legend: 'This is the first sign planted in the center of this field to mark the beginning of this magnificent temple in the year 1514.' The cathedral, completed in 1550, has much carved Mahogany woodwork, some of it considered the finest in the world, still in splendid condition after nearly four centuries in the tropics. Mahogany was early established as a ship-building wood and Cortez used it for the construction of ships for further voyages of discovery. . . . The first known European use of Mahogany was in the Escorial begun by Philip II of Spain in 1563 and completed in 1584. . . . The

Photo by F. H. Fischer

PLATE XXX. Hauling Mahogany logs from the forest near Vaca Falls, British Honduras, to a tributary of the Belize River for floating down to the seaport.

Photo by J. W. Gonggrijp

PLATE XXXI. Splitting out the small core of figured heartwood from a Letterwood tree (*Piratinera guianensis*) in Surinam.

earliest use of Mahogany known in England was in Nottingham castle built in 1680."

No one knows when Mahogany was first introduced into England, but it was probably used in shipbuilding long before it became fashionable for furniture, its identity concealed under the non-distinctive name of Cedar. In an account of the trees of Bermuda about 1619 (see *The World Displayed,* London, 1760, Vol. 4, Chap. 12), the native Cedar is described as "firmer and more durable than any of its kind we are acquainted with and answers in every respect to Oak timber. It is therefore used in shipbuilding." Certain rooms in Nottingham castle were wainscoted and floored in 1680 with "Cedar wood," as shown by the original bill for the timber, but contemporary evidence of the hardness and beauty of the woodwork leaves no room for doubt that the wood was Mahogany. It was probably to avoid the confusion with other kinds of Cedar (*Cedrela* and *Juniperus*) that the English settlers began to use the name Mahogany, presumably a term of native origin. Its first recorded use is as "Mohogeny" in Ogilby's *America,* in 1671. Various spellings were subsequently used—Mohogony, Mohogany, Muhagnee, Mehogeny, Mehogenny, Mahagoni, Mahoginy, and Mahogany—the last mentioned appearing for the first time about 1724. The French name for Mahogany is Acajou, and this apparently owes its origin to the early practice of coating the ends of the logs with resin from the Acajou or Cashew tree (*Anacardium occidentale* L.).

The earliest mentioned of Mahogany ("Mohagony wood") in an English newspaper appears to be in an advertisement in the *London Gazette,* February 22 to 25, 1702, regarding the sale of the cargoes of two prize ships. The first reference in the statistics of imports filed at the Public Records Office is for the year "Xmas 1699 to Xmas 1700" and pertains to a small lot of "Mohogony wood" from Jamaica. In an authoritative paper on *Early Imports of Mahogany for Furniture* (*The Connoisseur* [London] October and December 1934), R. W. Symonds says:

"Taking into consideration all the available evidence, I think it is permissible to state that Mahogany was employed in England from 1715 onwards for the making of tables, sometimes of gate-legged construction, but usually with straight round legs terminating in club feet or with the plain cabriole-shaped legs. Tables such as these were made in considerable numbers by many firms of London joiners and cabinet-makers and also by provincial furniture makers who lived in towns where a supply of imported Mahogany was available. Previous to 1715, Mahogany tables were only made sporadically owing to the cabinet-makers not being able to obtain a regular supply of the wood. . . . Mahogany overcame the difficulty of making table tops, owing to the large widths of the planks of this wood in comparison to Walnut. It was for this reason that Mahogany became at once popular with cabinet-makers. On its introduction, numerous new types of tables were designed, the construction of which would not have been possible in Walnut. These tables were not only made for the wealthy classes, but large numbers were produced of a plain character for the less well-to-do householder. Evidence in support of this last statement is to be found in the very large quantity of flap and tripod tables of a plain design that have survived. The number of the latter, however, has considerably decreased within recent years owing to the obnoxious habit of the furniture faker of carving up the plain example so that he can pass it off to the unwary collector, at a high rate of profit, as a period piece with the claw-and-ball feet. . . . It would appear from the contemporary information, cited from the statistics of imports, Sheraton's *Cabinet Dictionary,* and the *History of Jamaica,* that the first Mahogany to be imported into England in the early eighteenth century was Jamaican and, afterwards, Cuban. In the third quarter of the same century, Honduras Mahogany was imported. The reason for the cessation of any particular variety of Mahogany was because the trees near the coast having been felled, the traders sought another supply which was cheaper, owing to its being more easily procurable. It was not so much a question of seeking wood of fine quality, otherwise exporters would have gone to the trouble and expense of transporting the better quality timber from the interior."

The following description of the Mahogany tree and of the different kinds of its timber appears in the *History of Jamaica*, 1774. "This graceful and valuable tree, which furnishes a constant share toward the annual exports from the island, grew formerly in great abundance along the coast; but, having been almost exterminated from those parts in process of time, it is at present found chiefly in the woodland, mountainous recesses, where vast quantities of it still remain, particularly in the uncultivated districts of Clarendon, and the leeward parishes. It thrives in most soils, but varies in its grain and texture. What grows in rocky ground is of small diameter, but proportionally of closer grain, heavier weight, and more beautifully veined. What is produced in low, rich, and moist lands is larger in dimensions, more light and porous, and of a paler complexion. This constitutes the difference between the Jamaica wood and that which is collected from the coast of Cuba and the Spanish Main; the former is mostly found on rocky eminences; the latter is cut in swampy soils, near the seacoast. The superior value of the Jamaica wood, for beauty of coloring, firmness, and durability, may therefore be easily accounted for; but, as a large quantity of balks and planks is brought from the Spanish American coasts to this island, to be shipped from thence to Great Britain, the dealers are apt to confound all under the name of Jamaica wood, which in some measure hurts the credit of this staple production. The tree grows tall and straight, rising often sixty feet from the spur to the limbs; the foliage is a beautiful deep green; and the appearance, made by the whole tree, so elegant, that none would be more ornamental for an avenue or to decorate a plantation. It generally bears a great number of *capsulae* in the season. The flowers are of a reddish or saffron color; and the fruit, of an oval form, about the size of a turkey's egg. It is easily propagated from the seeds and grows rapidly. Some of them have reached to a monstrous size, exceeding one hundred feet in height, and proportionably bulky. One was cut, a few years since, in St. Elizabeth's, which measured twelve feet in diameter, and cleared to the proprietor about £500 currency. . . . We may imagine the plenty of it in former times here when it used to be cut up for beams, joists, plank, and even shingles. But it is now grown scarce within ten or twelve miles from the seacoast, and must every year become still scarcer, and consequently dearer, unless nurseries, or plantations, are formed of it in places where the carriage is more convenient for the market. In felling these trees, the most beautiful part is commonly left behind. The Negro workmen raise a scaffolding, of four or five feet elevation above the ground, and hack off the trunk, which they cut up into balks. The part below, extending to the root, is not only of the largest diameter, but of closer texture than the other parts, most elegantly diversified with shades or clouds, or dotted, like ermine, with black spots; it takes the highest polish, with a singular lustre, so firm as even to reflect objects like a mirror. This part is only to be come at by digging below the spur to the depth of two or three feet and cutting it through; which is so laborious an operation that few attempt it, except they are uncommonly curious in their choice of the wood, or to serve a particular order. Yet I apprehend it might be found to answer the trouble and expense, if sent for a trial to the British market, as it could not fail of being approved of beyond any other wood, or even tortoise-shell which it most resembles."

Sheraton gives an account of Jamaican Mahogany in his *Cabinet Dictionary* (1803), and is chiefly concerned with three types, designated as Cuba, Spanish, and Honduras Mahogany. Cuba wood is "a kind of Mahogany somewhat harder than Honduras wood, but with no figure in the grain. It is inferior to Spanish wood, through probably the Cuba and Spanish Mahogany are the same, as the island of Cuba is a Spanish colony. . . . That, however, which is generally distinguished by Spanish Mahogany is finer than that

MAP 7. Range of Mahogany in South America.

called Cuba, which is pale, straight-grained, and some of it only a bastard kind of Mahogany. It is generally used for chair wood, for which some of it will do very well." Regarding the Honduras variety, he says: "From this province is imported the principal kind of Mahogany in use amongst cabinet-makers, which generally bears the name of Honduras Mahogany, and sometimes Bay-wood, from the bay or arm of the sea which runs up to it. The difference between Honduras and Spanish wood is easily perceived by judges, but not by others unskilled in wood. The marks of the former are, as to size, its length and width, which generally run much more than in the latter wood. . . . The grain of Honduras wood is of a different quality from that of Cuba, which is close and hard, without black speckles, and of a rosy hue, and sometimes strongly figured; but Honduras wood is of an open nature, with black or grey spots, and frequently of a more flashy figure than Spanish. The best quality of Honduras wood is known by its being free from chalky and black speckles, and when the color is inclined to a dark gold hue. The common sort of it looks brisk at a distance and of a lively pale red, but, on close inspection, is of an open and close grain, and of a spongy appearance."

There have been various changes in the Mahogany industry since Sheraton's time, but the qualities of the timber and the reasons given for the differences remain about the same. The West Indian trade has greatly diminished, but commercial lots of logs are still exported from Cuba and the Dominican Republic and have an important place in the furniture trade. The use of tractors in Central America has made possible the logging of timber which previously was considered inaccessible, thereby adding materially to the area of merchantable forest, especially in the British Honduras region (Plate XXX).

The comparatively recent discovery of commercial quantities of *Swietenia* in the upper Amazon region is of great importance to the industry, as it opens up a vast, though still undetermined, area of virgin forest containing Mahogany. For the circumstances attending this discovery and the initial development of the trade the authors are indebted to J. W. Massey, agent at Iquitos, Peru, for Booth & Company, Ltd., of London. He says that in 1921, when business was poor and shippers at Iquitos were looking for new articles to export, the firm of Israel & Company, Ltd., made a trial shipment of several kinds of hardwood logs to W. R. Grace & Company, New York. The results were disappointing, as the manufacturers reported that most of the woods were too difficult to saw. Some time later, while looking over some of the logs in a storage yard, Arthur Rushforth, a timber buyer, came across several that appeared to him to be genuine Mahogany. Upon learning their source, he went to Iquitos in June 1923 to investigate. He found that a shipment of 300 logs, believed to be Andiroba (*Carapa*), had been made three weeks before his arrival to C. Hernandez e Hijos in Hamburg. A second lot of 250 tons, originally intended for the same destination, was purchased by Mr. Rushforth and consigned to New York as Andiroba. The true identity of the wood was subsequently established and commercial exploitation begun in 1924.

The known and probable distribution of Mahogany in South America is approximately as follows: In Venezuela, according to a memorandum by Llewelyn Williams, the center of its natural range is that area known as Turén, south of Acarigua, in the State of Portuguesa, extending southward to the Río Portuguesa, eastward to a point between Aqua Blanca and Cojedes, and westward at least as far as Barinas. The tree is commonly cultivated for ornamental purposes around Caracas, and in the valley of the Aragua and elsewhere. In Colombia, it is mostly in the upper valleys of the Magdalena and Cauca Rivers. According to M. Acosta Solís (*Tropical Woods* 60: 52), *Swietenia* is a component of the virgin forests in northwestern Ecuador. In Peru, Mahogany occurs east of the divide at elevations of 400 to 4500 feet and is in a belt of forest varying in width up to several hundred miles and extending from southern Ecuador through the Ucayali basin to the headwaters of the Tambo and Urubamba Rivers in the south (see *Tropi-*

cal Woods 30: 31). Eastward across the border in Brazil, there are commercial stands of Mahogany in the upper reaches of the Juruá and Purús Rivers, and large quantities of logs are floated down to Manáos for export either in the round or after manufacture into lumber. There are definite reports of the occurrence of Mahogany in Bolivia in the upper reaches of the San Miguel, the Río Mamoré, and the Río Branco; also on the Río Machado, a tributary of the Madeira. Apparently there is no genuine Mahogany in any part of the basins of the Rio Negro and Putumayo, or in the lower portions of the Juruá and Purús, or below the falls of the Madeira. The discovery above the waterfalls of the Tapajoz and Xingú Rivers of old Indian canoes made of Mahogany implies an extension of the Bolivian belt across Matto Grosso, Brazil. While from the foregoing it appears that only fragmentary knowledge of the range of *Swietenia* in South America exists, enough is known to reassure anyone who fears that the supply of genuine, first-growth Mahogany timber is nearing exhaustion.

Little systematic effort is being made to provide for a future supply of the timber except in British Honduras where silviculture of Mahogany has been under way since the appointment of a Conservator of Forests in 1922. Three methods are being employed: (*a*) Underbrushing through the selected area to favor existing regeneration and to form a seeding felling. (*b*) Favoring existing regeneration in cut-over areas by following up the Mahogany operation and improving, by underbrushing, the regeneration occurring around the stumps. It was found that while regeneration is usually abundant around stumps during the first two years after the felling of the trees, it rapidly disappears thereafter as a result of competition in the bush. The abundant regeneration is thus saved by improvement during the first two years after the exploitation of the area, and the old trees are replaced by a large stock of seedlings which have every chance of coming to maturity. (*c*) "Taungya," or the planting of shifting-cultivation areas with Mahogany. This has been inaugurated among the Maya Indians of the south with excellent results. The natives do all the seed collecting, nursery work, and transplanting of seedlings and receive free rent of their land in return. The first method is intensive and can be used on small estates for the establishment of a dense stock of Mahogany. The second is extensive and serves to replace the former Mahogany stock by large groups of regeneration over wide areas; the stock is not only replaced but greatly increased. The third converts worthless second-growth forest into blocks of what eventually will be almost pure stands of Mahogany.

In his notes on the silviculture of Trinidad trees (see *Tropical Woods* 27: 28), R. C. Marshall gives the following information regarding the two principal species of *Swietenia*. Honduras Mahogany, *S. macrophylla*, is a virtually evergreen tree, not very exacting as to soil condition and able to grow in fairly moist sites. The flowers appear in the rainy season, and the fruit ripens in the dry season. Germination is good and seedling growth is rapid. The seedlings are taprooted but are readily transplanted. Young trees are fairly tolerant of shade, but conditions for optimum growth call for full overhead light combined with side protection. It is too subject to disease and insect attack to be generally recommended for planting in Trinidad and pure crops of it should be avoided. It does not coppice well and young trees are rather sensitive to mechanical injury. As for West Indian Mahogany, *S. mahagoni*, which is believed not to be indigenous to Trinidad or Tobago, he says: "Trees in adjoining plots at St. Clair, aged approximately 30 years, gave an average girth of just over two feet for West Indian Mahogany, compared with over three feet for the Honduras variety. The height growth also was proportionately less. Owing to its slower rate of growth it has been but little planted of recent years and practically no information is available as to its silvicultural requirements. Information is needed as to whether it suffers as badly from disease as the faster-growing variety. If not, it may yet come into prominence."

Heartwood reddish, pinkish, salmon-colored, or yellowish when fresh, deepening with age to deep rich red or brown; surface of newly sawed lumber turns dark red upon exposure to the sun; sapwood yellowish or nearly colorless. Luster typically high, golden. Odor and taste absent or not distinctive. Variable in density from rather light, soft, and tough to heavy, hard, and brittle; sp. gr. (air-dry) 0.40 to 0.85; mostly between 0.50 to 0.60 for *S. macrophylla;* weight 25 to 53 lbs. per cu. ft.; texture rather fine to coarse; grain straight to roey, wavy, or curly, often producing highly attractive figure; technical properties high to excellent.

COMMON NAMES: Mahogany—Cuban, Honduras, Mexican, Panama, Peruvian, Spanish, West Indian, etc. (Eng., trade); caoba (Span., general); acajou (Fr.); mogno (Port.); mogano (Ital.); mahonie (Dutch); madeira, redwood (Florida, Bah.); caobilla, caoba de caracolillo, c. de clavo, c. de ramazón, c. hembra, c. lisa, c. macho (Cuba); chiculte, cóbano, flor de venadillo, gateado, palo zopilote, punab, rosadillo, tzopilotl, venadillo, zopilocuahuitl, zopilote, z. colorado, z. negro, zopilotl, zopilozontecomacuahuitl (Mex.); orura (Venez.); aguano (Peru).

Trichilia is closely related to *Guarea* and comprises about 230 species of shrubs and small to large trees widely distributed throughout tropical America and sparingly in western Africa and Madagascar. The leaves are commonly odd-pinnate, sometimes digitately compound or reduced to a single leaflet (*e.g., T. Karstenii* C. DC. = *Odontandra Karstenii* Tr. & Pl.), the leaflets entire or spiny-toothed and frequently with pellucid dots; the small greenish or yellowish flowers are borne in terminal or axillary panicles; the fruit is generally a small 3-valved, 3-seeded capsule, each cell containing a single seed surrounded by a red and showy aril.

The best known tree of the genus in central and southern Brazil and Argentina is the Catiguá, typified by *Trichilia catigua* A. Juss. It is usually less than 45 feet high, with a short trunk rarely over 20 inches in diameter. The bark is the source of tannin, dyestuff, and ingredients of insecticides and medicines. The reddish or flesh-colored wood is moderately hard, easy to work, and used for about the same purposes as Birch (*Betula*) in northern countries.

The largest known representative of this genus is the Brazilian Pimentiera or Pau Rosa Branca, *Trichilia alta* Blake. H. M. Curran, who was the first to collect it, says: "This is a common tree of the Bahia coast forest where it grows in association with Jequitibá (*Couratari*), Pau d'Alho (*Gallesia*), Araça de Agua (*Terminalia*), etc. It occurs singly or in small groups, never in pure stands. Heights of 125 feet and diameters up to three feet are common, while the clear lengths are usually 40 to 50 feet. The trunks are of good timber form and have a thin black or slate-gray bark. The wood is not durable in contact with the ground and is little used, being probably unknown in the local markets. Tests made in the United States prove that it machines well, does not warp or check badly, and has good possibilities for flooring and as a substitute for Hickory." Tests on the mechanical properties made at the University of Michigan gave the following results: Sp. gr. (oven-dry) 0.71; weight (8.6 per cent moisture), 48.7 lbs. per cu. ft. *Crushing strength parallel to grain,* 7770 lbs. per sq. in. *Bending:* Modulus of elasticity, 1,840,750 lbs. per sq. in.; fiber stress at elastic limit, 11,990 lbs. per sq. in.

The Pracuúba da Terra Firme of Amazonian Brazil, *Trichilia LeCointei* Ducke, is described (*Arch. Jard. Bot. Rio de Janeiro* 3: 192) as a medium-sized to large tree, with a fine-textured, hard and compact, mildly scented timber susceptible of a high polish. Freshly felled wood is pinkish throughout but, upon drying, the heartwood deepens in color to bright reddish brown, very distinct from the sapwood. The timber is employed locally in construction and for the hafts of harpoons.

There are numerous species in northern South America, the West Indies, Mexico, and Central America, the two with the greatest range being *Trichilia havanensis* Jacq. and *T. hirta* L. Their timber is utilized in a small way for tool handles, broomsticks, and interior construction.

They are not likely to contribute to the export trade.

Heartwood mostly light reddish brown; distinct but not always sharply demarcated from the thick whitish or roseate sapwood; color of both becoming darker upon exposure. Luster medium. Odor and taste absent or not distinctive in dry material. Density widely variable; sp. gr. (air-dry) 0.55 to 0.80; weight 34 to 50 lbs. per cu. ft.; texture mostly medium; grain generally straight; working properties fair to good; durability rather low.

COMMON NAMES: Bariaco, cabo de hacha, guabán, jujubán, siguaraya (Cuba); broomstick, cabo de hacha, caracolillo, gaeta, guaita, guayavacón, jobillo, molinillo, palo de Anastasio, ramoncillo, retamo (P.R.); almendrillo, chicharrón (Dom. R.); bois arada, b. diou marron, dombou, Marie Jeanne, mombin bâtard (Haiti); acurel, obi (Trin.); cabo de hacha, cauache, choben-ché, cucharillo, estribillo, garbancillo, garrapatilla, ichbabach, ixbahach, kulimziz, limoncillo, xkulinsis, xpukusikil (Mex.); bastard lime, red cedar, sisím (Br. H.); barrehorno, limoncillo (Hond.); barredero, barrehorno, canelillo, canjuro, cedrillo, cola de pavo, jocotillo, ojo de muñeca, pimientillo (Salv.); matapiojo (Nic.); cedro cóbano, c. dulce, c. macho, urruca (C.R.); alfaje, alfajillo (Pan.); affaja, bagre, mangle blanco, m. dulce, manglesito, yayo, y. blanco, y. colorado (Col.); canalete, cazabito, cedrillo, cerezo macho, hayo blanco, marfil, mata-pollos, pan de trigo, pata de paloma (Venez.); ulu (Br. G.); caá-tigoa, camboatá, carrapeta, catiguá, c. graúdo, c. miúdo (Braz.); caá-vorobei, catiguá, c. blanca, c. colorado, c. puita, fe de gozo, guamirí, guatambú-y, mangacitara, palo anis, tapé-riguá (Arg.).

MENISPERMACEAE

THE Moonseed family is composed of about 55 genera and 150 species of twining, scandent, or rarely erect shrubs and a few small trees of pantropical distribution. The leaves are alternate, palmate or peltate, entire or lobate, and without stipules; the small unisexual flowers are typically in axillary fascicles or cymes; the fruit is a berry-like drupe, with a single usually crescent-shaped seed. The fruits and bitter roots of a few species are used in medicine; some are narcotic and highly poisonous. The true "parreira brava" of pharmacopeas is obtained from the roots of *Chondodendron platyphyllum* Miers, a climbing shrub of eastern Brazil, commonly known there as Abutua Grande; it was the object of much scientific investigation by European chemists and physiologists beginning 260 years ago. One of the ingredients of the arrow poison, curare, is said to have been obtained from species of *Cocculus* and *Abuta* in the Amazon region. The tropical American trees of the family belong to the genera *Abuta* and *Hyperbaena* and the largest individuals reported are 30 feet tall and about six inches in diameter. Their woods, described below, have no important uses.

Color pale olive or olive-brown throughout. Not highly lustrous. Without distinctive odor or taste. Hard, heavy, tough and strong; texture uneven; grain irregular; durability low. Of laminated (anomalous) structure, the included phloem and conjunctive tissue forming conspicuous concentric bands apparently terminating seasonal growths.

Pores rather small to minute; few in some species, very numerous in others; with a more or less pronounced tendency to tangential arrangement. Vessels with simple perforations; without spiral thickenings; pitting fine, alternate. Rays nearly all broad, composing a third to a half of the cross section; wider near the bands of conjunctive tissue; heterogeneous, many of the cells square; ray-vessel pit-pairs small, half bordered. Wood parenchyma diffuse or finely reticulate. Phloem in islands filling space between rays at margin of conjunctive parenchyma; layer of sclerotic cells sometimes present in conjunctive tissue and across rays. Wood fibers with very numerous small but distinctly bordered pits. Ripple marks absent. No gum ducts seen.

COMMON NAMES: *Abuta:* Pareira brava blanc (Fr. G.); abutua, baga de caboclo, butuá, caá-peba, café-rana, cipo amargoso, eko, icú, pani, parreira brava, uva de gentio, u. secca (Braz.); atinupa, caimitillo, sanango, trompetero-sanango (Peru). *Hyperbaena:* Bois coq (Haiti); bronquito, chicharrón, c. amarillo, c. de farallón, chicharroncillo, huevo de gallo, mabinga, pegoje, picha jutía (Cuba); knock-me-back, tkansik (Br. H.); grenadilla (Guat.); chaparrón, cuero del diablo, huevo del diablo, umineishte (Salv.); naranjo de monte (C.R.).

MONIMIACEAE

Composing this family are 32 genera and perhaps 400 species of evergreen aromatic shrubs and trees widely dispersed in tropical and subtropical regions, especially of the southern hemisphere. The leaves are typically opposite, simple, coriaceous, and punctate with pellucid glands; the small to medium-sized flowers are borne in cymes or racemes, or rarely solitary; the one-seeded drupaceous fruits are inserted on the cup-shaped receptacle which, in some genera, incloses them, forming pseudofruits. The species are most numerous in Malaysia and Oceania and the best timber trees occur in Australia, where they are generally known as Sassafras and used for the same purposes as Yellow Poplar (*Liriodendron*) in the United States. There are three genera in Madagascar and neighboring islands, two in tropical continental Africa, and eight in Latin America. Of the American genera, the five with wood specimens available for study are *Boldea* (*Peumus*), *Bracteanthus, Laurelia, Mollinedia,* and *Siparuna*. None is of more than local importance.

Heartwood typically light olive-brown, but ranging from yellow to blackish brown; not always clearly differentiated from the sapwood. Not highly lustrous. A few species aromatic. Sp. gr. (air-dry) 0.43 to 0.95, mostly 0.55 to 0.75, equivalent to 34 to 47 lbs. per cu. ft.; texture fine to medium; grain usually straight; working properties good; durability fair.

Growth rings usually not well defined. Pores rather few to very numerous, mostly very small, invisible without lens and sometimes indistinct with it; solitary and in short to long radial multiples; vessels fairly evenly distributed with either simple or multiple (scalariform) perforations or both together; spirals absent, except in *Boldea;* tyloses common; intervascular pitting fine to coarse, alternate, opposite, or scalariform. Rays typically heterogeneous; in some genera all comparatively narrow, in others mostly coarse and conspicuous; ray-vessel pitting variable, sometimes distinctly scalariform; oil cells sporadic in certain eastern species, but absent in the American. Wood parenchyma mostly diffuse or metatracheal; distinct without lens in *Bracteanthus* only. Wood fibers with simple or small bordered pits; septate in part, except in *Bracteanthus* and *Siparuna*. Ripple marks absent. No gum ducts seen.

Boldea, with a single species, *B. boldus* (Mol.) Looser (= *Peumus boldus* Mol. = *Boldea fragrans* C. Gay), is a strongly aromatic tree, occasionally 40 to 60 feet high, with rough leathery evergreen leaves, growing on dry sunny hillsides in Chile, where it is known as Boldo or Boldu. The seed kernels are edible, the leaves are used for seasoning food, and the bark is a source of tannin. The olive-brown wood is aromatic, hard, of fine and uniform texture, easy to work, finishes smoothly, and is resistant to decay. Apparently its only uses are for charcoal and minor structural purposes.

Bracteanthus *glycycarpus* Ducke, the sole species, is a large and apparently rare tree 65 to 100 feet tall, discovered in the lower Amazon region by Dr. Adolpho Ducke in 1926. All of the vegetative parts of the plants have a fetid odor like that of the local species of *Siparuna*. It has thin coarsely fibrous bark, opposite simple leaves, large floral bracts, and fig-like fruits. The timber is not utilized, though suitable for heavy construction.

Heartwood deep olive-brown, sharply defined; sapwood pale olive. Luster medium. Scent distinctive, but mild and not unpleasant; vaguely suggesting Black Walnut (*Juglans nigra*). Hard and heavy; sp. gr. (air-dry) 0.83 to 0.96; weight 52 to 60 lbs. per cu. ft.; texture medium; feel harsh; grain straight; not easy to cut but finishing smoothly; looks durable.

Common names: Caá-pitiú grande, piranheira de terra firme (Braz.).

Laurelia. There are two species, one in Chile, the other in New Zealand. The latter, *L. novae-zelandiae* A. Cunn., known locally as Pukatea, is a large timber tree of the lowlands of North Island and the extreme northern part of South Island. *L. aromatica* Juss., the Laurel of Chile, occurs in mixed forest or small nearly pure stands from Aconcagua to Llanquihue and Chiloé, but the principal supplies remaining are in Malleco and Cautín. The tree is usually not over 45 or 50 feet high, with a straight smooth-barked trunk generally not over 24 inches, sometimes up to three feet, in diameter. The leaves serve as a condiment and the fruits as a substitute for nutmeg. The greenish yellow timber is of about the consistency of Red Gum (*Liquidambar*) and is the most commonly used native timber for cheap furniture and boxes, and to less extent in the interiors of buildings. Its industrial importance has been greatly reduced through the destruction of the more readily accessible forests and the utilization of the land for agriculture.

Mollinedia includes about 75 species of small trees and shrubs, of which a few are Australian, the others occurring scatteringly from southern Mexico to southern Brazil. The yellowish woods are of medium density, uniform-textured, tough and strong, easy to work, and having attractive silver grain on radial surface. The timber is rarely used because of the small size and scarcity of the trees, but *M. Schottiana* DC., called Capixim in southeastern Brazil, is employed for barrel hoops and sieve rims.

COMMON NAMES: Limoncillo, quizarrá (C.R.); amoquí-ey (Peru); capixim (Braz.).

Siparuna, with many species of strongly scented, sometimes malodorous, shrubs and little trees, is of common occurrence in mixed hardwood forests from southern Mexico to Peru and southern Brazil. The leaves of some species are used medicinally. The rather soft to hard and strong, yellowish or olive-brown wood is of good technical properties, but is available only in small sizes.

COMMON NAMES: Chuche, hierba de la conchuda, h. del talaje, limoncillo (Mex.); cerbatana (Guat.); hierba de pasmo (Pan.); arakarri wadilikoro, banjabodjamaro, injakoppie, jakopie, jarakopie, koesariejepoe, koffiewato, manjabo-hororodikoro, menirie dan bojaroe, pjawabanwewe, pjoewapawewe, proewabinowewe, sokonèprobero, tebepau, tipoporin ieriakopie, waroro, wokere apoekoesja (Sur.); munuridang (Br. G.); vulnéraire (Fr. G.); guayusa, pera de monte (Ec.); curuinsi-sacha, isula caspi, i. micuna, macusaro, pampaorgano mashan (Peru); caá-pitiú, c.-p. fedorento, capituí, herva cidreira do campo, limoeira brava (Braz.).

MORACEAE

THE Mulberry family, with about 70 genera and more than 1000 species of unarmed or thorny trees and shrubs and a few herbs, generally with milky juice, is of cosmopolitan distribution, though most abundant in tropical and subtropical regions. The leaves are alternate, stipulate, small to very large, entire, toothed, or lobed, thin to thick and leathery, smooth or as rough as sandpaper; the flowers are unisexual, the two sexes borne on the same or on different individuals; the fruits are indehiscent, one-seeded, free or aggregated, and variously designated as figs, berries, cherries, nuts, or oranges.

The largest genus by far is *Ficus*, but of its several hundred species the only two that are important commercially are *F. carica* L., whose fruits are the edible figs, and *F. elastica* Roxb., a common hot-house plant which in its native habitat is the source of India rubber. The Mulberry trees (*Morus*) of the north temperate zone have edible fruit and it is upon their leaves that silkworms are fed. The foliage of some species of *Brosimum* and *Trophis* is useful as green fodder for livestock. From the inner bark of several genera the Indians make clothing, blankets, and matting. The family supplies few commercial timbers. The best for heavy, durable construction is the West African Iroko, *Chlorophora ex-*

celsa Benth.; the related American species, *C. tinctoria* (L.) Gaud., supplies a well-known yellow dyewood called Fustic. The Guiana Letterwood or Snakewood, with its peculiar hieroglyphic markings, is the heartwood principally of *Piratinera guianensis* Aubl. The beautiful Satiné of French Guiana is the heartwood of *Brosimum paraense* Huber. The Osage Orange, with a very limited natural range in the United States, has been widely planted for farm hedges and the wood is noted for its strength and elasticity, its high resistance to decay, and its low shrinkage. There are about 30 genera represented in the New World but only those mentioned are of any commercial value.

The family as a whole is well defined botanically, but the internal classification is in many instances unsatisfactory, principally because of the difficulty of obtaining ample and complete herbarium specimens. The following descriptions of the wood are based upon authentic specimens of American species of 27 genera. For several genera the amount of material is inadequate for thorough treatment. Wood samples from type trees cannot always be relied upon, because the botanists who proposed the new species are sometimes uncertain as to the genera to which they should be referred. Until more and better material becomes available the generic concepts in closely related groups will continue to be hazy.

Sapwood thin, heartwood abundant, bright yellow, turning brown upon exposure, in *Bagassa, Clarisia, Chlorophora, Maclura, Morus,* and possibly *Sorocea;* sapwood very thick, heartwood sparingly developed, bright red in *Brosimum,* brown with black markings in *Helicostylis, Piratinera, Trophis, Trymatococcus,* and perhaps *Pseudolmedia;* sapwood very thick, heartwood absent or not clearly differentiated, in the others. Luster medium to high. Without distinctive odor or taste. Density widely variable from low or fairly so in *Castilla, Cecropia, Coussapoa, Ficus, Poulsenia,* and *Pourouma,* to exceptionally high in the heartwood of *Piratinera* and *Trymatococcus;* texture fine to coarse; grain very straight to uneven; working properties generally good; durability high for distinctively colored heartwood, otherwise low. Commercial importance of the timbers not likely to increase.

Maclura and *Morus* distinctly ring-porous, the late-wood pores clustered; tyloses thin-walled; small vessels with spiral thickenings. Other woods diffuse-porous; pores typically rather small; solitary and in short radial multiples, well distributed; vessels without spiral thickenings; tyloses thin-walled to thick-walled, sclerotic in heartwood of some or all specimens studied of *Brosimum, Helianthostylis, Piratinera, Pseudolmedia,* and *Trymatoccus.* Vessel perforations exclusively simple throughout family; intervascular pitting alternate, fine to coarse. Rays usually not over 5, sometimes up to 12, cells wide and ranging from few to 150, commonly not over 40, cells in height; uniseriates very few to numerous, composed mostly or entirely of square or upright cells; multiseriates commonly with 1 or 2, sometimes up to 4 (occasionally up to 10), marginal rows of square or upright cells; sclerotic cells common in densest woods; crystals often present, sometimes abundant; latex tubes occasional to frequent in *Acanthosphaera, Anonocarpus, Brosimopsis, Brosimum, Castilla, Helicostylis, Naucleopsis, Noyera, Ogcodeia, Olmedia, Olmedioperebea, Olmediophaena, Perebea, Pseudolmedia,* and *Trymatococcus;* pits to vessels medium and oval to large and elongated, the latter often in more or less scalariform arrangement. Wood parenchyma moderately to very abundant; typically vasicentric in *Bagassa, Maclura, Morus, Olmediophaena maxima,* and *Olmedioperebea;* vasicentric and short aliform, with or without occasional concentric bands in *Cecropia, Helicostylis, Naucleopsis, Ogcodeia, Perebea, Poulsenia,* and *Pourouma;* long and narrowly aliform in *Brosimopsis, Brosimum, Helianthostylis, Piratinera,* and *Trymatococcus;* aliform and confluent into irregular, often anastomosing bands in *Castilla, Coussapoa, Noyera,* and *Pseudolmedia,* and sometimes in *Brosimum, Helicostylis,* and *Ogcodeia;* in concentric bands more or less independent of the pores in *Anonocarpus, Clarisia, Ficus, Olmedia aspera, Sorocea,* and *Trophis;* crystals sometimes present; sclerotic cells common in densest woods. Wood fibers with thin to very thick, sometimes gelatinous, walls; septate in *Acanthosphaera, Castilla, Helicostylis, Naucleopsis, Noyera, Ogcodeia, Olmedia, Olmedioperebea, Olmediophaena,* and *Pseudolmedia;* pits small, but with distinct borders in *Brosimopsis* and *Brosimum* (in part), very small and simple or indistinctly bordered in the others. Ripple marks absent. No gum ducts seen. For anatomy of the different genera see *Tropical Woods* 61: 13–54.

Acanthosphaera, with two species of small to medium-sized unarmed laticiferous trees, is often included in the genus *Ogcodeia,* but according to Ducke (*Arquiv. Serv. Flor.* 1: 1: 20) it is worthy of generic rank. *A. Ulei* Warb. is rather common in the upper Amazon region of Brazil and eastern Peru; it bears small edible fruits. *A. amara* Ducke occurs further eastward along the Amazon. Its specific epithet refers to the very bitter latex which is used in local medicine as a febrifuge and is responsible for the name Quina, the usual designation for *Quassia amara* L. The following description is based upon one sample each of the two species. No heartwood present in specimens; sapwood light grayish brown. Luster medium. Odorless and tasteless. Moderately hard and heavy; texture medium; grain straight; working properties good. Probably of no commercial possibilities.

Common names: Bálsamo, muiratinga da terra firme, quina (Braz.).

Anonocarpus *amazonicus* Ducke, the only species, is a medium-sized to large, unarmed, laticiferous tree related to *Batocarpus,* and occurs in the Amazon region of Brazil and Peru. The generic name refers to the resemblance of the succulent fruit to that of certain species of *Anona.* L. Williams, who collected a specimen along the lower Huallaga, Loreto, Peru, where it is known as Machunasti, says (*Woods of northeastern Peru,* p. 73) that the tree is often 90 feet tall, with a spreading crown and fairly straight, cylindrical bole with small buttresses and clear of branches for as much as 65 feet. The timber is used for making canoes.

No colored heartwood seen; sapwood grayish or yellowish, streaked with parenchyma. Luster rather high. Moderately hard, heavy, tough, and strong; texture coarse; grain straight to irregular; easy to work, finishing smoothly; sapwood perishable in contact with the soil. Presumably of no commercial possibilities.

Bagassa *guianensis* Aubl., the only well-defined species, is a large forest tree of rather infrequent occurrence in the Guianas and Brazilian Amazon region. The fibrous bark contains an abundance of latex which is very attractive to certain beetles. The heart-shaped leaves are entire or 3-lobed. The fruits, which are about the size of an orange, are astringent but have an agreeable flavor and are comestible. The timber is of good quality and is used locally for general construction, carpentry, and furniture. It apparently is too scarce to become important in the export trade.

Heartwood lustrous golden brown, becoming russet upon exposure; sharply defined from the pale yellow or nearly white sapwood. Without distinctive odor or taste. Moderately heavy, tough, and strong; sp. gr. (air-dry) 0.70 to 0.80; weight 44 to 50 lbs. per cu. ft.; texture coarse; grain usually roey; easy to saw, difficult to split radially, finishes smoothly, holds its place well when manufactured; is fairly resistant to decay.

Common names: Gele bagasse (Sur.); bagasse, b. jaune, bois bagasse (Fr. G.); amapá-rana, bagaceira, tatajuba (Braz.).

Brosimopsis, with five species of medium-sized to large, unarmed trees, is apparently limited to Brazil. *B. lactescens* Sp. Moore occurs in Matto Grosso. The three strictly Amazon species are *B. acutifolia* (Hub.) Ducke of the lower region, *B. amplifolia* Ducke of the central part, and *B. obovata* Ducke of the upper part. They are known as Mururé, and the latex, which contains an alkaloid called murerina, is used in local medicine as an anti-syphilitic. The wood is said to be yellowish with brownish stripes, and without dark-colored heartwood.

The most widely distributed species, and the only one represented in the Yale wood collections, is *Brosimopsis oblongifolia* Ducke, which grows on non-inundated lands throughout most of the Brazilian rain forests. H. M. Curran collected it in the Rio Grongogy basin, Bahia, where it is called Leiteira. It is a fairly common constituent of the upper story of the forest, attaining a maximum height of 150 feet, with a trunk three feet or more in diameter and free of branches for 50 to 60 feet. Although the felling and hewing of the green timber was accomplished without difficulty, the log obtained by Mr. Curran has proved almost impossible to saw when dry, owing to the large amount of gritty substance (silica) in the vessels. It is used locally to a minor extent as hewed timbers for interior construction.

Wood throughout grayish or oatmeal-colored, with a yellowish hue; apparently all sapwood. Luster rather high. Without distinctive odor or taste. Moderately hard, heavy, tough, and strong; sp. gr. (air-dry) 0.75; weight 47 lbs. per cu. ft.; texture coarse, feel harsh; grain straight; easy to cut with a knife, but very difficult to saw, finishes smoothly; is not resistant to decay.

COMMON NAMES: Leiteira, mercurio vegetal, moruré, murerú, mururé (Braz.).

Brosimum, with numerous species of medium-sized to very large unarmed laticiferous trees, is of general distribution throughout tropical America. The generic term is here used in a restricted sense and the species with brown or brown and black heartwood are for convenience treated as a separate genus, *Piratinera.* The species of *Brosimum* appear to be of two types, namely those without distinct heartwood and those with rich red or red and yellow heartwood. The first group includes *B. Alicastrum* Sw., *B. amplicoma* Ducke, *B. columbianum* Blake, *B. costaricanum* Liebm., and *B. terrabanum* Pittier. The second group is typified by *B. paraense* Huber, known as Muirapiranga (red wood) in Brazilian Amazonia and as Satiné Rubané or Bois de Feroles in French Guiana. Closely related to *B. paraense,* and perhaps only a variety of it, is *B. caloxylon* Standl., the Bloodwood Cacique of Panama. The taxonomic value of such a distinction is doubtful, however, as one specimen of *B. utile* (H.B.K.) Pittier has, near an injury, a thin layer of bright red heartwood very similar to that of *B. paraense.* Perhaps further investigation will reveal that other species may develop a small amount of heartwood, either normally or traumatically. A parallel case is that of *Pterocarpus;* the Asiatic and African species are the source of Padauk timber, whereas the American trees are virtually all sapwood, with only traces of richly colored heartwood formed near wounds.

Brosimum Alicastrum Sw. is a tall tree known in Jamaica as Breadnut as the cooked seeds are edible and have a nutlike flavor. The species is sparingly represented in Cuba and also grows on the mainland in Yucatán, Mexico, and northern British Honduras. Closely related to it are two Central American species, *B. costaricanum* Liebm. and *B. terrabanum* Pittier, which perhaps should be considered only as varieties, as Standley says (*Flora of Costa Rica* 1: 380) that they "are much alike in all their characters" and (*Forests and flora of British Honduras,* p. 110) "it is rather doubtful whether this species [*B. terrabanum*] is distinct from *B. Alicastrum.*" The green foliage and the fruits of all three forms are locally important as fodder for livestock, especially during the dry season. The timber, which is hard, strong, and resilient but subject to decay, is used commercially to a small extent for veneer and locally for making tool handles and pack saddles, and for fuel. The latex is sometimes an adulterant of chicle.

Representatives of the *Alicastrum* group in northern South America are *Brosimum latifolium* Standl. of Ecuador and *B. Dugandii* Standl. and *B. columbianum* Blake (= *Helicostylis bolivarensis* Pittier, published about a year earlier from the same type collection, Curran 304, Yale 1531). This last species serves to illustrate the difficulties confronting a wood anatomist working in this family. Pittier stated in 1921 (see *Timbers of Tropical America,* footnote on p. 127) that he thought the tree should be referred to *Pseudolmedia* rather than to *Helicostylis.* Blake, in a letter of April 20, 1938, to the senior author, says: "I was not able to find any staminate flowers when I described the plant, and neither was Pittier. To be a *Brosimum,* as you say you think it is, the plant should have staminate flowers on the same receptacle with the pistillate. So far as the material shows, it does not have these, and I do not feel like transferring Pittier's name to *Brosimum* without support from flower structure, in view of the very close generic relationships in this group." The wood structure agrees closely with that of the *Brosimum Alicastrum* complex.

The most noted *Brosimum* is the Venezuelan Cow Tree, *B. utile* (H.B.K.) Pittier, discovered at the beginning of the nineteenth century by Alexander von Humboldt who describes it in the following manner (*Personal narrative of travels to the equinoctial regions of America during the years 1799–1804*. Eng. ed. London, 1852, Vol. II, pp. 47–48):

"For many weeks we have heard a great deal of a tree whose juice is a nourishing milk. The tree itself is called the Cow Tree and we were assured that the negroes on the farm, who are in the habit of drinking large quantities of this vegetable milk, consider it as highly nutritive; an assertion which startled us the more, as almost all lactescent vegetable fluids are acrid, bitter, and more or less poisonous. Experience, however, proved to us during our residence at Bárbula that the virtues of the Cow Tree, or Palo de Vaca, have not been exaggerated. This fine tree bears the general aspect of the Star-apple tree (*Chrysophyllum cainito*); its oblong, pointed, coriaceous, and alternate leaves are about ten inches long and marked with lateral nerves, which are parallel and project beneath. The flower we had no opportunity of seeing; the fruit is somewhat fleshy and contains one or two kernels. Incisions made in the trunk of the tree are followed by a profuse flow of gluey and thickish milk, destitute of acridity and exhaling a very agreeable balsamic odor. It was offered to us in calabashes, and though we drank large quantities of it, both at night before going to bed and again early in the morning, we experienced no uncomfortable effects. The viscidity of this milk alone renders it rather unpleasant to those who are unaccustomed to it. The negroes and free people who work in the plantations use it by soaking in it bread made from maize, manioc, aropa, and cassava; and the superintendent of the farm assured us that the slaves become visibly fatter during the season when the Palo de Vaca yields most milk. When exposed to the air this fluid displays on its surface, probably by the absorption of the atmospheric oxygen, membranes of a highly animal nature, yellowish and thready, like those of cheese; which, when separated from the more watery liquid, are nearly as elastic as those of caoutchouc, but in process of time exhibit the same tendency to putrefaction as gelatine. The people give the name of cheese to the curd which thus separates when brought into contact with the air and say that a space of five or six days suffices to turn it sour, as I found to be the case in some small quantities that I brought to Valencia. The milk itself, kept in a corked bottle, had deposited a small portion of coagulum and, far from becoming fetid, continued to exhale balsamic scent. When mingled with cold water the fresh fluid coagulated with difficulty, but contact with nitric acid produced the separation of the viscous membranes."

The range of *Brosimum utile* extends along the Atlantic coast to Costa Rica. The tree attains a height of 80 feet with an erect trunk about 20 inches in diameter and covered with a thick, gray, smoothish bark. The leaves are large and leathery, with prominent pinnate venation, very distinct from those of the *Alicastrum* group. The latex is apparently harmless to drink, if taken in small quantities, but analysis has shown (see *Contr. U.S. Nat. Herb.* 20: 3: 104) that its composition is 57 per cent water, 37 per cent wax, and about 5 per cent gum and sugar, and hence is better suited for making chewing gum than as a food. (Other American Cow Trees are species of two genera, *Couma*, fam. Apocynaceae, and *Mimusops*, fam. Sapotaceae.) The timber, which is virtually all sapwood, is considerably less dense than that of *B. Alicastrum* and apparently is not utilized. In the thin strip of bright red heartwood about the hollow center in one Colombian sample (Yale 29517; A. E. Lawrance 765) the vessels and the wood and ray parenchyma cells are filled with garnet-colored deposits contrasting with the golden luster of the fiber background.

Closely related to *Brosimum utile* are at least three Amazonian species: *B. potabile*

Ducke, locally known as Amapá Doce, is said to have a thick, yellowish white sapwood and a slender core of dense, beautiful brownish red heartwood. *B. parinarioides* Ducke, called Mureré-rana and Amapá-rana, attains a height of 140 feet and has a moderately hard wood, yellowish white throughout, except in very large trunks which may develop hollow pipes of richly colored heartwood. *B. amplicoma* Ducke, called Caucho Macho, is especially noteworthy because of its great height and the wide extension of its crown, which perhaps even exceeds that of *B. potabile* (see *Tropical Woods* 31: 10–11, 22–23).

There is a small group of closely related species of *Brosimum* with sufficient red heartwood to be of value. Chief of these is *B. paraense* Huber (= *Ferolia guianensis* Aubl.) of the Amazon basin. Ducke says (*Tropical Woods* 51: 15–16) that it is "the true Muirapiranga of the timber dealers of Pará; at Manáos, however, it is more often called Páo Rainha (queen wood). It is a rather tall tree, widely distributed through the Brazilian states of Pará and Amazonas and perhaps the whole hylaea, growing in the upland rain forest, more frequently on sandy soils than on clay." Other Brazilian names are Condurú or Gondurú or Condurú de Sangue; in northeastern Peru the usual designation is Palo de Sangre. The wood best known commercially emanates from French Guiana and is usually called Satiné or Satiné Rubané. Karl Schmieg says (*Tropical Woods* 9: 2): "It has a wonderful satiny luster and there are different hues and colors varying from gray to a rich strawberry red overlaid with a golden sheen. It is obtainable only in small logs, which rarely exceed eight inches in diameter, and is straight-grained, fine-textured, hard, and heavy, receives a beautiful polish, and takes glue very well. The Muirapiranga of the lower Amazon and the Palo de Sangre of the Peruvian Amazon are like the Satiné in many ways, but the samples I have seen lack the rich coloration of the French Guiana wood."

At least two other species have been described in the *paraense* group. According to Ducke (*loc. cit.*, p. 16), "*Brosimum angustifolium* Ducke is a tall tree of the Brevos Islands in the mouth of the Amazon; also found in the lower Xingú. Its wood, which is more yellowish brown than red, is seldom found in the timber commerce of Pará and the dealers there consider it as an inferior quality of Muirapiranga." One of the rarest and most highly prized woods of Panama is known as Cacique. A fragment (Yale 6747) collected by Alvin G. Cox in Bocas del Toro in 1923 was recognized by the senior author as being closely related to *B. paraense* (see *Timbers of Tropical America*, p. 139). In 1928, G. Proctor Cooper obtained sterile herbarium material of the tree and P. C. Standley named it *B. caloxylon*. "Although known at present only from sterile material, it seems advisable to give a name to this important tree, the first of this group to be reported from North America. *Brosimum caloxylon* is related to *B. paraense* Huber, which has much shorter leaf acuminations and broader leaves." (See *Tropical Woods* 17: 11.) Cooper gives the following account of tree and wood (*Tropical Woods* 14: 1–3):

"The name Cacique (meaning Indian chieftain) is applied to more than one wood of western Panama, but the most highly prized and perhaps the rarest is the one best described by the term Bloodwood Cacique. The wood, as known to the Indians, is not obtained from living trees, but from old trunks which have lain partially buried in the putrescent forest floor until all but a core of heartwood has long since disappeared under the combined attack of insects and decay. The Bloodwood Cacique is held in highest esteem by the Indians in Panama. It is believed to have both medicinal and supernatural properties and has been the favorite remedy of the tribal medicine man. A small piece or a chip placed over a wound is said to relieve the pain and quickly stop the bleeding, while if placed behind the ears and secured by a cord it will overcome pains in the head. The fine chips or shavings are placed in rum, making a bitter drink which is considered an excellent remedy for malarial fevers. The mountain tribes still believe that bodily ills are in some way connected with evil spirits entering the afflicted person, and the use of Cacique may be considered as a charm against the bad effects of the spirits. The old chiefs used the wood as a symbol of office and authority and it is said they carved out very weird designs from the old logs found half buried in the ground. . . . The tree apparently does not get very large, probably not over a foot in diameter and 40 feet in height. The bark is smooth and has a white latex. The buds are very long-pointed. The thick sapwood is white and sharply demarcated from the heart which is red, with black streaks and a golden luster. In one of the trees felled there was a central core about an inch and a half thick that was of golden yellow color with streaks of light red. The wood from the old logs on the ground is of a considerably darker color than the fresh material, but retains its satiny luster."

The woods of the various species of *Brosimum* available for study have the same type of structure, though differing in details the significance of which cannot now be determined. Heartwood (so far as known) in various shades of rich lustrous red and yellow as stated above; sharply demarcated from the thick, yellowish white sapwood. Odorless and tasteless. Heartwood very hard and heavy; sp. gr. (air-dry) 0.90 to 1.05; weight 56 to 66 lbs. per cu. ft.; sapwood lighter, variable; sp. gr. (air-dry) 0.60 to 0.85; weight 38 to 53 lbs. per cu. ft.; texture rather fine to coarse; grain straight to variable; not difficult to work, finishing smoothly, taking a high polish. Heartwood of very limited commercial possibilities because of its small size and scarcity; sapwood suitable for veneers and miscellaneous purposes not requiring resistance to decay.

Common names: *Brosimum Alicastrum* group: Breadnut (Jam.); guáimaro, ramón de Méjico (Cuba); moussara (Trin.); apomo, ash, capomo, Juandiego, nazareno, ojite, ojoche blanco, ojochillo, oox, ox, oxitl, oxotzin, ramón (Mex.); breadnut, capomo, ramoon, masico (Br. H.); masicarón, masico (Guat., Hond.); ojuste, ujuste (Salv.); mesica, ojoche (Nic.); feguó (C.R.); guaimaro, g. comestible, guayamero, manata, mare, mondongo, pasita (Col.); tillo (Ec.); muiratinga, murure (Braz.); *B. utile* group: Cow tree, milk tree (Eng.); mastate (C.R.); cuqua, palo de leche, sandy (Pan.); avichurí (Col.); barimiso, palo de vaca, vacuno (Venez.); amapá doce, a. rana, caucho macho, murere-rana (Braz.); *B. paraense* group: Bloodwood, Brazil redwood, cardinal wood (Eng.); cacique (Pan.); ajeersi, oolemeriballi, polisthout, sokonéballi, warimiaballi (Sur.); bois baroit, b. de Cayenne, b. de lettre rouge, b. de feroles, satiné, s. gris, s. rubané (Fr. G.); amapá-rana, condurú, c. de sangue, cundúrú, gondurú, meurapiranga, moirapiranga, muirapiranga, mureré-rana, pau rainha, uanta (Braz.); palo de sangre (Peru).

Castilla (or *Castilloa*), with about 10 closely related species of unarmed, deciduous trees, is distributed from Mexico throughout Central America and northern South America to parts of the Brazilian Amazon region. The leaves are simple, large, coarse-textured, and usually densely hairy all over. The fruits are rather large, scaly, pubescent, and, in some instances, contain a soft, sweetish, edible pulp. About the only value of the trees is in their latex, the source of Castilloa rubber, now of minor importance as nearly all of the commercial rubber is produced by plantations of *Hevea* (family Euphorbiaceae).

The natural habitat of the trees is clearings and natural openings in the virgin forest and sparsely wooded alluvial flats, often in association with Cecropias. On rich soil growth is rapid and the trees attain great size, sometimes over 150 feet high and five feet in diameter; open-grown trees develop short, stout boles and large, coarse crowns. A peculiarity of the trees is the development of long, slender pseudo-branches or deciduous twigs bearing very large, showy leaves. Large forest trees are often buttressed by the spurs of the wide-spreading superficial roots. The best known and for long the only described species is *Castilla elastica* Cerv. Central American forms, *e.g., C. fallax* Cook, do not produce commercial rubber and are termed Caucho Macho or Hule Macho, literally male rubber, though the Spanish term "macho" when applied to plants often signifies some important departure from the type rather than any sexual difference. The Amazon species, *C. Ulei* Warb., is known locally as Caucho, whereas the tree (*Hevea*) producing Pará rubber is called Seringueira. The bark of *Castilla* is finely laminated and is used by the Sumo Indians in Nicaragua for making soft vegetable cloth, through a process of soaking and beating. The timber is rarely used as it is unattractive in appearance and lacks strength and durability.

Wood whitish when fresh, turning yellowish brown; apparently all sapwood. Luster medium to rather high. Odorless and tasteless. Rather light in weight, but firm and tough; texture coarse; grain fairly straight; easy to cut, saws rather woolly; perishable in contact with the ground.

COMMON NAMES: Rubber tree (English); caucho, hule, ule (Span.); árbol de ule, kiikche, yaxha (Mex.); cauchillo, ule macho, ule-ule (Cent. Am.); gsi-krá, serú, sini, soró, tseni, tsini (C.R.); caucho negro (Ec., Peru); caucho (Braz.).

Cecropia, with numerous closely related species of typically small unarmed deciduous trees, is distributed throughout tropical America and in many localities constitutes one of the most conspicuous and characteristic elements of the vegetation on lowland flats, often forming almost pure stands on old clearings. The trees grow quickly and are short-lived. At maturity they are commonly less than 40 feet high, with slender, upright, whitish trunks 8 to 12 inches in diameter, occasionally up to 60 feet tall and 24 inches through. The stems are hollow between the nodes, and the upper part and the branches are generally inhabited by small ants which inflict severe bites when the tree is molested. The very large long-stalked, peltate, deeply lobed, rough leaves are usually whitish beneath and being clustered at the ends of the few, coarse, candelabra-like branches produce a very striking effect. The bark supplies fiber for oakum and native cordage and latex for medicinal purposes. The trunks are sometimes used for making rafts or buoying heavy logs, and the wood is employed to some extent for match sticks, paper pulp, boxboards, charcoal powder, and kindling.

Wood whitish when fresh, becoming pale brown or oatmeal-colored upon exposure; apparently all sapwood. Fairly lustrous. Light and soft, but tough and strong for its weight; sp. gr. (air-dry) 0.35 to 0.45; weight 19 to 28 lbs. per cu. ft.; texture coarse; grain generally straight; not always easy to finish smoothly, holds nails firmly; is perishable in contact with the soil. Of no interest to the export trade.

COMMON NAMES: Shake wood, trumpet (Jam.); yagrumo (Cuba); yagrumo hembra (P.R.); bois trompette (Haiti); trompette (Guad.); bois canon (Granada); chancarro, coilótapalo, coilotopalo, guarima, guarumbo, guarumo, ix-coch, saruma, trompeta, xcoochlé (Mex.); igarata, trumpet (Br. G.); guarumo (Cent. Am., gen.); ikú, ko-krá, kokua-krá, kur, serúnguo, srung-dó, xkur (C.R.); guarumo, g. morado, orumo, yarumo (Col.); guarura (Venez.); congo pump, float wood, pumpwood, trumpet tree, wanasoro (Br. G.); boessi papaja, bosch papaya (Sur.); ambaita, bois canon, b. trompette, coulekin (Fr. G.); ambaiba, a. tinga, arvore da preguiça, embauba, imbauba, i. branca, i. da matta, i. verde, imbaubão, umbauba (Braz.); cetico, c. de oyada, imbauba, pungára, setico, tacona, tacuma, tacuna, umbauba (Peru); umbaubeira (Parag.); ambahú, ambaiba, ambatí, ambay, palo de lija (Arg.).

Chlorophora, with a few species of medium-sized to large, spiny or unarmed trees, is widely distributed in tropical West Africa and tropical America. *C. excelsa* B. & H. f. is an important African tree; the timber is esteemed locally for general construction and considerable quantities are exported to Europe; its most common trade name is Iroko (see *Tropical Woods* 28: 4–10).

In America the genus is usually considered as monotypic, but in addition to *Chlorophora tinctoria* (L.) Gaud., one or two other forms are generally recognized either as varieties or sometimes as distinct species. The northern form, which grows in coastal lowlands of southern Mexico, Central America, the West Indies, and northern South America, where it is commonly called Mora, yields the well-known dyewood, Fustic, which has been an article of commerce since the middle of the seventeenth century. The tree rarely exceeds 65 feet in height, but may have a trunk diameter of 30 inches. The light brown, laticiferous bark is smooth on young stems and branches, which are sometimes spiny. The timber is exported in the form of short logs with the thin sapwood hewed off and vary in diameter from a few inches to two feet. The coloring principle, maclurin, is readily soluble in water and the natural color imparted is a fairly permanent dull yellowish brown or khaki. The use of mordants not only adds to the permanency of the dye, but permits a considerable range in shade; mixture with other dyewoods yields compound shades such as drab, fawn, and olive. Fustic has been largely superseded by aniline dyes but a remnant of the trade persists and probably will continue to do so indefinitely.

The southern form is a timber tree. H. M. Curran says that, though nowhere abundant, it is a constant factor in the forests of Misiones (Argentina), Paraguay, and southern Brazil. It is tall and symmetrical, with a rather fine-textured and comparatively smooth brown or gray bark and a rather narrow crown. At maturity the trunks are usually between 18 and 24 inches in diameter, with clear lengths of 40 to 50 feet. The timber reaches the market in the form of squared logs and is used where a hard, durable wood is needed, but not as a dyewood. It is favorably known to the wheelwright and carpenter and also makes good fence posts, props, and fuel.

Heartwood typically golden yellow when fresh, changing upon exposure to brown or russet, sometimes with a reddish tinge; sharply demarcated from the nearly white sapwood. Luster high. Without distinctive odor or taste. Mostly very hard, heavy, tough, and strong; sp. gr. (air-dry) 0.85 to 0.99; weight 53 to 62 lbs. per cu. ft.; texture medium to coarse; grain variable, often interwoven; not very difficult to work, taking a lustrous polish; is highly resistant to decay.

COMMON NAMES: Dyer's mulberry, fustic, f. wood, old fustic (Eng., trade); bois jaune, fastique, fustique (Fr.); Fiselholz, Fisetholz, Fustik, Fustikholz, Gelbholz (Germ.); geelhout, stockvischhout (Dutch); fustete, mora (Span.); jataiba (Port.); fustetto vecchio, legno giallo (Ital.); red fustic, snook (Jam.); fresno de America, mora del país (Cuba); mora macho, palo amarillo (Dom. R.); paaloe doesji (Curaçao); bois d'orange, palo naranjo (Trin.); barossa, moradilla, moral, m. amarilla, m. de clavo, m. liso, palo moral, yaga-huil (Mex.); palo negro (Guat.); jocomico, madrial, pinabete, quebracho de cerro (Hond.); brasil, palo de mora, tsukrá (C.R.); macano (Pan.); avinje, cauchillo, dinde, majagua gallina, morita, palo amarillo, p. de mora, palomora (Col.); charo, mora amarillo, mora lisa, moral (Venez.); moral, m. bobo (Ec.); insira, i. caspi, limulana, ynsira (Peru); amarillo (Boliv.); amoeira brava, a. de espinho, limão-rana, l.-r. amarello, moratana, runa, tatajiba, tatajuba, t. de espinho, tatayiba, tayuva (Braz.); mora amarilla, m. blanca, m. colorada, tatayibá, t. puitá, t. saiyú (Arg.).

Clarisia, with eight poorly distinguished species of unarmed small to large trees and shrubs, is distributed from southern Mexico to southeastern Brazil, but not in the West Indies, according to Lanjouw (*Rec. Trav. Bot. Néerlandais* 33: 254–276).

Clarisia mexicana (Liebm.) Lanj. is a tree, sometimes 100 feet tall and 24 inches in diameter, in southern Mexico and Guatemala. *C. urophylla* (Donn. Sm.) Lanj. is said to be a related species in the forest about Puerto Sierra, Honduras. *C. colombiana* (Rusby) Lanj. is a large tree of northern Colombia, called Ache or Aji. *C. biflora* R. & P. is a tree 50 feet high in northern Peru, where it is called Piamich and Yasmich. *C. mattogrossensis* Lanj. is a tree 70 feet high in Matto Grosso, Brazil. *C. Spruceana* Lanj. is described as a small tree in Venezuela. *C. ilicifolia* (Spreng.) Lanj. & Rossb. is distributed from the Guianas through the Amazon region to Rio de Janeiro, Brazil, and varies in size from shrubs to moderately large trees; the vernacular names in Brazil are Bainha d'Espada and Diconroque.

The best known species is *Clarisia racemosa* R. & P., a large tree widely distributed in Brazil and extending into northeastern Peru; its northern limits have not been determined. In eastern Brazil it is generally known as Oiticica or Oity, names applied also to fruit trees, *Licania rigida* Benth. and others, of the Chrysobalanoideae group of the Rosaceae. According to H. M. Curran, it is very common in the rain forests of Bahia where it grows scatteringly or in small clumps in association with Pau d'Alho (*Gallesia*), Massaranduba (*Mimusops*), Araça de Agua (*Terminalia*), and Jequitibá (*Cariniana*). It attains a height of 130 feet, with a well-formed nonbuttressed trunk sometimes 36 inches in diameter and free of branches for 50 to 60 feet. The bark is covered with small corky warts. The timber is well known to the rural population and is used for general construction and carpentry; shipments of lumber were made to Germany in 1938. Objections to it are that the irregular grain makes it difficult to finish smoothly with hand tools and that it is not adapted for use in places favorable to decay. It is common in the Amazon basin, where it is generally called Guariuba, and grows to large size, mostly on lands that are not subject to inundation. The laticiferous bark of the stem is gray and warty on the surface, but intensely red within as are also the roots.

In the Yale collections there are numerous specimens of the wood of *Clarisia,* mostly from Brazil, but including a few from Venezuela, Colombia, Panama, and northeastern Peru. They are all so much alike that they might be the produce of a single species. Heartwood bright yellow, becoming brown or russet, but retaining a golden luster, upon exposure; sharply defined from the thin, white sapwood. Dry specimens without distinctive odor or taste. Rather light and soft to moderately heavy; sp. gr. (air-dry) 0.50 to 0.65; weight 31 to 40 lbs. per cu. ft.; texture medium to coarse; grain variable, often decidedly roey; easy to cut, but cross-grained material requires sharp tools to produce a smooth surface; fresh wood saws woolly; holds its place well when manufactured; not very durable in contact with the soil; quarter-sawed roe-grained lumber has attractive figure.

Common names: Árbol del pan (Mex.); ache, agi, aji, dinde (Col.); pellejo de Indio (Venez.); capinuri, guariúba, huariuba, imauba, killo-muena, machunaste, mashonaste, maxunasti, piamich, tulpay, yasmich (Peru); bainha d'espada, diconroque, guariúba, janitá, oiti, oiticica, o. amarella, quariúba (Braz.).

Coussapoa, with numerous species of unarmed laticiferous trees and shrubs, is widely distributed in continental tropical America from Mexico to Brazil, Bolivia, and Peru. The plants are usually epiphytic, at least when young, and frequently also are more or less scandent; some of them develop into large trees. The stems are hollow and the leaves are large, sometimes 20 inches long, entire, leathery, prominently pinnately veined, and light-colored on the under side. The fiber of the inner bark is used by the natives for making coarse baskets and bags. The timber is not utilized.

Heartwood absent or not clearly differentiated from the pale brown or pinkish sapwood. Luster medium to rather high. Odorless and tasteless. Light in weight but comparatively firm, tough, and strong; texture coarse; grain straight to irregular; easily worked, perishable in contact with the soil. An unattractive wood without commercial possibilities.

Common names: Caraco (Col.); yele (Br. G.); coassapoua, coussapoui (Fr. G.); matapalo, m. colorado (Ec.); higo de monte (Boliv.); chichillica, uvilla, xenácocaspi (Peru); apuhy, a. grande, apuî, caimbé, c. rana, cibueiro, gummileira (Braz.).

Ficus, with several hundred named species of small to very large trees and shrubs, many of them epiphytic, at least in youth, is distributed over the warmer regions of the world, being most abundant in the East Indies and Africa. The best known plants are the common cultivated fig, *F. carica* L., and the Rubber plant of the florists, *F. elastica* Roxb., native of India and Malaysia. The latter species is a gigantic evergreen tree in its native habitat and its latex is the source of the India rubber of commerce. The famous Banyan, *F. bengalensis* L., with its numerous prop roots, may spread over an acre of ground. The Peepul or Pipal tree, *F. religiosa* L., another East Indian tree, is planted in villages and held sacred by the Hindus and Buddhists. The Sycamore of Biblical literature, whose wood was used by the ancient Egyptians for mummy cases, is *Ficus Sycomorus* L., not to be confused with the English Sycamore, which is a kind of Maple (*Acer*), or the American Sycamore (*Platanus*), which is closely related to the English Plane-tree.

According to Standley (*Contr. U.S. Nat. Herb.* 20: 1: 1. 1917), there are 33 species of *Ficus* in the West Indies, about 50 in South America, and 41 in Mexico and Central America. "It is a remarkable fact that only two of these are common to the West Indies and Central America, and one of them is doubtfully indigenous to the latter region." *Ficus* is a genus that lends itself readily to the making of herbarium species. Standley says (*loc. cit.*, p. 2): "While many of our [Mexican and Central American] species are constant in their characters, some are so variable that two collections may seem to represent quite distinct species until intermediate specimens are examined. Some characters which have been used to separate species are now found to be unreliable. Consequently, it is not improbable that certain species here recognized will have to be reduced when still more ample collections are obtained."

Some of the Fig trees develop normally and attain a height of 150 feet with a long trunk four to five feet in diameter above the buttresses and free of branches for 60 feet. Many others begin their growth in the tops of other trees and develop vine-like aerial roots which anastomose and grow around the host tree and eventually kill it, though a palm may survive for years with its trunk encased within a Strangler Fig or Matopalo, as they are commonly called. A Strangler Fig tree may finally assume the appearance of an ordinary tree, but during the course of its development it presents many fantastic or hideous forms.

In pre-conquest times the Aztecs of Mexico made extensive use of the bark of Fig trees for the preparation of paper for their records and correspondence. Large trunks are sometimes used for making dugout canoes, but they are not very lasting. At present, the trees are considered as weeds in the tropical American forest, as their wood is likely to rot or at least discolor before it can be dried. In localities where timber is scarce it is sometimes used for making boxes and for light interior construction.

The woods of *Ficus* are light-colored and of fairly uniform structure, being composed of alternating bands of wood fibers and soft parenchyma, the comparative thickness of each varying considerably in different species and specimens. This laminated structure is especially distinct if the soft layers have turned dark brown, as is so frequently the case with museum specimens. Heartwood not clearly differentiated. Luster low to medium. Odorless and tasteless. Light and soft to moderately hard and heavy, tough and strong for their weight; texture coarse to medium; grain variable; easy to work, finishes smoothly, holds nails firmly. Not likely to become of commercial importance, though suitable for many common purposes if lumber is kiln-dried before deteriorating.

COMMON NAMES: Fig, f. tree, wild fig (Eng.); higo, higón, higuerón, jagüeicillo, jagüey, j. de lavar, j. de peladero, j. hembra, j. macho, pinipiní (Cuba); hamo, hijo (Dom. R.); figuier, f. blanc, f. canelle, f. rouge (Haiti); álamo, amacostic, amacuahuitl, amate, a. amarillo, a. blanco, a. prieto, amatillo, amatl, amazquitl, amezquite, anabá, cabra-higo, camachín, capulín grande, chalote, chamachina, coamichín, comuchín, coobó, copó, copoy, cozahuique, higo loxe chico, h. l. grande, higuerón, higuito, hoeiamatl, itzmatl, jalamate, kopó, macahuite, matapalo, m. liso, nacapuli, ojite, palo chilamate, p. María, p. de coco, sabali, salate, s. bronco, samatito, tepeamate, tepeamatl, tescalama, tescalamate, texcalamate, texcalamatl, tlilamatl, xalama, zalate (Mex.); amate, copulamate, capulín amate, matapalo (Guat.); higo, higuero, higuillo (Hond.); betsúr, chilamate, detsí, gu-tsa, higuerón, h. colorado, higuito, káua-krá, keba-krá, krop, palo de agua, p. de sal, psií-krá, sigín, sotacaballo (C.R.); higo, higuerón, igo, matapalo, níspero blanco (Pan.); copé, copei, c. de tierra fría, guaimarito, higuerón, h. blanco, h. negro, jagúey, matapalo, nisperillo, pibijai, pivijay (Col.); abrazapalo, apamate, araguato, chopo, chuare blanco, higo, higuerón, higuerote, higuito, jagüey, matopalo, murcielaguero, sío (Venez.); dukalaballi, kumakaballi (Br. G.); doekoelia, doukaliebalié (Sur.); figuier grand bois (Fr. G.); labiata, matapalo, m. blanco (Ec.); capinúri, consájo, coucho renaco, huito, huitoc, ojé, renáco, renaquillo (Peru); apuhy, cachinguba, camelleira, caxinguba, coajinguva, cuaxinguba, gamelleira, guaxinguba, lombrigueira, mata-pau, uapuí, uapuim-assú (Braz.); figueira, higuerón, yatitá (Par.); agarro-palo, guajinguba, guapohy, guapoy, higuera blanca, h. del agua, h. morada, higuerón, h. bravo, ibapohy, ibapoy, i. caágui, i. morotí, i. say, ivapoy (Arg.).

Helianthostylis, with two closely related species of small unarmed laticiferous trees, is apparently confined to non-inundated land in the central Amazon region of Brazil. They are dioecious through reduction, the pollen-bearing staminate flowers being in heads with long-styled rudimentary pistils, whereas the fertile pistil is surrounded by staminate flowers without pollen. The twigs are slender and slightly zigzag; the smooth leathery entire leaves are 3 to 8 inches long, abruptly acuminate, the anastomosing veins depressed above and very prominent beneath; the yellowish green fruits are about the size of an ordinary marble. The timber is not utilized.

Helianthostylis paraensis Ducke was discovered in the forest along the Rio Tapajoz (see *Archiv. Inst. Biol. Vegetal* 2: 1: 30. 1935). The only wood sample available for study is of *H. Sprucei* Baill., a similar tree of infrequent occurrence along the Rio Negro and near Manáos, collected with fertile herbarium material by Adolpho Ducke (Yale 23642).

Heartwood absent or not distinguishable from the yellowish white sapwood. Luster rather high. Odorless and tasteless. Hard, heavy, tough, and strong; texture fine and uniform; grain straight; not difficult to work, finishing very smoothly. Of no commercial importance.

Helicostylis, with eight species of unarmed, laticiferous medium-sized to large, sometimes deciduous trees, is distributed from the Amazon region of Brazil through northern South America to Costa Rica. It is similar to *Perebea,* but the long and slender stigmas are somewhat contorted. The timber is rarely utilized.

The most northern species is *Helicostylis urophylla* Standl. of Costa Rica; it apparently is the only small tree in the group, and Standley states (*Flora of Costa Rica,* p. 389) that "it is possible that perfect material will show that this plant is better referable to the genus *Olmedia.*" Two species described from Panama are *H. latifolia* Pitt., a tree 60 to 80 feet tall and 12 to 16 inches in diameter, growing in the hill forests of Canal Zone and southern Darién, and *H. montana* Pitt., a deciduous tree 80 to 100 feet high and 36 to 40 inches in diameter, in the mountains of San Felix valley. *H. bolivarensis* Pitt. of Colombia appears to be misplaced generically, and Pittier stated in 1921 (see *Timbers of Tropical America,* p. 127, footnote) that "it is more likely to be a *Pseudolmedia.*" A year later the same type specimen (H. M. Curran 304) was described by S. F. Blake as *Brosimum columbianum* (*Proc. Biol. Soc. Wash.* 35: 179).

In the Amazon basin there are five named species, all described as medium-sized trees. *Helicostylis asperifolia* Ducke is distinguished by the roughness on both sides of its leaves, which feel like a fine grade of sandpaper (for description of the tree and wood, see *Tropical Woods* 31: 11, 23). The most widely distributed species is *H. tomentosa* (P. & E.) Rusby, which occurs from Bahia, Brazil, through the Amazon region to northeastern Peru and the Guianas. The heartwood is so slow in developing that trunks several inches in diameter may be all sapwood. Such timber, being perishable in contact with the soil, is not very serviceable in the moist tropics, though it could be used for the same purposes as Hickory (*Carya*) in the United States.

In the Yale collections there are 11 authentic samples of four species. Heartwood dark brown, somewhat streaked or variegated with black and yellow; sharply demarcated from the sapwood, which is nearly white when fresh, later becoming golden. Luster medium in heartwood, high in sapwood. Very hard, heavy, tough, and strong; texture medium; grain straight to roey; not very difficult to work, taking a high natural polish; heartwood appears highly durable.

COMMON NAMES: Bi, birsk, feguó, kabákrá, ojoche (C.R.); berbá, choybá, querendo (Pan.); sukune (Br. G.); basri letri, b. pauletoe, berekoro, boelikoro, kapitin hoedoe, man letter, moejepau lattoé, oenbatapo, ombatapo, paida, pauletoe aledin, perokoti, poevinga, sokoné fieroberoe, s. koereroe, sabana letri, toekoesie paida, tokolo apolikiri, tokora apolinore, wamiriaballi (Sur.); aimpem, inaré, jaboty, muiratinga (Braz.).

Maclura (or *Toxylon*), with a single species, *M. pomifera* (Raf.) Schneider, is a thorny, bushy, laticiferous tree of very limited natural distribution centering in the valley of the Red River in Oklahoma, but widely planted elsewhere in the United States, especially in the prairie regions, for hedge fences. It is commonly known as Osage Orange or Bodark, the latter a corruption of Bois d'Arc which refers to the use of the wood by the Indians for making bows. Owing to its strength and low shrinkage its principal use in manufacture has been for the rims of wagon wheels, particularly those for service in dry sandy countries. Other special uses are spokes, insulator pins, tree-nails, and to a minor extent as a dyewood. It is considered the most durable of all North American timbers and, inasmuch as the sapwood is very thin, even small-sized stems will give long service as stakes and posts.

Heartwood golden yellow, sometimes with reddish streaks, when fresh, becoming russet-brown upon exposure; sharply demarcated from the thin white sapwood. Luster high. Odor and taste absent or not distinctive. Very hard, heavy, tough, strong, and resilient; sp. gr. (air-dry) 0.85 to 0.90; weight 53 to 56 lbs. per cu. ft.; rather difficult to work, finishes very smoothly, holds its place remarkably well when manufactured, being comparatively inert to changes in atmospheric humidity. Tests by the U.S. Forest Service give the following results (U.S. Dept. Agr. Misc. Pub. No. 46, pp. 8–9): *Weight per cu. ft.*: green, 62 lbs., air-dry, 56 lbs., kiln-dry, 54 lbs. *Shrinkage* (in volume from green to oven-dry condition): 8.9 per cent, or only slightly more than Sugar Pine (*Pinus Lambertiana*) (8.4 per cent). Relative strength compared to White Oak (*Quercus alba* L.): *Bending*, 1.58; *crushing*, endwise 1.65, crosswise 3.11; *stiffness*, 1.01; *hardness*, 1.94; *shock resistance*, 2.85.

COMMON NAMES: Bodark, bodock, bois d'arc, bow-wood, hedge, mock orange, Osage, O. orange, O. apple tree, yellow wood (U.S.A.).

Morus, with several species of small to rather large, unarmed, laticiferous, typically deciduous trees and shrubs, is widely distributed in the temperate regions of the northern hemisphere and sparingly in the tropics. The fruit, known in English as the Mulberry and in Spanish as Mora, is edible and is composed of numerous small, juicy, 1-seeded drupes. The leaves are the principal food of silkworms. The lumber is of good quality and durable, but is not extensively utilized because of the scarcity and small size of the available timber. The smaller sizes are suitable for fence posts, railway crossties, larger logs for furniture and cabinet work.

The most widely known species are the Black Mulberry, *Morus nigra* L., native to Persia, and the White Mulberry, *M. alba* L., of China. Both are widely planted for shade and fruit trees. There are four American species. The Red Mulberry, *M. rubra* L., occurs throughout most of the eastern United States and southern Ontario, attaining its best development in the lower Ohio River valley and the foothills of the southern Appalachian Mountains. Though usually rather small, it sometimes attains a height of 60 to 70 feet, with a dense crown and a short trunk having a maximum diameter of four feet.

The Mexican Mulberry, *Morus microphylla* Buckl., is a small tree, 15 to 25 feet high and 10 to 15 inches in diameter, or sometimes shrubby, in western Oklahoma and Texas, southern New Mexico and Arizona, and northern Mexico. The wood is used to a limited extent by Mexican carpenters and formerly by the Indians in Texas for making bows. The species may be only a form of *M. celtidifolia* H.B.K., a tree rarely 30 feet high occurring in the uplands from Mexico through Central America and Colombia to Ecuador and Peru. The timber is said to be valued in Ecuador for building purposes.

The following description is based upon samples of *Morus rubra* and *M. microphylla* obtained in the United States. Heartwood orange-yellow, becoming russet to dark brown upon exposure; sharply demarcated from the thin, white or yellowish sapwood. Luster medium to rather high. Without distinctive odor or taste, at least when dry. Moderately light to rather heavy; sp. gr. (air-dry) 0.60 to 0.75; weight 37 to 47 lbs. per cu. ft.; of medium to coarse texture; straight-grained; easy to work, taking a high polish, holds its place well when manufactured; is very resistant to decay.

COMMON NAMES: Mulberry (black, red, Mexican), m. tree, murier sauvage (U.S.A.); hamdek-kiup, moral, tzitsi, yagabiyozaa (Mex.); brasil (C.R.); mora (Ec., Peru).

Naucleopsis. Four species of small to medium-sized unarmed Amazonian trees, with bitter latex, have been described, but the only specimen in the Yale collections (Yale 31959; Ducke 232) is of *N. macrophylla* from near Manáos. The tree occurs in the understory of the forest on noninundated land in the central Amazon region. The bark contains a green watery juice that is very bitter, and on that account the tree is called Quina. The thick, smooth, leathery, entire leaves are frequently more than a foot long and three inches wide. The timber apparently is not utilized.

Heartwood light yellowish brown; not clearly differentiated from the nearly white sapwood. Luster medium. Odorless and tasteless. Of medium density, but firm and strong; texture rather fine; grain straight; easy to work, saws somewhat woolly, but can be finished smoothly; probably not resistant to decay. Presumably of no commercial possibilities as the sample is not attractive in appearance and the trees seem to be scarce.

COMMON NAMES: Muiratinga, quina (Braz.).

Noyera, with two or three species of unarmed shrubs or small to medium-sized trees, occurs in the Guianas and Brazil. The best known is *N. mollis* (Poepp.) Ducke, a tree usually small but sometimes 65 feet high, fairly common in the second-growth forests of the middle and upper Amazon region. The rather large, leathery, distinctly pinnately veined leaves are rough on the upper surface and typically covered with velvety brown pubescence below. The bark contains an abundance of grayish yellow latex. The timber is not utilized.

Heartwood absent or not distinguishable from the yellowish or brownish sapwood. Luster medium. Odorless and tasteless. Hard, heavy, tough, and strong; texture medium; grain straight; not difficult to work, finishing smoothly; is not resistant to decay. Has no commercial possibilities.

COMMON NAMES: Caucho-rana, lechecaspi (Peru); caucho-rana, iawara-po, muiratinga da terra firme, pama-caruara (Braz.).

Ogcodeia, with a dozen species of unarmed laticiferous small or occasionally medium-sized trees, is distributed from Nicaragua to Bolivia, mostly in the Amazon basin. There is no information about the uses, if any, of the timber. The most northern species is *Ogcodeia naga* (Pitt.) Mildbr. of Costa Rica, Nicaragua, and Honduras. According to Standley (*Flora of Costa Rica* 1: 390), it is a forest tree 25 to 35 feet high. "The milky sap that exudes when the trunk is tapped has the appearance of cow's milk and is reported to have a similar flavor." *O. guianensis*

Mildbr. attains a height of 50 feet along the Cuyuni River in British Guiana; its latex is creamy white and very sticky. The species found in eastern Peru are: *O. tamamuri* Macbr., local name Tamamúri; *O. ternstroemiflora* Mildbr., 15 feet high and 8 inches in diameter, called Laná or Saná; and *O. Tessmannii* Mildbr., 25 feet high.

The following description is based on one wood sample each of four species. Heartwood (?) pale brown; not clearly differentiated from the lighter colored sapwood. Luster medium to rather high. Odorless and tasteless. Moderately hard and heavy; texture medium; grain straight; readily worked, finishing smoothly; presumably not durable in contact with the ground. Of no commercial possibilities, apapparently.

COMMON NAMES: Concha de Indio, majagua de Indio (Hond.); naga (C.R.); laná, naccho-huasca, saná, tamamúri (Peru).

Olmedia *aspera* R. & P., the only definitely known species, is a small to medium-sized unarmed laticiferous tree in Ecuador, Bolivia, eastern Peru, and the Brazilian Amazon country to Pará. It is well marked by its rather large, remotely and obscurely toothed, very scabrous leaves which feel like coarse sandpaper. By soaking and beating the inner bark the Indians make serviceable mats or blankets. The laminated wood resembles that of *Ficus* and is not utilized.

Heartwood absent from available specimens; sapwood grayish, of rather "stringy" appearance. Luster medium. Odorless and tasteless. Rather light in weight, but firm and tough; texture coarse; grain straight; easy to work, finishing smoothly; perishable in contact with the soil. Of no commercial possibilities.

COMMON NAMES: Llanchama, minchipata (Peru).

Olmedioperebea. The type of this genus, which was proposed by Adolpho Ducke in 1922 (*Archiv. Jard. Bot. Rio de* 3: 33), is *O. sclerophylla* Ducke, an unarmed tree 80 to 115 feet tall, growing on non-flooded land in the middle Amazon region of Brazil, where it shares the name Muiratinga with other white-wooded moraceous trees. The leaves are 8 to 12 inches long and 4 to 6 inches wide and are very hard. The fruit has a strong musk-like scent. The bark contains a yellowish gray latex. The timber is not utilized.

Heartwood absent or not clearly differentiated from the yellowish white sapwood. Fairly lustrous. Odorless and tasteless. Rather hard and heavy, tough and strong; texture coarse; grain straight; not difficult to work, finishing smoothly; is not resistant to decay. Of no commercial possibilities.

According to Ducke (*Arquiv. Serv. Flor.* 1: 1: 14; 1939), there is a second species, *Olmedioperebea calophylla* (P. & E.) Ducke, a rather small tree widely distributed in the non-inundated Amazon forest. The receptacles of the fruits, called "sorvinha" in São Paulo de Olivença, are juicy and comestible. The herbarium specimens appear very distinct from those of *O. sclerophylla* and there are likewise notable differences in the woods. Heartwood brown, with narrow black stripes; distinct from the sapwood. Density high to very high; texture medium; grain irregular; difficult to work; durability probably high.

Olmediophaena, with three species closely related to *Olmedia* and *Pseudolmedia,* occurs in Colombia (*O. coriacea* Karst.) and the Amazon region. *O. obliqua* (Huber) Ducke is a shrub or little tree of Pará and Amazonas, Brazil. *O. maxima* Ducke is one of the tallest trees on the periodically overflowed lands in the middle Amazon region, attaining a height of 130 feet or more and attracting attention because of its dense dark green foliage and its slender white or red-splotched trunk. Further eastward it becomes smaller and less conspicuous in the forest. It is the true Muiratinga, a Tupi Indian name meaning white wood. The copious grayish yellow latex is used in native medicine. Heartwood not available; sapwood pale grayish brown. Luster medium. Odorless and tasteless. Hard and moderately heavy; texture medium; grain somewhat irregular; not very difficult to work. Presumably without commercial possibilities.

COMMON NAMES: Capinury, muiratinga (Braz.).

Paraclarisia *amazonica* Ducke, the only species, is a small to medium-sized lactescent tree related to *Clarisia* and *Sorocea*. It occurs on periodically inundated land along the Amazon River in Brazil, where it is commonly known as Janitá, a name also given to *Clarisia ilicifolia*. (See *Arquiv. Serv. Flor.* 1: 1: 2.) The following description is based upon a single specimen (Yale 23655) collected by Adolpho Ducke below Santarem.

Heartwood not available; sapwood pale grayish brown, somewhat "stringy" in appearance. Luster rather high. Odorless and tasteless. Hard and heavy; texture rather coarse; grain somewhat irregular; not difficult to work. Probably of no commercial possibilities.

Perebea, with about 15 described but mostly poorly known species of small to medium-sized unarmed laticiferous trees, occurs throughout the Amazon basin and northward to Costa Rica. Some of the plants bear considerable resemblance to *Castilla*. So far as known, the members of this genus supply no economic products and their timber is not utilized. The wood material available for this study consists of authentic samples of five species, namely, *Perebea chimiqua* Macbr. of northeastern Peru (type); *P. glabrata* Standl. from Panama (type); *P. concinna* Standl. and *P. Tessmannii* Mildbr. from Brazil; and *P. castilloides* Pitt. from Panama (Yale 12142; Cooper 523). The last was originally determined as *Castilla panamensis* Cook, an error discovered in studying the wood (which is also unlike that of the other specimens of *Perebea*), thus adding strength to Pittier's expression of doubt (*Contr. U.S. Nat. Herb.* 13: 12: 438) whether it really belongs to *Perebea*.

Heartwood absent or indistinguishable from the yellowish white sapwood, which may acquire a pinkish tinge. Luster rather high, golden. Odorless and tasteless. Moderately hard, heavy, tough and strong; sp. gr. (air-dry) about 0.70; weight 44 lbs. per cu. ft.; texture coarse, feel harsh; grain straight; easy to split, saws rather woolly when fresh, takes a smooth finish; is not resistant to decay. Of no commercial possibilities.

COMMON NAMES: Caucho, ule, wild rubber (Pan.); kapiteinhout, zonderhart (Sur.); aberemou, vive éperou (Fr. G.); chimiqua, sipáruna (Peru); caucho-rana, muiratinga (Braz.).

Piratinera, best known as the source of Letterwood or Snakewood, includes several closely related species of medium-sized to large, unarmed, laticiferous trees, distributed throughout the Amazon region and northern South America, with one species (*P. panamensis* Pitt.) reaching southern Mexico. It was formerly customary to consider *Piratinera* Aublet (1775) synonymous with *Brosimum* Swartz (1788), and the latter name was made a *nomen conservandum* by the International Botanical Congress at Vienna in 1905. In 1918, however, Pittier (*Contr. U.S. Nat. Herb.* 20: 3: 96–100) presented evidence to show that the two genera are distinct. His conclusions were accepted, at least tentatively, by Blake who says (*Journ. Wash. Acad. of Sci.* 12: 17: 393) that "the floral characters brought forward by Pittier are sufficient to justify the separation of the two genera," and by Standley who believes (*Tropical Woods* 17: 9) that "until ampler herbarium material is available for study, it seems best to recognize *Piratinera* as a valid genus." The opinion reached from a study of the woods is that the differences between *Brosimum* and *Piratinera* are fully as pronounced (or no less obscure) than those separating some of the other genera in this family. So far as known, all species of *Piratinera* have dense woods and develop a considerable amount of heartwood which is a rather dull reddish brown with vertical or radial markings of black pigment, whereas in *Brosimum* (*sens. str.*) such specimens of heartwood as have been studied are remarkable for their rich coloration and silky, golden luster. From the point of view of their anatomy the two genera might well be merged, but for practical reasons it seems better to treat them as though they were distinct.

Letterwood has been an article of export from the Guianas since the time of their European settlement. Evidence of the early importance of the industry is contained in article seven of the terms of capitulation of March 6, 1667, by which the British Lieutenant General Byam gave over the Colony of Surinam to the Dutch Admiral Abraham Orijnssen. It is there specified that the English settlers should have the privilege of cutting Letterwood on the river. It is said that much of the original supply came from what was known as "tabukas," or heartwood parts of old fallen trees from which the sapwood had disappeared through the action of insects and decay.

The timber owes its value to the irregular black markings which have a fancied resemblance to hieroglyphs or letters and give to the tangential surface of the heartwood a peculiar appearance which suggests the spotted skin of certain snakes. These markings occur in all species of *Piratinera,* so far as known, but they are not a constant character. They extend radially and vary greatly in size and number not only in different specimens but also in the same piece. In tracing them along a radius it is found that they are much branched and anastomosed so that the figure they produce on the surface is never the same at different depths. The dark areas owe their existence, not to structural differences of any kind, but to variations in the color of the gummy deposits filling all of the cell cavities, the darkest parts being at the margins rather than in the center of the patches. In some trees, without distinction as to species, the heartwood is marked only with vertical black bands and these may develop at intervals in otherwise speckled wood. The Negroes who make a business of cutting the timber in the Guianas cannot tell if a standing tree has figured or plain wood. Those with plain wood are considered false or bastard kinds, but present knowledge indicates that these differences in trees are individual rather than specific or even varietal. The darkening of the cell contents is presumably the result of oxidation, but a satisfactory explanation of its radial distribution cannot now be given.

The principal species supplying commercial Letterwood is *Piratinera guianensis* Aubl. (= *Brosimum guianensis* [Aubl.] Huber = *B. Aubletii* P. & E. according to Macbride, but not according to Pittier). It often attains a height of 80 feet, with a long cylindrical bole sometimes over 30 inches in diameter above the root swelling and covered with a smooth bark containing a thick, sticky, white latex. Heartwood is slow in forming and a 15-inch log may have a core of heartwood only 1 to 4 inches thick (Plate XXXI). Trees vary in this respect and in Surinam certain creeks have a reputation among Letterwood hunters of producing timber with much heartwood, while in other localities equally large trees may not be worth felling.

The timber is exported in the form of small logs or sticks about seven feet long and from two to eight inches in diameter, with all sapwood removed. It is sold by weight and is one of the most expensive woods in the trade. It is very strong, though brittle, and was formerly used by the South American Indians for bows. Its principal use now is for walking sticks, for which purpose it is considered one of the finest of all woods, though it is difficult to bend without buckling and is also likely to split. Other uses are umbrella handles, drum sticks, fishing rod butts, and miscellaneous fancy articles. It is occasionally employed in violin bows, but is not very highly esteemed for this purpose. The larger sticks are sometimes sawed into thin veneers for cabinet work. Owing to the difficulty of finding a specimen of much size that is finely figured throughout, it is a common practice to stain patches of plain wood to simulate the natural markings. The sapwood, comprising by far the greater bulk of the tree, is left in the forest. It is hard, tough, and strong, and would possibly make good handles for implements and tools.

Heartwood brown or reddish brown with irregular radial black markings or with black vertical stripes alone or in conjunction with the speckles; distinctness of markings reduced as the color of background deepens upon exposure; sapwood thick, yellowish white, the line of demarcation often irregular and not very sharp. Luster medium. Odorless and tasteless. Heartwood extremely hard and heavy, strong but brittle; sp. gr. (air-dry) 1.20 to 1.365 (see *Tropical Woods* 6: 6); weight 75 to 84 lbs. per cu. ft.; texture fine and uniform; grain straight; hard to cut, splits readily, is inclined to be splintery, takes a very smooth finish and high natural polish; heartwood very resistant to decay, sapwood perishable. Of limited commercial importance because of the scarcity and small size of suitable timber.

Common names: Leopard wood, letterwood, snakewood, speckled wood, tortoiseshell wood (Eng.); bois de lettres (Fr.); Buchstabenholz, Letterholz, Muskatholz, Schlangenholz, Tigerholz (Germ.); letterhout (Dutch); legno serpente, l. tigre (Ital.); cacique carey (Pan.); palo de oro (Venez.); cangica, bourracourra, boutous, burokoro, burracurra, paira, tibicusi, tibikushi (Br. G.); basri letri, belokoro, bepaulétoe, boelekolle, gespikkeld letterhout, kapeweri letri, koelero, koereroe, kolero, letri, man letri, m. letterhout, moejepauletoe, paida, poevinga, roode letterhout, sokoné biberoe, tianalin wéwé, tokoro apolli marie, wekere paida (Sur.); bois d'amourette, b. de lettre gris, b. de l. moucheté, piratiner, piratinere (Fr. G.); barrueh, gomelleira preta, gateado, gatiá, moirapinima, muirapenima, muirapinima, pau rainha de listras, p. tartarugo (Braz.).

Poulsenia *armata* (Miq.) Standl., the single species, is a medium-sized to large, laticiferous tree, sometimes 100 feet tall with buttressed trunk, occurring from Vera Cruz, Mexico, through Central America and Colombia to Ecuador and Bolivia. It is readily distinguished from all other American members of the family by the numerous prickles investing the large stipules and the twigs. The ripe fruit heads, which are somewhat suggestive of small chirimoyas (*Anona cherimolia* Mill.), are edible and are sometimes sold in the markets of Vera Cruz (see *Tropical Woods* 33: 4–5). According to J. M. Duque (*Notizbl. Bot. Gart. Berlin-Dahlem* 13: 119: 496), the tree attains large dimensions in the virgin forest at an elevation of about 5500 feet along the Río Cali in Colombia and the timber is preferred for construction because of its reputed resistance to fire.

The inner bark of mature trees is very thick and composed of many layers of strong, interlaced fibers; it has long been used by the aborigines for making hammocks, blankets, mats, and clothing. According to Eduard Conzemius (see *Tropical Woods* 33: 32), bark cloth is still made by the Sumu Indians of Honduras and Nicaragua from the Tunu tree, which is believed to be either *Poulsenia armata* or *Ogcodeia naga* (Pitt.) Mildbr. The bark is soaked in water for a few days after which the sticky gum or milk adhering to it is scraped off. The bark is then dried in the sun and kept in the hut until the women find time to pound it into cloth. As it becomes hard and shrinks considerably, it has to be submerged in a neighboring stream for a short time before the pounding begins. The latter operation is performed on a small log with the aid of a wooden mallet made from the stems of two species of small palms. The bark extends gradually upon being pounded and becomes soft and flexible. After being washed and dried it is ready for use, and has a brownish color. A similar cloth, almost white in color and of superior quality, is obtained by the same process from the inner bark of a species of *Ficus* and likewise from the Rubber tree (*Castilla*), but in both cases manufacture of the cloth is more laborious.

Heartwood absent or not clearly distinguishable from the yellowish white sapwood, which becomes brownish or oatmeal colored upon exposure. Luster rather high. Odorless and tasteless. Rather light in weight, but firm and tough; texture coarse; grain straight to irregular; saws woolly when fresh; easy to cut but rather difficult to finish smoothly; is perishable in contact with the ground. Presumably of no commercial possibilities.

Common names: Ababábite, carnero, chirimoya (Mex.); tumu (Hond., Nic.); cocuá, maragua, mastate, namagua (Pan.); corbón, cucúa (Col.); majagua (Ec.).

Pourouma, with numerous species of small, medium-sized, or occasionally large unarmed trees, is widely distributed in tropical America from British Honduras to Peru and Brazil. The leaves are entire or palmately lobed or divided, the two forms sometimes appearing on the same tree. The pith is large and septate. The egg-shaped fruits are borne in cymes and in some species contain a juicy edible grape-like pulp. The light and perishable timber apparently has no special uses.

The northernmost species is *Pourouma aspera* Tréc., growing from northern South America to southern British Honduras. In eastern Nicaragua, according to notes supplied by F. C. Englesing (*Tropical Woods* 17: 35), it is a slender tree 85 to 110 feet high, with a cylindrical and somewhat arcuate trunk sometimes 24 inches in diameter, growing on low hills. The bark is smooth and mottled in various shades of brown, mauve, and gray. The stump of a freshly felled tree exudes a quantity of watery sap. Branches are few and ascending, forming a crown suggesting a candelabrum. Clustered at the ends of the twigs are palmately lobed leaves which are so scabrous that they are used by the Indians for sandpaper.

Most of the species are Amazonian. *Pourouma cecropiaefolia* Mart. has large radiately parted leaves that are velvety and gray or white beneath, suggesting *Cecropia*, though not peltate as in that genus. In at least one species, *C. myrmecophila* Ducke, the base of the petiole is enlarged and inhabited by very small, but pugnacious ants. A few species are cultivated in eastern Brazil for their fruits, but it is difficult to keep the trees low enough for convenient harvesting.

Heartwood absent or not clearly differentiated from the white sapwood, which becomes brownish. Luster rather high. Odorless and tasteless. Light in weight, but rather firm and tough; texture coarse; feel woolly; grain fairly straight; requires sharp tools for smooth working; poorly resistant to decay. Of no commercial promise.

Common names: Trumpet (Br. H.); guarumo macho, yabal (Nic.); guarumo, g. de montaña (C.R.); mangabe (Pan.); cormi, serpe, sirpe (Col.); buruma (Br. G.); pourouma (Fr. G.); ambauba mansa, amandier, cucúra, mapaty, sacha-uvilla, ubilla, uvilla (Peru); cucúra, imbauba de cheiro, i. de vinho, i. puruma, mapaty, tararanga, t. branca, t. preta, t. vermelha (Braz.).

Pseudolmedia, with several species of unarmed, laticiferous shrubs and small to medium-sized or rarely large trees, is distributed as follows: *P. spuria* (Sw.) Gris. grows to a maximum height of about 50 feet in Jamaica, Cuba, Puerto Rico, the island of Haiti, and British Honduras. *P. oxyphyllaria* Donn. Sm., a similar tree, occurs from Vera Cruz, Mexico, to Panama; during the dry season the branches are sometimes used for fodder for oxen. *P. mollis* Standl. has been described for Salvador. *P. Eggersii* Standl. of Ecuador is said to be "a tall erect tree with white wood used for inside construction" (see *Tropical Woods* 42: 27). There are several Peruvian species, and *P. multinervis* Mildbr. is said to reach a height of about 100 feet, with a long clear stem 36 inches in diameter above the root spurs which are about six feet high (see *Notizbl. Bot. Gart. Berlin-Dahlem* 10: 92: 189–190). There are a few species in Bolivia and British Guiana, and four or more in Brazil.

The following description applies particularly to *Pseudolmedia spuria*. Heartwood reddish brown, not very sharply demarcated from the thick sapwood which becomes grayish or pinkish brown. Luster medium. Odorless and tasteless. Very hard, heavy, tough, and strong; texture medium coarse; feel harsh; grain variable; not very difficult to work, finishing smoothly; reputed to be perishable in contact with the ground as most of the timber is sapwood. Of no commercial possibilities.

Common names: Bastard breadnut, milkwood (Jam.); macagua, m. amarilla (Cuba); negra lora (P.R.); macao (Dom. R.); bois mérese, longue barbe, mérisse (Haiti); cherry, manax (Br. H.); tepeujuhste (Salv.); ojoche (C.R.); vara piedra (Col.); guión (Ec.); chimicua, itaúba amarilla, loro-micunan (Peru); muiratinga, murere (Braz.).

Sorocea, with several species of unarmed laticiferous shrubs and small trees rarely 50 feet high and 18 inches in diameter, is distributed from Central America to northern Argentina and Uruguay. *S. affinis* Hemsl., a shrub or a tree up to 20 feet tall with entire or sinuate leaves and small red fruit in racemes, is known only in Panama. *S. colombiana* Standl., of about the same size, was discovered by H. M. Curran in the Department of Bolívar, Colombia; *S. Hirtella* Mildbr. is a shrub or a tree sometimes 40 feet high and eight inches in diameter growing in uplands in eastern Peru; *S. Briquetii* Macbr. and *S. opima* Macbr. are shrubs up to 12 feet high in the same general locality. *S. saxicola* Hassl. of Paraguay and Argentina is said to be a handsome tree 50 feet tall and 20 inches in diameter, with large dark green leaves and supplying a dark yellow tough and strong wood formerly utilized in making wheel hubs. The most widely distributed species is *S. ilicifolia* Miq., which has leaves with spine-tipped serrations. It is a shrub or a tree sometimes 40 feet high and 12 inches in diameter, occurring along the Amazon from above the estuary into Peru and southward into Paraguay and northern Argentina. Its elastic wood is used in Argentina for making barrel hoops.

Heartwood (present only in the sample of *Sorocea stenophylla*) light brown with orange hue, fading gradually into the thin nearly white sapwood. Luster medium to fairly high. Odorless and tasteless. Moderately hard and heavy; texture medium; grain straight; easy to work, finishing smoothly; durability doubtful. Of no commercial possibilities.

COMMON NAMES: Vara blanca (Col.); araçary (Braz.); ibirá-hú, María-molle, ñandipá-mí, ñ.-rá, soroco (Arg.); cincho (Urug.).

Trophis, with a few closely related and doubtfully distinct species of shrubs and small to rather large unarmed laticiferous trees, is distributed throughout the West Indies, southern Mexico, Central America, and the Andean region of South America into Peru. The leaves are rather large, entire or toothed, smooth or rough; the staminate flowers are in long catkins, the pistillate (on separate trees) in short, few-flowered spikes or racemes, thus differing from *Chlorophora* which has the pistils in small heads; the fruit is a small drupe with a large seed and scant, edible flesh.

The best known species is *Trophis racemosa* (L.) Urb. The timber is employed to a limited extent locally, but the chief value is in the green foliage which is widely used as fodder for horses and oxen, hence the general Spanish name Ramón (browse), anglicized to Ramoon.

Heartwood dark brown, with parenchyma markings suggesting Elm (*Ulmus*); sharply demarcated from the thick lighter-colored sapwood. Fairly lustrous. Odorless and tasteless. Hard, heavy, tough, and strong; texture medium; grain straight to irregular; not difficult to work, finishing smoothly; heartwood fairly durable.

COMMON NAMES: Ramón (Span., gen.); ramoon (Eng.); ramón de bestias, r. de caballos (Cuba); ramón de bestial (Dom. R.); chacox, confitura, huanchal, leche María, ramón de Castilla, ramoncillo (Mex.); white ramoon (Br. H.); chulujuhste, ojushte, pilijuhste, raspa lengua, ujushte (Salv.); San Ramón (Hond.); cafecillo (Nic.); gallote, lechoso, morillo, ojoche macho (Pan.); gigantón, guáimaro, g. lechoso, pan y cacao (Col.); charo, lechero, marfil (Venez.); cuchara-caspi, sinchi-caspi, urpai-machinga (Peru).

Trymatococcus, with four species of small to large unarmed laticiferous trees, is limited to the Amazon basin (see *Tropical Woods* 43: 34–35). *T. amazonicus* P. & E. is a shrubby little tree in northeastern Peru and northwestern Brazil. *T. turbinatus* (Baill.) Ducke is a small tree 20 feet high and 12 inches in diameter in Amazonas. *T. paraensis* Ducke is a large tree, sometimes 100 feet tall, with whitish wood; the latex is reported to be good for rheumatism. *T. oligandrus* (Benoist) Lanj. is a large tree in French Guiana and Surinam. Apparently the trees are poorly known and often confused with *Brosimum*. The timber is not utilized.

Heartwood absent from specimens; sapwood yellowish. Luster rather high. Odorless and tasteless. Very hard, heavy, tough, and strong; texture medium; grain straight; durability of heartwood unknown.

COMMON NAMES: Beloekoro, joekoeipio, letterhout (Sur.).

MYOPORACEAE

THIS family comprises five genera and about 100 species of shrubs or rarely trees. The main regions of distribution are in Australia and neighboring islands, but there is one species each in China, Japan, Hawaii, Mauritius, Africa, and the West Indies. The leaves are typically alternate, simple, entire, and without stipules; the flowers are axillary, solitary, or fascicled; the fruit is a 2-celled drupe with 2 to 10 seeds.

The useful products are few. *Eremophila Mitchelli* Benth., a small scrubby Australian tree, has a dense pleasantly scented heartwood noted for its durability and sometimes sold as a substitute for Sandalwood. *Myoporum platycarpum* R. Br., also of Australia, exudes a resin which is used to a limited extent for sealing wax. *M. sandwicensis* (DC.) A. Gray of Hawaii sometimes attains a height of 60 feet and a diameter of 36 inches; it is known locally as Naio and formerly supplied a low-grade Sandalwood after the exhaustion of the supply of Sandal trees (*Santalum*).

Bontia *daphnoides* L., the only species, is a small tree or a shrub of limited natural distribution in the West Indies and cultivated for ornament there and in northern South America. There are no known uses for the wood. The following description is based upon one small sample each from Curaçao and Dominican Republic. Heartwood light grayish brown, not clearly demarcated from the sapwood. Luster medium. With faint spicy scent, but without distinctive taste. Hard and heavy; fine-textured; grain fairly straight; carves easily and finishes very smoothly.

Growth rings indistinct. Pores small, not visible without lens; numerous but not crowded; mostly in short, infrequently rather long, radial multiples; well distributed, though with local tendencies to zonate arrangement of larger pores in early wood. Vessels thick-walled; perforations simple; pitting fine, alternate. Rays 1 or 2, sometimes 3, cells wide and up to 30, generally less than 20, cells high; vertical fusions common; heterogeneous; uniseriates and margins (2 to 6 cells high) of multiseriates composed of square or upright cells; ray-vessel pit pairs small, half-bordered, resembling intervascular in surface view. Wood parenchyma very sparingly paratracheal. Wood fibers with rather thick walls and small indistinctly bordered pits. Ripple marks of local occurrence and irregular. Gum ducts absent.

COMMON NAMES: Olive (Bah.); aceituna americana, olivo bastardo (Cuba); white alling (P.R.); olivo (Dom. R.); olivier bâtard (Haiti).

MYRICACEAE

Myrica, the only genus, includes about 35 species of shrubs and little trees of common occurrence in many of the temperate and tropical regions of the world. The resinous dotted often fragrant leaves are alternate, simple, sometimes pinnately lobed, and without stipules; the small, unisexual, monoecious or dioecious flowers are borne in short scaly spikes; the fruit is a small drupe, often covered with grains of wax. The most widely distributed of the American species is the common Bayberry, *Myrica cerifera* L., whose range includes the southeastern United States, the larger islands of the West Indies, and parts of Mexico and Central America. Though usually a slender shrub growing in lowland thickets, it sometimes develops into a small tree, 15 to 25 feet high. The wax from the fruit is used for making candles, which burn with a balsamic fragrance. A scarcely distinguishable species in the uplands of Mexico, often in high mountains, is *M. mexicana* Willd. Its wax is used for candles and also, of course, for medicinal purposes.

One of the principal sources of Myrtle wax in South America is *Myrica arguta* H.B.K. of Venezuela and Colombia. According to M. T. Dawe (*Journey through the western part of Colombia*, London, 1919, p. 9), the shrub or tree occurs in abundance in ravines near Manezales, Colombia, where it is called Laurel or Olivo, and stands of it are known as "oliveras." From the photographic illustration it is seen that the trees have small open crowns and slender crooked light-colored stems. The wax (cera de laurel) "is obtained by submerging the berries in a fiber bag in boiling water and subsequently pressing. The mixture of wax and water is then run into cold water, and when the wax solidifies it is collected and re-melted into moulds of the size and shape desired." From tests made at the Imperial Institute, London, it appears that the Colombian wax has the same general characteristics of commercial Myrtle wax from South Africa.

Heartwood dull reddish brown; without sharp demarcation from the lighter-colored sapwood. Odor and taste not distinctive. Moderately heavy and hard; texture very fine and uniform; grain variable, durability probably fair to good. Of no special uses, principally because of the small size and scarcity of the trees.

Growth rings present. Pores very small to minute, not visible without lens; nearly all solitary; numerous to abundant; well distributed without pattern. Vessels mostly with scalariform perforation plates having several bars; simple perforations may also be present, sometimes predominating; no spiral thickenings seen; intervascular pitting, if present, rather fine and opposite with tendency to scalariform. Rays mostly 1 to 3, sometimes 4, cells wide and few to 30, rarely to 50, cells high; heterogeneous; pits to vessels very small and rounded or narrowly elliptical. Wood parenchyma sparse to fairly well developed; diffuse or finely reticulate and barely visible with lens; crystals sometimes present. Wood fibers with moderately thick walls and very numerous small bordered pits. Ripple marks and gum ducts absent.

COMMON NAMES: Bayberry, sweet fern, s. gale, wax myrtle (U.S.A.); bayberry, mickle berry, waxberry, wild tea (Bah.); arraigán (Cuba); arrayán, bayberry, cerero, waxberry (P.R.); palo de cera (Dom. R.); árbol de cera, chacolol, chilpanxohuitl, huancanalá (Mex.); bayberry, teabark, teabox (Br. H.); cero vegetal (Hond.); laurel, olivo (Col.); encinillo, palomero, torcaz (Venez.); huacán timbú, laurel, tuppassaire, ssaire (Peru).

MYRISTICACEAE

THE Nutmeg family, according to the original classification, consisted of a single tropical genus, *Myristica,* with several subgenera, but some authorities now recognize 16 genera, as follows: *Gymnacranthera, Horsfieldia, Knema,* and *Myristica* in eastern Asia and Oceania; *Brochoneura, Cephalosphaera, Coelocaryon, Mauloutchia, Pycnanthus, Scyphocephalium,* and *Staudtia* in Madagascar and continental Africa; *Compsoneura, Dialyanthera, Iryanthera, Osteophloeum,* and *Virola* in tropical America. Regarding the geographical distribution in the New World, A. C. Smith says (*Brittonia* 2: 5: 396):

"The American species of Myristicaceae are strictly tropical, being found primarily at low elevations in an area circumscribed by southern Mexico and the lesser Antilles on the north and lowland Bolivia, Matto Grosso, and Santa Catharina on the south. A few species ascend the foothills of the Andes to an elevation of 1000 or even 1500 meters, but the great majority of species occurs only on the low tropical plains. The center of distribution of the American species is the western portion of the great Amazonian plain. In the Brazilian State of Amazonas, all of the five genera and 44 species have thus far been found of a total of 73 species in the family. In the adjacent portion of Peru the given genera are also represented, with 25 species. From this center the family is less richly represented toward the limits of its range, only a single species being known from Mexico (*Compsoneura Sprucei*), the Lesser Antilles (*Virola surinamensis*), and the Brazilian States of Paraná and Santa Catharina (*Virola oleifera*). *Virola* has the broadest distribution among American genera as well as the most species and apparently the most individuals. This genus is found throughout the range of the family except in Mexico, although it rarely occurs in the Andes at such high elevations as *Dialyanthera* and *Compsoneura*. These two genera appear well suited to the forests of low mountain regions. *Iryanthera,* like *Virola* and the monotypic *Osteophloeum,* is primarily a genus of the lowland forests, but it has a more restricted range than *Virola*."

The species are small to large trees, rarely shrubs, with alternate, simple, frequently large and aromatic leaves and small flowers. The fruits of *Myristica fragrans* Houtt., native of the Moluccas and widely planted, are the source of two well-known spices—nutmeg, the ruminated endosperm of the seed, and mace, the lacinate fleshy aril. The seeds of the other genera are of similar appearance and rich in vegetable tallow and wax, and are sometimes utilized for making soap and candles. The bark of some species is tapped for an astringent sap, one of the commercial kinos, usually bright yellow at first but turning red as it hardens. Although the trees are often of common occurrence, with well-formed trunks and easily worked wood, the timber has never been extensively used, mainly because of its liability to insect attack and its lack of resistance to decay.

Sapwood grayish or pinkish brown, often merging gradually into heartwood and darkening upon exposure; in some species the heartwood is dark red or chocolate-brown and sharply defined. No distinctive odor or taste. Density widely variable; sp. gr. (air-dry) 0.35 to 0.77, in some species 0.87 to 1.01; wt. per cu. ft. mostly between 22 and 48 lbs., occasionally 54 to 63 lbs. Texture uniform, fine to medium-coarse; grain generally straight, but sometimes irregular.

Growth rings present or absent. Pores small or medium-sized; not very numerous; solitary and in small multiples, well distributed without pattern. Vessels with predominantly simple to exclusively multiple perforations (scalariform, reticulate, or compound); spirals absent; intervascular pitting rather fine, mostly alternate, sometimes scalariform. Rays commonly 1 or 2, sometimes 3 to 6, cells wide, and not very high; heterogeneous; oil cells sometimes present; tanniniferous tubes characteristic of family though infrequent in some specimens; ray-vessel pitting mostly coarse and scalariform. Wood parenchyma rather well developed; paratracheal and, in certain species, metatracheal; sometimes terminal and distinct without lens. Wood fibers septate in part; pits small, simple or with an inconspicuous border. Ripple marks absent. No gum ducts seen.

Compsoneura. Of the several species of this genus, the most widely distributed in *C. Sprucei* (A. DC.) Warb., a large-leaved tree sometimes 40 feet tall with a trunk 10 inches in diameter, occurring in the mixed hardwood forests of southern Mexico, Central America, and the Amazon basin. The bark is smooth and the exuding sap turns blood-red. The wood is without distinctive color (in available specimens), of rather low to medium density (sp. gr. 0.45 to 0.65), straight-grained, even-textured, easily worked, not durable. The timber is not utilized.

COMMON NAMES: Sangre (Hond.); wild coffee (Pan.); senimoro-ey (Peru).

Dialyanthera. The most widely distributed of the six known species is *D. otoba* (H. & B.) Warb., a fairly common tree 60 to 90 feet high with an erect bole sometimes 24 inches in diameter, growing in uplands in Costa Rica, Panama, Colombia, Venezuela, and eastern Peru. The fat obtained by boiling and pressing the seeds is employed in Colombia as a remedy for parasites in animals. The timber is pinkish brown, light and soft (sp. gr. 0.38 to 0.50), straight-grained, very easy to work, holds nails firmly, but is not durable. It is utilized locally to a limited extent for boxes and interior construction.

COMMON NAMES: Fruta dorado, sebo (C.R.); bogamani verde, saba, white cedar (Pan.); otoba (Col.); otivo, otoba, otova (Venez.); coco (Ec.).

Iryanthera. According to Ducke (*Journ. Wash. Acad. Sci.* 26: 215), this very natural genus "is apparently restricted to the Amazonian hylaea (including the Guianas and the northwestern part of the State of Maranhão), where it is represented by a rather considerable number of species, though much less abundant in individuals than is *Virola;* it is one of the most characteristic elements of the hylaea flora. All species grow in upland virgin forest, where they prefer the neighborhood of small streamlets. All are known by the vernacular name Ucuhúba-rana (false Ucuhúba), those which furnish wood of good quality also as Punán." Most of the 20 species are medium-sized or small trees, but some of them are sometimes 100 feet tall, with a long thick cylindrical bole.

Heartwood medium to dark brown, sometimes deepening upon exposure to reddish or purplish, the color uniform or more or less striped; has a waxy appearance and feel; sharply demarcated from the oatmeal-colored sapwood, which is 2 to 4 inches thick in large trees and composes entire trunks of the smaller ones. Luster fairly high. No distinctive odor or taste. Wood rather light and soft to hard and heavy (sp. gr. 0.39 to 0.87), but always easy to work, the heartwood finishing with a high natural gloss; texture uniform, medium; grain mostly straight; dark-colored, waxy specimens appear durable. Suitable for cabinet making.

COMMON NAMES: Bémoonba, killakewa, kirakawa, kirikowa, k. fierberoe, k. wadelikoro, pajoelidan, pajoelie, pajoerilan, pajoririan, peritjaloipio, poeloe-moto, sewanna, soewana, srébébé, swamma, swanna, tjirikawa, wepetano-waroesie, wokomorokotore (Guianas); mouchigo rouge (Fr. G.); apuna, cacauhy, iquê, punán, ucuhúbarana, ucuúba vermelha (Braz.); cumala, c. del alturas (Peru).

Osteophloeum *platyspermum* (A. DC.) Warb., the only species, is, according to Ducke (*loc. cit.*, p. 215), "one of the most frequent and widely distributed Myristacaceae of the upland rain forest of Amazonian Brazil, from the mouths of the Amazon and the neighborhood of the capital of Pará to São Paulo de Olivença, not far from the Peruvian frontiers." It attains a height of 130 feet or more, but only the largest specimens develop heartwood. The timber is rather soft to moderately hard (sp. gr. 0.56 to 0.64), the sapwood yellowish or oatmeal-colored, the heartwood reddish brown and rather waxy. There are no recorded uses, but the lumber is suitable for box boards and various kinds of interior construction where resistance to insects and decay is not a factor.

COMMON NAMES: Coninga, sebo caspi (Col.); punán, ucuhúba-rana, ucuúba-rana (Braz.).

Virola. There are about 38 species of small, medium-sized, or large trees, common in the mixed hardwood forests of Central and South America. The most widely distributed species is *V. surinamensis* (Rol.) Warb. which, according to Ducke (*loc. cit.*, p. 258), "grows in some of the Lesser Antilles, Trinidad, the Guianas, southern Venezuela, and the northernmost part of the Brazilian State of Amazonas, the coastal region of Pará including the whole Amazon estuary, the northern part of Maranhão, and northeastern Ceará. . . . It is extremely abundant in the low islands of the great estuary, inundable by the Atlantic tide; in some of these it represents the majority of the rather large trees up to 20 meters high. The enormous quantities of Ucuhúba seeds yearly exported from Pará, or those consumed in industries, come from this species." Another well-known species, *Virola sebifera* Aubl., is (*idem*, p. 255) "common from the Guianas through the whole State of Pará to the State of Maranhão . . . It grows principally in secondary forest and in rather dry woods of the 'campos' regions. . . . The geographic area of *V. sebifera* includes the whole hylaea, the central part of Matto Grosso, the State of Goyoz, and the northern half of the State of São Paulo." The Central American species are typified by *V. Koschnyi* Warb. (= *V. merendonis* Pittier), ranging from British Honduras and Guatemala to Panama. The seeds of all species are rich in oil, used for making candles and soap, and are attractive because of the contrast between the shining brown surface and the white, pink, or red lace-like aril.

The woods of the different species are much alike, being pale brown when freshly cut but becoming deep reddish brown, often with a purplish hue; the sapwood is lighter and may be sharply demarcated or merge gradually into the heart. They are mostly of moderate density; sp. gr. 0.60 to 0.75; weight 27 to 47 lbs. per cu. ft.; medium-textured; ordinarily straight-grained. Tests made at the Imperial Institute, London, on timber (Banak) from British Honduras gave the following results: It cuts easily with hand and power saws, and a good surface is obtainable with jack and smoothing planes, both along and across the grain. Good clear boles are readily obtained with brad awl, gimlet, center bit and twist drill, without tendency to split. Nails and screws can be driven into the wood easily without splitting it, and hold fairly well. It cuts with facility in a mortising machine and works easily with gouge and chisel. The wood turns satisfactorily, though the fibers tear slightly, and a good finish is obtainable. It absorbs glue well, can be stained without difficulty, and gives satisfactory results in polishing and varnishing. It holds its place when manufactured, does not warp or check, and is very free from knots and other defects. In the following results of tests all values are in pounds per square inch. *Transverse bending* (*central loading*): Modulus of rupture, 9025; modulus of elasticity, 1,706,000; fiber stress at elastic limit, 6662. *Compression along the grain* (*8-in. specimen*): Crushing strength, 5345; modulus of elasticity, 1,735,000; fiber stress at elastic limit, 4990. *Compression across the grain:* Load at elastic limit, 3850; fiber stress at elastic limit, 954.

Attempts to establish a market for *Virola* timbers have not so far been highly successful, though trial lots have proved satisfactory for veneers and solid lumber for general utility purposes. One difficulty relates to the logging operations, for as soon as the trees are felled they are subject to the attack of small beetles, locally known as pinworms, which bore deeply into the wood. Apparently the only effective method of preventing this damage is to remove the logs from the forest as quickly as possible and store them under water. Dry wood is not subject to pinworm attacks and should give excellent service in temperate climates.

COMMON NAMES: Banak, mahban (trade, U.S.A.); muscadier fou (Guad.); acajou, anakin, cajuco, cayuco, wild nutmeg (Trin.); banak, bastard banak, b. cedar (Br. H.); sangre, palo de sangre (Guat., Hond., Nic.) sangredrago (Nic.); fruta dorado, ira rosa (C.R.); bogabani, bogamani, copidijo, fruta dorado, gorgoran, malagueta de montaña, tabegua (Pan.); camaticaro, c. blanco, c. rojo, cedrillo, cuajo (Venez.); arbre à suif, baboen, baboenhoedoe, baboenhout, baboentrie, babun-hudu, bali, dalli, dari, dayopa, guiaguia, guinguamadou, g. de monte, hill dalli, jea, jeamadou, kilikowa, mattoe moenba, moonba, mouchigo, m. rouge, moussigo, muscadier à suif, ouarouchi, pintrie, Saint-Jean rouge, tarosiepjo, virola, voirouchi, wallololo, waroesie, waroesierjan, warokoroballi, warokotie, yayamadou (Guianas); cacao de monte (Ec.); arvore de sebo, becuiba, bicuiba, b. assú, b. branca, b. cheirosa, b. mirim, b. vermelha, bicuhyba, mucuhyba, paricá, piquibucu, ucuhúba, ucuúba, u. branca, u. vermelha (Braz.); caupuri, cumala, c. blanca, c. caspi, ucufe-ey, cumula (Peru).

MYRSINACEAE

THIS family comprises about 32 genera and 900 species of shrubs, small or rarely large trees, and a few herbs, and is widely disseminated throughout tropical and subtropical regions. The leaves are alternate or pseudo-verticillate, simple, punctate or lineolate, and without stipules; the flowers are small, racemose, paniculate, or fascicled; the fruit is a berry or a drupe. In tropical America there are 10 arborescent genera and about 350 species. Although their woods are of good quality and have an attractive Oak-like figure when quartersawed, they are sparingly utilized locally and are of no commercial importance because of the small sizes ordinarily obtainable. The following general description is based on specimens of *Ardisia, Conomorpha, Cybianthus, Geissanthus, Grammadenia, Parathesis, Rapanea, Stylogyne,* and *Wallenia.*

Heartwood chestnut or pinkish brown, not sharply demarcated from the sapwood. Luster rather high. Without distinctive odor or taste. Usually of moderate density, sometimes heavy and hard; texture medium to rather coarse; feel slightly waxy; grain straight to irregular; fairly easy to work, though tending to flake on quarter-sawed lumber; finishes smoothly and takes a high natural polish; some species said to be fairly resistant to decay.

Growth rings usually indistinct. Pores sometimes barely visible but usually minute; variable in abundance but not crowded; occurring mostly in small multiples or short radial series, sometimes clustered, without definite pattern though with occasional tendency to diagonal or radial arrangement; infrequently in contact with the rays. Vessels with simple perforations; without spiral thickenings, though commonly with fine spiral striations; pitting very fine to fine, alternate. Rays virtually all multiseriate, variable in width from few to many cells, and moderately to very high; showing conspicuously on radial surface, where they are darker than the background; complexes of resinous cells, with walls of adjacent cells very thin or broken down so as to produce cysts, common in *Ardisia, Conomorpha, Grammadenia, Parathesis, Rapanea* (Plates LVI, 3 and 4; LVII, 1), and *Wallenia,* appearing as orange-colored specks under lens or (tangential section) as intercellular canals; heterogeneous, with most of the cells large and square or upright; lateral limit of rays not always clearly demarcated from the fibers on cross section; crystals common to numerous; gum abundant; pits to vessels very fine. Wood parenchyma very sparingly paratracheal and diffuse; not distinct with lens. Wood fibers with rather thin to thick walls; commonly septate; pits numerous, minute, simple. Ripple marks and gum ducts absent.

Ardisia (including *Icacorea*), with about 235 species, mostly Asiatic, is represented in America by about 60 species of shrubs and trees having a combined range covering the West Indies, southern Florida, part of Mexico, Central America, and South America to eastern Brazil and northern Peru. The trees are rarely over 25 or 30 feet tall and 8 to 10 inches in diameter. The timber is occasionally employed locally for furniture, cabinet work, and general construction.

COMMON NAMES: Dogberry, marlberry (Fla., B.W.I.); tapa camino (Cuba); bois de tremble, quatre chemins (Haiti); arrayán, camaca, capulín, c. de mayo, c. de tejón, c. manso, c. silvestre, capulincillo, chico correoso, huitumbio, laurel, l. de la sierra, laurelillo, mangle, morita, negrito, pimientilla, shka-na-tau, sirasil, xook-num (Mex.); cucuyul, uva de monte (Hond.); cerecilla, cerecita, cerezo silvestre, cotomate, uva (Salv.); cujia, uvita (Nic.); fruta de pava, guastomate, murta, sotacaballo, tucuico, tucuiquillo, uruca (C.R.); uvito (Pan.); huesito, tacaloa (Col.); mamey de monte (Ec.); capororoquinha, icacoré-caatinga (Braz.).

Conomorpha, with about 40 species of shrubs and little trees, ranges from the Lesser Antilles to Brazil and Peru. The moderately hard, rather fine-textured timber is of no importance because of the small sizes available.

COMMON NAMES: Olivo (Col.); koenaporang, teteroema (Sur.); caimito macho (Peru); garapapunta (Braz.).

Cybianthus, with about 35 species of little trees, is widely but sparsely distributed in the Amazon basin, with a few extensions to southeastern Brazil and northward to Trinidad. *C. detergens* Mart., of eastern and southern Brazil, is said to supply small timber for beams, frames, and minor interior work of houses, as well as for fuel and charcoal.

COMMON NAMES: Capororoca-assú, jacaré do matto (Braz.).

Geissanthus, with 25 or more species of shrubs and trees, is limited in its distribution to Colombia, Venezuela, Ecuador, Bolivia, and Peru. The only specimen available (Yale 28550; A. Rimbach 201) is of *G. ecuadorensis* Mez which occurs in the mountains of Ecuador where it is known as Tarqui. It is a medium-sized forest tree and supplies some good timber for local construction.

Grammadenia, with about a dozen species of shrubs, grows in the West Indies, Central America, and northwestern South America. At least two species are epiphytic.

COMMON NAMES: Aguacatillo de terra firme (Col.); cupis (Venez.).

Parathesis, with about 20 species, inhabits the West Indies, Mexico, Central America, Colombia, Venezuela, Ecuador, Peru, and Bolivia. The woods range in consistency from comparatively light and soft to heavy and hard, and in appearance from Beech (*Fagus*) to Oak (*Quercus*).

COMMON NAMES: Agracejo de sabana, cofa (Cuba); rasca garganta, seca garganta (P.R.); japelón (Dom. R.); raisin marron (Haiti); chico arrayán, ma-ku-lai (Mex.); chac manga, red mangrove, uva (Br. H.); camaco, chimiche, manchador (Guat.); cuya (Hond.); cugía (Nic.); tucuiquillo (C.R.); black cherry (Pan.).

Rapanea. The range of this genus, which contains nearly 150 species of shrubs and trees, includes nearly all the warm regions of the world. The Cape Beech or Beukenhout of South Africa, *Rapanea melanophleos* Mez, reaches a height of 50 feet and a diameter of over two feet and supplies a highly figured wood in local demand for making furniture and vehicles. The Muttonwood of East Australia, *R. variabilis* Mez, is a tree 45 to 50 feet high and a foot or more in diameter, with an Oak-figured wood suitable for furniture and cabinet work, though it is procurable only in small sizes.

Of the numerous American species, one, *Rapanea guianensis* Aubl., reaches its northern limit in southern Florida, where it is only a shrub, though elsewhere in its range, which includes the West Indies and northern South America, it is sometimes a medium-sized tree. *R. duidae* Gleason does not belong in this family but is one of the Sapotaceae, perhaps *Chrysophyllum*. The various Brazilian species are commonly known as Caporococa and some of the trees are from 45 to 60 feet high and up to three feet in diameter, but their timber is little used except for charcoal. Trees of this genus are common in Argentina where they are known generally as Canelón, also Capororoca or some variant of that name. They are sometimes 100 feet in height with trunks two to four feet through. The bark is said to be rich in tannin. The timber, which weighs about 35 lbs. per cu. ft., is used to a considerable extent for general carpentry, cabinet work, crates and boxes, house construction, and for fuel. The anatomy is very similar to that of *Parathesis*.

COMMON NAMES: Badula, camagüilla, mameyuelo (Cuba); arrayán, a. bobo, badula, cabra, cucubáno, memeyuelo (P.R.); bois plomb, b. savanne (Haiti); bois arrade, b. cassant, b. fourni, coca ravet (Fr. W.I.); manglillo (Trin.); amatillo (Salv.); ratón, ratincillo, sierra (C.R.); cucharo, cucubaro, espadero, huesito (Col.); mangle de montaña, manteco blanco (Venez.); dakara, konaparan, mannie botieie (Sur.); azeitona do matto, canellão, capororoca, c. branca, c. commum, c. de folha, c. mineira, jacaré do matto, jomirim, manque rosa, sobro (Braz.); muille, samal (Ec.); camesito, lluthulluthu, lucuma, sangre de drago (Peru); canelón, lanza blanca (Urug.); caá berá, c. poraca, c. pororo, c. pororoca, canelón, c. blanco, c. capororoca, c. colorado, lanza blanca, palo de San Antonio, pororoca, yaruma (Arg.).

Stylogyne, with about 40 species of shrubs and small trees, occurs throughout the whole of tropical America. Of the several species represented in the Yale collections, the largest is *S. latifolia* A. C. Smith, a spreading tree 35 feet tall in British Guiana. *S. amplifolia* Macbr. of the Peruvian Amazon region rarely exceeds 20 feet.

COMMON NAMES: Guastomate (C.R.); pajaweroe (Sur.); puca-varilla (Peru).

Wallenia, with about 20 species of trees and shrubs, is confined to the West Indies. *W. bumelioides* Gris. and *W. laurifolia* (Jacq.) Sw. are small Cuban trees of common occurrence in marshy areas. The timber, which is moderately hard and rather coarse-textured, is used locally to a minor extent in general construction and carpentry. *W. pendula* (Urb.) Mez is a shrub or less often a tree up to 35 feet high, endemic to Puerto Rico; its wood is of medium density and rather fine texture.

COMMON NAMES: Agrecejo, caimoní, camagua, c. macho, camagüilla, camao, carmoni, casmagua, caumao, cúrbano macho, guacamar, lustillo (Cuba); jacanillo, memeyuelo, quiebra-hacha, uva (P.R.); caimón, caimoní (Dom. R.); bois crepaud, petit raisin, raisin marron (Haiti); baibaiba (Mart.).

Weigeltia, with over 20 species of shrubs and trees, extends from the Lesser Antilles to southern Brazil. The wood has not been studied.

MYRTACEAE

THE Myrtle family comprises more than 70 genera and 3000 species of aromatic trees and shrubs of world-wide distribution, though most abundant in tropical and subtropical regions. The leaves are typically opposite, entire, pellucid-punctate, and persistent; the fruits are of two types, namely, berries or drupes, and dry capsules. Well-known representatives of the first group (Myrtoideae) are the Myrtle, *Myrtus communis* L., an aromatic shrub native to northern Europe and widely cultivated in gardens; the Pimento or Allspice tree, *Pimenta officinalis* Lindl., of the West Indies; *Eugenia aromatica* Baill., of the Molucca Islands, a tree whose dried flower buds are the cloves of the spice trade; the Rose Apple, *E. jambos* L., indigenous to southeastern Asia and cultivated throughout the tropics as a shade tree and for its delicately flavored fruit; and the Guava, *Psidium guajava* Raddi, a tropical American species cultivated for its fruit which is used largely for making jelly and jam.

The most important timber trees are of the second group (Leptospermoideae), chiefly Australasian. There are about 500 named species of *Eucalyptus* in Australia, varying in size from small trees to the largest of all hardwoods and exhibiting an equally great range in the appearance and properties of their woods. Various kinds are grown elsewhere for shade and ornament and also on a commercial scale for their timber. Good results have been obtained in southern Brazil with Eucalypt plantations for railway crossties. The most important genus in New Zealand is *Metrosideros*, the Ironwood or Rata.

There are about 20 arborescent genera and several hundred species of Myrtaceae growing naturally in Latin America. With one exception, *Tepualia stipularis* (Barn.) Gris., all belong to the Myrtoideae, and their woods are typified by the genus *Eugenia*. The trees are generally small or medium-sized and their principal value is in their fruits and miscellaneous minor products. The heartwood is variable in color, sometimes grayish brown with dark streaks, more often light to dark reddish brown, occasionally purplish; sapwood pale brown or pinkish, with either gradual or abrupt transition. The woods as a whole are hard, heavy, tough, strong, fine-textured, and finish smoothly, but they have a decided tendency to warp in drying and, being responsive to humidity changes, do not keep their place very well when manufactured. They make excellent fuel. Other common uses are tool handles, implement frames, and miscellaneous purposes requiring strong and tenacious material in small sizes or in the round. Sometimes the timber is good enough for general construction and selected lumber is employed to a limited extent in making furniture and vehicles. The woods are not noted for their durability in contact with the ground, but there are exceptions, particularly the dark-colored heartwood of old trees.

Growth rings often present, resulting from local variations in porosity and abundance of parenchyma. Pores mostly solitary; sub-circular; thick-walled; very small to minute, rarely visible without lens; numerous to very few, the number often decreasing during seasonal growth; irregularly distributed without definite pattern, though frequently tending to radial or diagonal arrangement. Perforations simple, except in *Myrceugenia apiculata* (DC.) Ndz. which has many-barred scalariform plates; spiral thickenings absent except in *M. Schulzii* Johow. and *M. Fernanduziana* (H. & A.) Berg; yellowish or brown gum often abundant; white deposits common in *Pimenta;* pits vestured. Rays mostly uniseriate or biseriate, occasionally 3 or 4 cells wide and up to 40, though generally less than 25, cells high, except in *Campomanesia* where they are 1 to 5 cells broad and up to 70, usually less than 40, cells high; decidedly heterogeneous, the multiseriate third made up of procumbent cells, the uniseriate parts composed of tall upright cells, except in *Campomanesia* where the rays are nearly homogeneous; disjunctive cells common; gum deposits abundant; pits to vessels very small to minute, except in *Myrceugenia* where they are small, circular and opposite or elongated and parallel. Wood parenchyma usually abundant; diffuse and in short or broken lines or zonate in irregular bands 2 to 8 cells wide; also weakly aliform in *Myrtus luma* Barn.; gum deposits abundant; crystalliferous strands sometimes present. Wood fibers with medium to very thick walls; pits numerous, rather large, distinctly bordered. Vasicentric tracheids commonly present. Ripple marks and gum ducts absent.

Amomis *caryophyllata* (Jacq.) Krug & Urban, the only species, is an aromatic tree reaching a maximum height of 50 feet and a diameter of 24 inches, native to the West Indies and northern South America and cultivated in the East Indies. The leaves and twigs yield by distillation an important essential oil, known as oil of bay, the distinctive ingredient of bay rum. According to Britton and Wilson (*Scientific Survey of Porto Rico and the Virgin Islands* 6: 27), "a superior kind of this oil is obtained on St. Jan, where there are extensive forests of the tree, obtained for the most part by clearing away other trees and bushes thus permitting the Bay tree to grow from seedlings without much cultivation. Much oil is also obtained from wild trees in Porto Rico, but little or none in St. Croix, St. Thomas, or Tortola. The species consists of many races differing mainly in the amount and quality of the oil contained, but also in shape, size, and color of the leaves and shape of the fruit. . . . The dark wood is strong, very hard, tough, mottled, and durable, with a specific gravity of about 0.90; it is utilized for rollers, sills, posts, and to some extent for carpentry."

Common names: Bayberry tree, bay rum tree, wild cinnamon, w. clove (Jam.); ausú, auzú, guayavita, limoncillo, malagueta (P.R., Virg. Is.); pimienta de Tabasco (Cuba); bois d'Inde français (Haiti); giroflier (Fr. W.I.); bayboom, beerum (Sur.).

Blepharocalyx, with numerous little-known species of shrubs and small or rarely large trees, is distributed from Colombia to Chile. The tallest tree appears to be *B. gigantea* Lillo, var. *montana* Lillo of Argentina. It attains a height of 100 feet, with a well-formed trunk sometimes 40 inches in diameter. The wood is said to be of a variegated coffee color, fine-textured, strong, and weighing about 50 lbs. per cu. ft. Its usefulness is impaired by the difficulty of seasoning it without checking and warping. There are no samples available for this study.

Common names: Vassourinha (Braz.); cocha-molle, horco-molle, mirta, multa (Arg.); arrayán, multa, murta (Urug.).

Britoa, with about 10 species of shrubs and small trees, is limited in its distribution to eastern Brazil and Argentina. *B. Sellowiana* Berg, which is known as Sete Casacas (Portuguese) and Siete Capotes (Spanish) in allusion to the shedding of the outer bark, is a source of comestible fruit, tanbark, and some timber for fuel and carving. *B. acida* Berg, native of the lower Amazon and planted elsewhere for its fruit, is called Araça do Pará. The fruits are edible raw or when dried or preserved. The dense wood is utilized for implement handles, articles of turnery, and charcoal. The genus is not represented in the Yale collections.

Common names: Araça do Pará, sete casacas (Braz.); marmelero, ñandú-a-puisá, siete capotes (Arg.).

Calycolpus, with 15 species of shrubs and little trees, is sparingly distributed from the West Indies and Central America to Peru and southern Brazil. The genus does not supply any timber of value. The only authentic wood sample available (Yale 9424) is of *Calycolpus glaber* (Benth.) Berg which was collected in British Guiana by A. C. Persaud. Heartwood uniform reddish brown, merging gradually into the lighter-colored sapwood. Luster medium. Without distinctive odor or taste. Hard, heavy, tough, and strong; texture fine and uniform; grain roey; rather difficult to cut but takes a smooth, glossy finish; is probably fairly durable. Presumably of no commercial possibilities.

Common names: Bamboo guava (Trin.); guayabillo (C.R., Pan.); wild guava (Br. G.).

Calyptranthes, with about 75 species of aromatic shrubs and small or rarely medium-sized trees, is widely distributed in tropical America, with two species reaching southern Florida. The plants have little value for any purpose. Heartwood light reddish brown, often with a grayish, bluish, or violet tinge; not sharply demarcated from the lighter-colored sapwood. Luster low to medium. Faintly scented. Moderately to extremely heavy and strong; texture fine. Utilized to a limited extent locally for fuel and for purposes requiring very tough and resilient timber in small sizes.

Common names: Spicewood, white spicewood (Florida); myrtle-of-the-river, spicewood, white stopper (Bah.); bastard greenheart, mountain bay, red rod, rod, rodwood (Jam.); clavellina, guaicaje, mije, miji, ramón amarillo (Cuba); hoja menuda, limoncillo, l. de monte (P.R.); arrayán, limoncillo (Dom. R.); tetet negresse (Guad.); debasse (Trin.); bastard turtle bone, Indio desnudo, walk-naked (Br. H.); murta (C.R.); guayabo prieto (Col.); canilla de venado, guayabo sortijo, sortijero (Venez.); kakreow, yawari-balli (Br. G.); aracasinho, braza viva, craveiro do campo, c. da terra, cravo do campo, pitangueira de cachorro (Braz.).

Campomanesia, with many species of shrubs and small trees usually less than 30, rarely up to 60, feet high, occurs throughout South America, though best represented in southern Brazil and Uruguay. The usefulness of the plants resides primarily in their edible fruits, although the wood finds local applications for tool handles, cart construction, articles of turnery, and fuel. Heartwood grayish brown, slightly variegated with red or purple; not sharply demarcated from the lighter-colored sapwood. Luster medium. Moderately hard and heavy, tough, and strong; sp. gr. (air-dry) 0.70 to 0.85; weight 44 to 53 lbs. per cu. ft.; texture rather fine, grain variable; not very difficult to work, finishing smoothly.

Common names: Bao, guayabito, guayabo anselmo (Col.); bois Caraibe, Carib wood, cometure (Trin.); araça lima, palillo (Peru); guabiróba, g. branca, guabirobeira, palilho, uvalha (Braz.); guabirá, g. guazú, g. puitá, guirobá, ibá-virá (Arg.); arazati rastrero, guabirobera del campo (Urug.).

Eugenia, with more than 500 species of trees and shrubs, is abundantly represented in tropical and subtropical America and Asia and rather sparingly in Australia and Africa. The fruits of many species are edible. Some of the trees are of commercial size for timber, but utilization is local and limited. The uses include interior heavy construction, implement frames, tool handles, turnery and carving, small cabinet work, and especially fuel, as the wood burns with much heat and little smoke. The variation in the appearance, properties, and structure of the woods of *Eugenia* is about as great as between those of the different genera in the section Myrtoideae.

Heartwood variable in color, light to dark grayish, pinkish, or reddish brown, often with a tinge of purple or violet, sometimes with black streaks; usually merging rather gradually into the sapwood. Luster commonly low. Frequently with mild and pleasant scent, but without distinctive taste. Moderately to very hard, heavy, tough, and strong; sp. gr. (air-dry) mostly between 0.80 and 1.05; weight 50 to 65 lbs. per cu. ft.; texture medium to fine; grain irregular; not easy to work, but finishes very smoothly; is inclined to warp and split in drying; resistance to decay poor to good (for deeply colored specimens).

Common names: Naked wood, spiceberry, stopper—gurgeon, red, Spanish, white (Florida); black wattle, ironwood, rodwood, stopper (red, Spanish, white), Surinam cherry, wattle (B.W.I.); camecará, guairage, guairaje, guairajillo, manzanilla, mije, pimienta, p. cimarrona, sumacará (Cuba); guasavera, guayabacón, guayabota, limoncillo, murta (P.R.); arrayán, escobón, guasara, limoncillo (Dom. R.); bois d'ine, b. mulâtre, b. myrte, brignolle, cerise marron, malaguette, petit bois d'Inde, z'os douvant (Haiti); casse hache, goyavier montagne (Guad.); guava—cherry, coffee, papery leaf, quinam, serette, wabin, wild (Trin.); arrayancillo, capulín, escobillo, guacuco, guayabillo, palo agrio, pimientilla, reyán, sacloob, xich-huhil, yagalán (Mex.); blossom berry, cacho de venado, female boy job, granada cimarrona, Indio desnudo, vainha de espada, walknaked (Br. H.); escobo, guacuco, pimientilla (Guat.); cacho de chivo, chamiso, escobo, guacoquito, guacuco, guayacán negro, naranjuelo, pinito cimarrón (Salv.); fierillo (Hond.); cacique, carro caliente, coquito, escobo, guayabillo, murta, turrú, turrusi, turú (C.R.); cacique blanco, coralillo, paico, sequarra (Pan.); arrayán, a. de tierra fría, cabeza de loro, carajito icotea, guayabo cascudo, g. colorado hoja menuda, g. de león, g. icotea, g. murciélago, g. prieto, manzanita de rosa prieta, raijón, vara real (Col.); guayabito, g. de monte, guayabo, g. blanco, g. de hierro, g. liso, g. negro, g. paují, naranjillo (Venez.); baniaballi, kakirio, savannah pepper, wariru-bime, wild guava (Br. G.); kersenboom, Surinaamsche kers (Sur.); bois goyave, cerise carée, goyabo amarillo, g. blanche, goyavier, iva-catinga (Fr. G.); arrayán (Ec.); cámu-cámu, puca-rupinia, unca (Peru); araça do ogapó, baitinga, cabelludo, cagaiteira, cambusarana, cambuhy, c. pitanga, c. verdadeiro, c. vermelho, cerejo, c. do Rio Grande, ginja, goiaba, grumichaba, grumixameira, grumixava, guabijú, guabillo, guabira-guassú, guiabirana do igapó, ibipitanga, ibiruba, jaboticabá, jambolão, lacre forte, mama de cachorro, murteira, ñhanica, nianica, pichuma, pelada, pimiento, pinna, pitanga, pitangatúba, pitangueira, p. da matta, p. preta, p. vermelha, pitomba, tatú, t. grande, t. pequeno, ubaia branca, ubatinga, ubipitanga, uvaia, uvalha, u. do campo, vassouri, vatinga (Braz.); arazá, arraguay, arraiján, arrayán, a. mato, a. negro, a. n. minor, arraicán, cambuy, ceresa, guabiyu, g. blanco, guaviyú, guayabo, guïli, g. blanco, horco-mato, ibá biyú, i.-hay, i.-h. amargo, i.-h.-mí, i.-h. morotí, i.-mbopy, i.-poroiti, i.-viyú, ibiráyepiro, igbá-jai, igua-jay, ivuaporoiti, lapachillo, manapiri, mato, m. blanco, m. colorado, ñangapary, ñangapiru, ñangapiry, peló-rebí, peró-rebí, pitanga, ubajay, vaporoti, vapuritín (Arg.); arrayán, arazá de monte, batinga, guabijú, guayabo blanco, guïli, g. colorado, mato, m. blanco, m. lapachillo, murta, murtiño-manso, iwahay (Urug.); guaviyú, nyangapirú (Par.); arrayán falso, cheken (Chile).

Feijoa (or *Orthrostemon*) *Sellowiana* Berg, the sole species, is a shrub or little tree less than 25 feet high in Uruguay and southern Brazil. The wood has not been studied.

COMMON NAMES: Arrayán, guayabo de la sierra (Urug.).

Marlierea (including *Eugeniopsis* and *Rubachia*) comprises about 50 species of shrubs and small trees growing from northern South America to southeastern Brazil. The best known species is the true Cambucá, *M. edulis* Ndz., a straggling tree 25 feet high native to Rio de Janeiro and cultivated in other Brazilian states for its large succulent apricot-like fruit of local commercial importance. The bark is rich in tannin and the dense fine-textured tough wood is used for implement handles and in small cabinet work. *M. tomentosa* Camb., another well-known Brazilian fruit tree, has a cherry-like fruit.

COMMON NAMES: Kwako (Br. G.); cambucá, c. verdadeiro, cambucáseiro, guaporanga, guapuronga (Braz.).

Gomidesia, often included in *Myrcia,* comprises 40 or more species of shrubs and little trees, mostly Brazilian. The species with the widest range is *G. Lindeniana* Berg, which grows from the West Indies to Brazil; its maximum height is about 30 feet. *G. reticulata* Berg of eastern Brazil supplies some small timber for use in native house framing and for fuel. The wood, judging from a single specimen (Yale 21421), is of the general type of *Eugenia.*

COMMON NAMES: Cieneguillo (P.R.); mangue do brejo, jaboticabeira branca (Braz.).

Myrceugenia, with about 15 species of shrubs and small trees, is distributed from Chile through Patagonia (Argentina) to Uruguay and southeastern Brazil. The best known species is *M. apiculata* (DC.) Ndz., the Arrayán of Chile and Patagonia. In the latter region it is rarely over 25 feet tall and 14 inches in diameter, but is larger in Chile and supplies some lumber of the general type of the Antarctic Beeches (*Nothofagus*) and used locally for the same purposes. The only other species represented in the Yale collections are *M. fernandeziana* (Hook. & Arn.) Berg and *M. Schulzei* Johow, both collected by Carl Skottsberg in Juán Fernández Islands, and *M. pitra* Berg, collected by E. L. Bernath in Chile. Their anatomy differs in several important details from that of the other members of the family studied.

Heartwood pale reddish brown; not sharply demarcated from the lighter-colored sapwood. Luster rather low. Odor and taste absent or not distinctive. Moderately hard and heavy, tough and strong; has about the consistency of Red Gum (*Liquidambar*); sp. gr. (air-dry) 0.55; weight 34 lbs. per cu. ft.; texture fine and uniform, grain fairly straight; easy to work, finishing very smoothly; appears fairly durable.

COMMON NAMES: Arrayán (Chile, Patagonia); pitra (Juán Fernández).

Myrcia, with about 500 species of shrubs and small to medium-sized trees, is of general distribution from the West Indies and Mexico to southern Brazil, Uruguay, and Chile. The plants are of little commercial value, but supply some edible fruits, dyes and tannin from the bark, and timber for fuel and miscellaneous purposes. The woods are similar in appearance and structure to those of *Eugenia*.

COMMON NAMES: Comecará, hoja menuda, pimentillo, pimienta cimarrona (Cuba); ausú, cienguillo, guayabacón, hoja menuda, punch berry, rama menuda (P.R.); huesito (Dom. R.); guepois (Guad.); yagalán (Mex.); pimiento, turro (Guat.); turro (C.R.); pimiento (Pan.); arrayán, arrayancito, guayabo macho, perilejo (Col.); guamufate, murto, oruma, sarura (Venez.); ibbi-banaru (Br. G.); couměté, mine (Sur.); bois goyave, goyavier (Fr. G.); macanúnu, orno-pichana, quinilla negra, rupinia (Peru); araça ferro, camboim, cambucá, cambuhy, cerejeira, cumaté, c. rana, cumaty, guapurunga, hume-caá, jaboticabeira, j. brava, murta cabelluda, oitchi, pedra hume, p. h. caá, pitanga colorada, p. miuda, uapixuna (Braz.); pitanguera de monte (Urug.); cuchá-guazú, pitanga colorada (Arg.)

Myrciaria, with about 60 species of shrubs and little trees, has its center of distribution in southeastern Brazil, with extensions to the West Indies. A few of the plants yield edible fruits, of which the best known is the Jaboticaba, *M. jaboticaba* (Vell.) Berg, native of southern Brazil from Rio Grande do Sul to Minas Geraes, and cultivated there and elsewhere. The Jaboticaba fruit is grape-like, with a tough and rather thick skin inclosing a juicy pulp of excellent flavor. It is borne abundantly in small clusters directly from the thin bark of the trunk and branches. It is common on the local markets and is very popular. The woods of these and other species are of the same general type as *Eugenia* and have the same applications.

COMMON NAMES: Camboimsinho, cambucá, cambuy, c. bala, c. preto, jaboticaba, jaboticabeira, j. do matto, jabuticaba, ubaia branca (Braz.); guayabo colorado, jaboticaba (Urug.).

Myrrhinium, with a few species, occurs from Colombia to Argentina and southern Brazil. Best known is *M. atropurpureum* Schott, a tree rarely over 25 feet high and 10 inches in diameter. The wood is hard and tough, weighs nearly 70 lbs. per cu. ft., and is fine-textured. From the single sample (Yale 32114) available for this study, it differs from most of the other genera in having only very fine rays, which are uniseriate or biseriate and up to 20, usually less than 10, cells high; otherwise the structure is similar to that of *Eugenia*. The attractively variegated timber has the local uses common to the American members of the family.

COMMON NAMES: Maitín, palo de lata, p. de hierro (Arg.); murtilla, palo fierro, pioje de chancho (Urug.).

Myrtus, with many species, mostly shrubs, is widely distributed over the world. The best known plant is the common Myrtle, *M. communis* L., of southern Europe and western Asia and often grown elsewhere for ornamental purposes; wreathes of its leaves were worn by the Athenian magistrates and by victors in the original Olympic games. Apparently the largest tree, which at best is only medium-sized, is the Luma, *M. luma* Barn., of Chile. Its pale brownish wood is similar in anatomy to that of *Eugenia.* Simmons says of it (*Lumber Markets of the West and North Coasts of South America,* p. 20): "From its properties and uses the Luma may be properly called the Hickory of Chile. Noted for its hardness, strength, weight, and flexibility, it is called on most for tool handles, vehicle parts, gymnasium apparatus, etc. The wood burns long and with an intense heat and is therefore appreciated above any other tree for fuel purposes. Steamboats in south Chile demand it for this purpose in large quantities."

COMMON NAMES: Arrayán (Mex.); arrayán, mirto, m. rosado (Guat.); arrayán, guayabito (Col.); guayabo arrayán (Venez.); cambuy, c. amarello, c. roxo, c. verdadeiro (Braz.); arazá, a. rostrero, mirto (Urug.); luma, patahua (Arg.); hurapo, luma (Chile).

Paivaea *Langsdorffii* Berg, the only species, is a little tree of limited distribution in São Paulo, Brazil, where it is known as Cambucy. It was appreciated in colonial times for its berries which were used to flavor rum. The wood is said to be beautiful and esteemed for small articles of turnery and cabinet work, but is too rare to be of any economic importance.

Pimenta *officinalis* Lindl., the sole species, is a very fragrant tree, sometimes 40 feet high, native to Jamaica, Cuba, southern Mexico, parts of Central America, and introduced through cultivation in other tropical countries. Fawcett and Rendle say (*Flora of Jamaica* 5: 326): "Wood, flowers, fruit, and leaves are aromatic. The berries while still green and unripe are gathered and dried for export, chiefly used as a condiment in cookery; they have the warm spicy taste of cloves and an aromatic odor resembling a mixture of cinnamon, cloves, and nutmegs, hence called allspice; they are used medicinally for their aromatic, carminative, and stimulant properties. The wood is tough and close-grained, used for cart shafts; saplings are made into walking sticks." Heartwood reddish brown, not sharply demarcated from the lighter-colored sapwood. Luster medium. Very hard and heavy; fine-textured; finishes very smoothly.

COMMON NAMES: Allspice tree, pimento (Jam.); pimienta (Cuba); malaguette, poivre Jamaïque (Haiti); bois d'Inde (Guad.); malagueta, patololote, pimentón, p. gorda, pimiento, xocoxochitl (Mex.); allspice, pimento (Br. H.); pimienta gorda (Guat.); pimiento oloroso (Nic.); Jamaíca (C.R.); malagueto, pepita de especie, pimiento (Venez.).

Psidiopsis *Moritziana* Berg, the only species, is a shrub or small tree growing at high elevations in the Venezuelan Andes. Its fruit is aromatic and sweet. The wood has not been studied.

COMMON NAMES: Cínaro, gumán, jumangue, sínaro (Venez.).

Psidium, with more than 100 species of shrubs and small to medium-sized trees, is distributed throughout tropical America. The best known plant is the Guava tree, *P. guajava* L., which produces one of the most widely known of tropical fruits. There are numerous species in Brazil and some of them yield timber of local utility. The wood is of the *Eugenia* type.

COMMON NAMES: Guayabo (Span.); guava (Anglicized, for both fruit and tree); goyavier (French); goyabier (Port.); guayabo agrio, cotorrero (Cuba); arrayán, coloc, pachi, pichi, pichiché (Mex.); cak, ch'amxuy, guayabillo, guayabo de agua, g. de monte, patá (Guat.); cás, guayabillo, guisaro, g. dulce, kás, kás-kra, kurib-krá, otera, seuí, suib-kra, sorí, sure (C.R.); guayabo de agua (Pan.); guayabito pirú (Col.); guayabo agrio, g. amarillo, g. casero (Venez.); kakirio, wild guava (Br. G.); water guave, watra gouaba, wilde guave (Sur.); goyavier porte-pois, pela (Fr. G.); araçá, a. da praca, a. do campo, a. felpudo, a. guazú, a.-gy, a.-ibá, a. mirim, a. de praya, a. peba, a. pera, a. piranga, a.

rana, a. r. branca, a. r. de concha, a. piranga, a. tuba, goyaba rana, goyabeira brava, g. do matto, g. preta, g. guazú, g. vermelha, guabiroba, guerapiranga, ubaçá, u. do campo, uba-caba (Braz.); arazá, a. guazú, a.-hú, a. puitá, a. saiyú, arrayán, arrayana, guacoco, guayabillo, guaybito rastrero, guayabo amarillo, g. colorado, g. negro (Arg.); arozá, a. charrúa, guayabo amarillo (Urug.); sahuintu (Peru).

Tepualia, the only American genus of the section Leptospermoideae, includes a single species, *T. stipularis* Gris., closely related to *Metrosideros*. It is a small tree, occasionally 35 feet high and a foot in diameter, of common occurrence in swamps of southern Chile and Patagonia (Arg.). Heartwood deep pink, merging gradually into the brownish sapwood. Luster rather high. Scentless and tasteless. Very hard, heavy, tough, and strong; texture very fine and uniform; grain straight; not very difficult to work, taking a glossy finish; probably durable. Presumably of no commercial possibilities.

Common names: Tepú (Chile); tepú, trepual (Arg.).

NYCTAGINACEAE

The Nyctaginaceae (sometimes termed the Allioniaceae), with 30 or more genera and about 300 species, are mostly herbs in temperate climates, but often shrubs, woody climbers, or small to medium-sized trees in the tropics. The leaves are alternate or opposite, simple, and without stipules; the flowers are small and inconspicuous, though sometimes surrounded by showy brightly colored bracts; the fruit is indehiscent and often glandular. Some of the plants are ornamental, particularly *Bougainvillea,* a woody vine native to South America but cultivated in most tropical countries. Other American genera containing woody species are *Neea, Pisonia, Pisoniella,* and *Torrubia* of the tribe Pisonieae; *Colignonia,* constituting the tribe Colignonieae; *Andradaea, Leucaster, Ramisia,* and *Reichenbachia* making up the tribe Leucastereae. The timbers are sometimes used locally but are of little or no commercial importance.

Heartwood yellowish or brown; not always clearly differentiated from the sapwood. Luster low. Without distinctive odor or taste when dry. Variable from light, soft, and spongy to moderately heavy and hard; sp. gr. (air-dry) 0.40 to 0.80; weight 25 to 50 lbs. per cu. ft.; texture medium to very coarse; grain irregular, more or less interwoven; durability low.

Structure anomalous. Vessels with simple perforations; without spirals; pitting fine, alternate. Normal rays fine, often minute. Normal wood parenchyma usually sparingly developed, but sometimes in widely spaced metatracheal lines. Raphides often present in conjunctive tissue. Wood fibers with simple pits. Ripple marks present in a few species of *Pisonia;* mostly indistinct.

In *Bougainvillea* the pores are radially arranged, the largest visible without lens; conjunctive tissue in very irregular anastomosing bands and fine to coarse rays. In *Colignonia* (Yale 16909) the pores are smaller, fewer, and not always radially arranged; conjunctive tissue in fairly regular bands, suggesting Phytolaccaceae. In *Reichenbachia colombiana* Standl. (Yale 32385), the only member of its section available for study, the wood is harder and finer-textured than the others; pores minute, numerous, in radial rows; conjunctive tissue in distinct bands and narrow rays. In *Neea, Pisonia,* and *Torrubia* the pores are medium-sized to small and in short radial rows or clusters, each terminating at an island of phloem, the whole when magnified giving the fanciful appearance of a mushroom; islands numerous and typically arranged in fairly regular diagonal rows (Plate LVIII, 3).

Neea, with over 70 described species of unarmed shrubs and small trees, occurs throughout tropical America. The flowers are small and cymose, the fruits are fleshy. There are no recorded uses for the woods, except occasionally for fuel.

Common names: Corail grandes feuilles (Haiti); palo ixtludo (Mex.); frutilla, puruma, p. de agua dulce, sangre de chucho, teñidor (Salv.); yana muco (Col.); cazabito, pica-paloma (Venez.); caparosa do campo, c. silvestre, herva caparosa, joão molle, Maria molle (Braz.); cumala, intuto-caspi, mullo-caspi, puca huayo, shula, topamaca blanca, yana-muco, yntutu-caspi (Peru).

Pisonia and **Torrubia** are not always considered separate genera, the latter genus being included with *Pisonia* in Heimerl's classification (*Pflanzenfamilien,* 2nd ed., 16c, p. 127). The principal distinctions are that the plants of *Torrubia,* which are all tropical American, are unarmed, and the fruit is juicy and without glands, whereas the other species, some of which belong to the eastern tropics, are armed with spines and have dry fruits with stalked glands along the angles. The largest tree known is the Zapallo of Argentina, *P. zapallo* Gris., which is said to reach a height of 60 to 70 feet and a diameter up to three feet; its soft, spongy but tenacious wood is used locally for sabots, slack cooperage, boxes, and fuel. Two West Indian species extend into southern Florida, and one of them, the Blolly, *P. obtusata* Jacq. or *Torrubia longifolia* (Heim.) Britt., is a spreading tree occasionally 30 to 50 feet high, with an erect or inclining trunk 15 to 20 inches thick. The woods of all species examined are much alike and are not likely to be used except when better timber is scarce. In the following lists of common names the two genera are kept separate, but some of those attributed to *Pisonia* probably belong to the other group.

Common names: *Pisonia:* Catclaw, cockspur, goodbread, hold-back, prickly mampoo, pull-back, salt (B.W.I.); uña de gato, vaca buey, zarza (Cuba); corcho, escambrón, lobloly, palo bobo, prickly mampoo, uña de gato, water mampoo (P.R.); uña de gato (Dom. R.); bois cassave, croc chien, palo perico (Haiti); amourette, croc-á-chien, mapoa gris (Guad.); laabra, mahoebaari, mappoo, masjiboeri, moesti sámbo, rondeisji (Dutch W.I.); bainoro prieto, beeb, coma de uña, espino y camote, garabato, g. blanco, g. prieto, garambullo, gu-ichi-gu, huele de noche, uña de gato, u. del diablo, vainoro prieto, zapo (Mex.); clavo, huele de noche (Guat.); beeb, uña de gato (Br. H.); cargalero (Hond.); cagalero, c. negro, cirzón, crucito, espuela del diablo, guaco (Salv.); espino negro (Nic.); uña de tigre (C.R.); adorate, aruñagato, bejuco de agua, buen amigo, pegapega, uña de gato (Col.); casabe, casare, more, pacurero (Venez.); mafo (Br. G.); cabolena, cebolino, cipo molle, cumichá, espora de gallo, joão dormendo, j. dormido, j. molle, pau de urubú, p. lepra, tapaciriba (Braz.); caspi zapallo, ombú-rá, palo bobo, yaguápindá, yuquerí buzu, y. ruzú, y. ruzú-rá, y.-sí, zapallo, z. caspi (Arg.). *Torrubia:* Beefwood, blolly, corkwood, pigeon wood, pork wood (Fla., B.W.I.); black mampoo, corcho, majagua que mona (P.R.); barrehorno, botijo, hilacho, vacabuey, zarza sin espinas (Cuba); muñeco (Dom. R.); estribo, sapo, sillo (Col.); amarillo pesjua, cazabito, pacurero, pecurero, salado, tinterero (Venez.); clavo-caspi (Peru).

NYSSACEAE

The Nyssaceae, sometimes included in the Cornaceae, comprise three genera and eight species of trees and shrubs. *Camptotheca* and *Davidia,* with one species each, are in Asia, while *Nyssa* has four species in eastern United States and two in eastern and southern Asia. All have simple, alternate, deciduous leaves and fine-textured woods of general utility.

Nyssa. The four American species are separable into two equal groups which are sometimes designated the Black Gums (*N. sylvatica* and *N. biflora*) and the Tupelos (*N. aquatica* and *N. ogeche,* the last a rare and local species of no commercial importance). The leaves are alternate, simple and mostly crowded at the ends of the branches; the minute greenish white flowers are polygamo-dioecious; the fruit is a drupe with thin flesh and a bony ribbed or winged stone. In the Black Gums the pistillate flowers have a disk-like calyx and are borne in small clusters and the fruits are blue, while in the Tupelos the calyx is cup-shaped, the flowers are solitary, and the olive-like fruits are red or purple. There are no pronounced distinctions between the woods of the two groups, though it is usually possible to recognize the different species.

Nyssa sylvatica Marsh., the common Black Gum, also known as Sour Gum, Pepperidge, and sometimes as Tupelo, is a medium-sized to large tree, occasionally on best sites attaining a height of over 100 feet and a diameter up to five feet. It is widely distributed throughout the eastern half of the United States from Maine and Michigan to Florida and Texas. It prefers moist situations and is a common associate of Red Maple and Black Ash, but adapts itself to various sites and occurs scatteringly in upland forests with such hardwoods as Red and White Oaks, Sugar Maple, and Beech. The principal sources of Black Gum lumber are the Appalachian Mountains, Arkansas, Texas, and the central and lower Mississippi valley. Swamp Black Gum, *N. biflora* Walt., is confined mostly to the coast region from Maryland to northern Florida and westward to eastern Texas, reaching its best development in the lower Mississippi valley. It inhabits swamps and the margins of ponds and streams along with Cypress (*Taxodium*), Tupelo, and lowland species of Oaks. Trees growing in the wettest sites are often much enlarged at the base of the trunk. It is estimated that the total stand of timber of these two species is about 30 billion board feet, of which one-third is *N. biflora*.

The principal source of Tupelo lumber is *Nyssa aquatica* L., a tree 80 to 100 feet tall, with a straight trunk sometimes four feet in diameter above the greatly enlarged tapering base. It is one of the most characteristic trees of the southern swamps where it grows in mixture with Cypress, Red Gum (*Liquidambar*), and Swamp White Oak. Writing in 1906, Herman von Schrenk says (*Southern Lumberman* 51: 603: 41): "Of the so-called 'inferior' woods there is probably none which has jumped from obscurity to the front rank as rapidly as Tupelo. Ten years ago this wood was unknown to the American market. Seven years ago the first car of Tupelo was shipped to London, from Southern Louisiana, and some six months later a second car was shipped, by a St. Louis manufacturer; today this same manufacturer is shipping 100 cars to the firm that ordered the first car in 1899." The average annual production of Tupelo lumber in Louisiana alone from 1916 to 1930 was about 100 million board feet. The total stand of Tupelo and Black Gum is estimated to be about 15 billion board feet.

The woods of the different species of *Nyssa* are so nearly alike in appearance and properties that ordinarily no attempt is made to separate them in localities where they occur in mixture in lowland forests, but Black Gum growing on uplands is usually considerably harder, heavier, and tougher than the others. The color is nearly white in the thick sapwood, with gradual transition to yellowish or brownish or somewhat streaked in the heartwood. Wood from the swollen parts of the trunk is very light, soft, and brittle; that from the main stem varies in weight (air-dry) from 25 to 47 lbs. per cu. ft., the average for Tupelo being about 32 lbs., that of the Black Gums about 35 lbs. The texture is fine and uniform; the grain is typically irregular and interlocked, making the wood tough to split and likely to warp and twist in seasoning unless carefully handled. Working properties are otherwise good, the finished surface being smooth and bright. Natural resistance to decay is low. The principal uses are for boxes and crates, factory flooring, moldings, woodenware, rollers, and veneers, especially for making plywood, boxes, and crates; the timber is also used for railway crossties (treated) and paper pulp.

Growth rings barely distinct, being demarcated by a narrow band of radially compressed and sometimes thicker-walled fibers. Pores small (nearly medium-sized in part in *Nyssa ogeche*); fairly to very numerous (*N. sylvatica*); solitary and in short to long radial multiples, evenly distributed. Vessels with long scalariform or scalariform-reticulate perforation plates having many fine bars; without spiral thickenings, except rarely in tips of members; pitting opposite to scalariform, the opposite pits small and rectangular. Rays heterogeneous; 1 or 2, occasionally 3, cells wide in *N. aquatica* and *N. sylvatica*, sometimes 4 or 5 cells wide in *N. biflora* and *N. ogeche;* uniseriate rays and ray margins composed of few to 40 rows of mostly square and upright cells; vertical fusions common; pits to vessels small, opposite. Wood parenchyma rather sparse; mostly diffuse, sometimes in short metatracheal rows, occasionally paratracheal; somewhat more plentiful in *N. ogeche* than in the others; crystals common, but rarely occupying entire strands. Wood fibers mostly thin-walled, but moderately thick-walled in *N. sylvatica;* pits with slit-like, often extended, apertures and distinct circular borders. Ripple marks absent. No gum ducts seen.

COMMON NAMES: *Nyssa aquatica:* Bay poplar, tupelo (trade); gum (bowl, cotton, ladle, papaw, sour, swamp, tupelo), olive-tree, tupelo (U.S.A.). *N. biflora:* Gum (black, lowland, l. black, swamp black), tupelo (U.S.A.). *N. ogeche:* Gopher plum, lime-tree, sour tupelo, tupelo (U.S.A.). *N. sylvatica:* Gum (black, sour, tupelo, yellow), pepperidge (U.S.A.).

OCHNACEAE

THIS widely distributed tropical family is composed of 22 genera and about 375 species of trees and shrubs, rarely half shrubs and herbaceous plants. The leaves are alternate, typically simple, and often closely pinninerved; stipules are present, often laciniate; the flowers are mostly in racemes or panicles; the fruit is various, drupaceous or capsular. Many of the plants have handsome flowers, and the seeds of some yield oil for soap-making or medicinal purposes, but the only genus of commercial importance for its timber is *Lophira,* with two species in tropical West Africa, supplying very dense durable dark red wood known as Bongossi, African Oak, or Red Ironwood. In tropical America there are over 200 species of 10 woody genera, but only a few are medium-sized to rather large trees, the others ranging down to low shrubs.

Wood dull brown or reddish brown throughout. Odorless and tasteless. Moderately to decidedly hard and heavy; texture medium-fine to coarse; feel rather harsh; grain straight to irregular; probably fairly resistant to decay. If obtainable in appropriate sizes and quantity it would be useful for general carpentry and heavy construction, but there is no likelihood that it will ever be a factor in the export trade.

Growth rings present or absent. Pores mostly very small to minute, but medium-sized in *Blastemanthus;* numerous, sometimes crowded; occurring singly and in small multiples, or sometimes in short radial rows or in clusters, uniformly distributed without definite pattern. Vessels with simple perforations, though a tendency to develop scalariform perforation plates has been reported; no spiral thickenings seen; pits very small to minute, the apertures often coalesced; pits vestured in the Exalbuminosae (*e.g., Ouratea* and *Elvasia*) only. Rays uniseriate and biseriate in *Blastemanthus* and sometimes in *Wallacea;* of two sizes in the others, the multiseriate 3 to 5 cells wide and rather low, or in some species more than 100 cells high, though usually not conspicuous because of lack of contrast with the background; followed out on cross section the wider rays sometimes appear to divide and coalesce; decidedly heterogeneous; pits to vessels numerous, small to very small; gum deposits common; large thick-walled crystalliferous cells, sometimes spherical, often present (*e.g., Elvasia* and *Ouratea*). Wood parenchyma sparsely to abundantly developed; paratracheal, metatracheal, and diffuse; crystals in short series, the length of an ordinary member of a strand, common in some genera (*e.g., Cespedesia*). Wood fibers with rather thick to very thick walls; sometimes septate; pits usually very numerous, distinctly to indistinctly bordered, occasionally few and minute. Ripple marks and gum ducts absent.

Blastemanthus, with five species of trees, is confined to British Guiana and the upper Amazon region. The only specimen at hand (Yale 3382) is of *B. grandiflorus* Spruce, a little tree collected by Adolpho Ducke on the Rio Negro, Amazonas, Brazil. The timber apparently has no special uses. Heartwood dark reddish brown, with a purplish hue; merging gradually into the thin sapwood. Luster low. Without distinctive odor or taste. Very heavy, hard, and strong; texture medium-fine; grain fairly straight; not easy to work, but finishing with a smooth, hard surface; durability probably high. Of no commercial possibilities.

Cespedesia, with five species of small to medium-sized trees, occasionally up to 65 feet tall and 24 inches in diameter, is distributed in western tropical America from Panama to Peru. *C. macrophylla* Seem. grows gregariously on hillsides at some distance from the sea in eastern Panama and is very attractive when in bloom, as great panicles of yellow flowers and clusters of large green leaves are massed at the ends of stout branchlets of the dense flattened parasol-like crown. The timber of *Cespedesia* is coarser-textured than others of the family and bears a superficial resemblance to some of the Philippine Dipterocarps of the Red Lauan type. It is not available in sufficient quantity to be commercially important. Heartwood reddish brown with a purplish tinge; not sharply demarcated from the light brown sapwood. Luster rather high. Odorless and tasteless. Moderately heavy, but firm and strong; texture medium to coarse; grain straight; working properties good; durability probably fairly high.

COMMON NAMES: Amarrón caspi, a. c. grande (Ec.); malafaia (Braz.).

Elvasia. There are six species of small trees in the Amazon basin. The only specimen available (Yale 21000) is of *E. calophyllea* DC. collected by Adolpho Ducke in Brazil, where it is known as Apapá. Heartwood rather dull brown tinged with red; merging gradually into the pale brown sapwood. Odorless and tasteless. Rather hard and heavy; texture medium; grain somewhat irregular; has about the consistency of Applewood (*Malus*); working properties good; durability probably fairly high. Has no commercial possibilities.

Ouratea, with about 200 species of small to large trees and shrubs, occurs in tropical Africa, the Indo-Malayan region, and throughout tropical America. The bark is rich in tannin. The wood is of good quality, but the American species are too small to furnish timber of more than local utility. Color dull grayish brown throughout. Odorless and tasteless. Moderately to decidedly heavy, hard, and strong; texture medium to rather fine; grain straight; working properties good; durability probably rather high.

COMMON NAMES: Cabbage bark (Jam.); gonfia lustrosa (P.R.); arete, contraguao, cordón de soldado, guanabilla, g. de monte, g. del Pinar, g. de sabana, nabaco, orilla de orroyo, rascabarriga, serrucho (Cuba); cinco negritos, zapotillo de la costa (Mex.); bastard sapodilla, billbird patter, tcanlol, xcanlol (Br. H.); coyolillo, naranjillo (Salv.); wild pigeon plum (Pan.); avououyra, oura-ara (Guianas); angelim, batiputá, cajú bravo, cajueiro bravo, jabotapitá, mangue do matto, pau de serra (Braz.).

Tyleria, with four described species of shrubs and small trees, occurs on Mount Duida in southern Venezuela. According to Tate (*Bull. Torrey Bot. Club* 58: 392), "the Tylerias, on account of their blunt stems, terminal clusters of leaves, and large flowers, are perhaps the most arresting group of trees on the summit of Duida, *T. grandiflora* in particular being so widespread as to occur most everywhere. . . . . *T. floribunda* and *T. spathulata* were seen about Central Camp, usually in forest about 20 feet high. . . . *T. linearis* has become specialized for growing along the banks of streams." The following description is based upon one specimen (Yale 16187; Tate 540) of the type of *Tyleria floribunda* Gleason. Color dull reddish brown throughout. Odorless and tasteless. Very heavy, hard and strong; texture fine and uniform; grain irregular; not difficult to work, finishing smoothly; is probably fairly durable. Of no commercial possibilities.

Wallacea, with two species of small trees, is found in Brazilian Amazonia. According to Adolpho Ducke (*Tropical Woods* 43: 22), *W. multiflora* Ducke "grows on the banks of the Curicuriáry, tributary of the upper Rio Negro, beyond the second cataract. *Wallacea insignis,* discovered by Richard Spruce at the Rio Uaupés near the village of Panuré (now Ipanoré), has also been collected at the headwaters of the Rio Tarumá, near Manáos, and on the banks of the Rio Aruan confluent of the Arapiuns, a left tributary of the lower Tapajoz." Wood of a dull brown color throughout. Without distinctive odor or taste. Hard, heavy, and strong; texture medium; grain straight; easily worked, finishing very smoothly; durability probably high. Apparently without commercial possibilities because of the scarcity and small size of the trees.

OLACACEAE

THE Olax family consists of 22 genera and about 260 species of trees, shrubs, and a few woody vines; some of the plants are semi-parasitic. The leaves are alternate, simple, and (except in *Agonandra*) without stipules; the small perfect flowers are borne in axillary cymes, racemes, or fascicles, or are sometimes solitary; the fruit is a one-seeded drupe, sometimes partly or completely enveloped by the accrescent calyx. *Schoepfia* and *Ximenia* are pantropical; *Aptandra, Heisteria,* and *Ptychopetalum* are limited to the tropics of the New World and Africa; there are eight strictly American genera, namely, *Agonandra, Brachynema, Cathedra, Chaunochiton, Eganthus, Liriosma, Minquartia* (including *Endusa*), and *Tetrastylidium.*

Woods yellowish, yellowish brown, or olive; not highly lustrous. Without distinctive odor and taste, except *Ximenia,* the heartwood of which is mildly fragrant, at least when fresh. Moderately to decidedly hard and heavy; texture medium to coarse; feel rather harsh; generally not difficult to work, finishing smoothly; durability variable.

Growth rings usually present, but not always clearly defined. Pores small to minute, or occasionally medium-sized; numerous; mostly solitary in *Agonandra, Heisteria, Liriosma,* and *Ximenia,* but in short to long radial multiples or series, suggesting certain Euphorbiaceae, in the others. Vessels with simple perforations except in *Minquartia* and *Heisteria;* spiral thickenings absent, but spiral striations sometimes present; tyloses common; intervascular pitting alternate, typically fine to very fine, but coarse in *Minquartia;* some of the pits to ray and wood parenchyma cells may be very large, the complements of the same size or several in a cluster that may easily be broken through in sectioning. Fibriform vessel members sometimes associated with ordinary vessels (*e.g., Chaunochiton*). Rays 1 to 5, mostly 1 or 2, cells wide; few to 25, sometimes up to 50, rarely (*Heisteria*) up to 100, cells high; moderately to decidedly heterogeneous; frequently very coarse-celled; small crystals common; ray-vessel pitting variable from very fine to very coarse, 2-sized in most cases, tending to scalariform in *Heisteria* and *Minquartia.* Wood parenchyma usually abundantly developed; typically reticulate, suggesting many Euphorbiaceae; paratracheal, aliform and confluent in *Schoepfia,* the cells in horizontal seriation; crystals sometimes present. Wood fibers with medium to very thick walls; pits variable from minute and simple or indistinctly bordered (*e.g., Cathedra, Chaunochiton, Minquartia, Ptychopetalum, Schoepfia*) to large and distinctly bordered (*e.g., Heisteria, Ximenia*). Ripple marks absent, except in parenchyma layers and patches in *Schoepfia,* where they are very fine, but distinct with lens. No gum ducts seen.

Agonandra, with eight named species, is of infrequent occurrence from southern Mexico to northern Argentina and southern Brazil. The trees are all small, mostly 20 to 30 feet high, with a rather stout, sometimes fluted, trunk 10 to 15 inches in diameter. The timber is of good quality, but because of its small size and scarcity it is unknown on the market and the only recorded special uses are for chair frames and wheel spokes. Heartwood orange-yellow, with gradual transition to the pale yellow sapwood. Not highly lustrous. Odorless and tasteless. Very hard, heavy, compact, and strong; texture fine; grain usually straight; not difficult to work, finishes very smoothly; the more deeply colored material durable.

COMMON NAMES: Granadillo, margarita, maromero, palo del golpe, ravienta cabra, suelda con suelda (Mex.); caimancillo, hoja menuda (Col.); amarellão, marfim, pau d'alho do campo, p. marfim, p. m. do cerrado, p. m. verdadeiro, tatú (Braz.); meloncillo, pata, sombra de toro (Arg.).

Aptandra. There is one species in tropical West Africa and three in the Amazon basin, all small trees of no importance for their timber. Wood yellow throughout. Luster medium. Odorless and tasteless. Hard, heavy, tough, and strong; texture medium; feel rather harsh; grain straight; not difficult to work, finishing smoothly; durability low. Suitable for about the same purposes as Maple (*Acer*).

COMMON NAMES: Castanha de cotia, quinquio, sapucainha (Braz.); pamashto, trompo-huayo (Peru).

Cathedra, with five species of trees, is sparingly distributed in Brazil. The bark is reddish; the leaves are leathery; the small flowers are borne in compact clusters in or near the axils of the leaves; the fruits have a thin fleshy exocarp and are inclosed by the accrescent calyx. The only wood sample examined was collected by B. A. Krukoff in the basin of the Rio Purus, Acre Territory. Color yellowish brown or pale olive throughout. Luster low. Without distinctive odor and taste. Very hard, heavy, and strong; texture medium; feel harsh; grain straight; splits readily, with a splintery fracture; rather difficult to saw, but can be finished very smoothly; durability probably good. Presumably of no commercial possibilities, but suitable for heavy construction.

Chaunochiton, with two or three species of medium-sized to large trees, is of infrequent occurrence from Central America to the Amazon basin. The timber apparently is not utilized for any special purposes. Wood yellow throughout. Luster medium. Without distinctive scent and taste when dry, but fresh material is said to have a strong odor of hydrocyanic acid. Moderately heavy and hard; texture rather fine; grain straight; very easily worked, being readily carved or turned; durability low.

Heisteria. Of the 40 or more species, three occur in tropical West Africa, the others in tropical America, ranging from southern Mexico to Peru and the Amazon region of Brazil. They are small trees, occasionally 35 feet high and six to eight inches in diameter, but *H. caloneura* Sleumer is said to attain a height of 100 feet in the basin of the Rio Madeira, Brazil. Heartwood light brown or reddish, merging into the sapwood; yellowish brown in *Heisteria Duckei* Sleumer and *H. flexuosa* Mart. Luster medium. Odorless and tasteless. Moderately to decidedly heavy, hard, and strong; texture mostly fine; grain straight; working properties good; durability fairly high in reddish material. Probably of no commercial possibilities.

COMMON NAMES: Pate macho (Guat., Hond.); cresta de gallo, sombrerito (Salv.); manglillo (C.R.); ajicillo, naranjillo colorado (Pan.); cascarilla negra, huesito negro (Col.); atrete (Ec.); chuchuhuasha, cotoma, masacey, huangana-caspi, huapa-caspi, huarmi-chuchuhuasha, moena, muena, platina caspi (Peru); klikli wete (Sur.).

Liriosma, with 14 species of shrubs and small trees, is limited in its distribution to the Amazon basin. The plants apparently are poorly known, the only recorded vernacular name being Senimoro-ey (Peru). The timber is not utilized. Wood yellow throughout. Luster medium. Odorless and tasteless. Moderately hard, heavy, and strong; texture fine to medium; feel harsh; grain straight; not difficult to work, taking a smooth finish; is perishable when exposed to decay. A general purpose material without commercial possibilities.

Minquartia. Two species are recognized, namely, *M. guianensis* Aubl., ranging from Nicaragua to Ecuador, the Guianas, and Brazil, and *M. punctata* (Radlk.) Sleumer of the upper Amazon region. The woods are alike in appearance and properties. Regarding the first species in Panama, where it is commonly known to English-speaking people as Manwood or Black Manwood, G. Proctor Cooper says (*Tropical Woods* 14: 4):

"The tree is found in the hot lowland forests on both the Caribbean and Pacific watersheds, but it grows only on well-drained slopes and not on the low flat ground back of the Mangrove. Occasionally it is found on islands where the land rises abruptly from the sea. . . . The tree has low buttresses and the bole is somewhat squared or fluted and twisted, especially at the base. It grows to be three feet in diameter and occasionally 100 feet tall. The core often is hollowed out by ants, leaving a shell about a foot thick. . . . Black Manwood is highly valued for its durability and strength. It is fairly easy to cut with an ax or machete, but the interlocked grain makes it difficult to split. The texture is very fine and the wood takes a high lustrous polish. The sapwood is from 1 to 2 inches thick and yellow-tan in color, but the heart is a deep chocolate or olive-brown. . . . As Black Manwood is scattered and difficult to locate in the bush it can never be considered as a commercial possibility. It should find use for walking sticks and special articles of turnery and inlay."

The uses of *Minquartia* timber are mainly for railroad and tram crossties, telephone poles, fence and house posts, and similar purposes requiring great strength and resistance to decay. Few woods equal its reputation for durability in contact with the ground, but the scarcity of the trees and their poor timber form prevent commercial exploitation. (See *Tropical Woods* 8: 10.)

COMMON NAMES: Manwood, plátano (Nic.); manú, palo de piedra (C.R.); black manwood, criollo, manwood, palo criollo, urari, urodibe (Pan.); jewalidanni, mincoa (Br. G.); arata or aratta, aratahoe-hoe, arratawerie, baggie-baggie, konthout, kontoe-hoedoe, makka, tomopio, wanania (Sur.); bois agouti, b. incorruptible, mincouart (Fr. G.); acaiquára, a. da varzea, a. do igapó, acapú, acaricoara, acariguara, araciúba, acary (Braz.); pechiche (Ec.); huacapú (Peru).

Ptychopetalum. Seven species have been described, five occurring in tropical Africa and two in the Amazon basin. The latter are slender trees less than 40 feet tall, with dark blue, plum-like fruits. Their principal use is for their roots, which supply a drug known as muira-puama, exported from Pará and Manáos. Wood brownish yellow throughout. Luster medium. Odorless and tasteless. Very hard, heavy, and strong; texture medium; feel harsh; grain straight; not difficult to work, taking a smooth finish; poorly resistant to decay. Presumably of no commercial possibilities.

Schoepfia. There are eight species in eastern and southeastern Asia and about 26 in Latin America. The combined range of the latter includes the West Indies, Mexico, Central America, and tropical South America. They are shrubs or small trees of no importance. Wood brownish yellow or pale olive throughout; appears waxy. Luster rather high. Without distinctive odor or taste. Moderately hard and heavy; texture rather fine; grain straight to irregular; very easy to work, taking a lustrous natural polish; durability probably rather low.

COMMON NAMES: White beefwood, whitewood (Bah.); boniatillo, cerillo, mije blanco (Cuba); palo fierro, tecolitillo (Mex.); sombra de armado (Hond.).

Ximenia. Of the 10 or more species, the best known and most widely distributed is *X. americana* L., a spiny semiparasitic shrub or small tree rarely 30 feet high, inhabiting tropical and subtropical regions throughout the world, often forming coastal thickets. Its bark is used locally as a source of tannin; the plum-like fruits are edible and have the scent of hydrocyanic acid; the seeds are rich in oil. The wood has been used to a limited extent as a substitute for Sandalwood (*Santalum*) to which it appears to be related. (See *Kew Bulletin* 1935, pp. 175–177.) Heartwood reddish yellow or orange-brown; rather sharply demarcated from the yellow sapwood. Luster fairly high. With mild fragrant scent, but without distinctive taste. Very hard and heavy; sp. gr. (air-dry) 0.95; weight 59 lbs. per cu. ft.; texture fine and uniform; grain straight to irregular; not difficult to work, taking a high natural polish; durability rather high.

COMMON NAMES: False sandalwood, hog plum, mountain plum, Spanish plum, tallow nut, t. wood, wild lime, w. olive, yellow sanders (Fla., B.W.I.); ciruelillo, ciruelo, c. cimarrón, jía manzanilla, yaná (Cuba); croc (Dom. R.); cerise de mer, macaby (Haiti); oranger des falaises, o. de montagne, prunier épineaux (Guad.); seaside plum (Trin.); ciruelillo, kic-ché, xkuk-ché (Mex.); manzanillo (Guat., Hond., Salv.); cagalera (Hond.); pepe nance (Salv.); chocomico (Nic.); teu-krá (C.R.); caimito de monte, espino de brujo, limoncillo (Col.); manzana guayabo, m.

del diablo, ligrito (Venez.); hevmassoli (Guianas); ambuy, ameixa, a. de espinho, ameixieira do Brasil, a. de espinhos, ameizero, espinheiro da ameixa (Braz.); albaricoque, albericoquillo, a. del campo, albarillo del campo, pata, p. de monte (Arg.).

OLEACEAE

THE Olive family comprises about 25 genera and over 400 species of trees, shrubs, and a few herbs, widely distributed in temperate and tropical regions, chiefly in the northern hemisphere. Some of the best known genera are *Olea,* the source of olives; *Syringa,* the Lilac; *Forsythia,* a decorative shrub with yellow flowers; *Ligustrum,* the Privet, a common hedge plant; *Jasminum,* the Jasmine, with fragrantly scented flowers; and *Fraxinus,* the Ash, the only important timber tree in the family. The following description of the woods, though based on American species, covers almost the entire range of variation in the family.

Heartwood light to chestnut brown, sometimes waxy and attractively variegated; often sharply demarcated from the white, pale brown, or pinkish sapwood. Mild odor and taste occasionally present, but not describable. Very hard and heavy to moderately so, occasionally rather light and soft; texture very fine and uniform to very coarse and uneven; grain straight to irregular; not difficult to work, finishing smoothly; deeply colored heartwood durable.

Growth rings usually present, barely visible to very distinct; ring-porous structure in *Fraxinus* and *Chionanthus.* Pores often thick-walled; greatly variable in size and distribution; large in early wood or *Fraxinus,* small to minute in the others; in the diffuse-porous woods and in late wood of the others they are solitary and, more often, in pairs or short radial rows, sometimes also in small clusters; in flame-like arrangement or zig-zag patches in *Osmanthus* and *Chionanthus.* Vessels with simple perforations; spiral thickenings present in *Chionanthus, Osmanthus,* and *Menodora;* pits medium-sized to minute, alternate, vestured in *Forestiera.* Rays all uniseriate, the cells upright, in *Menodora;* 1 to 3 cells wide and usually not over 25 to 30 cells high, except in vertical fusions, in the others; mostly heterogeneous, with a stratum of procumbent cells and margins of square or upright cells, but sometimes (*Fraxinus* and *Chionanthus*) homogeneous, at least in part; pits to vessels numerous, very small to minute. Wood parenchyma sparingly to abundantly developed; paratracheal in all genera; aliform and irregularly confluent, esp. in late wood, in *Fraxinus;* frequently in uniseriate to multiseriate bands, occasionally doubled or tripled, at margin of growth rings; in irregularly to closely spaced bands in *Linociera* (in part) and *Osmanthus;* strands commonly short-celled, occasionally with heavily sculptured and disjunctive radial walls; fusiform cells common in early wood of *Fraxinus.* Wood fibers with rather thin to very thick and gelatinous walls; rarely septate; pits typically very small, simple or indistinctly bordered, but sometimes (*Chionanthus, Menodora,* and *Osmanthus*) distinctly bordered; fine spiral thickenings present in *Menodora.* Gum ducts absent. Ripple marks of local occurrence in some specimens of *Fraxinus.*

Chionanthus, with two species of shrubs and small trees, occurs in northern and central China and middle and southern United States. The American species, *C. virginica* L., commonly called Fringe-tree or Old Man's Beard, is a tree, sometimes 30 feet high and 10 inches in diameter, found along streams in the Gulf states and northward into Arkansas and the southern Appalachians, and often cultivated for ornament. The leaves are simple, opposite, and deciduous; the fragrant flowers are borne in loose panicles in the axils of the upper leaves of the previous year; the fruit is a dark blue or blackish thin-fleshed drupe. The bark is used medicinally. There are no special uses for the pale brown hard and heavy wood.

Forestiera, with about 15 species, is distributed from southern United States through Mexico and Central America to Paraguay and through the West Indies to Brazil. The simple, opposite leaves are either deciduous or persistent; the minute flowers are borne in axillary panicles; the fruit is drupaceous. Most of the species are shrubby, but *F. acuminata* Poir. is sometimes 50 feet high and 10 inches in diameter in eastern Louisiana. Elsewhere the maximum heights reported for trees of this genus are 25 to 30 feet. Heartwood brown, absent from most specimens; sapwood thick, white. Luster rather high. Density medium to rather high; texture uniform, medium to fine; of good technical properties, but little used because of its small size and scarcity.

COMMON NAMES: Privet, swamp privet (U.S.A.); hueso blanco, yanilla blanca (Cuba); ink bush (Bah.); acebuche, lantisco, lantrisco, lentisco, mimbre, panalero, pico de pájaro (Mex.).

Fraxinus, with about 70 species of deciduous trees and shrubs, is widely distributed in the temperate regions of the northern hemisphere, with limited extensions into the tropics. The leaves are opposite and unequally pinnate, with 3 to 11 leaflets, or rarely reduced to a single leaflet; the terminal buds are compressed, obtuse, and much larger than the laterals; the green or white flowers are borne in fascicles or panicles; the fruit is a samara, winged at the apex. There are 18 species in the United States, and several of these and others to a total of 15, grow in the mountains of Mexico; there are also two species in Cuba. The Mexican and Cuban timber is used only locally.

According to W. D. Sterrett (U.S. Dept. Agr. Bull. No. 523, p. 2), about 98 per cent of the Ash lumber produced in the United States is from three species, namely, White Ash (*Fraxinus americana* L.), Black Ash (*F. nigra* Marsh.), and Green Ash (*F. pennsylvanica* Marsh., var. *lanceolata* Sarg.). The species supplying the other 2 per cent are Oregon Ash (*F. oregona* Nutt.), Blue Ash (*F. quadrangulata* Michx.), Biltmore Ash (*F. biltmoreana* Beadl.), Pumpkin Ash (*F. profunda* Bush.), and Red Ash (*F. pennsylvanica* Marsh.). "In the lumber trade Ash lumber is often not distinguished as to kinds, all species being sold under the common name of Ash. Much is sold under the name White Ash to distinguish it from Brown Ash (also known as Black Ash, *F. nigra*), which has mechanical properties quite different from those of White Ash but the same general appearance and structure and a more handsome grain. Lumber cut from all species, however, is often sold as White Ash. The terms Green, Red, and Biltmore Ash are not used at all in the lumber trade. Old-growth Ash from continuously wet river-bottom land is often called Pumpkin Ash because it is soft and brittle. The term is applied chiefly to Pumpkin Ash (*F. profunda*) and Green Ash (*F. lanceolata*). The terms Black and Blue Ash are often used locally to designate standing Ash timber, but do not necessarily refer to the species botanically known as *F. nigra* and *F. quadrangulata*. The term Oregon Ash is seldom used in trade on the Pacific coast."

All of the commercial species are in the eastern half of North America, except *Fraxinus oregona*, which occurs on rich soil along streams in the Pacific coast region from southern British Columbia to southern California, but attains its best development in southwestern Oregon. White Ash (*F. americana*) is a large tree, sometimes 120 feet tall with a well-formed bole six feet in diameter, widely distributed throughout eastern United States and southeastern Canada, being at its best in the central hardwood region. Red Ash (*F. pennsylvanica*) and Green Ash (var. *lanceolata*, sometimes considered a distinct species) are smaller trees and their range is much greater than that of White Ash, the principal sources of timber being in the Mississippi valley and southeastern states. Black Ash (*F. nigra*) is more northern and inhabits deep swamps and lowlands in southeastern Canada, New England, and the Lake states.

Most of the present supply of Ash is second-growth, chiefly in small timber tracts and woodlots attached to farms, and sapwood of fairly rapid uniform growth is preferred where strong timber is desired. The principal uses include handles of tools (such as shovels, hoes, and rakes rather than axes and hammers, where Hickory is preferable because of its greater resistance to shock), agricultural implements, butter tubs, bent work, athletic goods, baseball bats, and oars. Considerable quantities of lumber are consumed in the manufacture of kitchen furniture and interior trim.

Heartwood brown to dark brown in *F. nigra;* grayish brown, sometimes with a distinct reddish tinge, particularly in the late wood, in the others; sapwood nearly colorless, rather thin and sharply demarcated in old forest trees, thick and less clearly differentiated in second-growth. Luster medium to high. Without distinctive scent or taste. Light, soft, and brittle (Pumpkin Ash) to hard, heavy, tough, and strong, but mostly of medium density; sp. gr. of typical species (air-dry) 0.50 to 0.75, mostly between 0.55 and 0.65; weight 31 to 47, mostly 34 to 41, lbs. per cu. ft.; texture coarse; grain usually straight, sometimes curly; working properties excellent; durability of heartwood poor to medium (Black Ash); sapwood perishable in contact with soil, and subject to attack of powder-post beetles.

COMMON NAMES: *Fraxinus nigra:* Ash—basket, black, brown, hoop, splinter, swamp, water (U.S.A.). *Other species:* Ash—blue, cane, green, pop, pumpkin, red, water, white (U.S.A.); búfano (Cuba); baretta china, demettza, escobilla, fresnillo, fresno, f. asemillado, paramu, plumero, yaga guillaa, y. nisse (Mex.).

Haenanthus, a genus endemic to the Greater Antilles, is sometimes credited with six species, but according to Knoblauch (*Repert. Spec. Nov. Reg. Veg.* 34: 139–142) only two are clearly valid, namely, *H. incrassatus* (Sw.) Gris. of Jamaica and *H. salicifolius* Gris. (with three varieties) in Cuba, Puerto Rico, and the island of Haiti. The leaves are simple, opposite, and persistent; the yellow flowers are borne in axillary panicles; the fruit is drupaceous. The wood of the second species, a small tree known in Cuba as Caney, is brown, more or less variegated, and somewhat waxy in appearance and feel. It is very hard and heavy and takes a high polish, but the fibers are gelatinous and tough and tend to pull out in working. Durability is presumably high. The timber is not utilized because of its small size.

Linociera, often considered only a section of *Mayepea,* is represented in the West Indies, southern Mexico, Central America and Colombia by several species of shrubs and small to medium-sized trees. The leaves are simple, entire, opposite, and persistent; the flowers, which are comparatively large, are panicled; the fruit is a small oblong drupe. *L. panamensis* Standl. is a tree 40 to 50 feet high and 12 to 15 inches in diameter; its wood is of about the consistency of White Birch (*Betula alba* L.). Two or three other species occur in Central America, but the timber is not utilized. *L. areolata* Lundell is about 35 feet tall and six inches in diameter in moist *Liquidambar* forest in southern Mexico. The West Indian species are mostly small, occasionally medium-sized, and some of them have a small core of attractively variegated, brownish, waxy, very hard and heavy heartwood that is prized for small carved objects and articles of turnery.

COMMON NAMES: Bayito, coronel, dominguito, guaney, g. de corazón, g. de negro, jocotea, pico de gallo, ramón de costa (Cuba); avispillo, huesillo, hueso blanco (P.R.); lirio, tarana (Dom. R.); cayepon (Haiti); wild cocoplum (Br. H.).

Menodora, with about 15 species of low shrubs and herbs, occurs in the subtropical regions of both hemispheres. The leaves are simple, opposite to alternate; the yellow flowers are solitary or corymbose; the fruit is a 2-celled capsule. The wood of *M. scabra* Gray is described because its structure differs in several details from that of other members of the family.

Growth rings indistinct. Pores very small; very numerous (up to 300 or more per sq. mm.); mostly solitary, with some short radial multiples. Vessels with fine spiral thickenings; pits rather small, often in a single vertical row. Rays all uniseriate, mostly low but sometimes up to 25 cells high; all cells upright or square, often tall; pits to vessels small, circular. Wood parenchyma very sparingly paratracheal. Wood fibers with thick walls and fine spiral thickenings; pits numerous, with small borders.

Osmanthus, with 10 species of shrubs and small or rarely medium-sized evergreen trees, occurs in eastern Asia, the Himalayas, Polynesia, southern United States, and

Photo by Llewelyn Williams.

PLATE XXXII. Typical savanna in the lower Caura, Bolívar, Venezuela, where the dominant woody species are Chaparro (*Curatella americana*) and Alcornoque (*Bowdichia virgilioides*).

Photo by B. E. Dahlgren.

PLATE XXXIII. Forest along the Amazon River, Brazil.

central Mexico. The leaves are simple and opposite; the very small fragrant white or yellow flowers are borne in short axillary racemes; the fruit is a thin-fleshed succulent drupe. *O. fragrans* Lour. of China is a favorite garden plant because of its sweet-smelling flowers. The only American species, *O. americanus* (L.) B. & H., attains its largest size in rich hammocks and along streams of the coastal region of the south Atlantic and Gulf states, the maximum size reported being about 70 feet in height and a foot in diameter. The light brown, rather hard and heavy, fine-textured wood is of good quality, but is of no commercial importance.

COMMON NAMES: Devilwood, wild olive (U.S.A.).

ONAGRACEAE

THE Evening Primrose family, sometimes called the Oenotheraceae, comprises about 40 genera and several hundred species of aquatic and terrestrial herbs, erect, prostrate, epiphytic, or scandent shrubs, and a few small trees, widely disseminated but most abundant in the temperate and subtropical regions of the New World. The leaves are simple, entire or dentate, opposite or alternate, and without stipules; the flowers are mostly solitary, often showy; the fruit is a capsule or a nut; seeds sometimes winged (*Hauyea*). The principal uses of the plants are for ornament. Arborescent members of the family are few. Some species of *Hauya* of Mexico and Central America are said to attain a height of 25 to 50 feet, but there are no specimens available for study. *Fuchsia arborescens* Sims is occasionally a tree 20 feet tall in Mexico. The other woody representatives are shrubby. The following description of the wood is based upon a few samples of three genera, namely, *Fuchsia, Jussiaea,* and *Oenothera*.

Wood brownish to nearly white throughout specimens. Without distinctive scent or taste. Luster medium. Density rather low to medium; highest in *Fuchsia magellanica* Lam. (Yale 34062, collected by E. L. Bernath in southern Chile); texture fine and uniform; grain straight; presumably poorly resistant to decay.

Growth rings frequently present, but usually indistinct. Pores small to medium-sized (80 to 140μ); fairly numerous; solitary and in short radial multiples and little clusters; well distributed or with some tendency to zonate arrangement in early wood. Vessels with simple perforations; tyloses present in *Fuchsia;* pits vestured, large to very large (9 to 15μ), sometimes elongated. Rays 1 to 3, sometimes to 6, cells wide and up to 20, sometimes to 60, cells high; decidedly heterogeneous, with most of the cells square or tall upright, appearing irregularly flattened-hexagonal on tangential section; large crystals numerous in *Jussiaea latifolia* Benth. (Yale 17288, collected by L. Williams in eastern Peru); pits to vessels all or in part large, elongated, and in more or less scalariform arrangement. Wood parenchyma sparingly paratracheal. Wood fibers septate in *Fuchsia* and *Jussiaea;* pits numerous, small, simple or indistinctly bordered. Ripple marks absent. No gum ducts seen.

COMMON NAMES: *Fuchsia:* Adelaida, aretillo, atexúchil, chorros, flor de arete (Mex.); saca-tinta (Salv.); achiotillo, fusia (C.R.); aljaba, chilco (Arg.); chilco (Chile). *Oenothera:* Evening primrose (U.S.A.).

PALMACEAE

THE palms were referred to by Humboldt as "the princes of the vegetable kingdom," and they may indeed be regarded as one of nature's greatest gifts to man, particularly in the tropics, where they grow most abundantly and luxuriantly. People appreciate them not only for their beauty and shade, but also for the varied and valuable products they yield. The roots, trunks, leaves, flowers, and fruits variously supply food, drink, medicine, or means of providing shelter as well as such technical products as tannins, dyes, sugars, waxes, oils, fats, fiber, timber, ivory substitute, and other materials of more or less importance.

There are 87 genera and nearly 1200 species of American palms listed as presumably valid in Dahlgren's *Index of American palms* (1936). Their geographical distribution extends from 36° north to 38° south latitude. They are most abundant at the equator, as regards both number of species and of individuals, decreasing to the north and south and also in ascending the slopes of the high mountains, though a few species are said to flourish on the elevated paramos in Colombia, Ecuador, Peru, and Bolivia, at altitudes of 14,000 feet, which is well above the timber line. Notwithstanding the fact that the palms as a group are scattered over such a wide range of elevation, they find their optimum development near sea level as in the Amazon Valley. In general they prefer a moist soil and do not appear to suffer in ground that is distinctly saline, although many species grow also in the arid interior and on elevated plateaus of tropical America.

The leaves are crowded at the summit of the stems and some of them surpass in size those of all other forms of vegetation. In shape they vary from the general type of the pinnate frond of the Coconut (*Cocos*) or of the Cohune Palm (*Orbignya cohune* [Mart.] Dahlgren), which is sometimes 40 feet long, to the palmate type of entire leaf not longer than a man's hand. In whatever form or size these leaves develop, they are serviceable throughout tropical America for thatching and siding of native huts, as well as for various other purposes. The wax film on the leaves of *Copernicia cerifera* (Arr. Cam.) Mart. is the basis for the Carnaúba wax industry in Brazil (see "The Carnauba Palm and its wax," *Bull. Pan American Union,* January 1939).

The fruits, popularly known as nuts, vary in size from that of a pea to that of the Coconut, and are generally borne in clusters in the axils of the lower and older leaves or fronds. Some of the fruits are edible in the unprepared state, and most of them contain an oil which can be expressed and used for food. A kind of vegetable ivory, used for buttons and small articles of turnery and carving, is obtained from the seeds of *Phytelephas*.

The fundamental plan of all the well-known palms is strikingly similar, although the general appearance, size, and products of the various species differ markedly. Some attain great size, as for instance, the Royal Palm (*Roystonea*) of Cuba and the Wax Palm (*Ceroxylon*) of the Andes; others are small and delicate as in the case of *Geonoma* spp., in which stems are sometimes entirely wanting above the ground or are less than three feet high. Of those that develop trunks not all are straight and erect, some assuming the form of a vine like the rattans of the Far East; the Jacitára (*Desmoncus*) of Brazil is the best example of this type, and its trunks are scarcely as thick as a lead pencil and sometimes more than 100 feet long. In all palms, however, the outer portion of the trunk consists of fibro-vascular bundles imbedded in a mass of parenchyma tissue. Near the surface these bundles, which are horn-like and strong, are very numerous, but they become progressively fewer toward the center. In some species this outer shell of the trunk is relatively thin, with numerous fine bundles crowded densely together, while others have a thicker but less compact layer, with much larger and intertwining strands. The stems with the densest layers and smallest bundle strands supply material of beautiful figure and susceptible of very high polish and is accordingly in demand for the manufacture of walking sticks, umbrella handles, and fishing rods. The wood of one of these, the Black Palm, *Astrocaryum Standleyanum* Bailey, of Panama, is shown in Plate XXXVII, 2. Thin cross sections of palm stems are occasionally employed in marquetry. Whole stems are sometimes used for piling, posts, and house frames, while split sections of the outer shell of large logs serve for flooring and for constructing small bridges and crude buildings.

PAPAVERACEAE

The Poppy family comprises 42 genera and about 850 species of annual or perennial herbs with colored juice, very rarely shrubs or small trees, widely distributed in the north temperate zone and subtropics. The Poppies are noted for their large and brilliant flowers, and *Papaver somniferum* L. of eastern countries is the source of opium. Representatives of two American genera, *Bocconia* and *Dendromecon*, are arborescent.

Bocconia, with nine species of herbs, shrubs, and little trees, has a wide range in tropical America. The stems are simple or sparingly branched and the sap is yellow or reddish. The alternate simple membranous leaves are often very large; stipules are absent; the flowers are small and borne in large panicles; the fruit is a small few-seeded capsule, dehiscent to the base. *Bocconia arborea* S. Wats. of southern Mexico, Guatemala, and Salvador attains a maximum height of about 30 feet and a trunk diameter of 24 inches. *B. frutescens* L. is more widely distributed, occurring infrequently in the West Indies, Central America, and in South America to Argentina and Uruguay. Both have corky finely ridged bark and a large orange-colored pith that eventually becomes hollow. The principal local uses are as sources of a yellow dye and of alkaloids from the latex for anesthesia and other medicinal purposes.

Heartwood not seen; sapwood greenish yellow, sometimes with pinkish streaks. Luster medium. Without distinctive odor or taste. Of fairly light weight, but firm; texture rather coarse; grain straight; easy to work; probably not durable when exposed to decay. Of no commercial possibilities.

Growth rings poorly defined. Pores medium-sized to small, the largest near limit of vision; not very numerous; solitary and, more often, in small multiples, fairly evenly scattered without pattern. Vessels with simple perforations; no spiral thickenings seen; intervascular pitting coarse, alternate. Rays nearly all coarse, comprising about one-third of the cross section; conspicuous on radial surface; mostly 5 to 8, sometimes up to 12, cells wide and up to 50, occasionally to 150, cells high; decidedly heterogeneous, most of the cells square or upright without definite stratum of procumbent cells; sheath cells present; cells mostly hexagonal as seen on tangential section, or those in the interior much smaller and rounded; walls thin; ray-vessel pitting coarse and irregular, sometimes tending to scalariform. Wood parenchyma sparingly paratracheal. Wood fibers with numerous very small irregularly disposed simple pits. Ripple marks absent. Large open radial channels sometimes present.

COMMON NAMES: Palo amarillo, p. amargo, p. de pan cimarrón, yagrumita (Cuba); celandine, pan cimarrón, parrot weed (P.R.); celandine, John-crow bush, parrot weed (Jam.); bois codine, b. coq (Haiti); yagrumo macho (Dom. R); ahuacachilli, árbol de Judas, calderón, chicalote, c. de árbol, clacojegüite, cococxihuitl, cocojegüite, cuatlataya, cuau-chilli, enguande, engüemba, gordolobo, guachichil, guachilli, inguande, llora-sangre, mala mujer, mano de león, palmillo, palo amarillo, p. del diablo, p. de Judas, sauco, tlacoxihuitl, totolinyzochtl (Mex.); brasil, sangre de toro, tiñecanasta (Salv.); guacamayo, tabaquillo (C.R.); albarracín, azafrán, curarador, mata-chande, sarcillejo, sarno, trompeto, zarcilejo (Col.); celedonia, lechosa, sangre de drago, sanéculo (Venez.); sancho amargo, suncho amargo (Arg., Urug.); árbol Lillo (Arg.); mbiyuícaá (Par.); palo amarillo, yanalí (Peru).

Dendromecon, with about 20 species of shrubs and shrubby trees, occurs in dry regions of California, particularly in the southern part of the state, and extends into Baja California, Mexico. The stems are smooth and branched and the sap is watery and not colored. The leaves are alternate, simple, long and narrow, and without stipules; the rather large yellow flowers are solitary and terminal; the fruit is a slender 2-valved many-seeded capsule. *D. arborea* Greene and *D. rhamnoides* Greene both attain a maximum height of about 20 feet and a trunk diameter of a foot on dry ridges on Santa Catalina Island. The only specimen available (Yale 21697) is of *D. rigida* Benth. collected by Dr. Irma E. Webber in southern California, where it is known as Bush Poppy. It is said to grow as large as the other species, but has a leaning habit and requires support to reach a height of 20 feet.

Heartwood absent; sapwood yellow. Luster medium. Without distinctive odor or taste. Moderately dense, but hard and brittle; texture moderately fine; grain straight; easy to cut; is probably not durable. Has no commercial possibilities.

Growth rings present, owing to more or less ring-porous structure. Pores small to minute, not distinct without lens; numerous; the largest more or less zonate, the others clustered and arranged diagonally or zig-zag; minute pores associated with small ones and resembling parenchyma under lens. Vessels with simple perforations; spiral thickenings present; pitting medium, alternate. Rays nearly all multiseriate, mostly 4 cells wide, and high, sometimes up to 300 cells; decidedly heterogeneous, most of the cells square or upright, without definite stratum of procumbent cells; walls thin; sheath cells common; pits to other ray cells small and exceedingly numerous; pits to vessels medium-sized, rounded. Wood parenchyma very sparse. Wood fibers with rather thick walls and very numerous, small, simple or indistinctly bordered pits. Ripple marks absent. No gum ducts seen.

PHYTOLACCACEAE

The Pokeweed family is composed of about 17 genera and over 100 species of herbs, shrubs, climbers, and a few trees, widely distributed, but occurring for the most part in tropical and subtropical America. The leaves are alternate simple and entire; stipules are absent or minute; the small flowers are in simple or compound terminal or axillary racemes; the fruit is fleshy or dry; the seeds are sometimes arillate. The plants yield no important products and the trees are of no commercial value for their timber. The four best known genera with arborescent representatives are *Gallesia, Phytolacca, Rhabdodendron,* and *Seguieria*. Their woods (with the exception of *Rhabdodendron macrophyllum*) are of anomalous structure, being composed of soft to moderately hard layers of xylem, alternating with thinner and anastomosing laminations of phloem and conjunctive tissue, as in *Avicennia*. Since the phloem layers are rather pithy and perishable, the woody parts tend to separate like the layers of an onion.

Heartwood brownish, grayish, or yellowish, merging gradually into the sapwood. With or without distinctive taste; some species with garlic odor when fresh. Texture coarse to medium; grain irregular; often difficult to split radially; perishable when exposed to decay. Of no commercial possibilities.

Pores rather few and large to numerous and small to minute; without special arrangement, though tending to form radial rows because of the numerous rays. Vessels with simple perforations; without spiral thickenings; pitting coarse or fine (*Rhabdodendron* and *Seguieria*). Rays of two kinds; narrow and low ordinary rays and large and coarse bridges of conjunctive tissue; heterogeneous; ray-vessel pitting very coarse to fine (*Rhabdodendron* and *Seguieria*). Wood parenchyma sparingly paratracheal. Wood fibers with very small simple pits. Ripple marks and gum ducts absent. Raphides have been found in the conjunctive tissue of *Phytolacca dodecandra* L'Hér.

Gallesia. Of the two described species of this genus, the only one well known is *Gallesia integrifolia* (Spreng.) Harms, a common tree of rapid growth in the dense coastal rain forests of the Bahia-Victoria region. It is sometimes 125 feet tall, with a characteristically fluted, contorted, usually hollow trunk having a maximum diameter of 10 feet. It is very tenacious of life and sprouts so freely and persistently that it becomes a weed in the second-growth forest. When the trees are in blossom, and more especially when they are being felled, the whole forest reeks of garlic, hence the local name Pau d'Alho (garlic wood). The timber is not used in the southern part of its range where better kinds are abundant, but to the north where the forests have been depleted it finds a place in rough construction. It is perishable in situations favorable for decay. The wood ashes are rich in potash useful for fertilizer and soap-making. The wood loses most of its odor on drying, but retains an unpleasant taste. The white, spongy, anastomosing layers of conjunctive tissue are conspicuous and at their inner edge are the phloem strands in a peripheral row and resembling intercellular cavities or gum ducts. Sp. gr. (air-dry) 0.58; weight about 36 lbs. per cu. ft. Texture coarse and uneven; grain irregular; easy to cut, but difficult to finish smoothly because of the laminated structure. The other species, *Gallesia ovata*

Schmidt, grows in Peru, but there is little information available concerning it.

COMMON NAMES: Guararema, ibirarema, pau d'alho (Braz.).

Phytolacca is a widely dispersed genus of about 35 species, mostly herbaceous plants, of which the common American Pokeweed is a good example. The only large tree is the Ombú or Umbú, *Phytolacca dioica* L., growing wild in Argentina, Uruguay, Paraguay, and southern Brazil and Peru, and naturalized in many parts of the Old World. It grows rapidly, is not attacked by locusts or ants, will withstand hurricanes and prolonged heat and drouth, and provides excellent shade in regions where other trees will not grow. Regarding its occurrence and importance in Argentina, Walter L. Swindon writes in the *American Weekly of Buenos Aires* (3: 31: 5, Nov. 21, 1925) as follows:

"The country dwellers of Argentina have woven stories, legends, and songs about the Ombú until today it has become an object of traditional respect such as possibly nothing else in the interior inspires. . . . Probably the criollo affection for it is to be found in the fact that it serves as a traveller's guide, by which name it is often known, indicating as it does the situation of some native ranch or more pretentious dwelling where one is certain of receiving unbounded hospitality. . . . The Ombú has been planted, by unknown hands, in almost every part of Argentina, though it is found more generally in the Province of Corrientes and along the banks of the rivers, principally the Paraná, from Buenos Aires to Misiones. . . . The Ombú is unquestionably a magnificent tree, if tree it can be termed. Its peculiar characteristic is the extraordinary growth of the base of its trunk which divides into numerous outcrops of irregular shape having the appearance of bulky roots of exaggerated proportions, sometimes reaching as much as thirty meters in diameter, but the branches extend for 15 to 18 meters, above and around, and, being covered with a thick green foliage, thus protect its base from the sun's rays, affording cool, ample, and refreshing shade on the hottest day where no other sheltering tree will be found within hundreds of miles. Other characteristics of this remarkable tree are its spongy soft wood which is formed by numerous layers of loose fibrous tissues hardly touching each other. Over 400 separate layers have been counted in a trunk measuring one and a quarter meters in diameter. Sometimes these inside layers dry up and grow outwards, leaving large holes in the interior of the tree. The trunk and branches of the Ombú contain up to 80 per cent of water. On account of its structure and lack of solidity this plant may be considered as a gigantic herb instead of a tree. Except for its shade and picturesqueness the Ombú really possesses very few virtues. Its wood will not burn, but an infusion of its leaves acts as a purgative, and its fruit contains a soapy juice composed of salts of lime and potash."

COMMON NAMES: Bella sombra, belombra, ombú, umbú (Arg.).

Rhabdodendron, with two species of unarmed shrubs or small trees, is limited to the Amazon region of Brazil. The alternate leaves are large entire and glandular-dotted; the flowers have a concave receptacle, with almost obliterated calyx, five white petals, about 45 stamens, and a free, 1-celled ovary with lateral, basally inserted style, and are borne abundantly in axillary panicled racemes; the fruit is a small drupe with thin leathery exocarp and thin woody endocarp.

Rhabdodendron macrophyllum (Spruce) Huber has wood of normal structure. Color pale brownish throughout specimen (Yale 23639; Ducke 177). Luster medium. Odor and taste not distinctive. Hard, heavy, tough, and strong; of about the consistency of Sugar Maple (*Acer saccharum* Marsh.); working properties good; presumably poorly resistant to decay. Of no commercial possibilities.

Rhabdodendron amazonicum (Benth.) Huber exhibits considerable variation in the size, shape, and texture of the leaves and in the nature of the inflorescence. The wood resembles that of *Gallesia* but is odorless. It is conspicuously unlike that of *Rhabdodendron macrophyllum,* being of anomalous structure of the concentric type, that is, having successive bundles of xylem and phloem, repeating the structure of the young stem, separated by tangential bands of conjunctive parenchyma and interfascicular rays (see *Tropical Woods* 50: 22).

COMMON NAMES: Cachaceiro, muiracaua (Braz.).

Seguieria, with about 30 closely related species, mostly scandent shrubs with curved thorns, is distributed throughout most of tropical and subtropical America. *S. paraguayensis* is said to be a tree occasionally over 50 feet tall and 20 inches in diameter. *S. americana* L. is common in northern Colombia, where it is sometimes a tree 12 to 15 feet high with a trunk four to six inches in diameter. The fruits of this genus and *Gallesia* are winged and resemble those of Maple (*Acer*) except that they are not in pairs. The stem has a thin striated bark and a rather large solid yellow pith. The yellowish and slightly odorous wood has about the consistency of Maple (*Acer*).

COMMON NAMES: Barbera, barbero (Col.); cipo de alho (Braz.); espina del infierno, mboy-raí, nuatí-hú, verdaza, yoá-hu-y, yoá-ví (Arg.).

PICRODENDRACEAE

THIS family was proposed by John K. Small (*Jour. N.Y. Bot. Gard.* 18: 212; 180–186; August 1917) for the West Indian genus *Picrodendron,* which has been variously referred to Juglandaceae, Anacardiaceae, Sapindaceae, and Simarubaceae. There are three closely related and doubtfully distinct species of small unarmed deciduous trees. The leaves are alternate and digitately trifoliolate; the flowers are small, apetalous, and dioecious, the female stalked, solitary, and axillary, the male clustered or in spikes or aments; the fruit is a rounded orange-colored stalked drupe about an inch long when ripe, the thin fleshy exocarp containing a very bitter juice, the endocarp woody and brittle and usually containing a single seed with corrugated endosperm.

The type species is *Picrodendron baccatum* (L.) Krug & Urb., commonly known as the Jamaica Walnut. According to Small (*loc. cit.*), the first reference to it is in Han Sloane's catalogue of Jamaican plants published in 1695, where it is noted as a trifoliolate Walnut, with fruit the size of a nutmeg, growing along the Río Cobre and abundant in the savanna about St. Jago de la Vega. Subsequently he describes the tree more fully and states that it grows to a height of 20 feet, with a gray trunk "as thick as one's thigh." Fawcett & Rendle (*Flora of Jamaica* 4: 273) give the height as 25 to 40 feet. No uses for the plant are reported.

Small says (*loc. cit.*, p. 184): "A second species of *Picrodendron* grows at low elevations near the coasts in Cuba and the Bahamas, differing from the Jamaican tree in having blunt leaflets and ovoid or oval and smaller fruits. It was first described by A. Richard in 1845, and figured at the same time. The specimen studied by Richard was collected in the Vuelta de Abajo (Pinar del Río), Cuba. He referred the tree to the genus *Schmidelia* [*S. macrocarpa*] of the Soapberry family (Sapindaceae), but it is not of that relationship. In 1893, Krug and Urban described the Bahamian tree as a variety of the Jamaican species, *Picrodendron baccatum,* var. *bahamense,* but erroneously regarded the Cuban species of Richard as the same as the tree of Jamaica. Dr. Britton in 1906 first corrected the Cuban and Bahamian trees, as a species distinct from the Jamaica Walnut, under the name *Picrodendron macrocarpum* (A. Richard) Britton." Britton and Millspaugh state (*The Bahama Flora,* p. 103), that it is "a tree up to 20 m. high, or sometimes a shrub, in rocky coppices on seven of the islands of the archipelago; erroneously called Olive." In Cuba, according to Roig (*Diccionario Botanico,* p. 705), the species grows in deltas and coastal swamps, mostly at the rear of the Mangrove formations, and attains a height of about 20 feet. The very dense, black wood is used in naval construction and in turnery.

Picrodendron also occurs in the Island of Haiti. Small says: "The characters examined by those who determined the specimens led them to associate the plant with the Cuban species. An examination of the specimens in hand indicates that this Santo Domingan tree is different from the plants of the neighboring islands and may be named . . . *Picrodendron medium* Small, sp. nov. . . . This tree is interesting in that its leaflets resemble those of the Jamaican plant in outline and venation, whereas the fruits have the thick-walled stone (endocarp) of the Cuban plant. It is known to the natives of Santo Domingo as Manzanillo and, like its relatives, inhabits very dry regions."

The wood samples available for study are from Cuba (3) and Haiti (1). They are similar in structure and do not indicate close affinity to any of the four families to which the genus has been referred. Heartwood dark olive or olive-brown to nearly black, with a waxy appearance; distinct but not sharply demarcated from the yellowish sapwood. Luster medium. Without distinctive odor, but with bitter taste, as in Simarubaceae. Very hard, heavy, tough, and strong; texture medium; grain fairly straight; rather difficult to saw, but easy to cut with a knife, finishing smoothly and taking a high natural polish; presumably very resistant to decay. A good wood, but without commercial importance because of its small size and scarcity.

Growth rings poorly defined. Pores medium-sized to minute, the larger ones barely visible to indistinct without lens; rather numerous; occurring singly and in multiples of few to several pores each, rather well distributed. Vessels with simple perforations; without spiral thickenings; gum deposits abundant; white substance sometimes present; pitting alternate. Rays 1 or 2, occasionally 3, cells wide and up to 30 cells high; decidedly heterogeneous; gum deposits abundant in heartwood; pits to vessels medium-sized. Wood parenchyma abundant, but barely visible without lens; very irregularly arranged; narrowly vasicentric, short aliform, irregularly confluent, and in rather widely spaced narrow concentric bands, sometimes apparently terminal; also diffuse; crystalliferous strands common. Wood fibers with very thick walls and small simple or indistinctly bordered pits. Ripple marks absent. No gum ducts seen.

COMMON NAMES: Jamaica walnut (Jam.); blackwood (Bah.); aceituna, guayo, mangle negro, llana, llanilla, roblecillo, yana prieta, yanilla, y. prieta (Cuba); manzanillo (Dom. R.); gris-gris noir, simarouba (Haiti).

PIPERACEAE

THE Pepper family comprises several genera and hundreds of species of herbs and shrubs, erect or scandent, and a few small trees, mainly tropical. The only genus considered here is *Piper*.

Piper. A great many species have been described but most of them are separated by very small differences. The leaves are alternate entire and stipulate; the small greenish flowers are sessile in very dense spikes or sometimes in racemes; the fruit is a small berry. The plants are more or less aromatic and are the source of medicines and condiments. The leaves of the Betel Pepper, *Piper betle* L., are chewed by the natives of the Pacific Islands. The principal commercial products are black pepper from *P. nigrum* L. and cubeb berries from *P. cubeba* L., both East Indian plants. The family does not supply useful timber.

Wood pale brownish throughout; marked with conspicuous darker-colored rays on radial surface. Luster medium to rather high. Without distinctive odor or taste. Of medium density, but firm and tough; texture coarse; grain generally straight; inclined to split badly in drying; easy to work; probably perishable when exposed to insects and decay. Of no commercial possibilities.

Growth rings absent or poorly defined. Pores medium-sized to minute; few to fairly numerous; solitary and in small multiples; rarely in contact with the rays (Plate LI, 1). Vessels predominately with simple perforations, but with some tendency to form scalariform plates; without spiral thickenings; pitting very fine, alternate. Rays variable in width, but all wide to very wide and exceptionally high, extending the full length of the internodes (Plate LI, 2); heterogeneous, most of the cells square or upright; large oil cells sometimes present; pits to vessels uncommon, very small. Wood parenchyma sparingly paratracheal; pits to vessels small, often much elongated and in scalariform arrangement. Wood fibers storied in some species; pits numerous, simple or indistinctly bordered. Radial intercellular canals, visible without lens, present in the nodes.

Common names: Black jointer, joint wood, jointy benna, pepper elder (B. W.I.); cordoncillo (Sp. Am., gen.); bayuyo, caisimón, guayuyo, pimiento de costa, platanillo, p. de Cuba, p. de monte (Cuba); baquiña, basquiña, higüillo, h. de limón, h. oloroso (P.R.); anisillo, guayuyo, g. blanco (Dom. R.); bois major, collet de Notre Dame, feuilles sirop, noyau pays, sureau, s. plantain (Haiti); achiotlín, achotlín, acoyo, acuyo, cocolmeca, hachogue, hierba santa, hoja de aján, h. de anís, h. santa, maculan, mano de zopilote, mático, momo, pie de guicharo, rabo de zorra, Santa María, santillo de comer, s. de culebra, s. montés, soldadillo, tianipaquelite, tlanepaquelitl, tripa de zopilote, xmacolan, xmakulan, yaaxpehelché, yaxtehc-ché (Mex.); Spanish elder (Br. H.); hoja de jute, juniapre (Guat.); matarro, ombligo (Hond.); candelillo, chile, resfriado, Santa María (Salv.); alcotán, anisillo, caña de muela, estrella, hinojillo, monca blanca, Santa María (C.R.); cañotillo, gusanillo, hinojo (Pan.); cordoncillo de tierra fría, c. gigante, c. verda (Col.); anisillo, cordoncillo negro (Venez.); matico, mocco-mocco, sumpi (Peru); aguaxima, betre aromatica, caá-peba, caá-peua, canella de jacamin, catajé, corimbó-uassú, herva de soldado, jamborandi, jambú-rana, jambuy, malvaisco, malvarisco, matico, nhandi, nhandú, pimiento do matto (Braz.).

PLATANACEAE

Platanus, the only genus, includes about 11 species of large trees, three of them in southern Europe and Asia, eight in temperate North America. The bark exfoliates from the branches and young stems in large irregular plates. The large deciduous leaves are alternate, simple, palmately nerved and lobed, and the dilated petiole incloses the young bud; the stipules are laterally united into a short tube surrounding the twig and leaving a narrow ringlike scar in falling; the fruits are small achenes surrounded at the base by hairs and borne in dense heads.

The best known trees are the Oriental Plane, *Platanus acerifolia* Willd. or a hybrid of *P. orientalis* L., commonly planted for shade in parks and along streets, and the Sycamore, *P. occidentalis* L., the largest hardwood tree in the United States. The latter species is common along streams and lakes throughout most of the eastern half of North America, being at its best in the lower Ohio and Mississippi valleys where it occasionally attains a height of 170 feet and a trunk diameter of 14 feet. The lumber is used for boxes and crates, slack cooperage, furniture and fixtures, plywood, butchers' blocks, trunk slats, brush backs, and woodenware. Quartersawed material is attractively figured and is often known in the furniture trade as Lacewood. Two species, *P. Wrightii* S. Wats. and *P. racemosa* Nutt., occur in southwestern United States and northern Mexico; five others are found only in Mexico, and one, *P. chiapensis* Standl., extends from Chiapas into northern Quiché, Guatemala. Along water courses they attain a height of 125 feet and a trunk diameter of four to six feet, and have large spreading crowns. They are more useful for shade than for lumber and the Mexican timber is of minor commercial importance, though used locally for general carpentry, wooden dishes and spoons, and fuel. The woods of all species are much alike in structure and properties.

Heartwood light brown to pinkish, with prominent figure on radial surface from rays and ribbon grain; often merging gradually into the nearly white sapwood. Odorless and tasteless. Not very heavy, but firm, tough, and strong; sp. gr. (air-dry) about 0.50; weight 32 lbs. per cu. ft.; texture medium; grain usually irregular; fairly easy to work, finishing smoothly, but likely to warp unless carefully seasoned; perishable in contact with the ground.

Growth rings present. Pores small, not visible without lens; very numerous, crowded; well distributed without pattern. Vessels with both simple and multiple perforations, the scalariform plates of the latter having few to many narrow bars; spiral thickenings absent; intervascular pitting alternate to scalariform. Rays occasionally low and uniseriate, but mostly 4 to 8 cells wide and up to 100 (sometimes 300) cells high; homogeneous to somewhat heterogeneous; pits to vessels oval, often elongated and parallel. Wood parenchyma sparingly developed, mostly in short metatracheal lines and diffuse; not distinct with lens. Wood fibers with moderately thick walls and small distinctly bordered pits. Ripple marks and gum ducts absent.

COMMON NAMES: Buttonwood, lacewood, plane tree, sycamore (U.S.A.); álamo, a. blanco, aya (Mex.).

POLYGALACEAE

THE Milkwort family comprises about 11 genera and 1000 species of herbs, shrubs, climbers, or rarely trees, extensively distributed in tropical and temperate regions. The simple leaves are alternate, rarely opposite or whorled, and without stipules; the flowers are often brightly colored and have the general shape of a pea blossom (Leguminosae-Papilionaceae); the fruit is a capsule, drupe, or samara; the seeds are usually pubescent and arillate. Of the seven genera with distinctly woody plants in tropical America, the only real tree species are *Phlebotaenia Cowellii* Britton, of rare occurrence in Puerto Rico where it grows to a height of 50 or 60 feet and a diameter of 10 inches, and *Polygala scleroxylon* Ducke which attains about the same dimensions in the Amazon region of Brazil (see *Tropical Woods* 50: 35).

The climbers (*e.g.*, *Bredemeyera*, *Moutabea*, and *Securidaca*) have stems with anomalous wood structure (Plate LVIII, 1), that is, with successive layers of xylem and phloem separated by concentric or anastomosing bands of conjunctive parenchyma (see *Tropical Woods* 50: 24). Pores sometimes very small and scattered, more often large in part and numerous; rounded in outline; rarely in contact radially; distribution uniform to very irregular. Vascular pitting medium to rather coarse; perforations simple. Normal rays 1 or 2, occasionally 3, cells wide and few to several cells high; coarser rays (bridges of conjunctive tissue) sometimes present; pits to vessels oval to elongated. Wood parenchyma sparingly vasicentric to finely metatracheal. Wood fibers with rather thin to thick walls; pits bordered. Ripple marks absent. No gum ducts seen.

The woods of the trees and upright shrubs have normal structure. In *Polygala scleroxylon* the heartwood is canary yellow and rather sharply demarcated from the white sapwood. Heartwood not seen in the others; sapwood yellowish. Luster medium. Odorless and tasteless. Density medium to high; texture fine and uniform; grain straight to variable; working properties good. The wood of *P. scleroxylon,* if available in sufficient quantity, might serve some of the purposes of Boxwood.

Growth rings present or absent. Pores small to minute; usually rounded, isolated, and scattered, but angular and in short radial rows in *Monnina*. Vessels without spirals; perforations simple; intervascular pitting rather fine, alternate in *Monnina,* not observed in others. Rays mostly 1 or 2 cells wide, up to 5 cells in *Monnina,* and few to many cells high; decidedly heterogeneous; pits to vessels small. Wood parenchyma sparingly paratracheal in *Badeira* and *Polygala;* irregularly diffuse and sometimes terminal in *Monnina;* abundantly paratracheal and often aliform and confluent in *Phlebotaenia*. Wood fibers with medium to thick and gelatinous walls; pits very numerous, conspicuous, bordered. Ripple marks absent; no gum ducts seen.

COMMON NAMES: *Monnina:* Corneta, palo blanco (Venez.). *Moutabea:* Graine macaque (Fr. G.); aymontabou, caimito do monte, gogo de guariba, grãos de macaco (Braz.). *Phlebotaenia:* Cocuyo blanco (Cuba); carocolillo, palo de tortuga, violet tree (P.R.). *Securidaca* (*Elsota*): Enrededara de hacha, flor de la cruz, maravedi (Cuba); bejuco de sopla, b. zalzilla, jaboncillo (P.R.); mabelie (Haiti); flor de arrayán (Mex.); bejuco de purgación, corralmeca, tamagás (Salv.); penda morada (Col.); bejuco mulato, cascarón majomo, requena (Venez.); motelo-huasca (Peru); cumandá-y (Braz.).

POLYGONACEAE

The Buckwheat family includes about 40 genera and possibly 1000 species, mostly herbaceous plants (*e.g.*, sorrel, buckwheat, dock, rhubarb, and smartweed) in the temperate zone, but with some shrubs, woody climbers, and small to medium-sized trees in tropical regions. Trees are found in seven genera, namely, *Coccoloba* (*Coccolobis*), *Gymnopodium* (*Millspaughia*), *Neomillspaughia, Podopterus, Ruprechtia, Symmeria,* and *Triplaris,* all of which are confined to tropical America, with the exception of *Symmeria* whose single species occurs also in West Africa. The leaves are simple, alternate or sometimes opposite or whorled, and usually with sheathing united stipules (ocreae) which in falling leave a ring-like scar; the flowers are usually small, without petals, the inflorescence often racemose or spicate; the fruit is a 3-sided achene, usually surrounded by the persistent calyx. The stems are solid except in *Triplaris.* The timbers are of good technical quality, but because of the small sizes available are used only locally and then mostly for fuel.

Heartwood, so far as known, pinkish or reddish, occasionally dark reddish brown; sapwood pinkish or yellowish, sometimes sharply demarcated. Luster ranging from low to silky. Without distinctive odor or taste. Density variable from about 0.65 (*Triplaris*) to greater than 1.00; texture uniform, medium to fine; grain straight to irregular; easy to fairly difficult to work, finishing smoothly; dark heartwood durable.

Growth rings present or absent; never conspicuous. Pores mostly small to very small, sometimes medium-sized; rounded; rather few to numerous, but not crowded laterally; occurring singly and, more often, in pairs or short to rather long radial rows, and sometimes more or less clustered. Vessels with simple perforations; no spiral thickenings seen; pits very small to rather large, alternate, vestured; vessel members sometimes diagonally arranged. Rays all multiseriate and high in *Symmeria;* otherwise uniseriate or biseriate, mostly low but sometimes up to 30, occasionally to 50, cells high; usually homogeneous, sometimes weakly to decidedly heterogeneous; pits to vessels very small (*Gymnopodium, Podopterus,* and *Symmeria*), rather small and rounded, or fairly large and irregular (*Ruprechtia*). Wood parenchyma sparingly paratracheal; also diffuse in *Coccoloba* and *Neomillspaughia* with all strands apparently composed of short crystalliferous cells which may be square, vertically flattened, or rounded, and vary in size from about that of the wood fibers to large enough to be seen readily with a hand lens and appearing under the compound microscope as small, angular pores. Wood fibers with thin to thick walls and numerous small simple pits; commonly septate, and in *Podopterus, Ruprechtia,* and *Triplaris* many of them are finely chambered and contain starch in sapwood and small calcium oxalate crystals in heartwood; these chambered fibers may occur in irregular concentric layers appearing under lens as indistinct bands of parenchyma. Ripple marks absent. No gum ducts seen. Gum deposits abundant in all cells of the dark-colored heartwood.

Coccoloba. There are about 180 species in the West Indies, southern Florida, Mexico, Central America, and in South America to Paraguay. Most of them are erect or climbing shrubs and small trees, but a few attain rather large size on favorable sites. The leaves are sometimes very large; the flowers are fascicled within small bracts, the fascicles spicate; the perianth is often very fleshy in fruit.

The Pigeon Plum, *Coccoloba laurifolia* Jacq., is one of the largest and most abundant of the tropical trees of the seacoast of southern Florida where it attains a height of 60 to 70 feet, with a diameter up to two feet. It is common also in the West Indies and along the northern coast of South America. More widely distributed and better known is the Sea Grape, *C. uvifera* L., which occurs on sandy or rocky sea beaches of Florida, the West Indies, northern and eastern South America, both coasts of Central America, and down to Peru. In Cuba and Jamaica, where it is at its best, it also grows in the moist woods of the interior at elevations of 500 feet or more. At the limits of its range and in poor situations it is only a shrub or a low densely branched tree, but in better sites it is from 25 to 60 feet high and 18 to 24 inches in diameter. It has large rounded leaves that are very thick and leathery, and bears clusters of purple grape-like fruits that are good to eat, although the stone is large in proportion to the flesh. The bark yields, upon excision, an astringent red juice which is the source of the West Indian kino. This product, known also as gum kino, American kino, American extract of rhatany, and false rhatany extract, was formerly an article of trade, but the kino of commerce is now supplied by West Africa and parts of the East Indies.

The heartwood of *Coccoloba Tuerckheimii* D. Sm., a small Central American tree, is of a beautiful rich red color when freshly cut, but becomes chocolate-brown upon exposure. *C. armata* Gris. of Cuba has a very hard fine-textured red wood which differs from other species of *Coccoloba* examined (except possibly *C. caracasana* Meissn.) in that the crystalliferous parenchyma strands are replaced by chambered wood fibers as in *Ruprechtia*. *C. mollis* Casar. of Surinam and Brazil has a straight trunk supported by convex stilt-roots sometimes three feet high. *C. latifolia* Lam. is used in Brazil for making barrel hoops. Some of the larger trees, especially in the West Indies, supply a limited amount of wood for cabinet work, furniture, and durable construction, but there is no likelihood of the timber from any source entering the export trade. Owing to similarity of name, *Coccoloba* is occasionally confused with Cocobolo (*Dalbergia retusa* Hemsl.), a commercial timber from the Pacific coast of Central America.

Common names: Black grape, bow pigeon, crabwood, hopwood, horsewood, mangrove grape, pigeon plum, p. wood, red light, sea grape, seaside grape, s. plum, stavewood, tie-tongue, wild grape (Fla., B.W.I.); calambreña, cucubano, gateado, glateado, moralón, ortegón, puckhout, uva, u. del mar, uvero, uvilla (P.R.); chicharrón de sabana, hicaquillo, manati, moco de guanajo, uva caleta, uverillo, uvero, u. macho, uvilla, yarúa (Cuba); carga agua, guarape, saona, uva, uvilla (Dom. R.); gamelle, raisin la mer, r. marron (Haiti); bois baguette, b. rouge montagne, raisinier à grappes, Saint Jean (Fr. W.I.); cuchape, uvero de monte (Trin.); bob, bochiche, boob, carnero, c. de la costa, Juan Perez, manzano, niiche, palo de carnero, quiabaro, roble de la costa, tamulero, tepecohuite, tocó prieto, uva, u. de la mar, u. de la playa, uvero, xbobche (Mex.); bobche, grenada, iril, niiche, uva, wild grape (Br. H.); irayol de montaña (Guat.); cordoncillo, papaturro, rabo de león, tapatamal, tolondrón, uva, u. de la playa (Hond.); irire, juril, papalón, paparrón, papaturro (Salv.); carro caliente, papaturro, uva de la playa (C.R.); hueso, papaturro blanco, uva de la playa, uvero (Pan.); ají, cabeza de león, calenturo, cardo santo, cimarrón, corallero, guara, hueso de negro, Juan garrote, J. g. preto, maíz cocido, murta, palo bagre, pastelillo, simarrón, tacaloa, uvero, u. macho (Col.); arahueke, camare, cumare blanco, dreifi, kamalia, mangle de falda, quisanda, uvero, u. de la playa, u. macho (Venez.); baauwlifi, bradiliefi, brailifi, bredlief, droifi, druif, matoela, mattora, mierenhoedoe, venbatapo, patoela, pattoera, zeedruif (Sur.); apixuna, arco de pipa, cabaçú, cáuassú, guajabara, pichuna or pixuna, tinteira (Braz.); añalque, coccoloba, cunchu-caspi, eseri-ey, liquanco, nemoño-o, palo meta-caspi, purguá, quisanda, tangarana, t. mashau (Peru); duraznillo morado, ibararó-morotí, viraró blanco (Arg.).

Gymnopodium, with three species, appears to be limited to southern Mexico, British Honduras, and Guatemala. The largest tree reported is 40 feet high with a trunk eight inches in diameter. Usually the plants are shrubby, with many stiff branches; the fruits are small and winged. The bark is shreddy, suggesting a grapevine (*Vitis*). The rather hard and heavy, fine-textured, pinkish brown wood is not utilized.

Common names: Tzitzilché, zactcitsilché (Mex.); bastard logwood, cruceto (Br. H.).

Neomillspaughia includes a small group of shrubs and little trees occurring in Mexico and Central America. A specimen of Amarro Jabón, *N. paniculata* (Donn.

Smith) Blake, collected by the senior author near the Aguán River, Maloa District, Honduras, had very large thin slightly velvety leaves, deeply notched at both ends; the small dry 3-winged fruits are in large terminal panicles suggesting certain species of dock (*Rumex*). The moderately hard and heavy, strong, fine-textured wood is not utilized.

COMMON NAMES: Tsaitsa, xtzacitza, zacitsa (Mex.); amarro jabón (Hond.).

Podopterus, with three species of deciduous shrubs and small trees, occurs in southern Mexico and Guatemala. The best known species is *P. mexicanus* H. & B., a spinose plant sometimes 20 feet high, common in Yucatán where it is called Putsmucuy, a Maya name alluding to the sharp spine extending beyond the clusters of greenish brown flowers. According to Standley (*Flora of Yucatan,* p. 254), "the flowers yield a large amount of clear transparent honey of excellent flavor." The wood is not utilized. The following description is based upon a small stem of *Podopterus mexicanus* from Mexico (G. F. Gaumer 23206). Heartwood absent; sapwood pale brownish. Fairly lustrous. Odorless and tasteless. Of about the consistency of White Birch (*Betula papyrifera* Marsh.); texture fine and uniform; grain straight; works easily, finishing very smoothly; durability of heartwood unknown. Suitable for small articles of turnery.

Ruprechtia includes about 30 species of trees and shrubs occurring in southern Mexico, Central America, and northern and subtropical South America. Apparently the trees are of largest size in Argentina, where heights of 60 to 75 feet and trunk diameters up to 24 inches are reported, but the species there bear some of the same local names as the leguminous *Pterogyne nitens* Tul., the source of an attractive cabinet wood, and some of the published statements about *Ruprechtia* probably apply to the other timber. The wood is pinkish or light reddish brown, sometimes streaked, and has about the consistency of Birch (*Betula*), though occasionally rather soft. It is not resistant to decay, but is suitable for interior construction and carpentry, the better grades for plain furniture and flooring.

COMMON NAMES: Chachalaco, palo colorado, p. fierro, sangre de toro (Mex.); carreto, sangre de toro (Guat.); bolaó, guayabo volador, volador (Col.); cabritón, mazamorra, ojito, ubito (Venez.); maicharo-ey, tangarana, t. blanca (Peru); guayuvirá, virarú (Braz.); duraznillo, d. blanco, higuerita, ibirá hembra, i. pita-bi, i. pi-hú, i. pitá-y, i. puitá, i. puitá-y-rá, i. pyitá, ibiraró-mi, ivaró, lata, manzano del campo, manzanillo, marmelero, mata negra, palo borroso, p. de lanza, p. estaco, rama negra, sacha manzana, sarandí negro, viraró, virarú, v. colorado (Arg.); viraró (Urug.).

Symmeria *paniculata* Benth., the only species, is a shrubby tree known to occur in Senegal and Sierra Leone, West Africa, and in Colombia, the Guianas, and the lower Amazon region of Brazil. It has solid twigs and is without true ocreae, though the base of the leaf encircles the stem and leaves the characteristic ring scar; the fruits are in panicles and the enlarged calyx lobes give them the appearance of Beechnuts (seeds of *Fagus*). According to Curran (*Tropical Woods* 19: 29), it is a low-growing much-branched tree, sometimes 20 feet high and eight inches in trunk diameter, growing in almost pure stands at the edge of the forest along fresh-water streams near the confluence of the Magdalena and Cauca Rivers in northern Colombia. It is there known as Mangle, the name more commonly applied to the trees of the Mangrove formations fringing salt and brackish water. The only local use for the timber is for fuel. In Brazil the tree is found in marshy places along the margins of fresh-water streams and lakes, and provides a limited amount of material for interior construction. The dark reddish brown wood, which is lustrous in proper light and bears a superficial resemblance to *Tovomita* (Guttiferae), is hard, heavy and strong, of medium texture, and irregular grain. It has no commercial possibilities.

COMMON NAMES: Mangle (Col.); acaráuassú, mangue-rana (Braz.).

Triplaris. The dozen or more species of this genus are shrubs or slender trees of fairly common occurrence in the lowlands of southern Mexico, Central America, and most of South America, sometimes forming nearly pure stands on abandoned clearings. Most of them have hollow stems which are inhabited by small but vicious ants which emerge quickly and rain down upon anyone molesting the tree. When in flower and fruit the trees form masses of brilliant color that render them highly conspicuous. The nutlet-like fruits fall with the calyx at maturity and gyrate through the air like shuttlecocks. The wood is pinkish and lustrous and considerably softer than that of any other members of the family; sp. gr. 0.50 to 0.62; weight 32 to 39 lbs. per cu. ft. It is of medium texture, easy to work, finishes smoothly, is not resistant to decay. Though not much used because of the small size of the trees and the annoyance of cutting them, the timber is suitable for many of the same purposes as Pine (*Pinus*), namely, interior construction, joinery, and boxes.

Common names: Palo mulato (Mex.); canilla de mula, gallito, mulato, palo mulato (Salv.); hormigo, hormiguero, tobaco (C.R.); guayabo hormiguero, palo hormiguero, vara santa (Pan.); palo santo, uvero, vara santa, v. s. overa (Col.); barba de mono, barrabás, chupón, María barrabal, palo María (Venez.); kada-burichi, long John, sapahaki-apolli (Br. G.); don-hoedoe, dreitin, drytimehout, jakoma, jekoena, mierahoedoe, mierenboom, mierenhout, mirahoedoe, mirahoudoe, tasie, tassi (Sur.); bois fourmi (Fr. G.); fernansánchez (Ec.); mishu-quiro, palo santo, tangarana, t. blanca (Peru); formigueira, tachy, t. preto da varzea (Braz.).

PROTEACEAE

In this family are about 50 genera and 1000 species of trees and shrubs, mostly South African and Australasian, a few Latin American. The leaves are typically alternate and simple or variously divided, sometimes dimorphous; stipules are absent; the flowers have a colored corolla-like calyx and are borne in racemes or heads; the fruit is various—nut, drupe, follicle, or capsule; the seeds often are winged. Many of the plants are highly ornamental, but the economic products are limited to some edible nuts and a few fancy timbers. The best known tree is the Australian Silky Oak, *Grevillea robusta* A. Cunn., which has been planted in tropical and subtropical regions the world over as it will thrive in localities too dry for most trees; it is distinguished by fern-like leaves, silky pubescent beneath. Other species of *Grevillea* produce denser and more deeply colored woods, one of which is known as Beefwood. *Macadamia ternifolia* F. Muell. is cultivated in parts of tropical America on account of its edible nuts which are known as "nuez de Queensland." Two interesting South African species are the Silvertree, *Leucodendron argentum* R. Br., valued for decorative purposes, and the Zuckerbos (sugar bush), *Protea mellifera* Thunb., whose flowers contain a large quantity of watery nectar which the natives collect and boil down to make a cough syrup. The New World species are of seven genera, and the following description is based upon American specimens of five of them.

Heartwood variable in color from grayish or brownish to dark purplish red, usually attractively figured by the rays; sapwood brownish, grayish, or oatmeal-colored. Luster often silky. Without distinctive odor or taste when dry. Consistency ranges from that of Basswood (*Tilia*) through Beech (*Betula*) to the densest Oak (*Quercus*). Working qualities good, though careful seasoning is required to prevent splitting; deeply colored material resistant to decay.

Growth rings present or absent. Pores variable in size from readily visible to minute, in number from few to abundant; typically associated with parenchyma in festoons between broad rays (Plate XXXVIII); in *Roupala*, appearing suspended from each parenchyma band; in *Panopsis*, rather few, irregularly distributed, appearing nearly independent of the narrow closely spaced parenchyma lines; in *Embothrium*, mostly in a continuous tangential row in each parenchyma band; in *Guevina* and *Lomatia*, minute, very numerous, and composing bands or festoons of variable widths and spacing, the parenchyma obscure. Vessels nearly always with simple perforations, the smaller having wide rims; scalariform plates sometimes present in small vessels; spiral thickenings in *Guevina,* fine striations frequent in the others; gum plugs common; pitting fine to very fine, alternate. Rays of two sizes; the uniseriates few to rather numerous, very low, and composed of square or upright, usually irregularly shaped, cells; the multiseriates conspicuous and homogeneous or nearly so; sclerotic cells, with or without crystals, sometimes present; ray-vessel pitting fine to very fine, half-bordered. Wood parenchyma generally abundantly developed in numerous rather narrow scalloped bands in association with the pores or metatracheal in part; convex side of scallops toward center of tree. Wood fibers with small bordered pits, the apertures slit-like and extended. Ripple marks absent. No gum ducts seen in American woods (vertical traumatic ducts sometimes present in *Banksia, Cardwellia,* and *Grevillea*).

Embothrium, with several species of shrubs and little trees, occurs in the Andean region of South America from Ecuador to southern Chile, and also in Australia. The branches are long and Willow-like; the leaves are leathery and entire; the showy reddish or roseate flowers are in terminal racemes; the fruit is a woody capsule containing winged seeds. The timber is of good quality and of attractive appearance, but is of minor importance because of the scarcity and small size of the trees. It is sometimes employed locally for furniture and articles of turnery.

Wood light grayish brown, very strongly marked with prominent brown rays which give a speckled appearance to the tangential surface and show as large flakes on the radial. Luster high. Odorless and tasteless. Moderately heavy and hard; texture medium; grain straight; working properties good; durability probably low.

Common names: Catas, chappa, cocániro, llamas, machinparrani, mastimpanrani, picahua, salta-perico, tsacpá, zacpá (Peru); ciruelillo, notro (Chile); laurel (Tierra del Fuego).

Guevina *avellana* Molino, the only species, is a small to medium-sized tree of fairly common occurrence in the provinces of Cautín, Llanquihue, and Chilöe, Chile. The leaves are unequally pinnate, with serrate leaflets; the white flowers are borne in long axillary racemes. The tree is called Avellano because the coral-red edible fruits resemble hazelnuts (avellanas). The tree occasionally attains a height of 65 feet and a diameter of 30 inches, but ordinarily the mature sizes are about half those given. The timber is well known in the local markets and is used for furniture, picture frames, articles of turnery, and, in rural districts, for shingles.

Heartwood pale brown with a pinkish hue; merging gradually into the oatmeal-colored sapwood; prominently marked by the brown moderately high rays and exhibiting an Elm-like figure on tangential surface, owing to the lower luster of the bands of pores and parenchyma. Luster high. Scentless and tasteless. Rather light in weight, but firm and strong; texture medium to rather coarse; grain slightly irregular; very easily worked, finishing attractively; not very durable in contact with the ground.

Lomatia. There are several species in Australia and Tasmania and three or four in the southern Andes Mountains. The Radal, *L. hirsuta* (Lam.) Diels, occurs in Chile, Patagonia, and southern Peru, being at its best in the Chilean provinces of Cautín and Valdivia. The trees are low-branching, the short trunks 16 to 30 inches in diameter. Rural settlers use the dark brown sap of the bark to dye their ponchos. The heartwood is brown with darker stripes suggesting Walnut (*Juglans*) and is highly esteemed locally for furniture. The natural color is sometimes attractively variegated with blue or green imparted by fungi. *L. dentata* R. Br., a tree usually not over 25 feet tall and a foot in diameter, grows scatteringly in the forest from Linares to Osorno in Chile. Its light-colored wood has the general appearance of Sycamore (*Platanus*) and is well suited for making furniture. *L. ferruginea* (Cav.) R. Br. occurs around Valdivia and on the Chonos Islands, Chile.

The wood of the American species of *Lomatia,* so far as studied, is similar in structure to that of *Guevina,* the principal differences being as follows: Pores smaller (35 to 70μ). Vessels without spiral thickenings. Rays less conspicuous and having a larger proportion of the smaller sizes; larger rays also less frequently interrupted by the wood fibers.

COMMON NAMES: *Lomatia dentata:* Avellanillo, guarda-fuego, piñol (Chile); *L. ferruginea:* Fuinque, piune, romarillo (Chile); *L. hirsuta:* Andaga, garo. raral, shiapash (Peru); nogal, n. silvestre, palo negro, radal, ralral, raral (Chile, Arg.).

Panopsis, with several species of small to medium-sized trees, is widely distributed in continental tropical America. The leaves are entire, alternate or verticillate; the flowers are very small and racemose; the fruit is a hard one-seeded drupe. The best known species is *Panopsis rubescens* (Pohl) Pittier, which occurs rather sparingly throughout most of the Amazon basin. It is usually small, but sometimes attains a height of 50 feet. The timber of the different species examined is much the same in structure and properties, though varying somewhat in density and texture. It is used locally for cabinet work and fancy articles, but is too scarce to be important in the trade.

Heartwood pinkish brown to reddish brown, with prominent ray markings; rather sharply demarcated from the oatmeal-colored sapwood. Luster high. Odorless and tasteless. Light and soft to moderately heavy and hard; texture medium to very coarse; grain straight to irregular; easily worked, the finer-textured material finishing very attractively; durability probably rather high.

COMMON NAMES: Yolombo, colombo, zambo cedro-hembra (Col.); miao de lapa, solitario, tigua, yagüero, y. corococo (Venez.); mahoballi (Br. G.); mahoballie (Sur.); aderno, cedro bordado, loro jaya, louro faia, malheira, pau concha clara, p. malheira (Braz.).

Roupala, with many species of trees and shrubs, a few Australasian, is widely distributed in tropical America from Mexico to Peru and Argentina, though most abundantly represented in Brazil. The living leaves and branches give off a skunk-like scent when broken, but the dry wood is odorless. The leaves are usually simple but sometimes are also lobed or pinnate on the same tree. Most of the trees are small, but a few of them, *e.g., R. brasiliensis* Klotzsch, are up to 100 feet high with a long bole 24 to 30 inches in diameter. The Oak-like timber is used to some extent locally for furniture and construction, but is not exported.

Heartwood brown to dark reddish brown, sometimes with a purplish hue; distinct but not always sharply demarcated from the pale brownish sapwood; ray markings conspicuous. Luster medium to high; some specimens with an oily appearance. Without distinctive odor or taste when dry. Density variable, but mostly hard, heavy, and strong; sp. gr. (air-dry) 0.80 to 1.10; weight 50 to 79 lbs. per cu. ft.; texture typically coarse; feel harsh; grain straight to irregular; rather difficult to work, but can be finished smoothly and attractively; subject to collapse in drying; durability high.

COMMON NAMES: Palo de zorillo (Mex.); chancho, zorillo (Salv.); danto hediondo, ratón, ratoncillo (C.R.); azufre, carnefiambre, berraco, verraco (Col.); carne asada, chaparro, jigua, mapurite, mondingo, yaquero (Venez.); beefwood, guatapaná (Trin.); coco de monte (Ec.); aderno, carne de vacca, carvalho, c. catucaem, caxicaem, cochicahen, cutucaem, cutuahem, patuquiry, pau concha (Braz.); arellán, ingaina, paco-paco de la sierra, yngaina (Peru); mborebí-caá-guazú (Arg.).

QUIINACEAE

THIS family comprises about 35 species of trees and shrubs of three genera, namely, *Lacunaria, Quiina,* and *Touroulia.* They have their center of distribution in the Amazon basin, but there are extensions to southeastern Brazil and northward to the West Indies and Central America. The leaves are opposite or whorled, simple or rarely pinnately lobed, and provided with stipules; the flowers are borne in axillary panicles or racemes; the fruit is a berry with 1 to 4 tomentose seeds. The family is not the source of any valuable products. The timbers, which bear a superficial resemblance to Massaranduba (*Mimusops Huberi* Ducke), would be useful if they were larger and more abundant. The following description applies equally well to the available wood specimens of *Lacunaria* and *Quiina,* as the range of variation within a genus appears to be as great as that between the two genera.

Heartwood dark olive or reddish brown, sometimes with nearly black streaks; distinct from the lighter-colored sapwood, though without sharp line of demarcation. Luster rather low. Without distinctive odor or taste. Very heavy, hard, tough, and strong; sp. gr. (air-dry) up to 1.10; weight up to 69 lbs. per cu. ft.; texture rather fine; grain fairly straight; difficult to saw, but finishes very smoothly; is probably highly durable.

Growth rings sometimes present. Pores small to medium-sized (60 to 130μ), the largest barely visible without lens; numerous; often irregularly distributed, occurring singly and in small multiples or in short radial or diagonal rows, without definite pattern; in some specimens the pores, though close together, are not in contact radially, while in others they are. Vessels with simple perforations; without spiral thickenings; pits numerous, minute (all less than 4μ), alternate to subopposite, frequently with spirally coalescent apertures; gum plugs common. Rays very numerous; not distinct without lens on cross section, rather high in part on radial surface but not conspicuous because of lack of color contrast with the background; decidedly heterogeneous; of two sizes, the larger 2 to 4, occasionally 5 or 6, cells wide in median part and few to 100 or more cells high; walls thick and abundantly pitted; pits to vessels numerous, minute, of same appearance in face view as the intervascular; gum deposits abundant; no crystals seen. Wood parenchyma developed in varying amount, even in same specimen; sparse to abundant; paratracheal, sometimes confluent; also diffuse or in short tangential or diagonal lines; no crystalliferous strands observed. Wood fibers with very thick walls and minute cavities; non-septate; pits small, very numerous in both radial and tangential walls, distinctly bordered. Ripple marks and gum ducts absent.

Lacunaria, so named because of the latex-filled cavities in the fleshy pericarp of the *Theobroma*-like fruits, includes about a dozen species of small to medium-sized trees, mostly in the lower Amazon region. *L. Jenmani* (Radlk.) Ducke is known in Brazil as Moela de Muetum.

Quiina, with about 22 species of trees and shrubs, is best represented in the Amazon basin, but the geographical range includes parts of southeastern Brazil, northeastern Peru, the Guianas, Trinidad, Cuba, and Central America to southern British Honduras. The trees are not well known and there are no special uses for the timber.

COMMON NAMES: Hicaquillo de las pozas (Cuba); blackheart?, himirimia? (Br. G.); juravá-rana, quiina-rana, quina-rana (Braz.).

Touroulia, a little known genus of trees in northern Brazil and the Guianas, has only one unquestioned species, namely, *T. guianensis* Aubl., which differs from the other Quiinaceae in having pinnate leaves in the adult stage. The wood has not been studied.

RHAMNACEAE

THE Buckthorn family comprises about 50 genera and 500 species of erect or climbing shrubs and small to moderately large trees, often spiny, generally distributed in both temperate and tropical regions of the world, many of them occupying dry situations. The leaves are alternate or opposite, simple, and usually with very small stipules; the greenish flowers are typically small and borne in axillary cymes or umbels; the fruit is usually a more or less fleshy drupe, but sometimes 3-lobed and separating into nutlets. The economic products are few and are chiefly medicinal, although some species bear small fruits that are comestible and others are local sources of dyestuffs and soap substitutes. One of the best known plants is the European Buckthorn, *Rhamnus cathartica* L., which is often planted for hedges in Europe and eastern North America.

The woods, which are virtually unknown to the world markets, exhibit a wide range in appearance, structure, and properties. At one extreme in density is the West African *Maesopsis* with a light and soft timber about as easy to work as Spanish Cedar (*Cedrela*), while at the other is *Krugiodendron* of the West Indies with one of the heaviest and hardest woods known. The colors include various shades of yellow, orange, red, brown, and olive, sometimes striped with black. A rare wood noted for its peculiar color is the Red or Pink Ivory of northern Natal (see *Tropical Woods* 13: 4); the scientific name is *Rhamnus Zeyheri* Sond., but the species seems out of place in that genus.

The American trees represent about 17 genera, but the woods of only 13, including one large liana, are described here. For the most part the woods are attractively colored; without distinctive odor or taste when dry; very hard, heavy, and fine-textured; seasoning without much splitting or warping; moderately difficult to work, but taking a glossy polish; highly resistant to decay. Their utilization is limited because of the scarcity of trees large enough to supply timber.

Growth rings present; woods more or less distinctly ring-porous in *Ceanothus, Colletia, Discaria,* and *Rhamnus.* Pores medium-sized in *Colubrina* (in part) and *Sarcomphalus,* large in part in lianas (*e.g., Ampelozizyphus*), small to very small in the others; few in *Colubrina, Doerpfeldia,* and *Sarcomphalus,* fairly numerous in the others; generally solitary and in short to long radial multiples, but in flame-like or dendritic pattern in *Rhamnus* (*Eurhamnus*) and *Ceanothus* and in diagonal zig-zag, or ulmiform arrangement in *Colletia* and *Discaria.* Vessels with exclusively simple perforations; spiral thickenings present in *Colubrina, Doerpfeldia,* and *Sarcomphalus;* spiral striations observed in *Sageretia;* tyloses absent; gum deposits common, often as plugs near perforations; pits very small to minute in *Krugiodendron, Reynosia,* and *Sageretia,* small to medium-sized in the others. Rays typically 1 to 3 cells wide; not over 2 cells wide in *Ceanothus macrocarpus* Nutt., *Karwinskia* (in part), and *Krugiodendron;* up to 6 and 8 cells wide in *Discaria* and *Colletia,* respectively; distinctly two-sized, uniseriate or biseriate and very large and suggesting those of *Quercus* in *Ampelozizyphus;* height variable, less than 25 cells in *Krugiodendron,* occasionally up to 50, rarely to 100 or more, in the others; more or less distinctly heterogeneous with many square cells, sometimes homogeneous in part (*e.g., Doerpfeldia* and *Sarcomphalus*); crystals common; gum deposits frequently abundant; pits to vessels similar in size to the vascular. Wood parenchyma sparse to very abundant, mostly sparingly paratracheal; finely reticulate in *Sarcomphalus* and *Zizyphus* (in part); short to long aliform and confluent into short tangential to broken, wavy, or fairly uniform concentric bands in *Colubrina, Doerpfeldia, Reynosia,* and *Zizyphus* (in part); sometimes also in a terminal row or narrow band; pith flecks common; crystalliferous strands observed in certain species of *Colletia, Colubrina, Condalia, Doerpfeldia, Karwinskia, Rhamnidium, Sageretia,* and *Sarcomphalus.* Wood fibers typically small; walls moderately to extremely thick and gelatinous, the two types sometimes present in same growth rings; pits very small, simple. Ripple marks absent. No gum ducts seen. For anatomy of the different genera see *Tropical Woods* 58: 7-23.

Ampelozizyphus *amazonicus* Ducke, the only species, is a scandent shrub of the lower Amazon region of Brazil, where it is known as Saracura-mira, and the hinterlands of British Guiana, where the Waiwai Indians use the fragrant bark in place of soap. The leaves are large, leathery, prominently 3-nerved; the small green flowers are borne in axillary cymes; the fruit is a 3-lobed, 3-seeded, elastically dehiscent capsule. Wood brownish throughout, resembling Oak (*Quercus*). Luster medium. Odorless and tasteless; fresh inner bark is said to have the odor of methyl salicylate. Hard and heavy; texture coarse, grain irregular. Narrow wedges of bark penetrate rather deeply into the wood.

Ceanothus, with about 30 species of shrubs and small trees rarely over 30 feet high and 14 inches in diameter, is confined to the temperate and subtropical regions of North America, being most abundantly represented in California. The branches are often spinose; the alternate or opposite, usually 3-nerved leaves are persistent in the arborescent species; the little white, blue, or pink flowers are generally borne in axillary or terminal cymes or panicles; the fruit is small, dry, 3-lobed, and dehiscent.

Standley (*Trees and shrubs of Mexico*, p. 720) says: "Some of the species are showy when in flower. The blue-flowered ones are known on the Pacific coast of the United States as California Lilac. *Ceanothus americanus* L., of the United States, is known as New Jersey Tea. The astringent roots contain over 6 per cent of tannin, and an alkaloid, ceanothine. They have been used in the treatment of syphilis and are said to have purgative properties. The leaves were used by the Indians to make a beverage like tea, and during the Revolutionary War they were employed along the Atlantic coast as a substitute for Chinese tea. . . . The fresh flowers of some, and probably of all, of the species, when rubbed in water, give a cleansing lather which is a good substitute for soap." The arborescent species with the most southern range is *C. coeruleus* Lag., of southern Mexico and Guatemala; it is usually less than 25 feet high and its only recorded uses are medicinal. *Ceanothus* supplies no commercial timber.

Wood reddish brown; distinct but not very sharply demarcated from the whitish sapwood. Luster rather low. Without distinctive odor or taste. Hard and heavy; texture fine; grain variable; not difficult to work, finishing very smoothly; durability rather high.

COMMON NAMES: Blue blossom, b. myrtle, California lilac, deer brush, Jersey tea, mahala mat, New Jersey tea, red-root snow brush, white thorn, wild lilac (U.S.A.); chaquira, chaquirilla, cuaicuastle, huichagorare, palo colorado, sayolistle, tlaxistle, tnu-yoocó (Mex.).

Colletia, with numerous species of thorny shrubs and little trees, is limited to southern South America. Apparently the only uses are in local medicine. Wood of *C. spinosa* Lam. yellowish. Luster medium. Odorless and tasteless. Hard, heavy, tough, and strong; texture fine; grain fairly straight.

COMMON NAMES: Quina (Braz.); espina de la cruz (Urug.); barba de tigre, currú, curú-mamuel, espina-cruz, liaka, quina, q. del campo, tola (Arg.); crucero, yaquil (Chile); cjaru, r'ocke (Peru).

Colubrina, with about 25 species of unarmed evergreen shrubs and trees, is sparingly represented in the Old World and widely distributed in tropical America. The leaves are pinnately veined or 3-nerved, sometimes with glands on the under surface; the small yellowish flowers are fasciculate in the leaf axils; the fruit is a globose elastically dehiscent capsule. The shiny brown or black seeds are used for making necklaces and the bitter bark is employed in native medicine as a tonic and febrifuge. The timber of a few species is locally important.

The Saguaragy or Sobragy of southeastern Brazil, *Colubrina rufa* Reiss., is a tree of moderate height, but with trunks sometimes 32 inches in diameter, supplying an orange-colored or somewhat variegated heartwood that is fine-textured, strong, and durable. It is used for cabinet work, vehicles, shipbuilding, bridges, and is considered first class for railway crossties and fence posts.

Three arborescent species have their northern limit in southern Florida and one densely branched shrub crosses the border from Mexico into Texas. Best known is the Naked-wood or Soldier-wood, *Colubrina reclinata* Brongn., which grows in nearly pure stands on Umbrella Key, where it attains a height of 50 to 60 feet with a thick trunk divided by numerous deep irregular furrows like a tangled mass of serpents and covered with a thin orange-brown bark exfoliating like River Birch (*Betula nigra* L.). The species is also common in the West Indies. The rich velvety-lustered dark brown or olive-brown wood is little used except for fuel and a few minor purposes, but is suitable for small cabinet work and articles of turnery. Some of the common names recorded for the next species seem more appropriate for this one.

Colubrina ferruginosa Brongn. is widely distributed in the West Indies and also occurs in southern Mexico and Salvador;

it is said to reach its best development in Martinique, growing to a height of 75 feet and a diameter of two feet or more. In Haiti the species, called Bois Pelé, grows naturally in the foothills of the high mountains near the end of the southwest peninsula. According to W. R. Barbour (*Tropical Woods* 6: 13), the natives "make a practice of growing Bois Pelé near their homes, usually in a lot 50 to 100 feet square behind the house. The seeds are planted about six feet apart each way and the trees grow very tall and straight with clean boles and little taper. They do best on a rich well-watered soil, such as is suitable for bananas, and apparently are free from insect attacks and disease. It requires about ten years to grow a tree with a usable length of 40 feet or more and a diameter (breast high) of about six inches. These poles are used solely for rafters and ridgepoles for the wattle-and-daub huts or 'cayes.' The wood is hard, very stiff, works fairly well, is not badly attacked by termites, and lasts a long time when not exposed."

Heartwood generally orange or bright red with a yellow hue, somewhat striped or variegated; rich brown in *C. reclinata;* distinct but not always sharply demarcated from the whitish or yellowish sapwood. Luster medium to high. Without distinctive odor or taste. Moderately to very heavy and hard; sp. gr. (air-dry) 0.75 to 0.95; weight 47 to 59 lbs. per cu. ft.; texture medium to fine (*C. reclinata* being the densest and finest-textured); grain straight to variable; not difficult to work, finishing very smoothly and attractively; is highly durable. Not likely to become commercially important.

Common names: *Colubrina ferruginosa:* Bitters, snake bark (Bah.); black velvet, greenheart, mountain ebony, snakewood, wild ebony (Jam.); bijáguara, carbonero de costa, fuego (Cuba); abejuelo, achiotillo, avelluelo, greenheart, quitarán, ratón, snake bark, soap-tree (P.R.); bois conleuvre, b. de fer, b. de serpent (Mart.); corazón de paloma (Dom. R.); bois fer blanc, b. pelé (Haiti); guayul, manzanita, pimiento-ché, yax-puken, yax-pukim (Mex.); chaquirio or chaquiro (Salv.); *C. reclinata:* Naked-wood, soldier-wood (Florida); bijáguara, carbonero, c. de costa, ébano, guaciriano, jayajabico (Cuba); maabee, mabí, naked wood, smooth snake bark (P.R.); palo amargo (Dom. R.); bois de fer, b. mabí (Haiti). *Other species:* Wild coffee (Florida); aleznilla, café cimarrón (Mex.); pichy pang, Spanish elm, wild coffee (Pan.); saguaragy, sobragy, sobrasil (Braz.).

Condalia, with about 10 species of shrubs and trees rarely 30 feet high, is of common occurrence, often forming chaparral, in southwestern United States, northern Mexico, and southern South America. The branches are stiff and spinescent; the small leathery leaves are deciduous; the little greenish white flowers are solitary or clustered in the leaf axils; the fruit is a thin-fleshed edible little drupe. The bark of the roots of some species is used as a substitute for soap and for medicinal purposes. The genus supplies no timber of value.

The largest species is *Condalia obovata* Hook., which occurs in western Texas and Nuevo León and Tamaulipas, Mexico. It is often only a spiny shrub covering large areas with dense thickets, but on the high sandy banks of the lower Río Grande and its tributaries it is a tree sometimes 30 feet high and 6 to 8 inches in diameter. The dense wood makes excellent fuel and is said to be the source of a blue dye. The few specimens of *Condalia* in the Yale collections are of doubtful authenticity and differ too much in their appearance and anatomy to justify description.

Common names: Bluewood, logwood, purple haw (Texas); abrojo, barchatas, bindó, brasil, capul negro, capulín, chamís, chaparro prieto, clepe, crucillo, garambullo, garropata, mazquitillo, tecomblate (Mex.); piquillín (Arg.).

Discaria, with numerous species of typically spiny shrubs and little trees, has two representatives in Oceania, the others occurring in southern South America, especially Patagonia and Chile. The best known is *D. febrifuga* Mart., the Quina do Campo of Brazil, so named because the bitter bark, especially of the roots, is used like quinine as a febrifuge; it is said to yield a red dye also; the wood is a source of fuel and charcoal.

The following description is based upon two Chilean specimens of Chacay, one (Yale 5560) determined as *D. discolor* (Hook.) P. Dusén, the other (Yale 34054 of *D. serratifolia* (Vent.) B. & H. f. collected by E. L. Bernath in the Province of Malleco. Heartwood brownish, with Elm-like appearance; not sharply demarcated from the yellowish sapwood. Luster medium. Odorless and tasteless. Hard, heavy, tough, and strong; texture medium; grain straight to irregular; not difficult to work, but rather fibrous; durability presumably low. Of no commercial possibilities.

COMMON NAMES: Brusca, quina do campo (Braz.); bruquilla, chacái (Arg.); quina del campo (Urug.); chacay (Chile).

Doerpfeldia *cubensis* Urb., the only species, is a little tree growing in eastern Cuba, where it is called Bruja Negra and Hueso de Tortuga. The stiff branches bear many short leaf spurs; the leaves are small, obovate, entire, and leathery; the fruit is a small drupe. Heartwood probably dark reddish brown, judging from small wound areas; sapwood pale yellow. Luster medium. Odorless and tasteless. Very hard, heavy, and strong; texture fine and uniform; grain straight; not very difficult to work, taking a glossy polish. Of no known uses.

Karwinskia, with several species of unarmed shrubs and small to medium-sized trees, has its center of distribution in Mexico, with extensions into southwestern United States, northern Central America, and Haiti. The leaves are subopposite, pinnate-nerved, and pellucid-punctate; the small flowers are axillary, solitary or in cymes or umbels; the fruit is drupaceous.

The most widely distributed species is *Karwinskia Humboldtiana* (R. & S.) Zucc., ranging from western Texas to Yucatán and Oaxaca, Mexico. Standley (*Trees and shrubs of Mexico,* p. 717) says: "The fruit is sweet and edible, but the stones are harmful if swallowed. In people, especially children, paralysis, particularly of the lower limbs, is caused by eating the stones, and similar effects are said to be produced in pigs and chickens. . . . The seeds are oily, and they contain some principle which paralyzes the motor nerves. They are employed in Mexico as an anticonvulsive, particularly in the case of tetanus. An infusion or decoction of the leaves and roots is used locally for fevers." The species is usually a shrub or small tree, but on María Magdalena Islands, off the Tepic coast, where it is known as Tempisque, it occasionally develops a trunk 50 feet long and 24 inches in diameter. The dull red, very dense, fine-textured timber is considered excellent for railway crossties, but most of the trees there have irregular boles. Another species in southwestern Mexico is *K. latifolia* Standl., commonly called Margarita. It also grows on María Magdalena Island, and boles of mature trees are 14 to 18 inches in diameter and 20 feet to the first large branches. The timber is similar to that of the other species, but is often streaked with black, and defective wood may be almost wholly black; it is noted for its durability. The Salvador species, locally known as Güiligüiste or Huilihuiste, is *K. Calderonii* Standl. According to Standley and Calderon (*Flora de El Salvador,* p. 141), it is a common tree, sometimes 40 feet high, and supplies timber of excellent quality for making railway crossties, hubs of wheels, weavers' shuttles, mortars and pestles, bowling balls, and fuel.

Heartwood dull red or reddish brown, deepening upon exposure; uniform or streaked with black, which may predominate; distinct but not sharply demarcated from the yellowish sapwood. Without distinctive odor or taste. Very hard, heavy, and strong; sp. gr. (air-dry) 1.05 to 1.20; weight 65 to 75 lbs. per cu. ft.; texture fine and uniform; grain fairly straight; rather difficult to work, but finishing very smoothly; seems to season readily without bad checking and to hold its place when manufactured. A good timber in its class, but too scarce to be of value for export.

COMMON NAMES: Cacachilla, c. china, c. silvestre, cachila, capulincillo, c. cimarrón, coyotilla, frutillo, margarita, m. del cerro, negrito, palo negrito, tlalcapolin, tullidor, tullidora (Mex.); güiligüiste, huilihuiste (Salv.).

Krugiodendron *ferreum* (Vahl) Urban, the only species, is an evergreen tree often shrubby but sometimes 30 to 50 feet tall and 15 to 20 inches in diameter, growing in southern Florida, the West Indies from the Bahamas to St. Vincent, and in Yucatán, Mexico, and northern British Honduras. The leaves are subopposite; the small yellow-green flowers are borne in short axillary clusters; the fruit is a small, round, black, thin-fleshed drupe. The wood, which is sparingly utilized, is noted for being one of the densest in the world.

Heartwood orange to dark brown, usually more or less streaked; has a waxy appearance; sharply demarcated from the yellowish sapwood. Luster fairly high. Odor and taste absent or not distinctive. Exceedingly dense, horn-like, and strong; sp. gr. (air-dry) 1.34 to 1.42; weight 84 to 89 lbs. per cu. ft.; texture very fine and uniform; grain generally straight; difficult to cut, fairly easily split, takes a high polish; is very resistant to decay. Of no commercial possibilities.

COMMON NAMES: Black ironwood (Florida, Jam.); acero, carey de costa, coronel, palo diablo (Cuba); bariaco, espejuelo, palo de hierro (P.R.); palo de hierro (Dom. R.); bois de fer (Haiti, Guad.); chimtoc, quiebrahacha (Mex.); axe-master (Br. H.).

Reynosia, with several species of unarmed evergreen shrubs and small trees, has its center of distribution in the West Indies, with one extension into southern Florida. The leaves are mostly opposite short-petioled and leathery; the minute yellowish green flowers are borne in axillary clusters; the fruit is a thin-fleshed drupe. The northernmost species is *R. septentrionalis* Urb., a shrub or a tree sometimes 30 feet tall and eight inches in diameter, common on scrublands of the Bahamas and also along the coast and islands of southern Florida, where it is known as Red Ironwood and Darling Plum. There are a few species in Cuba that get large enough to supply timber suitable for fence posts and railway crossties, but the supply is very limited.

Heartwood orange-brown, becoming reddish brown upon exposure; rather waxy looking; sometimes with blackish streaks; rather sharply demarcated from the thin yellowish sapwood. Luster rather low. Odorless and tasteless; exceedingly hard, heavy, and strong; texture fine and uniform; grain mostly straight; rather difficult to work, but taking a high natural polish; very resistant to decay. Apparently without commercial possibilities because of the small size and scarcity of the timber.

COMMON NAMES: Darling plum, red ironwood (Florida); almendrillo, a. de costa, almendro, brujilla, carey cocuyo, cocuyo de costa, membrillo, m. silvestre (Cuba); guama, chicharrón (P.R.); brillol, gallegalle (Haiti).

Rhamnidium, with a few species of unarmed shrubs and small to medium-sized trees, has its center of distribution in eastern Brazil. Some species have been described from the West Indies, but Urban has transferred three of them, namely, *R. cubense* Britt. & Wils., *R. jamaicense* Urb., and *R. reticulatum* Gris., to the genus *Auerodendron.* The wood material available is inadequate as a basis for an opinion on the proposed classification. One of the most characteristic features of *Rhamnidium* is the opposite entire leathery leaves having deeply depressed midrib and parallel lateral nerves with finely anastomosing veins between them. The twigs are slender; the very small flowers are borne in axillary clusters; the fruit is a globular drupe. The timber is little known because of the scarcity of the larger trees.

Rhamnidium caloneurum Standl. was discovered by G. Proctor Cooper in the region of Bocas del Toro, Panama: According to the collector's notes, it is a tall tree, with a long clear unbuttressed bole 12 inches in diameter. The fruits are greenish red, and the wood, when freshly cut, has a scent suggesting peanuts. *R. elaeocarpum* Reiss. occurs in eastern Brazil, where it is called Azeitona, and in Formosa, Argentina; it was also found in northeastern Peru by Llewelyn Williams, who says

(*Woods of northeastern Peru*, p. 299) that it is a shrub or a small tree about 18 feet in height, branching a few feet above the ground. *R. glabrum* Reiss., another Brazilian species, is generally described as a shrub or little tree, but according to J. G. Kuhlmann (*Revista Florestal* 2: 1: 54; *Tropical Woods* 32: 34) it will grow to a height of 65 feet, with a trunk 18 to 20 inches thick, on good soil in humid sites. The sapwood is comparatively thin; the heartwood is light red, deepening to fiery red upon exposure, fine-textured, durable, and well suited for cabinet work. Some experimental forest plantations of it have been made.

Heartwood bright red, orange, or orange-brown, sometimes with dark streaks; rather sharply demarcated from the yellowish sapwood. Luster medium. Odorless and tasteless when dry. Very hard, heavy, and strong; texture fine and uniform; grain fairly straight; appears to season readily without checking; not very difficult to work, taking a glossy polish; durability high. Of no possibilities for the export trade.

Common names: Almendrón charrasco, fruta del bién (Cuba); azeitona, tarumai (Braz.).

Rhamnus, with about 150 species of armed or unarmed shrubs and small to medium-sized trees, is widely distributed in the temperate and subtropical regions of the northern hemisphere, with a few representatives elsewhere. The bark is bitter; there is no terminal bud; the leaves are alternate, feather-veined, entire or toothed; the small flowers are borne in axillary clusters; the fruit is a succulent drupe. The principal uses of the plants are medicinal. The species can be grouped into two subgenera and the woods of the temperate zone, so far as studied, are separable into two classes that are more distinct than most of the other genera. In subgenus *Eurhamnus* the woods are diffuse-porous and the pores are very small to minute and arranged in conspicuous flame-like, zig-zag, or dendritic pattern (Plate XLIII, 3); in subgenus *Frangula* the woods are more or less ring-porous and the pores, though never large, diminish gradually in size during seasonal growth.

The center of distribution of the American species is in Mexico with a few as far south as Costa Rica and one in Venezuela, while several cross the border into southwestern United States. *Rhamnus crocea* Nutt. and its varieties are spinescent evergreen shrubs or little trees 25 to 30 feet high growing in southern California and Arizona, often forming thickets. This species belongs to the subgenus *Eurhamnus* and the wood is similar to that of the common European Buckthorn, *R. cathartica* L. There are in the United States two arborescent species belonging to the subgenus *Frangula;* both are unarmed and deciduous. The Indian Cherry, *R. caroliniana* Walt., is a tree sometimes 30 to 40 feet high, its slender trunk rarely over eight inches in diameter, widely distributed in the southeastern quarter of the United States, but at its best in southern Arkansas. It has no commercial uses.

The best known American species is the Cascara, *Rhamnus Purshiana* DC., a shrub-like tree rarely 40 feet tall and 20 inches in diameter, usually separating 10 or 15 feet from the ground into numerous stout, generally upright stems. It is indigenous to the northwest United States and British Columbia, making its best growth in coniferous forests on rich bottomland in the Puget Sound region. The winter buds lack scales but are covered with rusty brown hairs. The leaves are finely serrate and have a deeply depressed midrib and parallel lateral veins; the flowers are greenish yellow; the fruits are small black drupes. The interior of freshly cut bark is bright yellow, turning brown upon exposure to sunlight, and has a bitter taste. The bark has long been an article of commerce, being the source of an official drug called "cascara sagrada," which is used in medicine as a laxative. The trees are stripped during the dry season and the stumps allowed to coppice. The bark is dried on wires in the shade and shipped in 100-pound bags or bales to the dealers who allow it to age a year or more before using. An average-sized tree will yield about 10 pounds of dried bark. The supply has been seriously depleted in Washington and Oregon, but is still plentiful across the Canadian border. The bark of *R. californica* Esch. is sometimes used as a substitute for that of the genuine Cascara.

Wood yellow, orange, or orange-brown; distinct but often not sharply demarcated from the whitish or pale olive sapwood. Luster medium. Odor and taste absent or not distinctive. Density fairly low to medium in *Frangula,* rather high in *Eurhamnus;* texture medium; grain straight to irregular; easy to work, finishing smoothly; durability fair to good. Of no commercial importance because of the small sizes available.

COMMON NAMES: *Rhamnus caroliniana:* Bog birch, brittle wood, buckthorn (alder, Carolina, yellow), elbow brush, Indian cherry, polecat tree, p. wood, stink berry, s. cherry, s. wood, yellow wood (U.S.A.); *R. Purshiana:* Bayberry, bitter bark, bear berry, b. wood, cascara, c. buckthorn, c. sagrada, chittern, chittim, coffee berry, c. bush, c. tree, pigeon berry, wahoo, wild coffee, yellow wood (U.S.A., Can.). *Other species:* Buckthorn, California holly (U.S.A.); capulincillo, tlalcapollin (Mex.); duraznillo (C.R.); anoncito, arraclán (Col.); zamorito (Venez.).

Sageretia, with several species of shrubs and little trees, often with spinose branchlets, has its center of distribution in Asia, but there are a few American representatives with a combined range extending from Florida and southwestern United States through Mexico and Central America to Peru. Some of the species have edible drupaceous fruits, and the leaves of *S. theezans* (L.) Brongn. are used in China as a substitute for tea. The only American species represented in the Yale collections is of *S. Wrightii* S. Wats., collected in the mountains of Arizona by S. B. Detwiler (Yale 26687). It is a shrub with spreading branches that take root and hold the soil on steep slopes.

Heartwood rich purplish brown, more or less streaked or variegated; sharply demarcated from the nearly white sapwood. Luster rather high. Odorless and tasteless. Very hard, heavy, and strong; texture fine and uniform; takes a high polish. Of no commercial possibilities.

Sarcomphalus, with about eight species of spinescent or unarmed shrubs and trees up to 45 feet tall, is limited to the West Indies. The leathery leaves are palmately or pinnately veined; the little flowers are borne in terminal panicles; the fruit is a small, nearly dry drupe. The species attaining the largest size is *S. laurinus* Gris. of Jamaica. According to Fawcett and Rendle (*Flora of Jamaica* 5: 67), it is sometimes 45 feet high, with a trunk up to 30 inches in diameter, and its dark-colored, hard, fine-textured wood, called Bastard Lignum-vitae, "is looked upon as one of the best timber woods in the island." The following description is based on specimens of *S. crenatus* Urb., *S. domingensis* Krug & Urb., and *S. reticulatus* (Vahl) Urb. from Haiti and Dominican Republic. On a basis of wood anatomy, *Sarcomphalus* should include a group of species now included in *Zizyphus.*

Heartwood orange, sometimes with blackish brown streaks; distinct but not always sharply demarcated from the yellowish sapwood. Odorless and tasteless. Hard, heavy, tough, and strong; texture fine to medium, uniform; grain fairly straight; not difficult to work, taking a glossy finish; durability probably high. A good timber, but without commercial possibilities because of the scarcity of large trees.

COMMON NAMES: Bastard lignum-vitae (Jam.); azofaifa de costa, a. de playa, bruja (Cuba); cacao rojo, espejuelo (P.R.); saona, sopaipo (Dom. R.); coquemolle (Haiti).

Zizyphus, with about 60 species of armed or unarmed, mostly deciduous, erect or vine-like shrubs and small to medium-sized trees, occurs in temperate and tropical regions of both hemispheres. The stipular prickles, when present, are mostly short, straight or recurved. The leaves are alternate to opposite, usually leathery, serrate or entire, and with 3 or 5 prominent nerves extending from the base; the small greenish flowers are borne in axillary clusters or cymes; the fruit is drupaceous, dry or fleshy and comestible. The best known species is the Jujube, *Z. jujuba* Lam., a small spiny tree native of the Old World but widely cultivated in the tropics generally for its mealy-fleshed fruit from which is obtained the jujube paste used in confectionery. *Zizyphus* is not an important source of timber as the larger trees are scarce, but the woods are of good quality and are used locally to a minor extent.

From the standpoint of wood structure the genus lacks homogeneity. The American woods are readily separable into at least two groups and both are distinct from the Old World species so far as studied. One group is characterized by finely reticulate and narrowly terminal parenchyma as in *Sarcomphalus*. Here belong *Zizyphus angolito* Standl., a beautiful tree 40 to 50 feet tall and 20 to 30 inches in diameter in northern Colombia (see *Tropical Woods* 32: 20); *Z. cyclocardia* Blake, a shrub or little tree of northern Venezuela; *Z. havanensis* H.B.K., a Cuban shrub; *Z. guatemalensis* Hemsl., a small tree of the interior of Guatemala; and *Z. sonorensis* S. Wats., a Mexican shrub or a tree up to 40 feet high, with small red fruits used locally as a substitute for soap in washing clothes. Only one of the specimens contains heartwood, which is dark brown.

In the second group the parenchyma is distinctly aliform to confluent into tangential to broken or regular concentric bands 2 to 5 cells wide and 2 to 5 pore-widths apart; sometimes the pores are imbedded, but more often they are free on one side. Included here are the Argentine Mistol, *Zizyphus mistol* Gris., a tree up to 50 feet in height and 24 inches in diameter, with bright red to reddish brown wood; *Z. rhodoxylon* Urb., a small tree of Haiti and Dominican Republic with dull reddish brown wood; the Jamaican Cogwood or Greenheart, *Z. chloroxylon* (L.) Oliv., a tree 30 to 60 feet high, usually with a low-branched trunk, the inner bark bright red. The last species is now scarce, but formerly its very dense and strong, greenish yellow or olive-colored wood was considered the best in Jamaica for use in the coffee and sugar factories for solid framework, cogs, and rollers. In this, as in most of the other species, the heartwood is durable, but slow in forming.

Heartwood typically red or reddish brown; sometimes pale to rather dark olive; distinct, but usually not sharply demarcated from the thick yellowish sapwood. Luster low to medium. Odor and taste not distinctive. Density high; sp. gr. (air-dry) 0.90 to 1.12; weight 56 to 70 lbs. per cu. ft.; texture fine and uniform; grain fairly straight; not very difficult to work, finishing very smoothly. Of no possibilities for the export trade.

COMMON NAMES: Cogwood, greenheart, Jamaica laurel (Jam.); cocuya (Cuba); hojancha prieta, yagua (Dom. R.); casser hache (Haiti); jujubier du pays (Mart.); amapole, a. dulce, confite, nanche de la costa, uayum, uayumke (Mex.); mocoso (Guat.); angolito, azufaifa, mondongüito (Col.); cacagüillo, cana, chicha, chichiboa, jacyuari, mamón de venao, mayo, naranjillo, nigua, nigüito (Venez.); joazeiro, j. grande, juá (Braz.); mistol, m. cuaresmillo, sacha-mistol (Arg.).

RHIZOPHORACEAE

THE Mangrove family, with 14 to 16 genera and about 100 species of shrubs and trees, mostly of small size, is widely distributed in tropical and subtropical regions. The Rhizophoreae, or Mangroves proper (*Rhizophora, Brugiera, Ceriops,* and *Kandelia*), are confined to coastal mud flats and estuaries where the water is saline or brackish, and where, together with certain members of other families (*e.g., Avicennia* and *Aegiceras*), they form associations commonly known as Mangrove swamps. The trees protect the land from erosion, and as they gradually advance in the shallow water, silt and debris are deposited about their roots and new land is formed. The trees are of local importance as sources of tanbark, charcoal, and some structural timber, but their chief service to man is as builders and protectors of land upon which more valuable vegetation can develop. The other members of the family, as it is now constituted, grow in the interior or at least far enough from the coast not to be affected by sea water. They differ from the

Photo by B. E. Dahlgren.

PLATE XXXIV. Mangrove tree (*Rhizophora mangle*) near Shark River, Florida. Oysters attached to the stilt-like roots are exposed at low tide.

Photo by U.S. Forest Service.

PLATE XXXV. Stand of Northern Black Cottonwood (*Populus trichocarpa*) on the Columbia National Forest, Washington. The trees are about 150 feet tall and 36 inches in diameter.

Rhizophoreae not only in habitat, but in many other ways, including the structure of their wood. As shown by Marco (*Tropical Woods* 44: 1–20), the genera are divisible into two fairly homogeneous groups, the Gynotrocheae and the Macarisieae with two unclassified. The American members of the family are of three genera, namely, *Rhizophora* (Rhizophoreae) and *Cassipourea* and *Sterigmapetalum* (Macarisieae). The leaves are opposite, simple, leathery, and evergreen; the caducous stipules are between the petioles; the flowers often have detate or fringed petals and are borne in axillary clusters; the fruit of *Rhizophora* is one-celled, one-seeded, and indehiscent, whereas that of the other two is septicidally dehiscent and the several seeds are winged. Their timbers are of little or no importance in the export trade.

Cassipourea, with about 65 described species of shrubs and small to rather large trees, occurs in tropical Africa and America. *C. alba* Gris. of the West Indies is sometimes 30 feet tall in moist situations, and the wood is used for fuel and house poles. *C. latifolia* Alston occurs in the understory of the forest in Trinidad and attains a height of 50 to 60 feet and a trunk diameter of 8 to 12 inches; the inner bark has a faint smell of garlic. *C. podantha* Standley is a small Central American tree, occasionally 25 or 30 feet tall and six inches in diameter; the wood is used locally for house framing. *C. elliptica* Poir and *C. guianensis* Aubl. occur in the Guianas and the lower or middle Amazon region; they are small trees or shrubs of no economic importance.

Heartwood pale reddish brown; sapwood yellowish. Odorless and tasteless. Moderately hard, heavy, tough, and strong; sp. gr. (air-dry) 0.75 to 0.80; weight 47 to 50 lbs. per cu. ft.; fine-textured; usually straight-grained; not difficult to work, finishing very smoothly; probably not very resistant to decay.

Growth rings often present, but not always distinct. Pores small to very small, not visible without lens; angular in outline; very numerous; mostly solitary, sometimes in pairs or short rows; fairly well distributed without pattern. Vessels with both simple and multiple perforations, the scalariform plates of the latter with 3 to 40 bars; spiral thickenings absent; tyloses sometimes present, occasionally sclerotic; tangential intervascular pitting infrequent, alternate or tending to scalariform. Rays nearly all uniseriate or biseriate and few to 65 cells high; decidedly heterogeneous; cells frequently disjunctive; crystals common to abundant; ray-vessel pitting coarse to very coarse, often scalariform. Wood parenchyma invisible without lens; sparingly paratracheal and in numerous short or broken tangential lines and diffuse; cells often disjunctive; pits to vessels large, elongated, scalariformly arranged. Wood fibers with very thick, often gelatinous, walls; non-septate; pits numerous, small, distinctly bordered. Ripple marks and gum ducts absent.

Common names: Cuco (Cuba); palo de gongoli, p. de toro, p. de oreja (P.R.); palo Robinson (Dom. R.); bois l'ail, garlic wood (Trin.); bois l'ill (Dom.); water wood (Br. H.); naranjo, schimés (Guat.); goatwood (Pan.); laranjorana (Braz.).

Rhizophora. Two of the three species of this genus occur only in the eastern hemisphere, the other, *Rhizophora mangle* L., the Red Mangrove, encircles the globe. They are all much alike, being evergreen shrubs or small to large trees with the stem supported on stilt-like roots which often form impenetrable thickets along low muddy seashores and fringe the streams as far inland as the water is brackish. Aerial roots are frequently developed from the lateral branches and extend down into the silt (Plate XXXIV). The seed germinates while still attached to the tree, developing a plummet-like radicle about a foot long which eventually falls into the oozy mud all ready to put forth leaves and roots. Some are carried away by the ebbing tide and float upright until grounded, so that every newly formed river bar or island is soon populated with Mangrove seedlings. The bark contains 20 to 30 per cent of tannin which is used locally and in some regions is an important article of commerce. In harvesting the bark the wood is usually wasted, but in certain localities there is considerable demand for it for fuel and fencing, and the larger logs are sawed into construction lumber. On favorable sites, for example in the Orinoco delta and formerly about Lake Maracaibo in Venezuela, there are, or were, pure forests of Red Mangrove trees 100 feet high with trunks 18 to 24, sometimes up to 36, inches in diameter, free of branches for 30 to 40 feet, and supported 6 to 10 feet above the ground by a great mass of roots. Considerable quantities of Mangrove are consumed in Ecuador because of its accessibility in places where other timber is scarce. The logs are usually stored in water and sawed only as needed. The lumber is used for rafters, beams, and joists of buildings, for knees and ribs of boats, and also for posts, piling, railway crossties, charcoal, and firewood. There are large areas of Mangrove swamps in the Amazon estuary, but neither the bark nor the wood is utilized extensively. The range of the species extends to southern Florida and Baja California.

Heartwood light red, deepening to dark red or reddish brown, sometimes purplish; uniform or more or less striped; rather sharply defined in old specimens from the yellowish, grayish, or pinkish sapwood, which often is thick. Not highly lustrous. Odorless and without distinctive taste. Very hard, heavy, tough, and strong; sp. gr. (air-dry) 0.95 to 1.12; weight 60 to 70 lbs. per cu. ft.; texture fine; grain straight to very irregular; dry wood difficult to cut, but finishes well and takes a high polish; resistant to decay but not to marine borers.

Growth rings present or absent, usually indistinct. Pores very small to minute, not distinct without lens; thick-walled, with circular or rounded outline; moderately numerous; occurring singly and, less commonly, in pairs, short rows, or small clusters, well distributed without pattern. Vessels with scalariform perforation plates, the bars thick, few to several (4 to 10); spiral thickenings absent; tyloses often abundant; pitting finely scalariform. Rays infrequently uniseriate or biseriate, mostly 3 to 6 cells wide and of variable heights up to 100 cells or more; inconspicuous on cross section, distinct on radial; heterogeneous, at least in part; crystals and gum deposits abundant; ray-vessel pitting very coarse, often unilaterally compound. Wood parenchyma sparingly vasicentric; cells often disjunctive; pits to vessels large, often elongated and in scalariform arrangement. Wood fibers with thick, often gelatinous walls; non-septate; gum deposits common; pits numerous, very small, simple. Ripple marks absent. No gum ducts seen.

Common names: Mangrove, red mangrove (Eng.); mangle, m. colorado (Span.); paletuvier, p. rouge (French); mangue, m. vermelho (Port.); mangle zapatero (P.R.); manglier (Mart.); manggel tan (Curaçao); candelón, mangle dulce, m. tinto, tab-ché, tap-ché (Mex.); mangle gateador (C.R.); mangle rojo (Col.); purgua (Venez.); duizendbeenboom, wortelboom, zwamp mangro (Sur.); apareiba, guapereiba, mangarabeira, mangue bravo, m. do brejo, m. preto, m. sapateiro, m. verdadeiro, m. vermelho, ratimbo (Braz.); mangle geli (Ec.).

Sterigmapetalum *obovatum* Kuhlmann, the only species, is a medium-sized to large tree in high forest on non-inundated lands near Manáos, Brazil. No local uses for the timber are known. The following description is based upon a single specimen (Yale 2069; Ducke 10).

Wood light grayish brown throughout. Not highly lustrous. Without distinctive odor or taste. Very hard, heavy, and strong; sp. gr. (air-dry) 0.98; weight 61 lbs. per cu. ft.; texture medium; feel harsh; grain somewhat irregular; not difficult to work, finishing smoothly; probably not resistant to decay.

Growth rings apparently absent. Pores small to medium-sized, the largest near limit of vision; round or oval; virtually all solitary; fairly numerous, but not crowded; well distributed without pattern, though with some tendency to diagonal arrangement. Vessels with both simple and multiple perforations, the scalariform plates having 7 to 20 thin bars; spiral thickenings absent. Rays barely visible on cross section, inconspicuous on radial; 1 to 4, mostly 3, cells wide, and up to 40, sometimes to 80, cells high; decidedly heterogeneous; the marginal cells large, upright or square, the procumbent cells low and in a definite stratum; no crystals observed; pits to vessels variable in size and form, usually very large and elongated, with tendency to scalariform arrangement. Wood parenchyma fairly abundant, not distinct without lens; unilaterally paratracheal (outside only) and short aliform, sometimes confluent (Plate XLVI, 3); also sparingly diffuse; pits to vessels large and elongated. Wood fibers with thick, often gelatinous, walls; pits numerous, rather large, distinctly bordered. Ripple marks absent. No gum ducts seen.

ROSACEAE

The Rose family, with about 90 genera and 2000 species of armed or unarmed herbs, shrubs, and small to occasionally large trees, is of cosmopolitan distribution, though chiefly in temperate regions. The leaves are alternate or rarely opposite, simple or compound, sometimes with glandular teeth; stipules are usually present in pairs, sometimes adnate to the petiole; the flowers typically have five petals inserted with the numerous stamens on the edge of a disk lining the calyx tube; the fruit is various. The only strictly tropical group is the Chrysobalanoideae. The family is of little value for timber but of great economic importance for fruit, such as the apple, quince, pear, almond, apricot, peach, plum, cherry, strawberry, blackberry, and raspberry, and for ornamental plants, particularly the rose. Exclusive of the Chrysobalanoideae, which are considered separately, there are 14 genera with arborescent species in the New World. The only important timber tree is Black Cherry, *Prunus serotina* Ehrh.

Heartwood in various shades of red or brown; sapwood yellowish or flesh-colored. Luster medium to high. Odor and taste usually not distinctive. Density medium to high; texture typically fine; grain straight to irregular; working properties and durability variable, frequently good, sometimes excellent.

Growth rings usually distinct. Pores small to minute; rather few to very numerous; generally more numerous and larger in early wood; larger pores in an initial band in *Cowania,* often widely spaced in a single row in *Cercocarpus, Lyonothamnus, Quillaja,* and *Vauquelinia;* pores typically solitary, but in multiples or radial series in *Prunus,* and producing flame-like pattern in *Prunus ilicifolia* (Nutt.) Walp. Vessels nearly always with simple, round or short-oval perforations; perforation plates with few bars or reticulate openings sometimes present (*e.g., Polylepis* and *Sorbus*); spiral thickenings observed in some or all available specimens of all genera except *Cowania, Crataegus, Polylepis,* and *Quillaja.* Rays generally 1 to 4, frequently only 1 or 2, cells wide and mostly less than 25, sometimes up to 40, cells high; narrow, but up to 120 cells high in *Kageneckia;* conspicuous in some species of *Prunus,* being up to 8 or 10 cells wide in *P. serotina;* homogeneous to heterogeneous; pits to vessels generally rather small, rounded. Wood parenchyma generally not visible without lens and not always distinct with it; mostly in fine uniseriate lines and diffuse, frequently reticulate; most abundantly developed, sometimes in visible bands, in *Prunus.* Wood fibers with numerous distinctly bordered pits in all instances except some species of *Prunus;* spiral thickenings present in *Amelanchier, Cercocarpus, Heteromeles, Lyonothamnus, Osteomeles,* and *Vauquelinia.* Ripple marks absent. Vertical traumatic gum ducts common in some species of *Prunus.*

Amelanchier, with several species of unarmed deciduous shrubs and small or rarely medium-sized trees, is sparingly represented in China, Japan, and the Mediterranean region, and more abundantly in the temperate and mountainous regions of North America from Newfoundland and Alaska to Oaxaca, Mexico. Some of the plants are cultivated in gardens for the beauty of their early and conspicuous flowers and occasionally for their small edible fruits. They are practically valueless as a source of timber. Best known and largest of the American species is the Service Berry or Shad Bush, *A. canadensis* Med., of eastern United States and Canada. Though usually a small and slender tree or in parts of its range only a shrub, it occasionally attains a height of 50 to 70 feet with a trunk 12 to 18 inches in diameter.

Heartwood brown or reddish brown, usually absent from small specimens; sapwood thick, slightly brownish; appearance of lumber usually marred by numerous brown lines (pith flecks). Luster medium. Odor and taste absent or not distinctive. Hard, heavy, compact, tough, and strong; sp. gr. (air-dry) 0.85; weight 53 lbs. per cu. ft.; texture fine and uniform; grain straight to irregular; rather easily worked, taking a good polish; dark heartwood durable. Of no commercial possibilities because of the scarcity and small size of the trees.

Common names: Currant tree, Indian cherry, June berry, May cherry, pigeon berry, sarvice berry, saskatoon, service berry, shad berry, s. blow, s. bush (U.S.A.); madronillo, membrillito, membrillo, m. cimarrón, tlaxisqui, tlaxistle, tomistlacati (Mex.).

Cercocarpus, with about 20 named species of unarmed, often contorted shrubs and little trees rarely over 25 feet high, is confined to the dry interior and mountainous region of western North America from Oregon and Montana to Oaxaca, Mexico. The fruit resembles an oat grain with a long feathery tail. Because of the rich color of their heartwood the plants are commonly known in the United States as Mountain Mahogany. The largest species is *C. ledifolius* Nutt., a round-topped slightly aromatic tree rarely 40 feet tall and 30 inches in diameter. Owing to the usually small size and poor form of the trunk, the chief use of the wood is for fuel; a very limited amount is made into domestic articles of turnery and carving.

Heartwood cherry red, reddish brown, or chocolate-brown, often with intermingling of lighter and darker shades; sharply demarcated from the thin yellowish sapwood. Luster medium. Without distinctive odor or taste. Very hard, heavy, compact, and strong; sapwood tough, heartwood generally brittle; weight (air-dry) about 65 lbs. per cu. ft.; texture fine; grain irregular; rather difficult to work, but taking a high polish; very resistant to decay. Suitable for brush backs, knife handles, and small turnery.

Common names: Hardtack, mountain mahogany (U.S.A.); lentisco, ramón, zunuiña (Mex.).

Cowania, with three or four species of unarmed shrubs, rarely little trees, is limited to the dry interior region of the United States and Mexico. The viscid leaves are dentate, pinnatifid, or entire; the fruit is a cluster of achenes, each with a long feathery tail as in *Cercocarpus.* The largest species is *C. mexicana* D. Don, which, at its best development, in Arizona, is a tree 20 to 25 feet high, with an erect trunk six to eight inches in diameter and covered with shaggy grayish brown bark separating freely into thin fibrous layers. The inner bark was formerly used by the Indians for weaving into clothing, sandals, mats, and ropes. The only wood sample available (Yale 40428) was collected on the rim of the Grand Canyon in Arizona by W. F. Opdyke. Heartwood brown; sharply demarcated from the thin yellowish white sapwood. Luster medium. Odorless and tasteless. Moderately hard and heavy; texture fine and uniform; grain straight; easy to work, finishing very smoothly. Would have many uses if available in larger sizes.

Common names: Cliff rose, quinine bush (U.S.A.); chivatillo, romerillo cimarrón, romero cedro (Mex.).

Crataegus, with about 300 named species of armed shrubs and small trees, is widely distributed in north temperate regions, very sparingly in the South American Andes. The usual English name is Hawthorn. Sudworth says (*Check list of the forest trees of the United States,* 1927, p. 135): "With the exception of the genus *Eucalyptus,* doubtless *Crataegus* is the most difficult group of woody plants. The distinctions relied upon to separate many of the species include the color and number of the stamens, form and color of the fruit, the number and markings of its seeds, etc., all of which require carefully collected sets of specimens taken from the same individual plant from the early flowering to fruiting stages. Most of these characteristics are quite beyond the lay student of trees, while they are exceedingly difficult for technically trained observers. Perhaps less than 25 of our native Hawthorns could be accurately distinguished by the layman. Later studies may show also that a good many of the Hawthorns now regarded as species are actually of hybrid origin. Years of study and painstaking investigation in growing plants from seed will be required before this most perplexing group can be fully understood." Some of the species are cultivated for ornamental purposes, and the fruits of a few kinds are esteemed for jellies and preserves. The wood is not of commercial value, but is used locally to a limited extent for miscellaneous purposes requiring strength, toughness, and fine texture.

Heartwood brown or reddish brown, more or less variegated; distinct but not always sharply demarcated from the thick flesh-colored sapwood; commonly streaked with brown pith-flecks. Luster rather low. Odor and taste absent or not distinctive. Hard, heavy, compact, tough, and strong; sp. gr. (air-dry) 0.65 to 0.80; weight 41 to 50 lbs. per cu. ft.; texture very fine and uniform; grain variable; not very difficult to work, finishing very smoothly; heartwood fairly durable.

COMMON NAMES: Albespine, haw (apple, black, crimson, green, May, parsley, pear, red, scarlet, yellow), hawthorn (various kinds), senellier, thorn (black, cockspur, pear, pin, prairie, Washington, white), t. apple, t. bush, t. plum, turkey apple (U.S.A.); dopini, dopri, manzanita tejocotera, tejocote, tejocotl, vipeni (Mex.); manzanilla (Guat.); huagra-manzana, manzanita (Ec.).

Heteromeles, with a single species, *H. arbutifolia* Roem. or *H. salicifolia* (Presl) Abrams, is an unarmed shrub in the chaparral belt of the California coast ranges and Sierras southward from Mendicino and Shasta Counties into northern Baja California, Mexico; on the islands off the southern California coast, especially on Santa Catalina, it becomes a tree 15 to 30 feet high, with a very short trunk 10 to 18 inches through at the base and supporting a crown composed of many upright branches. The smooth bright red or yellow berries have a dry astringent taste, and branches covered with them are used for Christmas decorations. The wood is suitable for small ornamental turnery, but owing to its scarcity it is not utilized.

Heartwood light to dark reddish brown; rather sharply demarcated from the thin, flesh-colored sapwood. Luster fairly high. Odor and taste absent or not distinctive. Hard, heavy, and strong; sp. gr. (air-dry) 0.96; weight 60 lbs. per cu. ft.; texture fine and uniform; grain variable; not difficult to work, taking a high polish; is resistant to decay.

COMMON NAMES: California holly, Christmas berry, tollón, toyón (Calif.).

Kageneckia, with three species of unarmed evergreen shrubs and little trees less than 25 feet high, is confined to southern Chile and Argentina. Its fruit is a dehiscent 5-parted capsule containing numerous small winged seeds. Apparently the plants have no special uses. Heartwood pale orange-brown; not sharply demarcated from the lighter-colored sapwood. Luster fairly high. Odorless and tasteless. Hard, heavy, tough, and strong; texture fine and uniform; grain straight; durability doubtful.

COMMON NAMES: Bollén, huayo (Chile); durazno de la sierra, sacha-durazno (Arg.).

Lyonothamnus *floribundus* Gray, the only species, is an unarmed evergreen shrub or a bushy tree rarely 25 to 35 feet high, forming small thickets on dry sites on Santa Cruz, Santa Catalina, and a few other islands off the coast of southern California. Young stems are glossy red, old ones are covered with a dark red laminated bark which separates into long thin strips suggesting a large grape vine; the twigs are glossy red. The leaves are of two types, simple and fern-like; the fruit consists of very small dehiscent capsules. Owing to the scarcity of the larger stems the wood is not utilized. Heartwood reddish brown with a yellowish tinge; not sharply demarcated from the lighter-colored sapwood. Luster medium. Odorless and tasteless. Hard, heavy, tough, and strong; texture fine and uniform; grain fairly straight. Not difficult to work, finishing very smoothly.

COMMON NAMES: Ironwood—Catalina, Santa Cruz (Calif.).

Malus, the apple genus, with numerous species of mostly unarmed small to medium-sized trees, is widely distributed in the north temperate region. It is frequently included, along with *Sorbus,* in *Pyrus* which, in the restricted sense, is the pear genus. The horticultural varieties of apples are derived from *M. pumila* Mill. of southeastern Europe and central Asia. The species has become widely naturalized in northeastern North America and is the only member of the genus supplying commercial timber in the United States; the amount is small and uncertain and the principal use is for handsaw handles. There are nine native species of *Malus* in the United States and Canada, all eastern except one, *M. rivularis* (Dougl.) Roem., which ranges from California to Alaska. The tart yellow and red fruits, crab apples, are used for making jelly and preserves. Some of the trees grow 25 to 30 feet high and 10 to 12 inches in diameter, but the timber is used only locally for tool handles, levers, and small domestic articles.

Heartwood reddish brown; distinct but not always sharply demarcated from the yellowish sapwood. Luster medium to low. Without distinctive odor or taste. Hard, moderately heavy, tough, and strong; sp. gr. (air-dry) 0.70 to 0.80; weight 44 to 50 lbs. per cu. ft.; texture fine and uniform; grain straight to irregular; easy to work, finishing smoothly; holds its place well when manufactured.

COMMON NAMES: Crab, crab apple (various kinds), c. tree, garland tree (U.S.A.).

Osteomeles, with several species of unarmed or thorny evergreen shrubs and small trees, is sparingly represented in Polynesia and China and more abundantly in the Andean region of South America from Venezuela to Peru. The Pujín of the eastern Cordillera of Ecuador is *O. latifolia* H.B.K. It is common at elevations of 9000 to 11,000 feet and develops a very thick trunk which divides near the ground into several short stems supporting a broad crown. It is unarmed; the rather large finely serrate leathery leaves are rusty brown on the under side; the small bright red fleshy fruits are like those of *Crataegus*. The wood is used locally to some extent for heavy construction. *O. glabrata* H.B.K. of the same locality has a similar fruit and is called Caisha-pujín (thorny Pujín) because the branchlets are spine-tipped. It is a shrub or little tree with small glossy leathery leaves and rather disagreeably scented flowers with white petals and red anthers. The wood is used for plow handles and whip stocks. This species also occurs in the Andes and coast Cordillera of Venezuela, where it is known as Mortiño; the trunks are sometimes over 30 feet long, according to Pittier (*Plantas usuales de Venezuela*, p. 301), but the timber is not utilized. *O. obtusifolia* Kunth is a shrubby thorny tree common at elevations of 6500 to 7500 feet in the Province of Loja, Ecuador, where its vernacular name is Quiqui. Heartwood (of two species) reddish brown; not sharply demarcated from the lighter-colored sapwood. Luster medium. Odorless and tasteless. Hard, heavy, tough, and strong; texture fine; grain fairly straight; easy to work, taking a high polish.

COMMON NAMES: Membrillito, mortiño, tiboi (Venez.); caisha-pujín, pujín, quiqui (Ec.).

Polylepis, with several species of unarmed evergreen shrubs and little trees, grows at high elevations in the Andean region of South America from Venezuela to Argentina. Rimbach says (*Tropical Woods* 31: 3) that in places on the outer slope of the western Cordillera in Ecuador the upper limit of arborescent vegetation is formed by pure stands of *Escallonia, Polylepis,* and *Gynoxis,* bordering, sometimes abruptly, upon the paramo, the grass and herb mat between timberline and perpetual snow. The principal species there is *P. lanuginosa* H.B.K., called Motilón, Panza, or Quiñua. It is generally less than 15 feet high, has a very crooked irregularly branched trunk, with the bark separating into many red papery layers. The small leaves are odd-pinnate, with 5 to 7 leathery leaflets, glossy green above and whitish woolly beneath, clustered at the jacketed nodes; the small flowers are borne in slender pendant racemes. The wood is not utilized. The Venezuelan species, known as Coloradito or Colorado, is *P. sericea* Wedd.; it also grows at the highest limits of vegetation. The Queñua or Tabaquillo of Argentina, *P. racemosa* R. & P., is a shrub or a tree up to 25 feet in height and 10 inches in diameter. Heartwood brown or reddish brown; not sharply distinguished from the lighter-colored sapwood. Luster medium. Odorless and tasteless. Hard, heavy, and strong (*P. sericea*) to rather light and brittle (*P. lanuginosa*); texture fine; grain irregular; easy to work, finishing very smoothly.

Common names: Coloradito, colorado (Venez.); motilón, panza, guiñoar, quiñua (Ec.); kehuiña (Boliv.); queñua, tabaquillo (Arg.).

Prunus, in the broad sense of the generic term, includes more than 100 species of thorny or unarmed shrubs and small or occasionally large trees, chiefly valuable for their fruits, which include the almond, apricot, peach, plum, and cherry. This large group is sometimes subdivided into several smaller genera, and from the standpoint of their fruits there is good reason for doing so. The woods exhibit greater variation than is to be found in any other genus of the Rosaceae, but they have not been studied fully enough to determine the taxonomic significance of the differences observed.

The American species are of three general types, namely, Plum trees, Cherry trees, and Cherry Laurels. In the first group the fruits are comparatively large and slightly 2-lobed, often covered with a glaucous bloom; the flesh is pulpy and the stone is flattened. The trees, which usually are thorny, are native to the north temperate zone and many species and varieties are cultivated. There are about a dozen American species, of which the best known is *Prunus americana* Marsh., a spreading tree, rarely 35 feet high, with a range covering the eastern two-thirds of the United States and southern Canada, and merging into *P. mexicana* S. Wats. in northern Mexico. The brown or reddish brown wood is not utilized; its anatomy is of the general type of the second group.

Common names: Black sloe, plum (Chickasaw, goose, hog, horse, red, sloe, wild, w. goose, yellow), sloe (U.S.A.); ciruelo (Mex.).

In the second or true cherry group the fruits are smaller fleshy drupes with round seeds and without bloom. There are several species of unarmed shrubs and trees in temperate North America and one, the Wild Black Cherry (*P. serotina*), is native to the entire eastern half of the United States and parts of Canada and Mexico. Regarding its occurrence in western South America, Popenoe says (*Cont. U.S. Nat. Herb.* 24: 5: 114–115; 1924): "It has generally been considered by botanists that this species is indigenous in Ecuador, but there is little evidence to substantiate this belief. It is thoroughly naturalized in several regions, but it is known in all of them under the name [Capulí] taken from the Nahuatl tongue of Mexico, and history records its introduction from Mexico into other parts of South America after the Conquest. It is most probable that the species was not known south of Central America in pre-Columbian times. . . . The botany of this interesting and valuable fruit tree has long been in confusion. Recently Blake (*Journal of Heredity* 13: 51–62; 1922) has gone over the available material in the herbaria at Washington and has reached the conclusion that Capulí or Capulín, grown from Mexico to Peru and Chile, is a cultivated form of the northern Black Cherry, *Prunus serotina* Ehrh., which occurs in a wild state from Nova Scotia to Mexico. Other botanists have considered it to be distinct, and it is often mentioned in literature under the name *Prunus salicifolia* H.B.K. or *Prunus capuli* Cav." A study of the wood tends to confirm Blake's opinion and also Standley's conclusion (*Trees and shrubs of Mexico*, p. 340) that *P. virens* (Woot. & Standl.) Standl. is "perhaps not sufficiently distinct from *P. capuli*" to be given specific rank.

Prunus serotina is the only American species of commercial value for its timber. It is often shrubby in poor situations and especially at the northern limit of its range, but on rich moist soil in the Appalachian region it attains a height of 100 feet or more, with a long well-formed trunk four to five feet in diameter. Like other members of the genus, it has a bitter aromatic bark and leaves; the bark of the branches and roots is used medicinally, the active ingredient being hydrocyanic acid. The ripe black somewhat astringent fruit is used for flavoring alcoholic liquors; near Abato,

Ecuador, according to Popenoe (*loc. cit.*), "there is a famous tree whose fruit is large, very juicy, and as sweet and pleasant as the best European cherries." The lumber was formerly prized for cabinet work and fine furniture in the United States, but its principal use now is for making blocks for electrotypes.

Heartwood brownish, often with greenish tinge, deepening upon exposure to rich reddish brown with a golden luster; sharply demarcated from the yellowish sapwood. Scent mildly aromatic; taste not distinctive. Of medium density, firm and strong; average sp. gr. (air-dry) about 0.58; weight 36 lbs. per cu. ft.; texture fine and uniform; grain generally straight; easy to work, finishing very smoothly; holds its place exceptionally well when manufactured; is moderately resistant to decay.

Common names: *P. serotina:* Cherry—black, rum, whiskey, wild, w. black (U.S.A.); capollín, capulí, capulín, detze, ghohto, pa-kshmuk, tnunday, xeugua (Mex.); cerezo (Guat.); cerezo de los Andes, mují (Venez.); capulí, capulín (Ec.). *Other species:* Cherry—bird, bitter, choke, fire, pigeon, pin, quinine, wild, w. red (U.S.A.).

In the third group, which includes the Cherry Laurels (*Laurocerasus*), there are several species of unarmed evergreen shrubs and trees native to both temperate and tropical regions. Their fruit is the shape of a cherry, but the flesh is thin and not pulpy, and the stone is thin-shelled.

Prunus ilicifolia (Nutt.) Walp. of California and Baja California is usually a chaparral shrub, but in sheltered canyons grows to a height of 25 feet, with a trunk a foot in diameter. Its leaves are spine-tipped like Holly (*Ilex*), but there is a variety (var. *integrifolia* Sudw.) with entire leaves. The very dense, tough, harsh-textured, brownish wood exhibits the flame-like pore-pattern characteristic of *P. Laurocerasus* L. of Europe.

Common names: Hollyleaf cherry, islay, mountain evergreen cherry, Spanish wild cherry (Calif.); isláy, ysláy (Mex.).

The most widely distributed species of this group is *Prunus myrtifolia* (L.) Urb., a small to medium-sized slender tree rarely 40 feet high, occurring from southern Florida through the West Indies to southern Brazil and northwestern Argentina. The timber is sparingly utilized. The heartwood is light clear red; not sharply demarcated from the flesh-colored sapwood, which turns orange or bronze on the surface as in *Alnus*.

Common names: West Indian cherry (Florida); ant's wood, wild cassada (Jam.); almendrillo, cuajaní hembra, cuajanincillo (Cuba); membrillito (Dom. R.); amandier à petites feuilles (Haiti); amandier des bois (Mart.); noyau (Guad.); almendro (Venez.); marmelo bravo, m. do matto, virarú (Braz.); caá ró, persiguero bravo (Arg.); duraznero de monte, d. bravo, rama negra, tarumán (Urug.).

Prunus occidentalis Sw. is a West Indian tree similar to the preceding, but with larger leaves and somewhat greater size, being occasionally 65 feet tall and 30 inches in diameter. Macfadyen says (*Flora of Jamaica*): "It is from the kernels of the drupe that the celebrated liqueur, the Noyau of Martinique, is prepared. They yield a flavor much superior to that of the peach, being rich, oily, and nutty, combined with that of prussic acid. The timber of the tree is of a red color resembling Cedar and is very hard and durable and, from its taking a fine polish, makes a beautiful flooring for houses, but it is not adapted for outdoor work or where it is exposed to the weather, for in such situations it very soon rots." In Cuba the bark and leaves are used medicinally and the wood is employed for posts, poles, implement frames, and railway crossties.

Heartwood rich dark reddish brown, often somewhat variegated; not sharply demarcated from the flesh-colored sapwood. Luster medium to high. Without distinctive scent or taste when dry. Very hard, heavy, tough, and strong; sp. gr. (air-dry) 0.90 to 1.05; weight 56 to 66 lbs. per cu. ft.; texture medium to rather coarse; grain variable; not difficult to work, finishing with a lustrous natural polish. Probably without commercial possibilities.

Common names: Pruan, prune tree (Jam.); almendrillo, almendrón (P.R.); almendro, cuajaní, c. macho (Cuba); membrillo (Dom. R.); amandier (Haiti); amandier, noyeau (Dom.). *Other species:* Cherry laurel, evergreen cherry, laurel cherry, laury mundy, mock olive, m. orange, wild orange, w. peach (U.S.A.); carretero, cerezo, cortapico, huaparín, iza, wasiki (Mex.); congrejillo (Salv.); cacique macho, mariquita (C.R.); duraznillo del cerro, palo de luz (Arg.).

Quillaja, with a few species of unarmed evergreen shrubs and small trees, is confined to southern Brazil, northeastern Argentina, Uruguay, and Chile. The leathery leaves are simple and the fruit is a dehiscent capsule containing numerous small winged seeds. The best known species are *Q. brasiliensis* Mart. and *Q. Saponaria* Mol. The bark contains saponin, which is useful in cleansing fine textiles, and tannin, which serves as an astringent in medicine. The timber is sparingly utilized, presumably because of its small size and scarcity. The only wood specimen available (Yale 23783; I. E. Webber 4) is from a little tree of *Q. Saponaria* which was planted for ornamental purposes in Berkeley, California. Heartwood absent; according to Pio Corrêa (*Flora do Brazil,* p. 70) it is black in old trees; sapwood yellowish. Luster medium. Of medium density; texture fine; easy to cut. Of no commercial possibilities.

COMMON NAMES: Timbouva? (Braz.); lava cabeza, palo de jabón, quillay (Urug.); palo de jabón (Arg.); quillay (Chile).

Sorbus, with numerous species of unarmed shrubs and small trees, is well represented in Europe and eastern Asia, sparingly so in North America. The European Rowan, *S. Aucuparia* L., is often cultivated as an ornamental tree in Canada and New England. The only arborescent American species is *S. americana* Marsh. which has a transcontinental range in Canada and northern United States. It has large pinnate leaves and inhabits uplands, hence the common name, Mountain Ash. It is rarely 30 feet high, with a smooth trunk sometimes 12 inches in diameter. The small yellow or orange-red fruits are borne in conspicuous flat cymes or clusters and add to the attractiveness of the tree for planting. The wood is not utilized. Heartwood brown; not sharply demarcated from the lighter-colored sapwood. Luster medium. Odorless and tasteless. Of medium density and weight, not very strong; texture fine; grain straight; very easy to work, finishing very smoothly; durability low. Of no commercial possibilities.

COMMON NAMES: Dogberry, elder-leaved sumach, life-of-man, missy-moosey, mountain ash, m. sumach, peruve, roundwood, rowan tree (U.S.A.).

Vauquelinia, with a few species of unarmed shrubs and small trees rarely 25 feet high, is confined to southwestern United States and Mexico. The best known species are *V. californica* (Torr.) Sarg. and *V. corymbosa* Corrêa. The trees have slender trunks, stiff and crooked branches, and thin dark reddish brown bark. The fruit is a small woody dehiscent capsule with winged seeds. The wood and bark are occasionally used locally as the source of a yellow dye. The only wood sample available for this study (Yale 14429) is of *V. californica,* collected by S. B. Detwiler in Superstition Mountain, Arizona. Heartwood rich reddish brown or chocolate-colored, sometimes streaked with red; sharply demarcated from the flesh-colored sapwood. Luster medium. Without distinctive odor or taste. Very hard, heavy, compact and strong; texture fine; grain mostly irregular; not very difficult to work, taking a high polish. Of no commercial possibilities.

COMMON NAMES: Árbol prieto, guauyul, guayule, palo prieto, p. verde (Mex.).

ROSACEAE-CHRYSOBALANOIDEAE

THE Chrysobalanoideae comprise 12 genera and more than 200 species of unarmed evergreen shrubs and trees of pantropical distribution, though most abundant in tropical America. They have simple entire alternate leaves; the flowers differ from those of the other Rosaceae in having the style inserted near the base of the ovary; the fruit is a drupe, often very large. The group is sometimes combined with the Prunoideae to form a separate family, Amygdalaceae, but from the standpoint of the woods the Prunoideae belong with the Rosaceae and the Chrysobalanoideae seem distinct enough to be segregated into a family of their own. The six genera with arborescent species in America are *Chrysobalanus, Couepia, Hirtella, Licania (Moquilea)*, and *Parinarium.* Their woods are so similar, judging from the fairly plentiful material of them at hand, that the differences between genera appear no greater than the variation within a single genus. The timber is not of much commercial importance.

Heartwood light to dark reddish brown, often with a grayish hue; sapwood pale brown with a pinkish tinge or superficially yellowish brown; transition to heartwood gradual. Luster low or occasionally medium. Odorless and tasteless when dry, except for a somewhat rancid scent in *Parinarium barbatum* Ducke. Moderately to very hard and heavy; sp. gr. (air-dry) 0.75 to 1.10; weight 44 to 69, mostly over 55, lbs. per cu. ft.; texture medium to rather coarse; grain fairly straight to irregular; working properties fair, sometimes interfered with by presence of siliceous grit; usually susceptible to decay, but some species are said to be highly resistant to marine borers.

Growth rings generally present, owing to periodic wider spacing of parenchyma bands. Pores solitary; rather few; noticeably different in size in same specimen, the largest barely to readily visible without lens; irregularly distributed without definite pattern, though sometimes with tendency to diagonal arrangement. Vessels with simple perforations; without spiral thickenings; tyloses common to abundant, sometimes thick-walled, occasionally sclerotic; pits to wood fibers fairly numerous, rather small. Rays mostly uniseriate, sometimes biseriate; commonly less than 25, sometimes up to 50, rarely to 80, cells high; decidedly heterogeneous; gum deposits abundant in heartwood; cells thick-walled; pits to other parenchyma cells large; ray-vessel pit-pairs of two types: (1) rounded or oval, showing small apertures and wide borders, and (2) large to very large, oval to much elongated, generally in scalariform arrangement with long axis of pit vertical, and showing narrow or incomplete borders. Wood parenchyma abundantly developed; distinct with lens and occasionally visible without it; vasicentric, 2 to 4 cells wide; metatracheal in fine wavy lines or bands, 1 to 3 cells wide, of variable spacing but often about a pore-width apart; gum deposits common to abundant. Wood fibers with thick to very thick walls and numerous large bordered pits. Ripple marks absent. No gum ducts seen.

Chrysobalanus, with a few species of shrubs and small trees rarely over 30 feet high, is widely distributed along the coastal regions of tropical and subtropical America and western Africa. The best known species is the Coco Plum or Icaco, *C. icaco* L., which has a large edible fruit so wrinkled as to suggest the face of a monkey. The seeds are rich in oil which is sometimes expressed for industrial purposes; the bark, leaves, and roots are astringent. The wood is used for fuel and minor local purposes.

Common names: Coco plum, fat pork, gopher plum, pigeon plum, pork-fat apple (English); hicaco, icaco, jicaco (Span.); icaco cimarrón, i. común, i. de costa, i. de montaña, i. de playa, i. dulce, i. negro, i. peludo, icaquillo (Cuba); zicaque (Haiti); icaque, i. des bois (Guad.); xicaco (Mex.); kulimiro (Br. G.); prunier coton, p. de l'anse (Fr. G.); ajurú, guagerú, guajurú, uajurú (Braz.).

Couepia, with numerous species of shrubs and medium-sized to rather large trees, is sparingly represented in western Africa and abundantly in tropical America from southern Mexico to Brazil. The timber is of no commercial value.

Common names: Guayabito de tinta, pío, uspib, uspío, uzbib, zapote amarillo, zapotillo (Mex.); baboon cap, monkey cup (Br. H.); sunsapotillo, ulezapote, ulozapote, uluzapote, zapote bolo, zapotillo, z. amarillo, zunzapotillo (Salv.); munzap (Hond.); olosapo (C.R.); merecure, querebere (Venez.); anaura, doekoelia, doekoelie, ganmasagon, japopalli, japoparé, joenoepé, kaierieballi, kweebie (Sur.); castanha de gallinha, c. pendula, cupéra, macucú, marimary, mary-rana, oiticica, oity chiado, o. coró, pajurá, paracari, parinary, pirá-uchy, tucuribá, uchirana, uchy-rana, u.-r. grande, uixirana, umary-rana (Braz.); capricornia, mashusacha, parinári, p. de altura, p. de seniso, sacha-umari, uchpa-parinari (Peru).

Hirtella, with many species of shrubs and small to occasionally medium-sized trees, is chiefly a tropical American genus, though there are a few representatives in western Africa. The bark is a source of tannin, and the ashes, which are rich in silica, are sometimes used in the making of native pottery and dishes. The timber is employed locally for fuel, charcoal, and some heavy construction.

COMMON NAMES: Hicaquillo, icaco de aura, i. prieto, siguapa, teta de yegua (Cuba); icacillo, teta de burra, t. de b. cimarrón (P.R.); bois canari, cauto (Trin.); cajetillo, icaquillo, uayamche (Mex.); grenada, pigeon plum, wild coco plum, w. pigeon plum (Br. H.); icaco montes (Salv.); pasta (Hond.); pelo de Indio (Nic.); serén-gró (C.R.); camaroncillo, carapato, chicharrón (Pan.); guamo mestizo (Col.); oreja de león (Venez.); booka-booka, buku-buku, fat pork, fukurero-kauta, kauta, warracabradanni (Br. G.); bokko-bokkoton, boko-bokton, foengoe hoedoe, f. pau, kauro-bandikoro, koembotassi, miendjoe, vonkjout, wooitanokombetassi (Sur.); bois de gaulette (Fr. G.); ajurú, carapé-rana, comandatuba, macucú, m. rana (Braz.); quinilla, yaco shembillo (Peru).

Licania, with many species of shrubs and medium-sized to large trees, is widely distributed in tropical America, but is most abundant in Brazil and the Guianas. The timber is used to a considerable extent locally for charcoal and for heavy construction not in contact with the ground, as its resistance to decay is poor.

The most important species is the Oiticica, *Licania rigida* Benth., of northeastern Brazil. Its ovoid fruit contains a seed having an average weight of three grams and an oil content of about 60 per cent. The seeds have become a highly valuable article of commerce, as the oil is well adapted for use in paints and varnishes.

The two best known species in the Guianas are *Licania mollis* Benth. and *L. heteromorpha* Benth. The former is said to be plentiful in the hilly lands of British Guiana, where it has an average height of 80 feet and trunks large enough to produce timbers squaring 14 inches. The wood is of a brownish color and has an unpleasant odor when fresh. It is very hard, heavy, and strong, but is not used extensively on account of the difficulty in working it.

Licania platypus (Hemsl.) Pittier is a Central American forest tree of stately proportions. The fruit, known in British Honduras as the monkey apple, is often over six inches long and four inches through; its yellow juicy and somewhat fibrous flesh is of a slightly acid flavor and is considered by some superior to the zapote (*Achras*). Two other species supplying some timber for heavy construction in Central America are *L. arborea* Seem. and *L. hypoleuca* Benth. The latter occurs also in the forests of Colombia and Venezuela, and the range of the former extends along the dry coast of Mexico.

COMMON NAMES: Icaquillo (Cuba); bois gris (Grenada); gaulette rouge, icaque montagne (Guad.); bois gris, case (Trin.); caca de niño, cacahoanache, cacahoanantzin, cacahuananche, cacahuate, caña dulce, frailecillo, mesonzapote, mezontli, palo de fraile, quirindol, q. cacahoananche, totopostle, zapote amarillo, z. borracho, z. cabello (Mex.); chozo, monkey apple, pigeon plum (Br. H.); chozo, encino (Guat.); canilla de mula, jobo, roble, súngano, sunza, sunzapote (Salv.); borraco, urraco (Hond.); alcornoque, zapote (C.R.); camaroncillo, carbonero, rasca, raspa, sangre, urono, zapote (Peru); caña dulce, carbonero, garcero, quitosol (Col.); icaquito, mezela (Venez.); aruadanni, burada, coutabally, iron Mary, kairiballi, kautaballi, kiakia, konoko, koonookoo, kunoko, marishiballi, muri kautaballi, uni, u. kia-kia (Br. G.); alauna, anaura, bongro, bosoho, foengoe, f. pau, hegron anaura, hoogland tapoeripa, iengibarki, japopalli, jappoparé, kaieriballi, kauston, kauta, koepésini, kokoho, kwata, kwebie, kwepie, kwepirian, man fengoe, m. gorro-gorro, m. vonkhout, marenballi wadilikoro, marisiballi, m. kibelekoberoe, m. talaroe, oni-knia-knia, pau avangolja, p. saandri, sabanna iengi-barki, soerorme umbakeloiré, solo-solo-bakelotje, soro-soro-unbakaloiré, sponsehoedoe, vonkhout, wekeroe koepésinirian (Sur.); anaoura, bois gaulette rouge, caligni, couépi, gris-gris, g.-g. coumaté, g.-g. rouge (Fr. G.); ajurú, anauerá, caraipé, c. tariian, caripé, c. rana, copuda, c. miuda, cutimandioca, guariuba, macé de fogo, macucú, milho cozido preto, oitiseiro, oiticica, oity grande, pajurá-rana, piachirana, pintadinho, turiuva (Braz.).

Moquilea, with numerous species of shrubs and trees, has the same range as *Licania* and the two genera are not always kept separate. It is most abundant and best known in Brazil. The Oity, *Moquilea tomentosa* Benth., is extensively planted in parks and along streets, particularly in eastern and southern Brazil; the crowns are kept in shape by frequent trimming. The tree is common also in the forest, and there has a short thick bole 24 to 36 inches in diameter. The timber is heavy, hard, and strong, and is used in civil and naval construction and carpentry, and also for fence posts, piling, and railway crossties.

There are several Brazilian species, *e.g.*, *Moquilea utilis* Hook., commonly called Caripé or Caraipé, which attain good proportions, but their timber is not highly esteemed, being hard to work and of low durability in contact with the ground. The siliceous bark, however, is of great usefulness to the natives in the manufacture of fireproof pottery, comprising every kind of cooking utensil. In relating his experiences on the Amazon, Spruce says of this pottery (*Notes of a botanist on the Amazon and Andes* 1: 12): "It was made of equal parts of a fine clay, found in the beds of igarapés, and of calcined Caraipé bark; but in other places where I have seen the manufacture of pottery carried on (and there is no Indian's house in the Amazon valley where it is not familiar) a much smaller proportion of the bark was used. The property which renders the bark available for this purpose is the great quantity of silex contained in it. In the best sorts, such as I afterwards saw on the river Uaupés, the crystals of silex may be observed with a lens even in the fresh bark; and the burnt bark turns out a flinty mass (with a very slight residuum of light ash, which may be blown away), so that for mixing with clay it requires to be reduced to powder with a pestle and mortar. The bark I saw at Caripi is, however, much less siliceous and, when burnt, may be broken up with the fingers."

COMMON NAMES: Sonzapote (C.R.); perquetano (Col.); kauta (Br. G.); couépi, maho (Fr. G.); camacary, caraipé, caripé, guayti, macucú, oiti, oiticica, oity, o. coroya, o. de praia, turiuva (Braz.).

Parinarium, with 75 or more species, mostly shrubs or small trees, though a few attain rather large proportions, is very widely distributed in the tropical and subtropical regions of the southern hemisphere. The wood of the Luísin of the Philippines, *P. corymbosum* Miq., is not durable in contact with the ground or exposed to the weather, but it is rarely attacked by insects and is one of the most resistant of all woods to the attacks of teredo. It is accordingly used for salt-water piling, though it is necessary to protect the parts above water with artificial preservatives, such as thick coatings of paint or tar. The Foengoe or Yonkhout of Surinam, believed to be a species of *Parinarium,* although the names are applied to species of other genera of the family, has properties very similar to the Luísin. The high resistance of the wood to marine borers is ascribed to the presence of silica bodies in the cell cavities.

The best known American species is *Parinarium campestre* Aubl. of the Guianas, Trinidad, and northern Brazil. It is a common tree in British Guiana, where it attains an average height of about 75 feet, with a diameter of 24 to 30 inches. The timber, which is hard and heavy and of a uniform grayish brown color with a pinkish hue, is little used, but deserves a trial for salt-water piling with the tops treated as stated above. According to Ducke (see *Tropical Woods* 44: 43), *P. montanum* (Aubl. in part) Huber is known in Amazonia by the names Pajurá and Paranary; it bears edible fruits. *P. Rodolphi* Huber, called Parinary and Paranary, has fruits which are scarcely edible. The name Pajurá is given to other trees, including *Couepia bracteosa* Benth., and to *Lucuma speciosa* Ducke of the Sapotaceae.

COMMON NAMES: Cautoro (Trin.); perefuetano (Col.); clavellino negro, tostado (Venez.); boohoorada, burada, buhurada (Br. G.); beherada, behoerada, beurata, boeirata, buirata, foengoe, f. hoedoe, f. pau, koebésini, koepesiini, vonkhout (Sur.); nefle, ouroucou-merepa, parinari (Fr. G.); caraipé-rana, farinha secca, macucú, pajurá, p. de matta, paranary, parinary (Braz.); uchpa-umarí (Peru).

RUBIACEAE

THE Rubiaceae comprise more than 350 genera and 6000 species of annual or perennial herbs, erect or climbing shrubs, and small, medium-sized, or rarely large trees of cosmopolitan distribution but most abundant in the tropics. Many of the plants are armed with spines; the leaves are opposite or verticillate, simple, entire, and usually provided with stipules which occasionally are leaf-like; the flowers are sometimes showy; the fruit is a capsule, berry, or drupe.

The family is the source of some important commercial products, notably coffee (*Coffea*) and quinine (*Cinchona*); others, of less value, are madder, an orange-red dye from the roots of *Rubia;* ipecac, a drug from the roots of *Cephaelis;* gambier, a tanning agent from the leaves of a species of *Uncaria;* and various decorative plants, notably the Gardenia. There are few economic timbers. Some of the largest trees grow in West Africa; the Badi or Bilinga (*Sarcomphalus*) has a durable yellow wood useful for construction and veneers, and species of *Mitragyna, Morinda,* and *Pausinystalia* supply local needs for building material. A Siamese species of *Gardenia* is the source of limited amounts of small timber suitable for most of the purposes of Boxwood.

There are arborescent species of about 70 genera in tropical America, but with a single exception they do not contribute to the world's timber market. The Cuban Degame (*Calycophyllum*) is well known to makers of archery bows under the trade name of Lemonwood; another species, the Palo Blanco of Argentina, appears suitable for weavers' shuttles. Some of the other light-colored, fine-textured, tough and strong woods may eventually find a foreign market, particularly for articles of turnery. The following description is based on specimens of 58 genera.

Heartwood usually not distinctly colored, being whitish, yellowish, or brownish, sometimes with a purplish tinge or streaks; bright yellow, orange, or yellow-brown, frequently with a greenish hue and dark reddish brown streaks in *Chimarrhis, Hamelia,* and *Morinda;* uniform brown in *Pinckneya* and *Sickingia;* brown with dark striping in *Erithalis;* yellowish to purplish brown with greenish yellow streaks in *Exostema;* sapwood white to brownish, becoming rose or deep red in *Sickingia* and likely to turn bluish in *Genipa.* Luster medium to rather low. Odor lacking or not distinctive in dry specimens; taste slightly bitter in the yellow group. Density usually medium, sometimes rather to very high (maximum in *Erithalis* and *Exostema*); texture fine (*e.g., Exostema*) to coarse (*e.g., Henriquezia, Pinckneya,* and *Platycarpum*), mostly rather fine and uniform; grain variable, but commonly fairly straight; working properties good; resistance to decay poor except in deeply colored heartwood.

Growth rings usually present but often indistinct; ring-porous structure in *Cephalanthus occidentalis* L. and *Pinckneya pubens* Michx. Pores typically small to very small, numerous to very numerous; mostly in small multiples, but nearly all solitary in *Coutarea, Erithalis, Exostema, Gleasonia, Ixora, Kotchubaea, Ladenbergia, Randia, Remijia, Retiniphyllum,* and *Sphinctanthus;* fairly uniformly distributed without distinctive pattern in most instances; arranged in irregular tangential or broken concentric bands in part in *Henriquezia* and *Platycarpum,* and in radial, diagonal, or somewhat tangential lines or bands in a few others, *e.g., Genipa, Gleasonia,* and *Ladenbergia.* Vessels with simple perforations; reticulate-scalariform or otherwise malformed plates occasionally present also; no spiral thickenings seen; pits vestured, mostly small to minute, alternate; finely scalariform pitting in part in *Cephaelis;* tyloses sometimes present (*e.g., Henriquezia* and *Kotchubaea*); gum deposits common to abundant. Rays often 2-sized; uniseriate, and up to 5, sometimes to 10, cells wide, and few to nearly 200, commonly less than 50, cells high; all uniseriate or partly biseriate in *Alibertia, Bertiera, Casasia, Cephalanthus, Coutarea, Erithalis, Exostema, Gleasonia, Henriquezia, Kotchubea, Platycarpum, Retiniphyllum,* and *Tocoyena;* rays 6 or more cells wide in *Coussarea, Elaeagia, Faramea, Ferdinandusa, Macrocnemum,* and *Morinda,* occasional in *Capirona, Chimarrhis, Hamelia, Rudgea,* and *Warscewiczia;* 1 to 5 cells wide in the others, with considerable range in different species of same genus; vertically fused rays common; heterogeneous, often decidedly so, with high margins of square or upright cells; by-pass vessel members common; sheath cells numerous in *Cephaelis, Faramea, Morinda,* and *Pentagonia;* large crystals sometimes present (*e.g., Guettarda*); raphides seen in *Faramea, Morinda,* and *Psychotria;* ray-vessel pitting fine to very fine, frequently unilaterally compound. Wood parenchyma very sparingly paratracheal in most genera; sometimes diffuse to reticulate; rarely in narrow tangential or broken concentric bands; aliform and confluent in *Henriquezia* and *Platycarpum;* large crystals occasionally present (*e.g., Ixora*); raphides noted in *Morinda.* Wood fibers often septate; walls medium to very thick and gelatinous; pits with slit-like extended apertures and very small to medium-sized borders, being largest in *Henriquezia* and *Platycarpum.* Ripple marks absent. No gum ducts seen.

Alibertia, with about 25 species of shrubs and little trees, is widely distributed in tropical America. The best known species is *A. edulis* A. Rich., so-named because of its comestible fruit. The rather hard and heavy, fine-textured, grayish brown wood is not utilized for any special purpose.

COMMON NAMES: Guayabita del Pinar, pitajoní, p. hembra (Cuba); costarica (Mex.); wild guava (Br. H.); guayaba de monte (Guat.); torolillo (Salv.); lirio (Hond.); madroño, m. de comer, trompillo (C.R.); lagartillo, madroño, m. de comer, trompito, trompo (Pan.); perita (Col.); goyave noir (Fr. G.); nigua (Boliv.); apuruhy, marmelinho do campo, puruhy, p. grande, p. pequeno, p. sinho, uapuruhy (Braz.).

Alseis, with about eight species of small to medium-sized trees, is of infrequent occurrence from Mexico to Brazil and Bolivia. The wood is light yellowish brown, moderately hard and heavy, rather fine-textured, and easily worked.

COMMON NAMES: Cacao-ché, hasché, jasché (Mex.); wild mamee (Br. H.); totumillo (Venez.); mishu-quiro, palo blanco (Peru); serrilha (Braz.).

Amaioua, with about six species of shrubs and small trees, occurs in Cuba, Panama, and through the Guianas to southeastern Brazil. The brownish, rather hard, heavy, strong, and fine-textured wood is used for making tool handles and sometimes in interior construction.

COMMON NAMES: Café cimarrón, cafetillo, c. cimarrón, palo café, pitajoní, p. cimarrón, p. macho (Cuba); camayung (Trin.); madroño (Pan.); kumara-mara (Br. G.); boaka vienga, kwariroman konokodikoro, omi marmadosoe, tapoeroe erere, t. vienga, tomoto mokjin, womimarmadosoe (Sur.); graine à tatou (Fr. G.); amaiua, goyabeira preta, marmalada brava, purahy grande, p. g. da matta (Braz.).

Anisomeris, with about 30 species of shrubs and little trees, often with spinose branchlets, is distributed from Guatemala to Peru. The wood is pale yellowish brown, moderately heavy and fine-textured. No special uses known.

COMMON NAMES: Clavo (Guat.); cunshicashan, cunshu-huacran, rifari (Peru); espino blanco (Boliv.).

Antirrhoea, with about 30 species of shrubs and trees up to 50 feet in height, occurs in eastern Asia to tropical Australia and in the New World in the West Indies and Panama. Some botanists segregate the American species into a distinct group, *Stenostomum,* but according to Standley (*North American Flora* 32: 263), "there appears to be no necessity for such a treatment." The yellowish, hard, fine-textured wood is used locally for the same purposes as Maple (*Acer*).

COMMON NAMES: Gold-spoon, pegwood, pigeon wood, Susan wood (Jam.); caobilla de costa, llorón, Juan Perez, vera, virgueta naranjo (Cuba); aquilón, boje, palo de quina, quina, tortuguillo (P.R.); candela, zapaliso (Pan.).

Balmea *Stormae* Martínez, the only species, is a shrub or small tree sometimes 30 feet tall and six to eight inches in diameter, growing on stony and arid soil near Uruapan, Michoacán, Mexico, where it is known as Ayuque. The bark has a grayish parchment-like epidermis that scales off readily. The broadly oval, deciduous leaves are clustered near the ends of the branches; the triangular stipules are interpetiolary; the deep red or violet, night-fragrant, short-lived flowers are pendulous in terminal cymes; the fruit is an erect dehiscent capsule with numerous imbricated winged seeds. Wood whitish or slightly yellowish throughout. Fairly lustrous. Without distinctive odor or taste. Light in weight but firm, having about the consistency of White Pine (*Pinus strobus* L.); texture medium fine and uniform; grain straight; very easy to work; poorly resistant to decay and subject to sapstain. Presumably of no commercial importance because of the small size of the stems.

Bathysa, with a few species of shrubs and trees up to 50 feet in height, has its center of distribution in eastern Brazil. The pale brown, medium-textured wood has no special uses.

COMMON NAMES: Autuparana, cauassú, quina do matto, quina de Santa Catharina, q. de serra, q. do matto, q. do Paraná (Braz.).

Bertiera, with about 35 species of shrubs and small trees, occurs in tropical Africa and South America. The only American specimen available is of *B. guianensis* Aubl. The wood is pinkish white, of medium density, and of fine and uniform texture. No uses known.

COMMON NAMES: Mullaca grange, ruichaey (Peru).

Bothriospora *corymbosa* Hook. f., the only species, is a small to medium-sized Amazonian tree, sometimes up to 70 feet tall and 16 inches in diameter according to L. Williams (*Woods of northeastern Peru*, p. 455). It is credited in literature with being highly poisonous, but Ducke (*Arch. Inst. Pesq. Agron. Pernambuco* 1: 20: 22) says that this reputation is due to confusing this species with a euphorbiaceous shrub, *Euphorbia capansa* Ducke. The wood is yellowish brown, with darker streaks of denser fiber layers; moderately hard and heavy, medium-textured, easy to work. No uses reported.

COMMON NAMES: Sardinheira (Braz.); junuisco-ey, quinilla (Peru).

Calycophyllum, with five species of medium-sized to large trees, is of general occurrence throughout tropical America. The bark is characteristic, being smooth, shiny, reddish or brown, and continually peeling off in shreddy strips.

The Palo Blanco of Argentina, *Calycophyllum multiflorum* Gris., is common in the forests of Formosa, Jujuy, and Salta, where it attains a height of 65 to 70 feet and a diameter of 20 to 30 inches. It also occurs in Paraguay and Matto Grosso, Brazil. The timber is used in making vehicles, implements, cogs of wheels, shoe lasts, turned articles, and similar purposes. The following results of an examination of the timber at the U.S. Forest Products Laboratory at Madison, Wisconsin, are reported by Arthur Koehler (*Tropical Woods* 14: 19): "In density this species ranks above Hickory [*Carya*] and Dogwood [*Cornus florida* L.]. The wide sapwood is a pale and dingy yellowish brown and the heartwood is light olive-brown. The wood has an exceedingly fine and uniform texture, which, together with its great hardness and straight grain, should make it an excellent material for certain exacting purposes such as rules, shuttles, shoe lasts, flooring, wood pulleys, and some of the other purposes for which boxwood is used." Shrinkage in drying from green to oven-dry, in percentage of green dimension, was found to be: volumetric, 11.2; radial, 4.0; tangential, 6.8.

According to Ducke (*Tropical Woods* 49: 1–4), there are three species of *Calycophyllum* in the Amazon region, all known as Pau Mulato in reference to the bark, though this common name is also applied to *Capirona* and sometimes to trees of other families. The Pau Mulato da Terra Firme is *C. acreanum* Ducke, a rather tall tree in the upland rain forest along the Rio Acre, while Pau Mulato da Catinga is *C. obovatum* Ducke, a medium-sized tree which "may be considered one of the characteristic elements of the flora of the catingas (localities of much rainfall and no well-defined dry season) of the upper Rio Negro, where it is not rare in slightly swampy places." These two species have no commercial value at present, and the name Pau Mulato commonly refers to *C. Spruceanum* Benth., a moderately large tree "very common in upper Amazonia and along the whole Amazon River, chiefly in inundable lowlands with fertile clay soil. It is used locally for timber and firewood. When exported (chiefly to Ceará and Rio de Janeiro, where it is employed in furniture) the timber is known to the trade as Pau Marfim (ivory wood), but that name must not be confused with the Pau Marfim of the drier upland forest of the lower Amazon, which is *Agonandra brasiliensis* Miers (fam. Opiliaceae) and is not exported." (For description of the wood anatomy see illustrated paper by F. R. Milanez, *Arch. Inst. Biol. Veg.* 3: 1: 111–129. 1936.) Practical tests of Pau Mulato for archery bows have not proved very satisfactory.

The species supplying the Lemonwood of the bow-makers in the United States is *Calycophyllum candidissimum* (Vahl) DC., a small to medium-sized tree 40 to 65 feet high and 8 to 20 inches in diameter occurring in Cuba and from southern Mexico through Central America to Colombia and Venezuela, merging into *C. obovatum* on the upper Rio Negro. It is conspicuous when in bloom because of the large white calyx lobes. In some parts of its range the trees are abundant and may form small nearly pure stands, but so far all of the timber known to the export trade comes from Cuba. It is shipped in the form of slender logs, commonly called Degame spars. The wood is similar in strength, toughness, and resilience to Lancewood (*Oxandra*) and Hickory (*Carya*), and is used locally in making vehicles, agricultural implements, tool handles, various articles of turnery, and the frames of buildings.

Heartwood brownish, sometimes more or less variegated; not sharply demarcated from the thick and nearly colorless sapwood. Luster rather low. Without distinctive odor and taste. Hard, heavy, tough, and strong; sp. gr. (air-dry) 0.80 to 0.85; weight 50 to 53 lbs. per cu. ft.; texture fine and uniform; grain straight to very irregular; not difficult to work, does not split readily, takes a glossy polish; holds its place well when manufactured; is not very resistant to decay.

COMMON NAMES: *Calycophyllum candidissimum:* Degame, d. lancewood, degamme, lemonwood (U.S.A., trade); dagame (Cuba); camarón, palo camarón (Mex.); madroño, salamo (Cent. Am., gen.); chulub, ucá (Guat.); solano, urraco (Hond.); surrá (C.R.); alazano, guayabo alazano, harino (Pan.); alazano, guayabo colorado, g. joveroso (Col.); araguato, betún (Venez.). *C. multiflorum:* Ibirá-morotí, morotibí, palo blanco (Arg.); palo blanco (Par.). *C. Spruceanum:* Corusicaá (Ec.); capirona (Peru); bayabochi (Boliv.); pau mulato, p. marfim (Braz.).

Capirona, with four species of medium-sized trees related to *Calycophyllum*, occurs in Surinam, Colombia, eastern Peru, and the Amazon region of Brazil. The best known species, *C. decorticans* Spruce, is described by Ducke (*Tropical Woods* 49: 3) as "a magnificent upland forest tree, with dark red flowers and scarlet calyx appendages." According to Llewelyn Williams (*Woods of northeastern Peru,* p. 456), it is a slender forest tree about 40 feet tall, with a spreading crown and a straight cylindrical trunk free of branches for three-fourths the height; the bark is dark purplish brown or almost black; the timber is used locally for general carpentry and fuel.

Heartwood pale brown with a pinkish tinge; fairly distinct but not clearly demarcated from the yellowish sapwood. Luster medium. Odorless and tasteless. Moderately hard and heavy; texture rather fine, uniform; grain straight; easy to work, finishing very smoothly; probably perishable in contact with the soil.

Common names: Akatombe, akatomno, akegoemio, akepemio, akhorok kalidan, akikada hoeballi, kantasie hoedoe, koraraballi, krasi-krasi hoedoe, mapiri, moentene, moentenehe, talipi, tarepi, toelala hoedoe (Sur.); capirona, c. negra, palo colorado (Peru); mulatorana, pau mulato, p. m. da terra firme (Braz.).

Casasia includes about eight species of West Indian shrubs and little trees. The yellowish or brownish, hard and strong, fine-textured wood is used to a minor extent for small tool handles.

Common names: Seven-year apple (Florida); wild pomegranate (Jam.); cocuyo, guayacancillo de loma, jagua amarilla, j. azul, j. de costa, jicarita, palo cabra, p. de la mar, p. de las bolas (Cuba).

Cephaelis is a pantropical genus with about 125 species of herbs, shrubs, and little trees closely related to *Palicourea* and *Psychotria*. Its only economic uses are medicinal. The wood is light-colored, of medium density, fine-textured, straight-grained, and very easy to work, but is available only in small sizes.

Common names: Ipecacuana cimarrona (Cuba); raicilla (Pan.); kamejoeroe, koejatta enaka, paipayodapari, pakira ponapiri (Sur.); ipeca annelé (Fr. G.); couve do matto, ipéca verdadeira, ipé-caá-coena, ipecacuanha verdadeira, poaya verdadeira (Braz.); awa, picho sisa, sufia, usiyapuiño, yaku-sisa blanca (Peru).

Cephalanthus, with about 15 species of shrubs and small trees, is sparingly represented in Asia and Africa, more abundantly in America. The best known species is *C. occidentalis* L., which inhabits swamps and low wet borders of streams throughout the eastern half of the United States and adjacent Canada, also occurring in New Mexico, Arizona, California, Mexico, northern Central America, and Cuba. Though usually a shrub in the northern part of its range, in southern Arkansas, eastern Texas, and parts of Mexico it attains a height of 50 feet with a straight trunk 12 inches in diameter. There are other species in South America, but no wood samples are available for study.

Heartwood light pinkish brown; distinct but not sharply demarcated from the yellowish sapwood. Luster medium. Odorless and tasteless. Moderately hard and heavy; texture medium to rather coarse; grain straight; easily worked; poorly resistant to decay. Of no commercial value because of the scarcity of the larger trees.

Common names: Button bush, b. willow, crooked wood (U.S.A.); aroma de ciénaga, a. de laguna, clavellina de ciénaga, c. de mota (Cuba); jazmín, j. blanco, mimbre, uvero (Mex.); botoncillo (Hond.); sarandí blanco, s. colorado, s. molle, s. negro (Arg., Urug.).

Chimarrhis, with about a dozen species of small to medium-sized trees, is widely distributed in tropical South America, southern Central America, and sparingly in the West Indies. The orange-colored, rather coarse-textured, moderately heavy wood is not difficult to work and is probably durable, but apparently it has no special uses. The wood of *Pseudochimarrhis barbata* Ducke is similar in all respects.

Common names: Cera (Cuba); bois riviere, resolu (Fr. W.I.); yema de huevo (C.R.); jagua amarilla, plátano, yema de huevo (Pan.); lomo de caimán (Col.); lengua de vaca (Venez.); alipo-menango, arrawerie, jean-jean hoedoe, mabarabe, takajoe (Sur.); tahuari, tuvara, yacu-caspi (Peru); pau de remo (Braz.).

Chomelia, with about six American species of spiny or unarmed shrubs and little and often crooked trees, is distributed from Guatemala to Brazil and Peru. The yellowish or brownish wood is moderately hard, fine-textured, easy to work, and suitable for small turned articles.

Common names: Clavo (Guat.); malacahuite (C.R.); chocolatillo (Col.); cassicus bloem, jawohe sarapan, kinoto-potele (Sur.); limao-rana da varzea, papaterra (Braz.); anzuelo casha, garras de gato (Peru).

Cinchona, with several species of shrubs and small or rarely medium-sized trees, is limited in natural distribution to the South American Andes, with the exception of *C. pubescens* Vahl, which extends as far north as Costa Rica. The genus is noted for the alkaloids obtained from the bark, particularly quinine, the universal remedy for malaria.

According to an article in the *Bulletin of the Imperial Institute* 37: 1: 18–31. 1939), "the classification of the group has presented considerable difficulty to botanists, owing to the facility with which the species hybridize with one another, giving rise to many different forms. Four species only have been cultivated to any extent as a source of alkaloids. These are: *Cinchona Ledgeriana* Moens ex Trimen . . . from which Ledger Bark is obtained; *C. succirubra* Pavon ex Klotsch, the source of Red Bark; and finally *C. calisaya* Wedd. and *C. officinalis* Linn., yielding Yellow Bark and Crown Bark or Loxa respectively. . . . At the present time practically the entire supply of Cinchona bark in commerce is obtained from *C. Ledgeriana* and *C. succirubra*. . . . It is found that in cultivation the different species have similar climatic requirements. Summarized briefly, the most suitable conditions are a tropical climate and an elevation of from 3000 to 6000 ft.; a fairly high average temperature with relatively small range of variation, high atmospheric humidity, high rainfall well distributed throughout the year, a light well-drained soil rich in organic matter, and a sloping situation sheltered from the wind. . . . It is estimated that the Netherlands East Indies now produce more than nine-tenths of the world's supply of Cinchona bark. This extraordinary strong position has been attained, not only as a result of favorable climatic conditions, but also through the systematic thoroughness with which the industry has been carried on from the start. . . . The exports of bark in 1936 amounted to 19,978,463 lbs. (£1,026,012) and of quinine 6,774,646 oz. (£411,584) . . . 1938, bark 15,337,801 lbs. (£919,721), quinine 6,430,407 oz. (£395,112)." India is the only producer of Cinchona bark in the British Empire but it "still has to import large quantities of quinine in order to meet the requirements of its sufferers from malaria who are estimated to amount to 100,000,000 at any particular time." Although synthetic febrifuges are being manufactured, they "have not yet become generally available at prices that would compete with quinine and, moreover, it would seem that the use of such drugs is attended with some risks, for which reason they cannot be distributed generally for use without medical supervision."

Although the year 1638 is generally accepted as the first date in the history of quinine, Dr. George T. Moore, President of the Shaw Botanical Gardens, has established through old volumes in the Garden's library that the first recognized use occurred in 1630. In that year, he found, the bark cured the malaria of Juan Lopez Canizares, Spanish corregidor of Loxa. Traditionally, it was the corregidor who recommended the treatment to the Countess of Chincón, wife of the Governor of Peru, whose cure at Lima in 1638 gave the Cinchona tree its name. The story of the natives, who used it for an unknown period, is that a Cinchona tree fell into a pool of water, and that a native, drinking the water, found his fever cured. Knowledge of its value was disseminated throughout Europe by the Jesuits, so that the "sacred bark" became known also as "Jesuit's bark."

COMMON NAMES: Quina, q. árbol (Venez.); cascarilla, quina (Ec.); caruacarua, cargua-cargua, capirona amarilla, c. del bajo, c. negra, cascarilla, ichu cascarilla, motosolo, palo blanco, quepo cascarilla, quina (Peru); mameluco, quina verdadeira (Braz.).

Cosmocalyx *spectabilis* Standl., the only species, is a tree 15 to 25 feet high, occurring in Michoacán and Yucatán, Mexico. The only specimen available (Yale 29766) was collected in Yucatán by R. S. Flores. The wood closely resembles *Sickingia*. The vernacular name is given as Chacte-coc.

Coussarea, with about 75 species of shrubs and small trees rarely over 30 feet high, is widely distributed in tropical South America, extending northward to Costa Rica. The brownish, rather fine-textured, moderately dense wood is easy to work, but has no special uses.

COMMON NAMES: Flor de muerto, manzano negro (Col.); boesi koffie, koeroewakoe erepare, koffie balli, k. waton, korowako erepalli, mapoeri erepare, mattoe koffie, popokai briengie, witte feoeta (Sur.); chonchuela, chorchulla, ginsira-caspi, motelo micuna, ñucnu-huito, supi-caspi (Peru); mborebí-caá, yerba de anta (Arg.).

Coutarea, with about eight species of shrubs and trees, occurs from southern Mexico to Argentina and southern Brazil. The most widely distributed species, with a range as great as for the whole genus, is *C. hexandra* (Jacq.) Schum. This is usually a small tree, but according to Llewelyn Williams (*Woods of northeastern Peru,* p. 465), it attains a height of 120 feet, with a straight cylindrical trunk 20 to 36 inches in diameter and free of branches up to 30 feet. "The dense wood is esteemed locally for house posts, furniture, and general carpentry." The bark is bitter and is often employed in local medicine as a febrifuge.

Heartwood pinkish brown, sometimes superficially pink; distinct but not sharply demarcated from the sapwood. Without odor, but sometimes with bitter taste. Hard, heavy, and strong; of fine and uniform texture; straight-grained; easy to work, taking a glossy polish.

COMMON NAMES: Cabalkax, campanilla, caparche, copalchí, copalquín, falsa quina, pailuch, palo amargo, quina, San Juan (Mex.); quina, q. blanca, quinita, zalas (Salv.); amargo, mariangola, mediagola (Col.); cabrito negro, campanilla (Venez.); kwatta thé, leletie, pakeeli, pakeri, tatobaballi (Sur.); huacamayo-caspi (Peru); murta do matto, quina, q. do Pernambuco, q. do Piauhy, quina-quina (Braz.); cascarilla, mbavihpuitá (Arg.).

Dialypetalanthus *fuscescens* Kuhlm., the only species, is a tree 10 to 25 feet high in the lower Amazon region. The yellowish or grayish brown, hard and heavy, fine-textured, easily worked wood is not utilized because of the small size and scarcity of the trees.

Duroia is a South American genus with about 25 species of shrubs and small to medium-sized trees rarely over 45 feet high and 10 inches in diameter. The pinkish white or brownish, moderately dense, fine-textured wood has no special uses.

COMMON NAMES: Turma de mico (Venez.); kumara-mara, maru-mara (Br. G.); angomoelé, assokarri, atakalli, atakara, bofroe marmadosoe, bosch marmeldoos, fierobero-koemaramara, jallakalloe-oepoepo, jarakara-oepoepo, koemaramara, marmadosoe, marmeldoosje, mekoepoepo-atakarie, siengie hoedoe, tanimme asakalli, tokotoro kjin wewe, waniaballi (Sur.); palo del diablo, quinilla, pamparemo-caspi, sacharuna-caspi, supai-caspi, s.-quinilla (Peru); caá-jussara, gurupea, folha de comichão, puruhy, p. grande (Braz.).

Elaeagia is a South American genus with about 10 species of trees. The best known species is *E. Mariae* Wedd. of Peru and Bolivia, which attains rather large size and is the source of a resin or wax, called aceite María, used medicinally and also for making candles. The wood is pale brown with a pinkish tinge, of medium density, rather fine texture, and easy to work. No particular uses are recorded.

Erithalis is a West Indian genus with six species of shrubs and occasionally trees up to 30 feet high. The best known species is *E. fruticosa* L. It has a distinct heartwood which is light brown with dark streaks, very dense, extremely hard, fine-textured, and takes a glossy finish, but because of the small sizes available, it is only suitable for small articles of turnery.

COMMON NAMES: Cuaba prieta, cuabilla, jayajabico, rompe machete, yayajabico, víbona (Cuba); black torch, jayajabico (P.R.); bois chandelle, b. c. noir (Fr. W.I.).

Exostema, with about 30 species of small trees and shrubs, has its center of distribution in the West Indies, with a few representatives on the mainland from Mexico to Peru. The best known is *E. caribaeum* (Jacq.) R. & S., a coastal tree rarely 25 feet tall and 12 inches in diameter, growing in many of the Antilles, southern Florida, and from southern Mexico along the Pacific coast to Costa Rica. The bark was formerly used as a febrifuge. The attractive wood is saffron-brown richly variegated with purple or green, very dense, fine-textured, taking a lustrous polish and locally esteemed for small cabinet work and articles of turnery. Its use is now very limited because of the scarcity of the larger sizes.

COMMON NAMES: Princewood (Florida); Caribee-bark tree, Jamaica Jesuit's bark, maroon lance (Jam.); agracejo carbonero, carey de costa, cera amarilla, cerillo, clavellina del río, cuabilla, lirio santana, macagua de costa, vigueta (Cuba); alvarillo, cuero de sapo, princewood, yellow torch (P.R.); pini-pini (Dom. R.); chandelle anglais, quinquina pays (Haiti); bois tabac, quinquina caraibe, q. piton (Fr. W.I.); falsa quina, huesillo, sabac-ché (Mex.); puca yanta (Peru).

Faramea, with about 150 species of shrubs and trees generally less than 25, rarely up to 50, feet high, is distributed throughout tropical America. The whitish or yellowish, moderately hard, fine-textured, readily worked wood is used locally for carving and turnery, tool handles, and small interior construction.

COMMON NAMES: Wild coffee, w. jessamine (Jam.); café cimarrón, cafetillo, jujano, nabaco, palo de toro (Cuba); cafetillo, palo de toro (P.R.); café marron (Dom.); hueso, ma-la (Mex.); cafecillo (Salv.); bonewood, huesillo, huesito (Pan.); clavo (Col.); cafecillo danta, danta orejona, jazmín de estrella (Venez.); charichuela, iru-huaco, itulli-caspi, sisíncaca, situlli-caspi, uchpa-caspi, uchu sanango (Peru); café-rana, carvoeiro (Braz.).

Ferdinandusa is a South American genus with about 20 species, mostly small to medium-sized trees, occasionally up to 65 feet high, but sometimes upright or scandent shrubs. The only wood samples available are of *F. rudgeoides* (Benth.) Wedd. from British Guiana and the Amazon region of Brazil. Color uniformly brownish yellow throughout. Luster golden. Without distinctive odor or taste. Very hard, heavy, and strong, but somewhat brittle; texture rather coarse; grain straight to irregular; not difficult to work, but finishing smoothly and presenting attractive appearance when quarter-sawed. Suitable for furniture, but of doubtful commercial possibilities.

COMMON NAMES: Acahú, acahúa acaú, acaúa (Braz.); guacamayo, louro-micuna (Peru).

Genipa is a small genus of medium-sized to large trees of general distribution throughout tropical America. *G. americana* L. occurs from the West Indies and Mexico to Argentina. The typical form is glabrous, but the commoner form in continental North America is characterized by abundant dense pubescence of the branches and lower leaf surfaces, and is sometimes recognized as a distinct species, *G. caruto* H.B.K., sometimes only as a variety, *G. americana,* var. *caruto* (H.B.K.) Schum. Apparently the woods are not distinguishable. The tree is common in the lowlands of the West Indies and the usual name there, as also in Mexico, is Jagua. The maximum size is about 70 feet in height, with a trunk 24 to 40 feet long and 24 inches through, but commercial logs are mostly 15 to 18 inches in diameter. The wood is used for boxes, chests, barrel hoops, sieve rims, shoe lasts, tool handles, plow beams, and many other purposes requiring strength and resilience combined with fine and uniform texture, but not highly resistant to decay. In Panama the tree is rarely over 50 feet tall and 14 inches in diameter and occurs generally in open forest or at the edge of savannas. The wood is used for general interior construction and is always preferred for the side of the door and door casing bearing the hinges.

According to H. M. Curran, the tree occurs sparingly in the virgin forest of eastern Brazil, rarely averaging more than one per acre, but is abundant in and about towns and villages, since it is planted as a shade tree or comes up spontaneously and is protected because of its fruit. It is usually rather short-boled, a clear length of 50 feet being the maximum, and the diameter ranges up to three feet. The lumber is well known and is suitable for a great many purposes, particularly for bent work, but the wood usually contains so much moisture that it is rather difficult to season properly without the use of dry-kilns. It should prove a satisfactory substitute for Ash (*Fraxinus*). The fruit, which is about two inches in diameter, has a leathery shell inclosing a dark pulp that is edible though not very palatable. The juice, transparent at first, soon turns black and leaves an indelible stain on everything it touches. In Hans Staden's account of his captivity among the cannibal tribes of Tupinambá Indians in eastern Brazil some 385 years ago is a note to the effect that the savages painted themselves with the juice of the apple-like fruit of the Junipappeeywa tree, which has been identified as *Genipa americana* (see *Tropical Woods* 23: 5). This aboriginal use is still continued among natives in various parts of tropical America, particularly in preparation for ceremonial rites.

Heartwood brownish or grayish brown, sometimes with a pinkish or yellowish hue; uniform or streaked; merging gradually into the sapwood, which sometimes resembles old ivory or may show a natural blue-black stain just below the surface. Luster medium. Odorless and tasteless. Hard, heavy, tough, strong, and resilient; sp. gr. (air-dry) 0.73 to 0.85; weight 45 to 53 lbs. per cu. ft.; texture rather fine; grain straight to irregular; not difficult to work, finishing smoothly; not very resistant to decay. A good timber, but its consumption will probably continue to be local.

Common names: Jagua (Sp. Am., general); gêne-pas, genipayer (Haiti); juniper (Trin.); jagua azul, j. blanca, maluca, xagua (Mex.); irayol, i. de loma (Guat.); genipap (Hond.); irayol, tambor, tiñedientes (Salv.); gigualtí, tapaculo, yigualtí (Nic.); brir, guayatil (C.R.); guayatil blanco, jagua amarilla, j. blanca, j. de montaña, j. negra (Pan.); angelina, angelito, jago, piginio amarillo (Col.); caruto, guaricha (Venez.); arasaloe, lana, oeman tapoeripa, sawa, taparoepa, tapoeloepa, tapoeripa, taproepa (Sur.); jagua dulce (Ec.); huitoc, palo colorado, vitu, yakuhuito (Peru); ñandipá (Arg.); genipapeiro, genipapinho, genipapo, g. do matto (Braz.).

Gleasonia, with three species of small trees, occurs in the northern Amazon region and the hinterlands of Venezuela. Wood throughout rather dull pale brown with a pinkish tinge, usually more or less streaked. Very hard and heavy; sp. gr. (air-dry) 0.92; weight 57 lbs. per cu. ft.; texture rather fine, uniform; grain straight; not easy to work, but finishing smoothly.

Gonzalagunia, with about 13 species of erect or straggling shrubs and a few trees up to 30, very rarely to 60, feet high, is widely but sparingly distributed in tropical America. The wood is not readily distinguished from some of the other light-colored, fine-textured kinds with narrow rays and very little parenchyma.

Common names: Lab-lab (Jam.); palo de semillero (Cuba); cachito de renado (Guat.); nigüita (Pan.); bochata, mullaca (Peru).

Guettarda, with about 125 species of small to medium-sized trees and erect or climbing, sometimes thorny, shrubs occurs throughout tropical America, and there is one representative along the coast in the East Indies. The yellowish brown to gray wood is often tinged or streaked with green or brown. It is moderately to very hard, fine-textured, and easy to work, and is employed locally in building native huts, making handles, and for other small uses.

Common names: Silver wood, velvet seed (Jam.); carapacho, chicharrón de monte, cigüilla, contraguao, cuero, c. duro, guayabillo, guayabito, guayanejo, hueso, jagüilla de monte, vigueta (Cuba); blackberry, palo de cucubano, serrasuela, velvetberry (P.R.); false ironwood (Virg. Is.); guayabón (Dom. R.); calle noir, goavier marron (Haiti); kiich-che (Mex.); glassy wood, prickle wood (Br. H.); mosqueta de montaña (C.R.); espino, guayabo (Peru); fruta de pava, guayabo prieto, huesito (Col.); cruceto (Venez.); yakki (Br. G.);

popokai makra (Sur.); garrabato (Peru); cuentrillo, jazmín de monte, níspero cimarrón, n. silvestre, peludilla, peludiña (Arg.); cuentrillo, níspero cimarrón, peludilla, peludiña, tala, vellodiño, vellosiño, velludo (Urug.); angelica, a. do matto, velludinha (Braz.).

Hamelia is a tropical American genus with about 40 species of shrubs and little trees of general distribution. Some of the plants are cultivated for their conspicuous yellow or scarlet flowers. The wood varies in color from yellowish brown with a greenish tinge to orange, sometimes with purplish streaks. It is of medium density and hardness, fine-textured, and easy to cut, but is rarely utilized because of the small sizes obtainable.

COMMON NAMES: Bonasí, palo de coral, ponasí (Cuba); balsamillo, balsamo (P.R.); buzunuco (Dom. R.); corail, c. rouge (Haiti); aguacatillo, cacanapazue, campanilla, chacloco, chactoc, coralillo, kanan, neanan, pañete, sangre de toro, xkaná, xkanán (Mex.); red fowl (Br. H.); chahmah, chichipín, clavito, flor de cangrejo, sisipinse, sikunkhen (Guat.); chichipín, clavillo, coloradillo, coral, coralillo (Hond.); chichipince, coralillo, doncella, flor de baño, sancocho, sisipinse, xuchil paltimatia, zambumbia (Salv.); canilla de venado, coralillo, chupamiel, papamiel (Nic.); añileto, azulillo, coralillo, palo camarón, zorillo, z. colorado, z. real (C.R.); guayabo, uvero (Pan.); bensenuco, coralito, recadito, tinto (Col.); coralillo, coralito (Venez.); herva de rato (Braz.); juto blanco, usiya-ey (Peru).

Henriquezia, with a few species of medium-sized to large trees, occurs in the north-central Amazon basin. Two species are represented by authentic wood specimens in the Yale collections, namely, *H. macrophylla* Ducke and *H. verticillata* Benth., known locally as Macaco-patrona. Heartwood pale to rather dark brown, with an orange hue; distinct in *H. macrophylla,* but not sharply demarcated from the sapwood. Without distinctive scent or taste. Very hard, heavy, and strong, but brittle; texture coarse; feel harsh; grain straight; rather difficult to work, but can be finished smoothly; durability doubtful. Apparently without possibilities for export.

Hoffmania, with a large number of closely related species of herbs, shrubs, and little trees, ranges from Mexico to the southern Andes and is most abundantly represented in Costa Rica. The only specimen at hand (Yale 38369; H. E. Stork 4153) is of *H. ramonensis* Standl., a little Costa Rican tree rarely 20 feet high and five inches in diameter. The light and soft, rather fine-textured, straight-grained wood is egg-yolk yellow throughout. It is not utilized.

Holtonia *myriantha* Standl., the only species, is a Colombian tree sometimes 50 feet high and 12 to 24 inches in diameter (see *Tropical Woods* 30: 37–38). The gray and brownish streaked wood is hard, heavy, strong, fine-textured, and easy to work. No uses are recorded.

COMMON NAMES: Blanquito, huesito de tierra fría (Col.).

Isertia, with about 20 species of shrubs and trees rarely 50 feet high, is distributed from Cuba and Guatemala to Peru and Brazil. Some of the plants are cultivated for their showy red flowers. The brownish streaked wood is moderately hard and heavy, fine-textured, and easy to work, but is little used.

COMMON NAMES: Bois fer (Trin.); canelito, fruta de mono, f. de murciélago, huevo de mono (Pan.); jaboncillo (Col.); mamayahooka (Br. G.); kamadan balli, k. ibibero, loto hoedoe, melamaroelan, tamoenin wewe, witte kjinkjin hoedoe (Sur.); asaquiro, isico-ey (Peru); coraleira (Braz.).

Ixora is a pantropical genus with about 350 species of shrubs and small trees, mostly in the Old World. Many are planted for ornament. The wood of the American species is pale brownish or yellowish with a tinge of green, more or less streaked; of medium to rather high density, fine-textured, and easy to work, but is little used. (The Hackia of British Guiana, sometimes confused in literature with *Ixora ferrea*

Benth., is a species of *Tabebuia,* fam. Bignoniaceae.)

Common names: Café cimarrón, cafetillo cimarrón, lengua de vaca (Cuba); dajao, palo de dajao (P.R.); bois pignette (Dom.); amaco (Guat.); palo de María (C.R.); oqüito (Pan.); kuraraballi (Br. G.); pajawaroelang, wesepi (Sur.); chimicúa (Peru).

Kotchubaea, with four species of shrubs and small trees, occurs in the Amazon region of Brazil. Heartwood dull olive-gray to olive-brown; fairly distinct but not sharply demarcated from the lighter sapwood. Without distinctive scent or taste. Very hard, heavy, tough, and strong; texture rather coarse; feel harsh; grain fairly straight; not easy to work; splits readily, finishes smoothly; durability doubtful. Apparently not utilized and presumably without commercial possibilities.

Ladenbergia, with about 40 species of shrubs and medium-sized to large trees, is widely distributed in continental tropical America from Costa Rica to Bolivia and Brazil. The genus is rather closely related to *Cinchona* and the bitter principle in the bark is used medicinally, though apparently it is without value as a febrifuge. The wood, which is of the color of old ivory with a slight olive hue, is rather dense, of fine to medium texture, rather harsh feel, and not very difficult to work. It is used locally to some extent for interior construction, turnery, and fuel.

Common names: Agujilla, quina (C.R.); azahar, cascarilla boba, c. roja, requesón (Col.); quina, quina-quina (Venez.); azahar hembra, a. macho, cascarilla, c. amarilla, c. azaharito, c. bobo, c. flor de azahar, cargua-cargua, c.-c. chica, c.-c. grande, carua (Peru); carua, cascarilla de mula (Boliv.); quina do Rio (Braz.).

Laugeria, with four species of shrubs and small or rarely medium-sized trees having very resinous branchlets, is limited in natural range to the West Indies. The best known and most widely distributed species, the only one represented in the Yale collections, is *L. resinosa* Vahl, sometimes referred to *Antirrhoea* or *Guettarda.* It occurs in thickets from the Bahamas to Trinidad and seldom exceeds 20 feet in height; the twigs are stout and densely leafy. There are no special uses for the plant and the only vernacular name known to the authors is Aquilón in Puerto Rico. Heartwood uniform medium reddish brown, fading gradually into the sapwood. Luster moderate. Without distinctive odor or taste. Moderately heavy, hard; fine-textured; straight-grained; easily worked, finishing smoothly.

Machaonia, with about 25 species of spiny or unarmed shrubs and little trees, is distributed from the West Indies and Mexico to Argentina. The hard, fine-textured, yellowish or pale olive-colored wood has no special uses.

Common names: Espino, e. de hojas chicas (Cuba); roseta (P.R.); huele de noche del campo, kampocolche, kuchel (Mex.); palito blanco (Venez.); limão-rana-sinho, poaya da praia, p. do rio (Braz.); arazá-ñuatí, guayabo espinudo (Arg.).

Macrocnemum, with about 15 species of shrubs and trees, occurs from the West Indies and Costa Rica to Brazil. *M. jamaicense* L., a tree sometimes 40 feet high, is considered a fairly good substitute for Boxwood (*Buxus*) in Cuba. The only specimen of the genus available is of *M. roseum* (R. & P.) Wedd. of Peru (Yale 18636; Williams 4957), which is said to attain a height of 80 feet, with a trunk 12 to 20 inches in diameter. Wood yellowish or brownish, more or less streaked. Taste slightly bitter. Moderately dense, strong; fine-textured; easy to work. The only uses are for fuel and occasionally in general construction.

Common names: White thorn (Jam.); boj indígena (Cuba); palo cuadrado (C.R.); asmonich, cascarilla bruta, c. parda, palo de San Juan, shamoja (Peru); acariquára (Braz.).

Morinda includes about 80 species of erect or scandent shrubs and small to rather large trees, chiefly East Indian and Malayan, a few in tropical Africa and America. The American species are all small, the trees rarely 30 feet high and six inches in diameter. The roots yield a red or orange dye. There are no special uses for the wood. Heartwood distinctively colored, orange with a greenish hue, somewhat streaked or variegated. Luster medium. Moderately hard and heavy; rather fine-textured; straight-grained; readily worked; is probably durable.

COMMON NAMES: Safran (Haiti); hooyoc, hoyoc, jojoc, palo de peine, xhoyoc, xoyencab (Mex.); turkey victuals, wild egg (Br. H.); piñuela, yema de huevo (Guat.); conchade huevo (Hond.); quidive, yema de huevo (Col.).

Palicourea, with about 200 species of shrubs and little trees, occurs throughout most of tropical America. The ivory-colored or grayish brown woods are of medium density, fine-textured, and easy to work, but are sparingly used because of the small sizes obtainable.

COMMON NAMES: Wild coffee, w. lime (Jam.); puerco hueso, taburete, tapa camino (Cuba); balsamo real, cachimbo, tafetán, wild cappel (P.R.); tafetán (Dom. R.); zorillo amarillo (C.R.); amargo, malibú muyo pavo, saca ojo (Col.); café de monte, cafecito, chaparrillo, chaparro bobo, c. cacho de venado (Venez.); kamadanni (Br. G.); bakioe wiwirie, bofroe kosabatiekie, boschvlier, diabitta, kamadan, marpjoeli ekerepore, m. kerapale, m. kierabelle, pahewe awhamoejihoenda, pangapanga, wajamoe sansamoeroe (Sur.); quillo-sisa (Peru); totó-caá (Par.); bota, cha de bugre, congonha de gentio, cotócotó, genipapo rosa, herva de rato, tangaracá-assú (Braz.).

Pentagonia, with about a dozen species of branched or unbranched shrubs and little trees noted for the large size of their leaves, occurs sparingly from Costa Rica to the upper Amazon. The pale grayish brown wood is not utilized.

Pinckneya *pubens* Michx., the only species, is a rare tree 20 to 30 feet high and eight to ten inches in diameter of infrequent occurrence in wet lands along streams of the coast region from South Carolina to Georgia and Florida. Heartwood chestnut-brown; not sharply demarcated from the sapwood. Luster medium. Odorless and tasteless. Rather light and soft, brittle; texture coarse; grain straight; very easy to work. Resembles Sassafras.

COMMON NAMES: Fever-tree, Florida quinine-bark, Georgia bark (U.S.A.).

Pittoniotis *trichantha* Gris., the sole species, is a small tree rarely over 30 feet high and 12 inches in diameter described from Panama, where it is known as Candelo, but extending into northern Colombia. The wood is pale yellowish brown with a tinge of olive; moderately hard, fine-textured, and easy to work, but apparently it is not utilized.

Platycarpum *orinocense* H. & B., the only species, is a large tree of the Orinoco River basin. No authentic material is available, but a fruiting specimen provisionally identified as this genus was collected with a wood sample (Yale 31964) by A. Ducke along the lower Rio Madeira, Amazonas, Brazil, and as the wood structure is of the type of the closely related genus *Henriquezia*, it is believed to be correctly determined. Heartwood light reddish brown, merging gradually into the sapwood. Luster rather low. Odorless and tasteless. Very hard, heavy, and strong, but somewhat brittle; texture coarse; feel harsh; grain straight; not easy to work, but finishing smoothly. Apparently without commercial possibilities.

Pogonopus, with three or four species of shrubs and small trees sometimes 30 feet high, is distributed sparingly from Salvador to Argentina. The plants are noted for the brilliance of their flowers. The only species in North America is *P. speciosus* (Jacq.) K. Schum. Its wood is brownish yellow, fine-textured, and slightly bitter.

COMMON NAMES: Chorcha de galla (Salv.); quina (Boliv.); quina morada (Arg.).

Posoqueria, with about 10 species of shrubs and small to medium-sized trees, is widely distributed in continental tropical America. The yellowish or gray, fine-textured, moderately dense wood is used locally for tool handles, turnery, and small articles of joinery. The unpleasant scent of the fresh wood is not noticeable in dry material.

COMMON NAMES: Palo chino, p. de peine, p. de p. blanco (Mex.); chintonrol, mountain guava, snake seed (Br. H.); chintonrol (Guat.); cachito (Hond.); jazmín de árbol, jicarillo (Nic.); fruta de mono, guayabo de mico (C.R.); boca vieja, borajó, fruta de mono, f. de murciélago, huevo de mono, monkey apple, mosquito wood (Pan.); cachaco, c. de monte, guayabo de mico, jazmín de mico, malibú, manzana amarilla (Col.); kamadan (Br. G.); aimiala posowepo, atalitjaka, boesi-mammi, dagoekoko, dagoeston, hondenpitten, koejakinerépare, koekoelitjiballi, koeroeritjiballi, kororitiballi, spikrietiekie, tamaara (Sur.); ucu-llúcuy (Peru); açucena do matto, aymara, fruta de macaco, papaterra, pau de macaco (Braz.).

Psychotria (including *Mapouria*) is a pantropical genus of about 1200 species of low to tall shrubs and comparatively few small trees. Some of the plants are used medicinally. The wood is yellowish or brownish with an olive or purplish hue, moderately hard, fine-textured, and easy to work, but is not available in large enough sizes to be useful.

COMMON NAMES: Wild coffee (Florida); café cimarrón, dagame cimarrón, lengua de vaca, palo moro, piñita, taburete de flor blanca, tapa comino (Cuba); cachimbo (P.R.); bois latelle, café marron (Haiti); cancerillo, pochitoco, yaxcanan (Mex.); casada, dead-man's bones, white wood (Br. H.); hueso de finado (Guat.); chinche de laguna, hierba del pájaro, h. del sapo (Salv.); cocobilito, garricillo, huesito, raicilla, r. macho (Pan.); amarga, anoncito de terra fría, chupa-chupa, esmeralda, enredadera de monte, fruta de pava, gotorero, huesito amarillo (Col.); aku-owmu, kill-cow bird's foot (Br. G.); akami ekoenale, a. enoeloe, akokoapa, boelinali oeneberadikoro, boschvlier, manjala kopie, manmantrie, poelinari, wasiliwo koeroe (Sur.); asier à l'asthme, simira (Fr. G.); ipecacuana falsa (Boliv.); brocha sisa, chirapa shacha, cucha-caspi, gidoro-ey, mullaca, paufil chaqui, shuturi, sonia, topamaqui, trompetero-caspi, ucumi micuna, yaku shuturi (Peru); café roxo do matto, capança, corta-asthma, dourada, douradinha, herva de rato, ipecacuanha preta, japiimcaá, matadeira, tangaraca (Braz.).

Randia (including *Basanacantha*) is a pantropical genus of about 350 species of unarmed or spinose, upright or clambering shrubs and small trees. Some species are cultivated on account of their beautiful and fragrant flowers. The wood, which is whitish, yellowish, or brownish, of medium density, fine-textured, and easily worked, is of little value because of its scarcity in usable sizes.

COMMON NAMES: Box briar, indigo berry, ink berry, prickle bush (Jam.); agalla de costa, café cimarrón, espuela de caballero, pitajoni bravo, p. espinoso, yamaguey, y. de costa (Cuba); box brier, cabai nagte, cambrón, dogwood, escambrón, ink berry, palo de cotorra, tintillo (P.R.); azota criollo (Dom. R.); bois sadine, croc à chien (Haiti); arbol de las cruces, caax, canalkax, canastilla, capulín corona, chapote, crucecilla, crucero, c. blanco, cruceto, cruzeta, espino cruz, huile de noche, kax, limoncillo, mehenkax, nanche, palo de la cruz, papache, p. picudo, papachillo, pechcitam, sapuchi de la sierra, tomatillo, xcaax, xpechcitam, zapotillo (Mex.); rosetillo (Guat.); cagalero, crucetilla, rosetillo (Hond.); caca de mico, crucitilla, crucito, jicarillo, tintero, torolillo (Salv.); crucillo, espino blanco, horquetilla, mostrenco (C.R.); carretillo, chocolatico, corallero, maíz tostado, mariangola, pinta pava (Col.); cruceto, c. negro, c. real, quipito hediondo, sajadito (Venez.); bird seed (Br. G.); aremie kieran, boesie droifi, jamahi sarapan, marmeldoosje (Sur.); chingai, clavo-caspi, espino, espuela casha, jugu-ey, millucassa, ñupchucri, umruyo (Peru); açucena, a. estrella, estrella, e. do norte, fructa decacharro, jasmin do matto, limão do matto, l. rana, l. sinho, mororó, papaterra, sucena (Braz.); ñandipá-guàzú, nuati-curuzú, palo de Santo Domingo, tista-tista (Arg.).

Remijia, with about 18 species of shrubs and small to large trees, is widely distributed in South America. The only uses apparently are medicinal. The wood, which is yellowish, streaked or tinged with olive, is hard, fine-textured, readily worked, and well adapted for carving and turnery.

COMMON NAMES: Azahar, cascarilla, quina (Col.); asar lisa, cascarilla, collar lisa (Peru); quina (Braz.).

Retiniphyllum, with about 15 species of shrubs and little trees, is sparingly distributed in Venezuela, the Guianas, and the Amazon region of Brazil and eastern Peru. The wood, which is brownish with a tinge of pink, is of medium density, fine texture, and good working qualities.

COMMON NAME: Kaia-kaia-danni (Br. G.).

Rondeletia is a tropical American genus of about 125 species of shrubs and slender trees commonly less than 20, occasionally up to 40, feet high, occurring for the most part in the West Indies, Mexico, and Central America, with a few extensions through the Andes to Peru. The yellowish or brownish, fine-textured wood is apparently not utilized.

COMMON NAMES: Bomitey, caobilla de sabana, clavellina, cocuyo de sabana, cuje de caballo, encospe, hatillo, nisperillo, vigueta blanca, v. de sabana, v. naranjo (Cuba); cordobancillo (P.R.); hierba de la muchachita, huele de noche (Mex.); candelillo (Hond.); bouquet de la reina (Salv.); quina, teresa (C.R.); candelo (Pan.); voiguio-ey (Peru).

Rudgea, with about 125 species of shrubs and small trees, is widely distributed throughout continental tropical America, but is most abundantly represented in Brazil. Apparently the only uses are medicinal. The wood is light-colored and fine-textured.

COMMON NAMES: Ashes wood, bois tatoo, kapol (Trin.); cafecillo, fruta de paloma, guaricha, totumillo (Venez.); agoeston, kamabhallie, kamakha-ie, koenaporang, kwalelang, maipjorie kera porre, tajaboedoe (Sur.); saúco de monte (Boliv.); amanga, pichico runto, sanango de bajo (Peru); congonha do gentio, mulatinho, pau de boia (Braz.).

Sickingia, with about 15 species of shrubs and small to medium-sized trees rarely up to 65 feet high, occurs in continental tropical America from southern Mexico to southern Brazil and Paraguay. One of its outstanding features is the bright red color the fresh wood assumes upon exposure.

The northernmost species, *Sickingia salvadorensis* Standl. of Salvador, Guatemala, British Honduras, and southern Mexico, is a tree 15 to 50 feet tall, supplying some timber for local construction and a red dye employed for coloring hammocks and other native articles. (See *Revista Botánica* [San Salvador] 1: 2: 34–42; 1923.) There are two species in Panama. The upland form, with small leaves, is *S. Klugei* Standl.; it occurs also in northern Colombia and Venezuela. The large-leaved tree, growing in wet lands along the Atlantic coast, is *S. Maxonii* Standl.; it crosses the border into Costa Rica. G. Proctor Cooper, who collected the specimens necessary for the proper classification of this species, says (*Tropical Woods* 14: 3): "In the forests it is straight and slender, about 50 feet high and a foot through, but when growing in the open, as along the creeks, the bole is short and thick, sometimes two feet in diameter. Occasionally the bole divides near the ground, producing several small trunks. It is a common tree in clearings because the Indians leave it standing on account of its supposed medicinal properties. The leaves are broad and very large, sometimes 30 inches long on new shoots. . . . The bark, which is smooth except for a few furrows and warts, is thick and brittle and contains a deep-red sap. It is used in native medicine as a purgative and, in the form of tea, as a febrifuge. When the tree is first cut the wood is light brown or tan, but in an hour or less the entire surface becomes blood-red. If a log is barked and allowed to dry quickly the color will not strike in and when the dry log is sawed the tan color of the interior will remain. If, however, the fresh unbarked log is left on the ground for some time the wood will become pinkish red, perhaps throughout. The red color bleaches to faint pink or even to yellow under prolonged exposure to strong light."

The principal species of the Amazon basin is *Sickingia tinctoria* (H.B.K.) K. Schum. Its flowers are noted for their fragrance. The wood is used to some extent for making spoons and bowls and sometimes in general construction. Small lots of logs, under the name of Arariba Rosa, entered the New York market during World War I for use as a dyewood. The species also grows in the upper Orinoco in Venezuela where it is called Paraguatán, a name perhaps more commonly applied to *S. erythroxylon* Willd. of the Caracas region. There are a few species in southern South America. The Arariba Vermelha, *S. rubra* (Mart.) K. Schum., grows in the coastal forests of Brazil from Bahia to Rio de Janeiro and Minas Geraes, and is sometimes 65 feet tall and 20 inches in diameter. The red dye from the bark was well known to the aboriginal inhabitants, and the richly colored wood is used for carpentry, interior construction, and fancy articles. The Arariba Branca of Rio de Janeiro is said to be *S. viridifolia* (Sald. & Fr. Allem.) K. Schum., but the published description of the wood as dingy white or orange yellow suggests some genus other than *Sickingia*.

Heartwood rather dark brown, usually poorly developed; sapwood yellowish or grayish when fresh, usually becoming red, pink, roseate, or violet rose throughout or more often variegated, and often fading to yellowish brown on the surface, suggesting certain species of *Aspidosperma* (fam. Apocynaceae). Luster medium to low. Odorless when dry; taste slightly bitter. Density widely variable in different species; sp. gr. (air-dry) 0.60 to 0.90; weight 37 to 56 lbs. per cu. ft.; texture rather fine, uniform; grain straight to irregular; working properties good; resistance to decay probably poor. Denser material appears worthy of trial for weavers' shuttles.

Common names: Chickavanté, chuchemuch, nazareno (Mex.); John crow wood, redwood, rosita (Br. H.); brasil, campeche, drago, palo colorado, quina, sangre de chucho (Salv.); alcarreto, guayatil, g. colorado, jagua de montaña, palo colorado, red india ink (Pan.); brasilete, carmesí, coralito, palo brasil, piginio (Col.); aguacatire, aguatire, araguato, carreto, comodoro, cucharo, guatán, lacre, palo cucharo, paraguatá, paraguatán, p. carreto, quina roja (Venez.); huacamain-caspi, machu-sacha, palo rosado, puca-quiro (Peru); arareua, arariba, a. branca, a. rosa, a. vermelha, arariua, asarauba, canella samambaia, iriba, pau brasil, p. d'arara, quina vermelha (Braz.).

Sommera, with eight species of shrubs and little trees rarely 25 feet high, occurs from Mexico to Peru and the Brazilian Amazon region. The wood of *S. sabiceoides* K. Schum., known in eastern Peru as Varilla, is brownish and fine-textured.

Sphinctanthus is a South American genus with six species of herbs, shrubs, and little trees not over 15 feet high. The wood of *S. maculatus* Spruce, collected by Llewelyn Williams in northeastern Peru, is light-colored, fine-textured, and of the type of *Randia*.

Stachyarrhena, with five species of shrubs and slender trees rarely 30 feet high, occurs in Panama and Brazil. The wood is pinkish brown, more or less streaked; hard, heavy, fine-textured, straight-grained, and easily worked. In Panama, where it is called Cabazuelo, the poles are used for rafters and poles in buildings.

Tocoyena is a tropical American genus of about 15 species of shrubs and small trees, widely but sparsely distributed, mostly in South America but with one species in Cuba and two in southern Central America. The yellowish or pinkish, moderately dense, fine-textured wood has no special uses.

Common names: Remo-caspi (Peru); genipapim, genipapinho preto, genipapo bravo, g. do campo (Braz.).

Warscewiczia, with seven species of large shrubs and small to medium-sized trees, is mostly limited to South America, but one species, *W. coccinea* (Vahl) Klotzsch, extends into Trinidad and Costa Rica, and is noted for the brilliance of its bright red inflorescence. Heartwood brownish, with a tinge of olive or pink; usually not very distinct from the sapwood. Luster medium. Without distinctive odor and taste when dry. Hard and heavy to only moderately so; texture rather fine; grain straight to irregular; working properties good; durability probably low. Little used because of its scarcity.

Common names: Wakamy, wild poinsettia (Trin.); barba de gallo (Col.); jayacoma, puca-sisa, quinilla, rafeicoño, r.-ey, shambosisa, tayacona (Peru); amor dobrado, curacy, c. mira, geniparana, picapau, rabo de arara (Braz.).

RUTACEAE

This family, best known as the source of Citrus fruits and Satinwood, consists of about 140 genera and 1600 species of aromatic trees, shrubs, and a few herbs of cosmopolitan distribution, though most abundant in the warm regions of the world. The plants are often armed with spines and usually have glands in the bark, leaves, and fruit. The leaves are opposite or alternate, or sometimes whorled, simple or digitately or pinnately compound, and without stipules; the flowers are perfect or unisexual, large or small, the stamens as many or twice as many as the petals; the fruit is a follicle, capsule, samara, drupe, or berry.

In the New World there are about 400 species of shrubs and small to medium-sized, rarely large, trees, representing 44 genera. The only ones supplying commercial wood are *Amyris, Balfourodendron, Citrus, Esenbeckia*(?), *Euxylophora,* and *Zanthoxylum.* Only two species of the entire family, namely, *Zanthoxylum flavum* Vahl, the West Indian Satinwood, and *Chloroxylon Swietenia* DC., the Ceylon or East Indian Satinwood, are well known to the timber trade of the world. All of the timbers have good technical properties, but with very few exceptions they are too small or too scarce to be of more than local utility.

Color of wood predominantly yellow, whitish, brownish, or greenish, often without much contrast between heartwood and sapwood; exceptions are *Ptelea* (chestnut-brown), *Plethadenia* (very dark brown), *Ravenia spectabilis* Engl. (roseate brown), and a few species of *Zanthoxylum* typified by *Z. monophyllum* Lam. (coppery brown). Luster medium to very high. Odor distinctive in some species of *Amyris* and *Zanthoxylum;* taste mildly bitter in *Hortia,* decidedly bitter in *Ptelea,* otherwise not distinctive. Density medium to high; texture fine to moderately coarse; grain straight to irregular; working properties generally excellent; resistance to decay variable, often low.

Growth rings usually visible; ring-porous structure more or less pronounced in *Choisya, Cneoridium, Erythrochiton, Ptelea, Thamnosma, Zanthoxylum americanum* L., and *Z. Clava-Herculis* L. Pores sometimes visible, but mostly indistinct or invisible without lens; commonly thick-walled; variable in abundance; mostly in short, sometimes long, radial multiples. Vessels with exclusively simple perforations except in *Adiscanthus;* spiral thickenings present in *Choisya, Cneoridium, Ptelea,* and *Thamnosma;* gum deposits common to very abundant; pitting fine to very fine. Rays all uniseriate in *Amyris* and *Choisya;* 1 or 2, occasionally 3, cells wide in *Adiscanthus, Cneoridium, Erythrochiton, Helietta, Sohnreyia,* and certain species of *Zanthoxylum;* 1 to 3, 4, 5, or 6 cells wide in the others; frequently less than 25, nearly always less than 60, rarely up to 100, cells high; homogeneous or weakly heterogeneous in *Amyris, Casimiroa, Citrus, Cneoridium, Dictyoloma, Euxylophora, Helietta* (in part), *Hortia, Nycticalanthus, Raputia, Sohnreyia,* and *Zanthoxylum;* decidedly heterogeneous in *Choisya, Erythrochiton, Esenbeckia, Metrodorea,* and *Ravenia;* less distinctly heterogeneous in the others; crystals common; enlarged, thin-walled oil cells present in *Euxylophora;* ray-vessel pitting mostly fine to very fine, often unilaterally compound. Wood parenchyma very sparse to very abundant; forms represented are paratracheal, aliform, confluent, and apparently demarcating seasonal growths; crystals common; numerous large bundles of raphides present in the diffuse parenchyma of *Raputia magnifica* Engl. Wood fibers with medium to very thick and gelatinous walls; pits small to very small, simple or indistinctly bordered. Ripple marks sometimes present in *Esenbeckia.* Small to very small, vertical, traumatic gum ducts of sporadic occurrence in *Balfourodendron, Citrus, Esenbeckia, Euxylophora, Helietta, Metrodorea, Pilocarpus, Ravenia,* and *Zanthoxylum.* For anatomy of the different genera see *Tropical Woods* 64: 3–27.

Adiscanthus *fusciflorus* Ducke, the only species, is an unarmed shrub or little tree, sometimes 15 feet high, in the undergrowth of somewhat open forests on rather swampy but non-inundated land in the central Amazon region of Brazil. The simple, alternate leaves are smooth, leathery, gland-dotted, entire, the larger ones being 14 to 20 inches long and gradually tapering from the base to a maximum width of about 3 inches; the flowers, which are dark purple outside and white and woolly within, are borne in clusters at the ends of the finely speckled or cross-striped peduncles a foot or more in length; the fruit consists of 1 to 5 dehiscent, 1-seeded, little capsules, each shaped like a clam shell. There are no known uses for the plant. Heartwood absent or not distinguishable from the pale brownish yellow sapwood. Luster medium. Odorless and tasteless. Hard, heavy, tough, and strong; texture moderately fine; grain straight; easy to work, finishing very smoothly; not durable. Of no commercial possibilities.

Amyris, with about 20 species of unarmed shrubs and small to medium-sized trees, occurs from the southern boundary of the United States (Florida and Texas) through the West Indies and Middle America to northern South America. The alternate to opposite leaves are compound, unifoliolate, trifoliolate, or imparipinnate, the leaflets with very numerous pellucid dots, the petioles sometimes winged; the small white flowers are borne in axillary or terminal panicles; the fruit is a small, oily, aromatic, black or reddish drupe.

The type of the genus is *Amyris balsamifera* L., and so closely related to it that they might well be considered only varieties are two other species, namely, *A. elemifera* L. and *A. sylvatica* Jacq. Though often only shrubs, all three forms attain tree size, occasionally 40 to 50 feet high and a foot or more in diameter. The resinous timber is of excellent quality but too small and scarce to be of value for lumber; owing to its resinous nature it is used locally for fuel, torches, and small cabinet work, and to a limited extent commercially as a source of ethereal oil. The only country now exporting the timber appears to be Venezuela. Pittier says (*Bol. Min. de Rel. Ext.* 7: 346. 1930) that the wood of Quigua or Tigua, *A. balsamifera,* contains a resin of the elemi type which yields upon distillation about 30 per cent of ethereal oil; the resin is also the source of an alkaloid, *amirina,* of the formula $C_{30}H_{49}OH$. The wood is shipped in small quantities to Germany and the United States in the form of branches and short logs, in mixture with Candil, *A. simplicifolia* Karst. The first imports into the United States were under the name of West Indian or Venezuelan Sandalwood, but this was later changed to Amyris Wood.

Heartwood yellowish brown deepening upon exposure; more or less streaked; has an oily appearance; sharply demarcated, at least in old specimens, from the thin, nearly white sapwood. Luster medium to high. Odor mildly fragrant or sometimes rather unpleasant; taste somewhat spicy resinous. Very hard, heavy, and strong but brittle; sp. gr. (air-dry) 0.90 to 1.10; weight 62 to 68 lbs. per cu. ft.; texture fine and uniform; grain variable; very easy to work, takes a lustrous polish, holds its place well when manufactured; very oily material highly resistant to decay.

Common names: Torch, torchwood (Florida); candle wood (black, white), rosewood, sandalwood, torchwood (Jam.); cuaba, c. amarilla, c. a. de costa, c. blanca, c. de costa, c. de la maestra, c. de monte, c. de sabana, cuabilla, incienso, i. de costa, palo de incienso, p. de roble, p. de resina, sasafrás del país (Cuba); palo de tea, puerco, tea, torchwood (P.R.); guaconejo (Dom. R.); bois chandelle, chandelle blanc

(Haiti); bossoea (Curaçao); limoncillo, ocotillo blanco, palo de gas, tojtanyuc (Mex.); waika pine (Br. Hond.); melón, ocotillo, roldán (Salv.); chilillo, pimienta (Hond.); marfil, palo de marfil, naranjito, ulanda (Col.); candil, c. de montaña, c. de playa, quigua, tigua (Venez.); seca olorosa (Ec.).

Balfourodendron *Riedelianum* Engl., the sole species, is an unarmed tree, rarely up to 80 feet high and 30 inches in diameter, usually much smaller, occurring in the State of São Paulo, Brazil, northern and central Paraguay, and in Misiones and Corrientes, Argentina. The leaves are large, digitately compound, with three pellucid-punctate leaflets; the small flowers are borne abundantly in terminal panicles; the fruit is a capsule about an inch long with 3 or 4 greenish leathery netted-veined wings. The wood, which has about the consistency of Sugar Maple (*Acer saccharum* Marsh.), is of excellent quality and is highly esteemed in Argentina, where the largest sizes are obtainable, for turnery, tool handles, oars, agricultural implements, interior construction, and furniture.

Heartwood apparently absent or not readily distinguished from the nearly white or pale yellowish brown sapwood; sometimes with a slight greenish tinge. Luster medium. Without distinctive odor or taste. Hard, heavy, strong, and elastic; sp. gr. (air-dry) 0.75 to 0.83; weight 47 to 52 lbs. per cu. ft.; texture fine and uniform; grain generally straight; not difficult to work, splits readily, takes a high polish; is not resistant to decay.

Common names: Marfim, pau liso, p. marfim, pequia marfim (Braz.); guatambú blanco, g. morotí, ibirá-ñeté (Arg.).

Casimiroa, with a few closely related and doubtfully distinct species of unarmed shrubs and trees, is limited in natural distribution to Mexico and Central America to Costa Rica. The leaves are alternate, digitately compound, usually with 3 or 5 leathery pellucid-dotted leaflets; the small white or greenish yellow flowers are borne in axillary racemes; the fruit is a 2–5-celled drupe with a single seed in each cell. The timber is not utilized for any special purpose.

The typical species is *Casimiroa edulis* Llave & Lex., a medium-sized tree, commonly called Zapote Blanco or White Sapote. Standley says (*Trees and shrubs of Mexico,* p. 527): "The White Sapote (including also *Casimiroa Sapota* Oerst.) is a well-known tree in Mexico, but is little grown outside that country. It is cultivated in the West Indies and has been introduced into southern California. The fruit varies in size and quality. It somewhat resembles an apple, and the best varieties are as large as a good-sized orange. The tender yellowish skin is thin, like that of an apple; the pulp is soft and cream-colored, of delicate texture, with a pleasant sweet flavor. The fruit ripens in July and August. It is much eaten in Mexico and is commonly sold in the markets. The fruits are popularly believed to induce sleep if eaten in quantity, and to calm rheumatic pains. The bark, leaves, and especially the seeds are said to contain a glucoside, casimirosine, which has a hypnotic and sedative effect upon cerebral centers. A small dose, it is stated, produces, at the end of an hour, deep sleep which lasts four to six hours. This principle, obtained chiefly from the seeds, has been used by Mexican physicians." The Central American species is usually referred to *Casimiroa tetrameria* Millsp., but according to Standley (*loc. cit.,* p. 526), it may be only a pubescent form of *C. edulis.*

Wood whitish throughout. Luster medium. Odorless and tasteless. Density medium; texture rather coarse; grain straight; easy to work, finishing very smoothly; not durable. Of no commercial possibilities.

Common names: Mango tarango, sapote blanco (Cuba, intr.); chapote, coaxmuttza, cochitzapotl, hyuy, iztactzopotl, matasano, yuy, zapote, z. blanco, z. de rata (Mex.); white sapote (Br. H.); matasano (Cent. Am., general); abache (Guat.).

Choisya, with seven species of unarmed aromatic shrubs or little trees, is limited in distribution to the southwestern United States and Mexico. The leaves are opposite or nearly so, digitately compound, with 3 to 13 oblong to linear leaflets; the white flowers are borne in showy axillary cymes near the ends of the branches; the fruit consists of 3 to 5 leathery 2-valved carpels. Some species are often cultivated in gardens for decorative purposes. The stems are too small to be utilized. Heartwood not seen; sapwood yellowish white. Luster medium. Odorless and tasteless. Hard, heavy, and strong; texture fine and uniform. Of no commercial interest.

COMMON NAMES: Zorillo (New Mex.); clavillo, clavo de olor, flor de clavo, hierba de clavo (Mex.).

Citrus is an Asiatic genus, with about a dozen species of aromatic shrubs and small trees, mostly with thorny branches, widely planted throughout the warmer regions of the world, and in many places thoroughly naturalized. The principal fruits are of the following species: Citron, *C. medica* L.; lemon, *C. Limonia* Osbeck; lime, *C. aurantiifolia* (Chr.) Swingle; common or sweet orange, *C. sinensis* (L.) Osbeck; sour orange, *C. Aurantium* L.; tangerine, *C. nobilis* Lour.; grapefruit, *C. paradisi* Macf.; and shaddock, *C. grandis* Osbeck. The woods of the different species are similar and enter the market in limited quantity in the form of small and generally irregular logs or sticks under the name of Orangewood. It is used for making small articles of turnery and carving, fancy boxes and novelties, inlays, and particularly for manicure sticks. Other woods used for the latter purpose and under the name of Orangewood are Venezuelan "Boxwood," *Gossypiospermum praecox* (Gris.) P. Wils., European Spindle-tree, *Evonymus europaeus* L., and Black Gum, *Nyssa sylvatica* Marsh. (see *Tropical Woods* 25: 28). The so-called Lemonwood of the archery bowmakers is the Cuban Degame or Dagama, *Calycophyllum candidissimum* (Vahl) DC., family Rubiaceae.

Wood light clear yellow throughout. Luster high. Odorless and tasteless. Hard, heavy, and strong; texture fine and uniform; grain variable; fairly easy to work, taking a glossy polish; holds its shape well when manufactured; is perishable in contact with the soil.

Cneoridium *dumosum* (Nutt.) Hook. f., the only species, is a densely branched unarmed strongly scented shrub of southern California and northern Baja California, Mexico. The linear gland-dotted leaves are opposite or fascicled on short branches; the flowers are very small and white; the fruit is a little reddish brown capsule containing 1 or 2 seeds. The plant apparently is of no utility.

Dictyoloma, with two species of unarmed shrubs or small trees, is limited in distribution to eastern Brazil and eastern Peru. The leaves are doubly pinnate; the small flowers are borne in many-flowered umbel-like racemes; the fruit is a 5-celled dehiscent capsule containing a few small kidney-shaped seeds, each with a cleft membranous wing. The Brazilian species, *D. incanescens* DC., grows in dry localities and rarely exceeds 20 feet in height. *D. peruviana* Planch. is a shrubby tree in the uplands of northeastern Peru. The crushed fresh leaves are said to be used locally as a substitute for soap and to stupefy fish. The wood apparently has no special uses.

Heartwood pale yellow; not sharply demarcated from the white sapwood. Luster high. Without distinctive odor or taste. Of rather light weight, but firm; texture fine; grain straight; very easy to work, taking a lustrous polish; not durable. Of no commercial possibilities.

COMMON NAMES: Barbasco negro, huaman-samane (Peru).

Erythrochiton, with five species of unarmed shrubs or little trees, is widely but sparingly distributed in tropical America. The leaves, which are clustered at the ends of the branches, have three leaflets in *E. trifoliatus* Pilger, but only one in the others; the showy white or roseate flowers are in long-pedunculate cymes or clusters except in *E. hypophyllanthus* Planch. & Linden, where they are borne singly on the dorsal midrib of the leaves; the fruit is a 5-parted capsule. The wood is not used for any special purpose. The only species in North America is *E. Lindeni* (Baill.) Hemsl., a shrub or little tree ranging from southern Mexico along the Pacific coast to Costa Rica. The most widely distributed species in South America is *E. brasiliense* Nees & Mart., a shrubby tree in southern and northern Brazil and eastern Bolivia and Peru. It is sometimes cultivated for ornamental purposes and the bark of the root is used in local medicine as a vermifuge. Heartwood not seen; sapwood nearly white. Luster medium. Odorless and tasteless. Hard and heavy; texture rather fine; grain irregular; easy to work; not durable. Has no commercial possibilities.

COMMON NAME: Sabia miuda (Braz.).

Esenbeckia, with about 30 species of unarmed shrubs and small to medium-sized trees, is widely distributed in America from Texas and northern Mexico to Chile, Argentina, and southern Brazil, though poorly represented in the Amazon basin. The leaves are simple, unifoliolate, or palmately compound with 3 to 5 leaflets; the yellowish flowers are borne in large terminal or axillary panicles; the fruit is an angled, usually rough or spiny, woody capsule with 4 or 5 elastically dehiscent carpels.

Best known of the northern species is *Esenbeckia pentaphylla* (Macfad.) Gris., a medium-sized tree occurring in Jamaica. A closely related species, *E. belizensis* Lundell, grows in Yucatán, British Honduras, and eastern Guatemala. The smooth greenish bark has irregular vertical rows of light-colored lenticels. The yellowish, fine-textured wood is suitable for tool handles and turned articles but is not used except for fuel.

There are numerous species in Brazil. *Esenbeckia febrifuga* A. Juss., a tree sometimes 45 feet high and 20 inches in diameter, occurs in the southeastern parts of that country and also in Paraguay and Misiones, Argentina, in much of its range being only a shrub. The bark is used as a substitute for quinine in treating fevers and the wood is a local favorite for making wooden spoons and dishes and other articles of carving and turnery. Other species, such as *E. leiocarpa* Engl. and *E. grandiflora* Mart., provide some timber for fence posts, implement frames, spokes of wagon wheels, and similar purposes requiring strength, durability, and fine texture.

Esenbeckia alata (Karst. & Tr.) Tr. & Pl. is a tree 30 to 40 feet high with a straight trunk 12 inches in diameter, of infrequent occurrence in northern Colombia. The bark is smooth and gray; the leaves and fruits are mildly aromatic. The timber is not utilized. *E. atata* Pittier occurs in the thorn forest of the dry coastal region of Venezuela and attains a maximum height of 50 feet and a diameter of 16 inches. It is claimed that small quantities of the timber were formerly exported to Germany and the United States as a substitute for Boxwood (*Buxus*), but the trees are now too scarce to be of commercial value. (See Record & Garratt's *Boxwoods*, Bull. 14, Yale School of Forestry, pp. 68–71.)

The following description is based on wood samples of eight species of *Esenbeckia*. Heartwood light clear yellow; not always sharply demarcated from the nearly white sapwood. Odorless and tasteless. Luster high. Very hard, heavy, and strong; sp. gr. (air-dry) 0.95 to 1.12; weight 59 to 70 lbs. per cu. ft.; texture very fine and uniform; grain straight to very irregular; easy to work, taking a lustrous polish; some heartwood specimens apparently durable. An excellent wood, but apparently without commercial possibilities.

COMMON NAMES: Wild orange (Jam.); gasparee—red, white, yellow (Trin.); hokab, jopoy, palillo, palo amarillo, yax-hocob (Mex.); verde lucero (Br. H.); angustora del Brasil, quina del Brasil (Col.); anacoa, ata-ata, atata (Venez.); apogitaguára, canella de cotia, guarantá, guaratan, larangeira, l. do matto, mamonhinho, mendanha, pau de veado, p. duro, quina do matto, tres folhas vermelhas (Braz.); apoitá-guará (Par.); ibirá-obí-guazú (Arg.).

Euxylophora *paraensis* Huber, the only species, is a large unarmed tree sometimes 130 feet tall, confined to non-inundated lands of lower Amazon region in the State of Pará, Brazil. The alternate, simple, entire, leathery leaves have a deeply grooved petiole and a blade ranging up to 10 inches long and 4 inches wide; the creamy white fragrantly scented flowers are borne in large terminal cymose panicles; the fruit is a

5-parted capsule with each of the dehiscent carpels containing two shiny black seeds.

Huber, who described the genus and species, says of it (translated from *Bol. Mus. Goeldi* 6: 184–185; 1909): "The most important representative of this family [Rutaceae] in Pará, from the standpoint of the utilization of its timber, is the Pau Amarello, a large tree of the forests on *terra firme* in the eastern and southern part of the state. Only in the last few years has it been possible for me to get flowers and fruits of this tree, emanating from the Augusto Montenegro Experiment Station where it is fairly common though not abundant. Examination of this material revealed the unexpected fact that the Pau Amarello is a representative of a new genus of the tribe Cusparieae, to which I have given the name *Euxylophora* in token of the excellent wood that it produces. . . . The Pau Amarello is one of the timbers most used in Pará and if found in greater abundance (I hear that in some places, for example along the upper Capim and in certain stretches of the Tocantins, it is of very frequent occurrence) it would have even more varied applications. On account of the large size of the boles and the great homogeneity of the lovely light yellow heartwood, it can be used for doors, tables, etc., which require wide planks. Its most frequent application is in the beautiful floors made of alternate boards of Acapú [*Vouacapoua*] and Pau Amarello which add so much to the attractiveness of the houses in Pará. In furniture, also, the juxtaposition of Pau Amarello, especially curly-grained pieces which are sometimes called Pau Setim, to dark-colored wood such as Acapú, Pau Santo [*Zollernia*], Jacarandá [*Dalbergia Spruceana* Benth.], Macacaúba [*Platymiscium*], etc., produces a most beautiful effect. It is probable this timber, were it possible to cultivate it (which unfortunately, judging from our experiments, appears not to be a very easy thing to do), could be exported on a large scale and meet with an unlimited market. As conditions now exist, Pau Amarello is a rather costly wood even here and probably could not be exported with profit."

The timber is known in the New York market, sometimes as Brazilian Satinwood or Sateen, but it lacks attractive figure, high luster, and other special properties essential for fine cabinet work. It has been used to some extent for the backs of hair brushes. A New York manufacturer attempted to use this timber for making Mah Jongg tiles faced with pyralin which is applied in a heated hydraulic press and requires that the wood be very dry. According to a report to the senior author, "wood experimentally dried to 4 per cent of moisture was so brittle that a board would shatter upon being dropped on a hard floor." Apparently the most satisfactory use for the timber is in a type of flooring and furniture for which the local demand is equal to the supply.

Heartwood bright clear yellow, deepening upon exposure; not sharply defined from the yellowish white sapwood. Luster fairly high. Without distinctive odor or taste. Rather hard and heavy; sp. gr. (air-dry) about 0.81; weight 51 lbs. per cu. ft.; texture medium and uniform; grain straight to irregular; not very difficult to work, taking a high polish; probably of low resistance to decay.

COMMON NAMES: Brazilian boxwood, B. satinwood, canary wood, sateen wood (U.S. trade); limão-rana, pau amarello, p. setim, pequia setim (Braz.).

Helietta, with five or six species of unarmed shrubs and small trees, occurs in tropical and subtropical America from Texas and northern Mexico to Paraguay. The leaves are compound, with 3 leaflets; the small flowers are borne in terminal or axillary panicles; the fruit consists of 3 or 4 indehiscent winged carpels, suggesting a group of Maple (*Acer*) seeds.

The only North American species is *Helietta parvifolia* (A. Gray) Benth., a shrub or little tree up to 25 feet high, forming thickets of considerable extent in the valley of the Rio Grande, but becoming somewhat larger on the limestone ridge of the Sierra Madre in Nueva León, Mexico. The only West Indian species is *H. glaucescens* Urb. of eastern Cuba. *H. Plaeana* Tul. is a little tree of Colombia and Venezuela, and Pittier suggests (*Bol. Min. Rel. Ext.* 7:

344) that it may have industrial possibilities as a substitute for Boxwood (*Buxus*). The species with southernmost range is *H. longifoliata* Britt., a tree usually less than 35 feet high and 12 inches in diameter, occasionally considerably larger. The timber is used locally for tool handles and various articles of turnery and appears suitable for making weaving shuttles.

Wood yellowish or nearly white throughout. Lustrous. Odorless and tasteless. Very hard, heavy, tough, and strong; sp. gr. (air-dry) 0.90 to 0.95; weight 56 to 61 lbs. per cu. ft.; texture fine and uniform; grain irregular; easy to work, taking a lustrous polish; probably poorly resistant to decay.

COMMON NAMES: Baretta, barreta, barretta (Texas, Mex.); gipato (Col.); caritivá, caritivano, caritivar, marfil (Venez.); canella do veado (Braz.); canela de venado, hira-oby, ibirá-obí (Arg.).

Hortia, with six species of unarmed trees and shrubs, occurs in tropical south and central Brazil and the Amazon basin. The simple alternate leathery leaves are sometimes 36 inches long and 6 inches wide, with the margin rolled toward the base; the small unscented red or roseate flowers are borne in large cymose panicles; the fruit is a drupe. *H. regia* Sandw. is the species known to occur in British Guiana, where it is one of the most distinctive trees in the forest because of its very long and narrow leaves. The fruit is yellow, juicy, and edible. The largest species is *H. excelsa* Ducke, a tree sometimes 100 feet tall in the humid high forest near Garupá in the Brazilian Amazon region. The inner bark has an odor suggesting wine made from sugar cane and this is imparted to the fresh wood. Closely related to it, but not so large a tree, is *H. superba,* also with scented bark; the only wood sample available for study is from the type of this species (see *Tropical Woods* 43: 21).

Color yellowish throughout. Fairly lustrous. Odorless when dry, but with slightly bitter taste. Very hard, heavy, tough, and strong; texture medium; grain straight; not difficult to work, easy to split, takes a glossy polish; is poorly resistant to decay. A good wood of the general type of Maple (*Acer*), but apparently of no commercial possibilities because of its scarcity.

COMMON NAMES: Bush orange, powistail tree, warunama (Br. G.); cachaceiro, pau amarello (Braz.).

Metrodorea, with five or six species of unarmed shrubs and small to medium-sized trees, is apparently confined to Brazil. The leaves are opposite, compound, with 2 or 3 leaflets; the flowers are small and borne in mostly terminal panicles; the fruit is a rather large, irregular, woody, 5-angled capsule. *M. pubescens* St. Hil. is a medium-sized tree of common occurrence on the plains of southeastern Brazil, and is said to supply a very dense, yellowish white timber of rather poor quality used for interior construction, but not suitable for flooring; its bark is medicinal. Another tree of the same general region is *M. nigra* St. Hil.; its wood is used to a limited extent for carving and carpentry. *M. flavida* K. Krause occurs in the central Amazon region and attains a height of about 60 feet; apparently the timber has no special uses.

Wood light yellow or yellowish white throughout. Luster high. Odor and taste absent or not distinctive. Hard, heavy, tough, and strong; sp. gr. (air-dry) 0.90 to 1.10; weight 56 to 72 lbs. per cu. ft.; texture fine; grain fairly straight; not very difficult to work, finishing very smoothly; poorly resistant to decay. Presumably of no commercial possibilities.

COMMON NAMES: Caputuna, caputuva, cataguá, c. bianca, c. rajada, chupa-ferre, laranjeira do matto, limoeiro do matto, tembletaru (Braz.).

Nycticalanthus *speciosus* Ducke, the only species, is an unarmed night-blooming shrub discovered in 1930 near Manáos, Brazil. The 3-foliolate leaves are very large; the large and conspicuous white flowers are borne in cymose panicles and suggest certain Bauhinias; the fruit is capsular. The plant is related to *Spiranthera*. The wood is yellowish white; odorless and tasteless; hard, heavy, strong, and fine-textured.

Peltostigma, with three described species of small unarmed trees, occurs in Jamaica and on the mainland from southern Mexico to Costa Rica. The best known and perhaps the only valid species is *P. pteleoides* (Hook.) Walp., a balsam-scented tree sometimes 25 feet high having the range of the genus. The leaves are alternate and digitately compound with 3 to 5 leaflets; the large fragrant white or creamy flowers are borne in axillary cymes; the fruit is capsular, with 6 to 10 cocci. The timber is not utilized, presumably because of its small size and scarcity. The following description is based upon a single Costa Rican specimen (Yale 38377; H. E. Stork 4161). Wood nearly white throughout stem (4 inches in diameter). Luster medium. Odorless and tasteless. Hard, heavy, tough, and strong, having about the consistency of Sugar Maple (*Acer saccharum* Marsh.); texture medium; grain fairly straight; not difficult to work, taking a smooth finish; probably perishable in contact with the ground.

COMMON NAME: Candle wood (Jam.).

Pilocarpus, with about 20 species of unarmed shrubs and little trees, is widely distributed in tropical America, but is only sparingly represented in the Amazon basin and apparently not at all in Central America. The leaves, which are simple or odd-pinnate with 3 to 9 leaflets, are congested at the ends of the branches and are alternate, opposite, or whorled; the small flowers are borne in terminal or axillary racemes or spikes; the fruit is a capsule with 1 to 5 dehiscent 1-seeded carpels. The principal value of the plants is medicinal, as the leaves contain an alkaloid, pilocarpine, well known to the drug trade. The woods have no special uses.

The West Indian species is *Pilocarpus racemosus* Vahl; it also occurs in Mexico, though the form there has been named *P. longipes* Rose. It rarely exceeds 20 feet in height. The Venezuelan shrub, *P. Alvaradoi* Pittier, does not appear essentially different from *P. racemosus*. Most of the species are native to eastern and southern Brazil, Paraguay, and Misiones, Argentina, the best known being *P. pennatifolius* Lem. and variety *Selloanus* (Engl.) Hessl. The following description is based on three authentic specimens of the foregoing species and variety from Mexico, Venezuela, and Argentina.

Wood lustrous yellowish or yellowish brown; sometimes with fairly distinct heartwood. Odorless and tasteless. Hard, heavy, and strong; texture fine; grain variable; rather easily worked, taking a high polish; poorly resistant to decay.

COMMON NAMES: Kokob-ché (Mex.); borrachero, mata sarna, sarna (Venez.); jaborandy (Braz.); pau de cotia (Urug.); caá-tay-guazú, ibirá-tay (Par.); ibirá-tai, jaborandí, yaguarandí (Arg.).

Plethadenia, with two closely related species of unarmed shrubs or little trees, is of limited distribution in the West Indies. The leaves are evenly pinnate, with narrowly winged rachis and 2 to 4 pairs of small glandular and punctate leaflets; the small white flowers are borne in axillary cymes; the capsular fruit is composed of four 1-seeded carpels. *P. granulata* (Krug & Urb.) Urb. occurs in dry uplands in the Dominican Republic, and *P. cubensis* Urb. is found in similar situations in eastern Cuba. The plants appear to be rare and no common names for them are recorded. Wood deep reddish brown throughout small branches, at least in dry material, becoming blackish brown with an oily appearance toward the center of older stems. Luster medium. Odor mildly fragrant; taste not distinctive. Very hard, heavy, and strong; texture fine; grain fairly straight; takes a high natural polish; probably durable. Of no commercial possibilities.

Ptelea, in a conservative sense, consists of a single species, *P. trifoliata* L., an unarmed shrub or a small tree rarely 25 feet high, with a smooth-barked trunk sometimes eight inches in diameter; it is widely distributed in North America from southern Canada throughout the eastern half of the United States and southward from New Mexico and California to Tamaulipas and Oaxaca, Mexico. The species exhibits much variation, and more than 50 segregates have been described and named, but according to Standley (*Trees and shrubs of Mexico*, p. 531) "it seems impossible to divide the specimens into groups by any constant character." The alternate leaves are digitately compound, with 3 or rarely 5 leaflets which are either glabrous or pubescent, and variable in size and form; the greenish white flowers are borne in compound cymes; the fruit is a wafer-like samara nearly an inch across or rarely wingless. The bark and foliage are strong-scented. Standley says (*loc. cit.*): "In the United States, where the plant is known as Hoptree and Wafer Ash, the fruit has been employed as a substitute for hops. The root has a bitter, pungent, and slightly acrid but not disagreeable taste and a somewhat aromatic odor. It contains the alkaloid berberine. It has been employed as a remedy for dyspepsia and as a mild tonic." The tree is often planted for ornamental purposes in parks and gardens. There are no special uses for the wood.

Heartwood chestnut-brown; rather sharply demarcated from the yellowish sapwood. Luster medium. Odor absent; taste of heartwood very bitter. Hard, heavy, tough, and strong; sp. gr. (air-dry) 0.85 to 0.90; weight 53 to 58 lbs. per cu. ft.; texture medium; grain straight to irregular; not difficult to work, finishing very smoothly; fairly durable. Of no commercial possibilities.

Common names: Ague bark, hop tree, wafer ash (U.S.A.); cola de zorillo (Mex.).

Raputia, with eight species of unarmed shrubs and small trees, is distributed from Venezuela through the lower Amazon region to southeastern Brazil. The 1–7-foliolate leaves are alternate or opposite and clustered at the ends of the branchlets; the flowers are large and racemose; the fruit is a dehiscent capsule composed of five 2-seeded carpels. The largest tree is *R. magnifica* Engl., with a range extending from Rio de Janeiro to São Paulo. It is known as Arapóca and is said to be up to 50 or 60 feet high with a trunk 20 to 25 feet long and 16 to 32 inches in diameter. The timber is used locally for joinery and general construction. The following description applies to this species, as it is the only one represented in the Yale collections.

Wood yellowish throughout; fairly uniform except for occasional fine brownish streaks. Luster rather high. Without distinctive odor or taste. Moderately hard and heavy; having about the consistency of Sugar Maple (*Acer saccharum* Marsh.); texture rather fine, uniform; grain fairly straight; easily worked, taking a very smooth finish; appears suitable for turnery and carving.

Common names: Amarellinho, arapóca, a. amarella, a. branca, a. de cheiro, gemma de ova, guatayapóca, gurataiapóca, pau amarello (Braz.).

Ravenia, with 10 species of unarmed shrubs and small to rarely medium-sized trees, occurs from the West Indies to southeastern Brazil, with a single species in Central America. The leaves are opposite large and simple or palmately compound with three rather small leaflets; the white, roseate, or red flowers are large and showy; the fruit is a capsule with 1 to 5 dehiscent carpels. There are apparently no special uses for the plants except for decorative purposes.

The largest species is the Tortugo Prieto, *Ravenia Urbani* Engl., of Puerto Rico, where in the high forests of the eastern mountains it is said to attain a height sometimes of 50 feet. Its wood has not been studied. *R. rosea* Standl. is a shrub or a small branchy tree not exceeding 25 feet in height and four inches in diameter, discovered by F. C. Englesing in Nicaragua and said by him to be abundant in dense shade over the whole area between the Rawawas and Ocongwas Rivers. The trunk is usually straight and cylindrical, the bark smooth and almost black except for the numerous small brown lenticels. The plant has no known uses and even the Indians have no name for it despite the fact that it bears rather conspicuous deeply rose-colored flowers from September to early February. Wood yellowish white throughout. Luster rather high. Without distinctive odor or taste. Hard, moderately heavy, tough, and strong; has about the consistency of Maple (*Acer*); easy to work, finishing very smoothly; is presumably low in resistance to decay. *R. spectabilis* (Lindl.) Planch. is a shrub or little tree growing in Haiti, Guadeloupe, and Cuba; it is known in Cuba as Lemonia and possibly also as Arraiján. Heartwood pinkish or roseate brown; distinct and rather sharply demarcated from the thin brownish sapwood. Luster medium. Odorless and tasteless. Very hard and heavy; texture fine; grain fairly straight; appears durable.

Sohnreyia *excelsa* Krause, the only species, is an unarmed tree 50 to 65 feet high, with the aspect of a palm, growing in the lower and central Amazon regions of Brazil. The pinnate leaves are sometimes seven feet in length, with many leaflets about an inch wide and 8 to 12 inches long; the flowers are small and racemose; the winged 2-seeded fruit resembles that of *Ptelea*. The plant is highly ornamental but supplies no important products.

Wood lustrous golden yellow throughout specimen. Scentless and tasteless. Of moderate density, firm and strong; texture medium and uniform; grain straight; easy to work, taking a satiny polish; probably not very resistant to decay. An excellent wood suitable for many purposes but not available in sufficient quantity to be of economic value.

Thamnosma, with six species of strong-scented shrubs and half-shrubs, occurs in South Africa and in southwestern United States and northern Mexico. The leaves are simple alternate and entire, sometimes reduced to scales; the flowers are small and racemose; the fruit is a leathery 2-lobed capsule opening at the apex. The only authentic specimen available (Yale 26658) is of *T. montana* Torr. & Frém. collected by S. B. Detwiler in Arizona, where it is known as Cordoncillo. Wood uniform pale yellow throughout. Luster medium. Odorless and tasteless. Rather hard and moderately heavy; texture fine; grain straight. Apparently of no commercial possibilities because of the small size of the plants.

Zanthoxylum and **Fagara** were treated as distinct genera by Linneaus and this view was adopted by Engler and many other botanists. Considered separately, *Zanthoxylum* includes about 15 species of shrubs and trees of the north temperate zone, all in eastern Asia with the exception of *Z. americanum* Mill., the Prickly Ash of eastern North America, while over 200 pantropical species of shrubs and trees are referred to *Fagara*. Prominent American authorities, including Sargent, Sudworth, Wilson, and Standley, consider *Fagara* a synonym for the older name *Zanthoxylum* (or *Xanthoxylum*). The trees are mostly small or medium-sized, in comparatively few instances exceeding 50 feet in height, with trunk diameters of 12 to 18, rarely 24, inches. In many species the branches are armed with stipular spines or prickles, and parts of the stems may be covered with corky knobs. The leaves are alternate, even-pinnate, odd-pinnate, or rarely unifoliolate, the leaflets entire or crenulate, the rachis often grooved, sometimes winged, unarmed or prickly; the flowers are small white, yellowish, or greenish and borne mostly in clustered axillary racemes or terminal panicles; the fruit consists of 1 to 5 follicles. The bark is aromatic and that of the roots is sometimes used in medicine as a stimulant and tonic.

The most northern species is *Zanthoxylum americanum* Mill., a much-branched spiny shrub or a tree sometimes 20 feet high, often forming dense thickets, from Quebec and Ontario to North Dakota and southward to Georgia, Alabama, and Oklahoma. The wood is definitely ring-porous. The prickly shrub or little tree, *Z. Clava-Herculis* L., commonly known as Hercules Club, Toothache-tree, and also Prickly Ash, grows from Virginia to Florida and westward into Texas. The bark is used as a remedy for toothache and rheumatism. There are about 25 species in Mexico and Central America and about 25 more in the West Indies and northern South America. The genus is poorly represented in the Amazon basin, but there are numerous species in southern South America. A few are large enough to supply some timber for local use, but are not likely to contribute to the export trade.

The only important commercial timber

is the West Indian Satinwood. Although more than one species may contribute to the supply, the principal one is *Zanthoxylum flavum* Vahl. This is an unarmed tree sometimes 40 feet high and 20 inches in diameter, usually much smaller and often reduced to a shrub, with a range including Dominican Republic, Puerto Rico, the Bahamas, Bermuda, and southern Florida. The wood is hard and heavy (sp. gr. about 0.90; weight 56 lbs. per cu. ft.), fine-textured, often with a beautiful, wavy grain, and is of a creamy or golden yellow color, with an oily appearance and feel, and when freshly worked gives off a pronounced scent of coconut. It has been known to the fine-furniture trade for a long time and is used for all kinds of cabinet work, inlays, and marquetry; it also is employed in turnery and for making the backs of hand mirrors and hair brushes. Antique Satinwood furniture is almost exclusively of this species, but in later times Ceylon or East Indian Satinwood (*Chloroxylon Swietenia* DC.) came into general use. The latter is generally more highly figured, paler in color, and somewhat harder than the other; it differs structurally in having distinct ripple marks and in the definite radial arrangement of the pores.

Another kind of West Indian Satinwood is exported in small quantities from Dominican Republic where it is known as Pino Macho. The tree has been provisionally identified as *Zanthoxylum elephantiasis* Macfad. It grows also in Cuba and Jamaica, and sparingly on the mainland from Vera Cruz, Mexico, to Panama. The wood differs from that of *Z. flavum* in being coarser-textured, less heavy and solid, and is slightly dull brown rather than lustrous golden yellow, and lacks the coconut scent. Some of it is beautifully figured and gives a very pleasing effect when made into furniture. It has been used successfully to a minor extent in New York and England, mostly under the name of Concha Satinwood.

Heartwood typically yellowish, with a more or less pronounced greenish hue, becoming brown superficially upon exposure; not clearly differentiated from the sapwood. In a few species, notably *Zanthoxylum monophyllum* Lam., the heartwood is coppery brown and sharply demarcated from the lemon-yellow sapwood. Luster usually high, suggesting some of the Lauraceae. Without distinctive odor and taste, except in *Z. flavum* which is coconut-scented. Light and soft to heavy and hard, but mostly in the class of Birch (*Betula*) and Yellow Poplar (*Liriodendron*) and suitable for the same purposes; working properties excellent; durability fair to poor.

Common names: *Zanthoxylum elephantiasis:* Concha satinwood (trade); satinwood, yellow sanders (Jam.); ayúa varía (Cuba); pino macho (Dom. R.); lagartillo (C.R.). *Z. flavum:* West Indian satinwood (trade); satinwood, yellow sanders, y. wood (Florida, B.W.I.); aceitillo (Cuba, P.R.); ayúa amarilla, limoncillo cimarrón (Nassau); espinillo (Dom. R.); satiné jaune (Fr. W.I.); calibori (Venez.). *Z. monophylum:* Carubio, mapurito, mopurito, rubia, yellow prickle (P.R.); malacapa (Hond.); lagarto amarillo, l. negro (C.R.); bosúa, bosuda, bosuga, paneque (Venez.). *Other species:* Bastard ironwood, colima, correosa, doctor's club, frêne piquant, Hercules club, pepperwood, pillenterry, prickly ash, sea ash, sting-tongue, tear-blanket, toothache tree, wait-a-bit, wild lime, w. orange, yellow wood (U.S.A.); Caesar wood, licca tree, lignum-rorum, prickly yellow, rosewood, saven tree, sevin tree, suarra wood, yellow Hercules, y. sanders (Jam.); aceitillo, ayúa, a. blanca, a. varía, ayuda, a. blanca, a. hembra, a. macho, a. varía, bayúa, b. lisa, bayuda, chivo, espino, limón café, limoncillo, mate arbol, niaragato, tomeguín, uña de gato, zorillo (Cuba); ayúa, carubio, cenizo, espino, e. rubial (P.R.); bois piné, piné, p. jaune (Haiti); arbre à pian, bois espinaux jaune, b. jaune, espin de bobo, espineaux blanc, e. jaune, e. rouge (Fr. W.I.); bosoo, l'epinet, stinking l'epinet (Trin.); colima, limoncillo, palo de ropa, p. mulato, sinanche, tamcazche, tancazche, uña de gato, vole, xic-ché, zorrillo (Mex.); prickly yellow—alligator-toothed, bastard, black, smooth-barked (Br. H.); ceibillo, lagarto, l. amarillo (Guat.); cedro espino, pochote, rosillo, salitrero (Salv.); cedrillo, cedro espino, chincho, lagarto amarillo (Hond.); ku-krá, lagartillo, lagarto amarillo, l. blanco, l. colorado, l. negro, limoncillo, norí, zorrillo (C.R.); acabú, alcabú, arcabú, ruda (Pan.); amamor, araña-gato, barbasco, carey vegetal, matijón, palo carey, uña de gato (Col.); araña-gato, bosúa, bosuda, bosuga, mapurite, m. blanco, ñaragato, panequa, uña de gato (Venez.); sada (Br. G.); awarratalla, awarratarra, boeboeraballi, boeroeaballi, geel steckel boom, hapau, he-he, he-makoa, karidan, k. hariraroe, karimadan kakhekora, kawarratarra, mana, palakoea, pieterjarie, pritijari, pritijarie, toepoeroe-proroem (Sur.); bois piquant, cacatier, clavalier des Antilles (Fr. G.); arruda do matto, betaru amarello, chupa ferro, coentrilho, cupim, curaturá, espinheiro, espinho de vintem, laranjeira do matto, laranjinha, limãozinha, mammica de cachorra, m. de cadella, m. de porca, mamminha de porca, mamuda, marupá-rana, tamanqueira, t. de espinho, t. da terra firme, t. da varzea, tambataruga, tembetarú, t. de espinho, temetarú, tinguaciba (Braz.); alcanfor-sacha, culantrillo, espino, hualaja, quillo-casha, shapillejo (Peru); coentrillo, culentrillo, mamica de cadena, mamilla de cadena, tembetarí, tembeterí, teta de cadena (Urug.); cochudo, coco, c. de cordoba, cuentrillo, culantrillo, curatúrá, c. morotí, mamica da cadella, naranjillo, sacha limón, saúco hediondo, tembetarí, t. blanco, t.-hú, t.-mí, t.-morotí, t.-puiá, t.-puitá, t.-saiyú, tembetary, t.-guazú (Arg.)

SABIACEAE

An unimportant family of three or four genera and about 150 species of trees and upright or climbing shrubs occurring in the Far East and tropical America. The leaves are alternate, simple or unequally pinnate, without stipules; the small flowers are borne in compound axillary or terminal racemes; the dry or drupaceous fruit is one-seeded, the embryo large, the cotyledons contorted, and the radicle curved. The species of *Sabia* are lianas or scandent shrubs in eastern Asia; they have coarse-textured, broad-rayed woods. *Ophiocaryon,* with two species, is sparsely distributed in the Amazon basin. *Meliosma* occurs in Asia and tropical America.

Meliosma, with about 130 species of shrubs and trees, is most abundantly represented in the Indo-Malayan region. There are about 30 species in tropical America, their range including the West Indies, southern Mexico, Central America, the lower Amazon, and eastern Brazil. Many of the Old World species have pinnate leaves, but all of those in America have simple leaves except *M. alba* (Schl.) Walp. of Mexico, which will be treated separately.

Wood pale brown, becoming orange superficially. Luster medium. Without distinctive odor or taste. Rather light in weight, but firm and tough; texture moderately coarse; grain variable; easy to work; poorly resistant to decay. Has rather attractive figure on radial surface, but the trees are too small or infrequent to supply commercial timber.

Growth rings absent or poorly defined. Pores rather thin-walled, rounded in outline; indistinct to barely visible without lens; 90 to 150μ in tang. diam.; moderately numerous, often in pairs, distributed without definite pattern. Vessels with multiple perforations; scalariform plates with several to many widely spaced bars, frequently reticulate; foraminate perforations sometimes present, especially in *Meliosma panamensis;* intervascular pitting medium, alternate. Rays distinct on cross section, usually conspicuous on radial surfaces, being darker than background; 1 to 4, sometimes 5, cells wide and up to 50, occasionally over 100, cells high; the uniseriates and biseriates few; decidedly heterogeneous, with most of the cells square or upright in *M. glabrata* and *M. panamensis,* but with definite strata of procumbent cells in *M. Herbertii;* pits to vessels medium-sized in *M. glabrata,* but very large and irregular in the other two. Wood parenchyma sparingly paratracheal; not distinct with lens. Wood fibers thin-walled; septate in part, particularly near outer limit of growth rings; pits numerous, with small borders and slit-like extended apertures. Ripple marks absent. No gum ducts seen.

Common names: Aguacatillo, algarrobo, arroyo, cacaillo, cacao bobo, ciralillo, guayaroto, serillo (P.R.); cacao bobo (Dom. R.); gros grain (Grenada); gounelle (Haiti); ira (C.R.).

Meliosma alba (Schl.) Walp. is a Mexican tree of limited distribution in eastern

Mexico. The only specimen available (Yale 37935) was collected by C. H. Muller at an altitude of 6000 feet near Villa Santiago, Nuevo León, with sterile herbarium material determined by Paul C. Standley. The buds are naked yellowish and tomentose; the unequally pinnate leaves are alternate and resemble those of Hickory (*Carya*); the small flowers appear in June and the fruit is a small one-seeded drupe. The tree is said to be restricted to very moist arroyos and attains a height of 50 feet and a diameter of 40 inches. The timber has about the consistency and appearance of Soft Maple (*Acer rubrum* L.) and is highly esteemed locally for making furniture.

Growth rings fairly distinct. Pores small to medium-sized (up to 125μ), not crowded; usually more numerous in outer part of growth ring; often in short radial multiples. Perforations simple, frequently irregular in outline; vascular pits large (up to 12μ), oval, alternate, not crowded; single row of pits common around perforation. Rays 1 to 3 cells wide (very few uniseriate) and up to 30 cells high; homogeneous; cells thick-walled and abundantly pitted; ray-vessel pit-pairs mostly confined to marginal cells, medium-sized, in surface view similar to intervascular in shape but with narrower border. Wood parenchyma terminal and in several narrow bands 1 or 2 cells wide, becoming more closely spaced near periphery of growth ring; also sparingly paratracheal in early wood. Wood fibers rather thin-walled; septate, at least in part; pits numerous, with slit-like apertures and very small borders. Ripple marks absent. No gum ducts seen.

COMMON NAMES: Ayón, palo blanco (Mex.).

Ophiocaryon. There are two species, sometimes considered as belonging to separate genera. *O. heterophyllum* (Benth.) Urb. is a shrub or tree of the upper Amazon. The leaves are usually pinnate, but those on the flowering twigs are reduced to a single leaflet. *O. paradoxum* Schomb. is said to be a tall tree in British Guiana, where it is known as Snake-nut because of the serpentine appearance of radicle in the seed. Both of these species are apparently rare and there are no wood samples of either of them in the Yale collections.

SALICACEAE

THE Willow family consists of two genera, *Salix* and *Populus*, each with a great many species of deciduous trees and shrubs of wide distribution over the world, but mostly in the north temperate zone. The buds of *Populus* have numerous scales, those of *Salix* only one. The leaves are alternate simple and stipular; the flowers are borne in unisexual aments from buds in axils of the leaves of the previous year, the male and female nearly always on different plants; the fruit is a small dehiscent capsule with numerous silky tufted seeds. The Willows (*Salix*) and the Aspens, Poplars, or Cottonwoods (*Populus*) are among the commonest and best known woody plants, especially along streams, and some of them are highly important sources of timber. Approximately 200,000 acres of the delta of the Río Paraná in Argentina are planted with Poplar and Willow, yielding annually about ten million square meters of timber for making boxes.

Heartwood brownish; usually distinct but not always sharply demarcated from the thick white sapwood. Luster low to silky. Without distinctive scent or taste when dry. Sp. gr. (air-dry) 0.32 to 0.48; weight 20 to 30 lbs. per cu. ft.; texture very fine to rather coarse; easy to work, though often sawing woolly; tough and strong for its weight; perishable in contact with the ground.

Growth rings present. Pores very small to large enough to be seen without lens; numerous, crowded, often gradually diminishing in size during a season's growth, sometimes tending to concentric or oblique arrangement. Vessels with simple perforations; without spirals; tyloses common; pitting alternate. Rays often not distinct even with lens on cross section; uniseriate and usually less than 20 cells high; homogeneous in *Populus;* heterogeneous in *Salix*, the marginal cells large and squarish; cells thin-walled; pits to vessels rather large, nearly circular. Wood parenchyma terminal in 1 to 3 rows. Wood fibers thin-walled; pits rather few, minute, rounded, simple. Ripple marks absent. No gum ducts seen.

Populus, with about 35 species of fast-growing trees, is of general distribution throughout the north temperate zone, often forming extensive forests in the extreme northern parts of its range. The usual English name for the tree is Poplar, but in America those with smooth greenish or whitish bark are commonly called Aspens or Popple, while the larger trees with rough deeply furrowed grayish bark are known as Cottonwoods. In the American timber trade the lumber is sold as Cottonwood and the pulpwood as Poplar, without distinction as to species. The name Yellow Poplar usually refers to *Liriodendron* (fam. Magnoliaceae).

There are two species and a few varieties of Aspen. *Populus tremuloides* Michx., usually a small tree less than 40 feet high but sometimes much larger, occurs from Labrador, Hudson's Bay, and Alaska southward through Canada and most of the United States, except the south Atlantic and Gulf states, to northern Mexico. The species closely resembles and for a long time was supposed to be the same as the European Aspen, *P. tremula* L. The Large-tooth Aspen, *P. grandidentata* Michx., is similar to the preceding, but its range is less extensive, being mostly in the north-eastern part of the United States and south-eastern Canada. The woods of the Aspens are practically identical and are distinguished from Cottonwoods by their finer texture and higher luster. The total stand of timber in the United States is estimated to be over ten billion board feet, of which only about 2 per cent is large enough for lumber. The principal use of the wood is for making book paper; other purposes are excelsior, match sticks, and pails and boxes for containing food.

Of the thirteen species of Cottonwood, the four most important commercially are as follows: Eastern Cottonwood or Carolina Poplar, *P. deltoides* Marsh., is of irregular or poorly known distribution throughout the eastern half of the United States and often planted for windbreaks, shade, and pulpwood; estimated stand of saw timber about three billion board feet. A form or variety called Yellow Cottonwood is recognized in the Middle West and has yellowish wood highly esteemed for turnery, house siding, and various other purposes for which *Liriodendron* is suitable. Swamp or River Cottonwood, *P. heterophylla* L., is most abundant in the south Atlantic and Gulf regions and the Mississippi valley; estimated stand about two billion board feet. Black Cottonwood, *P. trichocarpa* Hook. (Plate XXXV), the largest deciduous tree of the Pacific coast, grows from southern Alaska to southern California; estimated stand in the United States over a billion board feet. Balsam Poplar or Balm of Gilead, *P. balsamifera* L., occurs in Canada and Alaska, and along the northern border of the United States; at its best it is a very large tree, sometimes six feet in diameter. The principal uses of Cottonwood lumber are for crates and boxes, as it has a clean appearance, suitable for printing and stenciling, is odorless and tasteless when dry, light in weight but tough and strong, and nails without splitting. Other uses are planing-mill products, woodenware, plywood, excelsior, and for pulp, especially by the soda process for book paper.

Ten species of *Populus* extend into Mexico, two or three of them as far south as the capital, where they occur mostly along streams. On the best sites at the lower elevations, the Cottonwoods, locally known as Alamo, sometimes attain a height of 100 feet and a trunk diameter up to six feet. The Aspen, there called Alamillo, is usually a small slender tree rarely 50 feet tall and 24 inches in diameter. The timber has few uses except where other wood is scarce. Various European and North American species of *Populus* are cultivated in South America, particularly in Argentina and Chile, both for shade and in plantations for lumber for boxes, general carpentry, light construction, fuel, and in a small way for paper pulp.

Common names: Aspen, baumier, cottonwood, balm of Gilead, liard, tremble (Can.); aspen, balm, b. of Gilead, cotton gum, c. tree, cottonwood, langues des femmes, liar, poplar (black, Carolina, necklace, Norway, trembling, white, yellow, etc.), popple, quakenasp, quaking asp, tacamahac, trembling aspen, whitewood (U.S.A.); alamo, alamillo, chopo, guaribo, güérigo, olmo (Mex.).

Salix. More than 200 species and many varieties of Willow have been described, mostly of temperate regions, but ranging from the tropics to the arctic circle. They are chiefly shrubs and small poorly formed trees, but a few of them attain large dimensions. The pliable young shoots of some species are employed in making baskets and wicker furniture, and mats to protect river banks. Well-known uses of the wood of European trees are cricket bats, artificial limbs, and charcoal for black powder. The largest American tree is the Black Willow, *S. nigra* Marsh., which grows throughout the entire eastern half of the United States and adjacent parts of Canada and Mexico, but makes its best development in rich bottomlands of the lower Ohio and Mississippi valleys. The wood is used to a limited extent for boxes, crates, core stock for veneers, slack cooperage, excelsior, and charcoal. There are about 17 species in Mexico, and one of the largest and most widely dispersed is *S. taxifolia* H.B.K., a tree sometimes 50 feet high, with a wide-spreading open crown and a rather crooked trunk 12 to 18 inches in diameter. The most generally distributed tropical American species is *S. chilensis* Molino or *S. Humboldtiana* Willd. It is common along streams throughout much of Central and South America, varying in size from a shrub to a rather tall tree with crooked trunk and open crown. It is often planted along streams and irrigation ditches to prevent erosion. The wood is used to a limited extent locally for light construction, boxes, and fuel. The bark is rich in tannin and both bark and leaves contain a bitter principle (salicin) to which various therapeutic properties are ascribed.

COMMON NAMES: Willow (Eng.); sauce (Span.); ahuejote, c'os, hoo-cuy, huejocote, huexotl, jaray, saus blanco, sauz, taraix, taray, t. del río, tepehuexote, tocoy, xitzas, yaga-grieza, yutnu-nuu (Mex.); chorão, oeirana, salgueiro, s. do matto (Braz.); sauce colorado, s. común, s. criollo (Arg.); sarandí (Par.).

SANTALACEAE

THE Sandalwood family consists of 29 genera and about 400 species of herbs, shrubs, and small trees, sometimes parasitic, of very wide distribution in tropical and temperate regions. The leaves are alternate or opposite, entire, sometimes scale-like; stipules are absent; the flowers often are greenish; the fruit is indehiscent, nut-like or drupaceous. The best known and most important genus is *Santalum,* of the East Indies, Polynesia, and Australasia. The use of the fragrant wood and oil of the Indian species, *S. album* L., is of great antiquity. Sandalwood of other species was discovered in Hawaii about 1790 and an important trade with China developed, being at its height between 1810 and 1825. The temporary prosperity enabled the first king to lead his people from barbarism to civilization in a single generation, but at the same time it was almost their undoing, for it led to extravagance, oppression, and famine which came to an end in 1840 with the almost complete exhaustion of a great natural resource. Plantations of the trees are now being made, but the seedlings perish before the end of their first year unless their roots attach themselves by means of spongy sucking appendages (haustoria) to the roots of young trees of other kinds which serve as hosts.

There are four genera of Santalaceae native to South America, namely, *Myoschilos* in Chile and Patagonia; *Acanthosyris* in Bolivia, Brazil, and Argentina; *Jodina* in Argentina, Uruguay, and Brazil; and *Cervantesia* in the mountains of Colombia, Ecuador, and Peru. They are at best only little trees, and their unscented timber is of no commercial importance. Owing to their many dissimilarities, the woods will be described separately.

Acanthosyris. There are two or three species in southern South America, the largest being a tree occasionally nearly 40 feet high and 16 inches in trunk diameter in northern Argentina; it is noted for its sweet fruit.

Heartwood yellowish brown, sometimes with pinkish streaks; sapwood lighter. Without distinctive odor or taste. Moderately hard and heavy; has about the consistency of Maple (*Acer*); texture medium; feel harsh; grain irregular; easy to work, finishing smoothly, but is inclined to warp; durability probably low. Presumably without commercial possibilities.

Growth rings fairly distinct. Pores numerous; small to very small, not individually distinct without lens; solitary and, more often, in short to rather long radial rows or multiples occupying most of the space between the coarse rays; well distributed without pattern, though tending to tangential arrangement in early wood. Vessels with simple perforations; no spiral thickenings seen; pitting rather coarse, alternate to opposite. Rays distinct on cross section, conspicuous on radial; numerous, 1 or 2 pore-widths apart; nearly all multiseriate, up to 6 cells wide and 60 cells high; homogeneous to heterogeneous; pits to vessels variable from small and rounded to very large and irregular. Wood parenchyma in terminal bands, also sparingly paratracheal and diffuse; pits to vessels large, gash-like. Wood fibers with thick walls and rather few, very small, indistinctly bordered pits. Ripple marks absent. No gum ducts seen.

COMMON NAMES: Sombra de touro (Braz.); cabo de lanza, guá-hé, ibá-hé-é, ibará-hú, quebrachillo, sacha-pera, sombra de toro hembra, yvá-hehé (Arg.); quebrachillo (Urug.); ybá-hehé (Par.).

Cervantesia, with five species of small to perhaps medium-sized trees, is distributed from northern Colombia to Ecuador and Peru. The only wood samples available are of *C. colombiana* A. C. Smith, which were collected by Armando Dugand G. in the Department of Atlántico, Colombia, where the tree is known as Matamaíz (see *Tropical Woods* 51: 12–14).

Sapwood yellowish, with a golden luster in proper light; heartwood not seen, but probably light brown (judging from knots). Without distinctive odor or taste. Hard and heavy; sp. gr. (air-dry) 0.68 to 0.80; weight 42.5 to 50 lbs. per cu. ft.; rather coarse-textured and harsh; fairly straight-grained; not very difficult to work, finishing smoothly; is probably not resistant to decay.

Growth rings sometimes distinct, owing to wider spacing or more orderly arrangement of parenchyma. Pores small, near limit of vision; fairly numerous; occurring mostly in well-distributed radial multiples of 2 to 4. Vessels with simple perforations; without spiral thickenings; intervascular pitting fine, alternate. Rays visible on cross section, very distinct on radial; sometimes as wide as the pores and spaced 1 to 2 pore-widths apart; 1 to 6 cells wide and variable in height to over 100 cells; homogeneous to heterogeneous, most of the cells being large and rather short, some of them square or occasionally upright; pits to vessels small to very large and irregular; large crystals of calcium oxalate common. Wood parenchyma abundant, but variable; sparingly developed about some of the pores, surrounding others and frequently confluent into irregular broken to continuous bands, producing a sort of ulmiform pattern as seen with the unaided eye. Wood fibers not in radial rows; walls medium to thick; pits indistinctly bordered. Ripple marks and gum ducts absent. One specimen (Yale 27095; Dugand 563) differs from the others in having parenchyma mostly limited to the vicinity of the pores and only occasionally forming bands. The density is much greater; sp. gr. (air-dry) 0.91; weight 57 lbs. per cu. ft.

Jodina (or *Iodina*), with one species, *J. rhombifolia* Hook. & Arn., is a tall shrub or a tree sometimes 25 feet high and 14 inches in diameter in Argentina, characterized by its rhombus-shaped leaves with three bristle tips. Oil from the seeds is employed medicinally and the timber is used to a minor extent locally, mostly for fuel.

Heartwood pale yellow; sapwood white. Fairly lustrous. Without distinctive odor or taste. Hard and heavy; sp. gr. (air-dry) 0.77; weight about 48 lbs. per cu. ft.; texture medium-fine; grain irregular; working properties fair; finishes smoothly, but is inclined to warp; durability presumably low. Not likely to be of any importance for export.

Growth rings present or absent. Pores very small to minute; very numerous; arranged in concentric bands or in very irregular diagonal or zigzag patches suggesting certain species of *Rhamnus,* both types of arrangement occurring in same specimen. Vessels with simple perforations; spiral thickenings distinct; pitting rather coarse, alternate. Rays nearly all multiseriate, up to 10 cells wide and up to 50, occasionally 75 to 100, cells high; somewhat heterogeneous, the cells all small, most of them procumbent, some square; pits to vessels small to large and irregular. Wood parenchyma sparingly paratracheal. Wood fibers with thick walls and rather few very small indistinctly bordered pits. Ripple marks absent.

COMMON NAMES: Sombra de touro (Braz.); peje, quebrachillo, quebracho flojo, quinchilín, quinchirín, sombra de toro, s. de t. macho (Arg.); sombra de toro (Urug.).

Myoschilos *oblongus* Ruiz & Pav., the only species, is a large shrub of Chile and Patagonia, usually growing in association with *Nothofagus pumilis* Poepp. & Endl. The roots and leaves are used medicinally. The nearly white moderately hard and uniform-textured wood has about the consistency of White Birch (*Betula alba* L.), and is suitable for small articles of turnery.

Growth rings distinct, as wood is nearly ring-porous. Pores minute, scarcely distinct with lens; in an indefinite band in early wood and extending radially in irregular zones. Vessels with simple perforations; no spiral thickenings seen; intervascular pitting rather coarse, tending to scalariform, the vessel walls usually so narrow as to accommodate only 1 or 2 rows. Rays 1 to 3 cells wide and up to 15, sometimes over 30, cells high; heterogeneous, many of the cells square, all of them small; pits to vessels large, rounded or long-oval, 1 or 2 to each cross-field. Wood parenchyma not abundant; diffuse and in short tangential lines; pits to vessels mostly large and elongated. Wood fibers with rather thick walls and distinctly bordered pits. Ripple marks absent. No gum ducts seen.

COMMON NAMES: Codocoypu, orocoipu, senna (Chile); heidelbeere, lengua-myrtula (Arg.).

SAPINDACEAE

THE Soapberry family (excluding the Hippocastanaceae) consists of more than 120 genera and about 2000 species of trees, shrubs, and climbers, widely distributed in tropical and warm regions, especially in the Old World. The leaves are typically alternate, pinnate or trifoliolate, and without stipules; the flowers are generally small, variously arranged, but often in axillary or terminal panicles; fruit capsular or indehiscent, baccate, drupaceous, leathery, or composed of samaras; the seeds are often arillate. There are arborescent representatives of about 25 genera in tropical and subtropical America, but the entire list does not include an important timber tree.

Heartwood yellow to pinkish or reddish brown; usually not sharply differentiated from the sapwood. Luster typically medium. Odor and taste not distinctive. Moderately to decidedly hard and heavy; texture rather coarse to fine; grain variable; some specimens easy to work, others difficult; deeply colored kinds resistant to decay, others perishable.

Growth rings usually present, generally because of variation in the thickness of fiber walls or terminal parenchyma, or both; ring-porous structure common in *Sapindus,* local tendencies observed in some of the others (*e.g., Hypelate*). Pores in the diffuse-porous species (except in the lianas) typically small and indistinct or invisible without lens; solitary and in short to rather long radial multiples and sometimes in small clusters also; distributed without definite pattern, though with tendencies to radial and diagonal arrangement in certain instances (*e.g., Athyana, Cupania,* and *Hypelate*). Vessels with exclusively simple perforations, although tendencies to the formation of scalariform plates with few bars have been reported for the vessels near the pith in some species; gum deposits often abundant, frequently in the form of plugs near the ends of the vessel members; tyloses present in *Sapindus;* spiral thickenings absent, but fine striations common in *Athyana, Cupania, Dodonaea, Exothea, Hypelate, Sapindus,* and *Thouinia;* pitting alternate, very fine to coarse. Rays sometimes uniseriate or biseriate, more often up to 3, occasionally to 6, cells wide; commonly less than 25, sometimes up to 50, rarely to 90, cells high; typically homogeneous, with more or less tendency to heterogeneous, though rarely with distinctly upright cells; cells typically small (tang. section); no crystals seen; gum deposits often abundant; ray-vessel pit-pairs half-bordered and similar in size to the intervascular, except when unilaterally compound. Wood parenchyma of two general types: sparingly paratracheal or narrowly vasicentric and sparsely to abundantly diffuse and sometimes finely terminal; or (*e.g.*, in *Diplokeleba, Dipterodendron, Melicocca, Pseudima, Sapindus,* and *Talisia*) abundant in coarse-celled diagonal to concentric bands, sometimes composing a third or more of the cross section, suggesting certain Leguminosae; crystals frequent to abundant. Wood fibers commonly septate; walls medium to very thick in different specimens and sometimes varying noticeably in the same growth ring; pits small, simple or indistinctly bordered. Ripple marks absent. Small vertical traumatic gum ducts observed in one specimen of *Diplokeleba*. The lianas are of either normal or anomalous structure, those of the latter class having mostly compound or divided stems (*e.g.*, *Paullinia* and *Serjania*).

Allophylus, a pantropical genus, comprises more than 150 species of shrubs and small trees, of which about half are American. The woods are not utilized for any special purposes and have no commercial possibilities. Heartwood absent from available specimens or not distinguishable from the pale brownish sapwood. Luster medium. Odorless and tasteless. Hard and heavy to moderately so; texture rather fine; grain fairly straight; easily worked; probably of low resistance to decay.

Common names: Palo de caja (Cuba); palo blanco, quiebrahacha (P.R.); cucharita prieta, tres palabras (Dom. R.); bois nègre, chic-chic, Marguerite, tres feuilles, trois paroles (Haiti); bikbach, ixbahach, palo de caja (Mex.); bastard axemaster, cherry (Br. H.); huesillo (Guat.); esquitillo, huesillo (C.R.); San Pedro (Col.); quinilla colorado, shimbillo, shitari-runtucaspi, supai ocote, tortuga-caspi, yurac tortilla-caspi (Peru); vaccum (Braz.); chalchal, c. de gallina, frutilla, frutita, pitanga (Urug.); chalchal, cocu, granadillo, picazurembiu (Arg.).

Athyana *weinmannifolia* Radlk., the only species, is a small tree, sometimes 30 feet high, growing in Paraguay and northern Argentina. The dense fine-textured wood is said to be used locally to a minor extent for objects of turnery.

Common names: Quebrachillo, tarco (Arg.).

Cupania, with about 45 species of erect shrubs and small to medium-sized trees, is generally distributed throughout tropical America. The pale pinkish brown wood is sometimes used locally for poles, interior construction, and for fuel, but has no commercial possibilities.

Common names: Loblolly tree (Jam.); guara, g. blanca, g. colorada, g. de costa, g. de ley, g. macho, g̀uárana, g. de puerco, g. hembra, g. macho, guárano, zahumaya (Cuba); guara (P.R.); guárana (Dom. R.); bois de satanier (Haiti); huanchal, sacpom (Mex.); grande Betty, red copal (Br. H.); carboncillo, carbón colorado, cola de pava (Guat.); cedrillo, cola de pava, huesito, miacaguite (Salv.); bilabila, cola de pava (Nic.); cantarillo, carne asada, cascua, huesillo, manteco, pozolillo (C.R.); candelillo, gorgojero, gorgojo (Pan.); culo de Indio, guacharaco (Col.); guara, guamo Matias, lengua de vaca (Venez.); gatotie, gauwetie, gauwetrie, gauwiti, gauwtrie, hariraro, ietjoeroe tonorebjo, joeliballi, khalemeroe koelisiri, koelisiri, koemété, sabana tranga, s. t. hoedoe, tamoene tororebjo, tonorebjo, tonorooibjo, witte koeliesierie, w. tonorepio (Sur.); juapina (Peru); almesca ussu, camboatan, pau de arapuce (Braz.); cambuatá, corpus, ramo blanco, r. colorado (Urug.); corpus, ibirá-corpus, jacarandá-mí, ramo, r. colorado (Arg.).

Diatenopteryx *sorbifolia* Radlk., the only species, is a tree sometimes 80 feet tall and 24 inches in diameter in Paraguay, northeastern Argentina, and southern Brazil. The dense, strong, fine-textured, roseate wood is used locally for carpentry.

Common names: Ibirá pi-hú, María preta (Arg.).

Diplokeleba *floribunda* N. E. Brown, the sole species, is a small tree, occasionally 30 feet high and 20 inches in diameter, growing in northern Argentina. Wood of available specimens is yellowish throughout, with fine white parenchyma markings; not highly lustrous. Odorless, but with somewhat bitter taste. Hard and heavy; texture rather fine; grain irregular; not difficult to work, finishing rather attractively; is poorly resistant to decay. It is used locally to a minor extent for furniture.

COMMON NAMES: Palo blanco, urunday blanco (Arg.).

Dipterodendron, apparently with only one distinct species, *D. costaricense* Radlk., is a small to medium-sized tree sometimes 65 feet high, growing in the Pacific coast region of Costa Rica and Panama. It has fern-like foliage and is closely related to *Dilodendron;* in fact a form, *Dipterodendron elegans* Radlk., was originally named *Dilodendron bipinnatum* Radlk., var. *elegans* Radlk. The moderately dense, strong, readily worked, pinkish or brownish timber is sparingly utilized because of the scarcity of the larger sizes.

COMMON NAMES: Gallinazo, iguano, loro (C.R.); harino, jarino (Pan.).

Dodonaea comprises more than 50 species of shrubs and trees, nearly all Australian. One variable species, *D. viscosa* (L.) Jacq., occurs in tropical and subtropical regions throughout the world, often forming coastal thickets. It is a shrub or a little tree rarely over 15 feet high, the simple or pinnate leaves and the young shoots usually viscid. Standley says (*Trees and shrubs of Mexico,* p. 705): "The leaves are bitter and in various regions are used for fevers, colic, gout, rheumatism, and venereal diseases. The bark is employed in the preparation of astringent baths and fomentations, and the decoction of the wood is reported to have febrifuge properties. The seeds are said to be edible. In Australia the fruits, known as 'native hops,' were formerly much used as a substitute for true hops (*Humulus lupulus* L.) in making yeast and beer. The wood is described as brown, close-grained, and hard, and in India it has been utilized for engraving, turning, tool handles, and walking sticks." (See Gamble's *Manual of Indian Timbers,* p. 203.)

COMMON NAMES: Candlewood, dogwood (Bah.); switch sorrel (Jam.); chamiso, gitarán (P.R.); manglier petites feuilles, pativier (Haiti); bois couche, b. madam (Guad.); chapulizle, chapuliztoli, cuerco de cabra, granadina, grenadina, hierba de la cucaracha, jarilla munditos, ocotillo, pirimu, varal (Mex.); chulita (Guat., Salv.); ayuelo, hayuelo (Col.); granadillo, hayo, hayuelo, sen (Venez.); vassoura vermelha (Braz.); chamiso (Arg.) basura, chirca de monte (Urug.); chamana (Peru).

Exothea, with a single distinct species, *E. paniculata* Radlk., is a small to medium-sized tree rarely 50 feet high and 15 inches in diameter, occurring in the West Indies, southern Florida, southern Mexico, and northern Central America. The hard fine-textured reddish brown wood is used locally for articles of turnery, rural construction, piling, and fence posts.

COMMON NAMES: Butterbough, inkwood, ironwood (Florida); wild genip, w. guinep (Jam.); anoncillo cimarrón, guamacá, mulato, yaicuaje (Cuba); cuerno de buey (Dom. R.); bois couleuvre, b. mûlet (Haiti); copalillo, yaicua, yaicuage (Mex.); palo de cuiliote (Salv.).

Hypelate *trifoliata* Sw., the only species, is a shrub or a tree sometimes 40 feet high and 18 inches in diameter, limited to the West Indies and southern Florida. The hard brownish yellow fairly durable wood finds various domestic applications, but is of no commercial importance.

COMMON NAMES: White ironwood (Florida, Bah.); cerillo, cocuyo, cogote de toro, cuaba de ingenio, cuabilla, raspadura, vera amarilla (Cuba); cigua (P.R.); chandelle marron, gallipeau (Haiti).

Matayba (including *Ratonia*), with about 45 species of shrubs and small to medium-sized trees occasionally 60 feet high and 24 inches in diameter, is widely distributed throughout tropical America. The hard, strong, fine-textured, brownish or pinkish brown wood is fairly easy to work, but is not utilized for any special purposes and is of no commercial value.

COMMON NAMES: Bastard mahogany, coby, comanancy, cromanty, redwood, wanika (Jam.); macorí, macurije (Cuba); doncella, negra lora (P.R.); guara, ratón

(Dom. R.); bois graines noires, chataignier, satanier (Haiti); boy job, mabehu (Br. H.); coyolillo (Guat.); carbón (Hond.); culo de indio, lijo (Col.); yayo, zapatero (Venez.); kuleshiri (Br. G.); maca-apa-ipow (Fr. G.); gatotie, guawetie, gauwietie, gauwitie, gauwtrie, koelisiri karemeroe, k. khalemeroe, k. khonéméroe, koenatjeppi, pritijari, sabana blakkatranga-hoedoe, s. trange-hoedoe, siengabosoe, toepoeroe tomorebjo, tonoloipio, tonorebjo, t. toepoeroe, zwarte tonorebjo (Sur.); hiruhuaco-caspi (Peru); sama negra (Boliv.); cascudinho, fructa de pombo, pau d'espeto (Braz.); cambó-atá, yaguá-rataí (Arg., Urug.).

Melicocca, with two species of medium-sized to large trees, is apparently limited in natural distribution to South America. The southern species, *M. lepidopetala* Radlk., occurs in northern Argentina and adjacent regions of Bolivia and Paraguay. It attains a height of 35 feet and a diameter of 20 inches and is frequently planted for shade and ornament. The fruit is edible. The moderately hard yellowish white wood is noted for its flexibility, but it is very scarce. The northern species, *M. bijuga* L., is native to the Guianas, Venezuela, Colombia, and parts of Central America, and has become naturalized through planting in the West Indies and elsewhere. The salmon-colored pulp of the fruit is pleasantly flavored and the roasted seeds are edible. The tree is medium-sized to large and the timber is of good quality for interior work, but is not resistant to decay; the supply is too limited to be of commercial importance.

COMMON NAMES: Genip, guinep (B. W.I.); anoncillo, mamoncillo (Cuba); quenepa (P.R.); quénêpe (Haiti); knepier, quienet, quienette (Guad.); mamón de Cartagena (Salv.); mamón, m. de Cartagena (C.R.); mamón (Pan.); mamón, m. casero, m. cotoplix, m. de castilla, mamoncillo (Col.); macao, maco, mamón, mauco, muco (Venez.); knippa, knippelboom, knippen, tackelboom (Sur.); ibá-pobó, ibá-poó (Arg.).

Paullinia, with about 150 species of shrubs, usually climbing by means of 2-branched tendrils, is widely distributed throughout tropical and subtropical America, with a single species, *P. pinnata* L., occurring also in tropical Africa. The Cupana of Venezuela and Colombia and the Guaraná of northern Brazil is *P. cupana* H.B.K.; the roasted seeds are used in the preparation of a beverage having the stimulating properties of coffee. The crushed plants of various species of *Paullinia* and related genera are often thrown in streams to stupefy fish. The supple stems are used as a substitute for rope. The stems of some species are simple while others are compound, a common form of this anomaly consisting of a central woody body with one to several outer ones attached to it. (See Solereder's *Systematic anatomy of the dicotyledons,* p. 234.)

COMMON NAMES: Bread-and-cheese, supple jack (B.W.I.); bejuco de costilla (P.R.); azucarito, bejuco colorado, b. de vieja, b. matancero (Cuba); liane carée (Haiti); barbasco, bejuco costillón, b. de agua, b. vaquero, bejuquillo, cuaumecate, colorín, kexak, macalte-ik, panoquera, pico de guiloche, salatxiu (Mex.); fish poison, tie-tie (Br. Hond.); barbasco, bejuco cuadrado, b. de barbasco, chilmecate, nistamal, nistamalillo, palo de mimbre, pozolillo (Salv.); campalaca, pate (Hond.); chilmecate (Nic.); hoja de pájaro (C.R.); azucarerito, barbasco (Col.); azucarito, bejuco de mulato, cupana (Venez.); feifi fienga, jesikoesji, kamasoeli, koetoepoe, vijfvinger (Sur.); liane carrée (Fr. G.); cipó cruapé vermelho, c. timbó, cururú-apé, guaraná, timbó, t. de peixe (Braz.); caí-escalera-rá, isipó moroté, i. timbó, timbó (Arg.); arbol de la lecheguana (Urug.).

Pseudima, with only one well-known species, *P. frutescens* (Aubl.) Radlk., is a slender unbranched tree rarely 30 feet tall, with very large pinnate leaves clustered at the ends of the stems and with conspicuous terminal panicles of small white flowers, occurring in the Guianas and Amazonian Brazil. The shiny black inner shell of the fruit is peeled and dried and used for beads by natives of British Guiana. Wood yellowish white throughout. Luster medium. Odorless and tasteless. Moderately heavy and hard; texture medium; grain irregular; easy to work; probably perishable in contact with the soil. Without commercial possibilities.

COMMON NAMES: Camaá, fructo de anel, pitombeira, uarana (Braz.).

Sapindus, with 13 species of small to medium-sized trees, is widely distributed in tropical and subtropical regions. The only American representative is *S. Saponaria* L., known to English-speaking people as Soapberry because the fleshy parts of the fruits contain saponin and when macerated with water produce suds like soap and are frequently used for washing clothes. The seeds serve for ornamental purposes, and when pounded and thrown in the water will stupefy fish. The tree grows naturally or as an escape from cultivation from southern United States to Argentina. It is a variable species and some of the forms have been given specific names, *e.g., S. marginatus* Willd. and *S. Drummondii* Hook. & Arn. The timber is of little use except for fuel.

Wood yellowish or brownish, without very distinct heartwood; streaked or veined with white parenchyma markings. Luster rather low. Odorless and tasteless. Hard, heavy, and strong; sp. gr. (air-dry) about 0.88; weight 55 lbs. per cu. ft.; texture coarse; grain usually irregular; working properties rather poor; durability low. Of no commercial possibilities.

COMMON NAMES: Chinaberry, false dogwood, soapberry, wild China, w. Chinaberry (U.S.A.); soapberry tree (Jam.); jaboncillo (Cuba, P.R.); bois canique, b. savonette, graine canique, savonette pays (Haiti); savonetier, savonnier, savonette (Fr. W.I.); amole, a. de bolita, bibí, boliche, cholulo, devanador, gualulo, jaboncillo, jabón-che, matamuchacho, palo blanco, p. de cuentas, pipal, pipe, tehoitzli, tehuistle, tehuixtle, tehuiztle, xoken-cab, yamole, yamolli, zihom, zubul (Mex.); soapseed tree (Br. H.); pacún (Salv.); jaboncillo (Col.); parapara, pepo, zapatero (Venez.); sopo serie (Sur.); jaboncillo (Ec.); boliche, espingua, latápi, sullucu (Peru); saboeiro, saboneteiro (Braz.); casita, c.-rá, jaboncillo, palo jabón, quillái (Arg.).

Serjania, with about 200 species of large and small woody vines related to *Paullinia,* occurs throughout tropical and subtropical America. Slender tough stems are used as a local substitute for twine or rope and for making coarse baskets; crushed parts of the plant are thrown in the water to stupefy fish. Some of the stems are of normal structure, others are compound, divided, fluted, or otherwise anomalous.

COMMON NAMES: Fowl-foot (Bah.); mountain supple jack (Jam.); basket wood, bejuco de corrales, b. de costilla, black withe (P.R.); bejuco casero, b. colorado, b. cuadrante, b. de corrales, b. de vara, yaguajiro blanco (Cuba); bejuco de costella (Dom. R.); liane persil (Haiti); barbasco, bui, buiche, buyak, carretilla, cuamecatl, cuauhmecatl, cuaumecate, diente de culebra, kexak, nueve hojas, quamecatl, quirote culebra, sierrilla, tlatlanquaya, turizo (Mex.); barbasco, b. cuadrado, bejuco cuadrado, comemano (Salv.); crespillo (Hond.); barbasco, bejuco espinoso, juriso, tierizo (C.R.); esquinero, raya (Col.); bejuco colorado, b. de corral, b. de zarcillo, b. moreno, zarcillo (Venez.); abo, driekantie, koetoepoe, moetoepoe (Sur.); cipó-timbó, timbó (Braz.); isipó de agua, isipó-y (Arg.).

Talisia, with about 45 species of small to medium-sized trees, is mostly limited to South America, with a few representatives in Central America and southern Mexico. The best known species is *T. olivaeformis* (H.B.K.) Radlk., a tree sometimes 60 feet high, frequently planted for shade and for its edible fruits. The timber of *Talisia* has few applications; that of *T. cerasina* Radlk., the Pitomba of the Brazilian Amazon region, is used for making toothpicks.

Heartwood brownish with a tinge of pink; not sharply differentiated from the yellowish sapwood. Luster rather low. Odorless and tasteless. Hard, heavy, and strong; texture medium; grain fairly straight; not difficult to work; poorly resistant to decay. Apparently without commercial possibilities.

COMMON NAMES: Cotoperis (Cuba, int.); canip, kenep, kinep, coloc, guayo, mamón de mico, uayamcox, uayum (Mex.); tapaljocote (Salv.); dantisca (C.R.); cotopris, mamón cotopris, m. cotoprix, m. cutuplis, m. de mico (Col.); cotoperís, cotoprís, cotuplís, ramón cutuplís (Venez.); black moraballi, moraballi, wayam-cosh (Br. G.); hodemè, karaba, karababalli, makka krappa, sodimè, soepoepi, sondimè, tama kakalemoe, t. kalemoe, t. kolemoe, tonoloipio, zodèmè (Sur.); olho de boi, pitomba, pitombeira (Braz.); yaguana-ta-y (Arg.).

Thouinia, with about 15 species of shrubs and small trees, is limited to the West Indies, southern Mexico, and northern Central America. The hard and strong, fine-textured, pinkish wood is of good quality, but is not available in large enough quantity or sizes to be of economic value.

COMMON NAMES: Hard-bark, naked wood, quicksilver bush (Bah.); canelillo, copalillo, negra Cuba (Cuba); ceboroquillo, guava, quiebrahacha, serra-suela (P.R.); bois couré, b. la fièvre, b. poivre, malaguetle, marron (Haiti); kanchunup (Mex.); canjura, cortez amarillo, huesito, h. blanco, tegüecito (Salv.).

Thouinidium, with a few species of erect shrubs and slender trees, is confined to the West Indies, southern Mexico, and Central America. The only specimen available (Yale 1198; J. G. Ortega 33) is of *T. decandrum* (H. & B.) Radlk. from southwestern Mexico. The smooth-barked tree is rarely over 30 feet tall; the evenly pinnate leaves have 6 to 14 serrate leathery leaflets; the small flowers are borne in terminal panicles; the fruit consists of 2 or 3 laterally compressed samaras. The sapwood, which has about the consistency of Sugar Maple (*Acer saccharum* Marsh.), is whitish throughout, but there are indications that the heartwood may be brownish.

COMMON NAMES: Cucharillo, palo caimán (Cuba); bois brûle, gros peau (Haiti); charapo, panalillo, perico (Mex.); cola de pava, plumón, zorilla (Salv.).

Toulicia, with 14 species of small to rather large trees, is widely but sparsely distributed in South America from Venezuela and the Guianas to Brazil. The leaves are equally pinnate, with several to many leaflets; the flowers are borne in large panicles; the fruit is composed of three large-winged samaras. The wood of *T. guianensis* Aubl. is said to be used for torches in French Guiana, hence the name Bois Flambeau.

The only specimen at hand (Yale 22051) is of *T. bullata* Radlk. from the Rio Tapajoz in Brazil. Heartwood apparently absent from sample; sapwood badly stained, but presumably nearly white when fresh. Luster medium. Hard, strong, and moderately heavy; texture rather fine except for a few prominent vessel lines; grain straight; not difficult to work. Presumably of no commercial possibilities because of its scarcity.

COMMON NAMES: Bois flambeau, toulici (Fr. G.); karababalli, tatoe, toelisi (Sur.); pitambaina (Braz.).

Ungnadia *speciosa* Endl., the only species, is a shrub or a tree sometimes 30 feet high with a thin-barked trunk six to eight inches in diameter growing near streams and in canyons from eastern Texas and southern New Mexico in the United States to Chihuahua, Coahuila, and Nuevo León, Mexico. The flowers are sweet-scented, and the seeds are of a pleasant flavor but have powerful emetic properties and are reputed to be poisonous. The tree is occasionally cultivated for ornamental purposes in warm climates. The wood is not utilized.

Heartwood pale brown with a reddish tinge; not sharply differentiated from the sapwood. Luster medium. Odorless and tasteless. Moderately hard and heavy, having about the consistency of Yellow Birch (*Betula lutea* L.); texture fine; grain straight; easy to work, but probably not highly resistant to decay. Of no commercial possibilities.

COMMON NAMES: Buckeye—Mexican, Spanish, Texas (U.S.A.); monilla, monillo (Mex.).

Vouarana *guianensis* Aubl., the sole species, is a medium-sized tree, closely related to *Cupania,* occurring sparingly in the Guianas and Amazonian Brazil. The grayish brown, very hard and heavy, rather fine-textured, cross-grained timber is not easy to work and apparently is not utilized locally for any special purposes.

COMMON NAME: Bigi tiengimonnie (Sur.).

SAPOTACEAE

THE Sapota family includes several hundred species of small to very large trees and some shrubs and is represented in nearly all tropical and subtropical regions. More than 200 genera have been described, but most of them have been reduced to synonymy or to the status of sections of larger genera. Important contributions to the classification have been published recently, but there is still difference of opinion regarding the validity of some genera and the limits of others. The authors are convinced that a systematic study of the woods would be of great assistance in solving some of the taxonomic problems, but the material now available is insufficient for conclusive results. (See *Tropical Woods* 73: 1.)

The American woods are here treated under 16 generic designations, namely, *Achras, Bumelia, Calocarpum, Chromolucuma, Chrysophyllum, Dipholis, Ecclinusa, Henoonia, Labatia, Lucuma-Pouteria, Manilkara* or *Mimusops, Micropholis, Paralabatia, Pradosia, Sarcaulus,* and *Sideroxylon,* but according to Baehni's classification (*Candollea* 7: 394–508. 1938) the number would be reduced to eight. Some of the groups are maintained because of apparently valid differences in the woods, the others for convenience pending further reorganization of the family, since aggregation will be a simpler process than segregation.

The trees are characterized by a milky latex, often copious and in some instances of high commercial value, for example, chicle gum from *Achras* and balata from *Manilkara.* The leaves are alternate or rarely opposite, simple, entire or rarely dentate, and pinnately veined, the texture and color often profoundly affected by age; stipules are sometimes present, but caducous. The flowers are generally clustered in leaf axils, or above leaf scars, rarely cauliflorous; they are typically white or greenish, but highly colored in *Chromolucuma* and *Pradosia.* The fruit is indehiscent, the pericarp usually fleshy, with one to several glossy seeds; usually edible and in some instances succulent and highly esteemed, *e.g.,* the Sapodilla (*Achras*) and the Zapote (*Calocarpum*). The genus most valuable for timber is *Manilkara,* the source of Beefwood or Bulletwood of the Guianas and Massaranduba of Brazil.

Heartwood yellow or orange in *Bumelia, Henoonia,* and *Sideroxylon;* dark red or reddish brown in *Achras, Manilkara,* and some specimens of *Dipholis* and *Lucuma-Pouteria;* grayish or pinkish brown in the others; with oily appearance and feel in certain West Indian species of *Manilkara;* sapwood whitish and very distinct though often not sharply demarcated. Luster low to medium. Without distinctive odor; taste sometimes astringent or somewhat bitter, being sweet followed by astringent in *Pradosia.* Density variable from 0.65 to 1.25, mostly between 0.85 and 1.05; weight (airdry) 41 to 78, mostly 53 to 65, lbs. per cu. ft.; texture typically medium, uniform; grain generally straight, sometimes wavy or otherwise irregular; working properties variable, but usually good; densest woods rather splintery; durability low to very high.

Growth rings usually present. Pores small to minute, usually not individually distinct to the unaided eye, though the groups may be; mostly in small multiples which are sometimes fairly uniformly distributed (*e.g., Chromolucuma* and *Pradosia*), more often in radial or oblique series, sometimes in flame-like groups (*e.g., Bumelia, Henoonia,* and *Paralabatia*). Tyloses usually abundant, sclerotic in part in *Bumelia, Chromolucuma, Ecclinusa, Lucuma-Pouteria, Micropholis,* and *Paralabatia;* fine spirals present in at least some of the vessels in *Bumelia, Henoonia,* and *Paralabatia;* vessel perforations exclusively simple, commonly with a wide rim; intervascular pitting fine to very fine, alternate to sub-opposite. Rays all uniseriate in *Chromolucuma;* in part 2, occasionally 3, rarely 4, cells wide in the others; heights variable up to 60 cells, generally much lower; weakly heterogeneous, in part at least, in *Bumelia, Calocarpum, Chromolucuma,* and *Henoonia,* decidedly heterogeneous in the others; cell walls thick, rarely sclerotic; gum deposits abundant; pits to vessels uniformly small and subcircular or oval in *Henoonia*, in part large, short to long oval or boomerang-shaped, sometimes in scalariform arrangement, in the others; pit-pairs half-bordered or unilaterally compound. Wood parenchyma abundantly developed, but usually not distinct without lens; finely or coarsely reticulate, at least in part, in *Bumelia, Calocarpum, Chrysophyllum, Dipholis, Henoonia,* and *Sideroxylon;* typically in numerous, narrow, uniform to wavy or somewhat broken concentric bands, mostly 1 to 3 pore-widths apart, in the others; cells mostly large, rarely with sclerotic walls; crystals common to abundant in some instances (*e.g., Achras, Manilkara,* and *Sideroxylon*), but generally absent or sparse. Wood fibers with moderately to extremely thick walls, the latter often gelatinous; pits small, simple or indistinctly bordered. Vasicentric or sparingly diffuse tracheids present in *Bumelia, Henoonia,* and *Paralabatia*. Ripple marks absent. No gum ducts seen. For anatomy of the different genera see *Tropical Woods* 59: 23–51.

Achras comprises three closely related species of medium-sized to large trees apparently limited in natural distribution to southern Mexico, Central America, and northern Colombia. The southernmost form is *A. calcicola* Pittier, a deciduous tree sometimes 80 feet tall and three feet in diameter, of common occurrence in Panama and Atlántico, Colombia, and known as Níspero or Níspero de Monte. The leaves are large and leathery and the bark contains a copious white latex.

Achras chicle Pittier is known to occur in southern Mexico, British Honduras, Guatemala, and Salvador. It attains a height of 125 feet and a trunk diameter of three feet or more. It might be inferred from the specific name that this species is the principal source of chicle gum, but its latex is difficult to coagulate and is considered inferior to that of the Sapodilla, and in British Honduras the product is called "crown gum." In Guatemala the tree is known as Zapotillo and in Salvador as Níspero de Montaña.

The best known and most important tree is the Sapodilla, *Achras Zapota* L., a native of the Yucatán Peninsula and of indeterminate range southward in Central America, and cultivated generally throughout tropical regions. The glossy, leathery, evergreen leaves are clustered at the ends of the branchlets, as are also the small white flowers. The fruit is a rough and brownish berry, variable in size and form, the skin thin, the flesh reddish or brownish, melting and somewhat milky, of very sweet agreeable flavor, and containing few to several rather small seeds; it is considered one of the best of the tropical American fruits. The usual name in Central America is Níspero; in Mexico, Zapote, Chicozapote, and Zapotillo.

Achras Zapota is important commercially as the principal source of chicle, which until rather recently was the basis for nearly all chewing gum. At the peak of the industry, 1927 to 1929, the production was about twelve million pounds annually, but now it is very much less. The tree is at its best on the calcareous marl and disintegrated limestone over a large area including the Yucatán Peninsula of Mexico, the Petén province of northeastern Guatemala, and the northern half of British Honduras. It is estimated that the Sapodilla forest there contains at least one hundred million trees. It has been suggested as possible that when the ancient Mayas, who valued the tree highly, made agricultural clearings they spared the Sapodilla trees, which thus obtained an advantage over other vegetation when the areas were abandoned. Other factors in its favor are its tolerance of shade, its ability to reproduce prolifically, and its longevity. The native tappers, or chicleros, recognize three varieties, based on the color of the latex and the characters of the bark. Zapote Colorado, the commonest form, has a reddish latex and the bark has continuous fissures. Zapote Blanco has a whitish latex, often tinged with pink, the bark is easier to cut in tapping as the fissures are shallower and the inner part is not so fibrous. Zapote Morado is indistinguishable from the Zapote Colorado, except that the latex is darker and slightly purplish. Standley says (*Tropical Woods* 31: 40): "Careful study of the collected material leads one to suspect that such forms, although perhaps recognizable in the forest, never can be separated by characters of specific importance." The following information on

Sapodilla tapping is from a paper by H. M. Heyder (*Empire Forestry Journal* 9: 1: 107–113; abstract in *Tropical Woods* 24: 35–38):

"The tapping season is during the wet months of the year, roughly from October to March, and it begins after the period of heaviest rain, which usually comes about mid-September. Tapping depends greatly upon climatic conditions and a dry year implies a very scanty yield of chicle. . . . The method of tapping Sapodilla differs considerably from methods used in rubber tapping, and is more analogous to the tapping of gutta-percha. There is no continuous flow as in the case of rubber, and the healing of tapping cuts and replacement of latex are extremely slow. After one day's tapping the tree is usually allowed to rest for a period of three years or more, according to the area of bark which has been cut. The method which is used generally in Central America is to make zig-zag cuts in the bark, about eighteen inches apart, all the way up the tree from about two feet above the ground to the first branch. The zig-zag pattern of the cuts originates from the fact that it can easily be made with the machete [a keen-edged cutlass with a 28-inch blade] which every native carries in the forest in Central America. . . . Where the zig-zag cuts have been made for more than two-thirds the way round the stem, or where the cuts have been made too deeply, as frequently happens, the cambium is killed, the bark loosens, and the tree slowly dies. A large percentage of the mature and middle-aged Sapodilla now standing in the forests is in a moribund condition due to these causes. Tapping is generally done . . . between 6 a.m. and 11 a.m., as the air is then still and humid in the forest. The latex coagulates very rapidly on exposure to sun or drying wind, and even without these adverse factors it generally ceases to flow within four to six hours from the time of cutting, so that the chicleros are usually back in their camp soon after midday with the result of their morning's work. Rain does not interfere with tapping, as the extra water can easily be evaporated from the latex. . . . In cooking chicle, a large open cauldron holding about 40 gallons is used, and a small wood fire is placed below it. The chicle bubbles up, giving off a cloud of steam. All through the cooking process, a man stirs the chicle with a paddle to prevent it from scorching against the sides of the cauldron. When the moisture has been much reduced and the chicle has become a viscous mass which can hardly be moved with the paddle, it is dumped out of the cauldron on to a piece of canvas, previously rubbed with soap to prevent sticking, and there moulded into an oblong or oval block of about 20 lbs. weight. The blocks are set aside to harden for a few days and then packed into sacks, loaded on mules, and taken to the nearest river bank, whence they are despatched by boat to the export depot."

The woods of the different species of *Achras* are much alike, and very similar to those of *Manilkara*. The timber is noted for its strength and durability and was used extensively in early Mayan construction for lintels and supporting beams. It still serves the same purposes, as well as for railway crossties, heavy flooring, and tool handles, but is not exported.

Heartwood dark reddish or reddish brown; distinct but not sharply demarcated from the pinkish sapwood. Luster rather low. Without distinctive odor or taste. Very hard and heavy; sp. gr. (air-dry) about 1.09; weight 68 lbs. per cu. ft.; texture rather fine; grain fairly straight; not easy to work and has a tendency to splinter, but can be finished smoothly; highly resistant to decay.

Common names: Chicle tree, naseberry, neesberry, nisberry, sapodilla (English); níspero, sapodilla (Span. Am., gen.); sapote, zapote (Cuba); nisperillo (Dom. R.); sapotille (Haiti); sapotier, sapotille, sapotiller, sapotillier (Fr. W.I.); mispel, m. boom, mispoe, sapatija (Dutch W.I.); chicle zapote, chicozapote, chiczapotl, guendaxiña, peruétano, tzaput, tzicozapotl, xicozapotl, ya, zapote, z. blanco, z. chico, z. colorado, z. de abejas, z. morado, zapotillo, zaya (Mex.); chicle macho, chiquibul, sapodilla (Br. H.); muy, zapotillo (Guat., Hond.); muyozapot, níspero de montaña (Salv.); ibán (Nic.); korób, níspero tierno (C.R.); níspero de monte (Col.); níspero de montaña (Venez.); níspero quitense (Ec.); sapoti, sapotilha (Braz.).

Bumelia, with about 50 species of deciduous or evergreen lactescent shrubs and small to medium-sized trees, is generally distributed from southern United States to Argentina, except for most of the Amazon basin. The plants are often armed with simple or branched spines; the small leaves are commonly clustered on spur-like branchlets; the little white or greenish flowers are fascicled in the leaf axils or along old branches; the fruits are small one-seeded and with dry or succulent and edible flesh. The yellowish timber is very hard tough and strong, but it is sparingly utilized because the trunks are generally short and poorly formed.

There are 12 species in the United States, with a combined range from Florida to Arizona and extending up the Mississippi valley to near the mouth of the Ohio River. The largest of these is *Bumelia lanuginosa* Pers., which in the Texas coast region attains a height of 50 feet and a trunk diameter up to 24 inches. There are several species in the West Indies and one of them, *B. obovata* (Lam.) A. DC., is said to extend through the high plateaus and western watersheds of Mexico and along the Pacific coast of Central America to northern Venezuela. It is sometimes 50 feet high, but more often smaller and branched near the base of the trunk. The wood is used for making household implements and tool handles.

Of the southern species the largest is *Bumelia obtusifolia* Roem. & Schulte, with a wide range centering in northern Argentina. It grows to a height of 65 feet or more with a trunk sometimes 36 inches thick. The yellowish wood takes a lustrous polish and is used for handles and wheelwright work, but is too dense for general carpentry and not durable enough for exposed construction. It is good for fuel, though several other kinds of native timbers are superior to it.

The following description is based upon samples of 12 species. Wood yellowish throughout, usually with white parenchyma markings, sometimes with tinge of green or pink. Luster medium. Odor not distinctive, taste often rather bitter. Mostly very hard and heavy; sp. gr. (air-dry) 0.80 to 1.01; weight 50 to 63 lbs. per cu. ft.; texture medium to rather coarse; grain variable; not easy to work, but can be finished very smoothly; not highly resistant to decay. Of no commercial importance.

Common names: Antswood, black haw, buckthorn, chittimwood, downward plum, gum elastic, ironwood, saffron plum, sloewood (U.S.A.); cocuyo, jiquí de costa, j. espinoso, sapote espinoso (Cuba); boxwood, breakbill (B.W.I.); bois de buis, b. de fer, petit buis (Fr. W.I.); bebelama, coma, c. resinera, hastoch, huicicialtemetl, mulché, putzmucuy, tempeschitle, tempesquistle, tempextle, tempiste, tempixquiztli, tempixtle, tempizquixtli, tempizquiztle, yantsin-tsu, zapotillo bravo (Mex.); ávalo (Guat.); ispundio, limoncillo, zapotillo de peña (Salv.); espino blanco (C.R.); caimitillo, limoncillo (Pan.); doncello, espino de brujo, guamachito, pacito de montaña, p. de monte (Col.); barba de tigre, igüí, malarmo, patillo, paují (Venez.); abio, rompe gibão (Braz.); cabo de lanza, guaraniná, guayaibí-rai, horco-molle, ibirá-hú, i.-ñirá, lanza colorada, molle de monte, m. negro (Arg.).

Calocarpum includes two closely related species of large deciduous laticiferous trees native to Central America and widely planted in tropical regions for the fruit. The leaves are very large thin and clustered at the ends of the twigs; the flowers are borne on the bare branches below the leaves. The timber is not of commercial importance as the trees are protected on account of their edible fruit.

The best known species is the Zapote, *Calocarpum Sapota* (Jacq.) Merrill (= *C. mammosum* [L.] Pierre), a tree sometimes 100 feet tall with an elongated crown or more often with a short trunk and spreading crown. It belongs to the class of semicultivated plants and the limits of its natural distribution are unknown, but it is generally considered a native of Central America. Pittier says (*Contr. U.S. Nat. Herb.* 18: 2: 83): "The fruit has a thick mesocarp of a reddish or pinkish color, and a little sticky on account of the latex it contains. The flavor is sweetish, with a peculiar squashy strain, quite delectable if we believe some Spanish authors, but not generally to the taste of foreigners. This strain might, however, be removed or improved by appropriate selection and culture. That same mesocarp can also be turned into an excellent marmalade, or into jelly, and although the fruit does not yet seem to have met with any great favor in our markets, it is not altogether without importance among tropical fruits. The seed contains a large, oily almond, which has a strong smell and a bitter taste. . . . In Costa Rica . . . the whole almond, finely ground, is made into an excellent confection. Moreover, . . . it seems to have been extensively used, and is still used on a small scale, in conjunction with cacao, in the preparation of the current beverage of the natives of Central America. It is called 'sapuyul.' . . . As a historical memorandum, we may also mention that during the first half of the nineteenth century the same seed was still used in Costa Rica in lieu of the present iron to smooth starched white linen."

Calocarpum viride Pittier is similar in appearance to the other species but differing, according to Pittier (*loc. cit.*, p. 85), "by the smaller leaves, downy and white beneath, . . . and, above all, the comparatively small green and thin-skinned fruit and the smaller, ovate seed. . . . The fruit is superior in quality to the common Zapote, the flesh not being so fibrous and being free from the squashy flavor that characterizes the latter." Englesing says (*Tropical Woods* 17: 34) that in Nicaragua it is "a large tree, 100 feet high, with straight and cylindrical trunk sometimes three feet in diameter above the buttresses (six feet) and free of branches for 60 feet, of frequent occurrence among the low hills near the Kukalaya River, attaining its best development in rich well-drained soil. Bark near the base scales off and leaves the trunk smooth, while that higher up is shallowly furrowed between confluent ridges; color greenish gray, sparsely mottled with lighter gray patches higher up. Branches heavy and ascending, with numerous branchlets and twigs at upper end which, with their dark green leaves, form a sort of crown for each branch. White sticky latex exudes copiously from wounded bark and leaves."

The woods of the two species are much alike. They are light-colored when first cut, changing to light brown or buff, eventually acquiring a slight reddish tinge; heartwood and sapwood not clearly differentiated. Luster rather low. Odor absent, taste sometimes slightly bitter. Hard, heavy, tough and strong; texture medium; grain usually straight; working properties fair to good; durability doubtful. Not likely to be of commercial importance.

Common names: Mamee apple, m. sapote, marmalade fruit, sapote (B.W.I.); mamey colorado, m. sapote (Cuba); sapotier (Haiti); grosse zapotte, zapotte, z. à crême (Fr. W.I.); atzapotiquahuitl, chacal haaz, haaz, lava-zapote, mamey, m. colorado, tezonzapote, tezonzapotl, tsapas sabani, tzapotl, zapote, z. colorado, z. mamey (Mex.); mamee apple, m. zapote, zapote (Br. H.); chul, chul-ul, ingerto, sal-tul, tulul, zapote, ingerto (Guat.); zapote (Cent. Am., gen.); zapotillo (Hond.); bko, fiú, komkrá, kurók, zapote blanco, zapotillo, z. calenturiente (C.R.); mamey, oabo (Pan.); zapote (Col., Venez., Ec.).

Chromolucuma *rubriflora* Ducke, the only species, is a large laticiferous tree sometimes 100 feet tall, with high buttresses, growing in lowland forests in the east-central Amazon region of Brazil, where it is known as Abiurana and Maiá. According to Baehni (*loc. cit.*, p. 429) the genus should be included with *Pouteria*. Ducke, who proposed the genus, says (*Archiv. Jard. Bot. Rio de Janeiro* 4: 101) that the tree is one of the most remarkable of the family, so abundantly represented in Brazil, by reason of its foliaceous stipules, its very large leaves, its very long and slender peduncles, and especially the color of the flowers which gradually changes from yellowish green in the bud through distinct yellow and orange to bright red in the adult calyx. With the exception of *Pradosia lactescens* (Vell.) Radlk., which has reddish violet-brown flowers, the characteristic floral colors in the family are green, white, or rusty-brown. Heartwood light brown with reddish or orange tinge; not clearly differentiated from the sapwood. Luster medium. Without distinctive odor or taste. Hard and heavy; texture medium; grain straight; rather difficult to work; durability doubtful.

Chrysophyllum, with about 150 species of small to large evergreen trees and shrubs, is widely distributed in tropical and subtropical regions, especially tropical America. The bark contains a milky latex; the leaves are glabrous and shiny above, densely silky hairy beneath; the small flowers are clustered in the axils of the leaves or above old leaf scars, rarely cauliflorous; the fruit is variable as to size, form, consistency, and number of seeds.

The best known species is the Caimito or Star-apple, *Chrysophyllum cainito* L., a tree 35 to 65, rarely up to 100, feet high, native in the West Indies and possibly also in eastern Central America. It is widely planted as a shade tree and also for its succulent edible fruit which is of the size and shape of a small apple and contains several compressed brown seeds arranged star-like about a central axis. The foliage is bright blue-green above and coppery beneath, affording attractive contrast when stirred by the wind.

The most widely distributed of the North American species is *Chrysophyllum oliviforme* L., a small to medium-sized tree with a straight and slender trunk, occurring in southern Florida, the Bahamas, Cuba, Puerto Rico, Jamaica, southern Mexico, Salvador, British Honduras, Guatemala, and Honduras. The Mexican form was described as a distinct species, *C. mexicanum* Brandeg., but Standley says (*Tropical Woods* 31: 42): "Careful examination of a large number of sheets of this plant, as represented in Mexico and Central America, shows that it cannot be separated from the West Indian *C. oliviforme*." It is closely related to *C. argenteum* Jacq. of Costa Rica and Panama, also of the West Indies.

There are several species in the Guianas and the Brazilian Amazon region, the best known perhaps being *Chrysophyllum sericeum* A. DC. The Kokoritiballi of British Guiana is *C. ambelaniifolium* Sandw. The largest of the Argentine species is the Picazú-rembiú, *C. ebenaceum* Mart., which is sometimes 75 feet high and 20 inches in diameter in Misiones. The Aguay, *C. lucumifolium* Gris., has a wider distribution; its fruit is savory and the latex, called balata, is used like gutta-percha.

The timbers of the various species of *Chrysophyllum* are employed locally in limited amounts for general construction and carpentry, and the darker-colored kinds are suitable for exposed works. They are not likely to become of importance in the export trade. Heartwood variable in color from pale brown or pinkish to rather dark brown, with gradual transition to the sapwood. Luster rather low. Odor and taste absent or not distinctive. Hard and heavy, though variable in different species; sp. gr. (air-dry) 0.65 to 0.90; weight 41 to 56 lbs. per cu. ft.; texture medium; grain fairly straight; not difficult to work, finishing smoothly; durability fair to good.

Common names: Satinleaf (Florida); caimitillo, caimito, c. blanco, c. cimarrón, c. morado, macanabo (Cuba); caimitillo, caimito, c. de perro, c. verde, cainit, lechecillo, teta de burro (P.R.); caimito, c. blanco cimarrón, c. cimarrón, carabana (Dom. R.); caimite, c. marron (Haiti); caimitier, cainitier, pomme surrette (Fr. W.I.); canela, caimito, cayumito, chiceh, palo de canela, zapote caimito (Mex.); chique, damsel, star-apple, wild star-apple (Br. H.); guayabillo, zapoyillo (Salv.); caimito, c. cimarrón (C.R.); caimito serrano (Venez.); kokoritiballi (Br. G.); apra, atakamara, borowéballi, boschkoffie, jorromeran, kwatta bobbi, laurierkers, loekrie hoedoe, peprebolletrie, riemhout, sterappel, takamala (Sur.); caimitier, cainitier, macoucou (Fr. G.); ajará, caimitero, caimito, camiquié, guajará branco, massaranduba-rana, sorva do Perú (Braz.); balata blanca, b. de manchal (Peru); aguay, a. blanco, blanquillo colorado, carapún, carne de vaca, chalchal, lanza blanca, olivorá, picazú-rembiú (Arg.).

Dipholis, with about a dozen species of small to medium-sized, unarmed, lactescent trees, occurs in southern Florida, West Indies, southern Mexico, and Central America. The buds are naked; the leaves are persistent; the flowers are usually fragrant and borne in clusters on the axils of the leaves or above leaf scars; the fruit is plum-like. The timber is strong and fairly durable and finds local uses in general construction.

The best known species, with the range of the genus, is *Dipholis salicifolia* (L.) A. DC., a slender tree usually less than 50, rarely up to 75, feet tall and 20 inches in diameter. It is commonly known as Bustic or Cassada in Florida and the British West Indies. The bark is rich in milky latex, which is a minor source of chicle gum in British Honduras and Guatemala.

Dipholis Stevensonii Standl. is a large tree, common in the Mopán region of British Honduras, where it is called Faisán or Zapote Faisán. It resembles *Calocarpum* and was originally confused with *Calocarpum viride* Pittier (see *Forests and flora of British Honduras*, p. 313; *Tropical Woods* 11: 21–22; 53: 43). It is distinguished from the other Central American Sapotaceae by the rust-like tomentum of the leaves. The latex is used for chicle, called chicle faisán. The chicleros recognize two varieties of tree, Red Faisán, whose chicle is as good as that of *Achras Sapota* L., and White Faisán, with higher yield but somewhat lower quality of latex.

According to Charles Baehni (*Candollea* 7: 434. 1938), *Dipholis* is not distinct enough from *Bumelia* to be entitled to generic rank. From a preliminary study of the woods, however, it appears that the separation is justifiable. Heartwood brownish to reddish brown, with gradual transition to the sapwood. Not highly lustrous. Odor not distinctive, taste somewhat bitter. Very hard and heavy; sp. gr. (air-dry) 0.90 to 1.00; weight 56 to 62 lbs. per cu. ft.; texture medium fine; grain fairly straight; not difficult to work, finishing smoothly; durability fair to good.

COMMON NAMES: Bustic, cassada (Florida); barberry bully, bullet (black, cherry, mountain, red, white), galimenta, gallamenta (Jam.); almendrillo, almendro, a. silvestre, Carolina, cullá, cuyá, jocuma, j. blanca, juba, j. prieta, jubilla, sangre de doncella (Cuba); almendrón, bustic, espejuelo, tabloncillo, varital (P.R.); caya colorado, c. de loma, c. rubia (Dom. R.); acomât rouge, bois d'Inde, sapotillier marron (Haiti); acomât bâtard, balata batard (Fr. W.I.); mijico, xac-chum (Mex.); cháchiga, faisán—red, white, zapote (Br. H.); níspero amarillo, tempisque (C.R.).

Ecclinusa includes about a dozen closely related species of small to large unarmed lactiferous trees apparently limited in distribution to the Guianas, the Brazilian Amazon region, and southward to Rio de Janeiro. The leathery leaves have well-developed stipules; the flowers are sessile and in axillary clusters; the small globular fruits contain several exalbuminous seeds. The timber is of good quality and is used to a limited extent locally for interior construction and carpentry.

Ecclinusa sanguinolenta (Pierre) Engl. occurs in French Guiana, where it is known as Balata Pommier and Bois Cochon. According to Bertin (*Les bois de la Guyane française*, pp. 67–69), it has a long trunk 28 to 40 inches in diameter at the base and free of branches for more than 75 feet. The timber has about the density of Oak (*Quercus*) and has good working properties, but is not very durable when exposed to decay. Another Guiana species is *E. guianensis* Eyma, generally known as Bartaballi or Barataballi. It attains a height of about 100 feet and the latex is used to adulterate balata (*Mimusops*).

Ecclinusa ramiflora Mart. is a small to medium-sized tree in the uplands of the lower Amazon and southward along the coast. Much like it, but smaller and sometimes shrubby, is *E. abbreviata* Ducke in the forests about Manáos. *E. balata* Ducke is a medium-sized to rather large tree of the central and western Amazon. Ducke says (*Tropical Woods* 31: 20) that it "produces almost all the inferior quality of balata of the Brazilian Amazon. This product contains only about 30 per cent of gutta (according to Le Cointe, Director of the Museu Commercial do Pará), but the total value of its exports greatly exceeds that of the superior balata derived from *Mimusops bidentata* A. DC., which, in Brazil, occurs only in the relatively narrow region along the frontier of the Guianas. Enormous quantities of this inferior balata have been exported, especially from Manáos, but this industry is destined to disappear because all the trees in accessible regions are being felled to obtain the latex."

The woods of the few species studied are similar in structure and properties. Heartwood light brown, with a pinkish or grayish tinge, and sometimes with widely spaced brown streaks; not sharply differentiated from the sapwood. Luster medium. Odorless; taste mildly bitter. Hard and heavy to moderately so; texture medium; grain fairly straight; easy to work, finishes very smoothly; probably poorly resistant to decay and insect attacks. Not likely to become important for export.

COMMON NAMES: Barataballi (Br. G.); ajowo, araatawere, baalata, bataballi, barataballi, battamballi, kodiebie joesie, malobbi, mattamatta wèwè, poeromotto, wasepoekoe (Sur.); balata, b. pommier, b. rouge, b. saignant, bois cochon, wapo (Fr. G.); abiurana, balata, coquirana, c. itaúba, c. malenta, ucuquirana (Braz.).

Henoonia consists of two species of shrubs apparently endemic in Cuba. The leaves are small and leathery; the little flowers are borne singly or clustered in the axils of the leaves; the fruits have a single exalbuminous seed. The only wood sample available for this study is of *H. angustifolia* Urb. (Yale 16160) collected with flowering herbarium material in eastern Cuba by G. C. Bucher and determined by J. T. Roig. The original species, *H. myrtifolia* Gris., has been referred by some botanists to the Solanaceae, but the structure of *H. angustifolia* shows close affinity to the Sapotaceae (see *Tropical Woods* 58: 3 and 75: 4). Color light yellow throughout specimen. Not highly lustrous. Without distinctive odor; taste slightly bitter. Hard, heavy, and strong; texture fine and uniform; grain fairly straight; takes a glossy polish; is probably not resistant to decay. Of no commercial importance.

COMMON NAMES: Rascabarriga, yareicillo (Cuba).

Labatia is a poorly defined genus with several tropical American species generally included with *Pouteria*. Apparently they are all small unarmed laticiferous trees with hairy fruits. There are two species in the island of Haiti, at least three in Cuba, and three in Panama. The latter resemble *Lucuma* and all of them were originally referred to that genus (see *Tropical Woods* 4: 8; 31: 43). The woods bear a close resemblance to those of *Paralabatia*, and apparently are not utilized for any special purposes.

COMMON NAMES: Chicharrón, guayabillo de la maestra, sapote culebra de costa, sapotillo (Cuba).

Lucuma and **Pouteria.** The many species in this group are small to large unarmed lactescent trees, mostly tropical American. The leaves are without stipules and with varying nervation; the flowers are borne in clusters in leaf axils or above leaf scars; the fruits, which vary in form consistency and number of seeds, are usually edible and in some instances highly esteemed. The timber is of good quality for strong construction and the more deeply colored kinds are highly resistant to decay.

Heartwood grayish brown to reddish brown, not sharply demarcated and usually not clearly differentiated from the lighter colored sapwood. Luster rather low. Without distinctive odor; taste sometimes astringent. Moderately to very heavy and hard; sp. gr. (air-dry) 0.70 to 1.10; weight 44 to 69 lbs. per cu. ft.; texture mostly medium, uniform; grain typically straight, occasionally wavy; working properties and durability variable.

COMMON NAMES: Canisté, canistel, sapote culebra, siguapa (Cuba); ácana, jácana (P.R.); egg fruit (Bah.); jaune d'oeuf (Haiti); pain d'espice, penny piece (Trin.); atzapolquahuitl, atzapotl, choch, comíngalo, cozticzapotl, huicón, kanizte, palo de calentura, p. huicón, zapote amarillo, z. blanco, z. borracho, z. de niño

(Mex.); mamey cerera, m. cerilla, m. ciruela, silly young (Br. H.); silión, zapotillo (Guat.); chicazapote, guaicume, güicume, pan de la vida, ulozapote (Salv.); silión (Hond., Nic.); canistel, mamón, siguapa, zapotillo (C.R.); ingerto de montaña, mamecillo (Pan.); caimo blanco, manzano morado, sapote macho (Col.); lechosillo, soberbio, temare, vaquetero (Venez.); assapoko, a. balli, bakupar, barata, kokeritiballi, limonaballi, male bullet tree, moraballo, sororo-borieng (Br. G.); abenbele, a. njambokka, ajapoekoe, akawasiba, aretoboma, aroomé, asepoko konoko, a. moraballi, atakamara, a. balli, basaa botie-ie, basra kokoni-oedoe, basterd bolletrie, biesatokon, boeloewéballi, boesi koesoewé, bosch koesoewé, djoe bolletrie, d. botrie, eukele, hariraro asepokoe, h. moraballi, iawé hepapaja, iengie hoedoe, janboka, jansnijder, jawahe paikoelia, j. papaje, kiemboto, kienboto, kodibiesie, kodibiosi-balli, koesariejeppo, koesiri paratare, k. pialatara, kokonihoedoe, koni-koni waata, konoko, k. balli, kororietje-balli, kwatta bobi, laurierkers, lemoepoe, logoesoe fehoeta, lohoedoe, mabijara, mamanten, mambiara, mapijara, mapilan, mapiran, m. warian, m. hariroro, m. khalé-meroe, mapirian, mapiwalan, moraballi, m. diamaro, njamboka, oeroemerian, olemelan, oro oromé, orromé, piento botrie, riemhout, remoe epe, saffoe-ka, sagwenki hoedoe, sueparatarie, tamoené koesali epo, t. paraata, t. paratare, tapoekoe, tepjori pjorikin, tipopolipoli kinwewe, toepoewe, toewonoele, topie a baakawan, t. njambokka, warrossieran, wasse poekoe, wesse poekoe, w. p. kessipoeloe, weti apokwi, witte djoe bolletrie, wokko moloko telle, wokoeloe allowe, zwart riemhout (Sur.); balata indien, b. singe rouge, jaune d'oeuf, pomme de pin, wapi, wapo (Fr. G.); abi, abia, abiu, a. grande, a. rana, a. r. grande, a. r. gutta, a. r. mucura, bapeba, b. assú, chauá, cutiti, cutitiribá, c. grande, c. rana, frutão, goititurubá, grão de gallo, guajará, guapeba, guapebeira, maparajuba, massaranduba branca, mata-olho, mucurí, muirápixi, oitituruba, pajurá, pariri, uajará (Braz.); balata mapa, caimito, huanganacaspi, locma, lucmo, lucumo, quinilla blanca, pucuna-caspi, sacha-caimito, uchpaquinilla, urcu-cumala (Peru); aguay-mí, mata-ojo (Par.); mata-ojo (Urug.); aguay, a. guazú, mata-ojo (Arg.).

Manilkara or **Mimusops.** Insofar as the American species are concerned, these two names are synonymous. In the sense now used by many botanists, they are distinct genera, *Manilkara* being a segregate from *Mimusops* (*sens lat.*). Considered in this way, *Mimusops* is restricted to the tropics of the Old World, while *Manilkara*, with about 50 species, occurs in tropical and subtropical regions of both hemispheres. From present knowledge of the woods, the whole group might well be included in a single genus.

The plants vary in size from little more than shrubs to massive trees 150 feet tall and over six feet in diameter. One of the best known species of *Mimusops* (*sens str.*) is *M. Heckelii* (A. Chev.) Hutch. & Dalz., the West African tree supplying the timber called African Cherry or Cherry Mahogany. The American timbers known to the export trade are the Beefwood or Bulletwood of British Guiana (*M. bidentata*), the Massaranduba of Brazil (*M. Huberi*), and the Almique, Acana, or Donsella from Cuba (*M. jaimiqui*). The product of greatest value is the coagulated latex, called balata, which is obtained from a few species, mostly South American. The fruits are edible but not highly esteemed.

The range of the American species includes southern Florida, the West Indies, southern Central America, and South America to the Peruvian Amazon and southeastern Brazil. The most northern representative is the Wild Dilly or Wild Sapodilla of the southern Florida Keys, *Manilkara emarginata* (L.) Britt. & Wils., a tree sometimes 35 feet tall and 12 inches in diameter; it also occurs in the Bahamas, Puerto Rica, and Cuba, although Roig (*Dic. Bot. Nom. Vulg. Cubanos*, p. 839) lists the name as a synonym for *Mimusops Wrightiana* Pierre.

The Ácana, Aimiquí, or Jaimiquí, of Cuba, *Manilkara jaimiqui* (C. Wr.) Dubard (= *Mimusops jaimiqui* C. Wright), is a large tree supplying one of the best timbers in the island for heavy and durable construction of all kinds. The wood is of a deep rich red color, with an oily appearance and feel; because of its hardness it is difficult to nail without splitting, and railway crossties made of it must be bored before spiking. It takes a beautiful polish and is often employed for articles of turnery. Limited quantities have been sold in New York under the native name and also as Doncella and Almique. The last is based upon the Cuban Almiquí, and G. C. Bucher of Santiago de Cuba informs the authors that "Ácana and Almiquí are synonyms, the names being interchangeable in this district." Roig, however, says (*loc. cit.*, p. 35) that Almiquí is *Mimusops discolor* Ekman, but that the common name is confused with Aimiquí or Jaimiquí, *Manilkara jaimiqui*.

Two Central American species have been described. Pittier says (*Contr. U.S. Nat. Herb.* 13: 12: 465–466) that *Manilkara spectabilis* (Pitt.) Standl. (= *Mimusops spectabilis* Pitt.), which he discovered near Port Limón, Costa Rica, is a very large tree with a straight trunk often reaching a height of 150 feet. "This noble tree is conspicuous by its towering proportions among the constituents of the littoral forests of the Atlantic coast. . . . It is one of the Nísperos of the Costa Ricans and the Jamaican immigrants call it Bully Tree. . . . The wood is hard, heavy, and dark and, being very resistant to water and wet soil, is in great demand for railroad ties." *Manilkara darienensis* (Pitt.) Standl. (= Mimu*sops darienensis* Pitt.) was found by Pittier on hills back of Puerto Obaldía, San Blas coast, Panama. He says (*Contr. U.S. Nat. Herb.* 18: 6: 249–250) that it is 130 to 165 feet tall, with an unbuttressed trunk often over five feet in diameter, and "has great economic importance as the source of the Panama balata or gutta-percha and the wood is also very valuable. From an ecological standpoint the tree, which is very abundant in the hilly hinterland of the Caribbean coast, is to be considered as a characteristic element of the rain forests of the eastern part of the Isthmus."

As a source of balata the most important species appears to be *Manilkara bidentata* (A. DC.) A. Chev. (= *Mimusops bidentata* A. DC.), a large tree of the Guianas, Venezuela, and the upper and northwestern Amazon region. This tree has often been referred to as *Mimusops balata* (Aubl.) Gaertn., but this name belongs to a tree which was introduced into French Guiana from Mauritius and known only in cultivation (see *Tropical Woods* 58: 24). The native tree is commonly called Pamashto or Quinilla in eastern Peru, Balata in Brazil and French Guiana, and Purgio in Venezuela. In British Guiana the usual names are Bullet, Bully, or Balata tree, while the timber is called Beefwood, Bulletwood, and sometimes Horseflesh. Eyma says (*loc. cit.*, p. 208): "Collectors' notes invariably give the vernacular name of this species in Surinam as Bolletrie; in a few instances the name Balata is added, and all labels containing Indian names have Borowé as the Arowaccan and Parata as the Caribbean name. The negro name is Botrie. The Dutch name Paardevleeschout (horseflesh wood, from the red color of the wood) is sometimes encountered in literature but never on collectors' labels. This appears to be the only species from which commercial balata is procured in Surinam." The yield of latex per tree is from 2 to 5 pints, equivalent to 1 to 3 pounds of dry balata, the yield varying roughly with the thickness of the bark.

There are at least five other species in the lower Amazon region of Brazil, all large to very large trees, known as Massaranduba or Maparajuba. Their value resides in their timber rather than their latex. The best for durable construction is the Massaranduba Verdadeiro (true Massaranduba), *Manilkara Huberi* (Ducke) Standl. (= *Mimusops Huberi* Ducke). Eyma says (*loc. cit.*, p. 209): "This species is conspicuous for its leaves, which are light-colored at first, often beautifully orange or yellow beneath, with distinct darker nervation, the ground color becoming paler to white with age. The species has hitherto been recorded for the greater part of the state of Pará and so I do not hesitate to refer to it some sterile specimens from the interior of Surinam." This is the Milk-tree or Cow-tree which attracted the attention of Richard Spruce during his visit to Pará in 1849 because it secreted an abundance of drinkable milk (see *Notes of a botanist on the Amazon and Andes,* Vol. I, pp. 50–52). He "made trial of the milk, fresh from the tree, both alone and mingled with coffee; its consistency is that of good cream and its taste perfectly creamy and agreeable. It retains its fluidity for weeks, but acquires an unpleasant odor. It is extremely viscid and can with difficulty be removed from the hands or whatever else it touches, a property which renders it an excellent substitute for glue but a rather unsafe article of diet."

Massaranduba in small quantities has been imported into the United States for a long time, but no particular attention was directed to it until timber became available as a result of clearing the site of the Ford rubber plantations along the Tapajoz River. Roswell S. Cheves, of the Day Lumber Company, Springfield, Massachusetts, supplies the following information regarding practical tests made in 1936. Because of insufficient resilience it is not as satisfactory as Maple (*Acer saccharum* Marsh.) for drop-forge hammer bars or as a substitute for Hickory (*Carya*) in picker-sticks of weavers' looms. It gives good service as flooring in industrial plants and machine shops where resistance to trucking and other kinds of hard usage is essential; also bench tops, stair treads, truck bodies, and in paper mills for beaters, agitator bars, and jordan and bed plate filling. In comparative tests of Massaranduba and Oak (*Quercus*) for beater filling under normal working conditions in two paper factories, the Oak failed completely before the Massaranduba showed any appreciable wear. New York timber dealers and manufacturers report that a fine dust arises during sawing which is rather peppery and irritating to the mucous membrane, and also that the lumber must be handled carefully to avoid splinters.

There are two species in Ceará, Piauhy, and Maranhão, Brazil, namely, *Mimusops rufula* Miq. and *M. triflora* Fr. Allem. (= *M. cearensis* Huber); the latter is a small tree, sometimes shrubby. In the coastal forests from Bahia to Rio de Janeiro there are five species, all called Massaranduba, and H. M. Curran says that the timber "is much used for railway ties and heavy durable outside construction. The trees occur scattered through the hardwood forests of the coast but are nowhere abundant and the available supply near transportation has been nearly exhausted. They are of large size, 100 feet high and three or four feet in diameter, with a dark brown or blackish bark of medium thickness, and heavy, dark green leaves 6 to 8 inches long. The heartwood is extremely hard and of a deep red color; the sapwood is usually narrow and is almost white when freshly cut."

The woods of the several American species of this genus studied are similar in structure. Heartwood red or reddish brown, deepening upon exposure; with oily appearance and feel in some West Indian species; distinct but usually not sharply demarcated from the whitish sapwood. Luster low. Without pronounced odor or taste. Generally extremely hard, heavy, and strong; sp. gr. (air-dry) 0.90 to 1.20; weight 56 to 75 lbs. per cu. ft.; easy to moderately difficult to work, finishing very smoothly; highly durable. An excellent timber likely to increase in importance in the export trade.

COMMON NAMES: Beefwood, bulletwood, doncella, massaranduba, red lancewood (trade); wild dilly, w. sapodilla (Florida); bulletwood, sapodilla bulletwood (Jam.); ácana, a. de costa, aimiquí, almiquí, balata, jaimiquí (Cuba); ácana, ausubo, bullet tree, b. wood, mameyuelo, sapota de costa (P.R.); ausubo, balata, nisperillo (Dom. R.); bois huile, sapotille, s. marron (Haiti); níspero (C.R., Pan.); balata, níspero, trapichero (Col.); ácana, balata, cochinillo, massarandú, níspero, pendare, pulvio, purgo, purgüey, purguo (Venez.); balata, beefwood, bulletwood, burueh, koboru (Br. G.); bad bolletrie, badwood, balata, b. boom, basterd bolletrie, boeletrie, borowé, botrie, b. ie, brosse balata, b. bolletrie, paardevleeschout, valsche bolletrie (Sur.); balata, b. franc, b. rouge (Fr. G.); apraüá, balata, chauá, maparajuba, massaranduba (Braz.); balata, palata rosada, pamashto, quinilla, q. colorado (Peru).

Micropholis, usually considered as a section of *Sideroxylon,* includes more than 30 species of small to very large trees, mostly South American. The Gumbijava of southeastern Brazil, *M. Gardnerianum* (A. DC.) Pierre, is usually a medium-sized tree, but is said to attain a height of 100 feet or more in the Serra do Mar. The yellowish or reddish timber is easy to work and is employed locally for carpentry, interior construction, chairs, and packing cases, but is not durable enough for service in exposed positions. There are numerous species in the Amazon region and at least three in the Guianas. Perhaps the largest and best known of the latter is the Balata Franc of French Guiana, *M. Melinoniana* Pierre, which, according to Bertin (*Les bois de la Guyane française,* pp. 55–56), is a fairly common tree with a long slender trunk sometimes 80 to 90 feet long and 36 to 40 inches in diameter above the moderately high buttress. The wood is yellowish or roseate when fresh, but becomes light brown with a tinge of red. The timber, though little used, is considered of excellent quality for interior construction and carriage work.

The following description is based on specimens of six species from Brazil and the Guianas. The woods are fairly uniform in appearance and structure and are readily separated from those of the *Mastichodendron* section of *Sideroxylon* because their prevailing color is brown instead of yellow and the wood parenchyma is in concentric bands instead of reticulate. Heartwood grayish brown, usually with a reddish tinge, sometimes (*M. paraensis* [Huber] Eyma) with a yellowish green hue; sapwood lighter, but not clearly differentiated. Luster medium. Without distinctive odor or taste. Hard and heavy to decidedly so; sp. gr. (air-dry) 0.80 to 1.05; weight 50 to 66 lbs. per cu. ft.; texture medium, uniform; grain straight; working properties fair to good.

Common names: Sapotillo árbol (Cuba); caimitillo (P.R.); chupón colorado, hácano (Venez.); moraballi (Br. G.); asépokoballi, awapau, bobi waata, koesiri balatarie, k. paratare, koni-koni hoedoe, lo-hoedoe, riemhout, remoe époe, seloe boerwin, serere boerwin, wasépoekoe, wokowokoeloe (Sur.); balata blanc, bois crapaud (Fr. G.); apixuna, balata rosada, caramury da varzea, gipy, gumbijava, gumbixama, gumbixava, mangaba-rana, preguiceira, rosadinha (Braz.); barilla de agua, quinilla (Peru); ibirá-camby (Arg.).

Paralabatia, with three species of small to medium-sized unarmed evergreen trees, is limited in distribution to the Greater Antilles. *P. dictyoneura* (Gris.) Pierre occurs in the mountains of Cuba, where it is known as Cocuyo in common with various other trees whose bark, wood, or fruit has a color suggesting the glowworm (cocuyo). *P. Fuertesii* Urb. is a shrubby tree of Dominican Republic. *P. portoricensis* Britt. & Wils. is a tree 50 feet or more in height growing on limestone hills in northern Puerto Rico.

The woods of the first two species, the only ones available, are yellowish brown or somewhat roseate, without clear differentiation between heartwood and sapwood. Luster medium. Without distinctive odor or taste. Very hard, heavy, tough, and strong; texture medium; grain fairly straight; not very difficult to work but inclined to be splintery; finishes very smoothly; probably not highly durable.

Common names: Cocuyo, jubilla (Cuba); cuero de puerco, cuyá (Dom. R.); caracolet (Haiti).

Pradosia is a Brazilian genus with a few species of small to large unarmed lactescent trees generally characterized by a sweet taste to the bark and the wood. The best known tree is the Buranhem or Casca Doce, *Pradosia glycyphloea* (Casar) Kuhlmann, of the coastal forests of Brazil from São Paulo northward. In Bahia it is said to attain a height of 100 to 125 feet, with a long straight unbuttressed trunk up to four feet in diameter. The bark is smooth and the reddish inner part contains a small amount of latex and is very sweet, though astringent because of the high tannin content. The flowers of this species are very small, green, and borne in clusters of 2 to

6 in the axils of the leaves. A specimen of the wood of Buranhem was collected near Bahia with sterile botanical material by H. M. Curran (Yale 4964; Curran 24). The leaves resemble those illustrated by Kuhlmann for *P. glycyphloea*. The wood has a sweetish and astringent taste and is similar in structure to authentic specimens of other species of *Pradosia*. *Pradosia lactescens* (Vell.) Kuhlmann (*Archiv. Jard. Bot. Rio de Janeiro* 5: 205) is always a small tree whose bark, instead of being sweet-tasting, is bitter and of a detestable flavor. The small flowers are of a vinaceous violet color and are borne in dense clusters all along the trunk and larger branches. It is commonly known as Bacurí or Bacupari, names generally applied to species of *Rheedia* (fam. Guttiferae) which it resembles. No wood samples are available for study.

Pradosia inophylla (Mart. ex Miq.) Ducke is a shrub or little tree called Abihy near Manáos and Pau Doce near Faro. *P. pedicellata* Ducke is similar to the preceding, but generally taller, occasionally a large tree; in Gurupá it shares with other Sapotaceae the name of Ajarahy. *P. Huberi* Ducke attains a height of 100 to 130 feet and occurs on inundated lands in the estuary; because of the resemblance of its trunk to that of the Pracuúba (*Dimorphandra paraensis* Ducke) it is often called Pracuúba Doce and Pracuúba de Leite. *P. praealta* Ducke is found in dense forests on non-inundated land near Belem do Pará and is considered the largest of all sapotaceous trees, attaining a height of 160 feet and having an erect cylindrical trunk with tabular buttresses 10 to 15 feet high. (See *Tropical Woods* 71: 7–18.)

The Amazonian species apparently have few uses, except for the sweet and comestible fruits. The bark is rather low in tannin content and the wood is not highly resistant to decay. The Buranhem of the coastal region supplies bark rich in tannin (up to 30 per cent) which is employed commercially. The timber lacks color and figure but is considered excellent for the frames of vehicles and farming implements, oars, and heavy interior construction. It appears well suited for tool handles, wheelwright and bent work, and cooperage. It is slow in drying and the thick sapwood is likely to stain during the process.

Wood of a dull grayish brown color throughout, usually uniform, but sometimes with rather vague streaks of yellowish to purplish brown. Without noticeable odor; taste highly distinctive, being typically sweet at first, then bitter or astringent. Very hard, heavy, tough, and strong; sp. gr. (air-dry) 0.95 to 1.10; weight 59 to 70 lbs. per cu. ft.; texture medium, uniform; grain generally straight; not very difficult to work, finishing smoothly; poorly resistant to decay. Will probably never be important for export.

COMMON NAMES: Abihy, ajaray, bacupari, bacuri, buraem, burahem, buranhe, buranhem, casca doce, guaranhem, guranhem, imyricem, mericeem, muiracehima, paracuhuba doce, p. de leite, pau doce, pracuúba doce, p. de leite (Braz.).

Sarcaulus. There is only one, imperfectly known, species, *S. brasiliensis* (A. DC.) Eyma (= *Chrysophyllum brasiliense* A. DC.), a lactescent tree occurring in the Guianas and Amazonian Brazil. The leaves, which are 4 to 9 inches long and about a third as wide, are glabrous, somewhat leathery, and show the nervation prominently on the under side; the flowers have a fleshy globular corolla and are borne in fascicles in or just above the axils of the leaves; the fruit is unknown. There are no recorded uses for the timber. The following description is based upon one sample (Krukoff 5318) collected by B. A. Krukoff in the basin of the Rio Purús, Acre Territory, Brazil. Specimen dull brownish, more or less streaked, throughout; probably all sapwood. Without distinctive odor or taste. Hard, heavy, and strong; texture rather fine; grain fairly straight; working properties good; durability probably low. Apparently without commercial possibilities.

Sideroxylon, in a broad sense, includes a large number of trees of wide distribution in tropical and subtropical regions. Pittier says (*Contr. U.S. Nat. Herb.* 13: 12: 458): "The genus *Sideroxylon* was established by Dillenius, the type, *S. inerme*

L., being a tree of the Cape Colony in South Africa. . . . Time after time unfortunate additions increased the genus and caused the original definition to be repeatedly altered. These additions not only included several Bumelias and a few other species belonging to closely related genera of the Sapotaceae, but also a *Scleroxylon* (Celastraceae), a *Myrsine*, and an *Olinia* (Oliniaceae)."

Botanists have divided the genus into sections, and one of these, following Eyma, is here treated as a separate genus, *Micropholis*. Remaining in the Yale collections are specimens of several species determined as *Sideroxylon* which are alike in general appearance and properties, particularly in their yellow or orange color. They differ, however, in the arrangement of parenchyma; in one group, section *Mastichodendron*, it is consistently reticulate, and crystalliferous strands are common, whereas it is in fairly uniform concentric arrangement and no crystalliferous strands were seen in an unnamed group including *S. colombianum* Standl., the Joveroso, Mamón de Leche, or Mamón de Tigre of northern Colombia (see *Tropical Woods* 22: 13) and *S. Meyeri* Standl., the Zapotillo of British Honduras and Campeche. The wood of the Silly Young of British Honduras, *S. amygdalinum* Standl. (= *Lucuma amygdalina* Standl. = *Bumelia laurifolia* Standl.), is brown instead of yellow and its parenchyma is in concentric bands, thus suggesting *Lucuma* rather than *Sideroxylon* or *Bumelia*. (See *Tropical Woods* 31: 45.)

The section *Mastichodendron* comprises several closely related species limited in distribution to the West Indies, southern Florida, Mexico, and Central America. The trees have large boles, sometimes 50 inches in diameter, but are usually not over 60 to 70 feet high, though occasionally up to 100 feet. The buds are naked; the leaves are mostly clustered near the ends of the branchlets; the very small greenish yellow ill-smelling flowers are borne in dense clusters on the old wood or in the leaf axils; the fruit is a small olive-like drupe, edible but not highly esteemed. The best known species is *Sideroxylon foetidissimum* Jacq. (= *S. Mastichodendron* Jacq.), commonly known to English-speaking people as Mastic; it grows in southern Florida, the Bahamas, and many of the Antilles. It appears at its best in Cuba, where it is called Jocuma, and the timber is used in heavy construction of all kinds, vehicles, ox-yokes, fence posts, railway crossties, and to a small extent in furniture.

The following description is based upon specimens of *Sideroxylon foetidissimum*, *S. angustifolium*, and *S. tempisque*. Heartwood lemon to orange, not clearly differentiated from the yellow sapwood. Luster medium. Without distinctive odor; taste somewhat bitter. Very hard, heavy, tough, and strong; sp. gr. (air-dry) about 1.05; weight about 66 lbs. per cu. ft.; texture medium; grain straight to variable; not easy to work, but finishes smoothly; requires care in drying to prevent splitting; durability fair. Not likely to be important for export.

COMMON NAMES: Mastic, m. bully, wild olive (Florida, B.W.I.); caguaní, ébano amarillo, jocuma, j. amarillo, lechero (Cuba); tabloncillo, tortugo amarillo (P.R.); caya amarillo (Dom. R.); acomat (Haiti); acoma, coopey, topee (Trin.); acoma bâtard (Guad.); capire, capiri, caracolillo, cosahuico, dzoi, ébano amarillo, huacux, subul, tempisque, tempixque, tempixtle, totozapotl, zapote de ave, zoy (Mex.); tempisque (C. Am., gen.); cream tree (Br. H.); kobak (Guat.); saquaia (Salv.).

SAURAUIACEAE

Saurauia, the only genus, has been referred by different botanists to the Actinidiaceae, Clethraceae, Dilleniaceae, and Theaceae. The leaves are alternate, simple, mostly serrate, often rough, and have prominent parallel nervation; stipules are absent; the rather small flowers are paniculate; the fruit is typically a pulpy berry with small seeds. About 250 species have been described, but the differences between them are often slight. They occur in tropical and subtropical Asia and America, and are shrubs or small or rarely medium-sized trees of no commercial importance. The American representatives are distributed, often in the mountains, from southern Mexico through Central America and western South America to Chile; they are absent from the West Indies and very infrequent in the Guianas and Brazil.

Wood pale reddish brown throughout; not attractive. Luster rather low. Odorless and tasteless. Rather light, but firm, tough, and strong; sp. gr. (air-dry) 0.58; weight 36 lbs. per cu. ft.; texture medium; grain straight; easy to cut, saws finely woolly, is rather hairy under the plane; probably not durable. Useful locally for general carpentry and interior construction.

Growth rings absent or poorly defined. Pores numerous; very small, not distinct without lens; virtually all solitary; well distributed without special pattern. Vessels with many-barred scalariform perforation plates; spirals occasionally present in tips of members; intervascular pitting, when present, opposite and rather fine. Rays near limit of vision on cross section, distinct to conspicuous on radial surface; of two sizes, uniseriate and multiseriate (up to 5 or 6 cells), and few to many, sometimes over 100, cells high; decidedly heterogeneous; gum deposits abundant; bundles of raphides present in some species; pits to vessels rather small, circular or oval. Wood parenchyma diffuse or in short tangential lines; occasionally rather abundantly developed; faintly visible with lens; no crystalliferous strands observed. Wood fibers with thick or rather thin walls; pits numerous, distinctly bordered. Ripple marks absent. No gum ducts seen.

COMMON NAMES: Almendrillo, mameyito, m. blanco, pipicho (Mex.); sapocillo, zapotillo (Hond.); alais, capulín, c. montés, c. de montaña, cresta de gallo (Salv.); mocó (C.R.); frio (Pan.); azuceno (Col.); mocó, moquillo (Venez.); carrón (Ec.).

SCROPHULARIACEAE

THIS family comprises about 180 genera and hundreds of widely dispersed species of herbs, erect or scandent shrubs, and a few, typically small, trees. The leaves are simple and opposite or less commonly alternate or verticillate, and without stipules; the flowers are usually irregular, often showy; the fruit is a many-seeded capsule. Some of the best known plants are Foxglove (*Digitalis*), Mullen (*Verbascum*), Snapdragon (*Antirrhinum*), and Toadflax (*Linaria*). The only important tree is *Paulownia tomentosa* (Thunb.) Baill., native to eastern Asia but naturalized in waste places and thickets in southeastern United States; its wood has about the consistency of *Catalpa*.

Very few of the American genera are arborescent. *Ghiesbreghtia grandiflora* A. Gray of Guatemala and Chiapas, Mexico, is said to attain a height of 25 feet, but no specimens are available for study. *Dermatocalyx parviflorus* Oerst. of Central America is a scandent fleshy-leaved shrub sometimes climbing to a height of 80 feet and having a basal diameter of six inches; there are questionable reports that it is sometimes a small tree. *Monttea Schickendantzii* Hieron. of Argentina is a shrub or a tree 15 feet high and six inches in diameter. The woody stems of many of the smaller plants are unusual in that they have no true rays. The following description applies to Yale specimens of *Dermatocalyx, Monttea,* and *Pentstemon.*

Heartwood brown; sharply demarcated from the brownish to nearly white sapwood. Luster rather low. Odorless and tasteless. Hard, heavy, and strong; texture medium in *Dermatocalyx,* very fine in the others; easy to work, finishing smoothly; probably not very durable. Heartwood of *Monttea* suitable for small articles of turnery.

Growth rings present; ring-porous structure in *Pentstemon.* Pores minute in *Monttea* (30μ), minute to small in *Pentstemon* (50 to 70μ in early wood), small to large in *Dermatocalyx* (up to 270μ); irregularly distributed in *Dermatocalyx,* well distributed and solitary in *Monttea.* Vessels with simple perforations; fine spiral thickenings present in *Monttea* and *Pentstemon;* intervascular pitting fine (6μ), alternate. Rays (when present) 1 or 2, sometimes up to 5, cells wide and up to 20, occasionally to 50, cells high; very small crystals of various shapes present in *Dermatocalyx* and *Monttea;* rather to decidedly heterogeneous; pits to vessels small or minute. Wood parenchyma in fine irregularly spaced bands in *Dermatocalyx,* absent or very sparse in the others. Wood fibers with spiral thickenings in *Dermatocalyx* and *Monttea;* pits large and conspicuous and distinctly bordered in *Monttea,* smaller and less distinctly bordered in the other two; septa present in some of the fibers of *Dermatocalyx*. Ripple marks absent. No gum ducts seen.

Common names: *Dermatocalyx:* Colmillo (C.R.). *Monttea:* Palo de sebo, tintillo (Arg.).

SIMARUBACEAE

This heterogeneous family comprises about 30 genera and 200 species of shrubs and small to large trees widely distributed over the world, though mostly tropical. The American arborescent genera are *Aeschrion, Castela, Holacantha, Picrolemma, Quassia* (also African), *Recchia, Simaba,* and *Simaruba* of the subfamily Simaruboideae; *Alvaradoa,* subfamily Alvaradoideae; and *Picramnia,* subfamily Picramnioideae. Two genera often included in this family, namely, *Suriana* and *Picrodendron,* are considered as constituting the monotypic families Surianiaceae and Picrodendraceae, respectively. The classification would be simplified still further by excluding *Alvaradoa* and *Picramnia.*

The only tree well known in the temperate zone is the Asiatic Ailanthus or Tree of Heaven, *Ailanthus altissima* (Mill.) Swingle (= *A. glandulosa* Desf.). It was first planted in England in 1751 and in the United States in 1784. Illick says (*The Ailanthus tree in Pennsylvania,* pp. 8, 29): "Soon after its introduction into America the Ailanthus sprang into popularity as a street tree. Its rapid and luxuriant growth impressed the early nurserymen, and its ability to thrive in very unfavorable situations, particularly in industrial centers, is responsible for its wide use. . . . Extensive field studies show that no other tree competes so aggressively with the native tree growth. . . . Unless some major use is found for its wood it will become a serious forest weed. . . . Preliminary investigations show that it has merit as a pulpwood."

The principal uses of the tropical American Simarubaceae are medicinal and depend upon a bitter principle which is present in all parts of the plants of the Simaruboideae group. Two genera, *Quassia* and *Eschrion,* are well known to the drug trade as the source of Lignum Quassiae or Bitterwood. The only lumber-producing genus is *Simaruba;* limited amounts of the timber are exported from Brazil under the name of Marupá and used for some of the same purposes of Basswood (*Tilia*) and Whitewood (*Liriodendron*).

The following description of the wood applies particularly to the American Simaruboideae, excepting *Recchia,* which has not been studied. Color nearly white to canary or greenish yellow throughout. Luster mostly high. Odor absent; taste mildly to decidedly bitter. Of medium to low density; texture fine to rather coarse; grain generally straight; very easy to work; reputedly resistant to insects, but not to decay.

Growth rings usually present, but not always distinct. (*Ailanthus* is ring-porous.) Pores variable in size and abundance; mostly small, indistinct to invisible without lens; fairly numerous; solitary and in short to long multiples and little groups; well distributed without pattern except in *Holacantha* and less distinctly in *Castela.* Vessels with exclusively simple perforations; tyloses absent; spiral thickenings present in *Castela* and *Holacantha.* Rays uniseriate and biseriate in *Quassia* and *Simaba,* 1 to 6 cells wide in the others, with much variation in height in different species and same specimen, maximum about 90 cells; homogeneous or nearly so to distinctly heterogeneous; crystals common; pits to vessels sometimes rather large in part (*e.g., Simaruba*), but mostly small to minute. Wood parenchyma more or less abundant; vasicentric, aliform, confluent, or in concentric bands, sometimes apparently terminal; crystals sometimes present, particularly in *Simaruba.* Wood fibers very thick-walled in *Castela* and *Holacantha,* mostly rather thin-walled in the others; pits small simple and slit-like or with very small border and narrow, extended aperture. Vasicentric tracheids present in *Castela.* Ripple marks characterize *Aeschrion, Castela,* and *Simaruba;* fairly uniform to irregular; 50 to 90 per inch. Small vertical gum ducts of more or less common occurrence in *Castela* and *Simaruba.* The fact that *Castela* and *Holacantha* are desert plants presumably accounts for the exceptional features noted for their woods.

Alvaradoa and *Picramnia* differ from the

Simaruboideae in the following particulars: Color brownish; heartwood rather sharply demarcated in *Picramnia.* Taste not bitter in *Alvaradoa.* Density medium to rather high; texture fine to medium. Pores small to minute; tending to definite radial arrangement. Vessels without spiral thickenings; pitting very fine. Rays mostly 1 to 3, sometimes up to 7, cells wide and generally less than 40 cells high; definitely heterogeneous; not storied; pits to vessels very small or minute. Wood parenchyma absent or very sparse. Septate wood fibers in parenchyma-like bands in *Alvaradoa,* generally distributed in *Picramnia.* Vasicentric tracheids present in *Alvaradoa.* Ripple marks absent. No gum ducts seen.

Aeschrion (or *Picraena,* or *Picrasma* in part), with five species of unarmed trees, occurs in the West Indies and southern South America. All parts of the plants have a bitter taste. The leaves are large and odd-pinnate, suggesting Ash (*Fraxinus*), hence the English name Bitter Ash; the small greenish flowers are borne in axillary corymbs; the fruit is a small globose thin-fleshed drupe, borne singly or two or three together.

The South American species is *Aeschrion crenata* Vell., a little tree rarely 20 feet high, with large brittle branches and a slender dark-colored trunk only a few inches thick. The only uses are in domestic medicine. (For detailed description of the tree and wood, see *Lilloa* 1: 272–282.) The best known species is the Jamaica Quassia or Bitterwood, *A. excelsa* (Sw.) Kuntze. Mature trees are generally from 40 to 60, rarely up to 80, feet tall, with a straight smooth white-barked trunk. The range of the species includes the Island of Haiti, Puerto Rico, and the Lesser Antilles, but the commercial supply of the wood is chiefly of Jamaican origin, and is in the form of logs 3 to 10 feet long and 3 to 10 inches in diameter, often knotty, and with the bark on. The uses are the same as those of the Surinam Quassia (*Quassia amara* L.).

The woods of *A. crenata* and *A. excelsa* are similar. Heartwood not clearly differentiated from the yellowish sapwood. Luster high. Odor lacking; taste very bitter. Light in weight, but firm; sp. gr. (air-dry) about 0.50; weight about 31 lbs. per cu. ft.; texture medium; feel soft; grain straight to irregular; very easy to work; resistant to insects but not to decay.

Common names: Bitterwood, Jamaica quassia, quassia, West Indian bitterwood (Eng.); bois amer, b. de quassia de la Jamaïque (Fr.); Jamaica Quassiaholz (Germ.); lignum quassiae jamaicensis (Pharm.); bitter ash, b. wood (B.W.I.); bois amer, b. de St. Martin, coache, peste à poux, quachi, simarouba (Fr. W.I.); goric (Haiti); palo amargo, quina brava (Arg.).

Alvaradoa, with four or five species of unarmed shrubs and little trees rarely over 25 feet high, occurs in the West Indies, southern Florida, Mexico, Central America, Bolivia, and Argentina. The best known and most widely distributed species, with a range almost the same as the genus as a whole, is *A. amorphoides* Lieb. The leaves are odd-pinnate, with numerous narrow leaflets; the small green or yellowish white flowers are borne in long racemes; the fruits are purplish yellow lanceolate samaras borne in drooping racemes which render the plant highly decorative. The juice of the bark is bitter. There are apparently no special uses for the timber.

Heartwood absent or not clearly differentiated from the brownish sapwood. Luster rather high. Odorless and tasteless. Density medium (*A. amorphoides*) to fairly high (*A. arborescens*), the wood of the latter very tough and strong; texture medium; grain variable; not difficult to work, finishing smoothly; durability probably low. Of no commercial possibilities.

Common names: Aroma blanca, tamarindillo (Cuba); abbé marron (Haiti); belcinic-ché, besinic-ché, palo de hormigas, pié de gallo, suitsinic-ché, xbesinic-ché (Mex.); plumajillo (Guat.); palo de sobo, plumajillo (Salv.); zorra (Hond.); pichi blanco (Arg.).

Castela (including *Castelaria* and *Neocastela*), with ten closely related species of thorny shrubs and little trees, is of common occurrence in dry regions of the West Indies, southwestern United States, Mexico, Galapagos Islands, and northern and southern South America. The branchlets are spinose or spur-like; the leaves are very small and simple; the red or purplish little flowers are solitary or clustered in the axils of the leaves; the fruit is a small drupe. The plants are too small to supply commercial timber. The most northern species is *C. texana* (Torr. & Gray) Rose, a densely branched shrub in the chaparral of southwestern Texas and northern Mexico. The juice of the bark is bitter and astringent and finds some employment in local medicine. *C. Nicholsoni* Hook. f. is characteristic of the xerophilous thorn formations of northern Venezuela. *C. coccinea* Gris. of Argentina is perhaps the largest species, though attaining a height of less than 20 feet and a diameter of eight inches. The bark is used medicinally.

The only wood samples available are of *C. Nicholsoni* from the Lesser Antilles, but the structure is similar to that of *C. coccinea,* according to O'Donell (*Lilloa* 1: 268–272). Heartwood absent or not distinct from the yellow sapwood. Luster rather high. Odorless, but with bitter taste. Hard, heavy, and strong; sp. gr. (air-dry) about 0.85; texture fine; grain variable; finishes very smoothly.

COMMON NAMES: Abrojo de costa (Cuba); rupa wit (Curaçao); chaparro amargoso (Texas); amargoso, bisbirinda, palo amargoso (Mex.); urupagüita (Venez.); molle sigle (Urug.); granadillo meloncillo, mistol de zorro, molle negro, quillai, quillay, sacha melón (Arg.).

Holacantha *Emoryi* A. Gray, the only species, is a shrub or much-branched little tree 10 feet high, with green or brownish spinose branchlets and reduced scale-like leaves, growing in the desert along the boundary of southwestern United States and Mexico. Apparently there are no uses for the plant. Heartwood absent from available material or not distinguishable from the yellow sapwood. Luster rather high. Odor absent; taste bitter. Moderately hard and heavy; texture fine.

COMMON NAMES: Crucifixion thorn (Eng.); corona de Cristo (Span.).

Picramnia, with about 40 closely related species of unarmed shrubs and small or very rarely medium-sized trees, is widely distributed in tropical and subtropical America, though apparently rare in the Brazilian Amazon region. The leaves are rather large and odd-pinnate with few to several leaflets; the flowers are small, the male in little clusters, the female in panicles; the fruits are red berries, often highly ornamental. The leaves and bark are very bitter. The only species reaching the United States is *P. pentandra* Sw., a shrub or a tree occasionally 20 feet high, with a range extending from southern Florida through the West Indies to Colombia. *P. antidesma* Sw. occurs in the West Indies, southern Mexico, and Central America. The bark was formerly exported to Europe, particularly from Jamaica, for use in medicines. The largest tree reported is *P. lineata* Macbr. which, according to Llewelyn Williams (*Woods of northeastern Peru,* p. 229), attains a height of 50 feet in the Peruvian Amazon region, with a straight trunk 7 to 15 inches in diameter; the only use for the timber is occasionally for fuel.

Heartwood brown; rather sharply demarcated from the lighter colored sapwood. Luster medium to high. Without odor, but with bitter taste. Rather heavy, hard, tough, and strong; sp. gr. (air-dry) about 0.75; texture fine; grain generally straight; not difficult to work, finishing very smoothly; durability doubtful. Of no commercial possibilities.

COMMON NAMES: Bitter bush (Florida); bitter bush, snake root (Bah.); macary bitter, majoe bitter (Jam.); aguedita, brasilete bastardo, b. falso, marigoncillo, palo amargo, guina del país (Cuba); bitter bush, guarema, hueso (P.R.); palo de pez, p. peje (Dom. R.); bois poison, vaillant garçon (Haiti); cáscara amarga, chilillo (Mex.); aceitunito, coralillo (Salv.); caregre, sartalillo (C.R.); mangle amarillo (Col.); café-rana (Braz.); sanipanga (Peru); cedrillo (Arg.).

Picrolemma, with three species of unarmed shrubs or little trees, is confined to the Amazon region of Brazil and Peru. The leaves are odd-pinnate, with numerous large leaflets; the small fragrant salmon-colored flowers are borne in axillary racemes; the fruit is a red drupe resembling coffee, hence the local name Café-rana (false coffee). The twigs have hollow cavities which are inhabited by ants. The roots contain a bitter principle and those of *P. pseudocoffea* Ducke are exported from Pará to Rio de Janeiro where they are used in local medicine as a febrifuge and tonic (see *Arch. Jard. Bot. Rio de Janeiro* 4: 196–198; 5: 144–145). Wood whitish throughout. Luster medium. Unscented, but with a very bitter taste which is sweetish at first. Rather light in weight and brittle; texture medium; grain fairly straight.

Quassia includes two species, namely, *Q. africana* Baill. of tropical West Africa, and *Q. amara* L. with a natural range extending from northern South America through the West Indies, Central America, and southern Mexico, and widely extended by planting. The plants are unarmed shrubs or small trees rarely over 25 feet high, and all parts contain a neutral bitter principle, quassin, as intense and lasting to the taste as quinine. The leaves of the *Q. amara* have a winged rachis and are odd-pinnate, with five leaflets; the rather large crimson flowers are borne in conspicuous terminal racemes; the fruit consists of five spreading black drupes about half an inch long.

The wood has been an item of commerce from Surinam since about the middle of the 18th century, being known as Surinam Quassia or Bitterwood. It was introduced into the London Pharmacopoeia in 1788, but in the edition of 1809 was superseded by Jamaica Bitterwood (*Aeschrion*), which was obtainable in larger sizes. The bitter principle of both kinds of wood is readily soluble in cold water and for many years there was some demand for cups, called bitter cups, turned from the wood. The drug is employed as a tonic and vermifuge and formerly as a febrifuge; it is used as an insecticide, particularly for spraying hops, and as a substitute for hops in ale and beer and as an ingredient of certain proprietary medicines.

Heartwood absent or not clearly differentiated from the white or yellowish sapwood. Luster rather high. Odorless, but with pronounced bitter taste. Light and soft; sp. gr. (air-dry) about 0.50; weight 31 lbs. per cu. ft.; texture fine; grain straight; very easily worked, finishing smoothly; resistant to insects but not to decay.

COMMON NAMES: Bitterwood, quassia, Surinam bitterwood, S. quassia (Eng.); lignum quassiae verum (Pharm.); Bitterholz, echtes Quassiaholz, Fliegenholz (Germ.); quasi, quasia (Cuba); cuasia, tinco (Mex.); padilla, pensilero (Salv.); guabo, hombre grande (C.R.); cruseta, guavito, g. amargo, puesilde (Pan.); acuasia, cruceto morado, cuasia (Col.); cuasia, palo Isidoro (Venez.); kabodan, kwassi, k. bita, kwassie hout (Sur.); bois amer, b. cayan, b. de frêne, b. de quassie, coachi, quina de Cayenne, quinquina de Cayenne (Fr. G., Fr. W.I.); pau de Surinam, quasia, quina (Braz.).

Recchia (or *Rigiostachys*), with two or three species of unarmed little trees usually less than 20 feet high, is apparently confined to the mountains of southwestern Mexico. The leaves are odd-pinnate, with 5 to 11 large leaflets; the fragrant yellow flowers are borne in large terminal panicles; the fruit consists of 1 to 3 drupes with thin flesh and a hard stone. *R. mexicana* Moc. & Sessé is a tree 15 to 20 feet tall in Oaxaca, where it is known as Corazón Bonito; its wood is hard and valuable, according to Standley (*Trees and shrubs of Mexico*, p. 541). The genus is not represented in the Yale wood collections.

Simaba, with about 20 species of unarmed trees, shrubs, and half shrubs, is widely distributed in tropical South America, with one species extending into Costa Rica. The leaves are mostly large and odd-pinnate, with few to numerous leaflets; the flowers are borne in small or large panicles; the fruit is a drupe. The plants have a bitter juice.

The species with the greatest range is *Simaba cedron* Planch., a slender tree rarely 25 feet high, bearing at the top a cluster of leaves which sometimes are a yard long and give the plant a palm-like appearance. It is distributed from the Amazon basin through Colombia and Panama to Costa Rica. The cotyledons of the seeds are used medicinally in the treatment of fevers and snake bites. They are known in pharmacy as "semen cedronis." Most of the members of this genus are small, but *S. paraensis* Ducke of the State of Pará, Brazil, is said to be a medium-sized to large tree; it is also unique in having ill-smelling flowers. There are at least four species in southeastern Brazil which go by the common name of Calunga; the bark and the roots are used locally as a substitute for *Quassia*. The only authentic wood sample available (Yale 21616) is of *S. guianensis* Aubl. from Rio Tapajoz, Brazil. This species is a little tree or a shrub common along clear-water streams and lakes in the Amazon basin. It has scented greenish white flowers and bright red fruits with a strongly acid juice.

Wood yellowish white throughout when fresh, turning bright yellow with a greenish tinge. Fairly lustrous. Odorless, but with bitter taste. Rather light and soft; texture medium; grain straight; very easily worked, finishing smoothly; probably resistant to insects but not to decay. Of no commercial possibilities for lumber.

COMMON NAMES: Cedrón (C.R.); amargo (Col., Venez.); cajú-rana, calumba, calunga, parahyba, mirim, pau paratudo, pitombeira, simaruba mirim (Braz.).

Simaruba (or *Simarouba*), with nine species of unarmed shrubs and small, medium-sized, or large trees, is widely distributed in tropical America. The leaves are odd-pinnate, with few to numerous leathery leaflets; the rather small flowers are borne in much-branched panicles; the typically clustered drupaceous fruits have a thin pulp and a hard stone. The bark is bitter and is used medicinally. The light easily worked timber is suitable for many of the same purposes as soft Pine and that of several species is commercially important.

The species extending farthest north is *Simaruba glauca* DC., a medium-sized to large tree growing in southern Florida, the West Indies, Yucatán, Central America, and part of South America. It is highly ornamental. The wood is used in Jamaica for making match sticks. *S. Tulae* Urb. is a Puerto Rican tree sometimes 60 feet high and 20 inches in diameter, locally known as Aceitillo, the name also given to the West Indian Satinwood, *Zanthoxylon flavum* Vahl. The frequently encountered statement that the Satinwood of Puerto Rico is *Simaruba Tulae* is erroneous.

The Pau Parahyba of the coastal forests of eastern Brazil is a well-formed timber tree believed to be *Simaruba versicolor* St. Hil. The lumber is used for making boxes and for general carpentry and interior construction. Tests at the U.S. Forest Products Laboratory on material collected by H. M. Curran near Bahia gave the following results: Sp. gr. (oven-dry) 0.39; weight (8 per cent moisture) 25 lbs. per cu. ft. *Maximum crushing strength parallel to the grain*, 4120 lbs. per sq. in. *Stiffness and strength in bending* (lbs. per sq. in.): modulus of elasticity, 1,289,900; fiber stress at elastic limit, 7720. Fabrication tests made elsewhere (see *Wood Turning* 16: 7: 17, 19; April 1923) showed that the timber is not well suited for turning, owing to "a decided tendency to twist off and chew up in the machines."

The Marupá of the Amazon region of Brazil and the Simarupa of the Guianas appear to be *Simaruba amara* Aubl. The wood is used locally for house sheathing and boxes and is said to be immune to insect attack because of its bitterness, but there is evidence that this claim is not fully justified. Small shipments of Marupá lumber have been made to New York, its principal use being for interior trim to be painted. The stock must be protected from the weather to prevent blue stain. Logs are likely to crack open in sawing, and wide boards also have a tendency to split from end to end in drying. The Conservator of Forests of British Guiana (Annual Report, 1935) says that this tendency can be reduced by water-seasoning the logs and in keeping the width of the boards to eight inches or less. "The lumber can be air-seasoned in two months. It is rather responsive to atmospheric changes." The Guiana Match Factory, Ltd., of Georgetown, reports in a letter of May 27, 1937, that Simarupa wood has not proved satisfactory for making match splints owing to its brittleness and its wasteful tendency to splinter.

Authentic samples of *S. amara, S. glauca,* and *S. versicolor* are similar in appearance, structure, and properties. Heartwood not differentiated from the nearly white or straw-colored sapwood; uniform except for occasional oily streaks in some specimens. Luster rather high. Without odor, but with mildly bitter taste. Of light weight; comparatively firm and strong, though rather brittle; sp. gr. (air-dry) 0.40 to 0.50; weight 25 to 31 lbs. per cu. ft.; texture medium and uniform; grain usually very straight; working properties generally good; holds its place well when manufactured; does not warp or check badly; easy to paint, stain, or varnish; presents a clean surface for printing on boxes; is not suitable for use in exposed situations.

Common names: Bitterwood, paradise tree (Florida); bitter damson, b. dan, damsel, mountain damson, stave wood (Jam.); gavilán, palo blanco, roblecillo, simaruba (Cuba); aceitillo (P.R.); daguillo, olivo amargo (Dom. R.); bois blanc, b. frêne, frêne (Haiti); marouba (Grenada); gall tree (Barbados); pasa-ak, xpazakil (Mex.); negrito (Br. H.); jocote de mico (Guat.); aceituno, negrito (Hond.); aceituno, jucumico (Salv.); aceituno negrito (Nic.); olivo (C.R.); aceituno, olivo (Pan.); simaruba (Col.); cedro blanco, simaruba (Venez.); adoonsidero, aruba, maruba, simaruba, simarupa (Br. G.); adonisidoro, samalombo, samboera, siemaroepa, simaropa, simiaroepa, soemaloeba, soemaroepa, somaroepa, walkara (Sur.); acajou blanc, bois blanc, simarouba (Fr. G.); caixeta, c. branca, cajú-rana, malacacheta, marubá, marupá, m. rana, m. úba, pau parahyba, pitomba, pitombeiro de Marajo (Braz.).

SOLANACEAE

The Potato family comprises about 80 genera and 2000 species of unarmed or prickly herbs, erect and climbing shrubs, and small trees, generally distributed over the earth, but most abundant in the tropics. The leaves are alternate, simple or compound, and without stipules; the flowers have a tubular to rotate corolla and are usually in cymes; the fruit is a many-seeded berry or capsule. Included in the family are such well-known and highly important plants as the potato, tomato, eggplant, and tobacco; some others are grown for ornamental purposes and many are employed in native medicine. The comparatively few trees are confined to the tropics and are negligible as a commercial source of timber, as the wood has no special properties that would justify the use of the small sizes obtainable. The following description is based upon American specimens of *Acnistus, Athenaea, Capsicum, Cestrum, Cyphomandra, Datura* (incl. *Brugmansia*), *Espadaea, Grabowskia, Tochroma, Lycianthes, Lycium, Nicotiana, Sessea, Solandra,* and *Solanum. Duckeodendron* is considered separately.

Heartwood yellow or brownish; absent from most of the samples; sometimes sharply demarcated (*e.g., Grabowski* and *Lycium*); sapwood generally white, sometimes suggesting Holly (*Ilex*), but occasionally yellow. Luster low to medium. Consistency variable from light, soft, and rather spongy to moderately hard, heavy, and compact. Texture typically fine; grain rather irregular; material easy to work; tough and strong in relation to weight; is probably perishable in contact with the ground.

Growth rings often present, mostly indicated by local differences in density; woods grown in temperate climates may be ring-porous (*e.g.,* species of *Grabowski, Lycianthes,* and *Lycium*). Pores very small and not visible without lens to fairly large and distinct; rather few to numerous; in diffuse-porous woods, often in small clusters or short to rather long radial multiples which are well distributed without definite pattern, though occasionally tending to the formation of tangential rows; in ring-porous woods, the early-wood pores small, arranged in irregular bands or sometimes in broken rows, the late-wood pores very small to minute, and in *Grabowskia* and *Lycium* arranged in patches and festoons resembling parenchyma. Vessels with simple perforations; spiral thickenings generally absent except in ring-porous species;

course of vessels often very irregular (tangential section), suggesting Capparidaceae; tyloses absent or sparingly developed; intervascular pitting alternate, medium to very coarse. Rays greatly variable in size and appearance; sometimes all uniseriate and biseriate and less than 25 cells high, but more often of two sizes, the larger ones 3 to 5, occasionally 6 to 8, cells wide and up to 50, rarely more, cells high; heterogeneous, often conspicuously so; in softer specimens, cells frequently large and very irregular in form (tangential section); sheath cells sometimes present; pits to vessels medium-sized to large, rounded and alternate to elongated and in scalariform arrangement. Wood parenchyma usually sparingly developed, not visible with lens; vasicentric and diffuse; finely reticulate in *Espadaea;* greatly enlarged crystalliferous cells present in *Grabowskia.* Wood fibers generally with thin to moderately thick walls, but occasionally (*e.g., Grabowskia*) very thick-walled; sometimes septate; pits very small, simple or inconspicuously bordered. Ripple marks and gum ducts absent. Large open radial channels sometimes present (*e.g., Acnistus, Cestrum, Datura*). Pith often large and chambered.

Duckeodendron. The taxonomic position of this genus is uncertain. The single known species, *D. cestroides* Kuhlmann, is a large tree, sometimes 100 feet tall, discovered by Dr. Adolpho Ducke in 1923 along the Tapajoz River, Brazil, where it is known as Pupunha-rana. Owing to the absence of mature fruits and to the resemblance of the flowers to those of *Cestrum,* Kuhlmann referred the genus to the family Solanaceae, but on the basis of later collections transferred it to the Boraginaceae, and then after further study returned it to the Solanaceae, setting up for it a new tribe, the Duckeodendrineae, and stating that the only alternative is the creation of a new independent family, the Duckeodendraceae (see *Tropical Woods* 33: 7 and 38: 47). The timber is of good quality and suitable for the same uses as Yellow Poplar (*Liriodendron*) in the United States.

Wood uniform light clear yellow throughout, becoming brownish superficially upon exposure. Luster rather high. Without distinctive odor and taste, at least when dry. Of rather light weight but firm; texture uniform and rather fine; grain straight; very easy to work, finishing smoothly; holds its place well when manufactured; probably not highly resistant to decay. Similar in texture and properties to Marupá (*Simaruba*) and appears suited to the same purposes.

Growth rings, when present, limited by narrow band of parenchyma. Pores two-sized, the smallest often in irregular groups associated with the larger ones (150 to 350μ) which are mostly in radial series or rows with the interior pores not much flattened. Vessels with simple perforations, the rims distinct; without spiral thickenings; tyloses absent; pits rather small, alternate, the borders circular, the apertures included. Rays uniseriate or locally biseriate and up to 30 cells high; homogeneous or nearly so, the cells large, thin-walled, radially elongated; pits to vessels rather small, rounded, numerous; pits to fibers of same size and superficial appearance as those from one fiber to another; pits to wood parenchyma cells small, variable in form; no crystals observed. Wood parenchyma coarse-celled; in finely wavy, faintly visible, concentric bands 1 to 3 cells wide, the spacing variable but roughly equal to the width of one of the larger pores; a few cells in association with pores and pore groups, but not surrounding them; no crystals seen. Wood fibers rather thin-walled, non-septate; pits mostly in radial walls, fairly numerous, the borders circular and distinct, the apertures slit-like and slightly extended. Widely separated radial intercellular channels present. Ripple marks absent. The wood resembles some of the Apocynaceae (*e.g., Couma*) much more than it does any of the Solanaceae examined.

STAPHYLEACEAE

The Bladdernut family, with three genera and about 25 species of shrubs and small trees, occurs in eastern Asia and the Malay Archipelago and in North and South America. The leaves are opposite, compound, 1–9-foliolate; stipules and stipels are usually present; the small white perfect flowers are borne in terminal racemes or panicles; the fruit is a membranous inflated capsule or a drupe. The two genera represented in America are *Staphylea* and *Turpinia,* neither of any importance for timber.

Wood yellow (*Staphylea*) or light chocolate-brown throughout. Fairly lustrous. Sometimes with mild but indescribable scent and taste. Of rather light weight to medium density; firm and strong; texture fine to medium (*Turpinia*); grain mostly straight; easy to work, finishing smoothly; not highly durable. Of no commercial possibilities.

Growth rings usually present. Pores medium-sized to minute, the largest (*Turpinia*) near limit of vision; numerous; well distributed, occurring singly (*Staphylea*) or in short radial multiples or clusters (*Turpinia*). Vessels with many-barred scalariform perforation plates; spirals rarely present; pitting opposite, with tendency to scalariform. Rays decidedly heterogeneous; of two distinct sizes, the uniseriates with all cells upright or square, the multiseriates up to 6, sometimes to 10, cells wide (variable in different species) and up to 50, sometimes to 80, cells high, the body cells narrow and procumbent, the marginal ones square or upright and usually in only a few rows; pits to vessels oval to elongated, mostly opposite, those in *Turpinia* considerably larger than those in *Staphylea*. Wood parenchyma limited to a few cells about the vessels and sparingly diffuse; not distinct with lens. Wood fibers with medium-thick walls; pits bordered, numerous and conspicuous in both radial and tangential walls, especially in *Turpinia*. Ripple marks and gum ducts absent.

Staphylea, with about 13 species, mostly shrubs but sometimes trees 25 feet high, is distributed from middle Europe to the Far East and in North America from the Atlantic states and northern California through Mexico into Central America. The plants have attractive flowers and bladder-like seed pods, and are often cultivated for ornament. The only wood specimens available are of *Staphylea trifolia* L., the Bladdernut, a slender shrub with smooth striped bark, occurring throughout most of the eastern half of the United States and adjacent parts of Canada. Heartwood absent or not distinguishable from the pale yellow sapwood. Luster medium. Rather hard and heavy; texture fine; grain straight; easy to work, finishing very smoothly; presumably perishable when exposed to decay. Suitable for small articles of turnery.

Turpinia, with about a dozen species, inhabits tropical regions in the Far East and the West Indies, Mexico, Central America, and in South America to Ecuador and northern Peru. The largest and most widely distributed species is *T. paniculata* Vent., a tree often 25 to 40, occasionally up to 65, feet high and 12 to 20 inches in diameter. The timber is used to a very limited extent locally for general carpentry and construction, but has no commercial possibilities. The following description is based upon one sample each of *Turpinia carnosa* Spruce from Ecuador, *T. heterophylla* (R. & P.) Harms & Loes. from northeastern Peru, and *T. paniculata* Vent. from Cuba. Heartwood lustrous brown, suggesting Walnut; rather sharply demarcated from the lighter-colored sapwood. (The specimen of *T. heterophylla* is all sapwood and yellowish.) Rather light in weight, but firm and tenacious; texture medium fine; grain fairly straight; very easy to work, finishing smoothly; durability probably low.

Common names: Cassada wood, ironwood, mutton wood, wild cassada (Jam.); saúco cimarrón, serrucho (Cuba); capulín cimarrón (Mex.); avispillo, cedro hembra, saúco cimarrón (P.R.); cedrillo (Pan.); yana-mullaca (Peru).

STERCULIACEAE

This family comprises about 50 genera and over 700 species of herbs, upright or scandent shrubs, and small to large trees and is widely distributed throughout the tropics and warm regions of the world. The leaves are alternate, simple or lobed, stipulate, and often stellate-hairy; the flowers are panicled or cymose; the fruit is dry or fleshy, dehiscent or separating into follicles. The only products of great importance are chocolate and cocoa obtained from the seeds of the Cacao tree, *Theobroma cacao* L. The only timbers of more than local utility at present are the Abachi, Ayous, or Samba, *Triplochiton scleroxylon* K. Schum. (See *Tropical Woods* 18: 43 and 25: 3), and Mansonia, Pruno, or Apruno, *Mansonia altissima* A. Chev. (see *Tropical Woods* 35: 2 and 46: 63), both of West

Africa and considered by some botanists to constitute a separate family, the Triplochitonaceae. There are 13 genera containing woody plants in America, but the only really arborescent species are of six genera, namely, *Basiloxylon, Chiranthodendron, Guazuma, Fremontia* (*Fremontodendron*), *Sterculia,* and *Theobroma.* The following description is based upon specimens of all of them except *Chiranthodendron. Buettneria, Helicteres,* and *Waltheria* are treated separately.

Heartwood pinkish, pale brown, or, less commonly, dark reddish brown; sapwood yellowish or oatmeal-colored; quarter-sawed lumber usually with prominent figure suggesting Sycamore (*Platanus*). Luster low to fairly high. Without distinctive odor or taste. Woods variable from very light to very heavy, mostly of medium density; texture medium to very coarse; feel rather harsh; grain fairly straight; working properties variable from excellent to poor; durability generally low.

Growth rings often present. Pores rather large and distinct in *Basiloxylon* and *Sterculia,* barely visible to minute in the others, being distinctly 2-sized in *Fremontia,* which is more or less clearly ring-porous; in diffuse-porous woods, pores rather few to fairly numerous, occurring singly and in short radial multiples or, less often, in clusters, well distributed; latewood pores in *Fremontia* in diagonal or zigzag pattern. Vessels with simple perforations; without spiral thickenings, except in *Fremontia;* pits alternate, medium-sized (*Sterculia* and *Theobroma*), small (*Basiloxylon* and *Fremontia*), or minute (*Guazuma*). Rays heterogeneous; 2-sized, the largest observed being 6 by 75 cells in *Guazuma,* 10 by 100 cells in *Fremontia,* 10 by 200 in *Basiloxylon,* and 20 by over 300 in *Sterculia* and *Theobroma,* with much variation in different species and even in different parts of the same specimen; sheath cells usually abundant in *Basiloxylon* and *Sterculia,* sparingly developed or absent in the others; tile cells of about the same height as the procumbent cells characteristic of *Guazuma* and *Buettneria,* but absent from the others; rhombohedral crystals of calcium oxalate common, not imbedded; pits to vessels all minute in *Guazuma,* small and oval to lenticular and sometimes rather large and elongated in the others. Wood parenchyma abundant; in *Basiloxylon* and *Sterculia,* coarsely paratracheal and confluent into irregular, unevenly spaced, more or less broken concentric bands; in the others, finely reticulate, not visible without lens and not always distinct with it; number of cells per strand variable from 1 to 8, but generally 4, the cells usually not in horizontal seriation; crystals frequently present, mostly solitary and apparently free in ordinary cells. Wood fibers with thin to very thick walls; pits very small and numerous, simple or with vestigial borders. Ripple marks present, though often irregular and indistinct, in *Basiloxylon, Buettneria, Guazuma, Sterculia, Waltheria,* and some species of *Theobroma;* large rays not storied or occupying 2 to 8 tiers. Vertical traumatic gum ducts observed in *Basiloxylon, Fremontia, Sterculia,* and *Theobroma.* Bark finely laminated and containing V-shaped patches.

Basiloxylon *brasiliensis* (Fr. Allem.) K. Schum. is a tall tree of southeastern Brazil, where it is known as Farinha Seca. The leaves are simple and heart-shaped and the fruit is a tardily dehiscent capsule containing numerous seeds having thick, corky wings. The timber is of fairly good quality and finds numerous local uses, but is too scarce for export. The only specimen available (Yale 36073) is from Espirito Santo. Wood pale brownish throughout, with conspicuous ray flecks on radial surface. Luster silvery in proper lighting. Odorless and tasteless. Of medium density, but hard, tough, and strong; sp. gr. (air-dry) 0.74; weight 46 lbs. per cu. ft.; texture coarse; feel harsh; grain straight; easily worked, but difficult to finish smoothly; is probably perishable in contact with the soil.

Buettneria (or *Byttneria*), with about 60 species of shrubs, vines, and a few trees, occurs in the tropics of both hemispheres, but is most abundantly represented in the New World. Many of the plants are armed with prickles. The following description is of a single specimen (Yale 32794) of *B. geminifolia* Turcz., a spiny shrub collected by A. Rimbach at an elevation of about 8000 feet near Quito, Ecuador. Wood yellow throughout. Luster medium. Odorless and tasteless. Hard and rather heavy; texture fine and uniform; grain fairly straight; very easily worked and suitable for lower grades of Boxwood if available in large enough sizes.

COMMON NAMES: Arrendador, bejuco cenizo, tezak, varilla prieta, xtexak, zarza (Mex.); zarza, z. hueca (Hond., Salv.); uña de gato (C.R.); rabo de iguana, rangay, zarza (Pan.); zarza, z. hueco (Col., Venez.); wajamakándekele (Sur.).

Chiranthodendron *pentadactylon* Larr., the sole species, is a large tree in the forests of southern Mexico and Central America, but no wood specimens are available for study. Standley says of it (*Trees and shrubs of Mexico,* p. 796): "The Hand-flower tree is one of the most celebrated of Mexican plants and was well known to the early inhabitants. It is restricted in its distribution and for a long time the only tree known to the residents of the Valley of Mexico was one growing at Toluca. Even long after the Conquest this was believed to be the only tree of its kind, except for its progeny planted elsewhere in the valley. The stamens bear a striking resemblance to a hand and wrist with outspread fingers, and because of the remarkable form of the flowers the tree was viewed with veneration." The same writer (*Tropical Woods* 67: 16) says: "In Guatemala the Hand-tree, far from being rare, dominates the humid forest on some of the volcanoes. Particularly on Acatenango there is a broad belt of trees below the Pine forest of the highest slopes. Many of the trees are real giants, with massive trunks crowded together. Seedlings are scarce in spite of the abundance of fallen seed pods everywhere in the forests. The tree is widely dispersed from Sacatepéquez westward along the volcanoes, and grows here and there slightly north of the southern cordillera. It is believed that it had some religious significance among the ancient inhabitants, and today when land is cleared these trees are often left standing."

COMMON NAMES: Árbol de las manitas, macpalxochicuahuitl, manita de león, mano de león, teyaqua (Mex.); kanak mano de mico, palo de tayuyo (Guat.).

Fremontia *californica* Torr., the only species, is a low intricately branched shrub, often forming thickets in the region of the Mojave Desert, but becoming a tree 20 to 30 feet high with a trunk 12 to 14 inches in diameter in the western foothills of the Sierra Nevada Mountains of California. The range of the species includes the mountains and foothills of Baja California, Mexico. Heartwood attractively colored and figured, being reddish to olive-brown, variegated; sapwood sharply defined. Rather hard and heavy; sp. gr. (air-dry) about 0.75; of medium texture; easy to work, but requiring care in seasoning; suitable for small cabinet work and turnery.

COMMON NAMES: Flannel bush, mountain leatherwood, slippery elm (Calif.).

Guazuma, with five species, is of general distribution in Latin America. Best known and of widest range is *G. ulmifolia* Lam., usually a small to medium-sized tree, occasionally over 65 feet high and 24 inches in diameter, occurring in the West Indies, throughout most of Mexico, all of Central America and South America to Argentina and southern Brazil. The bast fiber of the young stems is employed to a limited extent in making rope. The pinkish wood varies from rather light to moderately heavy, but it is firm and strong and used locally for general carpentry and interior construction, slack cooperage, boxes and crates, tool handles, and for fuel and fine charcoal. The woods of all species are much alike and are readily distinguished from the other American Sterculeaceae, except *Buettneria,* by the tile cells in the rays and the extremely fine pitting of the vessels and the rays.

COMMON NAMES: Guácima, guácimo (Span. Am., gen.); bastard cedar, West Indian elm (B.W.I.); jackocalalu (St. Thom.); guácimo baba (Cuba); bois d'orme (Haiti); bois puant, orme des Antilles (Mart.); bois d'orme, b. zombre (Grenada); aquiche, bulines, cabal-pixoy, cuahulote, cuaulote, majahua de toro, palote negro, pixoi, pixoy, tablote, vácima, yaco granadillo (Mex.); bastard cedar, bay cedar (Br. H.); cablote, caulote (Guat.); caulote (Hond.); caca de mico, chicharrón, caulote, tapaculo (Salv.); diankrá, kudzir, serúru, shumgín, sungí, sun-

Photo by Llewelyn Williams.

PLATE XXXVI. A large Camoruco tree (*Sterculia apetala*) near Maracay, Aragua, Venezuela.

gín, surruru, udzir (C.R.); bastard cedar, guácimo de ternero (Pan.); guácimo colorado (Col.); guácimo blanco, g. dulce, g. macho (Venez.); wonan (Br. G.); anhuiba, cèdre, c. jaune, mahot baba (Fr. G.); ibixuma, motamba, mutamba (Braz.); bolaina, iumanasi, lluicho-vainilla, papayillo (Peru); cambá-acá, cambeza de negro, inga-hú, i. negro, marmelero (Arg.).

Helicteres, with about 45 species of shrubs and a few little trees, is widely distributed in the tropics except in Africa. The most distinguishing feature is the fruit, which consists of five woody spirally twisted follicles. The fiber of the bark is noted for its strength. The pale brownish, hard and heavy, fine-textured wood is not utilized.

COMMON NAMES: Blind-eye bush, cow bush, salz bush, wild salve (Bah.); screw tree (Jam.); cuernecillo, gato, huevo de gato (P.R.); majagüilla de costa, tapaculo (Cuba); jeucon (Haiti); capitanejo, guacimilla, majagüilla, tornillo (Mex.); monecillo, palo de capulín (Guat.); barreno, tornillo (Salv.); cola de chancho (Nic.); rabo de puerco (C.R.); guacimillo, guácimo torcido, torcidillo (Pan.); tornillo (Venez.); caá-juccara, malva branca, m. b. sedora, m. caá-jussara, rosca, saca-hapo (Braz.).

Sterculia, with about 60 species of trees and shrubs, is of general occurrence in tropical and subtropical regions, especially in the Far East. Some of the trees are valuable as a source of strong and coarse bast fiber used for tying tobacco and for making rope and bagging, and some species are planted for decorative purposes. The American species are comparatively few and their timbers are of three general types. The wood of *S. apetala* (Jacq.) Karst., a large widely distributed tree with great maple-like leaves, is generally very light, soft, spongy, coarse-textured, and fibrous, though occasionally of a little better quality; rays sometimes 20 cells wide and very high. It is from the Indian name for this tree that the Republic of Panama derives its name. *S. Recordiana* Standl., a rather large tree of Panama, has very coarse-textured and fibrous wood which, though light in weight, is comparatively hard, tough, and strong, as the fibers are rather thick-walled. *S. pruriens* (Aubl.) Schum. of the Guianas and northern Brazil has timber of fairly good quality, being fairly lustrous, of medium density, not very fibrous, easy to work, and finishing smoothly.

COMMON NAMES: Anacagüita guana (Cuba); anacagüite (P.R.); bastard mahoe (Jam.); mahoe (Trin.); bois de caca, mahot-cochon (Fr. W.I.); bellota (Mex.); castaño (Cent. Am., gen.); odobacri, panamá (Pan.); camajonduro, camajorú, camajurú (Col.); cacaguillo, cacagüito, cacaito, camoruco, cumaruco yaguero (Venez.); maho, manmaho (Br. G.); bofrohoedoe, jahoballi, kobèhè, koebèhè, koejethie, kroekroe-amète, maipjoerie-ietararè, manneko, patoelapo (Sur.); mahot cochon, tourou-tourou (Fr. G.); cacao blanco, c. común, c. de monte (Ec.); zapote silvestre (Peru); chichá, c.-caá, manduvi-guazú (Par.); axixa, boia, bracaiá, chicha, c. brava, capote, envireira, imbira quiaba, pau rei, tacacazeiro, tuxapá, unha d'anta (Braz.).

Theobroma, with about 15 species of small trees rarely more than 35 feet tall, is of wide distribution in tropical America, except the West Indies. *T. cacao* L., with a natural range from southern Mexico to Brazil, is the principal source of the chocolate and cocoa of commerce, though other species also are planted, and there are numerous minor varieties, differing in the form of the fruit. Standley says (*Trees and shrubs of Mexico*, p. 805, *et seq.*): "Of all the numerous vegetable productions of the New World and especially of Mexico, none is more celebrated than the product of the Cacao plant, which attained high favor in Europe immediately after the Conquest. . . . Cacao in its wild state was well known to the early inhabitants of Mexico, and it was also cultivated, to what extent is uncertain. The use of the drink made from the seeds was confined chiefly to the higher classes, but the use of the seeds as money was a matter of importance to all classes, for Cacao seeds were the basis of the Mexican financial system.

. . . At an early date Cacao was introduced into the Canaries and Philippines, and now is grown extensively in many parts of the Old World tropics. . . . The young trees must be grown under shade. Chocolate is the term applied to sweetened preparations of the roasted and ground Cacao seeds, with a large proportion of the original fat retained. Cocoa is prepared in the same way, but most of the fat is removed from it. . . . Cacao seeds contain 45 to 50 per cent of oil, or cocoa butter, which is much used in pharmacy in making ointments. They also contain an alkaloid, theobromine." *Theobroma* is not a source of commercial timber, although the wood is suitable for purposes requiring toughness and strength rather than attractive appearance or resistance to decay. Most specimens are sapwood, white or oatmeal-colored; the heartwood is brown to dark reddish brown.

COMMON NAMES: Cacao (Lat. Am., gen.); balamte, bizoya, cacahuacuahuitl, cacahoatl, cacahuatl, cacao blanco, cacaotlquahuitl, cacauquauitl, cahequa, caocauatzaua, chudechu, deghy, kako, pataste, patatle, pataxte, yagabizoya (Mex.); cucu, wild cacao (Br. H.); caco, cucuh, kicob, kicou, pataxte (Guat.); cushta (Salv.); bik, cacao, calabacillo, c. de ardillo, c. de mico, c. pataste, c. silvestre, carvu, dzugmanguá, erefa, kaokrá, kau, kaxutsia, ko, kráaku, kugín, nunisup, pataiste, pataste, saparon, scarvo, skarub, soró, teta negra, tsirú, tsirukurú, uérba, uirub (C.R.); cacao cimarrón, c. mani, wild cacao (Pan.); bacao (Col.); cacao de monte (Ec.); cacahuillo, cacao senisa, c. silvestre, cumala, cupuassú, macambo, majambo, uchpacacao (Peru); cabeça de urubú, cacao azul, c. quadrado, c. rana, cacao-u, cacao-y, cupú, cupuahy, cupuassú, cupuhy (Braz.).

Waltheria, with about 60 species of herbs, shrubs, and a few little trees, is widely distributed throughout warm regions and is most common in tropical America. The plants are used to some extent as a source of cordage and native medicines. The wood of *W. americana* L., a pantropical velvety mucilaginous shrub, is yellowish, of medium density, and rather fine-textured.

COMMON NAMES: Basora prieta, malvavisco (P.R.); malva blanca (Cuba); friega-plato (Dom. R.); hierba del soldado, malva, m. de monte, zacxiu (Mex.); escobilla, hierba del buey (Salv.); escobilla blanca (C.R.); bretónica macho (Venez.); kamferblad, malva, marfa (Sur.); malva branca, m. velludo (Braz.).

STYRACACEAE

THE Storax family, with seven or eight genera and over 100 species of shrubs and small to rather large trees, is represented in the Mediterranean region, eastern Asia and the Malay Archipelago, and North and South America. The leaves are alternate simple and often with stellate pubescence; stipules are absent; the flowers are in axillary or terminal racemes; the fruit is a drupe or a capsule, the calyx persistent. The timber is of little importance, although it is suitable for general carpentry and miscellaneous purposes requiring plain ordinary lumber. The principal products are aromatic resinous balsams used for perfume, incense, and medicine. The American species are of three genera, namely, *Halesia, Pamphilia,* and *Styrax.*

Heartwood brownish, sometimes with a reddish tinge; sapwood white. Luster mostly low. Without distinctive odor or taste. Rather light in weight, but firm, tough, and strong; texture fine and uniform; grain fairly straight; easy to work, finishing smoothly; not very resistant to decay.

Growth rings present or absent. Pores mostly small; numerous but not crowded; usually in multiples of 2 to 5, well distributed. Vessels with scalariform perforation plates; no spiral thickenings seen; pitting typically very fine. Rays mostly 2-sized, the larger 2 to 4, occasionally up to 7, cells wide; decidedly heterogeneous; pits to vessels very small. Wood parenchyma finely reticulate. Wood fibers with bordered pits. Ripple marks absent. Vertical traumatic gum ducts observed in *Styrax.*

Halesia (or *Mohrodendron*), with a few species of trees and shrubs, inhabits eastern China and the southeastern quarter of the

United States. Because of their small drooping white flowers the usual American name for the plants is Silverbell; some of the forms are cultivated for ornamental purposes; the fruit is a drupe with 2 to 4 thin lateral wings. The best known species is *H. carolina* L., with a range from the mountains of southern West Virginia to western Florida, western Kentucky, southern Illinois, central Alabama, and Georgia, and cultivated in eastern United States (to Massachusetts), California, and western and central Europe. It is often a shrub with wide-spreading stems, but sometimes a tree up to 40 feet in height and 18 inches in diameter. *H. diptera* Ellis, with 2-winged fruits, is a shrub or slender tree up to 30 feet high, growing in lowlands in the Gulf states. The largest tree is *H. monticola* Sarg., perhaps only a form of *H. carolina,* which attains a maximum height of 90 feet with a well-formed trunk 36 inches in diameter in the mountains of western North Carolina and eastern Tennessee. The timber of *Halesia* is of no commercial importance because of its scarcity. Heartwood pale brown; sharply demarcated from the wide white sapwood. Luster medium. Odorless and tasteless. Rather light in weight, but firm and tough; texture fine and uniform; grain fairly straight; working properties good; not highly resistant to decay.

COMMON NAMES: Bell tree, bellwood, boxelder, cowlicks, possumwood, silverbell, silver tree, snowdrop tree (U.S.A.).

Pamphilia, with two or three closely related species of shrubs and small trees, occurs in southeastern Brazil. The trees are called Benjoero and they yield tears of aromatic resin known locally as "incenso da America" and used for incense and for medicinal purposes. The wood is similar to that of *Styrax.*

Styrax, with about 100 species, has a wide range in warm and tropical countries, except in tropical and South Africa and in Australasia. *S. officinale* L., of southern Europe and Asia Minor, is the source of storax, and *S. benzoin* Dryand., of Malaysia, supplies benzoin. There are four species (one, *S. grandiflora* Ait., sometimes 40 feet high) in southern United States, nine in Mexico, three in Central America, a few in the West Indies, one (*S. guianensis* A. DC.) in the Amazon basin, and many below the Amazon region to Argentina. The resin obtained from the heartwood of several species is used as a substitute for storax and benzoin. The lumber is employed locally to a limited extent, but there are no special uses.

Heartwood pale brown with a reddish tinge; sapwood white. Luster low. Without distinctive odor and taste. Rather soft to moderately hard; of about the consistency of Red Gum (*Liquidambar*); sp. gr. (air-dry) 0.55 to 0.65; weight 33 to 40 lbs. per cu. ft.; fine-textured; easy to work; not highly durable. Of no possibilities for export.

COMMON NAMES: Mock orange, snowbell (U.S.A.); aceituno silvestre, azulejo de loma (Cuba); azahar de monte, capulín, chilacuate, hoja de jabón, levadura, ruín (Mex.); bracino, estoraque, quiquicirrí, resino (C.R.); sahumerio (Pan.); estoraque (Col.); estoraque, historaque, olivo (Venez.); almiscar, arvore do balsamo, beijoeiro, benjoeiro, canella póca, carne de vacca, cuia de brejo, estoraque do campo, e. da lisa, e. do matto, estoraqueiro, limoeiro do campo, pau de remo, pindahyba, pindaubuna, pindauvuna, pinduiba, pororoca (Braz.); aguay-guazú (Par.); carne de vaca, cohy, ibirá-cuaté, María molle (Arg.); utsupa cacao (Peru).

SURIANACEAE

Suriana *maritima* L., the only genus and species, is commonly included with the Simarubaceae. It is a shrub or occasionally a small tree up to 25 feet high and 10 inches in diameter, of rather infrequent occurrence along the seacoast throughout the tropics. The narrow simple alternate leaves are densely clustered on the velvety branches and hide the small yellow flowers; the achene-like little fruits have a thick embryo shaped like a horse-shoe. Apparently there are no special uses for any part of the plant.

Heartwood dark red or reddish brown,

merging into the lighter-colored sapwood. Luster rather low. Without distinctive odor or taste. Hard, heavy, tough, and strong; texture fine and uniform; grain somewhat irregular; not difficult to work, finishing very smoothly; appears durable. Suitable for small articles of turnery, but too small and scarce to be of any importance.

Growth rings visible, but poorly defined. Pores small to minute, not visible without lens; rather numerous; solitary and, more often, in multiples of 2 to 8, or sometimes in small clusters, well distributed. Vessels with simple perforations; intervascular pitting very fine. Rays uniseriate or locally biseriate; mostly only a few cells high, but occasionally vertically fused; decidedly heterogeneous, nearly all of the cells being square or upright; gum abundant; pits to vessels very small. Wood parenchyma narrowly vasicentric and diffuse; scarcely visible with lens. Wood fibers rather thick-walled; usually with gummy contents; pits numerous, very small, simple or indistinctly bordered. Ripple marks present; fairly regular; not distinct without lens; 100 to 130 per inch; all elements storied, though fused rays may occupy 2 to 5 tiers. No gum ducts seen.

COMMON NAMES: Bay cedar (Florida); bay cedar, tassel plant (Bah.); cuabilla, c. de costa (Cuba); giterón, temporana (P.R.); jobero (Dom. R.); crisse marine (Haiti); palo corra (Curaçao); pantsil, pantzil (Mex.); cucharo (Venez.).

SYMPLOCACEAE

Symplocos, the only genus, includes about 300 species of shrubs and small to medium-sized or rarely large trees, which inhabit the warmer parts of Asia, Australia, and America. The leaves are alternate simple and without stipules; the flowers are borne in dense or lax axillary spikes or racemes; the fruit is a drupe or a berry.

Of the New World species, the most northern is *Symplocos tinctoria* (L.) L'Hér., a slender tree in the undergrowth of high forests in southeastern United States. The leaves have a sweet taste in the autumn and are relished by browsing animals, on which account the tree is called Sweetleaf and Horse Sugar. There are several species in the West Indies, the most widely distributed being *S. martinicensis* Jacq., whose wood, though of good quality, is of very limited utility because of the scarcity and small size of the trees. There are nine species in Mexico and a few in Central America, the largest being *S. chiriquensis* Pittier of the middle belt of Panama, where it is said to attain a height of 90 feet and a trunk diameter of 24 inches; the wood is not used except for fuel. The numerous Brazilian species are of little worth for timber, but the leaves of several are used as an adulterant of mate (*Ilex*) for making tea, and the bark and leaves of some are the source of yellow dyes.

Wood white, yellowish, or brownish, without sharp distinction between heartwood and sapwood. Not highly lustrous. Odorless and tasteless. Consistency and density variable, but mostly like that of Birch (*Betula*); fine-textured; usually straight-grained; easy to work; not durable.

Growth rings present or absent. Pores small to very small; numerous; infrequently in direct contact radially; fairly well to irregularly distributed, without definite pattern. Vessels with many-barred scalariform perforation plates; spiral thickenings sometimes present; pitting, when present, mostly scalariform. Rays 1 to 5 cells wide, few to 50 cells high; heterogeneous; ray-vessel pitting distinctly scalariform. Wood parenchyma diffuse and sometimes finely reticulate; not distinct with lens. Wood fibers with medium walls and numerous distinctly bordered pits; fine spiral thickenings occasionally present. Ripple marks absent. No gum ducts seen. (*Symplocos tinctoria* is exceptional in having smaller pores and finer rays than other species, the vessel and ray pits are small and circular, and the rays are nearly homogeneous.)

COMMON NAMES: Horse sugar, sweetleaf, yellow wood (U.S.A.); azulejo del Pinar, a. de Rosario, a. de sábana, piñipiñi (Cuba); aceituna, a. blanca, a. cimarrona, níspero cimarrón, palo de cebra (P.R.); bois graine bleue, b. Martinique (Dom.); garrapata, garrapatilla, limoncillo (Mex.); chillador, florecilla (Salv.); amarellinho, caá-apoam, caúna, cipónima, congonha de caixeta, c. grande, herva caúna (Braz.); siete sangrías (Arg.).

THEACEAE

The Tea family, also known as the Ternstroemiaceae and Camelliaceae, consists of shrubs and small to large trees of wide distribution in the tropics, particularly in the Malayan Archipelago and Latin America, and to a limited extent in China, Japan, and the United States. The leaves are typically alternate, simple, entire or serrate, evergreen in the tropical species and clustered at the ends of the branches; stipules are absent; the white or pinkish flowers are mostly solitary or in axillary clusters and frequently are showy and highly scented, thus making the plants desirable for cultivation in gardens; the fruit is either capsular and dehiscent or leathery or woody and indehiscent; in some genera the seeds are winged, in some others they are shaped like a horseshoe. The outstanding member of the family is the Tea plant (*Camellia sinensis* L. or *Thea sinensis* [L.] O. Ktze.). Several of the Asiatic trees supply useful timber for local uses and *Laplacea Brenesii* Standl. is a source of structural lumber in Costa Rica.

According to the classification of Melchior (*Pflanzenfamilien*, 2nd ed., 21: 109–154), the family is divisible into five tribes containing a total of 23 genera and about 380 species. In the New World there are representatives of 10 of the genera, of which three (*Franklinia, Gordonia,* and *Stewartia*) are limited to the southeastern United States. The tropical American species are of the following genera: *Archytaea, Bonnetia, Eurya, Laplacea, Patascoya, Pelliciera,* and *Ternstroemia*. The following description includes all of them but *Patascoya* and *Pelliciera*.

Heartwood light to dark brown or red, fairly uniform to more or less variegated, but typically dull and unattractive; usually merging gradually into the brownish or pinkish sapwood. Without distinctive odor or taste. Density variable, some specimens hard and heavy, others (even in the same genus) of comparatively low density, but firm and tough; consistency about that of Red Gum (*Liquidambar*); texture uniform, fine to medium; grain variable, sometimes very irregular; working properties good, though some specimens show a tendency to warp badly in drying; durability low to fairly good. Commercial possibilities absent or poor.

Growth rings present or absent; often poorly defined. Pores very small to medium-sized, not distinct without lens; typically numerous, sometimes crowded, but rarely in contact radially; distribution uniform and without pattern, though the pores may be somewhat larger in the early wood of trees growing in temperate regions. Vessels with simple perforations only in *Archytaea* and *Bonnetia;* otherwise with scalariform perforation plates having 15 to 100 bars (bordered only at the ends, if at all) crowded together or rather widely spaced, sometimes anastomosing; spiral thickenings absent except in some of the overlapping tips of the members in *Franklinia, Gordonia,* and *Stewartia;* tyloses sometimes present. Rays all uniseriate or locally biseriate and less than 30 cells high in *Franklinia* and *Gordonia;* 2-sized in the others, the larger usually 2 or 3, occasionally 4, cells wide and less than 50 cells high, except in *Ternstroemia,* where they are frequently 4 to 6, sometimes to 8, cells wide and up to 100, rarely to 200, cells high; heterogeneous, often with most of the cells square or upright; crystals sometimes present; cells often very thick-walled in part and abundantly pitted; pits to vessels small to very large and irregular, frequently elongated and in scalariform arrangement. Wood parenchyma sparse to fairly abundant, not visible without lens; mostly diffuse to finely reticulate; crystals occasionally present. Wood fibers with thick to very thick walls and minute to moderately large lumina; pits numerous in both radial and tangential walls, the apertures lenticular or slit-like, the borders large and circular. Ripple marks absent. No gum ducts seen. For anatomy of the different genera see *Tropical Woods* 70: 25–32.

Archytaea, with two species of little trees or shrubs, occurs along river banks in northern Brazil and the Guiana and Venezuelan hinterlands. The leathery leaves are clustered near the ends of the branches; the reddish flowers are borne in axillary clusters; the fruit is a 5-celled septicidally dehiscent capsule. The following description is based on a small specimen of *A. multiflora* Benth. (Yale 40408; Pinkus 48) collected at an altitude of 4200 feet in the Mt. Roraima district of Venezuela. Color (of sapwood) pale brown. Luster medium. Odorless and tasteless. Hard, heavy, and strong; texture fine and uniform; grain straight; durability unknown. Of no commercial possibilities.

COMMON NAME: Hitchiaballi (Br. G.).

Bonnetia, with nine species of shrubs and small trees, occurs along the seacoast and river banks in South America, particularly eastern Brazil and the Amazon basin. The leathery leaves are entire and finely feather-veined; the scented roseate flowers are borne singly or in short racemes in the leaf axils; the fruit is a 3-celled, septicidally dehiscent capsule; the seeds are slender and winged. Heartwood red; sapwood pinkish, sometimes with sulphur-yellow coloration. Luster low. Odorless and tasteless. Mostly hard, heavy, and strong; texture fine and uniform; grain fairly straight; not difficult to work, finishing smoothly; appears durable. Presumably without commercial possibilities.

COMMON NAMES: Cascarilla, c. legitimo (Peru).

Eurya, with about 80 species of shrubs and trees, is widely distributed throughout the Asiatic and American tropics. Four subgenera are recognized and some botanists have given them generic rank. The two with American species are *Cleyera* and *Freziera.*

Of the first, the only one represented in the Yale collections, is *E. theoides* (Sw.) Blume, a tree occurring in the West Indies, southern Mexico, and Central America. Its leaves are thick, crenulate and persistent; the flowers are small and yellowish; the fruit is small green and indehiscent; the seeds are horse-shoe shaped. Regarding a specimen (Yale 38382) from Palmira, Province of Alajuela, Costa Rica, the collector, Austin Smith, says the species is a characteristic and beautiful tree of that locality (elevation 7000 feet), growing in half shade on clay loam and sometimes attaining a height of 50 feet with an indented trunk two feet in diameter at the base, the bark grayish and corrugated. The wood is hard, heavy, brittle, and fine-textured, the heartwood dull brown, more or less streaked. The grain is irregular and the working properties are not very good.

In the *Freziera* group the available material represents six species of trees with a combined range in uplands from Cuba and Costa Rica to Ecuador and Bolivia. Some of them are said to be 90 feet tall and nearly three feet in basal diameter. The leaves are finely serrate, sometimes silky pubescent on the under surface; the flowers are small and borne in axillary clusters; the small dry fruits are indehiscent; the seeds are reniform. The woods are dull brownish or reddish brown throughout (in dry specimens), and ranging in density from hard and heavy to moderately so, having about the consistency of Red Gum. The timber is of the general utility class but apparently is not utilized. Some of the foreign species are said to be employed in general construction, furniture, and shipbuilding.

COMMON NAMES: Wild damson (Jam.); teta prieta (P.R.); capulincillo (Mex.); barratillo, durazno de monte (Guat.); coral, tito (C.R.); sajinillito, sajinillo (Pan.); avispo, cerezo de monte, motilón (Col.); huiscaparum (Ec.).

Franklinia *alatamaha* Bartr., the sole species, is a small tree discovered on October 1, 1765, along the Altamaha River near Fort Barrington, McIntosh County, Georgia, by John Bartram and his son William, both noted American botanists. In 1777, William Bartram found the plant growing abundantly in the same locality and collected living specimens and seeds for introduction into the Bartram garden at Philadelphia. The species has not been found in a wild state since 1790, despite many searches for it, and apparently it exists now only in cultivation. Its leaves are deciduous, turning scarlet in autumn; the large white flowers appear in September from the axils of the crowded upper leaves; the fruit is a globose woody capsule, the five valves splitting (loculicidally) downward for about half their length and separating (septicidally) upward from the base for an equal distance; the seeds are not winged. Heartwood brownish; fairly distinct from the yellowish or whitish sapwood. Luster medium. Odorless and tasteless. Hard, moderately heavy, strong, suggesting White Birch; texture very fine and uniform; grain straight; easy to work, finishing very smoothly. Of no commercial possibilities.

Gordonia includes about 30 species of trees and shrubs, but they are all Asiatic except *G. Lasianthus* (L.) Ellis, which grows in swampy lands along the coast region of the United States from southern Virginia to Florida, thence westward to the Mississippi River, and inland to Augusta, Georgia. This species is an evergreen tree, sometimes up to 80 feet high and 20 inches in diameter, with thick reddish brown heavily furrowed bark; often much smaller and sometimes reduced to a shrub. The serrate leaves are 4 to 5 inches long, narrowed at the base, and finely serrate; the large pungently fragrant flowers are borne singly on long pedicels; the fruit is a loculicidally dehiscent capsule splitting from above but not from the base; the squarish dotted seeds are winged. Heartwood pinkish; not sharply demarcated from the pale brownish sapwood. Luster medium. Odorless and tasteless. Rather light in weight, but firm and tough; texture fine and uniform; grain straight; working properties excellent; durability rather low. Has occasionally been used locally for making furniture but has no commercial importance.

COMMON NAMES: Bay, black laurel, holly bay, swamp laurel, tan bay (U.S.A.).

Laplacea. Of the 30 species of this genus, eight occur in Malaysia, the others in tropical America. The American species are mostly small trees or shrubs, but several attain a maximum height of 100 feet and a basal diameter of 36 inches. The leaves are serrate or crenate in part, sometimes with silky pubescence beneath; the flowers, which are white or pinkish and rather showy, are solitary in the leaf axils; the fruit is a loculicidally dehiscent capsule with a persistent axis and winged seeds.

The Bloodwood or Ironwood of Jamaica, *Laplacea haematoxylon* (Sw.) G. Don, is a tree 25 to 40 feet high, said to have a handsome dark red, hard and heavy, fine-textured timber, durable under exposure and suitable for heavy construction and articles of turnery. The most widely distributed species is *L. semiserrata* (Mart. & Zucc.) Cambess., a medium-sized to large tree occasionally 100 feet tall, growing in uplands throughout most of continental tropical America from Costa Rica to Peru and southern Brazil. It is used for common lumber in eastern Peru, but there is no information as to its employment elsewhere, though it is suitable for about the same purposes as Red Gum. *L. Brenesii* Standl. is one of the best known local timbers on the market in the Cartago region of Costa Rica, where it is known as Campano. According to a letter from C. L. Lankester, it is abundant in the locality and is in demand for scantlings for house and mill construction, but does not make good boards because it warps badly in seasoning. The wood is brown or brownish, with little contrast between heartwood and sapwood. The texture is uniform, but appreciably coarser than that of all but one of the authentic specimens of the genus in the Yale collections.

COMMON NAMES: Bloodwood, ironwood (Jam.); almendro (Cuba); maricao, niño de cota (P.R.); nanche-ahuatosa (Mex.); campano, c. chile, ira colorada, llorón, yoro (C.R.); níspero macho de tierra fría, vara de león (Col.); florecillo, pedralejo (Venez.).

Patascoya *Stuebelii* (Hieron.) Urb., the only species, is a small tree apparently limited in distribution to the mountains of northern Colombia. The twigs are woolly; the leaves are small, cordate at the base, stiff, and finely serrate; the flowers are solitary and subtended by bracts; the fruit is unknown. The wood has not been studied.

Pelliciera *rhizophorae* Planch. & Trian., the sole species, is an evergreen tree 15 to 25 feet high growing in Mangrove swamps along the Pacific coast from Costa Rica to Colombia. The thick leathery oblong-lanceolate leaves are clustered near the ends of the branches; the large, solitary, sessile, white or pink flowers are subtended by two colored bracts which are as long as the petals; the fruit is a ribbed and beaked woody nut. Mature wood is not available.

COMMON NAMES: Mangle piñuela (C.R.); palo de sal (Pan.).

Stewartia (or *Stuartia*) includes six species, four of them in China and Japan, two in southeastern United States. They are deciduous trees or shrubs with membranous, usually serrate leaves; the white or roseate flowers are axillary and showy; the fruit is a woody, loculicidally dehiscent, two-seeded capsule. The American species are *S. Malachodendron* L. (= *S. virginica* Cav.), growing in woods and along hillsides from Virginia to Alabama and Florida, and *S. pentagyna* (Dunn) L'Hér. (= *Malachodendron pentagynum* [L'Hér.] Small), occurring along mountain streams from Kentucky to Georgia.

Ternstroemia, with about 85 species of trees and shrubs, is well represented in tropical Asia, very sparingly in Africa, and abundantly (60 species) in tropical America. The leaves are leathery, entire or crenate, persistent, subverticillate, often clustered on short twigs; the flowers small, mostly white, solitary, and scented; the fruit is coriaceous, indehiscent, usually beaked and containing a few horse-shoe shaped seeds which often are covered with scarlet papillae. The trees are usually small or of medium size, rarely 75 feet tall and 24 inches in diameter. The timber apparently is not utilized for any special purposes, although it appears suitable for furniture, as it has a fairly attractive silver grain on the radial surface. The following description is based on 16 specimens of eight species.

Heartwood in various shades of brown, merging gradually into the sapwood. Luster low. Without distinctive odor or taste. Of medium density, but hard and strong; texture medium, fairly uniform; grain variable; not difficult to work, finishing smoothly; durability doubtful.

COMMON NAMES: Scarlet seed, wild mammee sapota (Jam.); mamey del cura, palo colorado (P.R.); botoncillo (Dom. R.); hierba del cura, limoncillo, tepezapote, tepezapotl, trompillo (Mex.); trompillo (Salv.); manglillo (Pan.); carne asada, uva de orso (Venez.); kaiarima, mamusaru omirir (Br. G.).

THEOPHRASTACEAE

THIS unimportant family consists of four genera and about 75 species of shrubs and little trees growing mostly in dry situations in the West Indies, Mexico, Central America, and northern South America. The stiff simple persistent leaves are alternate, subopposite, or nearly verticillate; the flowers are borne in terminal or axillary panicles or racemes; the fruit is a many-seeded berry. The plants furnish no economic products.

Woods yellowish to bright yellow throughout. Not highly lustrous. Odorless and tasteless. Hard, heavy, and brittle; of fine to medium texture; grain straight to irregular; likely to split badly in drying; not difficult to work, taking a high polish; is attractively figured by the coarse rays; is poorly resistant to decay. Apparently of no commercial possibilities.

Growth rings sometimes distinct. Pores very small to minute, the smallest not distinct with lens; numerous to fairly so; mostly in small multiples or clusters, sometimes (*Jacquinia*) with larger pores tangentially arranged giving locally ring-porous appearance. Vessels with simple perforations; without spirals; pits minute, alternate. Rays all broad, 6 to 20 cells, and from 50 to 300 cells high; often low in proportion to their width; fairly homogeneous, though interspersed oddly shaped cells are common; pits to vessels minute, infrequent in *Clavija* and *Theophrasta* as the rays are not often in contact with the vessels. Wood parenchyma very sparingly paratracheal, not distinct with lens; sometimes apparently absent. Wood fibers with rather thick to very thick, often gelatinous, walls; pits numerous, very small, simple or indistinctly bordered. Ripple marks absent. No gum ducts seen.

Clavija, with about 35 species of shrubs and little trees, is distributed from Costa Rica to south-central Brazil and northern Paraguay and Peru. The leaves are very large, clustered at the end of the stem, sometimes thorny-serrate, and for this reason some species are grown in greenhouses for decorative purposes. The pores of the wood are very small, but visible with a lens. A specimen collected by Dr. Ducke in Amazonas (Yale 33813; Ducke 286) has high, Oak-like rays.

COMMON NAMES: Viborrana (C.R.); sapo (Pan.); huevo de ocotea, h. de morrocoyo (Col.); barabara, Cristóbal, olivo, pepita de San Cristóbal, San Cristóbal (Venez.); trompetero-sacha (Peru).

Deherainia is closely related to *Theophrasta* and has three species, one in Cuba, one ranging from Tabasco, Mexico, to Guatemala, and one known only from Chiapas, Mexico. The plants are spiny-leaved shrubs or little trees rarely 15 feet high and eight inches in diameter. The wood has not been studied.

COMMON NAMES: Chicharroncillo de paredón, contraquao cimarrón (Cuba).

Jacquinia, with about 35 species of evergreen shrubs and bushy trees, occurs in the West Indies, southern Florida, Mexico, Central America, and the coastal region of northern South America from Venezuela to Ecuador and northern Peru. The maximum sizes recorded are 35 feet in height and 12 inches in trunk diameter. The leaves are often punctate with pellucid glands. The hard-shelled fruits, which have the color and shape of miniature oranges, are used in various localities for stupefying fish. The wood is attractive because of the conspicuous rays, but is obtainable only in small billets which split badly in drying.

COMMON NAMES: Ironwood, Joe bush, J. wood, sea myrtle (Fla., B.W.I.); espuela de caballero, e. de rey, hueso, manajucillo (Cuba); azucare, barbasco, chirriador (P.R.); bois bandé (Haiti); chacsik, chaczinax, chilillo, deáse, flor de Mayo, f. de niño, guie-zee, hoatzinxochitl, hoitzxochitl, muyché, neucxochitl, palo de las animas, pinicua, rosadillo, San Juan, San Juanito, siche, sicijan, sixje, xochipaltic, zinkinax (Mex.); knock-me-back, tcansik (Br. H.); ducuche, luruche, mata-peje (Guat.); barbasco, carambolillo, crucillo, espino ruco, limoncillo, mata-pescada, mirra (Salv.); burriquita, siempre-viva (C.R.); barbasco, b. montañero, b. de púa, sarnisclo (Col.); barbasco, chilca, chirca, olivo, trompillo (Venez.); barbasco (Ec.).

Theophrasta, with two or three species of shrubs, grows in the Dominican Republic and Haiti. The wood is characterized by exceedingly small pores, not much larger than the fibers and not distinct under a lens.

COMMON NAMES: Guayabo de indio (Dom. R.); coque molle, petit coco (Haiti).

THYMELAEACEAE

THE Thyme family, with about 40 genera and 500 species, mostly shrubs and small trees, rarely herbs and large trees, is represented in nearly every part of the world, especially South Africa, Australia, and the Mediterranean region. The leaves are opposite or alternate, simple, and without stipules; the flowers are usually in axillary or terminal umbels, spikes, or racemes; the fruit is commonly a one-seeded drupe. A few of the plants are cultivated for decorative purposes and some are the source of drugs and poisons, but they are best known for the toughness of their bark, which in several genera is the source of a white, soft, strong fiber used locally for cordage. The only timber tree is *Aquilaria* of the Indo-Malayan region, particularly Assam, which supplies the Eaglewood or Agarwood of commerce, the Lignaloes of the ancients. This material is procured in the form of irregular masses of heartwood saturated with resin and an essential oil having the fragrance of honey. Agar oil is distilled from the soft and nearly colorless gum, but the hard brown material is so highly prized by the Parsees and Arabs that it is rarely distilled (see *Indian Forester* 53: 3: 158). In the New World there are 11 genera, all erect or climbing shrubs or little trees, mostly tropical. The following description applies to the woods of *Daphnopsis*, *Dirca*, *Lagetta*, *Lasiadenia*, *Ovidia*, and *Schoenobiblos*.

Color white or yellowish throughout.

Without distinctive odor or taste when dry. Very light and soft to moderately so (suggesting Malvaceae); texture fine and uniform; grain straight to irregular; very easy to work; not durable. Has no commercial possibilities.

Growth rings usually present; *Dirca* more or less ring-porous, the others diffuse-porous. Pores very small to minute; thick-walled; fairly numerous; sometimes solitary and in small multiples, sometimes in clusters or in radial or diagonal rows. Vessels with simple perforations; spiral thickenings present in *Dirca* and *Ovidia;* members short; pits small to rather large, alternate, vestured. Rays variable, often only 1 to 3 cells wide and few to 10 cells high, occasionally 3 to 5 cells wide and up to 50 cells high, rarely 8 cells wide and over 100 cells high; never very conspicuous because of lack of contrast with background; sometimes storied, wholly or in part; nearly homogeneous to decidedly heterogeneous; ray-vessel pit-pairs half-bordered and of same size as the intervascular; rays widening irregularly in the bark. Wood parenchyma sparingly to abundantly developed; paratracheal, sometimes aliform and confluent into long narrow bands; occasionally also terminal and diffuse. Wood fibers with very thin to rather thick walls; short and cambiform in very light wood (*e.g.*, *Dirca* and *Lasiadenia*); pits with small to rather large borders. Ripple marks present in *Daphnopsis* and *Schoenobiblos;* uniform and distinct, with all elements storied in some species of *Daphnopsis*. Large radial channels sometimes present in *Daphnopsis* and *Schoenobiblos*. Narrow widely spaced concentric layers of included phloem present in *Dirca*. (Several of the Old World genera have included phloem, island type.)

Daphnopsis, with about 40 species of shrubs and small trees sometimes 30 feet tall, is widely distributed in tropical America from the West Indies and Mexico to Peru, Argentina, and southern Brazil. The wood of some of the larger trees is said to be used locally for joinery and pirography. Color white throughout in two species studied, lemon-yellow in *D. macrophylla*. Luster medium. Odorless and tasteless. Light and soft to rather hard and heavy; texture medium fine; grain straight to finely roey (*D. macrophylla*); easy to work, the soft specimens sawing woolly; perishable when exposed to decay. Of no commercial possibilities.

Common names: Burn-nose bark (Jam.); guacacoa, juacacoa, torvisco de las Antillas (Cuba); cienegüillo, emajagüillo, mahout, majagua brava, m. de sierra, m. quemadora (P.R.); maho (Dutch W.I.); ahuejote, hoja de San Pedro (Mex.); chilillo (Guat.); mastate (C.R.); sapán, s. de venado (Ec.); embira branca, imbira branca (Braz.); envira (Urug.).

Dirca *palustris* L., the only distinct species, is a much-branched shrub sometimes over ten feet high and up to four inches in diameter at the base. It is widely distributed in eastern North America from New Brunswick and Ontario to Florida and Mississippi. It is noted for the extreme toughness and pliability of its bark which was used by the Indians for thongs, hence the name Leatherwood. Wood white or yellowish white throughout. Luster medium. Without distinctive odor or taste. Very light, spongy, and brittle; texture fine; grain fairly straight; perishable in contact with the ground.

Common names: Leatherwood, moosewood, wicopy (U.S.A.).

Lagetta, with a few closely related species of trees, mostly small, but sometimes 40 feet in height, occurs in Jamaica, Cuba, and the Island of Haiti. The best known species is *Lagetta lagetto* (W. Wr.) Nash, remarkable for its inner bark, which consists of many very thin laminations that can be separated readily and present a striking similarity to fine lace. It has long been used not only for cordage but also for articles of wearing apparel and adornment. It is said that the Governor of Jamaica presented Charles II with a cravat, frill, and a pair of ruffles made from this bark. The whitish moderately hard rather fine-textured wood has no special uses.

Common names: Gauze, Indian lace, lacebark, linen lacebark (Eng.); Spitzenbaum (Germ.); daguilla, lagetto (Span.); daguilla común, d. de loma, d. valenzuelana, dabile, guanilla (Cuba); palo de encaje (Dom. R.); bois dentelle (Haiti).

Lasiadenia *rupestris* Benth., the only species, is a shrub occurring in the Guianas and northern Amazon region. The only specimen at hand (Yale 33822) was collected by Adolpho Ducke along the Rio Negro. The wood is white, very soft and spongy, and without any known uses.

Linodendron is a Cuban genus with a few species, the best known being *L. Lagetta* Gris., a small tree or a shrub known as Guana, a name also applied to *Sterculia cubensis* Urb. The only use is for the fiber obtained from the bark. The wood has not been studied.

Ovidia *pillo-pillo* (Gay) Meissn. is the only well-known species of this small genus which is confined to Chile and Patagonia. It is a shrub or a little tree rarely over 15 feet tall and six inches in diameter, commonly called Pillo-pillo. The white soft fine-textured wood is not utilized for any special purposes.

Schoenobiblos. Four species have been described, one each from Trinidad, Colombia, Brazil, and Peru. All are small trees or shrubs. The best known is *S. daphnoides* Mart. & Zucc. of Rio Japura, Brazil. The only one represented in the Yale collections is from the type of *S. peruvianus* Standl., collected by Llewelyn Williams near Timbuchi, Peru, where it is known as Barbasco-caspi (Yale 17560). He says (*Woods of northeastern Peru,* p. 366) that it is an uncommon tree, up to 25 feet in height, with a short slender trunk and irregular crown. The yellowish white rather light but firm and strong wood has a pleasant odor and slightly bitter taste when fresh. It is not utilized.

TILIACEAE

THE Linden family comprises 38 genera and about 400 species of trees, shrubs, and herbs of general distribution. The leaves are alternate, simple, stipulate, and commonly deciduous; the flowers are usually in cymes or panicles; the fruit is capsular or nut-like. Herbaceous species of *Corchorus* are cultivated in the tropics for their coarse and strong fiber, known as jute or gunny, which is an important article of commerce. The best known and most important trees of this genus in the north temperate zone are Linden and Basswood (*Tilia*) which supply highly valuable soft white lumber of great utility. In tropical America there are tree species of 12 genera, of which 11 have been studied, namely, *Apeiba, Belotia, Carpodiptera, Christiania, Goethalsia, Heliocarpus, Luehea, Lueheopsis, Mollia, Mortoniodendron,* and *Tilia;* there are no specimens available of *Vasivaea* of the Amazon basin. Some of these tropical trees are large and supply timber for local use, but not for export.

Heartwood nearly white, pinkish, light to dark brown, yellowish, or olive; often not clearly differentiated from the sapwood. Fairly lustrous. Without distinctive odor or taste when dry. Variable from very hard, heavy, and strong (*e.g., Carpodiptera* and *Christiania*) to very light, soft, and spongy (*Heliocarpus* and *Belotia*); texture fine to coarse; working properties excellent to poor; mostly subject to stain and decay unless kept dry. The material of intermediate weight is tenacious and comparatively strong and can be nailed without splitting, hence is suitable for boxes, crating, shelving, moulding, excelsior, and slack cooperage.

Growth rings usually present, but often not distinct; may be terminated by parenchyma. Pores medium-sized to minute, the largest visible without lens; very numerous to rather few; occurring singly and in pairs or in short to rather long radial multiples or in small groups, well distributed without definite pattern. Vessels with simple perforations; spiral thickenings absent except in *Tilia;* intervascular pitting alternate, extremely fine (*Mollia* and *Mortoniodendron*) to medium; tyloses sometimes present. Rays distinctly heterogeneous to nearly homogeneous; of two sizes, the uniseriates mostly 15 to 20 cells high, the cells often square or upright, the multiseriates variable in width (up to 7 or 8 cells in some species) and height (maximum over 300 cells); lower rays in fairly definite horizontal seriation, the others occupying few to several stories; sheath cells sometimes present; tile cells of about the same height as the procumbent cells present in *Luehea, Lueheopsis, Mollia,* and

Mortoniodendron; thin-walled, empty cells of similar structure, but much larger, present in *Belotia;* crystals common; pits to vessels all very small to minute in *Apeiba, Goethalsia, Luehea, Lueheopsis, Mollia, Mortoniodendron,* and *Tilia,* but medium-sized, oval to long oval in *Belotia* and *Heliocarpus,* and distinctly 2-sized, minute and rather large elongated and tending to scalariform arrangement in *Carpodiptera* and *Christiania.* Wood parenchyma moderately to very abundant; special type composed of large and extremely thin-walled cells in layers or strands present in *Apeiba* and *Heliocarpus;* normal parenchyma paratracheal, metatracheal, diffuse, and sometimes terminal; aliform and confluent into narrow to rather wide tangential or concentric bands, distinct to unaided eye and suggesting certain Leguminosae, in *Carpodiptera* and *Christiania;* the same, only finer and not distinct without lens, in *Lueheopsis, Mollia,* and *Mortoniodendron;* diffuse and in short irregular metatracheal lines scarcely distinct with lens in *Goethalsia, Luehea,* and *Tilia;* diffuse and not distinct with lens in *Apeiba, Belotia,* and *Heliocarpus;* strands composed of 2 or 4 cells in *Heliocarpus,* 4 to 8 cells in the others, the cells not in horizontal seriation; crystals sometimes present. Wood fibers with very thick walls and small lumen in *Carpodiptera, Christiania,* and *Lueheopsis,* but with medium to very thin walls in the others; pits very small, often abundant, simple or indistinctly bordered. Ripple marks usually present though often irregular and not always distinct. Gum ducts absent; large widely separated radial channels observed in all specimens of *Heliocarpus.* Bark finely laminated and containing few to many, large to rather small, V-shaped patches.

Apeiba, with 12 species of trees and shrubs, has a wide range in tropical America, with the center of distribution in northern South America. The best known species is *A. tibourbou* Aubl., a small or medium-sized tree growing in the West Indies, southern Mexico, Central America, and in South America to Brazil and Peru. The timber of all species examined is much alike, being white, light and soft, and unsuited for lumber because of its laminated structure. The only special use known is for fishing rafts (jangadas) along the eastern coast of Brazil.

The wood is peculiar in having few to many arcs or concentric bands, 1 to 4 mm. thick, of lustrous, cottony material alternating with zones of firm and rather compact wood of ordinary structure (Plate L). Sometimes the soft tissue comprises the bulk of the stem, but more often its occurrence is sporadic. Examination under the microscope reveals that, with the exception of the vessels, all of the cells in the soft layers are large and very thin-walled and elongated in a radial direction (Plate L, 1). In *Heliocarpus,* which has somewhat similar structure, the thin-walled cells are not elongated radially, but are mostly cubical, and the rays often are not interrupted.

Common names: Mahault chardon, tambaol (Trin.); peine de mico (Mex., Cent. Am.); burillo (Nic.); burio, kurakrá, kutsingró, palo de burio (C.R.); comb wood, cortez, cortexo, cunsagarocri, peinecillo, wacinia (Pan.); contra, erizo, malagano (Col.); cabeza de negro, catigüire, erizo, heriso, peine de mico (Venez.); duru (Br. G.); alastioelan, araskioerian, bordaballi, borredaballi, boentaflabom, boesi soersakka, doro, kan-kan-pau, keesi-keesi-kankan, klikli, mekoejongaree, onikhiakiaballi, pikien-nigrekam-kam, pondjo-wattare, wekere-japoepare (Sur.); cortica, gargauba, jangada, pau de jangada, pente de macaco (Braz.); maqui-sapa, m.-s. ñaccha (Peru).

Belotia includes a few closely related species of small to medium-sized trees apparently limited in range to Cuba, Mexico, and Central America. The type of the genus is *B. grewiifolia* A. Rich. (see *Kew Bulletin* 1939, pp. 517–521). There are no special uses for the wood, which is white or pale brownish throughout. Not highly lustrous. Odorless and tasteless. Very light and soft, often spongy; of about the weight of Balsa (*Ochroma*), but lacking its firmness and strength; texture rather coarse; feel soft; grain straight; perishable in contact with the soil.

Common names: Guacimilla, majagüilla, m. blanco, m. macho (Cuba); corcho colorado, jonote coyolillo, manechaháu, patita, yaco de cal, y. de venado (Mex.); bastard polak, moho, narrow-leaved moho, white moho (Br. H.); mecate colorado (Guat.);

capulín, sirín de paloma (Hond.); capulín savanero (Nic.).

Carpodiptera comprises six species, three in the coastal region of East Africa, and three in the West Indies and Yucatán Peninsula; all are shrubs or little trees. The only authentic wood sample at hand (Yale 19317; Bucher 193) is of *C. cubensis* Gris. from eastern Cuba. The timber is said to be used locally for railway crossties. Heartwood brown, somewhat waxy; sapwood lighter, sharply demarcated. Luster low. Hard, heavy, and compact; fine-textured; easy to work, finishing with a high natural gloss; probably fairly durable. Presumably of no possibility for export.

COMMON NAMES: Majagua de Cuba, majagüilla (Cuba); bois d'ortie, b. d'o. rouge (Haiti); telcón (Mex.); mountain pear (Br. H.).

Christiania consists of two species, one endemic in Madagascar, the other, *C. africana* DC., growing in central and western tropical Africa and in northern Brazil, British Guiana, and British Honduras. The American representative is a medium-sized tree apparently of rare occurrence; the only local name recorded is Palo Mulato in British Honduras. Heartwood light to dark olive or reddish brown; somewhat waxy; more or less variegated and showing parenchyma lines and layers suggesting certain Leguminosae; usually not sharply demarcated from the yellowish sapwood. Very hard, heavy, tough, and strong; texture medium; feel rather harsh; not very easy to work, but finishing very smoothly with a high natural gloss; probably durable.

Goethalsia *meiantha* (D. Sm.) Burret, the only species, is a tree infrequently more than 50 feet high and 14 to 16 inches in diameter, in southern Central America and northern Colombia. (See *Tropical Woods* 15: 15; 40: 18; 42: 21, 40.) Wood white or grayish, with a slight tinge of pink; knots are brown. Luster low. Odorless and tasteless. Light and soft; sp. gr. (air-dry) 0.35; weight 22 lbs. per cu. ft.; texture medium; feel soft; grain straight; easy to work, but saws rather woolly; subject to sapstain and perishable in contact with the ground. Apparently without commercial possibilities.

COMMON NAME: Guácima blanca (Pan.).

Heliocarpus, with about 20 species of small to large and fast-growing trees, is best represented in Mexico and Central America, but the range extends to Argentina and southern Brazil. The bark yields a fiber used for tying dried tobacco and making coarse cordage. The wood is typically white, gray, or brownish, very light and soft, coarse-textured, stringy, and perishable in contact with the soil. Timber grown in southern South America is somewhat better and finds some local utility for interior construction and boxes.

COMMON NAMES: Cahaulagua, catena, coche, copal, cuahualagua, cuaulahuac, cuauloe, holol, huanate, jolocin, j. blanco, jolotzin, jonote, majagua, quauhalagua, tolotzin, zolotzin, yaga-guichi, zamo baboso (Mex.); false dragon's blood (Par.); moho —broadleaf, white, yellow (Br. H.); mahau, majao, m. blanco, mecate de agua (Hond.); calagua, calagual, calagüe, mozote, mozotillo (Salv.); burillo falso, pestaño mula (Nic.); burio, burillo, b. blanco, gsükrá, rusuragró, stsa, tsari (C.R.); majagüillo (Pan.); majagua melada (Col.); jangada brava (Braz.); llausaquiro (Peru); afata grande blanca, amor seco, apeivá, ibirá-piré-hú, lapachillo, tapicá-guazú (Arg.).

Luehea, with about 20 species of trees and shrubs, is widely distributed in tropical America. Some of the trees attain large dimensions, being often 80 to 100 feet tall, usually with an irregularly fluted trunk 24 to 36 inches in diameter. *L. speciosa* Willd. occurs in Cuba, southern Mexico, Central America, and northern South America. *L. candida* (DC.) Mart. extends from Mexico to Venezuela. Another common Central American species is *L. Seemannii* Tr. & Pl. The principal species in Argentina and southern Brazil is *L. divaricata* Mart.; its timber is used locally for interior construction, carpentry, common furniture, woodenware, brush backs, clogs and shoe soles, saddle frames, and pack saddles.

Heartwood brown or brownish, sometimes with a pinkish tinge and more or less streaked; not clearly differentiated from the sapwood. Luster medium to fairly high. Without distinctive odor or taste. Moderately hard and heavy, tough and strong; sp. gr. (air-dry) 0.55 to 0.67; weight 33 to 42 lbs. per cu. ft.; has about the consistency of Birch (*Betula*); texture medium; grain straight to finely roey; easy to work, finishing very smoothly; not highly resistant to decay.

COMMON NAMES: Guácima amarillo, g. barilla, g. varía (Cuba); algodoncillo, chacah, kazcat, pataxte, pataxtillo, patazte, pepe cacao (Mex.); caulote, mapola, tapasquit (Br. H.); llalo, tapasquit (Guat.); caulote blanco, guácimo colorado, llalo, yayo (Hond.); bonete, cabo de hacha, caulote, contamal, cotonrón, pataste, tepecaulote, terciopelo (Salv.); guácimo molinero (Nic.); guácimo macho, g. m. de montaña, g. molenillo (C.R.); guácimo, g. molenillo (Pan.); algodón de monte, a. montañero, malagano, tablón (Col.); guácimo blanco (Venez.); calzoncillo, c. panga (Peru); açoita-cavallo, estribeiro, ivatingy, uacima do campo (Braz.); caá-obetí (Par.); azota-caballo, caá-o-vetí, cabeti, ibá-tingui, sota caballo (Arg.).

Lueheopsis, with seven species of medium-sized to large trees, occurs in the Guianas and northern Brazil. *L. rosea* (Ducke) Burret is the largest tiliaceous tree along the lower Amazon, attaining a height of 125 feet. Other species are smaller, *L. violacea* Standl., for example, being only 35 feet high and 12 inches in diameter. There are no recorded uses for the timber, except for fuel. Wood yellowish or grayish brown throughout. Fairly lustrous. Hard, heavy, and strong; coarse-textured; easy to work, finishing very smoothly; probably perishable when exposed to decay.

COMMON NAMES: Koese-wiran (Sur.); arapapá (Braz.).

Mollia, with 16 species, is apparently limited in its range to the Amazon basin. The trees are small to medium-sized, the tallest reported being 70 feet. The timber is not utilized for any special purposes. Wood brown or pinkish brown throughout. Luster rather low. Odorless and tasteless. Hard, heavy, tough, and strong; sp. gr. (air-dry) 0.80; weight about 50 lbs. per cu. ft.; rather fine-textured; straight-grained; easy to work, finishing smoothly; is probably perishable when exposed to decay.

COMMON NAMES: Yawhooballi (Br. G.); sururú (Braz.); uchu huayo, u. mullaca (Peru).

Mortoniodendron, with two species of trees, appears to be limited to Central America. *M. guatemalense* Standl. & Steyerm. occurs in Guatemala, but the wood has not been studied. *M. anisophyllum* (Standl.) Standl. & Steyerm., is a tree about 50 feet tall and 18 inches in diameter discovered by G. Proctor Cooper in the region of Almirante, Bocas del Toro, Panama, in 1928. There are no recorded uses for the timber. The wood of the type tree (Yale 11945) is whitish throughout. Luster low. Odorless and tasteless. Light in weight but comparatively hard and tough; texture medium, uniform; grain irregular; easy to work; durability presumably poor. Apparently without commercial possibilities.

Tilia, with numerous closely related species and varieties of small to large trees, is widely distributed in eastern Asia, Europe, and North America. The European Linden, *T. europaea* L., is a good shade tree and is often planted for this purpose; its timber is considered exceptionally well suited for large carvings. There are about 15 species in Mexico, but there is little information concerning them; their woods have not been studied. For the United States, 18 species and 12 varieties have been described, the principal ones being *T. glabra* Vent. (= *Tilia americana* L.) and *T. heterophylla* Vent., northern and southern forms, respectively. The woods are not distinguishable by any means now known, and the name Basswood is applicable to all of them. The present stand of the timber in the United States is estimated to be in excess of eight billion board feet, half of this amount being accredited to the Lake Superior region. The timber is preferred for products made of wood in the natural condition, especially when a clean attractive appearance, light weight, and freedom from odor are essentials, as in food containers (boxes, tubs, pails, and veneer baskets). It is a favorite wood for apiarists' supplies (beehives and honey sections), slack cooperage heading, Venetian blinds, excelsior, woodenware, toys, novelties, and many other purposes requiring a light but strong, easily worked material.

Heartwood creamy white to brownish; not always clearly demarcated from the thick, nearly white sapwood. Luster medium. Odorless and virtually tasteless. Density low; sp. gr. (air-dry) 0.40 to 0.50; weight 25 to 32 lbs. per cu. ft.; texture rather fine, uniform; grain straight or sometimes curly; easy to season and manufacture; holds its place well; poorly resistant to decay.

COMMON NAMES: American linden, bass tree, basswood, bee tree, lein, lime tree, linden, linn, wahoo, wickup (U.S.A.); basswood, whitewood (Canada); cirimo, sirimo, tilo, tirimo, yaca (Mex.).

TRIGONIACEAE

AN unimportant family of three genera and about 40 species of small trees and scandent shrubs. *Euphronia* and *Trigonia* occur in tropical America, *Trigoniastrum*, with one species, in Malaysia. The leaves are alternate or opposite, simple and entire; stipules are usually present, sometimes connate, early deciduous; the small flowers are borne in racemes or panicles; the fruit is a 3-celled capsule, the valves separating from the central column; the seeds are covered with cottony hairs. Members of this family have been referred by various botanists to the Euphorbiaceae, Hippocrateaceae, Malpighiaceae, Ochnaceae, Polygalaceae, Rosaceae, and Sapindaceae.

Growth rings present. Pores thick-walled, rounded; small to rather large; fairly numerous; nearly all solitary; irregularly to fairly uniformly distributed, without pattern. Vessels with simple perforations; no spiral thickenings seen; intervascular pitting, when present, very fine and alternate. Rays decidedly heterogeneous; the uniseriate rays and ray margins composed of square and upright cells, the multiseriate parts consisting of procumbent cells; crystals common; ray-vessel pitting mostly unilaterally compound, the vascular pits very small, with slit-like apertures, the ray pits usually much elongated and vertically, diagonally, or radially inclined. Wood parenchyma abundant, variously arranged. Wood fibers in definite radial rows; walls thick; pits with distinct borders and extended slit-like apertures. Ripple marks absent. No gum ducts seen.

Euphronia (or *Lightia*) consists of three species of shrubs and small trees in the Amazon basin (see *Notizbl. Bot. Gart. Berlin-Dahlem* 12: 115: 699). The only specimens available for this study are of *E. licanoides* (Spruce) Hallier, a small tree with lilac flowers, collected by Adolpho Ducke on the Rio Curicuriary, a tributary of the Rio Negro, in Amazonas, Brazil. Heartwood brown; rather sharply demarcated from the grayish brown sapwood. Luster medium. Without distinctive odor or taste. Very hard, heavy, and strong; texture medium; grain fairly straight; not easy to work but finishing smoothly; durability doubtful. Apparently without commercial possibilities.

Trigonia, with about 30 species, mostly scandent shrubs, is widely distributed in continental tropical America from British Honduras to southern Brazil. The following description is based upon a specimen (Yale 23678) of an undetermined species collected by Adolpho Ducke near Manáos, Brazil. Wood clear yellow throughout. Luster rather high. Odorless and tasteless. Very hard, heavy, and strong; texture medium-fine; grain fairly straight; not difficult to work, finishing smoothly; durability probably low. Suitable for small articles of turnery, but without commercial possibilities because of the small size of the stems.

TURNERACEAE

AN unimportant family of seven or eight genera and about 100 species, chiefly herbs and low shrubs, occasionally large shrubs and small trees, best represented in tropical

America and Africa. They do not supply any useful timber. There are specimens of *Erblichia* and *Turnera* in the Yale collections and owing to various differences in the woods they will be considered separately.

Erblichia. The only generally accepted American species is *E. odorata* Seem., a tree 25 to 50 feet tall, of infrequent occurrence from southern Mexico to Panama. The leaves are simple, alternate, short-stalked, inconspicuously crenate, and without stipules; the large yellow or orange flowers are very showy; the fruit is a 1-celled capsule with three woody valves and many arillate seeds.

Wood pale brownish throughout specimens, except for dark brown areas about wounds. Luster low. Odorless and tasteless. Of medium density, but hard and tough; has about the consistency of Red Gum (*Liquidambar*); texture rather fine; grain straight to irregular; not difficult to work, finishing smoothly; cross-grained material inclined to warp badly; presumably low in resistance to decay. Of no commercial possibilities.

Growth rings absent or poorly defined. Pores small to medium-sized, not distinct without lens; occurring in scattered radial multiples of 2 to 6. Vessels with simple, wide-rimmed perforations; without spirals; pitting very fine, alternate. Rays inconspicuous; 1 to 5 cells wide and up to 40 cells high; decidedly heterogeneous; the multiseriates with stratum of procumbent cells; upright cells often disjunctive; vessel-ray pitting resembles intervascular. Wood parenchyma finely reticulate, barely visible under lens. Wood fibers with rather thick walls and numerous indistinctly bordered pits with slit-like, extended apertures. Ripple marks and gum ducts absent.

Common names: Axochitl, azuche, chamiso, jarro de oro, sanjuanero, suelda con suelda (Mex.); butterfly tree (Br. H.); candelaria, c. de montaña, canop, conop (Guat.); flor de fuego (Salv.).

Turnera, with about 50 species of herbs, shrubs, and a few trees, is very sparingly represented in the Indo-Malayan region, abundantly in the New World from the West Indies and northern Mexico to Argentina. The leaves are simple, serrate, often biglandular at the base; the yellow, white, or pink flowers are usually solitary; the capsule is thin-walled, with numerous arillate seeds. The best known shrub is the Damiana plant, *T. diffusa* Willd., which contains an aromatic volatile oil with a warm bitter camphor-like taste and is reported to have various medicinal virtues. The only wood sample available (Yale 22804; A. Rimbach 94) is of *T. Hindsiana* Benth. from Ecuador.

Heartwood yellowish, with a roseate hue, slightly darker than the thin yellow sapwood. Luster rather high. Taste somewhat mucilaginous; faint odor present. Of medium density and hardness; texture fine and uniform; grain straight; very easy to work, finishing smoothly and attractively; decay resistance probably poor. Of no commercial possibilities.

Growth rings poorly defined. Pores very small, invisible without lens; very numerous and rather crowded; solitary and in short to long multiples, in distinct radial arrangement because of the close spacing of the rays. Vessels in part with simple and in part with multiple perforations, the scalariform plates short and having numerous delicate bars; fine spiral thickenings present; pitting very fine, alternate. Rays very numerous; 1 to 4 cells wide and up to 100 or more cells high; decidedly heterogeneous, nearly all of the cells being square to tall-upright; pits to vessels very small and rounded to narrow-elongate and in scalariform arrangement. Wood parenchyma diffuse; cells often disjunctive. Wood fibers in definite radial rows; walls thick and often gelatinous; pits abundant, very small, distinctly bordered. Ripple marks and gum ducts absent.

Common names: Marilópe, marilópez (Cuba); marilópez, yellow alder (P.R.); oreganillo (Dom. R.); zombi nan bois (Haiti); amaranto, caléndula, clavel de oro, damiana, hierba de la pastora, h. del venado, marilópez, pastorcita, xmisibcoc (Mex.); escobillo, e. blanco, flor de San Lorenzo (Salv.); margarita de los campos, oreja de coyote, San Juan (Nic.); malva (Col.); cumaná (Venez.); damiana (Braz.).

ULMACEAE

THE Elm family comprises 15 genera and over 150 species of trees and shrubs of general distribution in temperate and tropical regions of the world. The leaves are alternate and simple, unequal and often oblique at the base; the paired stipules fall off very early; the flowers are clustered or sometimes solitary on the current or the previous season's growth; the fruit is a samara, nut, or drupe. The best known tree is the Elm (*Ulmus*), highly esteemed for shade and ornamental purposes and also for its very tough and strong wood. In temperate North America there are three genera, namely, *Celtis, Planera,* and *Ulmus;* in tropical and subtropical America there are six, namely, *Ampelocera, Celtis, Chaetoptelea, Momisia, Phyllostylon,* and *Trema.*

Heartwood pale brown to dark or grayish brown; sapwood yellowish or nearly white, not always clearly demarcated. Luster low to rather high. Mostly without distinctive odor or taste; dark brown specimens sometimes with scent of licorice. Density greatly variable, *Trema* being light and soft, *Planera* medium, *Ampelocera, Celtis,* and *Phyllostylon,* and some species of *Ulmus,* hard and heavy; texture very fine (*Ampelocera, Phyllostylon*), fine (*Planera*), medium (tropical species of *Celtis*), or coarse (*Chaetoptelea, Momisia, Ulmus*); working qualities generally good; durability mostly low.

Growth rings usually present, sometimes distinct; woods ring-porous in *Celtis* (in part) and *Ulmus,* and more or less so in *Chaetoptelea.* Pores ranging in size from minute to rather large; mostly in radial multiples of two to several pores each or in small clusters; rather few to very numerous; fairly uniformly distributed without definite pattern in diffuse-porous woods; in diagonal rows, or broken or continuous, mostly wavy, tangential or concentric bands or festoons in the late wood of ring-porous woods (Plate XXXIX, 1). Vessels with simple perforations; spirals present in *Celtis* (in part), *Planera, Ulmus,* and smallest vessels of *Chaetoptelea;* deposits of calcium carbonate frequent to abundant in both sapwood and heartwood of *Ampelocera* and *Phyllostylon;* intervascular pitting medium to rather coarse, alternate. Rays mostly homogeneous in *Chaetoptelea, Planera,* and *Ulmus,* more or less distinctly heterogeneous in the others; mostly uniseriate or biseriate in *Phyllostylon* and *Trema,* 1 to 4 cells wide in *Ampelocera* and *Momisia,* 1 to 6 in *Chaetoptelea, Planera,* and *Ulmus,* and 1 to 10 in *Celtis;* mostly less than 25, sometimes up to 60, cells high, except in *Momisia* where some exceed 200 cells; crystals sometimes present; pits to vessels small to medium-sized, round to oval, sometimes (esp. in *Trema*) in part elongated, irregular in form, with tendency to scalariform arrangement. Wood parenchyma paratracheal and not distinct with lens in *Celtis* (in part), *Ulmus,* and *Trema;* in numerous uniseriate lines or reticulate, visible with lens in *Planera;* usually abundantly developed and distinct in the others, being paratracheal, aliform, and irregularly confluent; crystals sometimes abundant. Wood fibers with thin to thick walls; pits very small, simple. Ripple marks present in *Ampelocera* and *Phyllostylon;* all elements storied except the taller rays. No gum ducts seen.

Ampelocera, with seven species of shrubs or small to medium-sized unarmed trees, is widely but sparingly distributed in tropical America. *A. Ruizii* Kl. is a shrub credited to Peru, though without definite locality. *A. verrucosa* Kuhlm. and *A. edentula* Kuhlm. are small to medium-sized trees of the lower Amazon region of Brazil. *A. glabra* Kuhlm. is another Brazilian tree, sometimes 65 feet high, with a ridged or somewhat sulcate trunk, occurring in Espirito Santo and commonly in Minas Geraes (for detailed description of the wood see *Arch. Inst. Biol. Veget.* 3: 2: 211–215). There are two Cuban species, namely, *A. cubensis* Gris. and *A. crenulata* Urb., both small trees with rather thick, rough, gray outer bark, and finely laminated inner part. *A. Hottlei* Standl. is known to occur in southern Mexico, British Honduras, Guatemala, Honduras, Panama, and northern Colombia (see *Tropical Woods* 51: 11). It is usually a small tree in the forest undergrowth, but sometimes attains a height of 60 feet, with a trunk diameter of 20 inches. No special uses are reported for the timber of any species.

Heartwood purplish brown with irregular

blackish streaks; sharply demarcated from the thick yellowish sapwood. Luster medium. Without distinctive odor and taste. Very hard, heavy, tough, and strong; texture rather fine, uniform; grain mostly irregular; working properties fair to excellent; durability probably fairly high. Heartwood apparently suitable for brush backs, but presumably not available in sufficient quantity to have commercial value.

Common names: Hueso, jatía blanca, purio (Cuba); bois blanc (Haiti); cautivo (Mex.); bullhoof (Br. H.); manteca (Hond.); achote de monte, arepito, casça de venado (Col.); mentira, vareteiro da matta virgem (Braz.).

Celtis, with about 60 species of trees and shrubs, is of wide distribution in temperate and tropical regions. There are six species and several varieties in the United States, the two best known being the Hackberry, *C. occidentalis* L., a small to large tree occurring rather commonly throughout the eastern half of the country and sparingly in the Rocky Mountains, and the Sugarberry, *C. laevigata* Willd. (= *C. mississippiensis* Bosc.), a closely related form in the southeastern states. Their lumber is usually sold in mixture with White Elm, also called Soft Elm in the trade. The wood of these species is distinctly ring-porous and the grayish heartwood is not clearly demarcated from the sapwood.

A variety of *Celtis laevigata* extends from Texas into Tamaulipas and San Luis Potosí in Mexico, and the trees sometimes reach a height of 125 feet with a trunk diameter of over three feet. The timber is considerably lighter and weaker than that of the northern Hackberry and its uses are purely local. There are at least three other species in Mexico, with a combined range extending from the United States to Oaxaca; they vary from medium-sized trees to low crooked shrubs, and supply no timber of value.

The principal species of *Celtis* in the West Indies is *C. trinervia* Lam. It is a slender tree, usually less than 50 feet tall, common in dry thickets. The wood has an olive-brown heart and yellowish sapwood and is hard to moderately hard, tough and strong; used in rural construction, tool handles, and farm implements.

There are several species in southern South America, the best known and most widely distributed being the Tala of Argentina, *Celtis tala* Gill., and its two varieties. At its best the Tala is about 50 feet high and 20 inches in diameter. Its yellow or grayish yellow wood is moderately hard and heavy, medium-textured, not resistant to decay. It is used in vehicle construction, tool handles, and miscellaneous purposes requiring a tough and strong timber.

Common names: Beaverwood, connu, hackberry, hack tree, nettle tree, one-berry, sugarberry (U.S.A.); aguadita, cagada de gallina, gallinosa, graciliano, guaseriano, guisacillo, hueso, h. blanco, jaboncillo, malva de sierra, ramón de costa, r. de sierra, quiebrahacha de costa (Cuba); anasillo, palo amargo (Dom. R.); bois raie, casser hache gris (Haiti); acibuche, cumbro, grangeno, muc, palo blanco, p. de águila, p. mulato, zitsmuc (Mex.); duraznillo (Salv.); gurupiá, jauii (Braz.); tala, t. ema, t. pispa, yoá-sí-y (Arg.); yuá-sí-y (Par.); talt (Urug.).

Chaetoptelea *mexicana* Liebm., the only species, is a small to large tree, sometimes 125 feet high and three feet in diameter, of limited occurrence in the uplands of southern Mexico, Costa Rica, and Panama. Some botanists include it in the genus *Ulmus,* but the fruit is not winged. The wood (Yale 3006; Pittier 2938, from Panama) has the dark brown color of *U. fulva,* and is used locally for railway crossties and for the frames and wheels of vehicles.

Common names: Cempoalehuatl, olmo, papalote (Mex.); mezcal, sauchino (Salv.); dzú, ira, tirra (C.R.); cenizo (Pan.).

Momisia, usually treated as only a section or subgenus of *Celtis,* includes several species of woody vines, shrubs, or small trees. Unlike the species of *Celtis* proper (which are unarmed erect plants), the branches bear stout, usually recurved, stipular spines for climbing or clambering. *Momisia pallida* Planch. often forms dense impenetrable thickets in Mexico and southwestern United States. The most widely distributed species is *M. iguanaea* (Jacq.) Rose & Standl. of the West Indies, southern Florida, western Texas, Mexico, Central America, and northern South America. Though usually classed as a clambering shrub, it is sometimes a tree 40 feet high and 12 inches in diameter, with long spreading branches. The wood, which is yellowish brown, streaked, rather hard and heavy, tough and strong, and coarse-textured, is used to some extent for fuel and fence posts.

COMMON NAMES: Uña de gato, zarza blanca (Cuba); azufaifo, cockspur (P.R.); gallito (Dom. R.); liane crabe (Haiti); garabato, g. blanco, grangeno, granjeno, g. huasteco, muc (Mex.); cagalero, c. blanco, uña de gato (Salv.); cagalero comestible (Nic.); barimiso, guacharagüera, marimiso (Venez.); meloncito blanco, palo blanco (Peru); tala trapadora (Arg.).

Phyllostylon *brasiliensis* Cap., the only species, is a small to medium-sized tree known to occur in Cuba, Dominican Republic, Haiti, Mexico, Guatemala, Nicaragua, Colombia, Venezuela, southern Brazil, Paraguay, and Argentina. It is said to be common and to attain a diameter of two feet in the southern part of Tamaulipas and the parts of San Luis Potosí and Higalgo that form what is commonly called La Huasteca, Mexico. It is widely distributed in northern Argentina and is sometimes 80 feet tall and 30 inches in diameter; the timber is used for agricultural implements and furniture.

Mr. G. C. Bucher of Santiago de Cuba supplies the following information about the species in eastern Cuba: "The Jatía, as it is called here, occurs on the south coast in this province, growing in almost solid stands on the coastal plains. The largest stand that I have observed is around the mouth of Guantánamo River. There is another large one at the head of the bay, and another in the lower Yeteras valley. It is very rarely in the interior of the island, that is, after passing the first range of hills from the coast. Its principal use here is for firewood for the sugar mills near the coast and it is not even considered a good wood for charcoal."

In Haiti, where it is known as Bois Blanc, *Phyllostylon* is one of the most common trees in the western and northern parts and is abundant in the flat and semiarid region south of Gonaives. When it grows in pure open stands it is of poor form, but in better sites in mixture with other species it grows taller, and occasional trunks are 18 inches in diameter and clear of branches for 30 feet. The timber is not highly valued by the Haitians and their use of it is limited to the smaller sizes for fence posts and house framing.

In the Dominican Republic the tree, known as Baitoa, is said to be common in the valleys of the Yaque del Norte and the Yaque del Sur and on the dry foothills and lower mountain slopes in that region. Usual heights are between 50 and 70 feet, the largest trees having irregular or fluted trunks 20 to 24 inches through. A trial shipment of logs to New York in 1917 resulted in a small trade in the timber as a substitute for Zapatero or Venezuelan Boxwood (*Gossypiospermum*) which at that time could not be obtained in sufficient quantity. The wood, under the name of San Domingo Boxwood, proved to be fairly satisfactory for knife scales and could be stained black or "ebonized" without much difficulty, but it was not considered good enough for the best grade of rulers and silk-weaving shuttles. A decided reduction in the available supply of the Venezuelan timber in 1937 resulted in renewed interest in the wood of *Phyllostylon*.

Heartwood lemon-yellow, sometimes with tinge of brown, occasionally with dark streaks; sapwood yellowish or nearly white. Sp. gr. (air-dry) 0.95; weight about 59 lbs. per cu. ft.; texture fine and uniform; grain fairly straight, but sometimes irregular; wood readily turned and carved, takes a high polish. (For further details see Yale School of Forestry Bulletin 14, *Boxwoods*.)

COMMON NAMES: San Domingo boxwood, West Indian boxwood (U.S.A., trade); jatía (Cuba); baitoa (Dom. R.); bois blanc (Haiti); cerón (Mex.); sabonero (Col.); cara tibama, membrillo (Venez.); pau branco, vareteiro (Braz.); palo de lanza blanco (Par.); ibirá-catú, palo amarillo, p. blanco, p. de lanza, p. de l. negro, tala grande, yoá-si-y-guazú (Arg.).

Planera *aquatica* Gmel., the only species, is a tree sometimes 40 feet tall with a short trunk rarely 20 inches in diameter growing in swamps and lowlands, sparingly along the southeastern coast of the United States and more abundantly in the lower Mississippi valley, especially western Louisiana and southern Arkansas. The wood is pale brown throughout and of about the consistency of Soft Maple and suitable for the same purposes. It has only minor local uses because of the scarcity of the larger trees.

COMMON NAMES: Planer tree, plene, water elm (U.S.A.).

Trema, with numerous closely related species of unarmed shrubs and small trees, is widely distributed in tropical and subtropical regions. The trees are fast-growing and short-lived, being often among the first to occupy burned-over areas or abandoned clearings. The best known and most widely distributed species in America is *T. micrantha* (L.) Blume, often a shrub but under the most favorable conditions growing 50 feet tall and 20 inches in diameter. Its range covers the West Indies, much of Mexico, Central America, and South America to Argentina and southern Brazil. The trees are sometimes employed to shade coffee and cocoa plantations, the supple twigs are woven into baskets, a tough fiber from the bark is made into cordage, and the wood is used to a limited extent to make fine charcoal for gunpowder.

COMMON NAMES: Bass cedar (Jam.); capulí cimarrón, guásima boba, guasimilla, g. cimarrón (Cuba); guacimillo, pain-in-back, palo de cabra (P.R.); anisillo, memiso (Dom. R.); bois de soie, mahaut piment (Haiti); kib, zonote (Mex.); bastard bay cedar, capulín, white capulín, white bay cedar (Br. H.); capulín negro (Hond.); capulín montes, capulincillo, churrusco (Salv.); capulín, juco, vara blanca (C.R.); capulín macho (Pan.); balsa montañero, guayuyo, majagua colorada, majagüito, venaco (Col.); carraspero, carrasposo, cuero de toro, lavandero, mosaquillo, nigüito (Venez.); aisegerina, atadijo, yana-caspi (Peru); coatindiva, crindeuva, pau de polvora (Braz.); afata grande colorado, camba-acá, capoeirero, ingá blanca, i. morotí, fruta de paloma, palo pólvora, tala blanca (Arg.).

Ulmus, with about 20 species of small to very large trees, is widely distributed throughout the north temperate zone, except in the western half of North America, and is extensively planted for shade in parks and along highways. Of the six species native to the United States, three are of commercial importance. *U. fulva* Michx., called Slippery Elm because of its mucilaginous inner bark, is of scattered occurrence over the entire eastern half of the country and southeastern Canada. The tree is also called Red Elm, while the timber is usually known as Gray Elm to the trade. The heartwood is dark brown and often has a distinctive scent suggesting licorice. The band of pores in the early wood is commonly wider than in the other Elms. The Cork, Rock, or Hickory Elm, *U. racemosa* Thomas, is of limited occurrence in New York, Ohio, Indiana, Kentucky, Illinois, Iowa, Missouri, and parts of adjoining states, being at its best in the Ohio valley. The timber is considered superior to the others because of its straight grain and greater density and strength. The Winged Elm or Wahoo, *U. alata* Michx., grows in the south-central states and usually is a spreading tree with a short stout often poorly formed bole. Second-growth timber is noted for its strength and toughness, but its uses are mostly local.

The American or White Elm, *Ulmus americana* L., is by far the most important species for timber production. Its natural range includes the entire eastern half of the United States, extending from the Great Plains to the Atlantic and from Canada to the Gulf of Mexico. "Though a tree of the primeval forest, it did not disappear with the clearing of the land for agriculture; rather it remained or emerged to cast its massive shadow on farmsteads and pastures and country roads or to form cathedral-like arches over village streets. It has been planted and cherished from earliest colonial days, commanding respect for its stature and pliant strength and stirring admiration for its graceful beauty, qualities rarely found in such harmonious combination."

Elm timber is noted for its toughness, pliability, and resistance to wear and shock. Large quantities of it are used in the slack cooperage industry for staves, heading, and hoops. Another important use is in the manufacture of baskets, especially for outside and inside rims and for bent handles; also for crates, frames of vehicles, agricultural implements, and certain types of furniture. The surface veneers of bedroom furniture are sometimes of Elm and the banded structure gives a fairly distinct figure to rotary-cut veneer or flat-sawn lumber.

Heartwood pale brown, sometimes dark brown (*Ulmus fulva*), often with a reddish tinge; sapwood grayish, not always sharply demarcated. Usually odorless and tasteless, but licorice-scented in *U. fulva*. Luster medium to low. Moderately to decidedly heavy, hard, tough, and strong; sp. gr. (air-dry) 0.60 to 0.85; weight 38 to 53 lbs. per cu. ft.; texture coarse; grain straight to irregular; usually difficult to split, easy to bend, usually saws woolly, but can be finished smoothly; durability fair to poor.

Common names: Elm (American, basket, cliff, cork, gray, hickory, red, rock, slippery, water, white, witch), wahoo, whahoo (U.S.A.).

URTICACEAE

The Nettle family comprises about 45 genera and 500 species of herbs, small to large shrubs, and little trees, and is of general distribution, though most abundant in the tropics. Many of the plants are covered with stinging hairs. The leaves are simple, alternate or opposite, and usually with stipules; the very small unisexual flowers are usually in cymes, but sometimes on a common enlarged receptacle; the fruit is a dry achene or a fleshy drupe. About the only useful product is tough bast fiber sometimes employed in cordage and coarse textiles. Although the family is considered to be closely related to the Ulmaceae and Moraceae, it furnishes no timber of value. There are arborescent representatives of several genera in tropical America, the principal ones being *Myriocarpa* and *Urera*.

Wood light to dark brown throughout dried specimens; often badly affected by decay and very fibrous. Luster low. Scent and taste absent or not distinctive. Light and soft; coarse-textured; perishable. Without commercial possibilities.

Growth rings usually absent. Pores few; small to rather large; occurring singly and in small multiples or groups, well distributed without definite pattern. Vessels with simple perforations; very rarely with spiral thickenings; tyloses common; intervascular pitting alternate, rather coarse. Rays mostly 4 to 6, sometimes up to 10, cells wide and 50 to over 200 cells high; decidedly heterogeneous, the cells thin-walled, coarse, irregular, frequently flattened-hexagonal (tangential section); sheath cells often present; crystal druses common in some species; pits to vessels large to very large, irregular in form and arrangement. Wood parenchyma often of two types: (1) with normal lignified walls; paratracheal, sometimes abundant and confluent, more often sparingly developed; druses common; (2) with thin unlignified walls occurring as islands or arcs, sometimes including parts of the rays, and upon disintegrating leaving the stem a loosely aggregated mass of fiber strands or layers, resembling certain woods with included phloem. Wood fibers with thin to moderately thick walls; frequently septate; pits numerous, minute, simple. Ripple marks sometimes present; all elements storied except the rays. No gum ducts seen.

Common names: *Myriocarpa:* Chaya, cholagogue Indio, mal hombre (Mex.); chichicastillo (Hond.); chichicaste colorado, picapica (Salv.); cow-itch (Pan.); tripa de pato (Col.); ishanga, ortinga, tigre-tigre (Peru). *Urera:* Chichicastre, chichicaste, jamo, ortiga (Cuba); ortiga, o. brava, o. colorado (P.R.); pringamoza (Dom. R.); maman guêpes (Haiti); chichicastle, chichicaxtli, chichicazlillo, la, laal, l. tzimin, laltsimin, mala mujer, mal hombre, ortiga, o. de caballo, quemador, tachinole (Mex.); cow-itch (Br. H.); chichicaste (Guat., Hond., Salv.); chichicaste negua, nigüilla (Salv.); madera de pega, ortiga, palo ortiga (Pan.); guaina, pringamosa, p. de montaña (Col.); pringamoza (Venez.); ishanga, i. del agua, ortinga (Peru); cansanção, caracasana, punú, urtega vermelha, urtigão (Braz.); urtigão (Urug.); ortiga brava, pinó-guazú (Arg.).

VACCINIACEAE

The Blueberry family, often included in the Ericaceae, is composed of about 25 genera and many species of shrubs and little trees, sometimes epiphytic and subscandent, mostly inhabiting mountainous regions of tropical Asia and America. The leaves are alternate simple and without stipules; the flowers are solitary, clustered, or racemose; the fruit is a berry or a drupe. Certain species are well known in the United States because of their edible fruit, namely, the Blueberry (*Vaccinium*), Huckleberry (*Gaylussacia*), and Cranberry (*Oxycoccus*). Most of the genera are represented in tropical America, but because of the small size of the plants they are valueless as a source of wood, which is pale brown to reddish, odorless and tasteless, rather hard and heavy, and of fine and uniform texture.

Growth rings indistinct. Pores small to minute; sometimes minute in young stems and becoming larger outward, though never distinct without lens; very numerous; fairly well distributed without definite pattern, occurring singly or crowded together. Vessels typically with scalariform perforation plates having few to many bars; sometimes with spiral thickenings; intervascular pitting with more or less distinct tendency to scalariform. Rays decidedly heterogeneous, many of the cells square or upright; of two distinct sizes, the larger visible without lens and sometimes rather conspicuous; pits to vessels oval to much elongated, small to large, typically in scalariform arrangement. Wood parenchyma very sparingly developed, not visible with lens; sometimes apparently absent. Wood fibers with moderately thick walls and very numerous distinctly bordered pits. Ripple marks absent. No gum ducts seen.

Common names: *Cavendishia:* Colmillo de perro (C.R.); uvito de monte (Col.). *Gaylussacia:* Blue tangle, dangleberry, huckleberry, whortleberry (U.S.A.). *Macleania:* Cacagüito (Venez.); hualicón lanudo (Ec.). *Oxycoccus:* Cranberry (U.S.A.). *Psammisia:* Colmillo (C.R.); hualicón llucho (Ec.). *Thibaudia:* Asnaludo, chaquiludo (Col.); coralito (Venez.); monte capulí (Peru). *Vaccinium:* Bilberry, blueberry, cowberry, deerberry, farkleberry (U.S.A.); axocopaconi, cahuichi, cahuitzi, madoñito, madroño chino (Mex.); arrayán (C.R.); mortiña (Col.).

VERBENACEAE

The Verbenaceae comprise about 80 genera and 3000 species of herbs, shrubs, vines, and tropical trees, and are of very wide geographical distribution. The leaves are mostly simple, sometimes (*e.g.*, *Vitex*) digitately compound, typically opposite, generally deciduous; the flowers, which often are highly colored, are borne in spikes, racemes, cymes, or panicles; the fruit is either dry, separating at maturity into 2 to 4 nutlets, or a drupe containing the nutlets.

The best known and most important timber tree is Teak, *Tectona grandis* L.f., native to southeastern Asia and Malaya and planted to a limited extent for forestry purposes in West Africa and tropical America (see *Phytologia* 1: 154–164). Commercial timbers are also supplied by species of *Vitex* in the Philippine Islands, India, and New Zealand.

The American arborescent species are of 15 genera. A few of the American timbers are of local utility for carpentry and construction, but the only genus having commercial possibilities is *Vitex*. The following description is based on 160 wood samples, representing 91 species of 13 genera.

Heartwood yellowish, olive, or light to dark brown, sometimes more or less variegated or streaked; often distinct, but generally not sharply differentiated from the sapwood; in some specimens the color is whitish or pale brownish throughout. Luster variable from low to high. Odor and taste usually not distinctive. Density moderate to high; texture fine to medium;

grain typically straight, sometimes wavy or roey; working properties good; resistance to decay poor to good.

Growth rings commonly present. Pores mostly medium-sized or small, the largest scarcely distinct without lens; commonly thick-walled; fairly numerous; occurring singly and in pairs, less frequently in small clusters; fairly evenly distributed in *Callicarpa, Clerodendrum, Cornutia, Duranta, Petitia,* and *Pseudocarpidium,* but in more or less concentric (ring-porous), tangential, or diagonal arrangement in the others. Vessels with simple perforations and rarely also with occasional scalariform plates; without spiral thickenings; tyloses common to abundant in *Citharexylum, Cornutia, Petitia, Pseudocarpidium,* and *Vitex;* deposits of calcium carbonate observed in some of the vessels of *Citharexylum* and *Lippia;* pitting alternate, fine to moderately so. Rays variable in width, sometimes all narrow, sometimes up to 7 cells wide (up to 12 cells wide in a species of *Petrea,* a liana); height usually less than 50, occasionally more than 100, cells; weakly to decidedly heterogeneous; crystals sometimes present; pits to vessels rounded and similar in face view to the intervascular and in part, in some genera, elongated and occasionally tending to scalariform arrangement. Wood parenchyma typically sparse and paratracheal, but abundant, aliform to confluent, in *Aegiphila;* sometimes finely terminal; pith flecks common in woods of low density. Wood fibers septate in many genera; pits numerous, small to medium-sized, more or less distinctly bordered. Ripple marks absent. No gum ducts seen. For anatomy of the different genera see *Tropical Woods* 65: 5–21.

Aegiphila, with about 140 species of low to high shrubs, small trees, and a few lianas, is of general distribution throughout tropical and subtropical America. The leaves are simple, deciduous, penni-nerved, generally opposite but sometimes ternate or verticillate; the flowers are small, of various colors, sometimes fragrant, and usually borne in terminal panicles or axillary cymes; the fruit is drupaceous, 4-lobed, the ectocarp usually dry but occasionally fleshy, juicy, and edible.

The largest trees reported for this genus are about 50 feet high and 10 inches in diameter, but most of the plants are of much lower stature, many of them only straggling bushes. They are of little economic value. Some are planted for ornament, some are reputed to have medicinal virtues, especially as a remedy for snake bite, and a few supply white, easily worked wood for making boxes, wooden shoes, and various small articles, and for interior framing.

Wood white or cream-colored throughout, becoming brownish upon exposure. Luster low to medium. Odorless and tasteless when dry. Light and soft to moderately heavy and hard; texture coarse to rather fine; grain straight; easy to work, tough and strong for its weight; is perishable in contact with the ground. Of no commercial possibilities.

COMMON NAMES: Goatwood, lard wood, spirit weed (B.W.I.); guairo santo, guaro (Cuba); capaíllo (P.R.); bois cabril, b. de bouc, b. de fer, b. sendu, sureau gros (Fr. W.I.); hulub (Mex.); palo de zope (Salv.); vara blanca (Hond.); tabaquillo, zorrillo (C.R.); Juan de la verdad, wild jasmine (Pan.); saúco de monte (Col.); contra-culebra, totumillo (Venez.); mora-cooballi, wanini (Br. G.); bois de golette, b. tobaco, manabo (Fr. G.); lulu, tutumbo (Ec.); chirapa-sacha, fetoró-ey, huaca, ucullucuy-sacha, utcus (Peru); camaá, camará, cambará, carindiba, contra-cobra, cipó pitomba, pau de moquem, tamanqueiro, uruá-rana (Braz.).

Callicarpa, with about 135 species of low bushes, erect or subscandent shrubs, and some typically small trees, is most abundantly represented in the East Indies and Oceania. The American species, about 30 in number, have their center of distribution in Cuba, with extensions into southern United States, Mexico, Central America, and northern South America. The stems of some species are prickly; the leaves are simple, deciduous, and mostly opposite, often glandular or with resinous dots, and fragrant; the flowers are generally very small, frequently fragrant, of various colors, and borne in axillary or super-axillary cymes; the drupaceous fruit is berry-like, with a fleshy exocarp usually very attractively colored. The plants have few uses except for decorative purposes.

Wood white or grayish brown throughout. Luster medium. Mostly of medium density and texture, straight-grained, easy to work; poorly resistant to decay. Of no commercial possibilities because of the small sizes available.

COMMON NAMES: Beauty berry, bunchberry, mulberry (Bermuda, French, or Spanish), sourbush, turkey berry (U.S.A.); filigrana (Cuba); capá rosa (P.R.); patzahumacachil, pukin, sacpukin, uvilla, zacpukim (Mex.); fruta de chacha (Guat.); vara del alcalde (Hond.); blackberry (Pan.).

Citharexylum (meaning fiddlewood), with about 60 species of shrubs and small to medium-sized or rarely large trees, is widely distributed throughout tropical and subtropical America, exclusive of the Amazon basin. The leaves are simple, sometimes glandular, entire or toothed, opposite or ternate; the flowers are small, often fragrant, and borne usually in axillary or terminal spikes or racemes; the fruit is a small berry-like drupe with thin juicy flesh, which in some species is edible.

The best known species is *Citharexylum fruticosum* L., a slender tree commonly less than 30 feet high, growing in southern Florida and many of the Antilles. It has been known for 250 years or more in the British West Indies as Fiddlewood, said to be a corruption of the name Bois Fidèle employed by the early French colonists in allusion to the strength and toughness of the timber and not, as is sometimes stated, because the wood was used for making musical instruments. Sargent (*Manual of the trees of North America,* p. 865) describes the wood as "heavy, exceedingly hard, strong, close-grained, clear bright red, with thin lighter colored sapwood," but there appears to be some mistake as to the color, for the specimens in the Yale collections have at most only a brownish heartwood which merges gradually into the sapwood. The tree is known in Cuba as Guayo Blanco, Palo Guitarro, and Roble Amarillo and, according to Roig (*Diccionario botanico de nombres vulgares cubanos,* p. 340), the wood is white, hard, and compact, suggesting Oak.

The largest species appears to be the Iguanero of Panama, *Citharexylum macranthum* Pittier, a forest tree up to 100 feet tall, with a straight trunk 20 inches in diameter. Pittier says (*Contr. U.S. Nat. Herb.* 18: 4: 169): "This tree departs from all hitherto described species of the genus by the ternate leaves and racemes of the floral branchlets, the unusually large corollas, the stamens inserted well in the lower half of the corolla tube, and the very short pistil. It shares with *C. macradenium* Greenm. the peculiarity of having very large glands at the base of the leaf blade. The core of the trunk is of a dirty yellow color. The wood is hard but tough, and little used." The Yale collections contain a wood specimen from one of the type trees (Yale 3020; Pittier 4199); its texture is coarser and the density much lower than in the West Indian species described above.

The Pombeira or Pau de Viola of southern Brazil, *Citharexylum cinereum* L., is said to be a common tree, sometimes of fairly large size, supplying some timber for general carpentry, crating, and slack cooperage. There are several other species in that general region. Best known in Misiones, Argentina, and in Uruguay is the prickly Tarumá or Tarumán, *C. montevidense* (Spreng.) Mold., sometimes 50 feet high, with a stout bole supplying some moderately soft easily worked lumber useful for boxboards and other purposes satisfied by a plain material of medium strength and poor resistance to decay.

Heartwood dull grayish brown or chestnut-brown, with a subdued golden luster. Odorless and tasteless. Moderately light and soft to very heavy and hard; texture fine to rather coarse; grain straight to roey or otherwise irregular; working properties good to fair; durability poor. Of no commercial possibilities.

COMMON NAMES: Fiddlewood (Florida, B.W.I.); anacahuite, canilla de venado, collarette, guairo santo, guayo blanco, mangle de sabana, palo guitarro, penda, roble amarillo, r. de olor, r. de dulce (Cuba); cucubano, higuerillo, pendula, Susanna, old woman's bitter (P.R.); penda, pendula (Dom. R.); café marrón, mala mujer (Haiti); cotelette (Fr. W.I.); cutlet, Susanna (Trin.); aceitunillo, cacachila, chachalaca, chacalpezle, comida de cuervo, coral, iximché, naranjillo, negrito, orcajuela, palo de violín, panochillo, revientacabra, roble, r. amarillo, sac-xitch-ché, saúco hediondo, takakché, tepesi (Mex.); birdseed, pigeon-feed (Br. H.); café gigante, coralillo, cordoncillo, moco de pava, sogüilla, uva (Salv.); dama (C.R.); corimiente, iguanero, wild cherry, w. lime (Pan.); palo guitarro, pendare, pendula, totumilla (Venez.); pau de viola, pombeira, tarumão de espinho (Braz.); aguay-guazú, coronillo colorado, tarumá con espinas, tarumán espinudo (Arg.); espina de los bañados, naranjo de bañado, tarumá, tarumán, t. de espinas (Urug.).

Clerodendrum is a pantropical genus with about 350 species and varieties of perennial herbs, erect or scandent shrubs, and small trees, rarely armed with spines. There are comparatively few species native to tropical America, but several are cultivated for their ornamental flowers and some have become naturalized. The leaves are deciduous, simple, opposite or whorled, entire or variously dentate, sometimes punctate; the small to large and often beautifully colored flowers are borne in terminal or axillary cymes or panicles; the fruit is drupaceous, separating at maturity into 2 to 4 nutlets. The genus is of no commercial importance as a source of timber. Wood yellowish or whitish throughout. Luster medium. Without distinctive odor or taste. Moderately heavy and hard; texture rather fine; grain straight; easy to work; presumably perishable in contact with the ground.

Common names (both native and cultivated species): Clavellina espinosa, guardia civil, hiel de gallina, h. de perro, jamaiquina, magüira cimarrona, mil flores, palo sabonero, ramo de novia, roble guayo (Cuba); Danish flag, flor de muerto, jasmín hediondo, Santo Domingo, wild Jessamine (P.R.); bocomelia, itzimte, metroceder (Mex.); muste (Guat.); ala de ángel, boca-amelia, bocamelia, martinica, ministeriosa, santa alda (Salv.); fucsia, milflor, verbena (C.R.); bleeding-heart, camella (Pan.); baraque (Venez.).

Cornutia, with about a dozen variable species and several varieties of unarmed pubescent or velvety odoriferous shrubs and small trees, is of general distribution in tropical America except in the Amazon basin. The branches are stout, 4-angled, and brittle; the leaves are simple, entire or toothed, opposite, and deciduous; the small bluish or purplish flowers are typically in large terminal panicles, occasionally in axillary cymes; the fruit is a small globose drupe. The genus supplies no timber of value. Wood pale brownish throughout. Luster medium. Odorless and tasteless. Moderately heavy and hard; texture rather fine; grain straight; easy to work; poorly resistant to decay.

Common names: Salvilla (Cuba); penda (Dom. R.); fiddlewood, purple fiddlewood (B.W.I.); bois cagne, b. care, b. cassau, b. cassave, b. cotelet quarré, b. de saban, b. de savane, mouri debout (Fr. W.I.); latche, pangagé, tzultesnuk, xoltexnuc (Mex.); matasano (Br. H.); flor lila, hoja de jope, h. de zope, latche (Guat.); cucaracho, zopilote (Hond.); azari, pavilla (C.R.); cuatro caras, palo cuadrado, murciélago, murciélago (Pan.); tuónculape (Col.); dona (Peru).

Duranta, with about a dozen species of shrubs and small trees, is of general distribution in tropical America except in the Amazon basin. The branches are often spiny and usually long and drooping or trailing; the entire or dentate leaves are opposite or verticillate; the small white, lilac, or purplish flowers are borne in terminal or axillary racemes; the fruit is a drupe included in the calyx and contains four nutlets. Some of the species are planted for ornament; the genus is not the source of commercial timber. Wood yellow or pale brownish throughout. Luster medium. Without distinctive odor; taste slightly bitter. Hard and heavy; texture very fine and uniform; grain straight; easily worked; poorly resistant to decay. Is suitable for small articles of turnery.

Common names: Celosa, c. cimarrona, violetina (Cuba); azota-caballo, cuento de oro, lila, lluvia, pigeon berry (P.R.); bois jambette, maïs bouilli (Haiti); espino blanco, hombo-coché, kampohoché, kanppocoché (Mex.); chulada, heliotropio (Salv.); pensamiento (Nic.); espino de paloma, varita de San José (Pan.); adonis, a. blanco, a. morado, garbancillo, guapante (Col.); fruta de paloma, limoncillo (Venez.); tala blanca (Arg.).

Lantana, with about 70 species of shrubs and undershrubs, rarely little trees, is widely distributed in tropical and subtropical regions of the world. Best known is *L. camara* L., usually a crooked much-branched shrub, sometimes arborescent but rarely over 15 feet high and three inches in diameter. It is usually armed with stout recurved prickles. On account of its rather showy flowers, the species has been introduced into most tropical countries and in some of them, as in Hawaii, has become a pest. The aromatic leaves are employed in native medicine. Heartwood absent from specimens; sapwood whitish or yellowish. Luster medium to rather high. Moderately hard and heavy; texture rather fine; grain fairly straight; easily worked. Of no commercial possibilities.

COMMON NAMES: Wild sage (Florida); doñanica, filigrana, f. cimarrona, f. de piña, f. olorosa, f. salvia, té de costa (Cuba); coriaquillo, c. de Santa María, poley cimarrón, sage (P.R.); bonbonnier, herbe á plomb, mélisse marron (Haiti); alantana, alfombrillo hediondo, confite, c. negro, confituria, corona del sol, hierba de Cristo, ikilhaxiu, lampana, lantana, matizidilla, mora, orégano, orozuz del país, palabra de caballero, p. de mujer, peonía negra, petekin, siete colores, sonora, s. roja, tarepe, tres colores, uña de gato, xo-hexnuc, zapotillo, zarzamora, zicilhaxiu (Mex.); cinco coloraditos, c. negritos, comida de paloma, corronchocho, orégano de monte, salvia santa (Guat.); cinco negritos, juanilama (Hond.); chiligüe, cinco coloraditos, c. negritos, icaquito, santo negrito (Salv.); cinco negritos, cuasquito oloroso, toltolquelite (Nic.); jaral, jarilla, santo negrito (C.R.); hierba zorra, pasarín, San Rafaelito, wild mint, w. sage (Pan.); sorrito (Col.); bubita negra, cariaquito, c. blanco, c. colorado, c. de sabana, flor de duende, f. de sangre, maiz-zorro, romanceta (Venez.); camará, c. de cheiro, cambará, chumbinho roxo, herva cidreira, h. sagrada (Braz.); caburá-caá, camará, cambará, salvia morada (Arg.).

Lippia, with more than 100 species of herbs, shrubs, and small trees, is sparingly represented in Africa and widely distributed in America, especially in the tropical and subtropical regions. The leaves are opposite or ternate, usually dentate, often aromatic, frequently velvety; the small bracted flowers are of various colors and borne in spikes or heads; the fruit is drupaceous but dry, finally separating into two nutlets. The leaves are sometimes used for seasoning food or for medicinal purposes. The genus is of no value for timber. Heartwood brownish; sometimes rather sharply demarcated from the lighter sapwood. Luster medium. Odorless and tasteless. Hard and heavy to only moderately so; texture fine; grain straight to irregular; not difficult to work; durability probably low.

COMMON NAMES: Cabradora, mint (U.S.A.); azulejo, filigrana, orozuz, o. de la tierra, salvia americana (Cuba); cape weed, cidrón, mejorana, orozuz, poley (P.R.); orégano, té Chino (Dom. R.); fleur la cigale, Marguerite (Haiti); agrito, altamisa, cabalyaxnic, canelilla, cedrón, damiana, epazote, epazotl, hierba buena, h. de la princesa, h. de mula, h. del Negro, h. dulce, h. Luisa, huele de noche, jaboncillo, jasminillo, jazmincillo, mirto, nacare, neuctixihuitl, orégano, orozuz, o. del país, palo de gusano, rosa de Castilla, salvia, s. poblano, sonora, tabaquillo, té de maceta, t. del país, t. del pan, tehuacán, topozana, vara blanca, v. dulce, xakilché, xoltenuuc, xtukuexiu (Mex.); cutujume, tatascamite, té cimarrón (Guat.); carbonero, corronchocho, lechuga de laguna, macahuite, orégano montés, salvia, s. santa, tamayagua, tatascame (Salv.); juanislama de monte, orozul (Nic.); caragre, juanilama, orégano, orozuz, salvia (C.R.); mastranto (Pan.); aloysia, cidrón, Luisa, oreganito macho, rosa-vieja, varilazo, vara de lazo, velita (Col.); amogre, orégano, oreganote, té negro, yerba Luisa (Venez.); cedroncillo, chichara-caspi, huicho-caspi, lauraimana (Peru); alecrim do campo, camará, capitão do matto, chá de frade, c. de pedestre, herva cidreira brava, h. c. do campo, murtinha italiana, pau lixa, salva do Brasil (Braz.); caá-yaguá, niño-rupá, poleo, tungaó-caá (Par.); azahar del campo, cama de niño, cedrón, c. del campo, niño-urupá, retamo (Arg.); azahar del campo, cama de niño, cedrón del campo, cidrilla, cidrón, favorita, garupá, Luisa, María Luisa, niño urupá, sauce limón, yerba cidrera, y. Luisa (Urug.).

Petitia, with two or three closely related species of unarmed trees and shrubs, is confined to the West Indies. Best known is *P. domingensis* Jacq. which attains a maximum height of 70 feet and a diameter of 24 inches, though usually it is much smaller and sometimes only a shrub. Its large opposite entire leaves are dark green above and rusty beneath; the small whitish fragrant flowers are borne in axillary panicles; the fruit is a small rounded black drupe. The timber is of excellent quality and attractive appearance, but the quantity now available is very limited. It is used locally for rollers in coffee-hulling mills, for making carts, and for posts, piling, and props. The figured material is suitable for fine furniture, brush backs, and turned articles.

Heartwood light brown, more or less variegated or sometimes with handsome dark-colored waxy striping; not sharply differentiated from the sapwood. Luster medium to high. Odorless and tasteless. Very hard, heavy, tough, and strong; sp. gr. (air-dry) about 0.95; weight 59 lbs. per cu. ft.; texture rather fine; grain straight to roey or finely wavy; not difficult to work, taking a glossy finish; moderately durable in contact with the ground. Of very limited commercial possibilities.

Common names: Bastard stopper (B.W.I.); guayo prieto, roble de olor, r. guayo (Cuba); capá amarillo, c. blanco, c. de sábana (P.R.); capá blanco (Dom. R.); bois d'sortie, chêne callebassie (Haiti).

Petrea (or *Petraea*), with about 35 species and varieties of shrubs, small trees, and woody vines, is widely distributed in tropical and subtropical America from the West Indies and northern Mexico through Central America to Peru, Bolivia, Paraguay, and southern Brazil. The leaves are deciduous, simple, opposite or whorled, sometimes very rough; the lilac, blue, or white flowers are borne in many-flowered drooping axillary or terminal racemes sometimes suggesting Wisteria; the fruit is a hard 2-celled drupe inclosed in a winged calyx. Many of the species are cultivated for ornament. The wood is not used because of the small size of the plants. The following description applies particularly to specimens of *Petrea arborea* H.B.K. from Venezuela.

Heartwood pale orange-brown, not sharply demarcated from the yellowish or light olive sapwood; with distinct ray flecks on radial surface; has about the consistency of Maple (*Acer*). Luster fairly high. Odorless and tasteless. Moderately hard and heavy; texture medium; grain straight to finely roey; working properties good; probably poorly resistant to decay. Has no commercial possibilities.

Common names: Purple wreath, queen's wreath (Eng.); liane St. Jean (Haiti); lilac (Trin.); bejuco del caballo, comecate azul, estrella azul, opp-tzimin, piocha viejo, soltero, topopostillo, tortilla tostado del caballo, soltero, yoxoptzimin (Mex.); colación, cuera de zapo, Santa Rita (Guat.); carbonero de monte (Hond.); adelfa, adolfina, lengua de vaca (Salv.); hoja chigüe (Nic.); choreque, raspa guacal (C.R.); bejuco de hajo, biurá, buirá, flor de la cruz, f. de mayo, Santa Lucía (Pan.); azulina, azulito, chaparrito, chaparro, chicharrón, copo morado, jazmín azul, mayadena mamoncillo, pluma de la reina, plumito (Col.); María, penitente, Santa Lucía (Venez.); hajauballi saléroe, parapo (Sur.); flôr de São Miguel, f. de viuva, touca de viuva, viuvinha (Braz.).

Pseudocarpidium, with a few species of shrubs and little trees, is limited in distribution to Cuba and the Bahamas. The leaves are opposite, simple, entire or dentate and spine-tipped; the small bluish flowers are in axillary panicles; the fruit is lobed, slightly fleshy or dry. The type of the genus is *P. Wrightii* Millsp., a gray-barked little tree sometimes 20 feet high, with Holly-like leaves. In Gibara, Cuba, it is known as Chicharrón, a name applied elsewhere in the island to various hard-wooded trees, especially Combretaceae. The genus is not a source of timber. Wood brownish yellow throughout, except near wounds, where it is olive-brown. Luster medium. Odorless and tasteless. Very hard, heavy, compact, tough, and strong; texture fine and uniform; grain straight to roey; not difficult to work, taking a high polish; resistance to decay probably low. Of no economic importance.

COMMON NAMES: Chicharrón, granadillo de costa, g. de Cuba, g. macho, yanilla blanca (Cuba).

Recordia is a Bolivian genus with a single known species, *R. boliviana* Moldenke. It is an unarmed shrub or small tree resembling *Citharexylum* in habit and general appearance, and apparently is endemic to the mountains of Santa Cruz. The leaves are opposite, finely serrate, densely pubescent below, mostly clustered at the tips of young twigs, and deciduous; the inflorescence is many-flowered racemose and terminal; the fruit is unknown. (See *Phytologia* 1: 4: 171–174, 1935.) There are no wood samples available for this study.

Rehdera, with three known species of unarmed shrubs or trees, occurs from Yucatán, Mexico, to Nicoya, Costa Rica. The leaves are opposite entire and triplinerved; the flowers are borne in short drooping axillary racemes; the drupaceous fruit is dry. The type of the genus is *R. trinervis* (Blake) Moldenke, a shrub or a tree sometimes 65 feet tall inhabiting low hills and river valley forests in southern Mexico, Guatemala, Salvador, Honduras, and Costa Rica. The wood has not been studied.

COMMON NAMES: Sacuisilché, saquilzciché (Mex.); palo blanco (Guat.); llayo (C.R.).

Rhaphithamnus, with two species of spiny shrubs and small trees less than 30 feet high, occurs in Chile and adjacent parts of Peru and Argentina. The leaves are deciduous, small, entire, and opposite or ternate; the sparsely flowered purplish racemes are axillary; the fruit is a very fleshy 2-seeded drupe. Wood pale grayish brown throughout; with slight greenish tinge in one specimen. Luster medium. Odorless and tasteless. Moderately heavy and hard; rather fine-textured; straight-grained; easily worked. Presumably poorly resistant to decay. Without commercial possibilities.

COMMON NAMES: Prickly myrtle (Eng.); amyán macho, arrayán de espino, a. espinudo, a. macho, espino, e. blanco, e. negro, guayún, Juan bueno, rapú, repú, r. mayún (Chile); arrayán macho (Arg.).

Vitex, with about 70 species of unarmed shrubs and small to large trees, is widely distributed throughout the tropics, with a few representatives in temperate regions. The opposite mostly deciduous leaves are digitately compound, with 1 to 7, mostly 3 or 5, entire leaflets; the white, blue, lilac, purplish, or red flowers are borne in axillary cymes or terminal panicles and in some species are very showy; the fruit is a drupe, in some instances edible.

The genus is the source of several valuable timbers. The Puriri of New Zealand, *Vitex lucens* T. Kirk, sometimes called New Zealand Teak, has been used so much for railway crossties, posts, and house blocks that the supply is almost exhausted. *V. altissima* L.f. is a large and important tree in southern and western India and Ceylon, finding numerous local uses where a strong and durable timber is required. The Philippine Molave, *V. parviflora* Juss., is widely distributed throughout the islands in second-growth and open primary forests, but the accessible supply of well-formed trees is now very limited; it is used for ship building, especially for keels and ribs, and in house construction for sills, floors, window and door frames, and balusters.

American species also supply some good timbers, but they are not known to the export trade. *Vitex divaricata* Sw. is a tree 30 to 60 feet high and from 20 to 30 inches in diameter in the West Indies and extending into northern Venezuela; the wood is used locally for framework of houses, general carpentry, and in cabinet work. There are several species in Central America. *V.*

Kuylenii Standl., known in eastern Guatemala as Barabás or Barbás, is a tree sometimes 90 feet tall with a well-formed bole 30 inches in diameter; the yellowish wood is moderately hard and heavy, works readily, and finishes smoothly, but is not utilized. In the same region is another species, probably *V. Cooperi* Standl., called Rajate Bién because of the ease with which the timber can be split; it is dull brown throughout or somewhat streaked, very easy to work, dries very slowly but without checking badly, and is employed locally for durable construction and general carpentry. A common tree in British Honduras and Yucatán is *V. Gaumeri* Greenm., sometimes 60 feet tall and 24 inches in diameter; it is very attractive when covered with panicles of small blue flowers. There are two or three other species in Mexico and at least two more in Panama, but they are of minor utility.

There are four or five species in Colombia and Venezuela, and the usual common name for them is Aceituno. Some are small trees but others attain heights of 75 to 100 feet and diameters of two to three feet. *Vitex columbiensis* Pitt. has a dark olive-brown wood of the type of Rajate Bién of Guatemala. *V. cymosa* Bert., which extends into the Amazon basin, is more yellowish and of a somewhat different texture, and some specimens are comparatively soft. Another Amazonian species is *V. spongiocarpa* Ducke, said to be the largest representative of the genus in Brazil. Ducke says (*Tropical Woods* 31: 21) that it is especially remarkable "on account of the voluminous spongy mesocarp of its fruits which, during the rainy season, float in the swamps of the upland forests, sometimes in abundance. The blossoming trees, crowned by their wide upward-raised inflorescences, suggest Teak (*Tectona grandis*) rather than the other Brazilian *Vitex* species." Its pale olive-brown heartwood merges gradually in the whitish sapwood; it is rather light, but firm and tough, saws somewhat woolly when fresh, and does not appear very durable. (For further description of the wood see *Tropical Woods* 31: 28–29.)

The southernmost species is *Vitex montevidensis* Cham., a tree sometimes 60 feet tall and 36 inches in diameter, growing in parts of Argentina and Uruguay. The fruit is edible and of an agreeable flavor. The heartwood of mature trees is of a dark olive-green color, easy to work, and durable; it is used locally in general construction, ship building, and joinery.

The following description is based on 46 specimens of 21 species of *Vitex* representing almost the entire American range of the genus. Heartwood exhibiting various shades of brown, deepening upon exposure; sometimes oily looking; usually not sharply demarcated from the whitish, yellow, or pale brown sapwood. Luster low to high. Odor and taste absent or not distinctive. Density variable; sp. gr. (air-dry) 0.60 to 0.80; weight 37 to 50 lbs. per cu. ft.; texture rather fine to moderately coarse; grain usually straight, sometimes wavy; working properties good to excellent; holds its place well when manufactured; durability variable. A good timber for general purposes not requiring attractive figure or color. Some samples suggest Satinwood (*Zanthoxylum*), others have a superficial resemblance to *Cordia*.

Common names: Black fiddlewood, lizard wood (B.W.I.); chicharrón, c. de costa, ofón, o. criollo, roble guayo, r. güiro (Cuba); higuerillo, pendulo blanco (P.R.); matta becerro, palo perrito (Dom. R.); bois de savanne, gri-gri (Haiti); bois lezard (Fr. W.I.); agua-malario, aguilote, ahuilote, atuto, coyotemate, negrito coyote, negro coyote, obalamo, tescialama, torete, ualama, uvalama, uvalano, valama, yaaxnic, yaxnic (Mex.); blue blossom, fiddlewood, monkey fiddle (Br. H.); barabás, barbás, jocote de mico, rajate bién (Guat.. Hond.); cuaja, cuajada, quajado, yellow manwood (Pan.); aceituno, a. blanco, fruta de gonzalo, peronilla (Col.); aceituno, guaretaro, totumillo, t. negro (Venez.); hackiaballi (Br. G.); pechiche (Ec.); tahuari (Peru); jaramantaia, mammeira, María preta, tarumá cheiroso, t. da matta, t. do alagado, t. do igapó, t. frondoso, t. grande do campo, t. silvestre, t. tuira, velame do campo (Braz.); tarumá (Par.); tarumá de ley (Urug.); tarumá, t. duro, t. guazú (Arg.).

VIOLACEAE

THE Violet family comprises 17 genera and 800 species of herbs, half-shrubs, erect or scandent shrubs, and some small or rarely medium-sized trees. The leaves are simple, stipulate, alternate or rarely opposite; the solitary to paniculate flowers usually have the lowermost petals larger and spurred; the fruit is an elastically dehiscent capsule or a berry. The family supplies no timber of commerce. *Viola,* with about 400 herbaceous species, is widely distributed in temperate regions, mostly of the northern hemisphere; *Hymenanthera* and *Melicytus* have their center of distribution in New Zealand; the others are typically tropical. Occurring only in tropical America are nine genera, namely, *Amphirrox* (shrubs), *Anchietea* and *Corynostylis* (climbers), *Gleospermum* (trees and shrubs), *Leonia* (trees), *Noisettia* (half-shrubs), *Paypayrola* and *Rinoreocarpus* (trees and shrubs); there are also species of two pantropical genera, *Hybanthus* (herbs and shrubs) and *Rinorea* (trees and shrubs).

Heartwood typically brownish in *Rinorea,* but absent or not clearly differentiated from the lemon-yellow or cream-colored sapwood in the others. Luster medium. Odorless and tasteless except in *Hybanthus,* which sometimes has a sweetish taste. Moderately hard and heavy, though somewhat variable in different genera; texture fine and uniform; grain usually straight; easy to work, finishing very smoothly; poorly resistant to decay and subject to sapstain if not dried properly. Suitable in part for some of the same purposes as Venezuelan Boxwood (*Gossypiospermum*).

Growth rings absent or poorly defined; sometimes indicated by slight differences in density. Pores typically angular; small to minute; variable in abundance, numerous but not crowded; apparently all solitary in *Anchietea,* but often in small multiples in the others; tendency to radial arrangement more or less noticeable, being most distinct in *Hybanthus.* Vessels with short to very long scalariform perforation plates except in *Anchietea* and *Hybanthus* which (like Old World genera *Agatea, Hymenanthera, Melicytus,* and *Viola*) have exclusively simple perforations; intervascular pitting variable from irregularly alternate to opposite or (in *Leonia* and *Paypayrola*) scalariform; tyloses sometimes present. Rays decidedly heterogeneous; of two sizes, uniseriate, and up to 4, sometimes to 8, cells wide; the multiseriate often high and conspicuous; lighter than the background, and frequently appearing double in part on cross section; uniseriate rays and ray margins composed of upright cells, multiseriate parts of square or short procumbent, frequently crystalliferous, cells; sheath and vascular by-pass cells common; pits to vessels typically large, oval to elongated, in definitely opposite or scalariform arrangement; pits small, rounded, not opposite in *Hybanthus.* Wood parenchyma apparently absent. Wood fibers with thick walls and rather large lumen; usually abundantly septate; pits numerous, indistinctly bordered. Ripple marks absent. No gum ducts seen.

Amphirrox, with five species of shrubs and trees rarely 35 feet tall and six inches in diameter, is widely dispersed throughout tropical South America and extends northward into the Atlantic coastal region of Costa Rica. The pale yellow wood shows the rays as prominent white flakes on the radial surface and as white lines on the tangential.

COMMON NAME: Farinha seca (Braz.).

Anchietea, with eight species of scandent shrubs, occurs mostly in southern tropical South America, with extensions through the Andes of Peru into Colombia.

COMMON NAMES: Cipo suma, paraguaya, periguaia, piraguara, piraguayo, piriquaia, puru-uara (Braz.); isipó-mí (Arg.).

Gloeospermum, with nine described species of shrubs and small trees, has its center of distribution in Colombia and northeastern Peru. The largest species reported is *G. dichotomum* (Rusby) Melch. According to Ramón Espina, who collected three specimens at an elevation of about 5000 feet in the Sierra Nevada de Santa Marta, Colombia, it is sometimes 50 feet tall and 14 inches in diameter. The bright yellow hard fine-textured wood looks like a good Boxwood substitute. Available samples of *G. sphaerocarpum* Tr. & Pl. from the Brazilian Amazon are pale buff and the technical properties are inferior to the other.

COMMON NAMES: Millua-caspi (Peru); farinha seca (Braz.).

Hybanthus, with about 75 species of herbs and shrubs, is widely distributed in tropical and subtropical regions of the world, a few species entering the temperate zone. Woody forms occur from the West Indies and Mexico to southern Brazil and Argentina. The roots and rhizomes of *H. ipecacuanha* (L.) Baill. of Brazil are used medicinally. The woods of *H. anomalus* (H.B.K.) Melch. of Colombia and *H. havanensis* Jacq. of Cuba are of the Boxwood type, but the stems are all small.

COMMON NAMES: Diente de perro, yerba de San Martín (Cuba); sacbacelcan, ta (Mex.); hierba del rosario (Salv.); palo negro (Nic.); mapola de monte (Col.); pachaga-pichinya (Peru); ipéca branca, i. de Marajó, ipeconha de flor branca, piraaya, poaia, poaya, p. branca, p. da praia, purga de praia, p. do campo (Braz.).

Leonia, with three species of trees, is confined to the Amazon basin. *L. glycycarpa* R. & P. is said by Llewelyn Williams (*Woods of northeastern Peru,* p. 348) to attain a height of 60 to 70 feet, with a slender trunk 10 to 24 inches in diameter; the timber is esteemed locally for house construction, general carpentry, and fuel. The wood is very similar to that of *L. cymosa* Mart. of Brazil. Color pale yellow or buff throughout. Luster medium. Odorless and tasteless. Density medium, texture fine and uniform; grain rather irregular, easy to work, finishing smoothly; subject to blue stain.

COMMON NAMES: Arenillo (Col.); ninacaspi, trapiarana, urcu-tamara, witu-malla (Peru); trapia-rana (Braz.).

Paypayrola, with seven species of shrubs and little trees, occurs in tropical South America from the Guianas to Rio de Janeiro, Brazil, and westward through the Amazon region to northeastern Peru. The cream-colored, rather light wood is easy to work and has an attractive silver grain, but is not available in sizes large enough to be utilized.

COMMON NAMES: Tornilla muena (Peru); manacá-rana (Braz.).

Rinorea (= *Alsodeia*) is a pantropical genus of more than 250 species of shrubs and small trees rarely 40 feet tall and eight inches in diameter. There are about 40 species in Latin America. The leaves are sometimes opposite or apparently so, and frequently mucilaginous; the small white or yellowish flowers are regular, with free petals and stamens; the 3-valved mucilaginous capsules have a faint odor of Slippery Elm (*Ulmus fulva* Michx.). According to Blake (*Contr. U.S. Nat. Herb.* 20: 13: 494), "No species of *Rinorea* is known to be of much economic importance, although a few are used by natives for one purpose or another. To this fact, as well as to a lack of striking features in habit or color, is due their comparative paucity of vernacular names. . . . *Rinorea physiphora* (Mart.) Baill. of the Rio de Janeiro region is known as Lobolobó, and the leaves are said to be boiled and eaten as greens by negroes, a use for which they seem to be singularly unsuited. St. Hilaire, however, was of the opinion that the species might be improved and made of some importance by cultivation. Baillon states that the astringent bark of some species is used as a febrifuge." The typically brownish, hard to moderately hard, fine to medium-textured wood is apparently not utilized for any special purpose.

COMMON NAMES: Castarita, frutillo (Mex.); wild coffee (Br. H.); cafecillo (Guat.); guayacillo, molenillo (Pan.); jazmín, pie de venado (Col.); cuspa, rabo de cachicamo, tabaquito (Venez.); conohorié, jasmin (Fr. G.); cafecillo, chiricsanango (Peru); baridikutshi, mamusaru (Br. G.); ajará, canella de jacamim, inambú-quiçaua, jacamim-renepeá, lobolobó (Braz.).

Rinoreocarpus *Ulei* (Melch.) Ducke, the sole species, is a tree sometimes 40 feet tall occurring in the middle and western Amazon region. Wood light buff throughout; silver grain distinct. Luster rather high. Odorless and tasteless. Has about the consistency of Soft Maple (*Acer rubrum* L.); texture fine; grain straight; working properties good.

VITACEAE

THE Grape family (sometimes called the Ampelidaceae) consists of about a dozen genera and several hundred species, typically nodose or jointed woody vines, widely distributed in tropical and temperate regions. The leaves are alternate or the lower sometimes opposite, simple or variously compound, often gland-dotted, and with stipules on the petioles or absent; the clusters of small flowers and the tendrils are opposite the leaves; the fruit is a berry, often pulpy, usually with four bony seeds. The American representatives are all vines, such as the Virginia Creeper or Woodbine and various kinds of Grapes; they are mostly slender, although some grapevines attain a diameter of eight to ten inches. The plants are of no value as a source of wood.

Vitis, with upward of 100 species and many horticultural varieties, is native to the northern hemisphere, especially the temperate regions. In the true Grapes (*Euvitis*) the bark is loose and shreddy, the tendrils are forked, and the nodes are solid, whereas in the Muscadine Grapes (*Muscadinia*) the bark adheres closely to the branches, the tendrils are simple, and the pith is continuous through the nodes. The heartwood is dull brown, rather hard, very porous and coarse-textured. The stems contain a copious supply of water and are the occasional source of a refreshing drink to a thirsty hunter or traveler.

Growth rings usually present. Pores numerous; of two sizes, very large and sometimes zonate, and very small to minute and mostly flattened in radial rows. Vessels typically with simple perforations; without spiral thickenings; pitting coarse, generally scalariform. Rays up to 15 cells wide and usually high, sometimes exceedingly so; nearly homogeneous; most of the cells slender and procumbent, some of them square; pits to vessels usually large, elongated and parallel; bundles of raphides sometimes present. Wood parenchyma paratracheal, not very abundant; pitting scalariform when in contact with vessels. Wood fibers thick-walled; usually septate; pits numerous, small, simple. Vasicentric tracheids with scalariform pitting sometimes associated with smallest vessels; delicate spiral thickenings may be present. Ripple marks and gum ducts absent.

VOCHYSIACEAE

A FAMILY of six genera (*Callisthene, Erisma, Erismadelphus, Qualea, Salvertia,* and *Vochysia*) and about 160 species of small to very large trees, some shrubs, and a few herbaceous plants. Their center of distribution is in Brazil, but the range extends to southern Mexico, and one genus, *Erismadelphus,* with one species of medium-sized trees, occurs in the Cameroons and Middle Congo of Africa. The plants are characterized by opposite or whorled simple leathery leaves, with or without stipules; the spurred flowers are in conspicuous racemes or panicles; the fruits are 3-valved, capsular or samaroid. Huber (*Bol. Mus. Goeldi* 6: 188) says that they are among the largest of the Amazonian trees and upon blossoming are transformed into gigantic bouquets of yellow (*Vochysia*), white or rose (*Qualea*), or red (*Vismia*). The bark of most genera contains a colorless liquid resin that thickens upon exposure to the air. The timbers are of little or no commercial importance at present, but a few of them are promising.

Heartwood variable in color from nearly white or gray to pink or reddish brown; with or without sharp demarcation from the sapwood. Luster sometimes golden or silvery in proper light, although the surface may appear dull or mealy because of the abundance of parenchyma. Without distinctive odor or taste when dry. Density variable, light and soft to heavy and hard, mostly medium; texture typically coarse; grain straight to roey; working properties usually good, but some specimens dull tools quickly; durability fair to poor.

Growth rings present or absent. Pores medium-sized to large, usually distinct without lens; rather few to numerous, but not crowded;

occurring singly and in small multiples, well distributed without pattern, though sometimes tending to diagonal arrangement. Vessels with simple perforations; without spirals; pits alternate, usually medium-sized, vestured. Rays variable from all fine to distinctly 2-sized, the larger ones 3 to 8 cells wide and 50 or (in *Vochysia*) sometimes 100 cells high; homogeneous to decidedly heterogeneous; pits to vessels large in *Erisma,* small to medium-sized in the others. Wood parenchyma greatly variable in amount and arrangement, but usually abundant; paratracheal, aliform, often confluent into wide bands; sometimes terminal also; frequently adding prominence to vessel lines or producing figure on tangential surface, as in certain Leguminosae. Wood fibers with thin to thick walls; pits small, simple. Ripple marks absent. Vertical gum ducts, gummosis type, common in three genera. Included phloem typical of *Erisma,* rare in *Qualea.*

Erisma. There are about 20 species in the Amazon basin, all trees, sometimes 125 feet tall. The seeds are rich in oil for industrial purposes and in the Amazon estuary considerable quantities are collected and exported under the name of "jaboty." Although the timber is obtainable in large dimensions it is not utilized. Heartwood reddish or purplish brown; sapwood grayish. Luster rather high in proper light. Odorless and tasteless. Moderately heavy, hard, tough, and strong; texture coarse; feel harsh; grain irregular; not difficult to work, but does not finish very smoothly; durability probably high.

Common names: Alaän-kopie, koesalijepo, kwalidan, kwarerian, letè-balli-bèlèro, paloeloeibjo, paroeroe pio, sengrie kwarrie, singrie kwarie, waáta aláan-kopie, warapa kwarè, w. kwa^r^èrè, witti-hoedoe (Sur.); bruto, caramurú, jabi araconha, jaboty, j. da terra firme, j. da varzea, quaruba de flores roxas (Braz.).

Qualea, with about 45 species, is common in the Amazon region and the Guianas, and some of the trees attain a height of nearly 200 feet. The timber varies widely in appearance and properties, and the kind that is used locally for dug-out canoes, carpentry, and interior construction bears some resemblance to Spanish Cedar (*Cedrela*). According to Bertin (*Les bois de la Guyane française et du Brésil,* Paris, 1920, p. 81), the Cèdre Gris of French Guiana, *Q. rosea* Aubl., is of good timber form, sometimes 100 feet tall, with an average trunk diameter of two feet and free of large branches for 60 to 75 feet. In the areas he examined, the species ranked eighth in abundance, making up a little more than 2 per cent of the total. The heartwood is pinkish brown with a golden luster, sometimes with an attractive roe grain; moderately hard; sp. gr. 0.60 to 0.70; weight about 37 to 44 lbs. per cu. ft.; easy to work. Local uses include joinery, flooring, interior trim, and furniture.

An Amazonian wood closely resembling the Cèdre Gris in luster and grain has been identified as *Qualea paraensis* Ducke. It is somewhat harder and is yellowish rather than pinkish. In both species the parenchyma is not banded, but vasicentric, aliform, and more or less confluent diagonally. A specimen of *Q. retusa* Warm., local name Umiry-rana, is reddish brown, coarse-textured, with a harsh feel, and is considerably harder and heavier than the preceding; it has an attractive wavy grain. The woods of *Qualea* and *Vochysia* are much alike and the range of variation within a genus is about as great as between genera, but the rays of *Qualea* are finer and show less prominently on radial surface. Moreover, some of the woods of *Vochysia* are lighter and softer than any of the other genus examined. Occasional lots of logs are exported. In Germany the timber is sometimes known as Brazilian Okume; the name Aburaq (from Quaruba spelled backwards) has also been coined for it.

Heartwood pinkish brown to reddish brown, occasionally olive-brown; sometimes sharply demarcated from the grayish or yellowish sapwood, sometimes with gradual transition. Luster golden in reddish kinds, medium in the others. Odorless and tasteless. Density medium to high; texture moderately to decidedly coarse; grain generally straight; working properties good.

Common names: Florecillo (Venez.); cèdre gris, couari cèdre, grignon fou, g. f. rouge (Fr. G.); be-kwai, bikiti, goenfoloe, gronfoeloe, iria-kopie, jeriakopi, jakoppitaparin, jokopi, khaléméroe, kharréméroe, koenore-èrèparè, kwa-ie, maoranaballi, mawaranaballi, meniridan, m. hororadi-korro, m. oewe-berabandikoro, meniriolan, miniridan-balli, saniki pisie, tiapotano, tomoené-irakopie, wassie-wassie, w.-w. kwarie, water-kwarie, watra-kwarie, wossie-jossie, woto-kwaleli, woto-kwarie (Sur.); ariauá, cutiuba, laba-laba, lacreiro, mandioqueira, marajuba, monteiro de costa, muiraúba da terra firme, m. da varzea, pau de mastro, p. mulato da terre firma, p. terra, quaruba, q. azul, umiry-rana (Braz.).

Salvertia *convallariaeodora* St. Hil., the sole species, is a small to medium-sized tree in eastern Brazil, being of frequent occurrence in the upland campos of the lower Amazon region. It is highly ornamental, as the foliage is attractive and the flowers are large, white, and fragrant. The only specimen available (Yale 35235) was collected by A. C. Brade in Minas Geraes and determined by Adolpho Ducke; it is of a stem a few inches in diameter with a large pink pith, and thick, deeply fissured bark. Heartwood not seen; said to be pale reddish brown in old trees; sapwood brownish yellow. Luster rather low. Odorless and tasteless. Moderately heavy, hard, and tough; texture coarse; grain irregular; not difficult to work, but fibrous and unattractive. Of no commercial possibilities.

COMMON NAMES: Colher de vaqueiro, folha larga, Gonçala Alves, mafuá, mafurá, pau de arára (Braz.).

Vochysia, with about 90 species of trees and shrubs, is represented throughout tropical America from southern Mexico to Peru, but most abundantly in the Guianas and Brazil. The trees vary in size from small to very large and in many places are of common occurrence. The timber has the same local uses as that of *Qualea;* the two genera are not always distinguished and in a given locality the same vernacular name is likely to be applied indiscriminately to two or more closely related species. The woods vary in appearance and technical properties, but generally the heartwood is pinkish or salmon, often highly lustrous. Rows of dark red gum ducts, similar to those in Spanish Cedar, are common.

Apparently the only attempt to develop an export trade in the timber has been in British Honduras, and trial shipments of about 25,000 board feet of logs were sent to the United States in 1924–25. The tree is common in the second-growth forest, composing the bulk of stands on the moist but not swampy sandy clay soils of the southern coast, but attaining its largest dimensions in the mixed hardwood and Cohune Palm forest. The trunk is well formed and is often free of branches for 50 to 65 feet. Two classes of timber are distinguished after felling, namely, White Yemeri and Red Yemeri. The former seasons more quickly than the other, but it is difficult to work when dry and, being perishable when exposed to decay, is only suitable for interior construction. Red Yemeri is more durable, is easy to work, and is suitable for plywood for general utility purposes, drawer bottoms, and panels. Quartersawed lumber, if free from "gum veins," has a rather attractive appearance and should find a place in the manufacture of the less expensive grades of furniture. Although some botanists claim to recognize two Central American species, namely *Vochysia guatemalensis* Donn. Smith, and *V. hondurensis* Sprague, the differences are slight and do not account for the variations of the timber which, presumably, are attributable to conditions of growth.

The following general description is based upon specimens of 14 species of *Vochysia* from virtually all parts of the range of the genus. The group is less heterogeneous than *Qualea.* Heartwood pinkish; distinct but not always sharply demarcated from the thick whitish or oatmeal-colored sapwood. Luster high, more or less golden. Odorless and tasteless. Density usually medium, sometimes rather low; strong in proportion to weight; texture medium to coarse; grain somewhat roey; usually readily worked, finishing rather attractively; durability fair.

COMMON NAMES: Corpo, corpus, puanchap (Mex.); emery, white mahogany, yemeri (Br. H.); San Juan, S. Pedrano (Guat., Hond.); yemeri (Nic.); palo de chancho (C.R.); mecri, yemeri macho (Pan.); arracache, avelluelo, corosito, flor amarillo, malambito, salado (Col.); tintín (Venez.); aeta-balli, eta-balli, ite-balli (Br. G.); bois agouti, b. préfontaine, grignon fou, g. gris, g. rose, g. rouge, pageolet, préfontaine, wachi-wachi blanc (Fr. G.); alaan kopie, be kwa-ie, eta-balli, goewanna kwarie, gwanna, g. kwarie, ietiballi harira-roe, kowalli, kowaré, kwaí, kaw-ie, kwalie, kwaré, kwarie, seteballi-korero, witte kwarie, wosjie-wosjie, wossie-wossie (Sur.); chimbuya (Ec.); chambo caspi, goma amarilla, timareo de altura (Peru); caizeta, canela santa, cedro-rana, cinzeiro, coariúba, congonha do campo, congonheiro, coxa de frango, muricí da serra, pau de cinza, p. de tucano, p. de vinho, p. de v. preto, p. terra, quaruba, q. branca, q. de flor pequana, q. vermelha, rabo de arara, r. de tucano, urucuca, vinheiro do matto (Braz.).

WINTERACEAE

THIS small family of shrubs and small trees is closely related to the Magnoliaceae and often included with them. The range is extensive and includes the Malay Archipelago, Oceania, and Central and South America. The woods, though of good quality, are not commercially important, but two genera, *Drimys* and *Zygogynum,* are remarkable among the Dicotyledons because they are without vessels. The only genera represented in the New World are *Drimys* and *Illicium.* Their woods are very unlike.

Drimys, with several species, mostly shrubs, occurs in Borneo, New Caledonia, Australia, New Zealand, and North and South America. *D. Winteri* Forst., with its varietal forms to which some botanists give the rank of species, has an extensive range from Vera Cruz and Oaxaca, Mexico, southward to the Straits of Magellan. While nowhere abundant, it is fairly common in southern Chile and is sometimes the dominant element of small low stands. Usually it is not over 30 feet tall with a short trunk a foot through, and frequently is little more than a shrub, but occasional specimens attain a height of 60 to 80 feet and a diameter up to 36 inches. Its fragrant white flowers are a source of perfume, but the tree is best known on account of its bark. Standley (*Trees and shrubs of Mexico,* p. 276) gives the following account of it:

"This plant, which furnishes the Winter's bark of commerce, was first obtained by Winter, who was captain of one of the ships which accompanied Sir Francis Drake's expedition of 1577. The three vessels of the fleet were struck by a storm in the southern ocean, and Winter's ship was driven to the Straits of Magellan where three weeks were spent with the object of improving the health of the crew. *Drimys* was one of the plants which attracted Winter's attention, and he used the bark for treating scurvy. Specimens of the bark were presented to the famous botanist Clusius, who gave it the name of Cortex Winteranus. It became a favorite remedy in Europe, but as it was difficult to obtain the drug from South America the bark of *Canella alba,* a West Indian tree, was often substituted for it. Winter's bark is little used at the present time except in domestic medicine in the regions where it is native. It is aromatic and pungent and has toxic and antiscorbutic properties. In Brazil it is used for dysentery and for gastric disturbances. In Costa Rica the bark is chewed for toothache. The powdered bark is sometimes employed in Mexico as a condiment."

The timber, which has the appearance of Beech (*Fagus*) and the consistency of Basswood (*Tilia*), is used locally for boxes, cases, interior woodwork, and similar purposes. Color pale brown or pinkish throughout. Without distinctive odor or taste. Sp. gr. (air-dry) 0.50; weight about 31 lbs. per cu. ft.; tough and strong for its weight; texture fine and uniform; grain usually straight; easy to work, finishes smoothly, holds nails well; is poorly resistant to decay.

Growth rings fairly distinct, but with little contrast in density between early and late wood. Vessels absent. Ground mass composed of long tracheids; pits in two or more rows in radial walls, the borders conspicuous and circular, the apertures lenticular; pits to ray cells smaller but distinctly bordered. Rays 1 to 7 cells wide and few to 100 cells high; of two classes, the larger suggesting *Fagus* and distinct on all sections; heterogeneous, most of the cells square or upright; pitting between ray cells copious, irregular. Wood parenchyma sparingly diffuse, not visible with lens. Ripple marks and gum ducts absent.

COMMON NAMES: Winter's-bark tree (Eng.); chachaca, chilillo, palo de Chile, p. picante (Mex.); muelo, quiebra-muelas (C.R.); aji canelo, canelo de páramo, cupis, quinón, palo de agi (Col.); arbol de agi (Venez.); casca d'anta, carne d'anta, paratudo, pau paratudo, melambo, canella amarga, c. de paramo, cataia, catara, capororoca (Braz.); canelo (Chile); canelo, palo de mambo (Arg.).

Illicium, with seven species of aromatic evergreen shrubs and small trees, occurs in eastern Asia and North America. The fruit is a whorl of hard, drupe-like, finally dehiscent, one-seeded follicles and is the source of star anise of China and Japan, used for flavoring food and cordials. There are two American species, *I. parviflorum* Michx., occurring along the coast from Georgia to Florida, and *I. floridanum* Ellis, occupying swamps from Florida to Louisiana and also found in Vera Cruz and Puebla, Mexico. Both are shrubs or little trees not over 10 feet tall and are reputed to be poisonous to livestock. They have no special uses and are too small to produce any timber. The following description is based on a single specimen (Yale 19948) of *I. floridanum* collected by Roland H. Harper in Florida.

Wood pale brownish throughout, showing pinkish rays clearly on radial surface. Luster medium. Without distinct odor or taste, although the inner bark is spicily scented. Moderately heavy, hard, compact, suggesting Sugar Maple (*Acer saccharum* Marsh.); texture fine and uniform; grain straight; easy to work, finishing very smoothly; durability probably low.

Growth rings present. Pores angular; very small, not visible without lens; numerous; solitary; well distributed without pattern. Vessels with long scalariform perforation plates having very numerous bars; no spiral thickenings seen. Rays very numerous; uniseriate and biseriate and few to 25 cells high; decidedly heterogeneous, with most of the cells square or upright, the procumbent cells short; walls rather thick and abundantly pitted; pits to vessels elongated and in scalariform arrangement. Wood parenchyma very sparingly paratracheal. Wood fibers in definite radial rows; walls thick; pits numerous, large, distinctly bordered. Ripple marks and gum ducts absent.

COMMON NAMES: Poison bay, sweet laurel (U.S.A.); ixcapantl, mata caballos (Mex.).

ZYGOPHYLLACEAE

THIS family is composed of about 25 genera and 200 species of herbs, shrubs, and a few trees, usually inhabiting dry regions, often near the seacoast, of both hemispheres. The principal woody genera in the New World are *Guaiacum, Bulnesia, Porlieria,* and *Larrea*. The branches are commonly jointed at the nodes. The leaves are opposite and abruptly pinnate, the leaflets entire and inequilateral; the stipules are paired, deciduous or persistent, and sometimes spinescent; the blue, purplish, or yellow flowers are solitary or in umbel-like clusters; the fruit is an angular or winged fleshy or dry capsule with few, sometimes arillate, seeds. The only well-known timber is the Lignum-vitae of commerce, which has been an article of trade for more than four centuries; for certain exacting purposes no satisfactory substitute for it has ever been discovered. The woods of the four genera are much alike.

Heartwood olive-brown or superficially green, more or less striped; sharply demarcated from the yellowish sapwood; inner sapwood often with distinct dark green vessel lines. Luster low. Odor mild and pleasant, most noticeable when specimen is warmed; taste slightly acrid. Nearly always extremely hard and heavy (*Bulnesia retama* less dense); sp. gr. (air-dry heartwood) 1.10 to 1.32; weight 69 to 83 lbs. per cu. ft.; dry sapwood will float in water; texture fine; feel oily or waxy; grain interwoven as a result of the change in direction of narrow alternate fiber layers, making wood finely roe-grained and very difficult to split radially; highly resistant to compression and wear, but brittle under impact; very durable.

Growth rings distinct to indistinct. Pores small to minute (occasionally medium-sized in *Guaiacum*); radially arranged in *Bulnesia;* scattered, often very irregularly, in the other genera. Vessels with simple perforations; without spiral thickenings; direction of vessels, as seen on tangential section, frequently very irregular; pitting fine, alternate. Rays usually 1 to 3 cells wide and less than 8 cells high, but 1 to 3 cells wide and up to 30 cells high in *Bulnesia retama* (Gill.) Gris.; homogeneous to heterogeneous; pits to vessels of same size as the vascular. Wood parenchyma sparingly to abundantly developed; paratracheal, sometimes aliform and occasionally confluent in bands of variable width; apotracheal as single cells or in patches; crystalliferous strands common. Wood fibers short; not in radial rows; very thick-walled and small-lumined; pits typically very small, with indistinct borders. Ripple marks present, regular, all elements storied; exceedingly fine, 220 to 280 per inch, not visible without lens. Gum ducts absent. Gum-resin abundant in all elements of heartwood; color dark brown, oxidizing to green.

Bulnesia. Eight species of this South American genus have been described, but only two of them are large enough to supply any timber of more than local utility, the others being mostly low-branching shrubs on the dry Andean foothills in Argentina. One of the latter group is the Retamo, *B. retama* (Gill.) Gris., a well-known Argentine plant which ordinarily is a shrub less than 15 feet tall, but sometimes develops into a short and stout tree 25 feet high and 10 to 20 inches in diameter. The slender branchlets, leafless most of the time and suggesting *Casuarina*, are crowded together in broom-like masses which are used by mountaineers for thatching their houses. The yellowish olive heartwood is often beautifully figured with blackish green markings; the sapwood is yellowish white. The density is not as high as that of the other species studied, rarely exceeding 0.90, or 56 lbs. per cu. ft. The timber is used locally for fence posts, tool handles, walking sticks, articles of turnery, and furniture. The structure appears intermediate between *Bulnesia* and *Larrea*.

Bulnesia Sarmienti Lorentz occurs in groups or patches in the dry regions in Argentina and Paraguay and attains a height of 50 or 60 feet and a diameter of 40 inches, though usually it is much smaller. It is called Palo Santo because the highly resinous heartwood is used for incense in churches. The timber entered the export trade about 1890 under the name of Palo Balsamo, and yields upon distillation 5 to 6 per cent of oil which is used in the perfume industry and is known to the trade as "oil of guaiac wood," "oleum ligni guaiaci," "Guajacholzöl," or "essence de bois gaïac."

The species producing the timber known as Vera or Verawood is *Bulnesia arborea* (Jacq.) Engl. which occurs in the coastal region of Colombia and Venezuela. It is a common tree of the dry foothills between Porto Cabello and Lake Maracaibo and, as it usually grows in places more favorable for its development than is the case with the others of this family, it is often comparatively slender, straight and of rather good timber form. Although occasionally 100 feet tall, it is usually less than half that height with a trunk 14 to 20 inches in diameter and free of branches for 15 to 20 feet. The timber is employed for the same purposes as Lignum-vitae (*Guaiacum*), though it is not considered its equal for bearings. The trade in the wood, never very extensive, is less now than formerly. The principal uses are for the steps and collars of water turbines and to some extent for brush backs; locally for railway crossties and fence posts.

Common names: *Bulnesia arborea:* Maracaibo lignum-vitae, venesia, vera, verawood (U.S.A., trade); gayac de Caracas (French); Veraholz (Germ.); guayacán, g. de bola, g. resino, palo sano (Col.); bera, berra, cuchivaro, palo sano, vera, v. aceituna, v. amarilla, v. azul, v. blanca (Venez.). *Other species:* Palo balsamo (U.S.A., trade); palo santo, retama, retamo (Arg.).

Guaiacum, by far the most important genus of the family because it supplies the Lignum-vitae of commerce, is well represented on all the West Indian islands, the coastal region of tropical North America, and the northern fringe of coast and adjacent islands of South America (Map 8). Most of the timber now comes from the west coast of Central America and from Cuba, Jamaica, Haiti, and Dominican Republic. Numerous species have been named, but there are only four that are really distinct.

Guaiacum Coulteri A. Gray (including *G. Palmeri* Vail and *G. Planchoni* A. Gray) is a shrub or a small tree sometimes 25 to 35 feet high and 16 to 28 inches in diameter, with crooked branches and smooth mottled exfoliating bark. It occurs throughout Sinaloa at elevations from near sea level to about 300 feet and all along the coast from Sonora to Oaxaca. Growth is slow, the trunk increasing in diameter at the rate of about four inches per decade. The timber is obtainable in pieces 12 inches square and about 11 feet long, but is only occasionally exported for the manufacture of pulleys and bearings. It is highly esteemed locally for heavy and durable construction and fuel, and splints of the resinous heartwood are used for tapers. The tree is recommended for ornamental planting on account of its dainty foliage, which is retained nearly the whole year, and also because of its attractive and profuse flowering, which continues two or three months.

Guaiacum guatemalense Planch. attains its largest size and importance in western Nicaragua. The timber has been on the market for many years, but for a while there was some prejudice against it because of the tendency of the logs to split at the ends. Improved methods of handling have overcome this objection and the wood is in favor, especially where large sizes are wanted. The largest log of which there appears to be any authentic record was 9 feet long, 36 inches in diameter, and weighed 4260 pounds. Usually the logs are 4 to 8 feet long and 9 to 24 inches in diameter, with a good percentage over 12 inches. They are straight, smooth, cylindrical, and free of natural defects. The bark is thin, of a dull gray color, irregularly flaky, sometimes pitted and coarsely granular, instead of being smooth and glossy as in *G. officinale*. The principal port of shipment is Corinto.

Guaiacum sanctum L. is an evergreen tree usually less than 30 feet high and 12 inches in diameter, though sometimes reaching considerably larger dimensions. It occurs throughout the West Indies, sparingly on some of the Florida Keys, and in Yucatán, probably Tabasco, and perhaps also in Vera Cruz, Mexico. The wood is of considerable commercial importance, though it is less highly esteemed, particularly in the larger sizes, than that of *G. officinale,* and is often referred to in the trade as Bastard Lignum-vitae.

Guaiacum officinale L. is a low or medium-sized tree with a short trunk often 10 to 12, occasionally 18 to 30, inches through. The range includes all of the West Indies and the northern coast of South America, particularly Venezuela and Colombia. It is also credited to the natural flora of Panama and Honduras. The timber enters the market in the form of bolts or logs 2 to 10 feet long and from 3 to 20 or more inches in diameter. The sapwood is usually thin, though if the logs have lain for a long time on the ground or in the water it may be entirely absent. It was formerly the practice to hew off the sapwood before shipment, at least for certain grades of wood, but logs are now shipped entire. There are decided differences in the bark characters and these serve as a ready means for distinguishing the several kinds, that of *G. officinale* suggesting Sycamore (*Platanus*).

The wood of *Guaiacum* has been an article of trade since 1508, when it was introduced to the medical profession of Europe as a specific for many of the most serious diseases of mankind. The name Lignum-vitae (wood of life) originated from the supposition that the material was possessed of extraordinary remedial powers. So great was the demand that for a time the wood sold for as much as seven gold crowns a pound. Numerous learned treatises were published and served to establish the reputation of the wood so firmly that it was nearly two centuries before it was seriously questioned. Both wood and resin are now medicinally obsolete except in certain proprietary decoctions.

The resin, known as "guaiac" or "guaiaci

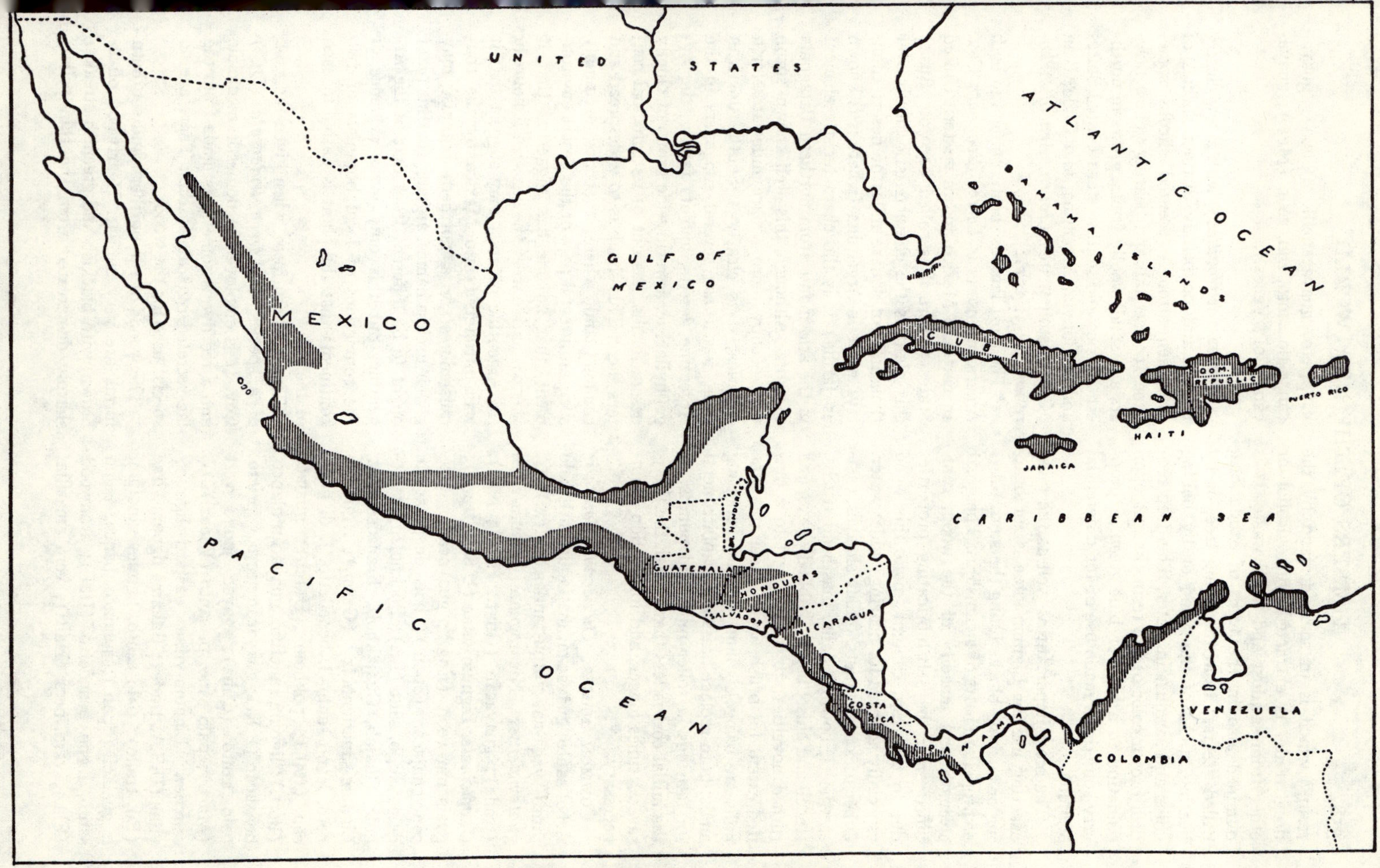

Map 8. Range of Lignum-vitae in tropical America.

resin," which is in some demand by the drug trade, is extracted with alcohol or ether from sawdust and wood waste. It is obtained also in the form of exudations, called "tears," from the living tree, or by heating sticks of the wood or by boiling chips and sawdust in water. The deep reddish brown color of the resin changes upon oxidation to a blue or blue-green, a property sometimes made use of for differential staining.

The most important as well as the most exacting use for Lignum-vitae is for bearing or bushing blocks lining the stern tubes of propeller shafts of steamships. The great strength and tenacity of the wood, combined with the self-lubricating properties due to the resin content, make this wood especially adapted for bearings under water. Other uses are mallets, pulley sheaves, caster wheels, bowling balls, masthead trucks, stencil and chisel blocks, cable dressers, and turned novelties; it is employed to a limited extent for brush backs. Steel and tube mills are using Lignum-vitae in increasing amounts to replace brass and babbitt metal for bearings in roller mills and pumps, as the initial cost is less than metal, the life is several times longer, and lubrication is unnecessary.

COMMON NAMES: *Guaiacum Coulteri:* Arbol santo, guayacán, matlaquahuitl, palo santo, yaga-na, yutnu-tandaa (Mex.). *G. guatemalense:* Nicaraguan lignum-vitae (trade); guayacán (Guat., Hond., Nic.). *G. officinale:* Lignum-vitae (U.S.A., trade); guaiacum-wood (Eng.); guayacán, guaiacán, guajacán (Span.); bois de gaïac, b. de gayac, b. sant (Fr.); Pockholz, Guaiacholz, Guajak, Guajakholz, Franzosenholz, Lignum-sanctumholz (Germ.); guaiaco nero, legno santo, l. benedetto, l. di guaicao (Ital.); pokhout (Dutch); guaiaco (Port.); lignum sanctum, guayacán blanco, hoaxacán (P.R.); guayacán negro, g. prieto, palo santo (Cuba); gayac jaune vert (Mart.); palo sano, p. santo (Venez.). *G. sanctum:* Lignum-vitae, bastard lignum-vitae (U.S.A., trade); Bahama lignum-vitae (Eu. trade); palo santo, lignum sanctum, l. guaiaci (Span.); guaiaco banco, legno santo, legno benedetto (Ital.); ironwood (Fla.); vera, bera (Haiti); vera amarilla, v. prieto, guayacancillo (Cuba); guayacán, palo santo, zon, zoon (Mex.); guayacán (Col., Venez.).

Larrea (or *Covillea*), with a few closely related and doubtfully distinct species of evergreen strongly scented shrubs sometimes 10 feet tall and bearing yellow flowers and small woolly fruits, grows in northern Mexico and southwestern United States and in Bolivia, Argentina, and Chile. Ivan M. Johnston writes (*Journ. Arnold Arboretum* 21: 3: 357):

"The most famous shrub common to both American deserts is *Larrea divaricata* Cav. In northern Mexico and in western United States (where it is called Creosote Bush) this plant is the characteristic shrub over thousands of square miles of desert country; while in Argentina (where it is known as Jarilla) it is the characteristic element in the arid monte from northern Patagonia to Salta. Although this well-known shrub may occupy more square miles in North America, it is clearly a South American type, for it has several congeners in the Argentine deserts, and its family, the Zygophyllaceae, a world-wide group of chiefly desert shrubs, has one of its principal centers there. This shrub is so widespread and common and, where it occurs, seemingly such an integral part of the desert environment, that no one has suggested that it was not native where he has seen it flourish. An Argentine botanist may suggest that it was introduced into Mexico by the conquistadores, or a northern botanist may give it useless names, arguing that his plant ought to be different from that of the Argentine, but this only gives evidence of the convictions of local botanists of two continents that the plant is native to each of them."

The northern form, when given the rank of a species, is *Larrea tridentata* (DC.) Cov. It is employed in domestic medicine, and a red dye is sometimes made from the lac secreted by scale insects. There are no special uses for the wood.

The woods of the northern and southern forms are very similar. Heartwood olive-brown, with blackish and greenish streaks; sharply demarcated from the thin yellow sapwood. Luster low. Faintly odorous; taste not distinctive. Very heavy, hard, and strong, but brittle; texture fine; feel rather waxy; grain interlocked; takes a very smooth finish; is highly resistant to decay. Suitable for small articles of turnery, but of no commercial possibilities because of the small size of the plants.

Common names: Creosote bush, hediondo (U.S.A.); falsa alcaparra, gobernadora, guamis, hediondilla, hediondo, hierba hedionda, huamis (Mex.); jarilla, j. crespa, j. del campo, j. del cerro, j. de la sierra, j. hembra (Arg.).

Porlieria is an unimportant genus of three or four species of shrubs and little trees inhabiting dry subtropical areas of Mexico and Texas and the corresponding zone in Argentina and Chile. The branches are short and stunted, and the stipules are mostly persistent, those of the South American species being spinescent. As in *Guaiacum,* the flowers are blue or purplish and the seeds have a scarlet or orange-colored aril. The northern species, *P. angustifolia* (Engelm.) Gray, is a shrub or small tree, sometimes 25 feet high and 10 inches in diameter, growing in southwestern Texas and northern Mexico. The bark is employed locally as a substitute for soap, particularly in washing woolens that are likely to fade. The principal use for the timber is for fence posts. *P. Lorentzii* Engl., growing on the Argentine steppes, is occasionally 25 feet tall and 10 inches in diameter. The lustrous yellow sapwood is said to make a fairly satisfactory substitute for Boxwood in turnery and carving, and it is used for making spoons, tobacco pipes, walking sticks, and whipstocks.

Heartwood dark brown, streaked with green; sharply demarcated from the yellowish sapwood. Luster rather low. Slightly fragrant; taste not distinctive. Very heavy, hard, and strong; sp. gr. (air-dry) about 1.15; weight 72 lbs. per cu. ft.; texture fine; feel oily; grain interlocked; finishes very smoothly; durability high.

Common names: Soap-bush (U.S.A.); guayacán (Mex.); turacasa (Peru); guayacán, guayaco, palo santo (Chile); chucharea, chuchupé, chucupé, cucharero, guayacán blanco (Arg.).

EXPLANATION OF THE WOOD DESCRIPTIONS

THE descriptions of the woods are based on specimens in the collections of the Yale School of Forestry, and all follow the same outline though varying in detail. The general properties are of the heartwood, except as otherwise stated. It should be borne in mind that such characteristics as color, odor, and taste may be materially affected by drying and long storage of the samples. Texture refers to the size of cellular elements and is described as to its fineness, coarseness, and evenness. Grain is a term denoting the directional arrangement of the fibers and other cells and is variously classified as straight, irregular, or interwoven. "Roe grain" is practically the same as "ribbon grain" and refers to the reversible light and dark striping seen on the radial surface of many woods; the stripes range in width from narrow (*e.g.*, Lignum-vitae) to wide (*e.g.*, Mahogany). Durability is used in the sense of the natural longevity of heartwood when exposed to conditions favorable for decay; in the absence of actual tests this property can be inferred from the physical characters, such as color, odor, and oiliness. The opinions expressed regarding the potential uses and commercial importance of a timber are based upon such information as is available at the moment and are not intended to be final. Species that produce wood of high quality but now valueless because of their scarcity are worthy of consideration for commercial planting.

In the descriptions of the minute anatomy the terminology follows that approved by the International Association of Wood Anatomists (*Tropical Woods* 36: 1–12; Dec. 1, 1933). The size classes for pores are also those approved by that organization (*Tropical Woods* 59: 51–52; Sept. 1, 1939). Other classifications are those of Record and Chattaway (*Tropical Woods* 57: 11–16; March 1, 1939) or, in a few instances, have been arbitrarily made. The general microscopical observations and comparisons were made at a low-power magnification of 60× and high power at 335×. Empirical estimates of comparative sizes of cells and pits were made at these magnifications. Some structures required magnifications up to 600× and occasional specific details were examined at much higher power. Microscope slides of stained and permanently mounted cross, radial, and tangential sections were available for all of the genera and a great many of the species, and many temporary mounts were made where the others were insufficient.

The following discussion of the anatomical descriptions is in the same general order as that followed in the wood descriptions in the text:

Growth rings, when present, result from one or more conditions, such as flattened late-wood fibers, terminal or initial parenchyma (Plate XLVII, 4), absence of pores from a narrow concentric zone (Plate XLIV, 1), greater abundance or larger size of early-wood pores (Plate XXXIX, 1), presence of bands of greater density or, rarely, of bands of distinctive coloration. The reason for the visibility of growth rings is either directly stated or can be inferred from other parts of the description. For woods of the north and south temperate zones, the term growth ring is usually synonymous with annual ring, but for tropical woods no correlation of the rings of growth with definite periods of time has as yet been determined. In the descriptions of tropical woods the term growth ring refers to concentric bands which resemble annual rings of woods grown in temperate regions.

They are frequently clearly demarcated and of diagnostic value.

Pores are described as to their visibility, size, abundance, distribution, arrangement, shape, and wall-thickness. The measurements given are of the tangential diameter of the largest pores, unless otherwise stated; medium-sized pores are barely visible to the unaided eye. The size classes used were: very small, up to 50μ; small, up to 100μ; medium-sized, 100 to 200μ; large, 200μ or over. Abundance of pores, based upon the appearance of porosity of the cross section, is dependent upon both the number and size of pores in a given area. Difficult as this combination is to express, it is, nevertheless, a useful and readily comprehended feature in its extremes. The terms used in the descriptions refer to the distance between pores or pore multiples as the gauge of porosity and the three groups, "scattered" (few and far between, Plate XXXIX, 2), "numerous but not crowded laterally" (Plate XLV), and "crowded laterally" (Plate XXXVIII), are extensively used. In some instances the pores were counted and classified as "very few" (not over 5 per sq. mm.), "few to numerous" (5 to 50 per sq. mm.), and "very numerous" (more than 50 per sq. mm.). In a very few diffuse-porous woods, pores of two distinct size classes occur together (Plate XLII, 3); usually, however, there is a fairly regular gradation from the largest to the smallest. A distinctive pore pattern provides one of the most useful features in identification. A definite ring-porous arrangement distinguishes numerous woods of the north temperate zone, but a semi-ring porous condition (*e.g., Juglans, Cedrela*) is generally less dependable (Plate XL, 3). Such patterns as "ulmiform" (*e.g., Ulmus, Patagonula, Fremontia*) and "flame-like" or "dendritic" (*e.g., Quercus, Rhamnus, Calophyllum*) are very distinctive and may occur either in diffuse-porous woods or in the late wood of those that are ring-porous (Plates XXXIX and XLIII). Tendency toward diagonal arrangement is very common, but the occurrence of all of the pores in a definite oblique or echelon pattern is sometimes distinctive. Patterns formed by radially aligned pores (long radial pore-multiples or pore-chains) and by the restricted area for pores between closely spaced coarse rays characterize some woods (Plates XLII and LI). Long radial multiples (adjoining tangential walls flattened) or chains (adjoining tangential walls unflattened) are themselves distinctive features even though the general pore arrangement is not radial. Clusters of pores are present in many woods and characterize a few. Short radial multiples (2 to 6 pores) are too common to have significance except when they appear to be much more numerous than the associated solitary pores. The expression pores "solitary and in small multiples" is frequently used to describe this very common association (Plate XLV). In a large group of woods the vessels are all solitary, or virtually so (Plate XLI, 2). The "solitary pores" classification does not preclude tangential pairs representing a cross section through the overlapping ends of two members of the same vessel. In woods in which the pores are virtually all solitary, intervascular pitting is absent or so rare that usually no mention is made of it in the description.

Vessels are described with reference primarily to their pits, perforations, spirals, striations, and contents, but other observed features are recorded if they are believed to be of diagnostic value. The size classifications of pits are: very small, less than 4μ; small, less than 7μ; medium-sized, 7–10μ; large, over 10μ; very large, over 15μ. The pattern of intervascular pitting (alternate, opposite, scalariform, or irregular) is given in the family or generic description. When used alone, the expression "pitting fine" implies alternate arrangement. Perforations are recorded as simple or multiple (scalariform, reticulate, or ephedrate), sometimes with additional information on width of perforation rim, thickness and number of bars, or length of plate. Spiral thickenings characterize some or all of the vessels of certain woods, mainly those from regions of temperate climate; in ring-porous woods they are often limited to the small vessels of the late wood. The inner apertures of alternate pits frequently coalesce and produce diagonal fissures which may be confused with spiral thickenings. Abundant fine striations characterize a few woods. Tyloses are common in the heartwood and inner part of the sapwood and frequently completely occlude the vessels; their walls are generally thin, but they may be thick and abundantly pitted, sometimes becoming sclerotic, as in *Hebepetalum* and *Piratinera*. Tyloses are occasionally highly distinctive; for example, they are known to occur in only eight of the many American genera of Leguminosae. Colored gums may fill the vessels or be localized as plates or plugs (*e.g.*, *Swietenia*). Lapachol compound occurs in the heartwood of *Avicennia* and some members of the Bignoniaceae, notably *Tabebuia*. Calcium carbonate is very common in dark-colored wound areas of dicotyledonous woods, but its occurrence in the vessels of the sapwood of certain Ulmaceae (*e.g.*, *Phyllostylon*) is an important diagnostic feature. Other deposits may be equally distinctive, for instance the bitter yellowish material in *Vatairea*.

Rays. Widths and maximum heights are given in number of cells as counted on the tangential section. The presence of two distinct size classes is a useful feature in identification; as seen with a lens on the cross section the larger rays are fairly uniformly spaced, with only much smaller rays between. Rays are considered conspicuous when they are the outstanding feature of the cross section. Extremes in ray abundance are noteworthy. Rays are said to be homogeneous when they are composed entirely of procumbent cells; heterogeneous, when some or all of the cells are square or upright. The degree of heterogeneity is frequently expressed, and departures from typical forms are described. Considerable attention is given to ray-vessel pitting because of its importance in identification; there are a few instances, however, where the rays are not in contact with the vessels, and the pitting between wood parenchyma cells and the vessels is described instead. Mention is made of the presence of crystals, latex tubes, intercellular canals, cysts, sheath cells, disjunctive cells, tile cells, oil cells, sclerotic cells, phloem bridges, abundant gum deposits, or other noteworthy features.

Wood parenchyma. The parenchyma pattern on cross section is described as seen with or without a hand lens or under the compound microscope, and the types are those given in Record and Chattaway's classification (*Tropical Woods* 57: 12–13). The term "reticulate" refers to the mesh-like appearance of very numerous, short, irregularly arranged rows of cells (Plate XLVII, 1); it is intermediate between abundantly diffuse and closely spaced concentric lines and characterizes woods in the Euphorbiaceae, Icacinaceae, Myrtaceae, and several other families. "Sparingly paratracheal" denotes the condition where a few cells are in contact with a pore but do not completely surround it; this type is usually indistinct or invisible with a hand lens. "Unilaterally paratracheal parenchyma" (appearing as caps or hoods, generally on the outer side of the pore) is difficult to determine accurately without the use of the compound microscope (Plate XLVI, 3). When parenchyma is in rather coarse, closely spaced bands there may be no ready way of deciding whether it is apotracheal or paratracheal (confluent); in such instances the pattern is usually described simply as "numerous concentric bands" (Plate XLVIII, 1). Among the microscopic details noted are crystals, crystalliferous strands, sclerotic cells, oil cells (Plate LVI, 1, 2), gum deposits, and apparent or actual septations, the last being rare (*e.g.*, Guttiferae). A typical crystalliferous strand is composed wholly of short or cubical cells, each containing a single crystal of calcium oxalate; sometimes the cells are distended. In other instances only a part of a strand may be crystalliferous. Parenchyma-vessel pitting is of the same general type as that between rays and vessels and usually is not described unless there is something peculiar about it or when the rays are not in contact with the vessels.

Wood fibers. The fibrous elements of dicotyledonous woods are all considered under the general term of "wood fibers," with the exception of "vasicentric tracheids" (which have typical vascular pitting) and "fibriform vessel members" (fiber-like cells with perforations). Two general classes of fibers are recognized, namely, those with distinctly bordered pits and those with simple or indistinctly bordered pits. Pits in the first group may be conspicuous under high-power magnification because of their large size or their great number, or both together. In some instances (*e.g.*, *Ilex* spp.) the fibers with distinctly bordered pits have spiral thickenings, but, as with vessels, this seems to be a temperate-zone feature. Fibers having simple or indistinctly bordered pits are likely to contain starch in the sapwood at proper season, particularly if wood parenchyma is sparingly developed. Such fibers often are septate. In some woods there are bands or patches composed of wide-lumined fibers which are rounded in outline and have wide interstitial spaces. Under the lens, such areas resemble wood parenchyma, though they are generally less sharply defined. If such fibers are not septate they are not always readily distinguishable from fusiform wood parenchyma cells, even under high magnification. Thickness of fiber wall is recorded for the extremes. Reference is made to a gelatinous condition when found, but this feature probably is of little if any diagnostic worth.

Ripple marks (Plate LIII), the result of storied structure, are recorded and, if the markings are regular enough, the average number of tiers per vertical inch is given. The number is not constant for a species and varies from 40 to 270 in woods of different families; only pronounced differences can be considered significant with our present knowledge. In certain woods some or all of the rays may occupy more than one tier or show no storied arrangement whatever; ripple marks, accordingly, may be indistinct except in parenchyma bands (Plate LIII, 4). Parenchyma cells, as well as the strands they compose, are sometimes in horizontal tiers, thus giving rise to secondary markings that are twice, rarely four times, as many as in the primary seriation (Plate LIII, 3).

Gum ducts in dicotyledonous woods (Plate LIV) are intercellular canals similar to the resin ducts of conifers. Both vertical and radial gum ducts occur, but in no instance did the authors find the two series in the same wood or even in the same family. Most of the vertical ducts are of traumatic origin and thus likely to be absent from a particular specimen. The radial ducts are of normal occurrence and hence more reliable as a diagnostic feature, but they vary in abundance and, when few in number, may be missed in section. Ducts are described with reference to their location, abundance, contents, and epithelial layer. Rays containing ducts are necessarily multiseriate, but they are not always fusiform (tangential section). Cysts (intercellular pockets, Plate LVI), phloem bridges (in certain woods having anomalous structure), and abnormally large and thin-walled cells may resemble ducts on tangential section. The so-called "radial channels" (Plate LV, 1, 2) are not gum ducts but intercellular spaces apparently resulting from the disintegration of parenchymatous tissue, probably of leaf traces. They are readily visible, sometimes conspicuous, and may be numerous or widely separated. Their presence provides an important diagnostic feature.

LISTS OF FAMILIES CLASSIFIED WITH REFERENCE TO SPECIAL PROPERTIES AND USES OF THEIR BARK, LEAVES, AND TIMBER

FIBROUS BARK USED FOR CORDAGE, ETC.

Anonaceae	Cupressaceae	Monimiaceae	Thymelaeaceae
Asclepiadaceae	Elaeocarpaceae	Moraceae	Tiliaceae
Bixaceae	Lecythidaceae	Rosaceae	Ulmaceae
Cochlospermaceae	Malvaceae	Sterculiaceae	Urticaceae

AROMATIC SUBSTANCES IN THE BARK

Anonaceae	Cupressaceae	Loganiaceae	Rhamnaceae
Apocynaceae	Ericaceae	Magnoliaceae	Rosaceae
Araliaceae	Euphorbiaceae	Meliaceae	Rutaceae
Betulaceae	Guttiferae	Menthaceae	Styracaceae
Bombacaceae	Hamamelidaceae	Monimiaceae	Turneraceae
Burseraceae	Humiriaceae	Myoporaceae	Winteraceae
Calycanthaceae	Juglandaceae	Myricaceae	Zygophyllaceae
Canellaceae	Lauraceae	Myristicaceae	
Chloranthaceae	Leguminosae	Piperaceae	

MALODOROUS SUBSTANCES IN THE BARK

Anacardiaceae	Capparidaceae	Lecythidaceae	Proteaceae
Anonaceae	Caprifoliaceae	Leguminosae	Simarubaceae
Aristolochiaceae	Celastraceae	Phytolaccaceae	Taxaceae

BITTER JUICES IN THE BARK

Acanthaceae	Capparidaceae	Hamamelidaceae	Rhamnaceae
Anacardiaceae	Caprifoliaceae	Leguminosae	Rubiaceae
Apocynaceae	Crossosomataceae	Lythraceae	Salicaceae
Berberidaceae	Cunoniaceae	Malpighiaceae	Simarubaceae
Burseraceae	Ebenaceae	Meliaceae	Thymelaeaceae
Buxaceae	Euphorbiaceae	Moraceae	Turneraceae
Cactaceae	Garryaceae	Papaveraceae	

ALKALOIDS AND GLUCOSIDES IN THE BARK

Apocynaceae	Coriariaceae	Leguminosae	Rubiaceae
Berberidaceae	Erythroxylaceae	Loganiaceae	Rutaceae
Buxaceae	Euphorbiaceae	Menispermaceae	Sapindaceae
Cactaceae	Garryaceae	Papaveraceae	Sterculiaceae
Compositae	Lauraceae	Rhamnaceae	

MILKY LATEX IN THE BARK

Anacardiaceae
Apocynaceae
Asclepiadaceae
Cactaceae
Campanulaceae
Caricaceae
Compositae
Convolvulaceae
Euphorbiaceae
Hippocrateaceae
Lobeliaceae
Malpighiaceae
Moraceae
Papaveraceae
Sapotaceae

RED OR YELLOW JUICES IN THE BARK

(r, red or reddish brown; y, yellow)

Anacardiaceae (y)
Bixaceae (y)
Burseraceae (r, y)
Cochlospermaceae (y)
Euphorbiaceae (r, y)
Guttiferae (y)
Koeberliniaceae (r)
Leguminosae (r, y)
Meliaceae (r)
Moraceae (r)
Myristicaceae (r, y)
Papaveraceae (y)
Polygonaceae (r)
Proteaceae (r)
Rubiaceae (r)
Vochysiaceae (y)
Zygophyllaceae (r)

DYESTUFFS IN ECONOMIC QUANTITIES

(b, bark; f, fruit; l, leaves; r, roots; w, wood)

Apocynaceae (f)
Berberidaceae (w)
Bixaceae (f)
Combretaceae (f)
Euphorbiaceae (b)
Hippocastanaceae (b)
Juglandaceae (b, f)
Julianiaceae (b)
Krameriaceae (r)
Leguminosae (b, f, l, w)
Lythraceae (b)
Moraceae (w)
Nyctaginaceae (f)
Papaveraceae (b)
Rubiaceae (f, r)
Symplocaceae (l)

TANNIN IN ECONOMIC QUANTITIES

(b, bark; f, fruit; l, leaves; w. wood)

Anacardiaceae (b, f, l, w)
Avicenniaceae (b)
Berberidaceae (w)
Betulaceae (b)
Burseraceae (b)
Combretaceae (b, f, l)
Cunoniaceae (b)
Dilleniaceae (b)
Ericaceae (b)
Eucryphiaceae (b)
Euphorbiaceae (b)
Fagaceae (b, f, l, w)
Flacourtiaceae (b, f)
Juglandaceae (b, f)
Julianiaceae (b)
Leguminosae (b, f)
Malpighiaceae (b)
Myrtaceae (b, w)
Ochnaceae (b)
Olacaceae (b)
Pinaceae (b)
Rhizophoraceae (b)
Rosaceae-Chrysobalanoideae (b)
Rubiaceae (b, l)
Salicaceae (b)
Sapotaceae (b)

SEEDS OF ECONOMIC IMPORTANCE FOR OIL, FAT, OR STARCH

Anacardiaceae
Apocynaceae
Araucariaceae
Bombacaceae
Buxaceae
Cactaceae
Caryocaraceae
Celastraceae
Chenopodiaceae
Combretaceae
Corylaceae
Euphorbiaceae
Fagaceae
Flacourtiaceae
Guttiferae
Hippocastanaceae
Hippocrateaceae
Icacinaceae
Juglandaceae
Lauraceae
Lecythidaceae
Leguminosae
Malvaceae
Meliaceae
Monimiaceae
Moraceae
Myristicaceae
Ochnaceae
Olacaceae
Oleaceae
Palmaceae
Pinaceae
Proteaceae
Rhamnaceae
Rosaceae-Chrysobalanoideae
Santalaceae
Sapindaceae
Sapotaceae
Sterculiaceae
Vochysiaceae

MOST IMPORTANT FAMILIES AND GENERA OF HARDWOODS IN TEMPERATE NORTH AMERICA

Aceraceae (*Acer*)
Betulaceae (*Betula*)
Fagaceae (*Castanea, Fagus, Quercus*)
Hamamelidaceae (*Liquidambar*)
Juglandaceae (*Carya, Juglans*)
Magnoliaceae (*Liriodendron*)
Nyssaceae (*Nyssa*)
Oleaceae (*Fraxinus*)
Platanaceae (*Platanus*)
Rosaceae (*Prunus*)
Salicaceae (*Populus, Salix*)
Tiliaceae (*Tilia*)
Ulmaceae (*Ulmus*)

FAMILIES AND GENERA OF THE HARDWOODS OF TROPICAL AMERICA BEST KNOWN TO FOREIGN COMMERCE

Anacardiaceae (*Schinopsis*)
Anonaceae (*Oxandra*)
Bignoniaceae (*Tabebuia*)
Bombacaceae (*Ochroma*)
Flacourtiaceae (*Gossypiospermum*)
Lauraceae (*Ocotea*)
Leguminosae (*Brya, Dalbergia, Haematoxylon*)
Meliaceae (*Carapa, Cedrela, Swietenia*)
Moraceae (*Brosimum, Chlorophora, Piratinera*)
Rubiaceae (*Calycophyllum*)
Rutaceae (*Zanthoxylum*)
Sapotaceae (*Manilkara*)
Zygophyllaceae (*Guaiacum*)

PRINCIPAL TROPICAL AMERICAN FAMILIES SUPPLYING HARDWOOD LUMBER SUITABLE FOR GENERAL CONSTRUCTION, CARPENTRY, AND FURNITURE

Anacardiaceae
Apocynaceae
Bignoniaceae
Bombacaceae
Boraginaceae
Burseraceae
Combretaceae
Euphorbiaceae
Guttiferae
Lauraceae
Lecythidaceae
Leguminosae
Meliaceae
Moraceae
Myristicaceae
Rosaceae
Sapotaceae
Simarubaceae
Verbenaceae
Vochysiaceae

VERY LIGHT AND SOFT WOODS

Anonaceae (*Anona*)
Apocynaceae (*Ambelania*)
Araliaceae
Bombacaceae
Cochlospermaceae
Compositae (*Tessaria*)
Cupressaceae
Euphorbiaceae
Hernandiaceae
Leguminosae
Leitneriaceae
Liliaceae (*Yucca*)
Malvaceae
Nyssaceae (*Nyssa*)
Solanaceae
Sterculiaceae
Thymelaeaceae
Tiliaceae

WOODS NOTED FOR THEIR NATURAL RESISTANCE TO DECAY

Anacardiaceae
Bignoniaceae
Boraginaceae
Combretaceae
Cupressaceae
Ebenaceae
Euphorbiaceae
Fagaceae
Lauraceae
Leguminosae
Meliaceae
Moraceae
Olacaceae
Podocarpaceae
Rhamnaceae
Sapotaceae
Taxaceae
Taxodiaceae
Zygophyllaceae

ATTRACTIVELY COLORED WOODS

Anacardiaceae
Anonaceae
Berberidaceae
Betulaceae
Bignoniaceae
Boraginaceae
Combretaceae
Cupressaceae
Dilleniaceae
Ebenaceae
Euphorbiaceae
Fagaceae
Guttiferae
Hamamelidaceae
Icacinaceae
Juglandaceae
Lauraceae
Leguminosae
Malvaceae
Meliaceae
Moraceae
Myristicaceae
Myrsinaceae
Oleaceae
Picrodendraceae
Platanaceae
Polygonaceae
Proteaceae
Rhamnaceae
Rosaceae
Rubiaceae
Rutaceae
Sapotaceae
Staphyleaceae
Surianaceae
Taxaceae
Taxodiaceae
Theophrastaceae
Verbenaceae
Violaceae
Vochysiaceae
Zygophyllaceae

WOODS WITH OILY OR WAXY APPEARANCE AND FEEL

Anacardiaceae	Erythroxylaceae	Malpighiaceae	Sapotaceae
Bignoniaceae	Euphorbiaceae	Myristicaceae	Taxodiaceae
Boraginaceae	Koeberliniaceae	Oleaceae	Tiliaceae
Burseraceae	Lauraceae	Picrodendraceae	Verbenaceae
Caryocaraceae	Lecythidaceae	Rhamnaceae	Zygophyllaceae
Dichapetalaceae	Leguminosae	Rutaceae	

WOODS WITH DISTINCTIVE ODOR

Anacardiaceae	Combretaceae	Melastomaceae	Rutaceae
Anonaceae	Coriariaceae	Meliaceae	Staphyleaceae
Bignoniaceae	Cupressaceae	Monimiaceae	Taxodiaceae
Bombacaceae	Hydrangeaceae	Myoporaceae	Thymelaeaceae
Boraginaceae	Juglandaceae	Myrtaceae	Turneraceae
Burseraceae	Lauraceae	Olacaceae	Ulmaceae
Canellaceae	Lecythidaceae	Oleaceae	Zygophyllaceae
Capparidaceae	Leguminosae	Phytolaccaceae	
Caprifoliaceae	Lythraceae	Pinaceae	
Caryocaraceae	Malvaceae	Rosaceae	

WOODS WITH DISTINCTIVE TASTE

Anacardiaceae	Lauraceae	Picrodendraceae	Staphyleaceae
Apocynaceae	Leguminosae	Pinaceae	Thymelaeaceae
Bignoniaceae	Loganiaceae	Rubiaceae	Turneraceae
Clethraceae	Meliaceae	Rutaceae	Verbenaceae
Cupressaceae	Myrtaceae	Sapindaceae	Violaceae
Fagaceae	Oleaceae	Sapotaceae	Zygophyllaceae
Juglandaceae	Phytolaccaceae	Simarubaceae	

WOODS YIELDING ETHEREAL OILS UPON DISTILLATION

Burseraceae	Olacaceae (*Ximenia*)	Zygophyllaceae (*Bulnesia Sarmientii*)
Lauraceae	Rutaceae (*Amyris*)	

WOODS SIMILAR TO BIRCH (*BETULA*)

Anacardiaceae (*Tapirira*)	Clethraceae	Polygonaceae (*Ruprechtia*)
Betulaceae	Combretaceae (*Terminalia*)	Santalaceae (*Myoschilos*)
Bignoniaceae (*Tabebuia insignis*)	Elaeocarpaceae (*Vallea*)	Symplocaceae
Burseraceae	Fagaceae (*Nothofagus*)	Tiliaceae (*Luehea*)
Canellaceae (*Capsicodendron*)	Lythraceae (*Physocalymma*)	
	Melastomaceae (*Calygonium*)	

WOODS SIMILAR TO BOXWOOD (*BUXUS*)

Apocynaceae (*Aspidosperma*)	Flacourtiaceae	Ulmaceae (*Phyllostylon*)
Buxaceae	Polygalaceae	Violaceae
Capparidaceae	Rubiaceae	Zygophyllaceae (*Porlieria*)
Celastraceae	Rutaceae	
Euphorbiaceae	Santalaceae	

WOODS SIMILAR TO MAPLE (*ACER*)

Acanthaceae (*Trichanthera*)
Aceraceae
Anacardiaceae (*Malosma*)
Apocynaceae (*Aspidosperma*)
Cornaceae
Ericaceae (*Oxydendrum*)
Malvaceae (*Tetrasida*)
Olacaceae (*Aptandra*)
Phytolaccaceae (*Rhabdodendron macrophyllum*)
Rubiaceae
Rutaceae
Sabiaceae (*Meliosma alba*)
Santalaceae
Sapindaceae
Ulmaceae (*Planera*)

WOODS SIMILAR TO RED GUM (*LIQUIDAMBAR*)

Aextoxicaceae
Cunoniaceae
Escalloniaceae
Eucryphiaceae
Fagaceae (*Nothofagus*)
Hamamelidaceae
Lythraceae
Malpighiaceae (*Byrsonima*)
Monimiaceae (*Laurelia*)
Myrtaceae (*Myrceugenia*)
Styracaceae
Theaceae
Turneraceae

POSSIBLE SUBSTITUTES FOR TEAK (*TECTONA*)

Bignoniaceae (*Paratecoma*)
Boraginaceae (*Cordia*)
Caryocaraceae (*Caryocar*)
Combretaceae (*Terminalia*)
Lauraceae (*Mezilaurus*)
Leguminosae (*Dicorynia*)
Verbenaceae (*Vitex*)

WOODS SIMILAR TO WALNUT (*JUGLANS*)

Anacardiaceae (*Metopium*)
Boraginaceae (*Cordia*)
Euphorbiaceae (*Hippomane*)
Juglandaceae
Lauraceae (*Phoebe porosa*)
Leguminosae (*Enterolobium*)
Staphyleaceae (*Turpinia*)

WOODS SIMILAR TO YELLOW POPLAR (*LIRIODENDRON*)

Araliaceae
Burseraceae
Hippocastanaceae
Magnoliaceae
Malvaceae
Monimiaceae
Simarubaceae
Solanaceae
Tiliaceae

WOODS SUITABLE FOR ARCHERY BOWS

Anonaceae (*Oxandra*)
Apocynaceae (*Aspidosperma*)
Boraginaceae (*Auxemma, Patagonula, Saccellium*)
Cupressaceae (*Juniperus*)
Leguminosae (*Apoplanesia, Robinia*)
Moraceae (*Maclura*)
Oleaceae (*Fraxinus*)
Rosaceae (*Cercocarpus*)
Rubiaceae (*Calycophyllum*)
Taxaceae (*Taxus*)
Ulmaceae (*Ulmus*)

WOODS SUITABLE FOR BENTWORK

Apocynaceae (*Aspidosperma*)
Boraginaceae (*Patagonula*)
Fagaceae (*Fagus, Quercus*)
Oleaceae (*Fraxinus*)
Rubiaceae (*Genipa*)
Ulmaceae (*Ulmus*)

WOODS USED FOR BOXES AND CRATES

Acanthaceae
Anacardiaceae
Araliaceae
Araucariaceae
Bombacaceae
Burseraceae
Cupressaceae
Euphorbiaceae
Hernandiaceae
Hippocastanaceae
Lauraceae
Leguminosae
Magnoliaceae
Malvaceae
Melastomaceae
Meliaceae
Moraceae
Myristicaceae
Nyctaginaceae
Nyssaceae
Pinaceae
Platanaceae
Polygonaceae
Salicaceae
Simarubaceae
Sterculiaceae
Tiliaceae
Verbenaceae
Winteraceae

WOODS USED FOR COOPERAGE

Aextoxicaceae
Apocynaceae
Bombacaceae
Boraginaceae
Burseraceae
Caryocaraceae
Fagaceae
Guttiferae
Hamamelidaceae
Hippocastanaceae
Lauraceae
Moraceae
Nyctaginaceae
Oleaceae
Platanaceae
Sapotaceae
Tiliaceae
Ulmaceae
Verbenaceae

WOODS USED FOR MAKING FISHING RODS

Anonaceae (*Oxandra*)
Bignoniaceae (*Tabebuia*)
Boraginaceae (*Patagonula*)
Lauraceae (*Ocotea*)
Moraceae (*Piratinera*)
Palmaceae (*Astrocaryum*)

WOODS USED FOR MATCH STICKS

Anacardiaceae (*Spondias*)
Apocynaceae (*Couma*)
Araliaceae
Betulaceae
Bignoniaceae (*Jacaranda*)
Bombacaceae
Burseraceae
Euphorbiaceae
Moraceae (*Cecropia*)
Pinaceae (*Pinus*)
Salicaceae
Simarubaceae

WOODS USED FOR TOOL HANDLES

Aceraceae
Anonaceae
Apocynaceae
Bignoniaceae
Boraginaceae
Corylaceae
Elaeocarpaceae
Eucryphiaceae
Euphorbiaceae
Fagaceae
Guttiferae
Juglandaceae
Leguminosae
Moraceae
Myrtaceae
Oleaceae
Rosaceae
Rubiaceae
Rutaceae
Sapotaceae
Ulmaceae
Verbenaceae
Zygophyllaceae

SOURCES OF INFORMATION AND MATERIAL

THE information in this book has come from innumerable sources. The principal publications consulted are listed by countries in the bibliography and there are frequent citations in the text. The authors have also been able to consult a wealth of unpublished reports, memoranda, field notes, and personal letters.

Principal Sources of the Wood Specimens

Canada: Prof. F. M. Knapp.

United States: Arnold Arboretum, Field Museum of Natural History, U.S. Forest Service, and the Morris K. Jesup collection; also Mr. Samuel B. Detwiler, Prof. Emanuel Fritz, Dr. Arthur H. Graves, Dr. R. M. Harper, Mr. Henry W. Hicock, Dr. Herbert F. Marco, Mr. D. N. Matthews, Mr. S. R. MacDonald, Mr. M. Scott, Mr. J. L. Stearns, Dr. Irma E. Webber, and members of the faculty of the Yale School of Forestry.

Cuba: Arnold Arboretum, Mr. G. C. Bucher, Sr. E. Duque Estrada, Sr. A. J. Fors, Prof. Henry S. Graves, Mr. H. C. Kluge, Rev. Bro. León, Prof. D. M. Matthews, Dr. Juan T. Roig, and Dr. H. N. Whitford.

Puerto Rico: Prof. I. W. Bailey, Mr. W. M. Barbour, Dr. N. L. Britton, Mr. W. D. Durland, Mr. L. R. Holdridge, and Mr. W. P. Kramer.

Virgin Islands: Mr. Arthur S. Fairchild and Prof. R. H. Woodworth.

Dominican Republic: Dr. W. L. Abbott, Mr. W. D. Durland, Mr. J. C. Scarff, Mr. J. G. Scarff, and Mr. W. N. Watkins.

Haiti: Mr. W. R. Barbour, Mr. E. C. Leonard, Mr. C. D. Mell, and Mr. L. R. Stadtmiller.

Trinidad: Mr. W. E. Broadway and Mr. W. J. Hutchinson.

Mexico: Mr. H. C. Benke, Prof. C. Conzatti, Dr. R. S. Flores, Mr. John A. Gamon, Mr. Tom Gill, Mr. H. C. Kluge, Prof. M. Martínez, Mr. C. D. Mell, Sr. Jesús Gonzalez Ortega, Mr. B. P. Reko, Dr. H. N. Whitford, and Mr. L. Williams.

British Honduras: Forestry Department, Mr. H. C. Kluge, Prof. C. L. Lundell, Prof. S. J. Record, Mr. W. A. Schipp, and Mr. H. W. Winzerling.

Guatemala: Sr. Carlos Galusser, Mr. H. Goulier, Mr. Henry Kuylen, Prof. S. J. Record, Sr. J. G. Salas, Mr. L. R. Stadtmiller, and Dr. H. N. Whitford.

Honduras: Dr. Ray Dawson, Mr. J. B. Edwards, Mr. Holger Johansen, Mr. P. H. Meyers, Prof. S. J. Record, and Mr. Paul C. Standley.

Salvador: Dr. Salvador Calderón.

Nicaragua: Mr. F. C. Englesing, Mr. F. H. Fischer, Mr. W. J. Hutchinson, and Mr. H. Bartlett Wells.

Costa Rica: Mr. F. Charles Clarke, Mr. G. Proctor Cooper, Rev. Ralph E. Danforth, Mr. C. D. Mell, Prof. W. R. Rowlee, Mr. Austin Smith, Mr. P. C. Standley, and Prof. H. E. Stork.

Panama: Mr. G. Proctor Cooper, Dr. Alvin G. Cox, Mr. F. H. Fischer, Mr. H. C. Kluge, Dr. H. Pittier, and Mr. Geo. M. Slater.

Colombia: Mr. H. M. Curran, Rev. Brother Daniel, Sr. Armando Dugand G., Sr. Ramón Espina, Sr. J. Giacometto, Mr. Henry Kuylen, Mr. A. E. Lawrance, Sr. José Miguel, Sr. J. Pinzón C., Prof. S. J. Record, and Dr. H. N. Whitford.

Venezuela: Dr. S. Aguerrevere, Mr. H. M. Curran, Sr. W. Gehriger, Mr. Miles Haman, Mr. C. D. Mell, Mr. George Newhall, Dr. H. Pittier, Dr. J. N. Rose, Sr. O. Stelling, Mr. G. H. Tate, Dr. H. N. Whitford, and Mr. L. Williams.

British Guiana: Forestry Department, Mr. Tom Gill, Mr. L. S. Hohenkerk, Mr. W. C. Joste, Mr. C. D. Mell, Mr. A. C. Persaud, Dr. A. C. Smith, University of Michigan, and Dr. H. N. Whitford.

Surinam: Dr. J. R. Beversluis, Mr. J. W. Gonggrijp, Mr. D. E. Headley, Mr. L. Junker, Mr. R. H. G. McArthur, and Dr. H. N. Whitford.

French Guiana: Administration Penetentiaire, Commandant A. Bertin, Mr. A. W. Betts, Prof. H. Lecomte, and Service des Bois Coloniaux.

Ecuador: Mr. R. M. Barber, Prof. W. H.

Brewer, New York Botanical Garden, and Dr. A. Rimbach.

Peru: Mr. Georges H. Barrel, Mr. L. Williams, and Mr. Felix Woytkowski.

Bolivia: Mr. B. A. Krukoff and Mr. R. S. Williams.

Paraguay: Goodell Pratt Co. (Greenfield, Mass.) and Mr. W. J. Hutchinson.

Uruguay: Dr. W. G. Herter.

Brazil: Companhia Ford Industrial do Brasil (through Field Museum), Mr. H. M. Curran, Dr. B. E. Dahlgren, Dr. Adolpho Ducke, Dr. F. C. Hoehne, Mr. Eugene Horn, Dr. J. Huber, Mr. W. J. Hutchinson, Mr. Charles Jackson, Mr. B. A. Krukoff, Dr. Paul LeCointe, Mr. Henry J. McCall, Dr. Edmondo Navarro, Sr. Paulo F. de Souza, Sr. Lino Tatto, Dr. Octavio Vecchi, and Dr. H. N. Whitford.

Argentina: Mr. H. M. Curran, Mr. W. J. Hutchinson, Mr. C. D. Mell, Sr. José F. Molfino, Oxford Imperial Forestry Institute, and Mr. Max Rothkugel.

Chile: Mr. E. L. Bernath, Prof. G. A. Garratt, Mr. W. J. Hutchinson, Mr. C. D. Mell, Dr. Carl Skottsberg, and Dr. G. R. Wieland.

Miscellaneous: Many timber dealers and manufacturers have contributed trade specimens and information concerning them. Special credit is due to the C. H. Pearson and Son Hardwood Company, 44 Whitehall Street, New York, and the I. T. Williams Company, 220 Eleventh Avenue, New York.

Determination of Herbarium Specimens

Arnold Arboretum (Jamaica Plain, Mass.): Dr. Ivan M. Johnston and Mr. Alfred Rehder.

Carleton College (Northfield, Minn.): Prof. H. E. Stork.

Field Museum of Natural History (Chicago): Dr. B. E. Dahlgren, Mr. J. Francis Macbride, Mr. Paul C. Standley, Dr. Julian A. Steyermark, and Mr. Llewelyn Williams.

Jardim Botanico do Rio de Janeiro (Brazil): Dr. Adolpho Ducke.

Missouri Botanical Garden (St. Louis): Dr. Robert E. Woodson, Jr., and Mr. Russell J. Seibert.

New York Botanical Garden: Dr. N. L. Britton, Dr. H. A. Gleason, Mr. B. A. Krukoff, Dr. Harold N. Moldenke, Dr. A. C. Smith, and Mr. Percy Wilson.

University of Michigan (Ann Arbor): Prof. C. L. Lundell.

U.S. Department of Agriculture (Washington, D.C.): Dr. Samuel F. Blake.

U.S. National Museum (Washington, D.C.): Mr. Ellsworth P. Killip, Mr. Emery C. Leonard, Dr. J. N. Rose, and Mr. William Watkins.

Venezuela: Dr. H. Pittier, Caracas.

Note: The two botanists who have rendered the greatest service to the authors are Mr. Paul C. Standley and Dr. Adolpho Ducke.

BIBLIOGRAPHY

GENERAL

American Midland Naturalist. Pub. bi-monthly by the University of Notre Dame, Notre Dame, Ind. 1906–.

BLAKE, S. F., and ALICE C. ATWOOD. *Geographical guide to the floras of the world.* Part I. Misc. Pub. No. 401, U.S. Dept. Agr., Washington, D.C. June 1942; pp. 336.

BOULGER, G. S. *Wood: A manual of the natural history and industrial applications of the timbers of commerce.* London: Edward Arnold, 1908; pp. 1–348; 43 text figs.; 48 plates.

Brittonia. Ser. pub. by N. Y. Botanical Garden, 1931–.

CHALK, L., and M. M. CHATTAWAY. Identification of woods with included phloem. *Tropical Woods* 50: 1–31; June 1, 1937.

CHATTAWAY, M. M. The wood anatomy of the family Sterculiaceae. *Phil. Trans., Royal Soc. London* (ser. B) 228: 554: 313–365; 3 plates; 34 text figs.; Nov. 22, 1937.

CHEVALIER, AUGUSTE. Notes historiques et souvenirs sur les acajous vrais. *Rev. Bot. App. & d'Agri. Tropicale* (Paris) 17: 194: 709–724; October 1937.

CLARKE, S. H. A comparison of certain properties of temperate and tropical timbers. *Tropical Woods* 52: 1–11; Dec. 1, 1937.

DADSWELL, H. E., and S. J. RECORD. Identification of woods with conspicuous rays. *Tropical Woods* 48: 1–30; Dec. 1, 1936.

DAHLGREN, B. E. *Index of American palms.* Chicago: Field Museum Bot. Ser. 14; 1–456; Apr. 30, 1936.

DIEHL, GEORGE A. A study of the Lecythidaceae. *Tropical Woods* 43: 1–15; Sept. 1, 1935.

ENGLER, A., and K. PRANTL. *Die natürlichen Pflanzenfamilien.* Leipzig: Wilhelm Engelmann. Vols. II–IV, 1889–1912; revised edition (incomplete) 1926–.

Field Museum Botanical Series. Pub. in parts at irreg. intervals by Field Museum of Natural History, Chicago, Ill., 1895–.

GARRATT, GEORGE A. Systematic anatomy of the woods of the Monimiaceae. *Tropical Woods* 39: 18–44; Sept. 1, 1934.

HALE, J. D. *The identification of woods commonly used in Canada.* Bull. No. 81, Canadian Forest Service, Ottawa, 1932; pp. 48; 2 text figs.; 14 plates.

HARRAR, ELLWOOD S. Some physical properties of modern cabinet woods. *Tropical Woods* 68: 1–11; 70: 1–15; 71: 26–32; Dec. 1, 1941, June 1, Sept. 1, 1942, resp.

HAYWARD, PHILLIPS A. *Wood—lumber and timbers.* New York: Chandler Cyclopedia, 930; pp. 521; 35 plates; 62 text figs.

HEIMSCH, CHARLES, JR. Comparative anatomy of the secondary xylem in the "Gruinales" and "Terebinthales" of Wettstein with reference to taxonomic grouping. *Lilloa* (Buenos Aires) 8: 83–198; 17 plates; 1942.

HESS, ROBERT W. Occurrence of raphides in wood. *Tropical Woods* 46: 22–31; June 1, 1936.

HOLLAND, J. H. *Overseas plant products.* London: John Bale, Sons & Curnone, Ltd., August 1937; pp. 279.

HOWARD, ALEXANDER L. *A manual of the timbers of the world, their characteristics and uses.* London: Macmillan & Co., 1934; pp. xxi + 672; 91 half-tones.

HUTCHINSON, J. *The families of flowering plants. I.* London: Macmillan & Co., 1926; pp. 328; 264 text figs.

Kew Bulletin of Miscellaneous Information. Pub. by Royal Botanic Gardens, Kew, England.

KOEHLER, ARTHUR. *Identification of furniture woods.* Misc. Circ. No. 66, U.S. Dept. Agr., Washington, D.C., 1926; pp. 78.

Lloydia. Pub. quarterly by Lloyd Library and Museum, Cincinnati, Ohio, 1938–.

MARCO, HERBERT. Systematic anatomy of the woods of the Rhizophoraceae. *Tropical Woods* 44: 1–20; Dec. 1, 1935.

MCLAUGHLIN, ROBERT P. Systematic anatomy of the woods of the Magnoliales. *Tropical Woods* 34: 3–39; June 1, 1933.

MEYER, HANS: *The book of wood names.* Hannover, Germany: M. & H. Schaper, 1933–1936; pp. xviii + 564.

North American Flora. Pub. in parts at irreg. intervals by New York Botanical Garden, 1905–.

PEIRCE, ALAN S. Anatomical interrelationships of the Taxodiaceae. *Tropical Woods* 46: 1–14; June 1, 1936.

PEIRCE, ALAN S. Systematic anatomy of the woods of the Cupressaceae. *Tropical Woods* 49: 5–21; March 1, 1937.

Phytologia. Serial, pub. at irreg. intervals by H. A. Gleason and Harold N. Moldenke, New York Botanical Garden, New York City; 1933–.

RECORD, SAMUEL J. Walnut woods—true and false. *Tropical Woods* 18: 4–29; June 1, 1929.

RECORD, SAMUEL J. Classifications of various anatomical features of dicotyledonous woods. *Tropical Woods* 47: 12–27; Sept. 1, 1936.

RECORD, SAMUEL J. The American woods of the orders Celastrales, Olacales, and Santales. *Tropical Woods* 53: 11–38; March 1, 1938.

RECORD, SAMUEL J. American woods of the family Anacardiaceae. *Tropical Woods* 60: 11–45; Dec. 1, 1939.

RECORD, SAMUEL J., and GEORGE A. GARRATT. *Boxwoods*. Bull. No. 14, Yale University School of Forestry, 1935; pp. 81; 8 plates, 2 text figs.

RECORD, SAMUEL J., and ROBERT W. HESS. American woods of the family Moraceae. *Tropical Woods* 61: 11–54; March 1, 1939.

SENEAR, FRANCIS EUGENE. Dermatitis due to woods. *Journ. Am. Medical Assn.* (Chicago) 101: 1527–1532; Nov. 11, 1933.

STONE, HERBERT. *The timbers of commerce and their identification*. London: Wm. Rider & Son, Ltd., 1904; pp. xxxviii + 290; 23 plates.

TIPPO, OSWOLD. Comparative anatomy of the Moraceae and their presumed allies. *Botanical Gazette* 100: 1: 1–99; 63 figs.; September 1938.

Tropical Woods. Serial, pub. quarterly by Yale University School of Forestry, 1925–.

WEBBER, IRMA E. Systematic anatomy of the woods of the Malvaceae. *Tropical Woods* 38: 15–36; June 1, 1934.

WEBBER, IRMA E. Systematic anatomy of the woods of the Simarubaceae. *American Journ. Botany* 23: 9: 577–587; 77 figs.; November 1936.

WEBBER, IRMA E. Systematic anatomy of the woods of the Burseraceae. *Lilloa* (Buenos Aires, Arg.) 6: 441–465; 4 plates; 1941.

ZON, RAPHAEL, and WILLIAM N. SPARHAWK. *Forest resources of the world*. New York: McGraw-Hill Book Co., 2 vols.; pp. 997; 16 maps in color.

UNITED STATES and CANADA

BENSON, LYMAN. The mesquites and screwbeans of the United States. *Am. Journ. Botany* 28: 9: 748–754; 5 figs.; November 1941.

BETTS, H. S. *Properties and uses of the southern pines*. Circ. No. 164, U.S. Forest Service, Washington, D.C., 1909; pp. 30.

BETTS, H. S. American woods. Leaflets pub. at irreg. intervals by U.S. Forest Service, Washington, D.C., 1928–.

BETTS, H. S. *The strength of North American woods*. Misc. pub. No. 46, U.S. Dept. Agr., Washington, D.C., 1929.

BOISEN, A. T., and J. A. NEWLIN. *The commercial hickories*. Bull. No. 80, U.S. Forest Service, Washington, D.C., 1910; pp. 64.

BROWN, H. P., and A. J. PANSHIN: *Identification of the commercial timbers of the United States*. New York: McGraw-Hill Book Co., 1934; pp. 223; 6 text figs.; 60 plates.

BROWN, H. P., and A. J. PANSHIN. *Commercial timbers of the United States: their structure, identification, properties, and uses*. New York: McGraw-Hill Book Co., 1940; pp. 554; 75 text figs.; 102 plates.

BRUSH, W. D. *Utilization of elm*. Bull. No. 683, U.S. Dept. Agr., Washington, D.C., 1916; pp. 43.

BRUSH, W. D. *Utilization of sycamore*. Bull. No. 884, U.S. Dept. Agr., Washington, D.C., 1920; pp. 24.

BRUSH, W. D. *Utilization of black walnut*. Bull. No. 909, U.S. Dept. Agr., Washington, D.C., 1921; pp. 89.

BRUSH, W. D. *Utilization of basswood*. Bull. No. 1007, U.S. Dept. Agr., Washington, D.C., 1922; pp. 64.

CARY, N. L. *Sitka spruce: its uses, growth, and management*. Bull. No. 1060, U.S. Dept. Agr., Washington, D.C., 1922; pp. 38.

CLINE, MCGARVEY, and J. B. KNAPP. *Properties and uses of Douglas fir*. Bull. No. 88, U.S. Forest Service, Washington, D.C., 1911; pp. 75.

CUNO, JOHN B. *Utilization of dogwood and persimmon*. Bull. No. 1436, U.S. Dept. Agr., Washington, D.C., 1926; pp. 43.

ELLIOTT, S. B. *The important timber trees of the United States*. Boston and New York: Houghton Mifflin Co., 1912; pp. 382.

FROTHINGHAM, E. H. *The eastern hemlock*. Bull. No. 102, U.S. Dept. Agr., Washington, D.C., 1915; pp. 43.

GARRATT, GEORGE A. *The mechanical properties of wood*. New York: John Wiley & Sons, Inc., 1931; pp. 1–276; 66 text figs.

GIBSON, HENRY H., and HU MAXWELL. *American forest trees.* Chicago, 1913; pp. 708; illustrated.

HALL, WILLIAM R., and HU MAXWELL. *Uses of commercial woods of the United States. I. Cedars, cypresses, and sequoias.* Bull. No. 95, U.S. Forest Service, Washington, D.C., 1911; pp. 62.

HALL, WILLIAM R., and HU MAXWELL. *Uses of commercial woods of the United States. II. Pines.* Bull. No. 99, U.S. Forest Service, Washington, D.C., 1911; pp. 96.

HARLOW, WILLIAM M., and ELLWOOD S. HARRAR. *Textbook of dendrology.* New York: McGraw-Hill Book Co., 1941; pp. 542; illustrated.

HOUGH, ROMEYN B. *American woods* (sections and text). Vols. 1–13; Lowville, N.Y., 1893–1913.

HOUGH, ROMEYN B. *Handbook of the trees of northern states and Canada east of the Rocky Mountains.* Lowville, N.Y., 1907; pp. 470; illustrated.

JEFFREY, EDWARD C. *The anatomy of woody plants.* Chicago: Univ. of Chicago Press, 1917; pp. 478.

JEPSON, WILLIS L. *The silva of California.* Berkeley: University Press, 1910; pp. 480; 10 text figs.; 3 maps; 85 plates.

JEPSON, WILLIS L., H. S. BETTS, and C. D. MELL. *California tanbark oak.* Bull. No. 75, U.S. Forest Service, Washington, D.C., 1911; pp. 34; 10 plates.

JOHNSON, H. M., E. J. HANZLIK, and W. H. GIBBONS. *Red alder in the Pacific Northwest.* Bull. No. 1437, U.S. Dept. Agr., Washington, D.C., 1926; pp. 46.

JOHNSON, R. P. A., and M. I. BRADNER. *Properties of western larch and their relation to the uses of the wood.* Tech. Bull. No. 285, U.S. Dept. Agr., Washington, D.C., 1932; pp. 93.

JOHNSON, R. P. A., and W. H. GIBBONS. *Properties of western hemlock and their relation to the uses of the wood.* Tech. Bull. No. 139, U.S. Dept. Agr., Washington, D.C., 1929; pp. 62.

KOEHLER, ARTHUR. *Guidebook for the identification of woods used for ties and timbers.* Pub. by U.S. Forest Service, Washington, D.C., 1917; pp. 79; 8 figs.; 31 plates.
wood. New York: McGraw-Hill Book Co., 1924; pp. 354; 129 text figs.

KOEHLER, ARTHUR. *The properties and uses of*

LAMB, GEORGE N. *Willows: their growth, use, and importance.* Bull. No. 316, U.S. Dept. Agr., Washington, D.C., 1915; pp. 52.

LUXFORD, R. F., and GEORGE B. TRAYER. *Wood handbook: basic information on wood as a material of construction with data on its use in design and specifications.* Pub. by U.S. Dept. Agr., Washington, D.C., 1940; pp. 326; 64 text figs.; 6 plates.

MARKWARDT, L. J., and T. R. C. WILSON. *Strength and related properties of woods grown in the United States.* Tech. Bull. No. 479, U.S. Dept. Agr., Washington, D.C., 1935; pp. 99; 39 text figs.; 21 tables.

MATOON, W. R. *Forest trees and forest regions of the United States.* Misc. Pub. No. 217, U.S. Dept. Agr., Washington, D.C., 1936; pp. 54; 11 figs.; 14 plates.

MATTOON, W. R. *Southern cypress.* Bull. No. 272, U.S. Dept. Agr., Washington, D.C., 1915; pp. 74.

MAXWELL, HU. *Uses of commercial woods of the United States: beech, birches, and maples.* Bull. No. 12, U.S. Dept. Agr., Washington, D.C., 1913; pp. 56.

MILLSPAUGH, CHARLES F., and LAWRENCE W. NUTTALL. *Flora of Santa Catalina Island (California).* Chicago: Field Museum Bot. Ser. 5: 1–413; 14 plates; 1 map; 1923.

MITCHELL, J. A. *Incense cedar.* Bull. No. 604, U.S. Dept. Agr., Washington, D.C., 1918; pp. 40.

MOHR, CHARLES, and FILIBERT ROTH. *The timber pines of the southern United States, together with a discussion of the structure of their wood.* Bull. No. 13 (rev. ed.), U.S. Div. Forestry, Washington, D.C., 1897; pp. 160; illustrated; range maps.

MUNNS, E. N. *The distribution of important forest trees of the United States.* Washington, D.C.: Misc. Pub. 287, U.S. Dept. Agr., April 1938; pp. 176; 170 maps.

NEWLIN, J. A., and THOMAS R. C. WILSON. *Mechanical properties of woods grown in the United States.* Bull. No. 556, U.S. Dept. Agr., Washington, D.C., 1917; pp. 47; 3 plates.

PENHALLOW, DAVID P. *A manual of the North American gymnosperms, exclusive of the Cycadales, but together with certain exotic species.* Boston: Ginn & Co., 1907; pp. 374; 48 text figs.; 55 plates.

RECORD, SAMUEL J. *Identification of the timbers of temperate North America.* New York: John Wiley & Sons, Inc., 1934; pp. 196; 7 plates; 47 text figs.

ROCHESTER, G. H. *The mechanical properties of Canadian woods.* Bull. No. 82, Canadian Forest Service, Ottawa, 1933; pp. 88.

SARGENT, CHARLES SPRAGUE. *Report on the forests of North America (exclusive of Mexico).* Vol. 9, Tenth Census, Washington, D.C., 1884.

SARGENT, CHARLES SPRAGUE. *The silva of North America: a description of the trees which grow naturally in North America, exclusive of Mexico.* Vols. 1–14, Boston: Houghton Mifflin Co., 1892–1902.

SARGENT, CHARLES SPRAGUE. *Manual of the trees of North America (exclusive of Mexico).* Boston: Houghton Mifflin Co., 1905–1933; pp. 910; 783 text figs.

SMALL, JOHN K. *Flora of the southeastern United States.* New York, 1903; pp. 1370.

STERNS, R. W. *Canadian Douglas fir; its mechanical and physical properties.* Bull. No. 60, Forest Branch, Dept. Int., Ottawa, Canada, 1918; pp. 84.

STERRETT, W. D. *The ashes: their characteristics and management.* Bull. No. 299, U.S. Dept. Agr., Washington, D.C., 1915; pp. 83.

SUDWORTH, GEORGE B. *Forest trees of the Pacific slope.* Washington, D.C.; U.S. Forest Service, 1908; pp. 441; 207 text figs.

SUDWORTH, GEORGE B. *Miscellaneous conifers of the Rocky Mountain region.* Bull. No. 680, U.S. Dept. Agr., Washington, D.C., 1908; pp. 45; 9 range maps, 13 plates.

SUDWORTH, GEORGE B. *The spruce and balsam fir trees of the Rocky Mountain region.* Bull. No. 327, U.S. Dept. Agr., Washington, D.C., 1910; pp. 43; 10 range maps; 25 plates.

SUDWORTH, GEORGE B. *The cypress and juniper trees of the Rocky Mountain region.* Bull. No. 207, U.S. Dept. Agr., Washington, D.C., 1915; pp. 36; 11 range maps; 26 plates.

SUDWORTH, GEORGE B. *The pine trees of the Rocky Mountain region.* Bull. No. 460, U.S. Dept. Agr., Washington, D.C., 1917; pp. 47; 14 range maps; 28 plates.

SUDWORTH, GEORGE B., and C. D. MELL. *The identification of important North American oak woods, based on a study of the secondary wood.* Bull. No. 102, U.S. Forest Service, Washington, D.C., 1911; pp. 56; 48 text figs.

SUDWORTH, GEORGE B., and C. D. MELL. *Distinguishing characteristics of North American gumwoods, based on the anatomy of the secondary wood.* Bull. No. 102, U.S. Forest Service, Washington, D.C., 1911; pp. 20; 9 text figs.

SUDWORTH, GEORGE B., and C. D. MELL. Identification of North American walnut woods. *Hardwood Record,* Chicago, Sept. 10 and 25, 1914.

LATIN AMERICA—GENERAL

Anais de Primeira Reunião Sul-Americana de Botanica, 12 a 19 de Outubre de 1938. 3 vols. Pub. Jardim Bot. do Rio de Janeiro, 1939–40.

BELCHER, R. S., et al. Cross-ties of foreign woods. *Proc. Am. Ry. Eng. Assn.* 26: 1033–1078; 1925.

Caribbean Forester. Pub. quarterly by Tropical Forest Exp. Sta., Rio Piedras, Puerto Rico, 1939–.

COCKRELL, ROBERT A. A comparative study of the wood structure of several South American species of *Strychnos. American Journ. Botany* 28: 1: 32–41; 2 plates; January 1941.

DUCKE, ADOLPHO. Notes on the wallaba trees (*Eperua* Aubl.). *Tropical Woods* 62: 21–28; June 1, 1940.

GARRATT, GEORGE A. Systematic anatomy of the woods of the Myristicaceae. *Tropical Woods* 35: 6–48; Sept. 1, 1933.

GARRATT, GEORGE A. Bearing of wood anatomy on the relationships of the Myrsinaceae. *Tropical Woods* 36: 20–44; Dec. 1, 1933.

GARRATT, GEORGE A. The wood of *Symphonia globulifera. Tropical Woods* 45: 1–15; March 1, 1936.

GILL, TOM. *Tropical forests of the Caribbean.* Washington, D.C., 1931; pp. xix + 318; 80 plates; 4 maps.

GREENE, EDWARD C., JR. Santa María: a neotropical timber of the genus *Calophyllum. Tropical Woods* 30: 9–16; June 1, 1932.

KOEHLER, ARTHUR. *The identification of true mahogany, certain so-called mahoganies, and some common substitutes.* Bull. No. 1050, U.S. Dept. Agr., Washington, D.C., 1922; pp. 18; 17 plates.

KRAMER, PAUL R. The woods of *Billa, Cashalia, Henoonia,* and *Juliania. Tropical Woods* 58: 1–5; June 1, 1939.

KYNOCH, WILLIAM, and NEWELL A. NORTON. *Mechanical properties of certain tropical woods, chiefly from South America.* Bull. No. 7, Univ. Mich. School of For. & Conservation, Ann Arbor, 1938; pp. 87.

MELL, C. D. *True mahogany.* Bull. No. 474, U.S. Dept. Agr., Washington, D.C., 1917; pp. 24; 4 text figs.; 3 plates.

MILANEZ, FERNANDO ROMANO. Estudo anatomico do lenho de trinta especies do genero *Aspidosperma. Physis* (Buenos Aires, Arg.) 15: 429–490; 49 photomicrographs; 1939.

NORTON, THOMAS H. *Tanning materials of Latin America.* Spl. Agts. Ser. No. 165, U.S. Dept. Com. (Washington, D.C.), 1918; pp. 32; 3 plates.

PANSHIN, A. J. Comparative anatomy of the woods of the Meliaceae, subfamily Swietenioideae. *American Journ. Botany* 20: 10: 663–668; 12 plates; December 1933.

PANSHIN, A. J. Wood anatomy of certain South

American rotenone-yielding plants. *American Journ. Botany* 24: 9: 587–591; 10 figs.; November 1937.

Payson, William Farquehar, et al. *Mahogany—antique and modern. A study of its history and use in the decorative arts.* New York: E. P. Dutton & Co., 1926; pp. xxii + 154; 250 pp. of half-tone plates.

Popenoe, Wilson. *Manual of tropical and subtropical fruits, excluding the banana, coconut, pineapple, citrus fruit, olive, and fig.* New York: The Macmillan Co., 1920; pp. 474; 24 states.

Record, Samuel J. *Lignum-vitae; a study of the woods of the Zygophyllaceae with reference to the true lignum-vitae of commerce—its sources, properties, uses, and substitutes.* Bull. No. 6, Yale University School of Forestry, 1921; pp. 48; 2 text figs.; 7 plates.

Record, Samuel J. American woods of the family Rhamnaceae. *Tropical Woods* 58: 6–24; June 1, 1939.

Record, Samuel J. American woods of the family Bombacaceae. *Tropical Woods* 59: 1–20; Sept. 1, 1939.

Record, Samuel J. American woods of the family Sapotaceae. *Tropical Woods* 59: 21–51; Sept. 1, 1939.

Record, Samuel J. Some new names for tropical American trees of the family Leguminosae. *Tropical Woods* 63: 1–6; Sept. 1, 1940.

Record, Samuel J. American timbers of the mahogany family. *Tropical Woods* 66: 7–34; June 1, 1941.

Record, Samuel J. American woods of the family Flacourtiaceae. *Tropical Woods* 68: 40–57; Dec. 1, 1941.

Record, Samuel J. American woods of the family Theaceae. *Tropical Woods* 70: 23–33; June 1, 1942.

Record, Samuel J. American timbers of the genera *Dalbergia* and *Machaerium. Tropical Woods* 72: 1–11; Dec. 1, 1942.

Record, Samuel J., and Robert W. Hess. Aids in identifying American timbers of the order Malvales. *Tropical Woods* 51: 1–10; Sept. 1, 1937.

Record, Samuel J., and Robert W. Hess. American timbers of the family Bignoniaceae. *Tropical Woods* 63: 9–38; Sept. 1, 1940.

Record, Samuel J., and Robert W. Hess. American woods of the family Rutaceae. *Tropical Woods* 64: 1–28; Dec. 1, 1940.

Record, Samuel J., and Robert W. Hess. American woods of the family Verbenaceae. *Tropical Woods* 65: 4–21; March 1, 1941.

Record, Samuel J., and Robert W. Hess. American woods of the family Boraginaceae. *Tropical Woods* 67: 19–33; Sept. 1, 1941.

Record, Samuel J., and Robert W. Hess. American timbers of the family Lauraceae. *Tropical Woods* 69: 7–33; March 1, 1942.

Record, Samuel J., and C. D. Mell. *Timbers of Tropical America.* New Haven: Yale University Press, 1924; pp. 610; 50 plates.

Revista Sudamericana de botánica. Organo oficial de la Asociación Sudamericana de Botánica, Montevideo, Uruguay, 1934–.

Smith, A. C. The American species of Myristicaceae. *Brittonia* (New York) 2: 393–510; 9 figs.; May 2, 1938.

Sudworth, George B., and C. D. Mell. *Fustic wood: its substitutes and adulterants.* Circ. No. 184, U.S. Forest Service, Washington, D.C., 1911; pp. 14; 4 figs.

Williams, Llewelyn. Studies of some tropical American woods. *Tropical Woods* 15: 14–24; Sept. 1, 1928.

Williams, Llewelyn. A study of the Caryocaraceae. *Tropical Woods* 42: 1–18; June 1, 1935.

WEST INDIES

Barker, Henry D., and William S. Dardeau. *Flore d'Haiti.* Pub. by Service Tech. Dept. Agr., Port-au-Prince, 1930; pp. 456.

Beard, J. S. The woodlands of Bermuda. *Empire Forestry Journal* (London) 19: 2: 258–263; 1 plate; July 1940.

Britton, N. L., and C. F. Millspaugh. *The Bahama flora.* New York, 1920; pp. 695.

Britton, N. L., and Percy Wilson. *Botany of Porto Rico and the Virgin Islands.* Vols. V & VI, Scientific Survey of Porto Rico and the Virgin Islands, N.Y. Academy of Sciences, 1924–1930; pp. 626 and 663 resp.

Brush, W. D., Louis S. Murphy, and C. D. Mell. *Trees of Porto Rico.* Appendix I, Bull. No. 354, U.S. Dept. Agr., Washington, D.C., 1916; pp. 56–97.

Chardón, Carlos E. Los pinares de la Republica Dominicana. *Caribbean Forester* 2: 120–131; April 1941.

Cook, Melville T., and H. A. Gleason. Ecological survey of the flora of Porto Rico. *Journ. Dept. Agr. Porto Rico* 12: 1–139; 75 plates; July 1928.

Cook, O. F., and G. N. Collins. Economic plants of Porto Rico. *Contrib. U.S. Nat. Herbarium* (Washington, D.C.) 8: 2: 57–269; plates 14–60; June 27, 1903.

Durland, W. D. The forests of Dominican Republic. *American Geographical Review* 12: 2: 206–222; April 1922.

Fawcett, William, and Alfred Barton Ren-

DLE. *Flora of Jamaica.* I–VII. London: The British Museum, 1926–36.

FORS, ALBERTO J. *Las maderas cubanos.* Pub. by Sec. Agr., Com. y Trabajo, Havana, 1929; pp. 98.

HOLDRIDGE, L. R. *Trees of Puerto Rico.* Volume I. Occ. paper No. 1, Trop. For. Exp. Sta.; Rio Piedras, March 1942; pp. 105; 50 plates.

MARSHALL, R. C. *Notes on the silviculture of the more important timber trees of Trinidad and Tobago, with information on the formation of woods.* Pub. by Forest Dept., Trinidad, 1930; pp. 50; 6 plates.

MARSHALL, R. C. The physiography and vegetation of Trinidad and Tobago. *Oxford Forestry Memoirs* 17: 1–56; 32 figs.; 1934.

MARSHALL, R. C. *Trees of Trinidad and Tobago.* Port-of-Spain: Govt. Printing Office, 1934; pp. 101; 20 plates.

MARSHALL, R. C. *Silviculture of the trees of Trinidad and Tobago, British West Indies.* London: Oxford University Press, 1939; pp. 296; 28 figs.; 16 plates.

MURPHY, L. S. *Forests of Porto Rico.* Bull. No. 354, U.S. Dept. Agr., Washington, D.C., 1916; pp. 55; 6 figs.; 12 plates; 1 map.

ROIG, JUAN T. *Diccionario botanico de nombres vulgares Cubanos.* Havana, 1928; pp. 897; 48 plates.

ROIG, JUAN T. *El cedro: estudio botanico y agricola.* Circ. No. 79, Estación Exp. Agronómica, Santiago de las Vegas, Cuba, May 1935; pp. 31; 7 plates.

ROIG, JUAN T. *Catalogo de maderas Cubanas.* Bol. No. 56, Estación Exp. Agron. Santiago de las Vegas, June 1935; pp. 77.

SCHIFFINO, JOSÉ. *Riqueza forestal Dominicana.* Santo Domingo: Imprinta "La Floresta," 1927; pp. 127.

SWABEY, CHRISTOPHER. *The principal timbers of Jamaica.* Bull. No. 29 (n.s.), Dept. Sci. & Agr., Kingston, 1941; pp. 37; 1 map.

MEXICO

BLANCO, CENOBIO E. Los pinos de Mexico. *Bol. Dept. Forestal y de Caza y Pesca* (Mexico, D.F.) 3: 11: 237–255; 9 plates; 1938.

MARTÍNEZ, JOSÉ GARCÍA. La exploitación de los pinos en Mexico. *Bol. Dept. Forestal y de Caza y Pesca* (Mexico, D.F.) 4: 12: 183–229; 1 map; 1938.

MARTÍNEZ, MAXIMINO. *Las plantas mas utiles que existen in la Republica Mexicana.* Mexico, D.F., 1928; pp. 381; 170 text figs.

MARTÍNEZ, MAXIMINO. *Las plantas medicinales de Mexico.* Mexico, D.F., 1933; pp. 644; many text figs.; 7 colored plates.

MARTÍNEZ, MAXIMINO. *Catalogo de nombres vulgares y cientificos de plantas Mexicanas.* Mexico, D.F., 1937; pp. 552.

Mexico Forestal. Pub. monthly by the Sociedad Forestal Mexicana, Calle del Eliseo 11, Mexico, D.F.

REHDER, ALFRED. The firs of Mexico and Guatemala. *Journ. Arnold Arboretum* 20: 3: 281–287; 1 fig.; July 1939.

ROBERTSON, C. C. The cultivation of Mexican pines in the Union of South Africa, with notes on the species and their original habitat. *Empire Forestry Journal* 11–13; 1932–1934; reprint, pp. 83; 1 map.

SAFFORD, W. E. Lignum nephriticum: its history and an account of the remarkable fluorescence of its infusion. *Smithsonian Report for 1915* (Washington, D.C.), 271–298; 6 text figs.; 7 plates.

SHAW, GEORGE RUSSELL. *The pines of Mexico.* Publ. No. 1, Arnold Arboretum (Jamaica Plain, Mass.), 1909; pp. 29; 1 map; 22 plates.

STANDLEY, PAUL C. *Flora of Yucatan.* Chicago: Field Museum Bot. Ser. 3: 3: 157–492; 1930.

STANDLEY, PAUL C. Trees and shrubs of Mexico. *Contr. U.S. Nat. Herb.* 23: 1–1721; October 1920 to November 1926.

WILLIAMS, LLEWELYN. Forest trees of the Isthmus of Tehuantepec, Mexico. *Tropical Woods* 53: 1–11; March 1, 1938.

CENTRAL AMERICA

COOPER, G. PROCTOR. Some interesting trees of western Panama. *Tropical Woods* 14: 1–8; June 1, 1928.

COOPER, G. PROCTOR. The forests of western Panama. *Tropical Woods* 16: 1–35; Dec. 1, 1928.

COOPER, G. PROCTOR. Espavé (*Anacardium rhinocarpus* DC.). *Tropical Woods* 22: 4–9; June 1, 1930.

DURLAND, WILLIAM D. The pine forests of Honduras. *Tropical Woods* 10: 8–9; June 1, 1927.

ENGLESING, F. C. Trees and shrubs collected in northeastern Nicaragua. *Tropical Woods* 17: 18–38; March 1, 1929.

FAHNESTOCK, GEORGE R., and GEORGE A. GARRATT. Nicaraguan pine. *Tropical Woods* 55: 1–16; Sept. 1, 1938.

Flora Salvadoreña. Pub. by Ministerio de Instrucción Publica de la Republica de El Salvador, 1926. 2 vols.; pp. 110 and 111, resp.; 100 plates, part in color.

HUMMEL, C. *Report on the forests of British Honduras with suggestions for a far reaching forest policy.* London: Crown Agents for the Colonies, 1925; pp. 122; 1 map.

KLUGE, H. C. Trees of the Bayano River watershed, Panama. *Tropical Woods* 5: 4–12; March 1, 1926.

LUNDELL, CYRUS L. *The vegetation of Petén.* Pub. No. 478, Carnegie Inst., Washington, D.C., June 16, 1937; pp. 244; 3 figs.; 39 plates.

PITTIER, H. *Ensayo sobre las plantas usuales de Costa Rica.* Washington, D.C.: H. L. & J. B. McQueen Co., 1908; pp. 176; 31 plates.

POPENOE, WILSON. *Cupressus Benthami:* a neglected opportunity? *Tropical Woods* 65: 1–4; March 1, 1941.

RECORD, SAMUEL J. Trees of Honduras. *Tropical Woods* 10: 10–47, June 1, 1927.

RECORD, SAMUEL J. Trees of Gualán, Guatemala. *Tropical Woods* 11: 10–18; Sept. 1, 1927.

RECORD, SAMUEL J., and GEORGE A. GARRATT. *Cocobolo.* Bull. No. 8, Yale School of Forestry, 1923; pp. 42; 2 figs.; 7 plates.

RECORD, SAMUEL J., and HENRY KUYLEN. Trees of the lower Río Motagua Valley, Guatemala. *Tropical Woods* 7: 10–29; Sept. 1, 1926.

Revista de Agricultura Tropical. Organo del Ministerio de Agricultura, San Salvador, El Salvador.

STANDLEY, PAUL C. An enumeration of the Sapotaceae of Central America. *Tropical Woods* 4: 1–11, December 1, 1925.

STANDLEY, PAUL C. Poisonous trees of Central America. *Tropical Woods* 9: 3–7; March 1, 1927.

STANDLEY, PAUL C. Flora of the Panama Canal Zone. *Contr. U.S. Nat. Herb.* (Washington) 27: 1–416; 67 plates; 7 text figs.; Jan. 31, 1928.

STANDLEY, PAUL C. A second list of the trees of Honduras. *Tropical Woods* 21: 9–41; March 1, 1930.

STANDLEY, PAUL C. *Flora of the Lancetilla Valley, Honduras.* Chicago: Field Museum Bot. Ser. 10: 1–418; 68 plates; Jan. 15, 1931.

STANDLEY, PAUL C. The flora of Barro Colorado Island, Panama. *Contr. Arnold Arboretum* (Jamaica Plain, Mass.) 5: 1–178; 21 plates; 1 map; Oct. 1, 1933.

STANDLEY, PAUL C. Additions to the trees of Honduras. *Tropical Woods* 37: 27–39; March 1, 1934.

STANDLEY, PAUL C. Flora of Costa Rica. *Field Museum Bot. Ser.* 18: 1–1616; 1937–1938.

STANDLEY, PAUL C. The forests of Guatemala. *Tropical Woods* 67: 1–18; Sept. 1, 1941.

STANDLEY, PAUL C., and SALVADOR CALDERÓN. *Lista preliminar de las plantas de El Salvador.* San Salvador, 1925; pp. 274.

STANDLEY, PAUL C., and SAMUEL J. RECORD. *The forests and flora of British Honduras.* Chicago: Field Museum Bot. Ser. 12: 1–432; 1936; 16 plates.

STEVENSON, DUNCAN, and NEIL S. Some secondary timbers of British Honduras. *Tropical Woods* 4: 12–16; December 1, 1925.

STEVENSON, NEIL S. The Honduras Rosewood. *Tropical Woods* 12: 1–3; Dec. 5, 1927.

STEVENSON, NEIL S. *Supply, consumption, and marketing of timber in British Honduras.* Belize: Govt. Printing Office, 1935; pp. 11.

YUNCKER, T. G. *Flora of the Aguán Valley and the coastal regions near La Ceiba, Honduras.* Chicago: Field Museum Bot. Ser. 9: 4: 245–346; March 22, 1940.

COLOMBIA and VENEZUELA

APOLINAR MARÍA, HERMANO. Vocabulario de terminos vulgares en historia natural colombiana. *Rev. Academia Colombiana,* Bogotá, Col., 1940–.

BRITTON, N. L., and E. P. KILLIP. Mimosaceae and Caesalpiniaceae of Colombia. *Annals N.Y. Academy of Science* 35: 101–208; 2 plates; April 1, 1936.

Caldasia. Pub. at irreg. intervals by Inst. Ciencias Naturales de la Universidad Nacional de Colombia, Bogotá, 1940–.

CUMMINGS, LEWIS A. The forests of Venezuela. *Tropical Woods* 18: 32–42; June 1, 1929.

CURRAN, HUGH M. The Lands of Loba, Colombia. *Tropical Woods* 19: 11–38; Sept. 1, 1929.

DUGAND, ARMANDO. The transition forests of Atlántico, Colombia. *Tropical Woods* 40: 1–14; Dec. 1, 1934.

DUGAND, ARMANDO. *Contribuciones a la historia natural Colombiana* (Barranquilla) 1: 1–13, March 25, 1938; 2: 1–2, Sept. 15, 1938; 3: 1–7, March 10, 1939.

ESPINA, RAMÓN, and JUAN GIACOMETTO. Trees of Sierra Nevada de Santa Marta. *Tropical Woods* 30: 17–37; June 1, 1932.

GLEASON, H. A. Botanical results of the Tyler-Duida expedition. *Bull. Torrey Botanical Club* (New York) 58: 277–506; plates 17–45; 1 map; Nov.–Dec. 1931.

PÉREZ ARBELÁEZ, ENRIQUE. *Plantas útiles de Colombia. I.* Bogotá: Imprinta Nacional, 1935; pp. 172; 116 text figs.

PITTIER, H. *Esbozo de las formaciones vegetales de Venezuela.* Caracas: Litografía del Comercio, 1920; pp. 44.

PITTIER, H. *Arboles y arbustos nuevos de Venezuela.* (Five parts.) Caracas, 1921–29.

PITTIER, H. *Exploraciones botanicas y otras en la cuenca de Maracaibo.* Caracas: Tipografía Mercantil, 1923; pp. 100; 13 plates.

PITTIER, H. *Manual de las plantas usuales de Venezuela.* Caracas: Litografía del Comercio, 1926; pp. 458; 42 plates.

PITTIER, H. *Arboles y arbustos del orden de las Leguminosas.* Caracas: Tipografía Americana, 1927–28; pp. 229.

PITTIER, H. *Suplemento a las plantas usuales de Venezuela.* Caracas, 1939; pp. 129.

PITTIER, H., and SAMUEL J. RECORD. La caoba venezolana. *Bol. Com. & Ind.* (Caracas) 18: 582–593; 3 text figs.; 1 plate; September 1921.

RECORD, SAMUEL J. The wood of the Venezuelan mahogany. *Bol. Com. & Ind.* (Caracas) 18: 577–581; September 1921.

RECORD, SAMUEL J., and HENRY KUYLEN. Santa Marta Valley, Colombia. *Tropical Woods* 23: 9–23; Sept. 1, 1930.

SPRAGUE, T. A., and L. A. BOODLE. West Indian boxwood (*Casearia praecox* Griseb.). *Kew Bulletin,* 1904; pp. 11–12.

STANDLEY, PAUL C. The Rubiaceae of Venezuela. Chicago: *Field Museum Bot. Ser.* 7: 4: 343–485; Oct. 12, 1931.

SUDWORTH, GEORGE B., and C. D. MELL. *"Colombian mahogany"* (Cariniana pyriformis): *its characteristics and its use as a substitute for true mahogany* (Swietenia mahagoni). Circ. 185, U.S. Forest Service, Washington, D.C., 1911; pp. 16; 11 text figs.

TAMAYO, FRANCISCO. Exploraciones botanicos en la Peninsula de Paraguaná, Estado Falcón. In *Bol. Soc. Venez. Cienc. Nat.* 47, 1941; pp. 90; 10 plates.

WHITFORD, H. N. Forest conditions in Colombia and Venezuela. *Bull. Pan American Union* (Washington, D.C.) 15: 4: 468–483; 7 plates; April 1918.

WILLIAMS, LLEWELYN. *Maderas económicas de Venezuela.* Bol. Técnico No. 2, Min. Agr. & Cria, Caracas, 1939; pp. 97.

WILLIAM, LLEWELYN. Botanical exploration in the middle and lower Caura, Venezuela. *Tropical Woods* 62: 1–20; June 1, 1940.

WILLIAMS, LLEWELYN. The Caura Valley and its forests. *Geographical Review* 31: 414–429; 10 text figs.; July 1941.

WILLIAMS, LLEWELYN. Forests of the Venezuelan Guiana. *Tropical Woods* 68: 13–40; Dec. 1, 1941.

WILLIAMS, LLEWELYN. Natural resources of Venezuela. *Chronica Botanica* (Waltham, Mass.) 7: 2: 75–77; March 1942.

THE GUIANAS

AITKEN, J. B. The wallabas of British Guiana. *Tropical Woods* 23: 1–4; Sept. 1, 1930.

BENOIST, R. Les bois de la Guyane française. *Archives de Botanique* 5: 1: 1–292; 10 figs.; 58 plates; 1931.

BERTIN, A. *Les bois de la Guyane française et du Brésil.* Pub. by Mission d'Études Forestières, Paris, 1920; pp. 315; 23 text figs.; 2 plates.

GONGGRIJP, J. W. Rozenhout, bois de rose femelle, uit Suriname. *Indische Mercuur* (Amsterdam), Apr. 23 and 30, 1920.

GONGGRIJP, J. W. Over Demerara greenheart, manbarklak en een nieuwe tegen paalworm bestande houtsort. *West Indië* (Paramaribo), September 1921.

GONGGRIJP, J. W. *Verslag van een dienstreis tot het onderzoeken van de mogelijkheid tot afzet van Surinaamsch hout vooral in Midden-Amerika.* Paramaribo: Bull. 46, Dept. Landbouw, Nijverheid en Handel, 1924; pp. 25.

GRAHAM, EDWARD H. Flora of the Kartabo region, British Guiana. *Annals Carnegie Museum* 22: 1: 17–292; 2 text figs.; 16 plates; Feb. 1, 1934.

HAMAN, MILES, and B. R. WOOD. The forests of British Guiana. *Tropical Woods* 15: 1–13; Sept. 1, 1928.

HOHENKIRK, L. S. British Guiana timbers. *Journ. Board Agr. Brit. Guiana* (Georgetown) 12: 3: 152–187; July 1919.

HOHENKIRK, L. S. Report on tests of woods from British Guiana by the Forest Products Laboratories of Canada. *Journ. Board Agr. Brit. Guiana* (Georgetown) 13: 2: 89–96; April 1920.

HUMPHREY, C. J. Tests on the durability of greenheart (*Nectandra Rodiaei* Schomb.). *Mycologia* 7: 204–209; 1 plate; July 1915.

JANSSONIUS, H. H. *Mikrographie einiger technisch wichtigen Holzarten aus Surinam.* Amsterdam: Johannes Müller, 1914; pp. 50; 16 text figs.

KRIBS, DAVID A. The Persaud Collection of British Guiana woods. *Tropical Woods* 13: 7–46; March 1, 1928.

MELL, C. D., and W. D. BRUSH. *Greenheart.* Circ. No. 211, U.S. Forest Service, Washington, D.C., 1913; pp. 12; 4 plates.

PFEIFFER, J. PH. *De houtsoorten van Suriname.* Afd. Handelsmuseum No. 6, Kon. Ver. Koloniaal Inst. te Amsterdam. Vol. I: pp. 505; 23 text figs.; 24 plates; 1926. Vol. II: pp. 269; 43 text figs.; 17 plates; 1927.

PULLE, A. *Flora of Surinam (Dutch Guyana).*

Vol. I–IV (incomplete). Afd. Handelsmuseum No. 11, Kon. Ver. Koloniaal Inst. te Amsterdam, 1932–38; pp. 1777.

RENAUD, MAURICE. Contribution a l'étude des bois de la Guyane française. *Bulletin de l'Agence Général des Colonies* (Paris) 25: 279: 970–1031, June 1932; 280: 1120–1177, July 1932; 281: 1265–1308, August 1932; 282: 1372–1434, September 1932; 22 plates.

STONE, HERBERT. Les bois utiles de la Guyane française. *Ann. Musée Colonial de Marseille* 25 (3rd ser.): 4: 2: 39–135; 5: 3: 3–159; 6: 2: 3–68; 8: 2: 1–98; 7 plates; 1916–1922.

STONE, HERBERT, and W. G. FREEMAN. *The timbers of British Guiana.* London: Crown Agts. for the Colonies, 1914; pp. 110; 1 plate.

WOOD, B. R. Greenheart. *Official Gazette* (Georgetown), July 17, 1926; pp. 4.

CHILE, ECUADOR, and PERU

BERNATH, E. L. Coniferous forest trees of Chile. *Tropical Woods* 52: 19–26; Dec. 1, 1937.

BERNATH, E. L. Arboles forestales coniferos de Chile. *Maderil* (Buenos Aires) 10: 117: 4–8; 4 photos; March 1938.

BERNATH, E. L. *Las hayas australes o antarticas de Chile.* Santiago de Chile, 1940; pp. 43; 1 map; 26 photos; 11 plates of drawings.

GREENHOUSE, SAMUEL. The culture of the balsa tree in Ecuador. *Journal of Forestry* 33: 10: 870–876; October 1935.

HARMS, H. Contribución al conocimiento de *Lysiloma guachapeli* (H.B.K.) Benth. *Rev. Colegio Nac. Vicente Rocafuerte* (Guayaquil, Ec.) 12: 27–31; 1930.

HERRERA, FORTUNATO L. *Catalogo alfabetico de los nombres vulgares y cientificos de plantas que existen en el Peru.* Lima: Univ. Mayor de San Marcos, 1939; pp. 363.

HERRERA, FORTUNATO L. *Contribución a flora del Departamento del Cuzco,* July 28, 1921; pp. 240.

HERRERA, FORTUNATO L. *Estudios sobre la flora del Departamento del Cuzco.* Vol. I, pp. 257, Lima, 1930; II, pp. 251, Cuzco, 1933.

HERRERA, FORTUNATO L. *Plantas que curan y plantas que matan de la flora del Cuzco (estudio folklórico).* Lima: Primer Congreso Nacional de Quimica, July 1938; pp. 201–264.

MACBRIDE, J. FRANCIS. *Flora of Peru.* Chicago: Field Museum Bot. Ser. Vol. 13, several parts (incomplete), 1936–.

POPENOE, WILSON. Economic fruit-bearing plants of Ecuador. *Contr. U.S. Nat. Herb.* (Washington, D.C.) 24: 5: 101–134; plates 34–49; 1924.

RIMBACH, AUGUST. The forests of Ecuador. *Tropical Woods* 31: 1–9; Sept. 1, 1932.

RUIZ, HIPÓLITO. *Travels of Ruiz, Pavón, and Dombey in Peru and Chile (1777–1788).* Translated by B. E. Dahlgren. Chicago: Field Museum Bot. Ser. 21: 1–372; March 28, 1940.

SOLÍS, M. ACOSTA. *Galapagos observado fitologicamente.* Quito: Universidad Central, 1937; pp. 78; 30 half-tones.

SOLÍS, M. ACOSTA. Principal timbers used in the Sierra del Ecuador. *Tropical Woods* 57: 1–6; March 1, 1939.

SOLÍS, M. ACOSTA. Productos forestales del Ecuador. *Maderil* (Buenos Aires, Arg.) 12: 133: 7–10, 27; July 1939.

WILLIAMS, LLEWELYN. Peruvian mahogany. *Tropical Woods* 31: 30–37, Sept. 1, 1932.

WILLIAMS, LLEWELYN. The forests of northeastern Peru. *Tropical Woods* 25: 5–12; March 1, 1931.

WILLIAMS, LLEWELYN. *Woods of Northeastern Peru.* Chicago: Field Museum Bot. Ser. 15: 1–587; 18 text figs.; 2 maps; Dec. 31, 1936.

ARGENTINA, PARAGUAY, and URUGUAY

CASTRO, EDUARDO B. *Las maderas Argentinas su importancia industrial.* Rosario, 1918; pp. 344.

CHEBATAROFF, J. Distribución de las Anacardiáceas del Uruguay. *Revista Sudamer. Bot.* (Montevideo) 5: 37–39; 2 figs.; November 1937.

COVAS, GUILLERMO. Las coniferas indígenas de la Republica Argentina. *Revista de la Facultad de Agronomia* (Buenos Aires) 21: 201–223; 2 plates; 1938.

DESCOLE, HORACIO R., and CARLOS A. O'DONELL. Estudios anatómicos en el leño de plantas Tucumanas. *Lilloa* (Buenos Aires) 1: 75–93; 8 plates.

DESCOLE, HORACIO R., et al. Revisión de las zigofilaceas Argentinas. *Lilloa* 5: 2: 257–352; 21 plates; 16 figs.; 2 maps; 1940.

DEVOTO, FRANCO E., and MAX ROTHKUGEL. *Informe sobre los bosques del Parque Nacional del Iguazú.* Buenos Aires: Ministerio de Agricultura, 1936; pp. 99; 50 half-tones.

DURLAND, WILLIAM D. The quebracho region of Argentina. *Geographical Review* (New York) 14: 2: 227–241; 2 maps; April 1924.

FIEBRIG-GERTZ, C. Guarang names of Paraguayan plants and animals. *Rev. Jard. Bot.*

y Mus. Hist. Nat. del Paraguay (Asunción) 2: 99–149; (1923) 1930.

FIEBRIG-GERTZ, C., and TEODORO ROJAS. Ensayo fitogeográfico sobre el Chaco Boreal. *Rev. Jard. Bot. y Mus. Hist. Nat. del Paraguay* (Asunción) 3: 3–87; 14 plates; 1933.

HAUMAN, LUCIEN, and LUIS H. IRIGOYEN. Catalogue des phanérogame de l'Argentine. Dicotylédones. I. *An. Mus. Nac. Hist. Nat. de Buenos Aires* 32: 1–314; 1923.

KERR, GEORGE A. *The quebracho forests of South America.* Washington, D.C.: Pan American Union, 1924; pp. 27; 6 figs.

KOUTCHE, VSEVOLOD, and JORGE N. F. CARMELICH. *Estudio forestal del caldén* (Prosopis algarrobilla *Griseb.*). Buenos Aires: Ministerio de Agricultura, 1936; pp. 22; 16 half-tones.

LATZINA, EDUARDO. Index de la flora dendrológica Argentina. *Lilloa* 1: 95–211; 14 plates; Dec. 27, 1937.

LATZINA, EDUARDO. Calorimetría de maderas pertenecientes a especies existentes in la Provincia de Tucumán. *Lilloa* 1: 235–248; Dec. 27, 1937.

LATZINA, EDUARDO. Neuva ensayos de fasificación de maderas del pais. *Lilloa* 2: 185–249; May 15, 1938.

LATZINA, EDUARDO. Ensayos de dureza, compacidad y porosidad efectuados con maderas del pais. *Lilloa* 2: 2: 353–412; 1938.

LATZINA, EDUARDO. Resultados preliminares de ensayos de resistencia a la compresión, tracción y flexión efectuados con maderas del pais. *Lilloa* 3: 81–209; Dec. 27, 1938.

LILLO, MIGUEL. *Segunda contribución al conocimiento de los árboles de la Argentina.* Buenos Aires: Univ. Nacional de Tucumán, 1924; pp. 55.

Lilloa (*Revista de botanica*). Ed. by Horacio R. Descole, Instituto "Miguel Lillo," Tucumán, Argentina, 1937–.

Maderil. Pub. monthly, official organ of Bolsa de Maderas de Buenos Aires, Arg.

MELL, C. D., and W. D. BRUSH. *Quebracho wood and its substitutes.* Circ. No. 202, U.S. Forest Service, Washington, D.C., 1912; pp. 12; 1 range map; 2 plates.

MEYER, TEODORO. Los árboles indigenas de importancia económica del Departamento de Resistencia (Chaco). *Revista Argentina de Agronómia* (Buenos Aires) 4: 3: 153–167; September 1937.

Nomina de las maderas del pais. Buenos Aires: Ministerio de Agricultura, 1932; pp. 8.

O'DONELL, CARLOS A. Anatomía comparada del leño de tres Simarubaceas Argentinas. *Lilloa* 1: 263–282; 3 plates; Dec. 27, 1937.

PALACIO, EMILIO. *Ensayos de resistencia de maderas Argentinas.* Buenos Aires: Revista Technica, 1903; pp. 64.

PARODI, DOMINGO. *Notas sobre algunas plantas usuales del Paraguay, de Corrientes y de Misiones.* Buenos Aires: Pablo E. Coni é Hijos, 1886; pp. 186.

PARODI, LORENZO R. Las plantas indigenas no alimenticas cultivadas en la Argentina. *Revista Argentina de Agronómia* (Buenos Aires) 1: 3: 165–212; 1 map; 1934.

Revista de Jardín Botánico y Museo de Historia Natural del Paraguay. I–III. Asunción, 1922–1933.

ROTHKUGEL, MAX. *Estudio del crecimiento y rendimiento del alamo criollo* (Populus nigra *var.* itálica *DuRoi*) *en el delta del Paraná.* Pub. by Min. Obras Publicas de la Provincia de Buenos Aires, 1938; pp. 44; 2 half-tones.

ROTHKUGEL, MAX. *Los bosques Patagónicos.* Buenos Aires: Ministerio de Agricultura, 1916; pp. 207; 21 maps; 68 half-tones; 10 plates of drawings.

SCHULZ, AUGUSTO G. *Las Bignoniáceas del* territorio del Chaco. *Lilloa* 5: 1: 131–158; 6 plates; 3 figs.; Feb. 14, 1940.

SPEGAZZINI, CARLOS, and CARLOS D. GIROLA. *Maderas.* Exposición Internacional de Agricultura de 1910; pp. 322–413.

TORTORELLI, LUCAS A. Technologia de maderas Argentinas. *Maderil* (Buenos Aires) 11: 121: 19–24; 7 figs.; July 1938.

TORTORELLI, LUCAS A. Maderas Argentinas. *Estudio xilologico y technologico de las principales especies arboreas del país.* Pub. by Univ. de Buenos Aires, July 1940; pp. 218; 49 plates; 5 maps.

VENTURI, SANTIAGO, and MIGUEL LILLO. *Contribución al conocimiento de los árboles de la Argentina.* Buenos Aires, 1910; pp. 127.

BRAZIL

Archivos do Instituto de Biologia Vegetal. Jardim Botanico, Rio de Janeiro, Brazil. Vol. I–IV; 1934–38.

Archivos do Jardim Botanico do Rio de Janeiro. Vol. I–VI; 1915–1933.

Arquivos de botanica do Estado de S. Paulo (n.s.) 1: 1–2; 1–52; 63 plates; 1938–39.

BABBITT, WILL C. Turnings from Brazilian hardwoods. *Wood Turning* (Milwaukee), April 1923.

BROTERO, FREDERICO A. *Estudo dos caractéres physicos e mecanicos das madeiras.* Bol. 8, Escola Polytechnica, São Paulo, 1932.

CURRAN, HUGH M. Forest Conditions in southeastern Bahia, Brazil. *Tropical Woods* 2: 6–9; March 1925.

De Assis Inglesias, Francisco. *Album floristico.* Rio de Janeiro: Serviço Florestal, 1940; pp. 134; 65 illustrations in color.

De Sampaio, A. J. *Nomes vulgares de plantas da Amazonia.* Rio de Janeiro, 1934; pp. 69.

De Sampaio, A. J. O problema florestal no Brazil em 1926. *Archivos do Museu Nacional* 28: 55–173; bibliography pp. 145–173; 1926.

Ducke, Adolpho. Les arbres producteurs de "balata." *Rev. Bot. App. & d'Agr. Tropicale* (Paris) 10: 111: 849–851; November 1930.

Ducke, Adolpho. Nova sapotacea productora de balata, da Amazonia brasileira. *Ann. Acad. Bras. Sci.* (Rio de Janeiro) 6: 4: 207–210; 1 plate; Dec. 31, 1934.

Ducke, Adolpho. Les genres *Coumarouna* Aubl. et *Taralea* Aubl. *Rev. Bot. App. & d'Agr. Tropicale* (Paris) 14: 154: 400–407; 1934.

Ducke, Adolpho. Notes on the vernacular names of trees from the Tapajoz River, Brazil. *Tropical Woods* 39: 11–16; Sept. 1, 1934.

Ducke, Adolpho. As espicies brasileiras de jatahy, jutahy ou jatobá (genero *Hymenaea* L.). *Ann. Acad. Bras. Sci.* (Rio de Janeiro) 7: 3: 203–211; 1 plate; Sept. 30, 1935.

Ducke, Adolpho. Notes on the species of *Hymenolobium,* giant trees of the Brazilian Amazonia. *Tropical Woods* 47: 1–10; Sept. 1, 1936.

Ducke, Adolpho. The páo mulato of Brazilian Amazonia. *Tropical Woods* 49: 1–4; March 1, 1937.

Ducke, Adolpho. The muirapirangas of Brazilian Amazonia. *Tropical Woods* 51: 15–18; Sept. 1, 1937.

Ducke, Adolpho. Notes on the purpleheart trees (*Peltogyne*) of Brazilian Amazonia. *Tropical Woods* 54: 1–7; June 1, 1938.

Ducke, Adolpho. Revision of the genus *Hevea,* mainly the Brazilian species. *Arq. Inst. Biol. Vegetal* (Rio de Janeiro) 2: 2: 217–346; 3 plates; December 1935. Revised and reprinted, 1939; pp. 32.

Ducke, Adolpho. *As Leguminosas da Amazonia Brasileira.* Rio de Janeiro: Serviço Florestal, Ministério da Agricultura, 1939; pp. 170.

Ducke, Adolpho. O genero *Anacardium* na Amazonia brasileira. *Ann. Acad. Bras. Sci.* 11: 1: 11–17; 3 plates; March 31, 1939.

Ducke, Adolpho. Notes on some highly aromatic Lauraceae of Brazilian Amazonia. *Tropical Woods* 60: 1–10; Dec. 1, 1939.

Ducke, Adolpho. Revision of the *Macrolobium* species of the Amazonia hylea. *Tropical Woods* 65: 21–31; March 1, 1941.

Hoehne, F. C. *A bracaatinga ou abaracaatinga.* São Paulo: Sec. Agr., Ind. e Com., May 1930; pp. 47; 15 figs.

Hoehne, F. C. Algo sobre a identidade botanica do "oiti" e do "pequiá" da Bahia. *Ann. Acad. Bras. Sci.* 4: 1–11; 2 plates; March 31, 1932.

Hoehne, F. C. *Flora Brasilica.* Pub. in parts at irreg. intervals by Sec. Agr., Ind. & Com. de São Paulo, 1940–.

Hoehne, F. C. *Plantas e substancias vegetais tóxicas e medicinais.* São Paulo, 1939; pp. 335; 252 text figs.; 26 plates in color.

Huber, J. Mattas e madeiras amazonicas. *Bol. Museu Goeldi* (Pará) 6: 91–225; 1909.

Le Cointe, Paul. *A Amazonia Brasileira. III. Arvores e plantas uteis (indigenas e acclimadas).* Belem-Pará: Livraria Classica, 1934; pp. 486; 14 plates.

Pereira, Huascar. *Diccionario das plantas uteis do Estado de S. Paulo.* São Paulo: Typographia Brasil de Rothschild & Co., 1929; pp. 779.

Pereira, Huascar. *The timber trees of the State of São Paulo, Brazil.* Pub. by Dept. Agr., Com. & Public Works; pp. 86; 27 plates.

Pereira, José Aranha. *Contribuição para a identificação micrographica das nossas madeiras.* Bull. No. 9, Escola Polytechnica de São Paulo, Brazil, 1933; pp. 165; 26 text figs.; 55 plates.

Pio Corrêa, M. *Diccionario das plantas uteis do Brasil e das exoticas cultivadas.* Rio de Janeiro: Ministerio da Agricultura, 1926, 1931. Vol. I, A–Cap, pp. 747; 657 text figs.; 106 plates. II, Car–E, pp. 707; 851 text figs.; 71 plates.

Pio Corrêa, M. *Flora do Brasil.* Rio de Janeiro: Typographia da Estatistica, 1909; pp. 154.

Rebouças, André e José. *Ensaio de indice geral das madeiras do Brasil.* Vol. I (A to C), Rio de Janeiro, 1877; pp. 435.

Resultados obtidos para algumas madeiras nacionaes. São Paulo, Brazil: Instituto de Pesquisas Technologicas, 1937; 1 large table.

Revista Florestal (Brasil). Orgão de defesa das florestas e assistencia as industrias de madeiras e sub-productos florestaes. Vol. I, Nos. 1–10; II, No. 1.; Rio de Janeiro, 1929–30.

Rodriguésia. Revista do Instituto de Biologia Vegetal, Jardim Botanico e Estação Biologica do Itatiaya, Nos. 1–11, 1935–37; Revista do Serviço Florestal, No. 12–, 1939–. Rio de Janeiro, Brazil.

Schmieg, Karl. Notes on new cabinet woods

from Brazil. *Tropical Woods* 5: 1–5; March 1, 1926.

Silveira Mello, Octavio. *A merindiba.* Rio de Janeiro: Ministerio da Agricultura, 1938; pp. 107; 35 text figs.

Standley, Paul C. Vernacular names of trees of the Tapajoz River, Brazil. *Tropical Woods* 29: 6–13; March 1, 1932.

Teixeira da Fonseca, Eurico. *Indicador de madeiras e plantas uteis do Brazil.* Rio de Janeiro: Villas-Boas & Co., 1922; pp. 343.

Teixeira da Fonseca, Eurico. *Oleos vegetaes Brasileiros (inclusive resinas, gommas, breus, ceras).* Rio de Janeiro, 1927; pp. 341.

Whitford, H. N. The structure and use of the Paraná pine forests of Brazil. *Journal of Forestry* 17: 2: 154–158; February 1019.

EXPLANATION OF THE PHOTOMICROGRAPHS

ALL of these photomicrographs were made by the junior author from specimens in the collections of the Yale School of Forestry. Unless otherwise indicated they are of cross sections and the top is toward the outside of the stem. Those marked 10 × were made with incident lighting and show parts of the wood as they appear under an ordinary hand lens magnifying ten diameters. Those of 20 × or more are of very thin, stained sections through which light was transmitted.

PLATE XXXVII

1. *Ephedra trifurca* Torr., Ephedraceae (Y. 14793). A gymnosperm having true vessels, ring-porous structure, and large rays, features ordinarily occurring only in angiosperms. 20 ×.

2. *Astrocaryum Standleyanum* Bailey, Palmaceae (Y. 7235). Showing very dense fibro-vascular bundles in the outer part of the stem of the Black Palm of Panama. 10 ×.

3. *Dracaena americana* D. Sm., Liliaceae. The fibro-vascular bundles are few and scattered near the center (below), but they are radially arranged (above) after the formation of a cambium layer. 20 ×.

PLATE XXXVIII

1. *Guevina avellana* Mol., Proteaceae (Y. 9548). On the outer (upper) side of each arc-like zone of pores is a narrow band of wood parenchyma. The coarse rays are dilated at the margins of the growth rings. 20 ×.

2. *Embothrium grandiflorum* Lam., Proteaceae (Y. 22819). The pores appear as though suspended from the hammock-like parenchyma bands. Most of the rays are broad. 20 ×.

3. *Roupala grossedentata* Pitt., Proteaceae (Y. 36189). The pores are mostly in short tangential rows and sometimes are imbedded in the parenchyma layers. Most of the rays are broad. 20 ×.

PLATE XXXIX

1. *Ulmus fulva* Michx., Ulmaceae (Y. 11871). The Slippery Elm has a wide band of large pores in the early wood (ring-porous). The many small pores in the late wood are in diagonal or wavy bands characteristic of most of the Ulmaceae, hence this pattern (which occurs in various other woods also) is described as "ulmiform." 20 ×.

2. *Carya glabra* Mill., Juglandaceae. The pores in the early wood are widely and irregularly spaced, those in the late wood few and small. Wood parenchyma is in numerous narrow, apotracheal bands in the late wood. 20 ×.

3. *Patagonula americana* L., Boraginaceae (Y. 23489). The numerous small pores are in irregularly ulmiform arrangement. Wood parenchyma is paratracheal, confluent, and diffuse. 20 ×.

PLATE XL

1. *Couma macrocarpa* Barb. Rodr., Apocynaceae (Y. 37137). Pores nearly all in short radial multiples. Tangential pairing results from overlapping vessel members. Wood parenchyma in indistinct reticulate arrangement. 20 ×.

2. *Jacaranda mimosifolia* D. Don, Bignoniaceae (Y. 23485). Pores mostly in small multiples fairly evenly distributed, but appearing in tangential or concentric arrangement because of the narrow bands of paratracheal parenchyma. 10 ×.

3. *Cedrela odorata* L., Meliaceae (Y. 35994). The growth rings in Spanish Cedar are demarcated by bands of wood parenchyma. The pores decrease gradually in size during seasonal growth, but there is a more or less pronounced tendency to ring-porous structure. 10 ×.

PLATE XLI

1. *Lyonia ferruginea* (Walt.) Nutt., Ericaceae (Y. 11538). Ring-porous wood. The pores are very numerous and virtually all solitary. 20 ×.

2. *Ouratea amplectens* Hutch. & Dalz., Ochnaceae (Y. 15096). Pores solitary, fairly evenly distributed. The rays are of two distinct sizes. 20 ×.

3. *Bumelia angustifolia* Nutt., Sapotaceae (Y. 6981). Pores in dendritic arrangement. Wood parenchyma abundant, diffuse and in loosely aggregated concentric bands. 20 ×.

PLATE XLII

1. *Amyris balsamifera* L., Rutaceae (Y. 9515). Pores in long multiples or radial chains. 20×.

2. *Monopteryx uaucu* Spruce, Leguminosae (Y. 21004). Long multiples associated with solitary pores. Wood parenchyma aliform and short-confluent. Some of the rays are very coarse. 20×.

3. *Banisteriopsis* sp., Malpighiaceae (Y. 9735). Long radial multiples of pores variable in size. 20×.

PLATE XLIII

1. *Calophyllum brasiliense* Camb., Guttiferae (Y. 10557). Pores solitary and in irregularly branched radial series. Wood parenchyma in narrow, broken, metatracheal bands. 20×.

2. *Manilkara bidentata* (A. DC.) A. Chev., Sapotaceae (Y. 18821). Pores in long multiples and radial series of short multiples. Wood parenchyma in numerous narrow, metatracheal bands. 20×.

3. *Rhamnus crocea* Nutt., Rhamnaceae (Y. 14798). Pores in flame-like pattern. 20×.

PLATE XLIV

1. *Guaiacum Coulteri* Gray, Zygophyllaceae (Y. 2944). Pores solitary and irregularly distributed. Wood parenchyma diffuse and aliform. A species of Lignum-vitae, noted for its density. 20×.

2. *Amburana cearensis* (Fr. Allem.) A. C. Smith, Leguminosae (Y. 23517). Wood parenchyma abundant, vasicentric and vasicentric-confluent. 10×.

3. *Platymiscium pinnatum* (Jacq.) Dugand, Leguminosae (Y. 8889). Wood parenchyma short aliform, locally confluent; also in narrow terminal bands. 10×.

4. *Ocotea Rodiaei* (Schomb.) Mez, Lauraceae (Y. 32858). Cross section of Greenheart showing unilaterally paratracheal or sparingly vasicentric parenchyma. 10×.

PLATE XLV

1. *Ormosia subsimplex* Benth., Leguminosae (Y. 37155). Wood parenchyma coarsely aliform and confluent. Some radial multiples of small, flattened pores are present. 10×.

2. *Cordia lomatoloba* Johnston, Boraginaceae (Y. 35943). Wood parenchyma confluent into tangential or concentric bands. Crystals large, visible with lens. 10×.

3. *Simaruba amara* Aubl., Simarubaceae (Y. 36590). Wood parenchyma narrowly aliform and confluent. 10×.

4. *Brosimum terrabanum* Pittier, Moraceae (Y. 7531). Wood parenchyma narrowly aliform and confluent into very numerous irregular bands. 10×.

PLATE XLVI

1. *Campsiandra angustifolia* Spruce, Leguminosae (Y. 36621). Wood parenchyma coarse, short aliform or locally confluent. 10×.

2. *Terminalia intermedia* (A. Rich.) Urb., Combretaceae (Y. 9216). Wood parenchyma irregularly aliform and locally confluent. 20×.

3. *Sterigmapetalum obovatum* Kuhlmann, Rhizophoraceae (Y. 20691). Wood parenchyma unilaterally paratracheal, with lateral extensions which sometimes are confluent. 20×.

4. *Brosimum Krukovii* Standl., Moraceae (Y. 36800). Wood parenchyma short to long aliform, sometimes confluent. 20×.

PLATE XLVII

1. *Adelia triloba* (Muell.) Hemsl., Euphorbiaceae (Y. 12047). Wood parenchyma in finely reticulate pattern. Growth rings are demarcated by concentric zones deficient in parenchyma. 20×.

2. *Cariniana pyriformis* Miers, Lecythidaceae (Y. 3968). Wood parenchyma in numerous narrow, concentric bands spaced about one pore-width apart. 20×.

3. *Bombax brevicuspe* Sprague, Bombacaceae (Y. 15285). Wood parenchyma composed of thin-walled cells and indistinctly reticulate, characteristic of many woods of the Malvales. 20×.

4. *Swietenia macrophylla* King, Meliaceae (Y. 34649). Widely spaced concentric bands of wood parenchyma, apparently demarcating growth rings, are characteristic of all species of *Swietenia* (Mahogany). Gum plugs are common in the vessels. 20×.

PLATE XLVIII

1. *Lonchocarpus latifolius* H.B.K., Leguminosae (Y. 20532). Wood parenchyma mostly paratracheal, in numerous coarse, irregular bands alternating with dense bands of wood fibers. 20×.

2. *Cymbopetalum costaricense* (D. Sm.) R. E. Fries, Anonaceae (Y. 12045). Wood parenchyma metatracheal, in numerous fine, fairly evenly spaced concentric bands characteristic of the family. 20×.

3. *Eschweilera decolorans* Sandw., Lecythidaceae (Y. 35451). Wood parenchyma metatracheal, in numerous narrow, concentric bands. Growth rings are produced by closer spacing in the late wood. Tyloses are abundant in the vessels. 20 ×.

4. *Lecythls grandiflora* Aubl., Lecythidaceae (Y. 5091). Wood parenchyma apotracheal, in narrow, closely and fairly uniformly spaced concentric bands. 20 ×.

PLATE XLIX

1. *Ficus aurea* Nutt., Moraceae. Wood parenchyma mostly apotracheal, in numerous coarse, concentric bands characteristic of the wood of Fig trees. 10 ×.

2. *Clitoria Hoffmanseggii* Benth., Leguminosae (Y. 36458). Wood parenchyma paratracheal, in very coarse concentric bands composing more than half of the ground mass. Pores few and of two distinct sizes, the small ones barely visible in photomicrograph. 10 ×.

3. *Erythrina Ulei* Harms, Leguminosae (Y. 21652). Wood parenchyma mostly apotracheal, in hammock-like bands between the coarse rays. 10 ×.

4. *Andira inermis* H.B.K., Leguminosae (Y. 6628). Wood parenchyma abundantly paratracheal, irregularly confluent. The contrast between the light-colored parenchyma and the dark fiber bands produces a figure characteristic of the Partridge woods. 10 ×.

PLATE L

1. *Apeiba tibourbou* Aubl., Tiliaceae (Y. 4063). Radial section showing one of the bands of unlignified parenchyma with the cells greatly elongated radially. 20 ×.

2. The same. Cross section showing three wide bands of unlignified parenchyma, which is soft and cottony in contrast with the normal wood. 10 ×.

PLATE LI

1. *Piper smilacifolium* C. DC., Piperaceae (Y. 10538). Cross section showing numerous pores between the very coarse medullary rays, but not in contact with them. 20 ×.

2. The same. Tangential section showing multiseriate rays continuous throughout the length of the section, as they are true medullary rays and their height is equal to the length of the internode. Uniseriate rays are absent. The wood fibers are storied. 20 ×.

3. *Ottoschulzia cubensis* (C. Wr.) Urb., Icacinaceae (Y. 9230). Cross section showing rays of two sizes, uniseriate and very coarse. Pores virtually all solitary. Wood parenchyma in irregular metatracheal lines. 20 ×.

PLATE LII

1. *Zuelania guidonia* (Sw.) Britt. & Millsp., Flacourtiaceae (Y. 34691). Rays narrow, very numerous, and less than one pore-width apart. 20 ×.

2. *Guevina avellana* Mol., Proteaceae (Y. 9548). Tangential section showing very coarse rays and some low uniseriate rays. 20 ×.

3. *Chrysochlamys membranacea* Tr. & Pl., Guttiferae (Y. 20968). Cross section showing uniseriate and multiseriate rays. The pores are rarely in contact with the large rays. 20 ×.

PLATE LIII

1. *Swietenia macrophylla* King, Meliaceae (Y. 2186). Tangential surface of Mahogany as seen under a hand lens, showing the rays and vessel members in horizontal seriation. This feature is sometimes absent or poorly developed. The vessels are filled with red gum (dark) or white substance (light). 10 ×.

2. *Bulnesia arborea* (Jacq.) Engl., Zygophyllaceae (Y. 380). Tangential section of the Venezuelan Vera, showing very low rays uniformly storied, the resulting ripple marks being more than 200 per vertical inch. This feature is characteristic of Lignum-vitae (*Guaiacum*) also. 20 ×.

3. *Lonchocarpus latifolius* H.B.K., Leguminosae (Y. 20532). Tangential section showing all elements in horizontal seriation. The individual cells of the 2-celled wood parenchyma strands are in secondary seriation, the resulting ripple marks in a parenchyma band being twice the number of those in the intervening fiber layers. 20 ×.

4. *Olneya tesota* Gray, Leguminosae (Y. 6698). Tangential section showing the wood parenchyma strands and their component cells storied, but the rays not in horizontal seriation. Ripple marks are accordingly not visible unless the plane of section is through a band of parenchyma. 50 ×.

PLATE LIV

1. *Eperua Schomburkgiana* Benth., Leguminosae (Y. 31971). Concentric rows of gum ducts apparently demarcating growth rings. The exudations produce brown stains on the surface of the wood. 10 ×.

2. *Vochysia biloba* Ducke, Vochysiaceae (Y. 37146). Tangential series of large gum ducts resulting from injury to the cambium. The wood parenchyma is confluent into irregular tangential or concentric bands. 10×.

3. *Buchenavia* sp., Combretaceae (Y. 10512). A tangential series of traumatic ducts is shown near bottom of the section. The wood parenchyma is aliform and irregularly confluent.

4. The same. Tangential section through the gum ducts showing how they are anastomosed. The two straight openings are vessels. 20×.

PLATE LV

1. *Himatanthus sucuuba* (Spruce) Woods., Apocynaceae (Y. 23465). Tangential surface showing a large radial channel as it appears under a hand lens. 10×.

2. *Alchornea sidaefolia* Baill., Euphorbiaceae (Y. 23811). Tangential surface showing three radial channels as they appear under a hand lens. 10×.

3. *Antonia ovata* Pohl, Loganiaceae (Y. 35488). Cross section showing numerous small islands of included phloem. 10×.

4. *Strychnos guianensis* (Aubl.) Baill., Loganiaceae (Y. 35716). Cross section showing fairly evenly distributed islands of included phloem which are much larger than the pores. 10×.

PLATE LVI

1. *Aniba roseadora* Ducke, Lauraceae (Y. 4435). The numerous oil cells in the Bois de Rose of French Guiana appear on cross section as very small pores. 20×.

2. The same. Radial section showing numerous vertically elongated oil cells in the wood parenchyma strands and a few in the margins of the rays. 20×.

3. *Rapanea leuconeura* (Mart.) Mez, Myrsinaceae (Y. 18952). Cross section showing gum cysts as irregular openings in the large rays. The pores are rarely in contact with these rays. 20×.

4. The same. Showing how the cysts appear in a tangential section. 20×.

PLATE LVII

1. *Rapanea leuconeura* (Mart.) Mez, Myrsinaceae (Y. 18952). Gum cysts in a large ray as they appear in a radial section after the breaking down of the cell walls. 20×.

2. *Seguieria americana* L., Phytolaccaceae (Y. 28508). Included phloem associated with concentric bands and broad rays of conjunctive tissue. 20×.

3. *Avicennia marina* (Forsk.) Vierh., Avicenniaceae. Included phloem associated with anastomosed bands of conjunctive tissue. The narrow rays are of normal structure. 20×.

PLATE LVIII

1. *Securidaca diversifolia* L., Polygalaceae (Y. 11116). Included phloem associated with coarse concentric bands of conjunctive tissue. The pores are comparatively few and large. 20×.

2. *Chamissoa altissima* (Jacq.) H.B.K., Amarantaceae (Y. 32380). Included phloem associated with irregularly concentric bands and wide rays of conjunctive tissue. 20×.

3. *Neea psychotrioides* D. Sm., Nyctaginaceae. The openings were left after the disintegration of the phloem strands and appear at the outer end of radial series of pores. 20×.

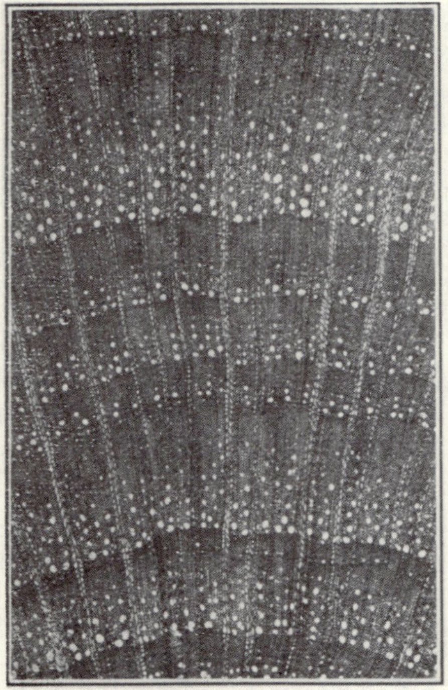

1. *Ephedra trifurca* Torr.

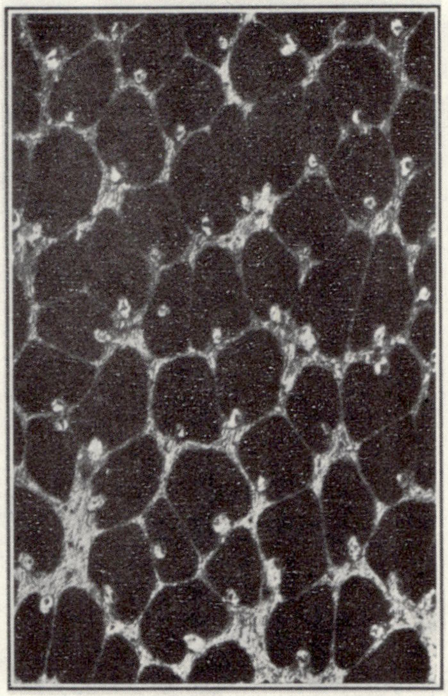

2. *Astrocaryum Standleyanum* Bailey.

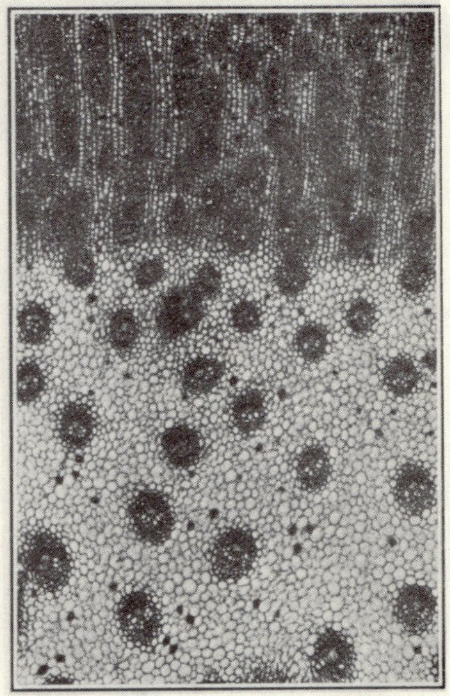

3. *Dracaena americana* D. Sm.

PLATE XXXVII. Gymnosperm with vessels, Black Palm wood, and stem of the Lily family.

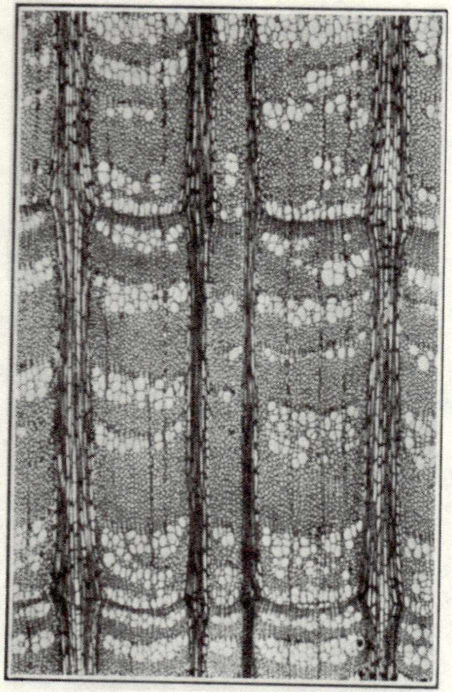

1. *Guevina avellana* Mol.

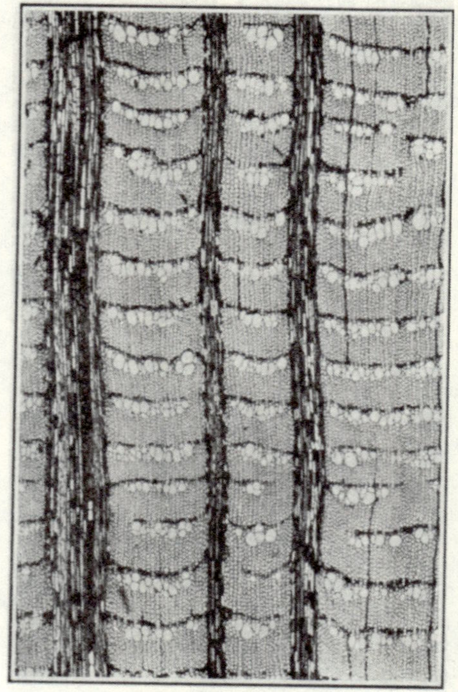

2. *Embothrium grandiflorum* Lam.

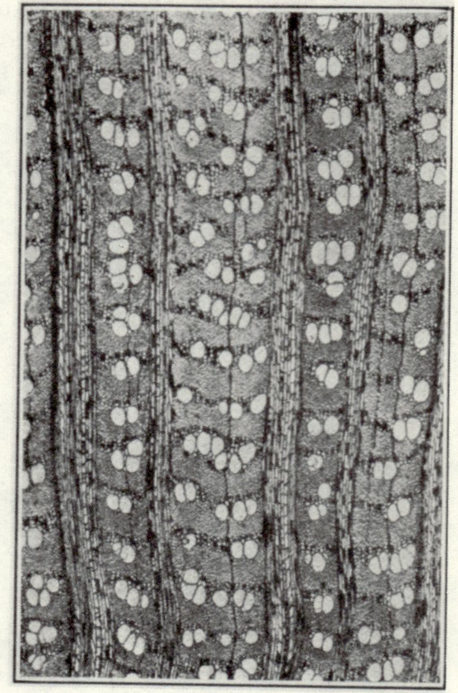

3. *Roupala grossedentata* Pitt.

PLATE XXXVIII. Pore and parenchyma arrangement and ray width in Proteaceae.

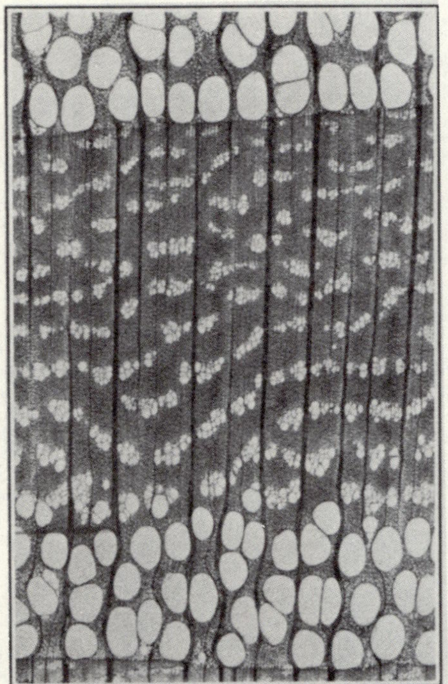

1. *Ulmus fulva* Michx.

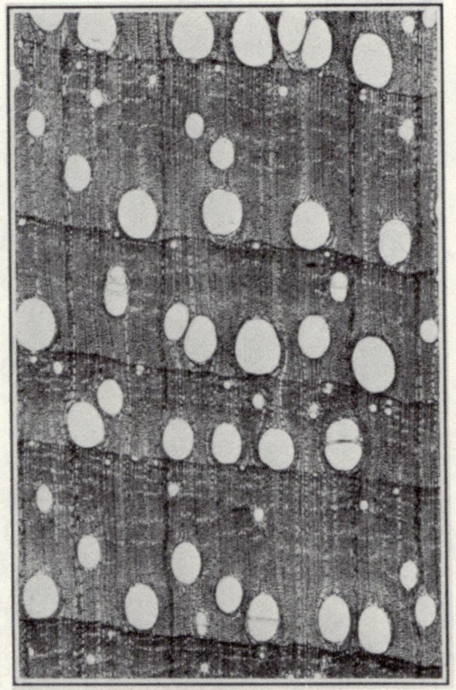

2. *Carya glabra* Mill.

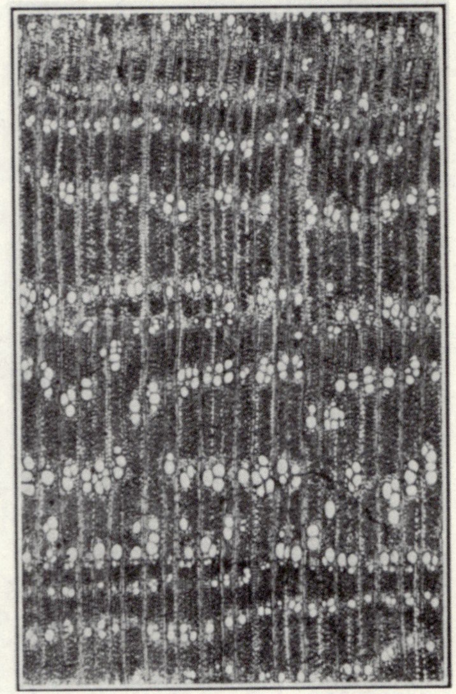

3. *Patagonula americana* L.

PLATE XXXIX. Types of pore arrangement in Elm, Hickory, and Guayabí.

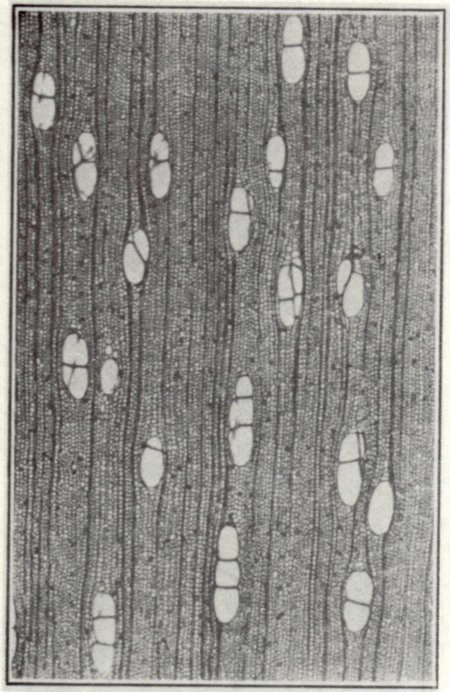

1. *Couma macrocarpa* Barb. Rodr.

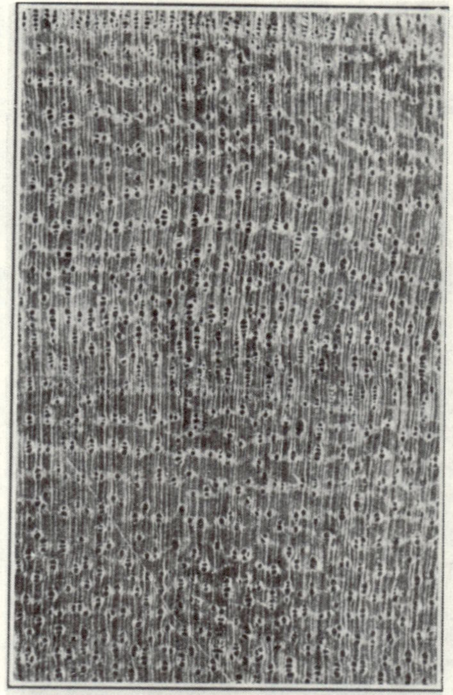

2. *Jacaranda mimosifolia* D. Don.

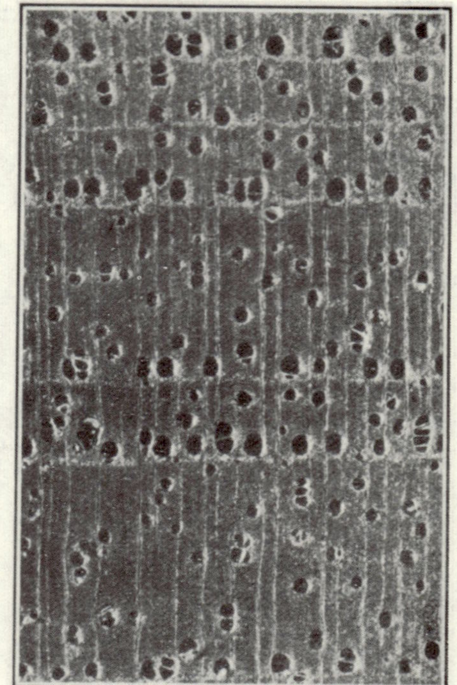

3. *Cedrela odorata* L.

PLATE XL. Three types of arrangement of pores and wood parenchyma.

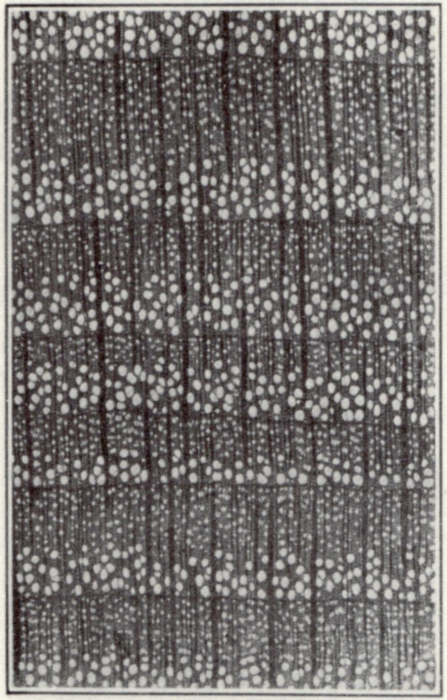

1. *Lyonia ferruginea* (Walt.) Nutt.

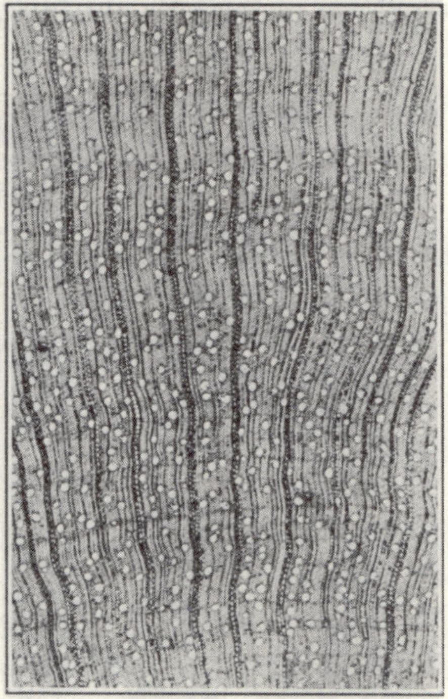

2. *Ouratea amplectens* Hutch. & Dalz.

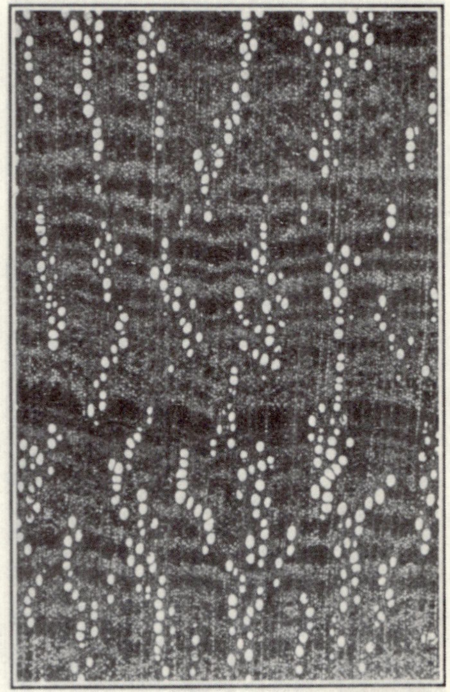

3. *Bumelia angustifolia* Nutt.

PLATE XLI. Three types of pore arrangement.

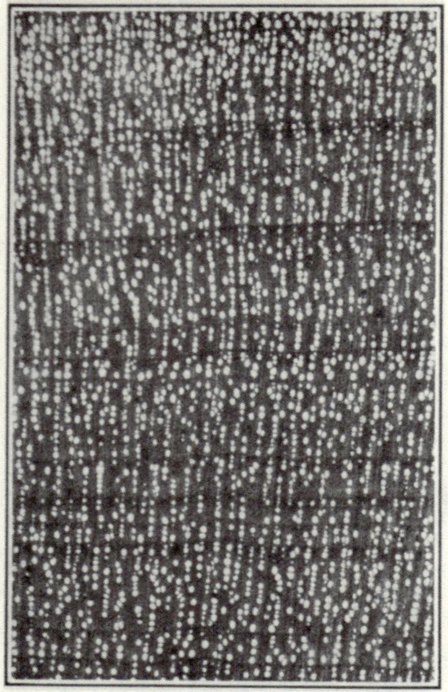

1. *Amyris balsamifera* L.

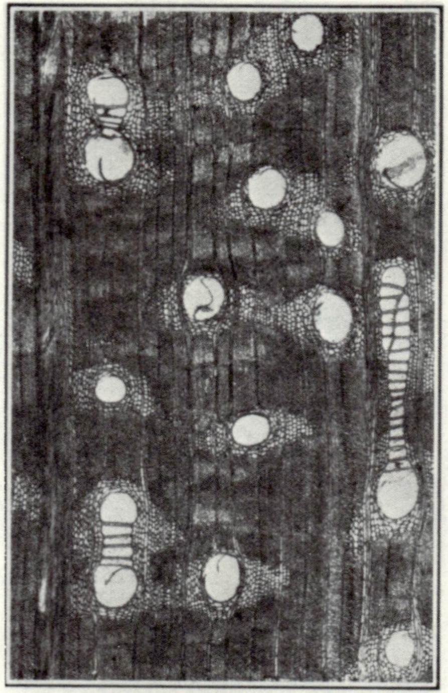

2. *Monopteryx uaucu* Spruce.

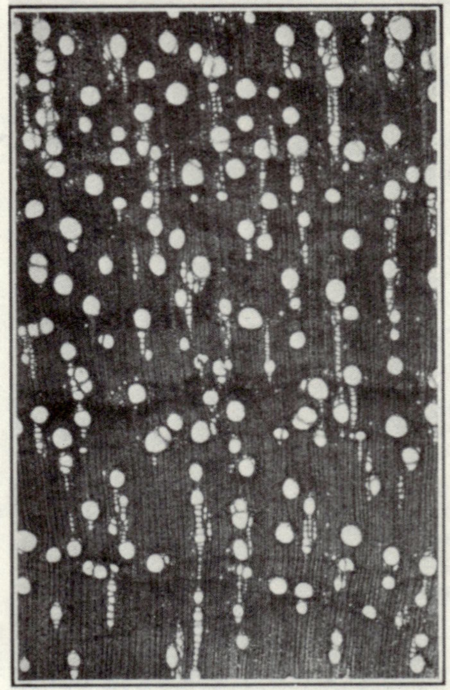

3. *Banisteriopsis* sp.

PLATE XLII. Pores in radial arrangement.

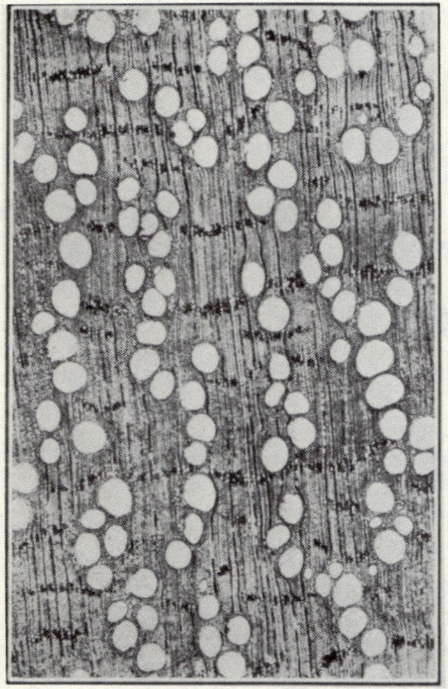

1. *Calophyllum brasiliense* Camb.

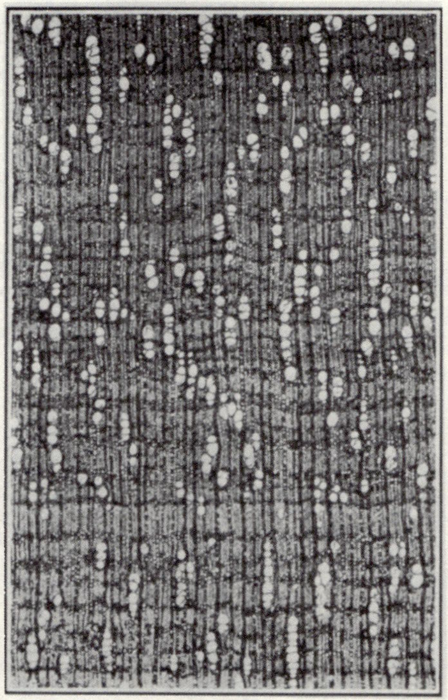

2. *Manilkara bidentata* (A.DC.) A. Chev.

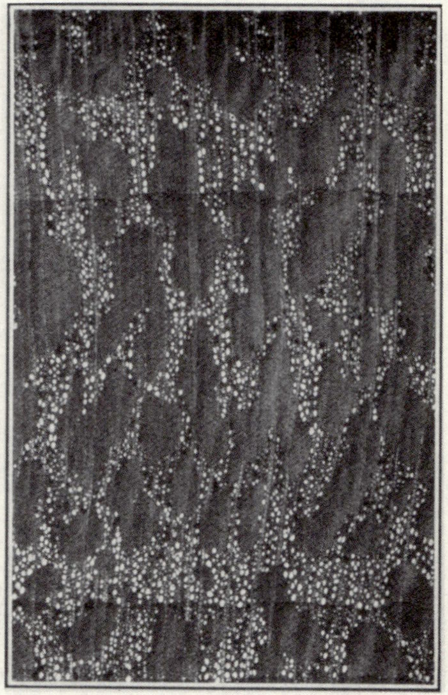

3. *Rhamnus crocea* Nutt.

PLATE XLIII. Pores in radial arrangement.

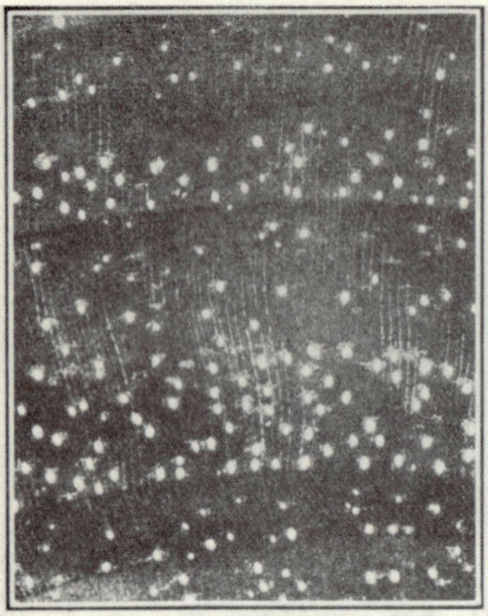

1. *Guaiacum Coulteri* Gray.

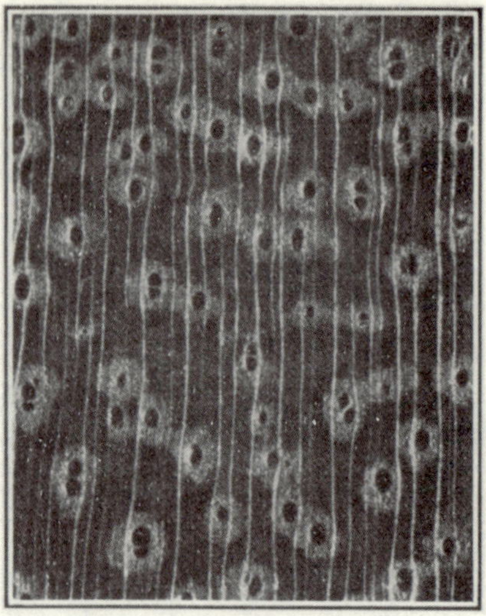

2. *Amburana cearensis* (Fr. Allem.) A. C. Sm.

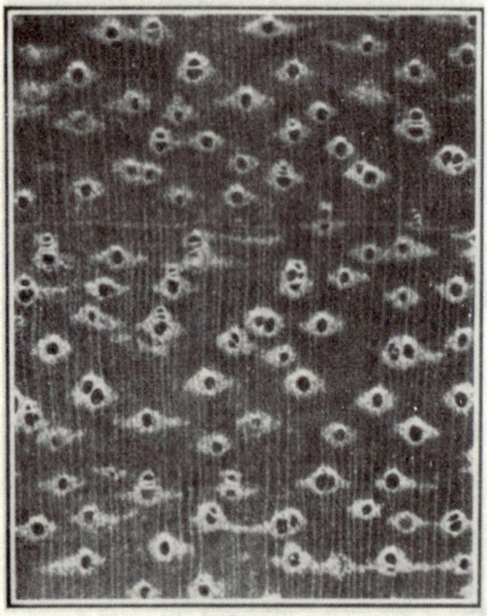

3. *Platymiscium pinnatum* (Jacq.) Dugand.

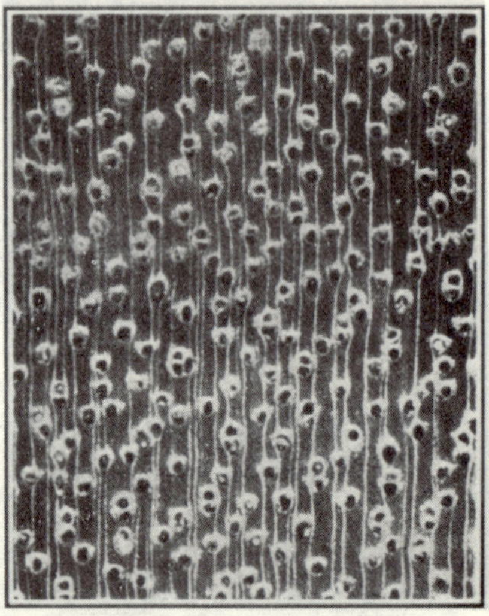

4. *Ocotea Rodiaei* (Schomb.) Mez.

PLATE XLIV. Types of vasicentric and aliform parenchyma.

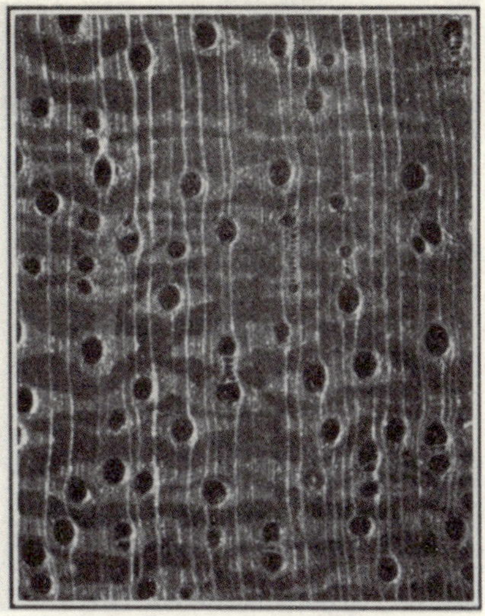

1. *Ormosia subsimplex* Benth.

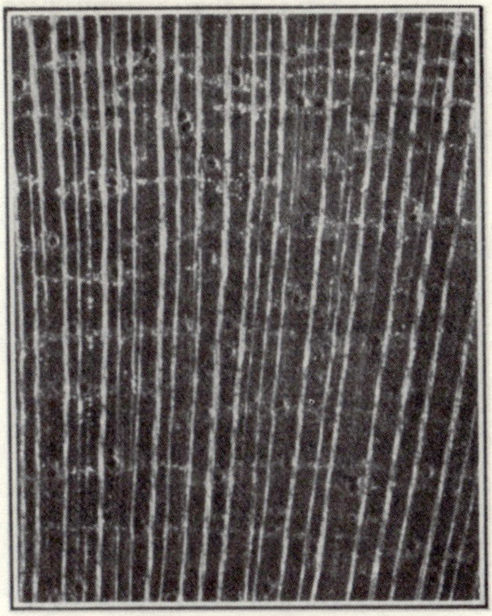

2. *Cordia lomatoloba* Johnston.

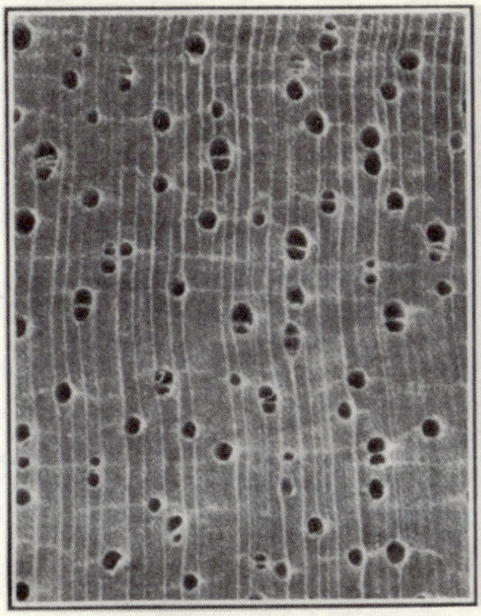

3. *Simaruba amara* Aubl.

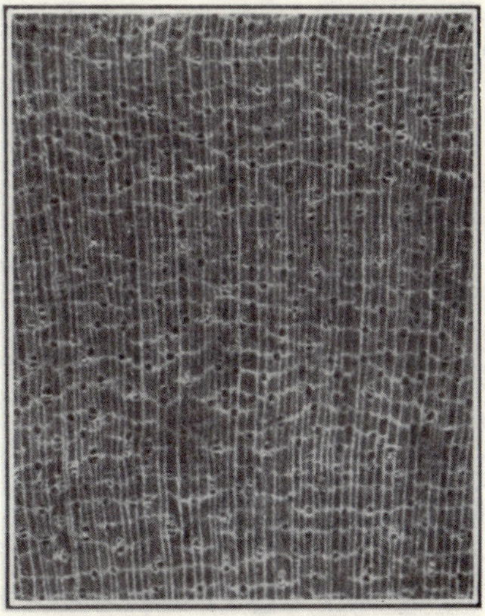

4. *Brosimum terrabanum* Pitt.

PLATE XLV. Types of aliform and confluent parenchyma.

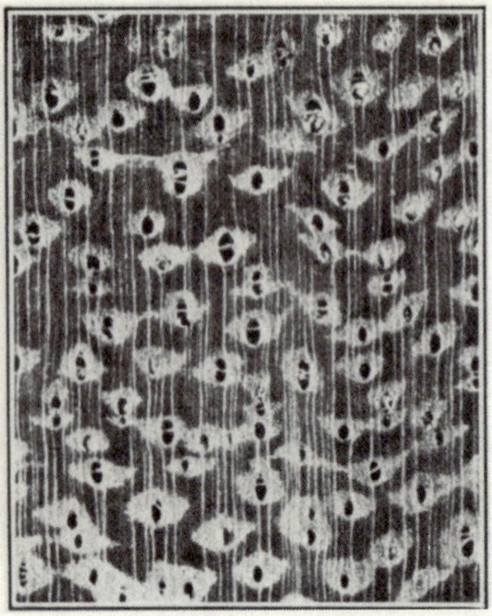

1. *Campsiandra angustifolia* Spruce.

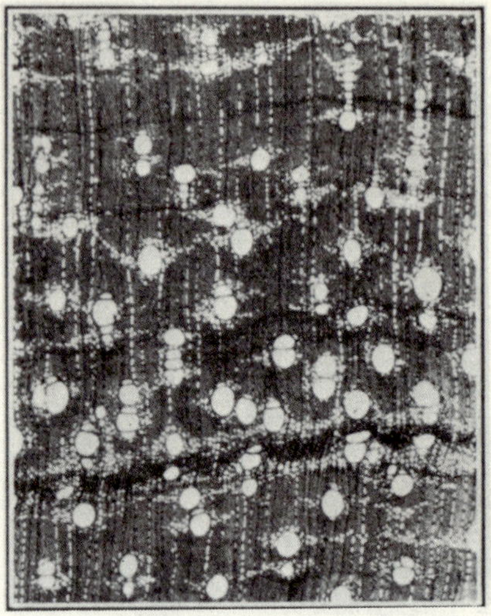

2. *Terminalia intermedia* (A. Rich.) Urb.

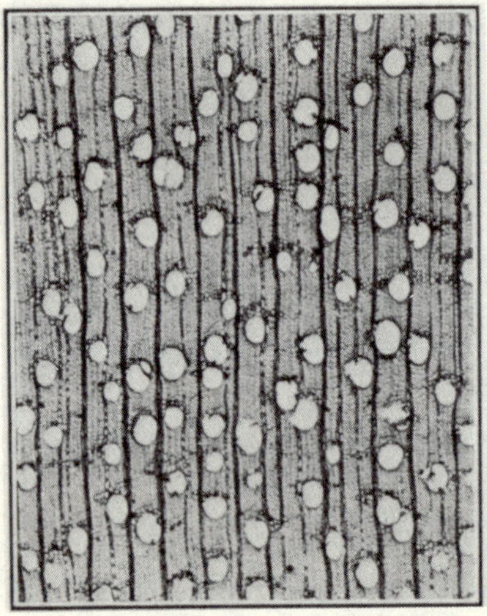

3. *Sterigmapetalum obovatum* Kuhlm.

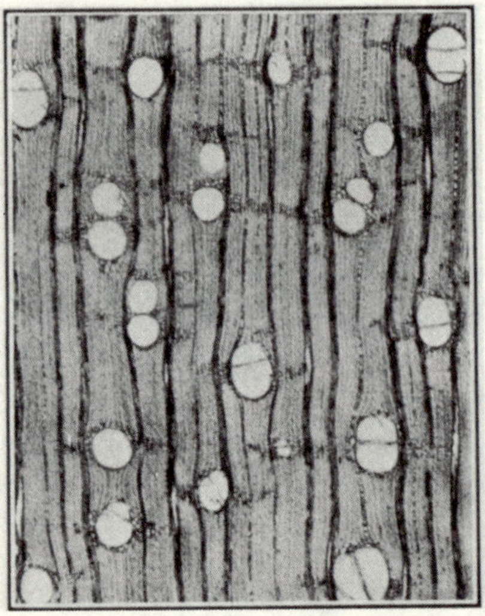

4. *Brosimum Krukovii* Standl.

PLATE XLVI. Types of aliform and confluent parenchyma.

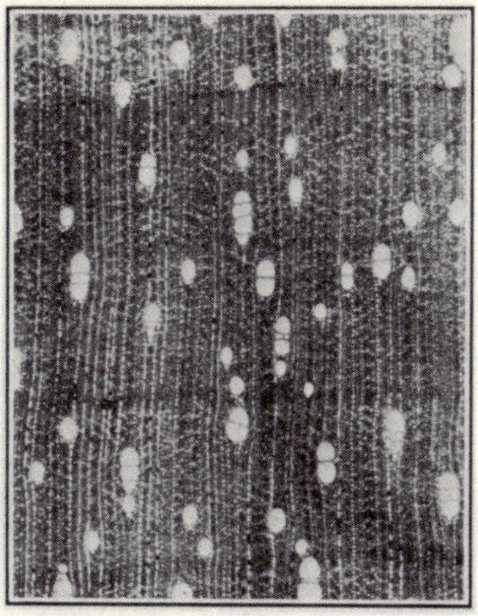

1. *Adelia triloba* (Muell.) Hemsl.

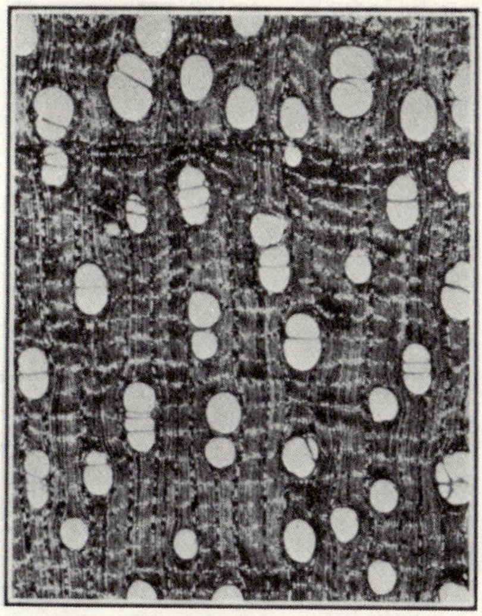

2. *Cariniana pyriformis* Miers.

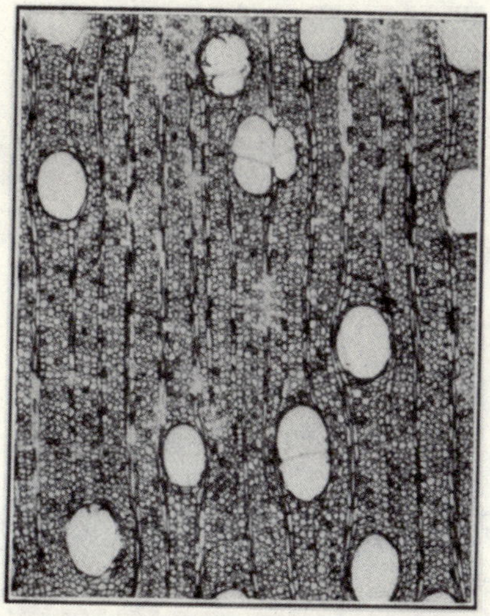

3. *Bombax brevicuspe* Sprague.

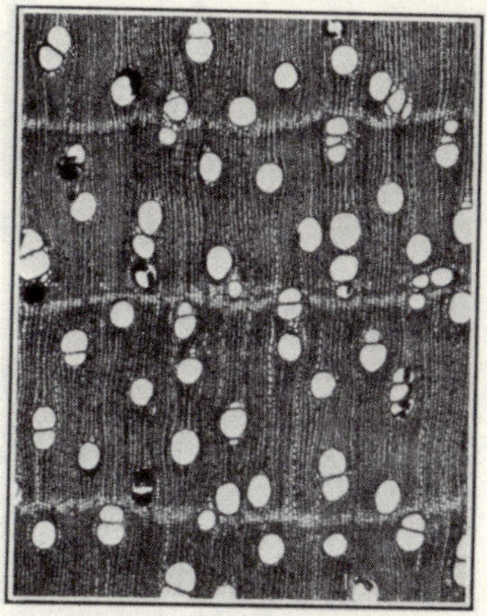

4. *Swietenia macrophylla* King.

PLATE XLVII. Different types of parenchyma distribution.

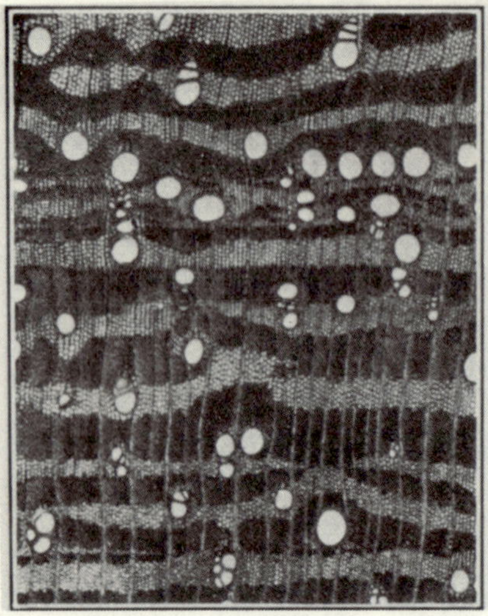

1. *Lonchocarpus latifolius* H.B.K.

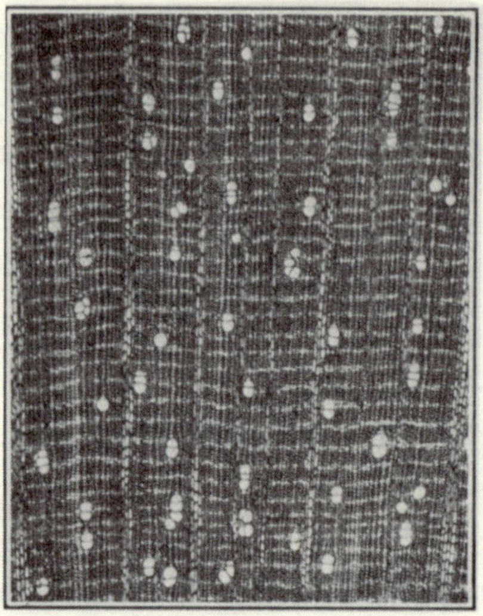

2. *Cymbopetalum costaricense* (D. Sm.) R. E. Fries.

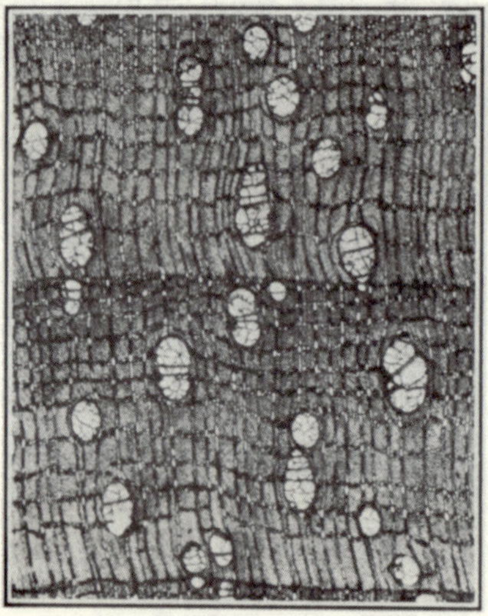

3. *Eschweilera decolorans* Sandw.

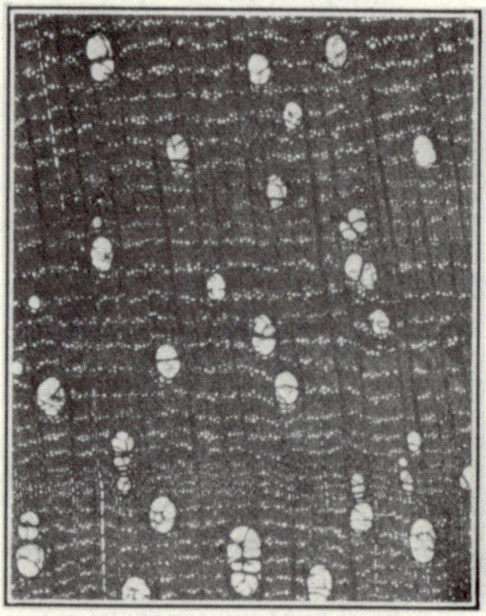

4. *Lecythis grandiflora* Aubl.

PLATE XLVIII. Types of banded wood parenchyma.

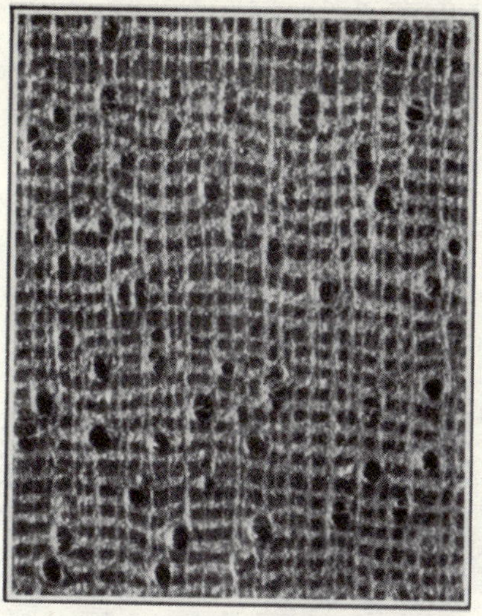

1. *Ficus aurea* Nutt.

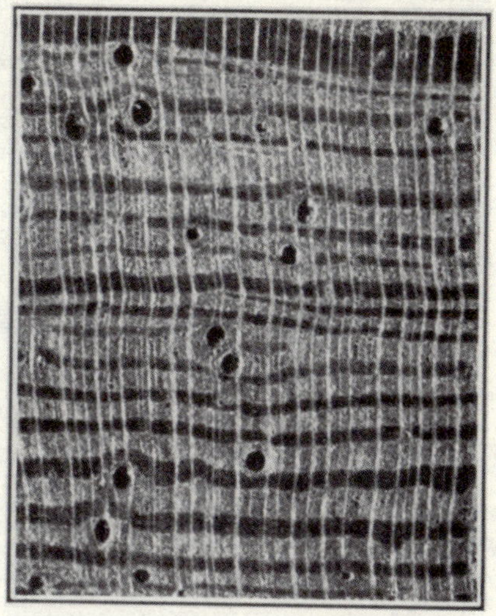

2. *Clitoria Hoffmanseggii* Benth.

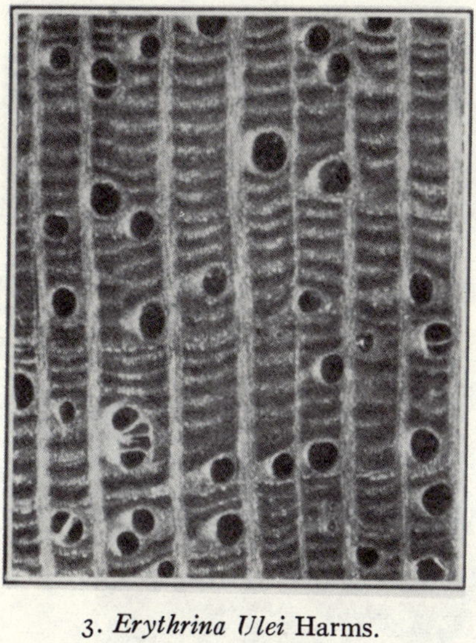

3. *Erythrina Ulei* Harms.

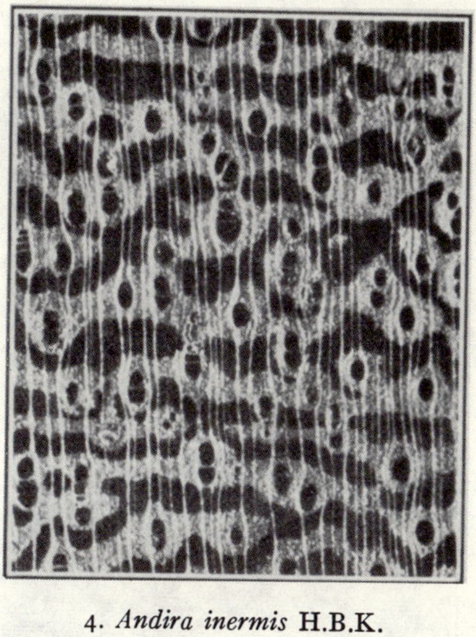

4. *Andira inermis* H.B.K.

PLATE XLIX. Types of banded wood parenchyma.

1. Radial section.

2. Cross section.

PLATE L. Concentric bands of unlignified tissue in *Apeiba*.

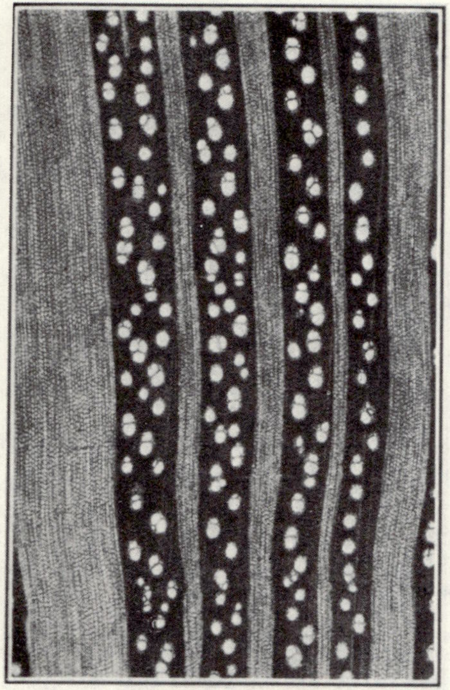

1. *Piper smilacifolium* C.DC.

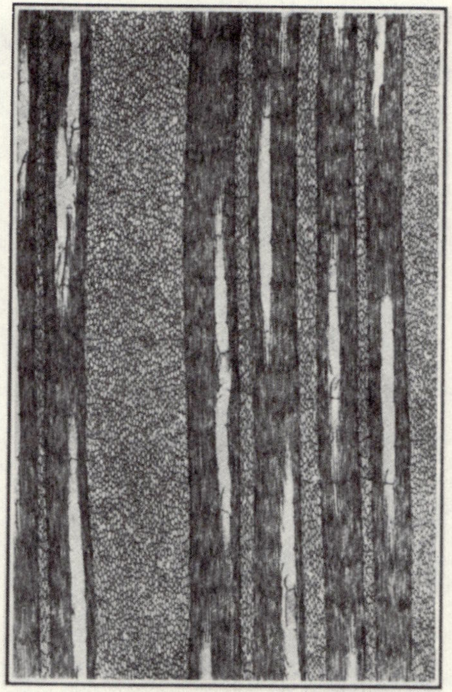

2. Same. Tangential section.

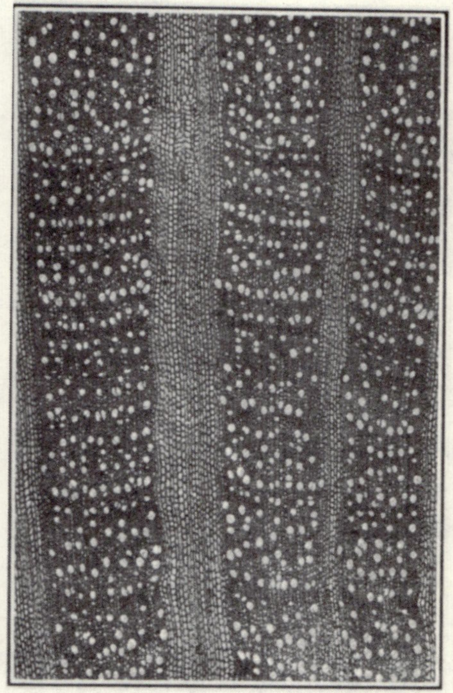

3. *Ottoschulzia cubensis* (C. Wr.) Urb.

PLATE LI. Woods of *Piper* and *Ottoschulzia* with very coarse rays.

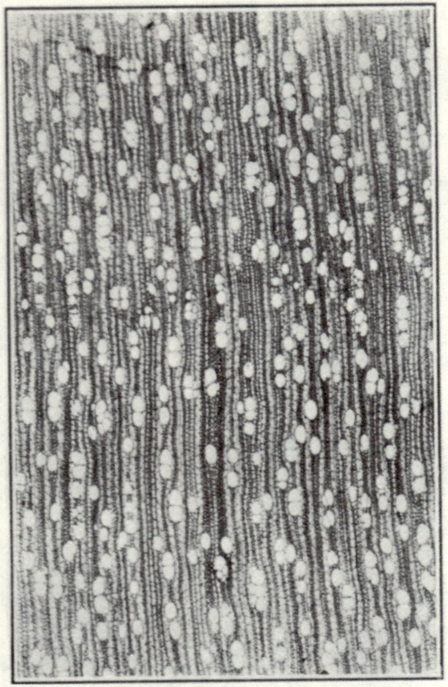

1. *Zuelania guidonia* (Sw.) Britt. & Millsp.

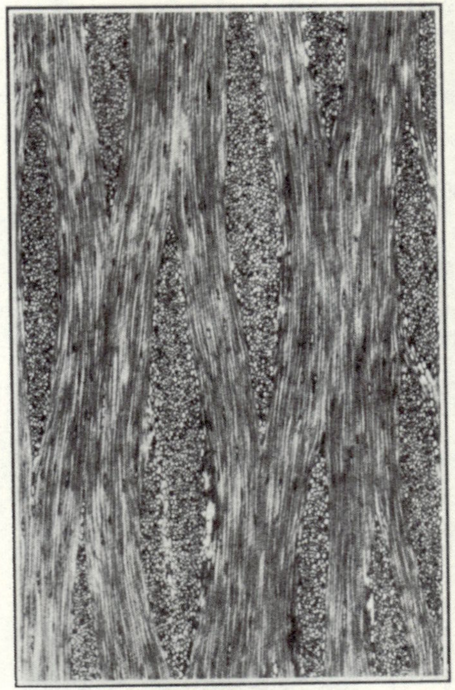

2. *Guevina avellana* Mol.

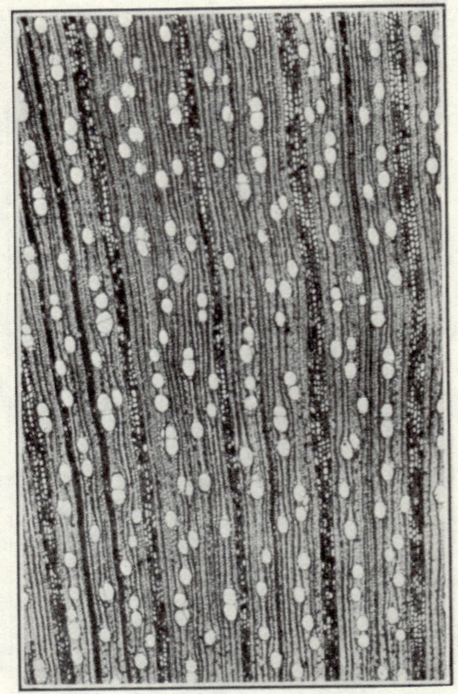

3. *Chrysochlamys membranacea* Tr. & Pl.

PLATE LII. Three types of rays.

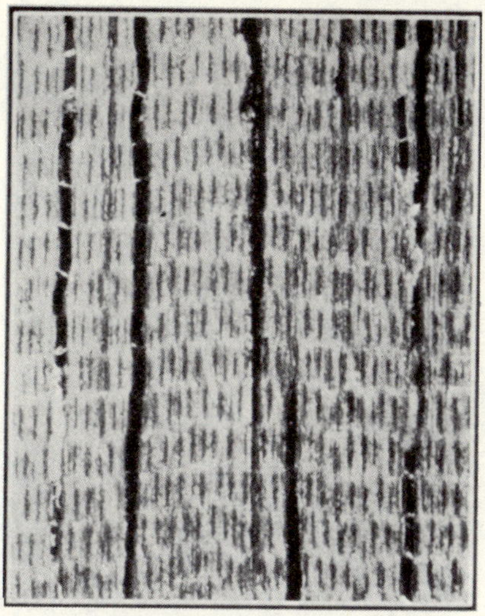

1. *Swietenia macrophylla King.*

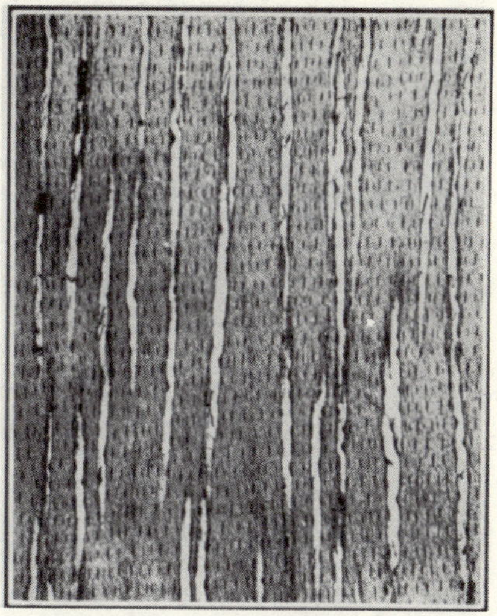

2. *Bulnesia arborea* (Jacq.) Engl.

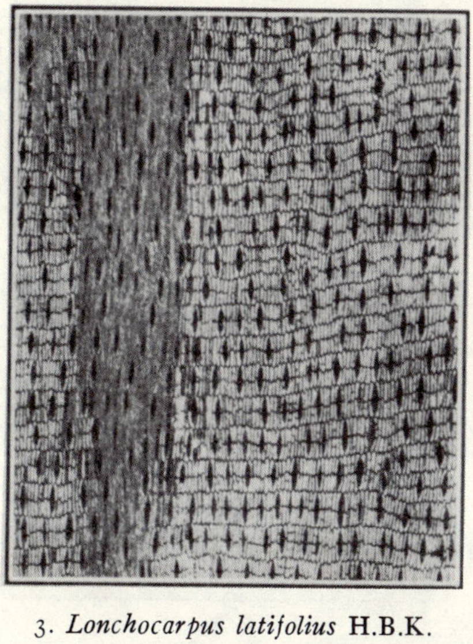

3. *Lonchocarpus latifolius* H.B.K.

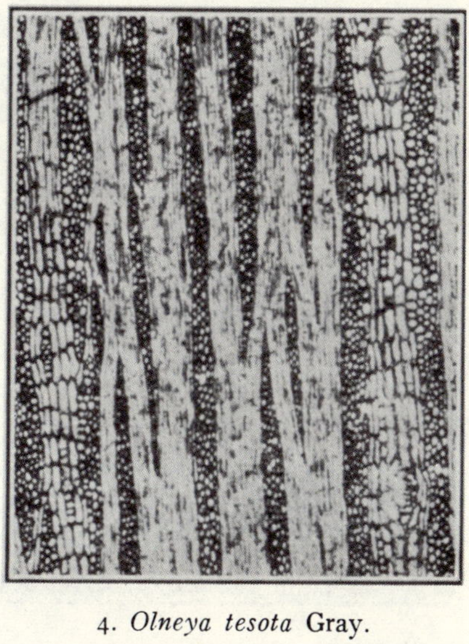

4. *Olneya tesota* Gray.

PLATE LIII. Different types of storied structure.

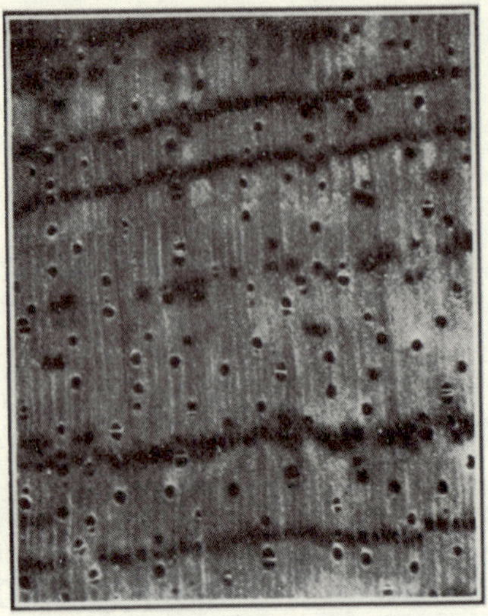

1. *Eperua Schomburgkiana* Benth.

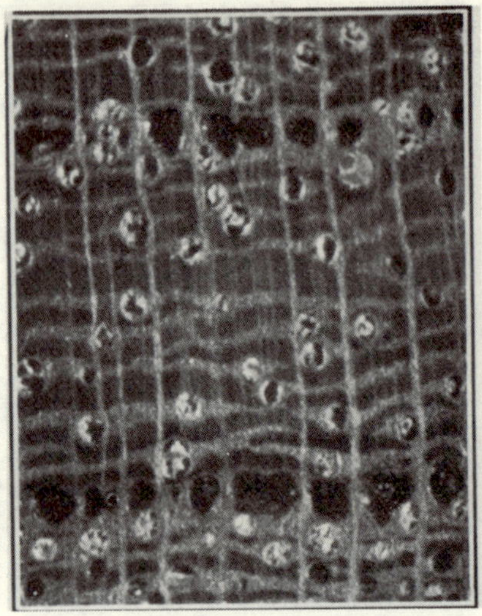

2. *Vochysia biloba* Ducke.

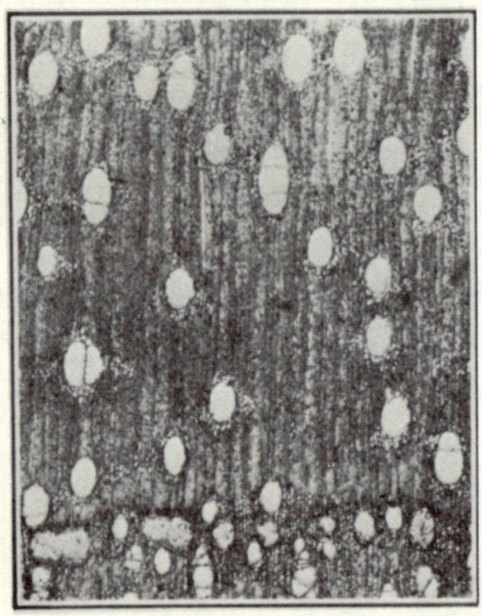

3. *Buchenavia* sp.

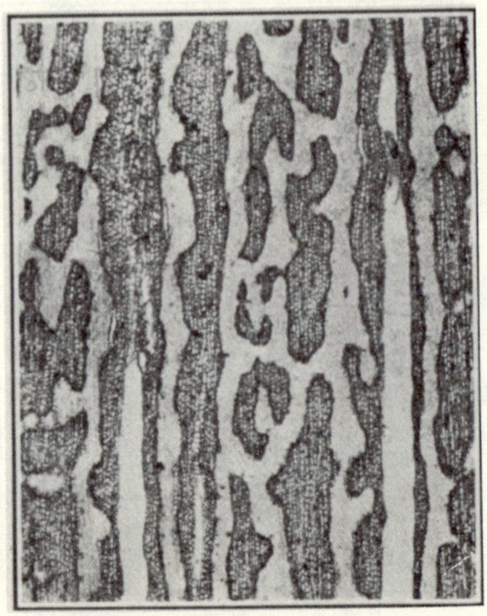

4. Same. Tangential section.

PLATE LIV. Gum ducts in dicotyledonous woods.

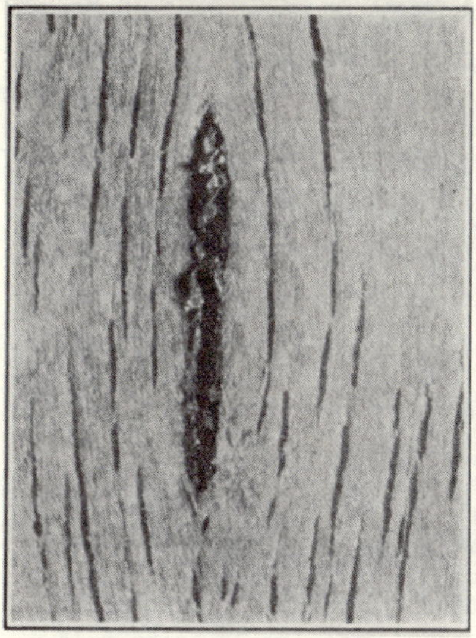

1. *Himatanthus sucuuba* (Spruce) Woods.

2. *Alchornea sidaefolia* Baill.

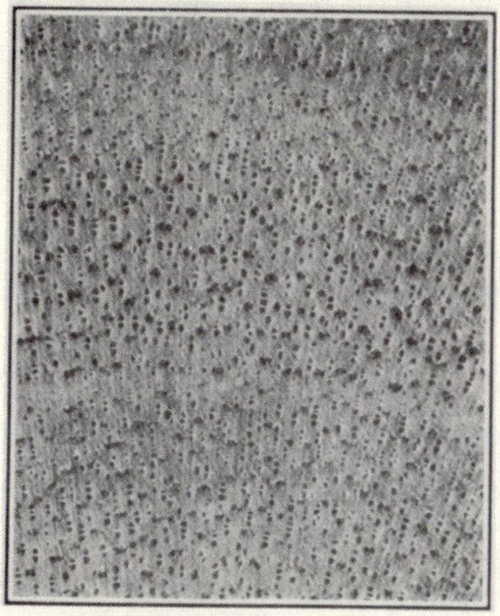

3. *Antonia ovata* Pohl.

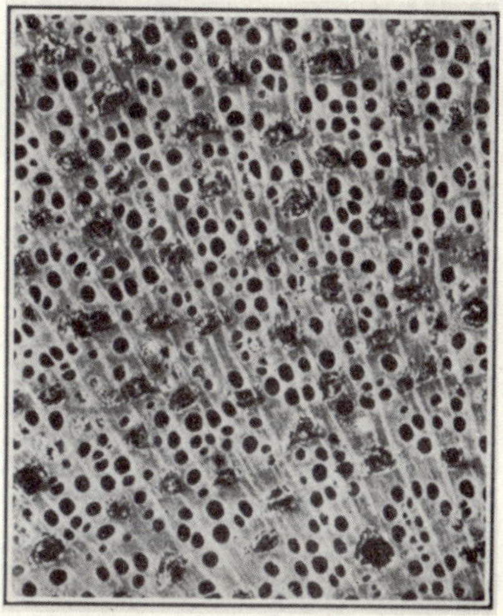

4. *Strychnos guianensis* (Aubl.) Baill.

PLATE LV. Large radial channels and included phloem strands.

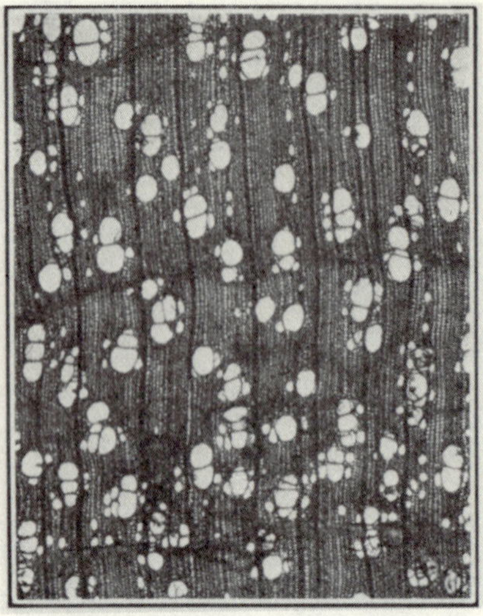

1. *Aniba roseadora* Ducke.

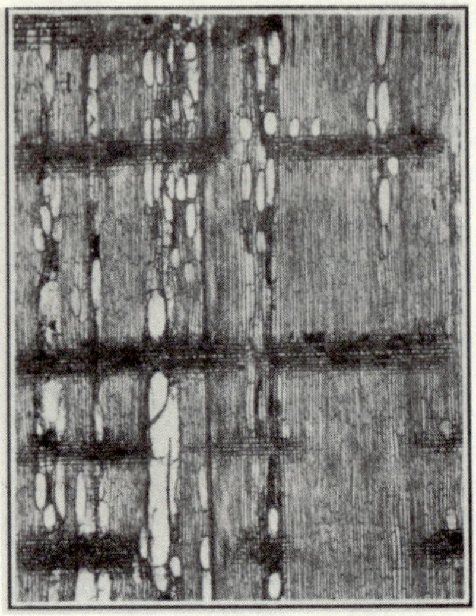

2. Same. Radial section.

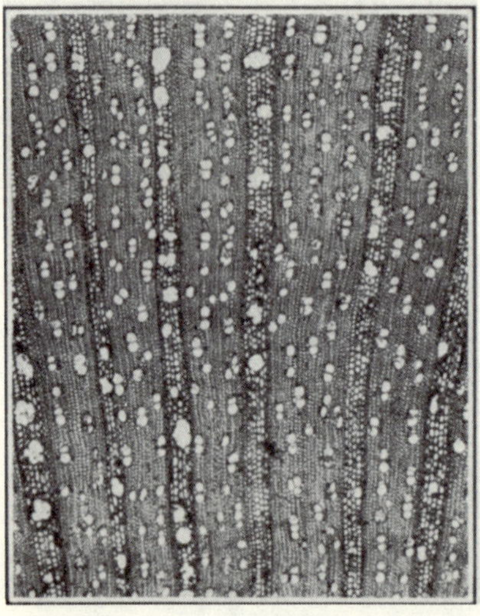

3. *Rapanea leuconeura* (Mart.) Mez.

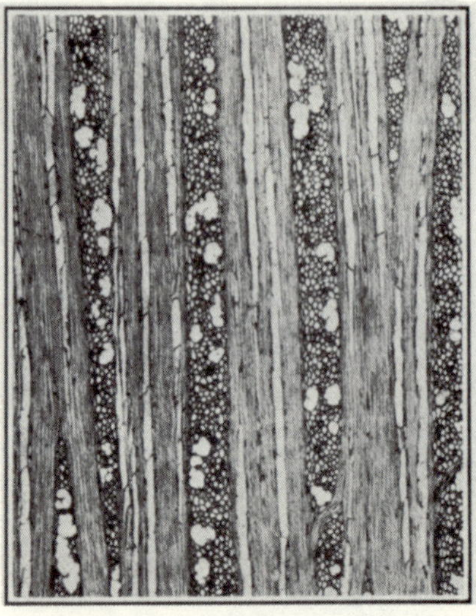

4. Same. Tangential section.

PLATE LVI. Oil cells and gum cysts.

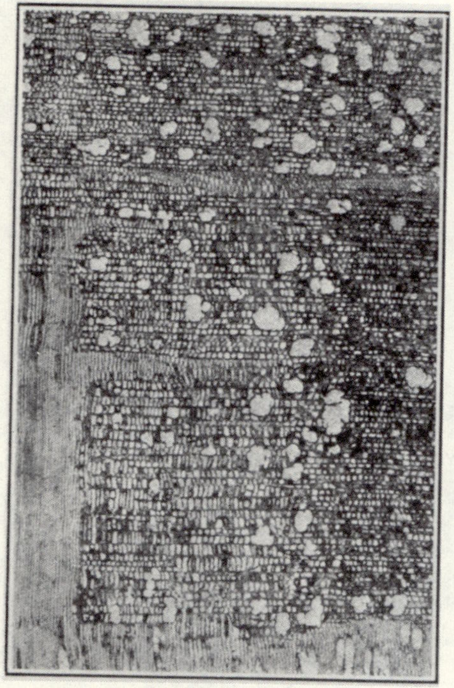

1. *Rapanea leuconeura* (Mart.) Mez.

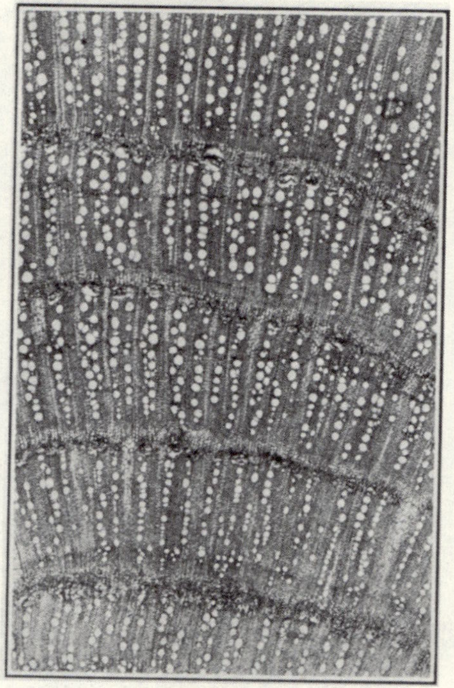

2. *Seguieria americana* L.

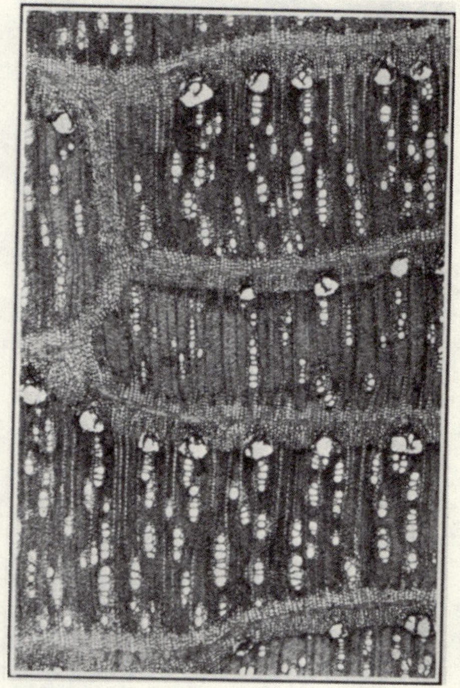

3. *Avicennia marina* (Forsk.) Vierh.

PLATE LVII. Gum cysts and included phloem bands.

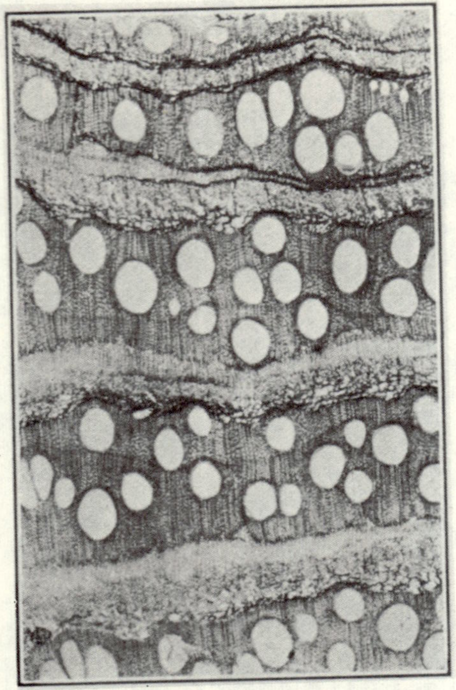

1. *Securidaca diversifolia* L.

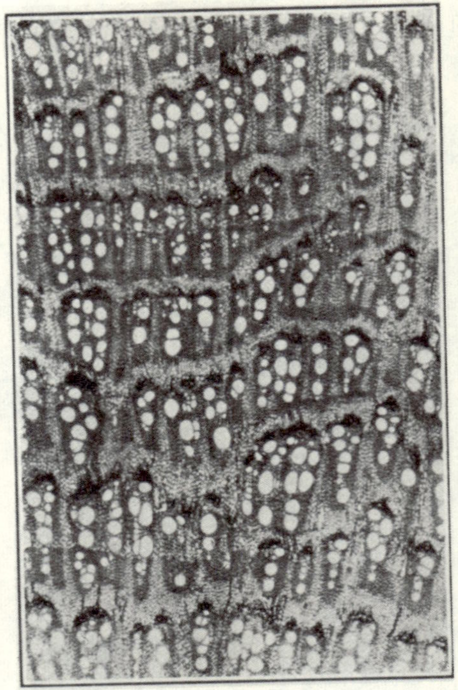

2. *Chamissoa altissima* (Jacq.) H.B.K.

3. *Neea psychotrioides* D. Sm.

PLATE LVIII. Woods of anomalous structure.

INDEX

Use and Abuse
of
America's Natural Resources

An Arno Press Collection

Ayres, Quincy Claude. **Soil Erosion and Its Control.** 1936

Barger, Harold and Sam H. Schurr. **The Mining Industries, 1899–1939.** 1944

Carman, Harry J., editor. **Jesse Buel:** Agricultural Reformer. 1947

Circular from the General Land Office Showing the Manner of Proceeding to Obtain Title to Public Lands. 1899

Fernow, Bernhard E. **Economics of Forestry.** 1902

Gannett, Henry, editor. **Report of the National Conservation Commission, February 1909.** Three volumes. 1909

Giddens, Paul H. **The Birth of the Oil Industry.** 1938

Greeley, William B. **Forests and Men.** 1951

Hornaday, William T. **Wild Life Conservation in Theory and Practice.** 1914

Ise, John. **The United States Forest Policy.** 1920

Ise, John. **The United States Oil Policy.** 1928

James, Harlean. **Romance of the National Parks.** 1939

Kemper, J. P. **Rebellious River.** 1949

Kinney, J. P **The Development of Forest Law in America.** *Including,* Forest Legislation in America Prior to March 4, 1789. 1917

Larson, Agnes M. **History of the White Pine Industry in Minnesota.** 1949

Liebig, Justus, von. **The Natural Lawss of Husbandry.** 1863

Lindley, Curtis H. **A Treatise on the American Law Relating to Mines and Mineral Lands.** Two volumes. 2nd edition. 1903

Lokken, Roscoe L. **Iowa**—Public Land Disposal. 1942

McGee, W. J., editor. **Proceedings of a Conference of Governors in the White House, May 13–15, 1908.** 1909

Mead, Elwood. **Irrigation Institutions.** 1903

Moreell, Ben. **Our Nation's Water Resources**—Policies and Politics. 1956

Murphy, Blakely M., editor. **Conservation of Oil & Gas**: A Legal History, 1948. 1949

Newell, Frederick Haynes. **Water Resources:** Present and Future Uses. 1920.

Nimmo, Joseph, Jr. **Report in Regard to the Range and Ranch Cattle Business of the United States.** 1885

Nixon, Edgar B., editor. **Franklin D. Roosevelt & Conservation, 1911–1945.** Two volumes. 1957

Peffer, E. Louise. **The Closing of the Public Domain.** 1951

Preliminary Report of the Inland Waterways Commission. 60th Congress, 1st Session, Senate Document No. 325. 1908

Puter, S. A. D. & Horace Stevens. **Looters of the Public Domain.** 1908

Record, Samuel J. & Robert W. Hess. **Timbers of the New World.** 1943

Report of the Public Lands Commission, with Appendix. 58th Congress, 3d Session, Senate Document No. 189. 1905

Report of the Public Lands Commission, Created by the Act of March 3, 1879. 46th Congress, 2d Session, House of Representatives Ex. Doc. No. 46. 1880

Resources for Freedom: A Report to the President by The President's Materials Policy Commission, Volumes I and IV. 1952. Two volumes in one.

Schoolcraft, Henry R. **A View of the Lead Mines of Missouri.** 1819

Supplementary Report of the Land Planning Committee to the National Resources Board, 1935–1942

Thompson, John Giffin. **The Rise and Decline of the Wheat Growing Industry in Wisconsin** (Reprinted from *Bulletin of the University of Wisconsin,* No. 292). 1909

Timmons, John F. & William G. Murray, editors. **Land Problems and Policies.** 1950

U.S. Department of Agriculture—Forest Service. **Timber Resources for America's Future:** Forest Resource Report No. 14. 1958

U.S. Department of Agriculture—Soil Conservation Service and Forest Service. **Headwaters Control and Use.** 1937

U.S. Department of Commerce and Labor—Bureau of Corporations. **The Lumber Industry,** Parts I, II, & III. 1913/1914

U.S. Department of the Interior. **Hearings before the Secretary of the Interior on Leasing of Oil Lands.** 1906

Whitaker, J. Russell & Edward A. Ackerman. **American Resources**: Their Management and Conservation. 1951

id